Lecture Notes in Physics

The Lecture Notes in Physics

The series Lecture Notes in Physics (LNP), founded in 1969, reports new developments in physics research and teaching – quickly and informally, but with a high quality and the explicit aim to summarize and communicate current knowledge in an accessible way. Books published in this series are conceived as bridging material between advanced graduate textbooks and the forefront of research and to serve three purposes:

- to be a compact and modern up-to-date source of reference on a well-defined topic
- to serve as an accessible introduction to the field to postgraduate students and nonspecialist researchers from related areas
- to be a source of advanced teaching material for specialized seminars, courses and schools

Both monographs and multi-author volumes will be considered for publication. Edited volumes should, however, consist of a very limited number of contributions only. Proceedings will not be considered for LNP.

Volumes published in LNP are disseminated both in print and in electronic formats, the electronic archive being available at springerlink.com. The series content is indexed, abstracted and referenced by many abstracting and information services, bibliographic networks, subscription agencies, library networks, and consortia.

Proposals should be sent to a member of the Editorial Board, or directly to the managing editor at Springer:

Christian Caron
Springer Heidelberg
Physics Editorial Department I
Tiergartenstrasse 17
69121 Heidelberg / Germany
christian.caron@springer.com

B. Coecke (Ed.)

New Structures for Physics

Springer

Bob Coecke
Parks Road
OX1 3QD Oxford
United Kingdom
coecke@comlab.ox.ac.uk

Coecke, B. (Ed.): *New Structures for Physics*, Lect. Notes Phys. 813 (Springer, Berlin Heidelberg 2011), DOI 10.1007/978-3-642-12821-9

Lecture Notes in Physics ISSN 0075-8450 e-ISSN 1616-6361
ISBN 978-3-642-12820-2 ISBN 978-3-642-12821-9 (eBook)
DOI 10.1007/978-3-642-12821-9
Springer Heidelberg Dordrecht London New York

Library of Congress Control Number: 2010930494

Cover design: Integra Software Services Pvt. Ltd., Pondicherry

Printed on acid-free paper

Springer is part of Springer Science+Business Media (www.springer.com)

Dedicated to the many bright young theoretical physicists that failed to escape the fate of having to work in institutions like banks.

Preface

New? In what sense? Surely I am not the only person who, after extensively justifying why certain mathematical structures naturally arise in physics, gets questions like: "this is all nice maths but what's the physics?" Meanwhile I figured out what this truly means: "I don't see any differential equations!" Okay, this is indeed a bit overstated. Nowadays any mathematical argument involving groups, when these are moreover referred to as "symmetry groups", stands a serious chance of being eligible for carrying the label "physics". But it hasn't always been like this. John Slater (cf. the Slater determinant in quantum chemistry) referred to the use of group theory in quantum physics by Weyl, Wigner et al. as der Gruppenpest, what translates as the "plague of groups". Even in 1975 he wrote [14]: "As soon as [my] paper became known, it was obvious that a great many other physicists were as 'disgusted' as I had been with the group-theoretical approach to the problem. As I heard later, there were remarks made such as 'Slater has slain the Gruppenpest'. I believe that no other piece of work I have done was so universally popular." Donkeys usually don't make the same mistake twice, . . .

. . . and, surely I am not the only person who, after extensively justifying why certain mathematical structures naturally arise in physics, gets questions like: "this is just the same thing in a different language!" Well, so was Copernicus' description of the planets as compared to Ptolemy's. Looking back at the facts, Ptolemy's description turned out to be more accurate, accounting even for relativistic effects. So was abandoning the view that the Earth was the centre of the universe and that planets move around it on hierarchies of epicycles a step backward? Of course not. Taking the superheavy object that the sun is to be a fixed point of reference unveiled the gravitational force as well as a critical glimpse of Newton's laws of motion, in terms of Galilei's visions and Kepler's work. Similarly, programming languages are not just a different way of writing down 0s and 1s, but also capture the flows of information within a computational process. Surely we wouldn't want the whole of mathematics to be written down entirely in terms of 0s and 1s; imagine researching physics in terms of nothing but 0s and 1s! All this just to say that language means structure and additional structure means additional content: using group theory is not just about using a different language but about identifying symmetry as a key ingredient of physics. The same goes for the structures that are discussed in this book: they all identify a key ingredient in physics that deserves our attention. They moreover identify this ingredient as present in a wide range of theories, including theories of information and computation.

The contributing research landscape. Once a subset of mathematics is accepted by the general physics community as relevant, many physicists seem to stop making a distinction between that piece of mathematics and the natural phenomenon this piece of mathematics aims to describe. For this reason, there is a high entrance fee for a mathematical structure to be awarded this privilege. But this also means that progress in physics does go hand-in-hand with the use of new mathematical structures. This book contains a number of such structures which recently have been finding their way into quantum information, foundations of general relativity, quantum foundations, and quantum gravity foundations. A surprising feature of many of these is that these structures are already heavily used in "Euro-style" computer science, and some were even crafted for this particular purpose. In general relativity "Scott domains" enable to reconstruct spacetime topology from the causal structure without making any reference to smoothness [10]. Dana Scott (= male) initially introduced these domains in the late 60s to provide semantics for the λ-calculus [13], which plays a key role both in the foundations of mathematics and in programming language foundations [19]. In quantum information monoidal categories [21] are becoming more prominent, for example, for the description of particular computational models such as topological quantum computing (see [11] for a survey), and measurement-based quantum computing (see for example [5, 6, 8]), in which the interaction between classical and quantum data is of key importance. Earlier it was already suggested that topological quantum field theories [2], which are functors between certain kinds of monoidal categories, could be relevant for a theory of quantum gravity [3, 7]. Again, these monoidal categories are of key importance in computer science, for example, they provide semantics for linear logic [20], a logic which is important in concurrency theory [18], the theoretical underpinning of mobile phone networks, internet protocols, cash machines etc.

At the n-category café John Baez suggested that a less opportunistic title for this volume would have been: "Structures you would already know about, had you been paying proper attention". While as title poetry this isn't great, he is of course right, and for more than one reason. John himself pointed to the fact that, for example, "Category theory has been important in algebraic topology ever since its interception in 1945. It's just taken a while for these structures to become part of the toolkit of the average mathematical physicist." He and Mike Stay have more examples on page 125 of their chapter entitled "Physics, topology, logic and computation: A Rosetta Stone" [4]. The other reason is the one I mentioned above: these structures are already heavily used in theoretical computer science, where the play the role of "logic of interaction" [1], "discrete (relativistic!) spacetime" [9, 12], among many other roles.

A personal appreciation. I started my research career in the late eighties in quantum foundations. If that didn't already guaranty academic suicide, I moreover studied hidden variable theories. After my PhD, in an attempt to save my career, I moved to the dying area of quantum logic within the retiring Geneva group led by Piron. Having become aware of my mistake I moved into pure mathematics, to category theory, an area hated by most non-category-theoretic-mathematicians, within the retiring category theory group at McGill University. The great surprise is that after all of this I am still standing, while many other scholars, far more brighter than I am, lost the battle. The worst carnage in terms of academic careers surely must have taken place in high energy physics [15, 16]. In quantum foundations the academic death-toll is less, but this mostly has to do with the the style quantum theory is taught in most places: "Don't think, just do!", resulting in not many people ending up in quantum foundations. The reason that I ended up surviving must be that although each of $|\text{quantum foundations}\rangle$, $|\text{quantum logic}\rangle$, $|\text{category theory}\rangle$ causes academic disaster, $|\text{quantum categorical logic foundations}\rangle$ proved to be some kind of a hit in European computer science circles where, surprisingly, "foundations" means "cool". In those circles structural research is indeed highly appreciated, the reason being that one simply can't do without. Meanwhile, the membership of our multidisciplinary group here at Oxford University Computing Laboratory [22] has grown to 30, which besides Samson Abramsky and myself now also includes Andreas Döring, and a zoo of DPhil (= Oxford PhD) students with backgrounds in theoretical physics, computer science, pure mathematics, philosophy, engineering, and even linguistics.

How did this all came about? In 2005 I organized an event called *Cats, Kets and Cloisters* (CKC) at Oxford University Computing Laboratory [23]. The event aimed to set the stage for an encounter of researchers studying mathematical structures in computer science, quantum foundations, pure mathematicians including specialists in logic, category theory and knot theory, and quantum informaticians. It in particular included twelve tutorial lectures by leading experts. The success of the conference what witnessed by the fact that since there was no budget to invite speakers, these twelve leading experts all covered there own expenses. Moreover, a chain of similar events [24–26] emerged after CKC, the most recent one being Categories, Quanta and Concepts (CQC) at the Perimeter Institute [27].

But a low in all this was the following. When asked by several PhD students were they could read about "this kind of stuff", there simply wasn't a satisfactory answer. This is were this volume kicks in: it collates a series of tutorials that do the job.

Contributions to this volume. We start with an ABC on monoidal category theory, by Abramsky and Tzevelekos, Baez and Stay, and Coecke and Paquette. These bulky contributions nicely complement each other, the first one being the lecture notes of the category theory course here at Oxford University Computing Laboratory, the second one exemplifying how the same structures arise in very different areas, and the third one establishing that monoidal categories have always been "out there" in physics. The "linear" feature of these categories is then further emphasized, in graphical realm by Selinger, and in computational realm by Haghverdi and Scott. In particular, Selinger's chapter is the first rock-solid comprehensive account on the topic of graphical calculi for monoidal categories, in which he fixes several caveats of the existing literature. Then follows a Blute-Panagaden double which applies the theory to formal distributions and formal Feynman diagrams. After that we have a living Legend, Jim Lambek, who exposes connections between particle physics and mathematical linguistics, an area which he pioneered in the 1950s. Next up is domain theory, starting with a tutorial overview by Martin, followed by a detailed account of the domain-theoretic structure on classical and quantum states by Coecke and Martin. This is then followed by a range of structures dealing with spacetime: first Martin and Panangaden's application of domain theory to general relativity, then Hiley's use of Clifford algebras, and finally Döring and Isham's use of topos theory in an 180 page long mega contribution. We end with applications of monoidal categories in quantum computational models, firstly a general account by Hines, which is followed by Panangaden and Paquette's survey of topological quantum computing.

Acknowledgments We in particular thank John Baez and the attendants of the n-category café for the "online public review process" of several chapters in this volume. Assistance in producing this volume was provided by the EC-FP6-STREP *Foundational Structures in Quantum Information and Computation* (QICS). We also acknowledge support from EPSRC Advanced Research Fellowship EP/D072786/1 entitled *The Structure of Quantum Information and its Ramifications for IT*.

Oxford, England Bob Coecke
August 2009

References

1. Abramsky, S., Gay, S.J., Nagarajan, R.: Interaction categories and the foundations of typed concurrent programming. In: Deductive Program Design: Proceedings of the 1994 Marktoberdorf International Summer School, NATO Advanced Science Institutes Series F, pp. 35–113. Springer, New York (1994)
2. Atiyah, M.: Topological quantum field theories. Publications Mathématique de l'Institut des Hautes Etudes Scientifiques **68**, 175–186 (1989)
3. Baez, J.C., Dolan, J.: Higher-dimensional algebra and topological quantum field theory. J. Math. Phys. **36**, 60736105. arXiv:q-alg/9503002 (1995)
4. Baez, J.C. Stay, M.: Physics, topology, logic and computation: A Rosetta Stone. In: Coecke, B. (ed.) New Structures for Physics, pp. 91–166, Springer Lecture Notes in Physics, New York. arXiv:0903.0340 (2009)
5. Coecke, B., Duncan, R.W.: Interacting quantum observables. In: Proceedings of the 35th International Colloquium on Automata, Languages and Programming (ICALP), pp. 298–310, Lecture Notes in Computer Science 5126. Springer, New York. arXiv:0906.4725 (2008)
6. Coecke, B., Paquette, E.O., Pavlovic, D.: Classical and quantum structuralism. In: Mackie, I., Gay, S. (eds.) Semantic Techniques for Quantum Computation. Cambridge University Press, Cambridge. arXiv:0904.1997 (2009)
7. Crane, L.: Clock and category: Is quantum gravity algebraic? J. Math. Phys. **36**, 6180–6193 (1995). arXiv:gr-qc/9504038
8. Duncan, R.W., Perdrix, S.: Graph states and the necessity of Euler decomposition. In: Proceedings of Computability in Europe: Mathematical Theory and Computational Practice (CiE'09), pp. 167–177. Lecture Notes in Computer Science 5635. Springer, New York (2009). arXiv:0902.0500

9. Lamport, L.: Time, clocks, and the ordering of events in a distributed system. Commun. ACM **21**, 558–565 (1978)
10. Martin, K., Panangaden, P.: A domain of spacetime intervals in general relativity. Commun. Math. Phys. **267**, 563–586 (2006)
11. Panangaden, P., Paquette, E.O.: A categorical presentation of quantum computation with anyons. In: Coecke, B. (ed.) New Structures for Physics, pp. 939–979. Springer Lecture Notes in Physics, New York (2009)
12. Petri, C.A.: State-transition structures in physics and in computation. Int. J. Theor. Phys. **12**, 979–992 (1982)
13. Scott, D.: Outline of a mathematical theory of computation. Technical Monograph PRG-2, Oxford University Computing Laboratory, Oxford (1970)
14. Slater, J.C.: Solid-State and Molecular Theory: A Scientific Biography. Wiley, New York (1975)
15. Smolin, L.: The Trouble with Physics. Houghton-Mifflin, Boston (2006)
16. Woit, P.: Not Even Wrong. Jonathan Cape, London (2006)
17. categorical logic: http://en.wikipedia.org/wiki/Categorical logic
18. concurrency: http://en.wikipedia.org/wiki/Concurrency (computer science)
19. lambda calculus: http://en.wikipedia.org/wiki/Lambda calculus
20. linear logic: http://en.wikipedia.org/wiki/Linear logic
21. monoidal category: http://en.wikipedia.org/wiki/Monoidal category
22. OUCL Quantum Group. http://web.comlab.ox.ac.uk/activities/quantum/
23. Cats, Kets and Cloisters (CKC) conference, organized by Bob Coecke, Oxford University Computing Laboratory, July 17–23, 2006. http://se10.comlab.ox.ac.uk:8080/FOCS/CKCinOXFORD en.html
24. Quantum Physics and Logic (QPL) workshop series, organized by Bob Coecke, Prakash Panangaden and Peter Selinger, started 2008. http://web.comlab.ox.ac.uk/people/Bob.Coecke/QPL 09.html
25. Categories, Logic and Foundations of Physics (CLAP) workshop series, organized by Bob Coecke and Andreas Döring, Imperial College and Oxford, started 2008. http://categorieslogicphysics.wikidot.com/
26. Informatic Phenomena (IP) workshop series, organized by Keye Martin and Mike Mislove, New Orleans, started 2008. http://www.math.tulane.edu/_mwm/WIP2009/Workshop on Informatic Phenomena.html
27. Categories, Quanta, Concepts (CQC), organized by Bob Coecke, Andreas Döring and Lucien Hardy, Perimeter Institute, June 1–5, 2009. http://www.perimeterinstitute.ca/en/Scientific/Conferences/Conferences/

Contents

Part I An ABC on Compositionality

1 Introduction to Categories and Categorical Logic 3
S. Abramsky and N. Tzevelekos
1.1 Introduction 3
1.2 Some Basic Constructions 13
1.3 Functors 23
1.4 Natural Transformations 30
1.5 Universality and Adjoints 37
1.6 The Curry–Howard Correspondence 51
1.7 Linearity 68
1.8 Monads and Comonads 79
References 94

2 Physics, Topology, Logic and Computation: A Rosetta Stone 95
J. Baez and M. Stay
2.1 Introduction 95
2.2 The Analogy Between Physics and Topology 98
2.3 Logic 132
2.4 Computation 149
2.5 Conclusions 166
References 168

3 Categories for the Practising Physicist 173
B. Coecke and É.O. Paquette
3.1 Prologue: Cooking with Vegetables 173
3.2 The 1D Case: New Arrows for Your Quiver 177
3.3 The 2D Case: Muscle Power 194
3.4 Quantum-Like Tensors 217
3.5 Classical-Like Tensors 232

3.6 Monoidal Functoriality, Naturality and TQFTs 271
3.7 Further Reading 281
References 283

Part II Manifestations of Linearity

4 A Survey of Graphical Languages for Monoidal Categories 289
P. Selinger
4.1 Introduction 289
4.2 Categories 292
4.3 Monoidal Categories 296
4.4 Autonomous Categories 307
4.5 Traced Categories 322
4.6 Products, Coproducts, and Biproducts 332
4.7 Dagger Categories 340
4.8 Bicategories 350
4.9 Beyond a Single Tensor Product 351
4.10 Summary 352
References 354

5 Geometry of Interaction and the Dynamics of Proof Reduction: A Tutorial 357
E. Haghverdi and P. Scott
5.1 Introduction 357
5.2 From Monoidal Categories to *-Autonomy 359
5.3 Linear Logic and Categorical Proof Theory 366
5.4 Traced Monoidal Categories 375
5.5 What is the Geometry of Interaction? 385
5.6 GoI Interpretation of MELL 389
5.7 Partial Trace and Abstract Orthogonality 398
5.8 Typed GoI for MELL in *-Categories 403
5.9 Concluding Remarks 412
Appendix 1: Graphical Representation of The Trace Axioms 412
Appendix 2: Comparing GoI Notation 414
References 414

Part III More Example Applications

6 Dagger Categories and Formal Distributions 421
R. Blute and P. Panangaden
6.1 Introduction 421
6.2 Dagger Categories and Nuclear Ideals 424
6.3 Distributions as Relations 425

6.4 Categories of Formal Distributions 428
6.5 Vertex Groups and Categories 433
6.6 Conclusion 435
References 435

7 Proof Nets as Formal Feynman Diagrams 437
R. Blute and P. Panangaden
7.1 Introduction 437
7.2 Functional Integrals in Quantum Field Theory 438
7.3 Linear Realizability Algebra 442
7.4 The ϕ-Calculus 445
7.5 Exponential Identities for Operators 453
7.6 Interpreting Proof Nets 456
7.7 Example Calculations 459
7.8 Conclusions 463
References 465

8 Compact Monoidal Categories from Linguistics to Physics 467
J. Lambek
8.1 Compact 2-Categories and Pregroups 468
8.2 Pregroups for Grammar 470
8.3 Free Compact 2-Categories 471
8.4 In Search of a Compact Feynman Category 475
8.5 A Pogroup for *QED* 477
8.6 From *QED* to the Standard Model 479
8.7 From 2-Categories to Bicategories 481
8.8 Other Operations in Bilinear Logic 485
8.9 Postscript 486
References 487

Part IV Informatic Geometry

9 Domain Theory and Measurement 491
K. Martin
9.1 Introduction 491
9.2 The Basic Elements 494
9.3 Fixed Points 508
9.4 Instances of Partiality 524
9.5 The Informatic Derivative 539
9.6 Forms of Process Evolution 559
9.7 Provocation 587
References 589

10 A Partial Order on Classical and Quantum States 593
B. Coecke and K. Martin
10.1 Introduction 593
10.2 Classical States 595
10.3 Quantum States 630
10.4 Synthesis 655
10.5 Applications 672
References 682

Part V Spatio-Temporal Geometry

11 Domain Theory and General Relativity 687
K. Martin and P. Panangaden
11.1 Introduction 687
11.2 Domains, Continuous Posets and Topology 689
11.3 The Causal Structure of Spacetime 691
11.4 Global Hyperbolicity 692
11.5 Spacetime from a Discrete Causal Set 694
11.6 Spacetime as a Domain 695
11.7 Time and Measurement 699
11.8 Spacetime Geometry from a Discrete Causal Set 701
11.9 Conclusions 702
References 703

12 Process, Distinction, Groupoids and Clifford Algebras: an Alternative View of the Quantum Formalism 705
B.J. Hiley
12.1 The Algebra of Process 705
12.2 Some Specific Algebras of Process 709
12.3 Connections with Other Mathematical Approaches 715
12.4 Some Radical New Ideas 719
12.5 The Conformal Clifford $\mathcal{C}_{2,4}$ 731
12.6 Connections in the Clifford Bundle 736
12.7 Expectation Values 740
12.8 The Symplectic Group 744
12.9 General Conclusions 749
References 750

13 "What is a Thing?": Topos Theory in the Foundations of Physics 753
A. Döring and C. Isham
13.1 Introduction 754
13.2 The Conceptual Background of Our Scheme 762
13.3 Propositional Languages and Theories of Physics 770

13.4 A Higher-Order, Typed Language for Physics 778
13.5 Quantum Propositions as Sub-objects of the Spectral Presheaf 788
13.6 Truth Values in Topos Physics 809
13.7 The de Groote Presheaves of Physical Quantities 827
13.8 The Presheaves $\underline{\mathrm{sp}(\hat{A})^{\succeq}}$, $\underline{\mathbb{R}^{\succeq}}$ and $\underline{\mathbb{R}^{\leftrightarrow}}$ 837
13.9 Extending the Quantity-Value Presheaf to an Abelian Group Object 858
13.10 The Role of Unitary Operators 863
13.11 The Category of Systems 871
13.12 Theories of Physics in a General Topos 886
13.13 The General Scheme Applied to Quantum Theory 894
13.14 Characteristic Properties of Σ_ϕ, $\mathcal{R}_\phi$ and $\mathbb{T}$, $\mathfrak{w}$ 904
13.15 Conclusion 911
Appendix 1: Some Theorems and Constructions Used in the Main Text .. 916
Appendix 2: A Short Introduction to the Relevant Parts of Topos Theory . 925
References 934

Part VI Geometry and Topology in Computation

14 Can a Quantum Computer Run the von Neumann Architecture? ... 941
P. Hines
14.1 Introduction 941
14.2 The von Neumann Architecture 942
14.3 Relevant Quantum Information Theory 944
14.4 Data/Code Interchangeability, and Evaluation 948
14.5 Evaluation in the One-Bit Computer 950
14.6 Implementing *Evaluation* by Unitary Operations? 950
14.7 Evaluation as Currying 953
14.8 Basic Category Theory 955
14.9 Categorical Closure and Hilbert Spaces 960
14.10 Abramsky and Coecke's Categorical Foundations for Quantum Mechanics 964
14.11 Evaluation by Teleportation, and the vN Architecture 966
14.12 Naming an Unknown Arrow 970
14.13 Other Aspects 972
Appendix 976
References 978

15 A Categorical Presentation of Quantum Computation with Anyons .. 983
P. Panangaden and É.O. Paquette
15.1 Introduction 983
15.2 Spin and Statistics 987
15.3 Anyons and Braids 989

15.4 The Algebra of Anyons: Modular Tensor Categories 992
15.5 An Example: Fibonacci Anyons 1014
15.6 Universal Quantum Computation with Fibonacci Anyons 1020
15.7 Conclusions 1022
References 1023

Index 1027

Part I
An ABC on Compositionality

Chapter 1
Introduction to Categories and Categorical Logic

S. Abramsky and N. Tzevelekos

Abstract The aim of these notes is to provide a succinct, accessible introduction to some of the basic ideas of category theory and categorical logic. The notes are based on a lecture course given at Oxford over the past few years. They contain numerous exercises, and hopefully will prove useful for self-study by those seeking a first introduction to the subject, with fairly minimal prerequisites. The coverage is by no means comprehensive, but should provide a good basis for further study; a guide to further reading is included.

The main prerequisite is a basic familiarity with the elements of discrete mathematics: sets, relations and functions. An Appendix contains a summary of what we will need, and it may be useful to review this first. In addition, some prior exposure to abstract algebra—vector spaces and linear maps, or groups and group homomorphisms—would be helpful.

1.1 Introduction

Why study categories—what are they good for? We can offer a range of answers for readers coming from different backgrounds:

- For **mathematicians**: category theory organises your previous mathematical experience in a new and powerful way, revealing new connections and structure, and allows you to "think bigger thoughts".
- For **computer scientists**: category theory gives a precise handle on important notions such as compositionality, abstraction, representation-independence, genericity and more. Otherwise put, it provides the fundamental mathematical structures underpinning many key programming concepts.

S. Abramsky (✉)
OUCL, University of Oxford, Oxford, UK
e-mail: samson@comlab.ox.ac.uk

N. Tzevelekos (✉)
OUCL, University of Oxford, Oxford, UK
e-mail: nikt@comlab.ox.ac.uk

Abramsky, S., Tzevelekos, N.: *Introduction to Categories and Categorical Logic*. Lect. Notes Phys. **813**, 3–94 (2011)
DOI 10.1007/978-3-642-12821-9_1

- For **logicians**: category theory gives a syntax-independent view of the fundamental structures of logic, and opens up new kinds of models and interpretations.
- For **philosophers**: category theory opens up a fresh approach to structuralist foundations of mathematics and science; and an alternative to the traditional focus on set theory.
- For **physicists**: category theory offers new ways of formulating physical theories in a structural form. There have *inter alia* been some striking recent applications to quantum information and computation.

1.1.1 From Elements To Arrows

Category theory can be seen as a "generalised theory of functions", where the focus is shifted from the pointwise, set-theoretic view of functions, to an abstract view of functions as *arrows*.

Let us briefly recall the arrow notation for functions between sets.[1] A function f with *domain* X and *codomain* Y is denoted by: $f : X \to Y$.

$$\text{Diagrammatic notation: } X \xrightarrow{f} Y\,.$$

The fundamental operation on functions is *composition*: if $f : X \to Y$ and $g : Y \to Z$, then we can define $g \circ f : X \to Z$ by $g \circ f(x) := g(f(x))$.[2] Note that, in order for the composition to be defined, the codomain of f must be the same as the domain of g.

$$\text{Diagrammatic notation: } X \xrightarrow{f} Y \xrightarrow{g} Z\,.$$

Moreover, for each set X there is an *identity function* on X, which is denoted by:

$$\mathsf{id}_X : X \longrightarrow X \qquad \mathsf{id}_X(x) := x\,.$$

These operations are governed by the *associativity law* and the *unit laws*. For $f : X \to Y$, $g : Y \to Z$, $h : Z \to W$:

$$(h \circ g) \circ f = h \circ (g \circ f)\,, \qquad f \circ \mathsf{id}_X = f = \mathsf{id}_Y \circ f\,.$$

Notice that these equations are formulated purely in terms of the algebraic operations on functions, without any reference to the elements of the sets X, Y, Z, W. We will refer to any concept pertaining to functions which can be defined purely in terms of composition and identities as *arrow-theoretic*. We will now take a first

[1] A review of basic ideas about sets, functions and relations, and some of the notation we will be using, is provided in Appendix A.

[2] We shall use the notation ":=" for "is defined to be" throughout these notes.

step towards learning to "think with arrows" by seeing how we can replace some familiar definitions couched in terms of elements by arrow-theoretic equivalents; this will lead us towards the notion of category.

We say that a function $f : X \longrightarrow Y$ is:

$$\begin{array}{ll} \textit{injective} & \text{if } \forall x, x' \in X.\, f(x) = f(x') \implies x = x', \\ \textit{surjective} & \text{if } \forall y \in Y.\, \exists x \in X.\, f(x) = y, \end{array}$$

$$\begin{array}{ll} \textit{monic} & \text{if } \forall g, h.\, f \circ g = f \circ h \implies g = h, \\ \textit{epic} & \text{if } \forall g, h.\, g \circ f = h \circ f \implies g = h. \end{array}$$

Note that injectivity and surjectivity are formulated in terms of elements, while epic and monic are arrow-theoretic.

Proposition 1 *Let $f : X \to Y$. Then,*

1. *f is injective iff f is monic.*
2. *f is surjective iff f is epic.*

Proof We show 1. Suppose $f : X \to Y$ is injective, and that $f \circ g = f \circ h$, where $g, h : Z \to X$. Then, for all $z \in Z$:

$$f(g(z)) = f \circ g(z) = f \circ h(z) = f(h(z)).$$

Since f is injective, this implies $g(z) = h(z)$. Hence we have shown that

$$\forall z \in Z.\, g(z) = h(z),$$

and so we can conclude that $g = h$. So f injective implies f monic.
For the converse, fix a one-element set $\mathbf{1} = \{\bullet\}$. Note that elements $x \in X$ are in 1–1 correspondence with functions $\bar{x} : \mathbf{1} \to X$, where $\bar{x}(\bullet) := x$. Moreover, if $f(x) = y$ then $\bar{y} = f \circ \bar{x}$. Writing injectivity in these terms, it amounts to the following.

$$\forall x, x' \in X.\, f \circ \bar{x} = f \circ \bar{x}' \implies \bar{x} = \bar{x}'$$

Thus we see that being injective is a *special case* of being monic. ∎

Exercise 2 Show that $f : X \to Y$ is surjective iff it is epic.

1.1.2 Categories Defined

Definition 3 A ***category*** $\mathcal{C}$ consists of:

- A collection $\mathsf{Ob}(\mathcal{C})$ of ***objects***. Objects are denoted by A, B, C, etc.
- A collection $\mathsf{Ar}(\mathcal{C})$ of ***arrows*** (or ***morphisms***). Arrows are denoted by f, g, h, etc.

- Mappings $\mathsf{dom}, \mathsf{cod} : \mathsf{Ar}(\mathcal{C}) \to \mathsf{Ob}(\mathcal{C})$, which assign to each arrow f its ***domain*** $\mathsf{dom}(f)$ and its ***codomain*** $\mathsf{cod}(f)$. An arrow f with domain A and codomain B is written $f : A \to B$. For each pair of objects A, B, we define the set

$$\mathcal{C}(A, B) := \{f \in \mathsf{Ar}(\mathcal{C}) \mid f : A \to B\}\,.$$

  We refer to $\mathcal{C}(A, B)$ as a ***hom-set***. Note that distinct hom-sets are *disjoint*.
- For any triple of objects A, B, C, a ***composition*** map

$$c_{A,B,C} : \mathcal{C}(A, B) \times \mathcal{C}(B, C) \longrightarrow \mathcal{C}(A, C)\,.$$

  $c_{A,B,C}(f, g)$ is written $g \circ f$ (or sometimes $f; g$). Diagrammatically:

$$A \xrightarrow{f} B \xrightarrow{g} C$$

- For each object A, an ***identity*** arrow $\mathsf{id}_A : A \to A$.

The above must satisfy the following axioms.

$$h \circ (g \circ f) = (h \circ g) \circ f\,, \qquad f \circ \mathsf{id}_A = f = \mathsf{id}_B \circ f\,.$$

whenever the domains and codomains of the arrows match appropriately so that the compositions are well-defined. ▲

1.1.3 Diagrams in Categories

Diagrammatic reasoning is an important tool in category theory. The basic cases are commuting triangles and squares. To say that the following triangle commutes

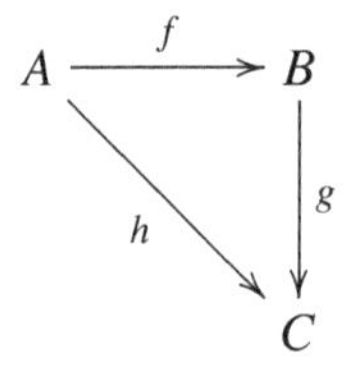

is exactly equivalent to asserting the equation $g \circ f = h$. Similarly, to say that the following square commutes

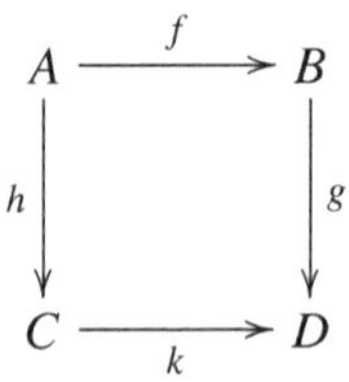

means exactly that $g \circ f = k \circ h$. For example, the equations

$$h \circ (g \circ f) = (h \circ g) \circ f\,, \qquad f \circ \mathsf{id}_A = f = \mathsf{id}_B \circ f\,,$$

can be expressed by saying that the following diagrams commute.

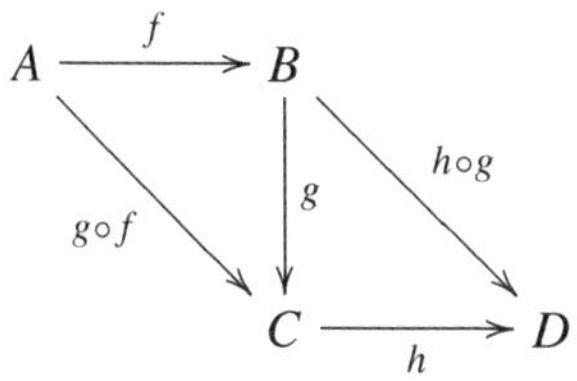

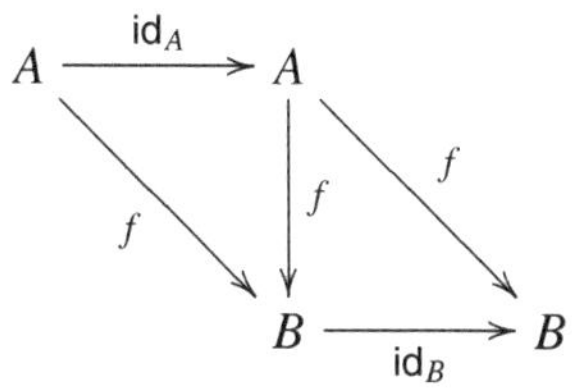

As these examples illustrate, most of the diagrams we shall use will be "pasted together" from triangles and squares: the commutation of the diagram as a whole will then reduce to the commutation of the constituent triangles and squares.

We turn to the general case. The formal definition is slightly cumbersome; we give it anyway for reference.

Definition 4 We define a ***graph*** to be a collection of ***vertices*** and ***directed edges***, where each edge $e : v \to w$ has a specified source vertex v and target vertex w. Thus graphs are like categories without composition and identities.[3] A ***diagram in a category*** $\mathcal{C}$ is a graph whose vertices are labelled with objects of $\mathcal{C}$ and whose edges are labelled with arrows of $\mathcal{C}$, such that, if $e : v \to w$ is labelled with $f : A \to B$, then we must have v labelled by A and w labelled by B. We say that such a diagram *commutes* if any two paths in it with common source and target, and at least one of which has length greater than 1, are equal. That is, given paths

$$A \xrightarrow{f_1} C_1 \xrightarrow{f_2} \cdots C_{n-1} \xrightarrow{f_n} B \qquad \text{and} \qquad A \xrightarrow{g_1} D_1 \xrightarrow{g_2} \cdots D_{m-1} \xrightarrow{g_m} B,$$

if $\max(n, m) > 1$ then

$$f_n \circ \cdots \circ f_1 = g_m \circ \cdots \circ g_1\,.$$

▲

To illustrate this definition, to say that the following diagram commutes

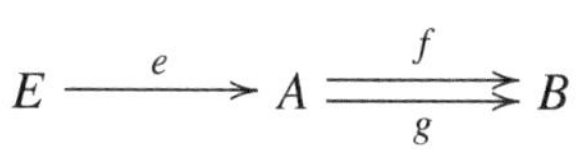

amounts to the assertion that $f \circ e = g \circ e$; it does *not* imply that $f = g$.

---

[3] This would be a "multigraph" in normal parlance, since multiple edges between a given pair of vertices are allowed.

1.1.4 Examples

Before we proceed to our first examples of categories, we shall present some background material on partial orders, monoids and topologies, which will provide running examples throughout these notes.

Partial orders

A *partial order* is a structure $(P, \leq)$ where P is a set and $\leq$ is a binary relation on P satisfying:

- $x \leq x$ (Reflexivity)
- $x \leq y \wedge y \leq x \;\Rightarrow\; x = y$ (Antisymmetry)
- $x \leq y \wedge y \leq z \;\Rightarrow\; x \leq z$ (Transitivity)

For example, $(\mathbb{R}, \leq)$ and $(\mathcal{P}(X), \subseteq)$ are partial orders, and so are strings with the sub-string relation.

If P, Q are partial orders, a map $h : P \to Q$ is a *partial order homomorphism* (or *monotone function*) if:

$$\forall x, y \in P.\, x \leq y \Longrightarrow h(x) \leq h(y)\,.$$

Note that homomorphisms are closed under composition, and that identity maps are homomorphisms.

Monoids

A *monoid* is a structure $(M, \cdot, 1)$ where M is a set,

$$\_\cdot\_ : M \times M \longrightarrow M$$

is a binary operation, and $1 \in M$, satisfying the following axioms.

$$(x \cdot y) \cdot z = x \cdot (y \cdot z)\,, \qquad 1 \cdot x = x = x \cdot 1\,.$$

For example, $(\mathbb{N}, +, 0)$ is a monoid, and so are strings with string-concatenation. Moreover, groups are special kinds of monoids.

If M, N are monoids, a map $h : M \to N$ is a *monoid homomorphism* if

$$\forall m_1, m_2 \in M.\, h(m_1 \cdot m_2) = h(m_1) \cdot h(m_2)\,, \qquad h(1) = 1\,.$$

Exercise 5 Suppose that G and H are groups (and hence monoids), and that $h : G \to H$ is a monoid homomorphism. Prove that h is a group homomorphism.

Topological spaces

A topological space is a pair (X, T_X) where X is a set, and T_X is a family of subsets of X such that

- $\varnothing, X \in T_X$,
- if $U, V \in T_X$ then $U \cap V \in T_X$,
- if $\{U_i\}_{i\in I}$ is any family in T_X, then $\bigcup_{i\in I} U_i \in T_X$.

A *continuous map* $f : (X, T_X) \to (Y, T_Y)$ is a function $f : X \to Y$ such that, for all $U \in T_Y$, $f^{-1}(U) \in T_X$.
Let us now see some first examples of categories.

- Any kind of mathematical structure, together with structure preserving functions, forms a category. E.g.
  - **Set** (sets and functions)
  - **Mon** (monoids and monoid homomorphisms)
  - **Grp** (groups and group homomorphisms)
  - **Vect**$_k$ (vector spaces over a field k, and linear maps)
  - **Pos** (partially ordered sets and monotone functions)
  - **Top** (topological spaces and continuous functions)
- **Rel**: objects are sets, arrows $R : X \to Y$ are *relations* $R \subseteq X \times Y$. Relational composition:

$$R; S(x, z) \iff \exists y.\, R(x, y) \wedge S(y, z)$$

- Let k be a field (for example, the real or complex numbers). Consider the following category **Mat**$_k$. The objects are natural numbers. A morphism $M : \mathrm{n} \to \mathrm{m}$ is an $\mathrm{n} \times \mathrm{m}$ matrix with entries in k. Composition is matrix multiplication, and the identity on n is the $\mathrm{n} \times \mathrm{n}$ diagonal matrix.
- ◇ Monoids are one-object categories. Arrows correspond to the elements of the monoid, with the monoid operation being arrow-composition and the monoid unit being the identity arrow.
- ◇ A category in which for each pair of objects A, B there is at most one morphism from A to B is the same thing as a ***preorder***, *i.e.* a reflexive and transitive relation.

Note that our first class of examples illustrate the idea of categories as *mathematical contexts*; settings in which various mathematical theories can be developed. Thus for example, **Top** is the context for general topology, **Grp** is the context for group theory, etc.

On the other hand, the last two examples illustrate that many important mathematical structures themselves appear as categories of particular kinds. The fact that two such different kinds of structures as monoids and posets should appear as extremal versions of categories is also rather striking.

This ability to capture mathematics both "in the large" and "in the small" is a first indication of the flexibility and power of categories.

Exercise 6 Check that **Mon**, $\mathbf{Vect}_k$, **Pos** and **Top** are indeed categories.

Exercise 7 Check carefully that monoids correspond exactly to one-object categories. Make sure you understand the difference between such a category and **Mon**. (For example: how many objects does **Mon** have?)

Exercise 8 Check carefully that preorders correspond exactly to categories in which each homset has at most one element. Make sure you understand the difference between such a category and **Pos**. (For example: how big can homsets in **Pos** be?)

1.1.5 First Notions

Many important mathematical notions can be expressed at the general level of categories.

Definition 9 Let $\mathcal{C}$ be a category. A morphism $f : X \to Y$ in $\mathcal{C}$ is:

- ***monic*** (or a ***monomorphism***) if $f \circ g = f \circ h \implies g = h$,
- ***epic*** (or an ***epimorphism***) if $g \circ f = h \circ f \implies g = h$.

An ***isomorphism*** in $\mathcal{C}$ is an arrow $i : A \to B$ such that there exists an arrow $j : B \to A$—the ***inverse*** of i—satisfying

$$j \circ i = \mathsf{id}_A , \qquad i \circ j = \mathsf{id}_B .$$

▲

We denote isomorphisms by $i : A \xrightarrow{\cong} B$, and write i^{-1} for the inverse of i. We say that A and B ***are isomorphic***, $A \cong B$, if there exists some $i : A \xrightarrow{\cong} B$.

Exercise 10 Show that the inverse, if it exists, is unique.

Exercise 11 Show that $\cong$ is an equivalence relation on the objects of a category.

As we saw previously, in **Set** monics are injections and epics are surjections. On the other hand, isomorphisms in **Set** correspond exactly to bijections, in **Grp** to group isomorphisms, in **Top** to homeomorphisms, in **Pos** to order isomorphisms, etc.

Exercise 12 Verify these claims.

Thus we have at one stroke captured the key notion of isomorphism in a form which applies to *all* mathematical contexts. This is a first taste of the level of generality which category theory naturally affords.

We have already identified monoids as one-object categories. We can now identify groups as *exactly those one-object categories in which every arrow is an isomorphism*. This also leads to a natural generalisation, of considerable importance in current mathematics: a ***groupoid*** is a category in which every morphism is an isomorphism.

Opposite Categories and Duality

The directionality of arrows within a category $\mathcal{C}$ can be reversed without breaking the conditions of being a category; this yields the notion of ***opposite category***.

Definition 13 Given a category $\mathcal{C}$, ***the opposite category*** $\mathcal{C}^{\mathsf{op}}$ is given by taking the same objects as $\mathcal{C}$, and

$$\mathcal{C}^{\mathsf{op}}(A, B) := \mathcal{C}(B, A).$$

Composition and identities are inherited from $\mathcal{C}$. ▲

Note that if we have

$$A \xrightarrow{f} B \xrightarrow{g} C$$

in $\mathcal{C}^{\mathsf{op}}$, this means

$$A \xleftarrow{f} B \xleftarrow{g} C$$

in $\mathcal{C}$, so composition $g \circ f$ in $\mathcal{C}^{\mathsf{op}}$ is defined as $f \circ g$ in $\mathcal{C}$!

Consideration of opposite categories leads to a ***principle of duality***: a statement S is true about $\mathcal{C}$ if and only if its dual (*i.e.* the one obtained from S by reversing all the arrows) is true about $\mathcal{C}^{\mathsf{op}}$. For example,

A morphism f is monic in $\mathcal{C}^{\mathsf{op}}$ if and only if it is epic in $\mathcal{C}$.

Indeed, f is monic in $\mathcal{C}^{\mathsf{op}}$ iff for all $g, h : C \to B$ in $\mathcal{C}^{\mathsf{op}}$,

$$f \circ g = f \circ h \implies g = h,$$

iff for all $g, h : B \to C$ in $\mathcal{C}$,

$$g \circ f = h \circ f \implies g = h,$$

iff f is epic in $\mathcal{C}$. We say that monic and epic are *dual notions*.

Exercise 14 If P is a preorder, for example $(\mathbb{R}, \leq)$, describe P^{op} explicitly.

Subcategories

Another way to obtain new categories from old ones is by restricting their objects or arrows.

Definition 15 Let $\mathcal{C}$ be a category. Suppose that we are given collections

$$\mathsf{Ob}(\mathcal{D}) \subseteq \mathsf{Ob}(\mathcal{C}), \quad \forall A, B \in \mathsf{Ob}(\mathcal{D}). \mathcal{D}(A, B) \subseteq \mathcal{C}(A, B).$$

We say that $\mathcal{D}$ is a ***subcategory*** of $\mathcal{C}$ if

$$A \in \mathsf{Ob}(\mathcal{D}) \Rightarrow \mathsf{id}_A \in \mathcal{D}(A, A), \quad f \in \mathcal{D}(A, B), g \in \mathcal{D}(B, C) \Rightarrow g \circ f \in \mathcal{D}(A, C),$$

and hence $\mathcal{D}$ itself is a category. In particular, $\mathcal{D}$ is:

- A ***full*** subcategory of $\mathcal{C}$ if for any $A, B \in \mathsf{Ob}(\mathcal{D})$, $\mathcal{D}(A, B) = \mathcal{C}(A, B)$.
- A ***lluf*** subcategory of $\mathcal{C}$ if $\mathsf{Ob}(\mathcal{D}) = \mathsf{Ob}(\mathcal{C})$. ▲

For example, **Grp** is a full subcategory of **Mon** (by Exercise 5), and **Set** is a lluf subcategory of **Rel**.

Simple cats

We close this section with some very basic examples of categories.

- $\mathbb{1}$ is the category with one object and one arrow, that is,

$$\mathbb{1} := \bullet \circlearrowleft$$

  where the arrow is necessarily $\mathsf{id}_\bullet$. Note that, although we say that $\mathbb{1}$ is *the* one-object/one-arrow category, there is by no means a *unique* such category. This is explained by the intuitively evident fact that any two such categories are isomorphic. (We will define what it means for categories to be isomorphic later.)
- In two-object categories, there is the one with two arrows, $\mathbb{2} := \bullet \quad \bullet$, and also:

$$\mathbb{2}_\rightarrow := \bullet \longrightarrow \bullet \quad , \quad \mathbb{2}_\rightrightarrows := \bullet \rightrightarrows \bullet \quad , \quad \mathbb{2}_\rightleftarrows := \bullet \rightleftarrows \bullet \quad \ldots$$

  Note that we have omitted identity arrows for economy. Categories with only identity arrows, like $\mathbb{1}$ and $\mathbb{2}$, are called *discrete categories*.

Exercise 16 How many categories $\mathcal{C}$ with $\mathsf{Ob}(\mathcal{C}) = \{\bullet\}$ are there? (Hint: what do such categories correspond to?)

1.1.6 Exercises

1. Consider the following properties of an arrow f in a category $\mathcal{C}$.

   - f is *split monic* if for some g, $g \circ f$ is an identity arrow.
   - f is *split epic* if for some g, $f \circ g$ is an identity arrow.

   (a) Prove that if f and g are arrows such that $g \circ f$ is monic, then f is monic.
   (b) Prove that, if f is split epic then it is epic.
   (c) Prove that, if f and $g \circ f$ are iso then g is iso.
   (d) Prove that, if f is monic and split epic then it is iso.

(e) In the category **Mon** of monoids and monoid homomorphisms, consider the inclusion map

$$i : (\mathbb{N}, +, 0) \longrightarrow (\mathbb{Z}, +, 0)$$

of natural numbers into the integers. Show that this arrow is both monic and epic. Is it an iso?

The **Axiom of Choice** in Set Theory states that, if $\{X_i\}_{i \in I}$ is a family of non-empty sets, we can form a set $X = \{x_i \mid i \in I\}$ where $x_i \in X_i$ for all $i \in I$.

(f) Show that in **Set** an arrow which is epic is split epic. Explain why this needs the Axiom of Choice.
(g) Is it always the case that an arrow which is epic is split epic? Either prove that it is, or give a counter-example.

2. Give a description of partial orders as categories of a special kind.

1.2 Some Basic Constructions

We shall now look at a number of basic constructions which appear throughout mathematics, and which acquire their proper general form in the language of categories.

1.2.1 Initial and Terminal Objects

A first such example is that of initial and terminal objects. While apparently trivial, they are actually both important and useful, as we shall see in the sequel.

Definition 17 An object I in a category $\mathcal{C}$ is ***initial*** if, for every object A, *there exists a unique arrow* from I to A, which we write $\iota_A : I \to A$.
A ***terminal*** object in $\mathcal{C}$ is an object T such that, for every object A, *there exists a unique arrow* from A to T, which we write $\tau_A : A \to T$. ▲

Note that initial and terminal objects are dual notions: T is terminal in $\mathcal{C}$ iff it is initial in $\mathcal{C}^{\mathsf{op}}$. We sometimes write **1** for the terminal object and **0** for the initial one. Note also the assertions of *unique existence* in the definitions. This is one of the *leitmotifs* of category theory; we shall encounter it again in a conceptually deeper form in Sect. 1.5.

Let us examine initial and terminal objects in our standard example categories.

- In **Set**, the empty set is an initial object while any one-element set $\{\bullet\}$ is terminal.
- In **Pos**, the poset $(\varnothing, \varnothing)$ is an initial object while $(\{\bullet\}, \{(\bullet, \bullet)\})$ is terminal.
- In **Top**, the space $(\varnothing, \{\varnothing\})$ is an initial object while $(\{\bullet\}, \{\varnothing, \{\bullet\}\})$ is terminal.
- In **Vect**$_k$, the one-element space $\{0\}$ is both initial and terminal.
- In a poset, seen as a category, an initial object is a least element, while a terminal object is a greatest element.

Exercise 18 Verify these claims. In each case, identify the canonical arrows.

Exercise 19 Identify the initial and terminal objects in **Rel**.

Exercise 20 Suppose that a monoid, viewed as a category, has either an initial or a terminal object. What must the monoid be?

We shall now establish a fundamental fact: initial and terminal objects are *unique up to (unique) isomorphism*. As we shall see, this is characteristic of all such "universal" definitions. For example, the apparent arbitrariness in the fact that any singleton set is a terminal object in **Set** is answered by the fact that what counts is the property of being terminal; and this suffices to ensure that any two concrete objects having this property must be isomorphic to each other.

The proof of the proposition, while elementary, is a first example of distinctively categorical reasoning.

Proposition 21 *If I and I' are initial objects in the category $\mathcal{C}$ then there exists a unique isomorphism $I \xrightarrow{\cong} I'$.*

Proof Since I is initial and I' is an object of $\mathcal{C}$, there is a unique arrow $\iota_{I'} : I \to I'$. We claim that $\iota_{I'}$ is an isomorphism.
Since I' is initial and I is an object in $\mathcal{C}$, there is an arrow $\iota'_I : I' \to I$. Thus we obtain $\iota_{I'}; \iota'_I : I \to I$, while we also have the identity morphism $\mathsf{id}_I : I \to I$. But I is initial and therefore there exists a *unique* arrow from I to I, which means that $\iota_{I'}; \iota'_I = \mathsf{id}_I$. Similarly, $\iota'_I; \iota_{I'} = \mathsf{id}_{I'}$, so $\iota_{I'}$ is indeed an isomorphism. ■

Hence, initial objects are "unique up to (unique) isomorphism", and we can (and do) speak of *the* initial object (if any such exists). Similarly for terminal objects.

Exercise 22 Let $\mathcal{C}$ be a category with an initial object $\mathbf{0}$. For any object A, show the following.

- If $A \cong \mathbf{0}$ then A is an initial object.
- If there exists a monomorphism $f : A \to \mathbf{0}$ then f is an iso, and hence A is initial.

1.2.2 Products and Coproducts

1.2.2.1 Products

We now consider one of the most common constructions in mathematics: the formation of "direct products". Once again, rather than giving a case-by-case construction of direct products in each mathematical context we encounter, we can express once and for all a general notion of product, meaningful in any category—and such that, if a product exists, it is characterised uniquely up to unique isomorphism, just as for initial and terminal objects. Given a particular mathematical context, *i.e.* a category, we can then verify whether on not the product exists in that category. The concrete construction appropriate to the context will enter only into the proof of *existence*;

all of the useful *properties* of the product follow from the general definition. Moreover, the categorical notion of product has a *normative* force; we can test whether a concrete construction works as intended by verifying that it satisfies the general definition.

In set theory, the cartesian product is defined in terms of the ordered pair:

$$X \times Y := \{(x, y) \mid x \in X \wedge y \in Y\}.$$

It turns out that ordered pairs can be *defined* in set theory, e.g. as

$$(x, y) := \{\{x, y\}, y\}.$$

Note that in no sense is such a definition canonical. The essential *properties* of ordered pairs are:

1. We can retrieve the first and second components x, y of the ordered pair (x, y), allowing *projection functions* to be defined:

$$\pi_1 : (x, y) \mapsto x, \qquad \pi_2 : (x, y) \mapsto y\,.$$

2. The information about first and second components completely determines the ordered pair:

$$(x_1, x_2) = (y_1, y_2) \iff x_1 = y_1 \wedge x_2 = y_2.$$

The categorical definition expresses these properties in arrow-theoretic terms, meaningful in any category.

Definition 23 Let A, B be objects in a category $\mathcal{C}$. An A,B–pairing is a triple (P, p_1, p_2) where P is an object, $p_1 : P \to A$ and $p_2 : P \to B$. A morphism of A,B–pairings

$$f : (P, p_1, p_2) \longrightarrow (Q, q_1, q_2)$$

is a morphism $f : P \to Q$ in $\mathcal{C}$ such that $q_1 \circ f = p_1$ and $q_2 \circ f = p_2$, *i.e.* the following diagram commutes.

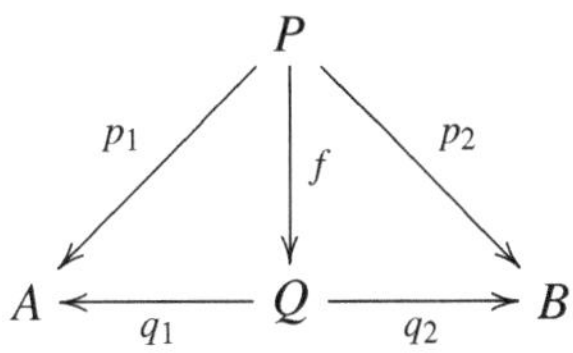

The A,B–pairings form a category $\mathbf{Pair}(A, B)$. We say that $(A \times B, \pi_1, \pi_2)$ is a ***product*** of A and B if it is *terminal* in $\mathbf{Pair}(A, B)$. ▲

Exercise 24 Verify that $\mathbf{Pair}(A, B)$ is a category.

Note that products are specified by triples $A \xleftarrow{\pi_1} A \times B \xrightarrow{\pi_2} B$, where π_i's are called *projections*. For economy (and if projections are obvious) we may say that $A \times B$ is the product of A and B. We say that $\mathcal{C}$ ***has (binary) products*** if each pair of objects A, B has a product in $\mathcal{C}$. A direct consequence of the definition, by Proposition 21, is that if products exist, they are unique up to (unique) isomorphism.

Unpacking the uniqueness condition from $\mathbf{Pair}(A, B)$ back to $\mathcal{C}$ we obtain a more concise definition of products which we use in practice.

Definition 25 (Equivalent definition of product) Let A, B be objects in a category $\mathcal{C}$. A product of A and B is an object $A \times B$ together with a pair of arrows $A \xleftarrow{\pi_1} A \times B \xrightarrow{\pi_2} B$ such that for every triple $A \xleftarrow{f} C \xrightarrow{g} B$ there exists a *unique* morphism

$$\langle f, g \rangle : C \longrightarrow A \times B$$

such that the following diagram commutes.

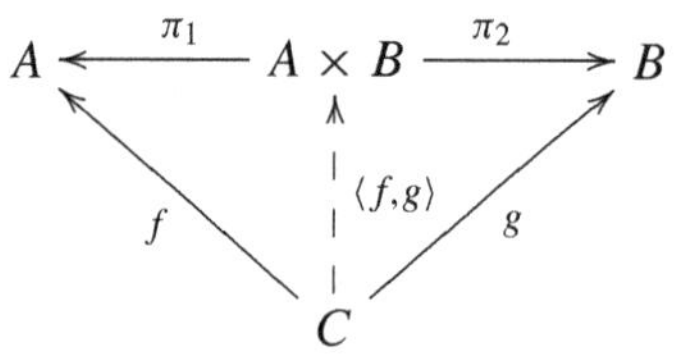

$$\left(\begin{array}{l} \pi_1 \circ \langle f, g \rangle = f \\ \pi_2 \circ \langle f, g \rangle = g \end{array} \right)$$

▲

We call $\langle f, g \rangle$ the *pairing* of f and g.

Note that the above diagram features a *dashed arrow*. Our intention with such diagrams is always to express the following idea: if the undashed part of the diagram commutes, then *there exists a unique arrow* (the dashed one) such that the whole diagram commutes. In any case, we shall always spell out the intended statement explicitly.

We look at how this definition works in our standard example categories.

- In **Set**, products are the usual cartesian products.
- In **Pos**, products are cartesian products with the pointwise order.
- In **Top**, products are cartesian products with the product topology.
- In $\mathbf{Vect}_k$, products are direct sums.
- In a poset, seen as a category, products are *greatest lower bounds*.

Exercise 26 Verify these claims.

The following proposition shows that the uniqueness of the pairing arrow can be specified purely equationally, by the equation:

$$\forall h : C \to A \times B.\ h = \langle \pi_1 \circ h, \pi_2 \circ h \rangle$$

Proposition 27 *For any triple* $A \xleftarrow{\pi_1} A \times B \xrightarrow{\pi_2} B$ *the following statements are equivalent.*

(I) *For any triple* $A \xleftarrow{f} C \xrightarrow{g} B$ *there exists a unique morphism* $\langle f, g\rangle : C \to A \times B$ *such that* $\pi_1 \circ \langle f, g\rangle = f$ *and* $\pi_2 \circ \langle f, g\rangle = g$.

(II) *For any triple* $A \xleftarrow{f} C \xrightarrow{g} B$ *there exists a morphism* $\langle f, g\rangle : C \to A \times B$ *such that* $\pi_1 \circ \langle f, g\rangle = f$ *and* $\pi_2 \circ \langle f, g\rangle = g$, *and moreover, for any* $h : C \to A \times B$, $h = \langle \pi_1 \circ h, \pi_2 \circ h\rangle$.

Proof For (I)⇒(II), take any $h : C \to A \times B$; we need to show $h = \langle \pi_1 \circ h, \pi_2 \circ h\rangle$. We have

$$A \xleftarrow{\pi_1 \circ h} C \xrightarrow{\pi_2 \circ h} B$$

and hence, by (I), there exists unique $k : C \to A \times B$ such that

$$\pi_1 \circ k = \pi_1 \circ h \quad \wedge \quad \pi_2 \circ k = \pi_2 \circ h \tag{$*$}$$

Note now that $(*)$ holds both for $k := h$ and $k := \langle \pi_1 \circ h, \pi_2 \circ h\rangle$, the latter because of (I). Hence, $h = \langle \pi_1 \circ h, \pi_2 \circ h\rangle$.

For (II)⇒(I), take any triple $A \xleftarrow{f} C \xrightarrow{g} B$. By (II), we have that there exists an arrow $\langle f, g\rangle : C \to A \times B$ such that $\pi_1 \circ \langle f, g\rangle = f$ and $\pi_2 \circ \langle f, g\rangle = g$. We need to show it is the unique such. Let $k : C \to A \times B$ s.t.

$$\pi_1 \circ k = f \quad \wedge \quad \pi_2 \circ k = g$$

Then, by (II),

$$k = \langle \pi_1 \circ k, \pi_2 \circ k\rangle = \langle f, g\rangle$$

as required. ■

In the following proposition we give some useful properties of products. First, let us introduce some notation for arrows: given $f_1 : A_1 \to B_1$, $f_2 : A_2 \to B_2$, define

$$f_1 \times f_2 := \langle f_1 \circ \pi_1, f_2 \circ \pi_2\rangle : A_1 \times A_2 \longrightarrow B_1 \times B_2.$$

Proposition 28 *For any* $f : A \to B$, $g : A \to C$, $h : A' \to A$, *and any* $p : B \to B'$, $q : C \to C'$,

- $\langle f, g\rangle \circ h = \langle f \circ h, g \circ h\rangle$,
- $(p \times q) \circ \langle f, g\rangle = \langle p \circ f, q \circ g\rangle$.

Proof For the first claim we have:

$$\langle f, g\rangle \circ h = \langle \pi_1 \circ (\langle f, g\rangle \circ h), \pi_2 \circ (\langle f, g\rangle \circ h)\rangle = \langle f \circ h, g \circ h\rangle.$$

And for the second:

$$\begin{aligned}(p \times q) \circ \langle f, g\rangle &= \langle p \circ \pi_1, q \circ \pi_2\rangle \circ \langle f, g\rangle \\ &= \langle p \circ \pi_1 \circ \langle f, g\rangle, q \circ \pi_2 \circ \langle f, g\rangle\rangle \\ &= \langle p \circ f, q \circ g\rangle.\end{aligned}$$

∎

General Products

The notion of products can be generalised to arbitrary arities as follows. A product for a family of objects $\{A_i\}_{i\in I}$ in a category $\mathcal{C}$ is an object P and morphisms

$$p_i : P \longrightarrow A_i \quad (i \in I)$$

such that, for all objects B and arrows

$$f_i : B \longrightarrow A_i \quad (i \in I)$$

there is a *unique* arrow

$$g : B \longrightarrow P$$

such that, for all $i \in I$, the following diagram commutes.

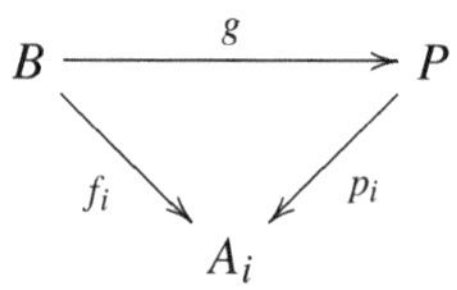

As before, if such a product exists, it is unique up to (unique) isomorphism. We write $P = \prod_{i\in I} A_i$ for the product object, and $g = \langle f_i \mid i \in I\rangle$ for the unique morphism in the definition.

Exercise 29 What is the product of the empty family?

Exercise 30 Show that if a category has binary and nullary products then it has all finite products.

1.2.2.2 Coproducts

We now investigate the dual notion to products: namely coproducts. Formally, coproducts in $\mathcal{C}$ are just products in $\mathcal{C}^{\mathsf{op}}$, interpreted back in $\mathcal{C}$. We spell out the definition.

Definition 31 Let A, B be objects in a category $\mathcal{C}$. A *coproduct* of A and B is an object $A + B$ together with a pair of arrows $A \xrightarrow{\mathsf{in}_1} A + B \xleftarrow{\mathsf{in}_2} B$ such that for every triple $A \xrightarrow{f} C \xleftarrow{g} B$ there exists a *unique* morphism

$$[f, g] : A + B \longrightarrow C$$

such that the following diagram commutes.

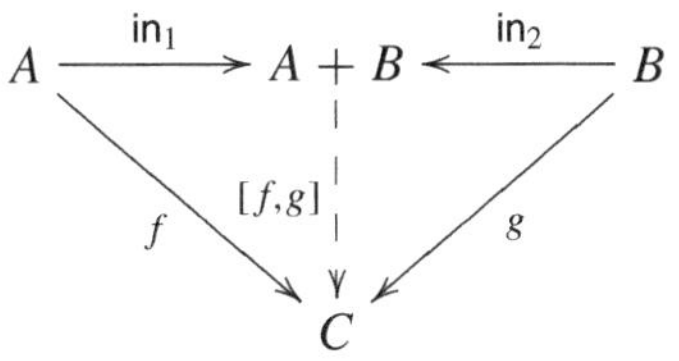

$$\begin{pmatrix} [f, g] \circ \mathsf{in}_1 = f \\ [f, g] \circ \mathsf{in}_2 = g \end{pmatrix}$$

▲

We call the in_i's *injections* and $[f, g]$ the *copairing* of f and g. As with pairings, uniqueness of copairings can be specified by an equation:

$$\forall h : A + B \to C.\, h = [h \circ \mathsf{in}_1, h \circ \mathsf{in}_2]$$

Coproducts in **Set**

This is given by *disjoint union* of sets, which can be defined concretely e.g. by

$$X + Y := \{1\} \times X \cup \{2\} \times Y.$$

We can define *injections*

$$X \xrightarrow{\mathsf{in}_1} X + Y \xleftarrow{\mathsf{in}_2} Y$$

$$\mathsf{in}_1(x) := (1, x)\,, \qquad \mathsf{in}_2(y) := (2, y)\,.$$

Also, given functions $f : X \longrightarrow Z$ and $g : Y \longrightarrow Z$, we can define

$$[f, g] : X + Y \longrightarrow Z$$

$$[f, g](1, x) := f(x)\,, \qquad [f, g](2, y) := g(y)\,.$$

Exercise 32 Check that this construction does yield coproducts in **Set**.

Note that this example suggests that coproducts allow for *definition by cases*.

Let us examine coproducts for some of our other standard examples.

- In **Pos**, disjoint unions (with the inherited orders) are coproducts.
- In **Top**, topological disjoint unions are coproducts.
- In $\mathbf{Vect}_k$, direct sums are coproducts.
- In a poset, *least upper bounds* are coproducts.

Exercise 33 Verify these claims.

Exercise 34 Dually to products, express coproducts as initial objects of a category **Copair**(A, B) of A,B–copairings.

1.2.3 Pullbacks and Equalisers

We shall consider two further constructions of interest: *pullbacks* and *equalisers*.

1.2.3.1 Pullbacks

Definition 35 Consider a pair of morphisms $A \xrightarrow{f} C \xleftarrow{g} B$. The ***pull-back*** of f along g is a pair $A \xleftarrow{p} D \xrightarrow{q} B$ such that $f \circ p = g \circ q$ and, for any pair $A \xleftarrow{p'} D' \xrightarrow{q'} B$ such that $f \circ p' = g \circ q'$, there exists a unique $h : D' \to D$ such that $p' = p \circ h$ and $q' = q \circ h$. Diagrammatically,

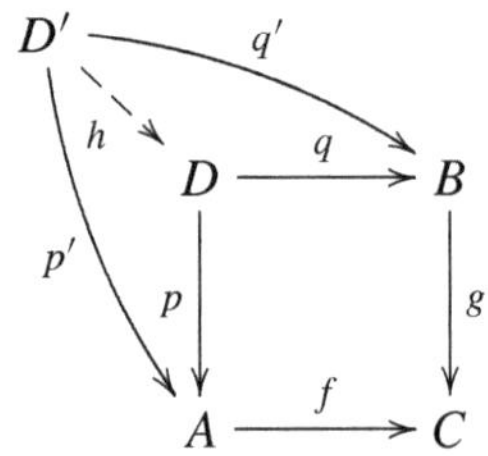

▲

Example 36

- In **Set** the pullback of $A \xrightarrow{f} C \xleftarrow{g} B$ is defined as a *subset of the cartesian product*:

$$A \times_C B = \{(a, b) \in A \times B \mid f(a) = g(b)\}.$$

  For example, consider a category $\mathcal{C}$ with

$$\mathsf{Ar}(\mathcal{C}) \xrightarrow{\mathsf{dom}} \mathsf{Ob}(\mathcal{C}) \xleftarrow{\mathsf{cod}} \mathsf{Ar}(\mathcal{C})\,.$$

  Then the pullback of dom along cod is the set of *composable morphisms*, *i.e.* pairs of morphisms (f, g) in $\mathcal{C}$ such that $f \circ g$ is well-defined.
- In **Set** again, subsets (*i.e.* inclusion maps) pull back to subsets:

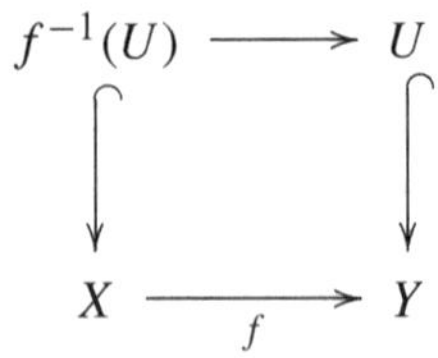

Exercise 37 Let $\mathcal{C}$ be a category with a terminal object $\mathbf{1}$. Show that, for any $A, B \in Ob(\mathcal{C})$, the pullback of $A \xrightarrow{\tau_A} \mathbf{1} \xleftarrow{\tau_B} B$ is the product of A and B, if it exists.

Just as for products, pullbacks can equivalently be described as terminal objects in suitable categories. Given a pair of morphisms $A \xrightarrow{f} C \xleftarrow{g} B$, we define an (f, g)*–cone* to be a triple (D, p, q) such that the following diagram commutes.

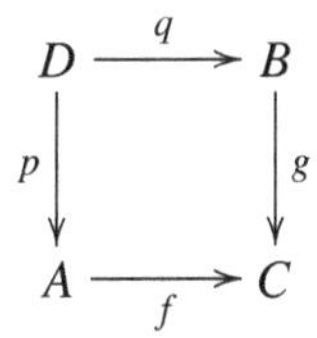

A morphism of (f, g)–cones $h : (D_1, p_1, q_1) \rightarrow (D_2, p_2, q_2)$ is a morphism $h : D_1 \rightarrow D_2$ such that the following diagram commutes.

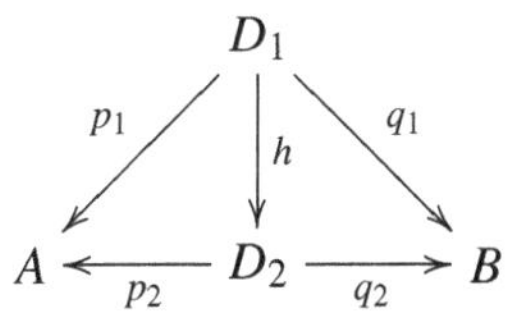

We can thus form a category $\mathbf{Cone}(f, g)$. A pull-back of f along g, if it exists, is exactly a terminal object of $\mathbf{Cone}(f, g)$. Once again, this shows the uniqueness of pullbacks up to unique isomorphism.

1.2.3.2 Equalisers

Definition 38 Consider a pair of parallel arrows $A \overset{f}{\underset{g}{\rightrightarrows}} B$. An ***equaliser*** of (f, g) is an arrow $e : E \rightarrow A$ such that $f \circ e = g \circ e$ and, for any arrow $h : D \rightarrow A$ such that $f \circ h = g \circ h$, there is a unique $\hat{h} : D \rightarrow E$ so that $h = e \circ \hat{h}$. Diagrammatically,

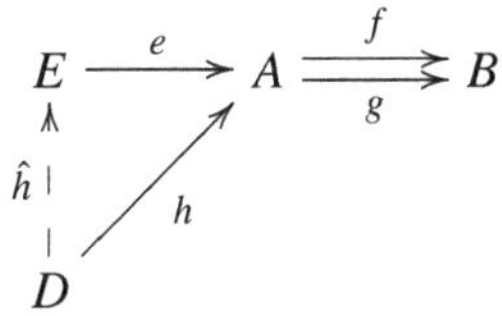

▲

As for products, uniqueness of the arrow from D to E can be expressed equationally:

$$\forall k : D \rightarrow E.\, \widehat{e \circ k} = k\,.$$

Exercise 39 Why is $\widehat{e \circ k}$ well-defined for any $k : D \rightarrow E$? Prove that the above equation is equivalent to the uniqueness requirement.

Example 40 In **Set**, the equaliser of f, g is given by the inclusion

$$\{x \in A \mid f(x) = g(x)\} \hookrightarrow A\,.$$

This allows *equationally defined subsets* to be defined as equalisers. For example, consider the pair of maps $\mathbb{R}^2 \underset{g}{\overset{f}{\rightrightarrows}} \mathbb{R}$, where

$$f : (x, y) \mapsto x^2 + y^2, \qquad g : (x, y) \mapsto 1\,.$$

Then, the equaliser is the unit circle as a subset of $\mathbb{R}^2$.

1.2.4 Limits and Colimits

The notions we have introduced so far are all special cases of a general notion of *limits* in categories, and the dual notion of *colimits* (Table 1.1).

Table 1.1 Examples of limits and colimits

Limits	Colimits
Terminal objects	Initial objects
Products	Coproducts
Pullbacks	Pushouts
Equalisers	Coequalisers

An important aspect of studying any kind of mathematical structure is to see what limits and colimits the category of such structures has. We shall return to these ideas shortly.

1.2.5 Exercises

1. Give an example of a category where some pair of objects lacks a product or coproduct.
2. (*Pullback lemma*) Consider the following commutative diagram.

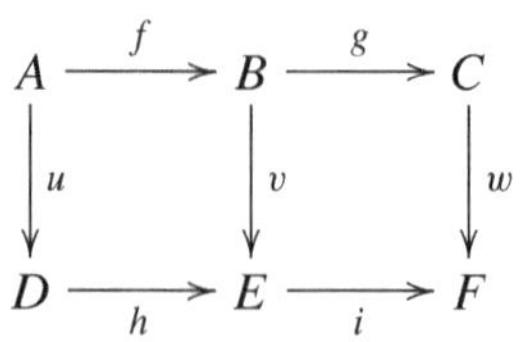

   Given that the right hand square $BCEF$ and the outer square $ACDF$ are pullbacks, prove that the left hand square $ABDE$ is a pullback.

3. Consider $A \xrightarrow{f} C \xleftarrow{g} B$ with pullback $A \xleftarrow{p} D \xrightarrow{q} B$. For each $A \xleftarrow{p'} D' \xrightarrow{q'} B'$ with $f \circ p' = g \circ q'$, let $\phi(p', q') : D' \to D$ be the arrow dictated by the pullback condition. Express uniqueness of $\phi(p', q')$ equationally.

1.3 Functors

Part of the "categorical philosophy" is:

Don't just look at the objects; take the morphisms into account too.

We can also apply this to categories!

1.3.1 Basics

A "morphism of categories" is a *functor*.

Definition 41 A ***functor*** $F : \mathcal{C} \to \mathcal{D}$ is given by:

- An object-map, assigning an object FA of $\mathcal{D}$ to every object A of $\mathcal{C}$.
- An arrow-map, assigning an arrow $Ff : FA \to FB$ of $\mathcal{D}$ to every arrow $f : A \to B$ of $\mathcal{C}$, in such a way that composition and identities are preserved:

$$F(g \circ f) = Fg \circ Ff\,, \qquad F\,\mathsf{id}_A = \mathsf{id}_{FA}.$$

▲

Note that we use the same symbol to denote the object- and arrow-maps; in practice, this never causes confusion. Since functors preserve domains and codomains of arrows, for each pair of objects A, B of $\mathcal{C}$, there is a well-defined map

$$F_{A,B} : \mathcal{C}(A, B) \to \mathcal{D}(FA, FB)\,.$$

The conditions expressing preservation of composition and identities are called *functoriality*.

Example 42 Let $(P, \leq)$, $(Q, \leq)$ be preorders (seen as categories). A functor $F : (P, \leq) \longrightarrow (Q, \leq)$ is specified by an object-map, say $F : P \to Q$, and an appropriate arrow-map. The arrow-map corresponds to the condition

$$\forall p_1, p_2 \in P.\, p_1 \leq p_2 \implies F(p_1) \leq F(p_2)\,,$$

i.e. to monotonicity of F. Moreover, the functoriality conditions are trivial since in the codomain $(Q, \leq)$ all hom-sets are singletons.
Hence, a functor between preorders is just a monotone map.

Example 43 Let $(M, \cdot, 1)$, $(N, \cdot, 1)$ be monoids. A functor $F : (M, \cdot, 1) \longrightarrow (N, \cdot, 1)$ is specified by a trivial object map (monoids are categories with a single object) and an arrow-map, say $F : M \to N$. The functoriality conditions correspond to

$$\forall m_1, m_2 \in M.\, F(m_1 \cdot m_2) = F(m_1) \cdot F(m_2)\,, \qquad F(1) = 1\,,$$

i.e. to F being a monoid homomorphism.
Hence, a functor between monoids is just a monoid homomorphism.

Other examples are the following.

- Inclusion of a sub-category, $\mathcal{C} \hookrightarrow \mathcal{D}$, is a functor (by taking the identity map for object- and arrow-map).
- The *covariant* powerset functor $\mathcal{P} : \mathbf{Set} \to \mathbf{Set}$:

$$X \mapsto \mathcal{P}(X)\,, \qquad (f : X \to Y) \mapsto \mathcal{P}(f) := S \mapsto \{f(x) \mid x \in S\}.$$

- $U : \mathbf{Mon} \to \mathbf{Set}$ is the "forgetful" or "underlying" functor which sends a monoid to its set of elements, "forgetting" the algebraic structure, and sends a homomorphism to the corresponding function between sets. There are similar forgetful functors for other categories of structured sets. Why are these trivial-looking functors useful?—We shall see!
- Group theory examples. The assignment of the commutator sub-group of a group extends to a functor from **Group** to **Group**; and the assignment of the quotient by this normal subgroup extends to a functor from **Group** to **AbGroup**. The assignment of the centraliser of a group does not!
- More sophisticated examples: e.g. *homology*. The basic idea of algebraic topology is that there are functorial assignments of algebraic objects (e.g. groups) to topological spaces, and variants of this idea ((co)homology theories) are pervasive throughout modern pure mathematics.

Functors "of several variables"

We can generalise the notion of a functor to a mapping from several domain categories to a codomain category. For this we need the following definition.

Definition 44 For categories $\mathcal{C}, \mathcal{D}$ define the ***product category*** $\mathcal{C} \times \mathcal{D}$ as follows. An object in $\mathcal{C} \times \mathcal{D}$ is a pair of objects from $\mathcal{C}$ and $\mathcal{D}$, and an arrow in $\mathcal{C} \times \mathcal{D}$ is a pair of arrows from $\mathcal{C}$ and $\mathcal{D}$. Identities and arrow composition are defined componentwise:

$$\mathsf{id}_{(A,B)} := (\mathsf{id}_A, \mathsf{id}_B)\,, \qquad (f, g) \circ (f', g') := (f \circ f', g \circ g')\,.$$

▲

A functor "of two variables", with domains $\mathcal{C}$ and $\mathcal{D}$, to $\mathcal{E}$ is simply a functor:

$$F : \mathcal{C} \times \mathcal{D} \longrightarrow \mathcal{E}\,.$$

For example, there are evident projection functors

$$\mathcal{C} \longleftarrow \mathcal{C} \times \mathcal{D} \longrightarrow \mathcal{D}.$$

1.3.2 Further Examples

1.3.2.1 Set-Valued Functors

Many important constructions arise as functors $F : \mathcal{C} \to \mathbf{Set}$. For example:

- If G is a group, a functor $F : G \to \mathbf{Set}$ is an *action of G on a set*.
- If P is a poset representing time, a functor $F : P \to \mathbf{Set}$ is a notion of *set varying through time*. This is related to *Kripke semantics*, and to *forcing arguments* in set theory.
- Recall that $\mathbb{2}_{\rightrightarrows}$ is the category $\bullet \rightrightarrows \bullet$. Then, functors $F : \mathbb{2}_{\rightrightarrows} \to \mathbf{Set}$ correspond to *directed graphs* understood as in Definition 4, *i.e.* as structures (V, E, s, t), where V is a set of vertices, E is a set of edges, and $s, t : E \to V$ specify the source and target vertices for each edge.

Let us examine the first example in more detail. For a group $(G, \cdot, 1)$, a functor $F : G \to \mathbf{Set}$ is specified by a set X (to which the unique object of G is mapped), and by an arrow-map sending each element m of G to an endofunction on X, say $m \bullet \_ : X \to X$. Then, functoriality amounts to the conditions

$$\forall m_1, m_2 \in G.\ F(m_1 \cdot m_2) = F(m_1) \circ F(m_2), \qquad F(1) = \mathsf{id}_X,$$

that is, for all $m_1, m_2 \in G$ and all $x \in X$,

$$(m_1 \cdot m_2) \bullet x = m_1 \bullet m_2 \bullet x, \qquad 1 \bullet x = x.$$

We therefore see that F defines an action of G on X.

Exercise 45 Verify that functors $F : \mathbb{2}_{\rightrightarrows} \to \mathbf{Set}$ correspond to directed graphs.

Example: Lists

Data-type constructors are functors. As a basic example, we consider lists. There is a functor

$$\mathsf{List} : \mathbf{Set} \longrightarrow \mathbf{Set}$$

which takes a set X to the set of all finite lists (sequences) of elements of X. List is functorial: its action on morphisms (*i.e.* functions, *i.e.* (functional) programs) is given by *maplist*:

$$\frac{f : X \longrightarrow Y}{\mathsf{List}(f) : \mathsf{List}(X) \longrightarrow \mathsf{List}(Y)}$$

$$\mathsf{List}(f)[x_1, \ldots, x_n] := [f(x_1), \ldots, f(x_n)]$$

We can upgrade List to a functor $\mathsf{MList} : \mathbf{Set} \to \mathbf{Mon}$ by mapping each set X to the monoid $(\mathsf{List}(X), *, \epsilon)$ and $f : X \to Y$ to $\mathsf{List}(f)$, as above. The monoid operation $* : \mathsf{List}(X) \times \mathsf{List}(X) \to \mathsf{List}(X)$ is list concatenation, and ϵ is the empty list. We call $\mathsf{MList}(X)$ the ***free monoid*** over X. This terminology will be justified in Chap. 5.

1.3.2.2 Products as Functors

If a category $\mathcal{C}$ has binary products, then there is automatically a functor

$$\_ \times \_ : \mathcal{C} \times \mathcal{C} \longrightarrow \mathcal{C}$$

which takes each pair (A, B) to the product $A \times B$, and each (f, g) to

$$f \times g := \langle f \circ \pi_1, g \circ \pi_2 \rangle .$$

Functoriality is shown as follows, using Proposition 28 and uniqueness of pairings in its equational form.

$$\begin{aligned}
(f \times g) \circ (f' \times g') &= (f \times g) \circ \langle f' \circ \pi_1, g' \circ \pi_2 \rangle = \langle f \circ f' \circ \pi_1, g \circ g' \circ \pi_2 \rangle \\
&= (f \circ f') \times (g \circ g'), \\
\mathsf{id}_A \times \mathsf{id}_B &= \langle \mathsf{id}_A \circ \pi_1, \mathsf{id}_B \circ \pi_2 \rangle = \langle \pi_1 \circ \mathsf{id}_{A \times B}, \pi_2 \circ \mathsf{id}_{A \times B} \rangle = \mathsf{id}_{A \times B} .
\end{aligned}$$

1.3.2.3 The Category of Categories

There is a category **Cat** whose objects are categories, and whose arrows are functors. Identities in **Cat** are given by identity functors:

$$\mathsf{Id}_{\mathcal{C}} : \mathcal{C} \longrightarrow \mathcal{C} := A \mapsto A, \; f \mapsto f.$$

Composition of functors is defined in the evident fashion. Note that if $F : \mathcal{C} \to \mathcal{D}$ and $G : \mathcal{D} \to \mathcal{E}$ then, for $f : A \to B$ in $\mathcal{C}$,

$$G \circ F(f) := G(F(f)) : G(F(A)) \longrightarrow G(F(B))$$

so the types work out. A category of categories sounds (and is) circular, but in practice is harmless: one usually makes some size restriction on the categories, and then **Cat** will be too "big" to be an object of itself. See Appendix A.

Note that product categories are products in **Cat**! For any pair of categories $\mathcal{C}, \mathcal{D}$, set

$$\mathcal{C} \xleftarrow{\pi_1} \mathcal{C} \times \mathcal{D} \xrightarrow{\pi_2} \mathcal{D}$$

where $\mathcal{C} \times \mathcal{D}$ the product category (defined previously) and π_i's the obvious projection functors. For any pair of functors $\mathcal{C} \xleftarrow{F} \mathcal{E} \xrightarrow{G} \mathcal{D}$, set

$$\langle F, G \rangle : \mathcal{E} \longrightarrow \mathcal{C} \times \mathcal{D} := A \mapsto (FA, GA),\ f \mapsto (Ff, Gf)\,.$$

It is easy to see that $\langle F, G \rangle$ is indeed a functor. Moreover, satisfaction of the product diagram and uniqueness are shown exactly as in **Set**.

1.3.3 Contravariance

By definition, the arrow-map of a functor F is *covariant*: it preserves the direction of arrows, so if $f : A \to B$ then $Ff : FA \to FB$. A *contravariant* functor G does exactly the opposite: it reverses arrow-direction, so if $f : A \to B$ then $Gf : GB \to GA$. A concise way to express contravariance is as follows.

Definition 46 Let $\mathcal{C}, \mathcal{D}$ be categories. A ***contravariant*** functor G from $\mathcal{C}$ to $\mathcal{D}$ is a functor $G : \mathcal{C}^{\mathsf{op}} \to \mathcal{D}$. (Equivalently, a functor $G : \mathcal{C} \to \mathcal{D}^{\mathsf{op}}$.) ▲

Explicitly, a contravariant functor G is given by an assignment of:

- an object GA in $\mathcal{D}$ to every object A in $\mathcal{C}$,
- an arrow $Gf : GB \to GA$ in $\mathcal{D}$ to every arrow $f : A \to B$ in $\mathcal{C}$, such that (notice the change of order in composition):

$$G(g \circ f) = Gf \circ Gg\,, \qquad G\mathsf{id}_A = \mathsf{id}_{GA}.$$

Note that functors of several variables can be covariant in some variables and contravariant in others, e.g.

$$F : \mathcal{C}^{\mathsf{op}} \times \mathcal{D} \longrightarrow \mathcal{E}\,.$$

Examples of Contravariant Functors

- The contravariant powerset functor, $\mathcal{P}^{\mathsf{op}} : \mathbf{Set}^{\mathsf{op}} \to \mathbf{Set}$, is given by:

$$\mathcal{P}^{\mathsf{op}}(X) := \mathcal{P}(X)\,.$$
$$\mathcal{P}^{\mathsf{op}}(f : X \to Y) : \mathcal{P}(Y) \longrightarrow \mathcal{P}(X) := T \mapsto \{x \in X \mid f(x) \in T\}\,.$$

- The dual space functor on vector spaces:

$$(\_)^* : \mathbf{Vect}_k^{\mathsf{op}} \longrightarrow \mathbf{Vect}_k := V \mapsto V^*.$$

Note that these are both examples of the following idea: send an object A into functions from A into some fixed object. For example, the powerset can be written as $\mathcal{P}(X) = 2^X$, where we think of a subset in terms of its characteristic function.

Hom-functors

We now consider some fundamental examples of **Set**-valued functors. Given a category $\mathcal{C}$ and an object A of $\mathcal{C}$, two functors to **Set** can be defined:

- The covariant Hom-functor at A,

$$\mathcal{C}(A, \_) : \mathcal{C} \longrightarrow \mathbf{Set},$$

  which is given by (recall that each $\mathcal{C}(A, B)$ is a set):

$$\mathcal{C}(A, \_)(B) := \mathcal{C}(A, B), \qquad \mathcal{C}(A, \_)(f : B \to C) := g \mapsto f \circ g.$$

  We usually write $\mathcal{C}(A, \_)(f)$ as $\mathcal{C}(A, f)$. Functoriality reduces directly to the basic category axioms: associativity of composition and the unit laws for the identity.
- There is also a contravariant Hom-functor,

$$\mathcal{C}(\_, A) : \mathcal{C}^{\mathsf{op}} \longrightarrow \mathbf{Set},$$

  given by:

$$\mathcal{C}(\_, A)(B) := \mathcal{C}(B, A), \qquad \mathcal{C}(\_, A)(h : C \to B) := g \mapsto g \circ h.$$

Generalising both of the above, we obtain a ***bivariant*** Hom-functor,

$$\mathcal{C}(\_, \_) : \mathcal{C}^{\mathsf{op}} \times \mathcal{C} \longrightarrow \mathbf{Set}.$$

Exercise 47 Spell out the definition of $\mathcal{C}(\_, \_) : \mathcal{C}^{\mathsf{op}} \times \mathcal{C} \longrightarrow \mathbf{Set}$. Verify carefully that it is a functor.

1.3.4 Properties of Functors

Definition 48 A functor $F : \mathcal{C} \to \mathcal{D}$ is said to be:

- ***faithful*** if each map $F_{A,B} : \mathcal{C}(A, B) \to \mathcal{D}(FA, FB)$ is injective;
- ***full*** if each map $F_{A,B} : \mathcal{C}(A, B) \to \mathcal{D}(FA, FB)$ is surjective;
- an ***embedding*** if F is full, faithful, and injective on objects;
- an ***equivalence*** if F is full, faithful, and *essentially surjective*: *i.e.* for every object B of $\mathcal{D}$ there is an object A of $\mathcal{C}$ such that $F(A) \cong B$;
- an ***isomorphism*** if there is a functor $G : \mathcal{D} \to \mathcal{C}$ such that

$$G \circ F = \mathsf{Id}_{\mathcal{C}}, \qquad F \circ G = \mathsf{Id}_{\mathcal{D}}.$$

▲

We say that categories $\mathcal{C}$ and $\mathcal{D}$ are isomorphic, $\mathcal{C} \cong \mathcal{D}$, if there is an isomorphism between them. Note that this is just the usual notion of isomorphism applied to **Cat**. Examples:

- The forgetful functor $U : \mathbf{Mon} \to \mathbf{Set}$ is faithful, but not full. For the latter, note that not all functions $f : M \to N$ yield an arrow $f : (M, \cdot, 1) \to (N, \cdot, 1)$. Similar properties hold for other forgetful functors.
- The free monoid functor $\mathsf{MList} : \mathbf{Set} \to \mathbf{Mon}$ is faithful, but not full.
- The product functor $\_ \times \_ : \mathcal{C} \times \mathcal{C} \longrightarrow \mathcal{C}$ is generally neither faithful nor full. For the latter, e.g. in **Set**, the function $f : \mathbb{N}^2 \to \mathbb{N}^2 := (m, n) \mapsto (n, n)$ cannot be expressed in the form $f_1 \times f_2$. Faithfulness of the functor is examined in Exercise 1.3.5(2).
- There is an equivalence between $\mathbf{FDVect}_k$ the category of finite dimensional vector spaces over the field k, and $\mathbf{Mat}_k$, the category of matrices with entries in k. Note that these categories are very far from isomorphic! This example is elaborated in Exercise 1.3.5(1).

Preservation and Reflection

Let P be a property of arrows. We say that a functor $F : \mathcal{C} \to \mathcal{D}$ *preserves* P if whenever f satisfies P, so does $F(f)$. We say that F *reflects* P if whenever $F(f)$ satisfies P, so does f. For example:

a. All functors preserve isomorphisms, split monics and split epics.
b. Faithful functors reflect monics and epics.
c. Full and faithful functors reflect isomorphisms.
d. Equivalences preserve monics and epics.
- The forgetful functor $U : \mathbf{Mon} \to \mathbf{Set}$ preserves products.

Let us show c; the rest are given as exercises below. So let $f : A \to B$ in $\mathcal{C}$ be such that Ff is an iso, that is, it has an inverse $g' : FB \to FA$. Then, by fullness, there exists some $g : B \to A$ so that $g' = Fg$. Thus,

$$F(g \circ f) = Fg \circ Ff = g' \circ Ff = \mathsf{id}_{FA} = F(\mathsf{id}_A)\,.$$

By faithfulness we obtain $g \circ f = \mathsf{id}_A$. Similarly, $f \circ g = \mathsf{id}_B$ and therefore f is an isomorphism.

Exercise 49 Show items a, b and d above.

Exercise 50 Show the following.

- Functors do not in general reflect monics or epics.
- Faithful functors do not in general reflect isomorphisms.
- Full and faithful functors do not in general preserve monics or epics.

1.3.5 Exercises

1. Consider the category $\mathbf{FDVect}_{\mathbb{R}}$ of finite dimensional vector spaces over $\mathbb{R}$, and $\mathbf{Mat}_{\mathbb{R}}$ of matrices over $\mathbb{R}$. Concretely, $\mathbf{Mat}_{\mathbb{R}}$ is defined as follows:

$$
\begin{aligned}
Ob(\mathbf{Mat}_{\mathbb{R}}) &:= \mathbb{N}\,, \\
\mathbf{Mat}_{\mathbb{R}}(n, m) &:= \{M \mid M \text{ is an } n \times m \text{ matrix with entries in } \mathbb{R}\}\,.
\end{aligned}
$$

   Thus, objects are natural numbers, and arrows $n \to m$ are $n \times m$ real matrices. Composition is matrix multiplication, and the identity on n is the $n \times n$ identity matrix.
   Now let $F : \mathbf{Mat}_{\mathbb{R}} \to \mathbf{FDVect}_{\mathbb{R}}$ be the functor taking each n to the vector space $\mathbb{R}^n$ and each $M : n \to m$ to the linear function

$$FM : \mathbb{R}^n \longrightarrow \mathbb{R}^m := (x_1, \dots, x_n) \mapsto [x_1, \dots, x_n]M$$

   with the $1 \times m$ matrix $[x_1, \dots, x_n]M$ considered as a vector in $\mathbb{R}^m$. Show that F is full, faithful and essentially surjective, and hence that $\mathbf{FDVect}_{\mathbb{R}}$ and $\mathbf{Mat}_{\mathbb{R}}$ are equivalent categories. Are they isomorphic?
2. Let $\mathcal{C}$ be a category with binary products such that, for each pair of objects A, B,

$$\mathcal{C}(A, B) \neq \emptyset. \tag{$*$}$$

   Show that the product functor $F : \mathcal{C} \times \mathcal{C} \to \mathcal{C}$ is faithful.
   Would F still be faithful in the absence of condition $(*)$?

1.4 Natural Transformations

> "Categories were only introduced to allow functors to be defined; functors were only introduced to allow natural transformations to be defined."

Just as categories have morphisms between them, namely functors, so functors have morphisms between them too—*natural transformations*.

1.4.1 Basics

Definition 51 Let $F, G : \mathcal{C} \to \mathcal{D}$ be functors. A ***natural transformation***

$$t : F \longrightarrow G$$

is a family of morphisms in $\mathcal{D}$ indexed by objects A of $\mathcal{C}$,

$$\{\, t_A : FA \longrightarrow GA \,\}_{A \in Ob(\mathcal{C})}$$

such that, for all $f : A \to B$, the following diagram commutes.

$$\begin{array}{ccc} FA & \xrightarrow{Ff} & FB \\ {\scriptstyle t_A}\downarrow & & \downarrow{\scriptstyle t_B} \\ GA & \xrightarrow[Gf]{} & GB \end{array}$$

This condition is known as *naturality*.
If each t_A is an isomorphism, we say that t is a ***natural isomorphism***:

$$t : F \overset{\cong}{\longrightarrow} G\,.$$

▲

Examples:

- Let Id be the identity functor on **Set**, and $\times \circ \langle \mathsf{Id}, \mathsf{Id} \rangle$ be the functor taking each set X to $X \times X$ and each function f to $f \times f$. Then, there is a natural transformation $\Delta : \mathsf{Id} \longrightarrow \times \circ \langle \mathsf{Id}, \mathsf{Id} \rangle$ given by:

$$\Delta_X : X \longrightarrow X \times X := x \mapsto (x, x)\,.$$

Naturality amounts to asserting that, for any function $f : X \to Y$, the following diagram commutes.

$$\begin{array}{ccc} X & \xrightarrow{f} & Y \\ {\scriptstyle \Delta_X}\downarrow & & \downarrow{\scriptstyle \Delta_Y} \\ X \times X & \xrightarrow[f \times f]{} & Y \times Y \end{array}$$

We call Δ the *diagonal* transformation on **Set**. In fact, it is the *only* natural transformation between these functors.

- The diagonal transformation can be defined for any category $\mathcal{C}$ with binary products by setting, for each object A in $\mathcal{C}$,

$$\Delta_A : A \longrightarrow A \times A := \langle \mathsf{id}_A, \mathsf{id}_A \rangle\,.$$

Projections also yield natural transformations. For example the arrows

$$\pi_{1(A,B)} : A \times B \longrightarrow A$$

specify a natural transformation $\pi_1 : \times \to \boldsymbol{\pi}_1$. Note that $\times, \boldsymbol{\pi}_1 : \mathcal{C} \times \mathcal{C} \to \mathcal{C}$ are the functors for product and first projection respectively.

- Let $\mathcal{C}$ be a category with terminal object T, and let $K_T : \mathcal{C} \to \mathcal{C}$ be the functor mapping all objects to T and all arrows to id_T. Then, the canonical arrows

$$\tau_A : A \longrightarrow T$$

specify a natural transformation $\tau : \mathsf{Id} \to K_T$ (where Id the identity functor on $\mathcal{C}$).

- Recall the functor $\mathsf{List} : \mathbf{Set} \to \mathbf{Set}$ which takes a set X to the set of finite lists with elements in X. We can define (amongst others) the following natural transformations,

$$\mathsf{reverse} : \mathsf{List} \longrightarrow \mathsf{List}\,, \quad \mathsf{unit} : \mathsf{Id} \longrightarrow \mathsf{List}\,, \quad \mathsf{flatten} : \mathsf{List} \circ \mathsf{List} \longrightarrow \mathsf{List}\,,$$

by setting, for each set X,

$$\begin{aligned} \mathsf{reverse}_X &: \mathsf{List}(X) \longrightarrow \mathsf{List}(X) := [x_1, \ldots, x_n] \mapsto [x_n, \ldots, x_1]\,, \\ \mathsf{unit}_X &: X \longrightarrow \mathsf{List}(X) := x \mapsto [x]\,, \\ \mathsf{flatten}_X &: \mathsf{List}(\mathsf{List}(X)) \longrightarrow \mathsf{List}(X) \\ &:= [\,[x_1^1, \ldots, x_{n_1}^1], \ldots, [x_1^k, \ldots, x_{n_k}^k]\,] \mapsto [x_1^1, \ldots\ldots, x_{n_k}^k]\,. \end{aligned}$$

- Consider the functor $P := \times \circ \langle U, U \rangle$ with $U : \mathbf{Mon} \to \mathbf{Set}$, *i.e.*

$$P : \mathbf{Mon} \longrightarrow \mathbf{Set} := (M, \cdot, 1) \mapsto M \times M,\ f \mapsto f \times f\,.$$

Then, the monoid operation yields a natural transformation $t : P \to U$ defined by:

$$t_{(M,\cdot,1)} : M \times M \longrightarrow M := (m, m') \mapsto m \cdot m'\,.$$

Naturality corresponds to asserting that, for any $f : (M, \cdot, 1) \to (N, \cdot, 1)$, the following diagram commutes,

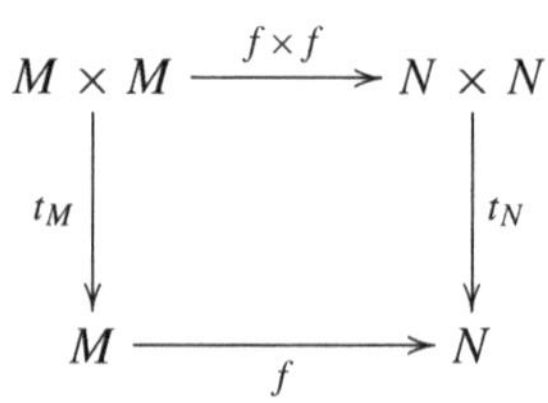

that is, for any $m_1, m_2 \in M$, $f(m_1) \cdot f(m_2) = f(m_1 \cdot m_2)$.

- If V is a finite dimensional vector space, then V is isomorphic to both its first dual V^* and to its second dual V^{**}.
  However, while it is naturally isomorphic to its second dual, there is no natural isomorphism to the first dual. This was actually the original example which motivated Eilenberg and Mac Lane to define the concept of natural transformation; here naturality captures *basis independence*.

Exercise 52 Verify naturality of diagonal transformations, projections and terminals for a category $\mathcal{C}$ with finite products.

Exercise 53 Prove that the diagonal is the only natural transformation $\mathsf{Id} \longrightarrow \times \circ \langle \mathsf{Id}, \mathsf{Id} \rangle$ on **Set**. Similarly, prove that the first projection is the only natural transformation $\times \to \pi_1$ on **Set**.

1.4.2 Further Examples

1.4.2.1 Natural Isomorphisms for Products

Let $\mathcal{C}$ be a category with finite products, *i.e.* binary products and a terminal object **1**. Then, we have the following canonical natural isomorphisms.

$$
\begin{aligned}
a_{A,B,C} &: A \times (B \times C) \xrightarrow{\cong} (A \times B) \times C\,, \\
s_{A,B} &: A \times B \xrightarrow{\cong} B \times A\,, \\
l_A &: \mathbf{1} \times A \xrightarrow{\cong} A\,, \\
r_A &: A \times \mathbf{1} \xrightarrow{\cong} A\,.
\end{aligned}
$$

The first two isomorphisms are meant to assert that the product is *associative* and *symmetric*, and the last two that **1** is its *unit*. In later sections we will see that these conditions form part of the definition of *symmetric monoidal categories*.

These natural isomorphisms are defined explicitly by:

$$
\begin{aligned}
a_{A,B,C} &:= \langle\langle \pi_1, \pi_1 \circ \pi_2 \rangle, \pi_2 \circ \pi_2 \rangle\,, \\
s_{A,B} &:= \langle \pi_2, \pi_1 \rangle\,, \\
l_A &:= \pi_2\,, \\
r_A &:= \pi_1\,.
\end{aligned}
$$

Since natural isomorphisms are a *self-dual* notion, similar natural isomorphisms can be defined if $\mathcal{C}$ has binary coproducts and an initial object.

Exercise 54 Verify that these families of arrows are natural isomorphisms.

1.4.2.2 Natural Transformations Between Hom-Functors

Let $f : A \to B$ in a category $\mathcal{C}$. Then, this induces a natural transformation

$$\mathcal{C}(f, \_) : \mathcal{C}(B, \_) \longrightarrow \mathcal{C}(A, \_)\,,$$
$$\mathcal{C}(f, \_)_C : \mathcal{C}(B, C) \longrightarrow \mathcal{C}(A, C) := (g : B \to C) \mapsto (g \circ f : A \to C)\,.$$

Note that $\mathcal{C}(f, \_)_C$ is the same as $\mathcal{C}(f, C)$, the result of applying the contravariant functor $\mathcal{C}(\_, C)$ to f. Hence, naturality amounts to asserting that, for each $h : C \to D$, the following diagram commutes.

$$\begin{array}{ccc} \mathcal{C}(B, C) & \xrightarrow{\mathcal{C}(B,h)} & \mathcal{C}(B, D) \\ {\scriptstyle \mathcal{C}(f,C)}\downarrow & & \downarrow{\scriptstyle \mathcal{C}(f,D)} \\ \mathcal{C}(A, C) & \xrightarrow[\mathcal{C}(A,h)]{} & \mathcal{C}(A, D) \end{array}$$

Starting from a $g : B \to C$, we compute:

$$\mathcal{C}(A, h)(\mathcal{C}(f, C)(g)) = h \circ (g \circ f) = (h \circ g) \circ f = \mathcal{C}(f, D)(\mathcal{C}(B, h)(g))\,.$$

The natural transformation $\mathcal{C}(\_, f) : \mathcal{C}(\_, A) \to \mathcal{C}(\_, B)$ is defined similarly.

Exercise 55 Define the natural transformation $\mathcal{C}(\_, f)$ and verify its naturality.

There is a remarkable result, the ***Yoneda Lemma***, which says that *every* natural transformation between Hom-functors comes from a (unique) arrow in $\mathcal{C}$ in the fashion described above.

Lemma 1 *Let A, B be objects in a category $\mathcal{C}$. For each natural transformation $t : \mathcal{C}(A, \_) \to \mathcal{C}(B, \_)$, there is a unique arrow $f : B \to A$ such that*

$$t = \mathcal{C}(f, \_)\,.$$

Proof Take any such A, B and t and let

$$f : B \longrightarrow A := t_A(\mathsf{id}_A)\,.$$

We want to show that $t = \mathcal{C}(f, \_)$. For any object C and any arrow $g : A \to C$, naturality of t means that the following commutes.

$$\begin{array}{ccc} \mathcal{C}(A, A) & \xrightarrow{\mathcal{C}(A,g)} & \mathcal{C}(A, C) \\ {\scriptstyle t_A}\downarrow & & \downarrow{\scriptstyle t_C} \\ \mathcal{C}(B, A) & \xrightarrow[\mathcal{C}(B,g)]{} & \mathcal{C}(B, C) \end{array}$$

Starting from id_A we have that:

$$t_C(\mathcal{C}(A, g)(\mathsf{id}_A)) = \mathcal{C}(B, g)(t_A(\mathsf{id}_A))\,, \quad \textit{i.e.}\ \ t_C(g) = g \circ f\,.$$

Hence, noting that $\mathcal{C}(f, C)(g) = g \circ f$, we obtain $t = \mathcal{C}(f, \_)$.
For uniqueness we have that, for any $f, f' : B \to A$, if $\mathcal{C}(f, \_) = \mathcal{C}(f', \_)$ then

$$f = \mathsf{id}_A \circ f = \mathcal{C}(f, A)(\mathsf{id}_A) = \mathcal{C}(f', A)(\mathsf{id}_A) = \mathsf{id}_A \circ f' = f'.$$

■

Exercise 56 Prove a similar result for contravariant hom-functors.

Alternative definition of equivalence

Another way of defining equivalence of categories is as follows.

Definition 57 We say that categories $\mathcal{C}$ and $\mathcal{D}$ are ***equivalent***, $\mathcal{C} \simeq \mathcal{D}$, if there are functors $F : \mathcal{C} \to \mathcal{D}$, $G : \mathcal{D} \to \mathcal{C}$ and natural isomorphisms

$$G \circ F \cong \mathsf{Id}_{\mathcal{C}}\,, \qquad F \circ G \cong \mathsf{Id}_{\mathcal{D}}\,.$$

▲

1.4.3 Functor Categories

Suppose we have functors $F, G, H : \mathcal{C} \to \mathcal{D}$ and natural transformations

$$t : F \longrightarrow G\,, \qquad u : G \longrightarrow H\,.$$

Then, we can compose these natural transformations, yielding $u \circ t : F \to H$:

$$(u \circ t)_A := FA \xrightarrow{t_A} GA \xrightarrow{u_A} HA.$$

Composition is associative, and has as identity the natural transformation

$$I_F : F \longrightarrow F := \{\, (I_F)_A := \mathsf{id}_A : FA \longrightarrow FA \,\}_A\,.$$

These observations lead us to the following.

Definition 58 For categories $\mathcal{C}, \mathcal{D}$ define the ***functor category*** $\mathsf{Func}(\mathcal{C}, \mathcal{D})$ by taking:

- Objects: functors $F : \mathcal{C} \to \mathcal{D}$.
- Arrows: natural transformations $t : F \to G$.

Composition and identities are given as above. ▲

Remark 59 We see that in the category **Cat** of categories and functors, each hom-set $\mathbf{Cat}(\mathcal{C}, \mathcal{D})$ *itself has the structure of a category*. In fact, **Cat** is the basic example of a "2-category", *i.e.* of a category where hom-sets are themselves categories.

Note that a natural isomorphism is precisely an isomorphism in the functor category. Let us proceed to some examples of functor categories.

- Recall that, for any group G, functors from G to **Set** are G-actions on sets. Then, $\mathsf{Func}(G, \mathbf{Set})$ is the category of G-actions on sets and *equivariant functions*: $f : X \to Y$ such that $f(m \bullet x) = m \bullet f(x)$.
- $\mathsf{Func}(2_{\rightrightarrows}, \mathbf{Set})$: Graphs and graph homomorphisms.
- If $F, G : P \to Q$ are monotone maps between posets, then $t : F \to G$ means that

$$\forall x \in P.\, Fx \leq Gx\,.$$

  Note that in this case naturality is trivial (hom-sets are singletons in Q).

Exercise 60 Verify the above descriptions of $\mathsf{Func}(G, \mathbf{Set})$ and $\mathsf{Func}(2_{\rightrightarrows}, \mathbf{Set})$.

Remark 61 The composition of natural transformations defined above is called *vertical composition*. The reason for this terminology is depicted below.

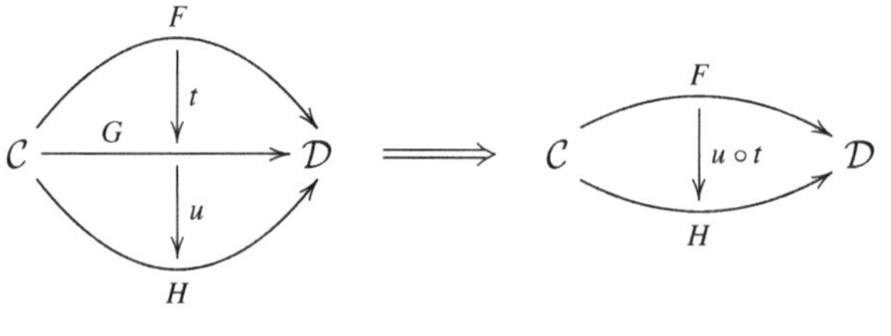

As expected, there is also a *horizontal composition*, which is given as follows.

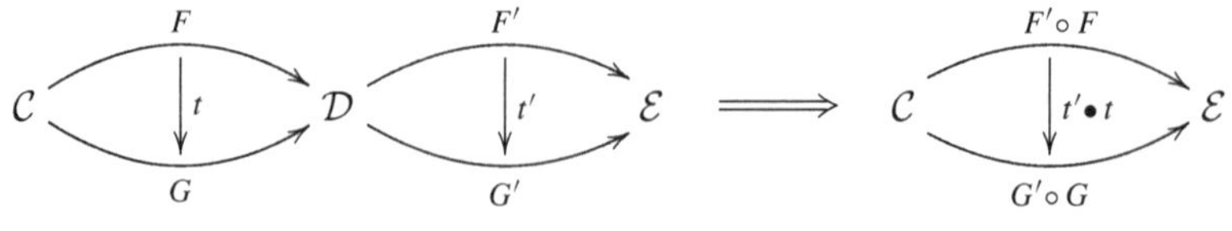

1.4.4 Exercises

1. By identifying the relevant functors, express pairing $\langle \_, \_ \rangle$ as a natural transformation. What does naturality correspond to explicitly?
2. Show that the two definitions of equivalence of categories, namely

   (a) $\mathcal{C}$ and $\mathcal{D}$ are equivalent if there is an equivalence $F : \mathcal{C} \to \mathcal{D}$ (definition 48),
   (b) $\mathcal{C}$ and $\mathcal{D}$ are equivalent if there are $F : \mathcal{C} \to \mathcal{D}$, $G : \mathcal{D} \to \mathcal{C}$, and isomorphisms $F \circ G \cong \mathsf{Id}_{\mathcal{D}}$, $G \circ F \cong \mathsf{Id}_{\mathcal{C}}$ (Definition 57), are: equivalent! Note that this will need the Axiom of Choice.

3. Define a relation on objects in a category $\mathcal{C}$ by: $A \cong B$ iff A and B are isomorphic.

   (a) Show that this relation is an equivalence relation.
   Define a *skeleton* of $\mathcal{C}$ to be the (full) subcategory obtained by choosing one object from each equivalence class of $\cong$ (note that this involves choices, and is not uniquely defined).
   (b) Show that $\mathcal{C}$ is equivalent to any skeleton.
   (c) Show that any two skeletons of $\mathcal{C}$ are isomorphic.
   (d) Give an example of a category whose objects form a proper class, but whose skeleton is finite.

4. Given a category $\mathcal{C}$, we can define a functor

$$y : \mathcal{C} \longrightarrow \mathsf{Func}(\mathcal{C}^{\mathsf{op}}, \mathbf{Set}) := A \mapsto \mathcal{C}(\_, A), \;\; f \mapsto \mathcal{C}(\_, f).$$

   Prove carefully that this is indeed a functor. Use exercise 56 to conclude that y is full and faithful. Prove that it is also injective on objects, and hence an embedding. It is known as the *Yoneda embedding*.
5. Define the horizontal composition $u \bullet t$ of natural transformations explicitly. Prove that it is associative.

1.5 Universality and Adjoints

There is a fundamental triad of categorical notions:

Functoriality, Naturality, Universality.

We have studied the first two notions explicitly. We have also seen many *examples* of universal definitions, notably the various notions of limits and colimits considered in Sect. 1.2. It is now time to consider universality in general; the proper formulation of this fundamental and pervasive notion is one of the major achievements of basic category theory.

Universality arises when we are interested in finding *canonical solutions* to problems of construction: that is, we are interested not just in the *existence* of a solution but in its *canonicity*. This canonicity should guarantee uniqueness, in the sense we have become familiar with: a canonical solution should be *unique up to (unique) isomorphism*.

The notion of canonicity has a simple interpretation in the case of posets, as an *extremal solution*: one that is the least or the greatest among all solutions. Such an extremal solution is obviously unique. For example, consider the problem of finding a lower bound of a pair of elements A, B in a poset P: a *greatest lower bound* of A and B is an extremal solution to this problem. As we have seen, this is the specialisation to posets of the problem of constructing a product:

- ⇝ A product of A, B in a poset is an element C such that $C \leq A$ and $C \leq B$, (C is a lower bound);
- ⇝ and for any other solution C', *i.e.* C' such that $C' \leq A$ and $C' \leq B$, we have $C' \leq C$. (C is a greatest lower bound.)

Because the ideas of universality and adjunctions have an appealingly simple form in posets, which is, moreover, useful in its own right, we will develop the ideas in that special case first, as a prelude to the general discussion for categories.

1.5.1 Adjunctions for Posets

Suppose $g : Q \to P$ is a monotone map between posets. Given $x \in P$, a ***g*-approximation of** x (from above) is an element $y \in Q$ such that $x \leq g(y)$.
A **best *g*-approximation of** x is an element $y \in Q$ such that

$$x \leq g(y) \wedge \forall z \in Q. (x \leq g(z) \implies y \leq z) .$$

If a best g-approximation exists then it is clearly unique.

1.5.1.1 Discussion

It is worth clarifying the notion of best g-approximation. If y is a best g-approximation to x, then in particular, by monotonicity of g, $g(y)$ is the least element of the set of all $g(z)$ where $z \in Q$ and $x \leq g(z)$. However, the property of being a best approximation is much *stronger* than the mere existence of a least element of this set. We are asking for y *itself* to be the least, *in* Q, among all elements z such that $x \leq g(z)$. Thus, even if g is *surjective*, so that for every x there is a $y \in Q$ such that $g(y) = x$, there need not exist a *best* g-approximation to x. This is exactly the issue of having a *canonical choice* of solution.

Exercise 62 Give an example of a surjective monotone map $g : Q \to P$ and an element $x \in P$ such that there is no best g-approximation to x in Q.

If such a best g-approximation $f(x)$ exists for all $x \in P$ then we have a function $f : P \to Q$ such that, for all $x \in P$, $z \in Q$:

$$x \leq g(z) \iff f(x) \leq z . \tag{1.1}$$

We say that f is the ***left adjoint*** of g, and g is the ***right adjoint*** of f. It is immediate from the definitions that the left adjoint of g, if it exists, is uniquely determined by g.

Proposition 63 *If such a function f exists, then it is monotone. Moreover,*

$$\mathsf{id}_P \leq g \circ f\,, \qquad f \circ g \leq \mathsf{id}_Q\,, \qquad f \circ g \circ f = f\,, \qquad g \circ f \circ g = g\,.$$

Proof If we take $z = f(x)$ in Eq. (1.1), then since $f(x) \leq f(x)$, $x \leq g \circ f(x)$. Similarly, taking $x = g(z)$ we obtain $f \circ g(z) \leq z$. Now, the ordering on functions $h, k : P \longrightarrow Q$ is the *pointwise order*:

$$h \leq k \iff \forall x \in P.\, h(x) \leq k(x).$$

This gives the first two equations.

Now, if $x \leq_P x'$ then $x \leq x' \leq g \circ f(x')$, so $f(x')$ is a g-approximation of x, and hence $f(x) \leq f(x')$. Thus, f is monotone.

Finally, using the fact that composition is monotone with respect to the pointwise order on functions, and the first two equations:

$$g = \mathsf{id}_P \circ g \leq g \circ f \circ g \leq g \circ \mathsf{id}_Q = g,$$

and hence $g = g \circ f \circ g$. The other equation is proved similarly. ∎

Examples:

- Consider the inclusion map

$$i : \mathbb{Z} \hookrightarrow \mathbb{R}\,.$$

This has both a left adjoint f^L and a right adjoint f^R, where $f^L, f^R : \mathbb{R} \to \mathbb{Z}$. For all $z \in \mathbb{Z}$, $r \in \mathbb{R}$:

$$z \leq f^R(r) \iff i(z) \leq r\,, \qquad f^L(r) \leq z \iff r \leq i(z)\,.$$

We see from these defining properties that the right adjoint maps a real r to the *greatest integer below it* (the extremal solution to finding an integer below a given real). This is the standard *floor function*.

Similarly, the left adjoint maps a real to the least integer above it yielding the *ceiling function*. Thus:

$$f^R(r) = \lfloor r \rfloor\,, \qquad f^L(r) = \lceil r \rceil\,.$$

- Consider a relation $R \subseteq X \times Y$. R induces a function:

$$f_R : \mathcal{P}(X) \longrightarrow \mathcal{P}(Y) := S \mapsto \{y \in Y \mid \exists x \in S.\, xRy\}\,.$$

This has a right adjoint $[R] : \mathcal{P}(Y) \longrightarrow \mathcal{P}(X)$:

$$S \subseteq [R]T \iff f_R(S) \subseteq T\,.$$

The definition of $[R]$ which satisfies this condition is:

$$[R]T := \{x \in X \mid \forall y \in Y.\, xRy \Rightarrow y \in T\}\,.$$

If we consider a set of *worlds* W with an *accessibility relation* $R \subseteq W \times W$ as in Kripke semantics for modal logic, we see that $[R]$ gives the usual Kripke semantics for the modal operator $\Box$, seen as a propositional operator mapping the set of worlds satisfied by a formula ϕ to the set of worlds satisfied by $\Box\phi$.

On the other hand, if we think of the relation R as the denotation of a (possibly non-deterministic) program, and T as a predicate on *states*, then $[R]T$ is exactly the *weakest precondition* $\mathbf{wp}(R, T)$. In *Dynamic Logic*, the two settings are combined, and we can write expressions such as $[R]T$ directly, where T will be (the denotation of) some formula, and R the relation corresponding to a program.

- Consider a function $f : X \to Y$. This induces a function:

$$f^{-1} : \mathcal{P}(Y) \longrightarrow \mathcal{P}(X) := T \mapsto \{x \in X \mid f(x) \in T\}\,.$$

This function f^{-1} has both a left adjoint $\exists(f) : \mathcal{P}(X) \longrightarrow \mathcal{P}(Y)$, and a right adjoint $\forall(f) : \mathcal{P}(X) \longrightarrow \mathcal{P}(Y)$. For all $S \subseteq X$, $T \subseteq Y$:

$$\exists(f)(S) \subseteq T \iff S \subseteq f^{-1}(T)\,, \qquad f^{-1}(T) \subseteq S \iff T \subseteq \forall(f)(S)\,.$$

How can we define $\forall(f)$ and $\exists(f)$ explicitly so as to fulfil these defining conditions? – As follows:

$$\exists(f)(S) := \{y \in Y \mid \exists x \in X.\, f(x) = y \wedge x \in S\}\,,$$
$$\forall(f)(S) := \{y \in Y \mid \forall x \in X.\, f(x) = y \Rightarrow x \in S\}\,.$$

If $R \subseteq X \times Y$, which we write in logical notation as $R(x, y)$, and we take the projection function $\pi_1 : X \times Y \longrightarrow X$, then:

$$\forall(\pi_1)(R) \equiv \forall y.\, R(x, y)\,, \qquad \exists(\pi_1)(R) \equiv \exists y.\, R(x, y)\,.$$

This extends to an algebraic form of the usual Tarski model-theoretic semantics for first-order logic, in which:

Quantifiers are Adjoints

1.5.1.2 Couniversality

We can dualise the discussion, so that starting with a monotone map $f : P \to Q$ and $y \in Q$, we can ask for the best P-approximation to y from below: $x \in P$ such that $f(x) \leq y$, and for all $z \in P$:

$$f(z) \leq y \iff z \leq x.$$

If such a best approximation $g(y)$ exists for all $y \in Q$, we obtain a monotone map $g : Q \to P$ such that g is right adjoint to f. From the symmetry of the definition, it is clear that:

$$f \text{ is the left adjoint of } g \iff g \text{ is the right adjoint of } f$$

and each determines the other uniquely.

1.5.2 Universal Arrows and Adjoints

Our discussion of best approximations for posets is lifted to general categories as follows.

Definition 64 Let $G : \mathcal{D} \to \mathcal{C}$ be a functor, and C an object of $\mathcal{C}$. A ***universal arrow from*** C ***to*** G is a pair (D, η) where D is an object of $\mathcal{D}$ and

$$\eta : C \longrightarrow G(D),$$

such that, for any object D' of $\mathcal{D}$ and morphism $f : C \to G(D')$, *there exists a unique* morphism $\hat{f} : D \to D'$ in $\mathcal{D}$ such that $f = G(\hat{f}) \circ \eta$.
Diagrammatically:

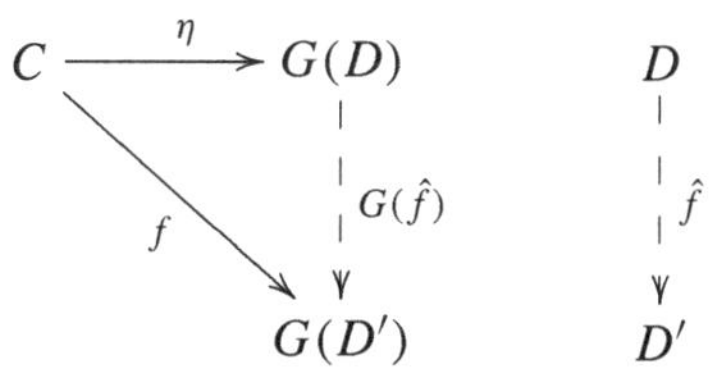

▲

As in previous cases, uniqueness can be given a purely equational specification:

$$\forall h : D \longrightarrow D'. \widehat{G(h) \circ \eta} = h. \tag{1.2}$$

Exercise 65 Show that if (D, η) and (D', η') are universal arrows from C to G then there is a unique isomorphism $D \cong D'$.

Exercise 66 Check that the equational specification of uniqueness (1.2) is valid.

Examples:

- Take $U : \mathbf{Mon} \to \mathbf{Set}$. Given a set X, the universal arrow is

$$\eta_X : X \longrightarrow U(\mathsf{MList}(X)) := x \mapsto [x] .$$

  Indeed, for any monoid $(M, \cdot, 1)$ and any function $f : X \to M$, set

$$\hat{f} : \mathsf{MList}(X) \longrightarrow (M, \cdot, 1) := [x_1, \ldots, x_n] \mapsto f(x_1) \cdot \cdots \cdot f(x_n) .$$

  It is easy to see that $\hat{f}$ is a monoid homomorphism, and that $U(\hat{f}) \circ \eta_X = f$. Moreover, for uniqueness we have that, for any $h : \mathsf{MList}(X) \to (M, \cdot, 1)$,

$$\begin{aligned} \widehat{U(h) \circ \eta_X} &= \widehat{x \mapsto h([x])} = [x_1, \ldots, x_n] \mapsto h([x_1]) \cdot \cdots \cdot h([x_n]) \\ &= [x_1, \ldots, x_n] \mapsto h([x_1] * \cdots * [x_n]) \\ &= [x_1, \ldots, x_n] \mapsto h([x_1, \ldots, x_n]) = h . \end{aligned}$$

- Let $K : \mathcal{C} \to \mathbb{1}$ be the unique functor to the one-object/one-arrow category. A universal arrow from the object of $\mathbb{1}$ to K corresponds to an initial object in $\mathcal{C}$. Indeed, such a universal arrow is given by an object I of $\mathcal{C}$ (and a trivial arrow in $\mathbb{1}$), such that for any A in $\mathcal{C}$ (and relevant arrow in $\mathbb{1}$) there exists a unique arrow from I to A (such that a trivial condition holds).
- Consider the functor $\langle \mathsf{Id}_{\mathcal{C}}, \mathsf{Id}_{\mathcal{C}} \rangle : \mathcal{C} \to \mathcal{C} \times \mathcal{C}$, taking each object A to (A, A) and each arrow f to (f, f). A universal arrow from an object (A, B) of $\mathcal{C} \times \mathcal{C}$ to $\langle \mathsf{Id}_{\mathcal{C}}, \mathsf{Id}_{\mathcal{C}} \rangle$ corresponds to a coproduct of A and B.

Exercise 67 Verify the description of coproducts as universal arrows.

As in the case of posets, a related notion to universal arrows is that of *adjunction*.

Definition 68 Let $\mathcal{C}, \mathcal{D}$ be categories. An ***adjunction*** from $\mathcal{C}$ to $\mathcal{D}$ is a triple (F, G, θ), where F and G are functors

$$\mathcal{C} \underset{G}{\overset{F}{\rightleftarrows}} \mathcal{D}$$

and θ is a family of bijections

$$\theta_{A,B} : \mathcal{C}(A, G(B)) \overset{\cong}{\longrightarrow} \mathcal{D}(F(A), B) ,$$

for each $A \in Ob(\mathcal{C})$ and $B \in Ob(\mathcal{D})$, **natural** in A and B.
We say that F is ***left adjoint*** to G, and G is ***right adjoint*** to F. ▲

Note that θ should be understood as the "witnessed" form—*i.e.* arrows instead of mere relations—of the defining condition for adjunctions in the case of posets:

$$x \leq g(y) \iff f(x) \leq y.$$

This is often displayed as a two-way 'inference rule':

$$\frac{A \longrightarrow GB}{FA \longrightarrow B}$$

Naturality of θ is expressed as follows: for any $f : A \to G(B)$ and any $g : A' \to A$, $h : B \to B'$,

$$\theta_{A',B}(f \circ g) = \theta_{A,B}(f) \circ F(g)\,,$$
$$\theta_{A,B'}(G(h) \circ f) = h \circ \theta_{A,B}(f)\,.$$

Note that f, g are in $\mathcal{C}$, and h is in $\mathcal{D}$. In one line:

$$\theta_{A',B'}(G(h) \circ f \circ g) = h \circ \theta_{A,B}(f) \circ F(g)\,.$$

Diagrammatically:

$$\begin{array}{ccccccc}
\mathcal{C}(A,GB') & \xleftarrow{\mathcal{C}(A,Gh)} & \mathcal{C}(A,GB) & \xrightarrow{\mathcal{C}(g,GB)} & \mathcal{C}(A',GB) & \mathcal{C}(A,GB) & \xrightarrow{\mathcal{C}(g,Gh)} & \mathcal{C}(A',GB') \\
\downarrow{\scriptstyle \theta_{A,B'}} & & \downarrow{\scriptstyle \theta_{A,B}} & & \downarrow{\scriptstyle \theta_{A',B}} & \downarrow{\scriptstyle \theta_{A,B}} & & \downarrow{\scriptstyle \theta_{A',B'}} \\
\mathcal{D}(FA,B') & \xleftarrow[\mathcal{D}(FA,h)]{} & \mathcal{D}(FA,B) & \xrightarrow[\mathcal{D}(Fg,B)]{} & \mathcal{D}(FA',B) & \mathcal{D}(FA,B) & \xrightarrow[\mathcal{D}(Fg,h)]{} & \mathcal{D}(FA',B')
\end{array}$$

Thus, θ is in fact a natural isomorphism

$$\theta : \mathcal{C}(\_\,, G(\_)) \overset{\cong}{\to} \mathcal{D}(F(\_), \_)\,,$$

where $\mathcal{C}(\_\,, G(\_)) : \mathcal{C}^{\mathsf{op}} \times \mathcal{D} \to \mathbf{Set}$ is the result of composing the bivariant hom-functor $\mathcal{C}(\_\,, \_)$ with $\mathsf{Id}_{\mathcal{C}^{\mathsf{op}}} \times G$, and $\mathcal{D}(F(\_), \_)$ is similar.

In the next propositions we show that universal arrows and adjunctions are equivalent notions.

Proposition 69 (Universals define adjunctions) *Let $G : \mathcal{D} \to \mathcal{C}$. If for every object C of $\mathcal{C}$ there exists a universal arrow $\eta_C : C \to G(F(C))$, then:*

1. *F uniquely extends to a functor $F : \mathcal{C} \to \mathcal{D}$ such that $\eta : \mathsf{Id}_{\mathcal{C}} \to G \circ F$ is a natural transformation.*
2. *F is uniquely determined by G (up to unique natural isomorphism), and vice versa.*
3. *For each pair of objects C of $\mathcal{C}$ and D of $\mathcal{D}$, there is a natural bijection:*

$$\theta_{C,D} : \mathcal{C}(C, G(D)) \cong \mathcal{D}(F(C), D)\,.$$

Proof For 1, we extend F to a functor as follows. Given $f : C \to C'$ in $\mathcal{C}$, we consider the composition

$$\eta_{C'} \circ f : C \longrightarrow GFC'.$$

By the universal property of η_C, there exists a unique arrow $Ff : FC \to FC'$ such that the following diagram commutes.

$$\begin{array}{ccc} C & \xrightarrow{\eta_C} & GFC \\ {\scriptstyle f}\downarrow & & \downarrow{\scriptstyle GFf} \\ C' & \xrightarrow[\eta_{C'}]{} & GFC' \end{array}$$

Note that the above is the naturality diagram for η on C, hence the arrow-map thus defined for F is the unique candidate that makes η a natural transformation.
It remains to verify the functoriality of F. To show that F preserves composition, consider $g : C' \to C''$. We have the following commutative diagram,

$$\begin{array}{ccccc} C & \xrightarrow{f} & C' & \xrightarrow{g} & C'' \\ {\scriptstyle \eta_C}\downarrow & & {\scriptstyle \eta_{C'}}\downarrow & & \downarrow{\scriptstyle \eta_{C''}} \\ GFC & \xrightarrow[GFf]{} & GFC' & \xrightarrow[GFg]{} & GFC'' \end{array}$$

from which it follows that

$$G(Fg \circ Ff) \circ \eta_C = GFg \circ GFf \circ \eta_C = \eta_{C''} \circ g \circ f\,,$$

$$\therefore\ F(g \circ f) = \widehat{\eta_{C''} \circ g \circ f} = G(\widehat{Fg \circ Ff}) \circ \eta_C = Fg \circ Ff\,,$$

where the last equality above holds because of (1.2). The verification that F preserves identities is similar.

For 2, we have that each FC is determined uniquely up to unique isomorphism, by the universal property, and once the object part of F is fixed, the arrow part is uniquely determined.

For 3, we need to define a natural isomorphism $\theta_{C,D} : \mathcal{C}(C, G(D)) \cong \mathcal{D}(F(C), D)$. Given $f : C \to GD$, $\theta_{C,D}(f)$ is defined to be the unique arrow $FC \to D$ such that the following commutes, as dictated by universality.

$$
\begin{array}{ccc}
C & \xrightarrow{\eta_C} & GFC \\
& \searrow^{f} & \Big\downarrow{\scriptstyle G(\theta_{C,D}(f))} \\
& & GD
\end{array}
$$

Suppose that $\theta_{C,D}(f) = \theta_{C,D}(g)$. Then

$$f = G(\theta_{C,D}(f)) \circ \eta_C = G(\theta_{C,D}(g)) \circ \eta_C = g\,.$$

Thus $\theta_{C,D}$ is injective. Moreover, given $h : FC \to D$, by the equational formulation of uniqueness (1.2) we have:

$$h = \theta_{C,D}(Gh \circ \eta_C)\,.$$

Thus $\theta_{C,D}$ is surjective. We are left to show naturality, *i.e.* that the following diagram commutes, for all $h : C' \to C$ and $g : D \to D'$.

$$
\begin{array}{ccc}
\mathcal{C}(C, GD) & \xrightarrow{\mathcal{C}(h,Gg)} & \mathcal{C}(C', GD') \\
{\scriptstyle \theta_{C,D}}\Big\downarrow & & \Big\downarrow{\scriptstyle \theta_{C',D'}} \\
\mathcal{D}(FC, D) & \xrightarrow[\mathcal{D}(Fh,g)]{} & \mathcal{D}(FC', D')
\end{array}
$$

We chase around the diagram, starting from $f : C \to GD$.

$$
\begin{aligned}
\mathcal{D}(Fh, g) \circ \theta_{C,D}(f) &= g \circ \theta_{C,D}(f) \circ Fh \\
\theta_{C',D'} \circ \mathcal{C}(h, Gg)(f) &= \theta_{C',D'}(Gg \circ f \circ h)
\end{aligned}
$$

Now:

$$
\begin{aligned}
g \circ \theta_{C,D}(f) \circ Fh &= \theta_{C',D'}(G(g \circ \theta_{C,D}(f) \circ Fh) \circ \eta_{C'}) && \text{by (1.2)} \\
&= \theta_{C',D'}(Gg \circ G(\theta_{C,D}(f)) \circ GFh \circ \eta_{C'}) && \text{functoriality of } G \\
&= \theta_{C',D'}(Gg \circ G(\theta_{C,D}(f)) \circ \eta_C \circ h) && \text{naturality of } \eta \\
&= \theta_{C',D'}(Gg \circ f \circ h) && \text{by (1.2).}
\end{aligned}
$$

■

Proposition 70 (Adjunctions define universals) *Let $G : \mathcal{D} \to \mathcal{C}$ be a functor, $D \in Ob(\mathcal{D})$ and $C \in Ob(\mathcal{C})$. If, for any $D' \in Ob(\mathcal{D})$, there is a bijection*

$$\phi_{D'} : \mathcal{C}(C, G(D')) \cong \mathcal{D}(D, D')$$

natural in D' then there is a universal arrow $\eta : C \to G(D)$.

Proof Take $\eta : C \to G(D) := \phi_D^{-1}(\mathsf{id}_D)$ and, for any $g : C \to G(D')$, take $\hat{g} : D \to D' := \phi_{D'}(g)$.
We have that

$$G(\hat{g}) \circ \eta = G(\hat{g}) \circ \phi_D^{-1}(\mathsf{id}_D) \overset{\text{nat}}{=} \phi_{D'}^{-1}(\hat{g}) = g\,.$$

Moreover, for any $h : D \to D'$,

$$\phi_{D'}(Gh \circ \eta) = \phi_{D'}(Gh \circ \phi_D^{-1}(\mathsf{id}_D)) \overset{\text{nat}}{=} \phi_{D'}(\phi_{D'}^{-1}(h)) = h\,,$$

where equalities labelled with "nat" hold because of naturality of ϕ. ∎

Corollary 71 *Let (F, G, θ) be an adjunction with $F : \mathcal{C} \to \mathcal{D}$. Then, for each $C \in Ob(\mathcal{C})$ there is a universal arrow $\eta : C \to G(F(C))$.* ∎

Equivalence of Universals and Adjoints

Thus we see that the following two situations are equivalent, in the sense that each determines the other uniquely.

- We are given a functor $G : \mathcal{D} \to \mathcal{C}$, and for each object C of $\mathcal{C}$ a universal arrow from C to G.
- We are given functors $F : \mathcal{C} \to \mathcal{D}$ and $G : \mathcal{D} \to \mathcal{C}$, and a natural bijection

$$\theta_{C,D} : \mathcal{C}(C, G(D)) \cong \mathcal{D}(F(C), D)\,.$$

Couniversal Arrows

Let $F : \mathcal{C} \to \mathcal{D}$ be a functor, and D an object of $\mathcal{D}$. A ***couniversal arrow from*** F ***to*** D is an object C of $\mathcal{C}$ and a morphism

$$\epsilon : F(C) \longrightarrow D$$

such that, for every object C' of $\mathcal{C}$ and morphism $g : F(C') \to D$, there exists a unique morphism $\bar{g} : C' \to C$ in $\mathcal{C}$ such that $g = \epsilon \circ F(\bar{g})$.
Diagrammatically:

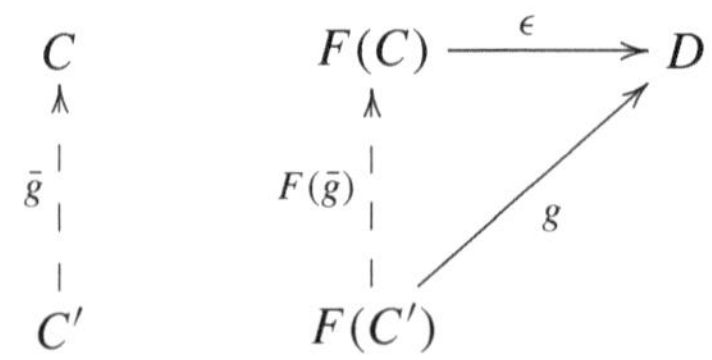

By exactly similar (but dual) reasoning to the previous propositions, an adjunction implies the existence of couniversal arrows, and the existence of the latter implies the existence of the adjunction. Hence,

$$\text{Universality} \equiv \text{Adjunctions} \equiv \text{Couniversality}\,.$$

Some examples of couniversal arrows:

- A terminal object in a category $\mathcal{C}$ is a couniversal arrow from the unique functor $K : \mathcal{C} \to \mathbb{1}$ to the unique object in $\mathbb{1}$.
- Let A, B be objects of $\mathcal{C}$. A product of A and B is a couniversal arrow from $\langle \mathsf{Id}_{\mathcal{C}}, \mathsf{Id}_{\mathcal{C}} \rangle : \mathcal{C} \to \mathcal{C} \times \mathcal{C}$ to (A, B).

1.5.3 Limits and Colimits

In the previous paragraph we described products $A \times B$ as couniversal arrows from the *diagonal functor* $\Delta : \mathcal{C} \to \mathcal{C} \times \mathcal{C}$ to (A, B). Δ is the functor assigning (A, A) to each object A, and (f, f) to each arrow f. Noting that $\mathcal{C} \times \mathcal{C} = \mathcal{C}^2$, where $\mathcal{C}^2$ is a *functor category*, this suggests an important generalisation.

Definition 72 Let $\mathcal{C}$ be a category and $\mathcal{I}$ be another category, thought of as an "index category". A *diagram of shape* $\mathcal{I}$ in $\mathcal{C}$ is just a functor $F : \mathcal{I} \to \mathcal{C}$. Consider the functor category $\mathcal{C}^{\mathcal{I}}$ with objects the functors from $\mathcal{I}$ to $\mathcal{C}$, and natural transformations as morphisms. There is a *diagonal functor*

$$\Delta : \mathcal{C} \longrightarrow \mathcal{C}^{\mathcal{I}},$$

taking each object C of $\mathcal{C}$ to the constant functor $K_C : \mathcal{I} \to \mathcal{C}$, which maps every object of $\mathcal{I}$ to C. A ***limit*** for the diagram F is a couniversal arrow from Δ to F. ▲

This concept of limit subsumes products (including infinite products), pullbacks, inverse limits, etc.

For example, take $\mathcal{I} := 2_{\rightrightarrows}$ (we have seen this before: $2_{\rightrightarrows} = \bullet \rightrightarrows \bullet$). A functor F from $\mathcal{I}$ to $\mathcal{C}$ corresponds to a diagram:

$$A \underset{g}{\overset{f}{\rightrightarrows}} B$$

A couniversal arrow from Δ to F corresponds to the following situation,

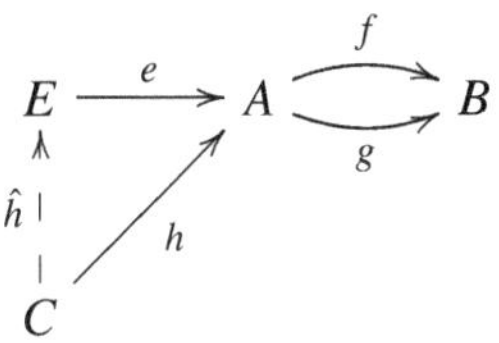

i.e. to an equaliser!

By dualising limits we obtain *colimits*. Some important examples are coproducts, coequalisers, pushouts and ω-colimits.

Exercise 73 Verify that pullbacks are limits by taking:

$$\mathcal{I} := \bullet \longrightarrow \bullet \longleftarrow \bullet$$

Limits as Terminal Objects

Consider $\Delta : \mathcal{C} \to \mathcal{C}^{\mathcal{I}}$ and $F : \mathcal{I} \to \mathcal{C}$. A *cone to* F is an object C of $\mathcal{C}$ and family of arrows γ,

$$\{\, \gamma_I : C \longrightarrow FI \,\}_{I \in Ob(\mathcal{I})}\,,$$

such that, for any $f : I \to J$, the following triangle commutes.

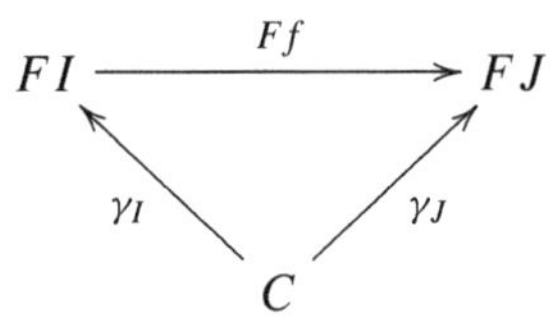

Thus a cone is exactly a natural transformation $\gamma : \Delta C \to F$. A morphism of cones ('mediating morphism') $(C, \gamma) \longrightarrow (D, \delta)$ is an arrow $g : C \to D$ such that each of the following triangles commutes.

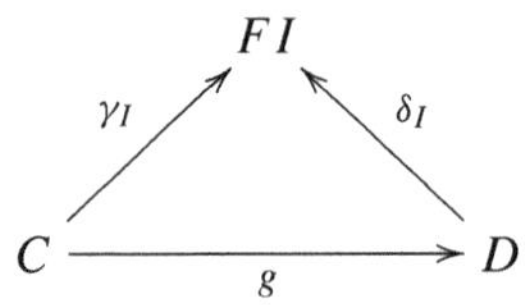

We obtain a category **Cone**(F) whose objects are cones to F and whose arrows are mediating morphisms. Then, a limit of F is a terminal object in **Cone**(F).

1.5.4 Exponentials

In **Set**, given sets A, B, we can form the set of functions $B^A :=$ **Set**(A, B), *which is again a set, i.e. an object of* ***Set***. This closure of **Set** under forming "function spaces" is one of its most important properties.

How can we axiomatise this situation? Once again, rather than asking what the elements of a function space *are*, we ask instead what we can *do* with them operationally. The answer is simple: apply functions to their arguments. That is, there is a map

$$\mathsf{ev}_{A,B} : B^A \times A \longrightarrow B \ \text{ such that } \ \mathsf{ev}_{A,B}(f, a) = f(a)\,.$$

We can think of the function as a "black box": we can feed it inputs and observe the outputs.

Evaluation has the following couniversal property. For any $g : C \times A \to B$, there is a unique map $\Lambda(g) : C \to B^A$ such that the following diagram commutes.

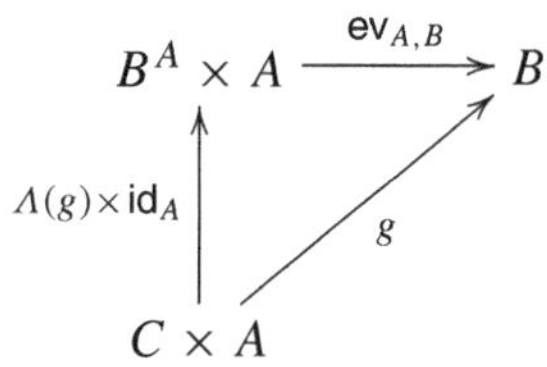

In **Set**, this is defined by:

$$\Lambda(g)(c) : A \longrightarrow B := a \mapsto g(c, a) .$$

This process of transforming a function of two arguments into a function-valued function of one argument is known as *currying*, after H. B. Curry. It is an algebraic form of *λ-abstraction.*

We are now led to the general definition of exponentials. Note that, for each object A of a category $\mathcal{C}$ with products, we can define a functor

$$\_ \times A : \mathcal{C} \longrightarrow \mathcal{C} .$$

Definition 74 Let $\mathcal{C}$ be a category with binary products. We say that $\mathcal{C}$ ***has exponentials*** if for all objects A and B of $\mathcal{C}$ there is a couniversal arrow from $\_ \times A$ to B, *i.e.* an object B^A of $\mathcal{C}$ and a morphism

$$\mathsf{ev}_{A,B} : B^A \times A \longrightarrow B$$

with the couniversal property: for every $g : C \times A \to B$, there is a unique morphism $\Lambda(g) : C \to B^A$ such that the following diagram commutes.

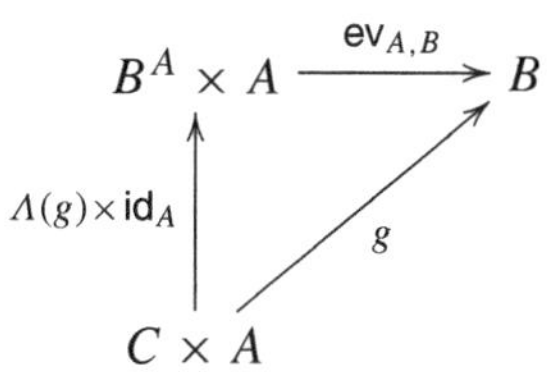

▲

As before, the couniversal property can be given in purely equational terms, as follows. For every $h : C \to B^A$,

$$\Lambda(\mathsf{ev}_{A,B} \circ h \times \mathsf{id}_A) = h .$$

Equivalently, $\mathcal{C}$ has exponentials if, for every object A, the functor $\_ \times A$ has a right adjoint, that is, there exists a functor $\_^A : \mathcal{C} \to \mathcal{C}$ and a bijection

$$\Lambda_{B,C} : \mathcal{C}(C \times A, B) \xrightarrow{\cong} \mathcal{C}(C, B^A)$$

natural in B, C. In that case, $\mathsf{ev}_{A,B} := \Lambda^{-1}(\mathsf{id}_{B^A})$.

Exercise 75 Derive $\_^A$ and Λ^{-1} of the above description from ev and Λ of definition 74.

Exercise 76 Show that $\mathcal{C}$ has exponentials iff, for every $A, B, C \in \mathsf{Ob}(\mathcal{C})$, there is an object B^A and a bijection

$$\theta_C : \mathcal{C}(C \times A, B) \xrightarrow{\cong} \mathcal{C}(C, B^A)$$

natural in C.

Notation 77 The notation B^A for exponential objects is standard in the category theory literature. For our purposes, however, it will be more convenient to write $A \Rightarrow B$.

Exponentials bring us to another fundamental notion, this time for understanding functional types, models of λ-calculus, and the structure of proofs.

Definition 78 A category with a terminal object, products and exponentials is called a ***Cartesian Closed Category (CCC)***. ▲

For example, **Set** is a CCC. Another class of examples are *Boolean algebras*, seen as categories:

- Products are given by conjunctions $A \wedge B$. We define exponentials as *implications*:

$$A \Rightarrow B := \neg A \vee B .$$

- Evaluation is just *Modus Ponens*,

$$(A \Rightarrow B) \wedge A \leq B$$

  while couniversality is the *Deduction Theorem*,

$$C \wedge A \leq B \iff C \leq A \Rightarrow B .$$

1.5.5 Exercises

1. Suppose that $U : \mathcal{C} \to \mathcal{D}$ has a left adjoint F_1, and $V : \mathcal{D} \to \mathcal{E}$ has a left adjoint F_2. Show that $V \circ U : \mathcal{C} \to \mathcal{E}$ has a left adjoint.
2. A *sup-lattice* is a poset P in which every subset $S \subseteq P$ has a supremum (least upper bound) $\bigvee S$. Let P, Q be sup-lattices, and $f : P \to Q$ be a monotone map.

(a) Show that if f has a right adjoint then f preserves least upper bounds:

$$f(\bigvee S) = \bigvee\{f(x) \mid x \in S\}.$$

(b) Show that if f preserves least upper bounds then it has a right adjoint g, given by:

$$g(y) = \bigvee\{x \in P \mid f(x) \leq y\}.$$

(c) Dualise to get a necessary and sufficient condition for the existence of left adjoints.

3. Let $F : \mathcal{C} \to \mathcal{D}$, $G : \mathcal{D} \to \mathcal{C}$ be functors such that F is left adjoint to G, with natural bijection $\theta_{C,D} : \mathcal{C}(C, GD) \xrightarrow{\cong} \mathcal{D}(FC, D)$. Show that there is a natural transformation $\varepsilon : F \circ G \to \mathsf{Id}_{\mathcal{D}}$, the ***counit*** of the adjunction.
Describe this counit explicitly in the case where the right adjoint is the forgetful functor U : **Mon** $\to$ **Set**.
4. Let $F : \mathcal{C} \to \mathcal{D}$ and $G : \mathcal{D} \to \mathcal{C}$ be functors, and assume F is left adjoint to G with natural bijection θ.

(a) Show that F preserves epimorphisms.
(b) Show that F is faithful if and only if, for every object A of $\mathcal{C}$, $\eta_A : A \to GF(A)$ is monic.
(c) Show that if, for each object A of $\mathcal{C}$, there is a morphism $s_A : GF(A) \to A$ such that $\eta_A \circ s_A = \mathsf{id}_{GF(A)}$ then F is full.

1.6 The Curry–Howard Correspondence

We shall now study a beautiful three-way connection between logic, computation and categories:

Table 1.2 The Curry–Howard correspondence

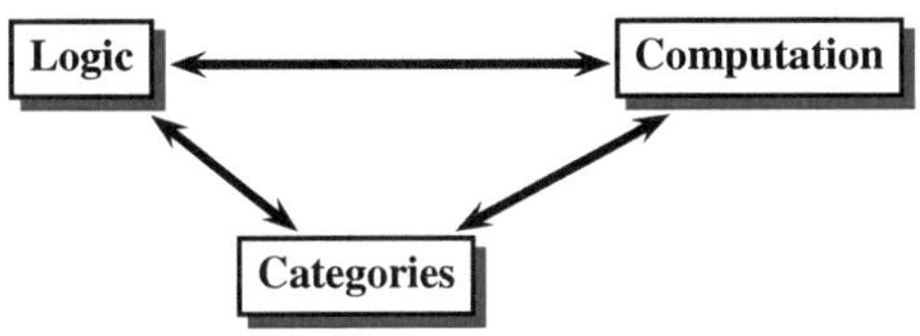

This connection has been known since the 1970s, and is widely used in Computer Science—it is also beginning to be used in Quantum Informatics! It is the upper link (Logic–Computation) that is usually attributed to Haskell B. Curry and William A. Howard, although the idea is related to the operational interpretation of intuitionistic logic given in various formulations by Brouwer, Heyting and Kolmogorov. The link to Categories is mainly due to the pioneering work of Joachim Lambek (Table 1.2).

1.6.1 Logic

Suppose we ask ourselves the question: *What is Logic about?* There are two main kinds of answer: one focuses on *Truth*, and the other on *Proof*. We focus on the latter, that is, on:

What follows from what

Traditional introductions to logic focus on Hilbert-style proof systems, that is, on generating the set of *theorems* of a system from a set of *axioms* by applying rules of inference (e.g. Modus Ponens).

A key step in logic took place in the 1930s with the advent of *Gentzen-style systems*. Instead of focusing on theorems, we look more generally and symmetrically at *What follows from what*: in these systems the primary focus is on *proofs from assumptions*. We will examine two such kinds of systems: Natural Deduction systems and Gentzen sequent calculi.

Definition 79 Consider the fragment of propositional logic with logical connectives $\wedge$ and $\supset$. The assertion that a formula A can be proved from assumptions $A_1, \ldots, A_n$ is expressed by a ***sequent***:

$$A_1, \ldots, A_n \vdash A$$

We use Γ, Δ to range over finite sets of formulas, and write Γ, A for $\Gamma \cup \{A\}$. Proofs are built using the proof rules of Table 1.3; the resulting proof system is called *the* ***Natural Deduction system for*** $\wedge,\supset$. ▲

For example, we have the following proof of $\supset$-transitivity.

$$\dfrac{\dfrac{\dfrac{\dfrac{}{A \supset B, B \supset C, A \vdash B \supset C}\,\text{Id} \qquad \dfrac{\dfrac{}{A \supset B, B \supset C, A \vdash A \supset B}\,\text{Id} \quad \dfrac{}{A \supset B, B \supset C, A \vdash A}\,\text{Id}}{A \supset B, B \supset C, A \vdash B}\,{\supset}\text{E}}{A \supset B, B \supset C, A \vdash C}\,{\supset}\text{E}}{A \supset B, B \supset C \vdash A \supset C}\,{\supset}\text{I}}{}$$

An important feature of Natural Deduction is the systematic pattern it exhibits in the structure of the inference rules. For each connective $\Box$ there are *introduction rules*,

Table 1.3 Natural deduction system for $\wedge,\supset$

Identity	Conjunction	Implication
$\dfrac{}{\Gamma, A \vdash A}\,\text{Id}$	$\dfrac{\Gamma \vdash A \quad \Gamma \vdash B}{\Gamma \vdash A \wedge B}\,\wedge\text{intro}$	$\dfrac{\Gamma, A \vdash B}{\Gamma \vdash A \supset B}\,{\supset}\text{intro}$
	$\dfrac{\Gamma \vdash A \wedge B}{\Gamma \vdash A}\,\wedge\text{elim}_1$	$\dfrac{\Gamma \vdash A \supset B \quad \Gamma \vdash A}{\Gamma \vdash B}\,{\supset}\text{elim}$
	$\dfrac{\Gamma \vdash A \wedge B}{\Gamma \vdash B}\,\wedge\text{elim}_2$	

which show how formulas $A \Box B$ can be derived, and *elimination rules*, which show how such formulas can be used to derive other formulas.

Admissibility

We say that a proof rule

$$\frac{\Gamma_1 \vdash A_1 \quad \cdots \quad \Gamma_n \vdash A_n}{\Delta \vdash B}$$

is ***admissible*** in Natural Deduction if, whenever there are proofs of $\Gamma_i \vdash A_i$ then there is also a proof of $\Delta \vdash B$. For example, the following *Cut rule* is admissible.

$$\frac{\Gamma \vdash A \quad A, \Delta \vdash B}{\Gamma, \Delta \vdash B}\ \mathsf{Cut}$$

Exercise 80 Show that the following rules are admissible in Natural Deduction.

1. The Weakening rule:

$$\frac{\Gamma \vdash B}{\Gamma, A \vdash B}$$

2. The Cut rule.

Our focus will be on ***Structural Proof Theory***, that is studying the "space of formal proofs" as a mathematical structure in its own right, rather than focussing only on

$$\text{Provability} \longleftrightarrow \text{Truth}$$

(*i.e.* the usual notions of "soundness and completeness"). One motivation for this approach comes from trying to understand and use the *computational content* of proofs, epitomised in the "Curry-Howard correspondence".

1.6.2 Computation

Our starting point in computation is the pure calculus of functions called the *λ-calculus*.

Definition 81 Assume a countably infinite set of ***variables***, ranged over by x, y, z and variants. ***λ-calculus terms***, ranged over by t, u, v etc, are constructed from the following inductive definition.

- Every variable x is a term.
- If t and u are terms, then $t\ u$ is a term (***application***).
- If x is a variable and t is a term, then $\lambda x.\, t$ is a term (***λ-abstraction***). ▲

The above definition can be given in the following compact form, which will be followed in similar definitions in the sequel.

$$\begin{aligned} \mathrm{VA} &\ni x, y, z, \ldots \\ \mathrm{TE} &\ni t, u, v \ ::= \ x \mid t\,u \mid \lambda x.t \end{aligned}$$

The computational content of the calculus is exhibited in the following examples. Note that the first example is not part of our formal syntax: it presupposes some *encoding* of numerals and successors (Table 1.4).

Table 1.4 Examples of λ-terms

$\lambda x.x+1$	successor function
$\lambda x.x$	identity function
$\lambda f.\lambda x.fx$	application
$\lambda f.\lambda x.f(fx)$	double application
$\lambda f.\lambda g.\lambda x.g(f(x))$	composition and application

What we also note above is the use of parentheses in order to disambiguate the structure of terms (*i.e.* the precedence of term constructors). To avoid notational clutter we also use the following conventions.

- Applications associate to the left. For example, $f\,x\,y$ stands for $(fx)\,y$.
- The scope of an abstractions goes as far to the right as possible. For example,

$$\lambda f.(\lambda x.f(xx))\,\lambda x.f(xx) \ \text{ stands for } \ \lambda f.((\lambda x.(f(xx)))(\lambda x.(f(xx)))).$$

The *free variables* of a term are those that are not bound by any λ; they can be seen as the *assumptions* of the term.

Definition 82 The set of ***free variables*** of a term t, $\mathsf{fv}(t)$, is given by:

$$\begin{aligned} \mathsf{fv}(x) &:= \{x\}, \\ \mathsf{fv}(t\,u) &:= \mathsf{fv}(t) \cup \mathsf{fv}(u), \\ \mathsf{fv}(\lambda x.t) &:= \mathsf{fv}(t) \setminus \{x\}. \end{aligned}$$

▲

The notation $\lambda x.t$ is meant to serve the purpose of expressing formally

the function that returns t on input x.

Thus, λ is a *binder*, that is, it binds the variable x in the "function" $\lambda x.t$, in the same way that e.g. $\int$ binds x in $\int f(x)\,dx$. This means that there should be no difference between $\lambda x.t$ and $\lambda x'.t'$, where t' is obtained from t by swapping x with some *fresh* variable x' (*i.e.* with some x' not appearing free in t). For example, the terms

$$\lambda x.x \quad \text{and} \quad \lambda x'.x'$$

should be "equal", as they both stand for the identity function. We formalise this by stipulating that

Terms are identified up to α-equivalence

where we say that two terms are α-equivalent iff they differ solely in the choice of variables appearing in *binding positions*. This is formally defined in two steps, as follows.

Definition 83 We define *variable-swapping* on terms recursively as follows.

$$\begin{aligned}(y\ x)\bullet z &:= \begin{cases} y & \text{if } z = x \\ x & \text{if } z = y \\ z & \text{otherwise}\end{cases}\\ (y\ x)\bullet t\,u &:= ((y\ x)\bullet t)((y\ x)\bullet u)\\ (y\ x)\bullet \lambda z.t &:= \lambda((y\ x)\bullet z).((y\ x)\bullet t)\end{aligned}$$

Then, ***α-equivalence***, $=_\alpha$, is the relation on terms defined inductively by:[4]

- $x =_\alpha x$,
- $t\,u =_\alpha t'\,u'$ if $t =_\alpha t'$ and $u =_\alpha u'$,
- $\lambda x.t =_\alpha \lambda x'.t'$ if, for all y not appearing in $t\,t'$, $(y\ x)\bullet t =_\alpha (y\ x')\bullet t'$. ▲

Equating terms modulo α-equivalence means that we work with $\mathrm{TE}/{=_\alpha}$ instead of TE. Henceforth, we will refer to elements of $\mathrm{TE}/{=_\alpha}$ as *terms*, and to elements of TE as ***raw terms***. Note that α-equivalence is meaningful only on raw terms.

Exercise 84 Prove the following α-equivalences.

$$\lambda x.x =_\alpha \lambda y.y\,, \quad \lambda x.\lambda y.\,xy =_\alpha \lambda y.\lambda x.\,yx\,, \quad x(\lambda x.x) =_\alpha x(\lambda y.y)\,.$$

Exercise 85 Show that, for all raw terms t, t' and variables x, x', if $t =_\alpha t'$ then $\mathsf{fv}(t) = \mathsf{fv}(t')$ and $(x\ x')\bullet t =_\alpha (x\ x')\bullet t'$.

Moreover, show that, for any $x, x' \notin \mathsf{fv}(t)$, $t =_\alpha (x\ x')\bullet t$. Hence infer that, for any $x' \notin \mathsf{fv}(t)$, $\lambda x.t =_\alpha \lambda x'.(x\ x')\bullet t$.

From the above exercise we obtain that fv and variable-swapping extend to terms (*i.e.* to $\mathrm{TE}/{=_\alpha}$) in a straightforward manner. Moreover, we have that, for any term t and any $x' \notin \mathsf{fv}(t)$,

$$\lambda x.t = \lambda x'.(x\ x')\bullet t\,.$$

Since λ-abstractions stand for functions, an application of a λ-abstraction on another term should result to a *substitution* of the latter inside the body of the abstraction.

---

[4] The last clause can be replaced by any of the following:

- ...if, for some y not appearing in $t\,t'$, $(y\ x)\bullet t =_\alpha (y\ x')\bullet t'$.
- ...if, for all y not appearing free in $t\,t'$, $(y\ x)\bullet t =_\alpha (y\ x')\bullet t'$.
- ...if, for some y not appearing free in $t\,t'$, $(y\ x)\bullet t =_\alpha (y\ x')\bullet t'$.

Definition 86 Define the ***substitution*** of a term t for a variable x inside a term inductively by:

$$y[t/x] := \begin{cases} t & \text{if } y = x \\ y & \text{if } y \neq x \end{cases}$$
$$(uv)[t/x] := (u[t/x])(v[t/x])$$
$$(\lambda z.u)[t/x] := \lambda z.\,(u[t/x]) \qquad (*)$$

where $(*)$ indicates the condition that $z \notin \mathsf{fv}(x\,t)$. ▲

Note that, due to identification of α-equivalent (raw) terms, it is always possible to rename bound variables so that condition $(*)$ be satisfied: for example,

$$(\lambda z.zx)[z/x] = (\lambda y.yx)[z/x] = \lambda y.yz$$

Exercise 87 Show that, for all λ-terms u, t, t' and variables x, x' such that $x' \notin \mathsf{fv}(u) \setminus \{x\}$,

$$u[t/x][t'/x'] = u[(t[t'/x'])/x]\,.$$

We proceed to the definition of β-reduction and β-conversion. These are relations defined on pairs of terms and express the computational content of the calculus.

Definition 88 We take ***β-reduction***, $\longrightarrow_\beta$, to be the relation defined by:

$$(\lambda x.t)\,u \longrightarrow_\beta t[u/x]\,.$$

This extends to arbitrary terms as follows. If $t \longrightarrow_\beta t'$ then:

$$t\,u \longrightarrow_\beta t'u\,, \quad u\,t \longrightarrow_\beta u\,t', \quad \lambda x.t \longrightarrow_\beta \lambda x.t'.$$

We take ***β-conversion***, $=_\beta$, to be the symmetric reflexive transitive closure of β-reduction, that is, the equivalence relation induced by:

$$(\lambda x.t)\,u =_\beta t[u/x]\,.$$

▲

With β-reduction we obtain a notion of "computational dynamics". For example:

$$\begin{aligned}(\lambda f.\,f(f\,y))(\lambda x.\,x+1) &\longrightarrow_\beta (\lambda x.\,x+1)((\lambda x.\,x+1)\,y)\\ &\longrightarrow_\beta ((\lambda x.\,x+1)\,y)+1 \longrightarrow_\beta (y+1)+1\end{aligned}$$

$$\begin{aligned}(\lambda f.\,f(f\,y))(\lambda x.\,x+1) &\longrightarrow_\beta (\lambda x.\,x+1)((\lambda x.\,x+1)\,y)\\ &\longrightarrow_\beta (\lambda x.\,x+1)(y+1) \longrightarrow_\beta (y+1)+1\end{aligned}$$

Note that in the sequel we will usually write β-reduction simply by "$\longrightarrow$".

1.6.3 Simply-Typed λ-Calculus

The "pure" λ-calculus we have discussed so far is *very* unconstrained. For example, it allows *self-application*, *i.e.* terms like xx are perfectly legal. On the one hand, this means that the calculus very expressive: for example, we can encode *recursion* by setting

$$\mathbf{Y} := \lambda f.(\lambda x.\, f(xx))\, \lambda x.f(xx)\,.$$

We have:

$$\mathbf{Y}t \longrightarrow (\lambda x.\, t(xx))\, \lambda x.\, t(xx) \longrightarrow t((\lambda x.\, t(xx))\, \lambda x.\, t(xx)) \longleftarrow t(\mathbf{Y}t)$$

However, self-application leads also to *divergences*. The most characteristic example is the following. Setting $\Omega := (\lambda x.xx)\, \lambda x.xx$, we have:

$$\Omega \longrightarrow \Omega \longrightarrow \Omega \longrightarrow \cdots$$

Historically, Curry extracted $\mathbf{Y}$ from an analysis of Russell's Paradox, so it should come as no surprise that it too leads to divergences: setting t' to be $\lambda x.\, t(xx)$,

$$\mathbf{Y}t \longrightarrow t't' \longrightarrow t(t't') \longrightarrow t(t(t't')) \longrightarrow \cdots$$

The solution is to introduce *types*. The original idea, due to Church following Russell, was that:

Types are there to stop you doing bad things

However, it has turned out that types constitute one of the most fruitful *positive* ideas in Computer Science, and provide one of the key disciplines of programming.

Definition 89 Let us assume a set of *base types*, ranged over by b. The ***simply-typed λ-calculus*** is defined as follows.

Type	$\mathrm{TY} \ni T, U ::= b \mid T \to U \mid T \times U$
Term	$\mathrm{TE} \ni t, u ::= x \mid t\, u \mid \lambda x.\, t \mid \langle t, u \rangle \mid \pi_1 u \mid \pi_2 u$
Typing context	$\Gamma ::= \varnothing \mid x : T, \Gamma \qquad$ (x does not appear in Γ)

A ***typing judgement*** is a triple of the form

$$\Gamma \vdash t : T\,,$$

Table 1.5 Simply-typed λ-calculus for ×, →

Variable	Product	Function
$\dfrac{}{\Gamma, x : T \vdash x : T}$	$\dfrac{\Gamma \vdash t : T \qquad \Gamma \vdash u : U}{\Gamma \vdash \langle t, u \rangle : T \times U}$	$\dfrac{\Gamma, x : U \vdash t : T}{\Gamma \vdash \lambda x. t : U \to T}$
	$\dfrac{\Gamma \vdash v : T \times U}{\Gamma \vdash \pi_1 v : T}$	$\dfrac{\Gamma \vdash t : U \to T \qquad \Gamma \vdash u : U}{\Gamma \vdash t\, u : T}$
	$\dfrac{\Gamma \vdash v : T \times U}{\Gamma \vdash \pi_2 v : U}$	

which is to be understood as the assertion that term t has the type T under the assumptions that x_1 has type $T_1, \ldots, x_k$ has type T_k, if $\Gamma = x_1 : T_1, \ldots, x_k : T_k$. A ***typed term*** is a term t accompanied with a type T and a context Γ, such that the judgement $\Gamma \vdash t : T$ is derivable by use of the *typing rules* of Table 1.5. ▲

Note that contexts are sets, and so $x : T, \Gamma$ stands for $\{x : T\} \cup \Gamma$ with x not appearing in Γ. As before, terms are identified up to α-equivalence.

From the definition of types we see that the simply-typed λ-calculus is a calculus of functions and products. For example:

$$
\begin{array}{ll}
b \to b \to b & \text{first-order function type} \\
(b \to b) \to b & \text{second-order function type}
\end{array}
$$

Exercise 90 Can you type the following terms?

$$\lambda x. xx\,, \quad \lambda f. (\lambda x. f(xx))(\lambda x. f(xx))\,.$$

Exercise 91 (Weakening & Cut) Show that Weakening and Cut are admissible in the typing system of the simply-typed λ-calculus:

$$\frac{\Gamma \vdash t : T}{\Gamma, x : U \vdash t : T}\ \mathsf{Weak} \qquad \frac{\Gamma \vdash t : T \qquad \Gamma, x : T \vdash u : U}{\Gamma \vdash u[t/x] : U}\ \mathsf{Cut}$$

We proceed to the rules for reduction and conversion. These are given as in the untyped case, with the addition of η-rules, which are essentially extensionality principles.

Definition 92 We define β-reduction, $\longrightarrow_\beta$, by the following rules, and let β-conversion, $=_\beta$, be its symmetric reflexive transitive closure.

$$
\begin{aligned}
(\lambda x. t)u &\longrightarrow_\beta t[u/x] \\
\pi_1 \langle t, u \rangle &\longrightarrow_\beta t \\
\pi_2 \langle t, u \rangle &\longrightarrow_\beta u
\end{aligned}
$$

Moreover, η-conversion, $=_\eta$, is the symmetric reflexive transitive relation obtained by the following rules,

$$t =_\eta \lambda x.tx \qquad x \notin \mathsf{fv}(t), \text{ at function types}$$
$$v =_\eta \langle \pi_1 v, \pi_2 v\rangle \qquad \text{at product types}$$

and λ-conversion, $=_\lambda$, is the transitive closure of $=_\beta \cup =_\eta$. ▲

Implicit in the above definition is the fact that η-rules relate typed terms. For example, $t =_\eta \lambda x.tx$ has as side condition that t be of function type, *i.e.* that t be a typed term $\Gamma \vdash t : T \to U$. Now, following our intuitive interpretation of arrows as functions, we can read this η-rule as:

t is the function returning $t(x)$ to every input x

Note that the above statement is in fact the couniversal property of currying in **Set**; we will see more on this in the next sections!

Exercise 93 (Subject Reduction) Show that, for any typed term $\Gamma \vdash t : T$, if $t \longrightarrow_\beta t'$ then $\Gamma \vdash t' : T$ is derivable.

Strong Normalisation

Term reduction results in a *normal form*: an *explicit but much longer expression* in which no more reductions are applicable. Formally, a λ-term is called a ***redex*** if it is in one of forms of the left-hand-side of the β-reduction rules, and therefore β-reduction can be applied to it. A term is in ***normal form*** if it contains no redexes. In the light of the correspondence presented in the next paragraph, a term in normal form corresponds to a proof in which all lemmas have been eliminated.

Fact 94 (SN) *For every term t, there is no infinite sequence of β-reductions:*

$$t \longrightarrow t_0 \longrightarrow t_1 \longrightarrow t_2 \longrightarrow \cdots$$

The above result states that *every* reduction sequence leads eventually to a term in normal form. Note, though, that reduction to normal form has enormous (*non-elementary*) complexity.

The Correspondence Between Logic and Computation

Comparing the following two systems,

Natural Deduction System for $\wedge,\supset$	vs	Simply-Typed λ-calculus for $\times,\to$

we notice that if we equate

$$\wedge \equiv \times$$
$$\supset \equiv \to$$

then they are the same! This is the Logic–Computation part of the Curry-Howard correspondence (sometimes: "Curry-Howard isomorphism"). It works on three levels (Table 1.6):

Table 1.6 Correspondence between logic and computation

Natural deduction system	Simply-typed λ-calculus
Formulas	Types
Proofs	Terms
Proof transformations	Term reductions

The view of proofs as containing computational content can also be detected in the *Brouwer-Heyting-Kolmogorov interpretation* of intuitionistic logic:

- A proof of an implication $A \supset B$ is a procedure which transforms any proof of A into a proof of B.
- A proof of $A \wedge B$ is a pair consisting of a proof of A and a proof of B.

These readings motivate identifying $A \wedge B$ with $A \times B$, and $A \supset B$ with $A \to B$. Moreover, these ideas have strong connections to computing. The λ-calculus is a "pure" version of functional programming languages such as Haskell and SML. So we get a reading of:

Proofs as Programs

1.6.4 Categories

We now have our link between Logic and Computation. We now proceed to complete the triangle of the Curry-Howard correspondence by showing the connection to Categories.

We establish the link from Logic (and Computation) to Categories. Let $\mathcal{C}$ be a cartesian closed category. We shall interpret formulas (or types) as objects of $\mathcal{C}$. A morphism $f : A \to B$ will then correspond to a proof of B from assumption A, *i.e.* a proof of $A \vdash B$ (a typed term $x : A \vdash t : B$). Note that the bare structure of a category only supports proofs from a single assumption. Since $\mathcal{C}$ has finite products, a proof of

$$A_1, \ldots, A_k \vdash A$$

will correspond to a morphism

$$f : A_1 \times \cdots \times A_k \longrightarrow A \,.$$

The correspondence is depicted in the Table 1.7.

Moreover, the rules for β- and η-conversion are all then *derivable* from the equations of cartesian closed categories. So cartesian closed categories are *models* of $\wedge,\supset$-logic at the level of proofs and proof-transformations, and of simply typed

Table 1.7 Correspondence between logic and categories

Axiom	$\dfrac{}{\Gamma, A \vdash A}\ \mathsf{Id}$	$\dfrac{}{\pi_2 : \Gamma \times A \longrightarrow A}$
Conjunction	$\dfrac{\Gamma \vdash A \qquad \Gamma \vdash B}{\Gamma \vdash A \wedge B}\ \wedge\mathsf{I}$	$\dfrac{f : \Gamma \longrightarrow A \qquad g : \Gamma \longrightarrow B}{\langle f, g \rangle : \Gamma \longrightarrow A \times B}$
	$\dfrac{\Gamma \vdash A \wedge B}{\Gamma \vdash A}\ \wedge\mathsf{E}_1$	$\dfrac{f : \Gamma \longrightarrow A \times B}{\pi_1 \circ f : \Gamma \longrightarrow A}$
	$\dfrac{\Gamma \vdash A \wedge B}{\Gamma \vdash B}\ \wedge\mathsf{E}_2$	$\dfrac{f : \Gamma \longrightarrow A \times B}{\pi_2 \circ f : \Gamma \longrightarrow B}$
Implication	$\dfrac{\Gamma, A \vdash B}{\Gamma \vdash A \supset B}\ \supset\mathsf{I}$	$\dfrac{f : \Gamma \times A \longrightarrow B}{\Lambda(f) : \Gamma \longrightarrow (A \Rightarrow B)}$
	$\dfrac{\Gamma \vdash A \supset B \quad \Gamma \vdash A}{\Gamma \vdash B}\ \supset\mathsf{E}$	$\dfrac{f : \Gamma \longrightarrow (A \Rightarrow B) \quad g : \Gamma \rightarrow A}{\mathsf{ev}_{A,B} \circ \langle f, g \rangle : \Gamma \longrightarrow B}$

λ-calculus at the level of terms and term-conversions. The connection to computation is examined in more detail below.

Remark 95 In our translation of Logic sequents there is an implicit ordering of assumptions: a *set* of assumptions is mapped to an assumption *product*,

$$\{A_1, \ldots, A_n\} \longmapsto A_1 \times \cdots \times A_n \,.$$

In practice, since for any permutation $A'_1, \ldots, A'_n$ of $A_1, \ldots, A_n$ we have

$$A_1 \times \cdots \times A_n \cong A'_1 \times \cdots \times A'_n \,,$$

such an ordering is harmless.

1.6.5 Categorical Semantics of Simply-Typed λ-Calculus

We translate the simply-typed λ-calculus into a cartesian closed category $\mathcal{C}$, so that to each typed term $x_1 : T_1, \ldots, x_k : T_k \vdash t : T$ corresponds an arrow

$$[\![t]\!] : [\![T_1]\!] \times \cdots \times [\![T_k]\!] \longrightarrow [\![T]\!] \,.$$

The translation if given by the function $[\![\_]\!]$ defined below ("semantic brackets").

Definition 96 (Semantic translation) Let $\mathcal{C}$ be a CCC and suppose we are given an assignment of an object $\tilde{b}$ to each base type b. Then, the translation is defined recursively on types by:

$$[\![b]\!] := \tilde{b}\,, \quad [\![T \times U]\!] := [\![T]\!] \times [\![U]\!]\,, \quad [\![T \rightarrow U]\!] := [\![T]\!] \Rightarrow [\![U]\!]\,,$$

and on typed terms by:

$$\frac{}{[\![\Gamma, x : T \vdash x : T]\!] := \pi_2 : [\![\Gamma]\!] \times [\![T]\!] \longrightarrow [\![T]\!]}$$

$$\frac{[\![\Gamma \vdash t : T \times U]\!] = f : [\![\Gamma]\!] \longrightarrow [\![T]\!] \times [\![U]\!]}{[\![\Gamma \vdash \pi_1 t : T]\!] := [\![\Gamma]\!] \xrightarrow{f} [\![T]\!] \times [\![U]\!] \xrightarrow{\pi_1} [\![T]\!]}$$

$$\frac{[\![\Gamma \vdash t : T]\!] = f : [\![\Gamma]\!] \longrightarrow [\![T]\!] \qquad [\![\Gamma \vdash u : U]\!] = g : [\![\Gamma]\!] \longrightarrow [\![U]\!]}{[\![\Gamma \vdash \langle t, u\rangle : T \times U]\!] := [\![\Gamma]\!] \xrightarrow{\langle f,g\rangle} [\![T]\!] \times [\![U]\!]}$$

$$\frac{[\![\Gamma, x : T \vdash t : U]\!] = f : [\![\Gamma]\!] \times [\![T]\!] \longrightarrow [\![U]\!]}{[\![\Gamma \vdash \lambda x.t : T \to U]\!] := \Lambda(f) : [\![\Gamma]\!] \longrightarrow ([\![T]\!] \Rightarrow [\![U]\!])}$$

$$\frac{[\![\Gamma \vdash t : T \to U]\!] = f \qquad [\![\Gamma \vdash u : T]\!] = g}{[\![\Gamma \vdash t\,u : U]\!] := [\![\Gamma]\!] \xrightarrow{\langle f,g\rangle} ([\![T]\!] \Rightarrow [\![U]\!]) \times [\![T]\!] \xrightarrow{\mathsf{ev}} [\![U]\!]}$$

▲

Our aim now is to verify that λ-conversion (induced by β- and η-rules) is *preserved* by the translation, *i.e.* that, for any t, u,

$$t =_\lambda u \implies [\![t]\!] = [\![u]\!] .$$

This would mean that our categorical semantics is *sound*.

Let us recall some structures from CCC's. Given $f_1 : D_1 \to E_1$, $f_2 : D_2 \to E_2$, we defined

$$f_1 \times f_2 = \langle f_1 \circ \pi_1, f_2 \circ \pi_2\rangle : D_1 \times D_2 \longrightarrow E_1 \times E_2 ,$$

and we showed that $(f_1 \times f_2) \circ \langle h_1, h_2\rangle = \langle f_1 \circ h_1, f_2 \circ h_2\rangle$. Moreover, exponentials are given by the following natural bijection.

$$\frac{f : D \times E \longrightarrow F}{\Lambda(f) : D \longrightarrow (E \Rightarrow F)}$$

Equivalently, recall the basic equation:

$$\mathsf{ev} \circ (\Lambda(f) \times \mathsf{id}_E) = f ,$$

where $\Lambda(f)$ is the unique arrow $D \to (E \Rightarrow F)$ satisfying this equation, with uniqueness being specified by:

$$\forall h : D \longrightarrow (E \Rightarrow F).\, \Lambda(\mathsf{ev} \circ (h \times \mathsf{id}_E)) = h .$$

Naturality of Λ is then proven as follows.

Proposition 97 *For any* $f : A \times B \to C$ *and* $g : A' \to A$,

$$\Lambda(f) \circ g = \Lambda(f \circ (g \times \mathsf{id}_B)) .$$

Proof

$$\begin{aligned} \Lambda(f) \circ g &= \Lambda(\mathsf{ev} \circ ((\Lambda(f) \circ g) \times \mathsf{id}_B)) \\ &= \Lambda(\mathsf{ev} \circ (\Lambda(f) \times \mathsf{id}_B) \circ (g \times \mathsf{id}_B))) = \Lambda(f \circ (g \times \mathsf{id}_B)) . \end{aligned}$$

■

Substitution Lemma

We consider a ***simultaneous substitution*** for all the free variables in a term.

Definition 98 Let $\Gamma = x_1 : T_1, \ldots, x_k : T_k$. Given typed terms

$$\Gamma \vdash t : T \quad \text{and} \quad \Gamma \vdash t_i : T_i\,,\ 1 \leq i \leq k\,,$$

we define $t[\boldsymbol{t}/\boldsymbol{x}] \equiv t[t_1/x_1, \ldots, t_k/x_k]$ recursively by:

$$\begin{aligned} x_i[\boldsymbol{t}/\boldsymbol{x}] &:= t_i \\ (\pi_i\, t)[\boldsymbol{t}/\boldsymbol{x}] &:= \pi_i(t[\boldsymbol{t}/\boldsymbol{x}]) \\ \langle t, u\rangle[\boldsymbol{t}/\boldsymbol{x}] &:= \langle t[\boldsymbol{t}/\boldsymbol{x}], u[\boldsymbol{t}/\boldsymbol{x}]\rangle \\ (t\, u)[\boldsymbol{t}/\boldsymbol{x}] &:= (t[\boldsymbol{t}/\boldsymbol{x}])(u[\boldsymbol{t}/\boldsymbol{x}]) \\ (\lambda x.\, t)[\boldsymbol{t}/\boldsymbol{x}] &:= \lambda x.\, t[\boldsymbol{t}, x/\boldsymbol{x}, x] . \end{aligned}$$

▲

Note that, in contrast to ordinary substitution, simultaneous substitution can be defined directly on raw terms, that is, prior to equating them modulo α-equivalence. Moreover, we can show that:

$$t[t_1/x_1, \ldots, t_k/x_k] = t[t_1/x_1] \cdots [t_k/x_k] .$$

We can now show the following *Substitution Lemma*.

Proposition 99 *For* $t, t_1, \ldots, t_k$ *as in the previous definition,*

$$[\![t[t_1/x_1, \ldots, t_k/x_k]]\!] = [\![t]\!] \circ \langle [\![t_1]\!], \ldots, [\![t_k]\!]\rangle .$$

Proof By induction on the structure of t.
(1) If $t = x_i$:

$$[\![x_i[\boldsymbol{t}/\boldsymbol{x}]]\!] = [\![t_i]\!] = \pi_i \circ \langle [\![t_1]\!], \ldots, [\![t_k]\!]\rangle = [\![x_i]\!] \circ \langle [\![t_1]\!], \ldots, [\![t_k]\!]\rangle .$$

(2) If $t = uv$ then, abbreviating $\langle [\![t_1]\!], \ldots, [\![t_k]\!]\rangle$ to $\langle [\![\boldsymbol{t}]\!]\rangle$ we have:

$$\begin{aligned}
[\![uv[\boldsymbol{t}/\boldsymbol{x}]]\!] &= [\![(u[\boldsymbol{t}/\boldsymbol{x}])(v[\boldsymbol{t}/\boldsymbol{x}])]\!] && \text{Defn of substitution}\\
&= \mathsf{ev} \circ \langle [\![u[\boldsymbol{t}/\boldsymbol{x}]]\!], [\![v[\boldsymbol{t}/\boldsymbol{x}]]\!]\rangle && \text{Defn of semantic function}\\
&= \mathsf{ev} \circ \langle [\![u]\!] \circ \langle [\![\boldsymbol{t}]\!]\rangle, [\![v]\!] \circ \langle [\![\boldsymbol{t}]\!]\rangle\rangle && \text{Induction hyp.}\\
&= \mathsf{ev} \circ \langle [\![u]\!], [\![v]\!]\rangle \circ \langle [\![\boldsymbol{t}]\!]\rangle && \text{Property of products}\\
&= [\![uv]\!] \circ \langle [\![\boldsymbol{t}]\!]\rangle && \text{Defn of semantic function}
\end{aligned}$$

(3) If $t = \lambda x.\, u$:

$$\begin{aligned}
[\![\lambda x.u[\boldsymbol{t}/\boldsymbol{x}]]\!] &= [\![\lambda x.(u[\boldsymbol{t}, x/\boldsymbol{x}, x])]\!] && \text{Defn. of substitution}\\
&= \Lambda([\![u[\boldsymbol{t}, x/\boldsymbol{x}, x]]\!]) && \text{Defn. of semantic function}\\
&= \Lambda([\![u]\!] \circ (\langle [\![\boldsymbol{t}]\!]\rangle \times \mathsf{id})) && \text{Induction hyp.}\\
&= \Lambda([\![u]\!]) \circ \langle [\![\boldsymbol{t}]\!]\rangle && \text{Prop. 97}\\
&= [\![\lambda x.u]\!] \circ \langle [\![\boldsymbol{t}]\!]\rangle && \text{Defn. of semantic function}
\end{aligned}$$

(4,5) The cases of projections and pairs are left as exercise. ■

Exercise 100 Complete the proof of the above proposition.

Validating the Conversion Rules

We can now show that the conversion rules of the λ-calculus are preserved by the translation, and hence the interpretation is sound. Observe the correspondence between η-rules and uniqueness (couniversality) principles.

- For β-conversion: $\big[(\lambda x.\, t)u = t[u/x]\,,\ \pi_1\langle t, u\rangle = t\,,\ \pi_2\langle t, u\rangle = u\big]$

$$\begin{aligned}
[\![(\lambda x.\, t)u]\!] &= \mathsf{ev} \circ \langle \Lambda([\![t]\!]), [\![u]\!]\rangle && \text{Defn. of semantics}\\
&= \mathsf{ev} \circ (\Lambda([\![t]\!]) \times \mathsf{id}) \circ \langle \mathsf{id}_{[\![\Gamma]\!]}, [\![u]\!]\rangle && \text{Property of } \times\\
&= [\![t]\!] \circ \langle \mathsf{id}_{[\![\Gamma]\!]}, [\![u]\!]\rangle && \text{Defn. of } \Lambda\\
&= [\![t[\boldsymbol{x}, u/\boldsymbol{x}, x]]\!] && \text{Substitution lemma.}
\end{aligned}$$

$$[\![\pi_1\langle t, u\rangle]\!] = \pi_1 \circ [\![\langle t, u\rangle]\!] = \pi_1 \circ \langle [\![t]\!], [\![u]\!]\rangle = [\![t]\!]\,.$$

- For η-conversion: $\big[t = \lambda x.\, tx\,,\ \langle \pi_1 t, \pi_2 t\rangle = t\big]$

$$[\![\lambda x.\, tx]\!] = \Lambda(\mathsf{ev} \circ ([\![t]\!] \times \mathsf{id})) = [\![t]\!] \qquad \text{Uniqueness equation } (\Rightarrow)$$

$$[\![\langle \pi_1 t, \pi_2 t\rangle]\!] = \langle \pi_1 \circ [\![t]\!], \pi_2 \circ [\![t]\!]\rangle = [\![t]\!] \qquad \text{Uniqueness equation } (\times)$$

1.6.6 Completeness?

It is the case that, in a general CCC $\mathcal{C}$, there may be equalities which are *not* reflected by the semantic translation, *i.e.*

$$[\![t]\!] = [\![u]\!] \quad \text{yet} \quad t \neq_\lambda u \,.$$

In the rest of this section, we show how to *construct* a CCC $\mathcal{C}_\lambda$ in which equalities between arrows correspond precisely to λ-conversions between terms. We call $\mathcal{C}_\lambda$ a *term model*, due to its dependence on the syntax.

Definition 101 We define a family of relations on variable-term pairs by setting $(x,t) \sim_{T,U} (y,u)$ if $x : T \vdash t : U$ and $y : T \vdash u : U$ are derivable and

$$t =_\lambda u[x/y] \,.$$

These are equivalence relations, so we set:

$$[(x,t)]_{T,U} := \{\, (y,u) \mid (x,t) \sim_{T,U} (y,u) \,\}.$$

Similarly, $(\centerdot, t) \sim_{\centerdot,U} (\centerdot, u)$ if $\vdash t : U$ and $\vdash u : U$ are derivable and $t =_\lambda u$. Moreover,

$$[(\centerdot,t)]_{\centerdot,U} := \{\, (\centerdot,u) \mid (\centerdot,t) \sim_{\centerdot,U} (\centerdot,u) \,\}\,.$$

▲

We denote $[(x,t)]_{T,U}$ simply by $[x,t]$, and $[(\centerdot,t)]_{\centerdot,U}$ simply by $[\centerdot,t]$ (these are *not* to be confused with copairings!). We proceed with $\mathcal{C}_\lambda$.

Definition 102 The category $\mathcal{C}_\lambda$ is defined as follows. We take as set of objects the set of λ-types augmented with a terminal object **1**:

$$Ob(\mathcal{C}_\lambda) := \{\mathbf{1}\} \cup \{\, \widetilde{T} \mid T \text{ a } \lambda\text{-type} \,\}$$

The homsets of $\mathcal{C}_\lambda$ contain equivalence relations on typed terms (definition 101), or terminal arrows τ:

$$\begin{aligned}
\mathcal{C}_\lambda(\widetilde{T}, \widetilde{U}) &:= \{\, [x,t] \mid x : T \vdash t : U \;\text{ is derivable} \,\} \\
\mathcal{C}_\lambda(\mathbf{1}, \widetilde{U}) &:= \{\, [\centerdot,t] \mid\; \vdash t : U \;\text{ is derivable} \,\} \\
\mathcal{C}_\lambda(A, \mathbf{1}) &:= \{\, \tau_A \,\}
\end{aligned}$$

The identities are:

$$\mathsf{id}_{\widetilde{T}} := [x,x]\,, \quad \mathsf{id}_{\mathbf{1}} := \tau_{\mathbf{1}}\,,$$

and arrow composition is defined by:

$$\begin{aligned}
[x,t]\circ[y,u] &:= [y, t[u/x]]\\
[x,t]\circ[\centerdot,u] &:= [\centerdot, t[u/x]]\\
[\centerdot,t]\circ\tau_A &:= \begin{cases}[y,t] & \text{if } A=\widetilde{U}\\ [\centerdot,t] & \text{if } A=\mathbf{1}\end{cases}\\
\tau_B\circ h &:= \tau_A \qquad\qquad (h\in\mathcal{C}_\lambda(A,B))
\end{aligned}$$

▲

Note that, for each variable x', any arrow $[x,t]:\widetilde{T}\to\widetilde{U}$ can be written in the form $[x',t']$, since $t=(t[x'/x])[x/x']$ and therefore $[x,t]=[x',t[x'/x]]$.

Proposition 103 *$\mathcal{C}_\lambda$ is a category.*

Proof It is not difficult to see that id's are identities. For associativity, we show the most interesting case (and leave the rest as an exercise):

$$[x,t]\circ([y,u]\circ[z,v]) = [x,t]\circ[z,u[v/y]] = [z,t[(u[v/y])/x]]\,,$$

$$([x,t]\circ[y,u])\circ[z,v] = [y,t[u/x]]\circ[z,v] = [z,t[u/x][v/y]]\,.$$

By Exercise 87, the above are equal. ■

Proposition 104 *$\mathcal{C}_\lambda$ has finite products.*

Proof Clearly, $\mathbf{1}$ is terminal with canonical arrows $\tau_A : A\to\mathbf{1}$. For (binary) products, $\mathbf{1}\times A = A\times\mathbf{1} = A$. Otherwise, define $\widetilde{T}\xleftarrow{\pi_1}\widetilde{T}\times\widetilde{U}\xrightarrow{\pi_2}\widetilde{U}$ by:

$$\begin{aligned}
\widetilde{T}\times\widetilde{U} &:= \widetilde{T\times U}\\
\pi_i &:= [x,\pi_i x] \quad i=1,2\,.
\end{aligned}$$

Given $\widetilde{T}\xleftarrow{[x,t]}\widetilde{V}\xrightarrow{[x,u]}\widetilde{U}$, take $\langle[x,t],[x,u]\rangle : \widetilde{V}\to\widetilde{T}\times\widetilde{U} := [x,\langle t,u\rangle]$. Then:

$$\begin{aligned}
\pi_1\circ\langle[x,t],[x,u]\rangle &= [y,\pi_1 y]\circ[x,\langle t,u\rangle] && \text{Definitions}\\
&= [x,\pi_1\langle t,u\rangle] && \text{Defn of composition}\\
&= [x,t] && \beta\text{-conversion}
\end{aligned}$$

Uniqueness left as exercise. The case of $\widetilde{T}\xleftarrow{[\centerdot,t]}\mathbf{1}\xrightarrow{[\centerdot,u]}\widetilde{U}$ is similar. ■

Proposition 105 *$\mathcal{C}_\lambda$ has exponentials.*

Proof We have that $\mathbf{1}\Rightarrow A = A$ and $A\Rightarrow\mathbf{1}=\mathbf{1}$, with obvious evaluation arrows. Otherwise,

$$\begin{aligned}
\widetilde{U}\Rightarrow\widetilde{T} &:= \widetilde{U\to T}\\
\mathsf{ev}_{\widetilde{U},\widetilde{T}} : (\widetilde{U}\Rightarrow\widetilde{T})\times\widetilde{U} &\longrightarrow \widetilde{T} := [x,(\pi_1 x)(\pi_2 x)]
\end{aligned}$$

Given $[x,t] : \widetilde{V} \times \widetilde{U} \to \widetilde{T}$, take $\Lambda([x,t]) := [x_1, \lambda x_2.t[\langle x_1, x_2\rangle/x]]$.
Then,

$$\begin{aligned}
\mathsf{ev} \circ \Lambda([x,t]) \times \mathsf{id} &= \mathsf{ev} \circ \langle \Lambda([x,t]) \circ \pi_1, \mathsf{id} \circ \pi_2\rangle \\
&= \mathsf{ev} \circ \langle [x_1, \lambda x_2.t[\langle x_1, x_2\rangle/x]] \circ [y, \pi_1 y], [y, \pi_2 y]\rangle \\
&= \mathsf{ev} \circ \langle [y, \lambda x_2.t[\langle \pi_1 y, x_2\rangle/x]], [y, \pi_2 y]\rangle \\
&= [z, (\pi_1 z)(\pi_2 z)] \circ [y, \underbrace{\langle \lambda x_2.t[\langle \pi_1 y, x_2\rangle/x], \pi_2 y\rangle}_{u}] \\
&= [y, (\pi_1 u)(\pi_2 u)] \stackrel{\beta}{=} [y, (\lambda x_2.t[\langle \pi_1 y, x_2\rangle/x])(\pi_2 y)] \\
&\stackrel{\beta}{=} [y, t[\langle \pi_1 y, \pi_2 y\rangle/x]] \stackrel{\eta}{=} [y, t[y/x]] = [x,t].
\end{aligned}$$

Uniqueness left as exercise. The case of $[x,t] : \mathbf{1} \times \widetilde{U} \to \widetilde{T}$ is similar. ∎

Exercise 106 Complete the proof of the previous propositions.

Hence, $\mathcal{C}_\lambda$ is a CCC and a sound model of the simply-typed λ-calculus. Moreover, applying our translation from the λ-calculus to a CCC (Definition 96) we can show that we have

$$[\![\Gamma \vdash t : T]\!] = [x,\ t[\pi_i x/x_i]_{i=1..n}]$$

where $\Gamma = \{x_1 : T_1, \ldots, x_n : T_n\}$, $x \notin \Gamma$ and $x : \prod_{i=1}^{n} T_i$. Then,

$$t =_\lambda u \quad \Longleftrightarrow \quad [\![\Gamma \vdash t : T]\!] = [\![\Gamma \vdash u : T]\!].$$

This means that our term model is *complete*.

1.6.7 Exercises

1. Give Natural Deduction proofs of the following sequents.

   - $\vdash (A \supset B) \supset ((B \supset C) \supset (A \supset C))$
   - $\vdash (A \supset (A \supset B)) \supset (A \supset B)$
   - $\vdash (C \supset A) \supset ((C \supset B) \supset (C \supset (A \wedge B)))$
   - $\vdash (A \supset (B \supset C)) \supset ((A \supset B) \supset (A \supset C))$

   In each case, give the corresponding λ-term and the corresponding arrow in a CCC $\mathcal{C}$.
2. For each of the following λ-terms, find a type for it. Try to find the "most general" type, built from "type variables" α, β etc. For example, the most general type for the identity $\lambda x.\,x$ is $\alpha \to \alpha$. In each case, give the derivation of the type for this term (where you may assume that types can be built up from type variables as well as base types).

   - $\lambda f.\,\lambda x.\,fx$
   - $\lambda x.\,\lambda y.\,\lambda z.\,x(yz)$

- $\lambda x.\lambda y.\lambda z.xzy$
- $\lambda x.\lambda y.xyy$
- $\lambda x.\lambda y.x$
- $\lambda x.\lambda y.\lambda z.xz(yz)$

Reflect a little on the methods you used to do this exercise. Could they be made algorithmic?

1.7 Linearity

In the system of Natural Deduction, implicit in our treatment of assumptions in sequents

$$A_1, \ldots, A_n \vdash A$$

is that we can use them as many times as we want (including not at all). In this section we will explore the field that is opened once we apply restrictions to this approach, and thus render our treatment of assumptions more *linear* (or *resource sensitive*).

1.7.1 Gentzen Sequent Calculus

In order to make the manipulation of assumptions more visible, we now represent the assumptions as a *list* (possibly with repetitions) rather than a set, and use explicit structural rules to control copying, deletion and interchange of assumptions.

Definition 107 The ***structural rules*** for Logic are given in the Table 1.8. ▲

Table 1.8 Structural rules for logic

$\dfrac{}{A \vdash A}$ Id	$\dfrac{\Gamma, A, B, \Delta \vdash C}{\Gamma, B, A, \Delta \vdash C}$ Exch
$\dfrac{\Gamma, A, A \vdash B}{\Gamma, A \vdash B}$ Contr	$\dfrac{\Gamma \vdash B}{\Gamma, A \vdash B}$ Weak

If we think of using proof rules *backwards* to reduce the task of proving a given sequent to various sub-tasks, then we see that the Contraction rule lets us *duplicate* premises, and the Weakening rule lets us *discard* them, while the Exchange rule merely lets us re-order them. The Identity axiom as given here is equivalent to the one with auxiliary premises given previously in the presence of Weakening.

The structural rules have clear categorical meanings in a category $\mathcal{C}$ with products. Recalling the *diagonal transformation* $\Delta_A := \langle \mathsf{id}_A, \mathsf{id}_A \rangle$ and the *symmetry transformation* $s_{A,B} := \langle \pi_2, \pi_1 \rangle$, the meanings are as follows.

$$\dfrac{\Gamma, A, B, \Delta \vdash C}{\Gamma, B, A, \Delta \vdash C}\,\mathsf{Exch} \qquad \dfrac{f : \Gamma \times A \times B \times \Delta \longrightarrow C}{f \circ (\mathsf{id}_\Gamma \times s_{A,B} \times \mathsf{id}_\Delta) : \Gamma \times B \times A \times \Delta \longrightarrow C}$$

$$\dfrac{\Gamma, A, A \vdash B}{\Gamma, A \vdash B}\,\mathsf{Contr} \qquad \dfrac{f : \Gamma \times A \times A \longrightarrow B}{f \circ (\mathsf{id}_\Gamma \times \Delta_A) : \Gamma \times A \longrightarrow B}$$

$$\dfrac{\Gamma \vdash B}{\Gamma, A \vdash B}\,\mathsf{Weak} \qquad \dfrac{f : \Gamma \longrightarrow B}{f \circ \pi_1 : \Gamma \times A \longrightarrow B}$$

In order to analyse Natural Deduction, Gentzen introduced **sequent calculi** based on **Left** and **Right** rules, instead of Introduction and Elimination rules. These kind of systems are more adequate for our discussion on linearity.

Definition 108 We define the ***Gentzen sequent calculus*** for $\wedge,\supset$ as the proof system obtained by the structural rules (Definition 107) and the rules in Table 1.9 for connectives. ▲

For example, the proof of $\supset$-transitivity is now given as follows.

$$\dfrac{\dfrac{\dfrac{\dfrac{\dfrac{}{A \vdash A}\,\mathsf{Id} \quad \dfrac{}{B \vdash B}\,\mathsf{Id}}{A, A \supset B \vdash B}\supset\mathsf{L}}{A \supset B, A \vdash B}\,\mathsf{Exch} \quad \dfrac{}{C \vdash C}\,\mathsf{Id}}{\dfrac{A \supset B, A, B \supset C \vdash C}{A \supset B, B \supset C, A \vdash C}\,\mathsf{Exch}}\supset\mathsf{L}$$

$$\dfrac{A \supset B, B \supset C, A \vdash C}{A \supset B, B \supset C \vdash A \supset C}\supset\mathsf{R}$$

Exercise 109 Show that the Gentzen-rules are admissible in Natural Deduction. Moreover, show that the Natural Deduction rules are admissible in the Gentzen sequent calculus.

The Cut rule allows the use of *lemmas* in proofs. It also yields a *dynamics of proofs* via *Cut Elimination*, that is, a dynamics of proof transformations towards the goal of eliminating the uses of the Cut rule in a proof, *i.e.* removing all lemmas and making the proof completely "explicit", meaning Cut-free. Such transformations are always possible, as is shown in the following seminal result of Gentzen (*Hauptsatz*).

Fact 110 (Cut Elimination) *The* Cut *rule is admissible in the Gentzen sequent calculus without* Cut.

Table 1.9 Gentzen sequent calculus for $\wedge,\supset$

Conjunction	Implication	Cut
$\dfrac{\Gamma \vdash A \quad \Delta \vdash B}{\Gamma, \Delta \vdash A \wedge B}\wedge\mathsf{R}$	$\dfrac{\Gamma, A \vdash B}{\Gamma \vdash A \supset B}\supset\mathsf{R}$	$\dfrac{\Gamma \vdash A \quad A, \Delta \vdash B}{\Gamma, \Delta \vdash B}\,\mathsf{Cut}$
$\dfrac{\Gamma, A, B \vdash C}{\Gamma, A \wedge B \vdash C}\wedge\mathsf{L}$	$\dfrac{\Gamma \vdash A \quad B, \Delta \vdash C}{\Gamma, A \supset B, \Delta \vdash C}\supset\mathsf{L}$	

1.7.2 Linear Logic

In the presence of the structural rules, the Gentzen sequent calculus is entirely equivalent to the Natural Deduction system we studied earlier. Nevertheless:

> What happens if we **drop** the Contraction and Weakening rules (but keep the Exchange rule)?

It turns out we can still make good sense of the resulting proofs, terms and categories, but now in the setting of a different, 'resource-sensitive' logic.

Definition 111 ***Multiplicative Linear logic*** is a variant of standard logic with *linear* logical connectives. The multiplicative connectives for conjunction and implication are $\otimes$ and $\multimap$. Proof sequents are of the form $\Gamma \vdash A$, where Γ is now a *multiset*. The proof rules for $\otimes,\multimap$-Linear Logic, given in Table 1.10, are the multiplicative versions of the Gentzen rules. ▲

Multiplicativity here means the use of *disjoint* (*i.e.* non-overlapping) contexts. The use of multisets allows us to omit explicit use of the Exchange rule in our proof system.

Note that the given system satisfies Cut-elimination, and this leans heavily on the $\multimap\mathsf{L}$ rule. We could have used instead the following rule,

$$\frac{\Gamma \vdash A \multimap B \qquad \Delta \vdash A}{\Gamma, \Delta \vdash B}\ \multimap\mathsf{E}$$

which is more intuitive computationally, but then cut-elimination would fail. Note, though, that:

$$\multimap\mathsf{L}\,,\ \mathsf{Cut}\,,\ \mathsf{Id}\ \equiv\ \multimap\mathsf{E}\,,\ \mathsf{Cut}\,,\ \mathsf{Id}\,.$$

This is shown as follows.

$$\frac{\dfrac{\Gamma \vdash A \qquad \dfrac{}{A \multimap B \vdash A \multimap B}\,\mathsf{Id}}{\Gamma, A \multimap B \vdash B}\,\multimap\mathsf{E} \qquad B, \Delta \vdash C}{\Gamma, A \multimap B, \Delta \vdash C}\,\mathsf{Cut} \qquad \frac{\Gamma \vdash A \multimap B \qquad \dfrac{\Delta \vdash A \qquad \dfrac{}{B \vdash B}\,\mathsf{Id}}{A \multimap B, \Delta \vdash B}\,\multimap\mathsf{L}}{\Gamma, \Delta \vdash B}\,\mathsf{Cut}$$

The resource-sensitive nature of Linear Logic is reflected in the following exercise.

Table 1.10 Rules for $\otimes,\multimap$-linear logic

Conjunction	Implication	Cut
$\dfrac{\Gamma \vdash A \qquad \Delta \vdash B}{\Gamma, \Delta \vdash A \otimes B}\,\otimes\mathsf{R}$	$\dfrac{\Gamma, A \vdash B}{\Gamma \vdash A \multimap B}\,\multimap\mathsf{R}$	$\dfrac{\Gamma \vdash A \qquad A, \Delta \vdash B}{\Gamma, \Delta \vdash B}\,\mathsf{Cut}$
$\dfrac{\Gamma, A, B \vdash C}{\Gamma, A \otimes B \vdash C}\,\otimes\mathsf{L}$	$\dfrac{\Gamma \vdash A \qquad B, \Delta \vdash C}{\Gamma, A \multimap B, \Delta \vdash C}\,\multimap\mathsf{L}$	

Exercise 112 Can you construct proofs in Linear Logic of the following sequents? (Hint: Use the Cut Elimination property.)

- $A \vdash A \otimes A$
- $(A \otimes A) \multimap B \vdash A \multimap B$
- $\vdash A \multimap (B \multimap A)$

Related to linear logic is the *linear λ-calculus*, which is a linear version of the simply-typed λ-calculus.

Definition 113 The ***linear λ-calculus*** is defined as follows.

Type	$\mathrm{TY} \ni T, U ::= b \mid T \multimap U \mid T \otimes U$
Term	$\mathrm{TE} \ni t, u ::= x \mid t\,u \mid \lambda x.t \mid t \otimes u \mid \mathsf{let}\ z\ \mathsf{be}\ x \otimes y\ \mathsf{in}\ t$
Typing context	$\Gamma ::= \varnothing \mid x : T, \Gamma$

Terms are typed by use of the typing rules of Table 1.11. Finally, the rules for β-reduction are:

$$\begin{aligned}(\lambda x.t)u &\longrightarrow_\beta t[u/x] \\ \mathsf{let}\ t \otimes u\ \mathsf{be}\ x \otimes y\ \mathsf{in}\ v &\longrightarrow_\beta v[t/x, u/y]\,.\end{aligned}$$

▲

Note here that, again, $x : T, \Gamma$ stands for $\{x : T\} \cup \Gamma$ with x not appearing in Γ. Note also that Cut-free proofs *always yield terms in normal form.*
Term formation is now highly constrained by the form of the typing judgements. In particular,

$$x_1 : A_1, \ldots, x_k : A_k \vdash t : A$$

now implies that each x_i occurs **exactly once** (free) in t.

Moreover, note that, for function application, instead of the rule on the LHS below, we could have used the more intuitive rule on the RHS.

$$\frac{\Gamma \vdash t : T \qquad x : U, \Delta \vdash u : V}{\Gamma, f : T \multimap U, \Delta \vdash u[ft/x] : V} \qquad\qquad \frac{\Gamma \vdash t : A \multimap B \qquad \Delta \vdash u : A}{\Gamma, \Delta \vdash t\,u : B}$$

Table 1.11 Linear λ-calculus for $\otimes$, $\multimap$

Variable cut	$\dfrac{}{x : T \vdash x : T}$	$\dfrac{\Gamma \vdash t : T \qquad x : T, \Delta \vdash u : U}{\Gamma, \Delta \vdash u[t/x] : U}$
Linear tensor	$\dfrac{\Gamma \vdash t : T \qquad \Delta \vdash u : U}{\Gamma, \Delta \vdash t \otimes u : T \otimes U}$	$\dfrac{\Gamma, x : T, y : U \vdash v : V}{\Gamma, z : T \otimes U \vdash \mathsf{let}\ z\ \mathsf{be}\ x \otimes y\ \mathsf{in}\ v : V}$
Linear function	$\dfrac{\Gamma, x : U \vdash t : T}{\Gamma \vdash \lambda x.t : U \multimap T}$	$\dfrac{\Gamma \vdash t : T \qquad x : U, \Delta \vdash u : V}{\Gamma, f : T \multimap U, \Delta \vdash u[ft/x] : V}$

As we did in the logic, we can show that the typing systems with one or the other rule are equivalent.

1.7.3 Linear Logic in Monoidal Categories

We proceed to give a categorical counterpart to linearity by providing a categorical interpretation of linear logic. Note that CCC's are no longer adequate for this task as they contain arrows

$$\Delta_A : A \longrightarrow A \times A\,, \quad \pi_1 : A \times B \longrightarrow A$$

which violate linearity. It turns out that the right setting is that of *symmetric monoidal closed categories*.

Definition 114 A ***monoidal category*** is a structure $(\mathcal{C}, \otimes, I, a, l, r)$ where:

- $\mathcal{C}$ is a category,
- $\otimes : \mathcal{C} \times \mathcal{C} \to \mathcal{C}$ is a functor (*tensor*),
- I is a distinguished object of $\mathcal{C}$ (*unit*),
- a, l, r are natural isomorphisms (*structural isos*) with components:

$$a_{A,B,C} : A \otimes (B \otimes C) \xrightarrow{\cong} (A \otimes B) \otimes C$$
$$l_A : I \otimes A \xrightarrow{\cong} A \qquad r_A : A \otimes I \xrightarrow{\cong} A$$

such that $l_I = r_I : I \otimes I \to I$ and the following diagrams commute.

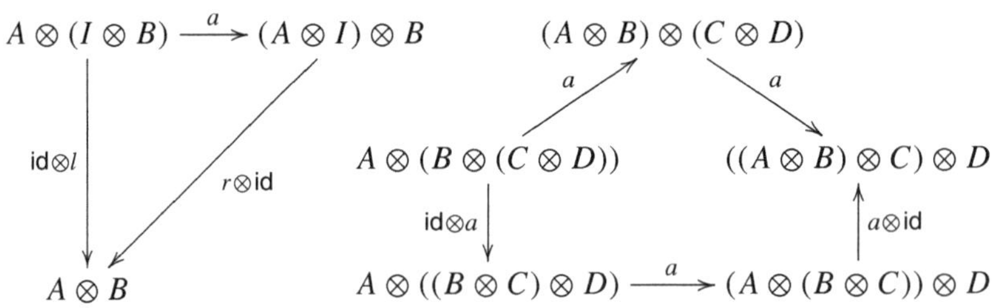

▲

The monoidal diagrams ensure ***coherence***, described by the slogan:

"... *'all' diagrams involving a, l and r must commute.*"

Examples:

- Both products and coproducts give rise to monoidal structures—which are the common denominator between them. (But in addition, products have *diagonals* and *projections*, and coproducts have *codiagonals* and *injections*.)

- $(\mathbb{N}, \leq, +, 0)$ is a monoidal category.
- **Rel**, the category of sets and relations, with cartesian product (which is *not* the categorical product).
- $\mathbf{Vect}_k$ with the tensor product.

Let us examine the example of **Rel** in some detail. We take $\otimes$ to be the cartesian product, which is defined on relations $R : X \to X'$ and $S : Y \to Y'$ as follows.

$$\forall (x, y) \in X \times Y, (x', y') \in X' \times Y'. (x, y) R \otimes S (x', y') \iff x R x' \wedge y S y' .$$

It is not difficult to show that this is indeed a functor. Note that, in the case that R, S are *functions*, $R \otimes S$ is the same as $R \times S$ in **Set**. Moreover, we take each $a_{A,B,C}$ to be the associativity function for products (in **Set**), which is an iso in **Set** and hence also in **Rel**. Finally, we take I to be the one-element set, and l_A, r_A to be the projection functions: their relational converses are their inverses in **Rel**. The monoidal diagrams commute simply because they commute in **Set**.

Exercise 115 Verify that $(\mathbb{N}, \leq, +, 0)$ and $\mathbf{Vect}_k$ are monoidal categories.

Tensors and Products

As we mentioned earlier, products are tensors with extra structure: natural diagonals and projections. This fact, which reflects *no-cloning* and *no-deleting* of Linear Logic, is shown as follows.

Proposition 116 *Let $\mathcal{C}$ be a monoidal category $(\mathcal{C}, \otimes, I, a, l, r)$. $\otimes$ induces a product structure iff there exist natural diagonals and projections, i.e. natural transformations given by arrows*

$$d_A : A \longrightarrow A \otimes A , \qquad p_{A,B} : A \times B \longrightarrow A , \qquad q_{A,B} : A \times B \longrightarrow B ,$$

such that the following diagrams commute.

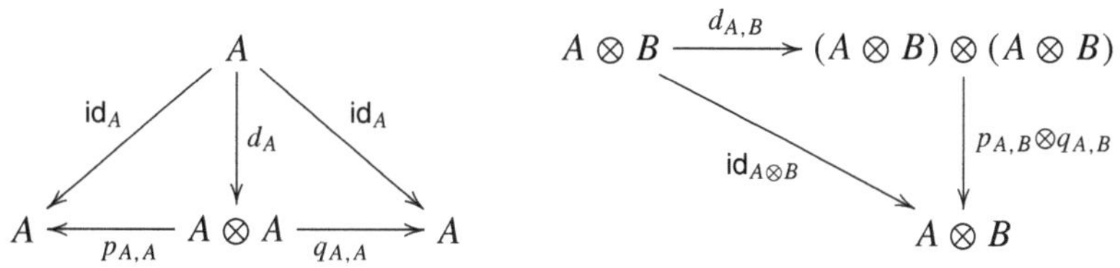

Proof The "only if" direction is straightforward. For the converse, let $\mathcal{C}$ be monoidal with natural projections and diagonals. Then, we take product pairs to be pairs of the form

$$A \xleftarrow{p_{A,B}} A \otimes B \xrightarrow{q_{A,B}} B .$$

Moreover, for any pair of arrows $B \xleftarrow{f} A \xrightarrow{g} C$, define

$$\langle f, g \rangle := A \xrightarrow{d_A} A \otimes A \xrightarrow{f \otimes g} B \otimes C\,.$$

Then the product diagram commutes. For example:

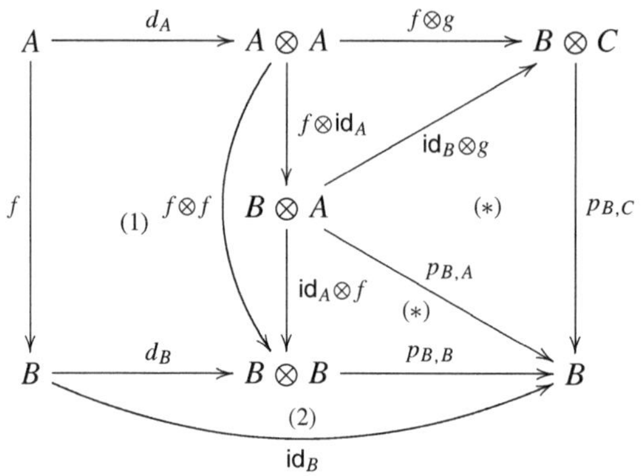

(∗) naturality of p
(1) naturality of d
(2) hypothesis

For uniqueness, if $h : A \to B \otimes C$ then the following diagram commutes,

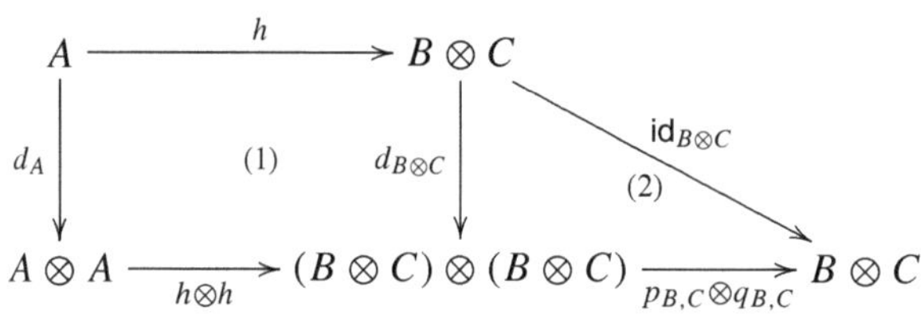

(1) naturality of d
(2) hypothesis

so $h = \langle \pi_1 \circ h, \pi_2 \circ h \rangle$. ∎

SMCC's

Linear Logic is interpreted in monoidal categories with two more pieces of structure: monoidal symmetry and closure. The former allows the Exchange rule to be interpreted, while the latter realises linear implication.

Definition 117 A ***symmetric monoidal category*** is a monoidal category $(\mathcal{C}, \otimes, I, a, l, r)$ with an additional natural isomorphism (*symmetry*),

$$s_{A,B} : A \otimes B \xrightarrow{\cong} B \otimes A$$

such that $s_{B,A} = s_{A,B}^{-1}$ and the following diagrams commute.

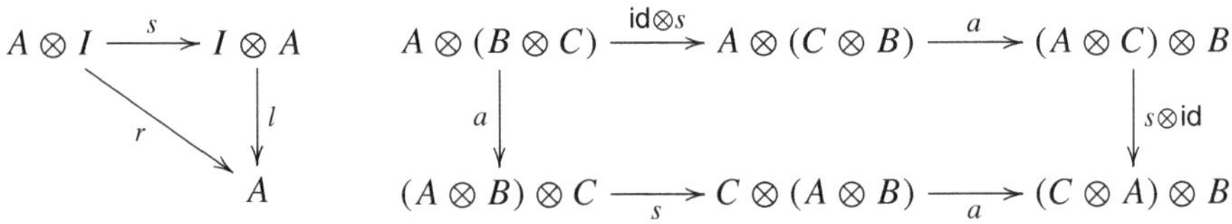

▲

Definition 118 A ***symmetric monoidal closed category (SMCC)*** is a symmetric monoidal category $(\mathcal{C}, \otimes, I, a, l, r, s)$ such that, for each object A, there is a couniversal arrow to the functor

$$\_ \otimes A : \mathcal{C} \longrightarrow \mathcal{C}\,.$$

That is, for all pairs A, B, there is an object $A \multimap B$ and a morphism

$$\mathsf{ev}_{A,B} : (A \multimap B) \otimes A \longrightarrow B$$

such that, for every morphism $f : C \otimes A \to B$, there is a *unique* morphism $\Lambda(f) : C \to (A \multimap B)$ such that

$$\mathsf{ev}_{A,B} \circ (\Lambda(f) \otimes \mathsf{id}_A) = f\,.$$

▲

Note that, although we use notation borrowed from CCC's (ev, Λ), these are *different* structures! Examples of symmetric monoidal closed categories are **Rel**, **Vect**$_k$, and (*a fortiori*) cartesian closed categories.

Exercise 119 Show that **Rel** is a symmetric monoidal closed category.

Linear Logic in SMCC's

Just as cartesian closed categories correspond to $\wedge,\supset$-logic (and simply-typed λ-calculus), so do symmetric monoidal closed categories correspond to $\otimes,\multimap$-logic (and linear λ-calculus).

So let $\mathcal{C}$ be a symmetric monoidal closed category. The interpretation of a linear sequent

$$A_1, \ldots, A_k \vdash A$$

will be a morphism

$$f : A_1 \otimes \cdots \otimes A_k \longrightarrow A\,.$$

Table 1.12 Categorical translation of $\otimes,\multimap$-linear logic

$\overline{A \vdash A}$	$\overline{\mathsf{id}_A : A \longrightarrow A}$
$\dfrac{\Gamma \vdash A \qquad A, \Delta \vdash B}{\Gamma, \Delta \vdash B}$	$\dfrac{f : \Gamma \longrightarrow A \qquad g : A \otimes \Delta \longrightarrow B}{g \circ (f \otimes \mathsf{id}_\Delta) : \Gamma \otimes \Delta \longrightarrow B}$
$\dfrac{\Gamma \vdash A \qquad \Delta \vdash B}{\Gamma, \Delta \vdash A \otimes B}$	$\dfrac{f : \Gamma \longrightarrow A \qquad g : \Delta \longrightarrow B}{f \otimes g : \Gamma \otimes \Delta \longrightarrow A \otimes B}$
$\dfrac{\Gamma, A, B \vdash C}{\Gamma, A \otimes B \vdash C}$	$\dfrac{f : (\Gamma \otimes A) \otimes B \longrightarrow C}{f \circ a_{\Gamma,A,B} : \Gamma \otimes (A \otimes B) \longrightarrow C}$
$\dfrac{\Gamma, A \vdash B}{\Gamma \vdash A \multimap B}$	$\dfrac{f : \Gamma \otimes A \longrightarrow B}{\Lambda(f) : \Gamma \longrightarrow (A \multimap B)}$
$\dfrac{\Gamma \vdash A \multimap B \qquad \Delta \vdash A}{\Gamma, \Delta \vdash B}$	$\dfrac{f : \Gamma \longrightarrow (A \multimap B) \qquad g : \Delta \longrightarrow A}{\mathsf{ev}_{A,B} \circ (f \otimes g) : \Gamma \otimes \Delta \longrightarrow B}$

To be precise in our interpretation, we will again treat contexts as lists of formulas, and explicitly interpret the Exchange rule by:

$$\frac{\Gamma, A, B, \Delta \vdash C}{\Gamma, B, A, \Delta \vdash C} \qquad \frac{f : \Gamma \otimes A \otimes B \otimes \Delta \longrightarrow C}{f \circ (\mathsf{id}_\Gamma \otimes s_{B,A} \otimes \mathsf{id}_\Delta) : \Gamma \otimes B \otimes A \otimes \Delta \longrightarrow C}$$

The rest of the rules are translated as follows (Table 1.12).
Note that, because of coherence in monoidal categories, we will not be scholastic with associativity arrows a in our translations and will usually omit them. For the same reason, consecutive applications of tensor will be written without specifying associativity, e.g. $A_1 \otimes \cdots \otimes A_n$.

Exercise 120 Let $\mathcal{C}$ be a symmetric monoidal closed category. Give the interpretation of the $\multimap$-left rule in $\mathcal{C}$:

$$\frac{\Gamma \vdash A \qquad B, \Delta \vdash C}{\Gamma, A \multimap B, \Delta \vdash C} \multimap\mathsf{L}$$

Exercise 121 Is it possible to translate $\otimes,\multimap$-logic into a CCC $\mathcal{C}$? Is this in accordance with linearity of $\otimes,\multimap$-logic?

1.7.4 Beyond the Multiplicatives

Linear Logic has three "levels" of connectives, each describing a different aspect of standard logic:

- The **multiplicatives**: e.g. $\otimes$, $\multimap$,
- The **additives**: *additive conjunction* & and *disjunction* $\oplus$,
- The **exponentials**, allowing controlled access to copying and discarding.

We focus on additive conjunction and the exponential "**!**", which will allow us to recover the 'expressive power' of standard $\wedge,\supset$-logic.

Definition 122 The logical connective for ***additive disjunction*** is &, and the related proof rules are the following.

$$\frac{\Gamma \vdash A \qquad \Gamma \vdash B}{\Gamma \vdash A \,\&\, B}\,\&\mathsf{R} \qquad \frac{\Gamma, A \vdash C}{\Gamma, A \,\&\, B \vdash C}\,\&\mathsf{L} \qquad \frac{\Gamma, B \vdash C}{\Gamma, A \,\&\, B \vdash C}\,\&\mathsf{L}$$

▲

So additive conjunction has proof rules that are identical to those of standard conjunction ($\wedge$). Note though that, since by linearity an argument of type $A \,\&\, B$ can only be used once, each use of a left rule for & makes a once-and-for-all *choice* of a projection. On the other hand, $A \otimes B$ represents a conjunction where *both* projections must be available.

Additive conjunction can be interpreted in any symmetric monoidal category with products, *i.e.* a category $\mathcal{C}$ with structure $(\otimes, \times)$ where $\otimes$ is a symmetric monoidal tensor and $\times$ is a product.

$$\frac{f : \Gamma \longrightarrow A \qquad g : \Gamma \longrightarrow B}{\langle f, g\rangle : \Gamma \longrightarrow A \times B} \qquad \frac{f : \Gamma \otimes A \longrightarrow C}{f \circ (\mathsf{id} \otimes \pi_1) : \Gamma \otimes (A \times B) \longrightarrow C}$$

Moreover, we can extend the linear λ-calculus with term constructors for additive conjunction as follows.

$$\frac{\Gamma \vdash t : A \qquad \Gamma \vdash u : B}{\Gamma \vdash \langle t, u\rangle : A \,\&\, B} \quad \frac{\Gamma, x : A \vdash t : C}{\Gamma, z : A \,\&\, B \vdash \mathbf{let}\, z = \langle x, -\rangle\, \mathbf{in}\, t : C}$$

$$\frac{\Gamma, B \vdash C}{\Gamma, z : A \,\&\, B \vdash \mathbf{let}\, z = \langle -, y\rangle\, \mathbf{in}\, t : C}$$

The β-reduction rules related to these constructs are:

$$\begin{aligned} \mathbf{let}\,\langle t, u\rangle = \langle x, -\rangle\, \mathbf{in}\, v &\longrightarrow_\beta v[t/x] \\ \mathbf{let}\,\langle t, u\rangle = \langle -, y\rangle\, \mathbf{in}\, v &\longrightarrow_\beta v[u/y]\,. \end{aligned}$$

Finally, we can gain back the lost structural rules, in *disciplined* versions, by introducing an exponential *bang* operator ! which is a kind of *modality* enabling formulas to participate in structural rules.

Definition 123 The logical connective for ***bang*** is **!**, and the related proof rules are the following.

$$\frac{\Gamma, A \vdash B}{\Gamma, !A \vdash B}\,!\mathsf{L} \qquad \frac{!\Gamma \vdash A}{!\Gamma \vdash !A}\,!\mathsf{R} \qquad \frac{\Gamma \vdash B}{\Gamma, !A \vdash B}\,\mathsf{Weak} \qquad \frac{\Gamma, !A, !A \vdash B}{\Gamma, !A \vdash B}\,\mathsf{Contr}$$

Note that $!\{A_1, \ldots, A_n\} := !A_1, \ldots, !A_n$. ▲

We can now see the discipline imposed on structural rules: in order for the rules to be applied, the participating formulas need to be tagged with a bang.

Interpreting Standard Logic

We are now in position to recover the standard logical connectives $\wedge$, $\supset$ within Linear Logic. If we interpret

$$\begin{aligned} A \supset B &:= \,!A \multimap B \\ A \wedge B &:= A \,\&\, B \end{aligned}$$

and each $\wedge,\supset$-sequent $\Gamma \vdash A$ as $!\Gamma \vdash A$, then each proof rule of the Gentzen system for $\wedge,\supset$ is admissible in the proof system of Linear Logic for $\otimes,\multimap,\&,!$.

Note in particular that the interpretation

$$A \supset B \;:=\; !A \multimap B$$

decomposes the fundamental notion of implication into finer notions—like "splitting the atom of logic"!

1.7.5 Exercises

1. Give proofs of the following sequents in Linear Logic.

   a) $\vdash A \multimap A$
   b) $A \multimap B, B \multimap C \vdash A \multimap C$
   c) $\vdash (A \multimap B \multimap C) \multimap (B \multimap A \multimap C)$
   d) $A \otimes (B \otimes C) \vdash (A \otimes B) \otimes C$
   e) $A \otimes B \vdash B \otimes A$

   For each of the proofs constructed give:

   - the corresponding linear λ-term,
   - its interpretation in **Rel**.

2. Consider a symmetric monoidal closed category $\mathcal{C}$.

   (a) Suppose the sequents $\Gamma_1 \vdash A$, $\Gamma_2 \vdash B$ and $A, B, \Delta \vdash C$ are provable and let their interpretations (*i.e.* the interpretations of their proofs) in $\mathcal{C}$ be $f_1 : \Gamma_1 \to A$, $f_2 : \Gamma_2 \to B$ and $g : A \otimes B \otimes \Delta \to C$ respectively. Find then the interpretations h_1, h_2 of the following proofs.

$$\dfrac{\dfrac{\dfrac{\vdots}{\Gamma_1 \vdash A} \quad \dfrac{\vdots}{\Gamma_2 \vdash B}}{\Gamma_1, \Gamma_2 \vdash A \otimes B}\,\otimes\mathsf{R} \qquad \dfrac{\dfrac{\vdots}{A, B, \Delta \vdash C}}{A \otimes B, \Delta \vdash C}\,\otimes\mathsf{L}}{\Gamma_1, \Gamma_2, \Delta \vdash C}\,\mathsf{Cut} \qquad\qquad \dfrac{\dfrac{\vdots}{\Gamma_2 \vdash B} \qquad \dfrac{\dfrac{\vdots}{\Gamma_1 \vdash A} \quad \dfrac{\vdots}{A, B, \Delta \vdash C}}{\Gamma_1, B, \Delta \vdash C}\,\mathsf{Cut}}{\Gamma_1, \Gamma_2, \Delta \vdash C}\,\mathsf{Cut}$$

   and show that $h_1 = h_2$.

(b) Suppose now $\mathcal{C}$ has *also* binary products, given by $\times$. Given that the sequents $\Gamma \vdash A$, $\Gamma \vdash B$ and $A, \Delta \vdash C$ are provable, and that their interpretations in $\mathcal{C}$ are $f_1 : \Gamma \to A$, $f_2 : \Gamma \to B$ and $g : A \otimes \Delta \to C$ respectively, find the interpretations h_1, h_2 of the following proofs.

$$\dfrac{\dfrac{\dfrac{\vdots}{\Gamma \vdash A} \quad \dfrac{\vdots}{\Gamma \vdash B}}{\Gamma \vdash A \,\&\, B}\,\&\mathsf{R} \qquad \dfrac{\dfrac{\vdots}{A, \Delta \vdash C}}{A \,\&\, B, \Delta \vdash C}\,\&\mathsf{L}}{\Gamma, \Delta \vdash C}\,\mathsf{Cut} \qquad\qquad \dfrac{\dfrac{\vdots}{\Gamma \vdash A} \quad \dfrac{\vdots}{A, \Delta \vdash C}}{\Gamma, \Delta \vdash C}\,\mathsf{Cut}$$

and show that $h_1 = h_2$.

3. Show that the condition $l_I = r_I$ in the definition of monoidal categories is redundant.
Moreover, show that the condition $\mathsf{id}_A \otimes l_B = a_{A,I,B} \circ r_A \otimes \mathsf{id}_B$ in the definition of symmetric monoidal categories is redundant.

1.8 Monads and Comonads

Recall that an adjunction is given by a triple $\langle F, G, \theta \rangle$, with $F : \mathcal{C} \to \mathcal{D}$ and $G : \mathcal{D} \to \mathcal{C}$ being functors, and θ a natural bijection between homsets. By composing the two functors we obtain endofunctors

$$G \circ F : \mathcal{C} \longrightarrow \mathcal{C}, \quad F \circ G : \mathcal{D} \longrightarrow \mathcal{D}.$$

These can be seen as *encapsulating the effect* of the adjunction inside their domain category. For example, if we consider the functors

$$\mathsf{MList} : \mathbf{Set} \longrightarrow \mathbf{Mon}, \quad U : \mathbf{Mon} \longrightarrow \mathbf{Set},$$

then $U \circ \mathsf{MList}$ encodes the free monoid construction inside **Set**.

The study of such endofunctors on their own right gave rise to the notions of *monad* and *comonad*, which we examine in this section.

1.8.1 Basics

Definition 124 A ***monad*** over a category $\mathcal{C}$ is a triple (T, η, μ) where T is an endofunctor on $\mathcal{C}$ and $\eta : \mathsf{Id}_{\mathcal{C}} \to T$, $\mu : T^2 \to T$ are natural transformations such that the following diagrams commute. (Note that $T^2 := T \circ T$, etc.).

$$\begin{array}{ccc} T^3A & \xrightarrow{\mu_{TA}} & T^2A \\ {\scriptstyle T\mu_A}\downarrow & & \downarrow{\scriptstyle \mu_A} \\ T^2A & \xrightarrow[\mu_A]{} & TA \end{array} \qquad \begin{array}{ccc} TA & \xrightarrow{\eta_{TA}} & T^2A \\ {\scriptstyle T\eta_A}\downarrow & \searrow^{\mathsf{id}_{TA}} & \downarrow{\scriptstyle \mu_A} \\ T^2A & \xrightarrow[\mu_A]{} & TA \end{array}$$

▲

We call η the *unit* of the monad, and μ its *multiplication*; the whole terminology comes from monoids. Let us now proceed to some examples.

- Let $\mathcal{C}$ be a category with coproducts and let E be an object in $\mathcal{C}$. We can define a monad (T, η, μ) of E-coproducts (computationally, E-exceptions) by taking $T : \mathcal{C} \to \mathcal{C}$ to be the functor $\_ + E$, and η, μ as follows.

$$T := A \mapsto A + E\,,\ f \mapsto f + \mathsf{id}_E$$
$$\eta_A := A \xrightarrow{\mathsf{in}_1} A + E$$
$$\mu_A := (A + E) + E \xrightarrow{[\mathsf{id}_{A+E},\, \mathsf{in}_2]} A + E$$

As an injection, η is a natural transformation. For μ, we can use the properties of the coproduct. For $f : A \to B$,

$$\begin{aligned} Tf \circ \mu_A &= Tf \circ [\mathsf{id}_{A+E}, \mathsf{in}_2] = [Tf \circ \mathsf{id}_{A+E}, Tf \circ \mathsf{in}_2] = [Tf, Tf \circ \mathsf{in}_2] \\ &= [Tf, (f + \mathsf{id}_E) \circ \mathsf{in}_2] = [Tf, \mathsf{in}_2] \\ &= [\mathsf{id}_{B+E} \circ Tf, \mathsf{in}_2 \circ \mathsf{id}_E] = [\mathsf{id}_{B+E}, \mathsf{in}_2] \circ (Tf + \mathsf{id}_E) \\ &= \mu_B \circ T^2 f\,. \end{aligned}$$

The monadic diagrams follow in a similar manner. For example,

$$\begin{aligned} \mu_A \circ \mu_{TA} &= \mu_A \circ [\mathsf{id}_{TA+E}, \mathsf{in}_2] = [\mu_A \circ \mathsf{id}_{TA+E}, \mu_A \circ \mathsf{in}_2] = [\mu_A, \mu_A \circ \mathsf{in}_2] \\ &= [\mu_A, [\mathsf{id}_{A+E}, \mathsf{in}_2] \circ \mathsf{in}_2] = [\mu_A, \mathsf{in}_2] \\ &= [\mathsf{id}_{A+E} \circ \mu_A, \mathsf{in}_2 \circ \mathsf{id}_E] = [\mathsf{id}_{A+E}, \mathsf{in}_2] \circ (\mu_A + \mathsf{id}_E) \\ &= \mu_A \circ T\mu_A\,. \end{aligned}$$

- Now let $\mathcal{C}$ be a cartesian closed category and let ξ be some object in $\mathcal{C}$. We can define a monad of ξ-side-effects by taking T to be the functor $\xi \Rightarrow (\_ \times \xi)$, and η, μ as follows.

$$T := A \mapsto \xi \Rightarrow (A \times \xi)\,,\ f \mapsto \xi \Rightarrow (f \times \mathsf{id}_\xi)$$
$$\eta_A := \Lambda(\, A \times \xi \xrightarrow{\mathsf{id}_{A\times\xi}} A \times \xi\,)$$
$$\mu_A := \Lambda(\, T(TA) \times \xi \xrightarrow{\mathsf{ev}_{\xi, TA\times\xi}} TA \times \xi \xrightarrow{\mathsf{ev}_{\xi, A\times\xi}} A \times \xi\,)$$

Naturality of η, μ follows from naturality of Λ: for any $f : A \to A'$,

$$\begin{aligned} Tf \circ \eta_A &= (\xi \Rightarrow f \times \mathsf{id}_\xi) \circ \Lambda(\mathsf{id}_{A\times\xi}) = \Lambda(f \times \mathsf{id}_\xi \circ \mathsf{id}_{A\times\xi}) \\ &= \Lambda(\mathsf{id}_{A'\times\xi} \circ f \times \mathsf{id}_\xi) = \Lambda(\mathsf{id}_{A'\times\xi}) \circ f = \eta_{A'} \circ f\,, \end{aligned}$$

$$\begin{aligned} \mu_{A'} \circ T^2 f &= \Lambda(\mathsf{ev}_{\xi,A'\times\xi} \circ \mathsf{ev}_{\xi,TA'\times\xi}) \circ T^2 f = \Lambda(\mathsf{ev}_{\xi,A'\times\xi} \circ \mathsf{ev}_{\xi,TA'\times\xi} \circ T^2 f \times \mathsf{id}_\xi) \\ &= \Lambda(\mathsf{ev}_{\xi,A'\times\xi} \circ Tf \times \mathsf{id}_\xi \circ \mathsf{ev}_{\xi,TA\times\xi}) = \Lambda(f \times \mathsf{id}_\xi \circ \mathsf{ev}_{\xi,A\times\xi} \circ \mathsf{ev}_{\xi,TA\times\xi}) \\ &= (\xi \Rightarrow f \times \mathsf{id}_\xi) \circ \Lambda(\mathsf{ev}_{\xi,A\times\xi} \circ \mathsf{ev}_{\xi,TA\times\xi}) = Tf \circ \mu_A\,. \end{aligned}$$

The monadic diagrams are shown in a similar manner.

- Our third example employs the functor U : **Mon** $\to$ **Set**. In particular, we take $T := U \circ \mathsf{MList}$ and η, μ as follows.

$$\begin{aligned} T &:= X \mapsto \bigcup\nolimits_{n\in\omega} \{[x_1, \dots, x_n] \mid x_1, \dots, x_n \in X\}\,, \\ &\quad\; f \mapsto ([x_1, \dots, x_n] \mapsto [f(x_1), \dots, f(x_n)])\,. \\ \eta_X &:= x \mapsto [x] \\ \mu_X &:= [[x_{11}, \dots, x_{1n_1}], \dots, [x_{k1}, \dots, x_{kn_k}]] \mapsto [x_{11}, \dots, x_{1n_1}, \dots, x_{k1}, \dots, x_{kn_k}] \end{aligned}$$

Naturality of η, μ is obvious—besides, η is the unit of the corresponding adjunction. The monadic diagrams are also straightforward: they correspond to the following equalities of mappings (we use $\boldsymbol{x}$ for $x_1, \dots, x_n$).

$$\begin{array}{ccc} [[[\boldsymbol{x}_{11}], \dots, [\boldsymbol{x}_{1n_1}]], \dots, [[\boldsymbol{x}_{k1}], \dots, [\boldsymbol{x}_{kn_k}]]] & \overset{\mu}{\longmapsto} & [[\boldsymbol{x}_{11}], \dots, [\boldsymbol{x}_{1n_1}], \dots, [\boldsymbol{x}_{k1}], \dots, [\boldsymbol{x}_{kn_k}]] \\ {\scriptstyle T\mu}\downarrow & & \downarrow{\scriptstyle \mu} \\ [[\boldsymbol{x}_{11}, \dots, \boldsymbol{x}_{1n_1}], \dots, [\boldsymbol{x}_{k1}, \dots, \boldsymbol{x}_{kn_k}]] & \underset{\mu}{\longmapsto} & [\boldsymbol{x}_{11}, \dots, \boldsymbol{x}_{1n_1}, \dots, \boldsymbol{x}_{k1}, \dots, \boldsymbol{x}_{kn_k}] \end{array}$$

$$\begin{array}{ccc} [x_1, \dots, x_n] & \overset{\eta}{\longmapsto} & [[x_1, \dots, x_n]] \\ {\scriptstyle T\eta}\downarrow & \overset{\mathsf{id}}{\searrow} & \downarrow{\scriptstyle \mu} \\ [[x_1], \dots, [x_n]] & \underset{\mu}{\longmapsto} & [x_1, \dots, x_n] \end{array}$$

Exercise 125 Show that the E-coproduct monad and the ξ-side-effect monads are indeed monads.

Our discussion on monads can be dualised, leading us to *comonads*.

Definition 126 A ***comonad*** over a category $\mathcal{C}$ is a triple (Q, ε, δ) where Q is an endofunctor on $\mathcal{C}$ and $\varepsilon : Q \to \mathsf{Id}_{\mathcal{C}}$, $\delta : Q \to Q^2$ are natural transformations such that the following diagrams commute.

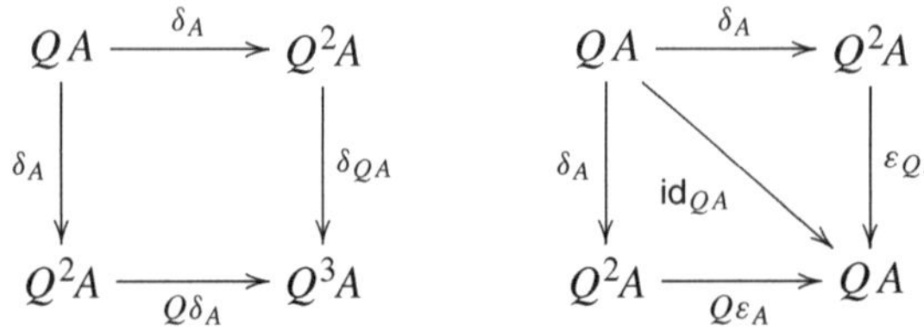

▲

ε is the *counit* of the comonad, and δ its *comultiplication*. Two of our examples from monads dualise to comonads.

- If $\mathcal{C}$ has finite products then, for any object S, we can define the S-product comonad with functor $Q := S \times \_$.
- We can form a comonad on **Mon** with functor $Q := \mathsf{MList} \circ U$ (and counit that of the corresponding adjunction).

Exercise 127 Give an explicit description of the comonad on **Mon** with functor $Q := \mathsf{MList} \circ U$ described above. Verify it is a comonad.

1.8.2 (Co)Monads of an Adjunction

In the previous section, we saw that an adjunction between **Mon** and **Set** yielded a monad on **Set** (and a comonad on **Mon**), with its unit being the unit of the adjunction. We now show that this observation generalises to any adjunction. Recall that an adjunction is specified by:

- a pair of functors $\mathcal{C} \underset{G}{\overset{F}{\rightleftarrows}} \mathcal{D}$,
- for each $A \in Ob(\mathcal{C})$, $B \in Ob(\mathcal{D})$, a bijection $\theta_{A,B} : \mathcal{C}(A, GB) \cong \mathcal{D}(FA, B)$ natural in A, B.

For such an adjunction we build a monad on $\mathcal{C}$: the functor of the monad is simply $T := G \circ F$, and unit and multiplication are defined by setting

$$\eta_A : A \longrightarrow GFA := \theta^{-1}_{A,FA}(\mathsf{id}_{FA}),$$
$$\mu_A : GFGFA \longrightarrow GFA := G(\theta_{GFA,FA}(\mathsf{id}_{GFA})).$$

Observe that η is the unit of the adjunction.

Proposition 128 *Let* (F, G, η) *be an adjunction. Then the triple* (T, η, μ) *defined above is a monad on* $\mathcal{C}$.

Proof Recall that naturality of θ means concretely that, for any $f : A \to GB$, $g : A' \to A$ and $h : B \to B'$,

$$\theta_{A',B'}(Gh \circ f \circ g) = h \circ \theta_{A,B}(f) \circ Fg.$$

η is the unit of the adjunction and hence natural. We show naturality of μ:

$$\begin{aligned} GFGFf \circ \mu_B &= G\theta_{GFB,FB}(\mathsf{id}_{GFB}) \circ GFGFf = G(\theta_{GFB,FB}(\mathsf{id}_{GFB}) \circ FGFf) \\ &\overset{\text{nat.}\theta}{=} G\theta_{GFA,FB}(\mathsf{id}_{GFB} \circ GFf) = G\theta_{GFA,FB}(GFf \circ \mathsf{id}_{GFA}) \\ &\overset{\text{nat.}\theta}{=} G(Ff \circ \theta_{GFA,FA}(\mathsf{id}_{GFA})) = GFf \circ \mu_A \,. \end{aligned}$$

The monoidal condition for μ also follows from naturality of θ:

$$\begin{aligned} \mu_A \circ \mu_{GFA} &= G(\theta(\mathsf{id}_{GFA}) \circ \theta(\mathsf{id}_{GFGFA})) \overset{\text{nat}}{=} G\theta(G\theta(\mathsf{id}_{GFA}) \circ \mathsf{id}_{GFGFA}) \\ &= G\theta(\mathsf{id}_{GFA} \circ G\theta(\mathsf{id}_{GFA})) \overset{\text{nat}}{=} G(\theta(\mathsf{id}_{GFA}) \circ FG\theta(\mathsf{id}_{GFA})) \\ &= \mu_A \circ GF\mu_A \,. \end{aligned}$$

Finally, for the η-μ conditions we also use the universality diagram for η and the uniqueness property (in equational form).

$$\begin{aligned} \mu_A \circ \eta_{GFA} &= G\theta_{GFA,FA}(\mathsf{id}_{GFA}) \circ \eta_{GFA} = \mathsf{id}_{GFA} \,, \\ \mu_A \circ GF\eta_{GFA} &= G\theta_{GFA,FA}(\mathsf{id}_{GFA}) \circ GF\eta_{GFA} = G(\theta_{GFA,FA}(\mathsf{id}_{GFA}) \circ F\eta_{GFA}) \\ &\overset{\text{nat}}{=} G\theta(\mathsf{id}_{GFA} \circ \eta_{GFA}) = G\theta(G\,\mathsf{id}_{FA} \circ \eta_{GFA}) = G\,\mathsf{id}_{FA} = GFA \,. \end{aligned}$$

■

Hence, every adjunction gives rise to a monad. It turns out that the converse is also true: *every* monad is described by means of an adjunction in this way. In particular, there are two canonical constructions of adjunctions from a given monad: the *Kleisli construction*, and the *Eilenberg-Moore construction*. These are in a sense minimal and maximal solutions to describing a monad via an adjunction. We describe the former one in the next section.

Finally, note that—because of the symmetric definition of adjunctions—the whole discussion can be dualised to comonads. That is, every adjunction gives rise to a comonad with counit that of the adjunction, and also every comonad can be derived from an adjunction in this manner.

1.8.3 The Kleisli Construction

The Kleisli construction starts from a monad (T, η, μ) on a category $\mathcal{C}$ and builds a category $\mathcal{C}_T$ of T-computations, as follows.

Definition 129 Let (T, η, μ) be a monad on a category $\mathcal{C}$. Construct the ***Kleisli category*** $\mathcal{C}_T$ by taking the same objects as $\mathcal{C}$, and by including an arrow $f_{.T} : A \to B$ in $\mathcal{C}_T$ for each $f : A \to TB$ in $\mathcal{C}$. That is,

$$\begin{aligned} Ob(\mathcal{C}_T) &:= Ob(\mathcal{C}) \,, \\ \mathcal{C}_T(A, B) &:= \{f_{.T} \mid f \in \mathcal{C}(A, TB)\} \,. \end{aligned}$$

The identity arrow for A in $\mathcal{C}_T$ is $\eta_{A.T}$, while the composite of $f_{.T} : A \to B$ and $g_{.T} : B \to C$ is $h_{.T}$, where:

$$h := A \xrightarrow{f} TB \xrightarrow{Tg} T^2C \xrightarrow{\mu_C} TC\,.$$

▲

The conditions for $\mathcal{C}_T$ being a category follow from the monadic conditions. For composition with identity, for any $f : A \to TB$,

$$f_{.T} \circ \eta_{A.T} = (\mu_B \circ Tf \circ \eta_A)_{.T} = (\mu_B \circ \eta_B \circ f)_{.T} = f_{.T}\,,$$
$$\eta_{B.T} \circ f_{.T} = (\mu_B \circ T\eta_B \circ f)_{.T} = f_{.T}\,.$$

For associativity of composition, for any $f : A \to TB$, $g : B \to TC$ and $h : C \to TD$,

$$\begin{aligned}(h_{.T} \circ g_{.T}) \circ f_{.T} &= (\mu_D \circ Th \circ g)_{.T} \circ f_{.T} = (\mu_D \circ T(\mu_D \circ Th \circ g) \circ f)_{.T}\\ &= (\mu_D \circ T\mu_D \circ T^2h \circ Tg \circ f)_{.T} = (\mu_D \circ \mu_{TD} \circ T^2h \circ Tg \circ f)_{.T}\\ &= (\mu_D \circ Th \circ \mu_C \circ Tg \circ f)_{.T} = h_{.T} \circ (g_{.T} \circ f_{.T})\,.\end{aligned}$$

Let us now proceed to build the adjunction between $\mathcal{C}$ and $\mathcal{C}_T$ that will eventually give us back the monad T. Construct the functors $F : \mathcal{C} \to \mathcal{C}_T$ and $G : \mathcal{C}_T \to \mathcal{C}$ as follows.

$$F := A \mapsto A\,,\ (f : A \to B) \mapsto ((\eta_B \circ f)_{.T} : A \to B)\,,$$
$$G := A \mapsto TA\,,\ (f_{.T} : A \to B) \mapsto (\mu_B \circ Tf : TA \to TB)\,.$$

Functoriality of F, G follows from the monad laws and the definition of $\mathcal{C}_T$. Moreover, for each $A, B \in Ob(\mathcal{C})$, construct the following bijection of arrows.

$$\theta_{A,B} : \mathcal{C}(A, TB) \xrightarrow{\cong} \mathcal{C}_T(A, B) := f \mapsto f_{.T}$$

To establish that (F, G, θ) is an adjunction we need only show that θ is natural in A, B. So take $f : A \to TB$, $g : A' \to A$ and $h_{.T} : B \to B'$. We then have:

$$\begin{aligned}\theta_{A',B'}(G(h_{.T}) \circ f \circ g) &= \theta_{A',B'}(\mu_{B'} \circ Th \circ f \circ g) = (\mu_{B'} \circ Th \circ f \circ g)_{.T}\\ &= h_{.T} \circ (f \circ g)_{.T} = h_{.T} \circ (\mu_B \circ Tf \circ \eta_A \circ g)_{.T}\\ &= h_{.T} \circ f_{.T} \circ (\eta_A \circ g)_{.T} = h_{.T} \circ \theta_{A,B}(f) \circ Fg\,.\end{aligned}$$

The final step in this section is to verify that the monad (T', η', μ') arising from this adjunction is the one we started from. The construction of T' follows the recipe given in the previous section, that is:

- $T' : \mathcal{C} \to \mathcal{C} := G \circ F$. Thus, T' maps each object A to TA, and each arrow $f : A \to B$ to $\mu_B \circ T\eta_A \circ Tf = Tf$.

- $\eta'_A : A \to TA := \theta^{-1}_{A,FA}(\mathsf{id}^{(\mathcal{C}_T)}_{FA}) = \theta^{-1}(\eta_{A.T}) = \eta_A$.
- $\mu'_A : T^2 A \to TA := G\theta_{GFA,FA}(\mathsf{id}^{(\mathcal{C})}_{GFA}) = G\theta(\mathsf{id}_{TA}) = \mu_A \circ T\,\mathsf{id}_{TA} = \mu_A$.

Thus, we have indeed obtained the initial (T, η, μ).

The Kleisli Construction on a Comonad

Dually to the Kleisli category of a monad we can construct the Kleisli category of a comonad[5]—and reobtain the comonad through an adjunction between the Kleisli category and the original one. Specifically, given a category $\mathcal{C}$ and a comonad (Q, ε, δ) on $\mathcal{C}$, we define the category $\mathcal{C}_Q$ as follows.

$$
\begin{aligned}
Ob(\mathcal{C}_Q) &:= Ob(\mathcal{C}) \\
\mathcal{C}_Q(A, B) &:= \{f_{.Q} \mid f \in \mathcal{C}(QA, B)\} \\
\mathsf{id}^{(\mathcal{C}_Q)}_A &:= \varepsilon_{A.Q} \\
g_{.Q} \circ f_{.Q} &:= (g \circ Qf \circ \delta_A)_{.Q}
\end{aligned}
$$

The Kleisli category of a comonad will be of use in the next sections, where comonads will be considered for modelling *bang* of Linear Logic. We end this section by showing a result that will be of use then.

Proposition 130 *Let $\mathcal{C}$ be a category and (Q, ε, δ) be a comonad on $\mathcal{C}$. If $\mathcal{C}$ has binary products then so does $\mathcal{C}_Q$.*

Proof Let A, B be objects in $\mathcal{C}, \mathcal{C}_Q$. We claim that their product in $\mathcal{C}_Q$ is given by $(A \times B, p_1, p_2)$, where

$$
p_1 := \big(Q(A \times B) \xrightarrow{\varepsilon} A \times B \xrightarrow{\pi_1} A\big)_{.Q}
$$

and similarly for p_2. Now, for each $f_{.Q} : C \to A$ and $g_{.Q} : C \to B$, setting $\langle f_{.Q}, g_{.Q}\rangle := \langle f, g\rangle_{.Q}$ we have:

$$
p_1 \circ \langle f_{.Q}, g_{.Q}\rangle = (\pi_1 \circ \varepsilon \circ Q\langle f, g\rangle \circ \delta)_{.Q} = (\pi_1 \circ \langle f, g\rangle \circ \varepsilon \circ \delta)_{.Q} = f_{.Q}\,,
$$

and similarly $p_2 \circ \langle f_{.Q}, g_{.Q}\rangle = g_{.Q}$. Finally, for any $h_{.Q} : C \to A \times B$,

$$
\begin{aligned}
\langle p_1 \circ h_{.Q}, p_2 \circ h_{.Q}\rangle &= \langle \pi_1 \circ \varepsilon \circ Qh \circ \delta, \pi_2 \circ \varepsilon \circ Qh \circ \delta\rangle_{.Q} = \langle \pi_1 \circ h, \pi_2 \circ h\rangle_{.Q} \\
&= h_{.Q}\,.
\end{aligned}
$$

∎

Exercise 131 Show that the Kleisli category $\mathcal{C}_Q$ of a comonad (Q, ε, δ) has a terminal object when $\mathcal{C}$ does.

---

[5] In some texts, this is called a *coKleisli* category.

1.8.4 Modelling of Linear Exponentials

In this section we employ comonads in order to model the exponential *bang* operator, $!$, of Linear Logic. Let us start by modelling a *weak bang* operator, $\hat{!}$, which involves solely the following proof rules.

$$\dfrac{\Gamma, A \vdash B}{\Gamma, \hat{!}A \vdash B}\,\hat{!}\mathsf{L} \qquad\qquad \dfrac{\hat{!}B \vdash A}{\hat{!}B \vdash \hat{!}A}\,\hat{!}\mathsf{R}$$

Observe that, compared to $!$, $\hat{!}$ is weak in its Right rule, and it also misses Contraction and Weakening.

Let us now assume as given a symmetric monoidal closed category $\mathcal{C}$ along with a comonad (Q, ε, δ) on $\mathcal{C}$. As seen previously, $\mathcal{C}$ is a model of $(\otimes\!-\!\circ)$-Linear Logic. Moreover, $(\mathcal{C}, Q)$ yields a model of $(\otimes\!-\!\circ\hat{!})$-Linear Logic by modelling each formula $\hat{!}A$ by QA (*i.e.* Q applied to the translation of A). The rules for weak bang are then interpreted as follows.

$$\dfrac{f : \Gamma \otimes A \longrightarrow B}{f \circ \mathsf{id}_\Gamma \otimes \varepsilon_A : \Gamma \otimes QA \longrightarrow B} \qquad\qquad \dfrac{f : QB \longrightarrow A}{Qf \circ \delta_B : QB \longrightarrow QA}$$

We know that arrow-equalities in $\mathcal{C}$ correspond to proof-transformations in the proof system. Thus, the comonadic law $\varepsilon_{QA} \circ \delta_A = \mathsf{id}_{QA} = Q\varepsilon_A \circ \delta_A$ corresponds to the following transformations.

$$\dfrac{\dfrac{\dfrac{}{\hat{!}A \vdash \hat{!}A}\,\mathsf{Id}}{\hat{!}A \vdash \hat{!}\hat{!}A}\,\hat{!}\mathsf{R} \qquad \dfrac{\dfrac{}{\hat{!}A \vdash \hat{!}A}\,\mathsf{Id}}{\hat{!}\hat{!}A \vdash \hat{!}A}\,\hat{!}\mathsf{L}}{\hat{!}A \vdash \hat{!}A}\,\mathsf{Cut} \quad\Longrightarrow\quad \dfrac{}{\hat{!}A \vdash \hat{!}A}\,\mathsf{Id} \quad\Longleftarrow\quad \dfrac{\dfrac{\dfrac{}{A \vdash A}\,\mathsf{Id}}{\hat{!}A \vdash A}\,\hat{!}\mathsf{L}}{\hat{!}A \vdash \hat{!}A}\,\hat{!}\mathsf{R}$$

Exercise 132 Find a proof-transformation corresponding to the comonadic law $\delta_{QA} \circ \delta_A = Q\delta_A \circ \delta_A$.

In order to extend our translation to the general $!\mathsf{R}$ rule, we need arrows in $\mathcal{C}$ of the form

$$Q^2A_1 \otimes \cdots \otimes Q^2A_n \longrightarrow Q(QA_1 \otimes \cdots \otimes QA_n).$$

Hence, we need to impose (a coherent) distributivity of the tensor—either binary ($\otimes$) or nullary (I)—over the comonad Q. This can be formalised by stipulating that Q be a *symmetric monoidal* endofunctor.

Definition 133 Let $(\mathcal{C}, \otimes, I, a, l, r, s)$ and $(\mathcal{C}', \otimes', I', a', l', r', s')$ be symmetric monoidal categories. A functor $F : \mathcal{C} \to \mathcal{C}'$ is called ***symmetric monoidal*** if there exist:

- a morphism $m_0 : I' \to F(I)$,

- a natural transformation $m_2 : F(\_) \otimes' F(\_) \to F(\_ \otimes \_)$,

such that the following diagrams commute.

$$\begin{array}{ccccc}
FA \otimes' (FB \otimes' FC) & \xrightarrow{\mathsf{id} \otimes' m_2} & FA \otimes' F(B \otimes C) & \xrightarrow{m_2} & F(A \otimes (B \otimes C)) \\
{\scriptstyle a'}\downarrow & & & & \downarrow{\scriptstyle Fa} \\
(FA \otimes' FB) \otimes' FC & \xrightarrow[m_2 \otimes' \mathsf{id}]{} & F(A \otimes B) \otimes' FC & \xrightarrow[m_2]{} & F((A \otimes B) \otimes C)
\end{array}$$

$$\begin{array}{ccc}
FA \otimes' I' & \xrightarrow{\mathsf{id} \otimes' m_0} & FA \otimes' FI \\
{\scriptstyle r'}\downarrow & & \downarrow{\scriptstyle m_2} \\
FA & \xleftarrow[Fr]{} & F(A \otimes I)
\end{array}
\qquad
\begin{array}{ccc}
FA \otimes' FB & \xrightarrow{m_2} & F(A \otimes B) \\
{\scriptstyle s'}\downarrow & & \downarrow{\scriptstyle Fs} \\
FB \otimes' FA & \xrightarrow[m_2]{} & F(B \otimes A)
\end{array}$$

We may write such an F as (F, m). Moreover, if $(F, m), (G, n) : \mathcal{C} \to \mathcal{C}'$ are (symmetric) monoidal functors then a natural transformation $\phi : F \to G$ is called ***monoidal*** whenever the following diagrams commute.

$$\begin{array}{ccc}
I' & \xrightarrow{m_0} & FI \\
& {\scriptstyle n_0}\searrow & \downarrow{\scriptstyle \phi} \\
& & GI
\end{array}
\qquad
\begin{array}{ccc}
FA \otimes' FB & \xrightarrow{m_2} & F(A \otimes B) \\
{\scriptstyle \phi \otimes' \phi}\downarrow & & \downarrow{\scriptstyle \phi} \\
GA \otimes' GB & \xrightarrow[n_2]{} & G(A \otimes B)
\end{array}$$

▲

For example, the identity functor is symmetric monoidal. Moreover, if F and G are symmetric monoidal functors then so is $G \circ F$. Other examples are the following.

- The constant endofunctor K_I, which maps each object to I and each arrow to id_I, is symmetric monoidal with structure maps:

$$m_0 : I \longrightarrow I := \mathsf{id}_I\,, \quad m_2 : I \otimes I \longrightarrow I := r_I\,.$$

- The endofunctor $\otimes \circ \langle \mathsf{Id}_{\mathcal{C}}, \mathsf{Id}_{\mathcal{C}} \rangle$, which maps each object A to $A \otimes A$ and each arrow f to $f \otimes f$, is symmetric monoidal with:

$$m_0 : I \longrightarrow I \otimes I := r_I^{-1}\,, \quad m_2 := (A \otimes A) \otimes (B \otimes B) \longrightarrow (A \otimes B) \otimes (A \otimes B)\,,$$

the latter given by use of structural transformations.

Exercise 134 Verify that if $F : \mathcal{C} \to \mathcal{D}$, $G : \mathcal{D} \to \mathcal{E}$ are symmetric monoidal functors then so is $G \circ F$.

Definition 135 A comonad (Q, ε, δ) on a SMCC $\mathcal{C}$ is called a ***monoidal comonad*** if Q is a symmetric monoidal functor, say (Q, m), and ε, δ are monoidal natural transformations. We write Q as $(Q, \varepsilon, \delta, m)$. ▲

Now let us assume $\mathcal{C}$ is a SMCC and $(Q, \varepsilon, \delta, m)$ is a monoidal comonad on $\mathcal{C}$. The coherence of m_2 with a, expressed by the first diagram of symmetric monoidal functors, allows us to generalise m_0 and m_2 to arbitrary arities and assume arrows:

$$m_n : QA_1 \otimes \cdots \otimes QA_n \longrightarrow Q(A_1 \otimes \cdots \otimes A_n) .$$

We can give the interpretation of the Right rule for bang as follows.

$$\frac{f : QB_1 \otimes \cdots \otimes QB_n \longrightarrow A}{Qf \circ m_n \circ (\delta_{B_1} \otimes \cdots \otimes \delta_{B_n}) : QB_1 \otimes \cdots \otimes QB_n \longrightarrow QA}$$

Contraction and Weakening

Our discussion on the categorical modelling of linear exponentials has only touched the issues of Right and Left rules. However, we also need adequate structure for translating Contraction and Weakening.

$$\frac{\Gamma, !A, !A \vdash B}{\Gamma, !A \vdash B}\ \mathsf{Contr} \qquad \frac{\Gamma \vdash B}{\Gamma, !A \vdash B}\ \mathsf{Weak}$$

For these rules we can use appropriate (monoidal) natural transformations. For Contraction, we stipulate a transformation with components $d_A : QA \to QA \otimes QA$, *i.e.*

$$d : Q \longrightarrow \otimes \circ \langle Q, Q \rangle .$$

For Weakening, a transformation with components $e_A : QA \to I$, *i.e.*

$$e : Q \longrightarrow K_I .$$

Although the above allow the categorical interpretation of the proof-rules, they do not necessarily preserve the intended proof-transformations. For that, we need to impose some further coherence conditions, which are epitomised in the following notion.

Definition 136 Let $\mathcal{C}$ be a SMCC. A monoidal comonad $(Q, \varepsilon, \delta, m)$ on $\mathcal{C}$ is called a ***linear exponential comonad*** if there exist monoidal natural transformations

$$d : Q \longrightarrow \otimes \circ \langle Q, Q \rangle , \quad e : Q \longrightarrow K_I ,$$

such that:

(a) for each object A, the triple (QA, d_A, e_A) is a *commutative comonoid in* $\mathcal{C}$, *i.e.* the following diagrams commute,

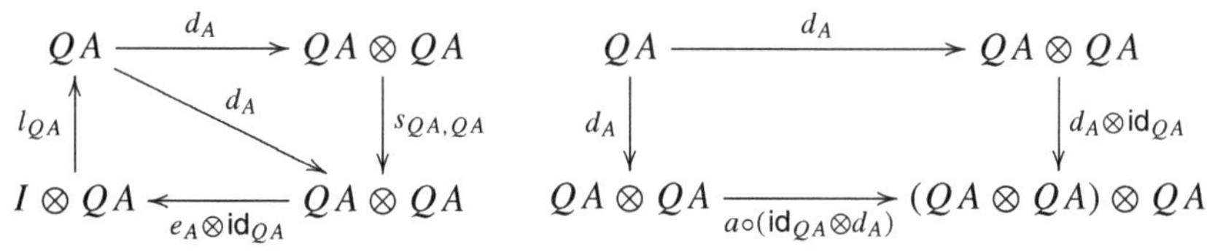

(b) for each object A, the following diagrams commute.

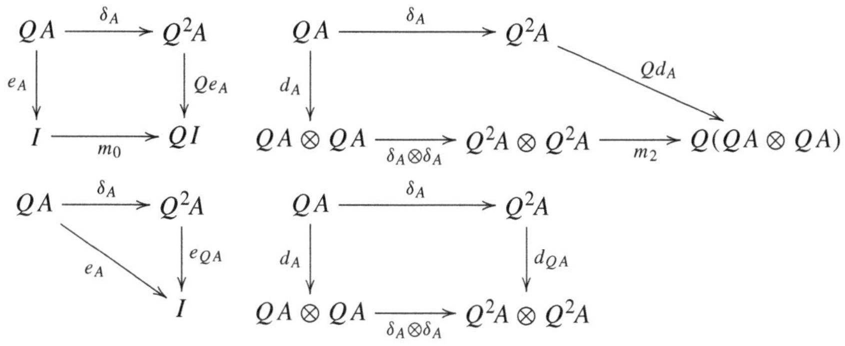

We write Q as $(Q, \varepsilon, \delta, m, d, e)$. ▲

Exercise 137 Express what it means concretely for d, e to be monoidal natural transformations.

Exercise 138 Give the categorical interpretation of Contraction and Weakening in a SMCC $\mathcal{C}$ with a linear exponential comonad.

Including Products

We now consider the fragment of Linear Logic which includes all four linear connectives we have seen thus far, *i.e.* $\otimes \multimap !\&$, and their respective proof rules (see definitions 122, 123). The categorical modelling of ($\otimes \multimap !\&$)-Linear Logic requires:

- a symmetric monoidal closed category $\mathcal{C}$,
- a linear exponential comonad $(Q, \varepsilon, \delta, m, d, e)$ on $\mathcal{C}$,
- finite products in $\mathcal{C}$.

The above structure is adequate for modelling the proof rules as we have seen previously. Moreover, it provides rich structure for the Kleisli category $\mathcal{C}_Q$. The next result and its proof demonstrate categorically the 'interpretation' of ordinary logic within Linear Logic given by:

$$A \Rightarrow B \equiv\, !A \multimap B\,.$$

Proposition 139 *Let $\mathcal{C}$ be a SMCC with finite products and let $(Q, \varepsilon, \delta, m, d, e)$ be a linear exponential comonad on $\mathcal{C}$. Then:*

(a) *The Kleisli category $\mathcal{C}_Q$ has finite products.*
(b) *There exists an isomorphism $i : Q\mathbf{1} \to I$ and a natural isomorphism $j : Q(\_\times\_) \to Q(\_) \otimes Q(\_)$.*
(c) *$\mathcal{C}_Q$ is cartesian closed, with the exponential of objects B, C being $QB \multimap C$.*

Proof Part (a) has been shown previously (Proposition 130, Exercise 131), and part (b) is left as exercise. For (c), we have the following isomorphisms:

$$\begin{aligned} \mathcal{C}_Q(A \times B, C) &= \mathcal{C}(Q(A \times B), C) && \text{definition of } \mathcal{C}_Q \\ &\cong \mathcal{C}(QA \otimes QB, C) && \text{part (b)} \\ &\cong \mathcal{C}(QA, QB \multimap C) && \text{monoidal closure of } \mathcal{C} \\ &= \mathcal{C}_Q(A, QB \multimap C) && \text{defn of } \mathcal{C}_Q. \end{aligned}$$

Concretely, we obtain $\theta_A : \mathcal{C}_Q(A \times B, C) \xrightarrow{\cong} \mathcal{C}_Q(A, QB \multimap C)$ by:

$$\begin{aligned} \theta_A &:= (f_{.Q} : A \times B \to C) \longmapsto (\Lambda(f \circ j^{-1}_{A,B}))_{.Q} \\ \theta^{-1}_A &:= (g_{.Q} : A \to QB \multimap C) \longmapsto (\Lambda^{-1}(g) \circ j_{A,B})_{.Q}\,. \end{aligned}$$

Clearly, θ_A is a bijection. In order to establish couniversality of the exponential, we need also show naturality in A (see Exercise 76). So take $f_{.Q} : A \times B \to C$ and $h_{.Q} : A' \to A$. Note first that the following commutes.

$$\begin{array}{ccccc} Q(A \times B) & \xrightarrow{\delta} & Q^2(A \times B) & \xrightarrow{Q\langle Q\pi_1, Q\pi_2\rangle} & Q(QA \times QB) \\ \downarrow{\scriptstyle j} & & & & \downarrow{\scriptstyle j} \\ QA \otimes QB & & \xrightarrow[\delta\otimes\delta]{\hspace{6em}} & & Q^2A \otimes Q^2B \end{array} \qquad (*)$$

Note also that, for any $h_{i.Q} : A'_i \to A_i$ in $\mathcal{C}_Q$, $i = 1, 2$, we have:

$$h_{1.Q} \times h_{2.Q} := \big(Q(A'_1 \times A'_2) \xrightarrow{\langle Q\pi_1, Q\pi_2\rangle} QA'_1 \times QA'_2 \xrightarrow{h_1 \times h_2} A_1 \times A_2\big)_{.Q}$$

Thus, noting that $\mathsf{id}^{(\mathcal{C}_Q)}_B = \varepsilon_{B.Q}$,

$$\begin{aligned} \theta_{A'}(f_{.Q} \circ h_{.Q} \times \mathsf{id}^{(\mathcal{C}_Q)}_B) &= \big(\Lambda(f \circ Q(h \times \varepsilon \circ \langle Q\pi_1, Q\pi_2\rangle) \circ \delta \circ j^{-1})\big)_{.Q} \\ &= \big(\Lambda(f \circ Q(h \times \varepsilon) \circ Q\langle Q\pi_1, Q\pi_2\rangle \circ \delta \circ j^{-1})\big)_{.Q} \\ &\overset{(*)}{=} \big(\Lambda(f \circ Q(h \times \varepsilon) \circ j^{-1} \circ \delta \otimes \delta)\big)_{.Q} \end{aligned}$$

$$
\begin{aligned}
&= \big(\Lambda(f \circ j^{-1} \circ Qh \otimes Q\varepsilon \circ \delta \otimes \delta)\big)_{.Q} \\
&= \big(\Lambda(f \circ j^{-1} \circ (Qh \circ \delta) \otimes \mathsf{id})\big)_{.Q} \\
&= \big(\Lambda(f \circ j^{-1}) \circ Qh \circ \delta\big)_{.Q} = \theta_A(f_{.Q}) \circ h_{.Q}
\end{aligned}
$$

as required. ■

Exercise 140 Show part (b) of Proposition 139. For the defined j, show commutativity of $(*)$.

1.8.5 Exercises

1. We say that a category $\mathcal{C}$ is *well-pointed* if it contains a terminal object $\mathbf{1}$ and, for any pair of arrows $f, g : A \to B$,

   $$f \neq g \implies \exists h : 1 \to A.\, f \circ h \neq g \circ h\,.$$

   Let now $\mathcal{C}$ be a well-pointed category with a terminal object $\mathbf{1}$ and binary coproducts, and consider the functor $G : \mathcal{C} \to \mathcal{C}$ given by:

   $$G := A \mapsto A + \mathbf{1}\,,\ f \mapsto f + \mathsf{id}_{\mathbf{1}}\,.$$

   If $\mathcal{C}(\mathbf{1}, \mathbf{1}+\mathbf{1}) = \{\mathsf{in}_1, \mathsf{in}_2\}$ with $\mathsf{in}_1 \neq \mathsf{in}_2$, show that if (G, η, μ) is a monad on $\mathcal{C}$ then, for each object A:

   $$\eta_A = A \xrightarrow{\mathsf{in}_1} A + \mathbf{1}\,, \quad \mu_A = (A+\mathbf{1}) + \mathbf{1} \xrightarrow{[\mathsf{id}_{A+\mathbf{1}}, \mathsf{in}_2]} A + \mathbf{1}\,.$$

2. Let $\mathcal{C}$ be a SMCC and let (Q, ε, δ) be a comonad on $\mathcal{C}$.

   (a) Suppose that the sequents $\hat{!}A \vdash B$ and $\hat{!}B \vdash C$ are provable and let $f : QA \to B$ and $g : QB \to C$ be their interpretations (*i.e.* the interpretations of their proofs) in $\mathcal{C}$. Find the interpretations of the sequent $\hat{!}A \vdash \hat{!}C$ which correspond to each of the following proofs and show that the two interpretations are equal.

   $$\dfrac{\dfrac{\dfrac{\vdots}{\hat{!}A \vdash B}}{\hat{!}A \vdash \hat{!}B}\,\hat{!}\mathsf{R} \qquad \dfrac{\dfrac{\vdots}{\hat{!}B \vdash C}}{\hat{!}B \vdash \hat{!}C}\,\hat{!}\mathsf{R}}{\hat{!}A \vdash \hat{!}C}\,\mathsf{Cut}
   \qquad\qquad
   \dfrac{\dfrac{\dfrac{\dfrac{\vdots}{\hat{!}A \vdash B}}{\hat{!}A \vdash \hat{!}B}\,\hat{!}\mathsf{R} \qquad \dfrac{\vdots}{\hat{!}B \vdash C}}{\hat{!}A \vdash C}\,\mathsf{Cut}}{\hat{!}A \vdash \hat{!}C}\,\hat{!}\mathsf{R}$$

   (b) Find the interpretations in $\mathcal{C}$ of the following proofs; are the interpretations equal?

$$\dfrac{\dfrac{\dfrac{}{\hat{!}A \vdash \hat{!}A}\,\mathsf{Id}}{\hat{!}\hat{!}A \vdash \hat{!}A}\,\hat{!}\mathsf{L}}{\hat{!}\hat{!}A \vdash \hat{!}\hat{!}A}\,\hat{!}\mathsf{R} \qquad\qquad \dfrac{\dfrac{\dfrac{}{\hat{!}A \vdash \hat{!}A}\,\mathsf{Id}}{\hat{!}A \vdash \hat{!}\hat{!}A}\,\hat{!}\mathsf{R}}{\hat{!}\hat{!}A \vdash \hat{!}\hat{!}A}\,\hat{!}\mathsf{L}$$

3. Show that a symmetric monoidal category $\mathcal{C}$ has finite products (given by $\otimes$, I, etc.) iff there are monoidal natural transformations

$$d : \mathsf{Id}_{\mathcal{C}} \longrightarrow \otimes \circ \langle \mathsf{Id}_{\mathcal{C}}, \mathsf{Id}_{\mathcal{C}} \rangle\,, \quad e : \mathsf{Id}_{\mathcal{C}} \longrightarrow K_I\,,$$

such that the following diagram commutes, for any $A \in Ob(\mathcal{C})$.

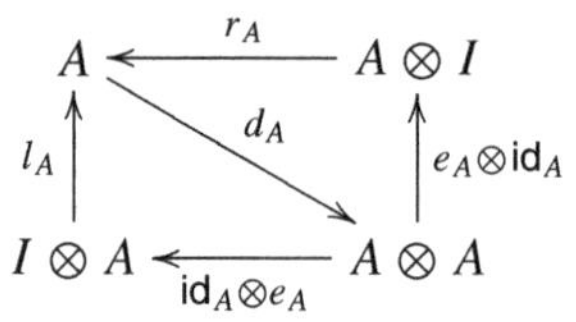

A Review of Sets, Functions and Relations

Our aim in this Appendix is to provide a brief review of notions we will assume in the notes. If the first paragraph is not familiar to you, you will need to acquire more background before being ready to read the notes.

Cartesian Products, Relations and Functions

Given sets X and Y, their cartesian product is

$$X \times Y = \{(x, y) \mid x \in X \wedge y \in Y\}\,.$$

A *relation* R from X to Y, written $R : X \to Y$, is a subset $R \subseteq X \times Y$. Given such a relation, we write $(x, y) \in R$, or equivalently $R(x, y)$. We compose relations as follows: if $R : X \to Y$ and $S : Y \to Z$, then for all $x \in X$ and $z \in Z$:

$$R; S(x, z) \equiv \exists y \in Y.\, R(x, y) \wedge S(y, z)\,.$$

A relation $f : X \to Y$ is a *function* if it satisfies the following two properties:

- (single-valuedness): if $(x, y) \in f$ and $(x, y') \in f$, then $y = y'$.
- (totality): for all $x \in X$, for some $y \in Y$, $(x, y) \in f$.

If f is a function, we write $f(x) = y$ or $f : x \mapsto y$ for $(x, y) \in f$. Function composition is written as follows: if $f : X \to Y$ and $g : Y \to Z$,

$$g \circ f(x) = g(f(x))\,.$$

It is easily checked that $g \circ f = f; g$, viewing functions as relations.

Equality of Functions

Two functions $f, g : X \to Y$ are *equal* if they are equal as relations, *i.e.* as sets of ordered pairs. Equivalently, but more conveniently, we can write:

$$f = g \iff \forall x \in X.\, f(x) = g(x)\,.$$

The right-to-left implication is the standard tool for proving equality of functions on sets. As we shall see, when we enter the world of category theory, which takes a more general view of "arrows" $f : X \to Y$, for most purposes we have to leave this familiar tool behind!

Making the Arrow Notation for Functions and Relations Unambiguous

Our definitions of functions and relations, as they stand, have an unfortunate ambiguity. Given a relation $R : X \to Y$, we cannot uniquely recover its "domain" X and "codomain" Y. In the case of a function, we can recover the domain, because of totality, but not the codomain.

Example Consider the set of ordered pairs $\{(n, n) \mid n \in \mathbb{N}\}$, where $\mathbb{N}$ is the set of natural numbers. Is this the identity function $\mathsf{id}_{\mathbb{N}} : \mathbb{N} \longrightarrow \mathbb{N}$, or the inclusion function $\mathsf{inc} : \mathbb{N} \hookrightarrow \mathbb{Z}$, where $\mathbb{Z}$ is the set of integers?

We wish to have unambiguous notions of domain and codomain for functions, and more generally relations. Thus we modify our official definition of a relation from X to Y to be an ordered triple (X, R, Y), where $R \subseteq X \times Y$. We then define composition of (X, R, Y) and (Y, S, Z) in the obvious fashion, as $(X, R; S, Z)$. We treat functions similarly. We shall not belabour this point in the notes, but it is implicit when we set up perhaps the most fundamental example of a category, namely the category of sets.

Size

We shall avoid explicit discussion of set-theoretical foundations in the text, but we include a few remarks for the interested reader. Occasionally, distinctions of set-theoretic size do matter in category theory. One example which does arise in the notes is when we consider **Cat**, the category of "all" categories. Does this category belong to (is it an object of) itself, at the risk of a Russell-type paradox? The way we avoid this is to impose some set-theoretic limitation of size on the categories gathered into **Cat**. **Cat** will then be too big to fit into itself. For example, we can limit **Cat** to those categories whose collections of objects and arrows form *sets* in the sense of some standard set theory such as ZFC. **Cat** will then be a proper class, and will not be an object of itself. One assumption we do make throughout the notes is that the categories we deal with are "locally small", *i.e.* that all hom-sets are indeed sets. Another place where some technical caveat would be in order is when we form functor categories. In practice, these issues never (well, hardly ever) cause problems, because of the strongly-typed nature of category theory. We leave the interested reader to delve further into these issues by consulting some of the standard texts.

B Guide to Further Reading

Of the many texts on category theory, we shall only mention a few, which may be particularly useful to someone who has read these notes and wishes to learn more.

The short text [10] is very nicely written and gently paced; it is probably a little easier going than these notes. A text which is written with a clarity and at a level which makes it ideal as a next step after these notes is [5]. A text particularly useful for its large number of exercises with solutions is [1].

Another very nicely written text, focussing on the connections between categories and logic, and especially topos theory, is [4], recently reissued by Dover Books. A classic text on categorical logic is [6]. A more advanced text on topos theory is [9].

The text [8] is a classic by one of the founders of category theory. It assumes considerable background knowledge of mathematics to fully appreciate its wide-ranging examples, but it provides invaluable coverage of the key topics.

A stimulating text on the correspondence between computation and logic is [3]; it is out of print, but available online. A more recent text on this topic is [11].

The 3-volume handbook [2] provides coverage of a broad range of topics in category theory. The book [7] is somewhat idiosyncratic in style, but offers insights by one of the key contributors to category theory.

References

1. Barr, M., Wells, C.: Category Theory for Computing Science, 3rd edn. Publications CRM, Montreal (1999)
2. Borceux, F.: *Handbook of Categorical Algebra Volumes 1–3.* Cambridge University Press, Cambridge (1994)
3. Girard, J.-Y., Taylor, P., Lafont, Y.: Proofs and Types. Cambridge University Press, Cambridge (1989)
4. Goldblatt, R.: Topoi, the Categorial Analysis of Logic. North-Holland, Amsterdam (1984). Reprinted by Dover Books, 2006.
5. Herrlich, H., Strecker, G.: Category Theory, 3rd edn. Heldermann, Berline (2007)
6. Lambek, J., Scott, P.J.: Introduction to Higher-Order Categorical Logic. Cambridge University Press, Cambridge (1986)
7. Lawvere, F., Schanuel, S.: Conceptual Mathematics: A First Introduction to Categories. Cambridge University Press, Cambridge (1997)
8. Mac Lane, S.: Categories for the Working Mathematician, 2nd edn. Springer, New York (1998)
9. Mac Lane, S., Moerdijk, I.: Sheaves in Geometry and Logic: A First Introduction to Topos Theory. Springer, New York (1994)
10. Pierce, B.: Basic Category Theory for Computer Scientists. MIT Press, Cambridge (1991)
11. Sørensen, M.H., Urzyczyn, P.: Lectures on the Curry-Howard Isomorphism. Elsevier, New York (2006)

Chapter 2
Physics, Topology, Logic and Computation: A Rosetta Stone

J. Baez and M. Stay

Abstract In physics, Feynman diagrams are used to reason about quantum processes. In the 1980s, it became clear that underlying these diagrams is a powerful analogy between quantum physics and topology. Namely, a linear operator behaves very much like a "cobordism": a manifold representing spacetime, going between two manifolds representing space. This led to a burst of work on topological quantum field theory and "quantum topology". But this was just the beginning: similar diagrams can be used to reason about logic, where they represent proofs, and computation, where they represent programs. With the rise of interest in quantum cryptography and quantum computation, it became clear that there is extensive network of analogies between physics, topology, logic and computation. In this expository paper, we make some of these analogies precise using the concept of "closed symmetric monoidal category". We assume no prior knowledge of category theory, proof theory or computer science.

2.1 Introduction

Category theory is a very general formalism, but there is a certain special way that physicists use categories which turns out to have close analogues in topology, logic and computation. A category has *objects* and *morphisms*, which represent *things* and *ways to go between things*. In physics, the objects are often *physical systems*, and the morphisms are *processes* turning a state of one physical system into a state of another system—perhaps the same one. In quantum physics we often formalize this by taking *Hilbert spaces* as objects, and *linear operators* as morphisms.

J. Baez (✉)
Department of Mathematics, University of California, Riverside, CA, USA
e-mail: baez@math.ucr.edu

M. Stay (✉)
Computer Science Department, University of Auckland and Google, Auckland, Newzealand
e-mail: stay@google.com

Baez, J., Stay, M.: *Physics, Topology, Logic and Computation: A Rosetta Stone*. Lect. Notes Phys. **813**, 95–172 (2011)
DOI 10.1007/978-3-642-12821-9_2

Sometime around 1949, Feynman [63] realized that in quantum field theory it is useful to draw linear operators as diagrams:

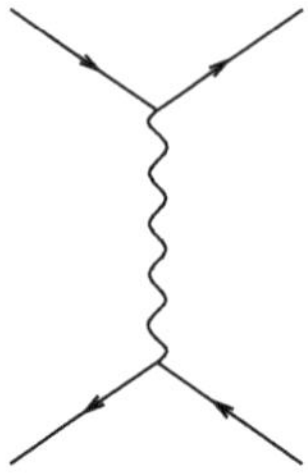

This lets us reason with them pictorially. We can warp a picture without changing the operator it stands for: all that matters is the topology, not the geometry. In the 1970s, Penrose realized that generalizations of Feynman diagrams arise throughout quantum theory, and might even lead to revisions in our understanding of spacetime [84–87]. In the 1980s, it became clear that underlying these diagrams is a powerful analogy between quantum physics and topology! Namely, a linear operator behaves very much like a "cobordism"—that is, an n-dimensional manifold going between manifolds of one dimension less:

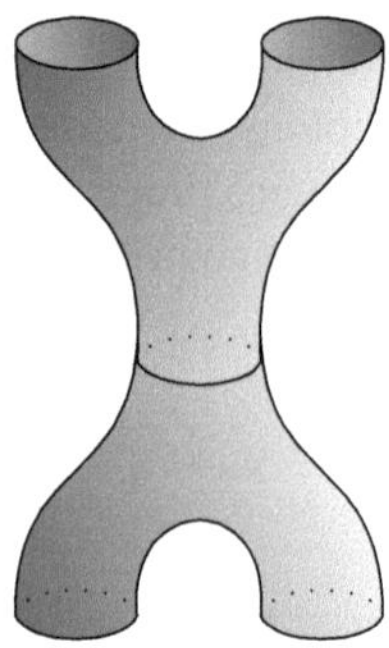

String theory exploits this analogy by replacing the Feynman diagrams of ordinary quantum field theory with 2-dimensional cobordisms, which represent the worldsheets traced out by strings with the passage of time. The analogy between operators and cobordisms is also important in loop quantum gravity and—most of all—the more purely mathematical discipline of "topological quantum field theory".

Meanwhile, quite separately, logicians had begun using categories where the objects represent *propositions* and the morphisms represent *proofs*. The idea is that a proof is a process going from one proposition (the hypothesis) to another (the conclusion). Later, computer scientists started using categories where the objects represent *data types* and the morphisms represent *programs*. They also started using "flow charts" to describe programs. Abstractly, these are very much like Feynman diagrams!

The logicians and computer scientists were never very far from each other. Indeed, the "Curry–Howard correspondence" relating proofs to programs has been well-known at least since the early 1970s, with roots stretching back earlier

[36, 37, 56]. But, it is only in the 1990s that the logicians and computer scientists bumped into the physicists and topologists. One reason is the rise of interest in quantum cryptography and quantum computation [29]. With this, people began to think of quantum processes as forms of information processing, and apply ideas from computer science. It was then realized that the loose analogy between flow charts and Feynman diagrams could be made more precise and powerful with the aid of category theory [3].

By now there is an extensive network of interlocking analogies between physics, topology, logic and computer science. They suggest that research in the area of common overlap is actually trying to build a new science: *a general science of systems and processes*. Building this science will be very difficult. There are good reasons for this, but also bad ones. One bad reason is that different fields use different terminology and notation.

The original Rosetta Stone, created in 196 BC, contains versions of the same text in three languages: demotic Egyptian, hieroglyphic script and classical Greek. Its rediscovery by Napoleon's soldiers let modern Egyptologists decipher the hieroglyphs. Eventually this led to a vast increase in our understanding of Egyptian culture.

At present, the deductive systems in mathematical logic look like hieroglyphs to most physicists. Similarly, quantum field theory is Greek to most computer scientists, and so on. So, there is a need for a new Rosetta Stone to aid researchers attempting to translate between fields. Table 2.1 shows our guess as to what this Rosetta Stone might look like.

Table 2.1 The Rosetta Stone (pocket version)

Category theory	Physics	Topology	Logic	Computation
Object	System	Manifold	Proposition	Data type
Morphism	Process	Cobordism	Proof	Program

The rest of this paper expands on this table by comparing how categories are used in physics, topology, logic, and computation. Unfortunately, these different fields focus on slightly different kinds of categories. Though most physicists don't know it, quantum physics has long made use of "compact symmetric monoidal categories". Knot theory uses "compact braided monoidal categories", which are slightly more general. However, it became clear in the 1990s that these more general gadgets are useful in physics too. Logic and computer science used to focus on "cartesian closed categories"—where "cartesian" can be seen, roughly, as an antonym of "quantum". However, thanks to work on linear logic and quantum computation, some logicians and computer scientists have dropped their insistence on cartesianness: now they study more general sorts of "closed symmetric monoidal categories".

In Sect. 2.2 we explain these concepts, how they illuminate the analogy between physics and topology, and how to work with them using string diagrams. We assume no prior knowledge of category theory, only a willingness to learn some. In Sect. 2.3 we explain how closed symmetric monoidal categories correspond to a small fragment of ordinary propositional logic, which also happens to be a fragment of Girard's "linear logic" [47]. In Sect. 2.4 we explain how closed symmetric monoidal categories correspond to a simple model of computation. Each of these sections starts with some background material. In Sect. 2.5, we conclude by presenting a larger version of the Rosetta Stone.

Our treatment of all four subjects—physics, topology, logic and computation—is bound to seem sketchy, narrowly focused and idiosyncratic to practitioners of these subjects. Our excuse is that we wish to emphasize certain analogies while saying no more than absolutely necessary. To make up for this, we include many references for those who wish to dig deeper.

2.2 The Analogy Between Physics and Topology

2.2.1 Background

Currently our best theories of physics are general relativity and the Standard Model of particle physics. The first describes gravity without taking quantum theory into account; the second describes all the other forces taking quantum theory into account, but ignores gravity. So, our world-view is deeply schizophrenic. The field where physicists struggle to solve this problem is called *quantum gravity*, since it is widely believed that the solution requires treating gravity in a way that takes quantum theory into account.

Table 2.2 Analogy between physics and topology

Physics	Topology
Hilbert space (system)	$(n-1)$-Dimensional manifold (space)
Operator between Hilbert spaces (process)	Cobordism between $(n-1)$-dimensional manifolds (spacetime)
Composition of operators	Composition of cobordisms
Identity operator	Identity cobordism

Nobody is sure how to do this, but there is a striking similarity between two of the main approaches: string theory and loop quantum gravity. Both rely on the analogy between physics and topology shown in Table 2.2. On the left we have a basic ingredient of quantum theory: the category Hilb whose objects are Hilbert spaces, used to describe physical *systems*, and whose morphisms are linear operators, used to describe physical *processes*. On the right we have a basic structure in differential topology: the category nCob. Here the objects are $(n{-}1)$-dimensional manifolds, used to describe *space*, and whose morphisms are n-dimensional cobordisms, used to describe *spacetime*.

As we shall see, Hilb and nCob share many structural features. Moreover, both are very different from the more familiar category Set, whose objects are sets and whose morphisms are functions. Elsewhere we have argued at great length that this is important for better understanding quantum gravity [11] and even the foundations of quantum theory [12]. The idea is that if Hilb is more like nCob than Set, maybe we should stop thinking of a quantum process as a function from one set of states to another. Instead, maybe we should think of it as resembling a "spacetime" going between spaces of dimension one less.

This idea sounds strange, but the simplest example is something very practical, used by physicists every day: a Feynman diagram. This is a 1-dimensional graph going between 0-dimensional collections of points, with edges and vertices labelled in certain ways. Feynman diagrams are topological entities, but they describe linear operators. String theory uses 2-dimensional cobordisms equipped with extra structure—string worldsheets—to do a similar job. Loop quantum gravity uses 2d generalizations of Feynman diagrams called "spin foams" [10]. Topological quantum field theory uses higher-dimensional cobordisms [14]. In each case, processes are described by morphisms in a special sort of category: a "compact symmetric monoidal category".

In what follows, we shall not dwell on puzzles from quantum theory or quantum gravity. Instead we take a different tack, simply explaining some basic concepts from category theory and showing how Set, Hilb, nCob and categories of tangles give examples. A recurring theme, however, is that Set is very different from the other examples.

To help the reader safely navigate the sea of jargon, here is a chart of the concepts we shall explain in this section:

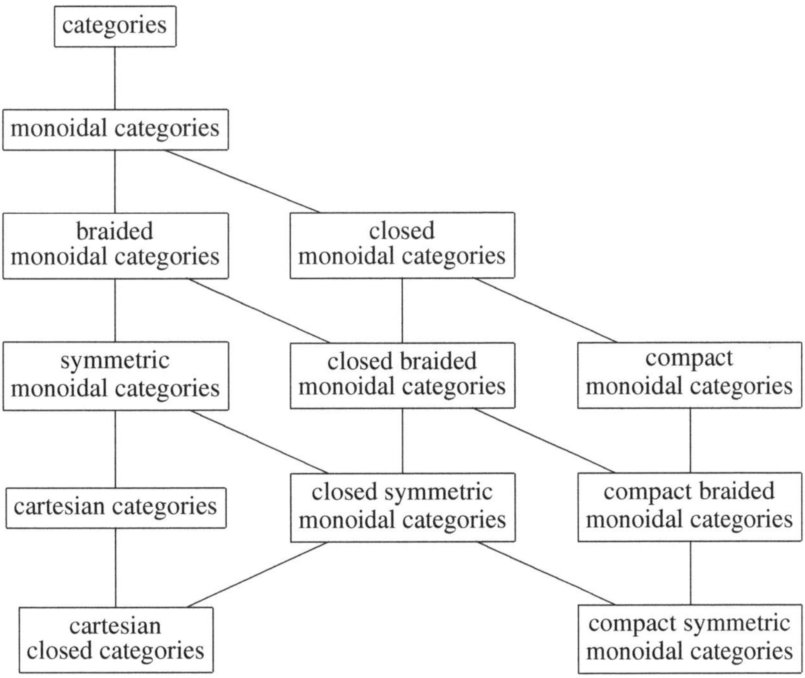

The category Set is cartesian closed, while Hilb and nCob are compact symmetric monoidal.

2.2.2 Categories

Category theory was born around 1945, with Eilenberg and Mac Lane [40] defining "categories", "functors" between categories, and "natural transformations" between functors. By now there are many introductions to the subject [35, 78, 81], including some available for free online [21, 50]. Nonetheless, we begin at the beginning:

Definition 1 A **category** C consists of:

- a collection of **objects**, where if X is an object of C we write $X \in C$, and
- for every pair of objects (X, Y), a set $\hom(X, Y)$ of **morphisms** from X to Y. We call this set $\hom(X, Y)$ a **homset**. If $f \in \hom(X, Y)$, then we write $f\colon X \to Y$.

such that:

- for every object X there is an **identity morphism** $1_X\colon X \to X$;
- morphisms are composable: given $f\colon X \to Y$ and $g\colon Y \to Z$, there is a **composite morphism** $gf\colon X \to Z$; sometimes also written $g \circ f$.
- an identity morphism is both a **left and a right unit** for composition: if $f\colon X \to Y$, then $f 1_X = f = 1_Y f$; and
- composition is **associative**: $(hg)f = h(gf)$ whenever either side is well-defined.

Definition 2 We say a morphism $f\colon X \to Y$ is an **isomorphism** if it has an inverse—that is, there exists another morphism $g\colon Y \to X$ such that $gf = 1_X$ and $fg = 1_Y$.

A category is the simplest framework where we can talk about systems (objects) and processes (morphisms). To visualize these, we can use "Feynman diagrams" of a very primitive sort. In applications to linear algebra, these diagrams are often called "spin networks", but category theorists call them "string diagrams", and that is the term we will use. The term "string" here has little to do with string theory: instead, the idea is that objects of our category label "strings" or "wires":

and morphisms $f\colon X \to Y$ label "black boxes" with an input wire of type X and an output wire of type Y:

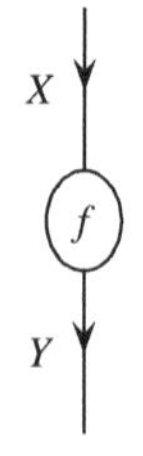

We compose two morphisms by connecting the output of one black box to the input of the next. So, the composite of $f: X \to Y$ and $g: Y \to Z$ looks like this:

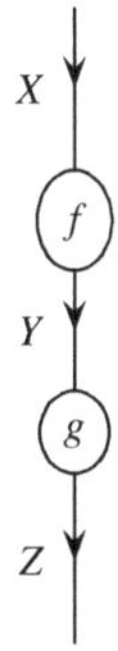

Associativity of composition is then implicit:

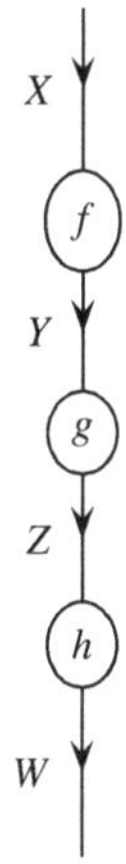

is our notation for both $h(gf)$ and $(hg)f$. Similarly, if we draw the identity morphism $1_X: X \to X$ as a piece of wire of type X:

then the left and right unit laws are also implicit.

There are countless examples of categories, but we will focus on four:

- Set: the category where objects are sets.
- Hilb: the category where objects are finite-dimensional Hilbert spaces.
- nCob: the category where morphisms are n-dimensional cobordisms.
- Tang_k: the category where morphisms are k-codimensional tangles.

As we shall see, all four are closed symmetric monoidal categories, at least when k is big enough. However, the most familiar of the lot, namely Set, is the odd man out: it is "cartesian".

Traditionally, mathematics has been founded on the category Set, where the objects are *sets* and the morphisms are *functions*. So, when we study systems and processes in physics, it is tempting to specify a system by giving its set of states, and a process by giving a function from states of one system to states of another.

However, in quantum physics we do something subtly different: we use categories where objects are *Hilbert spaces* and morphisms are *bounded linear operators*. We specify a system by giving a Hilbert space, but this Hilbert space is not really the set of states of the system: a state is actually a ray in Hilbert space. Similarly, a bounded linear operator is not precisely a function from states of one system to states of another.

In the day-to-day practice of quantum physics, what really matters is not sets of states and functions between them, but Hilbert space and operators. One of the virtues of category theory is that it frees us from the "Set-centric" view of traditional mathematics and lets us view quantum physics on its own terms. As we shall see, this sheds new light on the quandaries that have always plagued our understanding of the quantum realm [12].

To avoid technical issues that would take us far afield, we will take Hilb to be the category where objects are *finite-dimensional Hilbert spaces* and morphisms are *linear operators* (automatically bounded in this case). Finite-dimensional Hilbert spaces suffice for some purposes; infinite-dimensional ones are often important, but treating them correctly would require some significant extensions of the ideas we want to explain here.

In physics we also use categories where the objects represent choices of *space*, and the morphisms represent choices of *spacetime*. The simplest is nCob, where the objects are $(n-1)$*-dimensional manifolds*, and the morphisms are n*-dimensional cobordisms*. Glossing over some subtleties that a careful treatment would discuss [90], a cobordism $f: X \to Y$ is an n-dimensional manifold whose boundary is the disjoint union of the $(n-1)$-dimensional manifolds X and Y. Here are a couple of cobordisms in the case $n = 2$:

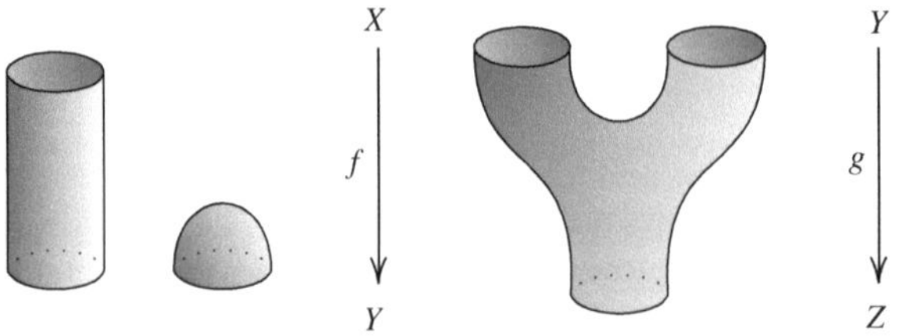

We compose them by gluing the "output" of one to the "input" of the other. So, in the above example $gf: X \to Z$ looks like this:

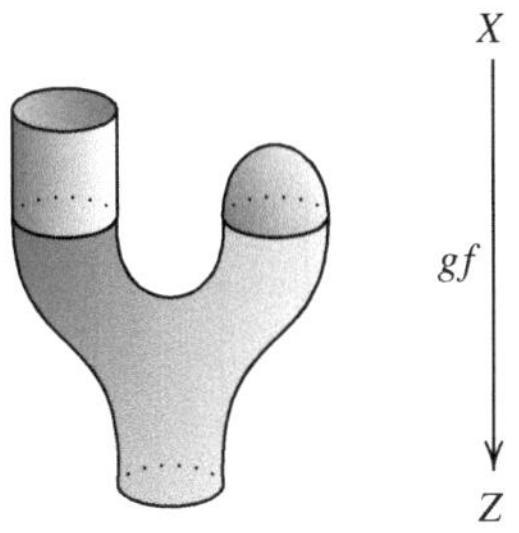

Another kind of category important in physics has objects representing *collections of particles*, and morphisms representing their *worldlines and interactions*. Feynman diagrams are the classic example, but in these diagrams the "edges" are not taken literally as particle trajectories. An example with closer ties to topology is Tang_k.

Very roughly speaking, an object in Tang_k is a collection of points in a k-dimensional cube, while a morphism is a "tangle": a collection of arcs and circles smoothly embedded in a $(k+1)$-dimensional cube, such that the circles lie in the interior of the cube, while the arcs touch the boundary of the cube only at its top and bottom, and only at their endpoints. A bit more precisely, tangles are "isotopy classes" of such embedded arcs and circles: this equivalence relation means that only the topology of the tangle matters, not its geometry. We compose tangles by attaching one cube to another top to bottom.

More precise definitions can be found in many sources, at least for $k = 2$, which gives tangles in a 3-dimensional cube [46, 64, 90, 99, 107, 111]. But since a picture is worth a thousand words, here is a picture of a morphism in Tang_2:

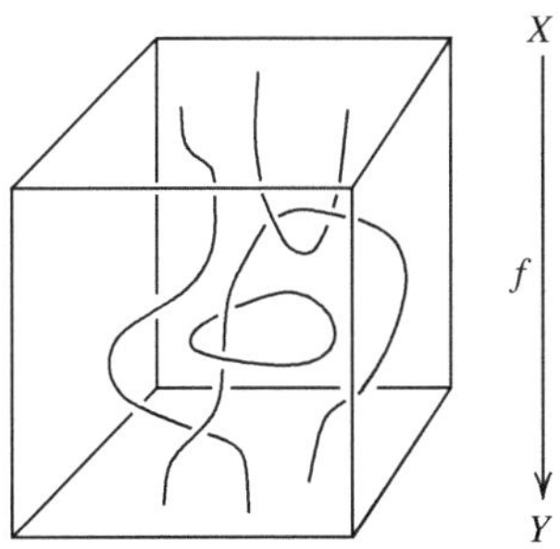

Note that we can think of a morphism in Tang_k as a 1-dimensional cobordism *embedded in a k-dimensional cube*. This is why Tang_k and $n\mathrm{Cob}$ behave similarly in some respects.

Here are two composable morphisms in Tang_1:

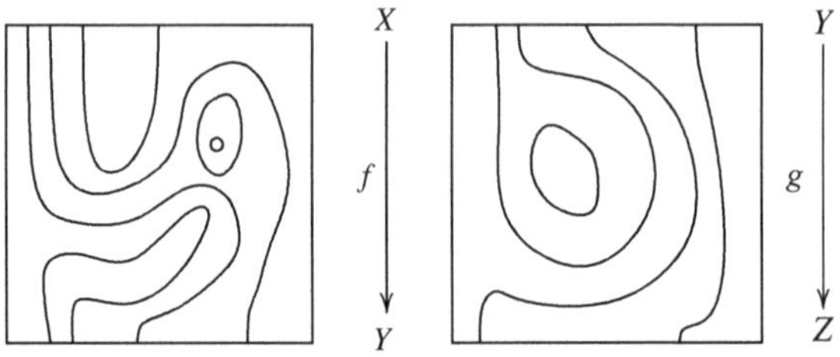

and here is their composite:

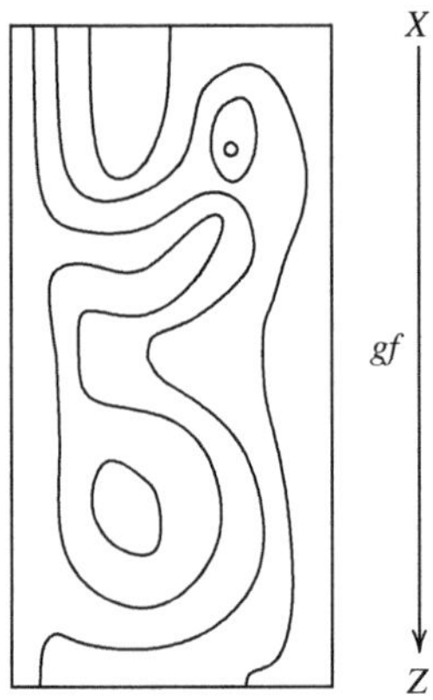

Since only the tangle's topology matters, we are free to squash this rectangle into a square if we want, but we do not need to.

It is often useful to consider tangles that are decorated in various ways. For example, in an "oriented" tangle, each arc and circle is equipped with an orientation. We can indicate this by drawing a little arrow on each curve in the tangle. In applications to physics, these curves represent worldlines of particles, and the arrows say whether each particle is going forwards or backwards in time, following Feynman's idea that antiparticles are particles going backwards in time. We can also consider "framed" tangles. Here each curve is replaced by a "ribbon". In applications to physics, this keeps track of how each particle twists. This is especially important for fermions, where a 2π twist acts nontrivially. Mathematically, the best-behaved tangles are both framed and oriented [14, 99], and these are what we should use to define Tang_k. The category $n\mathrm{Cob}$ also has a framed oriented version. However, these details will barely matter in what is to come.

It is difficult to do much with categories without discussing the maps between them. A map between categories is called a 'functor':

Definition 3 A **functor** $F: C \to D$ from a category C to a category D is a map sending:

- any object $X \in C$ to an object $F(X) \in D$,
- any morphism $f: X \to Y$ in C to a morphism $F(f): F(X) \to F(Y)$ in D,

in such a way that:

- F **preserves identities**: for any object $X \in C$, $F(1_X) = 1_{F(X)}$;
- F **preserves composition**: for any pair of morphisms $f: X \to Y$, $g: Y \to Z$ in C, $F(gf) = F(g)F(f)$.

In the sections to come, we will see that functors and natural transformations are useful for putting extra structure on categories. Here is a rather different use for functors: we can think of a functor $F: C \to D$ as giving a picture, or "representation", of C in D. The idea is that F can map objects and morphisms of some 'abstract' category C to objects and morphisms of a more "concrete" category D.

For example, consider an abstract group G. This is the same as a category with one object and with all morphisms invertible. The object is uninteresting, so we can just call it $\bullet$, but the morphisms are the elements of G, and we compose them by multiplying them. From this perspective, a **representation** of G on a finite-dimensional Hilbert space is the same as a functor $F: G \to$ Hilb. Similarly, an **action** of G on a set is the same as a functor $F: G \to$ Set. Both notions are ways of making an abstract group more concrete.

Ever since Lawvere's 1963 thesis on functorial semantics [75], the idea of functors as representations has become pervasive. However, the terminology varies from field to field. Following Lawvere, logicians often call the category C a "theory", and call the functor $F: C \to D$ a "model" of this theory. Other mathematicians might call F an "algebra" of the theory. In this work, the default choice of D is usually the category Set.

In physics, it is the functor $F: C \to D$ that is called the "theory". Here the default choice of D is either the category we are calling Hilb or a similar category of *infinite-dimensional* Hilbert spaces. For example, both "conformal field theories" [95] and topological quantum field theories [8, 9] can be seen as functors of this sort.

If we think of functors as models, natural transformations are maps between models:

Definition 4 Given two functors $F, F': C \to D$, a **natural transformation** $\alpha: F \Rightarrow F'$ assigns to every object X in C a morphism $\alpha_X: F(X) \to F'(X)$ such that for any morphism $f: X \to Y$ in C, the equation $\alpha_Y F(f) = F'(f)\alpha_X$ holds in D. In other words, this square commutes:

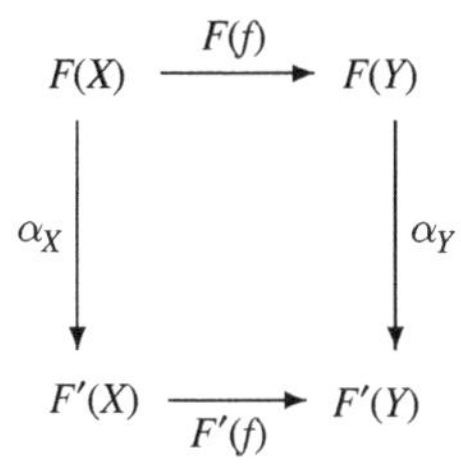

(Going across and then down equals going down and then across.)

Definition 5 A **natural isomorphism** between functors $F, F' : C \to D$ is a natural transformation $\alpha: F \Rightarrow F'$ such that α_X is an isomorphism for every $X \in C$.

For example, suppose $F, F' : G \to \mathrm{Hilb}$ are functors where G is a group, thought of as a category with one object, say $\bullet$. Then, as already mentioned, F and F' are secretly just representations of G on the Hilbert spaces $F(\bullet)$ and $F'(\bullet)$. A natural transformation $\alpha: F \Rightarrow F'$ is then the same as an **intertwining operator** from one representation to another: that is, a linear operator

$$A: F(\bullet) \to F'(\bullet)$$

satisfying

$$AF(g) = F'(g)A$$

for all group elements g.

2.2.3 Monoidal Categories

In physics, it is often useful to think of two systems sitting side by side as forming a single system. In topology, the disjoint union of two manifolds is again a manifold in its own right. In logic, the conjunction of two statement is again a statement. In programming we can combine two data types into a single "product type". The concept of "monoidal category" unifies all these examples in a single framework.

A monoidal category C has a functor $\otimes: C \times C \to C$ that takes two objects X and Y and puts them together to give a new object $X \otimes Y$. To make this precise, we need the cartesian product of categories:

Definition 6 The **cartesian product** $C \times C'$ of categories C and C' is the category where:

- an object is a pair (X, X') consisting of an object $X \in C$ and an object $X' \in C'$;
- a morphism from (X, X') to (Y, Y') is a pair (f, f') consisting of a morphism $f: X \to Y$ and a morphism $f': X' \to Y'$;
- composition is done componentwise: $(g, g')(f, f') = (gf, g'f')$;
- identity morphisms are defined componentwise: $1_{(X,X')} = (1_X, 1_{X'})$.

Mac Lane [77] defined monoidal categories in 1963. The subtlety of the definition lies in the fact that $(X \otimes Y) \otimes Z$ and $X \otimes (Y \otimes Z)$ are not usually equal. Instead, we should specify an isomorphism between them, called the "associator". Similarly, while a monoidal category has a "unit object" I, it is not usually true that $I \otimes X$ and $X \otimes I$ equal X. Instead, we should specify isomorphisms $I \otimes X \cong X$ and $X \otimes I \cong X$. To be manageable, all these isomorphisms must then satisfy certain equations:

Definition 7 A **monoidal category** consists of:

- a category C,
- a **tensor product** functor $\otimes: C \times C \to C$,

- a **unit object** $I \in C$,
- a natural isomorphism called the **associator**, assigning to each triple of objects $X, Y, Z \in C$ an isomorphism

$$a_{X,Y,Z} : (X \otimes Y) \otimes Z \xrightarrow{\sim} X \otimes (Y \otimes Z),$$

- natural isomorphisms called the **left** and **right unitors**, assigning to each object $X \in C$ isomorphisms

$$l_X : I \otimes X \xrightarrow{\sim} X$$
$$r_X : X \otimes I \xrightarrow{\sim} X,$$

such that:

- for all $X, Y \in C$ the **triangle equation** holds:

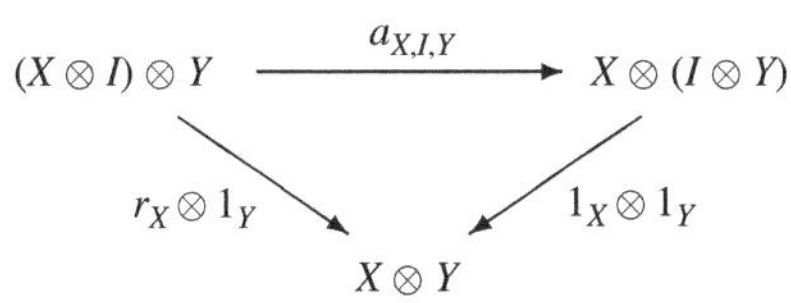

- for all $W, X, Y, Z \in C$, the **pentagon equation** holds:

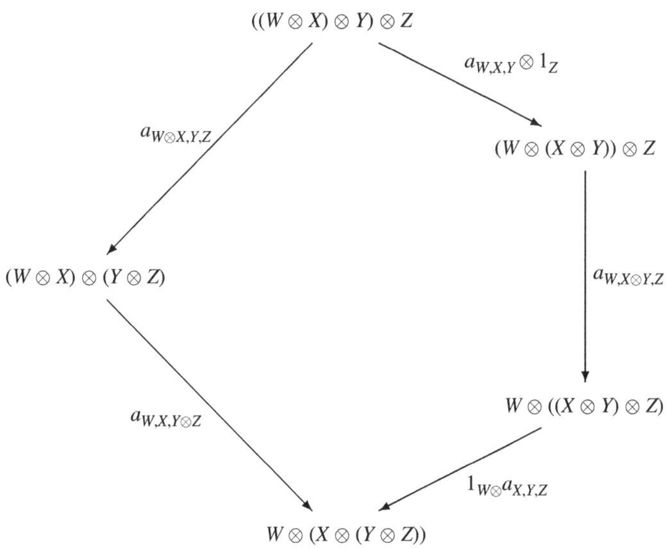

When we have a tensor product of four objects, there are five ways to parenthesize it, and at first glance the associator lets us build two isomorphisms from $W \otimes (X \otimes (Y \otimes Z))$ to $((W \otimes X) \otimes Y) \otimes Z$. But, the pentagon equation says these isomorphisms are equal. When we have tensor products of even more objects there are even more ways to parenthesize them, and even more isomorphisms between them built from the associator. However, Mac Lane showed that the pentagon identity implies these isomorphisms are all the same. Similarly, if we also assume the triangle equation, all isomorphisms with the same source and target built from the associator, left and right unit laws are equal.

In a monoidal category we can do processes in "parallel" as well as in "series". Doing processes in series is just composition of morphisms, which works in any category. But in a monoidal category we can also tensor morphisms $f: X \to Y$ and $f': X' \to Y'$ and obtain a "parallel process" $f \otimes f': X \otimes X' \to Y \otimes Y'$. We can draw this in various ways:

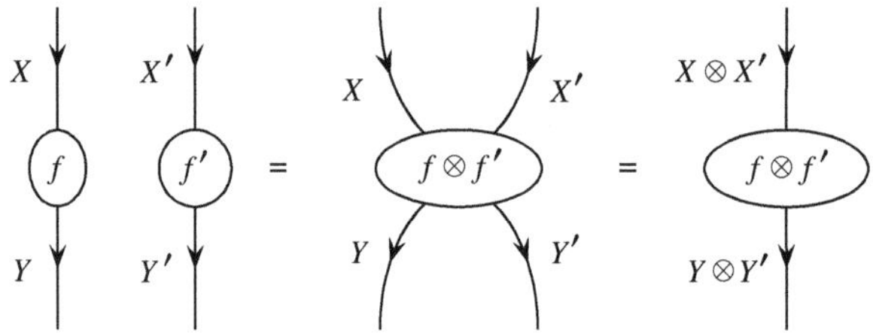

More generally, we can draw any morphism

$$f: X_1 \otimes \cdots \otimes X_n \to Y_1 \otimes \cdots \otimes Y_m$$

as a black box with n input wires and m output wires:

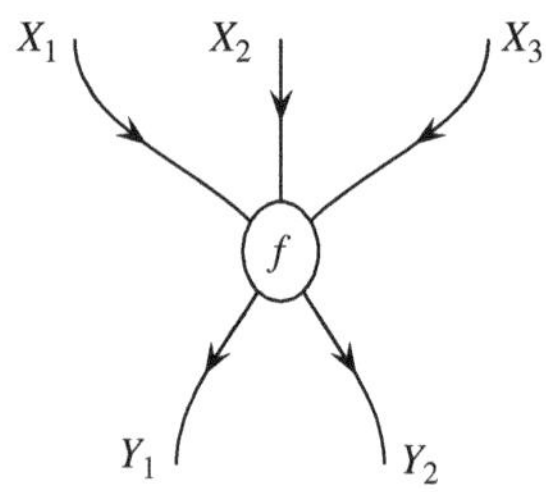

We draw the unit object I as a blank space. So, for example, we draw a morphism $f: I \to X$ as follows:

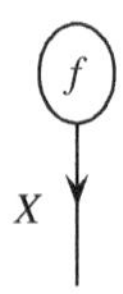

By composing and tensoring morphisms, we can build up elaborate pictures resembling Feynman diagrams:

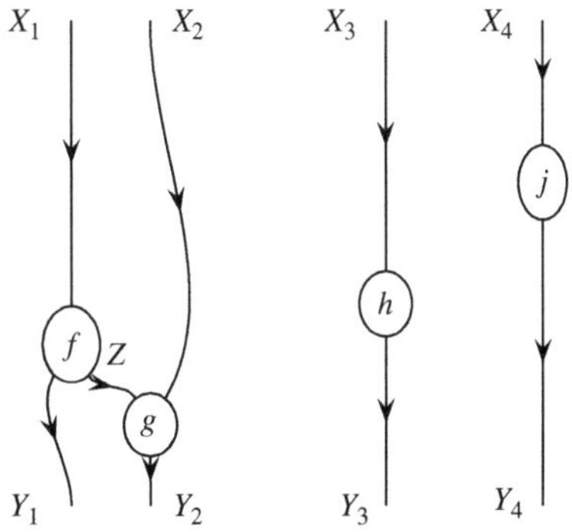

The laws governing a monoidal category allow us to neglect associators and unitors when drawing such pictures, without getting in trouble. The reason is that Mac Lane's Coherence Theorem says any monoidal category is "equivalent", in a suitable sense, to one where all associators and unitors are identity morphisms [77].

We can also deform the picture in a wide variety of ways without changing the morphism it describes. For example, the above morphism equals this one:

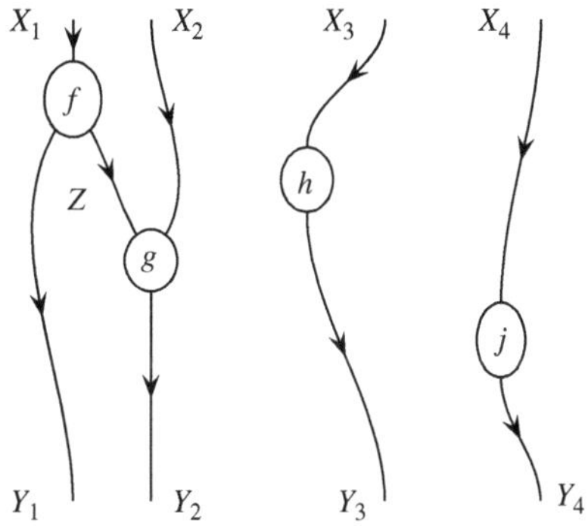

Everyone who uses string diagrams for calculations in monoidal categories starts by worrying about the rules of the game: *precisely how* can we deform these pictures without changing the morphisms they describe? Instead of stating the rules precisely—which gets a bit technical—we urge you to explore for yourself what is allowed and what is not. For example, show that we can slide black boxes up and down like this:

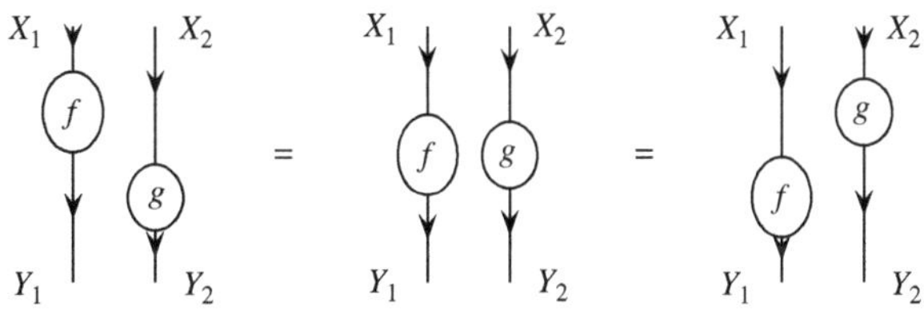

For a formal treatment of the rules governing string diagrams, try the original papers by Joyal and Street [59, 60] and the book by Yetter [111].

Now let us turn to examples. Here it is crucial to realize that the same category can often be equipped with different tensor products, resulting in different monoidal categories:

- There is a way to make Set into a monoidal category where $X \otimes Y$ is the cartesian product $X \times Y$ and the unit object is any one-element set. Note that this tensor product is not strictly associative, since $(x, (y, z)) \neq ((x, y), z)$, but there's a natural isomorphism $(X \times Y) \times Z \cong X \times (Y \times Z)$, and this is our associator. Similar considerations give the left and right unitors. In this monoidal category, the tensor product of $f: X \to Y$ and $f': X' \to Y'$ is the function

$$\begin{aligned} f \times f' : X \times X' &\to Y \times Y' \\ (x, x') &\mapsto (f(x), f'(x')). \end{aligned}$$

  There is also a way to make Set into a monoidal category where $X \otimes Y$ is the disjoint union of X and Y, which we shall denote by $X + Y$. Here the unit object is the empty set. Again, as indeed with all these examples, the associative law and left/right unit laws hold only up to natural isomorphism. In this monoidal category, the tensor product of $f: X \to Y$ and $f': X' \to Y'$ is the function

$$\begin{aligned} f + f' : X + X' &\to Y + Y' \\ x &\mapsto \begin{cases} f(x) \text{ if } x \in X, \\ f'(x) \text{ if } x \in X'. \end{cases} \end{aligned}$$

  However, in what follows, when we speak of Set as a monoidal category, we always use the cartesian product!
- There is a way to make Hilb into a monoidal category with the usual tensor product of Hilbert spaces: $\mathbb{C}^n \otimes \mathbb{C}^m \cong \mathbb{C}^{nm}$. In this case the unit object I can be taken to be a 1-dimensional Hilbert space, for example $\mathbb{C}$.
  There is also a way to make Hilb into a monoidal category where the tensor product is the direct sum: $\mathbb{C}^n \oplus \mathbb{C}^m \cong \mathbb{C}^{n+m}$. In this case the unit object is the zero-dimensional Hilbert space, $\{0\}$.
  However, in what follows, when we speak of Hilb as a monoidal category, we always use the usual tensor product!
- The tensor product of objects and morphisms in nCob is given by disjoint union. For example, the tensor product of these two morphisms:

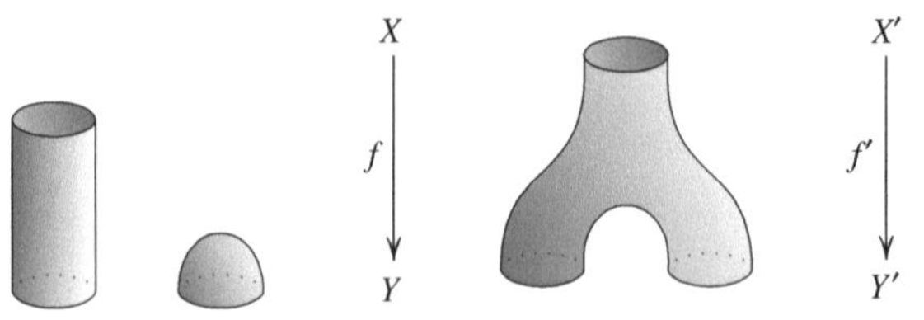

is this:

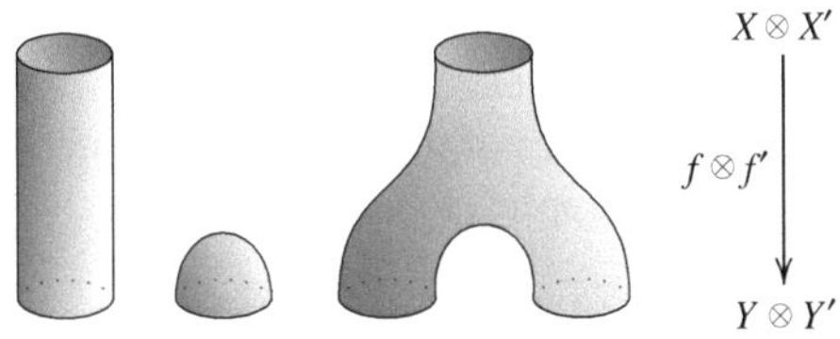

- The category Tang_k is monoidal when $k \geq 1$, where the the tensor product is given by disjoint union. For example, given these two tangles:

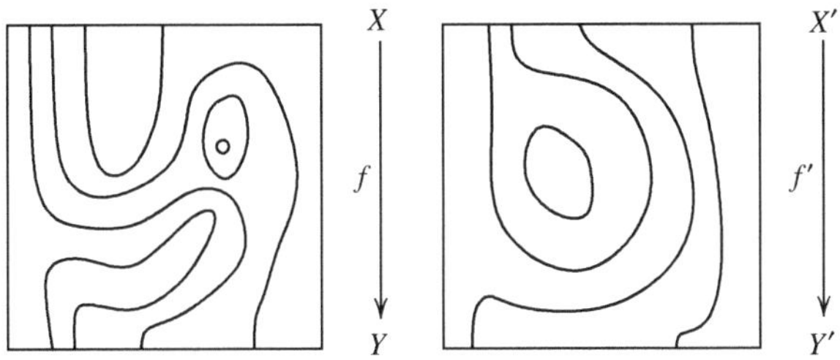

their tensor product looks like this:

The example of Set with its cartesian product is different from our other three main examples, because the cartesian product of sets $X \times X'$ comes equipped with functions called "projections" to the sets X and X':

$$X \xleftarrow{p} X \times X' \xrightarrow{p'} X'$$

Our other main examples lack this feature—though Hilb made into a monoidal category using $\oplus$ has projections. Also, every set has a unique function to the one-element set:

$$!_X : X \to I.$$

Again, our other main examples lack this feature, though Hilb made into a monoidal category using $\oplus$ has it. A fascinating feature of quantum mechanics is that we make Hilb into a monoidal category using $\otimes$ instead of $\oplus$, even though the latter approach would lead to a category more like Set.

We can isolate the special features of the cartesian product of sets and its projections, obtaining a definition that applies to any category:

Definition 8 Given objects X and X' in some category, we say an object $X \times X'$ equipped with morphisms

$$X \xleftarrow{p} X \times X' \xrightarrow{p'} X'$$

is a **cartesian product** (or simply **product**) of X and X' if for any object Q and morphisms

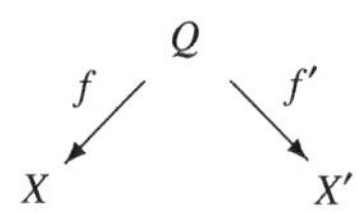

there exists a unique morphism $g\colon Q \to X \times X'$ making the following diagram commute:

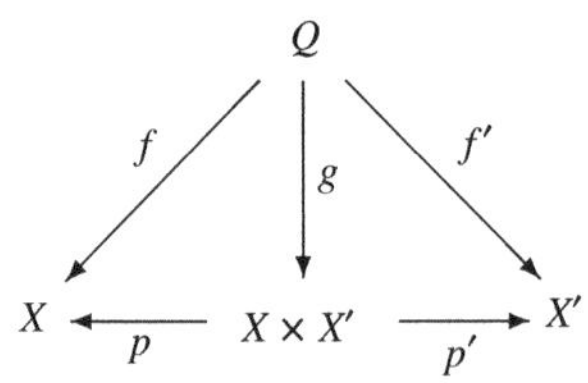

(That is, $f = pg$ and $f' = p'g$.) We say a category has **binary products** if every pair of objects has a product.

The product may not exist, and it may not be unique, but when it exists it is unique up to a canonical isomorphism. This justifies our speaking of "the" product of objects X and Y when it exists, and denoting it as $X \times Y$.

The definition of cartesian product, while absolutely fundamental, is a bit scary at first sight. To illustrate its power, let us do something with it: combine two morphisms $f\colon X \to Y$ and $f'\colon X' \to Y'$ into a single morphism

$$f \times f'\colon X \times X' \to Y \times Y'.$$

The definition of cartesian product says how to build a morphism of this sort out of a pair of morphisms: namely, morphisms from $X \times X'$ to Y and Y'. If we take these to be fp and $f'p'$, we obtain $f \times f'$:

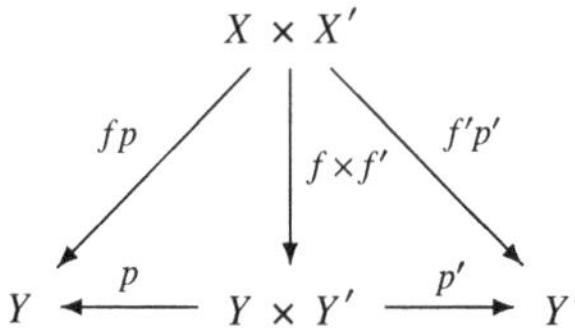

Next, let us isolate the special features of the one-element set:

Definition 9 An object 1 in a category C is **terminal** if for any object $Q \in C$ there exists a unique morphism from Q to 1, which we denote as $!_Q \colon Q \to 1$.

Again, a terminal object may not exist and may not be unique, but it is unique up to a canonical isomorphism. This is why we can speak of 'the' terminal object of a category, and denote it by a specific symbol, 1.

We have introduced the concept of binary products. One can also talk about n-ary products for other values of n, but a category with binary products has n-ary products for all $n \geq 1$, since we can construct these as iterated binary products. The case $n = 1$ is trivial, since the product of one object is just that object itself (up to canonical isomorphism). The remaining case is $n = 0$. The zero-ary product of objects, if it exists, is just the terminal object. So, we make the following definition:

Definition 10 A category has **finite products** if it has binary products and a terminal object.

A category with finite products can always be made into a monoidal category by choosing a specific product $X \times Y$ to be the tensor product $X \otimes Y$, and choosing a specific terminal object to be the unit object. It takes a bit of work to show this! A monoidal category of this form is called **cartesian**.

In a cartesian category, we can "duplicate and delete information". In general, the definition of cartesian products gives a way to take two morphisms $f_1 \colon Q \to X$ and $f_2 \colon Q \to Y$ and combine them into a single morphism from Q to $X \times Y$. If we take $Q = X = Y$ and take f_1 and f_2 to be the identity, we obtain the **diagonal** or **duplication** morphism:

$$\Delta_X \colon X \to X \times X.$$

In the category Set one can check that this maps any element $x \in X$ to the pair (x, x). In general, we can draw the diagonal as follows:

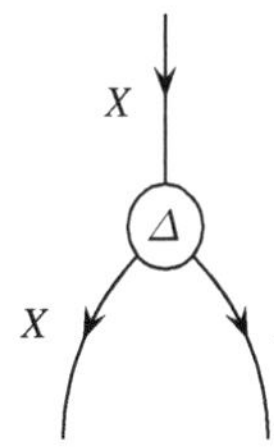

Similarly, we call the unique map to the terminal object

$$!_X : X \to 1$$

the **deletion** morphism, and draw it as follows:

Note that we draw the unit object as an empty space.

A fundamental fact about cartesian categories is that duplicating something and then deleting either copy is the same as doing nothing at all! In string diagrams, this says:

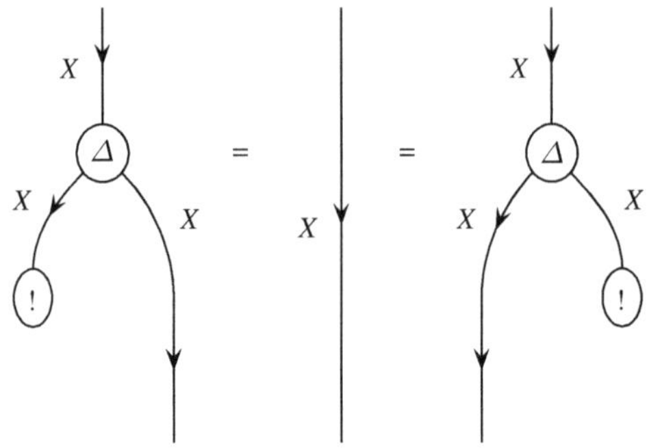

We leave the proof as an exercise for the reader.

Many of the puzzling features of quantum theory come from the noncartesianness of the usual tensor product in Hilb. For example, in a cartesian category, every morphism

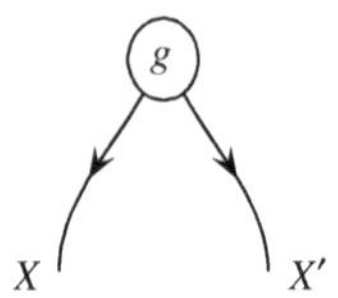

is actually of the form

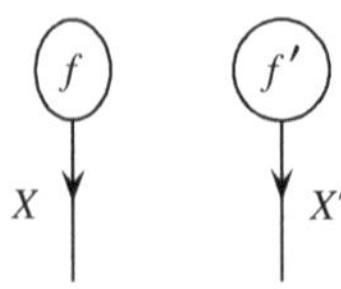

In the case of Set, this says that every point of the set $X \times X'$ comes from a point of X and a point of X'. In physics, this would say that every state g of the combined system $X \otimes X'$ is built by combining states of the systems X and X'. Bell's theorem [20] says that is *not* true in quantum theory. The reason is that quantum theory uses the noncartesian monoidal category Hilb!

Also, in quantum theory we *cannot* freely duplicate or delete information. Wootters and Zurek [110] proved a precise theorem to this effect, focused on duplication: the "no-cloning theorem". One can also prove a "no-deletion theorem". Again, these results rely on the noncartesian tensor product in Hilb.

2.2.4 Braided Monoidal Categories

In physics, there is often a process that lets us "switch" two systems by moving them around each other. In topology, there is a tangle that describes the process of switching two points:

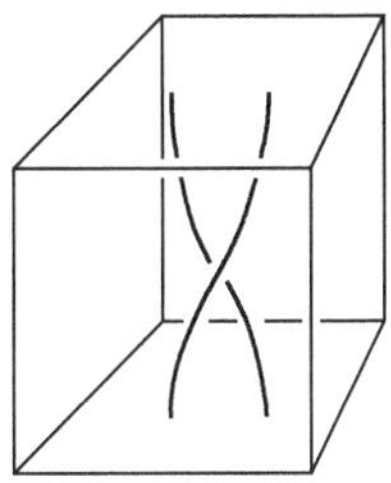

In logic, we can switch the order of two statements in a conjunction: the statement "X and Y" is isomorphic to "Y and X". In computation, there is a simple program that switches the order of two pieces of data. A monoidal category in which we can do this sort of thing is called "braided":

Definition 11 A **braided monoidal category** consists of:

- a monoidal category C,
- a natural isomorphism called the **braiding** that assigns to every pair of objects $X, Y \in C$ an isomorphism

$$b_{X,Y} : X \otimes Y \to Y \otimes X,$$

such that the **hexagon equations** hold:

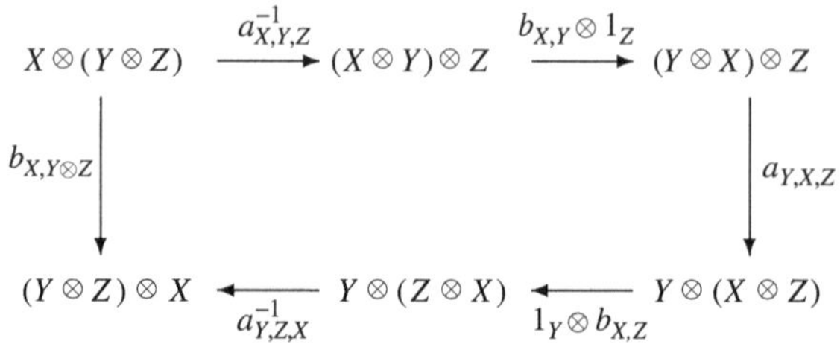

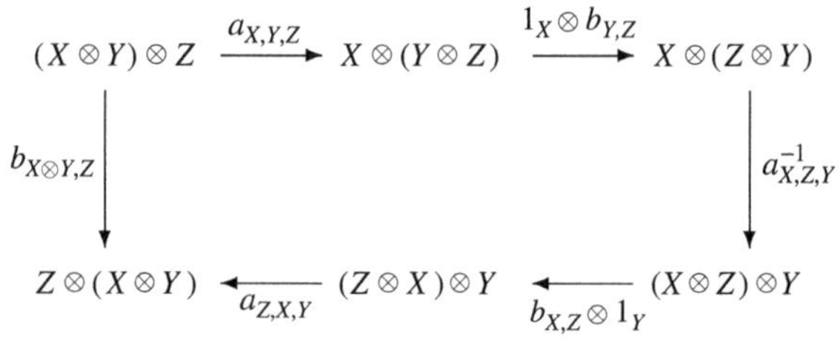

The first hexagon equation says that switching the object X past $Y \otimes Z$ all at once is the same as switching it past Y and then past Z (with some associators thrown in to move the parentheses). The second one is similar: it says switching $X \otimes Y$ past Z all at once is the same as doing it in two steps.

In string diagrams, we draw the braiding $b_{X,Y}: X \otimes Y \to Y \otimes X$ like this:

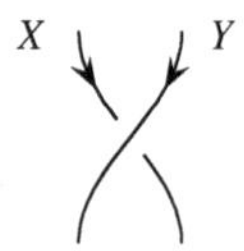

We draw its inverse $b^{-1}_{X,Y}$ like this:

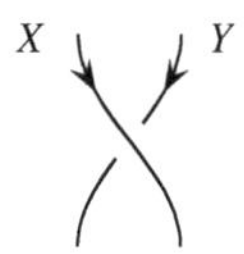

This is a nice notation, because it makes the equations saying that $b_{X,Y}$ and $b^{-1}_{X,Y}$ are inverses "topologically true":

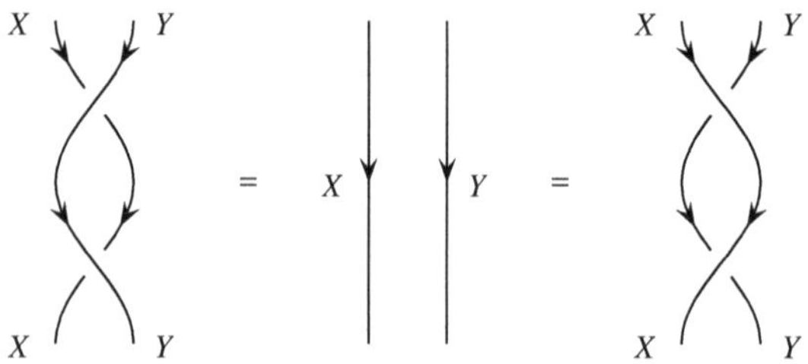

Here are the hexagon equations as string diagrams:

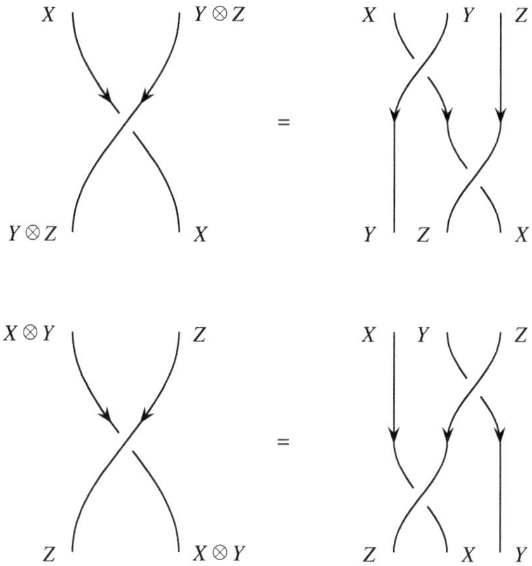

For practice, we urge you to prove the following equations:

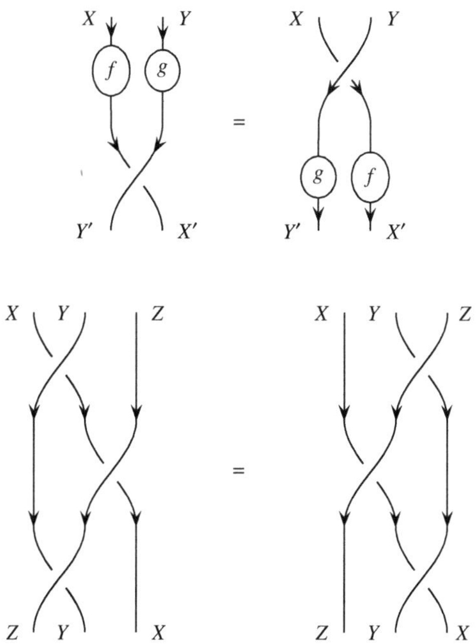

If you get stuck, here are some hints. The first equation follows from the naturality of the braiding. The second is called the **Yang–Baxter equation** and follows from a combination of naturality and the hexagon equations [61, 62].

Next, here are some examples. There can be many different ways to give a monoidal category a braiding, or none. However, most of our favorite examples come with well-known "standard" braidings:

- Any cartesian category automatically becomes braided, and in Set with its cartesian product, this standard braiding is given by:

$$\begin{aligned} b_{X,Y} : X \times Y &\to Y \times X \\ (x, y) &\mapsto (y, x). \end{aligned}$$

- In Hilb with its usual tensor product, the standard braiding is given by:

$$\begin{aligned} b_{X,Y} : X \otimes Y &\to Y \otimes X \\ x \otimes y &\mapsto y \otimes x. \end{aligned}$$

- The monoidal category nCob has a standard braiding where $b_{X,Y}$ is diffeomorphic to the disjoint union of cylinders $X \times [0, 1]$ and $Y \times [0, 1]$. For 2Cob this braiding looks as follows when X and Y are circles:

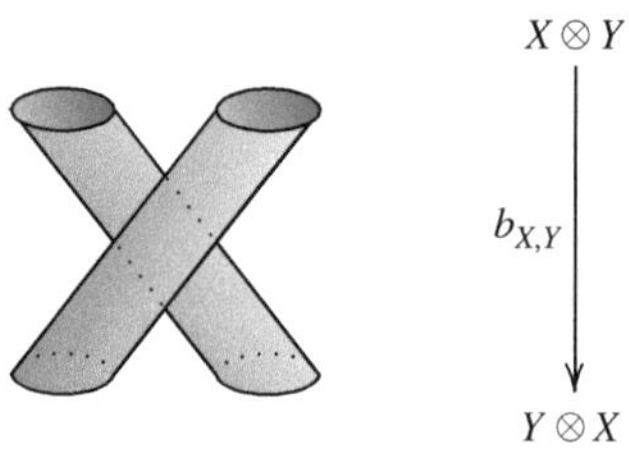

- The monoidal category Tang_k has a standard braiding when $k \geq 2$. For $k = 2$ this looks as follows when X and Y are each a single point:

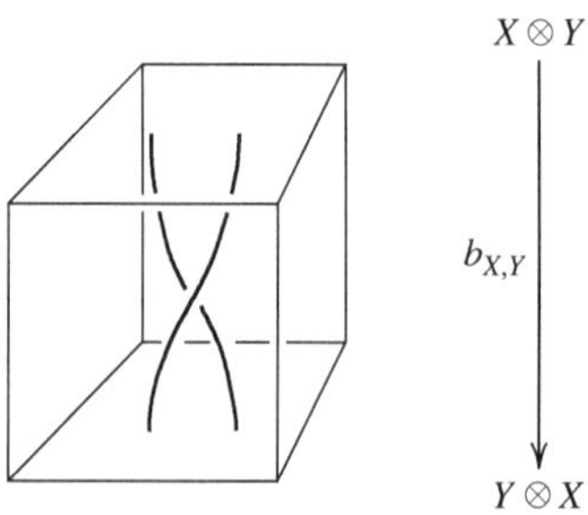

The example of Tang_k illustrates an important pattern. Tang_0 is just a category, because in 0-dimensional space we can only do processes in "series": that is, compose morphisms. Tang_1 is a monoidal category, because in 1-dimensional space we can also do processes in "parallel": that is, tensor morphisms. Tang_2 is a braided monoidal category, because in 2-dimensional space there is room to move one object around another. Next we shall see what happens when space has 3 or more dimensions!

2.2.5 Symmetric Monoidal Categories

Sometimes switching two objects and switching them again is the same as doing nothing at all. Indeed, this situation is very familiar. So, the first braided monoidal categories to be discovered were "symmetric" ones [77]:

Definition 12 A **symmetric monoidal category** is a braided monoidal category where the braiding satisfies $b_{X,Y} = b^{-1}_{Y,X}$.

So, in a symmetric monoidal category,

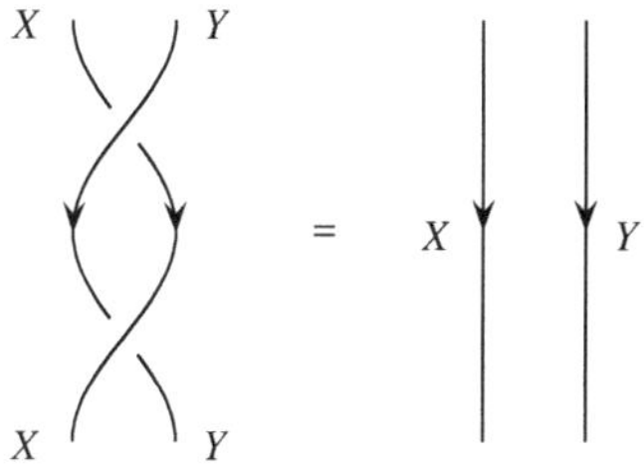

or equivalently:

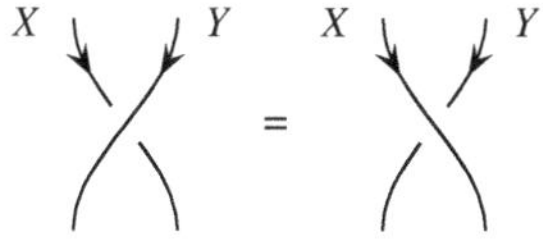

Any cartesian category automatically becomes a symmetric monoidal category, so Set is symmetric. It is also easy to check that Hilb, nCob are symmetric monoidal categories. So is Tang_k for $k \geq 3$.

Interestingly, Tang_k "stabilizes" at $k = 3$: increasing the value of k beyond this value merely gives a category equivalent to Tang_3. The reason is that we can already untie all knots in 4-dimensional space; adding extra dimensions has no real effect. In fact, Tang_k for $k \geq 3$ is equivalent to 1Cob. This is part of a conjectured larger pattern called the "Periodic Table" of n-categories [14]. A piece of this is shown in Table 2.3.

An n-category has not only morphisms going between objects, but 2-morphisms going between morphisms, 3-morphisms going between 2-morphisms and so on up

Table 2.3 The periodic table: conjectured descriptions of $(n + k)$-categories with only one j-morphism for $j < k$

	$n = 0$	$n = 1$	$n = 2$
$k = 0$	Sets	Categories	2-Categories
$k = 1$	Monoids	Monoidal categories	Monoidal 2-categories
$k = 2$	Commutative monoids	Braided monoidal categories	Braided monoidal 2-categories
$k = 3$	‘’	Symmetric monoidal categories	Sylleptic monoidal 2-categories
$k = 4$	‘’	‘’	Symmetric monoidal 2-categories
$k = 5$	‘’	‘’	‘’
$k = 6$	‘’	‘’	‘’

to n-morphisms. In topology we can use n-categories to describe tangled higher-dimensional surfaces [15], and in physics we can use them to describe not just particles but also strings and higher-dimensional membranes [14, 16]. The Rosetta Stone we are describing concerns only the $n = 1$ column of the Periodic Table. So, it is probably just a fragment of a larger, still buried n-categorical Rosetta Stone.

2.2.6 Closed Categories

In quantum mechanics, one can encode a linear operator $f: X \to Y$ into a quantum state using a technique called “gate teleportation” [51]. In topology, there is a way to take any tangle $f: X \to Y$ and bend the input back around to make it part of the output. In logic, we can take a proof that goes from some assumption X to some conclusion Y and turn it into a proof that goes from no assumptions to the conclusion “X implies Y”. In computer science, we can take any program that takes input of type X and produces output of type Y, and think of it as a piece of data of a new type: a “function type”. The underlying concept that unifies all these examples is the concept of a ‘closed category’.

Given objects X and Y in any category C, there is a *set* of morphisms from X to Y, denoted $\mathrm{hom}(X, Y)$. In a closed category there is also an *object* of morphisms from X to Y, which we denote by $X \multimap Y$. (Many other notations are also used.) In this situation we speak of an “internal hom”, since the object $X \multimap Y$ lives inside C, instead of “outside”, in the category of sets.

Closed categories were introduced in 1966, by Eilenberg and Kelly [41]. While these authors were able to define a closed structure for any category, it turns out that the internal hom is most easily understood for monoidal categories. The reason is that when our category has a tensor product, it is closed precisely when morphisms from $X \otimes Y$ to Z are in natural one-to-one correspondence with morphisms from Y to $X \multimap Z$. In other words, it is closed when we have a natural isomorphism

$$\begin{aligned} \hom(X \otimes Y, Z) &\cong \hom(Y, X \multimap Z) \\ f &\mapsto \tilde{f} \end{aligned}$$

For example, in the category Set, if we take $X \otimes Y$ to be the cartesian product $X \times Y$, then $X \multimap Z$ is just the set of functions from X to Z, and we have a one-to-one correspondence between

- functions f that eat elements of $X \times Y$ and spit out elements of Z

and

- functions $\tilde{f}$ that eat elements of Y and spit out functions from X to Z.

This correspondence goes as follows:

$$\tilde{f}(x)(y) = f(x, y).$$

Before considering other examples, we should make the definition of "closed monoidal category" completely precise. For this we must note that for any category C, there is a functor

$$\hom\colon C^{\mathrm{op}} \times C \to \mathrm{Set}.$$

Definition 13 The **opposite category** C^{op} of a category C has the same objects as C, but a morphism $f\colon x \to y$ in C^{op} is a morphism $f\colon y \to x$ in C, and the composite gf in C^{op} is the composite fg in C.

Definition 14 For any category C, the **hom functor**

$$\hom\colon C^{\mathrm{op}} \times C \to \mathrm{Set}$$

sends any object $(X, Y) \in C^{\mathrm{op}} \times C$ to the set $\hom(X, Y)$, and sends any morphism $(f, g) \in C^{\mathrm{op}} \times C$ to the function

$$\begin{aligned} \hom(f, g)\colon \hom(X, Y) &\to \hom(X', Y') \\ h &\mapsto ghf \end{aligned}$$

when $f\colon X' \to X$ and $g\colon Y \to Y'$ are morphisms in C.

Definition 15 A monoidal category C is **left closed** if there is an **internal hom** functor

$$\multimap\colon C^{\mathrm{op}} \times C \to C$$

together with a natural isomorphism c called **currying** that assigns to any objects $X, Y, Z \in C$ a bijection

$$c_{X,Y,Z}\colon \hom(X \otimes Y, Z) \xrightarrow{\sim} \hom(X, Y \multimap Z)$$

It is **right closed** if there is an internal hom functor as above and a natural isomorphism

$$c_{X,Y,Z} \colon \hom(X \otimes Y, Z) \xrightarrow{\sim} \hom(Y, X \multimap Z).$$

The term "currying" is mainly used in computer science, after the work of Curry [36, 37]. In the rest of this section we only consider *right* closed monoidal categories. Luckily, there is no real difference between left and right closed for a braided monoidal category, as the braiding gives an isomorphism $X \otimes Y \cong Y \otimes X$.

All our examples of monoidal categories are closed, but we shall see that, yet again, Set is different from the rest:

- The cartesian category Set is closed, where $X \multimap Y$ is just the set of functions from X to Y. In Set or any other cartesian closed category, the internal hom $X \multimap Y$ is usually denoted Y^X. To minimize the number of different notations and emphasize analogies between different contexts, we shall not do this: we shall always use $X \multimap Y$. To treat Set as *left* closed, we define the curried version of $f \colon X \times Y \to Z$ as above:

$$\tilde{f}(x)(y) = f(x, y).$$

  To treat it as *right* closed, we instead define it by

$$\tilde{f}(y)(x) = f(x, y).$$

  This looks a bit awkward, but it will be nice for string diagrams.
- The symmetric monoidal category Hilb with its usual tensor product is closed, where $X \multimap Y$ is the set of linear operators from X to Y, made into a Hilbert space in a standard way. In this case we have an isomorphism

$$X \multimap Y \cong X^* \otimes Y$$

  where X^* is the dual of the Hilbert space X, that is, the set of linear operators $f \colon X \to \mathbb{C}$, made into a Hilbert space in the usual way.
- The monoidal category Tang_k ($k \geq 1$) is closed. As with Hilb, we have

$$X \multimap Y \cong X^* \otimes Y$$

  where X^* is the orientation-reversed version of X.
- The symmetric monoidal category nCob is also closed; again

$$X \multimap Y \cong X^* \otimes Y$$

  where X^* is the $(n-1)$-manifold X with its orientation reversed.

Except for Set, all these examples are actually "compact". This basically means that $X \multimap Y$ is isomorphic to $X^* \otimes Y$, where X^* is some object called the "dual" of X. But to make this precise, we need to define the 'dual' of an object in an arbitrary monoidal category.

To do this, let us generalize from the case of Hilb. As already mentioned, each object $X \in$ Hilb has a dual X^* consisting of all linear operators $f : X \to I$, where the unit object I is just $\mathbb{C}$. There is thus a linear operator

$$\begin{array}{rcl} e_X : X \otimes X^* & \to & I \\ x \otimes f & \mapsto & f(x) \end{array}$$

called the **counit** of X. Furthermore, the space of all linear operators from X to $Y \in$ Hilb can be identified with $X^* \otimes Y$. So, there is also a linear operator called the **unit** of X:

$$\begin{array}{rcl} i_X : I & \to & X^* \otimes X \\ c & \mapsto & c\, 1_X \end{array}$$

sending any complex number c to the corresponding multiple of the identity operator.

The significance of the unit and counit become clearer if we borrow some ideas from Feynman. In physics, if X is the Hilbert space of internal states of some particle, X^* is the Hilbert space for the corresponding antiparticle. Feynman realized that it is enlightening to think of antiparticles as particles going backwards in time. So, we draw a wire labelled by X^* as a wire labelled by X, but with an arrow pointing 'backwards in time': that is, up instead of down:

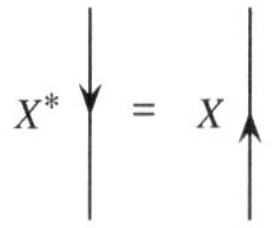

(Here we should admit that most physicists use the opposite convention, where time marches up the page. Since we read from top to bottom, we prefer to let time run down the page.)

If we draw X^* as X going backwards in time, we can draw the unit as a **cap**:

and the counit as a **cup**:

In Feynman diagrams, these describe the *creation* and *annihilation* of virtual particle-antiparticle pairs!

It then turns out that the unit and counit satisfy two equations, the **zig-zag equations**:

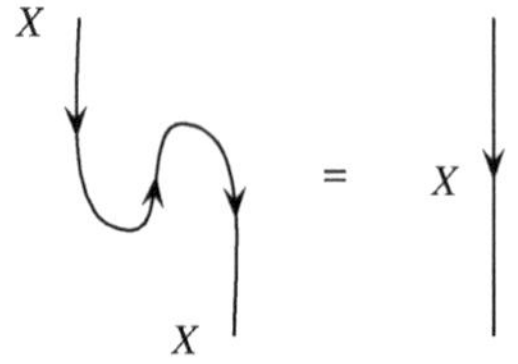

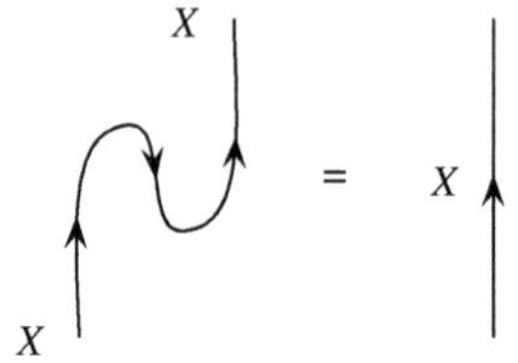

Verifying these is a fun exercise in linear algebra, which we leave to the reader. If we write these equations as commutative diagrams, we need to include some associators and unitors, and they become a bit intimidating:

$$\begin{array}{ccccc}
X \otimes I & \xrightarrow{1_X \otimes i_X} & X \otimes (X^* \otimes X) & \xrightarrow{a^{-1}_{X,X^*,X}} & (X \otimes X^*) \otimes X \\
\downarrow{\scriptstyle r_X} & & & & \downarrow{\scriptstyle e_X \otimes 1_X} \\
X & & \xleftarrow[\;l_X\;]{} & & I \otimes X
\end{array}$$

$$\begin{array}{ccccc}
I \otimes X^* & \xrightarrow{i_X \otimes 1_X} & (X^* \otimes X) \otimes X^* & \xrightarrow{a_{X^*,X,X^*}} & X^* \otimes (X \otimes X^*) \\
\downarrow{\scriptstyle l_X} & & & & \downarrow{\scriptstyle 1_{X^*} \otimes e_X} \\
X^* & & \xleftarrow[\;r_{X^*}\;]{} & & X^* \otimes I
\end{array}$$

But, they really just say that zig-zags in string diagrams can be straightened out.

This is particularly vivid in examples like Tang_k and $n\mathrm{Cob}$. For example, in 2Cob, taking X to be the circle, the unit looks like this:

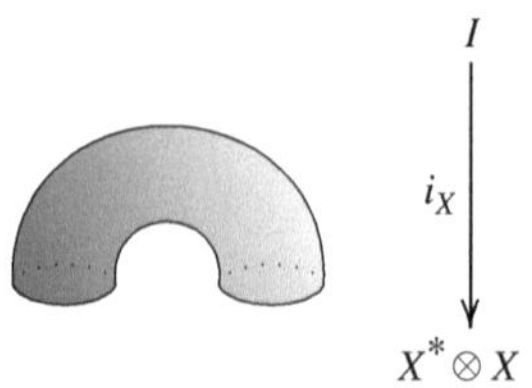

while the counit looks like this:

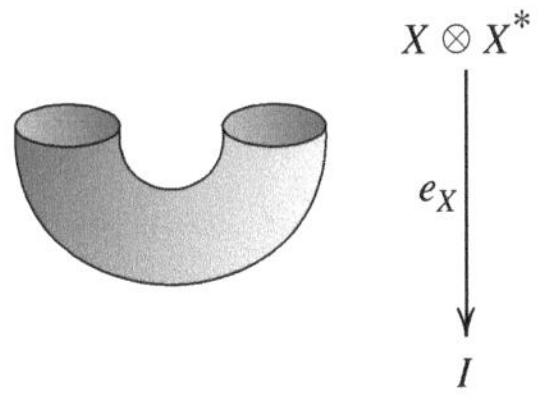

In this case, the zig-zag identities say we can straighten a wiggly piece of pipe.

Now we are ready for some definitions:

Definition 16 Given objects X^* and X in a monoidal category, we call X^* a **right dual** of X, and X a **left dual** of X^*, if there are morphisms

$$i_X \colon I \to X^* \otimes X$$

and

$$e_X \colon X \otimes X^* \to I,$$

called the **unit** and **counit** respectively, satisfying the zig-zag equations.

One can show that the left or right dual of an object is unique up to canonical isomorphism. So, we usually speak of "the" right or left dual of an object, when it exists.

Definition 17 A monoidal category C is **compact** if every object $X \in C$ has both a left dual and a right dual.

Often the term "autonomous" is used instead of "compact" here. Many authors reserve the term "compact" for the case where C is symmetric or at least braided; then left duals are the same as right duals, and things simplify [46]. To add to the confusion, compact symmetric monoidal categories are often called simply "compact closed categories".

A partial explanation for the last piece of terminology is that any compact monoidal category is automatically closed! For this, we define the internal hom on objects by

$$X \multimap Y = X^* \otimes Y.$$

We must then show that the $*$ operation extends naturally to a functor $* \colon C \to C$, so that $\multimap$ is actually a functor. Finally, we must check that there is a natural isomorphism

$$\hom(X \otimes Y, Z) \cong \hom(Y, X^* \otimes Z)$$

In terms of string diagrams, this isomorphism takes any morphism

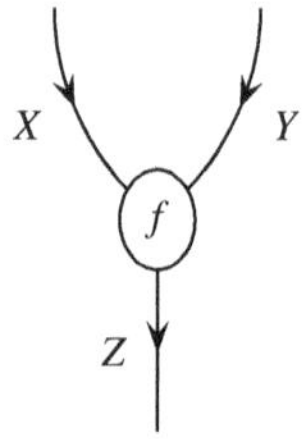

and bends back the input wire labelled X to make it an output:

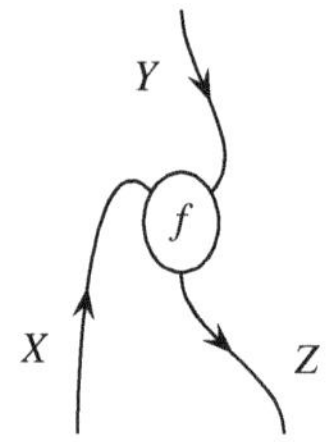

Now, in a compact monoidal category, we have:

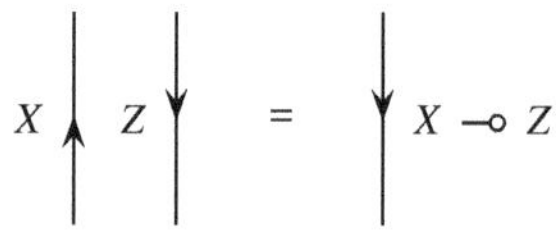

But in general, closed monoidal categories don't allow arrows pointing up! So for these, drawing the internal hom is more of a challenge. We can use the same style of notation as long as we add a decoration—a **clasp**—that binds two strings together:

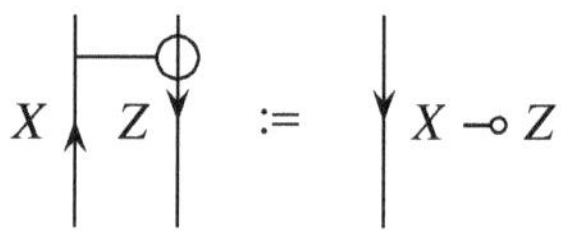

Only when our closed monoidal category happens to be compact can we eliminate the clasp.

Suppose we are working in a closed monoidal category. Since we draw a morphism $f: X \otimes Y \to Z$ like this:

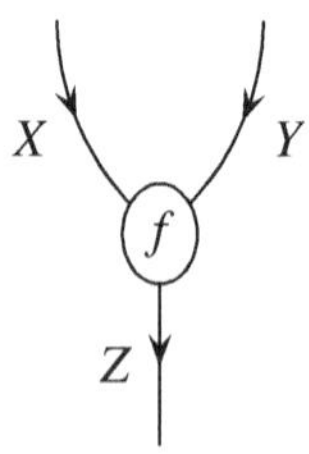

we can draw its curried version $\tilde{f}: Y \to X \multimap Z$ by bending down the input wire labelled X to make it part of the output:

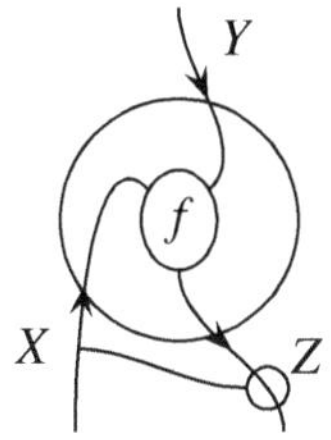

Note that where we bent back the wire labelled X, a cap like this appeared:

Closed monoidal categories don't really have a cap unless they are compact. So, we drew a **bubble** enclosing f and the cap, to keep us from doing any illegal manipulations. In the compact case, both the bubble and the clasp are unnecessary, so we can draw $\tilde{f}$ like this:

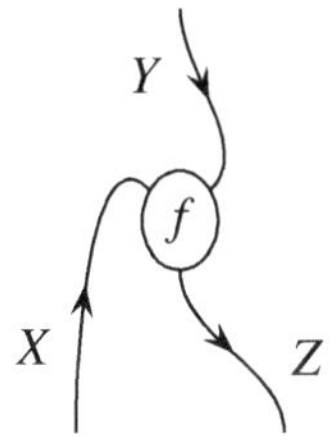

An important special case of currying gives the **name** of a morphism $f: X \to Y$,

$$\ulcorner f \urcorner : I \to X \multimap Y.$$

This is obtained by currying the morphism

$$f r_x : I \otimes X \to Y.$$

In string diagrams, we draw $\ulcorner f \urcorner$ as follows:

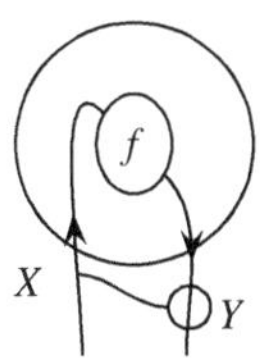

In the category Set, the unit object is the one-element set, 1. So, a morphism from this object to a set Q picks out a point of Q. In particular, the name $\ulcorner f \urcorner : 1 \to X \multimap Y$ picks out the element of $X \multimap Y$ corresponding to the function $f : X \to Y$. More generally, in any cartesian closed category the unit object is the terminal object 1, and a morphism from 1 to an object Q is called a **point** of Q. So, even in this case, we can say the name of a morphism $f : X \to Y$ is a point of $X \multimap Y$.

Something similar works for Hilb, though this example is compact rather than cartesian. In Hilb, the unit object I is just $\mathbb{C}$. So, a nonzero morphism from I to any Hilbert space Q picks out a nonzero vector in Q, which we can normalize to obtain a **state** in Q: that is, a unit vector. In particular, the name of a nonzero morphism $f : X \to Y$ gives a state of $X^* \otimes Y$. This method of encoding operators as states is the basis of "gate teleportation" [51].

Currying is a bijection, so we can also **uncurry**:

$$c^{-1}_{X,Y,Z} : \hom(Y, X \multimap Z) \xrightarrow{\sim} \hom(X \otimes Y, Z)$$
$$g \mapsto \underset{\sim}{g}.$$

Since we draw a morphism $g : Y \to X \multimap Z$ like this:

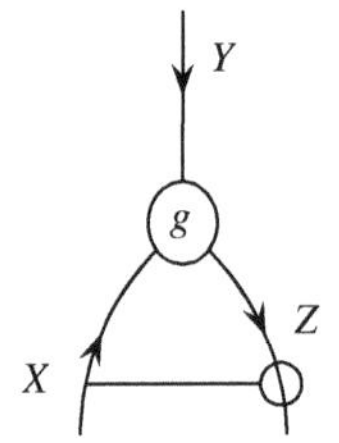

we draw its "uncurried" version $\underset{\sim}{g} : X \otimes Y \to Z$ by bending the output X up to become an input:

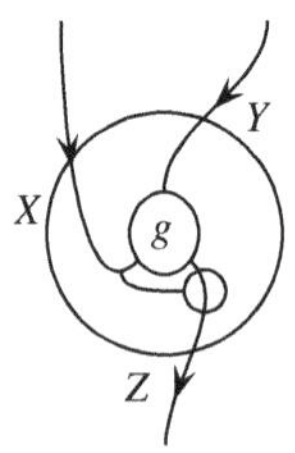

Again, we must put a bubble around the "cup" formed when we bend down the wire labelled Y, unless we are in a compact monoidal category.

A good example of uncurrying is the **evaluation** morphism:

$$\mathrm{ev}_{X,Y} : X \otimes (X \multimap Y) \to Y.$$

This is obtained by uncurrying the identity

$$1_{X \multimap Y} \colon (X \multimap Y) \to (X \multimap Y).$$

In Set, $\mathrm{ev}_{X,Y}$ takes any function from X to Y and evaluates it at any element of X to give an element of Y. In terms of string diagrams, the evaluation morphism looks like this:

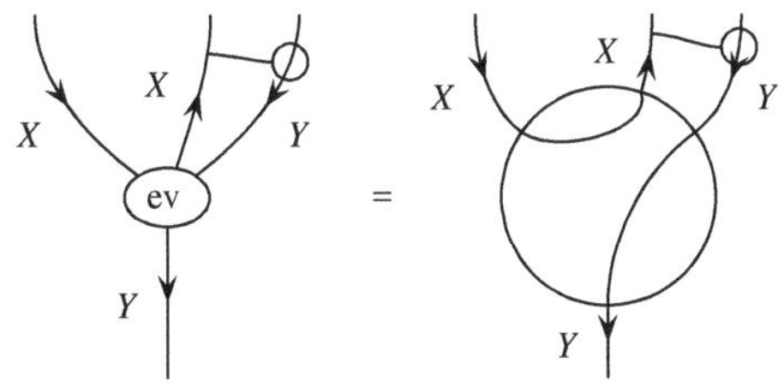

In any closed monoidal category, we can recover a morphism from its name using evaluation. More precisely, this diagram commutes:

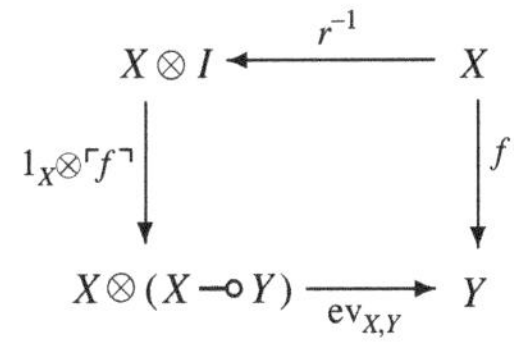

Or, in terms of string diagrams:

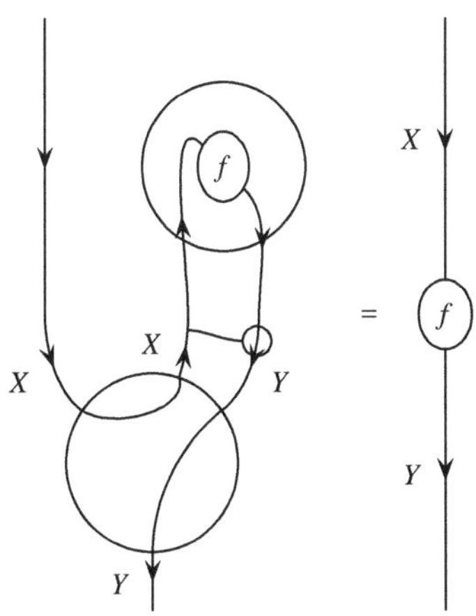

We leave the proof of this as an exercise. In general, one must use the naturality of currying. In the special case of a compact monoidal category, there is a nice picture proof! Simply pop the bubbles and remove the clasp:

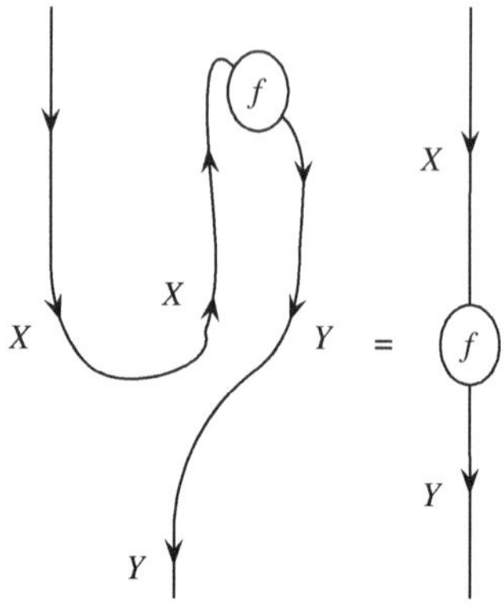

The result then follows from one of the zig-zag identities.

In our rapid introduction to string diagrams, we have not had time to illustrate how these diagrams become a powerful tool for solving concrete problems. So, here are some starting points for further study:

- Representations of Lie groups play a fundamental role in quantum physics, especially gauge field theory. Every Lie group has a compact symmetric monoidal category of finite-dimensional representations. In his book *Group Theory*, Cvitanovic [38] develops detailed string diagram descriptions of these representation categories for the classical Lie groups SU(n), SO(n), SU(n) and also the more exotic "exceptional" Lie groups. His book also illustrates how this technology can be used to simplify difficult calculations in gauge field theory.
- Quantum groups are a generalization of groups which show up in 2d and 3d physics. The big difference is that a quantum group has compact *braided* monoidal category of finite-dimensional representations. Kauffman's *Knots and Physics* [65] is an excellent introduction to how quantum groups show up in knot theory and physics; it is packed with string diagrams. For more details on quantum groups and braided monoidal categories, see the book by Kassel [64].
- Kauffman and Lins [66] have written a beautiful string diagram treatment of the category of representations of the simplest quantum group, $SU_q(2)$. They also use it to construct some famous 3-manifold invariants associated to 3d and 4d topological quantum field theories: the Witten–Reshetikhin–Turaev, Turaev–Viro and Crane–Yetter invariants. In this example, string diagrams are often called "q-deformed spin networks" [101]. For generalizations to other quantum groups, see the more advanced texts by Turaev [107] and by Bakalov and Kirillov [17]. The key ingredient is a special class of compact braided monoidal categories called "modular tensor categories".
- Kock [70] has written a nice introduction to 2d topological quantum field theories which uses diagrammatic methods to work with 2Cob.
- Abramsky, Coecke and collaborators [2–4, 31, 33, 34] have developed string diagrams as a tool for understanding quantum computation. The easiest introduction is Coecke's "Kindergarten quantum mechanics" [32].

2.2.7 Dagger Categories

Our discussion would be sadly incomplete without an important admission: *nothing we have done so far with Hilbert spaces used the inner product!* So, we have not yet touched on the essence of quantum theory.

Everything we have said about Hilb applies equally well to Vect: the category of finite-dimensional *vector spaces* and linear operators. Both Hilb and Vect are compact symmetric monoidal categories. In fact, these compact symmetric monoidal categories are "equivalent" in a certain precise sense [78].

So, what makes Hilb different? In terms of category theory, the special thing is that we can take the Hilbert space adjoint of any linear operator $f: X \to Y$ between finite-dimensional Hilbert spaces, getting an operator $f^{\dagger}: Y \to X$. This ability to 'reverse' morphisms makes Hilb into a 'dagger category':

Definition 18 A **dagger category** is a category C such that for any morphism $f: X \to Y$ in C there is a specified morphism $f^{\dagger}: Y \to X$ such that

$$(gf)^{\dagger} = f^{\dagger} g^{\dagger}$$

for every pair of composable morphisms f and g, and

$$(f^{\dagger})^{\dagger} = f$$

for every morphism f.

Equivalently, a dagger category is one equipped with a functor $\dagger: C \to C^{\mathrm{op}}$ that is the identity on objects and satisfies $(f^{\dagger})^{\dagger} = f$ for every morphism.

In fact, all our favorite examples of categories can be made into dagger categories, except for Set:

- There is no way to make Set into a dagger category, since there is a function from the empty set to the 1-element set, but none the other way around.
- The category Hilb becomes a dagger category as follows. Given any morphism $f: X \to Y$ in Hilb, there is a morphism $f^{\dagger}: Y \to X$, the **Hilbert space adjoint** of f, defined by

  $$\langle f^{\dagger}\psi, \phi\rangle = \langle \psi, f\phi\rangle$$

  for all $\phi \in X$, $\psi \in Y$.
- For any k, the category Tang_k becomes a dagger category where we obtain $f^{\dagger}: Y \to X$ by reflecting $f: X \to Y$ in the vertical direction, and then switching the direction of the little arrows denoting the orientations of arcs and circles.
- For any n, the category nCob becomes a dagger category where we obtain $f^{\dagger}: Y \to X$ by switching the input and output of $f: X \to Y$, and then switching the orientation of each connected component of f. Again, a picture speaks a thousand words:

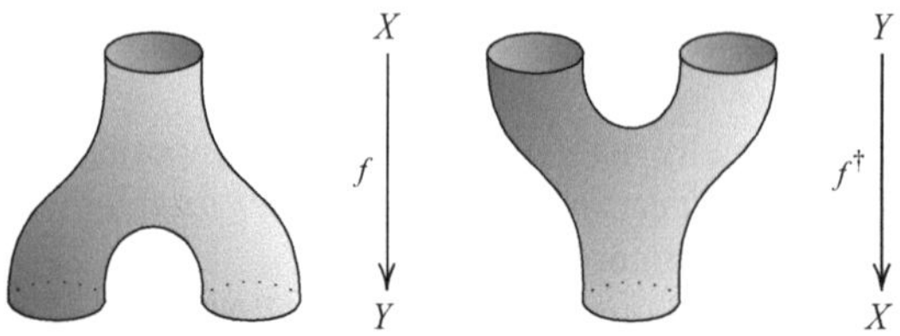

In applications to physics, this dagger operation amounts to "switching the future and the past".

In all the dagger categories above, the dagger structure interacts in a nice way with the monoidal structure and also, when it exists, the braiding. One can write a list of axioms characterizing how this works [2, 3, 97]. So, it seems that the ability to "reverse" morphisms is another way in which categories of a quantum flavor differ from the category of sets and functions. This has important implications for the foundations of quantum theory [12] and also for topological quantum field theory [14], where dagger categories seem to be part of larger story involving "n-categories with duals" [15]. However, this story is still poorly understood—there is much more work to be done.

2.3 Logic

2.3.1 Background

Symmetric monoidal closed categories show up not only in physics and topology, but also in logic. We would like to explain how. To set the stage, it seems worthwhile to sketch a few ideas from twentieth-century logic.

Modern logicians study many systems of reasoning beside ordinary classical logic. Of course, even classical logic comes in various degrees of strength. First there is the "propositional calculus", which allows us to reason with abstract propositions $X, Y, Z, \ldots$ and these logical connectives:

and	$\wedge$
or	$\vee$
implies	$\Rightarrow$
not	$\neg$
true	$\top$
false	$\bot$

Then there is the "predicate calculus", which also allows variables like $x, y, z, \ldots$, predicates like $P(x)$ and $Q(x, y, z)$, and the symbols "for all" ($\forall$) and "there exists" ($\exists$), which allow us to quantify over variables. There are also higher-order systems that allow us to quantify over predicates, and so on. To keep things simple, we mainly confine ourselves to the propositional calculus in what follows. But even here, there are many alternatives to the "classical" version!

The most-studied of these alternative systems are *weaker* than classical logic: they make it harder or even impossible to prove things we normally take for granted. One reason is that some logicians deny that certain familiar principles are actually valid. But there are also subtler reasons. One is that studying systems with rules of lesser strength allows for a fine-grained study of precisely which methods of reasoning are needed to prove which results. Another reason—the one that concerns us most here—is that dropping familiar rules and then adding them back in one at at time sheds light on the connection between logic and category theory.

For example, around 1907 Brouwer [53] began advocating "intuitionism". As part of this, he raised doubts about the law of excluded middle, which amounts to a rule saying that from $\neg\neg X$ we can deduce X. One problem with this principle is that proofs using it are not "constructive". For example, we may prove by contradiction that some equation has a solution, but still have no clue how to construct the solution. For Brouwer, this meant the principle was invalid.

Anyone who feels the law of excluded middle is invalid is duty-bound to study intuitionistic logic. But, there is another reason for studying this system. Namely: we do not really *lose* anything by dropping the law of excluded middle! Instead, we *gain* a fine-grained distinction: the distinction between a direct proof of X and a proof by contradiction, which yields merely $\neg\neg X$. If we do not care about this distinction we are free to ignore it, but there is no harm in having it around.

In the 1930's, this idea was made precise by Gödel [49] and Gentzen [104]. They showed that we can embed classical logic in intuitionistic logic. In fact, they found a map sending any formula X of the propositional calculus to a new formula X°, such that X is provable classically if and only if X° is provable intuitionistically. (More impressively, this map also works for the predicate calculus.)

Later, yet another reason for being interested in intuitionistic logic became apparent: its connection to category theory. In its very simplest form, this connection works as follows. Suppose we have a set of propositions $X, Y, Z, \ldots$ obeying the laws of the intuitionistic propositional calculus. We can create a category C where these propositions are objects and there is at most one morphism from any object X to any object Y: a single morphism when X implies Y, and none otherwise!

A category with at most one morphism from any object to any other is called a **preorder**. In the propositional calculus, we often treat two propositions as equal when they both imply each other. If we do this, we get a special sort of preorder: one where isomorphic objects are automatically equal. This special sort of preorder is called a **partially ordered set**, or **poset** for short. Posets abound in logic, precisely because they offer a simple framework for understanding implication.

If we start from a set of propositions obeying the intuitionistic propositional calculus, the resulting category C is better than a mere poset. It is also cartesian, with $X \wedge Y$ as the product of X and Y, and $\top$ as the terminal object! To see this, note that any proposition Q has a unique morphism to $X \wedge Y$ whenever it has morphisms to X and to Y. This is simply a fancy way of saying that Q implies $X \wedge Y$ when it implies X and implies Y. It is also easy to see that $\top$ is terminal: anything implies the truth.

Even better, the category C is cartesian closed, with $X \Rightarrow Y$ as the internal hom. The reason is that

$$X \wedge Y \text{ implies } Z \quad \text{iff} \quad Y \text{ implies } X \Rightarrow Z.$$

This automatically yields the basic property of the internal hom:

$$\hom(X \otimes Y, Z) \cong \hom(Y, X \multimap Z).$$

Indeed, if the reader is puzzled by the difference between "X implies Y" and $X \Rightarrow Y$, we can now explain this more clearly: the former involves the homset $\hom(X, Y)$ (which has one element when X implies Y and none otherwise), while the latter is the internal hom, an object in C.

So, C is a cartesian closed poset. But, it also has one more nice property, thanks to the presence of $\vee$ and $\bot$ We have seen that $\wedge$ and $\top$ make the category C cartesian; $\vee$ and $\bot$ satisfy exactly analogous rules, but with the implications turned around, so they make C^{op} cartesian.

And that is all! In particular, negation gives nothing more, since we can define $\neg X$ to be $X \Rightarrow \bot$, and all its intuitionistically valid properties then follow. So, the kind of category we get from the intuitionistic propositional calculus by taking propositions as objects and implications as morphisms is precisely a **Heyting algebra**: a cartesian closed poset C such that C^{op} is also cartesian.

Heyting, a student of Brouwer, introduced Heyting algebras in intuitionistic logic before categories were even invented. So, he used very different language to define them. But, the category-theoretic approach to Heyting algebras illustrates the connection between cartesian closed categories and logic. It also gives more evidence that dropping the law of excluded middle is an interesting thing to try.

Since we have explained the basics of cartesian closed categories, but not said what happens when the *opposite* of such a category is *also* cartesian, in the sections to come we will take a drastic step and limit our discussion of logic even further. We will neglect $\vee$ and $\bot$, and concentrate only on the fragment of the propositional calculus involving $\wedge$, $\top$ and $\Rightarrow$.

Even here, it turns out, there are interesting things to say—and interesting ways to modify the usual rules. This will be the main subject of the sections to come. But to set the stage, we need to say a bit about proof theory.

Proof theory is the branch of mathematical logic that treats proofs as mathematical entities worthy of study in their own right. It lets us dig deeper into the propositional calculus by studying not merely *whether or not* some assumption X implies some conclusion Y, but the whole *set of proofs* leading from X to Y. This amounts to studying not just posets (or preorders), but categories that allow many morphisms from one object to another.

In Hilbert's approach to proof, there were many axioms and just one rule to deduce new theorems: *modus ponens*, which says that from X and "X implies Y" we can deduce Y. Most of modern proof theory focuses on another approach, the "sequent calculus", due to Gentzen [104]. In this approach there are few axioms but many inference rules.

An excellent introduction to the sequent calculus is the book *Proofs and Types* by Girard, Lafont and Taylor, freely available online [48]. Here we shall content ourselves with some sketchy remarks. A "sequent" is something like this:

$$X_1, \ldots, X_m \vdash Y_1, \ldots, Y_n$$

where X_i and Y_i are propositions. We read this sequent as saying that *all* the propositions X_i, taken together, can be used to prove at least *one* of the propositions Y_i. This strange-sounding convention gives the sequent calculus a nice symmetry, as we shall soon see.

In the sequent calculus, an "inference rule" is something that produces new sequents from old. For example, here is the **left weakening** rule:

$$\frac{X_1, \ldots, X_m \vdash Y_1, \ldots, Y_n}{X_1, \ldots, X_m, A \vdash Y_1, \ldots, Y_n}$$

This says that from the sequent above the line we can get the sequent below the line: we can throw in the extra assumption A without harm. Thanks to the strange-sounding convention we mentioned, this rule has a mirror-image version called **right weakening**:

$$\frac{X_1, \ldots, X_m \vdash Y_1, \ldots, Y_n}{X_1, \ldots, X_m \vdash Y_1, \ldots, Y_n, A}$$

In fact, Gentzen's whole setup has this mirror symmetry! For example, his rule called **left contraction**:

$$\frac{X_1, \ldots, X_m, A, A \vdash Y_1, \ldots, Y_n}{X_1, \ldots, X_m, A \vdash Y_1, \ldots, Y_n}$$

has a mirror partner called **right contraction**:

$$\frac{X_1, \ldots, X_m \vdash Y_1, \ldots, Y_n, A, A}{X_1, \ldots, X_m \vdash Y_1, \ldots, Y_n, A}$$

Similarly, this rule for "and"

$$\frac{X_1, \ldots, X_m, A \vdash Y_1, \ldots, Y_n}{X_1, \ldots, X_m, A \wedge B \vdash Y_1, \ldots, Y_n}$$

has a mirror partner for "or":

$$\frac{X_1, \ldots, X_m \vdash Y_1, \ldots, Y_n, A}{X_1, \ldots, X_m \vdash Y_1, \ldots, Y_n, A \vee B}$$

Logicians now realize that this mirror symmetry can be understood in terms of the duality between a category and its opposite.

Gentzen used sequents to write inference rules for the classical propositional calculus, and also the classical predicate calculus. Now, in these forms of logic we have

$$X_1, \ldots, X_m \vdash Y_1, \ldots, Y_n$$

if and only if we have

$$X_1 \wedge \cdots \wedge X_m \vdash Y_1 \vee \cdots \vee Y_n.$$

So, why did Gentzen use sequents with a *list* of propositions on each side of the $\vdash$ symbol, instead just a single proposition? The reason is that this let him use only inference rules having the "subformula property". This says that every proposition in the sequent above the line appears as part of some proposition in the sequent below the line. So, a proof built from such inference rules becomes a "tree" where all the propositions further up the tree are subformulas of those below.

This idea has powerful consequences. For example, in 1936 Gentzen was able prove the consistency of Peano's axioms of arithmetic! His proof essentially used induction on trees (Readers familiar with Gödel's second incompleteness theorem should be reassured that this sort of induction cannot itself be carried out in Peano arithmetic.)

The most famous rule *lacking* the subformula property is the 'cut rule':

$$\frac{X_1, \ldots, X_m \vdash Y_1, \ldots, Y_k, A \qquad X_{m+1}, \ldots, X_n, A \vdash Y_{k+1}, \ldots, Y_\ell}{X_1, \ldots, X_n \vdash Y_1, \ldots, Y_\ell}$$

From the two sequents on top, the cut rule gives us the sequent below. Note that the intermediate step A does not appear in the sequent below. It is "cut out". So, the cut rule lacks the subformula property. But, one of Gentzen's great achievements was to show that any proof in the classical propositional (or even predicate) calculus that can be done *with* the cut rule can also be done *without* it. This is called 'cut elimination'.

Gentzen also wrote down inference rules suitable for the intuitionistic propositional and predicate calculi. These rules lack the mirror symmetry of the classical case. But in the 1980s, this symmetry was restored by Girard's invention of "linear logic" [47].

Linear logic lets us keep track of how many times we use a given premise to reach a given conclusion. To accomplish this, Girard introduced some new logical connectives! For starters, he introduced 'linear' connectives called $\otimes$ and $\multimap$, and a logical constant called I. These act a bit like $\wedge$, $\Rightarrow$ and $\top$. However, they satisfy rules corresponding to a symmetric monoidal category instead of a cartesian closed category. In particular, from X we can prove neither $X \otimes X$ nor I. So, we cannot freely "duplicate" and "delete" propositions using these new connectives. This is reflected in the fact that linear logic drops Gentzen's contraction and weakening rules.

By itself, this might seem unbearably restrictive. However, Girard *also* kept the connectives $\wedge$, $\Rightarrow$ and $\top$ in his system, still satisfying the usual rules. And, he introduced an operation called the "exponential", !, which takes a proposition X and turns it into an "arbitrary stock of copies of X". So, for example, from $!X$ we can prove 1, and X, and $X \otimes X$, and $X \otimes X \otimes X$, and so on.

Full-fledged linear logic has even more connectives than we have described here. It seems baroque and peculiar at first glance. It also comes in both classical and intuitionistic versions! But, *just as classical logic can be embedded in intuitionistic logic, intuitionistic logic can be embedded in intuitionistic linear logic* [47]. So, we do not lose any deductive power. Instead, we gain the ability to make even more fine-grained distinctions.

In what follows, we discuss the fragment of intuitionistic linear logic involving only $\otimes$, $\multimap$ and I. This is called "multiplicative intuititionistic linear logic" [52, 91]. It turns out to be the system of logic suitable for closed symmetric monoidal categories—nothing more or less.

2.3.2 Proofs as Morphisms

In Sect. 2.2 we described categories with various amounts of extra structure, starting from categories pure and simple, and working our way up to monoidal categories, braided monoidal categories, symmetric monoidal categories, and so on. Our treatment only scratched the surface of an enormously rich taxonomy. In fact, each kind of category with extra structure corresponds to a system of logic with its own inference rules!

To see this, we will think of *propositions* as *objects* in some category, and *proofs* as giving *morphisms*. Suppose X and Y are propositions. Then, we can think of a proof starting from the assumption X and leading to the conclusion Y as giving a morphism $f: X \to Y$. (In Sect. 2.3.3 we shall see that a morphism is actually an equivalence class of proofs—but for now let us gloss over this issue.)

Let us write $X \vdash Y$ when, starting from the assumption X, there is a proof leading to the conclusion Y. An inference rule is a way to get new proofs from old. For example, in almost every system of logic, if there is a proof leading from X to Y, and a proof leading from Y to Z, then there is a proof leading from X to Z. We write this inference rule as follows:

$$\frac{X \vdash Y \qquad Y \vdash Z}{X \vdash Z}$$

We can call this **cut rule**, since it lets us "cut out" the intermediate step Y. It is a special case of Gentzen's cut rule, mentioned in the previous section. It should remind us of composition of morphisms in a category: if we have a morphism $f: X \to Y$ and a morphism $g: Y \to Z$, we get a morphism $gf: X \to Z$.

Also, in almost every system of logic there is a proof leading from X to X. We can write this as an inference rule that starts with *nothing* and concludes the existence of a proof of X from X:

$$\frac{}{X \vdash X}$$

This rule should remind us of how every object in category has an identity morphism: for any object X, we automatically get a morphism $1_X\colon X \to X$. Indeed, this rule is sometimes called the **identity rule**.

If we pursue this line of thought, we can take the definition of a closed symmetric monoidal category and extract a collection of inference rules. Each rule is a way to get new morphisms from old in a closed symmetric monoidal category. There are various superficially different but ultimately equivalent ways to list these rules. Here is one:

$$\frac{}{X \vdash X}\,(\mathrm{i}) \qquad \frac{X \vdash Y \quad Y \vdash Z}{X \vdash Z}\,(\circ)$$

$$\frac{W \vdash X \quad Y \vdash Z}{W \otimes Y \vdash X \otimes Z}\,(\otimes) \qquad \frac{W \vdash (X \otimes Y) \otimes Z}{\overline{W \vdash X \otimes (Y \otimes Z)}}\,(\mathrm{a})$$

$$\frac{X \vdash I \otimes Y}{\overline{X \vdash Y}}\,(\mathrm{l}) \qquad \frac{X \vdash Y \otimes I}{\overline{X \vdash Y}}\,(\mathrm{r})$$

$$\frac{W \vdash X \otimes Y}{\overline{W \vdash Y \otimes X}}\,(\mathrm{b}) \qquad \frac{X \otimes Y \vdash Z}{\overline{Y \vdash X \multimap Z}}\,(\mathrm{c})$$

Double lines mean that the inverse rule also holds. We have given each rule a name, written to the right in parentheses. As already explained, rules (i) and ($\circ$) come from the presence of identity morphisms and composition in any category. Rules ($\otimes$), (a), (l), and (r) come from tensoring, the associator, and the left and right unitors in a monoidal category. Rule (b) comes from the braiding in a braided monoidal category, and rule (c) comes from currying in a closed monoidal category.

Now for the big question: *what does all this mean in terms of logic?* These rules describe a small fragment of the propositional calculus. To see this, we should read the connective $\otimes$ as "and", the connective $\multimap$ as "implies", and the proposition I as "true".

In this interpretation, rule (c) says we can turn a proof leading from the assumption "Y and X" to the conclusion Z into a proof leading from X to "Y implies Z". It also says we can do the reverse. This is true in classical, intuitionistic and linear logic, and so are all the other rules. Rules (a) and (b) say that "and" is associative and commutative. Rule (l) says that any proof leading from the assumption X to the conclusion "true and Y" can be converted to a proof leading from X to Y, and vice versa. Rule (r) is similar.

What do we do with these rules? We use them to build "deductions". Here is an easy example:

$$\dfrac{\dfrac{}{X \multimap Y \vdash X \multimap Y}\,(\mathrm{i})}{X \otimes (X \multimap Y) \vdash Y}\,(\mathrm{c}^{-1})$$

First we use the identity rule, and then the inverse of the currying rule. At the end, we obtain

$$X \otimes (X \multimap Y) \vdash Y.$$

This should remind us of the evaluation morphisms we have in a closed monoidal category:

$$\mathrm{ev}_{X,Y} \colon X \otimes (X \multimap Y) \to Y.$$

In terms of logic, the point is that we can prove Y from X and "X implies Y". This fact comes in handy so often that we may wish to abbreviate the above deduction as an extra inference rule—a rule derived from our basic list:

$$\dfrac{}{X \otimes (X \multimap Y) \vdash Y}\,(\mathrm{ev})$$

This rule is called **modus ponens**.

In general, a deduction is a tree built from inference rules. Branches arise when we use the $(\circ)$ or $(\otimes)$ rules. Here is an example:

$$\dfrac{\dfrac{\dfrac{}{(A \otimes B) \otimes C \vdash (A \otimes B) \otimes C}\,(\mathrm{i})}{(A \otimes B) \otimes C \vdash A \otimes (B \otimes C)}\,(\mathrm{a}) \qquad A \otimes (B \otimes C) \vdash D}{(A \otimes B) \otimes C \vdash D}\,(\circ)$$

Again we can abbreviate this deduction as a derived rule. In fact, this rule is reversible:

$$\dfrac{A \otimes (B \otimes C) \vdash D}{(A \otimes B) \otimes C \vdash D}\,(\alpha)$$

For a more substantial example, suppose we want to show

$$(X \multimap Y) \otimes (Y \multimap Z) \vdash X \multimap Z.$$

The deduction leading to this will not even fit on the page unless we use our abbreviations:

$$\dfrac{\dfrac{\dfrac{\dfrac{\dfrac{}{X \otimes (X \multimap Y) \vdash Y}\,(\mathrm{ev}) \qquad \dfrac{}{Y \multimap Z \vdash Y \multimap Z}\,(\mathrm{i})}{(X \otimes (X \multimap Y)) \otimes (Y \multimap Z) \vdash Y \otimes (Y \multimap Z)}\,(\otimes) \qquad \dfrac{}{Y \otimes (Y \multimap Z) \vdash Z}\,(\mathrm{ev})}{(X \otimes (X \multimap Y)) \otimes (Y \multimap Z) \vdash Z}}{X \otimes ((X \multimap Y) \otimes (Y \multimap Z)) \vdash Z}\,(\alpha^{-1})}{(X \multimap Y) \otimes (Y \multimap Z) \vdash X \multimap Z}\,(\mathrm{c})$$

Since each of the rules used in this deduction came from a way to get new morphisms from old in a closed monoidal category (we never used the braiding), it follows that in every such category we have **internal composition** morphisms:

$$\bullet_{X,Y,Z} : (X \multimap Y) \otimes (Y \multimap Z) \to X \multimap Z.$$

These play the same role for the internal hom that ordinary composition

$$\circ\colon \hom(X, Y) \times \hom(Y, Z) \to \hom(X, Z)$$

plays for the ordinary hom.

We can go ahead making further deductions in this system of logic, but the really interesting thing is what it omits. For starters, it omits the connective "or" and the proposition "false". It also omits two inference rules we normally take for granted—namely, **contraction**:

$$\frac{X \vdash Y}{X \vdash Y \otimes Y} \, (\Delta)$$

and **weakening**:

$$\frac{X \vdash Y}{X \vdash I} \, (!)$$

which are closely related to duplication and deletion in a cartesian category. Omitting these rules is a distinctive feature of linear logic [47]. The word "linear" should remind us of the category Hilb. As noted in Sect. 2.2.3, this category with its usual tensor product is noncartesian, so it does not permit duplication and deletion. But, what does omitting these rules mean *in terms of logic?*

Ordinary logic deals with propositions, so we have been thinking of the above system of logic in the same way. Linear logic deals not just with propositions, but also other resources—for example, physical things! Unlike propositions in ordinary logic, we typically can't duplicate or delete these other resources. In classical logic, if we know that a proposition X is true, we can use X as many or as few times as we like when trying to prove some proposition Y. But if we have a cup of milk, we can't use it to make cake and then use it again to make butter. Nor can we make it disappear without a trace: even if we pour it down the drain, it must go somewhere.

In fact, these ideas are familiar in chemistry. Consider the following resources:

$$\begin{aligned} H_2 &= \text{one molecule of hydrogen} \\ O_2 &= \text{one molecule of oxygen} \\ H_2O &= \text{one molecule of water} \end{aligned}$$

We can burn hydrogen, combining one molecule of oxygen with two of hydrogen to obtain two molecules of water. A category theorist might describe this reaction as a morphism:

$$f\colon O_2 \otimes (H_2 \otimes H_2) \to H_2O \otimes H_2O.$$

A linear logician might write:

$$O_2 \otimes (H_2 \otimes H_2) \vdash H_2O \otimes H_2O$$

to indicate the existence of such a morphism. But, we cannot duplicate or delete molecules, so for example

$$H_2 \nvdash H_2 \otimes H_2$$

and

$$H_2 \nvdash I$$

where I is the unit for the tensor product: not iodine, but "no molecules at all".

In short, ordinary chemical reactions are morphisms in a symmetric monoidal category where objects are collections of molecules. As chemists normally conceive of it, this category is not closed. So, it obeys an even more limited system of logic than the one we have been discussing, a system lacking the connective $\multimap$. To get a closed category—in fact a compact one—we need to remember one of the great discoveries of twentieth-century physics: *antimatter*. This lets us define $Y \multimap Z$ to be "anti-Y and Z":

$$Y \multimap Z = Y^* \otimes Z.$$

Then the currying rule holds:

$$\frac{Y \otimes X \vdash Z}{X \vdash Y^* \otimes Z}$$

Most chemists don't think about antimatter very often—but particle physicists do. They don't use the notation of linear logic or category theory, but they know perfectly well that since a neutrino and a neutron can collide and turn into a proton and an electron:

$$\nu \otimes n \vdash p \otimes e,$$

then a neutron can turn into a antineutrino together with a proton and an electron:

$$n \vdash \nu^* \otimes (p \otimes e).$$

This is an instance of the currying rule, rule (c).

2.3.3 Logical Theories from Categories

We have sketched how different systems of logic naturally arise from different types of categories. To illustrate this idea, we introduced a system of logic with inference rules coming from ways to get new morphisms from old in a *closed symmetric monoidal category*. One could substitute many other types of categories here, and get other systems of logic.

To tighten the connection between proof theory and category theory, we shall now describe a recipe to get a logical theory from any closed symmetric monoidal category. For this, we shall now use $X \vdash Y$ to denote the *set* of proofs—or actually, equivalence classes of proofs—leading from the assumption X to the conclusion Y. This is a change of viewpoint. Previously we would write $X \vdash Y$ when this set of proofs was nonempty; otherwise we would write $X \nvdash Y$. The advantage of treating $X \vdash Y$ as a set is that this set is precisely what a category theorist would call $\hom(X, Y)$: a homset in a category.

If we let $X \vdash Y$ stand for a homset, an inference rule becomes a function from a product of homsets to a single homset. For example, the cut rule

$$\frac{X \vdash Y \qquad Y \vdash Z}{X \vdash Z}\,(\circ)$$

becomes another way of talking about the composition function

$$\circ_{X,Y,Z} : \hom(X, Y) \times \hom(Y, Z) \to \hom(X, Z),$$

while the identity rule

$$\frac{}{X \vdash X}\,(\mathrm{i})$$

becomes another way of talking about the function

$$\mathrm{i}_X : 1 \to \hom(X, X)$$

that sends the single element of the set 1 to the identity morphism of X. (Note: the set 1 is a *zero-fold* product of homsets.)

Next, if we let inference rules be certain functions from products of homsets to homsets, deductions become more complicated functions of the same sort built from these basic ones. For example, this deduction:

$$\frac{\dfrac{\dfrac{}{X \otimes I \vdash X \otimes I}\,(\mathrm{i})}{X \otimes I \vdash X}\,(\mathrm{r}) \qquad \dfrac{}{Y \vdash Y}\,(\mathrm{i})}{(X \otimes I) \otimes Y \vdash X \otimes Y}\,(\otimes)$$

specifies a function from 1 to $\hom((X\otimes I)\otimes Y,\, X\otimes Y)$, built from the basic functions indicated by the labels at each step. This deduction:

$$\dfrac{\dfrac{\dfrac{}{(X\otimes I)\otimes Y \vdash (X\otimes I)\otimes Y}\,(\mathrm{i})}{(X\otimes I)\otimes Y \vdash X\otimes(I\otimes Y)}\,(\mathrm{a}) \qquad \dfrac{\dfrac{\dfrac{}{I\otimes Y\vdash I\otimes Y}\,(\mathrm{i})}{I\otimes Y\vdash Y}\,(\mathrm{r}) \quad \dfrac{}{X\vdash X}\,(\mathrm{i})}{X\otimes(I\otimes Y)\vdash X\otimes Y}\,(\otimes)}{(X\otimes I)\otimes Y\vdash X\otimes Y}\,(\circ)$$

gives another function from 1 to $\hom((X \otimes I) \otimes Y, X \otimes Y)$.

If we think of deductions as giving functions this way, the question arises when two such functions are equal. In the example just mentioned, the triangle equation in the definition of monoidal category (Definition 7):

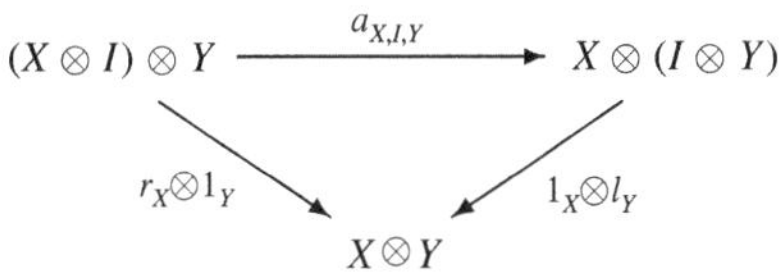

says these two functions *are* equal. Indeed, the triangle equation is precisely the statement that these two functions agree! (We leave this as an exercise for the reader.)

So: even though two deductions may look quite different, they may give the same function from a product of homsets to a homset if we demand that these are homsets in a closed symmetric monoidal category. This is why we think of $X \multimap Y$ as a set of *equivalence classes* of proofs, rather than proofs: it is forced on us by our desire to use category theory. We could get around this by using a 2-category with proofs as morphisms and "equivalences between proofs" as 2-morphisms [93, 94]. This would lead us further to the right in the Periodic Table (Table 2.3). But let us restrain ourselves and make some definitions formalizing what we have done so far.

From now on we shall call the objects $X, Y, \ldots$ "propositions", even though we have seen they may represent more general resources. Also, purely for the sake of brevity, we use the term "proof" to mean "equivalence class of proofs". The equivalence relation must be coarse enough to make the equations in the following definitions hold:

Definition 19 A **closed monoidal theory** consists of the following:

- A collection of **propositions**. The collection must contain a proposition I, and if X and Y are propositions, then so are $X \otimes Y$ and $X \multimap Y$.
- For every pair of propositions X, Y, a set $X \vdash Y$ of **proofs** leading from X to Y. If $f \in X \vdash Y$, then we write $f : X \to Y$.
- Certain functions, written as **inference rules**:

$$\frac{}{X \vdash X}\ (\mathrm{i}) \qquad \frac{X \vdash Y \qquad Y \vdash Z}{X \vdash Z}\ (\circ)$$

$$\frac{W \vdash X \qquad Y \vdash Z}{W \otimes Y \vdash X \otimes Z}\ (\otimes) \qquad \frac{W \vdash (X \otimes Y) \otimes Z}{W \vdash X \otimes (Y \otimes Z)}\ (\mathrm{a})$$

$$\frac{X \vdash I \otimes Y}{X \vdash Y}\ (\mathrm{l}) \qquad \frac{X \vdash Y \otimes I}{X \vdash Y}\ (\mathrm{r})$$

$$\frac{X \otimes Y \vdash Z}{Y \vdash X \multimap Z}\ (\mathrm{c})$$

A double line means that the function is invertible. So, for example, for each triple X, Y, Z we have a function

$$\circ_{X,Y,Z} \colon (X \vdash Y) \times (Y \vdash Z) \to (X \vdash Z)$$

and a bijection

$$c_{X,Y,Z} \colon (X \otimes Y \vdash Z) \to (Y \vdash X \multimap Z).$$

- Certain equations that must be obeyed by the inference rules. The inference rules $(\circ)$ and (i) must obey equations describing associativity and the left and right unit laws. Rule $(\otimes)$ must obey an equation saying it is a functor. Rules (a), (l), (r), and (c) must obey equations saying they are natural transformations. Rules (a), (l), (r) and $(\otimes)$ must also obey the triangle and pentagon equations.

Definition 20 A **closed braided monoidal theory** is a closed monoidal theory with this additional inference rule:

$$\frac{W \vdash X \otimes Y}{W \vdash Y \otimes X}\ (\mathrm{b})$$

We demand that this rule give a natural transformation satisfying the hexagon equations.

Definition 21 A **closed symmetric monoidal theory** is a closed braided monoidal theory where the rule (b) is its own inverse.

These are just the usual definitions of various kinds of closed category—monoidal, braided monoidal and symmetric monoidal—written in a new style. This new style lets us *build such categories from logical systems*. To do this, we take the objects to be propositions and the morphisms to be equivalence classes of proofs, where the equivalence relation is generated by the equations listed in the definitions above.

However, the full advantages of this style only appear when we dig deeper into proof theory, and generalize the expressions we have been considering:

$$X \vdash Y$$

to "sequents" like this:

$$X_1, \ldots, X_n \vdash Y.$$

Loosely, we can think of such a sequent as meaning

$$X_1 \otimes \cdots \otimes X_n \vdash Y.$$

The advantage of sequents is that they let us use inference rules that—except for the cut rule and the identity rule—have the "subformula property" mentioned near the end of Sect. 2.3.1.

Formulated in terms of these inference rules, the logic of closed symmetric monoidal categories goes by the name of "multiplicative intuitionistic linear logic", or MILL for short [52, 91]. There is a "cut elimination" theorem for MILL, which says that with a suitable choice of other inference rules, the cut rule becomes redundant: any proof that can be done with it can be done without it. This is remarkable, since the cut rule corresponds to *composition of morphisms* in a category. One consequence is that in the free symmetric monoidal closed category on any set of objects, the set of morphisms between any two objects is *finite*. There is also a decision procedure to tell when two morphisms are equal. For details, see Trimble's thesis [105] and the papers by Jay [58] and Soloviev [100]. Also see Kelly and Mac Lane's coherence theorem for closed symmetric monoidal categories [67], and the related theorem for compact symmetric monoidal categories [68].

MILL is just one of many closely related systems of logic. Most include extra features, but some *subtract* features. Here are just a few examples:

- Algebraic theories. In his famous thesis, Lawvere [75] defined an **algebraic theory** to be a cartesian category where every object is an n-fold cartesian power $X \times \cdots \times X$ $(n \geq 0)$ of a specific object X. He showed how such categories regarded as logical theories of a simple sort—the sort that had previously been studied in "universal algebra" [26]. This work initiated the categorical approach to logic which we have been sketching here. Crole's book [35] gives a gentle introduction to algebraic theories as well as some richer logical systems. More generally, we can think of any cartesian category as a generalized algebraic theory.
- Intuitionistic linear logic (ILL). ILL supplements MILL with the operations familiar from intuitionistic logic, as well as an operation ! turning any proposition (or resource) X into an "indefinite stock of copies of X". Again there is a nice category-theoretic interpretation. Bierman's thesis [24] gives a good overview, including a proof of cut elimination for ILL and a proof of the result, originally due to Girard, that intuitionistic logic can be be embedded in ILL.
- Linear logic (LL). For full-fledged linear logic, the online review article by Di Cosmo and Miller [39] is a good place to start. For more, try the original paper by Girard [47] and the book by Troelstra [106]. Blute and Scott's review article [25] serves as a Rosetta Stone for linear logic and category theory, and so do the lectures notes by Schalk [91].

- Intuitionistic Logic (IL). Lambek and Scott's classic book [73] is still an excellent introduction to intuitionistic logic and cartesian closed categories. The online review article by Moschovakis [83] contains many suggestions for further reading.

To conclude, let us say precisely what an "inference rule" amounts to in the setup we have described. We have said it gives a function from a product of homsets to a homset. While true, this is not the last word on the subject. After all, instead of treating the propositions appearing in an inference rule as *fixed*, we can treat them as *variable*. Then an inference rule is really a "schema" for getting new proofs from old. How do we formalize this idea?

First we must realize that $X \vdash Y$ is not just a set: it is a set *depending in a functorial way* on X and Y. As noted in Definition 14, there is a functor, the "hom functor"

$$\mathrm{hom}\colon C^{\mathrm{op}} \times C \to \mathrm{Set},$$

sending (X, Y) to the homset $\mathrm{hom}(X, Y) = X \vdash Y$. To look like logicians, let us write this functor as $\vdash$.

Viewed in this light, most of our inference rules are *natural transformations*. For example, rule (a) is a natural transformation between two functors from $C^{\mathrm{op}} \times C^3$ to Set, namely the functors

$$(W, X, Y, Z) \mapsto W \vdash (X \otimes Y) \otimes Z$$

and

$$(W, X, Y, Z) \mapsto W \vdash X \otimes (Y \otimes Z).$$

This natural transformation turns any proof

$$f\colon W \to (X \otimes Y) \otimes Z$$

into the proof

$$a_{X,Y,Z} f\colon W \to X \otimes (Y \otimes Z).$$

The fact that this transformation is *natural* means that it changes in a systematic way as we vary W, X, Y and Z. The commuting square in the definition of natural transformation, Definition 4, makes this precise.

Rules (l), (r), (b) and (c) give natural transformations in a very similar way. The $(\otimes)$ rule gives a natural transformation between two functors from $C^{\mathrm{op}} \times C \times C^{\mathrm{op}} \times C$ to Set, namely

$$(W, X, Y, Z) \mapsto (W \vdash X) \times (Y \vdash Z)$$

and

$$(W, X, Y, Z) \mapsto W \otimes Y \vdash X \otimes Z.$$

This natural transformation sends any element $(f, g) \in \hom(W, X) \times \hom(Y, Z)$ to $f \otimes g$.

The identity and cut rules are different: they *do not* give natural transformations, because the top line of these rules has a different number of variables than the bottom line! Rule (i) says that for each $X \in C$ there is a function

$$\mathrm{i}_X : 1 \to X \vdash X$$

picking out the identity morphism 1_X. What would it mean for this to be natural in X? Rule ($\circ$) says that for each triple $X, Y, Z \in C$ there is a function

$$\circ\colon (X \vdash Y) \times (Y \vdash Z) \to X \vdash Z.$$

What would it mean for this to be natural in X, Y and Z? The answer to both questions involves a generalization of natural transformations called "dinatural" transformations [77].

As noted in Definition 4, a natural transformation $\alpha\colon F \Rightarrow G$ between two functors $F, G\colon C \to D$ makes certain squares in D commute. If in fact $C = C_1^{\mathrm{op}} \times C_2$, then we actually obtain commuting cubes in D. Namely, the natural transformation α assigns to each object (X_1, X_2) a morphism α_{X_1,X_2} such that for any morphism $(f_1\colon Y_1 \to X_1, f_2\colon X_2 \to Y_2)$ in C, this cube commutes:

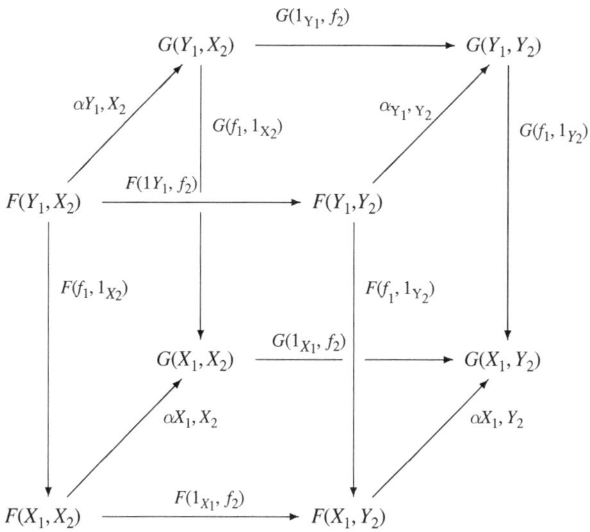

If $C_1 = C_2$, we can choose a single object X and a single morphism $f\colon X \to Y$ and use it in both slots. As shown in Fig. 2.1, there are then two paths from one corner of the cube to the antipodal corner that only involve α for repeated arguments: that is, $\alpha_{X,X}$ and $\alpha_{Y,Y}$, but not $\alpha_{X,Y}$ or $\alpha_{Y,X}$. These paths give a commuting hexagon.

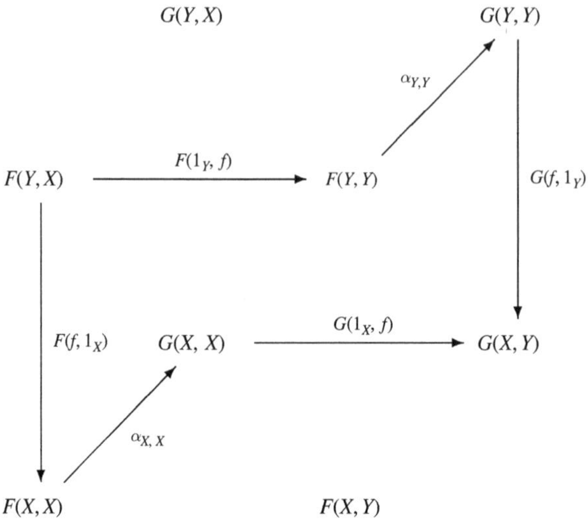

Fig. 2.1 A natural transformation between functors $F, G\colon C^{\mathrm{op}} \times C \to D$ gives a commuting cube in D for any morphism $f\colon X \to Y$, and there are two paths around the cube that only involve α for repeated arguments

This motivates the following:

Definition 22 A **dinatural transformation** $\alpha\colon F \Rightarrow G$ between functors $F, G\colon C^{\mathrm{op}} \times C \to D$ assigns to every object X in C a morphism $\alpha_X\colon F(X, X) \to G(X, X)$ in D such that for every morphism $f\colon X \to Y$ in C, the hexagon in Fig. 2.1 commutes.

In the case of the identity rule, this commuting hexagon follows from the fact that the identity morphism is a left and right unit for composition: see Fig. 2.2. For the cut rule, this commuting hexagon says that composition is associative: see Fig. 2.3.

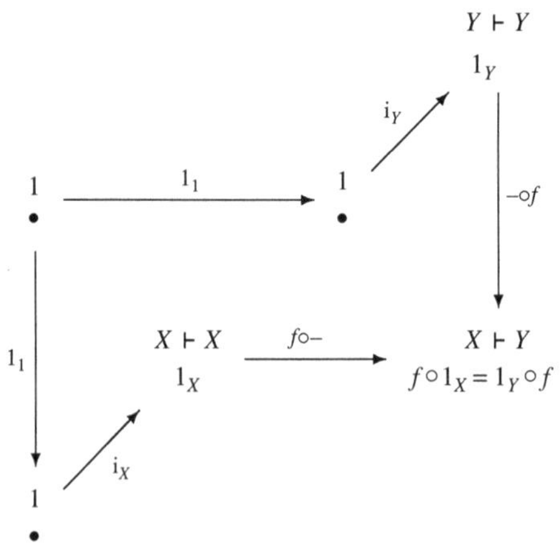

Fig. 2.2 Dinaturality of the (i) rule, where $f\colon X \to Y$. Here $\bullet \in 1$ denotes the one element of the one-element set

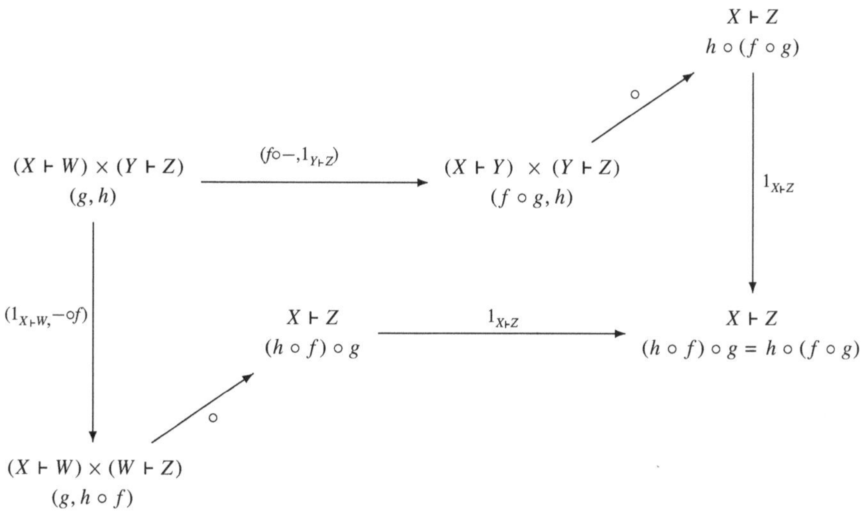

Fig. 2.3 Dinaturality of the cut rule, where $f: W \to Y$, $g: X \to W$, $h: Y \to Z$

So, in general, the sort of logical theory we are discussing involves:

- A *category* C of propositions and proofs.
- A *functor* $\vdash: C^{\mathrm{op}} \times C \to \mathrm{Set}$ sending any pair of propositions to the set of proofs leading from one to the other.
- A set of *dinatural transformations* describing inference rules.

2.4 Computation

2.4.1 Background

In the 1930s, while Turing was developing what are now called "Turing machines" as a model for computation, Church and his student Kleene were developing a different model, called the "lambda calculus" [30, 69]. While a Turing machine can be seen as an idealized, simplified model of computer *hardware*, the lambda calculus is more like a simple model of *software*.

By now the are many careful treatments of the lambda calculus in the literature, from Barendregt's magisterial tome [18] to the classic category-theoretic treatment of Lambek and Scott [73], to Hindley and Seldin's user-friendly introduction [55] and Selinger's elegant free online notes [96]. So, we shall content ourselves with a quick sketch.

Poetically speaking, the lambda calculus describes a universe where everything is a program and everything is data: *programs are data*. More prosaically, everything is a "λ-term", or "term" for short. These are defined inductively:

- **Variables:** there is a countable set of "variables" $x, y, z, \ldots$ which are all terms.
- **Application:** if f and t are terms, we can "apply" f to t and obtain a term $f(t)$.
- **Lambda-abstraction:** if x is a variable and t is a term, there is a term $(\lambda x.t)$.

Let us explain the meaning of application and lambda-abstraction. Application is simple. Since "programs are data", we can think of any term either as a program or a piece of data. Since we can apply programs to data and get new data, we can apply any term f to any other term t and get a new term $f(t)$.

Lambda-abstraction is more interesting. We think of $(\lambda x.t)$ as the program that, given x as input, returns t as output. For example, consider

$$(\lambda x.x(x)).$$

This program takes any program x as input and returns $x(x)$ as output. In other words, it applies any program to itself. So, we have

$$(\lambda x.x(x))(s) = s(s)$$

for any term s.

More generally, if we apply $(\lambda x.t)$ to any term s, we should get back t, but with s substituted for each free occurrence of the variable x. This fact is codified in a rule called **beta reduction**:

$$(\lambda x.t)(s) = t[s/x]$$

where $t[s/x]$ is the term we get by taking t and substituting s for each free occurrence of x. But beware: this rule is not an equation in the usual mathematical sense. Instead, it is a "rewrite rule": given the term on the left, we are allowed to rewrite it and get the term on the right. Starting with a term and repeatedly applying rewrite rules is how we take a program and let it run!

There are two other rewrite rules in the lambda calculus. If x is a variable and t is a term, the term

$$(\lambda x.t(x))$$

stands for the program that, given x as input, returns $t(x)$ as output. But this is just a fancy way of talking about the program t. So, the lambda calculus has a rewrite rule called **eta reduction**, saying

$$(\lambda x.t(x)) = t.$$

The third rewrite rule is **alpha conversion**. This allows us to replace a bound variable in a term by another variable. For example:

$$(\lambda x.x(x)) = (\lambda y.y(y))$$

since x is "bound" in the left-hand expression by its appearance in "λx". In other words, x is just a dummy variable; its name is irrelevant, so we can replace it with y. On the other hand,

$$(\lambda x.y(x)) \neq (\lambda x.z(x)).$$

We cannot replace the variable y by the variable z here, since this variable is "free", not bound. Some care must be taken to make the notions of free and bound variables precise, but we shall gloss over this issue, referring the reader to the references above for details.

The lambda calculus is a very simple formalism. Amazingly, starting from just this, Church and Kleene were able to build up Boolean logic, the natural numbers, the usual operations of arithmetic, and so on. For example, they defined "Church numerals" as follows:

$$\begin{aligned} \overline{0} &= (\lambda f.(\lambda x.x)) \\ \overline{1} &= (\lambda f.(\lambda x.f(x))) \\ \overline{2} &= (\lambda f.(\lambda x.f(f(x)))) \\ \overline{3} &= (\lambda f.(\lambda x.f(f(f(x))))) \end{aligned}$$

and so on. Note that f is a variable above. Thus, the Church numeral $\overline{n}$ is the program that "takes any program to the nth power": if you give it any program f as input, it returns the program that applies f n times to whatever input x it receives.

To get a feeling for how we can define arithmetic operations on Church numerals, consider

$$\lambda g.\overline{3}(\overline{2}(g)).$$

This program takes any program g, squares it, and then cubes the result. So, it raises g to the sixth power. This suggests that

$$\lambda g.\overline{3}(\overline{2}(g)) = \overline{6}.$$

Indeed this is true. If we treat the definitions of Church numerals as reversible rewrite rules, then we can start with the left side of the above equation and grind away using rewrite rules until we reach the right side:

$$\begin{aligned} (\lambda g.\overline{3}(\overline{2}(g))) &= (\lambda g.\overline{3}((\lambda f.(\lambda x.f(f(x))))(g))) && \text{def. of } \overline{2} \\ &= (\lambda g.\overline{3}(\lambda x.g(g(x)))) && \text{beta} \\ &= (\lambda g.(\lambda f.(\lambda x.f(f(f(x)))))(\lambda x.g(g(x)))) && \text{def. of } \overline{3} \\ &= (\lambda g.(\lambda x.(\lambda x.g(g(x)))((\lambda x.g(g(x)))((\lambda x.g(g(x)))(x))))) && \text{beta} \\ &= (\lambda g.(\lambda x.(\lambda x.g(g(x)))((\lambda g.g(g(x)))(g(g(x)))))) && \text{beta} \\ &= (\lambda g.(\lambda x.(\lambda x.g(g(x)))(g(g(g(g(x))))))) && \text{beta} \\ &= (\lambda g.(\lambda x.g(g(g(g(g(g(x)))))))) && \text{beta} \\ &= \overline{6} && \text{def. of } \overline{6} \end{aligned}$$

If this calculation seems mind-numbing, that is precisely the point: it resembles the inner workings of a computer. We see here how the lambda calculus can serve as a programming language, with each step of computation corresponding to a rewrite rule.

Of course, we got the answer $\overline{6}$ because $3 \times 2 = 6$. Generalizing from this example, we can define a program called "times" that multiplies Church numerals:

$$\text{times} = (\lambda a.(\lambda b.(\lambda x.a(b(x))))).$$

For example,

$$\text{times}(\overline{3})(\overline{2}) = \overline{6}.$$

The enterprising reader can dream up similar programs for the other basic operations of arithmetic. With more cleverness, Church and Kleene were able to write terms corresponding to more complicated functions. They eventually came to believe that *all* computable functions $f\colon \mathbb{N} \to \mathbb{N}$ can be defined in the lambda calculus.

Meanwhile, Gödel was developing another approach to computability, the theory of "recursive functions". Around 1936, Kleene proved that the functions definable in the lambda calculus were the same as Gödel's recursive functions. In 1937 Turing described his "Turing machines", and used these to give yet another definition of computable functions. This definition was later shown to agree with the other two. Thanks to this and other evidence, it is now widely accepted that the lambda calculus can define *any* function that can be computed by *any* systematic method. We say it is "Turing complete".

After this burst of theoretical work, it took a few decades for programmable computers to actually be built. It took even longer for computer scientists to profit from Church and Kleene's insights. This began around 1958, when McCarthy invented the programming language Lisp, based on the lambda calculus [80]. In 1965, an influential paper by Landin [74] pointed out a powerful analogy between the lambda calculus and the language ALGOL. These developments led to a renewed interest in the lambda calculus which continues to this day. By now, a number of computer languages are explicitly based on ideas from the lambda calculus. The most famous of these include Lisp, ML and Haskell. These languages, called "functional programming languages", are beloved by theoretical computer scientists for their conceptual clarity. In fact, for many years, everyone majoring in computer science at MIT has been required to take an introductory course that involves programming in Scheme, a dialect of Lisp. The cover of the textbook for this course [1] even has a big λ on the cover!

We should admit that languages of a different sort—"imperative programming languages"—are more popular among working programmers. Examples include FORTRAN, BASIC, and C. In imperative programming, a program is a series of instructions that tell the computer what to do. By constrast, in functional programming, a program simply describes a function. To run the program, we apply it to an input. So, as in the lambda calculus, "application" is a fundamental operation in functional programming. If we combine application with lambda abstraction, we obtain a language powerful enough to compute any computable function.

However, most functional programming languages are more regimented than the original lambda calculus. As we have seen, in the lambda calculus as originally developed by Church and Kleene, any term can be applied to any other. In real life, programming involves many kinds of data. For example, suppose we are writing a program that involves days of the week. It would not make sense to write

$$\mathrm{times}(\overline{3})(\mathrm{Tuesday})$$

because Tuesday is not a number. We might choose to represent Tuesday by a number in some program, but doubling that number doesn't have a good interpretation: is the first day of the week Sunday or Monday? Is the week indexed from zero or one? These are arbitrary choices that affect the result. We could let the programmer make the choices, but the resulting unstructured framework easily leads to mistakes.

It is better to treat data as coming in various "types", such as integers, floating-point numbers, alphanumeric strings, and so on. Thus, whenever we introduce a variable in a program, we should make a "type declaration" saying what type it is. For example, we might write:

$$\mathrm{Tuesday} : \mathrm{day}$$

This notation is used in Ada, Pascal and some other languages. Other notations are also in widespread use. Then, our system should have a "type checker" (usually part of the compiler) that complains if we try to apply a program to a piece of data of the wrong type.

Mathematically, this idea is formalized by a more sophisticated version of the lambda calculus: the "typed" lambda calculus, where every term has a type. This idea is also fundamental to category theory, where every morphism is like a black box with input and output wires of specified types:

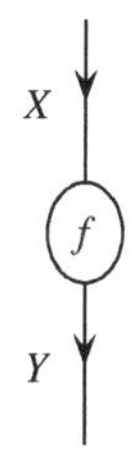

and it makes no sense to hook two black boxes together unless the output of the first has the same type as the input of the next:

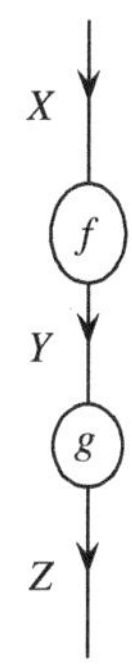

Indeed, there is a deep relation between the typed lambda calculus and cartesian closed categories. This was discovered by Lambek in 1980 [72]. Quite roughly speaking, a "typed lambda-theory" is a very simple functional programming language with a specified collection of basic data types from which other more complicated types can be built, and a specified collection of basic terms from which more complicated terms can be built. The data types of this language are *objects* in a cartesian closed category, while the programs—that is, terms—give *morphisms!*

Here we are being a bit sloppy. Recall from Sect. 2.3.3 that in logic we can build closed monoidal categories where the morphisms are equivalence classes of proofs. We need to take equivalence classes for the axioms of a closed monoidal category to hold. Similarly, to get closed monoidal categories from computer science, we need the morphisms to be equivalence classes of terms. Two terms count as equivalent if they differ by rewrite rules such as beta reduction, eta reduction and alpha conversion. As we have seen, these rewrites represent the steps whereby a program carries out its computation. For example, in the original "untyped" lambda calculus, the terms $\mathrm{times}(\overline{3})(\overline{2})$ and $\overline{6}$ differ by rewrite rules, but they give the same morphism. So, when we construct a cartesian closed category from a typed lambda-theory, we *neglect the actual process of computation*. To remedy this we should work with a cartesian closed 2-category which has:

- types as objects,
- terms as morphisms,
- equivalence classes of rewrites as 2-morphisms.

For details, see the work of Seely [93, 94], Hilken [54], and Melliès [79]. Someday this work will be part of the larger n-categorical Rosetta Stone mentioned at the end of Sect. 2.2.5.

In any event, Lambek showed that every typed lambda-theory gives a cartesian closed category—and conversely, every cartesian closed category gives a typed lambda-theory. This discovery led to a rich line of research blending category theory and computer science. There is no way we can summarize the resulting enormous body of work, though it constitutes a crucial aspect of the Rosetta Stone. Two good starting points for further reading are the textbook by Crole [35] and the online review article by Scott [89].

In what follows, our goal is more limited. First, in Sect. 2.4.2, we explain how every "typed lambda-theory" gives a cartesian closed category, and conversely. We follow the treatment of Lambek and Scott [73], in a somewhat simplified form. Then, in Sect. 2.4.3, we describe how every "linear type theory" gives a closed symmetric monoidal category, and conversely.

The idea here is roughly that a "linear type theory" is a programming language suitable for *both classical and quantum computation*. This language differs from the typed lambda calculus in that it forbids duplication and deletion of data except when expressly permitted. The reason is that while every object in a cartesian category comes equipped with "duplication" and "deletion" morphisms:

$$\Delta_X : X \to X \otimes X, \qquad !_X : X \to 1,$$

a symmetric monoidal category typically lacks these. As we saw in Sect. 2.2.3, a great example is the category Hilb with its usual tensor product. So, a programming language suitable for quantum computation should not assume we can duplicate all types of data [29, 110].

Various versions of "quantum" or "linear" lambda calculus have already been studied, for example by Benton, Bierman de Paiva and Hyland [22], Dorca and van Tonder [108], and Selinger and Valiron [98]. Abramsky and Tzevelekos sketch a version in their paper in this volume [6]. We instead explain the 'linear type theories' developed by Simon Ambler in his 1991 thesis [7].

2.4.2 The Typed Lambda Calculus

Like the original "untyped" lambda calculus explained above, the typed lambda calculus uses terms to represent both programs and data. However, now every term has a specific type. A program that inputs data of type X and outputs data of type Y is said to be of type $X \multimap Y$. So, we can only apply a term s to a term t of type X if s is of type $X \multimap Y$ for some Y. In this case $s(t)$ is a well-defined term of type Y. We call $X \multimap Y$ a **function type**.

Whenever we introduce a variable, we must declare its type. We write $t : X$ to mean that t is a term of type X. So, in lambda abstraction, we no longer simply write expressions like $(\lambda x \,.\, t)$. Instead, if x is a variable of type X, we write

$$(\lambda x : X \,.\, t).$$

For example, here is a simple program that takes a program of type $X \multimap X$ and "squares" it:

$$(\lambda f : X \multimap X \,.\, (\lambda x : X \,.\, f(f(x)))).$$

In the original lambda calculus, all programs take a single piece of data as input. In other words, they compute unary functions. This is no real limitation, since we can handle functions that take more than one argument using a trick called "currying", discussed in Sect. 2.2.6 This turns a function of several arguments into a function that takes the first argument and returns a function of the remaining arguments. We saw an example in the last section: the program "times". For example, $\mathrm{times}(\overline{3})$ is a program that multiplies by 3, so $\mathrm{times}(\overline{3})(\overline{2}) = \overline{6}$.

While making all programs compute unary functions is economical, it is not very kind to the programmer. So, in the typed lambda calculus we also introduce products: given types X and Y, there is a type $X \times Y$ called a **product type**. We can think of a datum of type $X \times Y$ as a pair consisting of a datum of type X and a datum of type Y. To make this intuition explicit, we insist that given terms $s : X$ and $t : Y$ there is a term $(s, t) : X \times Y$. We also insist that given a term $u : X \times Y$

there are terms $p(u) : X$ and $p'(u) : Y$, which we think of as the first and second components of the pair t. We also include rewrite rules saying:

$$\begin{aligned}(p(u), p'(u)) &= u \text{ for all } u : X \times Y,\\ p(s,t) &= s \text{ for all } s : X \text{ and } t : Y,\\ p'(s,t) &= t \text{ for all } s : X \text{ and } t : Y.\end{aligned}$$

Product types allow us to write programs that take more than one input. Even more importantly, they let us deal with programs that produce more than one output. For example, we might have a type called "integer". Then we might want a program that takes an integer and duplicates it:

$$\text{duplicate} : \text{integer} \multimap (\text{integer} \times \text{integer})$$

Such a program is easy to write:

$$\text{duplicate} = (\lambda x : \text{integer} \,.\, (x, x)).$$

Of course this a program we should *not* be allowed to write when duplicating information is forbidden, but in this section our considerations are all "classical", i.e., suited to cartesian closed categories.

The typed lambda calculus also has a special type called the "unit type", which we denote as 1. There is a single term of this type, which we denote as (). From the viewpoint of category theory, the need for this type is clear: a category with finite products must have not only binary products but also a terminal object (see Definition 10). For example, in the category Set, the terminal object can be taken as any one-element set, and () is the unique element of this set. It may be less clear why this type is useful in programming. One reason is that it lets us think of a constant of type X as a function of type $1 \multimap X$—that is, a "nullary" function, one that takes no arguments. There are some other reasons, but they go beyond the scope of this discussion. Suffice it to say that Haskell, Lisp and even widely used imperative languages such as C, C++ and Java include the unit type.

Having introduced the main ingredients of the typed lambda calculus, let us give a more formal treatment. As we shall see, a "typed lambda-theory" consists of types, terms and rewrite rules. From a typed lambda-theory we can get a cartesian closed category. The types will give objects, the terms will give morphisms, and the rewrite rules will give equations between morphisms.

First, the **types** are given inductively as follows:

- **Basic types:** There is an arbitarily chosen set of types called **basic types**.
- **Product types:** Given types X and Y, there is a type $X \times Y$.
- **Function types:** Given types X and Y, there is a type $X \multimap Y$.
- **Unit type:** There is a type 1.

There may be unexpected equations between types: for example we may have a type X satisfying $X \times X = X$. However, we demand that:

- If $X = X'$ and $Y = Y'$ then $X \times Y = X' \times Y'$.
- If $X = X'$ and $Y = Y'$ then $X \multimap Y = X' \multimap Y'$.

Next we define **terms**. Each term has a specific type, and if t is a term of type X we write $t : X$. The rules for building terms are as follows:

- **Basic terms:** For each type X there is a set of **basic terms** of type X.
- **Variables:** For each type X there is a countably infinite collection of terms of type X called **variables** of type X.
- **Application:** If $f : X \multimap Y$ and $t : X$ then there is a term $f(t)$ of type Y.
- **Lambda abstraction:** If x is a variable of type X and $t : Y$ then there is a term $(\lambda x : X \,.\, t)$ of type $X \multimap Y$.
- **Pairing:** If $s : X$ and $t : Y$ then there is a term (s, t) of type $X \times Y$.
- **Projection:** If $t : X \times X'$ then there is a term $p(t)$ of X and a term $p'(t)$ of type X'.
- **Unit term:** There is a term () of type 1.

Finally there are **rewrite rules** going between terms of the same type. Given any fixed set of variables $\mathcal{S}$, there will be rewrite rules between terms of the same type, all of whose free variables lie in the set $\mathcal{S}$. For our present purposes, we only need these rewrite rules to decide when two terms determine the same morphism in the cartesian closed category we shall build. So, what matters is not really the rewrite rules themselves, but the equivalence relation they generate. We write this equivalence relation as $s \sim_{\mathcal{S}} t$.

The relation $\sim_{\mathcal{S}}$ can be any equivalence relation satisfying the following list of rules. In what follows, $t[s/x]$ denotes the result of taking a term t and replacing every free occurence of the variable x by the term s. Also, when when we say 'term' without further qualification, we mean 'term all of whose free variables lie in the set $\mathcal{S}$'.

- **Type preservation:** If $t \sim_{\mathcal{S}} t'$ then t and t' must be terms of the same type, all of whose free variables lie in the set $\mathcal{S}$.
- **Beta reduction:** Suppose x is a variable of type X, s is a term of type X, and t is any term. If no free occurrence of a variable in s becomes bound in $t[s/x]$, then:

$$(\lambda x : X \,.\, t)(s) \sim_{\mathcal{S}} t[s/x].$$

- **Eta reduction:** Suppose the variable x does not appear in the term f. Then:

$$(\lambda x : X \,.\, f(x)) \sim_{\mathcal{S}} f.$$

- **Alpha conversion:** Suppose x and y are variables of type X, and no free occurrence of any variable in t becomes bound in $t[x/y]$. Then:

$$(\lambda x : X \,.\, t) \sim_{\mathcal{S}} (\lambda y : X \,.\, t[x/y]).$$

- **Application:** Suppose t and t' are terms of type X with $t \sim_{\mathcal{S}} t'$, and suppose that $f : X \multimap Y$. Then:

$$f(t) \sim_{\mathcal{S}} f(t').$$

- **Lambda abstraction:** Suppose t and t' are terms of type Y, all of whose free variables lie in the set $\mathcal{S} \cup \{x\}$. Suppose that $t \sim_{\mathcal{S}\cup\{x\}} t'$. Then:

$$(\lambda x : X . t) \sim_{\mathcal{S}} (\lambda x : X . t')$$

- **Pairing:** If u is a term of type $X \times Y$ then:

$$(p(u), p'(u)) \sim_{\mathcal{S}} u.$$

- **Projection:** if s is a term of type X and t is a term of type Y then:

$$\begin{aligned} p(s,t) &\sim_{\mathcal{S}} s \\ p'(s,t) &\sim_{\mathcal{S}} t. \end{aligned}$$

- **Unit term:** If t is a term of type 1 then:

$$t \sim_{\mathcal{S}} ().$$

Now we can describe Lambek's classic result relating typed lambda-theories to cartesian closed categories. From a typed lambda-theory we get a cartesian closed category C for which:

- The objects of C are the types.
- The morphisms $f : X \to Y$ of C are equivalence classes of pairs (x, t) consisting of a variable $x : X$ and a term $t : Y$ with no free variables except perhaps x. Here (x, t) is equivalent to (x, t') if and only if:

$$t \sim_{\{x\}} t'[x/x'].$$

- Given a morphism $f : X \to Y$ coming from a pair (x, t) and a morphism $g : Y \to Z$ coming from a pair (y, u) as above, the composite $gf : X \to Y$ comes from the pair $(x, u[t/y])$.

We can also reverse this process and get a typed lambda-theory from a cartesian closed category. In fact, Lambek and Scott nicely explain how to construct a category of category of cartesian closed categories and a category of typed-lambda theories. They construct functors going back and forth between these categories and show these functors are inverses up to natural isomorphism. We thus say these categories are "equivalent" [73].

2.4.3 Linear Type Theories

In his thesis [7], Ambler described how to generalize Lambek's classic result from cartesian closed categories to closed symmetric monoidal categories. To do this, he replaced typed lambda-theories with "linear type theories". A linear type theory can be seen as a programming language suitable for both classical and quantum computation. As we have seen, in a noncartesian category like Hilb, we cannot freely duplicate or delete information. So linear type theories must prevent duplication or deletion of data *except when it is expressly allowed.*

To achieve this, linear type theories must not allow us to write a program like this:

$$(\lambda x : X . (x, x)).$$

Even a program that "squares" another program, like this:

$$(\lambda f : X \multimap X . (\lambda x : X . f(f(x)))),$$

is not allowed, since it "reuses" the variable f. On the other hand, a program that composes two programs is allowed!

To impose these restrictions, linear type theories treat variables very differently than the typed lambda calculus. In fact, in a linear type theory, any term will contain a given variable at most *once*. But linear type theories depart even more dramatically from the typed lambda calculus in another way. They make no use of lambda abstraction! Instead, they use "combinators".

The idea of a combinator is very old: in fact, it predates the lambda calculus. Combinatory logic was born in a 1924 paper by Schönfinkel [92], and was rediscovered and extensively developed by Curry [36, 37] starting in 1927. In retrospect, we can see their work as a stripped-down version of the untyped lambda calculus that completely avoids the use of variables. Starting from a basic stock of terms called "combinators", the only way to build new ones is application: we can apply any term f to any term t and get a term $f(t)$.

To build a Turing-complete programming language in such an impoverished setup, we need a sufficient stock of combinators. Remarkably, it suffices to use three. In fact it is possible to use just *one* cleverly chosen combinator—but this tour de force is not particularly enlightening, so we shall describe a commonly used set of three. The first, called I, acts like the identity, since it comes with the rewrite rule:

$$I(a) = a$$

for every term a. The second, called K, gives a constant function $K(a)$ for each term a. In other words, it comes with a rewrite rule saying

$$K(a)(b) = a$$

for every term b. The third, called S, is the tricky one. It takes three terms, applies the first to the third, and applies the result to the second applied to the third:

$$S(a)(b)(c) = a(c)(b(c)).$$

Later it was seen that the combinator calculus can be embedded in the untyped lambda calculus as follows:

$$\begin{aligned} I &= (\lambda x.x) \\ K &= (\lambda x.(\lambda y.x)) \\ S &= (\lambda x.(\lambda y.(\lambda z.x(z)(y(z))))). \end{aligned}$$

The rewrite rules for these combinators then follow from rewrite rules in the lambda calculus. More surprisingly, any function computable using the lambda calculus can also be computed using just I, K and S! While we do not need this fact to understand linear type theories, we cannot resist sketching the proof, since it is a classic example of using combinators to avoid explicit use of lambda abstraction.

Note that all the variables in the lambda calculus formulas for I, K, and S are bound variables. More generally, in the lambda calculus we define a **combinator** to be a term in which all variables are bound variables. Two combinators c and d are **extensionally equivalent** if they give the same result on any input: that is, for any term t, we can apply lambda calculus rewrite rules to $c(t)$ and $d(t)$ in a way that leads to the same term. There is a process called "abstraction elimination" that takes any combinator in the lambda calculus and produces an extensionally equivalent one built from I, K, and S.

Abstraction elimination works by taking a term $t = (\lambda x.u)$ with a single lambda abstraction and rewriting it into the form $(\lambda x.f(x))$, where f has no instances of lambda abstraction. Then we can apply eta reduction, which says $(\lambda x.f(x)) = f$. This lets us rewrite t as a term f that does not involve lambda abstraction. We shall use the notation $[[u]]_x$ to mean "any term f satisfing $f(x) = u$".

There are three cases to consider; each case justifies the definition of one combinator:

1. $t = (\lambda x.x)$. We can rewrite this as $t = (\lambda x.I(x))$, so $t = [[x]]_x = I$.
2. $t = (\lambda x.u)$, where u does not depend on x. We can rewrite this as $t = (\lambda x. K(u)(x))$, so $t = [[u]]_x = K(u)$.
3. $t = (\lambda x.u(v))$, where u and v may depend on x. We can rewrite this as $t = (\lambda x.(([[u]]_x x)([[v]]_x x))$ or $t = (\lambda x.S([[u]]_x)([[v]]_x)(x))$, so $t = S([[u]]_x)([[v]]_x)$.

We can eliminate all use of lambda abstraction from any term by repeatedly using these three rules "from the inside out". To see how this works, consider the lambda term $t = (\lambda x.(\lambda y.y))$, which takes two inputs and returns the second. Using the rules above we have:

$$\begin{aligned}(\lambda x.(\lambda y.y)) &= (\lambda x.(\lambda y.[[y]]_y(y))) \\ &= (\lambda x.(\lambda y.I(y))) \\ &= (\lambda x.I) \\ &= (\lambda x.[[I]]_x(x)) \\ &= (\lambda x.K(I)(x) \\ &= K(I).\end{aligned}$$

We can check that it works as desired: $K(I)(x)(y) = I(y) = y$.

Now let us return to our main theme: linear type theories. Of the three combinators described above, only I is suitable for use in an arbitrary closed symmetric monoidal category. The reason is that K deletes data, while S duplicates it. We can see this directly from the rewrite rules they satisfy:

$$\begin{aligned}K(a)(b) &= a \\ S(a)(b)(c) &= a(c)(b(c)).\end{aligned}$$

Every linear type theory has a set of "basic combinators", which neither duplicate nor delete data. Since linear type theories generalize *typed* lambda-theories, these basic combinators are typed. Ambler writes them using notation resembling the notation for morphisms in category theory.

For example, given two types X and Y in a linear type theory, there is a **tensor product type** $X \otimes Y$. This is analogous to a product type in the typed lambda calculus. In particular, given a term s of type X and a term t of type Y, we can combine them to form a term of type $X \otimes Y$, which we now denote as $(s \otimes t)$. We reparenthesize iterated tensor products using the following basic combinator:

$$\mathrm{assoc}_{X,Y,Z} \colon (X \otimes Y) \otimes Z \to X \otimes (Y \otimes Z).$$

This combinator comes with the following rewrite rule:

$$\mathrm{assoc}_{X,Y,Z}((s \otimes t) \otimes u) = (s \otimes (t \otimes u))$$

for all terms $s : X, t : Y$ and $u : Z$.

Of course, the basic combinator $\mathrm{assoc}_{X,Y,Z}$ is just a mildly disguised version of the associator, familiar from category theory. Indeed, all the basic combinators come from natural or dinatural transformations implicit in the definition of "closed symmetric monoidal category". In addition to these, any given linear type theory also has combinators called "function symbols". These come from the morphisms particular to a given category. For example, suppose in some category the tensor product $X \otimes X$ is actually the cartesian product. Then the corresponding linear type theory should have a function symbol

$$\Delta_X \colon X \to X \otimes X$$

which lets us duplicate data of type X, together with function symbols

$$p: X \otimes X \to X, \qquad p': X \otimes X \to X$$

that project onto the first and second factors. To make sure these work as desired, we can include rewrite rules:

$$\begin{aligned} \Delta(s) &= (s \otimes s) \\ p(s \otimes t) &= s \\ p'(s \otimes t) &= t. \end{aligned}$$

So, while duplication and deletion of data is not a "built-in feature" of linear type theories, we can include it when desired.

Using combinators, we could try to design a programming language suitable for closed symmetric monoidal categories that completely avoid the use of variables. Ambler follows a different path. He retains variables in his formalism, but they play a very different—and much *simpler*—-role than they do in the lambda calculus. Their only role is to help decide which terms should count as equivalent. Furthermore, lambda abstraction plays no role in linear type theories, so the whole issue of free versus bound variables does not arise! In a sense, all variables are free. Moreover, every term contains any given variable at most once.

After these words of warning, we hope the reader is ready for a more formal treatment of linear type theories. A **linear type theory** has types, combinators, terms, and rewrite rules. The types will correspond to objects in a closed symmetric monoidal category, while equivalence classes of combinators will correspond to morphisms. Terms and rewrite rules are only used to define the equivalence relation.

First, the set of **types** is defined inductively as follows:

- **Basic types:** There is an arbitarily chosen set of types called **basic types**.
- **Product types:** Given types X and Y, there is a type $(X \otimes Y)$.
- **Function types:** Given types X and Y, there is a type $(X \multimap Y)$.
- **Trivial type:** There is a type I.

There may be equations between types, but we require that:

- If $X = X'$ and $Y = Y'$ then $X \otimes Y = X' \otimes Y'$.
- If $X = X'$ and $Y = Y'$ then $X \multimap Y = X' \multimap Y'$.

Second, a linear type theory has for each pair of types X and Y a set of **combinators** of the form $f: X \to Y$. These are defined by the following inductive rules:

- Given types X and Y there is an arbitrarily chosen set of combinators $f: X \to Y$ called **function symbols**.
- Given types X, Y, and Z we have the following combinators, called **basic combinators**:
  - $\mathrm{id}_X : X \to X$
  - $\mathrm{assoc}_{X,Y,Z} : (X \otimes Y) \otimes Z \to X \otimes (Y \otimes Z)$
  - $\mathrm{unassoc}_{X,Y,Z} : X \otimes (Y \otimes Z) \to (X \otimes Y) \otimes Z$

- $\text{braid}_{X,Y}\colon X \otimes Y \to Y \otimes X$
- $\text{left}_X\colon I \otimes X \to X$
- $\text{unleft}_X\colon X \to I \otimes X$
- $\text{right}_X\colon I \otimes X \to X$
- $\text{unright}_X\colon X \to I \otimes X$
- $\text{eval}_{X,Y}\colon X \otimes (X \multimap Y) \to Y$

- If $f\colon X \to Y$ and $g\colon Y \to Z$ are combinators, then $(g \circ f)\colon X \to Z$ is a combinator.
- If $f\colon X \to Y$ and $g\colon X' \to Y'$ are combinators, then $(f \otimes g)\colon X \otimes X' \to Y \otimes Y'$ is a combinator.
- If $f\colon X \otimes Y \to Z$ is a combinator, then we can **curry** f to obtain a combinator $\tilde{f}\colon Y \to (X \multimap Z)$.

It will generally cause no confusion if we leave out the subscripts on the basic combinators. For example, we may write simply "assoc" instead of $\text{assoc}_{X,Y,Z}$.

Third, a linear type theory has a set of **terms** of any given type. As usual, we write $t : X$ to say that t is a term of type X. Terms are defined inductively as follows:

- For each type X there is a countably infinite collection of **variables** of type X. If x is a variable of type X then $x : X$.
- There is a term 1 with $1 : I$.
- If $s : X$ and $t : Y$, then there is a term $(s \otimes t)$ with $(s \otimes t) : X \otimes Y$, as long as no variable appears in both s and t.
- If $f\colon X \to Y$ is a combinator and $t : X$ then there is a term $f(t)$ with $f(t) : X$.

Note that any given variable may appear at most once in a term.

Fourth and finally, a linear type theory has **rewrite rules** going between terms of the same type. As in our treatment of the typed lambda calculus, we only care here about the equivalence relation $\sim$ generated by these rewrite rules. This equivalence relation must have all the properties listed below. In what follows, we say a term is **basic** if it contains no combinators. Such a term is just an iterated tensor product of distinct variables, such as

$$(z \otimes ((x \otimes y) \otimes w)).$$

These are the properties that the equivalence relation $\sim$ must have:

- If $t \sim t'$ then t and t' must be terms of the same type, containing the same variables.
- The equivalence relation is **substitutive**:
  - Given terms $s \sim s'$, a variable x of type X, and terms $t \sim t'$ of type X whose variables appear in neither s nor s', then $s[t/x] \sim s'[t'/x]$.
  - Given a basic term t with the same type as a variable x, if none of the variables of t appear in the terms s or s', and $s[t/x] \sim s'[t/x]$, then $s \sim s'$.
- The equivalence relation is **extensional**: if $f : X \multimap Y$, $g : X \multimap Y$ and $\text{eval}(t \otimes f) = \text{eval}(t \otimes g)$ for all basic terms $t : X$, then $f = g$.

- We have:
  - $\mathrm{id}(s) \sim s$
  - $(g \circ f)(s) \sim g(f(s))$
  - $(f \otimes g)(s \otimes t) \sim (f(s) \otimes g(t))$
  - $\mathrm{assoc}((s \otimes t) \otimes u) \sim (s \otimes (t \otimes u))$
  - $\mathrm{unassoc}(s \otimes (t \otimes u)) \sim ((s \otimes t) \otimes u)$
  - $\mathrm{braid}(s \otimes t) \sim (t \otimes s)$
  - $\mathrm{left}(1 \otimes s) \sim s$
  - $\mathrm{unleft}(s) \sim (1 \otimes s)$
  - $\mathrm{right}(1 \otimes s) \sim s$
  - $\mathrm{unright}(s) \sim (1 \otimes s)$
  - $\mathrm{eval}(s \otimes \tilde{f}(t)) \sim f(s \otimes t)$

Note that terms can have variables appearing anywhere within them. For example, if x, y, z are variables of types X, Y and Z, and $f\colon Y \otimes Z \to W$ is a function symbol, then

$$\mathrm{braid}(x \otimes f(y \otimes z))$$

is a term of type $W \otimes X$. However, every term t is equivalent to a term of the form $\mathrm{cp}(t)(\mathrm{vp}(t))$, where $\mathrm{cp}(t)$ is the **combinator part** of t and $\mathrm{vp}(t)$ is a basic term called the **variable part** of t. For example, the above term is equivalent to

$$\mathrm{braid} \circ (\mathrm{id} \otimes (f \circ (\mathrm{id} \otimes \mathrm{id})))(x \otimes (y \otimes z)).$$

The combinator and variable parts can be computed inductively as follows:

- If x is a variable of type X, $\mathrm{cp}(x) = \mathrm{id}\colon X \to X$.
- $\mathrm{cp}(1) = \mathrm{id}\colon I \to I$.
- For any terms s and t, $\mathrm{cp}(s \otimes t) = \mathrm{cp}(s) \otimes \mathrm{cp}(t)$.
- For any term $s : X$ and any combinator $f : X \to Y$, $\mathrm{cp}(f(s)) = f \circ \mathrm{cp}(s)$.
- If x is a variable of type X, $\mathrm{vp}(x) = x$.
- $\mathrm{vp}(1) = 1$.
- For any terms s and t, $\mathrm{vp}(s \otimes t) = \mathrm{vp}(s) \otimes \mathrm{vp}(t)$.
- For any term $s : X$ and any combinator $f\colon X \to Y$, $\mathrm{vp}(f(s)) = \mathrm{vp}(s)$.

Now, suppose that we have a linear type theory. Ambler's first main result is this: there is a symmetric monoidal category where objects are types and morphisms are equivalence classes of combinators. The equivalence relation on combinators is defined as follows: two combinators $f, g\colon X \to Y$ are equivalent if and only if

$$f(t) \sim g(t)$$

for some basic term t of type X. In fact, Ambler shows that $f(t) \sim g(t)$ for *some* basic term $t : X$ if and only if $f(t) \sim g(t)$ for *all* such basic terms.

Ambler's second main result describes how we can build a linear type theory from any closed symmetric monoidal category, say C. Suppose C has composition $\Box$, tensor product $\bullet$, internal hom $\multimap$, and unit object ι. We let the basic types of our linear type theory be the objects of C. We take as equations between types those generated by:

- $\iota = I$
- $A \bullet B = A \otimes B$
- $A \multimap\!\bullet\, B = A \multimap B$

We let the function symbols be all the morphisms of C. We take as our equivalence relation on terms the smallest allowed equivalence relation such that:

- $1_A(x) \sim A$
- $(g \,\Box\, f)(x) \sim g(f(x))$
- $(f \bullet g)(x \otimes y) \sim (f(x) \otimes g(y))$
- $a_{A,B,C}((x \otimes y) \otimes z) \sim (x \otimes (y \otimes z))$
- $b_{A,B}(x \otimes y) \sim (y \otimes x)$
- $l_A(1 \otimes x) \sim x$
- $r_A(x \otimes 1) \sim x$
- $\mathrm{ev}_{A,B}(x \otimes \tilde{f}(y)) \sim f(x \otimes y)$

Then we define

- $\mathrm{id} = 1$
- $\mathrm{assoc} = a$
- $\mathrm{unassoc} = a^{-1}$
- $\mathrm{braid} = b$
- $\mathrm{left} = l$
- $\mathrm{unleft} = l^{-1}$
- $\mathrm{right} = r$
- $\mathrm{unleft} = r^{-1}$
- $\mathrm{eval} = \mathrm{ev}$
- $g \circ f = g \,\Box f$

and we're done!

Ambler also shows that this procedure is the "inverse" of his procedure for turning linear type theories into closed symmetric monoidal categories. More precisely, he describes a category of closed symmetric monoidal categories (which is well-known), and also a category of linear type theories. He constructs functors going back and forth between these, based on the procedures we have sketched, and shows that these functors are inverses up to natural isomorphism. So, these categories are "equivalent".

In this section we have focused on closed symmetric monoidal categories. What about closed categories that are just braided monoidal, or merely monoidal? While we have not checked the details, we suspect that programming languages suited to these kinds of categories can be obtained from Ambler's formalism by removing various features. To get the braided monoidal case, the obvious guess is to remove Ambler's rewrite rule for the 'braid' combinator and add two rewrite rules corresponding to the hexagon equations (see Sect. 2.2.4 for these). To get the monoidal case, the obvious guess is to completely remove the combinator "braid" and all rewrite rules involving it. In fact, Jay [57] gave a language suitable for closed monoidal categories in 1989; Ambler's work is based on this.

2.5 Conclusions

In this paper we sketched how category theory can serve to clarify the analogies between physics, topology, logic and computation. Each field has its own concept of "thing" (object) and "process" (morphism)—and these things and processes are organized into categories that share many common features. To keep our task manageable, we focused on those features that are present in every closed symmetric monoidal category. Table 2.4, an expanded version of the Rosetta Stone, shows some of the analogies we found.

Table 2.4 The Rosetta Stone (larger version)

Category Theory	Physics	Topology	Logic	Computation
Object X	Hilbert space X	Manifold X	Proposition X	Data type X
Morphism $f: X \to Y$	Operator $f: X \to Y$	Cobordism $f: X \to Y$	Proof $f: X \to Y$	Program $f: X \to Y$
Tensor product of objects: $X \otimes Y$	Hilbert space of joint system: $X \otimes Y$	Disjoint union of manifolds: $X \otimes Y$	Conjunction of propositions: $X \otimes Y$	Product of data types: $X \otimes Y$
Tensor product of morphisms: $f \otimes g$	Parallel processes: $f \otimes g$	Disjoint union of cobordisms: $f \otimes g$	Proofs carried out in parallel: $f \otimes g$	Programs executing in parallel: $f \otimes g$
Internal hom: $X \multimap Y$	Hilbert space of "anti-X and Y": $X^* \otimes Y$	Disjoint union of orientation-reversed X and Y: $X^* \otimes Y$	Conditional proposition: $X \multimap Y$	Function type: $X \multimap Y$

However, we only scratched the surface! There is much more to say about categories equipped with extra structure, and how we can use them to strengthen the ties between physics, topology, logic and computation—not to mention what happens when we go from categories to n-categories. But the real fun starts when we exploit these analogies to come up with new ideas and surprising connections. Here is an example.

In the late 1980s, Witten [109] realized that string theory was deeply connected to a 3d topological quantum field theory and thus the theory of knots and tangles [71]. This led to a huge explosion of work, which was ultimately distilled into a beautiful body of results focused on a certain class of compact braided monoidal categories called "modular tensor categories" [17, 107].

All this might seem of purely theoretical interest, were it not for the fact that superconducting thin films in magnetic fields seem to display an effect—the "fractional quantum Hall effect"—that can be nicely modelled with the help of such categories [102, 103]. In a nutshell, the idea is that excitations of these films can act like particles, called 'anyons'. When two anyons trade places, the result depends on how they go about it:

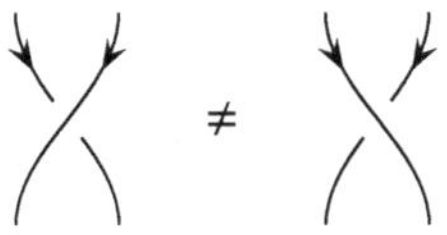

So, collections of anyons are described by objects in a braided monoidal category! The details depend on things like the strength of the magnetic field; the range of possibilities can be worked out with the help of modular tensor categories [82, 88].

So far this is all about physics and topology. Computation entered the game around 2000, when Freedman, Kitaev, Larsen and Wang [43–45] showed that certain systems of anyons could function as "universal quantum computers". This means that, in principle, arbitrary computations can be carried out by moving anyons around. Doing this *in practice* will be far from easy. However, Microsoft has set up a research unit called Project Q attempting to do just this. After all, a working quantum computer could have huge practical consequences.

But regardless of whether topological quantum computation ever becomes practical, the implications are marvelous. A simple diagram like this:

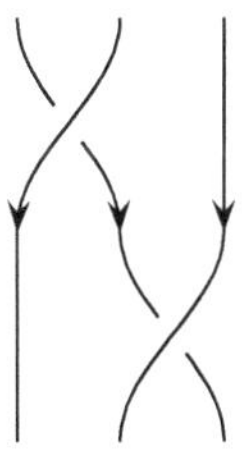

can now be seen as a *quantum process*, a *tangle*, a *computation*—or an abstract morphism in any braided monoidal category! This is just the sort of thing one would hope for in a general science of systems and processes.

Acknowledgments We owe a lot to participants of the seminar at UCR where some of this material was first presented: especially David Ellerman, Larry Harper, Tom Payne—and Derek Wise, who took notes [13]. This paper was also vastly improved by comments by Andrej Bauer, Tim Chevalier, Derek Elkins, Greg Friedman, Matt Hellige, Robin Houston, Theo Johnson–Freyd, Jürgen Koslowski, Todd Trimble, Dave Tweed, and other regulars at the n-Category Café. MS would like to thank Google for letting him devote 20% of his time to this research, and Ken Shirriff for helpful corrections. This work was supported by the National Science Foundation under Grant No. 0653646.

References

1. Abelson, H., Sussman, G.J., Sussman, J.: Structure and Interpretation of Computer Programs, MIT Press, Cambridge (1996). Available at http://mitpress.mit.edu/sicp/
2. Abramsky, S.: Abstract scalars, loops, and free traced and strongly compact closed categories. In: Proceedings of CALCO 2005, Lecture Notes in Computer Science 3629, Springer, Berlin, 2005, 1–31. Also available at http://web.comlab.ox.ac.uk/oucl/work/samson.abramsky/calco05.pdf
3. Abramsky, S., Coecke, B.: A categorical semantics of quantum protocols. Available at arXiv:quant-ph/0402130
4. Abramsky, S., Duncan, R.: A categorical quantum logic, to appear in Mathematical Structures in Computer Science (2006). Also available as arXiv:quant-ph/0512114
5. Abramsky, S., Haghverdi, E., Scott, P.: Geometry of interaction and linear combinatory algebras, Math. Struct. Comput. Sci. **12**, 625–665 (2002). Also available at http://citeseer.ist.psu.edu/491623.html
6. Abramsky, S., Tzevelokos, N.: Introduction to categories and categorical logic, in this volume. Also available at http://web.comlab.ox.ac.uk/people/Bob.Coecke/AbrNikos.pdf
7. Ambler, S.: First order logic in symmetric monoidal closed categories, Ph.D. thesis, U. of Edinburgh, 1991. Available at http://www.lfcs.inf.ed.ac.uk/reports/92/ECS-LFCS-92-194/
8. Atiyah, M.F.: Topological quantum field theories. Publ. Math. IHES Paris **68**, 175–186 (1989).
9. Atiyah, M.F.: The Geometry and Physics of Knots. Cambridge University Press, Cambridge (1990)
10. Baez, J.: An introduction to spin foam models of quantum gravity and *BF* theory. In: Gausterer, H., Grosse, H. (eds.) Geometry and Quantum Physics, pp. 25–93. Springer, Berlin (2000). Also available at arXiv:gr-qc/9905087
11. Baez, J.: Higher-dimensional algebra and Planck-scale physics. In: Callender, C., Huggett, N. (eds.) Physics Meets Philosophy at the Planck Length, pp. 177–195. Cambridge University Press, Cambridge (2001). Also available as arXiv:gr-qc/9902017
12. Baez, J.: Quantum quandaries: A category-theoretic perspective. In: French, S., Rickles, D., Saatsi, J. (eds.) Structural Foundations of Quantum Gravity, pp. 240–265. Oxford University Press, Oxford (2006). Also available as arXiv:quant-ph/0404040
13. Baez, J.: Classical versus quantum computation, U. C. Riverside seminar notes by D. Wise. Available at http://math.ucr.edu/home/baez/qg-fall2006/ and http://math.ucr.edu/home/baez/qg-winter2007
14. Baez, J., Dolan, J.: Higher-dimensional algebra and topological quantum field theory. J. Math. Phys. **36**, 6073–6105 (1995). Also available as arXiv:q-alg/9503002
15. Baez, J., Langford, L.: Higher-dimensional algebra IV: 2-tangles. Adv. Math. **180**, 705–764 (2003). Also available as arXiv:q-alg/9703033
16. Baez, J., Lauda, A.: A prehistory of n-categorical physics. In: Hans Halvorson (ed.) To Appear in Proceedings of Deep Beauty: Mathematical Innovation and the Search for an Underlying Intelligibility of the Quantum World, Princeton, October 3, 2007. Also available at http://math.ucr.edu/home/baez/history.pdf
17. Bakalov, B., Kirillov, A., Jr.: Lectures on Tensor Categories and Modular Functors, American Mathematical Society, Providence, Rhode Island, 2001. Preliminary version available at http://www.math.sunysb.edu/_kirillov/tensor/tensor.html
18. Barendregt, H.: The Lambda Calculus, Its Syntax and Semantics. North-Holland, Amsterdam (1984)
19. Barr, M., Wells, C.: Toposes, Triples and Theories. Springer, Berlin (1983). Revised and corrected version available at http://www.cwru.edu/artsci/math/wells/pub/ttt.html
20. Bell, J.S.: On the Einstein-Podolsky-Rosen paradox, Physics**1**, 195–200 (1964)
21. Bell, J.L.: The Development of Categorical Logic. Available at http://publish.uwo.ca/ _jbell/catlogprime.pdf

22. Benton, N., Bierman, G.M., de Paiva, V., Hyland, J.M.E.: Linear lambda-calculus and categorical models revisited, in Computer Science Logic (CSL'92), Selected Papers, Lecture Notes in Computer Science **702**, pp. 61–84. Springer, Berlin (1992). Also available at http://citeseer.ist.psu.edu/benton92linear.html
23. Benton, N., Bierman, G., de Paiva, V., Hyland, M.: Term Assignment for Intuitionistic Linear Logic, Technical Report 262, University of Cambridge Computer Laboratory, August 1992. Also available at http://citeseer.ist.psu.edu/1273.html
24. Bierman, G.: On Intuitionistic Linear Logic, PhD Thesis, Cambridge University. Available at http://research.microsoft.com/_gmb/Papers/thesis.pdf
25. Blute, R., Scott, P.: Category theory for linear logicians. In: Ehrhard, T., Girard, J.-Y., Ruet, P., Scott, P. (eds.) Linear Logic in Computer Science, pp. 3–64. Cambridge University Press, Cambridge (2004). Also available at http://www.site.uottawa.ca/_phil/ papers/catsurv.web.pdf
26. Burris, S.N., Sankappanavar, H.P.: A Course in Universal Algebra. Springer, Berlin (1981). Also available at http://www.math.uwaterloo.ca/_snburris/htdocs/ualg.html
27. Chari, V., Pressley, A.: A Guide to Quantum Groups. Cambridge University Press, Cambridge (1995)
28. Cheng, E., Lauda, A.: Higher-Dimensional Categories: An Illustrated Guidebook. Available at http://www.dpmms.cam.ac.uk/_elgc2/guidebook/
29. Nielsen, M.A., Chuang, I.L.: Quantum Computation and Quantum Information. Cambridge University Press, Cambridge (2000)
30. Church, A.: An unsolvable problem of elementary number theory. Am. J. Math. **58**, 345–363 (1936)
31. Coecke, B.: De-linearizing linearity: projective quantum axiomatics from strong compact closure. In: Proceedings of the 3rd International Workshop on Quantum Programming Languages (QPL 2005), pp. 49–72. Elsevier (2007). Also available as arXiv:quantph/0506134
32. Coecke, B.: Kindergarten quantum mechanics. To appear in Proceedings of QTRF-III. Also available as arXiv:quant-ph/0510032
33. Coecke, B., Paquette, E.O.: POVMs and Naimark's theorem without sums. To appear in Proceedings of the 4th International Workshop on Quantum Programming Languages (QPL 2006). Also available as arXiv:quant-ph/0608072
34. Coecke, B., Pavlovic, D.: Quantum measurements without sums. In: Chen, G., Kauffman, L., Lomonaco, S. (eds.) To appear in the Mathematics of Quantum Computation and Technology. Taylor and Francis. Also available as arXiv:quant-ph/0608035
35. Crole, R.L.: Categories for Types. Cambridge University Press, Cambridge (1993)
36. Curry, H.B., Feys, R.: Combinatory Logic, vol. I. North-Holland, Amsterdam (1958)
37. Curry, H.B., Findley, J.R., Selding, J.P.: Combinatory Logic, vol. II. North-Holland, Amsterdam (1972)
38. Cvitanovic, P.: Group Theory. Princeton University Press, Princeton (2003). Available at http://www.nbi.dk/GroupTheory/
39. Di Cosmo, R., Miller, D.: Linear logic, Stanford Encyclopedia of Philosophy. Available at http://plato.stanford.edu/entries/logic-linear/
40. Dorca, M., van Tonder, A.: Quantum computation, categorical semantics and linear logic. Available as arXiv:quant-ph/0312174
41. Eilenberg, S., Kelly, G.M.: Closed categories. In: Proceedings of the Conference on Categorical Algebra (La Jolla, 1965), pp. 421–562. Springer, Berlin (1966)
42. Eilenberg, S., Mac Lane, S.: General theory of natural equivalences. Trans. Am. Math. Soc. **58**, 231–294 (1945)
43. Freedman, M., Kitaev, A., Larsen, M., Wang, Z.: Topological quantum computation. Available as arXiv:quant-ph/0101025
44. Freedman, M., Kitaev, A., Wang, Z.: Simulation of topological field theories by quantum computers. Commun. Math. Phys. **227**, 587–603 (2002). Also available as arXiv:quantph/0001071

45. Freedman, M., Kitaev, A., Wang, Z.: A modular functor which is universal for quantum computation. Commun. Math. Phys. **227**, 605–622 (2002). Also available as arXiv:quantph/0001108
46. Freyd, P., Yetter, D.: Braided compact monoidal categories with applications to low dimensional topology. Adv. Math. 77, 156–182 (1989)
47. Girard, J.-Y.: Linear logic. Theor. Comp. Sci. **50**, 1–102 (1987). Also available at http://iml.univ-mrs.fr/ girard/linear.pdf
48. Girard, J.-Y., Lafont, Y., Taylor, P.: Proofs and Types. Cambridge University Press, Cambridge (1990). Also available at http://www.monad.me.uk/stable/Proofs%2BTypes.html
49. Gödel, K.: Zur intuitionistischen Arithmetik und Zahlentheorie. Ergebnisse eines mathematischen Kolloquiums**4**, 34–38 (1933)
50. Goldblatt, R.: Topoi: The Categorial Analysis of Logic. North-Holland, New York (1984). Also available at http://cdl.library.cornell.edu/cgi-bin/cul.math/docviewer?did=Gold010
51. Gottesman, D., Chuang, I.L.: Quantum teleportation is a universal computational primitive. Nature **402**, 390–393 (1999). Also available as arXiv:quant-ph/9908010
52. Hasegawa, M.: Logical predicates for intuitionistic linear type theories. In: Girard, J.-Y. (ed.) Typed Lambda Calculi and Applications: 4th International Conference, TLCA '99. Lecture Notes in Computer Science **1581**. Springer, Berlin (1999). Also available at http://citeseer.ist.psu.edu/187161.html
53. Heyting, A., Brouwer, L.E.J. (ed.): Collected Works 1: Philosophy and Foundations of Mathematics. Elsevier, Amsterdam (1975)
54. Hilken, B.: Towards a proof theory of rewriting: The simply-typed 2*-calculus. Theor. Comput. Sci. **170**, 407–444 (1996)
55. Hindley, J.R., Seldin, J.P.: Lambda-Calculus and Combinators: An Introduction. Cambridge University Press, Cambridge (2008)
56. Howard, W.A.: The formulae-as-types notion of constructions. In: Seldin, J.P., Hindley, J.R. (eds.) To Curry, H.B.: Essays on Combinatory Logic, Lambda Calculus and Formalism, pp. 479–490. Academic Press, New York (1980)
57. Jay, C.B.: Languages for monoidal categories. J. Pure Appl. Alg. **59**, 61–85 (1989)
58. Jay, C.B.: The structure of free closed categories. J. Pure Appl. Alg. **66**, 271–285 (1990)
59. Joyal, A., Street, R.: The geometry of tensor calculus I. Adv. Math. **88**, 55–113 (1991)
60. Joyal, A., Street, R.: The geometry of tensor calculus II. Available at http://www.math.mq.edu.au/ street/GTCII.pdf
61. Joyal, A., Street, R.: Braided monoidal categories, Macquarie Math Reports 860081 (1986). Available at http://rutherglen.ics.mq.edu.au/_street/JS86.pdf
62. Joyal, A., Street, R.: Braided tensor categories. Adv. Math.**102**, 20–78 (1993)
63. Kaiser, D.: Drawing Theories Apart: The Dispersion of Feynman Diagrams in Postwar Physics. University of Chicago Press, Chicago (2005)
64. Kassel, C.: Quantum Groups. Springer, Berlin (1995)
65. Kauffman, L.H.: Knots and Physics. World Scientific, Singapore (1991)
66. Kauffman, L.H., Lins, S.: Temperley–Lieb Recoupling Theory and Invariants of 3-Manifolds. Princeton University Press, Princeton (1994)
67. Kelly, G.M., Mac Lane, S.: Coherence in closed categories. J. Pure Appl. Alg. **1**, 97–140, 219 (1971)
68. Kelly, G.M., Laplaza, M.L.: Coherence for compact closed categories. J. Pure Appl. Alg.**19**, 193–213 (1980)
69. Kleene, S.: *-definability and recursiveness. Duke Math. J. **2**, 340–353 (1936)
70. Kock, J.: Frobenius Algebras and 2D Topological Quantum Field Theories, London Mathematical Society Student Texts **59**. Cambridge University Press, Cambridge (2004)
71. Kohno, T. (ed.): New Developments in the Theory of Knots. World Scientific, Singapore (1990)
72. Lambek, J.: From *-calculus to cartesian closed categories. In: Seldin, J.P., Hindley, J.R. (eds.) To Curry, H.B.: Essays on Combinatory Logic, Lambda Calculus and Formalism, pp. 375–402. Academic Press, New York (1980)

73. Lambek, J., Scott, P.J.: Introduction to Higher-order Categorical Logic. Cambridge University Press, Cambridge (1986)
74. Landin, P.: A correspondence between ALGOL 60 and Church's lambda-notation. Commun. ACM **8**, 89–101, 158–165 (1965)
75. Lawvere, F.W.: Functorial Semantics of Algebraic Theories, Ph.D. Dissertation, Columbia University (1963). Also available at http://www.tac.mta.ca/tac/reprints/articles/5/tr5abs.html
76. Leinster, T.: A survey of definitions of n-category. Theor. Appl. Cat. **10**, 1–70 (2002). Also available as arXiv:math/0107188
77. Mac Lane, S.: Natural associativity and commutativity. Rice Univ. Stud. **49** 28–46 (1963)
78. Mac Lane, S.: Categories for the Working Mathematician. Springer, Berlin (1998)
79. Melliès, P.: Axiomatic rewriting theory I: A diagrammatic standardisation theorem. In: Processes, Terms and Cycles: Steps on the Road to Infinity, Lecture Notes in Computer Science **3838**, pp. 554–638. Springer, New York (2005). Also available at http://www.pps.jussieu.fr/ mellies/papers/jwkfestschrift.pdf
80. McCarthy, J.: Recursive functions of symbolic expressions and their computation by machine, Part I. Comm. ACM **4**, 184–195 (1960). Also available at http://wwwformal.stanford.edu/jmc/recursive.html
81. McLarty, C.: Elementary Categories, Elementary Toposes. Clarendon Press, Oxford (1995)
82. Moore, G., Read, N.: Nonabelions in the fractional quantum Hall effect. Nucl. Phys. B **360**, 362–396 (1991)
83. Moschovakis, J.: Intuitionistic logic, Stanford Encyclopedia of Philosophy. Available at http://plato.stanford.edu/entries/logic-intuitionistic/
84. Penrose, R.: Applications of negative dimensional tensors. In: Welsh, D. (ed.) Combinatorial Mathematics and Its Applications, pp. 221–244. Academic Press, New York (1971)
85. Penrose, R.: Angular momentum: An approach to combinatorial space-time. In: Bastin, T. (ed.) Quantum Theory and Beyond, pp. 151–180. Cambridge University Press, Cambridge (1971)
86. Penrose, R.: On the nature of quantum geometry. In: Klauder J. (ed.) Magic Without Magic, pp. 333–354. Freeman, San Francisco (1972)
87. Penrose, R.: Combinatorial quantum theory and quantized directions. In: Hughston, L., Ward, R. (eds.) Advances in Twistor Theory, Pitman Advanced Publishing Program, pp. 301–317 (1979)
88. Rowell, E., Stong, R., Wang, Z.: On classification of modular tensor categories, available as arXiv:0712.1377
89. Scott, P.: Some aspects of categories in computer science. In: Hazewinkel, M. (ed.) Handbook of Algebra, vol. 2, Elsevier, Amsterdam (2000). Also available at http://www.site.uottawa.ca/phil/papers/handbook.ps
90. Sawin, S.: Links, quantum groups and TQFTs. Bull. Am. Math. Soc.**33**, 413–445 (1996). Also available as arXiv:q-alg/9506002
91. Schalk, A.: What is a categorical model for linear logic? Available at http://www.cs.man.ac.uk/_schalk/notes/llmodel.pdf
92. Schönfinkel, M.: Über die Bausteine der mathematischen Logik. Math. Ann. **92**, 305–316 (1924). Also available as On the building blocks of mathematical logic. Trans. Bauer-Mengelberg, S.: In: van Heijenoort, J. (ed.) A Source Book in Mathematical Logic, 1879–1931, pp. 355–366. Harvard University Press, Cambridge, Massachusetts (1967)
93. Seely, R.A.G.: Weak adjointness in proof theory, Applications of Sheaves, Lecture Notes in Mathematics **753**, pp. 697–701. Springer, Berlin. Also available at http://www.math.mcgill.ca/rags/WkAdj/adj.pdf
94. Seely, R.A.G.: Modeling computations: A 2-categorical framework. In: Proceedings of the Symposium Logic Computer Science 1987, Computer Society of the IEEE, pp. 65–71. Also available at http://www.math.mcgill.ca/rags/WkAdj/LICS.pdf
95. Segal, G.: The definition of a conformal field theory. In: Tillmann, U.L. (ed.) Topology, Geometry and Quantum Field Theory: Proceedings of the 2002 Oxford Symposium in Honour of the 60th Birthday of Graeme Segal. Cambridge University Press, Cambridge (2004)

96. Selinger, P.: Lecture notes on the lambda calculus. Available at http://www.mscs.dal.ca/_selinger/papers/#lambdanotes
97. Selinger, P.: Dagger compact closed categories and completely positive maps. In: Proceedings of the 3rd International Workshop on Quantum Programming Languages (QPL 2005), pp. 139–163. Elsevier, Amsterdam (2007). Also available at http://www.mscs.dal.ca/_selinger/papers/#dagger
98. Selinger, P., Valiron, B.: A lambda calculus for quantum computation with classical control. Math. Struct. Comp. Sci. **13**, 527–552 (2006). Also available at http://www.mathstat.dal.ca/_selinger/papers/#qlambda
99. Shum, M.-C.: Tortile tensor categories. J. Pure Appl. Alg. **93**, 57–110 (1994)
100. Soloviev, S.: Proof of a conjecture of S. Mac Lane. Ann. Pure Appl. Logic **90**, 101–162 (1997)
101. Smolin, L.: The future of spin networks. In: Hugget, S., Tod, P., Mason, L.J. (eds.) The Geometric Universe: Science, Geometry, and the Work of Roger Penrose. Oxford University Press, Oxford (1998). Also available as arXiv:gr-qc/9702030
102. Stern, A.: Anyons and the quantum Hall effect – A pedagogical review. Ann. Phys. **323**, 204–249 (2008). Available as arXiv:0711.4697
103. Stone, M. (ed.): Quantum Hall Effect. World Scientific, Singapore (1992)
104. Szabo, M.E. (ed.): Collected Papers of Gerhard Gentzen. North–Holland, Amsterdam (1969)
105. Trimble, T.: Linear Logic, Bimodules, and Full Coherence for Autonomous Categories, Ph.D. thesis, Rutgers University (1994)
106. Troelstra, A.S.: Lectures on Linear Logic, Center for the Study of Language and Information. Stanford, California (1992)
107. Turaev, V.G.: Quantum Invariants of Knots and 3-Manifolds. de Gruyter, Berlin (1994)
108. van Tonder, A.: A lambda calculus for quantum computation. SIAM J. Comput. **33**, 1109–1135 (2004). Also available as arXiv:quant-ph/0307150
109. Witten, E.: Quantum field theory and the Jones polynomial. Commun. Math. Phys. **121**, 351–399 (1989)
110. Wootters, W.K., Zurek, W.H.: A single quantum cannot be cloned. Nature **299**, 802–803 (1982)
111. Yetter, D.N.: Functorial Knot Theory: Categories of Tangles, Coherence, Categorical Deformations, and Topological Invariants. World Scientific, Singapore (2001)

Chapter 3
Categories for the Practising Physicist

B. Coecke and É.O. Paquette

Abstract In this chapter we survey some particular topics in category theory in a somewhat unconventional manner. Our main focus will be on monoidal categories, mostly symmetric ones, for which we propose a physical interpretation. Special attention is given to the category which has finite dimensional Hilbert spaces as objects, linear maps as morphisms, and the tensor product as its monoidal structure (**FdHilb**). We also provide a detailed discussion of the category which has sets as objects, relations as morphisms, and the cartesian product as its monoidal structure (**Rel**), and thirdly, categories with manifolds as objects and cobordisms between these as morphisms (**2Cob**). While sets, Hilbert spaces and manifolds do not share any non-trivial common structure, these three categories are in fact structurally very similar. Shared features are diagrammatic calculus, compact closed structure and particular kinds of internal comonoids which play an important role in each of them. The categories **FdHilb** and **Rel** moreover admit a categorical matrix calculus. Together these features guide us towards topological quantum field theories. We also discuss posetal categories, how group representations are in fact categorical constructs, and what strictification and coherence of monoidal categories is all about. In our attempt to complement the existing literature we omitted some very basic topics. For these we refer the reader to other available sources.

3.1 Prologue: Cooking with Vegetables

Consider a "raw potato". Conveniently, we refer to it as A. Raw potato A admits several *states* e.g. "dirty", "clean", "skinned", ... Since raw potatoes don't digest well we need to *process* A into "cooked potato" B. We refer to A and B as *kinds* or *types* of food. Also B admits several states e.g. "boiled", "fried", "baked with skin",

B. Coecke (✉)
OUCL, University of Oxford, Oxford, UK
e-mail: coecke@comlab.ox.ac.uk

É.O. Paquette (✉)
School of Computer Science, McGill University, Montreal, QC, Canada
e-mail: eopaquette@holon23.net

Coecke, B., Paquette, É.O.: *Categories for the Practising Physicist*. Lect. Notes Phys. **813**, 173–286 (2011)
DOI 10.1007/978-3-642-12821-9_3

"baked without skin",... Correspondingly, there are several ways to turn raw potato A into cooked potato B e.g. "boiling", "frying", "baking", to which we respectively refer as f, f' and f''. We make the fact that each of these cooking *processes* applies to raw potato A and produces cooked potato B explicit via *labelled arrows*:

$$A \xrightarrow{f} B \qquad A \xrightarrow{f'} B \qquad A \xrightarrow{f''} B\,.$$

Sequential composition. A plain cooked potato tastes dull so we'd like to process it into "spiced cooked potato" C. We refer to the composite process that consists of first "boiling" $A \xrightarrow{f} B$ *and then* "salting" $B \xrightarrow{g} C$ as

$$A \xrightarrow{g\circ f} C\,.$$

We refer to the trivial process of "doing nothing to vegetable X" as

$$X \xrightarrow{1_X} X\,.$$

Clearly we have $1_Y \circ \xi = \xi \circ 1_X = \xi$ for all processes $X \xrightarrow{\xi} Y$. Note that there is a slight subtlety here: we need to specify what we mean by equality of cooking processes. We will conceive two cooking processes $X \xrightarrow{\xi} Y$ and $X \xrightarrow{\zeta} Y$ as equal, and write $\xi = \zeta$, if the resulting effect on each of the states which X admits is the same. A stronger notion of equality arises when we also want some additional details of the processes to coincide e.g. the brand of the cooking pan that we use.

Let D be a "raw carrot". Note that it is indeed very important to explicitly distinguish our potato and our carrot and any other vegetable such as "lettuce" L in terms of their respective names A, D and L, since each admits distinct ways of processing. And also a cooked potato admits different ways of processing than a raw one, for example, while we can mash cooked potatoes, we can't mash raw ones. We denote all processes which turn raw potato A into cooked potato B by $\mathbf{C}(A, B)$. Consequently, we can repackage composition of cooking processes as a function

$$- \circ - : \mathbf{C}(X, Y) \times \mathbf{C}(Y, Z) \to \mathbf{C}(X, Z)\,.$$

Parallel composition. We want to turn "raw potato" A and "raw carrot" D into "carrot-potato mash" M. We refer to the fact that this requires (or consumes) both A *and* D as $A \otimes D$. Refer to 'frying the carrot' as $D \xrightarrow{h} E$. Then, by

$$A \otimes D \xrightarrow{f\otimes h} B \otimes E$$

we mean "boiling potato A" *while* "frying carrot D" and by

$$C \otimes F \xrightarrow{x} M$$

we mean "mashing spiced cooked potato C and spiced cooked carrot F".

Laws. The whole process from raw components A and D to "meal" M is

$$A \otimes D \xrightarrow{f \otimes h} B \otimes E \xrightarrow{g \otimes k} C \otimes F \xrightarrow{x} M \;=\; A \otimes D \xrightarrow{x \circ (g \otimes k) \circ (f \otimes h)} M\,,$$

where "peppering the carrot" is referred to as $E \xrightarrow{k} F$. We refer to the list of the operations that we apply, i.e. (f while h, g while k, x), as a *recipe*. Distinct recipes can yield the same meal. The reason for this is that the two operations "and then" (i.e. $- \circ -$) and "while" (i.e. $- \otimes -$) which we have at our disposal are not totally independent but *interact* in a certain way. This is exemplified by the equality

$$(1_B \otimes h) \circ (f \otimes 1_D) = (f \otimes 1_E) \circ (1_A \otimes h) \tag{3.1}$$

on cooking processes, which states that it makes no difference whether "we first boil the potato and then fry the carrot", or, "first fry the carrot and then boil the potato".

Equation (3.1) is in fact a generally valid equational law for cooking processes, which does not depend on specific properties of A, B, D, E, f nor h.

Of course, chefs do not perform computations involving Eq. (3.1), since their "intuition" accounts for its content. But, if we were to teach an android how to become a chef, which would require it/him/her to reason about recipes, then we would need to teach it/him/her the laws governing these recipes.

In fact, there is a more general law governing cooking processes from which Eq. (3.1) can be derived, namely,

$$(g \circ f) \otimes (k \circ h) = (g \otimes k) \circ (f \otimes h)\,. \tag{3.2}$$

That is, "boiling the potato and then salting it, while, frying the carrot and then peppering it", is equal to "boiling the potato while frying the carrot, and then, salting the potato while peppering the carrot".[1] A proof of the fact that Eq. (3.1) can be derived from Eq. (3.2) is in Proposition 2 below.

Logic. Equation (3.2) is indeed a *logical* statement. In particular, note the remarkable similarity, but at the same time also the essential difference, of Eq. (3.2) with the well-known *distributive law of classical logic*, which states that

$$A \text{ and } (B \text{ or } C) = (A \text{ and } B) \text{ or } (A \text{ and } C)\,. \tag{3.3}$$

For simple situations, if one possesses enough brainpower, "intuition" again accounts for this distributive law. Ot the other hand, it needs to be explicitly taught to androids, since this distributive law is key to the resolution method which is the standard implementation of artificial reasoning in AI and robotics [58]. Also for complicated sentences we ourselves will need to rely on this method too.

---

[1] In the light of the previous footnote, note here that this law applies to any reasonable notion of equality for processes.

The $(\circ, \otimes)$-logic is a *logic of interaction*. It applies to cooking processes, physical processes, biological processes, logical processes (i.e. proofs), or computer processes (i.e. programs). The theory of *monoidal categories*, the subject of this chapter, is the mathematical framework that accounts for the common structure of each of these *theories of processes*. The framework of monoidal categories moreover enables *modeling* and *axiomatising* (or "classify") the extra structure which certain families of processes may have. For example, how cooking processes differ from physical processes, and how quantum processes differ from classical processes.

Pictures. We mentioned that our intuition accounts for $(\circ, \otimes)$-logic. Wouldn't it be nice if there would be mathematical structures which also "automatically" (or "implicitly") account for the logical mechanisms which we intuitively perform? Well, these mathematical structures do exist. While they are only a fairly recent development, they are becoming more and more prominent in mathematics, including in important "Fields Medal awarding areas" such as algebraic topology and representation theory—see for example [53, and references therein]. Rather than being symbolic, these mathematical structures are purely graphical. Indeed, by far ***the*** coolest thing about monoidal categories is that they admit a purely pictorial calculus, and these pictures automatically account for the logical mechanisms which we intuitively perform. As pictures, both sides of Eq. (3.2) become:

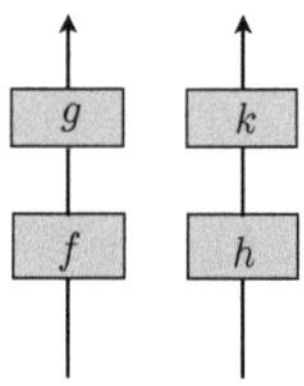

Hence Eq. (3.2) becomes an implicit salient feature of the graphical calculus and needs no explicit attention anymore. This, as we will see below, substantially simplifies many computations. To better understand in which manner these pictures simplify computations note that the differences between the two sides of Eq. (3.2) can be recovered by introducing "artificial" brackets within the two pictures:

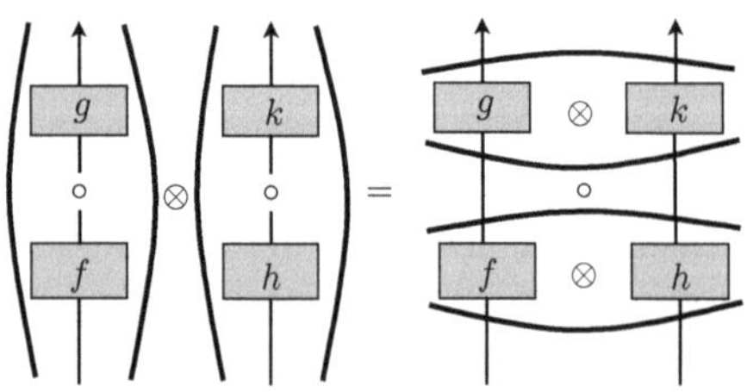

A detailed account on this graphical calculus is in Sect. 3.3.2.

In the remainder of this chapter we provide a formal tutorial on several kinds of monoidal categories that are relevant to physics. If you'd rather stick to the informal story of this prologue you might want to first take a bite of [20, 21].[2] Section 3.2 introduces categories and Sect. 3.3 introduces tensor structure. Section 3.4 studies quantum-like tensors and Sect. 3.5 studies classical-like tensors. Sect. 3.6 introduces mappings between monoidal categories (= monoidal functors), and natural transformations between these, which enable to concisely define topological quantum field theories. Section 3.7 suggests further reading.

3.2 The 1D Case: New Arrows for Your Quiver

The bulk of the previous section discussed the two manners in which we can compose processes, namely *sequentially* and *in parallel*, or more physically put, in time and in space. These are indeed the situations we truly care about in this chapter. Historically however, category theoreticians cared mostly about *one-dimensional fragments* of the *two-dimensional* monoidal categories. These one-dimensional fragments are (ordinary) *categories*, hence the name *category theory*. Some people will get rebuked by the terminology and particular syntactic language used in category theory—which can sound and look like unintelligible jargon—resulting in its unfortunate label of *generalised abstract nonsense*. The reader should realise that initially category theory was crafted as "a theory of mathematical structures". Hence substantial effort was made to avoid any reference to the underlying *concrete models*, resulting in its seemingly idiosyncratic format. The personalities involved in crafting category theory, however brilliant minds they had, also did not always help the cause of making category theory accessible to a broader community.

But this "theory of mathematical structures" view is not the only way to conceive category theory. As we argued above, and as is witnessed by its important use in computer science, in proof theory, and more recently also in quantum informatics and in quantum foundations, category theory is a theory which brings the notions of (*type* of) *system* and *process* to the forefront, two notions which are hard to cast within traditional monolithic mathematical structures.

We profoundly believe that the fact that the mainstream physics community has not yet acquired this (type of) systems/process structure as a primal part of its theories is merely accidental, and temporary, ... and will soon change.

3.2.1 Categories

We will use the following syntax to denote a function:

$$f : X \to Y :: x \mapsto y$$

[2] Paper [20] provided a conceptual template for setting up the content of this paper. However, here we go in more detail and provide more examples.

where X is the set of arguments, Y the set of possible values, and

$$x \mapsto y$$

means that argument x is mapped on value y.

Definition 1 A *category* $\mathbf{C}$ consists of

1. A family[3] $|\mathbf{C}|$ of *objects*;
2. For any $A, B \in |\mathbf{C}|$, a set $\mathbf{C}(A, B)$ of *morphisms*, the *hom-set*;
3. For any $A, B, C \in |\mathbf{C}|$, and any $f \in \mathbf{C}(A, B)$ and $g \in \mathbf{C}(B, C)$, a *composite* $g \circ f \in \mathbf{C}(A, C)$, i.e., for all $A, B, C \in |\mathbf{C}|$ there is a *composition operation*

$$- \circ - : \mathbf{C}(A, B) \times \mathbf{C}(B, C) \to \mathbf{C}(A, C) :: (f, g) \mapsto g \circ f\,,$$

   and this composition operation is *associative* and has *units*, that is,

   i. for any $f \in \mathbf{C}(A, B)$, $g \in \mathbf{C}(B, C)$ and $h \in \mathbf{C}(C, D)$ we have

$$h \circ (g \circ f) = (h \circ g) \circ f\,;$$

   ii. for any $A \in |\mathbf{C}|$, there exists a morphism $1_A \in \mathbf{C}(A, A)$, called the *identity*, which is such that for any $f \in \mathbf{C}(A, B)$ we have

$$f = f \circ 1_A = 1_B \circ f\,.$$

A shorthand for $f \in \mathbf{C}(A, B)$ is $A \xrightarrow{f} B$. As already mentioned above, this definition was proposed by Samuel Eilenberg and Saunders Mac Lane in 1945 as part of a framework which intended to unify a variety of mathematical constructions within different areas of mathematics [33]. Consequently, most of the examples of categories that one encounters in the literature encode mathematical structures: the objects will be examples of this mathematical structure and the morphisms will be the *structure-preserving* maps between these. This kind of categories is usually referred to as *concrete categories* [5]. We will also call them *concrete categorical models*.

3.2.2 Concrete Categories

Traditionally, mathematical structures are defined as a set equipped with some operations and some axioms, for instance:

---

[3] Typically, "family" will mean a class rather than a set. While for many constructions the *size* of $|\mathbf{C}|$ is important, it will not play a key role in this paper.

- A *group* is a set G with an associative binary operation $- \bullet - : G \times G \to G$ and with a two-sided identity $1 \in G$, relative to which each element is invertible, that is, for all $g \in G$ there exists $g^{-1} \in G$ such that $g \bullet g^{-1} = g^{-1} \bullet g = 1$.

Similarly we define rings and fields. Slightly more involved but in the same spirit:

- A *vector space* is a pair $(V, \mathbb{K})$, respectively a commutative group and a field, and these interact via the notion of scalar multiplication, i.e. a map $V \times \mathbb{K} \to V$ which is subject to a number of axioms.

It is to these operations and axioms that one usually refers to as *structure*. Functions on the underlying sets which preserve (at least part of) this structure are called *structure preserving maps*. Here are some examples of structure preserving maps:

- *group homomorphisms*, i.e. functions which preserve the group multiplication, from which it then also follows that the unit and inverses are preserved;
- *linear maps*, i.e. functions from a vector space to a vector space which preserve linear combinations of vectors.

Example 1 Let **Set** be the concrete category with:

1. all sets as objects,
2. all functions between sets as morphisms, that is, more precisely, if X and Y are sets and $f : X \to Y$ is a function between these sets, then $f \in \mathbf{Set}(X, Y)$,
3. ordinary composition of functions, that is, for $f : X \to Y$ and $g : Y \to Z$ we have $(g \circ f)(x) := g(f(x))$ for the composite $g \circ f : X \to Z$, and,
4. the obvious identities i.e. $1_X(x) := x$.

Set is indeed a category since:

- function composition is associative, and,
- for any function $f : X \to Y$ we have $(1_Y \circ f)(x) = f(x) = (f \circ 1_X)(x)$.

Example 2 $\mathbf{FdVect}_{\mathbb{K}}$ is the concrete category with:

1. finite dimensional vectors spaces over $\mathbb{K}$ as objects,
2. all linear maps between these vectors spaces as morphisms, and
3. ordinary composition of the underlying functions, and,
4. identity functions.

$\mathbf{FdVect}_{\mathbb{K}}$ is indeed category since:

- the composite of two linear maps is again a linear map, and,
- identity functions are linear maps.

Example 3 **Grp** is the concrete category with:

1. groups as objects,
2. group homomorphisms between these groups as morphisms, and,
3. ordinary function composition, and,
4. identity functions.

Grp is indeed category since:

- the composite of two group homomorphisms is a group homomorphism, and,
- identity functions are group homomorphisms.

Example 4 (elements) Above we explained that mathematical structures such as groups typically consist of a set with additional structure. In the case of a category we have a collection of objects, and for each pair of objects a set of morphisms. The "structure of a category" then consists of the composition operation on morphisms and the identities on objects. So there is no reference to what the individual objects actually are (e.g. a set, a vector space, or a group). Consequently, one would expect that when passing from a mathematical structure (cf. group) to the corresponding concrete category with these mathematical structures as objects (cf. **Grp**), one looses the object's "own" structure. But fortunately, this happens not to be the case. The fact that we consider structure preserving maps as morphisms will allow us to recover the mathematical structures that we started from. In particular, by only relying on categorical concepts we are still able to identify the "elements" of the objects.

For the set $X \in |\mathbf{Set}|$ and some chosen element $x \in X$ the function

$$e_x : \{*\} \to X :: * \mapsto x \,,$$

where $\{*\}$ is any one-element set, maps the unique element of $\{*\}$ onto the chosen element x. If X contains n elements, then there are n such functions each corresponding to the element on which $*$ is mapped. Hence the elements of the set X are now encoded as the set $\mathbf{Set}(\{*\}, X)$.

In a similar manner we can single out vectors in vectors spaces. For the vector space $V \in |\mathbf{FdVect}_{\mathbb{K}}|$ and some fixed vector $v \in V$ the linear map

$$e_v : \mathbb{K} \to V :: 1 \mapsto v \,,$$

where $\mathbb{K}$ is now the one-dimensional vector space over itself, maps the element $1 \in \mathbb{K}$ onto the chosen element v. Since e_v is linear, it is completely characterised by the image of the single element 1. Indeed, $e_v(\alpha) = e_v(\alpha \cdot 1) = \alpha \cdot e_v(1) = \alpha \cdot v$, that is, the element 1 is a basis for the one-dimensional vector space $\mathbb{K}$.

Example 5 **Pos** is the concrete category with:

1. partially ordered sets, that is, a set together with a reflexive, anti-symmetric and transitive relation, as objects,

2. order preserving maps, i.e. $x \leq y \Rightarrow f(x) \leq f(y)$, as morphisms, and,
3. ordinary function composition, and identity functions.

An extended version of this category is **Pre** where we consider arbitrary pre-ordered sets, that is, a set together with a reflexive and transitive relation.

Example 6 **Cat** is the concrete category with[4]:

1. categories as objects,
2. so-called *functors* between these as morphisms (see Sect. 3.2.6), and,
3. functor composition, and identity functors.

3.2.3 Real World Categories

But viewing category theory as some kind of *metatheory about mathematical structure* is not necessarily the most useful perspective for the sort of applications that we have in mind. Indeed, here are a few examples of the kind of categories we truly care about, and which are not categories with mathematical structures as objects and structure preserving maps as morphisms.

Example 7 The category **PhysProc** with

1. all physical systems $A, B, C, \ldots$ as objects,
2. all physical processes which take a physical system of type A into a physical system of type B as the morphisms of type $A \longrightarrow B$ (these processes typically require some finite amount of time to be completed), and,
3. sequential composition of these physical processes as composition, and the process which leaves system A invariant as the identity 1_A.

Note that in this case associativity of composition admits a physical interpretation: if we first have process f, then process g, and then process h, it doesn't matter whether we either consider $(g \circ f)$ as a single entity after which we apply h, or whether we consider $(h \circ g)$ as a single entity which we apply after f. Hence brackets constitute superfluous data that can be omitted i.e.

$$h \circ g \circ f := h \circ (g \circ f) = (h \circ g) \circ f\,.$$

Example 8 The category **PhysOpp** is an *operational variant* of the above where, rather than general physical systems such as stars, we focus on systems which can be manipulated in the lab, and rather than general processes, we consider the operations which the practising experimenter performs on these systems, for example, applying force-fields, performing measurements etc.

---

[4] In order to conceive **Cat** as a concrete category, the family of objects should be restricted to the so-called "small" categories i.e., categories for which the family of objects is a set.

Example 9 The category **QuantOpp** is a restriction of the above where we restrict ourselves to quantum systems and operations thereon. Special processes in **QuantOpp** are *preparation procedures*, or *states*. If Q denotes a qubit, then the type of a preparation procedure would be $\mathrm{I} \longrightarrow Q$ where I stands for "unspecified". Indeed, the point of a preparation procedure is to provide a qubit in a certain state, and the resources which we use to produce that state are typically not of relevance for the remainder of the experimental procedure. We can further specialise to either pure (or closed) quantum systems or mixed (or open) quantum systems, categories to which we respectively refer as **PurQuantOpp** and **MixQuantOpp**.

Obviously, Example 9 is related to the concrete category which has Hilbert spaces as objects and certain types of linear mappings (e.g. completely positive maps) as morphisms. The preparation procedures discussed above then correspond to "categorical elements" in the sense of Example 4. We discuss this correspondence below.

While to the sceptical reader the above examples still might not seem very useful yet, the next two ones, which are very similar, have become really important for Computer Science and Logic. They are the reason that, for example, University of Oxford Computing Laboratory offers category theory to its undergraduates.

Example 10 The category **Comp** with

1. all data types, e.g. Booleans, integers, reals, as objects,
2. all programs which take data of type A as their input and produce data of type B as their output as the morphisms of type $A \longrightarrow B$, and,
3. sequential composition of programs as composition, and the programs which output their input unaltered as identities.

Example 11 The category **Prf** with

1. all propositions as objects,
2. all proofs which conclude from proposition A that proposition B holds as the morphisms of type $A \longrightarrow B$, and,
3. concatenation (or chaining) of proofs as composition, and the tautologies "from A follows A" as identities.

Computer scientists particularly like category theory because it explicitly introduces the notion of *type*: an arrow $A \xrightarrow{f} B$ has type $A \longrightarrow B$. These types prevent silly mistakes when writing programs, e.g. the composition $g \circ f$ makes no sense for $C \xrightarrow{g} D$ because the output — called the *codomain* — of f doesn't match the input — called the *domain* — of g. Computer scientists would say:

"types don't match".

Similar categories **BioProc** and **ChemProc** can be build for organisms and biological processes, chemicals and chemical reactions, etc.[5] The recipe for producing these categories is obvious:

Name	**Objects**	**Morphisms**
some area of science	corresponding systems	corresponding processes

Composition boils down to "first f and then g *happens*" and identities are just "nothing *happens*". Somewhat more operationally put, composition is "first *do* f and then *do* g" and identities are just "*doing* nothing". The reason for providing both the "objectivist" (= passive) and "instrumentalist" (= active) perspective is that we both want to appeal to members of the theoretical physics community and members of the quantum information community. The first community typically doesn't like instrumentalism since it just doesn't seem to make sense in the context of theories such as cosmology; on the other hand, instrumentalism is as important to quantum informatics as it is to ordinary informatics. We leave it up to the reader to decide whether it should play a role in the interpretation of quantum theory.

3.2.4 Abstract Categorical Structures and Properties

One can treat categories as mathematical structures in their own right, just as groups and vector spaces are mathematical structures. In contrast with concrete categories, abstract categorical structures then arise by either endowing categories with more *structure* or by requiring them to satisfy certain *properties*.

We are of course aware that this is not a formal definition. Our sheepish excuse is that physicists rarely provide precise definitions. There is however a formal definition which can be found in [5]. We do provide one below in Example 24.

Example 12 A *monoid* $(M, \bullet, 1)$ is a set together with a binary associative operation

$$- \bullet - : M \times M \to M$$

which admits a unit—i.e. a "group without inverses". Equivalently, we can define a monoid as a category $\mathbf{M}$ with a single object $*$. Indeed, it suffices to identify

- the elements of the hom-set $\mathbf{M}(*, *)$ with those of M,
- the associative composition operation

[5] The first time the 1st author heard about categories was in a Philosophy of Science course, given by a biologist specialised in population dynamics, who discussed the importance of category theory in the influential work of Robert Rosen [59].

$$- \circ - : \mathbf{M}(*, *) \times \mathbf{M}(*, *) \to \mathbf{M}(*, *)$$

with the associative monoid multiplication $\bullet$, and
- the identity $1_* : * \to *$ with the unit 1.

Dually, in any category $\mathbf{C}$, for any $A \in |\mathbf{C}|$, the set $\mathbf{C}(A, A)$ is always a monoid.

Definition 2 Two objects $A, B \in |\mathbf{C}|$ are *isomorphic* if there exists morphisms $f \in \mathbf{C}(A, B)$ and $g \in \mathbf{C}(B, A)$ such that $g \circ f = 1_A$ and $f \circ g = 1_B$. The morphism f is called an *isomorphism* and $f^{-1} := g$ is called the *inverse* to f.

The notion of isomorphism known to the reader is the set-theoretical one, namely that of a bijection. We now show that in the concrete category **Set** the category-theoretical notion of isomorphism coincides with the notion of bijection. Given functions $f : X \to Y$ and $g : Y \to X$ satisfying $g(f(x)) = x$ for all $x \in X$ and $f(g(y)) = y$ for all $y \in Y$ we have:

- $f(x_1) = f(x_2) \Rightarrow g(f(x_1)) = g(f(x_2)) \Rightarrow x_1 = x_2$ so f is injective, and,
- for all $y \in Y$, setting $x := g(y)$, we have $f(x) = y$ so f is surjective,

so f is indeed a bijection. We leave it to the reader to verify that the converse also holds. For the other concrete categories mentioned above the categorical notion of isomorphism also coincides with the usual one.

Example 13 Since a group $(G, \bullet, 1)$ is a monoid with inverses it can now be equivalently defined as a category with one object in which each morphism is an isomorphism. More generally, a *groupoid* is a category in which each morphism has an inverse. For instance, the category **Bijec** which has sets as objects and bijections as morphisms is such a groupoid. So is **FdUnit** which has finite dimensional Hilbert spaces as objects and unitary operators as morphisms. Groupoids have important applications in mathematics, for example, in algebraic topology [17].

From this, we see that any group is an example of an abstract categorical structure. At the same time, all groups together, with structure preserving maps between them, constitute a concrete category. Still following? That categories allow several ways of representing mathematical structures might seem confusing at first, but it is a token of their versatility. While monoids correspond to categories with only one object, with groups as a special case, similarly, pre-orders are categories with very few morphisms, with partially ordered sets as a special case.

Example 14 Any preordered set $(P, \leq)$ can be seen as a category $\mathbf{P}$:

- The elements of P are the objects of $\mathbf{P}$,
- Whenever $a \leq b$ for $a, b \in P$ then there is a single morphism of type $a \longrightarrow b$, that is, $\mathbf{P}(a, b)$ is a singleton, and whenever $a \not\leq b$ then there is no morphism of type $a \longrightarrow b$, that is, $\mathbf{P}(a, b)$ is empty.
- Whenever there is pair of morphisms of types $a \longrightarrow b$ and $b \longrightarrow c$, that is, whenever $a \leq b$ and $b \leq c$, then transitivity of $\leq$ guarantees the existence of a unique morphism of type $a \longrightarrow c$, which we take to be the composite of the morphisms of type $a \longrightarrow b$ and $b \longrightarrow c$.

- Reflexivity guarantees the existence of a unique morphism of type $a \longrightarrow a$, which we take to be the identity on the object a.

Conversely, a category **C** of which the objects constitute a set, and in which there is at most one morphism of any type i.e., hom-sets are either singletons or empty, is in fact a preordered set. Concretely:

- The set $|\mathbf{C}|$ are the elements of the preordered set,
- We set $A \leq B$ if and only if $\mathbf{C}(A, B)$ is non-empty,
- Since **C** is a category, whenever there exist morphisms $f \in \mathbf{C}(A, B)$ and $g \in \mathbf{C}(B, C)$, that is, whenever both $\mathbf{C}(A, B)$ and $\mathbf{C}(B, C)$ are non-empty, then there exist a morphism $g \circ f \in \mathbf{C}(A, C)$, so $\mathbf{C}(A, C)$ is also non-empty. Hence, $A \leq B$ and $B \leq C$ yields $A \leq C$, so $\leq$ is transitive.
- Since $1_A \in \mathbf{C}(A, A)$ we also have $A \leq A$, so $\leq$ is reflexive.

Hence, preordered sets indeed constitute an abstract category: its defining property is that every hom-set contains at most one morphism. Such categories are sometimes called *thin categories*. Conversely, categories with non-trivial hom-sets are called *thick*. Partially ordered sets also constitute an abstract category, namely one in which:

- every hom-set contains at most one morphism;
- whenever two objects are isomorphic then they must be equal.

This second condition imposes anti-symmetry on the partial order.

Let $\{*\}$ and $\emptyset$ denote a singleton set and the empty set respectively. Then for any set $A \in |\mathbf{Set}|$, the set $\mathbf{Set}(A, \{*\})$ of all functions of type $A \to \{*\}$ is itself a singleton, since there is only one function which maps all $a \in A$ on $*$, the single element of $\{*\}$. This concept can be *dualised*. The set $\mathbf{Set}(\emptyset, A)$ of functions of type $\emptyset \to A$ is again a singleton consisting of the "empty function". Due to these special properties, we call $\{*\}$ and $\emptyset$ respectively the *terminal* object and the *initial* object in **Set**. All this can be generalised to arbitrary categories as follows:

Definition 3 An object $\top \in |\mathbf{C}|$ is *terminal* in **C** if, for any $A \in |\mathbf{C}|$, there is only one morphism of type $A \longrightarrow \top$. Dually, an object $\bot \in |\mathbf{C}|$ is *initial* in **C** if, for any $A \in |\mathbf{C}|$, there is only one morphism of type $\bot \longrightarrow A$.

Proposition 1 *If a category **C** has two initial objects then they are isomorphic. The same property holds for terminal objects.*

Indeed. Let $\bot$ and $\bot'$ both be initial objects in **C**. Since $\bot$ is initial, there is a unique morphism f such that $\mathbf{C}(\bot, \bot') = \{f\}$. Analogously, there is a unique morphism g such that $\mathbf{C}(\bot', \bot) = \{g\}$. Now, since **C** is a category and relying again on the fact that $\bot$ is initial, it follows that $g \circ f \in \mathbf{C}(\bot, \bot) = \{1_\bot\}$. Similarly, $g \circ f \in \mathbf{C}(\bot', \bot') = \{1_{\bot'}\}$. Hence, $\bot \simeq \bot'$ as claimed. Similarly we show that $\top \simeq \top'$.

Example 15 A partially ordered set P is *bounded* if there exist two elements $\top$ and $\bot$ such that for all $a \in P$ we have $\bot \leq a \leq \top$. Hence, when P is viewed as a category, this means that it has both a terminal and an initial object.

The next example of an abstract categorical structure is the most important one in this paper. Therefore, we state it as a definition. Among many (more important) things, it axiomatises "cooking with vegetables".

Definition 4 A *strict monoidal category* is a category for which:

1. objects come with monoid structure $(|\mathbf{C}|, \otimes, \mathrm{I})$ i.e, for all $A, B, C \in |\mathbf{C}|$,

$$A \otimes (B \otimes C) = (A \otimes B) \otimes C \quad \text{and} \quad \mathrm{I} \otimes A = A = A \otimes \mathrm{I}\,,$$

2. for all objects $A, B, C, D \in |\mathbf{C}|$ there exists an operation

$$- \otimes - : \mathbf{C}(A, B) \times \mathbf{C}(C, D) \to \mathbf{C}(A \otimes C, B \otimes D) :: (f, g) \mapsto f \otimes g$$

which is associative and has 1_I as its unit, that is,[6]

$$f \otimes (g \otimes h) = (f \otimes g) \otimes h \quad \text{and} \quad 1_\mathrm{I} \otimes f = f = f \otimes 1_\mathrm{I}\,,$$

3. for all morphisms f, g, h, k with matching types we have

$$(g \circ f) \otimes (k \circ h) = (g \otimes k) \circ (f \otimes h)\,, \tag{3.4}$$

4. for all objects $A, B \in |\mathbf{C}|$ we have

$$1_A \otimes 1_B = 1_{A \otimes B}\,. \tag{3.5}$$

As we will see in Sect. 3.6.1, the two equational constraints Eqs. (3.4) and (3.5) can be conceived as a single principle.

The symbol $\otimes$ is sometimes called the tensor. We will also use this terminology, since "tensor" is shorter than "monoidal product". However, the reader should not deduce from this that the above definition necessitates $\otimes$ to be anything like a tensor product, since this is not at all the case.

The categories of systems and processes discussed in Sect. 3.2.3 are all examples of strict monoidal categories. We already explained in Sect. 3.1 what $- \otimes -$ stands for: it enables dealing with situations where several systems are involved. To a certain extent $- \otimes -$ can be interpreted as a logical conjunction:

$$A \otimes B := \text{system } A \ \underline{\text{and}} \text{ system } B$$
$$f \otimes g := \text{process } f \ \underline{\text{and}} \text{ process } g\,.$$

There is however considerable care required with this view: while

$$A \wedge A = A\,,$$

[6] Note that this operation on morphisms is a *typed* variant of the notion of monoid.

in general

$$A \otimes A \neq A\,.$$

This is where the so-called *linear logic* [36, 61] kicks in, which is discussed in substantial detail in [4].

For the special object I we have

$$A \otimes \mathrm{I} = A = \mathrm{I} \otimes A$$

since it is the unit for the monoid. Hence, it refers to a system which leaves any system invariant when adjoined to it. In short, it stands for "unspecified", for "no system", or even for "nothing". We already made reference to it in Example 9 when discussing preparation procedures. Similarly, 1_{I} is the operation which "does nothing to nothing". The system I will allow us to encode a notion of state within arbitrary monoidal categories, and also a notion of number and probabilistic weight—see Example 27 below.

Example 16 Now, a monoid $(M, \bullet, 1)$ can also be conceived as a strict monoidal category in which all morphisms are identities. Indeed, take M to be the objects, $\bullet$ to be the tensor and 1 to be the unit for the tensor. By taking identities to be the only morphisms, we can equip these with the same monoid structure as the monoid structure on the objects. Hence it satisfies Eq. (3.5). By

$$(1_A \circ 1_A) \otimes (1_B \circ 1_B) = 1_A \otimes 1_B = 1_{A\otimes B} = 1_{A\otimes B} \circ 1_{A\otimes B} = (1_A \otimes 1_B) \circ (1_A \otimes 1_B)$$

eq. (3.2) is also satisfied.

3.2.5 Categories in Physics

In the previous section, we saw how groups and partial orders, both of massive importance for physics, are themselves abstract categorical structures.

- While there is no need to argue for the importance of group theory to physics here, it is worth mentioning that John Slater (cf. Slater determinant in quantum chemistry) referred to Weyl, Wigner and others' use of group theory in quantum physics as *der Gruppenpest*, what translates as the "plague of groups". Even in 1975 he wrote: As soon as [my] paper became known, it was obvious that a great many other physicists were as disgusted as I had been with the group-theoretical approach to the problem. As I heard later, there were remarks made such as "Slater has slain the Gruppenpest". I believe that no other piece of work I have done was so universally popular. Similarly, we may wonder whether it are the category theoreticians or their opponents which are the true aliens.
- Partial orders model spatio-temporal causal structure [56, 64]. Roughly speaking, if $a \leq b$ then events a and b are causally related, if $a < b$ then they are time-like

separated, and if a and b don't compare then they are space-like separated. This theme is discussed in great detail in [49].

- The degree of bipartite quantum entanglement gives rise to a preorder on bipartite quantum states [52]. The relevant preorder is Muirheads' majorization order [51]. However, multipartite quantum entanglement and mixed state quantum entanglement are not well understood yet. We strongly believe that category theory provides the key to the solution, in the following sense:

$$\frac{\text{bipartite entanglement}}{\text{some preorder}} = \frac{\text{multipartite entanglement}}{\text{some thick category}}$$

We also acknowledge the use of category theory in several involved subjects in mathematical physics ranging from topological quantum field theories (TQFTs) to proposals for a theory of quantum gravity; here the motivation to use category theory is of a mathematical nature. We discuss one such topic, namely TQFT, in Sect. 3.6.5.

But the particular perspective which we would like to promote here is *categories as physical theories*. Above we discussed three kinds of categories:

- *Concrete categories* have mathematical structures as objects, and structure preserving maps between these as morphisms.
- *Real world categories* have some notion of system as objects, and corresponding processes thereoff as morphisms.
- *Abstract categorical structures* are mathematical structures in their own right; they are defined in terms of additional structure and/or certain properties.

The real world categories constitute the *area* of our focus (e.g. quantum physics, proof theory, computation, organic chemistry, . . .), the concrete categories constitute the formal mathematical *models* for these (e.g., in the case of quantum physics, Hilbert spaces as objects, certain types of linear maps as morphisms, and the tensor product as the monoidal structure), while the abstract categorical structures constitute *axiomatisations* of these.

The latter is the obvious place to start when one is interested in comparing theories. We can study which axioms and/or structural properties give rise to certain physical phenomena, for example, which tensor structures give rise to teleportation (e.g. [2]), or to non-local quantum-like behavior [24]. Or, we can study which structural features distinguish classical from quantum theories (e.g. [27, 26]).

Quantum theory is subject to the so-called No-Cloning, No-Deleting and No-Broadcasting theorems [7, 54, 69], which impose key constraints on our capabilities to process quantum states. Expressing these clearly requires a formalism that allows to vary types from a single to multiple systems, as well as one which explicitly accommodates processes (cf. copying/deleting process). Monoidal categories provide the appropriate mathematical arena for this on-the-nose.

Example 17 Why does a tiger have stripes and a lion doesn't? One might expect that the explanation is written within the fundamental building blocks which these animals are made up from, so one could take a big knife and open the lion's and the tiger's bellies. One finds intestines, but these are the same for both animals. So maybe the answer is hidden in even smaller constituents. With a tiny knife we keep cutting and identify a smaller kind of building block, namely the cell. Again, there is no obvious difference between tigers and lions at this level. So we need to go even smaller. After a century of advancing "small knife technology" we discover DNA and this constituent truly reveals the difference. So yes, now we know why tigers have stripes and lions don't! Do we really? No, of course not. Following in the footsteps of Charles Darwin, your favorite nature channel would tell you that the explanation is given by a process of type

$$\textit{prey} \otimes \textit{predator} \otimes \textit{environment} \longrightarrow \textit{dead prey} \otimes \textit{eating predator}$$

which represents the successful challenge of a predator, operating within some environment, on some prey. Key to the success of such a challenge is the predator's camouflage. Sandy savanna is the lion's habitat while forests constitute the tiger's habitat, so their respective coat blends them within their natural habitat. Any (neo-)Darwinist biologist will tell you that the fact that this is encoded in the animal's DNA is not a cause, but rather a consequence, via the process of natural selection.

This example illustrates how monoidal categories enable to shift the focus from an *atomistic* or *reductionist* attitude to one where systems are studied in terms of their interactions with other systems, rather than in terms of their constituents. Clearly, in recent history, physics has solely focused on chopping down things into smaller things. Focussing on interactions might provide us with a complementary understanding of the fundamental theories of nature.

3.2.6 Structure Preserving Maps for Categories

The notion of structure preserving map between categories—which we referred to in Example 6—wasn't made explicit yet. These "maps which preserve categorical structure", the so-called *functors*, must preserve the *structure of a category*, that is, *composition* and *identities*. An example of a functor that might be known to the reader because of its applications in physics, is the linear representation of a group. A representation of a group G on a vector space V is a group homomorphism from G to $\mathrm{GL}(V)$, the general linear group on V, i.e., a map $\rho : G \to \mathrm{GL}(V)$ such that

$$\rho(g_1 \bullet g_2) = \rho(g_1) \circ \rho(g_2) \quad \text{for all} \quad g_1, g_2 \in G\,, \quad \text{and}\,, \quad \rho(1) = 1_V\,.$$

Consider G as a category $\mathbf{G}$ as in Example 13. We also have that $\mathrm{GL}(V) \subset \mathbf{FdVect}_{\mathbb{K}}(V, V)$ (cf. Example 2). Hence, a group representation ρ from G to $\mathrm{GL}(V)$ induces "something" from $\mathbf{G}$ to $\mathbf{FdVect}_{\mathbb{K}}$:

$$\rho : G \to \mathrm{GL}(V) \quad \rightsquigarrow \quad \mathbf{G} \xrightarrow{R_\rho} \mathbf{FdVect}_{\mathbb{K}}\,.$$

However, specifying $\mathbf{G} \xrightarrow{R_\rho} \mathbf{FdVect}_{\mathbb{K}}$ requires some care:

- Firstly, we need to specify that we are representing on the general linear group of the vector space $V \in \mathbf{FdVect}_{\mathbb{K}}$. We do this by mapping the unique object $*$ of $\mathbf{G}$ on V, thus defining a map from objects to objects

$$R_\rho : |\mathbf{G}| \to |\mathbf{FdVect}_{\mathbb{K}}| :: * \mapsto V\,.$$

- Secondly, we need to specify to which linear map in

$$\mathrm{GL}(R_\rho(*)) \subset \mathbf{FdVect}_{\mathbb{K}}(R_\rho(*), R_\rho(*))$$

  a group element

$$g \in \mathbf{G}(*, *) = G$$

  is mapped. This defines a map from a hom-set to a hom-set, namely

$$R_\rho : \mathbf{G}(*, *) \to \mathbf{FdVect}_{\mathbb{K}}(R_\rho(*), R_\rho(*)) :: g \mapsto \rho(g)\,.$$

  The fact that ρ is a group homomorphism implies in our category-theoretic context that R_ρ preserves composition of morphisms as well as identities, that is, R_ρ preserves the categorical structure.

Having this example in mind, we infer that a functor must consist not of a single but of *two* kinds of mappings: one map on the objects, and a family of maps on the hom-sets which preserve identities and composition.

Definition 5 Let $\mathbf{C}$ and $\mathbf{D}$ be categories. A *functor*

$$F : \mathbf{C} \longrightarrow \mathbf{D}$$

consists of:

1. A mapping

$$F : |\mathbf{C}| \to |\mathbf{D}| :: A \mapsto F(A)\,;$$

2. For any $A, B \in |\mathbf{C}|$, a mapping

$$F : \mathbf{C}(A, B) \to \mathbf{D}(F(A), F(B)) :: f \mapsto F(f)$$

   which preserves identities and composition, i.e.,

   i. for any $f \in \mathbf{C}(A, B)$ and $g \in \mathbf{C}(B, C)$ we have

$$F(g \circ f) = F(g) \circ F(f)\,,$$

ii. and, for any $A \in |\mathbf{C}|$ we have

$$F(1_A) = 1_{F(A)} .$$

Typically one drops the parentheses unless they are necessary. For instance, $F(A)$ and $F(f)$ will be denoted simply as FA and Ff.

Consider the category **PhysProc** of Example 7 and a concrete category **Mod** in which we wish to model these mathematically by assigning to each process a morphism in the concrete category **Mod**. Functoriality of

$$F : \mathbf{PhysProc} \longrightarrow \mathbf{Mod}$$

means that sequential composition of physical processes is mapped on composition of morphisms in **Mod**, and that void processes are mapped on the identity morphisms. From this, we see that functoriality is an obvious requirement when designing mathematical models for physical processes.

Example 18 Define the category $\mathbf{Mat}_{\mathbb{K}}$ with

1. the set of natural numbers $\mathbb{N}$ as objects,
2. all $m \times n$-matrices with entries in $\mathbb{K}$ as morphisms of type $n \longrightarrow m$, and
3. matrix composition, and identity matrices.

This example is closely related to Example 2. However, it strongly emphasizes that objects are but labels with no internal structure. Strictly speaking this is not a concrete category in the sense of Sect. 3.2.2. However, for all practical purposes, it can serve as well as a model as any other concrete category. Therefore, we can relax our conception of concrete categories to accommodate such models.

Assume now that for each vector space $V \in |\mathbf{FdVect}_{\mathbb{K}}|$, we pick a fixed basis. Then any linear function $f \in \mathbf{FdVect}_{\mathbb{K}}(V, W)$ admits a matrix in these bases. This "assigning of matrices" to linear maps is described by the functor

$$F : \mathbf{FdVect}_{\mathbb{K}} \longrightarrow \mathbf{Mat}_{\mathbb{K}}$$

which maps vector spaces on their respective dimension, and which maps linear maps on their matrices in the chosen bases. Importantly, note that it is the functor F which encodes the choices of bases, and not the categorical structure of $\mathbf{FdVect}_{\mathbb{K}}$.

Example 19 In $\mathbf{Mat}_{\mathbb{C}}$, if we map each natural number on itself and conjugate all the entries of each matrix we also obtain a functor.

We now introduce the concept of *duality* which we already hinted at above. Simply put, it means reversal of the arrows in a given category **C**. We illustrate this notion in term of an example. Transposition of matrices, just like a functor, is a mapping on both objects and morphisms which:

i. preserves objects and identities,
ii. reverses the direction of the morphisms since when the matrix M has type $n \longrightarrow m$, then the matrix M^T has type $m \longrightarrow n$, and

iii. preserves the composition 'up to this reversal of the arrows', i.e. for any pair of matrices N and M for which types match we have

$$(N \circ M)^T = M^T \circ N^T .$$

So transposition is a functor *up to reversal of the arrows*.

Definition 6 A *contravariant functor* $F : \mathbf{C} \to \mathbf{D}$ consists of the same data as a functor, it also preserves identities, but reverses composition that is:

$$F(g \circ f) = Ff \circ Fg ,$$

In contrast to contravariant functors, ordinary functors are often referred to as *covariant functors*.

Definition 7 The *opposite category* $\mathbf{C}^{op}$ of a category $\mathbf{C}$ is the category with

- the same objects as $\mathbf{C}$,
- in which morphisms are "reversed", that is,

$$f \in \mathbf{C}(A, B) \Leftrightarrow f \in \mathbf{C}^{op}(B, A) ,$$

  where to avoid confusion from now on we denote $f \in \mathbf{C}^{op}(B, A)$ by f^{op},
- identities in $\mathbf{C}^{op}$ are those of $\mathbf{C}$, and

$$f^{op} \circ g^{op} = (g \circ f)^{op} .$$

Contravariant functors of type $\mathbf{C} \to \mathbf{D}$ can now be defined as functors of type $\mathbf{C}^{op} \to \mathbf{D}$. Of course, the operation $(-)^{op}$ on categories is involutive: reversing the arrows twice is the same as doing nothing. The process of reversing the arrow is sometimes indicated by the prefix "co", indicating that the defining equations for those structures are the same as the defining equations for the original structure, but with arrows reversed.

Example 20 The *transpose* is the involutive contravariant functor

$$T : \mathbf{FdVect}_{\mathbb{K}}^{op} \to \mathbf{FdVect}_{\mathbb{K}}$$

which maps each vector space on the corresponding dual vector space, and which maps each linear map f on its transpose f^T.

Example 21 A Hilbert space is a vector space over $\mathbb{C}$ with an inner-product

$$\langle -, - \rangle : \mathcal{H} \times \mathcal{H} \to \mathbb{C} .$$

Let **FdHilb** be the category with finite dimensional Hilbert spaces as objects and with linear maps as morphisms. Of course, one could define other categories with

Hilbert spaces as objects, for example, the groupoid **FdUnit** of Example 13. But as we will see below in Sect. 3.3.3, the category **FdHilb** as defined here comes with enough extra structure to extract all unitary maps from it. Hence, **FdHilb** subsumes **FdUnit**. This extra structure comes as a functor, whose action is *taking the adjoint* or *hermitian transpose*. This is the contravariant functor

$$\dagger : \mathbf{FdHilb}^{op} \longrightarrow \mathbf{FdHilb}$$

which:

1. is *identity-on-object*, that is,

$$\dagger : |\mathbf{FdHilb}^{op}| \to |\mathbf{FdHilb}| :: \mathcal{H} \mapsto \mathcal{H} ,$$

2. and assigns morphisms to their adjoints, that is,

$$\dagger : \mathbf{FdHilb}^{op}(\mathcal{H}, \mathcal{K}) \to \mathbf{FdHilb}(\mathcal{K}, \mathcal{H}) :: f \mapsto f^{\dagger} .$$

Since for $f \in \mathbf{FdHilb}(\mathcal{H}, \mathcal{K})$ and $g \in \mathbf{FdHilb}(\mathcal{K}, \mathcal{L})$ we have:

$$1_{\mathcal{H}}^{\dagger} = 1_{\mathcal{H}} \qquad \text{and} \qquad (g \circ f)^{\dagger} = f^{\dagger} \circ g^{\dagger}$$

we indeed obtain an identity-on-object contravariant functor. This functor is moreover *involutive*, that is, for all morphisms f we have

$$f^{\dagger\dagger} = f .$$

While the morphisms of **FdHilb** do not reflect the inner-product structure, the latter is required to specify the adjoint. In turn, this adjoint will allow us to recover the inner-product in purely category-theoretic terms, as we shall see in Sect. 3.3.3.

Example 22 Define the category $\mathbf{Funct}_{\mathbf{C},\mathbf{D}}$ with

1. all functors from **C** to **D** as objects,
2. *natural transformations* between these as morphisms (cf. Sect. 3.6.2), and,
3. composition of natural transformations and corresponding identities.

Example 23 The defining equations of strict monoidal categories, that is,

$$(g \circ f) \otimes (k \circ h) = (g \otimes k) \circ (f \otimes h) \quad \text{and} \quad 1_A \otimes 1_B = 1_{A \otimes B} , \tag{3.6}$$

to which we from now on refer as *bifunctoriality*, is nothing but functoriality of a certain functor. We will discuss this in detail in Sect. 3.6.1.

Example 24 A *concrete category*, or even better, a **Set**-concrete category, is a category **C** together with a functor $U : \mathbf{C} \longrightarrow \mathbf{Set}$. The way in which we construct this functor for categories with mathematical structures as objects is by sending each object to the underlying set, and morphisms to the underlying functions. So we forget the extra structure the object has. Therefore the functor U is typically called *forgetful*. For example, the category **Grp** is a concrete category for the functor

$$U : \mathbf{Grp} \longrightarrow \mathbf{Set} :: \begin{cases} (G, \bullet, 1) \mapsto G \\ f \mapsto f \end{cases}$$

which "forgets" the group's multiplication and unit, and morphisms are mapped on their underlying functions. More generally, a **D**-concrete category is a category **C** with a functor $U : \mathbf{C} \longrightarrow \mathbf{D}$.

Example 25 The TQFTs of Sect. 3.6.5 are special kinds of functors.

3.3 The 2D Case: Muscle Power

We now genuinely start to study the interaction of the parallel and the sequential modes of composing systems, and operations thereon.

3.3.1 Strict Symmetric Monoidal Categories

The starting point of this Section is the notion of a strict monoidal category as given in Definition 4. Such categories enable us to give formal meaning to physical processes which involve several types, e.g. classical and quantum as the following example clearly demonstrates.

Example 26 Define **CQOpp** to be the strict monoidal category containing both classical and quantum systems, with operations thereon as morphisms, and with the obvious notion of monoidal tensor, that is, a physical analogue of the tensor for vegetables that we saw in the prologue. Concretely, by $A \otimes B$ we mean that we have *both A and B* available to operate on. Note in particular that at this stage of the discussion there are no Hilbert spaces involved, so $\otimes$ cannot stand for the tensor product, but this does not exclude that we may want to model it by the tensor product at a later stage. In this category, non-destructive (projective) measurements have type

$$Q \longrightarrow X \otimes Q$$

where Q is a quantum system and X is the classical data produced by the measurement. Obviously, the hom-sets

$$\mathbf{CQOpp}(Q, Q) \qquad \text{and} \qquad \mathbf{CQOpp}(X, X)$$

have a very different structure since $\mathbf{CQOpp}(Q, Q)$ stands for the operations we can perform on a quantum system while $\mathbf{CQOpp}(X, X)$ stands for the classical operations (e.g. classical computations) which we can perform on classical systems. But all of these now live within a single mathematical entity **CQOpp**.

The structure of a strict monoidal category does not yet capture certain important properties of cooking with vegetables. Denote the strict monoidal category constructed in the Prologue by **Cook**.

Clearly "boil the potato while fry the carrot" is very much the same thing as "fry the carrot while boil the potato". But we cannot just bluntly say that in the category **Cook** the equality

$$h \otimes f = f \otimes h$$

holds. By plain set theory, for this equality to be meaningful, the two morphisms $h \otimes f$ and $f \otimes h$ need to live in the same set. That is, respecting the structure of a category, within the same hom-set. So

$$A \otimes D \xrightarrow{f \otimes h} B \otimes F \qquad \text{and} \qquad D \otimes A \xrightarrow{h \otimes f} F \otimes B$$

need to have the same type, which implies that

$$A \otimes D = D \otimes A \qquad \text{and} \qquad B \otimes F = F \otimes B \tag{3.7}$$

must hold. But this completely blurs the distinction between a carrot and a potato. For example, we cannot distinguish anymore between "boil the potato while fry the carrot", which we denoted by

$$A \otimes D \xrightarrow{f \otimes h} B \otimes F\,,$$

and "fry the potato while boil the carrot", which given Eq. (3.7), we can write as

$$A \otimes D = D \otimes A \xrightarrow{h \otimes f} F \otimes B = B \otimes F\,.$$

So we basically threw out the child with the bath water.

The solution to this problem is to introduce an operation

$$\sigma_{A,D} : A \otimes D \longrightarrow D \otimes A$$

which *swaps* the role of the potato and the carrot relative to the monoidal tensor. The fact that "boil the potato while fry the carrot" is essentially the same thing as "fry the carrot while boil the potato" can now be expressed as

$$\sigma_{B,F} \circ (f \otimes h) = (h \otimes f) \circ \sigma_{A,D}\,.$$

In our "real world example" of cooking this operation can be interpreted as physically swapping the vegetables [21]. An equational law governing "swapping" is:

$$\sigma_{B,A} \circ \sigma_{A,B} = 1_{A \otimes B}\,.$$

Definition 8 A *strict symmetric monoidal category* is a strict monoidal category **C** which moreover comes with a family of isomorphisms

$$\left\{A \otimes B \xrightarrow{\sigma_{A,B}} B \otimes A \;\middle|\; A, B \in |\mathbf{C}|\right\}$$

called *symmetries*, and which are such that:

- for all $A, B \in |\mathbf{C}|$ we have $\sigma_{A,B}^{-1} = \sigma_{B,A}$, and
- for all $A, B, C, D \in |\mathbf{C}|$ and all f, g of appropriate type we have

$$\sigma_{C,D} \circ (f \otimes g) = (g \otimes f) \circ \sigma_{A,B} . \tag{3.8}$$

All Examples of Sect. 3.2.3 are strict symmetric monoidal categories for the obvious notion of symmetry in terms of "swapping".

We can rewrite Eq. (3.8) in a form which makes the types explicit:

$$\begin{array}{ccc} A \otimes B & \xrightarrow{\sigma_{A,B}} & B \otimes A \\ {\scriptstyle f\otimes g}\downarrow & & \downarrow{\scriptstyle g\otimes f} \\ C \otimes D & \xrightarrow[\sigma_{C,D}]{} & D \otimes C \end{array} \tag{3.9}$$

This representation is referred to as *commutative diagrams*.

Proposition 2 *In any strict monoidal category we have*

$$\begin{array}{ccc} A \otimes B & \xrightarrow{1_A\otimes g} & A \otimes D \\ {\scriptstyle f\otimes 1_B}\downarrow & & \downarrow{\scriptstyle f\otimes 1_D} \\ C \otimes B & \xrightarrow[1_C\otimes g]{} & C \otimes D \end{array} \tag{3.10}$$

Indeed, relying on bifunctoriality we have:

$$\begin{array}{c} (f \otimes 1_D) \circ (1_A \otimes g) = (f \circ 1_A) \otimes (1_D \circ g) \\ \| \\ f \otimes g \\ \| \\ (1_C \circ f) \otimes (g \circ 1_B) = (1_C \otimes g) \circ (f \otimes 1_B) . \end{array}$$

The reader can easily verify that, given a connective $-\otimes-$ defined both on objects and morphisms as in items 1 & 2 of Definition 4, the four equations

$$(f \circ 1_A) \otimes (1_D \circ g) = f \otimes g = (1_B \circ f) \otimes (g \circ 1_C) \tag{3.11}$$

$$(g \otimes 1_B) \circ (f \otimes 1_B) = (g \circ f) \otimes 1_B \tag{3.12}$$

$$(1_A \otimes g) \circ (1_A \otimes f) = 1_A \otimes (g \circ f)\,, \tag{3.13}$$

when varying over all objects $A, B, C, D \in |\mathbf{C}|$ and all morphisms f and g of appropriate type, are equivalent to the single equation

$$(g \circ f) \otimes (k \circ h) = (g \otimes k) \circ (f \otimes h) \tag{3.14}$$

when varying over f, g, h, k. Eqs. (3.12), (3.13) together with

$$1_A \otimes 1_B = 1_{A\otimes B}$$

is usually referred to as $-\otimes-$ being *functorial in both arguments*. They are indeed equivalent to the mappings on objects and morphisms

$$(1_A \otimes -) : \mathbf{C} \longrightarrow \mathbf{C} \qquad \text{and} \qquad (- \otimes 1_B) : \mathbf{C} \longrightarrow \mathbf{C}$$

both being functors, for all objects $A, B \in |\mathbf{C}|$ — their action on objects is

$$(1_A \otimes -) :: X \mapsto A \otimes X \qquad \text{and} \qquad (- \otimes 1_B) :: X \mapsto X \otimes B\,.$$

Hence, functoriality in both arguments is strictly weaker than bifunctoriality (cf. Example 23), since the latter also requires Eq. (3.11).

3.3.2 Graphical Calculus for Symmetric Monoidal Categories

The most attractive, and at the same time, also the most powerful feature of strict symmetric monoidal categories, is that they admit a purely diagrammatic calculus. Such a graphical language is subject to the following characteristics:

- The symbolic ingredients in the definition of strict symmetric monoidal structure, e.g. $\otimes$, $\circ$, A, I, f etc., or any other abstract categorical structure which refines it, all have a purely diagrammatic counterpart ;
- The corresponding axioms become very intuitive graphical manipulations ;
- And crucially, an equational statement is derivable in the graphical language if and only if it is symbolically derivable from the axioms of the theory.

For a more formal presentation of what we precisely mean by a graphical calculus we refer the reader to Peter Selinger's marvelous paper [63] in these volumes.

These diagrammatic calculi trace back to Penrose's work in the early 1970s, and have been given rigorous formal treatments in [35, 38, 39, 62]. Some examples of possible elaborations and corresponding applications of the graphical language presented in this paper are in [25, 26, 23, 45, 63, 65, 67, 68].

The graphical counterparts to the axioms are typically much simpler then their formal counterparts. For example, in the Prologue we mentioned that bifunctoriality becomes a tautology in this context. Therefore such a graphical language radically simplifies algebraic manipulations, and in many cases trivialises something very complicated. Also the physical interpretation of the axioms, something which is dear to the authors of this paper, becomes very direct.

The graphical counterparts to strict symmetric monoidal structure are:

- The *identity* 1_{I} is the empty picture (= it is not depicted).
- The *identity* 1_A for and object A different of I is depicted as

- A *morphism* $f : A \longrightarrow B$ is depicted as

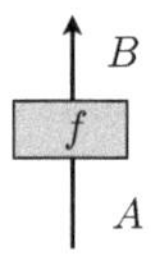

- The *composition* of morphisms $f : A \longrightarrow B$ and $g : B \longrightarrow C$ is depicted by locating g above f and by connecting the output of f to the input of g, i.e.

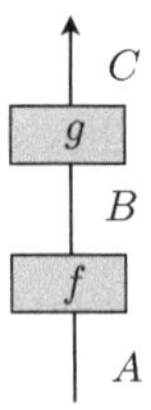

- The *tensor product* of morphisms $f : A \longrightarrow B$ and $g : C \longrightarrow D$ is depicted by aligning the graphical representation of f and g side by side in the order they occur within the expression $f \otimes g$, i.e.

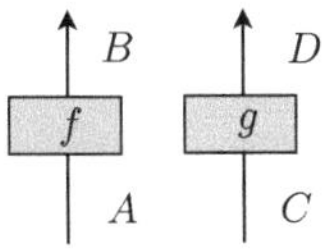

– *Symmetry*

$$\sigma_{AB} : A \otimes B \longrightarrow B \otimes A$$

is depicted as

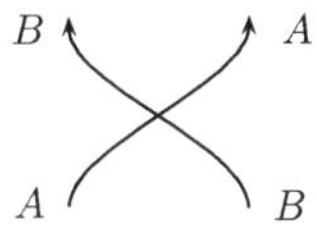

– Morphisms

$$\psi : \mathrm{I} \longrightarrow A \quad , \quad \phi : A \longrightarrow \mathrm{I} \quad \text{and} \quad s : \mathrm{I} \longrightarrow \mathrm{I}$$

are respectively depicted as

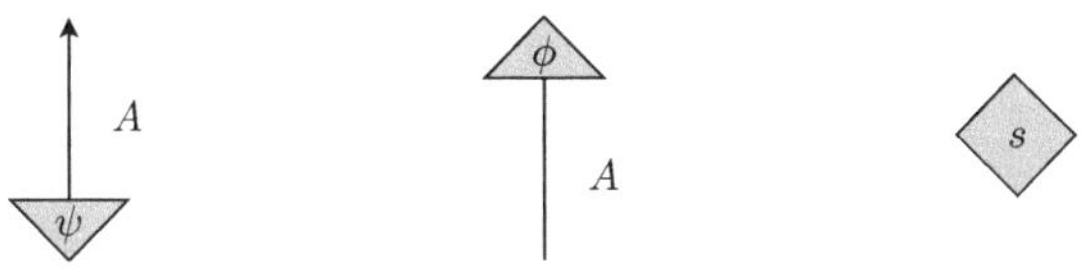

The diamond shape of the morphisms of type $\mathrm{I} \longrightarrow \mathrm{I}$ indicates that they arise when composing two triangles:

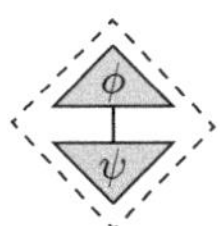

Example 27 In the category **QuantOpp** the triangles of respective types $\mathrm{I} \longrightarrow A$ and $A \longrightarrow \mathrm{I}$ represent states and effects, and the diamonds of type $\mathrm{I} \longrightarrow \mathrm{I}$ can be interpreted as probabilistic weights: they give the likeliness of a certain effect to occur when the system is in a certain state. In the usual quantum formalism these values are obtained when computing the Born rule or Luders' rule. In appropriate

categories, we find these exact values back as one of these diamonds, by composing a state and an effect [22, 63].

The equation

$$f \otimes g = (f \otimes 1_D) \circ (1_A \otimes g) = (1_B \otimes g) \circ (f \otimes 1_C) \tag{3.15}$$

established in Proposition 2 is depicted as:

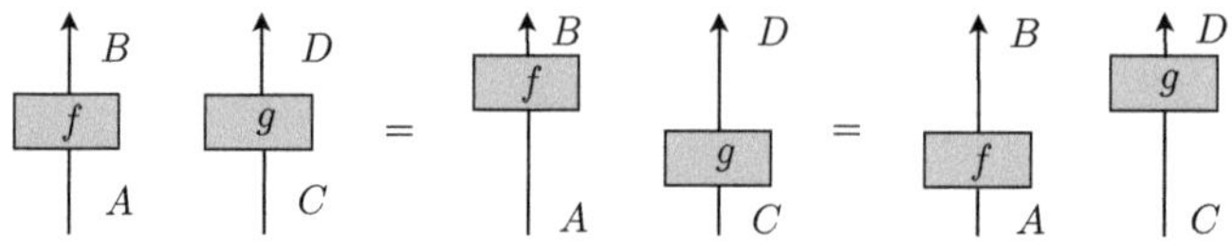

In words: we can "slide" boxes along their wires.

The first defining equation of symmetry, i.e. Eq. (3.9), depicts as:

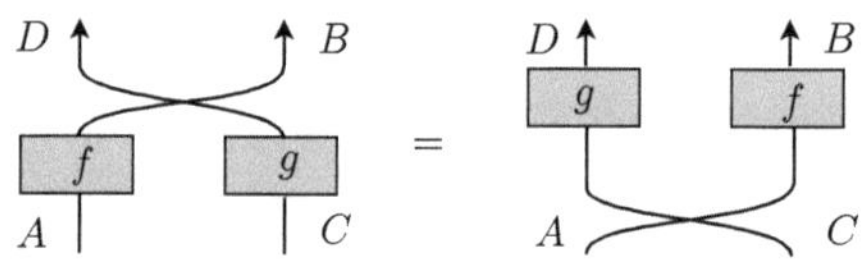

i.e., we can still "slide" boxes along crossings of wires. The equation

$$\sigma_{B,A} \circ \sigma_{A,B} = 1_{A,B}\,, \tag{3.16}$$

which when varying $A, B \in |\mathbf{C}|$ states that

$$\sigma_{A,B}^{-1} = \sigma_{B,A}\,,$$

depicts as

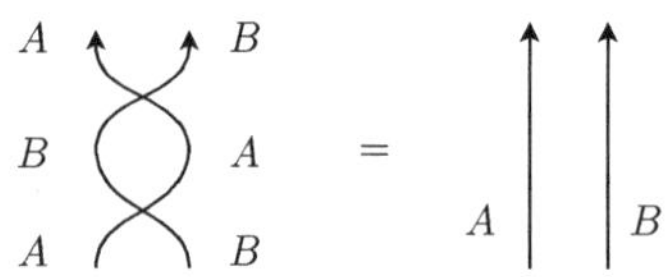

Suppose now that for any three arbitrary morphisms

$$f : A \longrightarrow A' \quad , \quad g : B \longrightarrow B' \qquad \text{and} \qquad h : C \longrightarrow C'$$

in any strict symmetric monoidal category, one intends to prove that

$$
\begin{aligned}
&(\sigma_{B',C'} \otimes f) \circ (g \otimes \sigma_{A,C'}) \circ (\sigma_{A,B} \otimes h) \\
&\quad = (h \otimes \sigma_{A',B'}) \circ (\sigma_{A',C} \otimes 1_{B'}) \circ (1_{A'} \otimes \sigma_{B',C}) \circ (f \otimes g \otimes 1_C)
\end{aligned}
$$

always holds. Then, the typical textbook proof proceeds by *diagram chasing*:

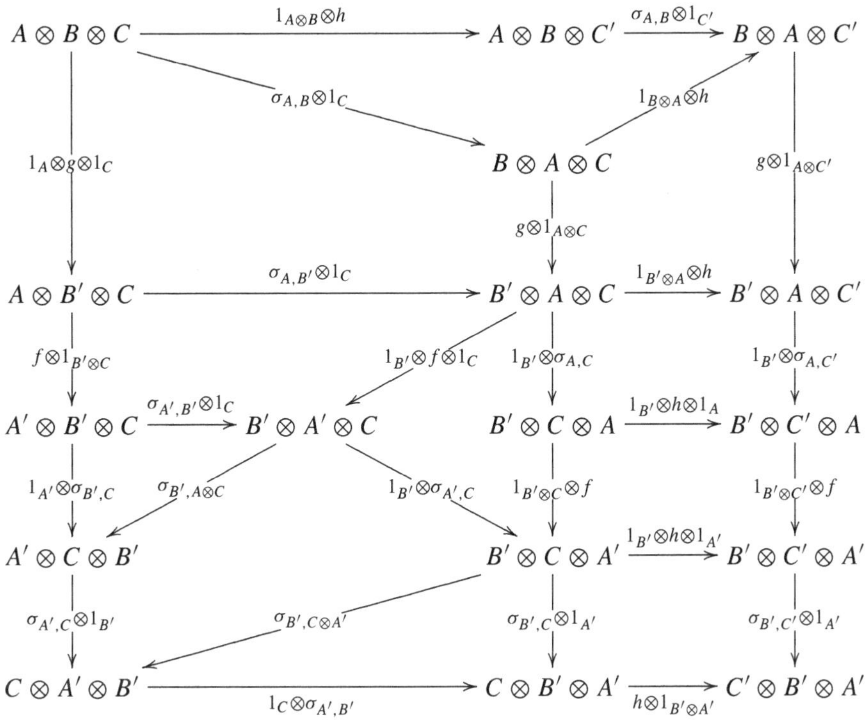

One needs to read this "dragon" as follows. The two outer paths both going from the left-upper-corner to the right-lower-corner represent the two sides of the equality we want to prove. Then, we do what category-theoreticians call diagram chasing, that is, "pasting" together several commutative diagrams, which connect one of the outer paths to the other. For example, the triangle at the top of the diagram expresses that

$$
(\sigma_{A,B} \otimes 1_{C'}) \circ (1_{A\otimes B} \otimes h) = (1_{B\otimes A} \otimes h) \circ (\sigma_{A,B} \otimes 1_C),
$$

that is, an instance of bifunctoriality. Using properties of strict symmetric monoidal categories, namely bifunctoriality and Eq. (3.9) expressed as commutative diagrams, we can pass from the outer path at the top and the right to the outer path on the left and the bottom. This is clearly a very tedious task and getting these diagrams into LaTeX becomes a time-consuming activity.

On the other hand, when using the graphical calculus, one immediately sees that

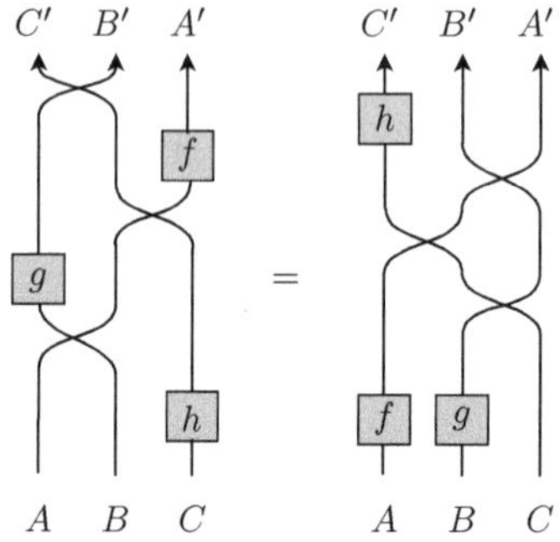

must hold. We pass from one picture to the other by sliding the boxes along wires and then by rearranging these wires. In terms of the underlying equations of strict symmetric monoidal structure, "sliding the boxes along wires" uses Eqs. (3.9) and (3.15), while "rearranging these wires" means that we used Eq. (3.9) as follows:

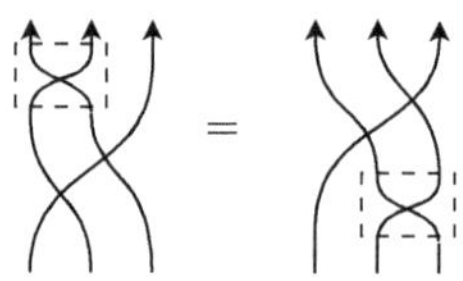

Indeed, since symmetry is a morphism it can be conceived as a box, and hence we can "slide it along wires".

In a broader historical perspective, we are somewhat unfair here. Writing equational reasoning down in terms of these commutative diagrams rather than long lists of equalities was an important step towards a better geometrical understanding of the structure of proofs.

3.3.3 Extended Dirac Notation

Definition 9 A *strict dagger monoidal category* **C** is a strict monoidal category equipped with an involutive identity-on-objects contravariant functor

$$\dagger : \mathbf{C}^{op} \longrightarrow \mathbf{C}\,,$$

that is,

- $A^\dagger = A$ for all $A \in |\mathbf{C}|$, and
- $f^{\dagger\dagger} = f$ for all morphisms f,

and this functor preserves the tensor, that is,

$$(f \otimes g)^\dagger = f^\dagger \otimes g^\dagger . \tag{3.17}$$

We will refer to $B \xrightarrow{f^\dagger} A$ as the *adjoint* to $A \xrightarrow{f} B$. A *strict dagger symmetric monoidal category* **C** is both a strict dagger monoidal category and a strict symmetric monoidal category such that

$$\sigma^\dagger_{A,B} = \sigma^{-1}_{A,B}.$$

Definition 10 [2] A morphism $U : A \longrightarrow B$ in a strict dagger monoidal category **C** is called *unitary* if its inverse and its adjoint coincide, that is, if

$$U^\dagger = U^{-1} .$$

Let $\psi, \phi : \mathrm{I} \longrightarrow A$ be "elements" in **C**. Their *inner-product* is the "scalar"

$$\langle \phi \mid \psi \rangle := \phi^\dagger \circ \psi : \mathrm{I} \longrightarrow \mathrm{I} .$$

So in any strict monoidal category we refer to morphisms of type

$$\mathrm{I} \longrightarrow A$$

as *elements* (cf. Example 4), to those of type

$$A \longrightarrow \mathrm{I}$$

as *co-elements*, and to those of type

$$\mathrm{I} \longrightarrow \mathrm{I}$$

as *scalars*. As already discussed in Example 27 in the category **QuantOpp** these corresponds respectively to states, effects and probabilistic weights.

Even at this abstract level, many familiar things follow from Definition 10. For example, we recover the defining property of adjoints for any dagger functor:

$$\begin{aligned}
\langle f^\dagger \circ \psi \mid \phi \rangle &= (f^\dagger \circ \psi)^\dagger \circ \phi \\
&= (\psi^\dagger \circ f) \circ \phi \\
&= \psi^\dagger \circ (f \circ \phi) \\
&= \langle \psi \mid f \circ \phi \rangle .
\end{aligned}$$

From this it follows that unitary morphisms preserve the inner-product:

$$\begin{aligned}\langle U \circ \psi \mid U \circ \phi \rangle &= \langle U^\dagger \circ (U \circ \psi) \mid \phi \rangle \\ &= \langle (U^\dagger \circ U) \circ \psi \mid \phi \rangle \\ &= \langle \psi \mid \phi \rangle .\end{aligned}$$

Importantly, the graphical calculus of the previous section extends to strict dagger symmetric monoidal categories. Following Selinger [63], we introduce an asymmetry in the graphical notation of the morphisms $A \xrightarrow{f} B$ as follows:

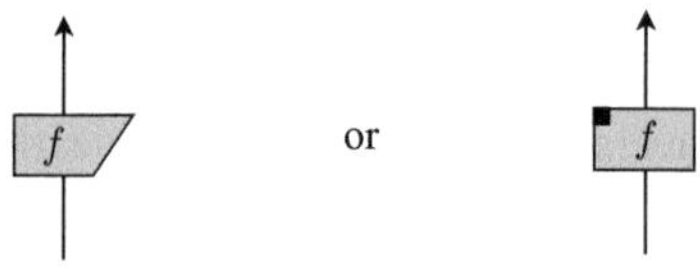

Then we depict the adjoint $B \xrightarrow{f^\dagger} A$ of $A \xrightarrow{f} B$ as follows:

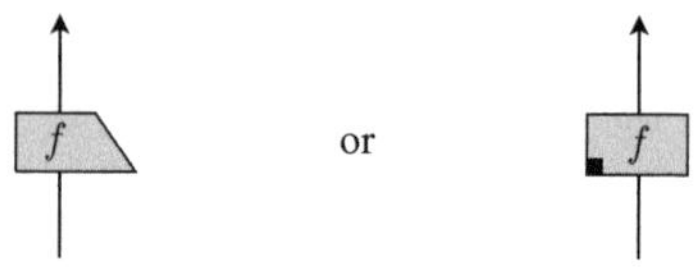

that is, we turn the box representing f upside-down. All this enables interpreting Dirac notation [31] in terms of strict dagger symmetric monoidal categories, and in particular, in terms of the corresponding graphical calculus:

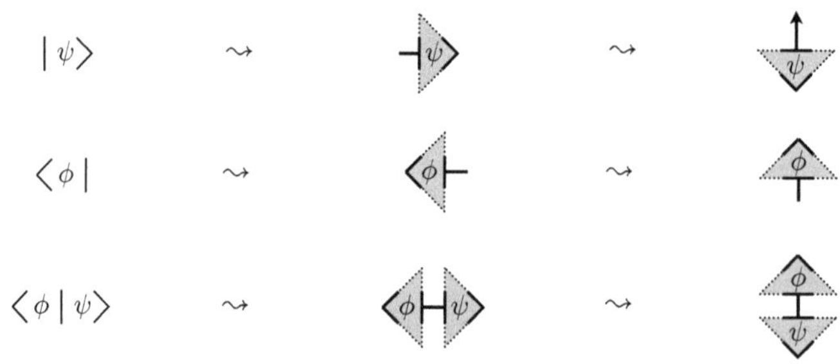

The latter notation merely requires closing the bra's and ket's and performing a 90° rotation.[7] Summarising we now have:

[7] This 90° rotation is merely a consequence of our convention to read pictures from bottom-to-top. Other authors obey different conventions e.g. top-to-bottom or left-to-right.

Dirac	matrix	strict †-SMC	picture
$\vert\psi\rangle$	$\begin{pmatrix}\psi_1\\ \vdots\\ \psi_n\end{pmatrix}$	$\mathrm{I} \xrightarrow{\psi} A$	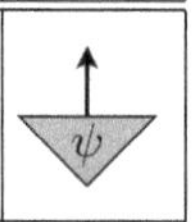
$\langle\phi\vert$	$(\bar{\phi}_1 \ldots \bar{\phi}_n)$	$A \xrightarrow{\phi} \mathrm{I}$	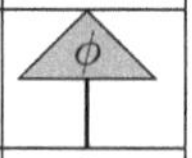
$\langle\phi\vert\psi\rangle$	$(\bar{\phi}_1 \ldots \bar{\phi}_n)\begin{pmatrix}\psi_1\\ \vdots\\ \psi_n\end{pmatrix}$	$\mathrm{I} \xrightarrow{\psi} A \xrightarrow{\phi^\dagger} \mathrm{I}$	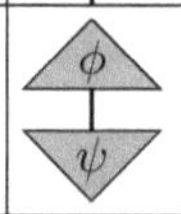
$\vert\psi\rangle\langle\phi\vert$	$\begin{pmatrix}\psi_1\\ \vdots\\ \psi_n\end{pmatrix}(\bar{\phi}_1 \ldots \bar{\phi}_n)$	$A \xrightarrow{\phi^\dagger} \mathrm{I} \xrightarrow{\psi} A$	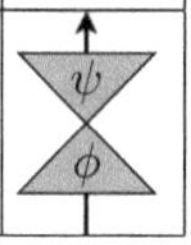

In particular, note that in the language of strict dagger symmetric monoidal categories both a bra-ket and a ket-bra are compositions of morphisms, namely $\phi^\dagger \circ \psi$ and $\psi \circ \phi^\dagger$ respectively. What the diagrammatic calculus adds to standard Dirac notation is a second dimension to accommodate the monoidal composition:

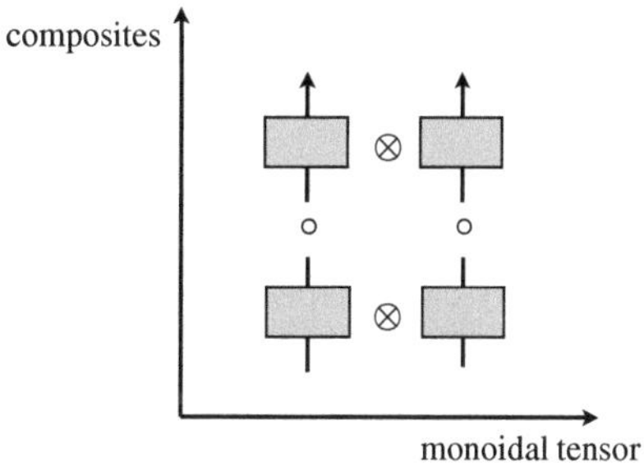

The advantages of this have already been made clear in the previous section and will even become clearer in Sect. 3.4.1.

Concerning the types of the morphisms in the third column of the above table, recall that in Example 4 we showed that the vectors in Hilbert spaces $\mathcal{H}$ can be faithfully represented by linear maps of type $\mathbb{C} \to \mathcal{H}$. Similarly, complex numbers $c \in \mathbb{C}$, that is, equivalently, vectors in the 'one-dimensional Hilbert space $\mathbb{C}$', can be faithfully represented by linear maps

$$s_c : \mathbb{C} \to \mathbb{C} :: 1 \mapsto c\,,$$

since by linearity the image of 1 fully specifies this map.

However, by making explicit reference to **FdHilb** and hence also by having matrices (morphisms in **FdHilb** expressed relative to some bases) in the above table, we are actually cheating. The fact that Hilbert spaces and linear maps are set-theoretic based mathematical structures has non-trivial "unpleasant" implications. In particular, while the $\otimes$-notation for the monoidal structure of strict monoidal categories insinuates that the tensor product would turn **FdHilb** into a strict symmetric monoidal category, this turns out not to be true in the "strict" sense of the word true.

3.3.4 The Set-Theoretic Verdict on Strictness

As outlined in Sect. 3.2.5, we "model" real world categories in terms of concrete categories. While the real world categories *are* indeed strict monoidal categories, their corresponding models typically *aren't*.

What goes wrong is the following: for set-theory based mathematical structures such as groups, topological spaces, partial orders and vector spaces, neither

$$A \otimes (B \otimes C) = (A \otimes B) \otimes C \quad \text{nor} \quad \mathrm{I} \otimes A = A = A \otimes \mathrm{I}$$

hold. This is due to the fact that for the underlying sets X, Y, Z we have that

$$(x, (y, z)) \neq ((x, y), z) \qquad \text{and} \qquad (*, x) \neq x \neq (x, *)$$

so, as a consequence, neither

$$X \times (Y \times Z) = (X \times Y) \times Z \quad \text{nor} \quad \{*\} \times X = X = X \times \{*\}$$

hold. We do have something very closely related to this, namely

$$X \times (Y \times Z) \simeq (X \times Y) \times Z \quad \text{and} \quad \{*\} \times X \simeq X \simeq X \times \{*\}\,.$$

That is, we have *isomorphisms* rather than *strict* equations. But these isomorphisms are not just ordinary isomorphisms but so-called *natural* isomorphisms. They are an instance of the more general *natural transformations* which we will discuss in Sect. 3.6.2.[8] Meanwhile we introduce a restricted version of this general notion of natural transformation, one which comes with a clear interpretation.

Consider a category **C** that comes with an operation on objects

$$-\otimes- : |\mathbf{C}| \times |\mathbf{C}| \to |\mathbf{C}| :: (A, B) \mapsto A \otimes B\,, \tag{3.18}$$

and with for all objects $A, B, C, D \in |\mathbf{C}|$ we also have an operation on hom-sets

$$-\otimes- : \mathbf{C}(A, B) \times \mathbf{C}(C, D) \to \mathbf{C}(A \otimes C, B \otimes D) :: (f, g) \mapsto f \otimes g\,. \tag{3.19}$$

[8] Naturality is one of the most important concepts of formal category theory. In fact, in the founding paper [33] Eilenberg and MacLane argue that *their* main motivation for introducing the notion of a category is to introduce the notion of a functor, and that *their* main motivation for introducing the notion of a functor is to introduce the notion of a natural transformation.

Let

$$\Lambda(x_1, \ldots, x_n, C_1, \ldots, C_m) \quad \text{and} \quad \Xi(x_1, \ldots, x_n, C_1, \ldots, C_m)$$

be two well-formed expressions built from:

- $- \otimes -$,
- brackets,
- variables $x_1, \ldots, x_n$,
- and constants $C_1, \ldots, C_m \in |\mathbf{C}|$.

Then a natural transformation is a family

$$\left\{ \Lambda(A_1, \ldots, A_n, C_1, \ldots, C_m) \xrightarrow{\xi_{A_1,\ldots,A_n}} \Xi(A_1, \ldots, A_n, C_1, \ldots, C_m) \mid A_1, \ldots, A_n \in \mathbf{C} \right\}$$

of morphisms which are such that for all objects $A_1, \ldots, A_n, B_1, \ldots, B_n \in |\mathbf{C}|$ and all morphisms $A_1 \xrightarrow{f_1} B_1, \ldots, A_n \xrightarrow{f_n} B_n$ we have:

$$\begin{array}{ccc}
\Lambda(A_1, \ldots, A_n, C_1, \ldots, C_m) & \xrightarrow{\xi_{A_1,\ldots,A_n}} & \Xi(A_1, \ldots, A_n, C_1, \ldots, C_m) \\
{\scriptstyle \Lambda(f_1,\ldots,f_n,1_{C_1},\ldots,1_{C_m})}\Big\downarrow & & \Big\downarrow{\scriptstyle \Xi(f_1,\ldots,f_n,1_{C_1},\ldots,1_{C_m})} \\
\Lambda(B_1, \ldots, B_n, C_1, \ldots, C_m) & \xrightarrow[\xi_{B_1,\ldots,B_n}]{} & \Xi(B_1, \ldots, B_n, C_1, \ldots, C_m)
\end{array}$$

A natural transformation is a *natural isomorphism* if, in addition, all these morphisms $\xi_{A_1,\ldots,A_n}$ are isomorphisms in the sense of Definition 2.

Examples of such well-formed expressions are

$$x \otimes (y \otimes z) \qquad \text{and} \qquad (x \otimes y) \otimes z$$

and the corresponding constraint on the morphims is

$$\begin{array}{ccc}
A \otimes (B \otimes C) & \xrightarrow{\alpha_{A,B,C}} & (A \otimes B) \otimes C \\
{\scriptstyle f\otimes(g\otimes h)}\Big\downarrow & & \Big\downarrow{\scriptstyle (f\otimes g)\otimes h} \\
A' \otimes (B' \otimes C') & \xrightarrow[\alpha_{A',B',C'}]{} & (A' \otimes B') \otimes C'
\end{array} \tag{3.20}$$

If Diagram (3.20) commutes for all $A, B, C, A', B', C', f, g, h$ and the morphisms

$$\alpha := \{\alpha_{A,B,C} \mid A, B, C \in \mathbf{C}\}$$

are all isomorphisms, then this natural isomorphism is called *associativity*. Its name refers to the fact that this natural isomorphism embodies a *weaker* form of the strict associative law $A \otimes (B \otimes C) = (A \otimes B) \otimes C$. A better name would actually be *re-bracketing*, since that is what it truly does: it is a morphism—which we like to think of as a *process*—which transforms type $A \otimes (B \otimes C)$ into type $(A \otimes B) \otimes C$. In other words, it provides a *formal witness* to the actual *processes of re-bracketing a mathematical expression*. The *naturality condition* in Diagram (3.20) formally states that re-bracketing *commutes* with any triple of operations f, g, h we apply to the systems, and hence it tells us that the process of re-bracketing *does not interfere* with any non-trivial processes f, g, h—almost as if it wasn't there.

Other important pairs of well-formed formal expressions are

$$x \quad \text{and} \quad c \otimes x \qquad\qquad x \quad \text{and} \quad x \otimes c$$

and, if I is taken to be the constant object, the corresponding naturality constraint is

$$\begin{array}{ccc} A & \xrightarrow{\lambda_A} & \mathrm{I} \otimes A \\ {\scriptstyle f}\downarrow & & \downarrow{\scriptstyle 1_\mathrm{I} \otimes f} \\ B & \xrightarrow[\lambda_B]{} & \mathrm{I} \otimes B \end{array} \qquad\qquad \begin{array}{ccc} A & \xrightarrow{\rho_A} & A \otimes \mathrm{I} \\ {\scriptstyle f}\downarrow & & \downarrow{\scriptstyle f \otimes 1_\mathrm{I}} \\ B & \xrightarrow[\rho_B]{} & B \otimes \mathrm{I} \end{array} \tag{3.21}$$

The natural isomorphisms λ and ρ in Diagrams (3.21) are called *left-* and *right unit*. In this case, a better name would have been *left-* and *right introduction* since they correspond to the process of introducing a new object relative to an existing one.

We encountered a fourth important example in Definition 8, namely

$$x \otimes y \qquad \text{and} \qquad y \otimes x\,,$$

for which Diagram (3.9) is the naturality condition. The isomorphism σ is called *symmetry* but a better name could have been *exchange* or *swapping*.

Example 28 The category **Set** has associativity, left- and right unit, and symmetry natural isomorphisms relative to the Cartesian product, with the singleton set $\{*\}$ as the monoidal unit. Explicitly, setting

$$f \times f' : X \times X' \to Y \times Y' :: (x, x') \mapsto (f(x), f'(x'))$$

for $f : X \to Y$ and $f' : X' \to Y'$, these natural isomorphisms are

$$\alpha_{X,Y,Z} : X \times (Y \times Z) \to (X \times Y) \times Z :: (x, (y, z)) \mapsto ((x, y), z)$$

$$\lambda_X : X \to \{*\} \times X :: x \mapsto (*, x) \qquad\qquad \rho_X : X \to X \times \{*\} :: x \mapsto (x, *)$$

$$\sigma_{X,Y} : X \times Y \to Y \times X :: (x, y) \mapsto (y, x)$$

The reader can easily verify that Diagrams (3.9), (3.20) and (3.21) all commute. Showing that bifunctoriality holds is somewhat more tedious.

Definition 11 A *monoidal category* consists of the following data:

1. a category $\mathbf{C}$,
2. an object $\mathrm{I} \in |\mathbf{C}|$,
3. a *bifunctor* $- \otimes -$, that is, an operation both on objects and on morphisms as in prescriptions (3.18) and (3.19) above, which moreover satisfies

$$(g \circ f) \otimes (k \circ h) = (g \otimes k) \circ (f \otimes h) \quad \text{and} \quad 1_A \otimes 1_B = 1_{A \otimes B}$$

   for all $A, B \in |\mathbf{C}|$ and all morphisms f, g, h, k of appropriate type, and
4. three natural isomorphisms

$$\alpha = \{A \otimes (B \otimes C) \xrightarrow{\alpha_{A,B,C}} (A \otimes B) \otimes C \mid A, B, C \in |\mathbf{C}|\},$$
$$\lambda = \{A \xrightarrow{\lambda_A} \mathrm{I} \otimes A \mid A \in |\mathbf{C}|\} \quad \text{and} \quad \rho = \{A \xrightarrow{\rho_A} A \otimes \mathrm{I} \mid A \in |\mathbf{C}|\},$$

   hence satisfying Eqs. (3.20) and (3.21), and such that the *Mac Lane pentagon*

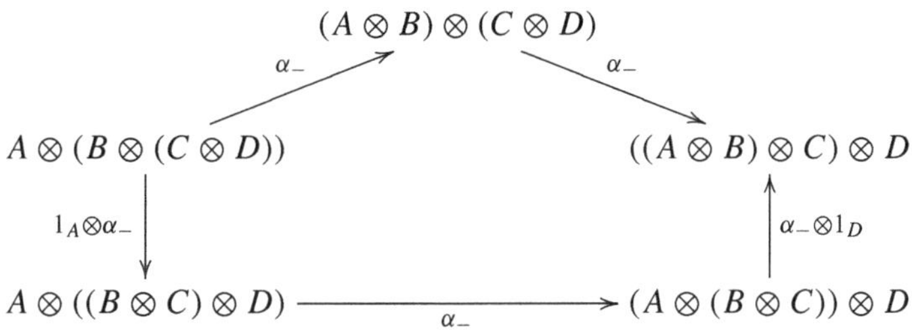

(3.22)

   commutes for all $A, B, C, D \in |\mathbf{C}|$, that also

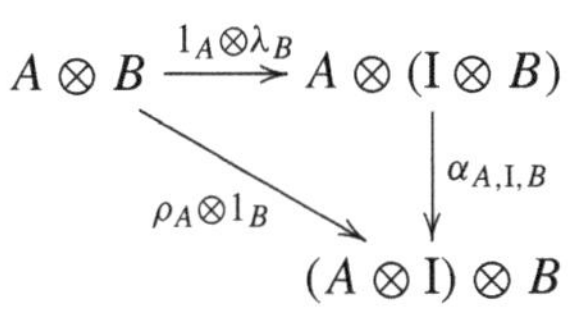

(3.23)

   commutes for all $A, B \in |\mathbf{C}|$, and that

$$\lambda_{\mathrm{I}} = \rho_{\mathrm{I}}\,. \tag{3.24}$$

A monoidal category is moreover *symmetric* if there is a fourth natural isomorphism

$$\sigma = \{A \otimes B \xrightarrow{\sigma_{A,B}} B \otimes A \mid A, B \in |\mathbf{C}|\},$$

satisfying Eq. (3.9), and such that

$$\begin{array}{ccc} A \otimes B & \xrightarrow{\sigma_{A,B}} & B \otimes A \\ & \searrow^{1_{A\otimes B}} & \downarrow \sigma_{B,A} \\ & & A \otimes B \end{array} \tag{3.25}$$

commutes for all $A, B \in |\mathbf{C}|$, that

$$\begin{array}{ccc} A & \xrightarrow{\lambda_A} & \mathrm{I} \otimes A \\ & \searrow^{\rho_A} & \downarrow \sigma_{\mathrm{I},A} \\ & & A \otimes \mathrm{I} \end{array} \tag{3.26}$$

commutes for all $A \in |\mathbf{C}|$, and that

$$\begin{array}{ccccc} A \otimes (B \otimes C) & \xrightarrow{\alpha_-} & (A \otimes B) \otimes C & \xrightarrow{\sigma_{(A\otimes B),C}} & C \otimes (A \otimes B) \\ \downarrow 1_A \otimes \sigma_{B,C} & & & & \downarrow \alpha_- \\ A \otimes (C \otimes B) & \xrightarrow[\alpha_-]{} & (A \otimes C) \otimes B & \xrightarrow[\sigma_{A,C} \otimes 1_B]{} & (C \otimes A) \otimes B \end{array} \tag{3.27}$$

commutes for all $A, B, C \in |\mathbf{C}|$.

The set-theoretic verdict on strictness is very hard! The punishment is grave: a definition which stretches over two pages, since we need to carry along associativity and unit natural isomorphisms, which, on top of that, are subject to a formal overdose of *coherence conditions*, that is, Eqs. (3.22), (3.23), (3.24), (3.25), (3.27). They embody rules which should be obeyed when natural ismorphisms interact with each other, in addition to the naturality conditions which state how natural isomorphisms interact with other morphisms in the category. For example, Eq. (3.26) tells us that if we introduce I on the left of A, and then swap I and A, that this should be the same as introducing I on the right of A. Equation (3.26) tells us that the two ways of re-bracketing the four variable expressions involved should be the same.

The idea behind coherence conditions is as follows: if for formal expressions $\Lambda(A_1, \ldots, A_n, C_1, \ldots, C_m)$ and $\Xi(A_1, \ldots, A_n, C_1, \ldots, C_m)$ there are two morphisms

$$\Lambda(A_1, \ldots, A_n, C_1, \ldots, C_m) \xrightarrow{f,g} \Xi(A_1, \ldots, A_n, C_1, \ldots, C_m)$$

which are obtained by composing the natural isomorphisms α, σ, λ, ρ and 1 both with $- \otimes -$ and $- \circ -$, then $f = g$ – identities are indeed natural isomorphisms, for the formal expressions $\Lambda(A) = \Xi(A) = A$. That Eqs. (3.22), (3.23), (3.24), (3.25), (3.27) suffice for this purpose is in itself remarkable. This is a the consequence of

MacLane's highly non-trivial coherence theorem for symmetric monoidal categories [50], which states that from this set of equations we can derive any other one.

If it wasn't for this theorem, things could have been even worse, potentially involving equations with an unbounded number of symbols.

Pffffffffffffffffffffff . . .

. . . sometimes miracles do happen:

Theorem 1 (Strictification [50] p.257) *Any monoidal category* **C** *is categorically equivalent, via a pair of strong monoidal functors* $G : \mathbf{C} \longrightarrow \mathbf{D}$ *and* $F : \mathbf{D} \longrightarrow \mathbf{C}$, *to a strict monoidal category* **D**.

The definitions of categorical equivalence and strong monoidal functor can be found below in Sect. 3.6.3. In words, what this means is that for practical purposes, arbitrary monoidal categories behave the same as strict monoidal categories. In particular, the connection between diagrammatic reasoning (incl. Dirac notation) and axiomatic reasoning for strict monoidal categories extends to arbitrary monoidal categories. The essence of the above theorem is that the unit and associativity isomorphims are so well-behaved that they don't affect this correspondence. In the graphical calculus, the associativity natural isomorphisms becomes implicit when we write

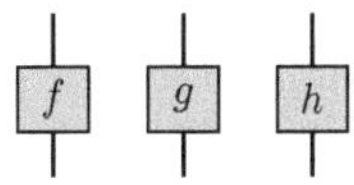

The absence of any brackets means that we can interpret this picture either as

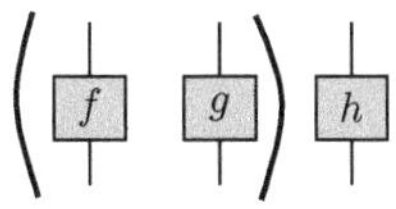

or

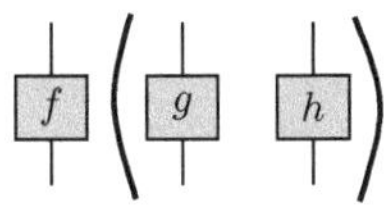

That is, it does not matter whether in first order we want to associate f with g, and then in second order this pair as a whole with h, or whether in first order we want to associate g with h, and then in second order this pair as a whole with f.

So things turn out not to be as bad as they looked at first sight!

Example 29 The category **Set** admits two important symmetric monoidal structures. We discussed the Cartesian product in Example 28. The other one is the *disjoint union*. Given two sets X and Y their disjoint union is the set

$$X + Y := \{(x, 1) \mid x \in X\} \cup \{(y, 2) \mid y \in Y\}.$$

This set can be thought of as the set of all elements both of X and Y, but where the elements of X are "coloured" with 1 while those of Y are "coloured" with 2. This guarantees that, when the same element occurs both in X and Y, it is twice accounted for in $X+Y$ since the "colours" 1 and 2 recall whether the elements in $X+Y$ either originated in X or in Y. As a consequence, the intersection of $\{(x,1) \mid x \in X\}$ and $\{(y,2) \mid y \in Y\}$ is empty, hence the name "disjoint" union.

For the disjoint union, we take the empty set $\emptyset$ as the monoidal unit and set

$$f+f' : X+X' \to Y+Y' :: \begin{cases} (x,1) \mapsto (f(x),1) \\ (x,2) \mapsto (f'(x),2) \end{cases}$$

for $f : X \to Y$ and $f' : X' \to Y'$. The natural isomorphisms of the symmetric monoidal structure are

$$\alpha_{X,Y,Z} : X+(Y+Z) \to (X+Y)+Z :: \begin{cases} (x,1) \mapsto ((x,1),1) \\ ((x,1),2) \mapsto ((x,2),1) \\ ((x,2),2) \mapsto (x,2) \end{cases}.$$

$$\lambda_X : X \to \emptyset + X :: x \mapsto (x,2) \qquad \rho_X : X \to X+\emptyset :: x \mapsto (x,1)$$

$$\sigma_{X,Y} : X+Y \to Y+X :: (x,i) \mapsto (x,3-i)$$

One again easily verifies that Diagrams (3.20), (3.21) and (3.9) all commute. Showing that bifunctoriality holds is again somewhat more tedious.

Example 30 The category $\mathbf{FdVect}_{\mathbb{K}}$ also admits two symmetric monoidal structures, provided respectively by the tensor product $\otimes$ and by the direct sum $\oplus$.

For the tensor product, the monoidal unit is the underlying field $\mathbb{K}$, while the natural isomorphisms of the monoidal structure are given by

$$\alpha_{V_1,V_2,V_3} : V_1 \otimes (V_2 \otimes V_3) \to (V_1 \otimes V_2) \otimes V_3 :: v' \otimes (v'' \otimes v''') \mapsto (v' \otimes v'') \otimes v'''$$

$$\lambda_V : V \to \mathbb{K} \otimes V :: v \mapsto 1 \otimes v \qquad \rho_V : V \to V \otimes \mathbb{K} :: v \mapsto v \otimes 1$$

$$\sigma_{V_1,V_2} : V_1 \otimes V_2 \to V_2 \otimes V_1 :: v' \otimes v'' \mapsto v'' \otimes v' .$$

Note that the inverse to λ_V is

$$\lambda_V^{-1} : \mathbb{K} \otimes V \to V :: k \otimes v \mapsto k \cdot v .$$

The "scalars" are provided by the field $\mathbb{K}$ itself, since it is in bijective correspondence with the linear maps from $\mathbb{K}$ to itself. We leave it to the reader to verify that this defines a monoidal structure.

On the other hand, the monoidal unit for the direct sum is the 0-dimensional vector space. Hence this monoidal structure only admits a single "scalar". The following subsection discusses scalars in more detail.

Definition 12 A *dagger monoidal category* $\mathbf{C}$ is a monoidal category which comes with an identity-on-objects contravariant involutive functor

$$\dagger : \mathbf{C}^{op} \longrightarrow \mathbf{C}$$

satisfying Eq. (3.17), and for which all unit and associativity natural isomorphisms are unitary. A *dagger symmetric monoidal category* **C** is both a dagger monoidal category and a symmetric monoidal category, in which the symmetry natural isomorphism is also unitary.

Example 31 The category **FdHilb** admits two dagger symmetric monoidal structures, respectively provided by the tensor product and by the direct sum. In both cases, the adjoint of Example 21 is the dagger functor.

Example 32 As we will see in great detail in Sects. 3.4.2 and 3.5.4, the category **Rel** which has sets as objects and relations as morphisms also admits two symmetric monoidal structures, just like **Set**: these are again the Cartesian product and the disjoint union. Moreover, **Rel** is dagger symmetric monoidal relative to both monoidal structures with the relational converse as the dagger functor. This is a first very important difference between **Rel** and **Set**, since the latter does not admit a dagger functor for either of the monoidal structures we identified on it.

Example 33 The category **2Cob** has 1-dimensional closed manifolds as objects, and 2-dimensional cobordisms between these as morphisms, it is dagger symmetric monoidal with the disjoint union of manifolds as its monoidal product and with the reversal of cobordisms as the dagger. This category will be discussed in great detail in Sect. 3.4.3.

Of course, in **FdHilb** the tensor product $\otimes$ and the direct sum $\oplus$ are very different monoidal structures as exemplified by the particular role each of these plays within quantum theory. In particular, as pointed out by Schrödinger in the 1930s [60], the tensor product description of compound quantum systems is what makes quantum physics so different from classical physics. We will refer to monoidal structures which are somewhat like $\otimes$ in **FdHilb** as *quantum-like*, and to those that are rather like $\oplus$ in **FdHilb** as *classical-like*. As we will see below, the quantum-like tensors allow for correlations between subsystems, so the joint state can in general not be decomposed into states of the individual subsystems. In contrast, the classical-like tensors can only describe "separated" systems, that is, the state of a joint system can always be faithfully represented by states of the individual subsystems.

The tensors considered in this paper have the following nature:

Category	**Classical-like**	**Quantum-like**	**Other** (see §3.4.3)
Set	$\times$		$+$
Rel	$+$	$\times$	
FdHilb	$\oplus$	$\otimes$	
nCob		$+$	

Observe the following remarkable facts:

- While $\times$ behaves "classical-like" in **Set**, it behaves "quantum-like" in **Rel**, and this despite the fact that **Rel** contains **Set** as a subcategory with the same objects as **Rel**, and which inherits its monoidal structures from **Rel**.

- There is a remarkable parallel between the role that the pair $(\oplus, \otimes)$ plays for **FdHilb** and the role that the pair $(+, \times)$ plays for **Rel**.
- In **nCob** the direct sum even becomes "quantum-like"—a point which has been strongly emphasized for a while by John Baez [9].

All of this clearly indicates that being either quantum-like and classical-like is something that involves not just the objects, but also the tensor and the morphism structure.

Sections 3.4 and 3.5 provide a detailed discussion of these two very distinct kinds of monoidal structures, which will shed more light on the above table.

To avoid confusion concerning which monoidal structure on a category we are considering, we may specify it e.g., (**FdHilb**, $\otimes$, $\mathbb{C}$).

3.3.5 Scalar Valuation and Multiples

In any monoidal category **C** the hom-set $\mathbb{S}_{\mathbf{C}} := \mathbf{C}(\mathrm{I}, \mathrm{I})$ is always a monoid with categorical composition as monoid multiplication. Therefore we call $\mathbb{S}_{\mathbf{C}}$ the *scalar monoid* of the monoidal category **C**. Such a monoid equips any monoidal category with explicit quantitative content. For instance, if **C** is dagger monoidal, scalars can be produced in terms of the inner-product of Definition 10.

The following is a fascinating fact discovered by Kelly and Laplaza in [41]: even for "non-symmetric" monoidal categories, the scalar monoid is always commutative. The proof is given by the following commutative diagram:

$$
\begin{array}{ccccccccc}
\mathrm{I} & \xrightarrow{\simeq} & \mathrm{I}\otimes\mathrm{I} & = & \mathrm{I}\otimes\mathrm{I} & = & \mathrm{I}\otimes\mathrm{I} & \xrightarrow{\simeq} & \mathrm{I} \\
{\scriptstyle t}\uparrow & & \uparrow{\scriptstyle 1_\mathrm{I}\otimes t} & & & & \uparrow{\scriptstyle s\otimes 1_\mathrm{I}} & & \uparrow{\scriptstyle s} \\
\mathrm{I} & \xrightarrow{\simeq} & \mathrm{I}\otimes\mathrm{I} & & \uparrow{\scriptstyle s\otimes t} & & \mathrm{I}\otimes\mathrm{I} & \xrightarrow{\simeq} & \mathrm{I} \\
{\scriptstyle s}\uparrow & & \uparrow{\scriptstyle s\otimes 1_\mathrm{I}} & & & & \uparrow{\scriptstyle 1_\mathrm{I}\otimes t} & & \uparrow{\scriptstyle t} \\
\mathrm{I} & \xrightarrow[\simeq]{} & \mathrm{I}\otimes\mathrm{I} & = & \mathrm{I}\otimes\mathrm{I} & = & \mathrm{I}\otimes\mathrm{I} & \xrightarrow[\simeq]{} & \mathrm{I}
\end{array}
\qquad (3.28)
$$

Equality of the two outer paths both going from the left-lower-corner to the right-upper-corner boils down to equality between:

- the outer left/upper path which consists of $t \circ s$, and the composite of isomorphism $\mathrm{I} \simeq \mathrm{I} \otimes \mathrm{I}$ with its inverse, so nothing but 1_I, giving all together $t \circ s$, and
- the outer lower/right path, giving all together $s \circ t$.

Their equality relies on bifunctoriality (cf. middle two rectangles) and naturality of the left- and right-unit isomorphisms (cf. the four squares).

Diagrammatically commutativity is subsumed by the fact that scalars do not have wires, and hence can 'move freely around in the picture':

This result has physical consequences. Above we argued that strict monoidal categories model physical systems and processes thereon. We now discovered that a strict monoidal category **C** always has a commutative endomorphism monoid $\mathbb{S}_{\mathbf{C}}$. So when varying quantum theory by changing the underlying field $\mathbb{K}$ of the vector space, we need to restrict ourselves to commutative fields, hence excluding things like "quaternionic quantum mechanics" [34].

Example 34 We already saw that the elements of $\mathbb{S}_{(\mathbf{FdHilb},\otimes,\mathbb{C})}$ are in bijective correspondence with those of $\mathbb{C}$, in short,

$$\mathbb{S}_{(\mathbf{FdHilb},\otimes,\mathbb{C})} \simeq \mathbb{C}\,.$$

In **Set** however, since there is only one function of type $\{*\} \to \{*\}$, namely the identity, $\mathbb{S}_{(\mathbf{Set},\times,\{*\})}$ is a singleton, in short,

$$\mathbb{S}_{(\mathbf{Set},\times,\{*\})} \simeq \{*\}\,.$$

Thus, the scalar structure on $(\mathbf{Set}, \times, \{*\})$ is trivial. On the other hand, in **Rel** there are two relations of type $\{*\} \to \{*\}$, the identity and the empty relation, so

$$\mathbb{S}_{(\mathbf{Rel},\times,\{*\})} \simeq \mathbb{B}\,,$$

where $\mathbb{B}$ are the Booleans. Hence, the scalar structure on $(\mathbf{Rel}, \times, \{*\})$ is non-trivial as it is that of Boolean logic. Operationally, we can interpret these two scalars as "possible" and "impossible" respectively. When rather considering $\oplus$ on **FdHilb** instead of $\otimes$ we again have a trivial scalar structure, since there is only one linear map from the 0-dimensional Hilbert space to itself. So

$$\mathbb{S}_{(\mathbf{FdHilb},\oplus,\mathbb{C})} \simeq \{*\}\,.$$

So scalars and scalar multiples are more closely related to the "multiplicative" tensor product structure than to the "additive" direct sum structure. We also have

$$\mathbb{S}_{(\mathbf{nCob},+,\emptyset)} \simeq \mathbb{N}\,.$$

In general, it is the quantum-like monoidal structures which admit non-trivial scalar structure. This might come as a surprise to the reader, given that for vector spaces one typically associates these scalars with linear combinations of vectors, which are very much "additive" in spirit.

The right half of commutative Diagram (3.28) states that

$$s \circ t = \mathrm{I} \xrightarrow{\simeq} \mathrm{I} \otimes \mathrm{I} \xrightarrow{s \otimes t} \mathrm{I} \otimes \mathrm{I} \xrightarrow{\simeq} \mathrm{I}\,.$$

We generalize this by defining *scalar multiples* of a morphism $A \xrightarrow{f} B$ as

$$s \bullet f := A \xrightarrow{\simeq} \mathrm{I} \otimes A \xrightarrow{s \otimes f} \mathrm{I} \otimes B \xrightarrow{\simeq} B\,.$$

These scalars satisfy the usual properties, namely

$$(t \bullet g) \circ (s \bullet f) = (t \circ s) \bullet (g \circ f) \tag{3.29}$$

and

$$(s \bullet f) \otimes (t \bullet g) = (s \circ t) \bullet (f \otimes g)\,, \tag{3.30}$$

cf. in matrix calculus we have

$$\left(y \begin{pmatrix} b_{11} & b_{12} \\ b_{21} & b_{22} \end{pmatrix}\right)\left(x \begin{pmatrix} a_{11} & a_{12} \\ a_{21} & a_{22} \end{pmatrix}\right) = yx \left(\begin{pmatrix} b_{11} & b_{12} \\ b_{21} & b_{22} \end{pmatrix}\begin{pmatrix} a_{11} & a_{12} \\ a_{21} & a_{22} \end{pmatrix}\right)$$

and

$$\left(x \begin{pmatrix} a_{11} & a_{12} \\ a_{21} & a_{22} \end{pmatrix}\right) \otimes \left(y \begin{pmatrix} b_{11} & b_{12} \\ b_{21} & b_{22} \end{pmatrix}\right) = xy \left(\begin{pmatrix} a_{11} & a_{12} \\ a_{21} & a_{22} \end{pmatrix} \otimes \begin{pmatrix} b_{11} & b_{12} \\ b_{21} & b_{22} \end{pmatrix}\right).$$

Diagrammatically these properties are again implicit and require 'artificial' brackets to be made explicit, for example, Eq. (3.29) is hidden as:

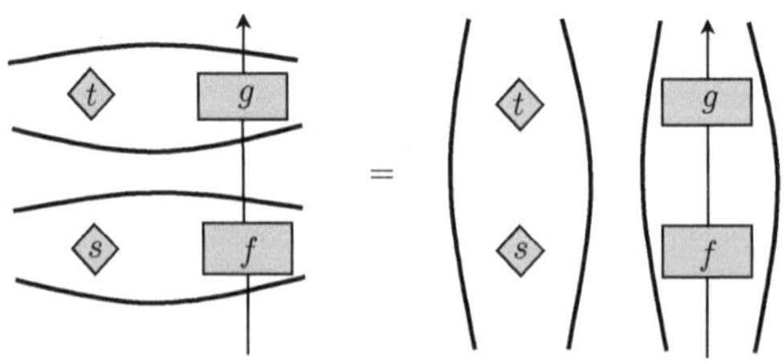

Of course, we could still prove these properties with commutative diagrams. For Eq. (3.29) the left-hand-side and the right-hand-side are respectively the top and the bottom path of the following diagram:

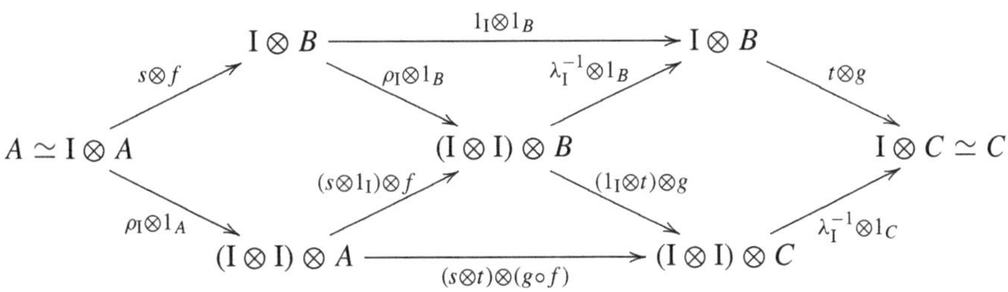

where we use the fact that $t \circ s = \lambda_I^{-1} \circ (s \otimes t) \circ \rho_I$. The diamond on the left commutes by naturality of ρ_I. The top triangle commutes because both paths are equal to $1_{I \otimes B}$ as $\lambda_I = \rho_I$. The bottom triangle commutes by Eq. (3.15). Finally, the right diamond commutes by naturality of λ_I.

3.4 Quantum-Like Tensors

So what makes $\otimes$ so different from $\oplus$ in the category **FdHilb**, what makes $\times$ so different in the categories **Rel** and **Set**, and what makes $\times$ so similar in the category **Rel** to $\otimes$ in the category **FdHilb**?

3.4.1 Compact Categories

Definition 13 A *compact (closed) category* **C** is a symmetric monoidal category in which every object $A \in |\mathbf{C}|$ comes with

1. another object A^*, the *dual* of A,
2. a pair of morphisms

$$\mathrm{I} \xrightarrow{\eta_A} A^* \otimes A \qquad \text{and} \qquad A \otimes A^* \xrightarrow{\epsilon_A} \mathrm{I},$$

   respectively called *unit* and *counit*,

which are such that the following two diagrams commute:

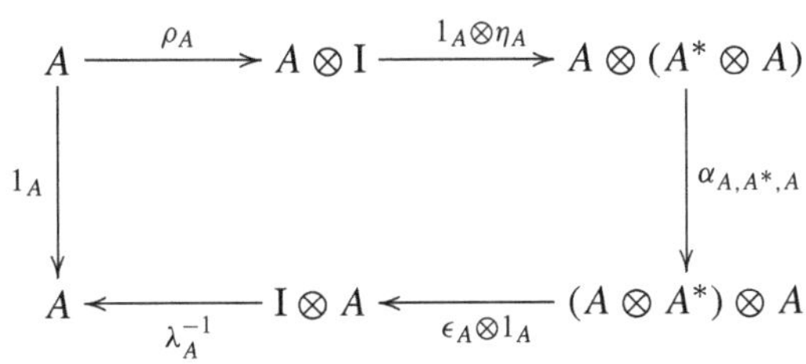

(3.31)

$$\begin{array}{ccccc} A^* & \xrightarrow{\lambda_{A^*}} & \mathrm{I}\otimes A^* & \xrightarrow{\eta_A\otimes 1_{A^*}} & (A^*\otimes A)\otimes A^* \\ \downarrow{\scriptstyle 1_{A^*}} & & & & \downarrow{\scriptstyle \alpha^{-1}_{A^*,A,A^*}} \\ A^* & \xleftarrow[\rho^{-1}_{A^*}]{} & A^*\otimes \mathrm{I} & \xleftarrow[1_{A^*}\otimes\epsilon_A]{} & A^*\otimes(A\otimes A^*) \end{array} \tag{3.32}$$

In the case that **C** is strict the above diagrams simplify to

$$\begin{array}{ccc} A & \xrightarrow{1_A} & A \\ \downarrow{\scriptstyle 1_A\otimes\eta_A} & \nearrow{\scriptstyle \epsilon_A\otimes 1_A} & \\ A\otimes A^*\otimes A & & \end{array} \qquad \begin{array}{ccc} A^* & \xrightarrow{\eta_A\otimes 1_{A^*}} & A^*\otimes A\otimes A^* \\ & \searrow{\scriptstyle 1_{A^*}} & \downarrow{\scriptstyle 1_{A^*}\otimes\epsilon_A} \\ & & A^* \end{array} \tag{3.33}$$

Definition 13 can also be expressed diagrammatically, provided we introduce some new graphical elements:

- As before A will be represented by an upward arrow:

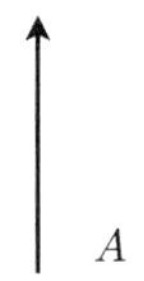

On the other hand, we depict A^*, the dual object to A, either by an upward arrow labelled by A^*, or by a downward arrow labelled A:

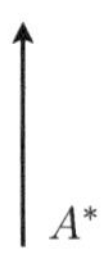

or

- The unit η_A and counit ϵ_A are respectively depicted as

- Commutation of the two diagrams now boils down to:

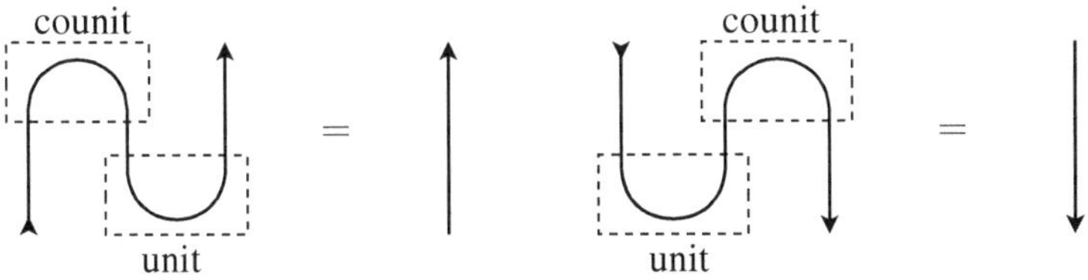

When expressed diagrammatically, these equational constraints admit the simple interpretation of "yanking a wire". While at first sight compactness of a category as stated in Definition 13 seems to be a somewhat ad hoc notion, this graphical interpretation establishes it as a very canonical one which extends the graphical calculus for symmetric monoidal categories with *cup*- and *cap*-shaped wires. As the following lemma shows, the equational constraints imply that we are allowed to 'slide' morphisms also along these cups and caps.

Lemma 1 *Given a morphism $f : A \longrightarrow B$ define its transpose to be*

$$f^* := (1_{A^*} \otimes \epsilon_B) \circ (1_{A^*} \otimes f \otimes 1_{B^*}) \circ (\eta_A \otimes 1_{B^*}) : B^* \longrightarrow A^* .$$

Diagrammatically, when depicting the morphism f as

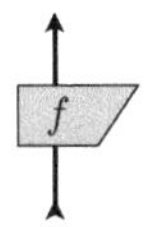

then its transpose is depicted as

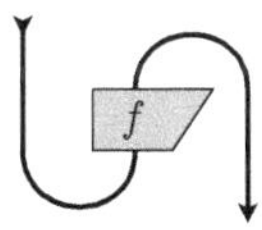

Anticipating what will follow, we abbreviate this notation for f^* to

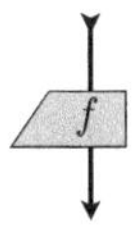

With this graphical notation we have:

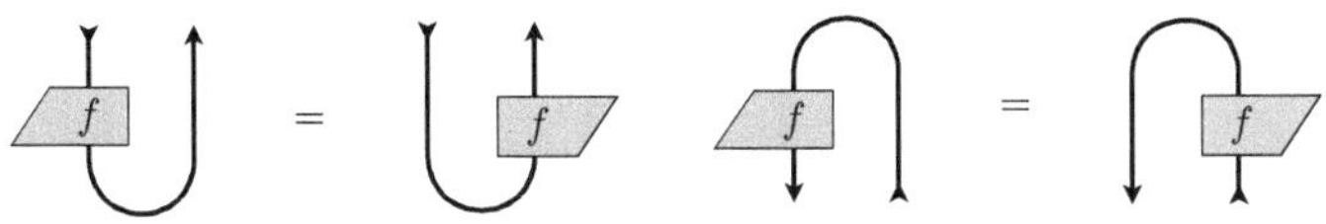

that is, we can "slide" morphisms along cup- and cap-shaped wires.

The proof of the first equality simply is

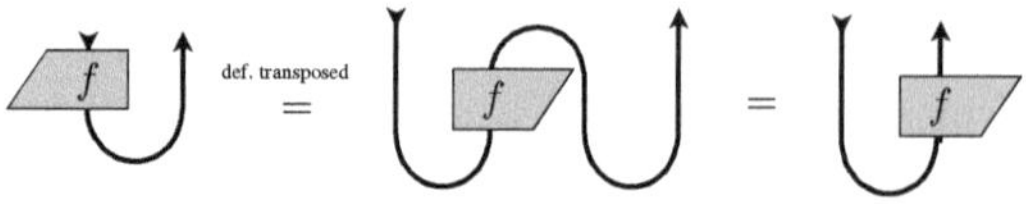

The proof for the second equality proceeds analogously.

Example 35 The category **FdVect**$_{\mathbb{K}}$ is compact. We take the usual linear algebraic dual space V^* to be V's dual object and the unit to be

$$\eta_V : \mathbb{K} \to V^* \otimes V :: 1 \mapsto \sum_{i=1}^{n} f_i \otimes e_i$$

where $\{e_i\}_{i=1}^n$ is a basis of V and $f_j \in V^*$ is the linear functional such that $f_j(e_i) = \delta_{i,j}$ for all $1 \leq i, j \leq n$. Finally, we take the counit to be

$$\epsilon_V : V \otimes V^* \to \mathbb{K} :: e_i \otimes f_j \mapsto f_j(e_i)\,.$$

We leave it to the reader to verify commutation of Diagrams 3.31 and 3.32. Two important points need to be made here:

- The linear maps η_V and ϵ_V do not depend on the choice of the basis $\{e_i\}_{i=1}^n$. It suffices to verify that there is a canonical isomorphism

$$\mathbf{FdVect}_{\mathbb{K}}(V, V) \xrightarrow{\simeq} \mathbf{FdVect}_{\mathbb{K}}(\mathbb{K}, V^* \otimes V)$$

  which does not depend on the choice of basis. The unit η_V is the image of 1_V under this isomorphism and since 1_V is independent of the choice of basis it follows that η_V does not depend on any choice of basis. The argument for ϵ_V proceeds analogously.
- There are other possible choices for η_V and ϵ_V which turn **FdVect**$_{\mathbb{K}}$ into a compact category. For example, if $f : V \to V$ is invertible then

$$\eta'_V := (1_{V^*} \otimes f) \circ \eta_V \qquad \text{and} \qquad \epsilon'_V := \epsilon_V \circ (f^{-1} \otimes 1_{V^*})$$

  make Diagrams (3.31) and (3.32) commute. Indeed, graphically we have:

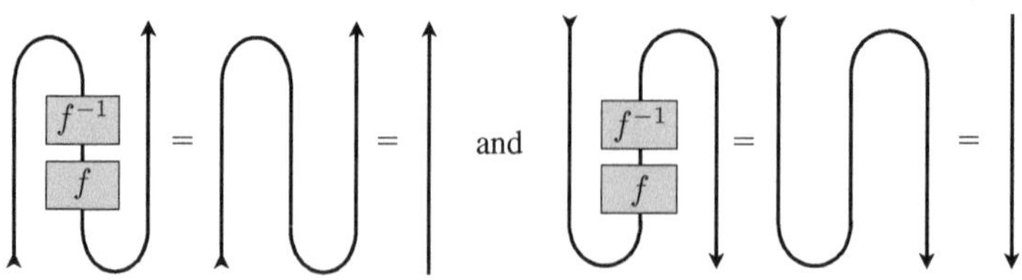

Example 36 The category **Rel** of sets and relations is also compact relative to the Cartesian product as we shall see in detail in Sect. 3.4.2.

Example 37 The category **QuantOpp** is compact. We can pick Bell-states as the units and the corresponding Bell-effects as counits. As shown in [2, 20], compactness is exactly what enables modeling protocols such as quantum teleportation:

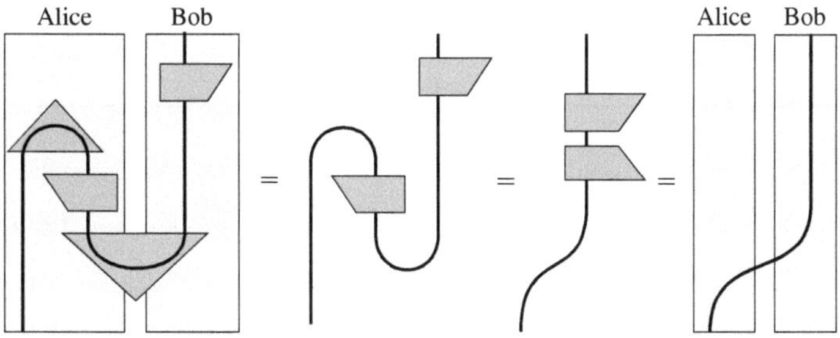

where the trapezoid is assumed to be unitary and hence, its adjoint coincides with its inverse. The classical information flow is (implicitly) encoded in the fact that the same trapezoid appears in the left-hand-side picture both at Alice's and Bob's side.

Given a morphism $f : A \longrightarrow B$ in a compact category, its *name*

$$\mathrm{I} \xrightarrow{\ulcorner f \urcorner} A^* \otimes B$$

and its *coname*

$$A \otimes B^* \xrightarrow{\llcorner f \lrcorner} \mathrm{I}$$

are defined by:

$$\begin{array}{ccc} A^* \otimes A & \xrightarrow{1_{A^*} \otimes f} & A^* \otimes B \\ {\scriptstyle \eta_A} \uparrow & \nearrow {\scriptstyle \ulcorner f \urcorner} & \\ \mathrm{I} & & \end{array} \qquad \text{and} \qquad \begin{array}{ccc} & & \mathrm{I} \\ & \nearrow {\scriptstyle \llcorner f \lrcorner} & \uparrow {\scriptstyle \epsilon_B} \\ A \otimes B^* & \xrightarrow[f \otimes 1_{B^*}]{} & B \otimes B^* \end{array}$$

Following [2] we can show that for $f : A \longrightarrow B$ and $g : B \longrightarrow C$

$$\lambda_C^{-1} \circ (\llcorner f \lrcorner \otimes 1_C) \circ (1_A \otimes \ulcorner g \urcorner) \circ \rho_A = g \circ f$$

always holds. The graphical proof is again trivial:

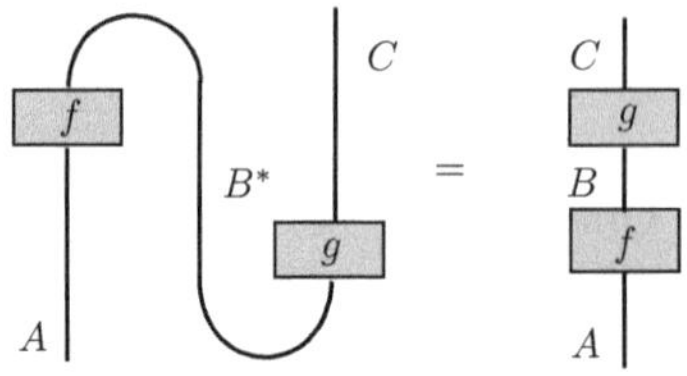

In contrast a (non-strict) symbolic proof goes as follows:

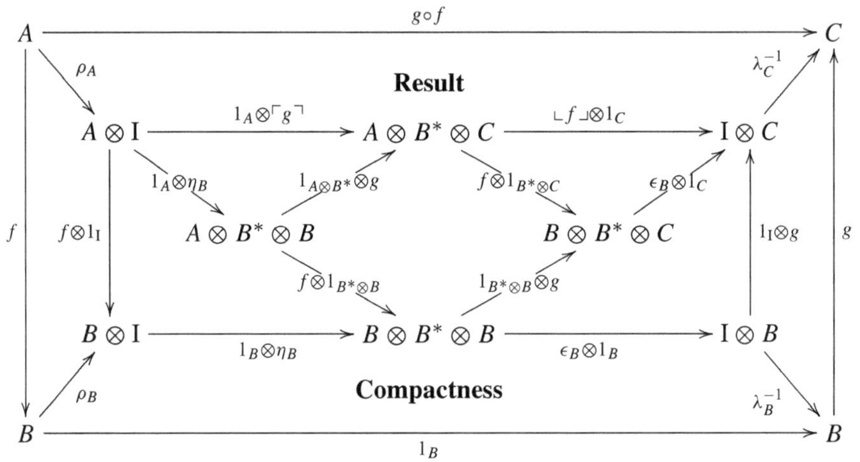

Both paths on the outside are equal to $g \circ f$. We want to show that the pentagon labelled "Result" commutes. To do this we will "unfold" arrows using equations which hold in compact categories in order to pass from the composite $g \circ f$ at the left/bottom/right to $\lambda_C^{-1} \circ (\llcorner f \lrcorner \otimes 1_C) \circ (1_A \otimes \ulcorner g \urcorner) \circ \rho_A$. This will transform the tautology $g \circ f = g \circ f$ into commutation of the pentagon labelled "Result". For instance, we use compactness to go from the identity arrow at the bottom of the diagram to the composite $\lambda_B^{-1} \circ (\epsilon_B \otimes 1_B) \circ (1_B \otimes \eta_B) \circ \rho_B$. The outer left and right trapezoids express naturality of ρ and λ. The remaining triangles/diamonds express bifunctoriality and the definitions of name/coname.

The scalar $\epsilon_A \circ \sigma_{A^*,A} \circ \eta_A : \mathbb{K} \to \mathbb{K}$ depicts as

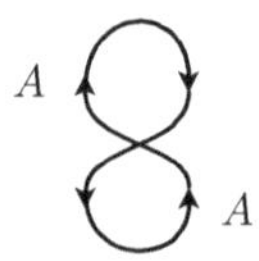

and when setting

it becomes an 'A-labelled circle'

Example 38 In $\mathbf{FdVect}_{\mathbb{K}}$ the V-labelled circle stands for the dimension of the vector space V. By the definitions of η_V and ϵ_V, the previous composite is equal to

$$\sum_{ij} f_j(e_i) = \sum_{ij} \delta_{i,j} = \sum_{i} 1 = \dim(V).$$

Definition 14 A *dagger compact category* $\mathbf{C}$ is both a compact category and a dagger symmetric monoidal category, such that for all $A \in |\mathbf{C}|$, $\epsilon_A = \eta_A^\dagger \circ \sigma_{A,A^*}$.

Example 39 The category **FdHilb** is dagger compact.

3.4.2 The Category of Relations

We now turn our attention to the category **Rel** of sets and relations, a category which we briefly encountered in previous sections. Perhaps surprisingly, **Rel** possesses more "quantum features" than the category **Set** of sets and functions. In particular, just like **FdHilb** it is a dagger compact category.

A *relation* $R : X \to Y$ between two sets X and Y is a subset of the set of all their ordered pairs, that is, $R \subseteq X \times Y$. Thus, given an element $(x, y) \in R$, we say that $x \in X$ *relates* to $y \in Y$, which we denote as xRy. The set

$$R := \{(x, y) \mid xRy\}$$

is also referred to as the *graph* of the relation.

Example 40 For the relation "strictly less than" or "<" on the natural numbers, we have that 2 relates to 5, which is denoted as $2 < 5$ or $(2, 5) \in \, < \, \subseteq \mathbb{N} \times \mathbb{N}$. For the relation "is a divisor of" or "|" on the natural numbers, we have $6|36$ or $(6, 36) \in \, | \subseteq \mathbb{N} \times \mathbb{N}$. Other examples are general preorders or equivalence relations.

Definition 15 The monoidal category **Rel** is defined as follows:

- The objects are sets.
- The morphisms are all relations $R : X \to Y$.
- For $R_1 : X \to Y$ and $R_2 : Y \to Z$ the composite $R_2 \circ R_1 \subseteq X \times Z$ is

$$R_2 \circ R_1 := \{(x, z) \mid \text{there exists a } y \in Y \text{ such that } xR_1y \text{ and } yR_2z\}.$$

Composition is easily seen to be associative. For $X \in |\mathbf{Rel}|$ we have

$$1_X := \{(x, x) \mid x \in X\}.$$

- The monoidal product of two sets is their Cartesian product, the unit for the monoidal structure is any singleton, and for two relations $R_1 : X_1 \to Y_1$ and $R_2 : X_2 \to Y_2$ the monoidal product $R_1 \times R_2 \subseteq X_1 \times X_2 \to Y_1 \times Y_2$ is

$$R_1 \times R_2 := \{((x, x'), (y, y')) \mid x R_1 y \text{ and } x' R_2 y'\} \subseteq (X_1 \times X_2) \times (Y_1 \times Y_2)\,.$$

We mentioned before that **Set** was contained in **Rel** as a "sub-monoidal category". In **Rel**, the left- and right-unit natural isomorphisms respectively are

$$\lambda_X := \{(x, (*, x)) \mid x \in X\} \quad \text{and} \quad \rho_X := \{(x, (x, *)) \mid x \in X\}\,,$$

and the associativity natural isomorphism is

$$\alpha_{X,Y,Z} := \{((x, (y, z)), ((x, y), z)) \mid x \in X,\, y \in Y \text{ and } z \in Z\}.$$

These relations are all single-valued, so they are also functions, and they are the same functions as the natural isomorphisms for the Cartesian product in **Set**. Let us verify the coherence conditions for them:

(i) The pentagon

$$\begin{array}{ccccc}
W \times (X \times (Y \times Z)) & \xrightarrow{\alpha_-} & (W \times X) \times (Y \times Z) & \xrightarrow{\alpha_-} & ((W \times X) \times Y) \times Z \\
{\scriptstyle 1\times\alpha_-}\downarrow & & & & \uparrow{\scriptstyle \alpha_-\times 1} \\
W \times ((X \times Y) \times Z) & & \xrightarrow[\alpha_-]{\hspace{6em}} & & (W \times (X \times Y)) \times Z
\end{array}$$

indeed commutes. The top part

$$\alpha_- \circ \alpha_- : W \times (X \times (Y \times Z)) \to ((W \times X) \times Y) \times Z$$

is by definition a subset of

$$(W \times (X \times (Y \times Z))) \times (((W \times X) \times Y) \times Z)\,.$$

Unfolding the definition of relational composition we obtain

$$\alpha_- \circ \alpha_- = \Big\{((w, (x, (y, z))), (((w'', x''), y''), z'')) \;\Big|\; \exists ((w', x'), (y', z')) \text{ s.t. } (w, (x, (y, z)))\alpha((w', x'), (y', z')) \text{ and } ((w', x'), (y', z'))\alpha(((w'', x''), y''), z'')\Big\}\,,$$

which by the definition of α simplifies to

$$\alpha_- \circ \alpha_- = \{((w, (x, (y, z))), (((w, x), y), z)) \mid w \in W, x \in X, y \in Y, z \in Z\}\,.$$

The bottom path yields the same result, hence making the pentagon commute. For the remaining diagrams we leave the details to the reader.

(ii) The triangle

$$\begin{array}{ccc} X \times Y & \xrightarrow{1_A \times \lambda_Y} & X \times (\{*\} \times Y) \\ & \searrow^{\rho_X \times 1_Y} & \downarrow \alpha_{X,\{*\},Y} \\ & & (X \times \{*\}) \times Y \end{array}$$

commutes as both paths are now equal to

$$\{((x, y), ((x, *), y)) \mid x \in X \text{ and } y \in Y\}.$$

As $\times$ is symmetric in **Set** we also expect **Rel** to be symmetric monoidal. For any X and $Y \in |\mathbf{Rel}|$, the natural isomorphism

$$\sigma_{X,Y} := \{((x, y), (y, x)) \mid x \in X \text{ and } y \in Y\}$$

also obeys the coherence conditions:

(i) The two triangles

$$\begin{array}{ccc} X \times Y & \xrightarrow{\sigma_{X,Y}} & Y \times X \\ & \searrow^{=} & \downarrow \sigma_{Y,X} \\ & & X \times Y \end{array} \quad , \quad \begin{array}{ccc} X & \xrightarrow{\lambda_X} & \{*\} \times X \\ & \searrow^{\rho_X} & \downarrow \sigma_{\{*\},X} \\ & & X \times \{*\} \end{array}$$

commute since both paths of the left triangle are equal to

$$\{((x, y), (x, y)) \mid x \in X \text{ and } y \in Y\},$$

while the paths of the right triangle are equal to

$$\{(x, (x, *)) \mid x \in X\}.$$

(ii) The hexagon

$$\begin{array}{ccccc} X \times (Y \times Z) & \xrightarrow{\alpha_-} & (X \times Y) \times Z & \xrightarrow{\sigma_{(X \times Y),Z}} & Z \times (X \times Y) \\ \downarrow 1_X \times \sigma_{Y,Z} & & & & \downarrow \alpha_- \\ X \times (Z \times Y) & \xrightarrow[\alpha_-]{} & (X \times Z) \times Y & \xrightarrow[\sigma_{X,Y} \times 1_Z]{} & (Z \times X) \times Y \end{array}$$

commutes since both paths are equal to

$$\{((x,(y,z)),((z,x),y)) \mid x \in X, y \in Y \text{ and } z \in Z\}.$$

So **Rel** is indeed a symmetric monoidal category as expected. **Rel** shares many common characteristics with **FdHilb**, one of them being a †-compact structure. Firstly, **Rel** is compact closed with self-dual objects that is, $X^* = X$ for any $X \in |\mathbf{Rel}|$. Moreover, for any $X \in |\mathbf{Rel}|$ let

$$\eta_X : \{*\} \to X \times X := \{(*, (x,x)) \mid x \in X\}$$

and

$$\epsilon_X : X \times X \to \{*\} := \{((x,x),*) \mid x \in X\}.$$

These morphisms make

$$\begin{array}{ccccc} X & \xrightarrow{\rho_X} & X \times \{*\} & \xrightarrow{1_X \times \eta_X} & X \times (X \times X) \\ {\scriptstyle 1_X}\downarrow & & & & \downarrow{\scriptstyle \alpha_-} \\ X & \xleftarrow[\lambda_X^{-1}]{} & \{*\} \times X & \xleftarrow[\epsilon_X \times 1_X]{} & (X \times X) \times X \end{array}$$

and its dual both commute. Indeed:

(a) The composite

$$(1_X \times \eta_X) \circ \rho_X : X \to X \times (X \times X)$$

is the set of tuples

$$\{(x,(x',(x'',x''')))\} \subseteq X \times (X \times (X \times X))$$

such that there exists an $(x'''', *) \in X \times \{*\}$ with

$$x\, \rho_X\, (x'''', *) \quad \text{and} \quad (x'''', *)\, (1_X \times \eta_X)\, (x', (x'', x''')).$$

By definition of ρ and 1_X, and of the product of relations, this entails that x, x'''' and x' are all equal. Moreover, by definition of η_X, and of the product of relations, we have that x'' and x''' are also equal. Thus,

$$(1_X \times \eta_X) \circ \rho_X := \{(x,(x,(x',x'))) \mid x, x' \in X\}.$$

(b) Hence the composite

$$\alpha \circ ((1_X \times \eta_X) \circ \rho) : X \to (X \times X) \times X$$

is

$$\alpha \circ ((1_X \times \eta_X) \circ \rho) = \{(x, ((x, x'), x') \mid x, x' \in X\}.$$

(c) The composite

$$(\epsilon_X \times 1_X) \circ (\alpha \circ (1_X \times \eta_X) \circ \rho) : X \to \{*\} \times X$$

is a set of tuples

$$\{(x, (*, x'))\} \subseteq X \times (\{*\} \times X)$$

such that there exists an $((x'', x'''), x'''') \in (X \times X) \times X$ with

$$x\ (\alpha \circ (1_X \times \eta_X) \circ \rho)\ ((x'', x'''), x'''') \text{ and } ((x'', x'''), x'''')\ (\epsilon_X \times 1_X)\ (*, x').$$

By the computation in (b) we have that $x = x''$ and $x''' = x''''$. By definition of ϵ_X, 1_X and the product of relations we have $x'' = x'''$ and $x'''' = x'$. All this together yields $x = x'' = x''' = x'''' = x'$ and hence

$$(\epsilon_X \otimes 1_X) \circ (\alpha \circ (1_X \otimes \eta_X) \circ \rho) = \{(x, (*, x)) \mid x \in X\}.$$

(d) Post-composing the previous composite with the natural isomorphism λ_X^{-1} yields a morphism of type $X \to X$, namely

$$\lambda_X^{-1} \circ (\epsilon_X \otimes 1_X) \circ \alpha \circ (1_X \otimes \eta_X) \circ \rho = \{(x, x) \mid x \in X\}$$

which is the identity relation as required.

Commutation of the dual diagram is done analogously. From this, we conclude that **Rel** is compact closed. The obvious candidate for the dagger

$$\dagger : \mathbf{Rel}^{op} \longrightarrow \mathbf{Rel}$$

is the relational converse. For relation $R : X \to Y$ its converse $R^{\cup} : Y \to X$ is

$$R^{\cup} := \{(y, x) \mid xRy\}.$$

We define the contravariant identity-on-objects involutive functor

$$\dagger : \mathbf{Rel} \longrightarrow \mathbf{Rel} :: R \mapsto R^{\cup}.$$

Note that the adjoint and the transpose coincide, that is,

$$R^* = (1_X \times \epsilon_Y) \circ (1_X \times R \times 1_Y) \circ (\eta_X \times 1_Y) = R^\dagger$$

which the reader may easily check. Finally, we verify that **Rel** is dagger compact:

- The category **Rel** is dagger monoidal:
  (i) From the definition of the monoidal product of two relations

  $$R_1 := \{(x, y) \mid xRy\} \quad \text{and} \quad R_2 := \{(x', y') \mid x'Ry'\}$$

  we have that

  $$(R_1 \times R_2)^\dagger = \{((y, y'), (x, x')) \mid xR_1y \text{ and } x'R_2y'\} = R_1^\dagger \times R_2^\dagger.$$

  (ii) The fact that $\alpha^\dagger = \alpha^{-1}$, $\lambda^\dagger = \lambda^{-1}$, $\rho^\dagger = \rho^{-1}$ and $\sigma^\dagger = \sigma^{-1}$ is trivial as the inverse of all these morphisms is the relational converse.

- The diagram

  $$\begin{array}{ccc} \{*\} & \xrightarrow{\epsilon_X^\dagger} & X \times X \\ & \searrow^{\eta_X} & \downarrow{\sigma_{X,X}} \\ & & X \times X \end{array}$$

  commutes since from

  $$\epsilon_X := \{((x, x), *) \mid x \in X\}$$

  follows

  $$\epsilon_X^\dagger := \{(*, (x, x)) \mid x \in X\}$$

  and hence $\sigma \circ \epsilon_X^\dagger = \epsilon_X^\dagger = \eta_X$.

So **Rel** is indeed a dagger compact category.

3.4.3 The Category of 2D Cobordisms

The category **2Cob** can be informally described as a category whose morphisms, so-called *cobordisms*, describe the "topological evolution" of manifolds of dimension $2 - 1 = 1$ through time. For instance, consider some snapshots of two circles which merge into a single circle, with time going upwards:

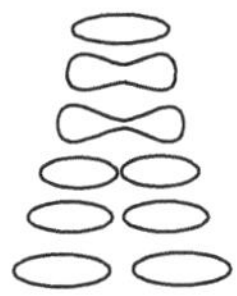

Passing to the continuum, the same process can be described by the cobordism

Thus, we take a cobordism to be a (compact) 2-dimensional manifold whose boundary is partitioned in two. We take these closed one-dimensional manifolds to be the domain and the codomain of the cobordism. Since we are only interested in the topology of the manifolds, each (co)domain consists of a finite number of closed strings.

Definition 16 The category **2Cob** is defined as follows:

- Each object is a finite number of closed strings. Hence each object can be equivalently represented by a natural number $n \in \mathbb{N}$:

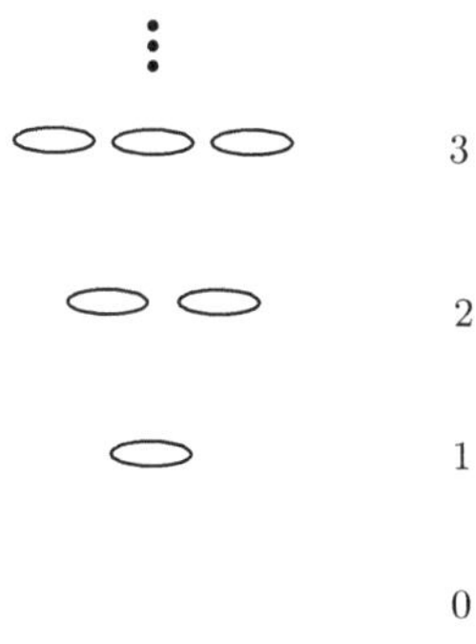

- Morphisms are cobordisms $M : n \to m$ taking $n \in \mathbb{N}$ (strings) to $m \in \mathbb{N}$ (strings), which are defined up to homeomorphic equivalence. Hence, if a cobordism can be continuously deformed into another cobordism, then these two cobordisms correspond to the same morphisms.
- For each object n, the identity $1_n : n \to n$ which is given by n parallel cylinders:

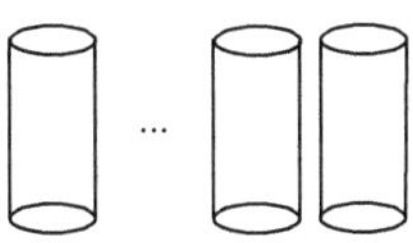

- Composition is given by "gluing" manifolds together, e.g.

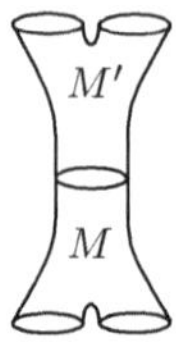

  is a composite

$$M' \circ M : 2 \to 2$$

  where the cobordism $M' : 1 \to 2$ is glued to $M : 2 \to 1$ along the object 1.
- The disjoint union of manifolds provides this category with a monoidal structure. For example, if $M : 1 \to 0$ and $M' : 2 \to 1$ are cobordisms, then the cobordism $M + M' : 1 + 2 \to 0 + 1$ depicts as:

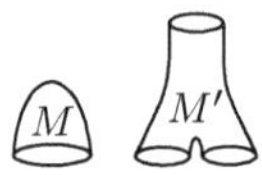

- The empty manifold 0 is the identity for the disjoint union.
- The *twist* cobordism provides symmetry. For example, the twist

$$T_{1,1} : 1 + 1 \to 1 + 1$$

  is depicted as

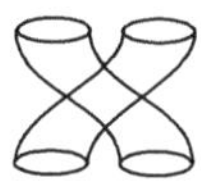

  The generalisation to

$$T_{n,m} : m + n \to n + m$$

  for any $m, n \in \mathbb{N}$ should be obvious.
- The unit and counit

$$\eta_1 : 0 \to 1 + 1 \qquad \text{and} \qquad \epsilon_1 : 1 + 1 \to 0$$

  of the compact structure on 1 are the cobordisms

and

We recover the equations of compactness as

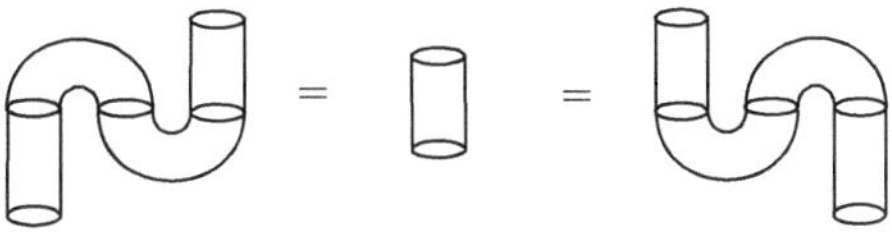

which hold since all cobordisms involved are homeomorphically equivalent. The generalisation of the units to arbitrary n is again obvious:

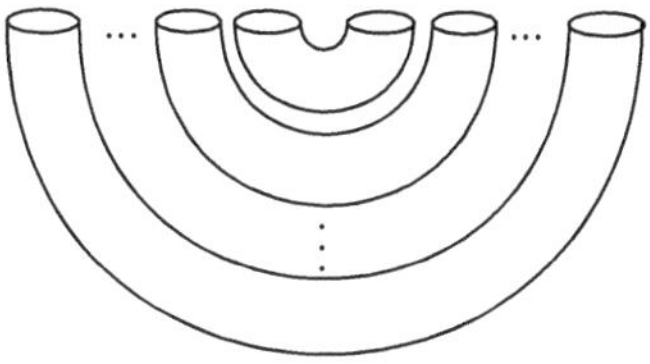

These together with corresponding counits are easily seen to always satisfy the equations of compactness.

- The dagger consists in 'flipping' the cobordisms, e.g. if $M : 2 \to 1$ is

then $M^{\dagger} : 1 \to 2$ is

Clearly the dagger is compatible with the disjoint union which makes **2Cob** a dagger monoidal category. It is also dagger compact since $\sigma_{1,1} \circ \epsilon_1^{\dagger}$ is

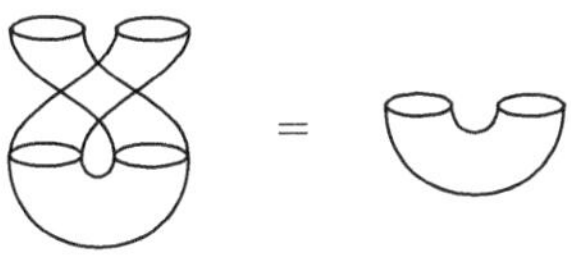

which is again easily seen to be true for arbitrary n.

Obviously, we have been very informal here. For a more elaborated discussion and technical details we refer the reader to [9, 10, 42, 66]. The key thing to remember is that there are important 'concrete' categories in which the morphisms are nothing like maps from the domain to the codomain.

Note also that we can conceive—again somewhat informally—the diagrammatic calculus of the previous sections as the result of contracting the diameter of the strings in **2Cob** to zero. These categories of cobordisms play a key role in *topological quantum field theory* (TQFT). We discuss this topic in Sect. 3.6.5.

3.5 Classical-Like Tensors

The tensors to which we referred as classical-like are not compact. Instead they do come with some other structure which, in all non-trivial cases, turns out to be incompatible with compactness [1]. In fact, this incompatibility is the abstract incarnation of the No-Cloning theorem which plays a key role in quantum information [30, 69].

3.5.1 Cartesian Categories

Consider the category **Set** with the Cartesian product as the monoidal tensor, as defined in Example 28. Given sets $A_1, A_2 \in |\mathbf{Set}|$, their Cartesian product $A_1 \times A_2$ consists of all pairs (x_1, x_2) with $x_1 \in A_1$ and $x_2 \in A_2$. The fact that Cartesian products consist of pairs is witnessed by the *projection* maps

$$\pi_1 : A_1 \times A_2 \to A_1 :: (x_1, x_2) \mapsto x_1 \quad \text{and} \quad \pi_2 : A_1 \times A_2 \to A_2 :: (x_1, x_2) \mapsto x_2 ,$$

which identify the respective components, together with the fact that, in turn, we can *pair* $x_1 = \pi_1(x_1, x_2) \in A_1$ and $x_2 = \pi_2(x_1, x_2) \in A_2$ back together into $(x_1, x_2) \in A_1 \times A_2$, merely by putting brackets around them. We would like to express this fact purely in category-theoretic terms. But both the projections and the pairing operation are expressed in terms of their action on elements, while categorical structure only recognises hom-sets, and not the internal structure of the underlying objects. Therefore, we consider the action of projections on hom-sets, namely

$$\pi_1 \circ - : \mathbf{Set}(C, A_1 \times A_2) \to \mathbf{Set}(C, A_1) :: f \mapsto \pi_1 \circ f$$

and

$$\pi_2 \circ - : \mathbf{Set}(C, A_1 \times A_2) \to \mathbf{Set}(C, A_2) :: f \mapsto \pi_2 \circ f ,$$

which we can combine into a single operation 'decompose'

$$dec_C^{A_1,A_2} : \mathbf{Set}(C, A_1 \times A_2) \to \mathbf{Set}(C, A_1) \times \mathbf{Set}(C, A_2) :: f \mapsto (\pi_1 \circ f, \pi_2 \circ f) ,$$

together with an operation ‘recombine’

$$rec_C^{A_1,A_2} : \mathbf{Set}(C, A_1) \times \mathbf{Set}(C, A_2) \to \mathbf{Set}(C, A_1 \times A_2) :: (f_1, f_2) \mapsto \langle f_1, f_2\rangle$$

where

$$\langle f_1, f_2\rangle : C \to A_1 \times A_2 :: c \mapsto (f_1(c), f_2(c)) .$$

In this form we have

$$dec_C^{A_1,A_2} \circ rec_C^{A_1,A_2} = 1_{\mathbf{Set}(C,A_1)\times\mathbf{Set}(C,A_2)}$$

and

$$rec_C^{A_1,A_2} \circ dec_C^{A_1,A_2} = 1_{\mathbf{Set}(C,A_1\times A_2)} ,$$

so $dec_C^{A_1,A_2}$ and $rec_C^{A_1,A_2}$ are now effectively each others inverses. In the light of Example 4, setting $C := \{*\}$, we obtain

$$\mathbf{Set}(\{*\}, A_1 \times A_2) \underset{rec_{\{*\}}^{A_1,A_2}}{\overset{dec_{\{*\}}^{A_1,A_2}}{\rightleftarrows}} \mathbf{Set}(\{*\}, A_1) \times \mathbf{Set}(\{*\}, A_2) ,$$

which corresponds to projecting and pairing elements exactly as in the discussion at the beginning of this section. All of this extends in abstract generality.

Definition 17 A *product* of A_1 and $A_2 \in |\mathbf{C}|$ is a triple which consists of another object $A_1 \times A_2 \in |\mathbf{C}|$ together with two morphisms

$$\pi_1 : A_1 \times A_2 \longrightarrow A_1 \quad \text{and} \quad \pi_2 : A_1 \times A_2 \longrightarrow A_2 ,$$

and which is such that for all $C \in |\mathbf{C}|$ the mapping

$$(\pi_1 \circ -, \pi_2 \circ -) : \mathbf{C}(C, A_1 \times A_2) \to \mathbf{C}(C, A_1) \times \mathbf{C}(C, A_2) \tag{3.34}$$

admits an inverse $\langle -, -\rangle_{C,A_1,A_2}$.

Below we omit the indices C, A_1, A_2 in $\langle -, -\rangle_{C,A_1,A_2}$.

Definition 18 (Cartesian category) A category $\mathbf{C}$ is *Cartesian* if any pair of objects $A, B \in |\mathbf{C}|$ admits a (not necessarily unique) product.

Proposition 3 *If a pair of objects admits two distinct products then the carrier objects are isomorphic in the category-theoretic sense of Definition 2.*

Indeed, suppose that A_1 and $A_2 \in |\mathbf{C}|$ have two products $A_1 \times A_2$ and $A_1 \boxtimes A_2$ with respective projections

$$\pi_i : A_1 \times A_2 \longrightarrow A_i \quad \text{and} \quad \pi'_j : A_1 \boxtimes A_2 \longrightarrow A_j .$$

Consider the pairs of morphisms

$$(\pi'_1, \pi'_2) \in \mathbf{C}(A_1 \boxtimes A_2, A_1) \times \mathbf{C}(A_1 \boxtimes A_2, A_2)$$

and

$$(\pi_1, \pi_2) \in \mathbf{C}(A_1 \times A_2, A_1) \times \mathbf{C}(A_1 \times A_2, A_2) .$$

By Definition 17 we can apply the respective inverses of

$$(\pi_1 \circ -, \pi_2 \circ -) \qquad \text{and} \qquad (\pi'_1 \circ -, \pi'_2 \circ -)$$

to these pairs, yielding morphisms in

$$\mathbf{C}(A_1 \boxtimes A_2, A_1 \times A_2) \quad \text{and} \quad \mathbf{C}(A_1 \times A_2, A_1 \boxtimes A_2) ,$$

say f and g respectively, for which we have

$$\pi'_1 = \pi_1 \circ f, \quad \pi'_2 = \pi_2 \circ f, \quad \pi_1 = \pi'_1 \circ g \quad \text{and} \quad \pi_2 = \pi'_2 \circ g.$$

Then, it follows that

$$(\pi'_1 \circ 1_{A_1 \boxtimes A_2}, \pi'_2 \circ 1_{A_1 \boxtimes A_2}) = (\pi_1 \circ f, \pi_2 \circ f) = (\pi'_1 \circ g \circ f, \pi'_2 \circ g \circ f) ,$$

and applying the inverse to $(\pi'_1 \circ -, \pi'_2 \circ -)$ now gives $1_{A_1 \boxtimes A_2} = g \circ f$. An analogue argument gives $f \circ g = 1_{A_1 \times A_2}$ so f is an isomorphism between the two objects $A_1 \times A_2$ and $A_1 \boxtimes A_2$ with g as its inverse.

The above definition of products in terms of "decomposing and recombining compound objects" is not the one that one usually finds in the literature.

Definition 19 A *product* of two objects A_1 and A_2 in a category $\mathbf{C}$ is a triple consisting of another object $A_1 \times A_2 \in |\mathbf{C}|$ together with two morphisms

$$\pi_1 : A_1 \times A_2 \longrightarrow A_1 \quad \text{and} \quad \pi_2 : A_1 \times A_2 \longrightarrow A_2 ,$$

and which is such that for any object $C \in |\mathbf{C}|$, and any pair of morphisms $C \xrightarrow{f_1} A_1$ and $C \xrightarrow{f_2} A_2$ in $\mathbf{C}$, there exists a unique morphism $C \xrightarrow{f} A_1 \times A_2$ such that

$$f_1 = \pi_1 \circ f \qquad \text{and} \qquad f_2 = \pi_2 \circ f .$$

We can concisely summarise this *universal* property by the commutative diagram

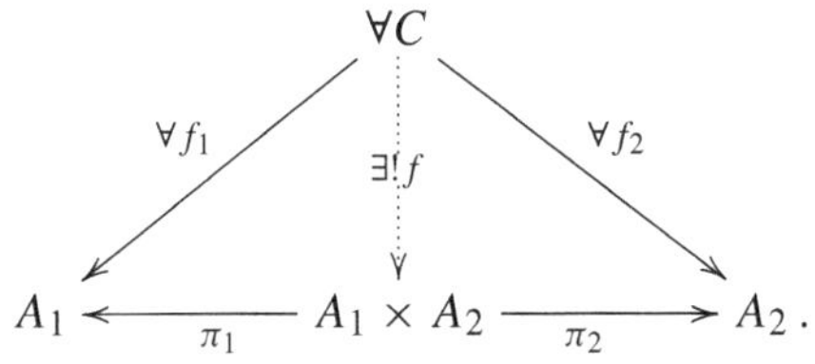

It is easy to see that this definition is equivalent to the previous one: the inverse $\langle -, - \rangle$ to $(\pi_1 \circ -, \pi_2 \circ -)$ provides for any pair (f_1, f_2) a unique morphism $f := \langle f_1, f_2 \rangle$ which is such that $(\pi_1 \circ f, \pi_2 \circ f) = (f_1, f_2)$. Conversely, uniqueness of $C \xrightarrow{f} A_1 \times A_2$ guarantees $(\pi_1 \circ -, \pi_2 \circ -)$ to have an inverse $\langle -, - \rangle$, which is obtained by setting $\langle f_1, f_2 \rangle := f$.

For more details on this definition, and the reason for its prominence in the literature, we refer to [4] and standard textbooks such as [5, 50].

Proposition 4 *If a category* $\mathbf{C}$ *is Cartesian, then each choice of a product for each pair of objects always defines a symmetric monoidal structure on* $\mathbf{C}$ *with* $A \otimes B := A \times B$*, and with the terminal object as the monoidal unit.*

Proving this requires work. First, for $f : A_1 \longrightarrow B_1$ and $g : A_2 \longrightarrow B_2$ let

$$f \times g : A_1 \times A_2 \longrightarrow B_1 \times B_2$$

be the unique morphism defined in terms of Definition 19 within

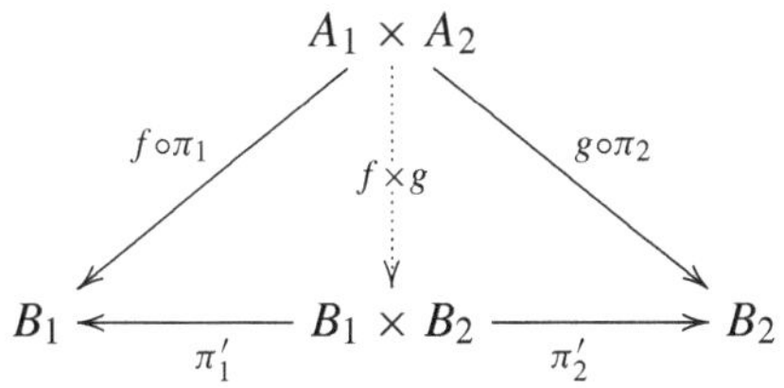

Then it immediately follows that the diagrams

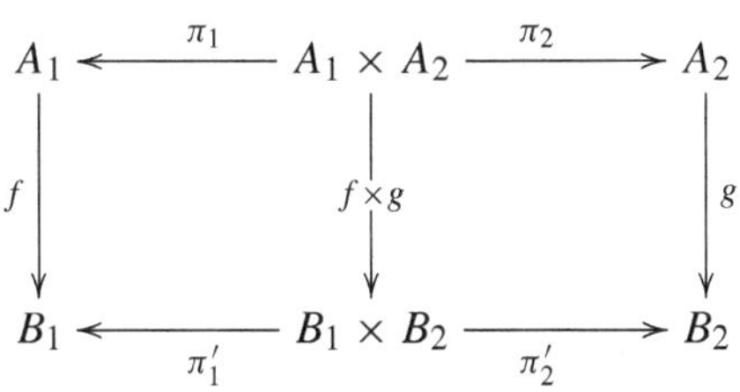

(3.35)

commute. From Definition 17 we know that for any h,

$$\langle \pi_1 \circ h, \pi_2 \circ h \rangle = h \,, \tag{3.36}$$

and, in particular, this entails

$$\langle \pi_1, \pi_2 \rangle = \langle \pi_1 \circ 1_{A_1 \times A_2}, \pi_2 \circ 1_{A_1 \times A_2} \rangle = 1_{A_1 \times A_2} . \tag{3.37}$$

Using Eq. (3.36) for $A \xrightarrow{f} B$, $B \xrightarrow{g} C$ and $B \xrightarrow{h} D$ we have

$$\begin{aligned} \langle g, h \rangle \circ f &= \langle \pi_1 \circ (\langle g, h \rangle \circ f), \pi_2 \circ (\langle g, h \rangle \circ f) \rangle \\ &= \langle (\pi_1 \circ \langle g, h \rangle) \circ f, (\pi_2 \circ \langle g, h \rangle) \circ f \rangle \\ &= \langle g \circ f, h \circ f \rangle \,. \end{aligned}$$

Using this, for $A \xrightarrow{f} B$, $A \xrightarrow{g} C$, $B \xrightarrow{h} D$ and $C \xrightarrow{k} E$, we have

$$\begin{aligned} (h \times k) \circ \langle f, g \rangle &= \langle h \circ \pi_1, k \circ \pi_2 \rangle' \circ \langle f, g \rangle \\ &= \langle h \circ \pi_1 \circ \langle f, g \rangle, k \circ \pi_2 \circ \langle f, g \rangle \rangle' \\ &= \langle h \circ f, k \circ g \rangle' \,, \end{aligned}$$

where $\langle -, - \rangle'$ is the pairing operation relative to $(\pi_1' \circ -, \pi_2' \circ -)$. In a similar manner the reader can verify that $- \times -$ is bifunctorial.

To support the claim in Proposition 4 we will now also construct the required natural isomorphisms, and leave verification of the coherence diagrams to the reader. Let $!_A$ be the unique morphism of type $A \longrightarrow \top$. Setting

$$\lambda_A := \langle !_A, 1_A \rangle : A \longrightarrow \top \times A$$

we have

$$\langle !_B, 1_B \rangle \circ f = \langle !_B \circ f, 1_B \circ f \rangle = \langle !_A, f \circ 1_A \rangle = (!_\top \times f) \circ \langle !_A, 1_A \rangle \,,$$

so we have established commutation of

$$\begin{array}{ccc} A & \xrightarrow{\lambda_A} & \top \times A \\ {\scriptstyle f}\downarrow & & \downarrow{\scriptstyle 1_\top \times f} \\ B & \xrightarrow[\lambda_B]{} & \top \times B \end{array}$$

that is, λ is natural. The components are moreover isomorphisms with π_2 as inverse. The fact that $\pi_2 \circ \lambda_A = 1_A$ holds by definition, and from

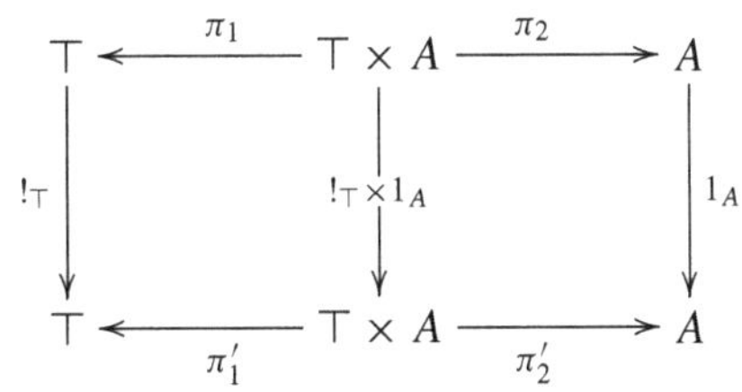

and the fact that by the terminality of $\top$ we have

$$!_{\top\times A} = !_\top \circ \pi_1 = !_A \circ \pi_2$$

it follows that

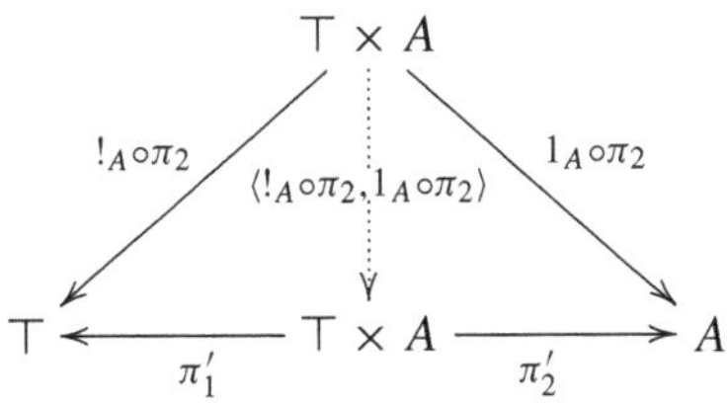

commutes, so by uniqueness, it follows that $\langle !_A \circ \pi_2, 1_A \circ \pi_2 \rangle = !_\top \times 1_A$, and hence

$$\langle !_A, 1_A \rangle \circ \pi_2 = \langle !_A \circ \pi_2, 1_A \circ \pi_2 \rangle = !_\top \times 1_A = 1_\top \times 1_A = 1_{\top\times A}\,.$$

Similarly the components $\rho_A := \langle 1_A, !_A \rangle$ also define a natural isomorphism.

For associativity, let us fix some notation for the projections:

$$A \xleftarrow{\pi_1} A \times (B \times C) \xrightarrow{\pi_2} B \times C \quad \text{and} \quad B \xleftarrow{\pi_1'} B \times C \xrightarrow{\pi_2'} C\,.$$

We define a morphism of type $A \times (B \times C) \to A \times B$ within

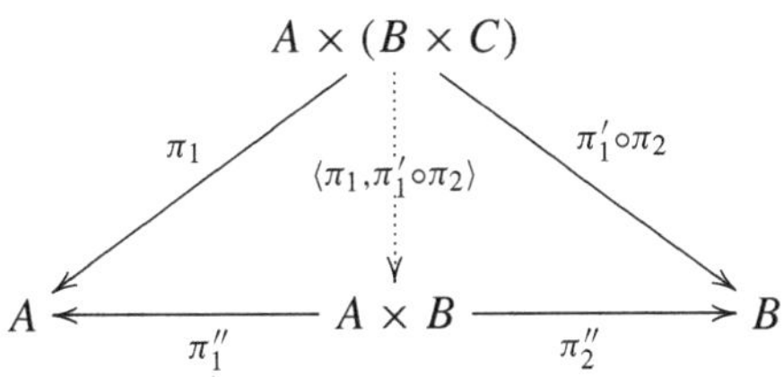

and we define $\alpha_{A,B,C}$ within

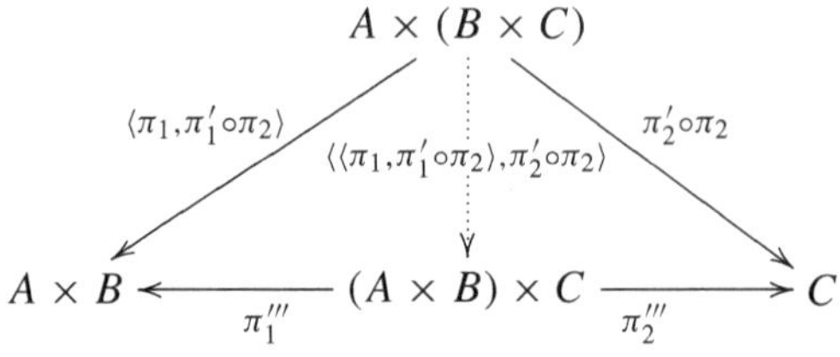

Naturality as well as the fact that the components are isomorphisms relies on uniqueness of the morphisms as defined above and is left to the reader.

For symmetry, the components $\sigma_{A,B} : A \times B \longrightarrow B \times A$ are defined within

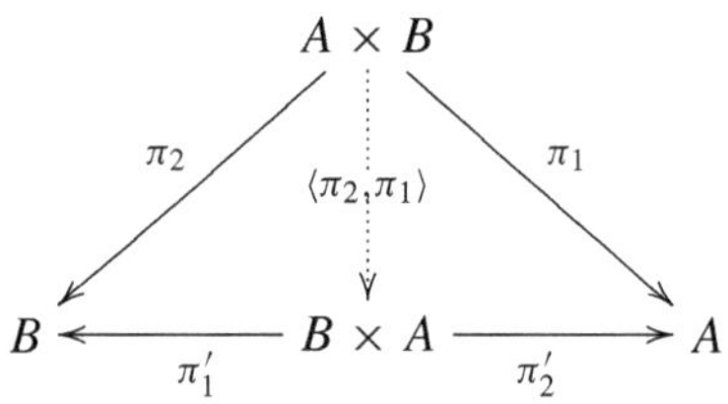

where again we leave verifications to the reader.

3.5.2 Copy-Ability and Delete-Ability

So how does all this translate in term of morphisms as physical processes? By a *uniform copying operation* or *diagonal* in a monoidal category **C** we mean a natural transformation

$$\Delta = \left\{ A \xrightarrow{\Delta_A} A \otimes A \;\middle|\; A \in |\mathbf{C}| \right\} .$$

The corresponding commutativity requirement

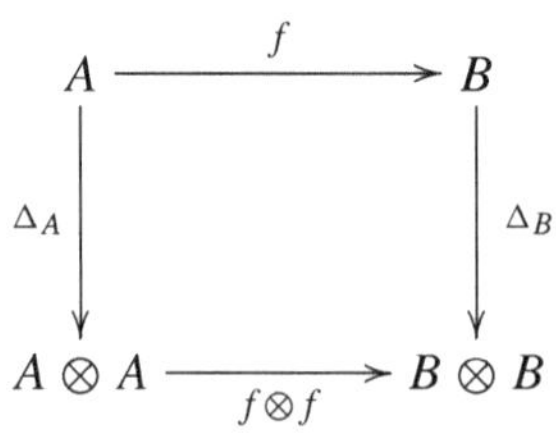

expresses that "when performing operation f on a system A and then copying it", is the same as "copying system A and then performing operation f on each copy". For example, correcting typos on a sheet of written paper and then Xeroxing it is the same as first Xeroxing it and then correcting the typos on each of the copies.

The category **Set** has

$$\left\{ \Delta_X : X \to X \times X :: x \mapsto (x, x) \;\middle|\; X \in |\mathbf{Set}| \right\}$$

as a uniform copying operation since we have commutation of

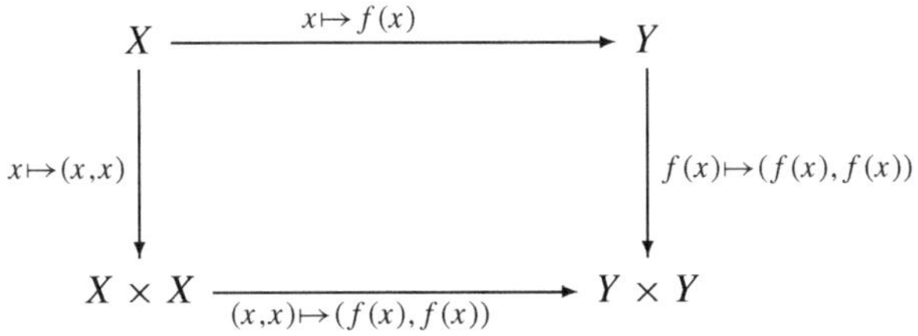

Example 41 Is there a uniform copying operation in **FdHilb**? We cannot just set

$$\Delta_{\mathcal{H}} : \mathcal{H} \to \mathcal{H} \otimes \mathcal{H} :: \psi \mapsto \psi \otimes \psi$$

since this map is not even linear. On the other hand, when for each Hilbert space $\mathcal{H}$ a basis $\{|i\rangle\}_i$ is specified, we can consider

$$\left\{ \Delta_{\mathcal{H}} : \mathcal{H} \to \mathcal{H} \otimes \mathcal{H} :: |i\rangle \mapsto |i\rangle \otimes |i\rangle \;\middle|\; \mathcal{H} \in |\mathbf{FdHilb}| \right\} .$$

But now the diagram

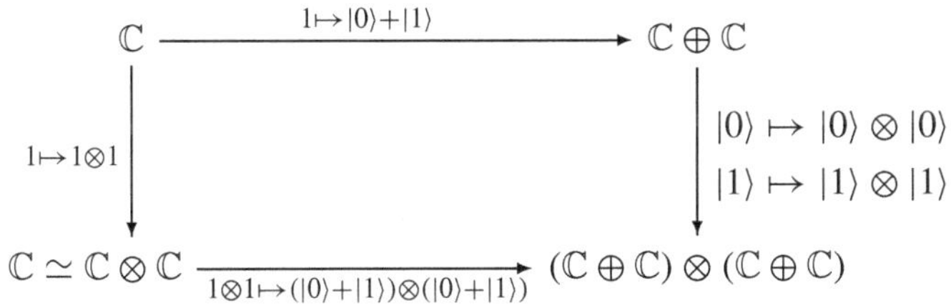

fails to commute, since via one path we obtain the (unnormalized) *Bell-state*

$$1 \mapsto |0\rangle \otimes |0\rangle + |1\rangle \otimes |1\rangle ,$$

while via the other path we obtain an (unnormalized) *disentangled state*

$$1 \mapsto (|0\rangle + |1\rangle) \otimes (|0\rangle + |1\rangle) .$$

This inability to define a uniform copying operation reflects the fact that we cannot copy (unknown) quantum states.

Example 42 Let us now turn our attention to **Rel** and, given that every function is also a relation, consider the family of functions which provided a uniform copying operation for **Set**. In more typical relational notation we have

$$\Delta_X := \{(x, (x, x)) \mid x \in X\} \subseteq X \times (X \times X)\,.$$

However, the diagram

$$\begin{array}{ccc}
\{*\} & \xrightarrow{\{(*,0),(*,1)\}} & \{0,1\} \\
{\scriptstyle \{(*,(*,*))\}}\big\downarrow & & \big\downarrow{\scriptstyle \{(0,(0,0)),(1,(1,1))\}} \\
\{(*,*)\} = \{*\} \times \{*\} & \xrightarrow[\{(*,0),(*,1)\}\times\{(*,0),(*,1)\}]{} & \{0,1\} \times \{0,1\}
\end{array}$$

fails to commute, since via one path we have

$$\{(*, (0, 0)), (*, (1, 1))\} = \{*\} \times \{(0, 0), (1, 1)\}\,,$$

while the other path yields

$$\{(*, (0, 0)), (*, (0, 1)), (*, (1, 0)), (*, (1, 1))\} = \{*\} \times (\{0, 1\} \times \{0, 1\})\,.$$

Note here in particular the similarity with the counterexample that we provided for the case of **FdHilb**, by identifying

$$\begin{array}{rcl}
|0\rangle \otimes |0\rangle + |1\rangle \otimes |1\rangle & \stackrel{\sim}{\longleftrightarrow} & \{(0, 0), (1, 1)\} \\
(|0\rangle + |1\rangle) \otimes (|0\rangle + |1\rangle) & \stackrel{\sim}{\longleftrightarrow} & \{0, 1\} \times \{0, 1\}\,.
\end{array}$$

Example 43 Similarly, the cobordism

is not a component of a uniform copying relation

$$\{\Delta_n : n \to n + n \mid n \in \mathbb{N}\}\,,$$

since in

$$\begin{array}{ccc}
0 & \xrightarrow{\Delta_0} & 0 + 0 \\
{\scriptstyle M}\big\downarrow & & \big\downarrow{\scriptstyle M+M} \\
1 & \xrightarrow[\Delta_1]{} & 1 + 1
\end{array}$$

where $M : 0 \to 1$ is

the upper path gives

while the lower path gives

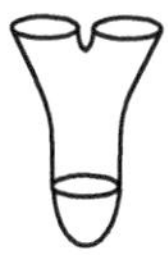

The category **Set** admits a uniform copying operation as a consequence of being Cartesian. We indeed have the following general result.

Proposition 5 *Each Cartesian category admits a uniform copying operation.*

Indeed, let

$$\Delta_A := \langle 1_A, 1_A \rangle$$

and let $A \xrightarrow{f} B$ be arbitrary. Then we have

$$\langle 1_B, 1_B \rangle \circ f = \langle 1_B \circ f, 1_B \circ f \rangle = \langle f \circ 1_A, f \circ 1_A \rangle = (f \times f) \circ \langle 1_A, 1_A \rangle\,,$$

so Δ is a natural transformation, and hence a uniform copying operation.

In fact, one can define Cartesian categories in terms of the existence of a uniform copying operation and a corresponding uniform deleting operation

$$\mathcal{E} = \left\{ A \xrightarrow{\mathcal{E}_A} \mathrm{I} \;\middle|\; A \in |\mathbf{C}| \right\}\,,$$

for which the naturality constraint now means that

$$\begin{array}{ccc} A & \xrightarrow{f} & B \\ {\scriptstyle \mathcal{E}_A}\downarrow & \swarrow {\scriptstyle \mathcal{E}_B} & \\ \mathrm{I} & & \end{array}$$

commutes. There are some additional constraints such as “first copying and then deleting results in the same as doing nothing”, and similar ones, which all together formally boil down to saying that for each object A in the category the triple $(A, \,\Delta_A, \mathcal{E}_A)$ has to be an *internal commutative comonoid*. We will define the concept of internal commutative comonoid below in Sect. 3.5.7.

Example 44 The fact that the diagonal in **Set** fails to be a diagonal in **Rel** seems to indicate that in **Rel** the Cartesian product does not provide a product in the sense of Definition 17. Consider

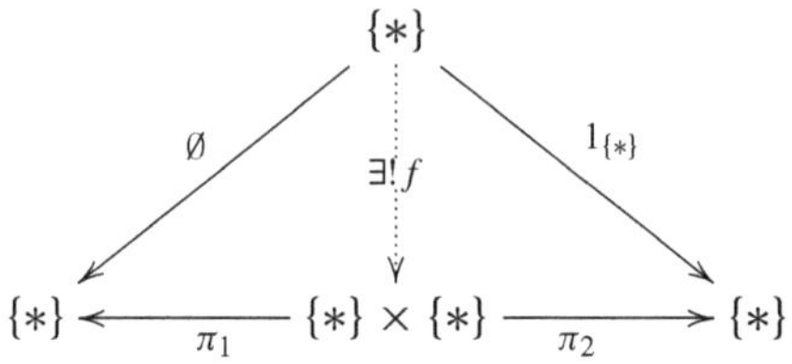

where $\emptyset$ stands for the empty relation. Since $\{*\} \times \{*\} = \{(*, *)\}$ is a singleton there are only two possible choices for π_1 and π_2, namely the empty relation and the singleton relation $\{((*, *), *)\} \subseteq \{(*, *)\} \times \{*\}$. Similarly there are also only two candidate relations to play the role of f. So since $\pi_1 \circ f = \emptyset$ either π_1 or f has to be $\emptyset$ and since $\pi_2 \circ f = 1_{\{*\}}$ neither π_2 nor f can be $\emptyset$. Thus π_1 has to be the empty relation and π_2 has to be the singleton relation. However, when considering

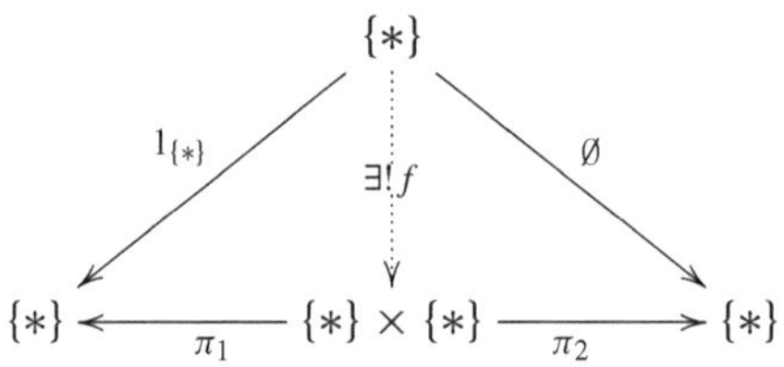

π_2 has to be the empty relation and π_1 has to be the singleton relation, so we have a contradiction. Key to all this is the fact that the empty relation is a relation, while it is not a function, or more generally, that relations need not be *total* (total = each argument is assigned to a value). On the other hand, when showing that the diagonal in **Set** was not a diagonal in **Rel** we relied on the multi-valuedness of the relation $\{(*, 0), (*, 1)\} \subseteq \{*\} \times \{0, 1\}$. Hence multi-valuedness of certain relations obstructs the existence of a natural diagonal in **Rel**, while the lack of totality of certain relations obstructs the existence of faithful projections in **Rel**, causing a break-down of the Cartesian structure of $\times$ in **Rel** as compared to the role it plays in **Set**.

3.5.3 Disjunction vs. Conjunction

As we saw in Sect. 3.5.1, the fact that in **Set** Cartesian products $X \times Y$ consist of pairs (x, y) of elements $x \in X$ and $y \in Y$ can be expressed in terms of a bijective correspondence

$$\mathbf{Set}(C, A_1 \times A_2) \simeq \mathbf{Set}(C, A_1) \times \mathbf{Set}(C, A_2)\,.$$

One can then naturally ask whether we also have that

$$\mathbf{Set}(A_1 \times A_2, C) \overset{?}{\simeq} \mathbf{Set}(A_1, C) \times \mathbf{Set}(A_2, C)\,.$$

The answer is no. But we do have

$$\mathbf{Set}(A_1 + A_2, C) \simeq \mathbf{Set}(A_1, C) \times \mathbf{Set}(A_2, C)\,.$$

where $A_1 + A_2$ is the disjoint union of two sets A_1 and A_2, that is, we repeat,

$$A_1 + A_2 := \{(x_1, 1) \mid x_1 \in A_1\} \cup \{(x_2, 2) \mid x_2 \in A_2\}\,.$$

This isomorphism now involves *injection* maps

$$\iota_1 : A_1 \to A_1 + A_2 :: x_1 \mapsto (x_1, 1) \quad \text{and} \quad \iota_2 : A_2 \to A_1 + A_2 :: x_2 \mapsto (x_2, 2)$$

They embed the elements of A_1 and A_2 within $A_1{+}A_2$. Their action on hom-sets is

$$\begin{aligned} - \circ \iota_1 &: \mathbf{Set}(A_1 + A_2, C) \to \mathbf{Set}(A_1, C) :: f \mapsto f \circ \iota_1 \\ - \circ \iota_2 &: \mathbf{Set}(A_1 + A_2, C) \to \mathbf{Set}(A_2, C) :: f \mapsto f \circ \iota_2\,, \end{aligned}$$

which converts a function that takes values on all elements that either live in A_1 *or* A_2, into two functions, one that takes values in A_1, and one that takes values in A_2. We can again recombine these two operations in a single one

$$codec_C^{A_1,A_2} : \mathbf{Set}(A_1 + A_2, C) \to \mathbf{Set}(A_1, C) \times \mathbf{Set}(A_2, C) :: f \mapsto (f \circ \iota_1, f \circ \iota_2)$$

which has an inverse, namely

$$corec_C^{A_1,A_2} : \mathbf{Set}(A_1, C) \times \mathbf{Set}(A_2, C) \to \mathbf{Set}(A_1 + A_2, C) :: (f_1, f_2) \mapsto [f_1, f_2]$$

where

$$[f_1, f_2] : A_1 + A_2 \to C :: \begin{cases} x \mapsto f_1(x) \textit{ iff } x \in A_1 \\ x \mapsto f_2(x) \textit{ iff } x \in A_2 \end{cases}.$$

The binary operation $[-,-]$ on functions now recombines two functions f_1 and f_2 into a single one. We have an isomorphism

$$\mathbf{Set}(A_1+A_2,C) \underset{corec_{\{*\}}^{A_1,A_2}}{\overset{codec_{\{*\}}^{A_1,A_2}}{\rightleftarrows}} \mathbf{Set}(A_1,C)\times\mathbf{Set}(A_2,C)\,.$$

Note that while $[f_1,f_2]$ produces an image either for the function f_1 <u>*or*</u> the function f_2, in contrast $\langle f_1,f_2\rangle$ produces an image both for the function f_1 <u>*and*</u> the function f_2. In operational terms, while the product allows to describe a pair of (classical) systems, the disjoint union allows to describe a situation where we have either of two systems. For example, it allows to describe the *branching structure* that arises as a consequence of non-determinism.

Definition 20 A *coproduct* of two objects A_1 and A_2 in a category $\mathbf{C}$ is a triple consisting of another object $A_1+A_2\in|\mathbf{C}|$ together with two morphisms

$$\iota_1:A_1\longrightarrow A_1+A_2 \quad\text{and}\quad \iota_2:A_2\longrightarrow A_1+A_2\,,$$

and which is such that for all $C\in|\mathbf{C}|$ the mapping

$$(-\circ\iota_1,-\circ\iota_2):\mathbf{C}(A_1+A_2,C)\to\mathbf{C}(A_1,C)\times\mathbf{C}(A_2,C)$$

admits an inverse. A category $\mathbf{C}$ *is co-Cartesian* if any pair of objects $A,B\in|\mathbf{C}|$ admits a (not necessarily unique) coproduct.

As in the case of products, we also have the following variant:

Definition 21 A *coproduct* of two objects A_1 and A_2 in a category $\mathbf{C}$ is a triple consisting of another object $A_1+A_2\in|\mathbf{C}|$ together with two morphisms

$$\iota_1:A_1\longrightarrow A_1+A_2 \quad\text{and}\quad \iota_2:A_2\longrightarrow A_1+A_2\,,$$

and which is such that for any object $C\in|\mathbf{C}|$, and any pair of morphisms $A_1\xrightarrow{f_1}C$ and $A_2\xrightarrow{f_2}C$ in $\mathbf{C}$, there exists a <u>unique</u> morphism $A_1+A_2\xrightarrow{f}C$ such that

$$f_1=f\circ\iota_1 \qquad\text{and}\qquad f_2=f\circ\iota_2\,.$$

We can again represent this in a commutative diagram:

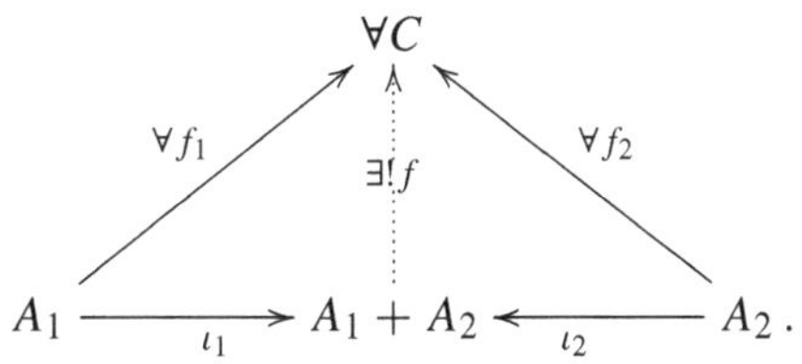

As a counterpart to the diagonal which we have in Cartesian categories we now have a *codiagonal*, with components

$$\nabla_A := [1_A, 1_A] : A + A \longrightarrow A\,.$$

Example 45 As explained in Example 14, we can think of a partially ordered set P as a category $\mathbf{P}$. In such a category products turn out to be *greatest lower bounds* or *meets*, and coproducts turn out to be *least upper bounds* or *joins*. The existence of an isomorphism

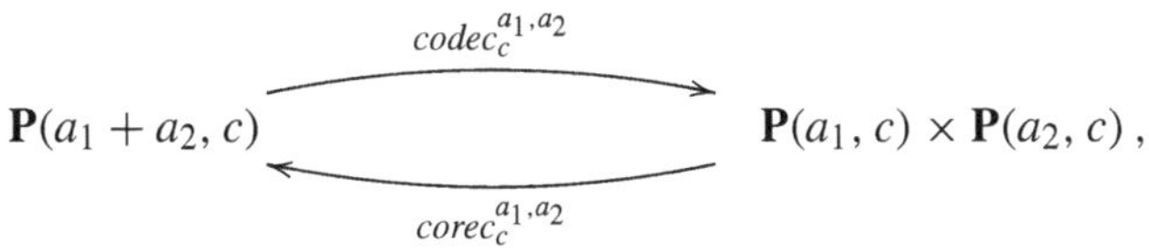

given that $\mathbf{P}(a_1 + a_2, c)$, $\mathbf{P}(a_1, c)$ and $\mathbf{P}(a_2, c)$ and hence also $\mathbf{P}(a_1, c) \times \mathbf{P}(a_2, c)$ are all either singletons or empty, means that $\mathbf{P}(a_1 + a_2, c)$ is non-empty if and only if $\mathbf{P}(a_1, c) \times \mathbf{P}(a_2, c)$ is non-empty, that is, if and only if both $\mathbf{P}(a_1, c)$ and $\mathbf{P}(a_2, c)$ are non-empty. Since non-emptiness of $\mathbf{P}(a, b)$ means that $a \leq b$, we indeed have

$$a_1 + a_2 \leq c \iff a_1 \leq c \;\&\; a_2 \leq c$$

so $a_1 + a_2$ is indeed the least upper bounds of a_1 and a_2. So Definition 21 provides us with a complementary but equivalent definition of least upper bounds. In

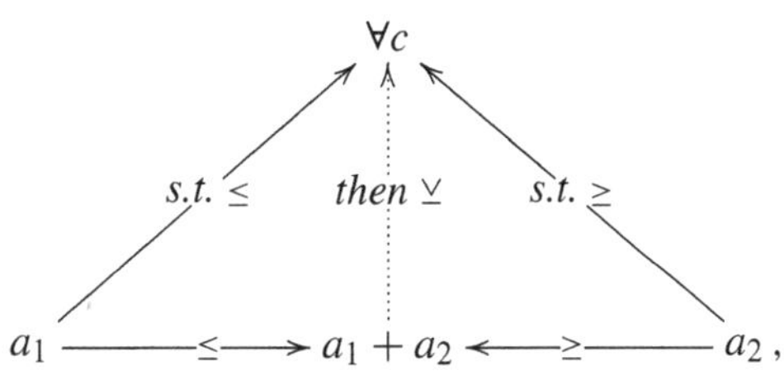

we now have that the existence of ι_1 and ι_2 assert that $a_1 \leq a_1 + a_2$ and $a_2 \leq a_1 + a_2$, so $a_1 + a_2$ is an upper bound for a_1 and a_2, and whenever there exists an element $c \in P$ which is such that both $a_1 \leq c$ and $a_2 \leq c$ hold, then we have that $a_1 + a_2 \leq c$, so $a_1 + a_2$ is indeed the least upper bound for a_1 and a_2.

Dually to what we did in a category with products, in a category with coproducts we can define sum morphisms $f + g$ in terms of commutation of

$$\begin{array}{ccccc} A_1 & \xrightarrow{\iota_1} & A_1 + A_2 & \xleftarrow{\iota_2} & A_2 \\ {\scriptstyle f}\downarrow & & \downarrow{\scriptstyle f+g} & & \downarrow{\scriptstyle g} \\ B_1 & \xrightarrow[\iota_1']{} & B_1 + B_2 & \xleftarrow[\iota_2']{} & B_2 \end{array}$$

and we have

$$h \circ [f, g] = [h \circ f, h \circ g] \quad \text{and} \quad [f, g] \circ (h + k) = [f \circ h, g \circ k] .$$

From this we can derive that coproducts provide a monoidal structure.

We already hinted at the fact that while a product can be interpreted as a conjunction, the coproduct can be interpreted as a disjunction. The distributive law

$$A \ and \ (B \ or \ C) = (A \ and \ B) \ or \ (A \ and \ C)$$

of classical logic incarnates in *categorical logic* as the existence of a natural isomorphism wich effectively 'distributes', namely

$$\{A \times (B + C) \xrightarrow{dist_{A,B,C}} (A \times B) + (A \times C) \mid A, B, C \in |\mathbf{C}|\} .$$

Shorter, we can write

$$A \times (B + C) \simeq (A \times B) + (A \times C) .$$

This of course requires the category to be both Cartesian and co-Cartesian.

Such an isomorphism does not always exist, as the following example illustrates.

Example 46 Let $\mathcal{H}$ be a Hilbert space and let $L(\mathcal{H})$ be the set of all of its (closed, in the infinite-dimensional case) subspaces, ordered by inclusion. Again this can be thought of as a category **L**. It has an initial object, namely the zero-dimensional subspace, and it has a terminal object, namely the whole Hilbert space itself. This category is Cartesian with intersection as product, and it is also co-Cartesian for

$$V + W := \bigcap\{X \in L(\mathcal{H}) \mid V, W \subseteq X\} ,$$

that is, the (closed) linear span of V and W. However, as observed by Birkhoff and von Neumann in [15], this lattice does not satisfy the distributive law. Take for example two vectors $\psi, \phi \in \mathcal{H}$ with $\phi \perp \psi$. Then we have

$$\mathrm{span}(\psi+\phi)\cap(\mathrm{span}(\psi)+\mathrm{span}(\phi)) = \mathrm{span}(\psi+\phi)\cap\mathrm{span}(\psi,\phi) = \mathrm{span}(\psi+\phi)\,,$$

while since

$$(\mathrm{span}(\psi+\phi)\cap\mathrm{span}(\psi)) \quad \text{and} \quad (\mathrm{span}(\psi+\phi)\cap\mathrm{span}(\phi))$$

only include the zero-vector 0, we have

$$(\mathrm{span}(\psi+\phi)\cap\mathrm{span}(\psi)) + (\mathrm{span}(\psi+\phi)\cap\mathrm{span}(\phi)) = 0\,,$$

and as a consequence

$$\begin{array}{c}\mathrm{span}(\psi+\phi)\cap(\mathrm{span}(\psi)+\mathrm{span}(\phi)) \\ \not\simeq \\ (\mathrm{span}(\psi+\phi)\cap\mathrm{span}(\psi)) + (\mathrm{span}(\psi+\phi)\cap\mathrm{span}(\phi))\,.\end{array}$$

Recall that an isomorphism consists of a pair of morphisms that are mutually inverse. So a natural isomorphism consists of a pair of natural transformations. In a category which is both Cartesian and co-Cartesian one of the two components of the distributivity natural isomorphism always exists, namely the natural transformation

$$\{(A\times B)+(A\times C) \xrightarrow{\theta_{A,B,C}} A\times(B+C) \mid A,B,C\in|\mathbf{C}|\}\,,$$

which we conveniently denote by

$$(A\times B)+(A\times C) \rightsquigarrow A\times(B+C)\,.$$

Indeed, by the assumption that the category is both Cartesian and co-Cartesian there exist unique morphisms f and g such that

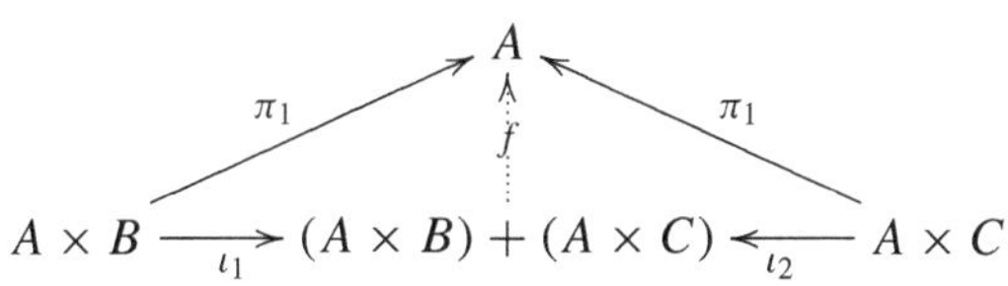

and

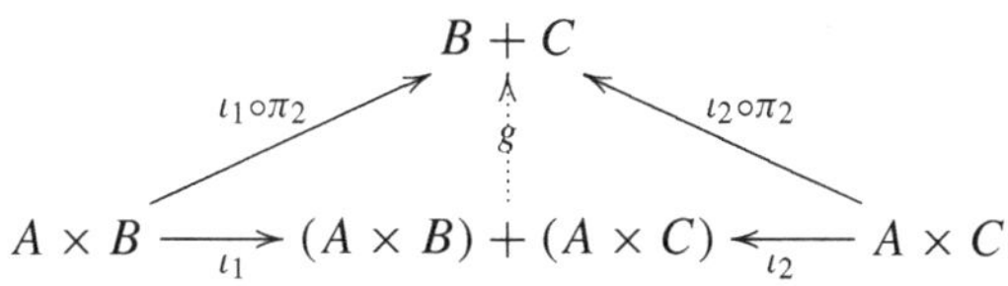

commute, namely $f := [\pi_1, \pi_1]$ and $g := [\iota_1\circ\pi_2, \iota_2\circ\pi_2]$, and hence there also exists a unique morphism $\theta_{A,B,C}$ such that

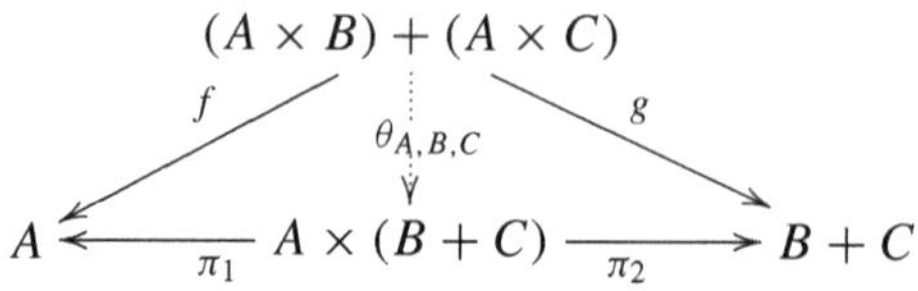

commutes, namely

$$\theta_{A,B,C} = \langle f, g \rangle = \langle [\pi_1, \pi_1], [\iota_1 \circ \pi_2, \iota_2 \circ \pi_2] \rangle \,.$$

The collection

$$\theta = \{\theta_{A,B,C} \mid A, B, C \in |\mathbf{C}|\}$$

is moreover a natural transformation since given

$$(f \times g) + (f \times h) : (A \times B) + (A \times C) \longrightarrow (A' \times B') + (A' \times C') \,,$$

using the various lemmas for products and coproducts, we have that

$$\begin{aligned}
&\langle [\pi'_1, \pi'_1], [\iota'_1 \circ \pi'_2, \iota'_2 \circ \pi'_2] \rangle \circ ((f \times g) + (f \times h)) \\
&\quad = \langle [\pi'_1, \pi'_1] \circ ((f \times g) + (f \times h)), [\iota'_1 \circ \pi'_2, \iota'_2 \circ \pi'_2] \circ ((f \times g) + (f \times h)) \rangle \\
&\quad = \langle [\pi'_1 \circ (f \times g), \pi'_1 \circ (f \times h)], [\iota'_1 \circ \pi'_2 \circ (f \times g), \iota'_2 \circ \pi'_2 \circ (f \times h)] \rangle \\
&\quad = \langle [f \circ \pi_1, f \circ \pi_1], [\iota'_1 \circ g \circ \pi_2, \iota'_2 \circ h \circ \pi_2] \rangle \\
&\quad = \langle f \circ [\pi_1, \pi_1], (g + h) \circ [\iota_1 \circ \pi_2, \iota_2 \circ \pi_2] \rangle \\
&\quad = (f \times (g + h)) \circ \langle [\pi_1, \pi_1], [\iota_1 \circ \pi_2, \iota_2 \circ \pi_2] \rangle \,,
\end{aligned}$$

which results in commutation of

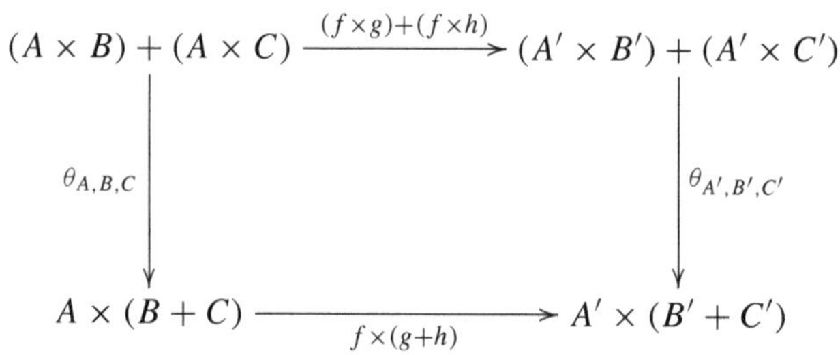

If this natural transformation is an isomorphism we have a *distributive category*.

Example 47 From the above it follows that in any lattice we have

$$(a \wedge b) + (a \wedge c) \leq a \wedge (b + c) \,.$$

Below we will see that also the so-called orthomodular law can be given a purely category-theoretic form, so Birkhoff-von Neumann style quantum logic can be entirely casted in purely category-theoretic terms.

3.5.4 Direct Sums

Example 48 The *direct sum* $V \oplus V'$ of two vector spaces V and V' is both a product and a coproduct in $\mathbf{FdVect}_{\mathbb{K}}$. Indeed, consider matrices

$$\pi_1 := (1_W | 0_{W,W'}) \qquad \text{and} \qquad \pi_2 := (0_{W',W} | 1_{W'}) ,$$

where 1_U denotes the identity on U and $0_{U,U'}$ is a matrix of 0's of dimension $\dim(U) \times \dim(U')$. Let $M : V \to W$ and $N : V \to W'$ also be represented as matrices. The unique matrix P which makes

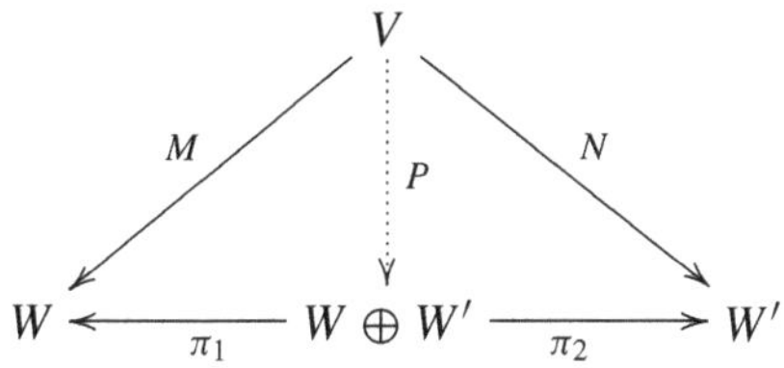

commute is

$$\left(\frac{M}{N}\right) .$$

Therefore $\oplus$ is a product. The dual is obtained by transposing the matrices in this diagram. Setting ι_i for the transpose of π_i the diagram

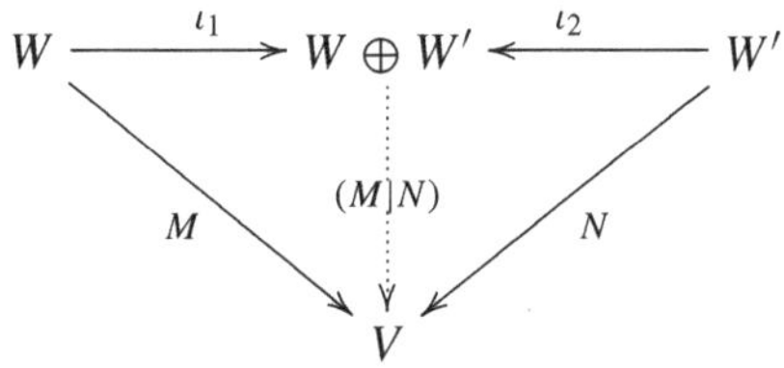

commutes. This shows that $W \oplus W'$ is indeed also a coproduct. Moreover, the zero-dimensional space is both initial and terminal.

Example 49 In the category **Rel** we can extend the disjoint union to morphisms. For any two relations $R_1 : X \to X'$ and $R_2 : Y \to Y'$ we set

$$R_1 + R_2 := \{((x, 1), (x', 1)) \mid x R_1 x'\} \cup \{((y, 2), (y', 2)) \mid y R_2 y'\} .$$

We define injection relations $\iota_1 : X \to X + Y$ and $\iota_2 : Y \to X + Y$ to be

$$\iota_1 := \{(x, (x, 1)) \mid x \in X\} \quad \text{and} \quad \iota_2 := \{(y, (y, 2)) \mid y \in Y\}$$

and the copairing relation $[R_1, R_2] : X + Y \to Z$ to be

$$[R_1, R_2] := \{((x, 1), z) \mid x R_1 z\} \cup \{((y, 2), z) \mid y R_2 z\}\,.$$

One easily verifies that all these data define a coproduct. We define projection relations as the relational converse of the injection relations, that is,

$$\pi_1 := \{((x, 1), x) \mid x \in X\} \quad \text{and} \quad \pi_2 := \{((y, 2), y) \mid y \in Y\}.$$

One easily verifies that this defines a product. So the diagrams expressing the product properties are converted into the diagrams expressing the coproduct properties by the relational converse. Since for any $X \in |\mathbf{Rel}|$ there is only one relation of type

$$\emptyset \to X \qquad \text{and} \qquad X \to \emptyset$$

it follows that the empty set is both initial and terminal. All of this makes the disjoint union within **Rel** very similar to the direct sum in $\mathbf{FdVect}_{\mathbb{K}}$.

Definition 22 A category **C** is *enriched in commutative monoids* if each hom-set $\mathbf{C}(A, B)$ is a commutative monoid

$$(\mathbf{C}(A, B)\,, +\,, 0_{A,B})\,,$$

and if for all $f \in \mathbf{C}(A, B)$, all $g_1, g_2 \in \mathbf{C}(B, C)$ and all $h \in \mathbf{C}(C, D)$ we have

$$\begin{aligned} (g_1 + g_2) \circ f &= (g_1 \circ f) + (g_2 \circ f) \qquad & 0_{B,C} \circ f &= 0_{A,C} \\ h \circ (g_1 + g_2) &= (h \circ g_1) + (h \circ g_2) \qquad & h \circ 0_{B,C} &= 0_{B,D}\,. \end{aligned}$$

Example 50 The category $\mathbf{FdVect}_{\mathbb{K}}$ is enriched in commutative monoids. The monoid operation is addition of linear maps and the unit is the zero linear map. Also the category **Rel** is enriched in commutative monoids. The monoid operation is the union of relations and the unit is the empty relation.

Definition 23 The *direct sum* or *biproduct* of two objects $A_1, A_2 \in |\mathbf{C}|$ is a quintuple consisting of another object $A_1 \oplus A_2 \in |\mathbf{C}|$ together with four morphisms

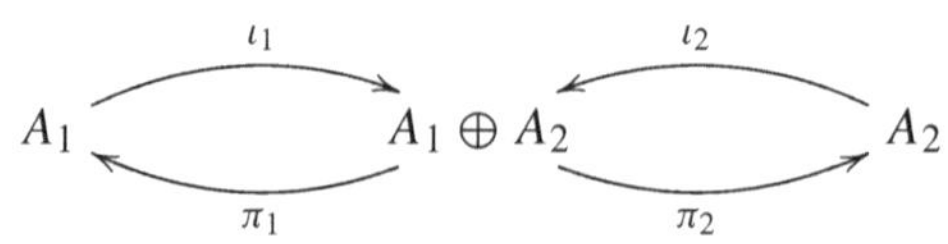

satisfying

$$\pi_1 \circ \iota_1 = 1_{A_1} \qquad \pi_2 \circ \iota_2 = 1_{A_2} \qquad \pi_2 \circ \iota_1 = 0_{A_1,A_2} \qquad \pi_1 \circ \iota_2 = 0_{A_2,A_1} \tag{3.38}$$

and

$$\iota_1 \circ \pi_1 + \iota_2 \circ \pi_2 = 1_{A_1 \oplus A_2} \,.$$

When setting

$$\delta_{ij} := \begin{cases} 1_{A_i} & i = j \\ 0_{A_j,A_i} & i \neq j \end{cases}$$

then Eq. (3.38) can be rewritten as

$$\pi_i \circ \iota_j = \delta_{ij} \,.$$

Note that Definition 3.38 does not explicitly require that $A_1 \oplus A_2$ is both a product and a coproduct. In particular, it does not make any reference to other objects C as the definitions of product and coproduct do.

Definition 24 A *zero object* is an object which is both initial and terminal.

If a category **C** has a zero object, then for each pair of objects $A, B \in |\mathbf{C}|$ we can construct a canonical *zero map* by relying on the uniqueness of morphism from the initial object to B and from A to the terminal object:

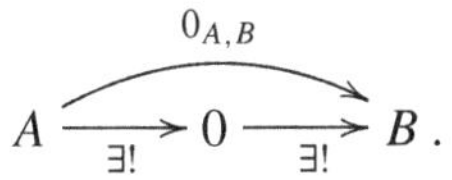

One can show that if a category with a zero object is enriched in commutative monoids, that these unique morphisms must be the units for the monoids.

Definition 25 *A* biproduct *category is a category with a zero object in which for any two objects A_1 and A_2 a biproduct $(A_1 \oplus A_2, \pi_1, \pi_2, \iota_1, \iota_2)$ is specified.*[9]

One can show that the above definition is equivalent to the following one, which does make explicit reference to products and coproducts [37].

---

[9] There is no particular reason why we ask for biproducts to be specified while in the case of Cartesian categories we only required existence. This is a matter of taste, whether one prefers "being Cartesian" or "being a biproduct category" to be conceived as a "property a category possesses" or "some extra structure it comes with". There are different "schools" of category theory which have strong arguments for either of these. Each of these have their virtues and therefore we decided to give an example of both.

Definition 26 Let **C** be both Cartesian and co-Cartesian with specified products and coproducts, and let $\bot$ and $\top$ respectively denote an initial and a terminal object of **C**. Then **C** is a *biproduct category* if:

1. the (unique) morphism $\bot \longrightarrow \top$ is an isomorphism,
2. setting

$$A_1 \xrightarrow{\quad\quad\quad\quad 0_{A_1,A_2} \quad\quad\quad\quad} A_2 \qquad := \qquad A_1 \longrightarrow 1 \overset{\simeq}{\rightleftarrows} 0 \longrightarrow A_2$$

   the morphism

$$[\langle 1_{A_1}, 0_{A_1,A_2}\rangle, \langle 0_{A_2,A_1}, 1_{A_2}\rangle] : A_1 + A_2 \longrightarrow A_1 \times A_2$$

   is an isomorphism for all objects $A_1,\, A_2 \in |\mathbf{C}|$.

In fact, any morphism

$$A_1 + A_2 \xrightarrow{f} B_1 \times B_2$$

is fully characterised by four 'component' morphisms, namely

$$f_{ij} := \pi_i \circ f \circ \iota_j \qquad \text{for} \qquad i = 1, 2\,,$$

since

$$f = [\langle f_{1,1}, f_{2,1}\rangle, \langle f_{1,2}, f_{2,2}\rangle]\,.$$

Indeed,

$$\begin{aligned}
&[\langle f_{1,1}, f_{2,1}\rangle, \langle f_{1,2}, f_{2,2}\rangle] \\
&\quad = [\langle \pi_1 \circ (f \circ \iota_1), \pi_2 \circ (f \circ \iota_1)\rangle, \langle \pi_1 \circ (f \circ \iota_2), \pi_2 \circ (f \circ \iota_2)\rangle] \\
&\quad = [f \circ \iota_1, f \circ \iota_2] \\
&\quad = f \circ [\iota_1, \iota_2] \\
&\quad = f\,.
\end{aligned}$$

Therefore it makes sense to think of f as the matrix

$$f = \begin{pmatrix} f_{1,1} & f_{1,2} \\ f_{2,1} & f_{2,2} \end{pmatrix}.$$

Using this, condition 2 in Definition 26 can now be stated as the requirement that

$$\begin{pmatrix} 1_{A_1} & 0_{A_2,A_1} \\ 0_{A_1,A_2} & 1_{A_2} \end{pmatrix}$$

is an isomorphism.

Example 51 In **FdVect**$_{\mathbb{K}}$ the direct sum $\oplus$ is a biproduct. We have

$$\pi_1 \circ \iota_1 = \pi_1 \circ \pi_1^T = (1_W | 0_{W,W'}) \left(\frac{1_W}{0_{W',W}}\right) = 1_W .$$

We also have

$$\pi_1 \circ \iota_2 = \pi_1 \circ \pi_2^T = (1_W | 0_{W,W'}) \left(\frac{0_{W,W'}}{1_{W'}}\right) = 0_{W',W} .$$

The two remaining equations are obtained in the same manner.

Example 52 In **Rel** the disjoint union $+$ is a biproduct. The morphism

$$\pi_1 \circ \iota_1 : X \to X + Y \to X$$

is a subset of $X \times X$. The composite of

$$\iota_1 = \{(x, (x, 1)) \mid x \in X\} \quad \text{and} \quad \pi_1 = \{((x, 1), x) \mid x \in X\}$$

is $\{(x, x) \mid x \in X\} = 1_X$. The morphism

$$\pi_1 \circ \iota_2 : Y \to X + Y \to X$$

is a subset of $X \times Y$, namely the set of pairs (x, y) such that there exists a $(x, z) \in \iota_2$ and $(z, x) \in \pi_1$. But there are no such elements z since the elements of X are labeled by 1 and those of Y by 2 within $X + Y$. Thus, we obtain the empty relation $0_{Y,X}$.

3.5.5 Categorical Matrix Calculus

By Definition 26 each biproduct category is Cartesian, hence by Proposition 4 it carries monoidal structure. We show now that from Definition 26 it indeed follows that each hom-set $\mathbf{C}(A, B)$ in a biproduct category $\mathbf{C}$ is a monoid, with

$$f + g := A \xrightarrow{\Delta_A} A \oplus A \xrightarrow{f \oplus g} B \oplus B \xrightarrow{\nabla_B} B$$

and with $0_{A,B}$ as the unit. Indeed, let $f : A \to B$ and consider

$$f + 0_{A,B} = A \xrightarrow{\Delta_A} A \oplus A \xrightarrow{f \oplus 0_{A,B}} B \oplus B \xrightarrow{\nabla_A} B .$$

The equality $f + 0_{A,B} = f$ can be shown via the commutation of

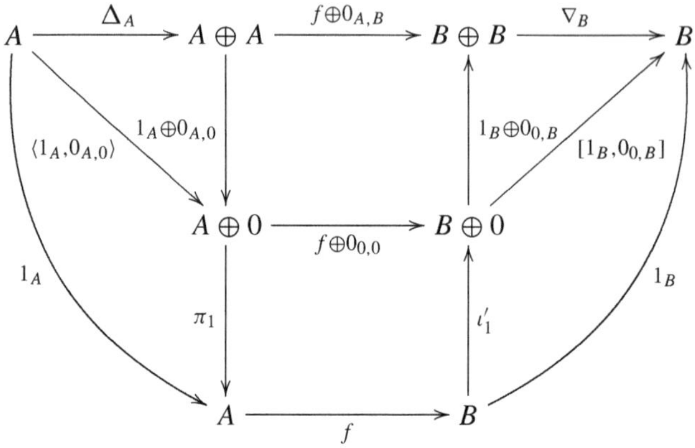

(3.39)

In the above diagram, all subdiagrams correspond to definitions, except for the square at the bottom. To show that it commutes, consider

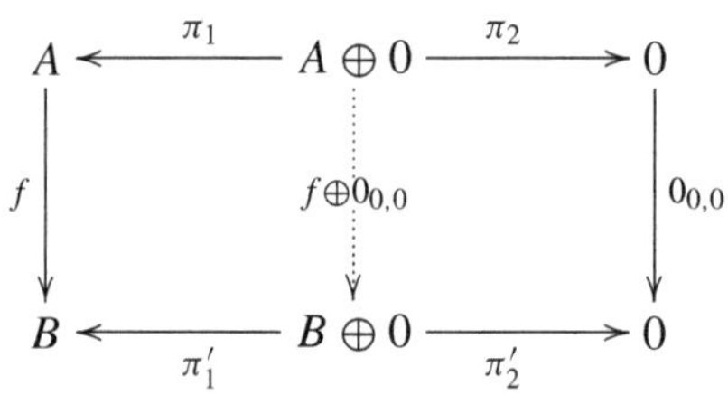

(3.40)

Since this is a product diagram, $f \oplus 0_{0,0}$ is the unique morphism making it commute. Moreover, the diagram

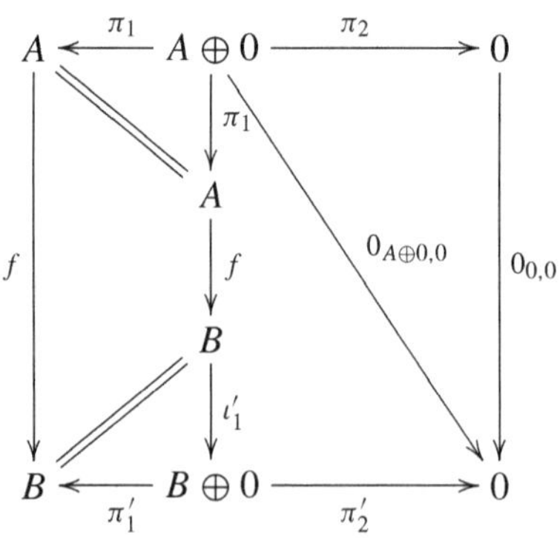

commutes, so it follows that $\iota'_1 \circ f \circ \pi_1$ also makes Diagram (3.40) commute. Thus,

$$f \oplus 0_{0,0} = \iota'_1 \circ f \circ \pi_1$$

by uniqueness, that is, the square at the bottom of Diagram (3.39) also commutes. To establish $0_{A,B} + f$ one proceeds similarly.

We also have to show that

$$(f + g) + h = f + (g + h)\,.$$

This is established in terms of commutation of the diagram

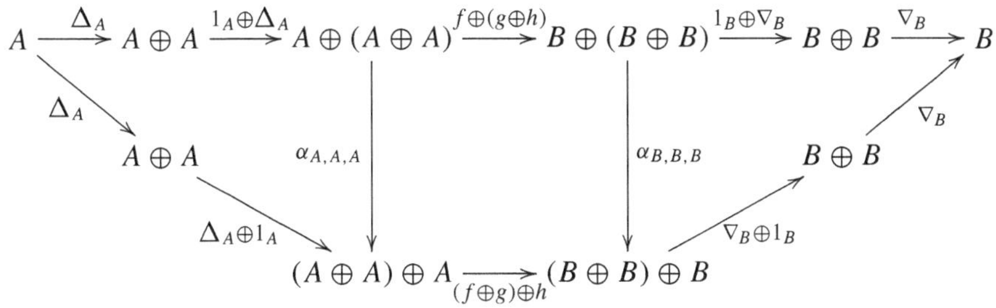

where $\alpha_{A,A,A}$ is defined as in Proposition 4. The central square commutes by definition. We now show that the left triangle also commutes. We have

$$\begin{aligned}
&\langle\langle \pi_1, \pi_1' \circ \pi_2\rangle, \pi_2' \circ \pi_2\rangle \circ (1_A \oplus \Delta_A) \circ \Delta_A \\
&= \langle\langle \pi_1, \pi_1' \circ \pi_2\rangle, \pi_2' \circ \pi_2\rangle \circ \langle 1_A, \langle 1_A, 1_A\rangle\rangle \\
&= \langle\langle \pi_1, \pi_1' \circ \pi_2\rangle \circ \langle 1_A, \langle 1_A, 1_A\rangle\rangle, \pi_2' \circ \pi_2 \circ \langle 1_A, \langle 1_A, 1_A\rangle\rangle\rangle \\
&= \langle\langle \pi_1 \circ \langle 1_A, \langle 1_A, 1_A\rangle\rangle, \pi_1' \circ \pi_2 \circ \langle 1_A, \langle 1_A, 1_A\rangle\rangle\rangle, \pi_2' \circ \pi_2 \circ \langle 1_A, \langle 1_A, 1_A\rangle\rangle\rangle \\
&= \langle\langle 1_A, 1_A\rangle, 1_A\rangle \\
&= (\Delta_A \oplus 1_A) \circ \Delta_A\,.
\end{aligned}$$

The right triangle is also easily seen to commute.

This addition moreover satisfies a distributive law, namely

$$(f + g) \circ h = (f \circ h) + (g \circ h) \quad \text{and} \quad h \circ (f + g) = (h \circ f) + (h \circ g)\,. \tag{3.41}$$

One usually refers to this additive structure on morphisms as *enrichment in monoids*. We leave it up to the reader to verify these distributive laws. A physicist-friendly introduction to *enriched category theory* suitable for the readers of this chapter is [16]. An inspiring paper which introduced the concept is [46].

We now show that from Definition 26 it also follows that for

$$\mathrm{Q}_i := \iota_i \circ \pi_i : A_1 \oplus A_2 \longrightarrow A_1 \oplus A_2$$

with $i = 1, 2$ we have

$$\sum_{i=1,2} \mathrm{Q}_i = 1_{A_1 \oplus A_2} \,. \tag{3.42}$$

Indeed, unfolding the definitions we have

$$\begin{aligned}\sum_{i=1,2} \mathrm{Q}_i &= \nabla_{A_1 \oplus A_2} \circ ((\iota_1 \circ \pi_1) \oplus (\iota_2 \circ \pi_2)) \circ \Delta_{A_1 \oplus A_2} \\ &= \nabla_{A_1 \oplus A_2} \circ ((\iota_1 \oplus \iota_2) \circ (\pi_1 \oplus \pi_2)) \circ \Delta_{A_1 \oplus A_2} \\ &= (\nabla_{A_1 \oplus A_2} \circ (\iota_1 \oplus \iota_2)) \circ ((\pi_1 \oplus \pi_2) \circ \Delta_{A_1 \oplus A_2})\end{aligned}$$

and using the fact that a biproduct of morphisms is at the same time a product of morphisms we obtain

$$(\pi_1 \oplus \pi_2) \circ \Delta_A = \langle \pi_1 \circ 1_{A_1 \oplus A_2}, \pi_2 \circ 1_{A_1 \oplus A_2} \rangle = \langle \pi_1, \pi_2 \rangle = 1_{A_1 \oplus A_2} \,.$$

Analogously, one obtains that

$$\nabla_A \circ (\iota_1 \oplus \iota_2) = 1_{A_1 \oplus A_2} \,,$$

and the composite of identities being again the identity, we proved the claim.

Definition 27 A *dagger biproduct category* is a category which is both a dagger symmetric monoidal category and a biproduct category for which the monoidal tensor and the biproduct coincide, and with $\iota_i = \pi_i^\dagger$ for all projections and injections.

These dagger biproduct categories were introduced in [2, 22, 62] in order to enable one to talk about quantum spectra in purely category-theoretic language. Let

$$A_1 \oplus A_2 \xrightarrow{\;U\;} B$$

be unitary in a dagger biproduct category. By the corresponding *projector spectrum* we mean the family $\{\mathrm{P}_i\}_i$ of *projectors*

$$\mathrm{P}_i^U := U \circ \mathrm{Q}_i \circ U^\dagger : B \longrightarrow B \,.$$

Proposition 6 *Binary projector spectra satisfy*

$$\sum_{i=1,2} P_i^U = 1_B \,.$$

This result easily extends to more general biproducts $A_1 \oplus \ldots \oplus A_n$, which can be defined in the obvious manner, and which allow us in addition to define

n-ary projector spectra too. In **FdHilb**, this n-ary generalisation of Proposition 6 corresponds to the fact that

$$\sum_{i=1}^{i=n} \mathrm{P}_i = 1_{\mathcal{H}} \qquad \text{where} \qquad \{\mathrm{P}_i\}_{i=1}^{i=n}$$

is the projector spectrum of an arbitrary self-adjoint operator. More details on this abstract view of quantum spectra are in [2, 22, 62].

Now, consider two biproducts $A_1 \oplus \ldots \oplus A_n$ and $B_1 \oplus \ldots \oplus B_m$ each with their respective injections and projections. As already indicated in the previous section, with each morphism

$$A_1 \oplus \ldots \oplus A_n \xrightarrow{f} B_1 \oplus \ldots \oplus B_m$$

we can associate a matrix

$$\begin{pmatrix} \pi_1 \circ f \circ \iota_1 & \ldots & \pi_1 \circ f \circ \iota_n \\ \vdots & \ddots & \vdots \\ \pi_m \circ f \circ \iota_1 & \ldots & \pi_m \circ f \circ \iota_n \end{pmatrix}.$$

Moreover, these matrices obey the usual matrix rules with respect to composition and the above defined summation. Indeed, for composition, the composite $g \circ f = h$ also has an associated matrix with entries

$$h_{ij} = \pi_i \circ (f \circ g) \circ \iota_j\,.$$

By Eq. (3.42) we have

$$\begin{aligned} h_{ij} &= \pi_i \circ (f \circ g) \circ \iota_j \\ &= \pi_i \circ (f \circ 1 \circ g) \circ \iota_j \\ &= \pi_i \circ \left(f \circ \left(\sum_r \iota'_r \circ \pi'_r \right) \circ g \right) \circ \iota_j \\ &= \sum_r \pi_i \circ f \circ \iota'_r \circ \pi'_r \circ g \circ \iota_j \\ &= \sum_r (\pi_i \circ f \circ \iota'_r) \circ (\pi'_r \circ g \circ \iota_j) \\ &= \sum_r f_{ir} \circ g_{rj} \end{aligned}$$

from which we recover matrix multiplication. For the sum, using the distributivity of the composition over the sum, one finds that for individual entries in $f + g$ we have

$$\begin{aligned}\pi_i \circ (f + g) \circ \iota_j &= (\pi_i \circ f + \pi_i \circ g) \circ \iota_j \\ &= \pi_i \circ f \circ \iota_j + \pi_i \circ g \circ \iota_j \\ &= f_{ij} + g_{ij}\end{aligned}$$

which indeed is the sum of matrices.

Example 53 We illustrate the concepts of this section for the category **Rel**. Somewhat unfortunately, the disjoint union bifunctor and the monoidal enrichment operation share the same notation $+$. But since their type are essentially different, i.e.

$$\text{tensor} + : \mathbf{Rel}(X, Y) \times \mathbf{Rel}(X', Y') \to \mathbf{Rel}(X + X', Y + Y')$$

and

$$\text{monoid} + : \mathbf{Rel}(X, Y) \times \mathbf{Rel}(X, Y) \to \mathbf{Rel}(X, Y)$$

respectively, this should not confuse the reader.

- The sum $R_1 + R_2 : X \to Y$ of two relations is, by definition, the composite

  $$X \xrightarrow{\Delta_X} X + X \xrightarrow{R_1+R_2} Y + Y \xrightarrow{\nabla_Y} Y\,.$$

  The relation Δ_X consists of all ordered pairs

  $$\{(x, (x, 1)) \mid x \in X\} \cup \{(x, (x, 2)) \mid x \in X\}\,.$$

  Thus the composite $(R_1 + R_2) \circ \Delta_X$ is then, by definition, the set

  $$\{(x, (y, 1)) \mid x R_1 y\} \cup \{(x', (y', 2)) \mid x' R_2 y'\}\,.$$

  Using the definition of copairing $\nabla_Y := [1_Y, 1_Y]$ we obtain

  $$\{(x, y) \mid x R_1 y\} \cup \{(x', y') \mid x' R_2 y'\}\,,$$

  that is,

  $$R_1 + R_2 = \{(x, y) \mid x R_1 y \ \underline{\text{or}}\ x R_2 y\}\,.$$

- Relations

  $$Q_X : X + Y \to X \to X + Y \quad \text{and} \quad Q_Y : X + Y \to Y \to X + Y$$

  are defined as $\iota_X \circ \pi_X$ and $\iota_Y \circ \pi_Y$ respectively, that is,

  $$Q_X = \{((x, 1), (x, 1)) \mid x \in X\} \quad \text{and} \quad Q_Y = \{((y, 2), (y, 2)) \mid y \in Y\}\,.$$

Using the definition of the sum we obtain

$$\begin{aligned} Q_X + Q_Y &= \{((x,1),(x,1)) \mid x \in X\} \cup \{((y,2),(y,2)) \mid y \in Y\} \\ &= \{((z,i),(z,i)) \mid (z,i) \in X + Y\} \\ &= 1_{X+Y} \end{aligned}$$

as required. It is easily seen that this generalises to an arbitrary number of terms in the biproduct.

- The matrix calculus in **Rel** is done over the semiring (= rig = ring without inverses) $\mathbb{B}$ of Booleans. Indeed, there are two relations between $\{*\}$ and itself, namely the empty relation and the identity relation. These will respectively be denoted by 0 and 1. The semiring operations arise from composing and adding these relations, which amounts to the semiring multiplication and the semiring addition respectively. By Eq. (3.41), we have distributivity, and we then easily see that we indeed get the Boolean semiring:

$$0 \cdot 0 = 0 \quad 0 \cdot 1 = 0 \quad 1 \cdot 1 = 1 \qquad 0 + 0 = 0 \quad 0 + 1 = 1 \quad 1 + 1 = 1$$

— contra the two-element field where we have $1 + 1 = 0$ — so the operations $- \cdot -$ and $- + -$ coincide with the Boolean logic operations:

$$\cdot \sim \wedge \qquad \text{and} \qquad + \sim \vee .$$

A relation $R : \{a, b\} \to \{c, d\}$ can now be represented by a 2×2 matrix, e.g.

$$R = \begin{pmatrix} 1 & 1 \\ 1 & 0 \end{pmatrix}$$

when aRc, bRc and aRd (and not bRd). Similarly, $R' : \{c, d\} \to \{e, f, g\}$ is

$$R' = \begin{pmatrix} 1 & 0 \\ 1 & 1 \\ 0 & 1 \end{pmatrix}$$

when cRe, cRf, dRf and dRg. Their composite

$$R' \circ R = \{(a,e),(a,f),(b,e),(b,f),(a,g)\}$$

can be computed by matrix multiplication:

$$\begin{pmatrix} 1 & 0 \\ 1 & 1 \\ 0 & 1 \end{pmatrix} \begin{pmatrix} 1 & 1 \\ 1 & 0 \end{pmatrix} = \begin{pmatrix} 1 & 1 \\ 1 & 1 \\ 1 & 0 \end{pmatrix} .$$

For a relation $R'' : \{a, b\} \to \{c, d\}$ represented by the matrix

$$\begin{pmatrix} 0 & 1 \\ 0 & 1 \end{pmatrix},$$

that is, $R'' = \{(b, c), (b, d)\}$, the sum $R + R''$ is given by

$$\{(a, c), (b, c), (a, d)\} \cup \{(b, c), (b, d)\} = \{(a, b), (a, c), (b, c), (b, d)\},$$

which indeed corresponds to the matrix sum

$$\begin{pmatrix} 1 & 1 \\ 1 & 0 \end{pmatrix} + \begin{pmatrix} 0 & 1 \\ 0 & 1 \end{pmatrix} = \begin{pmatrix} 1 & 1 \\ 1 & 1 \end{pmatrix}.$$

3.5.6 Quantum Tensors from Classical Tensors

Interesting categories such as **FdHilb** and **Rel** have both a classical-like and a quantum-like tensor. Obviously these two structures interact. For example, due to very general reasons we have distributivity natural isomorphisms

$$A \otimes (B \oplus C) \simeq (A \otimes B) \oplus (A \otimes C) \qquad \text{and} \qquad A \otimes 0 \simeq 0$$

both in the case of **FdHilb** and **Rel**. We can rely on so-called *closedness* of the $\otimes$-structure to prove this, something for which we refer to other sources. Another manner to establish this fact for the cases of **FdHilb** and **Rel**, is to observe that the $\otimes$-structure arises from the $\oplus$-structure.

Let $\mathbf{C}$ be a biproduct category and let $X \in \mathbf{C}$ be such that composition commutes in $\mathbf{C}(X, X)$. Define a new category $\mathbf{C}|X$ as follows:

- The objects of $\mathbf{C}|X$ are those objects of $\mathbf{C}$ which are of the form $X \oplus \ldots \oplus X$. We denote such an object consisting of n terms by $[n]$.
- For all $n, m \in \mathbb{N}$ we set $\mathbf{C}|X([n], [m]) := \mathbf{C}([n], [m])$.

Note that we can represent all morphisms in $\mathbf{C}([n], [m])$ by matrices, and hence also those in $\mathbf{C}|X([n], [m])$. Now we define a monoidal structure:

- $\mathrm{I} := X$
- $[n] \otimes [m] := [n \times m]$
- For all $f \in \mathbf{C}([n], [m])$ and $g \in \mathbf{C}([n'], [m'])$ we define

  $$f \otimes g \in \mathbf{C}|X([n] \otimes [n'], [m] \otimes [m'])$$

  to be the morphism with matrix entries

  $$(f \otimes g)_{(i,i'),(j,j')} := f_{i,j} \circ g_{i',j'}.$$

We leave it to the reader to verify that this provides $\mathbf{C}|X$ with a symmetric monoidal structure. Note that commutativity of $\mathbf{C}(X, X)$ is necessary, since otherwise we would be in contradiction with the fact that the scalar monoid in a monoidal category is always commutative — cf. Section 3.3.5. With these definitions we have:

$$[n] \otimes ([m] \oplus [k]) \simeq ([n] \otimes [m]) \oplus ([n] \otimes [k]) \qquad \text{and} \qquad [n] \otimes [0] \simeq [0]\,.$$

Indeed, note first that since $[n] = \underbrace{\mathrm{I} \oplus \cdots \oplus \mathrm{I}}_{n}$ we have

$$[n] \oplus [m] \simeq [n + m]$$

where $[n + m] = \underbrace{\mathrm{I} \oplus \cdots \oplus \mathrm{I}}_{n+m}$. Therefore,

$$\begin{aligned} [n] \otimes ([m] \oplus [k]) &\simeq [n] \otimes [m + k] \\ &\simeq [n \times (m + k)] \\ &= [(n \times m) + (n \times k)] \\ &\simeq [n \times m] \oplus [n \times k] \\ &\simeq ([n] \otimes [m]) \oplus ([n] \otimes [k]). \end{aligned}$$

Moreover,

$$\begin{aligned} [n] \otimes [0] &\simeq [n \times 0] \\ &= [0]. \end{aligned}$$

Example 54 In **FdHilb** there is one non-trivial object $\mathcal{H}$ such that $\mathbf{FdHilb}(\mathcal{H}, \mathcal{H})$ is commutative, namely $\mathbb{C}$. The category $\mathbf{FdHilb}|\mathbb{C}$ has Hilbert spaces of the form $\mathbb{C}^{\oplus n}$ with $n \in \mathbb{N}$ as objects, linear maps between these as morphisms, and the tensor product as the monoidal structure. This category is said to be *categorically equivalent* (a notion which we define later) to **FdHilb**. The only difference is that **FdHilb** contains for each $n \in \mathbb{N}$ many isomorphic Hilbert spaces of dimension n, while in $\mathbf{FdHilb}|\mathbb{C}$ there is exactly one Hilbert space of dimension n.

Example 55 In **Rel** it is the non-trivial object $\{*\}$ for which $\mathbf{Rel}(\{*\}, \{*\}) \simeq \mathbb{B}$ is commutative. We obtain a category with objects of the form

$$\{*\} + \ldots + \{*\}\,,$$

that is, a n-element set for each $n \in \mathbb{N}$, with relations between these as morphisms, and with the Cartesian product as the monoidal structure. Again we have that $\mathbf{Rel}|\mathbb{B}$ is categorically equivalent to **Rel**.

We can endow $\mathbf{C}|X$ with compact structure. Set:

- $[n]^* := [n]$
- Let $\eta_{[n]} \in \mathbf{C}|X(\mathrm{I}, [n]^* \otimes [n])$ be the morphism with matrix entries

$$(\eta_{[n]})_{(i,i),1} := 1_{\mathrm{I}} \qquad \text{and} \qquad (\eta_{[n]})_{(i,j\neq i),1} := 0_{\mathrm{I},\mathrm{I}} \,.$$

- Let $\epsilon_{[n]} \in \mathbf{C}|X([n] \otimes [n]^*, \mathrm{I})$ to be the morphism with matrix entries

$$(\epsilon_{[n]})_{1,(i,i)} := 1_{\mathrm{I}} \qquad \text{and} \qquad (\epsilon_{[n]})_{1,(i,j\neq i)} := 0_{\mathrm{I},\mathrm{I}} \,.$$

To see that this indeed defines a compact structure, observe that the identity of $[n]$ is

$$1_{[n]} = \delta_{i,j} := \begin{cases} 1_{\mathrm{I}} & \text{if } i = j \\ 0_{\mathrm{I},\mathrm{I}} & \text{otherwise} \end{cases} .$$

Using this, we find that

$$(1_{[n]} \otimes \eta_{[n]})_{(i,(j,k)),(l,1)} = \delta_{i,l} \circ \eta_{(j,k),1}$$

and

$$(\epsilon_{[n]} \otimes 1_{[n]})_{(1,i),((j,k),l)} = \epsilon_{1,(j,k)} \circ \delta_{i,l} .$$

We can now verify the equations of compactness by computing the composite—say e—of the two preceding morphisms using matrix calculus, i.e.

$$e_{(m,n)} = \sum_{j,k,l} (\epsilon_{[n]} \otimes 1_{[n]})_{(1,m),((j,k),l)} (1_{[n]} \otimes \eta_{[n]})_{(j,(k,l)),(n,1)} \,.$$

Note that the indexation over j, k and l has two different bracketings in the above sum. By definition of the identity, unit and counit, the term $e_{(m,n)}$ will be 1_I only if $j = k = l$, which entails that $e_{(m,n)} = \delta_{i,j}$, the identity. Since the objects are self-dual the other equation holds too.

Robin Houston proved a surprising result in [37] which to some extent is a converse to the above. It states that when a compact category is Cartesian (or co-Cartesian) then it also has direct sums.

3.5.7 Internal Classical Structures

In [2] unitary biproduct decompositions of the form

$$U : A \longrightarrow \underbrace{\mathrm{I} \oplus \ldots \oplus \mathrm{I}}_{n}$$

were used to encode the flow of classical data in quantum informatic protocols. In **FdHilb** such a map indeed singles out a basis. Explicitly, via the correspondence between vectors in Hilbert space $\mathcal{H}$ and linear maps of type $\mathbb{C} \to \mathcal{H}$, the linear maps

$$\{U^\dagger \circ \iota_i : \mathbb{C} \to \mathcal{H} \mid i = 1, \ldots, n\}$$

define a basis for $\mathcal{H}$, namely

$$\{|i\rangle := (U^\dagger \circ \iota_i)(1) \mid i = 1, \ldots, n\}\,.$$

These basis vectors are then identified with outcomes of measurements.

But there is another way to encode bases as morphisms in a category, one for which we only need to rely on the tensor structure, and hence we can stay in the diagrammatic realm of Sect. 3.3.2. If we have a basis

$$\mathcal{B} := \{|i\rangle \mid i = 1, \ldots, n\}$$

of a Hilbert space $\mathcal{H}$ then we can consider the linear maps

$$\delta : \mathcal{H} \to \mathcal{H} \otimes \mathcal{H} :: |i\rangle \mapsto |ii\rangle \quad \text{and} \quad \epsilon : \mathcal{H} \to \mathbb{C} :: |i\rangle \mapsto 1\,.$$

These two maps indeed faithfully encode the basis $\mathcal{B}$ since we can extract it back from them. It suffices to solve the equation

$$\delta(|\psi\rangle) = |\psi\rangle \otimes |\psi\rangle$$

in the unknown $|\psi\rangle$. Indeed, the only $|\psi\rangle$'s for which the right-hand-side is of the form $|\phi\rangle \otimes |\phi'\rangle$ are the basis vectors. For any other $\psi = \sum_i \alpha_i\, |i\rangle$ we have that

$$\delta(|\psi\rangle) = \sum_i \alpha_i\, |i\rangle \otimes |i\rangle\,,$$

that is, we obtain a genuinely *entangled* state.

The pair of maps (δ, ϵ) satisfies several properties e.g.

$$(\delta \otimes 1_{\mathcal{H}}) \circ \delta = (1_{\mathcal{H}} \otimes \delta) \circ \delta : \mathcal{H} \to \mathcal{H} \otimes \mathcal{H} \otimes \mathcal{H} :: |i\rangle \mapsto |iii\rangle$$

and

$$(\epsilon \otimes 1_{\mathcal{H}}) \circ \delta = (1_{\mathcal{H}} \otimes \epsilon) \circ \delta = 1_{\mathcal{H}} :: |i\rangle \mapsto |i\rangle$$

establishing it as an instance of the following concept in **FdHilb**:

Definition 28 Let $(\mathbf{C}, \otimes, \mathrm{I})$ be a monoidal category. An *internal comonoid* is an object $C \in |\mathbf{C}|$ together with a pair of morphims

$$C \otimes C \xleftarrow{\ \delta\ } C \xrightarrow{\ \epsilon\ } \mathrm{I}\,,$$

where δ is the *comultiplication* and ϵ the *comultiplicative unit*, which are such that

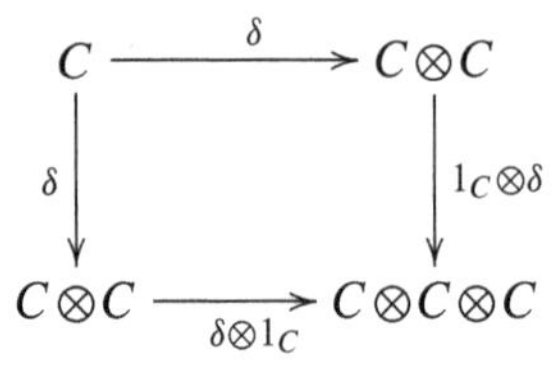

and

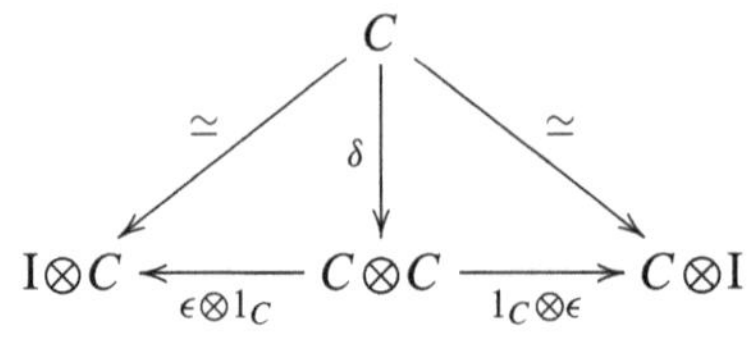

commute.

Example 56 The relations

$$\delta = \{(x, (x, x)) \mid x \in X\} \subseteq X \times (X \times X)$$

and

$$\epsilon = \{(x, *) \mid x \in X\} \subseteq X \times \{*\}$$

define an internal comonoid on X in **Rel** as the reader may verify. We could refer to these as the *copying* and *deleting* relations.

The notion of internal comonoid is dual to the notion of *internal monoid*.

Definition 29 Let $(\mathbf{C}, \otimes, \mathrm{I})$ be a monoidal category. An *internal monoid* is an object $M \in |\mathbf{C}|$ together with a pair of morphisms

$$M \otimes M \xrightarrow{\mu} M \xleftarrow{e} \mathrm{I},$$

where μ is the *multiplication* and e the *multiplicative unit*, which are such that

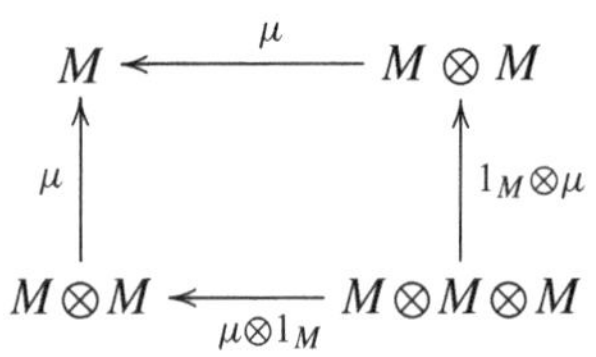

and

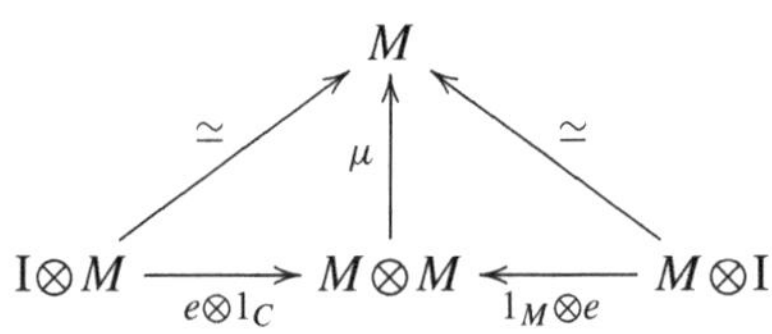

commute.

The origin of this name is the fact that monoids can equivalently be defined as internal monoids in **Set**. Since the notion of internal monoid applies to arbitrary monoidal categories, it generalises the usual notion of a monoid.

Example 57 A strict monoidal category can also be defined as an internal monoid in the category **Cat**, which has categories as objects, functors as morphisms and the product of categories as tensor—see Sect. 3.6.1 below. Proving this is slightly beyond the scope of this chapter but we invite the interested reader to do so.

We now show that internal monoids in **Set** are indeed ordinary monoids. Given such an internal monoid (X, μ, e) in **Set** with functions

$$\mu : X \times X \to X \quad \text{and} \quad e : \{*\} \to X,$$

we take the elements of the monoid to be those of X, the monoid operation to be

$$- \bullet - : X \times X \to X :: (x, y) \mapsto \mu(x, y)\,,$$

and the unit of the monoid to be $1 := e(*) \in X$. The condition

$$\begin{array}{ccc} X \times X \times X & \xrightarrow{1_X \times \mu} & X \times X \\ {\scriptstyle \mu \times 1_X}\downarrow & & \downarrow{\scriptstyle \mu} \\ X \times X & \xrightarrow[\mu]{} & X \end{array}$$

boils down to the fact that for all $x, y, z \in X$ we have

$$x \bullet (y \bullet z) = (x \bullet y) \bullet z\,,$$

that is, associativity of the monoid operation, and the condition

$$\begin{array}{ccccc} & & X & & \\ & \nearrow^{\simeq} & \uparrow{\scriptstyle \mu} & \nwarrow^{\simeq} & \\ \{*\} \times X & \xrightarrow[e \times 1_X]{} & X \times X & \xleftarrow[1_X \times e]{} & X \times \{*\} \end{array}$$

boils down to the fact that for all $x \in X$ we have

$$x \bullet 1 = 1 \bullet x = x\,,$$

that is, the element 1 is the unit of the monoid.

An internal definition of a group requires a bit more work.

Definition 30 Let **C** be a category with finite products and let $\top$ be the terminal object in **C**. An *internal group* is an internal monoid (G, μ, e) together with a morphism $\mathrm{inv} : G \longrightarrow G$ such that we have commutation of

$$\begin{array}{ccc} G & \xrightarrow{\;!\;} & \top \\ {\scriptstyle \langle 1_G, \mathrm{inv}\rangle}\downarrow & & \downarrow{\scriptstyle e} \\ G\times G & \xrightarrow{\;\mu\;} & G \\ {\scriptstyle \langle \mathrm{inv}, 1_G\rangle}\uparrow & & \uparrow{\scriptstyle e} \\ G & \xrightarrow[\;!\;]{} & \top \end{array}$$

The additional operation $\mathrm{inv} : G \longrightarrow G$ assigns the inverses to the elements of the group. We leave it to the reader to verify that internal groups in **Set** are indeed ordinary groups. When we rather consider groups in other categories, in particular those in categories of vector spaces, then one typically speaks about *Hopf algebras*, of which *quantum groups* are a special case. An excellent textbook on this topic is [65]. There are also lectures on this topic available on-line [19]. Also the notion of group homomorphism can be "internalized" in a category. We define a *group homomorphism* between two group objects $(G, \mu, e, \mathrm{inv})$ and $(G', \mu', e', \mathrm{inv}')$ to be a morphism $\phi : G \longrightarrow G'$ which commutes with all three structural morphisms, that is, the diagrams

$$\begin{array}{ccc} G\times G & \xrightarrow{\;\mu\;} & G \\ {\scriptstyle \phi\times\phi}\downarrow & & \downarrow{\scriptstyle \phi} \\ G'\times G' & \xrightarrow[\;\mu'\;]{} & G' \end{array} \quad , \quad \begin{array}{ccc} \top & \xrightarrow{\;e\;} & G \\ & {\scriptstyle e'}\searrow & \downarrow{\scriptstyle \phi} \\ & & G' \end{array} \quad \text{and} \quad \begin{array}{ccc} G & \xrightarrow{\;\mathrm{inv}\;} & G \\ {\scriptstyle \phi}\downarrow & & \downarrow{\scriptstyle \phi} \\ G' & \xrightarrow[\;\mathrm{inv}'\;]{} & G' \end{array}$$

all commute. Again, these diagrams generalise what we know about group homomorphisms, namely that they preserve multiplication, unit and inverses. The notion of (co)monoid homomorphism is defined analogously.

3.5.8 Diagrammatic Classicality

In a dagger monoidal category every internal comonoid

$$\left(X\,,\; X \xrightarrow{\;\delta\;} X\otimes X\,,\; X \xrightarrow{\;\epsilon\;} \mathrm{I}\right)$$

defines an internal monoid

$$\left(X\,,\; X\otimes X \xrightarrow{\;\delta^\dagger\;} X\,,\; \mathrm{I} \xrightarrow{\;\epsilon^\dagger\;} X\right).$$

This merely involves reversal of the arrows. We can easily see this in diagrammatic terms. We represent the comultiplication and its unit as follows:

$\delta :=$  $\epsilon :=$ 

Then, the corresponding requirements are:

Now, if we flip all of these upside-down we obtain a monoid:

$u :=$  $e :=$ 

with corresponding requirements:

A *dagger (co)monoid* is a (co)monoid satisfying all the preceding requirements.

The dagger comonoids in **FdHilb** and **Rel** which we have seen above both have some additional properties. For example, they are *commutative*:

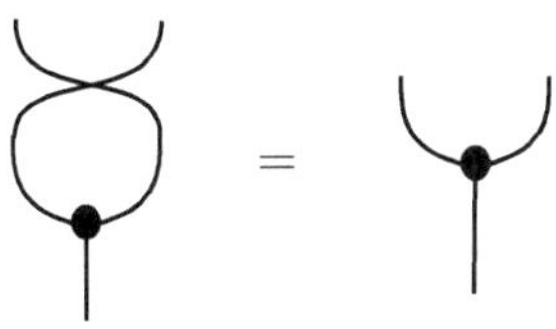

that is, symbolically,

$$\sigma_{X,X} \circ \delta = \delta \,.$$

The comultiplication is *isometric* or *special*:

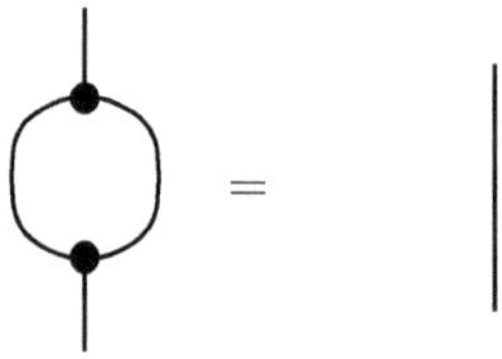

that is, symbolically,

$$\delta^\dagger \circ \delta = 1_X \,.$$

But by far, the most fascinating law which they obey are the *Frobenius equations*:

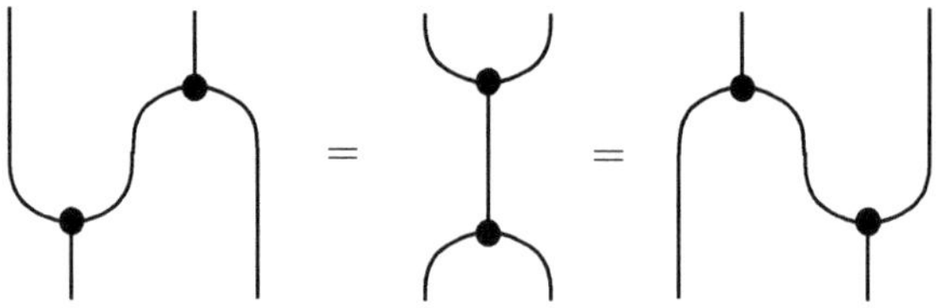

that is, symbolically,

$$(1_X \otimes \delta^\dagger) \circ (\delta \otimes 1_X) = \delta \circ \delta^\dagger = (\delta^\dagger \otimes 1_X) \circ (1_X \otimes \delta) \,.$$

For a commutative dagger comonoid these two equations are easily seen to be equivalent. We verify that these equations hold for the dagger comonoids in **FdHilb** and **Rel** discussed in the previous section.

In **FdHilb**, we have

$$\delta^\dagger : \mathcal{H} \otimes \mathcal{H} \to \mathcal{H} :: |ij\rangle \mapsto \delta_{ij} \cdot |i\rangle \quad \text{and} \quad \epsilon^\dagger : \mathbb{C} \to \mathcal{H} :: 1 \mapsto \sum_i |i\rangle$$

so

$$|ij\rangle \xmapsto{\delta \otimes 1_X} |iij\rangle \xmapsto{1_X \otimes \delta^\dagger} |i\rangle \otimes (\delta_{ij} \cdot |i\rangle) = \delta_{ij} \cdot |ii\rangle$$

and

$$|ij\rangle \xmapsto{\delta^\dagger} \delta_{ij} \cdot |i\rangle \xmapsto{\delta} \delta_{ij} \cdot |ii\rangle \,.$$

In **Rel** we have

$$\delta^\dagger = \{((x, x), x) \mid x \in X\} \subseteq (X \times X) \times X$$

and

$$\epsilon^\dagger = \{(*, x) \mid x \in X\} \subseteq \{*\} \times X$$

so we obtain

$$(1_X \otimes \delta^\dagger) \circ (\delta \otimes 1_X) = \delta \circ \delta^\dagger = \{((x, x), (x, x)) \mid x \in X\}.$$

One can show that the Frobenius equation together with isometry guarantees a normal form for any connected picture made up of dagger Frobenius (co)monoids, identities and symmetry, and which only depends on the number of input and output wires [25, 43]. As a result we can represent any such network as a "spider" e.g.,

"more complicated network" =

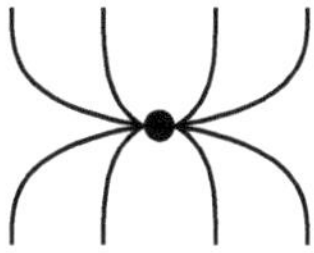

Hence commutative dagger special Frobenius comonoids turn out to be structures which come with a very simple graphical calculus, but at the same time they are of key importance to quantum theory, as is exemplified by this theorem [28]:

Theorem 2 *In* ***FdHilb*** *there is a bijective correspondence between dagger special Frobenius comonoids and orthonormal bases. Explicitly, each dagger special Frobenius comonoid in* ***FdHilb*** *is of the form*

$$\delta : \mathcal{H} \to \mathcal{H} \otimes \mathcal{H} :: |i\rangle \mapsto |ii\rangle \quad \textit{and} \quad \epsilon : \mathcal{H} \to \mathbb{C} :: |i\rangle \mapsto 1$$

relative to some orthonormal basis $\{|i\rangle\}_i$.

In the category **2Cob** we also encounter the Frobenius equation:

but the (co)monoids involved are not special, since the two cobordisms

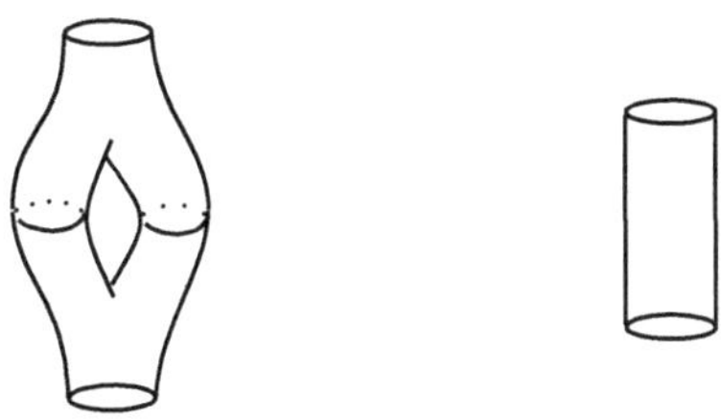

are not homeomorphic. Therefore a normal form in **2Cob** is of the form [42]

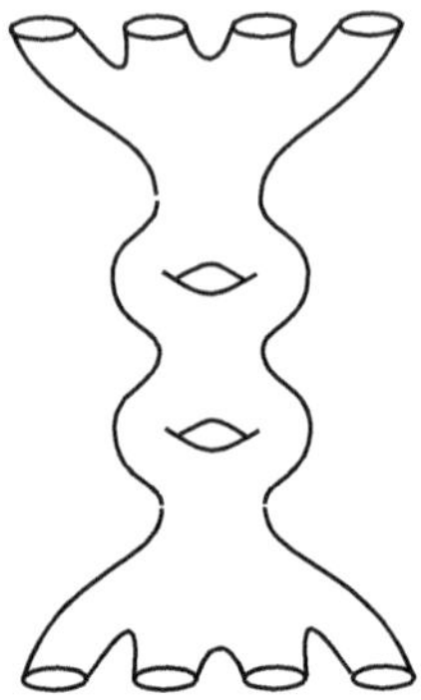

The commutative diagram in Definition 30 becomes

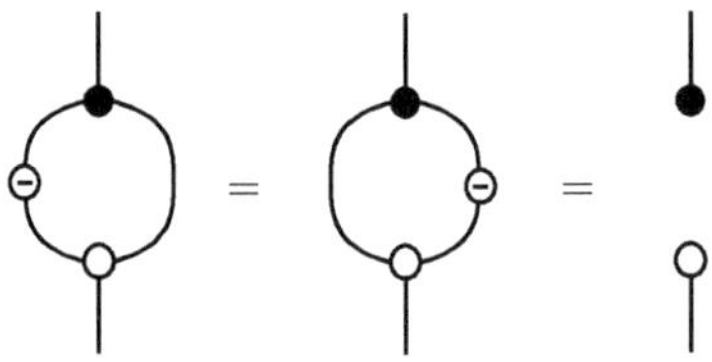

when setting

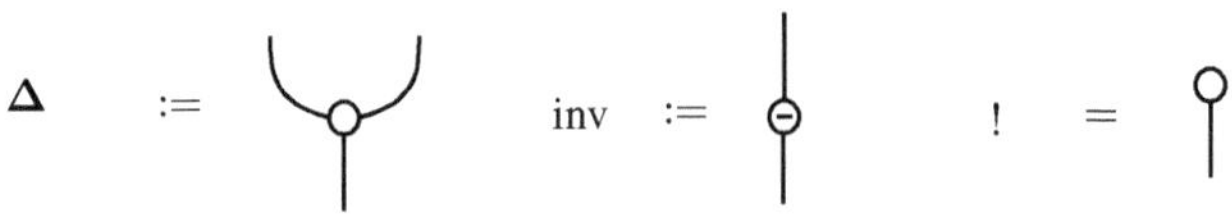

One refers to this equation typically as the *Hopf law*—cf. the Hopf algebras mentioned above. What also holds for these operations are the *bialgebra laws*:

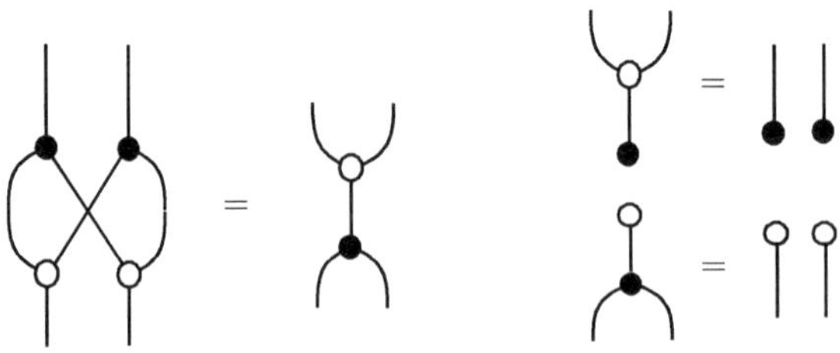

There's lots more to say on the connections between algebraic structures and these pictures. The reader may consult, for example, [42, 63, 65]. A great place to find some very well-explained introductions to this is John Baez' This Week's Finds in Mathematical Physics [8], for example, weeks 174, 224, 268.

3.6 Monoidal Functoriality, Naturality and TQFTs

In this section we provide the remaining bits of theory required to state the definition of a topological quantum field theory.

3.6.1 Bifunctors

The category **Cat** which has categories as objects and functors as morphisms also comes with a monoidal structure:

Definition 31 The *product* of categories $\mathbf{C}$ and $\mathbf{D}$ is a category $\mathbf{C} \times \mathbf{D}$:

1. objects are pairs (C, D) with $C \in |\mathbf{C}|$ and $D \in |\mathbf{D}|$,
2. morphisms are pairs $(f, g) : (C, D) \longrightarrow (C', D')$ where $f : C \longrightarrow C'$ is a morphism in $\mathbf{C}$ and $g : D \longrightarrow D'$ is a morphism in $\mathbf{D}$,
3. composition is componentwise, that is,

$$(f', g') \circ (f, g) = (f' \circ f, g' \circ g),$$

and the identities are pairs of identities.

This monoidal structure is Cartesian. The obvious projection functors

$$\mathbf{C} \xleftarrow{P_1} \mathbf{C} \times \mathbf{D} \xrightarrow{P_2} \mathbf{D}$$

provide the product structure:

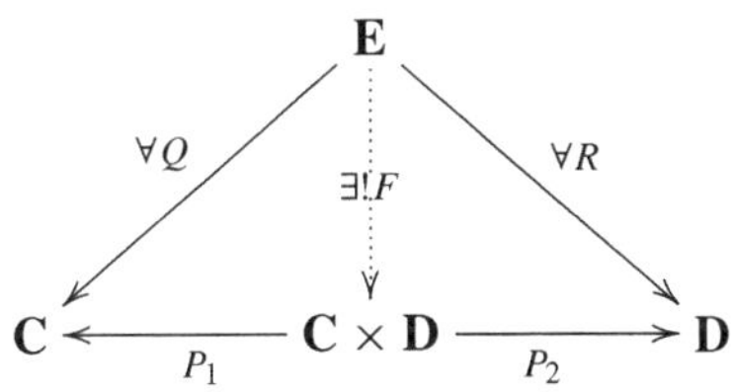

This notion of product allows for a very concise definition of *bifunctoriality*. A *bifunctor* is now nothing but an ordinary functor of type

$$F : \mathbf{C} \times \mathbf{D} \longrightarrow \mathbf{E}.$$

So, for instance, to say that a tensor is a bifunctor it now suffices to say that

$$- \otimes - : \mathbf{C} \times \mathbf{C} \longrightarrow \mathbf{C}$$

is a functor. Indeed, this implies that we have

$$\otimes(\varphi \circ \xi) = \otimes(\varphi) \circ \otimes(\xi) \quad \text{and} \quad \otimes(1_{\Xi}) = 1_{\otimes(\Xi)}$$

for all morphisms φ, ξ and all objects Ξ in $\mathbf{C} \times \mathbf{C}$, that is,

$$(g \circ f) \otimes (g' \circ f') = (g \otimes g') \circ (f \otimes f') \quad \text{and} \quad 1_A \otimes 1_B = 1_{A \otimes B}\,.$$

We give another example of bifunctor which is contravariant in the first variable and covariant in the second variable. This functor is key to the so-called *Yoneda Lemma*, which constitutes the core of many categorical constructs, for which we refer to the standard literature [50]. For all $A \in |\mathbf{C}|$ let

$$\mathbf{C}(A, -) : \mathbf{C} \longrightarrow \mathbf{Set}$$

be the functor which maps

1. each object $B \in |\mathbf{C}|$ to the set $\mathbf{C}(A, B) \in |\mathbf{Set}|$, and
2. each morphism $g : B \longrightarrow C$ to the function

$$\mathbf{C}(A, g) : \mathbf{C}(A, B) \to \mathbf{C}(A, C) :: f \mapsto g \circ f\,.$$

For all $C \in |\mathbf{C}|$ let

$$\mathbf{C}(-, C) : \mathbf{C}^{op} \longrightarrow \mathbf{Set}$$

be the functor which maps

1. each object $A \in |\mathbf{C}|$ to the set $\mathbf{C}(A, C) \in |\mathbf{Set}|$, and
2. each morphism $f : A \longrightarrow B$ to the function

$$\mathbf{C}(f, C) : \mathbf{C}(B, C) \to \mathbf{C}(A, C) :: g \mapsto g \circ f\,.$$

One verifies that given any pair $f : A \longrightarrow B$ and $h : C \longrightarrow D$ the diagram

$$\begin{array}{ccc}
\mathbf{C}(B, C) & \xrightarrow{\mathbf{C}(f,C)} & \mathbf{C}(A, C) \\
{\scriptstyle \mathbf{C}(B,h)}\downarrow & & \downarrow{\scriptstyle \mathbf{C}(A,h)} \\
\mathbf{C}(B, D) & \xrightarrow[\mathbf{C}(f,D)]{} & \mathbf{C}(A, D)
\end{array}$$

commutes, sending a morphism $g : B \longrightarrow C$ to the composite $h \circ g \circ f : A \longrightarrow D$. The bifunctor—also called *hom-functor* – which unifies the above two functors is

$$\mathbf{C}(-, -) : \mathbf{C}^{op} \times \mathbf{C} \longrightarrow \mathbf{Set}$$

which maps

1. each pair of objects $(A, B) \in |\mathbf{C}|$ to the set $\mathbf{C}(A, B) \in |\mathbf{Set}|$, and
2. each pair morphism $(f : A \longrightarrow B, h : C \longrightarrow D)$ to the function

$$\mathbf{C}(f, h) : \mathbf{C}(B, C) \to \mathbf{C}(A, D) :: g \mapsto h \circ g \circ f \,.$$

We can now identify

$$\mathbf{C}(A, -) := \mathbf{C}(1_A, -) \qquad \text{and} \qquad \mathbf{C}(-, A) := \mathbf{C}(-, 1_A)\,.$$

These functors are called *representable functors*. They enable us to represent objects and morphisms of any category as functors on the well-known category **Set**.

3.6.2 Naturality

We already encountered a fair number of examples of our restricted variant of natural isomorphisms, namely

$$\mathrm{I} \otimes A \simeq A \simeq A \otimes \mathrm{I} \quad , \quad A \otimes B \simeq B \otimes A \quad , \quad A \otimes (B \otimes C) \simeq (A \otimes B) \otimes C$$

and

$$A \times (B + C) \simeq (A \times B) + (A \times C)\,,$$

as well as some proper natural transformations, namely

$$A \rightsquigarrow A \times A \quad , \quad A + A \rightsquigarrow A \quad \text{and} \quad (A \times B) + (A \times C) \rightsquigarrow A \times (B + C)\,.$$

What makes all of these special is that all of the above expressions only involve objects of the category **C** without there being any reference to morphisms. This is not the case anymore for the general notion of natural transformations, which are in fact, structure preserving maps between functors.

Definition 32 Let $F, G : \mathbf{C} \longrightarrow \mathbf{D}$ be functors. A *natural transformation*

$$\tau : F \Rightarrow G$$

consists of a family of morphisms

$$\{\tau_A \in \mathbf{D}(FA, GA) \mid A \in |\mathbf{C}|\}$$

which are such that the diagram

$$\begin{array}{ccc} FA & \xrightarrow{\tau_A} & GA \\ {\scriptstyle Ff}\downarrow & & \downarrow{\scriptstyle Gf} \\ FB & \xrightarrow[\tau_B]{} & GB \end{array}$$

commutes for any $A, B \in |\mathbf{C}|$ and any $f \in \mathbf{C}(A, B)$.

Example 58 Given vector spaces V and W, then two group representations

$$\rho_1 : G \to \mathrm{GL}(V) \quad \text{and} \quad \rho_2 : G \to \mathrm{GL}(W)$$

are *equivalent* if there exists an isomorphism $\tau : V \to W$ so that for all $g \in G$,

$$\tau \circ \rho_1(g) = \rho_2(g) \circ \tau \,. \tag{3.43}$$

This isomorphism is a natural transformation. Indeed, taking the functorial point of view for the two representations above, we get two functors

$$\mathbf{G} \xrightarrow{R_{\rho_1}} \mathbf{FdVect}_{\mathbb{K}} \quad \text{and} \quad \mathbf{G} \xrightarrow{R_{\rho_2}} \mathbf{FdVect}_{\mathbb{K}}$$

where R_{ρ_1} maps $*$ on some vector space $R_{\rho_1}(*)$ and R_{ρ_2} maps $*$ on some vector space $R_{\rho_2}(*)$. Naturality means that the diagram

$$\begin{array}{ccc} R_{\rho_1}(*) & \xrightarrow{\tau_*} & R_{\rho_2}(*) \\ {\scriptstyle R_{\rho_1}g}\downarrow & & \downarrow{\scriptstyle R_{\rho_2}g} \\ R_{\rho_1}(*) & \xrightarrow[\tau_*]{} & R_{\rho_2}(*) \end{array}$$

commutes, which translates into Eq. (3.43).

Example 59 The family of canonical linear maps

$$\{\tau_V : V \to V^{**} \mid V \in \mathbf{FdVect}_{\mathbb{K}}\}$$

from a vector space to its double dual is a natural transformation

$$\tau : 1_{\mathbf{FdVect}_{\mathbb{K}}} \Rightarrow (-)^{**}$$

from the identity functor to the double dual functor. There is no natural transformation of type $1_{\mathbf{FdVect}_{\mathbb{K}}} \Rightarrow (-)^*$. Indeed, while each finite dimensional vector space is isomorphic with its dual, there is no "natural choice" of an isomorphism, since constructing one depends on a choice of basis.

The fact that for $\mathbf{FdVect}_{\mathbb{K}}$ naturality indeed means basis independence can immediately be seen from the definition of naturality. In

$$\begin{array}{ccc} FV & \xrightarrow{\tau_V} & GV \\ {\scriptstyle Ff}\downarrow & & \downarrow{\scriptstyle Gf} \\ FV & \xrightarrow[\tau_V]{} & GV \end{array}$$

the linear map $f : V \to V$ can be interpreted as a change of basis, and then the linear maps $Ff : FV \to FV$ and $Gf : GV \to GV$ apply this change of basis to the expressions FV and GV respectively. Commutation of the above diagram then means that it makes no difference whether we apply τ_V before the change of basis, or whether we apply it after the change of basis. Hence it asserts that τ_V is a basis independent construction.

3.6.3 Monoidal Functors and Monoidal Natural Transformations

A monoidal functor, unsurprisingly, is a functor between two monoidal categories that preserves the monoidal structure "coherently".

Definition 33 Let

$$(\mathbf{C}, \otimes, \mathrm{I}, \alpha_{\mathbf{C}}, \lambda_{\mathbf{C}}, \rho_{\mathbf{C}}) \quad \text{and} \quad (\mathbf{D}, \odot, \mathrm{J}, \alpha_{\mathbf{D}}, \lambda_{\mathbf{D}}, \rho_{\mathbf{D}})$$

be monoidal categories. Then a *monoidal functor* is a functor

$$F : \mathbf{C} \longrightarrow \mathbf{D}\,,$$

together with a natural transformation

$$\phi_{-,-} : (F-) \odot (F-) \Rightarrow F(- \otimes -)$$

with components

$$\{\phi_{A,B} : FA \odot FB \longrightarrow F(A \otimes B) \mid A, B \in |\mathbf{C}|\}\,,$$

and a morphism

$$\phi : \mathrm{J} \longrightarrow F\mathrm{I}\,,$$

which are such that for every $A, B, C \in |\mathbf{C}|$ the diagrams

$$\begin{array}{ccc}
(FA \odot FB) \odot FC & \xrightarrow{\alpha_{\mathbf{D}}^{-1}} & FA \odot (FB \odot FC) \\
\Big\downarrow{\scriptstyle \phi_{A,B} \odot 1_{FC}} & & \Big\downarrow{\scriptstyle 1_{FA} \odot \phi_{B,C}} \\
F(A \otimes B) \odot FC & & FA \odot F(B \otimes C) \\
\Big\downarrow{\scriptstyle \phi_{A \otimes B, C}} & & \Big\downarrow{\scriptstyle \phi_{A, B \otimes C}} \\
F((A \otimes B) \otimes C) & \xrightarrow[F\alpha_{\mathbf{C}}^{-1}]{} & F(A \otimes (B \otimes C))
\end{array}$$

and

$$\begin{array}{ccc}
FA \odot \mathrm{J} & \xrightarrow{1_{FA} \odot \phi} & FA \odot F\mathrm{I} \\
\Big\downarrow{\scriptstyle \rho_{\mathbf{D}}^{-1}} & & \Big\downarrow{\scriptstyle \phi_{A,\mathrm{I}}} \\
FA & \xleftarrow[F\rho_{\mathbf{C}}^{-1}]{} & F(A \otimes \mathrm{I})
\end{array}
\quad , \quad
\begin{array}{ccc}
\mathrm{J} \odot FB & \xrightarrow{\phi \odot 1_{FB}} & F\mathrm{I} \odot FB \\
\Big\downarrow{\scriptstyle \lambda_{\mathbf{D}}^{-1}} & & \Big\downarrow{\scriptstyle \phi_{\mathrm{I},B}} \\
FB & \xleftarrow[F\lambda_{\mathbf{C}}^{-1}]{} & F(\mathrm{I} \otimes B)
\end{array}$$

commute in **D**. Moreover, a monoidal functor between symmetric monoidal categories is *symmetric* if, in addition, for all $A, B \in |\mathbf{C}|$ the diagram

$$\begin{array}{ccc}
FA \odot FB & \xrightarrow{\sigma_{FA,FB}} & FB \odot FA \\
\Big\downarrow{\scriptstyle \phi_{A,B}} & & \Big\downarrow{\scriptstyle \phi_{B,A}} \\
F(A \otimes B) & \xrightarrow[F\sigma_{A,B}]{} & F(B \otimes A)
\end{array}$$

commutes in **D**. A monoidal functor is *strong* if the components of the natural transformation $\phi_{-,-}$ as well as the morphism ϕ are isomorphisms, and it is *strict* if they are identities. In this case the equational requirements simplify to

$$F(A \otimes B) = FA \odot FB \qquad \text{and} \qquad F\mathrm{I} = \mathrm{J}\,,$$

and

$$F\alpha_{\mathbf{C}} = \alpha_{\mathbf{D}} \quad , \quad F\lambda_{\mathbf{C}} = \lambda_{\mathbf{D}} \quad , \quad F\rho_{\mathbf{C}} = \rho_{\mathbf{D}} \quad \text{and} \quad F\sigma_{\mathbf{C}} = \sigma_{\mathbf{D}}\,.$$

Hence a strict monoidal functor between strict monoidal categories just means that the tensor is preserved by F.

Example 60 The functor $\dagger : \mathbf{C}^{op} \longrightarrow \mathbf{C}$ is a strict monoidal functor. In a compact category **C**, the functor $(-)^* : \mathbf{C}^{op} \longrightarrow \mathbf{C}$ which maps any object A on A^* and any morphism f on its transpose f^* is a strong monoidal functor.

Definition 34 A *monoidal natural transformation*

$$\theta : \big(F, \{\phi_{A,B} \mid A, B \in |\mathbf{C}|\}, \phi\big) \Rightarrow \big(G, \{\psi_{A,B} \mid A, B \in |\mathbf{C}|\}, \psi\big)$$

between two monoidal functors is a natural transformation such that

$$\begin{array}{ccc} FA \odot FB & \xrightarrow{\theta_A \odot \theta_B} & GA \odot GB \\ {\scriptstyle \phi_{A,B}}\downarrow & & \downarrow{\scriptstyle \psi_{A,B}} \\ F(A \otimes B) & \xrightarrow[\theta_{A\otimes B}]{} & G(A \otimes B) \end{array} \qquad \text{and} \qquad \begin{array}{ccc} & \mathrm{J} & \\ {\scriptstyle \phi}\swarrow & & \searrow{\scriptstyle \psi} \\ F\mathrm{I} & \xrightarrow[\theta_{\mathrm{I}}]{} & G\mathrm{I} \end{array}$$

commute. A monoidal natural transformation is *symmetric* if the two monoidal functors which constitute its domain and codomain are both symmetric.

3.6.4 Equivalence of Categories

In Example 6 we defined the category **Cat** which has categories as objects and functors as morphism. Definition 2 on isomorphic objects, when applied to this special category **Cat**, tells us that two categories **C** and **D** are isomorphic if there exists two functors $F : \mathbf{C} \longrightarrow \mathbf{D}$ and $G : \mathbf{D} \longrightarrow \mathbf{C}$ such that

$$G \circ F = 1_{\mathbf{C}} \qquad \text{and} \qquad F \circ G = 1_{\mathbf{D}} .$$

Thus, the functor F defines a bijection between the objects as well as between the hom-sets of **C** and **D**. However, many categories that are—for most practical purposes—equivalent are not isomorphic. For example,

- the category **FSet** which has all finite sets as objects, and functions between these sets as morphisms, and,
- a category which has for each $n \in \mathbb{N}$ exactly one set of that size as objects, and functions between these sets as morphisms.

Therefore, it is useful to define some properties for functors that are weaker than being isomorphisms. For instance, the two following definitions describe functors whose morphism assignments are injective and surjective respectively.

Definition 35 A functor $F : \mathbf{C} \longrightarrow \mathbf{D}$ is *faithful* if for any $A, B \in |\mathbf{C}|$ and any $f, g : A \longrightarrow B$ we have that

$$Ff = Fg : FA \longrightarrow FB \qquad \text{implies} \qquad f = g : A \longrightarrow B .$$

Definition 36 A functor $F : \mathbf{C} \longrightarrow \mathbf{D}$ is *full* if for any $A, B \in |\mathbf{C}|$ and for any $g : FA \to FB$ there exists an $f : A \longrightarrow B$ such that $Ff = g$.

A *subcategory* **D** of a category **C** is a collection of objects of **C** as well as a collection of morphisms of **C** such that

- for every morphism $f : A \longrightarrow B$ in **D**, both A and $B \in |\mathbf{D}|$,
- for every $A \in |\mathbf{D}|$, 1_A is in **D**, and
- for every pair of composable morphisms f and g in **D**, $g \circ f$ is in **D**.

These conditions entail that **D** is itself a category. Moreover, if **D** is a subcategory of **C**, the inclusion functor $F : \mathbf{D} \longrightarrow \mathbf{C}$ which maps every $A \in |\mathbf{D}|$ and $f \in \mathbf{D}$ to itself in **C** is automatically faithful. If in addition F is full, then we say that **D** is a *full subcategory* of **C**. A full and faithful functor is in general *not* an isomorphism, as we shall see in Theorem 3 below.

Definition 37 Two categories **C** and **D** are *equivalent* if there is a pair of functors $F : \mathbf{C} \longrightarrow \mathbf{D}$ and $G : \mathbf{D} \longrightarrow \mathbf{C}$ and natural isomorphisms

$$G \circ F \overset{\sim}{\Rightarrow} 1_{\mathbf{C}} \qquad \text{and} \qquad F \circ G \overset{\sim}{\Rightarrow} 1_{\mathbf{D}} \, .$$

An equivalence of categories is weaker than the notion of isomorphism of categories. It captures the essence of what we can do with categories without using concrete descriptions of objects: if two categories **C** and **D** are equivalent then any result following from the categorical structure in **C** remains true in **D**, and vice-versa.

Theorem 3 *[50, p. 93] A functor $F : \mathbf{C} \longrightarrow \mathbf{D}$ is an equivalence of categories if and only if it is both full and faithful, and if each object $B \in |\mathbf{D}|$ is isomorphic to an object FA for some $A \in |\mathbf{C}|$.*

Example 61 A *skeleton* **D** of a category **C** is any full subcategory of **C** such that each $A \in |\mathbf{C}|$ is isomorphic in **C** to exactly one $B \in |\mathbf{D}|$. An equivalence between a category **C** and one of its skeleton **D** is defined as follows:

1. As **D** is a full subcategory of **C**, there is an inclusion functor $F : \mathbf{D} \longrightarrow \mathbf{C}$.
2. By the definition of a skeleton, every $A \in |\mathbf{C}|$ is isomorphic to an $A' \in |\mathbf{D}|$, so we can set $GA := A'$ and pick an isomorphism $\tau_A : A \longrightarrow GA$.
3. From the preceding point, there is a unique way to define a functor $G : \mathbf{C} \longrightarrow \mathbf{D}$ such that we have $FG \overset{\sim}{\Rightarrow} 1_{\mathbf{C}}$ and $GF \overset{\sim}{\Rightarrow} 1_{\mathbf{D}}$.

Particular instances of this are:

- The two categories with sets as objects and functions as morphisms discussed at the beginning of this section.
- **FdHilb** is equivalent to the category with $\mathbb{C}^0, \mathbb{C}^1, \mathbb{C}^2, \cdots, \mathbb{C}^n, \cdots$ as objects and linear maps between these as morphisms. This category is isomorphic to the category $\mathbf{Mat}_{\mathbb{C}}$ of matrices with entries in $\mathbb{C}$ of Example 18.

3.6.5 Topological Quantum Field Theories

TQFTs are primarily used in condensed matter physics to describe, for instance, the fractional quantum Hall effect. Perhaps more accurately, TQFTs are quantum field theories that compute topological invariants. In the context of this paper, TQFTs are our main example of monoidal functors. Defining a TQFT as a monoidal functor is very elegant, however, the seemingly short definition that we will provide is packed with subtleties. In order to appreciate it to its full extent, we will first give the non-categorical axiomatics of a generic n-dimensional TQFTs as given in [66]. We then derive the categorical definition from it. The bulk of this section is taken from [42] to which the reader is referred for a more detailed discussion on the subject.

An *n-dimensional TQFT* is a rule $\mathcal{T}$ which associates to each closed oriented $(n-1)$-dimensional manifold Σ a vector space $\mathcal{T}(\Sigma)$ over the field $\mathbb{K}$, and to each oriented cobordism $M : \Sigma_0 \to \Sigma_1$ a linear map $\mathcal{T}(M) : \mathcal{T}(\Sigma_0) \to \mathcal{T}(\Sigma_1)$, subject to the following conditions:

1. if $M \simeq M'$ then $\mathcal{T}(M) = \mathcal{T}(M')$;
2. each cylinder $\Sigma \times [0, 1]$ is sent to the identity map of $\mathcal{T}(\Sigma)$;
3. If $M = M' \circ M''$ then

$$\mathcal{T}(M) = \mathcal{T}(M') \circ \mathcal{T}(M'') ;$$

4. the disjoint union $\Sigma = \Sigma' + \Sigma''$ is mapped to

$$\mathcal{T}(\Sigma) = \mathcal{T}(\Sigma') \otimes \mathcal{T}(\Sigma''),$$

   and the disjoint union $M = M' + M''$ is mapped to

$$\mathcal{T}(M) = \mathcal{T}(M') \otimes \mathcal{T}(M'') ;$$

5. the empty manifold $\Sigma = \emptyset$ is mapped to the ground field $\mathbb{K}$ and the empty cobordism is sent to the identity map on $\mathbb{K}$.

All of this can be written down in one line.

Definition 38 An *n-dimensional TQFT* is a symmetric monoidal functor

$$\mathcal{T} : (\mathbf{nCob}, +, \emptyset, T) \to (\mathbf{FdVect}_{\mathbb{K}}, \otimes, \mathbb{K}, \sigma)$$

where T are the "twist" cobordisms e.g. $T_1 =$ 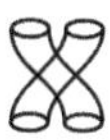 .

The rule that maps manifolds to vector spaces and cobordisms to linear maps gives the domain and the codomain of the functor. Condition 1 says that we consider homeomorphism classes of cobordisms. Conditions 2 and 3 spell out that the TQFT is a functor. Conditions 4 and 5 say that it is a monoidal functor.

We now construct such a functor. In the case of 2-dimensional quantum field theories, it turns out that this question can be answered with the material we introduced in the preceding sections.

We have the following result [42]:

Proposition 7 *The monoidal category **2Cob** is generated by*

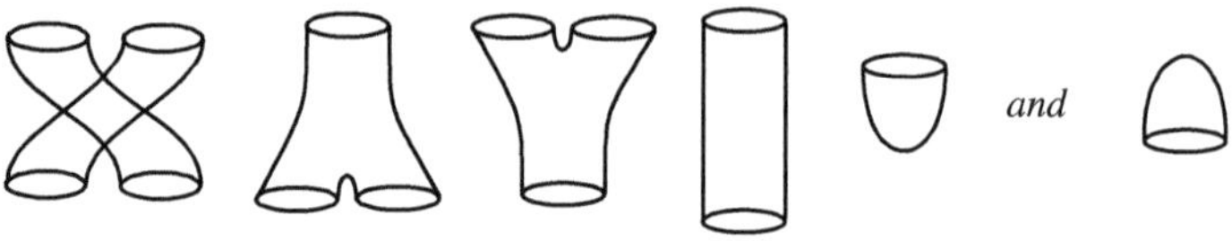

*That means, any cobordism in **2Cob** can be written in terms of these generators when using composition and tensor.*

Following the discussion of Sect. 3.5.7, it is easily seen that these generators satisfy the axioms of a Frobenius comonoid. Moreover, since $\mathcal{T}$ is a monoidal functor, it is sufficient to give the image of the generators of **2Cob** in order to specify it completely. Hence we can map this Frobenius comonoid in **2Cob** on a Frobenius comonoid in $\mathbf{FdVect}_{\mathbb{K}}$:

Objects: $n \mapsto \underbrace{V \otimes V \otimes \ldots \otimes V}_{n \text{ times}}$

Identity:  $\mapsto 1_V : V \to V$

Twist: 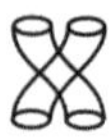 $\mapsto \sigma_{V,V} : V \otimes V \to V \otimes V$

e:  $\mapsto e : \mathbb{K} \to V$

μ:  $\mapsto \mu : V \otimes V \to V$

ϵ :  $\mapsto \epsilon : V \to \mathbb{K}$

δ :  $\mapsto \delta : V \to V \otimes V$

The converse is also true, that is, given a Frobenius comonoid on V, then we can define a TQFT with the preceding prescription, so there is a one-to-one correspondence between commutative Frobenius comonoids and 2-dimensional TQFTs. This is interesting in itself but we can go a step further.

We can now define the category $\mathbf{2TQFT}_{\mathbb{K}}$ of 2-dimensional TQFTs and symmetric monoidal natural transformations between them. Given two TQFTs $\mathcal{T}, \mathcal{T}' \in |\mathbf{2TQFT}_{\mathbb{K}}|$, then the components of the natural transformation θ must be—by the definition above—of the form

$$\theta_n : \underbrace{V \otimes V \otimes \ldots \otimes V}_{n \text{ times}} \to \underbrace{W \otimes W \otimes \ldots \otimes W}_{n \text{ times}} .$$

Since this natural transformation is monoidal, it is completely specified by the map $\theta_1 : V \to W$. The morphism $\theta_{\mathbb{K}}$ is the identity mapping from the trivial Frobenius comonoid on $\mathbb{K}$ to itself. Finally, naturality of θ means that the components must commute with the morphisms of **2Cob**. Since the latter can be decomposed into the generators listed in Proposition 7, we just have to consider these cobordisms. e.g.

$$\begin{array}{ccc} V \otimes V & \xrightarrow{\theta_2} & W \otimes W \\ {\scriptstyle \mu_V}\downarrow & & \downarrow{\scriptstyle \mu_W} \\ V & \xrightarrow[\theta_1]{} & W \end{array}$$

We can now define the category $\mathbf{CFC}_{\mathbb{K}}$ of commutative Frobenius comonoids and morphisms of Frobenius comonoids, that is, linear maps that are both comonoid homomorphisms and monoid homomorphisms.

Theorem 4 *[42] The category $\boldsymbol{2TQFT}_{\mathbb{K}}$ is equivalent to the category $\boldsymbol{CFC}_{\mathbb{K}}$.*

3.7 Further Reading

This concludes our tutorial of (a small fraction of) category theory. We particularly focussed on monoidal categories, given that we expect their role to grow within physics. We indicated how the monoidal structure encodes the nature of physical systems, e.g. classical versus quantum. Admittedly, the distinction as presented here requires substantial qualification, and by no means characterizes what quantum theory is truly about. A recent more elaborated categorical comparison of classical vs. quantum theories is in [24]. All of this is part of a novel vastly growing research area, and we hope that this chapter may help the interested reader to take a bite of it.

We end this chapter by pointing in the direction of other important categorical concepts, for which we refer the reader to other sources. A good place to start are the YouTube postings by the Catsters [18].

Adjoint functors are, at least from a mathematical perspective, the greatest achievement of category theory thus far: it essentially unifies all known mathematical constructs of a variety of areas of mathematics such as algebra, geometry, topology, analysis and combinatorics within a single mathematical concept.

The restriction of adjoint functors to posetal categories, that is, those discussed in Examples 14, 15, 45 and 46, is the concept of *Galois adjoints*. These play an important role in computer science when reasoning about *computational processes*. Let P be a partial order which represents the properties one wishes to attribute to the input data of a process, with "$a \leq b$" if and only if "whenever a holds, then b must hold too", and let Q be the partial order which represents the properties one wishes to attribute to the output data of that process. So the process is an order preserving map $f : P \to Q$. The order preserving map $g : Q \to P$, which maps a property b of the output to the "weakest" property (i.e. highest in the partial ordering) which the input data needs to satisfy in order to guarantee that the output satisfies b, is then the *left Galois adjoint* to f. One refers to $g(b)$ as the *weakest precondition*. Formally f is left Galois adjoint to g if and only if for all $a \in P$ and all $b \in Q$ we have

$$f(a) \leq b \Longleftrightarrow a \leq g(b)\,.$$

The *orthomodular law* of quantum logic [57], that is, in the light of Example 46, a weakening of the distributive law which $L(\mathcal{H})$ does satisfy, is an example of such an adjunction of processes, namely

$$\mathrm{P}_c(a) \leq b \Longleftrightarrow a \leq [c \rightarrow](b)$$

where:

- P_c is an order-theoretic generalization of the linear algebraic notion of an "orthogonal projector on subspace c", formally defined to be

$$\mathrm{P}_c : L \to L :: a \mapsto c \wedge (a \vee c^{\perp})\,,$$

  where $(-)^{\perp}$ stands for the orthocomplement;
- $[- \rightarrow](-)$ is referred to as *Sasaki hook*, or unfortunately, also sometimes referred to as "quantum implication", and is formally defined within

$$[c \rightarrow] : L \to L :: a \mapsto c^{\perp} \vee (a \wedge c)\,.$$

Heyting algebras, that is, the order-theoretic incarnation of intuitionistic logic, and which play an important role in the recent work by Doering and Isham [32] are, by definition, Galois adjoints now defined within

$$[c\wedge](a) \leq b \Longleftrightarrow a \leq [c \Rightarrow](b)\,.$$

So these Galois adjoints relate logical conjunction to logical implication.

The general notion of adjoint functors involves, instead of an "if and only if" between statements $f(a) \leq b$ and $a \leq g(b)$, a "natural equivalence" between homsets $\mathbf{D}(FA, B)$ and $\mathbf{C}(A, GB)$, where $F : \mathbf{C} \longrightarrow \mathbf{D}$ and $G : \mathbf{D} \longrightarrow \mathbf{C}$ are now functors. We refer to [4, 12] in these volumes for an account on adjoint functors and the role they play in logic. We also recommend [44] on this topic.

The composite $G \circ F : \mathbf{C} \longrightarrow \mathbf{C}$ of a pair of adjoint functors is a *monad*, and each monad arises in this manner. The posetal counterpart to this is a *closure operator*, of which the linear span in a vector space is an example.

The composite $F \circ G : \mathbf{D} \longrightarrow \mathbf{D}$ of a pair of adjoint functors is a *comonad*. Comonads are an instance of the research area of *coalgebra*, of which comonoids are also an instance. The study of coalgebraic structures has become increasingly important both in computer science and physics. These structures are very different from algebraic structures: while algebraic structures typically would take two pieces of data a and b as input, and produce the composite $a \bullet b$, coalgebraic structures would do the opposite, that is, take one piece of data as input and produce two pieces of data as output, cf. a copying operation. Another example of a coalgebraic concept is quantum measurement. Quantum measurements take a quantum state as input and produces another quantum state together with classical data [27].

There also is the area of *higher-dimensional category theory*, after which the n-category cafe is named [11]. Monoidal categories are a special case of *bicategories*, since we can compose the objects with the tensor, as well as the processes between these objects. There is currently much activity on the study of *n-categories*, that is, categories in which the hom-sets are themselves categories, and the hom-sets of these categories are again categories etc. Why would we be interested in that? If one is interested in processes then one should also be in modifying processes, and that is exactly what these higher dimensional categorical structures enable to model. An excellent book on higher-dimensional category theory is [48].

We end by recommending the other chapters in these volumes entitled New Structures for Physics, which, among many other things, contain complementary tutorials on category theory and its graphical calculus [4, 12, 63].

Acknowledgments We very much appreciated the feedback from the n-category cafe on a previous draft of this paper, by John Baez, Hendrik Boom, Dave Clarke, David Corfield and Aaron Lauda. We in particular thank Frank Valckenborgh for proofreading the final version.

References

1. Abramsky, S.: No-Cloning in categorical quantum mechanics. In: Mackie, I., Gay, S. (eds.) Semantic Techniques for Quantum Computation. Cambridge University Press, Cambridge (2008)
2. Abramsky, S., Coecke, B.: A Categorical semantics of quantum protocols. In: Proceedings of the 19th Annual IEEE Symposium on Logic in Computer Science (LiCS'04), IEEE Computer Science Press, 2004. arXiv:quant-ph/0402130 An updated & improved version appeared under the title Categorical quantum mechanics. In: Handbook of quantum logic and quantum structures, Elsevier. arXiv:0808.1023
3. Abramsky, S., Coecke, B.: Abstract physical traces. Theor. Appl. Categories **14**, 111–124 (2006). http://www.tac.mta.ca/tac/volumes/14/6/14–06abs.html
4. Abramsky, S., Tzevelekos, N.: Introduction to categories and categorical logic. In: Coecke, B. (ed.) New Structures for Physics. Springer Lecture Notes in Physics, New York (2009)
5. Adámek, J., Herrlich, H., Strecker, G.E.: Abstract and Concrete Categories – The Joy of Cats. Wiley, New York (1990). Freely available from http://katmat.math.uni-bremen.de/acc/acc.pdf
6. Asperti, A., Longo, G.: Categories, types, and Structures. An Introduction to Category theory for the Working Computer Scientist. MIT Press, New York (1991)
7. Barnum, H., Caves, C.M., Fuchs, C.A., Jozsa, R., Schumacher, B.: Noncommuting mixed states cannot be broadcast. Phys. Rev. Lett. **76**, 2818–2821 (1996). arXiv:quant-ph/9511010

8. Baez, J.C.: This Week's Finds in Mathematical Physics, 1993–2008. http://math.ucr.edu/home/baez/TWF.html
9. Baez, J.C.: Quantum quandaries: A category-theoretic perspective. In French, S. et al. (eds.) Structural Foundations of Quantum Gravity. Oxford University Press, Oxford (2004). arXiv:quant-ph/0404040
10. Baez, J.C., Dolan, J.: Higher-dimensional algebra and topological quantum field theory. J. Math. Phys. **36**, 60736105 (1995). arXiv:q-alg/9503002
11. Baez, J.C., Corfield, D., Schreiber, U.: The n-category cafe. Group blog. http://golem.ph.utexas.edu/category/
12. Baez, J.C., Stay, M.: Physics, topology, logic and computation: A Rosetta Stone. In: Coecke, B. (ed.) New Structures for Physics. Springer Lecture Notes in Physics, New York (2009). arXiv: 0903.0340
13. Barr, M., Wells, C.: Toposes, Triples and Theories. Springer, New York (1985). Republished in Reprints in Theory and Applications of Categories, No. 12, 1–287, 2005. http://www.tac.mta.ca/tac/reprints/articles/12/tr12abs.html
14. Bénabou, J.: Categories avec multiplication. Comptes Rendus des Séances de l'Académie des Sciences Paris **256**, 1887–1890 (1963)
15. Birkhoff, G., von Neumann, J.: The logic of quantum mechanics. Ann. Math. **37**, 823–843 (1936)
16. Borceux, F., Stubbe, I.: Short introduction to enriched categories. In: Coecke, B., Moore, D.J., Wilce, A. (eds.) Current Research in Operational Quantum Logic: Algebras, Categories and Languages, pp. 167–194. Fundamental Theories of Physics vol. 111. Springer, New York (2000)
17. Brown, R.: Topology: A Geometric Account of General Topology, Homotopy Types, and the Fundamental Groupoid. Halsted Press, New York (1988)
18. Cheng, E., Willerton, S.: The Catsters. YouTube videos. www.youtube.com/user/TheCatsters
19. Catsters: Group objects and Hopf algebras. See [18]
20. Coecke, B.: Kindergarten quantum mechanics. In: Khrennikov, A. (ed.) Quantum Theory: Reconsiderations of the Foundations III, pp. 81–98. AIP Press, New York (2005). arXiv:quant-ph/0510032v1
21. Coecke, B.: Introducing categories to the practicing physicist. In: What is Category Theory? pp. 45–74. Advanced Studies in Mathematics and Logic 30, Polimetrica Publishing, Italy (2006). arXiv:0808.1032
22. Coecke, B.: De-linearizing linearity: Projective quantum axiomatics from strong compact closure. Elect. Notes Theor. Comput. Sci. **170**, 47–72 (2007). arXiv:quant-ph/0506134
23. Coecke, B., Duncan, R.: Interacting quantum observables. In: Proceedings of the 35th International Colloquium on Automata, Languages and Programming, pp. 298–310, Lecture Notes in Computer Science, vol. 5126, Springer, New York (2008). An extended version is available under the title Interacting Quantum Observables: Categorical Algebra and Diagrammatics at arXiv:0906.4725
24. Coecke, B., Edwards, B., Spekkens, R.W.: The group-theoretic origin of non-locality for qubits. Oxford University Computing Laboratory Research Report PRG-RR-09–04 (2009) http://web.comlab.ox.ac.uk/publications/publication3026-abstract.html
25. Coecke, B., Paquette, E.O.: POVMs and Nairmarks theorem without sums, Electronic Notes in Theoretical Computer Science (to appear) (2008). arXiv:quant-ph/0608072
26. Coecke, B., Paquette, E.O., Pavlovic, D.: Classical and quantum structuralism. In: Mackie, I., Gay, S. (eds.) Semantic Techniques for Quantum Computation. To appear, Cambridge University Press, Cambridge (2008). arXiv:0904.1997
27. Coecke, B., Pavlovic, D.: Quantum measurements without sums. In: Chen, G., Kauffman, L., Lamonaco, S. (eds.) Mathematics of Quantum Computing and Technology, pp. 567–604. Taylor and Francis, Abington (2008). arXiv:quant-ph/0608035
28. Coecke, B., Pavlovic, D., Vicary, J.: A new description of orthogonal bases. Preprint (2008). arXiv:0810.0812

29. Deligne, P.: Catégories tannakiennes. In: The Grothendieck Festschrift Volume II, Progress in Mathematics 87, pp. 111–196. Birkhäuser, Boston (1990)
30. Dieks, D.G.B.J.: Communication by EPR devices. Phys. Lett. A **92**, 271–272 (1982)
31. Dirac, P.A.M.: The Principles of Quantum Mechanics, 3rd edn. Oxford University Press, Oxford (1947)
32. Doering, A., Isham, C.: 'What is a thing?': Topos theory in the foundations of physics. In: Coecke, B. (ed.) New Structures for Physics. Springer Lecture Notes in Physics, New York (2009). arXiv:0803.0417
33. Eilenberg, S., MacLane, S.: General theory of natural equivalences. Trans. Am. Math. Soc. **58**, 231–294 (1945)
34. Finkelstein, D.R., Jauch, J.M., Schiminovich, D., Speiser, D.: Foundations of quaternion quantum mechanics. J. Math. Phys. **3**, 207–220 (1962)
35. Freyd, P., Yetter, D.: Braided compact closed categories with applications to lowdimensional topology. Adv. Math. **77**, 156–182 (1989)
36. Girard, J.-Y.: Linear logic. Theor. Comput. Sci. **50**, 1–102 (1987)
37. Houston, R.: Finite products are biproducts in a compact closed category. J. Pure Appl. Algebra **212**, 394–400 (2008). arXiv:math/0604542
38. Joyal, A., Street, R.: The geometry of tensor calculus I. Adv. Math. **88**, 55–112 (1991)
39. Joyal, A., Street, R., Verity, D.: Traced monoidal categories. Proc. Cambridge Philoso. Soc. **119**, 447–468 (1996)
40. Kauffman, L.H.: Teleportation topology. Optics Spectroscopy **99**, 227–232 (2005). arXiv:quant-ph/0407224
41. Kelly, G.M., Laplaza, M.L.: Coherence for compact closed categories. J. Pure Appl. Algebra **19**, 193–213 (1980)
42. Kock, J.: Frobenius Algebras and 2D Topological Quantum Field Theories, London Mathematical Society, in Student Texts 59 (2004)
43. Lack, S.: Composing PROPs. Theory Appl. Categories **13**, 147–163 (2004)
44. Lambek, J., Scott, P.J.: Higher Order Categorical Logic. Cambridge University Press, Cambridge (1986)
45. Lauda, A.D., Pfeiffer, H.: State sum construction of two-dimensional open-closed Topological Quantum Field Theories. J. Knot Theory Ramifications **16**, 1121–1163 (2007)
46. Lawvere, F.W.: Metric spaces, generalized logic, and closed categories. Rendiconti del Seminario Matematico e Fisico di Milano **43**, 135–166 (1974)
47. Lawvere, F.W., Schanuel, S.H.: Conceptual Mathematics. Cambridge University Press, Cambridge (1997)
48. Leinster, T.: Higher Operads, Higher Categories, London Mathematical Society Lecture Note Series 298, Cambridge University Press, Cambridge (2004)
49. Martin, K., Panangaden, P.: Domain theory and general relativity. In: Coecke, B. (ed.) New Structures for Physics. Springer Lecture Notes in Physics, New York (2009)
50. Mac Lane, S.: Categories for the Working Mathematician, 2nd edn. Springer, New York (2000)
51. Muirhead, R.F.: Some methods applicable to identities and inequalities of symmetric algebraic functions of n letters. Proc. Edinburgh Math. Soc. **21**, 144157 (1903)
52. Nielsen, M.A.: Conditions for a class of entanglement transformations. Phys. Rev. Lett. **83**, 436–439 (1999)
53. Morrison, S.E.: A diagrammatic category for the representation theory of Uq(sln). PhD thesis, University of California at Berkeley, Berkeley (2007)
54. Pati, A.K., Braunstein, S.L.: Impossibility of deleting an unknown quantum state. Nature **404**, 164–165 (2000). arXiv:quant-ph/9911090
55. Penrose, R.: Applications of negative dimensional tensors. In: Combinatorial Mathematics and Its Applications, pp. 221–244, Academic Press, New York (1971)
56. Penrose, R.: Techniques of differential topology in relativity, Society for Industrial and Applied Mathematics (1972)
57. Piron, C.: Foundations of Quantum Physics. W. A. Benjamin, Reading (1976)

58. Robinson, J.A.: A machine-oriented logic based on the resolution principle. J. ACM **12**, 23–41 (1965)
59. Rosen, R.: Anticipatory Systems: Philosophical, Mathematical and Methodological Foundations. Pergamon Press, New York (1985)
60. Schrödinger, E.: Discussion of probability relations between separated systems. Proc. Cambridge Philosophical Soc. **31**, 555–563 (1935); **32**, 446–451 (1936)
61. Seely, R.A.G.: Linear logic, *-autonomous categories and cofree algebras. Contemporary Math. **92**, 371–382 (1998)
62. Selinger, P.: Dagger compact closed categories and completely positive maps. Electro. Notes Theor. Comput. Sci. **170**, 139–163 (2007)
63. Selinger, P.: A survey of graphical languages for monoidal categories. In: Coecke, B. (ed.) New Structures for Physics. Springer Lecture Notes in Physics, New York (2009)
64. Sorkin, R.: Spacetime and causal sets, In: J. D'Olivo (ed.) Relativity and Gravitation: Classical and Quantum. World Scientific, Singapore (1991)
65. Street, R.: Quantum Groups: A Path to Current Algebra. Cambridge University Press, Cambridge (2007)
66. Turaev, V.G.: Axioms for topological quantum field theories. In: Annales de la faculté des sciences de Toulouse 6e série **3**, 135–152 (1994)
67. Vicary, J.: A categorical framework for the quantum harmonic oscillator. Int. J. Theor. Phys. (to appear) (2008). arXiv: quant-ph/0706.0711
68. Yetter, D.N.: Functorial Knot Theory. Categories of Tangles, Coherence, Categorical Deformations, and Topological Invariants, World Scientific, Singapore (2001)
69. Wootters, W., Zurek, W.: A single quantum cannot be cloned. Nature **299**, 802–803 (1982)

Part II
Manifestations of Linearity

Chapter 4
A Survey of Graphical Languages for Monoidal Categories

P. Selinger

Abstract This article is intended as a reference guide to various notions of monoidal categories and their associated string diagrams. It is hoped that this will be useful not just to mathematicians, but also to physicists, computer scientists, and others who use diagrammatic reasoning. We have opted for a somewhat informal treatment of topological notions, and have omitted most proofs. Nevertheless, the exposition is sufficiently detailed to make it clear what is presently known, and to serve as a starting place for more in-depth study. Where possible, we provide pointers to more rigorous treatments in the literature. Where we include results that have only been proved in special cases, we indicate this in the form of caveats.

4.1 Introduction

There are many kinds of monoidal categories with additional structure—braided, rigid, pivotal, balanced, tortile, ribbon, autonomous, sovereign, spherical, traced, compact closed, $*$-autonomous, to name a few. Many of them have an associated graphical language of "string diagrams". The proliferation of different notions is often confusing to non-experts, and occasionally to experts as well. To add to the confusion, one concept often appears in the literature under multiple names (for example, "rigid" is the same as "autonomous", "sovereign" is the same as "pivotal", and "ribbon" is the same as "tortile").

In this survey, I attempt to give a systematic overview of the main notions and their associated graphical languages. My initial intention was to summarize, without proof, only the main definitions and coherence results that appear in the literature. However, it quickly became apparent that, in the interest of being systematic, I had to include some additional notions. This led to the sections on spacial categories, and planar and braided traced categories.

P. Selinger (✉)
Department of Mathematics and Statistics, Dalhousie University, Halifax, NS, Canada
e-mail: selinger@mathstat.dal.ca

Selinger, P.: *A Survey of Graphical Languages for Monoidal Categories*. Lect. Notes Phys. **813**, 289–355 (2011)
DOI 10.1007/978-3-642-12821-9_4

Historically, the terminology was often fixed for special cases before more general cases were considered. As a result, some concepts have a common name (such as “compact closed category”) where another name would have been more systematic (e.g. “symmetric autonomous category”). I have resisted the temptation to make major changes to the established terminology. However, I propose some minor tweaks that will hopefully not be disruptive. For example, I prefer “traced category”, which can be combined with various qualifying adjectives, to the longer and less flexible “traced monoidal category”.

Many of the coherence results are widely known, or at least presumed to be true, but some of them are not explicitly found in the literature. For those that can be attributed, I have attempted to do so, sometimes with a caveat if only special cases have been proved in the literature. For some easy results, I have provided proof sketches. Some unproven results have been included as conjectures.

While the results surveyed here are mathematically rigorous, I have shied away from giving the full technical details of the definitions of the graphical languages and their respective notions of equivalence of diagrams. Instead, I present the graphical languages somewhat informally, but in a way that will be sufficient for most applications. Where appropriate, full mathematical details can be found in the references.

Readers who want a quick overview of the different notions are encouraged to first consult the summary chart at the end of this article.

An updated version of this article will be maintained on the ArXiv, so I encourage readers to contact me with corrections, literature references, and updates.

Graphical Languages: An Evolution of Notation

The use of graphical notations for operator diagrams in physics goes back to Penrose [30]. Initially, such notations applied to multiplications and tensor products of linear operators, but it became gradually understood that they are applicable in more general situations.

To see how graphical languages arise from matrix multiplication, consider the following example. Let $M : A \to B$, $N : B \otimes C \to D$, and $P : D \to E$ be linear maps between finite dimensional vector spaces A, B, C, D, E. These maps can be combined in an obvious way to obtain a linear map $F : A \otimes C \to E$. In functional notation, the map F can be written

$$F = P \circ N \circ (M \otimes \mathrm{id}_C). \tag{4.1}$$

The same can be expressed as a summation over matrix indices, relative to some chosen basis of each space. In mathematical notation, suppose $M = (m_{j,i})$, $N = (n_{l,jk})$, $P = (p_{m,l})$, and $F = (f_{m,ik})$, where i, j, k, l, m range over basis vectors of the respective spaces. Then

$$f_{m,ik} = \sum_j \sum_l p_{m,l} n_{l,jk} m_{j,i}. \tag{4.2}$$

In physics, it is more common to write column indices as superscripts and row indices as subscripts. Moreover, one can drop the summation symbols by using Einstein's summation convention.

$$F_m^{ik} = P_m^l N_l^{jk} M_j^i. \tag{4.3}$$

In (4.2) and (4.3), the order of the factors in the multiplication is not relevant, as all the information is contained in the indices. Also note that, while the notation mentions the chosen bases, the result is of course basis independent. This is because indices occur in pairs of opposite variance (if on the same side of the equation) or equal variance (if on opposite sides of the equation). It was Penrose [30] who first pointed out that the notation is valid in many situations where the indices are purely formal symbols, and the maps may not even be between vector spaces.

Since the only non-trivial information in (4.3) is in the pairing of indices, it is natural to represent these pairings graphically by drawing a line between paired indices. Penrose [30] proposed to represent the maps M, N, P as boxes, each superscript as an incoming wire, and each subscript as an outgoing wire. Wires corresponding to the same index are connected. Thus, we obtain the graphical notation:

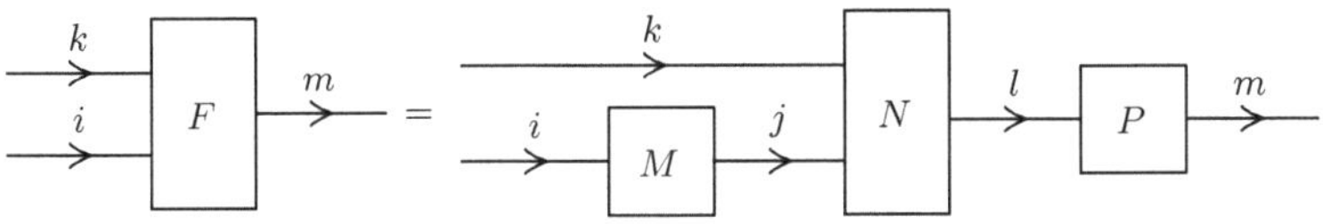

(4.4)

Finally, since the indices no longer serve any purpose, one may omit them from the notation. Instead, it is more useful to label each wire with the name of the corresponding space.

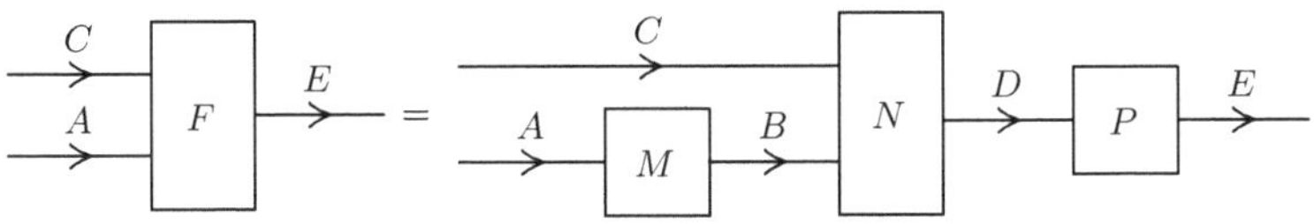

(4.5)

In the notation of monoidal categories, (4.5) can be expressed as a commutative diagram

$$\begin{array}{ccc} A \otimes C & \xrightarrow{F} & E \\ {\scriptstyle M\otimes \mathrm{id}_C}\downarrow & & \uparrow{\scriptstyle P} \\ B \otimes C & \xrightarrow{N} & D, \end{array} \tag{4.6}$$

or simply:

$$F = P \circ N \circ (M \otimes \mathrm{id}_C). \tag{4.7}$$

Thus, we have completed a full circle and arrived back at the notation (4.1) that we started with.

Organization of the Paper

In each of the remaining sections of this paper, we will consider a particular class of categories and its associated graphical language.

Acknowledgments

I would like to thank Gheorghe Ştefănescu and Ross Street for their help in locating hard-to-obtain references, and for providing some background information. Thanks to Fabio Gadducci, Chris Heunen, and Micah McCurdy for useful comments on an earlier draft.

4.2 Categories

We only give the most basic definitions of categories, functors, and natural transformations. For a gentler introduction, with more details and examples, see e.g. Mac Lane [29].

Definition 1 A *category* $\mathbf{C}$ consists of:

- a class $|\mathbf{C}|$ of *objects*, denoted $A, B, C, \ldots$;
- for each pair of objects A, B, a set $\hom_{\mathbf{C}}(A, B)$ of *morphisms*, which are denoted $f : A \to B$;
- *identity morphisms* $\mathrm{id}_A : A \to A$ and the operation of *composition*: if $f : A \to B$ and $g : B \to C$, then

$$g \circ f : A \to C,$$

subject to the three equations

$$\mathrm{id}_B \circ f = f, \qquad f \circ \mathrm{id}_A = f, \qquad (h \circ g) \circ f = h \circ (g \circ f)$$

for all $f : A \to B$, $g : B \to C$, and $h : C \to D$.

The terms "map" or "arrow" are often used interchangeably with "morphism".

Examples 2 Some examples of categories are: the category **Set** of sets (with functions as the morphisms); the category **Rel** of sets (with relations as the morphisms); the category **Vect** of vector spaces (with linear maps); the category **Hilb** of Hilbert spaces (with bounded linear maps); the category **UHilb** of Hilbert spaces (with unitary maps); the category **Top** of topological spaces (with continuous maps); the category **Cob** of n-dimensional oriented manifolds (with oriented cobordisms). Note that in each case, we need to specify not only the objects, but also the morphisms (and technically the composition and identities, although they are often clear from the context).

Categories also arise in other sciences, for example in logic (where the objects are propositions and the morphisms are proofs), and in computing (where the objects are data types and the morphisms are programs).

Many concepts associated with sets and functions, such as *inverse, monomorphism* (injective map), *idempotent*, *cartesian product*, etc., are definable in an arbitrary category.

Graphical Language

In the graphical language of categories, objects are represented as *wires* (also called *edges*) and morphisms are represented as *boxes* (also called *nodes*). An identity morphisms is represented as a continuing wire, and composition is represented by connecting the outgoing wire of one diagram to the incoming wire of another. This is shown in Table 4.1.

Table 4.1 The graphical language of categories

Object	A	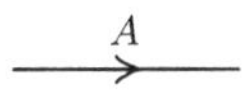
Morphism	$f : A \to B$	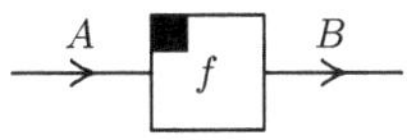
Identity	$\mathrm{id}_A : A \to A$	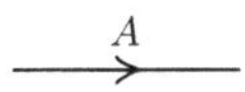
Composition	$t \circ s$	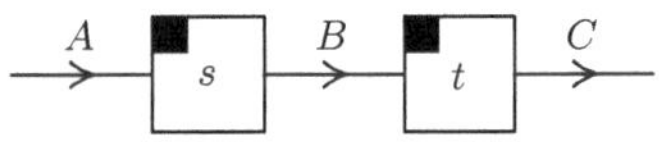

Coherence

Note that the three defining axioms of categories (e.g., $\mathrm{id}_B \circ f = f$) are automatically satisfied "up isomorphism" in the graphical language. This property is known as *soundness*. A converse of this statement is also true: every equation that holds in the graphical language is a consequence of the axioms. This property is called *completeness*. We refer to a soundness and completeness theorem as a *coherence theorem*.

Theorem 1 (Coherence for categories) *A well-formed equation between two morphism terms in the language of categories follows from the axioms of categories if and only if it holds in the graphical language up to isomorphism of diagrams.*

Hopefully it is obvious what is meant by *isomorphism of diagrams*: two diagrams are isomorphic if the boxes and wires of the first are in bijective correspondence with the boxes and wires of the second, preserving the connections between boxes and wires.

Admittedly, the above coherence theorem for categories is a triviality, and is not usually stated in this way. However, we have included it for sake of uniformity, and for comparison with the less trivial coherence theorems for monoidal categories in the following sections. The proof is straightforward, since by the associativity and unit axioms, each morphism term is uniquely equivalent to a term of the form

$$((f_n \circ \ldots) \circ f_2) \circ f_1$$

for $n \geq 0$, with corresponding diagram

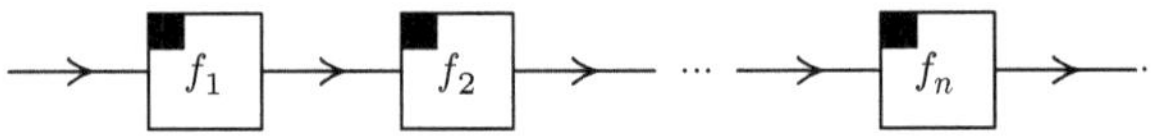

Remark 1 We have equipped wires with a left-to-right arrow, and boxes with a marking in the upper left corner. These markings are of no use at the moment, but will become important as we extend the language in the following sections.

4.2.1 Technicalities

Signatures, Variables, Terms, and Equations

So far, we have not been very precise about what the wires and boxes of a diagram are labeled with. We have also glossed over what was meant by "a well-formed equation between morphism terms in the language of categories". We now briefly explain these notions, without giving all the formal details. For a more precise mathematical treatment, see e.g. Joyal and Street [22].

The wires of a diagram are labeled with *object variables*, and the boxes are labeled with *morphism variables*. To understand what this means, consider the familiar language of arithmetic expressions. This language deals with *terms*, such as $(x+y+2)(x+3)$, which are built up from *variables*, such as x and y, *constants*, such as 2 and 3, by means of *operations*, such as addition and multiplication. Variables can be viewed in three different ways: first, they can be viewed as *symbols* that can be compared (e.g. the variable x occurs twice in the given term, and is different from the variable y). They can also be viewed as placeholders for arbitrary *numbers*, for example $x = 5$ and $y = 15$. Here x and y are allowed to represent different numbers or the same number; however, the two occurrences of x must denote the same number. Finally, variables can be viewed as placeholders for arbitrary *terms*, such as $x = a + b$ and $y = z^2$.

The formal language of category theory is similar, except that we require two sets of variables: object variables (for labeling wires) and morphism variables (for labeling boxes). We must also equip each morphism variable with a specified domain and codomain. The following definition makes this more precise.

Definition 3 A *simple (categorical) signature* Σ consists of a set Σ_0 of *object variables*, a set Σ_1 of *morphism variables*, and a pair of functions dom, cod : $\Sigma_1 \to \Sigma_0$. Object variables are usually written $A, B, C, \ldots$, morphism variables are usually written $f, g, h, \ldots$, and we write $f : A \to B$ if $\mathrm{dom}(f) = A$ and $\mathrm{cod}(f) = B$.

Given a simple signature, we can then build *morphism terms*, such as $f \circ (g \circ \mathrm{id}_A)$, which are built from morphism variables (such as f and g) and morphism constants (such as id_A), via operations (i.e., composition). Each term is recursively equipped with a domain and a codomain, and we must require compositions to respect the domain and codomain information. A term that obeys these rules is called *well-formed*. Finally, an equation between terms is called a *well-formed equation* if the left-hand side and right-hand side are well-formed terms that moreover have equal domains and equal codomains.

The graphical language is also relative to a given signature. The wires and boxes are labeled, respectively, with object variables and morphism variables from the signature, and the labeling must respect the domain and codomain information. This means that the wire entering (respectively, exiting) a box labeled f must be labeled by the domain (respectively, codomain) of f.

The above remark about the different roles of variables in arithmetic also holds for the diagrammatic language of categories. On the one hand, the labels can be viewed as formal symbols. This is the view used in the coherence theorem, where the formal labels are part of the definition of equivalence (in this case, isomorphism) of diagrams.

The labels can also be viewed as placeholders for specific objects and morphisms in an actual category. Such an assignment of objects and morphisms is called an *interpretation* of the given signature. More precisely, an interpretation i of a signature Σ in a category $\mathbf{C}$ consists of a function $i_0 : \Sigma_0 \to |\mathbf{C}|$, and for any $f \in \Sigma_1$ a morphism $i_1(f) : i_0(\mathrm{dom}\, f) \to i_0(\mathrm{cod}\, f)$. By a slight abuse of notation, we write $i : \Sigma \to \mathbf{C}$ for such an interpretation.

Finally, a morphism variable can be viewed as a placeholder for an arbitrary (possibly composite) diagram. We occasionally use this latter view in schematic drawings, such as the schematic representation of $t \circ s$ in Table 4.1. We then label a box with a morphism term, rather than a formal variable, and understand the box as a short-hand notation for a possibly composite diagram corresponding to that term.

Functors and Natural Transformations

Definition 4 Let $\mathbf{C}$ and $\mathbf{D}$ be categories. A *functor* $F : \mathbf{C} \to \mathbf{D}$ consists of a function $F : |\mathbf{C}| \to |\mathbf{D}|$, and for each pair of objects $A, B \in |\mathbf{C}|$, a function $F : \hom_{\mathbf{C}}(A, B) \to \hom_{\mathbf{D}}(FA, FB)$, satisfying $F(g \circ f) = F(g) \circ F(f)$ and $F(\mathrm{id}_A) = \mathrm{id}_{FA}$.

Definition 5 Let $\mathbf{C}$ and $\mathbf{D}$ be categories, and let $F, G : \mathbf{C} \to \mathbf{D}$ be functors. A *natural transformation* $\tau : F \to G$ consists of a family of morphisms $\tau_A : FA \to GA$, one for each object $A \in |\mathbf{C}|$, such that the following diagram commutes for all $f : A \to B$:

$$\begin{array}{ccc} FA & \xrightarrow{\tau_A} & GA \\ {\scriptstyle Ff}\downarrow & & \downarrow{\scriptstyle Gf} \\ FB & \xrightarrow{\tau_B} & GB. \end{array}$$

Coherence and Free Categories

Most coherence theorems are proved by characterizing the *free* categories of a certain kind.

Definition 6 We say that a category **C** is *free* over a signature Σ if it is equipped with an interpretation $i : \Sigma \to \mathbf{C}$, such that for any category **D** and interpretation $j : \Sigma \to \mathbf{D}$, there exists a unique functor $F : \mathbf{C} \to \mathbf{D}$ such that $j = F \circ i$.

Theorem 2 *The graphical language of categories over a signature Σ, with identities and composition as defined in Table 4.1, and up to isomorphism of diagrams, forms the free category over Σ.*

Theorem 1 is indeed a consequence of this theorem: by definition of freeness, an equation holds in all categories if and only if it holds in the free category. By the characterization of the free category, an equation holds in the free category if and only if it holds in the graphical language.

4.3 Monoidal Categories

In this section, we consider various notions of monoidal categories. We sometimes refer to these notions as "progressive", which means they have graphical languages where all arrows point left-to-right. This serves to distinguish them from "autonomous" notions, which will be discussed in Sect. 4.4, and "traced" notions, which will be discussed in Sect. 4.5.

4.3.1 (Planar) Monoidal Categories

A *monoidal category* (also sometimes called *tensor category*) is a category with an associative unital tensor product. More specifically:

Definition 7 [29, 23] A *monoidal category* is a category with the following additional structure:

- a new operation $A \otimes B$ on objects and a new object constant I;
- a new operation on morphisms: if $f : A \to C$ and $g : B \to D$, then

$$f \otimes g : A \otimes B \to C \otimes D;$$

- and isomorphisms

$$\begin{array}{ll} \alpha_{A,B,C}: & (A \otimes B) \otimes C \stackrel{\cong}{\to} A \otimes (B \otimes C), \\ \lambda_A: & I \otimes A \stackrel{\cong}{\to} A, \\ \rho_A: & A \otimes I \stackrel{\cong}{\to} A, \end{array}$$

subject to a number of equations:

- $\otimes$ is a bifunctor, which means $\mathrm{id}_A \otimes \mathrm{id}_B = \mathrm{id}_{A \otimes B}$ and $(k \otimes h) \circ (g \otimes f) = (k \circ g) \otimes (h \circ f)$;
- α, λ, and ρ are natural transformations, i.e., $(f \otimes (g \otimes h)) \circ \alpha_{A,B,C} = \alpha_{A',B',C'} \circ ((f \otimes g) \otimes h)$, $f \circ \lambda_A = \lambda_{A'} \circ (\mathrm{id}_I \otimes f)$, and $f \circ \rho_A = \rho_{A'} \circ (f \otimes \mathrm{id}_I)$;
- plus the following two coherence axioms, called the "pentagon axiom" and the "triangle axiom":

$$\begin{array}{ccccc} & (A \otimes (B \otimes C)) \otimes D & \xrightarrow{\alpha_{A,B\otimes C,D}} & A \otimes ((B \otimes C) \otimes D) & \\ {}^{\alpha_{A,B,C}\otimes D}\nearrow & & & & \searrow^{A\otimes\alpha_{B,C,D}} \\ ((A \otimes B) \otimes C) \otimes D & & & & A \otimes (B \otimes (C \otimes D)) \\ & \searrow_{\alpha_{A\otimes B,C,D}} & & \nearrow_{\alpha_{A,B,C\otimes D}} & \\ & & (A \otimes B) \otimes (C \otimes D) & & \end{array}$$

$$\begin{array}{ccc} (A \otimes I) \otimes B & \xrightarrow{\alpha_{A,I,B}} & A \otimes (I \otimes B) \\ & {}_{\rho_A \otimes \mathrm{id}_B}\searrow \quad \swarrow_{\mathrm{id}_A \otimes \lambda_B} & \\ & A \otimes B & \end{array}$$

When we specifically want to emphasize that a monoidal category is not assumed to be braided, symmetric, etc., we sometimes also refer to it as a *planar monoidal category*.

Examples 8 Examples of monoidal categories include: the category **Set** (of sets and functions), together with the cartesian product $\times$; the category **Set** together with the disjoint union operation $+$; the category **Rel** with either $\times$ or $+$; the category **Vect** (of vectors spaces and linear functions) with either $\oplus$ or $\otimes$; the category **Hilb** of Hilbert spaces with either $\oplus$ or $\otimes$; the categories **Top** and **Cob** with disjoint union $+$. Note that in each case, we need to specify a category and a tensor product (in general there are multiple choices). Technically, we should also specify associativity maps etc., but they are usually clear from the context.

Graphical Language

We extend the graphical language of categories as follows. A tensor product of objects is represented by writing the corresponding wires in parallel. The unit object

Table 4.2 The graphical language of monoidal categories

Tensor product	$S \otimes T$	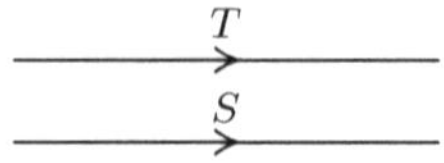
Unit object	I	(empty)
Morphism	$f : A_1 \otimes \ldots \otimes A_n \to B_1 \otimes \ldots \otimes B_m$	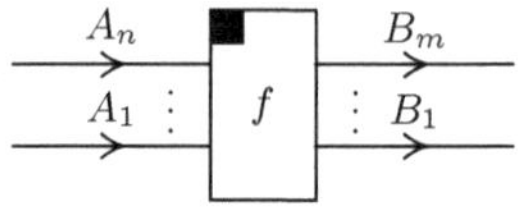
Tensor product	$s \otimes t$	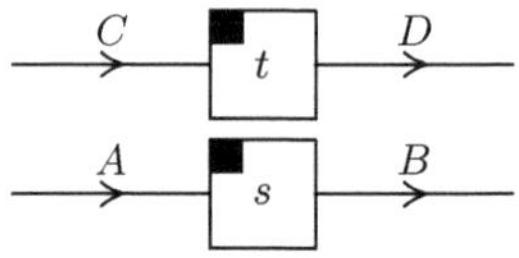

is represented by zero wires. A morphism variable $f : A_1 \otimes \ldots \otimes A_n \to B_1 \otimes \ldots \otimes B_m$ is represented as a box with n input wires and m output wires. A tensor product of morphisms is represented by stacking the corresponding diagrams. This is shown in Table 4.2.

Note that it is our convention to write tensor products in the bottom-to-top order. Similar conventions apply to objects as to morphisms: thus, a single wire is labeled by an *object variable* such as A, while a more general object such as $A \otimes B$ or I is represented by zero or more wires. For more details, see "Monoidal signatures" below.

Coherence

It is easy to check that the graphical language for monoidal categories is sound, up to deformation of diagrams in the plane. As an example, consider the following law, which is a consequence of bifunctoriality:

$$(\mathrm{id}_C \otimes g) \circ (f \otimes \mathrm{id}_B) = (f \otimes \mathrm{id}_D) \circ (\mathrm{id}_A \otimes g).$$

Translated into the graphical language, this becomes

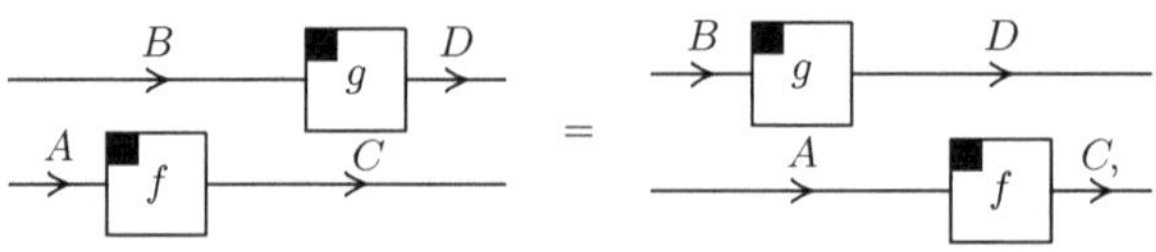

which obviously holds up to deformation of diagrams. We have the following coherence theorem:

Theorem 3 (Coherence for planar monoidal categories [21, Theorem. 1.5], [22, Theorem. 1.2]) *A well-formed equation between morphism terms in the language of monoidal categories follows from the axioms of monoidal categories if and only if it holds, up to planar isotopy, in the graphical language.*

Here, by "planar isotopy", we mean that two diagrams, drawn in a rectangle in the plane with incoming and outgoing wires attached to the boundaries of the rectangle, are equivalent if it is possible to transform one to the other by continuously moving around boxes in the rectangle, without allowing boxes or wires to cross each other or to be detached from the boundary of the rectangle during the moving. To make these notions mathematically precise, it is usually easier to represent morphism as points, rather than boxes. For precise definitions and a proof of the coherence theorem, see Joyal and Street [21, 22].

Caveat 9 Technically, Joyal and Street's proof in [21, 22] only applies to planar isotopies where each intermediate diagram during the deformation remains progressive, i.e., with all arrows oriented left-to-right. Joyal and Street call such an isotopy "recumbent". We conjecture that the result remains true if one allows arbitrary planar deformations. Similar caveats also apply to the coherence theorems for braided and balanced monoidal categories below.

The following is an example of two diagrams that are not isomorphic in the planar embedded sense:

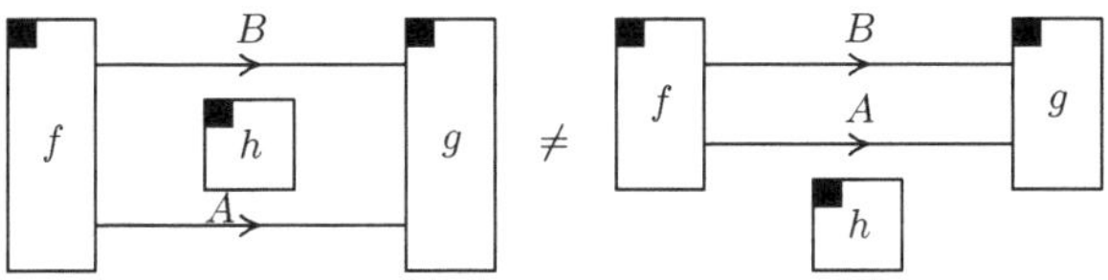

(4.1)

where $f : I \to A \otimes B$, $g : A \otimes B \to I$, and $h : I \to I$. And indeed, the corresponding equation $g \circ ((\rho_A \circ (\mathrm{id}_A \otimes h) \circ \rho_A^{-1}) \otimes \mathrm{id}_B) \circ f = g \circ ((\lambda_A \circ (h \otimes \mathrm{id}_A) \circ \lambda_A^{-1}) \otimes \mathrm{id}_B) \circ f$ *does not follow* from the axioms of monoidal categories. This is an easy consequence of soundness.

Note that because of the coherence theorem, it is not actually necessary to memorize the axioms of monoidal categories: indeed, one could use the coherence theorem as the *definition* of monoidal category! For practical purposes, reasoning in the graphical language is almost always easier than reasoning from the axioms. On the other hand, the graphical definition is not very useful when one has to check whether a given category is monoidal; in this case, checking finitely many axioms is easier.

Relationship to Traditional Coherence Theorems

Many category theorists are familiar with coherence theorems of the form "all diagrams of a certain type commute". Mac Lane's traditional coherence theorem for monoidal categories [28] is of this form. It states that all diagrams built from only α, λ, ρ, id, $\circ$, and $\otimes$ commute.

The coherence results in this paper are of a more general form (cf. Kelly [26, p. 107]). Here, the object is to characterize *all* formal equations that follow from a given set of axioms. We note that the traditional coherence theorem is an easy consequence of the general coherence result of Theorem 3: namely, if a given well-formed equation is built only from α, λ, ρ, id, $\circ$, and $\otimes$, then both the left-hand side and right-hand side denote identity diagrams in the graphical language. Therefore, by Theorem 3, the equation follows from the axioms of monoidal categories. Analogous remarks hold for all the coherence theorems of this article.

4.3.1.1 Technicalities

Monoidal Signatures

To be precise about the labels on diagrams of monoidal categories, and about the meaning of "well-formed equation" in the coherence theorem, we introduce the concept of a monoidal signature. This generalizes the simple signatures introduced in Sect. 4.2. Monoidal signatures were introduced under the name *tensor schemes* by Joyal and Street [21, 22]. We give a non-strict version of the definition.

Definition 10 ([22, Def. 1.4], [21, Def. 1.6]) Given a set Σ_0 of *object variables*, let Mon(Σ_0) denote the free ($\otimes$, I)-algebra generated by Σ_0, i.e., the set of *object terms* built from object variables and I via the operation $\otimes$. For example, if $A, B \in \Sigma_0$, then the term $(A \otimes B) \otimes (I \otimes A)$ is an element of Mon(Σ_0).

A *monoidal signature* consists of a set Σ_0 of object variables, a set Σ_1 of *morphism variables*, and a pair of functions dom, cod : $\Sigma_1 \to \mathrm{Mon}(\Sigma_0)$.

The concept of well-formed morphism terms and equations (in the language of monoidal categories) is defined relative to a given monoidal signature. In the graphical language, wires and boxes are labeled by object variables and morphism variables as before. An object term expands to zero or more parallel wires, by the rules of Table 4.2. As before, the labellings must respect the domain and codomain information, which now involves possibly multiple wires connected to a box. Just as we sometimes label a box by a morphism term in schematic drawings to denote a possibly composite diagram, we sometimes label a wire by an object term, such as S and T in Table 4.2. In this case, it is a short-hand notation for zero or more parallel wires.

Given a monoidal signature Σ and a monoidal category $\mathbf{C}$, an *interpretation* $i : \Sigma \to \mathbf{C}$ consists of an object function $i_0 : \Sigma_0 \to |\mathbf{C}|$, which then extends in a unique way to $\hat{i}_0 : \mathrm{Mon}(\Sigma_0) \to |\mathbf{C}|$ such that $\hat{i}_0(A \otimes B) = \hat{i}_0(A) \otimes \hat{i}_0(B)$ and $\hat{i}_0(I) = I$, and for any $f \in \Sigma_1$ a morphism $i_1(f) : i_0(\mathrm{dom}\, f) \to i_0(\mathrm{cod}\, f)$.

The remaining graphical languages in this Section 4.3 are all given relative to a monoidal signature.

Monoidal Functors and Natural Transformations

Definition 11 A *strong monoidal functor* (also sometimes called a *tensor functor*) between monoidal categories **C** and **D** is a functor $F : \mathbf{C} \to \mathbf{D}$, together with natural isomorphisms $\phi^2 : FA \otimes FB \to F(A \otimes B)$ and $\phi^0 : I \to FI$, such that the following diagrams commute:

$$
\begin{array}{ccccc}
(FA \otimes FB) \otimes FC & \xrightarrow{\phi^2 \otimes \mathrm{id}} & F(A \otimes B) \otimes FC & \xrightarrow{\phi^2} & F((A \otimes B) \otimes C) \\
\Big\downarrow{\scriptstyle \alpha} & & & & \Big\downarrow{\scriptstyle F(\alpha)} \\
FA \otimes (FB \otimes FC) & \xrightarrow{\mathrm{id} \otimes \phi^2} & FA \otimes F(B \otimes C) & \xrightarrow{\phi^2} & F(A \otimes (B \otimes C))
\end{array}
$$

$$
\begin{array}{ccc}
FA \otimes I & \xrightarrow{\rho} & FA \\
\Big\downarrow{\scriptstyle \mathrm{id} \otimes \phi^0} & & \Big\downarrow{\scriptstyle F(\rho)} \\
FA \otimes FI & \xrightarrow{\phi^2} & F(A \otimes I)
\end{array}
\qquad
\begin{array}{ccc}
I \otimes FA & \xrightarrow{\lambda} & FA \\
\Big\downarrow{\scriptstyle \phi^0 \otimes \mathrm{id}} & & \Big\downarrow{\scriptstyle F(\lambda)} \\
FI \otimes FA & \xrightarrow{\phi^2} & F(I \otimes A)
\end{array}
$$

Definition 12 Let **C** and **D** be monoidal categories, and let $F, G : \mathbf{C} \to \mathbf{D}$ be strong monoidal functors. A natural transformation $\tau : F \to G$ is called *monoidal* (or a *tensor transformation*) if the following two diagrams commute for all A, B:

$$
\begin{array}{ccc}
FA \otimes FB & \xrightarrow{\phi^2} & F(A \otimes B) \\
\Big\downarrow{\scriptstyle \tau_A \otimes \tau_B} & & \Big\downarrow{\scriptstyle \tau_{A \otimes B}} \\
GA \otimes GB & \xrightarrow{\phi^2} & G(A \otimes B)
\end{array}
$$

Coherence and Free Monoidal Categories

Similarly to what we stated for categories, the coherence theorem for monoidal categories is a consequence of a characterization of the free monoidal category. However, due to the extra coherence conditions in the definition of a strong monoidal functor, the definition of freeness is slightly more complicated.

Definition 13 A monoidal category **C** is a *free monoidal category* over a monoidal signature Σ if it is equipped with an interpretation $i : \Sigma \to \mathbf{C}$ such that for any monoidal category **D** and interpretation $j : \Sigma \to \mathbf{D}$, there exists a strong monoidal functor $F : \mathbf{C} \to \mathbf{D}$ such that $j = F \circ i$, and F is unique up to a unique monoidal natural isomorphism.

As before, the coherence theorem can be re-formulated as a freeness theorem.

Theorem 4 *The graphical language of monoidal categories over a monoidal signature Σ, with identities, composition, and tensor as defined in Tables 4.1 and 4.2, and up to planar isotopy of diagrams, forms a free monoidal category over Σ.*

Most of the coherence theorems (and conjectures) of this article can be similarly formulated in terms of freeness. An exception to this are the traced categories without braidings in Sects. 4.5.1, 4.5.2, 4.5.3, 4.5.4 and 4.7.5, as explained in Remark 15. From now on, we will only mention freeness when it is not entirely automatic, such as in Sect. 4.4.1.

4.3.2 Spacial Monoidal Categories

Definition 14 A monoidal category is *spacial* if it satisfies the additional axiom

$$\rho_A \circ (\mathrm{id}_A \otimes h) \circ \rho_A^{-1} = \lambda_A \circ (h \otimes \mathrm{id}_A) \circ \lambda_A^{-1}, \tag{4.2}$$

for all $h : I \to I$.

In the graphical language, this means that

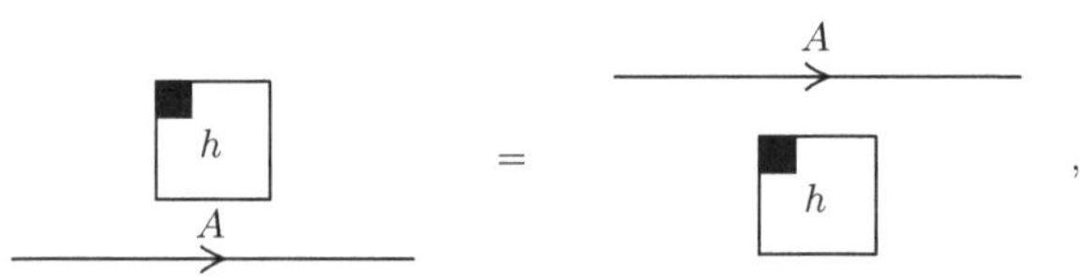

so in particular, it implies that the two terms in (4.1) are equal. The author does not know whether the concept of a spacial monoidal category appears in the literature, or if it does, under what name.

Graphical Language

The graphical language for spacial monoidal categories is the same as that for monoidal categories, except that planarity is dropped from the notion of diagram equivalence, i.e., diagrams are considered up to isomorphism. Obviously the axioms are sound; we conjecture that they are also complete.

Conjecture 1 (Coherence for spacial monoidal categories) A well-formed equation between morphism terms in the language of spacial monoidal categories follows from the axioms of spacial monoidal categories if and only if it holds, up to isomorphism of diagrams, in the graphical language.

Note that, in the case of planar diagrams, the notion of isomorphism of diagrams coincides with ambient isotopy in 3 dimensions. This explains the term "spacial".

4.3.3 Braided Monoidal Categories

Definition 15 [23] A *braiding* on a monoidal category is a natural family of isomorphisms $c_{A,B} : A \otimes B \to B \otimes A$, satisfying the following two "hexagon axioms":

$$\begin{array}{ccccc} & (B \otimes A) \otimes C & \xrightarrow{\alpha_{B,A,C}} & B \otimes (A \otimes C) & \\ {}^{c_{A,B}\otimes \mathrm{id}_C}\nearrow & & & & \searrow^{\mathrm{id}_B \otimes c_{A,C}} \\ (A \otimes B) \otimes C & & & & B \otimes (C \otimes A). \\ {}_{\alpha_{A,B,C}}\searrow & & & & \nearrow_{\alpha_{B,C,A}} \\ & A \otimes (B \otimes C) & \xrightarrow{c_{A,B\otimes C}} & (B \otimes C) \otimes A & \end{array}$$

$$\begin{array}{ccccc} & (B \otimes A) \otimes C & \xrightarrow{\alpha_{B,A,C}} & B \otimes (A \otimes C) & \\ {}^{c^{-1}_{B,A}\otimes \mathrm{id}_C}\nearrow & & & & \searrow^{\mathrm{id}_B \otimes c^{-1}_{C,A}} \\ (A \otimes B) \otimes C & & & & B \otimes (C \otimes A). \\ {}_{\alpha_{A,B,C}}\searrow & & & & \nearrow_{\alpha_{B,C,A}} \\ & A \otimes (B \otimes C) & \xrightarrow{c^{-1}_{B\otimes C,A}} & (B \otimes C) \otimes A & \end{array}$$

Note that every braided monoidal category is spacial; this follows from the naturality (in I) of $c_{A,I} : A \otimes I \to I \otimes A$.

A braided monoidal functor between braided monoidal categories is a monoidal functor that is compatible with the braiding in the following sense:

$$\begin{array}{ccc} FA \otimes FB & \xrightarrow{\phi^2} & F(A \otimes B) \\ {\scriptstyle c_{FA,FB}}\downarrow & & \downarrow{\scriptstyle Fc_{A,B}} \\ FB \otimes FA & \xrightarrow{\phi^2} & F(B \otimes A). \end{array}$$

Graphical Language

One extends the graphical language of monoidal categories with the *braiding*:

Braiding $c_{A,B}$

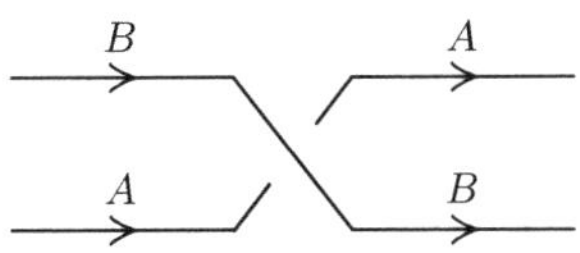

In general, if A and B are composite object terms, the braiding $c_{A,B}$ is represented as the appropriate number of wires crossing each other.

Note that the braiding satisfies $c_{A,B} \circ c^{-1}_{A,B} = \mathrm{id}_{A\otimes B}$, but not $c_{A,B} \circ c_{B,A} = \mathrm{id}_{A\otimes B}$. Graphically:

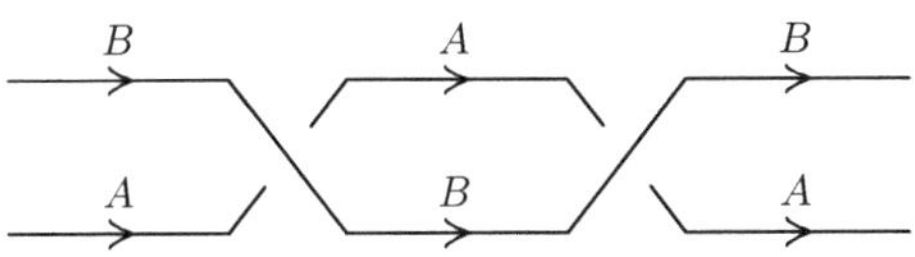

$= \mathrm{id}_{A\otimes B}$,

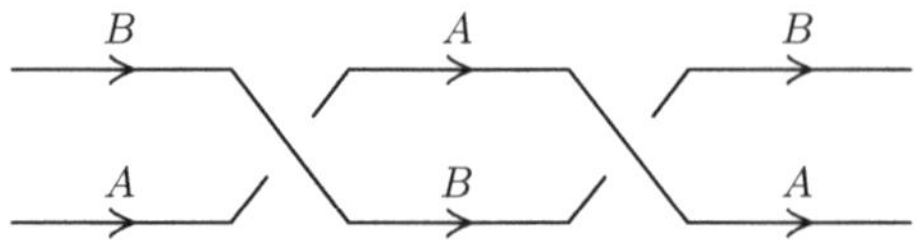

$\neq \mathrm{id}_{A\otimes B}.$

Example 1 The hexagon axiom translates into the following in the graphical language:

$$(\mathrm{id}_B \otimes c_{A,C}) \circ \alpha_{B,A,C} \circ (c_{A,B} \otimes \mathrm{id}_C) = \alpha_{B,C,A} \circ (c_{B,C\otimes A}) \circ \alpha_{A,B,C}$$

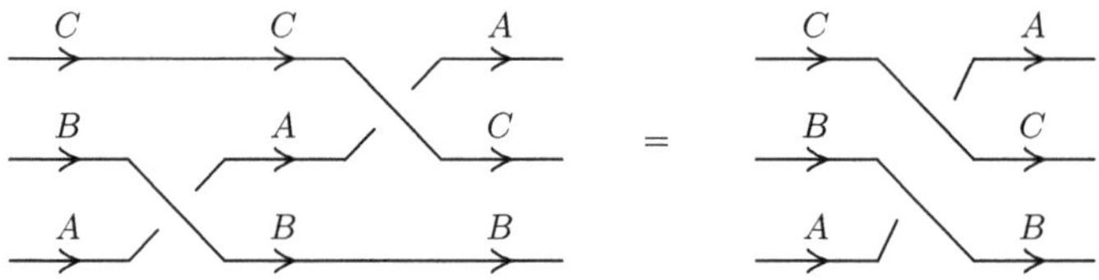

Example 2 The *Yang-Baxter equation* is the following equation, which is a consequence of the hexagon axiom and naturality:

$$(c_{B,C}\otimes\mathrm{id}_A)\circ(\mathrm{id}_B\otimes c_{A,C})\circ(c_{A,B}\otimes\mathrm{id}_C) = (\mathrm{id}_C\otimes c_{A,B})\circ(c_{A,C}\otimes\mathrm{id}_B)\circ(\mathrm{id}_A\otimes c_{B,C}).$$

In the graphical language, it becomes:

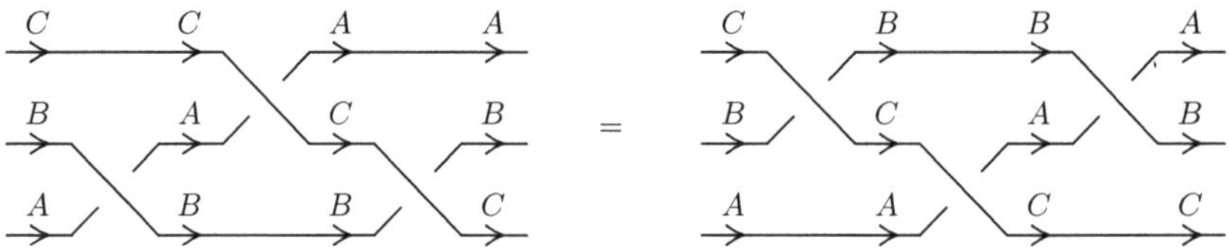

Theorem 5 (Coherence for braided monoidal categories [22, Theorem. 3.7]) *A well-formed equation between morphisms in the language of braided monoidal categories follows from the axioms of braided monoidal categories if and only if it holds in the graphical language up to isotopy in 3 dimensions.*

Here, by "isotopy in 3 dimensions", we mean that two diagrams, drawn in a 3-dimensional box with incoming and outgoing wires attached to the boundaries of the box, are isotopic if it is possible to transform one to the other by moving around nodes in the box, without allowing nodes or edges to cross each other or to be detached from the boundary during the moving. Also, the linear order of the edges entering and exiting each node must be respected. This is made more precise in Joyal and Street [22].

Caveat 16 The proof by Joyal and Street [22] is subject to some minor technical assumptions: graphs are assumed to be *smooth*, and the isotopies are progressive, with continuously changing tangent vectors.

4.3.4 Balanced Monoidal Categories

Definition 17 [23] A *twist* on a braided monoidal category is a natural family of isomorphisms $\theta_A : A \to A$, satisfying $\theta_I = \mathrm{id}_I$ and such that the following diagram commutes for all A, B:

$$\begin{array}{ccc} A \otimes B & \xrightarrow{c_{A,B}} & B \otimes A \\ {\scriptstyle \theta_{A\otimes B}}\downarrow & & \downarrow{\scriptstyle \theta_B \otimes \theta_A} \\ A \otimes B & \xleftarrow[c_{B,A}]{} & B \otimes A. \end{array} \tag{4.3}$$

A *balanced monoidal category* is a braided monoidal category with twist.

A balanced monoidal functor between balanced monoidal categories is a braided monoidal functor that is also compatible with the twist, i.e., such that $F(\theta_A) = \theta_{FA}$ for all A.

Graphical Language

The graphical language of balanced monoidal categories is similar to that of braided monoidal categories, except that morphisms are represented by flat ribbons, rather than 1-dimensional wires. A ribbon can be thought of as a pair of parallel wires that are infinitesimally close to each other, or as a wire that is equipped with a *framing* [22]. For example, the braiding looks like this:

$c_{A,B} =$  .

The twist map θ_A is represented as a 360-degree twist in a ribbon, or in several ribbons together, if A is a composite object term. This is easiest seen in the following illustration.

$\theta_A =$  , $\theta_{A\otimes B} =$ 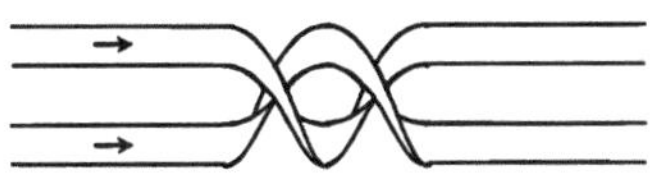 .

The meaning of (4.3) should then be obvious.

Theorem 6 (Coherence for Balanced Monoidal Categories [22, Theorem. 4.5]) *A well-formed equation between morphisms in the language of balanced monoidal categories follows from the axioms of balanced monoidal categories if and only if it holds in the graphical language up to framed isotopy in 3 dimensions.*

4.3.5 Symmetric Monoidal Categories

Definition 18 A *symmetric monoidal category* is a braided monoidal category where the braiding is self-inverse, i.e.:

$$c_{A,B} = c_{B,A}^{-1}$$

In this case, the braiding is called a *symmetry*.

Remark 2 Because of Eq. (4.3), a symmetric monoidal category can be equivalently defined as a balanced monoidal category in which $\theta_A = \mathrm{id}_A$ for all A.

Remark 3 The previous remark notwithstanding, there exist symmetric monoidal categories that possess a non-trivial twist (in addition to the trivial twist $\theta_A = \mathrm{id}_A$). Thus, in a balanced monoidal category, the symmetry condition $c_{A,B} = c_{B,A}^{-1}$ does not in general imply $\theta_A = \mathrm{id}_A$. In other words, a balanced monoidal category that is symmetric as a braided monoidal category is not necessarily symmetric as a balanced monoidal category. An example is the category of finite dimensional vector spaces and linear bijections, with $\theta_A(x) = nx$, where $n = \dim(A)$.

Examples 19 On the monoidal category (**Set**, $\times$) of sets with cartesian product, a symmetry is given by $c(x, y) = (y, x)$. On the category (**Vect**, $\otimes$) of vector spaces with tensor product, a symmetry is given by $c(x \otimes y) = y \otimes x$.

Graphical Language

The symmetry is graphically represented by a crossing:

Symmetry $c_{A,B}$

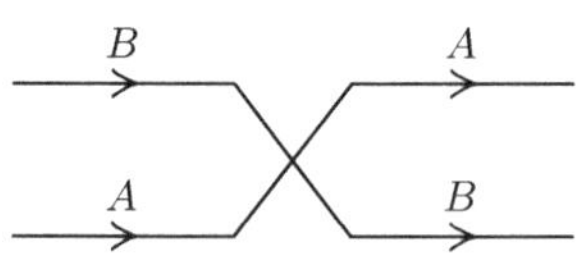

Theorem 7 (Coherence for symmetric monoidal categories [22, Theorem. 2.3]) *A well-formed equation between morphisms in the language of symmetric monoidal categories follows from the axioms of symmetric monoidal categories if and only if it holds, up to isomorphism of diagrams, in the graphical language.*

Note that the graphical language for symmetric monoidal categories is up to isomorphism of diagrams, without any reference to 2- or 3-dimensional structure. However, isomorphism of diagrams is equivalent to ambient isotopy in 4 dimensions, so we can still regard it as a geometric notion.

4.4 Autonomous Categories

Autonomous categories are monoidal categories in which the objects have *duals*. In terms of graphical language, this means that some wires are allowed to run from right to left.

4.4.1 (Planar) Autonomous Categories

Definition 20 [23] In a (without loss of generality strict) monoidal category, an *exact pairing* between two objects A and B is given by a pair of morphisms $\eta : I \to B \otimes A$ and $\epsilon : A \otimes B \to I$, such that the following two adjunction triangles commute:

$$\begin{array}{ccc} A & \xrightarrow{\mathrm{id}_A \otimes \eta} & A \otimes B \otimes A \\ & {\scriptstyle \mathrm{id}_A} \searrow & \downarrow {\scriptstyle \epsilon \otimes \mathrm{id}_A} \\ & & A, \end{array} \qquad \begin{array}{ccc} B & \xrightarrow{\eta \otimes \mathrm{id}_B} & B \otimes A \otimes B \\ & {\scriptstyle \mathrm{id}_B} \searrow & \downarrow {\scriptstyle \mathrm{id}_B \otimes \epsilon} \\ & & B. \end{array} \tag{4.1}$$

In such an exact pairing, B is called the *right dual* of A and A is called the *left dual* of B.

Remark 4 The maps η and ϵ determine each other uniquely, and they are respectively called the *unit* and the *counit* of the adjunction. Moreover, the triple (B, η, ϵ), if it exists, is uniquely determined by A up to isomorphism. The existence of duals is therefore a property of a monoidal category, rather than an additional structure on it. Moreover, every strong monoidal functor automatically preserves existing duals.

Definition 21 [20, 21, 23] A monoidal category is *right autonomous* if every object A has a right dual, which we then denote A^*. It is *left autonomous* if every object A has a left dual, which we then denote ${}^*A$. Finally, the category is *autonomous* if it is both right and left autonomous.

Remark 5 (Terminology) A [right, left, –] autonomous category is also called [right, left, –] rigid, see e.g. [32, p. 78]. Also, the term "autonomous" is sometimes used in the weaker sense of "monoidal closed". Although this latter usage is no longer common, it still lives on in the terminology "*-autonomous category" (Barr [4], see also Sect. 4.9).

If we wish to emphasize that an autonomous category is not necessarily symmetric or braided, we sometimes call it a *planar autonomous category*.

Graphical Language

If A is an object variable, the objects A^* and ${}^*A$ are both represented in the same way: by a wire labeled A running from right to left. The unit and counit are represented as half turns:

Dual $A^*, {}^*A$

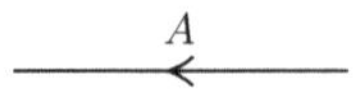

Unit $\eta_A : I \to A^* \otimes A$

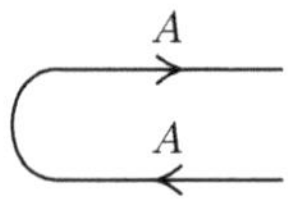

$\eta'_A : I \to A \otimes {}^*A$

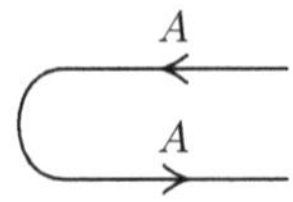

Counit $\epsilon_A : A \otimes A^* \to I$

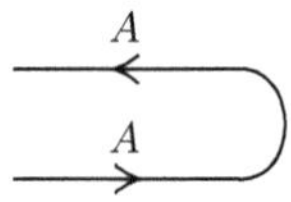

$\epsilon'_A : {}^*A \otimes A \to I$

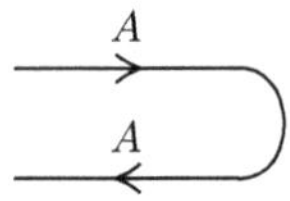

More generally, if A is a composite object represented by a number of wires, then A^* and ${}^*A$ are represented by the same set of wires running backward (rotated by 180 degrees), and the units and counits are represented as multiple wires turning.

Example 3 The two diagrams in (4.1), where $B = A^*$, translate into the graphical language as follows:

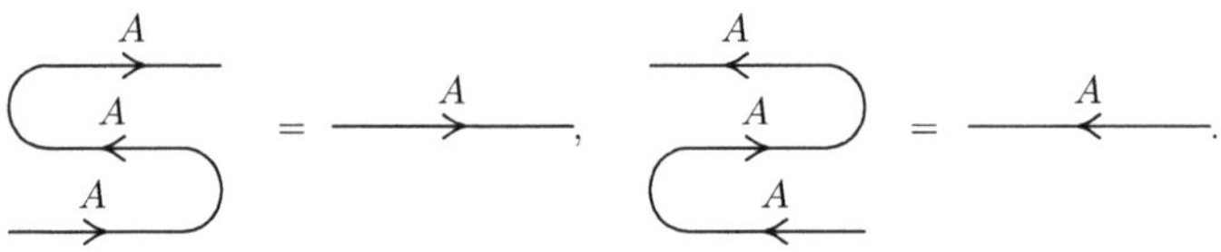

Example 4 For any morphism $f : A \to B$, it is possible to define morphisms $f^* : B^* \to A^*$ and ${}^*f : {}^*B \to {}^*A$, called the *adjoint mates* of f, as follows:

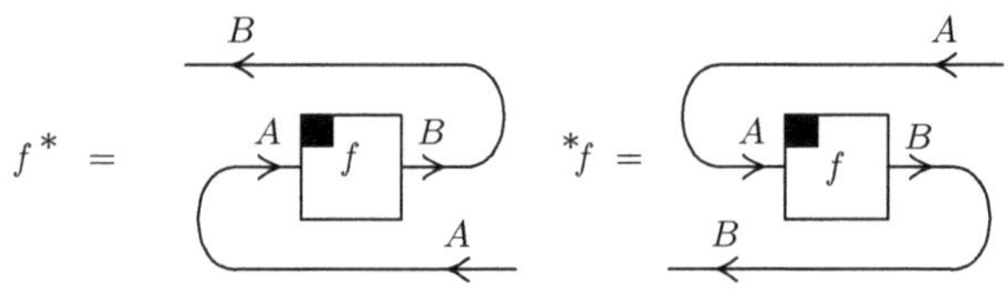

With these definitions, $(-)^*$ and ${}^*(-)$ become contravariant functors.

Theorem 8 (Coherence for planar autonomous categories [21, Theorem. 2.7]) *A well-formed equation between morphisms in the language of autonomous categories follows from the axioms of autonomous categories if and only if it holds in the graphical language up to planar isotopy.*

Here, the notion of planar isotopy is the same as before, except that the wires are of course no longer restricted to being oriented left-to-right during the deformation. However, the ability to turn wires upside down does not extend to boxes: the notion of isotopy for this theorem does not include the ability to rotate boxes. See Joyal and Street [21] for a more precise statement.

Caveat 22 The proof by Joyal and Street [21] assumes that the diagrams are piecewise linear.

Note that the same theorem applies to left autonomous, right autonomous, or autonomous categories. Indeed, each individual term in the language of autonomous categories involves only finitely many duals, and thus may be translated into a term of (say) left autonomous categories by replacing each object variable A by $A^{***\cdots *}$, for a sufficiently large, even number of $*$'s. The resulting term maps to the same diagram.

The same coherence theorem also holds for categories that are only right (or left) autonomous. This is a consequence of the following proposition.

Proposition 1 *Each right (or left) autonomous category can be fully embedded in an autonomous category.*

Proof Let $\mathbf{C}$ be a right autonomous category, and consider the strong monoidal functor $F : \mathbf{C} \to \mathbf{C}$ given by $F(A) = A^{**}$. This functor is full and faithful, and every object in the image of F has a left dual. Now let $\hat{\mathbf{C}}$ be the colimit (in the large category of right autonomous categories and strong monoidal functors) of the sequence

$$\mathbf{C} \xrightarrow{F} \mathbf{C} \xrightarrow{F} \mathbf{C} \xrightarrow{F} \cdots$$

Then $\hat{\mathbf{C}}$ is autonomous, and $\mathbf{C}$ is fully and faithfully embedded in $\hat{\mathbf{C}}$. The proof for left autonomous categories is analogous. □

Corollary 1 (Coherence for right (left) autonomous categories) *A well-formed equation between morphisms in the language of right (left) autonomous categories follows from the axioms of right (left) autonomous categories if and only if it holds in the graphical language up to planar isotopy.*

Proof It suffices to show that an equation (in the language of right autonomous categories) holds in all right autonomous categories if and only if it holds in all autonomous categories. The "only if" direction is trivial, since every autonomous category is right autonomous. For the opposite direction, suppose some equation holds in all autonomous categories, and let $\mathbf{C}$ be a right autonomous category. Then $\mathbf{C}$ can be faithfully embedded in an autonomous category $\hat{\mathbf{C}}$. By assumption, the equation holds in $\hat{\mathbf{C}}$, and therefore also in $\mathbf{C}$, since the embedding is faithful. □

4.4.1.1 Technicalities

Autonomous Signatures

The diagrams of autonomous categories, and the concept of well-formed equation in the coherence theorem, are defined relative to the notion of an autonomous signature. These were called *autonomous tensor schemes* by Joyal and Street [21]. We give a non-strict version of the definition.

Definition 23 [21, Def. 2.5] Given a set Σ_0 of *object variables*, let $\mathrm{Aut}(\Sigma_0)$ denote the free $(\otimes, I, {}^*(-), (-)^*)$-algebra generated by Σ_0, i.e., the set of *object terms* built from object variables and I via the operations $\otimes$, ${}^*(-)$, and $(-)^*$). For example, if $A, B \in \Sigma_0$, then the term $B^* \otimes ({}^{**}I \otimes A)^*$ is an element of $\mathrm{Aut}(\Sigma_0)$.

An *autonomous signature* consists of a set Σ_0 of object variables, a set Σ_1 of *morphism variables*, and a pair of functions $\mathrm{dom}, \mathrm{cod} : \Sigma_1 \to \mathrm{Aut}(\Sigma_0)$.

The concept of a *right autonomous signature* and *left autonomous signature* are defined analogously. The remaining graphical languages in this Section 4.4 are all given relative to an autonomous signature.

Functors and Natural Transformations of Autonomous Categories

Any strong monoidal functor preserves exact pairings: if $\eta : I \to B \otimes A$ and $\epsilon : A \otimes B \to I$ define an exact pairing, then so do

$$\hat{F}\eta : I \xrightarrow{\phi^0} FI \xrightarrow{F\eta} F(B \otimes A) \xrightarrow{(\phi^2)^{-1}} FB \otimes FA$$

and

$$\hat{F}\epsilon : FA \otimes FB \xrightarrow{\phi^2} F(A \otimes B) \xrightarrow{F\epsilon} FI \xrightarrow{(\phi^0)^{-1}} I.$$

In particular, if **C** and **D** are autonomous categories and $F : \mathbf{C} \to \mathbf{D}$ is a monoidal functor, by uniqueness of duals, there will be a unique induced natural isomorphism $F(A^*) \cong (FA)^*$ such that

$$\begin{array}{ccc} I & \xrightarrow{\hat{F}\eta_A} & F(A^*) \otimes FA \\ & \searrow^{\eta_{FA}} & \downarrow{\scriptstyle \cong \otimes \mathrm{id}} \\ & & (FA)^* \otimes FA \end{array} \qquad \text{and} \qquad \begin{array}{ccc} FA \otimes F(A^*) & \xrightarrow{\hat{F}\epsilon_A} & I, \\ \downarrow{\scriptstyle \mathrm{id} \otimes \cong} & \nearrow_{\epsilon_{FA}} & \\ FA \otimes (FA)^* & & \end{array}$$

and similarly for $F({}^*A) \cong {}^*(FA)$.

For natural transformations, we have the following lemma:

Lemma 1 (Saavedra Rivano [32, Prop. 5.2.3], see also [23, Prop. 7.1]) *Suppose $\tau : F \to G$ is a monoidal natural transformation between strong monoidal functors $F, G : \mathbf{C} \to \mathbf{D}$. If A has a right dual A^* in $\mathbf{C}$, then τ_{A^*} and $(\tau_A)^*$ are mutually inverse in $\mathbf{D}$ (up to the above canonical isomorphism), or more precisely:*

$$\begin{array}{ccc} F(A^*) & \xrightarrow{\tau_{A^*}} & G(A^*) \\ {\scriptstyle \cong}\downarrow & & \downarrow{\scriptstyle \cong} \\ (FA)^* & \xleftarrow[(\tau_A)^*]{} & (GA)^* \end{array}$$

In particular, if $\mathbf{C}$ *is autonomous, then any such monoidal natural transformation is invertible.*

Coherence and Free Autonomous Categories

The graphical language, as we have defined it above for autonomous categories, is sufficient for the purposes of Theorem 8. However, it does not characterize the free autonomous category over an autonomous signature as stated. For example, consider a signature with a single morphism variable $f : A \to A$. The problem is that there are clearly some diagrams, such as

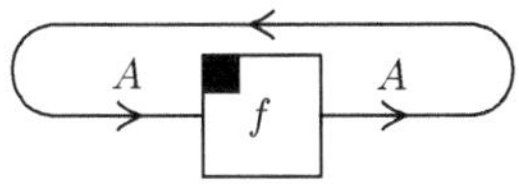

(4.2)

which are not translations of any well-formed term of autonomous categories. Indeed, for this diagram to correspond to a well-formed term, we would have to have e.g. $f : A^{**} \to A$ or $f : A \to {}^{**}A$.

Joyal and Street [21] characterize the free autonomous category by equipping each edge with a winding number. Effectively, the horizontal segments of edges are labeled with pairs (A, n), where A is an object variables and n is an integer winding number. Left-to-right segments have even winding numbers, right-to-left segments have odd winding numbers, and winding numbers increase by one on counterclockwise turns, and decrease by one on clockwise turns. The winding numbers on the input and output of each box, and on the global inputs and outputs, are restricted to be consistent with the domain and codomain information, where e.g. A^{**} corresponds to $(A, 2)$, and ${}^{***}B$ to $(B, -3)$. See [21] for precise details. Here is an example of a well-formed diagram of type $I \to B^{**} \otimes A$, where $g : I \to A \otimes B$:

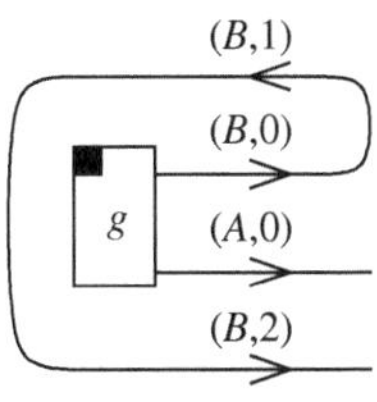

Theorem 9 *The graphical language (with winding numbers) of autonomous categories over an autonomous signature* Σ*, up to planar isotopy of diagrams, forms a free autonomous category over* Σ.

We remark that if a diagram of planar autonomous categories can be labeled with winding numbers, then this labeling is necessarily unique. In particular, for the purposes of Theorem 8, there is no harm in dropping the winding numbers, because by hypothesis, the theorem only considers diagrams that are the translation of well-formed terms, whose winding numbers can therefore uniquely reconstructed.

4.4.2 (Planar) Pivotal Categories

A pivotal category is an autonomous category with a suitable isomorphism $A \cong A^{**}$.

Definition 24 [15, 16, 19] A *pivotal category* is a right autonomous category equipped with a monoidal natural isomorphism $i_A : A \to A^{**}$.

Note that any pivotal category is immediately left autonomous, therefore autonomous. The requirement that i_A is a *monoidal* natural transformation here means that i_I is the canonical isomorphism $I \cong I^{**}$, and that the following diagram commutes, where the horizontal arrow is the canonical isomorphism derived from the autonomous structure:

$$\begin{array}{ccc} & A \otimes B & \\ {}^{i_A \otimes i_B}\swarrow & & \searrow^{i_{A\otimes B}} \\ A^{**} \otimes B^{**} & \xrightarrow{\cong} & (A \otimes B)^{**}. \end{array} \tag{4.3}$$

The following property, which is sometimes taken as part of the definition of pivotal categories [19, Def. 3.1.1], is a direct consequence of Saavedra Rivano's Lemma (Lemma 1).

Lemma 2 *In any pivotal category, the following diagram commutes:*

$$\begin{array}{ccc} A^* & \xrightarrow{i_{A^*}} & A^{***} \\ & {}_{id_{A^*}}\searrow & \downarrow{\scriptstyle i_A^*} \\ & & A^*. \end{array}$$

Remark 6 One can equivalently define a pivotal category as an autonomous category equipped with a monoidal natural isomorphism (of contravariant monoidal functors) $\phi : A^* \xrightarrow{\cong} {}^*A$. This was done by Freyd and Yetter [16]. Condition (S) of [16, Def. 4.1] is also a consequence of Saavedra Rivano's Lemma, and is therefore redundant.

Remark 7 (Terminology) Freyd and Yetter [16] also introduced the term *sovereign category* for a pivotal category.

A pivotal functor between pivotal categories is a monoidal functor that also satisfies

$$\begin{array}{ccc} FA & \xrightarrow{F(i_A)} & F(A^{**}) \\ & {}_{i_{FA}}\searrow & \downarrow{\scriptstyle \cong} \\ & & (FA)^{**}. \end{array}$$

Graphical Language

The graphical language for pivotal categories is the same as that for autonomous categories, where the isomorphism $i_A : A \to A^{**}$ is represented like an identity map. Of course, there are now additional diagrams that are the translation of well-formed terms. For example, when $f : A \to A$, then (4.2) is a well-formed diagram of pivotal categories, but not of autonomous categories. Indeed, in the case of pivotal categories, the problem of winding numbers (discussed before Theorem 9) disappears, as winding numbers are taken modulo 2, and hence add nothing beyond orientation.

Theorem 10 (Coherence for pivotal categories) *A well-formed equation between morphisms in the language of pivotal categories follows from the axioms of pivotal categories if and only if it holds in the graphical language up to planar isotopy, including rotation of boxes.*

Caveat 25 Only special cases of this theorem have been proved in the literature. Freyd and Yetter [16, Thm. 4.4] considered the case of the free pivotal category generated by a category. In our terminology, this means that they only considered diagrams for pivotal categories over *simple signatures*, rather than over *autonomous signatures*. In other words, they only considered boxes of the form

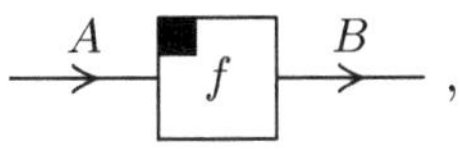

with exactly one input and one output. Joyal and Street's draft report [19] claims the general result but contains no proof.

The notion of planar isotopy for pivotal categories includes the ability to rotate boxes in the plane of the diagram. For example, the following two diagrams are isotopic in this sense:

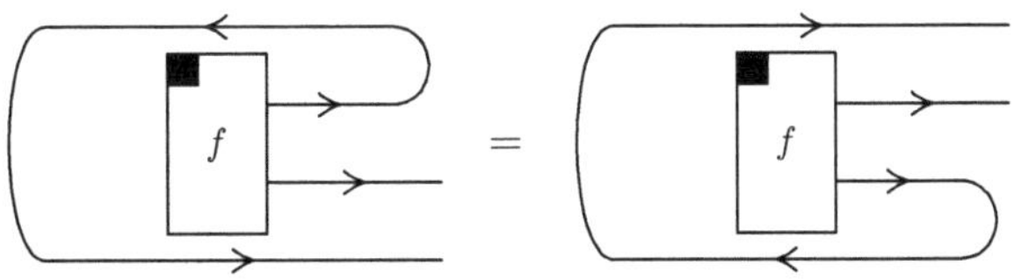

(4.4)

This also explains why we have marked a corner of each box. With the ability to rotate boxes, we need to keep track of their "natural" orientation, so that the diagrams from (4.4) can also be represented like this:

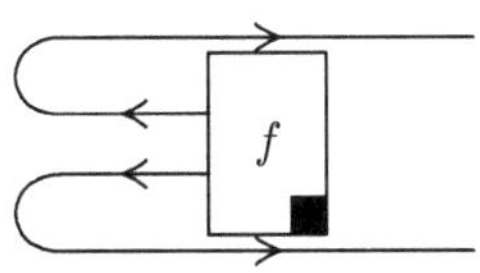

More generally, the adjoint mate of $f : A \to B$ can be represented by a rotated box:

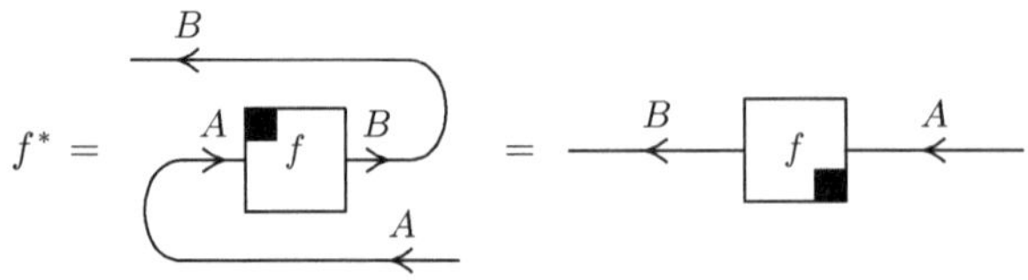

(4.5)

Also note that is f is a composite diagram, then the whole diagram may be rotated to obtain f^*.

4.4.3 Spherical Pivotal Categories

Definition 26 (Barrett and Westbury [5]) A pivotal category is *spherical* if for all objects A and morphisms $f : A \to A$,

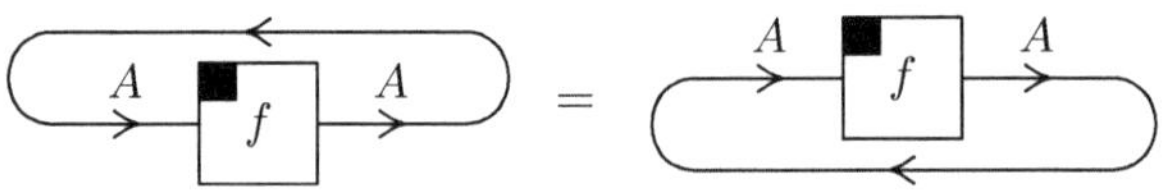

(4.6)

The intuition behind the "spherical" axioms is that diagrams should be embedded in a 2-sphere, rather than the plane. It is then obvious that the left-hand side of (4.6) can be continuously transformed into the right-hand side, namely by moving the loop across the back of the 2-sphere.

Failure of Coherence

The spherical axiom is not sound for the graphical language of diagrams embedded in the 2-sphere. The problem is that the notion of "diagram embedded in the 2-sphere" is not compatible with composition or tensor. The following is a consequence of the spherical axiom, but does not hold up to isotopy in the 2-sphere.

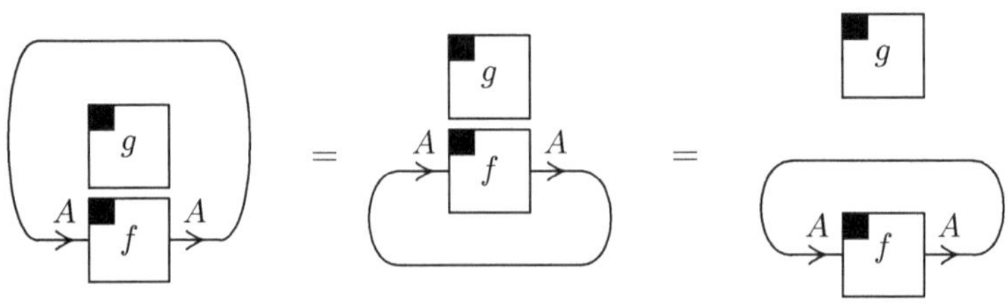

Note that this counterexample is similar to the spacial axiom (4.2), but does not quite imply it. If one adds the spacial axiom, as we are about to do, then any notion of isotopy is lost and equivalence of diagrams collapses to isomorphism.

4.4.4 Spacial Pivotal Categories

Definition 27 A pivotal category is *spacial* if it satisfies the spacial axiom (4.2) and the spherical axiom (4.6).

Graphical Language and Coherence

The graphical language for spacial pivotal categories is the same as that for planar pivotal categories, except that equivalence of diagrams is now taken up to isomorphism. Clearly, the axioms are sound for the graphical language. We conjecture that they are also complete.

Conjecture 2 (Coherence for spacial pivotal categories) A well-formed equation between morphisms in the language of spacial pivotal categories follows from the axioms of spacial pivotal categories if and only if it holds in the graphical language up to isomorphism.

4.4.5 Braided Autonomous Categories

An braided autonomous category is an autonomous category that is also braided (as a monoidal category). The notion of braided autonomous categories is not extremely natural, as the graphical language is only sound for a restricted form of isotopy called *regular isotopy*. Nevertheless, it is useful to collect some facts about braided autonomous categories.

Lemma 3 [23, Prop. 7.2] *A braided monoidal category is autonomous if and only if it is right autonomous.*

Proof If $\eta : I \to B \otimes A$ and $\epsilon : A \otimes B \to I$ form an exact pairing, then so do $c^{-1}_{A,B} \circ \eta : I \to A \otimes B$ and $\epsilon \circ c_{B,A} : B \otimes A \to I$. Therefore any right dual of A is also a left dual of A. □

In any braided autonomous category **C**, we can define a natural isomorphism $b_A : A^{**} \to A$. This follows from the proof of Lemma 3, using the fact that both A and A^{**} are right duals of A^*. More concretely, b_A and its inverse are defined by:

$$\begin{aligned} b_A &= A^{**} \xrightarrow{\eta_A \otimes \mathrm{id}} A^* \otimes A \otimes A^{**} \xrightarrow{\mathrm{id} \otimes c_{A,A^{**}}} A^* \otimes A^{**} \otimes A \xrightarrow{\epsilon_{A^*} \otimes \mathrm{id}} A, \\ b_A^{-1} &= A \xrightarrow{\mathrm{id} \otimes \eta_{A^*}} A \otimes A^{**} \otimes A^* \xrightarrow{c^{-1}_{A^{**},A} \otimes \mathrm{id}} A^{**} \otimes A \otimes A^* \xrightarrow{\mathrm{id} \otimes \epsilon_A} A^{**}. \end{aligned}$$

Here we have written, without loss of generality, as if **C** were strict monoidal. Graphically, b_A and its inverse look like this:

$b_A =$ 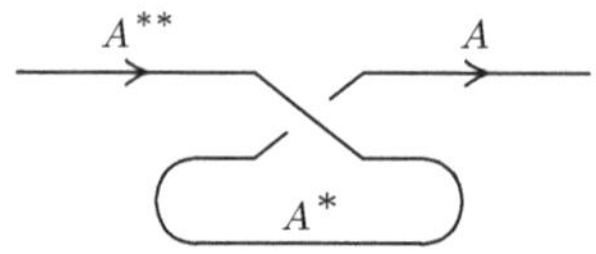

$b_A^{-1} =$ 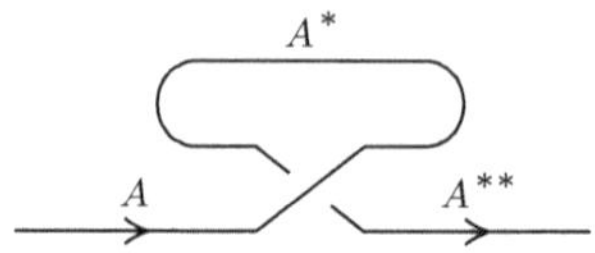

We must note that although b_A is a natural isomorphism, it is not canonical. In general, there exist infinitely many natural isomorphisms $A \cong A^{**}$. Also, b is not a *monoidal* natural transformation, and therefore does not define a pivotal structure on **C**. A general braided autonomous category is not pivotal.

Graphical Language and Coherence

The graphical language braided autonomous categories is obtained simply by adding braids to the graphical language of autonomous categories. However, the correct notion of equivalence of diagrams is neither planar isotopy (like for autonomous categories), nor 3-dimensional isotopy (like for braided monoidal categories), but an in-between notion called *regular isotopy* [25].

Table 4.3 Reidemeister moves and Λ-moves

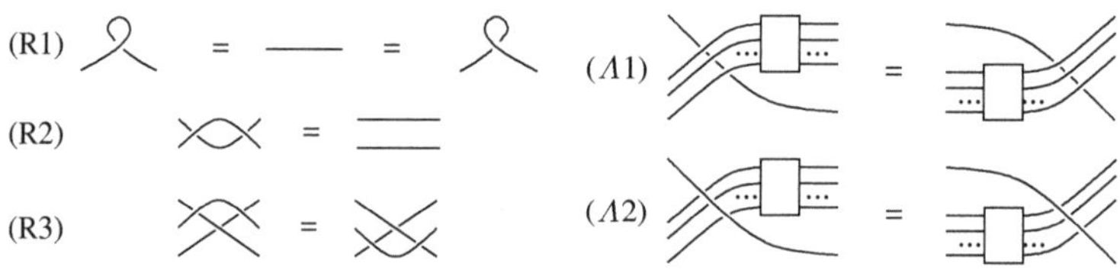

It is well-known that 3-dimensional isotopy of links and tangles is equivalent to planar isotopy of their (non-degenerate) projections onto a 2-dimensional plane, plus the three *Reidemeister moves* [31] shown as (R1)–(R3) in Table 4.3. To extend this to diagrams with nodes, one also has to add the moves (Λ1) and (Λ2).

Regular isotopy is defined to be the equivalence obtained by dropping Reidemeister move (R1). Note that regular isotopy is an equivalence on 2-dimensional representation of 3-dimensional diagrams (and not of 3-dimensional diagrams themselves).

Theorem 11 (Coherence for braided autonomous categories) *A well-formed equation between morphisms in the language of braided autonomous categories follows from the axioms of braided autonomous categories if and only if it holds in the graphical language up to regular isotopy.*

Caveat 28 Only special cases of this theorem have been proved in the literature. Freyd and Yetter [16, Thm. 3.8] proved this only for diagrams over a simple signature.

4.4.6 Braided Pivotal Categories

Lemma 4 (Deligne, see [43, Prop. 2.11]) *Let* $\mathbf{C}$ *be a braided autonomous category. Then giving a twist* $\theta_A : A \to A$ *on* $\mathbf{C}$ *(making* $\mathbf{C}$ *into a balanced category) is equivalent to giving a pivotal structure* $i_A : A \to A^{**}$ *(making* $\mathbf{C}$ *into a pivotal category).*

The lemma is remarkable because the concept of a braided autonomous category does not include any assumption relating the braided structure to the autonomous structure. Moreover, the axioms for a twist depend only on the braided structure, whereas the axioms for a pivotal structure depend only on the autonomous structure. Yet, they are equivalent if $\mathbf{C}$ is braided autonomous.

Proof of Lemma 4 Recall the natural isomorphism $b_A : A^{**} \to A$ that was defined in Sect. 4.4.5 for any braided autonomous category. Given a twist $\theta_A : A \to A$, we define a pivotal structure by

$$i_A = A \xrightarrow{\theta_A} A \xrightarrow{b_A^{-1}} A^{**}. \tag{4.7}$$

Conversely, given a pivotal structure $i_A : A \to A^{**}$, we define a twist by

$$\theta_A = A \xrightarrow{i_A} A^{**} \xrightarrow{b_A} A. \tag{4.8}$$

The two constructions are clearly each other's inverse. To verify their properties, it is obvious that i_A is a natural isomorphism if and only if θ_A is a natural isomorphism. Moreover, $\theta_I = \mathrm{id}$ iff $i_I = b_I^{-1}$, and b_I^{-1} is the canonical isomorphism $I \cong I^{**}$. What remains to be shown is that θ satisfies Eq. (4.3) if and only if i satisfies Eq. (4.3). However, this is a direct consequence of the following fact about b, which is easily verified:

$$\begin{array}{ccc}
A^{**} \otimes B^{**} & \xrightarrow{c_{A,B}} & B^{**} \otimes A^{**} \\
{\scriptstyle \cong}\downarrow & & \\
(A \otimes B)^{**} & & \downarrow{\scriptstyle b_B \otimes b_A} \\
{\scriptstyle b_{A\otimes B}}\downarrow & & \\
A \otimes B & \xleftarrow[c_{B,A}]{} & B \otimes A.
\end{array}$$

□

Corollary 2 *A braided pivotal category is the same thing as a balanced autonomous category.* □

Remark 8 While Lemma 4 establishes a one-to-one correspondence between twists and pivotal structures, the correspondence is not canonical. Indeed, instead of (4.7) and (4.8), we could have equally well used

$$i_A = A \xrightarrow{\theta_A^{-1}} A \xrightarrow{b'_A} A^{**} \tag{4.9}$$

and

$$\theta_A = A \xrightarrow{b'_A} A^{**} \xrightarrow{i_A^{-1}} A, \tag{4.10}$$

where

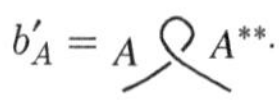

In fact, there are a countable number of such similar one-to-one correspondences, all induced by the existence of a monoidal natural transformation ${b'_A}^{-1} \circ i_A \circ b_A \circ i_A : A \to A$. They all coincide if and only if the category is tortile, as discussed in the next section.

Graphical Language and Coherence

The graphical language for braided pivotal categories is the same as the graphical language for pivotal categories, with the addition of braids. Equivalence of diagrams is up to regular isotopy, just as for braided autonomous categories (see Sect. 4.4.5).

Theorem 12 (Coherence for braided pivotal categories) *A well-formed equation between morphisms in the language of braided pivotal categories follows from the axioms of braided pivotal categories if and only if it holds in the graphical language up to regular isotopy.*

Caveat 29 Only special cases of this theorem have been proved in the literature. Freyd and Yetter [16, Thm. 4.4] proved this only for diagrams over a simple signature.

Remark 9 The equation

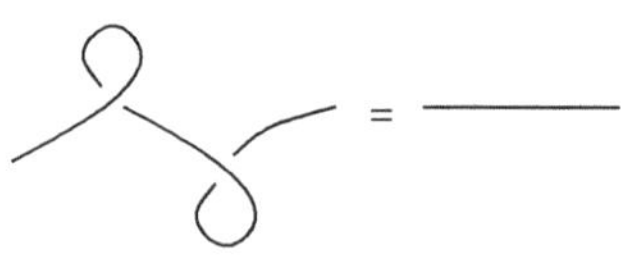

holds up to regular isotopy, as it can be proved using only the Reidemeister moves (R2) and (R3). It is therefore valid in braided pivotal categories (or even braided autonomous categories). On the other hand, the equation

holds up to isotopy, but not up to regular isotopy (because regular isotopy preserves total curvature, as pointed out by Freyd and Yetter [15, p. 169]). It is therefore not valid in braided pivotal categories. The use of regular isotopy does not seem natural, and this is precisely the reason why Joyal and Street introduced tortile categories, which we discuss in the next section.

Remark 10 A braided pivotal category is not in general spherical (and therefore also not spacial). Indeed, instead of the spherical axiom (4.6), only the following holds up to regular isotopy:

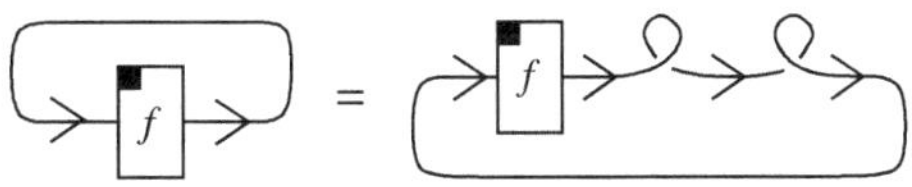

Along with Remark 8, this is further evidence that braided pivotal categories (and braided autonomous categories) are not "natural" notions.

4.4.7 Tortile Categories

Lemma 5 *Consider a braided pivotal category, which is equivalently balanced autonomous via (4.7) and (4.8). For any object A the following are equivalent:*

(a) $(\epsilon_{A^*} \otimes id_A) \circ (id_{A^*} \otimes c^{-1}_{A^{**},A}) \circ (\eta_A \otimes id_{A^{**}}) \circ i_A \circ (\epsilon_{A^*} \otimes id_A) \circ (id_{A^*} \otimes c_{A,A^{**}}) \circ (\eta_A \otimes id_{A^{**}}) \circ i_A = id_A$, *or graphically:*

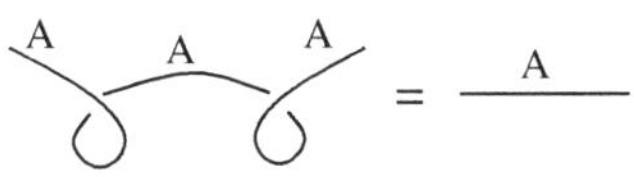

(b) $\theta_{A^*} = (\theta_A)^*$.

Proof The proof is a straightforward calculation, but it is best explained by the fact that the following hold in the graphical language:

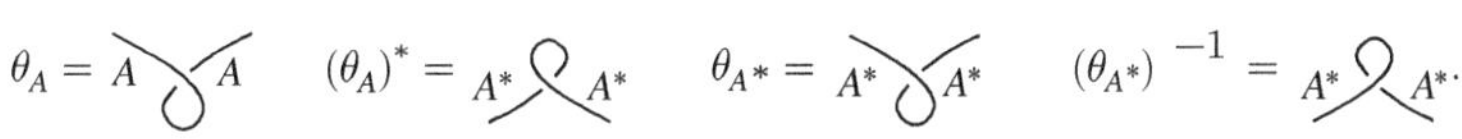

Therefore, the equation (b) is equivalent to

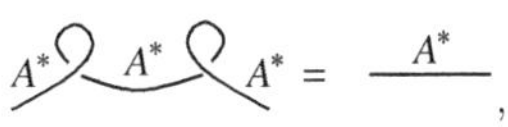

which is the adjoint mate of (a). □

Remark 11 The condition in Lemma 5(a) holds if and only if the two definitions of θ_A from (4.8) and (4.10) coincide.

Definition 30 [23] A *tortile category* is a braided pivotal category satisfying the condition of Lemma 5(a). Equivalently, a tortile category is a balanced autonomous category satisfying the condition of Lemma 5(b).

Remark 12 (Terminology) A tortile category is also sometimes called a *ribbon category*, see e.g. [42].

Graphical Language and Coherence

The graphical language for tortile categories is like the graphical language for braided pivotal categories, except that morphisms are represented by ribbons, rather than wires. These ribbons are just like the ones for balanced categories from Sect. 4.3.4. Units and counits are represented in the obvious way, for example

$\eta_A =$ 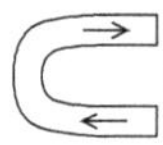 , $\epsilon_A =$ 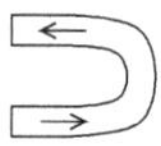 .

The twist map $\theta_A : A \to A$ can be represented in several equivalent ways:

$\theta_A =$  $=$ 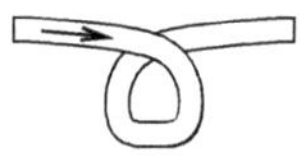 $=$  .

Note that these diagrams are equivalent up to framed 3-dimensional isotopy, and define the same morphism in a tortile category. (On the other hand, in a mere braided pivotal category, the latter two diagrams are not equal). Also note that the map b_A from Sect. 4.4.5 is also represented in the graphical language as

$b_A =$  ,

but this is of type $b_A : A^{**} \to A$, whereas $\theta_A : A \to A$. They differ, of course, only by an invisible pivotal map $i_A : A \to A^{**}$.

Theorem 13 (Coherence for tortile categories) *A well-formed equation between morphisms in the language of tortile categories follows from the axioms of tortile categories if and only if it holds in the graphical language up to framed 3-dimensional isotopy.*

Caveat 31 Only special cases of this theorem have been proved in the literature. Shum [34, Thm. 6.1] proved it for the case of the free tortile category generated by a category, i.e., for diagrams over a simple signature only.

4.4.8 Compact Closed Categories

A compact closed category is a tortile category that is symmetric (as a balanced monoidal category) in the sense of Sect. 4.3.5. Equivalently, because of Remark 2, a compact closed category is a tortile category in which $\theta_A = \mathrm{id}_A$ for all A.

The definition can be simplified. Notice that a right autonomous symmetric monoidal category is automatically autonomous (by Lemma 3), balanced (with $\theta_A = \mathrm{id}_A$) and therefore pivotal (by Lemma 4). Moreover, it is tortile (because $\theta_{A^*} = (\theta_A)^* = \mathrm{id}_{A^*}$). We can therefore define:

Definition 32 A *compact closed category* is a right autonomous symmetric monoidal category.

Remark 13 By analogy with Remark 3, it is possible for a compact closed category to possess a non-trivial twist (with the associated non-trivial pivotal structure), in addition to the trivial twist $\theta_A = \mathrm{id}_A$, making it into a tortile category. In other words, for a given tortile category, the symmetry condition $c_{A,B} = c_{B,A}^{-1}$ does not in general imply $\theta_A = \mathrm{id}_A$. However, it does imply $\theta_A^2 = \mathrm{id}_A$, as the following argument shows:

$\theta_A^2 =$ 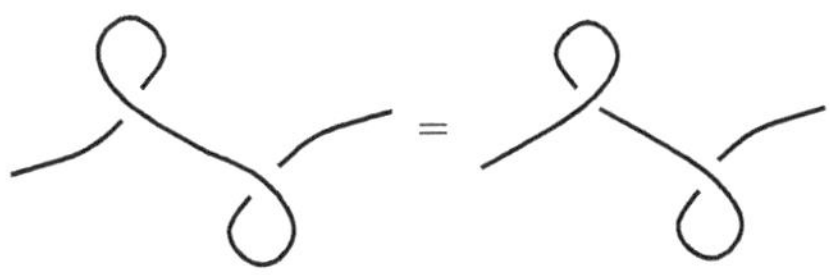 $= \mathrm{id}_A$.

To construct an example where $\theta \neq \mathrm{id}$, consider the category $\mathbf{C}$ of finite-dimensional real vector spaces and linear functions. Define an equivalence relation on objects by $A \sim B$ iff $\dim(A \otimes B)$ is a square. Then define a subcategory $\mathbf{C}\sim$ by

$$\hom_{\mathbf{C}_\sim}(A, B) = \begin{cases} \hom_{\mathbf{C}}(A, B) & \text{if } A \sim B, \\ \emptyset & \text{else.} \end{cases}$$

Then $\mathbf{C}_\sim$ is compact closed. Let $\mathbb{N}^+ = \{1, 2, 3, \ldots\}$ be the positive integers, and consider some multiplicative homomorphism $\phi : \mathbb{N}^+ \to \{-1, 1\}$. Any such homomorphism is determined by a sequence $a_1, a_2, \ldots \in \{-1, 1\}$ via

$$\phi(p_1^{n_1} p_2^{n_2} \cdots p_k^{n_k}) = a_1^{n_1} a_2^{n_2} \cdots a_k^{n_k},$$

where p_i is the ith prime number. Finally, define the twist map θ_A as multiplication by the scalar $\phi(\dim(A))$, or as id_A if A is 0-dimensional. With this twist, $\mathbf{C}_\sim$ is tortile. In fact, this shows that there exists a continuum of possible twists on $\mathbf{C}_\sim$.

Examples 33 The monoidal category $(\mathbf{Rel}, \times)$ is compact closed with $A^* = A$. The category $(\mathbf{FdVect}, \otimes)$ of finite dimensional vectors spaces is compact closed with A^* the dual space of A, and similarly for the category of finite dimensional Hilbert spaces $(\mathbf{FdHilb}, \otimes)$. The corresponding categories of possibly infinite dimensional spaces are not autonomous. (**Cob**, $+$) is compact closed with A^* equal to A with reversed orientation.

Graphical Language and Coherence

The graphical language for compact closed categories is like that of tortile categories, except that we remove the framing and twist maps, and use symmetries instead of braidings.

Theorem 14 (Coherence for compact closed categories) *A well-formed equation between morphisms in the language of compact closed categories follows from the axioms of compact closed categories if and only if it holds, up to isomorphism of diagrams, in the graphical language.*

Caveat 34 The special case of diagrams over a simple signature was proven by Kelly and Laplaza [27, Thm. 8.2]. The general case does not appear in the literature.

4.5 Traced Categories

The graphical languages considered in Sect. 4.3 were *progressive*, which means that all wires were oriented left-to-right. By contrast, the graphical languages of autonomous categories in Sect. 4.4 allow wires to be oriented left-to-right or right-to-left. We now turn out attention to an intermediate notion, namely *traced* categories.

Like autonomous graphical languages, traced graphical languages permit loops, but with a restriction: all wires must be directed left-to-right at their endpoints. In other words, traced diagrams are like autonomous diagrams, but are taken relative to a *monoidal signature* (see Sect. 4.3.1), rather than an *autonomous signature* (see Sect. 4.4.1). Table 4.4 shows a typical example of a traced diagram, and a typical example of an autonomous diagram that is not a traced diagram.

Table 4.4 (a) A traced diagram. (b) An autonomous diagram that is not traced

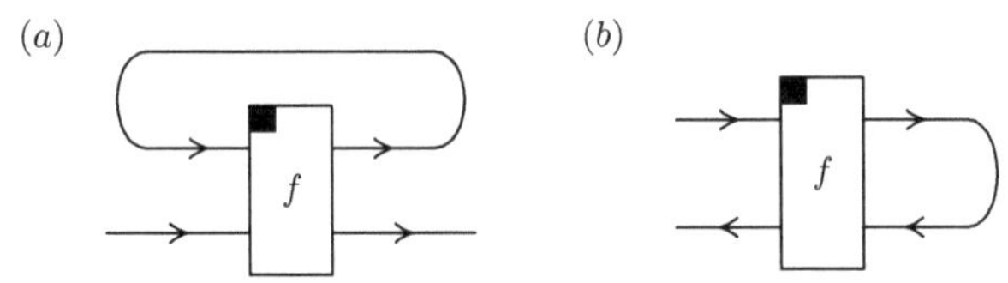

Logically, we should have considered traced categories before pivotal categories, because traced categories have less structure than pivotal categories (i.e., every pivotal category is traced, and not the other way around). However, many of the coherence theorems of this section are consequences of the corresponding theorems for pivotal categories, and therefore it made sense to present the pivotal notions first.

Symmetric traced categories and their graphical language (in the strict monoidal case, and with one additional axiom) were first introduced in the 1980s by Ştefănescu and Căzănescu under the name "biflow" [38, 10, 11]. Joyal, Street, and Verity later rediscovered this notion independently, generalized it to balanced monoidal categories, and proved the fundamental embedding theorem relating balanced traced categories to tortile categories [24].

Remark 14 Joyal, Street, and Verity use the term *traced monoidal category*. However, I prefer *traced category*, usually prefixed by an adjective such as planar, spacial, balanced, symmetric. The word "monoidal" is redundant, because one cannot have a traced structure without a monoidal structure. Also, by putting the adjective before the word "traced", rather than after it, we make it clear that the traced structure, and not just the underlying monoidal structure, if being modified.

4.5.1 Right Traced Categories

Definition 35 A *right trace* on a monoidal category is a family of operations

$$\mathrm{Tr}_{\mathrm{R}}^{X} : \hom(A \otimes X, B \otimes X) \to \hom(A, B),$$

satisfying the following four axioms. For notational convenience, we assume without loss of generality that the monoidal structure is strict.

(a) Tightening (naturality in A, B): $\mathrm{Tr}_{\mathrm{R}}^{X}((g \otimes \mathrm{id}_X) \circ f \circ (h \otimes \mathrm{id}_X)) = g \circ (\mathrm{Tr}_{\mathrm{R}}^{X} f) \circ h$;
(b) Sliding (dinaturality in X): $\mathrm{Tr}_{\mathrm{R}}^{Y}(f \circ (\mathrm{id}_A \otimes g)) = \mathrm{Tr}_{\mathrm{R}}^{X}((\mathrm{id}_B \otimes g) \circ f)$, where $f : A \otimes X \to B \otimes Y$ and $g : Y \to X$;
(c) Vanishing: $\mathrm{Tr}_{\mathrm{R}}^{I} f = f$ and $\mathrm{Tr}_{\mathrm{R}}^{X \otimes Y} f = \mathrm{Tr}_{\mathrm{R}}^{X}(\mathrm{Tr}_{\mathrm{R}}^{Y}(f))$;
(d) Strength. $\mathrm{Tr}_{\mathrm{R}}^{X}(g \otimes f) = g \otimes \mathrm{Tr}_{\mathrm{R}}^{X} f$.

A *(planar) right traced category* is a monoidal category equipped with a right trace.

These axioms are similar to those of Joyal, Street, and Verity [24], except that we have omitted the yanking axioms which does not apply in the planar case, and we have replaced the non-planar "superposing" axiom by the planar "strength" axiom. I do not know whether this set of planar axioms appears in the literature.

Graphical Language and Coherence

The right trace of a diagram $f : A \otimes X \to B \otimes X$ is graphically represented by drawing a loop from the output X to the input X, as follows:

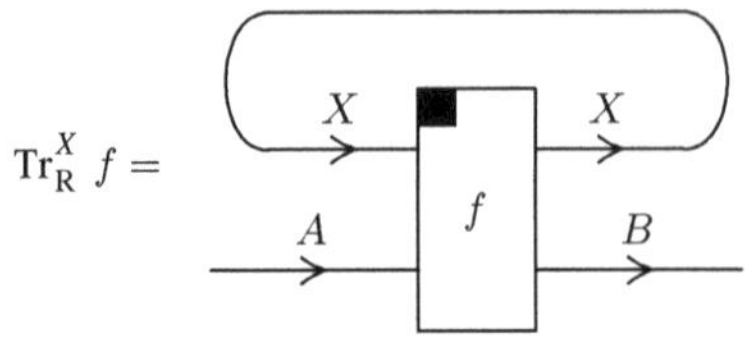

(4.1)

Note that in the graphical language of right traced categories, parts of wires can be oriented right-to-left, but each wire must be oriented left-to-right near the endpoints. The four axioms of right traced categories are illustrated in the graphical language in Table 4.5. The axioms of right traced categories are obviously sound for the graphical language, up to planar isotopy. We conjecture that they are also complete.

Table 4.5 The axioms of right traced categories

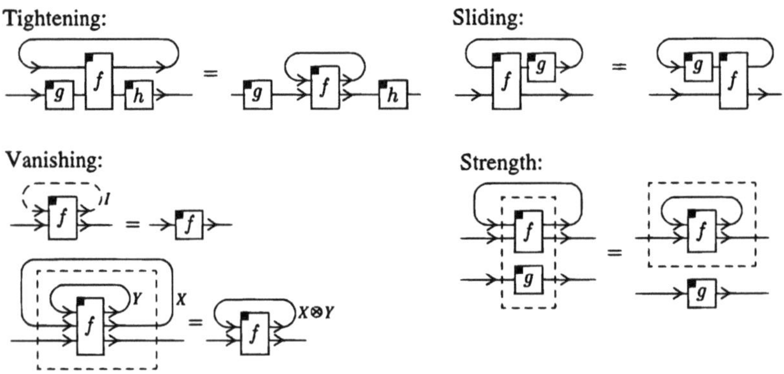

Conjecture 3 (Coherence for right traced categories) A well-formed equation between morphism terms in the language of right traced categories follows from the axioms of right traced categories if and only if it holds in the graphical language up planar isotopy.

This is a weak conjecture, in the sense that there is not much empirical evidence to support it, nor is there an obvious strategy for a proof. If this conjecture turns out to be false, the axioms for right traced categories should be amended until it becomes true.

The concept of a *left trace* is defined similarly as a family of operations

$$\mathrm{Tr}_{\mathrm{L}}^{X} : \hom(X \otimes A, X \otimes B) \to \hom(A, B),$$

satisfying symmetric axioms. A left trace is graphically depicted as follows:

$\mathrm{Tr}_{\mathrm{L}}^{X}\, g =$ 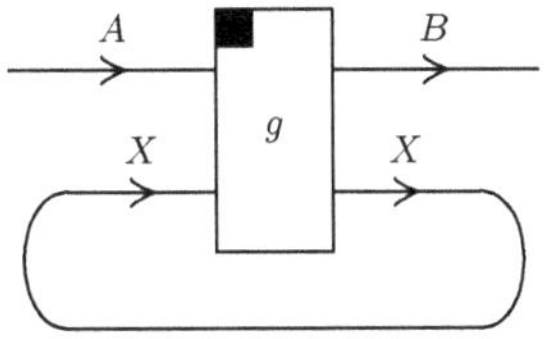

(4.2)

We say that a monoidal functor F *preserves right traces* if $F(\mathrm{Tr}_{\mathrm{R}}^{X} f) = \mathrm{Tr}_{\mathrm{R}}^{FX}((\phi^2)^{-1} \circ Ff \circ \phi^2)$, and similarly for left traces.

4.5.2 Planar Traced Categories

Definition 36 A *planar traced category* is a monoidal category equipped with a right trace and a left trace, such that the two traces satisfy three additional axioms:

(a) Interchange: $\mathrm{Tr}_{\mathrm{R}}^{X}(\mathrm{Tr}_{\mathrm{L}}^{Y} f) = \mathrm{Tr}_{\mathrm{L}}^{Y}(\mathrm{Tr}_{\mathrm{R}}^{X} f)$, for all $f : Y \otimes A \otimes X \to Y \otimes B \otimes X$;
(b) Left pivoting: $\mathrm{Tr}_{\mathrm{R}}^{B}(\mathrm{id}_B \otimes f) = \mathrm{Tr}_{\mathrm{L}}^{A}(f \otimes \mathrm{id}_A)$, for all $f : I \to A \otimes B$;
(c) Right pivoting: $\mathrm{Tr}_{\mathrm{R}}^{B}(\mathrm{id}_B \otimes f) = \mathrm{Tr}_{\mathrm{L}}^{A}(f \otimes \mathrm{id}_A)$, for all $f : A \otimes B \to I$.

Graphical Language and Coherence

The graphical language of planar traced categories consists of diagrams using the left and right trace together, modulo planar isotopy. The axioms of interchange, left pivoting, and right pivoting are shown graphically in Table 4.6. Compare also equation (4.4) on page 4.4. The axioms are clearly sound; we conjecture that they are also complete:

Table 4.6 Axioms relating left and right trace

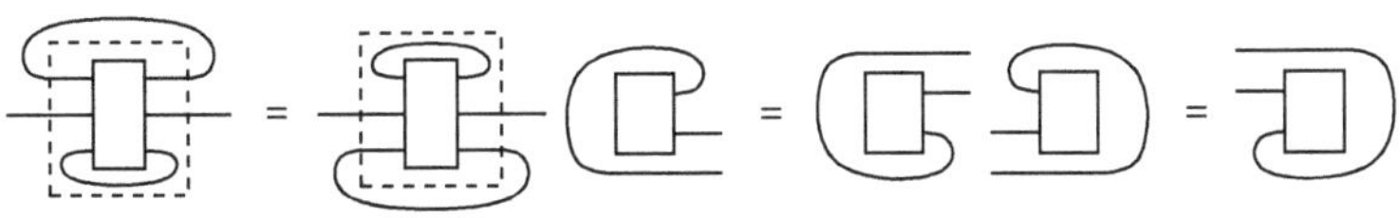

(a) interchange (b) left pivoting (c) right pivoting

Conjecture 4 (Coherence for planar traced categories) A well-formed equation between morphism terms in the language of planar traced categories follows from the axioms of planar traced categories if and only if it holds in the graphical language up planar isotopy.

As for right traced categories, this conjecture is weak. If it turns out to be false, then one should amend the axioms of planar traced categories accordingly.

Remark 15 Even if the conjecture is true, the graphical language does not in itself give an easy description of the free planar traced category. This is because there are diagrams, such as the following, that "look" planar traced, but are not actually the diagram of any planar traced term (not even up to planar isotopy).

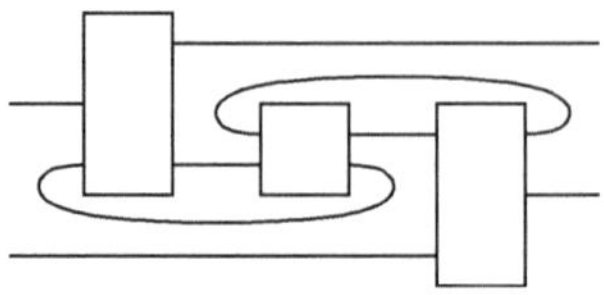

It is not obvious how to characterize the "planar traced" diagrams intrinsically, or how to extend the notion of planar traced categories to encompass all such diagrams.

Remark 16 An autonomous category is not necessarily traced. However, every pivotal category is planar traced with the obvious definitions of left and right trace:

$$\begin{aligned}\mathrm{Tr}_{\mathrm{R}}^{X} f &= (\mathrm{id}_B \otimes \epsilon_X) \circ \Big((f \circ (\mathrm{id}_A \otimes i_X^{-1})) \otimes \mathrm{id}_{X^*}\Big) \circ (\mathrm{id}_A \otimes \eta_{X^*})\,,\\ \mathrm{Tr}_{\mathrm{L}}^{X} f &= (\epsilon_{X^*} \otimes \mathrm{id}_B) \circ (\mathrm{id}_{X^*} \otimes ((i_X \otimes \mathrm{id}_B) \circ f)) \circ (\eta_X \otimes \mathrm{id}_A).\end{aligned}$$

In the graphical language, this looks just like the Diagrams (4.1) and (4.2). As a consequence, each diagram of planar traced categories can be regarded as a diagram of planar pivotal categories, but not the other way around.

4.5.3 Spherical Traced Categories

The concept of a spherical traced category is analogous to that of spherical pivotal categories from Sect. 4.4.3.

Definition 37 A planar traced category satisfies the *spherical axiom* if for all $f : A \to A$,

$$\mathrm{Tr}_{\mathrm{L}}^{A} f = \mathrm{Tr}_{\mathrm{R}}^{A} f, \tag{4.3}$$

or equivalently, in the graphical language:

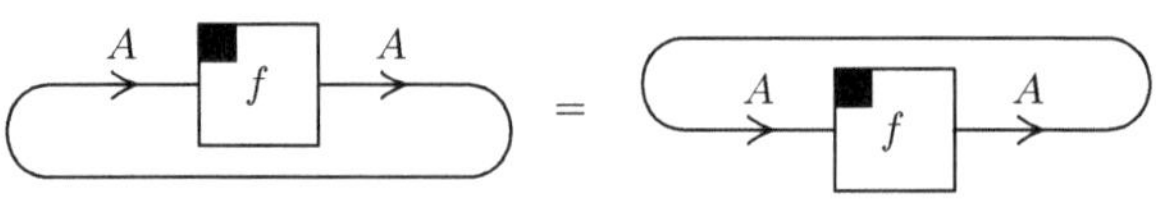

A *spherical traced category* is a planar traced category satisfying the spherical axiom.

Every spherical pivotal category is spherical traced.

Failure of Coherence

Just like for spherical pivotal categories, the graphical language of spherical traced categories is not coherent for any geometrically useful notion of equivalence of diagrams.

4.5.4 Spacial Traced Categories

Definition 38 A *spacial traced category* is a planar traced category if it satisfies the spacial axiom (4.2) and the spherical axiom (4.3)

Graphical Language and Coherence

The graphical language for spacial traced categories is the same as that for planar traced categories, except that equivalence of diagrams is now taken up to isomorphism.

Conjecture 5 (Coherence for spacial traced categories) A well-formed equation between morphism terms in the language of spacial traced categories follows from the axioms of spacial traced categories if and only if it holds, up to isomorphism of diagrams, in the graphical language.

Remark 17 Every spacial pivotal category is clearly spacial traced. I do not know whether conversely every spacial traced category can be faithfully embedded in a spacial pivotal category. If this is true, then Conjecture 5 follows from Conjecture 2.

4.5.5 Braided Traced Categories

Braided traced categories, like braided pivotal categories, are a somewhat unnatural notion, because coherence is only satisfied up to regular isotopy. (If one considers full isotopy, one obtains the more natural notion of balanced traced categories, which we will consider in the next section). Nevertheless, we include this section on braided traced categories, not least because it is the first traced notion for which we can actually prove a coherence theorem (modulo Caveat 29).

Definition 39 A *braided traced category* is a planar traced category with a braiding (as a monoidal category), such that

$$(\mathrm{Tr}_{\mathrm{L}}^{A}\, c_{A,A}) \circ (\mathrm{Tr}_{\mathrm{R}}^{A}\, c_{A,A}^{-1}) = \mathrm{id}_A, \tag{4.4}$$

or graphically:

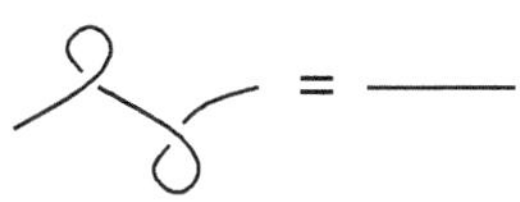

Lemma 6 (a) *The axiom (4.4) does not follow from the remaining axioms.*
(b) *In the presence of the remaining axioms, (4.4) is equivalent to*

$$(Tr_L^A c_{A,A}^{-1}) \circ (Tr_R^A c_{A,A}) = id_A, \tag{4.5}$$

or graphically:

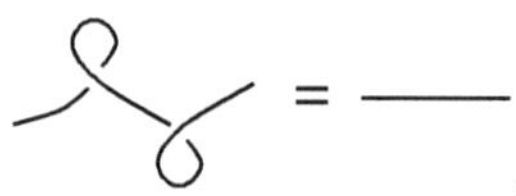

(c) *In the presence of the remaining axioms of braided traced categories, the left and right pivoting axioms are redundant.*

Proof (a) To see this, consider morphism terms in the language of braided traced categories with one object generator and no morphism generators. Define the *degree* of a term to the be tensor product of all traced-out objects, i.e., deg(id) = I, deg($f \circ g$) = deg(f) $\otimes$ deg(g), deg($\mathrm{Tr}_{\mathrm{R}}^X f$) = $X \otimes$ deg(f), etc. This is well-defined up to isomorphism. All the axioms of planar traced categories and braided categories respect degree; the only axioms where the left-hand side and right-hand side could potentially have different degree are sliding in Table 4.5 and pivoting in Table 4.6. However, in the absence of morphism generators, it is easy to show that all morphism terms are of the form $f : A \to B$ where $A \cong B$. Therefore, neither sliding nor pivoting change the degree (the latter because it is vacuous). Therefore degree is an invariant. On the other hand, (4.4) is not degree-preserving; therefore it cannot follow from the other axioms.

(b) The following graphical proof sketch can be turned into an algebraic proof:

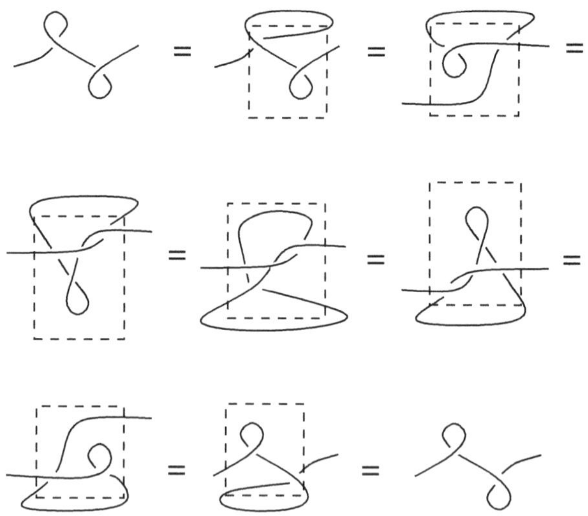

(c) Here is a proof sketch for the left pivoting axiom. Notably, the second to last step uses dinaturality (sliding).

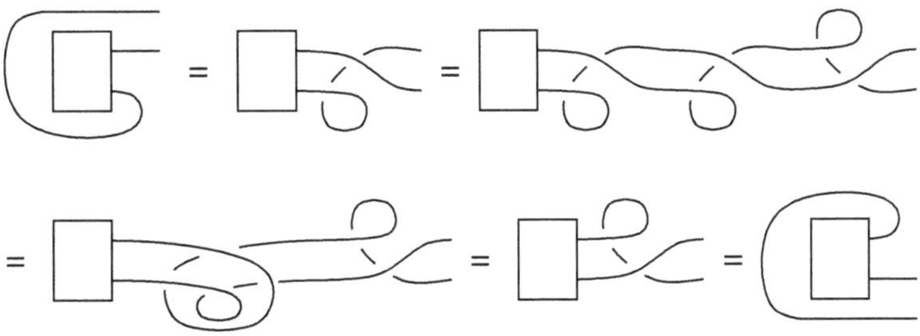

Remark 18 Each braided traced category possesses a balanced structure (as a braided monoidal category) given by $\theta_A = \mathrm{Tr}^A_{\mathrm{L}}\, c^{-1}_{A,A}$, with inverse $\theta^{-1}_A = \mathrm{Tr}^A_{\mathrm{R}}\, c_{A,A}$ (cf. (4.4)). However, this twist is not canonical; for example, another twist can be defined by $\theta'_A = \mathrm{Tr}^A_{\mathrm{R}}\, c_{A,A}$ with inverse $\theta'^{-1}_A = \mathrm{Tr}^A_{\mathrm{L}}\, c^{-1}_{A,A}$ (cf. (4.5)). In fact, there are countably many other possible twists. This is entirely analogous to Remark 8. The various twists coincide if and only if the yanking equation (4.6) holds, yielding a balanced traced category as discussed in Sect. 4.5.6 below.

We note that every braided pivotal category is braided traced, with the traced structure as given in Remark 16. Moreover, there is an embedding theorem giving a partial converse:

Theorem 15 (Representation of braided traced categories) *Every braided traced category **C** can by fully and faithfully embedded into a braided pivotal category Int(**C**), via a braided traced functor.*

Proof The proof exactly mimics the Int-construction of Joyal, Street, and Verity [24], except that we must replace the twist by , and be careful only to use the braided traced axioms. We omit the details, which are both lengthy and tedious. □

Remark 19 A braided traced category is not necessarily spherical (and therefore not spacial). This is analogous to Remark 10.

Graphical Language and Coherence

The graphical language for braided traced categories is obtained by adding braids to the graphical language of planar traced categories. Equivalence of diagrams is up to *regular isotopy* (see Sect. 4.4.5).

Theorem 16 (Coherence for braided traced categories) *A well-formed equation between morphisms in the language of braided traced categories follows from the axioms of braided traced categories if and only if it holds in the graphical language up to regular isotopy.*

Proof Soundness is easy to check by inspection of the axioms. Completeness is a consequence of Theorems 12 and 15. Namely, consider an equation in the language of braided traced categories that holds in the graphical language up to regular isotopy. The diagrams corresponding to the left-hand side and right-hand side of the equation can also be regarded as diagrams of braided pivotal categories, and since they are regularly isotopic, the equation holds in all braided pivotal categories by Theorem 12. Since any braided traced category **C** can be faithfully embedded in a braided pivotal category Int(**C**) by Theorem 15, an equation that holds in Int(**C**) must also hold in **C**. It follows that the equation in question holds in all braided traced categories **C**, and therefore, it is a consequence of the axioms. □

Caveat 40 Because of the dependence on Theorem 12, Caveat 29 also applies here.

4.5.6 Balanced Traced Categories

Definition 41 ([24]) A *balanced traced category* is a balanced monoidal category equipped with a right trace Tr, and satisfying the following *yanking axioms*:

$$\mathrm{Tr}^X(c_{X,X}) = \theta_X \qquad \text{and} \qquad \mathrm{Tr}^X(c^{-1}_{X,X}) = \theta^{-1}_X \tag{4.6}$$

Graphical Language and Coherence

The graphical language of balanced traced categories combines the ribbons and twists of balanced categories with the loops of traced categories. The trace is represented as expected:

$\mathrm{Tr}^X f =$ 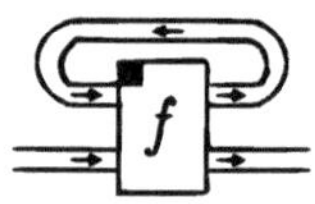 .

Note that there is no need to postulate a left trace, because a left trace is definable from the right trace and braidings as follows:

$\mathrm{Tr}^X_{\mathrm{L}} f =$ 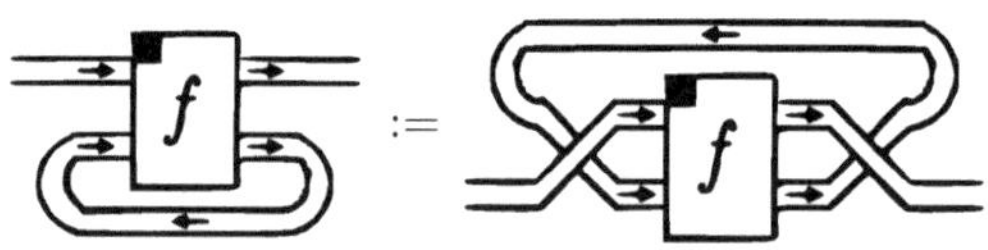

Remark 20 The defined left trace automatically satisfies interchange and the pivoting axioms (Table 4.6), as well as the spherical axiom (4.3) and the braided traced axiom (4.4). The spacial axiom (4.2) is satisfied by any braided monoidal category. Therefore, any balanced traced category is spacial traced and braided traced.

The graphical validity of the yanking axiom is easily verified using a shoe string:

Every tortile category is balanced traced, with the traced structure as given in Remark 16. Moreover, there is an embedding theorem:

Theorem 17 (Representation of balanced traced categories [24, Prop. 5.1]) *Every balanced traced category can be fully and faithfully embedded into a tortile category, via a balanced traced functor.*

Theorem 18 (Coherence for balanced traced categories) *A well-formed equation between morphisms in the language of balanced traced categories follows from the axioms of balanced traced categories if and only if it holds in the graphical language up to framed isotopy in 3 dimensions.*

Proof This follows from Theorems 13 and 17, by the exact same argument that was used in the proof of Theorem 16. □

Caveat 42 Because of the dependence on Theorem 13, Caveat 31 also applies here.

Remark 21 In any braided monoidal category with a right trace, the twist and its inverse are definable by Eq. (4.6). These maps are automatically natural and satisfy $\theta_I = \mathrm{id}_I$ and (4.3). However, they are not automatically inverse to each other. Therefore, a balanced traced category could be equivalently defined as a braided monoidal category with a right trace, satisfying

$$\mathrm{Tr}^X(c^{-1}_{X,X}) = \mathrm{Tr}^X(c_{X,X})^{-1}.$$

4.5.7 Symmetric Traced Categories

Definition 43 [11, 10, 24] A *symmetric traced category* is a symmetric monoidal category with a right trace Tr, satisfying the *symmetric yanking axiom*:

$$\mathrm{Tr}^X(c_{X,X}) = \mathrm{id}_X.$$

Remark 22 Because of Remark 2, a symmetric traced category can be equivalently defined as a balanced traced category in which $\theta_A = \mathrm{id}_A$ for all A.

Obviously every compact closed category is symmetric traced with the structure from Remark 16. Here, too, we have an embedding theorem:

Theorem 19 (Representation of symmetric traced categories [24]) *Every symmetric traced category can be fully and faithfully embedded into a compact closed category, via a symmetric traced functor.*

Example 5 [24] Consider the category **Rel** of sets and relations, with biproducts given by disjoint union $A+B$. Given a relation $R : A+X \to B+X$, define its trace $\mathrm{Tr}^X(R) : A \to B$ by $(a,b) \in \mathrm{Tr}^X(R)$ iff there exists $n \geq 0$ and $x_1, \ldots, x_n \in X$ such that $a\,R\,x_1\,R\,x_2\,R\,\ldots\,R\,x_n\,R\,b$. This defines a symmetric traced category which is not compact closed.

Graphical Language and Coherence

The graphical language is like that of planar traced categories, combined with the symmetry. A typical diagram looks like this:

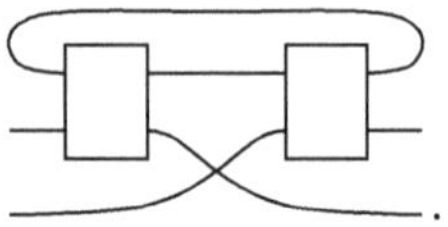

The notion of equivalence of diagrams is isomorphism.

Theorem 20 (Coherence for symmetric traced categories) *A well-formed equation between morphisms in the language of symmetric traced categories follows from the axioms of symmetric traced categories if and only if it holds in the graphical language up to isomorphism of diagrams.*

Proof A consequence of Theorems 14 and 19, as in Theorems 16 and 18.

Caveat 44 Because of the dependence on Theorem 14, Caveat 34 also applies here.

Remark 23 Strict symmetric traced categories, with the additional axiom

$$\mathrm{Tr}^X(\mathrm{id}_{A\otimes X}) = \mathrm{id}_A, \tag{4.7}$$

first appear in the work of Ştefănescu under the name "biflow". A precursor of the definition appears in [38], and the axioms were given their modern form by Căzănescu and Ştefănescu [10, 11]. The paper [38] also contains a detailed proof sketch of coherence, namely, that the graphical language, modulo isomorphism and the equation (4.7), forms the free biflow over a monoidal signature. This proof sketch remains valid with respect to the modern definition, provided that one assumes coherence for symmetric monoidal categories.

4.6 Products, Coproducts, and Biproducts

In this section, we consider graphical languages for monoidal categories where the monoidal structure is given by a categorical product, coproduct, or biproduct. The main difference with the graphical languages of "purely" monoidal categories from Sects. 4.3, 4.4 and 4.5 is that equivalence of diagrams must now be defined up to diagrammatic equations.

4.6.1 Products

Definition 45 In a category, a *product* of objects A and B is given by an object $A \times B$, together with morphisms $\pi_1 : A \times B \to A$ and $\pi_2 : A \times B \to B$, such that for all objects C and pairs of morphisms $f : C \to A$ and $g : C \to B$, there exists a unique morphism $h : C \to A \otimes B$ such that the following diagram commutes:

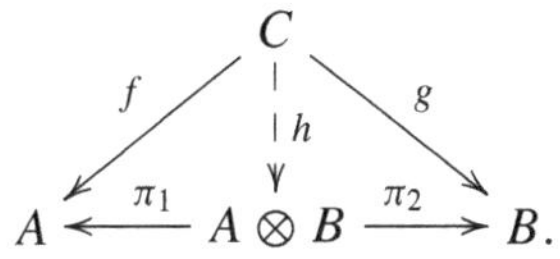

The unique morphism h is often written as $h = \langle f, g \rangle$. An object I is *terminal* if for all objects C, there exists a unique morphism $h : C \to I$. A *finite product category* (or *cartesian category*) is a category with a chosen terminal object, and a chosen product for each pair of objects.

Equivalently, a finite product category can be described as a symmetric monoidal category, together with natural families of *copy* and *erase* maps

$$\Delta_A : A \to A \otimes A, \qquad \Diamond_A : A \to I$$

subject to a number of axioms, shown in Table 4.7.

Table 4.7 The axioms for products

Naturality axioms:	
$\Delta_B \circ f = (f \otimes f) \circ \Delta_A$:	$A \to B \otimes B$
$\Diamond_B \circ f = \Diamond_A$:	$A \to I$
Commutative comonoid axioms:	
$(\mathrm{id}_A \otimes \Delta_A) \circ \Delta_A = (\Delta_A \otimes \mathrm{id}_A) \circ \Delta_A$:	$A \to A \otimes A \otimes A$
$(\mathrm{id}_A \otimes \Diamond_A) \circ \Delta_A = \rho_A^{-1}$:	$A \to A \otimes I$
$(\Diamond_A \otimes \mathrm{id}_A) \circ \Delta_A = \lambda_A^{-1}$:	$A \to I \otimes A$
$c_{A,A} \circ \Delta_A = \Delta_A$:	$A \to A \otimes A$
Coherence axioms:	
$\Delta_I = \lambda_I^{-1}$:	$I \to I \otimes I$
$(\mathrm{id}_A \otimes c_{B,A} \otimes \mathrm{id}_B) \circ \Delta_{A \otimes B} = \Delta_A \otimes \Delta_B$:	$A \otimes A \to B \otimes B \otimes A \otimes B$
$\Diamond_I = \mathrm{id}_I$:	$I \to I$
$\Diamond_{A \otimes B} = \lambda_I \circ (\Diamond_A \otimes \Diamond_B)$:	$A \otimes B \to I$

Graphical Language

We extend the graphical language of symmetric monoidal categories by adding the following representations of the copy and erase maps.

Copy $\Delta_A : A \to A \otimes A$

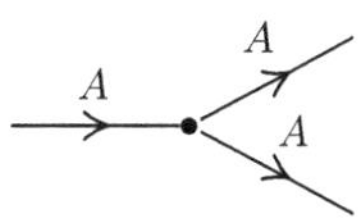

Erase $\Diamond_A : A \to I$

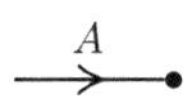

As usual, if A is a composite object term, a wire labeled A should be replaced by multiple parallel wires. Table 4.8 contains graphical representations of some of the axioms for finite product categories.

Table 4.8 Graphical representation of some product axioms

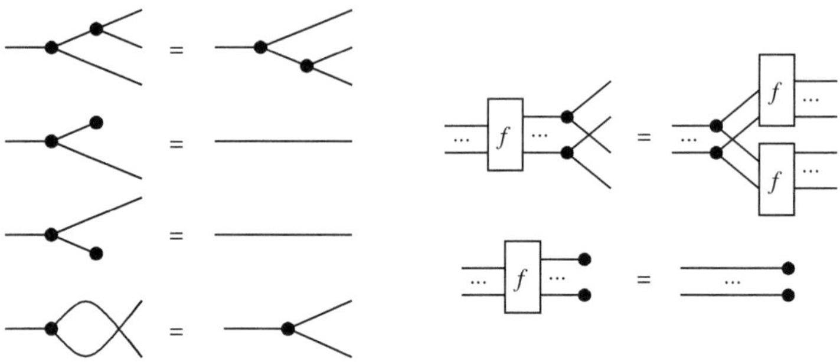

Commutative comonoid axioms Naturality

Note that the projections $\pi_1 : A \times B \to A$ and $\pi_2 : A \times B \to B$, and the pairing $h : C \to A \otimes B$ of $f : C \to A$ and $g : C \to B$, are represented graphically as follows:

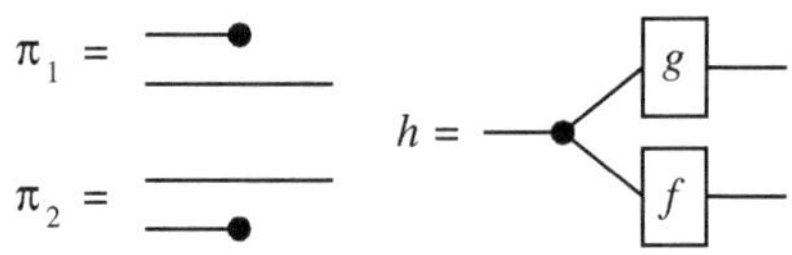

Coherence

As the equivalences in Table 4.8 demonstrate, coherence in the graphical language of finite product categories does not hold up to isomorphism or isotopy of diagrams. Rather, is holds up to *manipulations* of diagrams. So unlike the graphical languages considered in Sects. 4.2, 4.3, 4.4 and 4.5, we now have to consider axioms on diagrams.

Theorem 21 (Coherence for finite product categories) *A well-formed equation between morphism terms in the language of finite product categories follows from the axioms of finite product categories if and only if it holds in the graphical language, up to isomorphism of diagrams and the diagrammatic manipulations shown in Table 4.8.*

This theorem is a simple consequence of coherence for symmetric monoidal categories (Theorem 7), together with the fact that all the axioms of finite product categories, except those shown in Table 4.8, hold up to isomorphism of diagrams.

4.6.2 Coproducts

The definition of coproducts and initial objects is dual to that of products and terminal objects, i.e., it is obtained by reversing all the arrows in Sect. 4.6.1. Explicitly, an object 0 is *initial* if for all objects C, there exists a unique morphism $h : 0 \to C$. A *coproduct* of objects A, B is given by an object $A + B$, together with morphisms $\iota_1 : A \to A + B$ and $\iota_2 : B \to A + B$, such that for all objects C and pairs of morphisms $f : A \to C$ and $g : B \to C$, there exists a unique morphism $h : A + B \to C$ such that $h \circ \iota_1 = f$ and $h \circ \iota_2 = g$. One often writes $h = [f, g]$. A category with finite coproducts is also called a *co-cartesian category.*

Dualizing the presentation of Sect. 4.6.1, one can equivalently define a finite coproduct category as a symmetric monoidal category with natural families of *merge* and *initial* maps

$$\nabla_A : A \otimes A \to A, \qquad \Box_A : I \to A,$$

satisfying the duals of the axioms in Table 4.7.

Graphical Language

The graphical language of finite coproduct categories is obtained by dualizing that of finite product categories, with the duals of the axioms from Table 4.8.

Merge $\nabla_A : A \otimes A \to A$

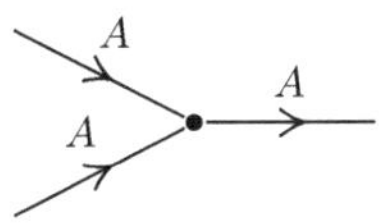

Initial $\Box_A : I \to A$

4.6.3 Biproducts

Definition 46 An object is called a *zero object* if it is initial and terminal. If $\mathbf{0}$ is a zero object, then there is a distinguished map $A \to \mathbf{0} \to B$ between any two objects, denoted $0_{A,B}$. A *biproduct* of objects A_1 and A_2 is given by an object $A_1 \oplus A_2$, together with morphisms $\iota_i : A_i \to A_1 \oplus A_2$ and $\pi_i : A_1 \oplus A_2 \to A_i$, for $i = 1, 2$, such that $A \oplus B$ is a product with π_1, π_2, a coproduct with ι_1, ι_2 and such that $\pi_i \circ \iota_j = \delta_{ij}$. Here $\delta_{ii} = \mathrm{id}_{A_i}$ and $\delta_{ij} = 0_{A_j,A_i}$ when $i \neq j$. We say that $\mathbf{C}$ is a *biproduct category* if it has a chosen zero object $\mathbf{0}$ and a chosen biproduct for any pair of objects.

Remark 24 The axiom $\pi_i \circ \iota_j = \delta_{ij}$ is equivalent to the assertion that the symmetric monoidal structure defined by $\oplus$ "as a product" coincides with the symmetric monoidal structure defined by $\oplus$ "as a coproduct". Therefore, a biproduct category is symmetric monoidal in a canonical way.

Equivalently, a biproduct category can be defined as a symmetric monoidal category, together with natural families of morphisms

$$\Delta_A : A \to A \otimes A, \qquad \Diamond_A : A \to I, \qquad \nabla_A : A \otimes A \to A, \qquad \Box_A : I \to A,$$

satisfying the axioms in Table 4.7, as well as their duals.

Graphical Language

The graphical language of biproducts is obtained by combining the graphical languages for products and coproducts. In this case, one has the equalities in Table 4.9, which are consequences of the naturality axioms in Table 4.8. Note that the axiom $\pi_i \circ \iota_j = \delta_{ij}$ holds automatically in the graphical language.

Table 4.9 Some biproduct laws

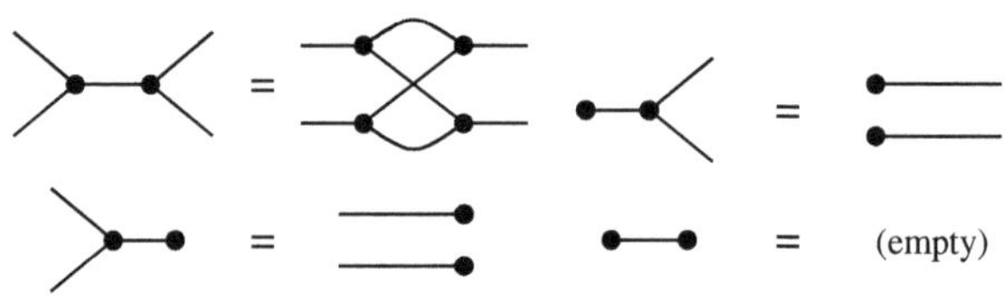

Theorem 22 (Coherence for biproduct categories) *A well-formed equation between morphism terms in the language of biproduct categories follows from the axioms of biproduct categories if and only if it holds in the graphical language, up to isomorphism of diagrams, the diagrammatic manipulations shown in Table 4.8, and their duals.*

This theorem is a simple consequence of coherence for symmetric monoidal categories, together with the fact that the axioms in Table 4.8 (and their duals) are exactly the graphical representations of the axioms in Table 4.7 (and their duals) that do not already hold up to graphical isomorphism.

4.6.4 Traced Product, Coproduct, and Biproduct Categories

It potentially makes sense to revisit each of the notions of Sects. 4.3, 4.4 and 4.5 and consider the case where the monoidal structure is given by a product, coproduct, or biproduct. Since products, coproducts, and biproducts are automatically symmetric, we do not need to consider the weaker notions (such as balanced, braided, etc).

Moreover, we do not need to consider any autonomous cases, because an autonomous category where the tensor is given by a product (or coproduct) is trivial. Indeed, for any objects A, B, the morphisms $f : A \to B$ are in one-to-one correspondence with morphism $A \otimes B^* \to I$. Since I is terminal, there is exactly one such morphism, and therefore there is a unique morphism between any two objects. Such a category is equivalent to the one-object one-morphism category.

Therefore, the only new notion from Sects. 4.3, 4.4 and 4.5 that admits non-trivial examples in the context of products, coproducts, or biproducts is that of a symmetric traced category.

Definition 47 A *traced product [coproduct, biproduct] category* is a symmetric traced category where the tensor is given by a categorical product [coproduct, biproduct].

Example 6 [24] The symmetric traced category (**Rel**, $+$) from Example 4.5.7 is a traced biproduct category.

Example 7 Consider the category $\mathbf{Set}_\perp$ whose objects are sets, and whose morphisms are partial functions, regarded as a subcategory of **Rel** from Example 4.6.4. In this category, the empty set 0 is a zero object, and the disjoint union operation $A + B$ defines a coproduct (but not a product). Trace is given as in Example 4.6.4. With these definitions, $\mathbf{Set}_\perp$ is a traced coproduct category.

Graphical Language

As expected, the graphical language of traced product [coproduct, biproduct] categories is given by adding a trace (as in Section 4.5) to the graphical language of finite product [finite coproduct, biproduct] categories.

Theorem 23 (Coherence for traced product [coproduct, biproduct] categories) *A well-formed equation between morphism terms in the language of traced product [coproduct, biproduct] categories follows from the respective axioms if and only if it holds in the graphical language, up to isomorphism of diagrams, and the diagrammatic manipulations shown in Table 4.8 and/or their duals (as appropriate).*

Remark 25 In computer science, traces arise naturally in the context of *data flow* (as fixed points), and in the context of *control flow* (as iteration). The two situations correspond to traced product categories and traced coproduct categories, respectively. The duality between data flow and control flow was first described by Bainbridge [3]. The following are typical examples of a data flow diagram (on the left) and a control flow diagram (on the right). The data flow diagram represents the fixed point expression $y = (3 + x)(x + y)$, parametric on an input x. The control flow diagram represents a generic "while loop". Note that data flow diagrams have a notion of "copying" data, whereas control flow diagrams have a dual notion of "merging" control paths.

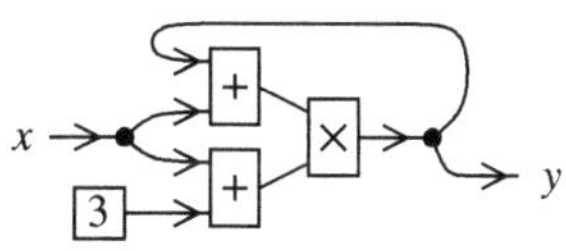

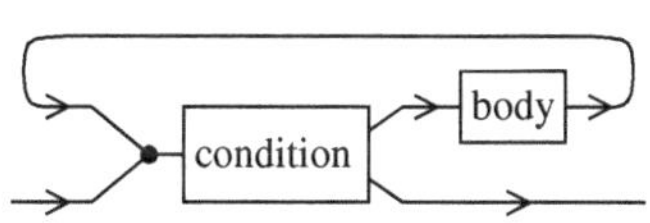

Proposition 2 (Căzănescu and Ştefănescu [10, 11]) *In a category with finite coproducts, giving a trace is equivalent to giving an iteration operator. Here, an iteration operator is a family of operations*

$$iter^X : hom(X, A + X) \to hom(X, A),$$

natural in A and dinatural in X, satisfying

1. *Iteration:* $iter(f) = [id_A, iter(f)] \circ f$, *for all* $f : X \to A + X$;
2. *Diagonal property:* $iter(iter(f)) = iter((id_A + [id_X, id_X]) \circ f)$, *for all* $f : X \to A + X + X$.

Dually, on a finite product category, giving a trace is equivalent to a fixed point operator $fix^X : hom(A \times X, X) \to hom(A, X)$.

This makes precise the intuitive idea that in the presence of coproducts, the while loop in Remark 25 is sufficient for constructing arbitrary traces.

Remark 26 In the presence of the other axioms, the diagonal property is equivalent to the so-called Bekič Lemma:

$$\text{iter}[f, g] = [\text{id}_A, \text{iter}([\text{id}_{A+X}, \text{iter}(g)] \circ f)] \circ [\text{in}_2, \text{iter}(g)],$$

for all $f : X \to A + X + Y$ and $g : Y \to A + X + Y$ [36, Prop. B.1].

Remark 27 Iteration operators in the sense of Proposition 2 were first defined, using different but equivalent axioms, by Căzănescu and Ungureanu [12, 9], under the name "algebraic theory with iterate".

Proposition 3 *[11] In a category with finite biproducts, giving a trace is equivalent to giving a repetition operation, i.e., a family of operators*

$$* : hom(A, A) \to hom(A, A)$$

satisfying

1. $f^* = \text{id} + ff^*$,
2. $(f + g)^* = (f^*g)^*f^*$.
3. $(fg)^*f = f(gf)^*$ (dinaturality).

Here, $f + g$ *denotes the morphism* $\nabla_A \circ (f \oplus g) \circ \Delta_A : A \to A$, *for* $f, g : A \to A$.

4.6.5 Uniformity and Regular Trees

Definition 48 Suppose we are given a traced category with a distinguished subclass of morphisms called the *strict* morphisms. Then the trace is called *uniform* if for all $f : A \otimes X \to B \otimes X$, $g : A \otimes Y \to B \otimes Y$, and strict $h : X \to Y$, the following implication holds:

$$(\mathrm{id}_B \otimes h) \circ f = g \circ (\mathrm{id}_A \otimes h) \quad \Rightarrow \quad \mathrm{Tr}^X(f) = \mathrm{Tr}^Y(g).$$

Equivalently, in pictures:

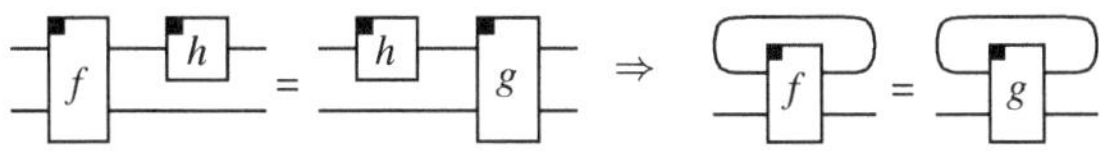

whenever h is strict. Note that uniformity is not an equational property.

Proposition 4 *[11] A traced coproduct category is uniformly traced if and only if for all* $f : X \to A + X$, $g : Y \to A + Y$, *and strict* $h : X \to Y$,

$$(id_A + h) \circ f = g \circ h \quad \Rightarrow \quad iter^X(f) = iter^Y(g) \circ h.$$

Moreover, a traced biproduct category is uniformly traced if and only if for all $f : X \to X$, $g : Y \to Y$, *and strict* $h : X \to Y$,

$$h \circ f = g \circ h \quad \Rightarrow \quad h \circ f^* = g^* \circ h.$$

In the particular case where the class of strict morphisms is taken to be the smallest co-cartesian subcategory containing all objects, Ştefănescu [36, 35] proved that the free uniformly traced coproduct category over a monoidal signature is given by the graphical language of traced coproduct categories, modulo a suitable notion of simulation equivalence on diagrams. This simulation equivalence is easiest to describe in the case where all morphism variables are of input arity 1. In this case, two diagrams are simulation equivalent if and only if they have the same infinite tree unwinding. There is also an analogous result for biproducts. We refer the reader to [36, 37, 40] for full details.

The following is an example of an equation that holds up to infinite tree unwinding, but fails in general traced coproduct categories:

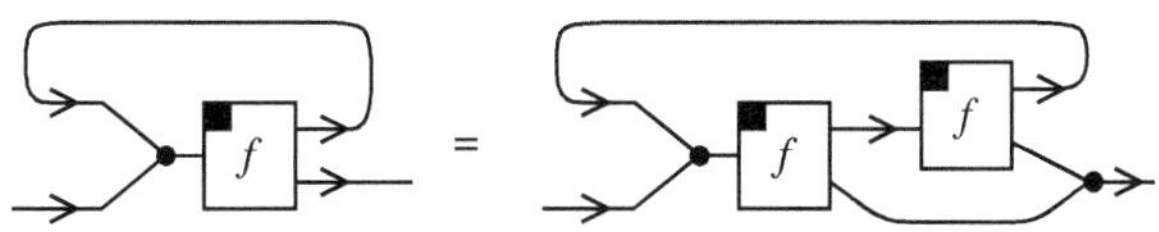

(4.1)

Ésik's "iteration theories" [14] are a direct equational axiomatization of such infinite tree unwindings. They include an iteration operator as in Proposition 2, but with an infinite family of additional properties, such as (4.1).

4.6.6 Cartesian Center

Sometimes it is useful to consider notions that are weaker than product categories, yet still have copy and erase maps $\Delta_A : A \to A \otimes A$ and $\Diamond_A : A \to I$. For example, it is common to drop the naturality axioms, while retaining the commutative comonoid and coherence axioms (see Tables 4.7 and 4.8). An equivalent way to describe such a category is as a symmetric monoidal category with (faithful) *cartesian center* [18], i.e., a symmetric monoidal category with a symmetric monoidal subcategory that contains all the objects and is cartesian. Similar ideas have occurred, with varying degrees of explicitness, in the literature on flowcharts, see e.g. [12, 7, 39].

Similarly, if one omits naturality from the axioms for coproducts, one obtains categories with a co-cartesian center. A weakened version of biproducts is obtained by combining the axioms of cartesian center and co-cartesian center. In this case, one requires the operations $\Delta, \Diamond, \nabla, \Box$ to be natural with respect to one another, yielding the properties from Table 4.9. More generally, one may require any subset of the operations Δ, $\Diamond$, ∇, $\Box$ to exist, and a further subset to be natural transformations. As the reader may imagine, this leads to a nearly endless number of categorical notions and corresponding graphical languages; see e.g. [39, 40].

4.7 Dagger Categories

The concept of a dagger category (also called *involutive category* or **-category* in the literature) is motivated by the category of Hilbert spaces, where each morphism $f : A \to B$ has an *adjoint* $f^\dagger : B \to A$.

Definition 49 A *dagger category* is a category $\mathbf{C}$ together with an involutive, identity-on-objects, contravariant functor $\dagger : \mathbf{C} \to \mathbf{C}$.

Concretely, this means that to every morphism $f : A \to B$, one associates a morphism $f^\dagger : B \to A$, called the *adjoint* of f, such that for all $f : A \to B$ and $g : B \to C$:

$$\begin{aligned} \mathrm{id}_A^\dagger &= \mathrm{id}_A && : A \to A, \\ (g \circ f)^\dagger &= f^\dagger \circ g^\dagger && : C \to A, \\ f^{\dagger\dagger} &= f && : A \to B, \end{aligned}$$

Example 8 The category **Hilb** of Hilbert spaces and bounded linear maps is a dagger category, where $f^\dagger : B \to A$ is given by the usual adjointness property of linear algebra, i.e., $\langle f^\dagger x \mid y\rangle = \langle x \mid fy\rangle$ for all $x \in B$ and $y \in A$.

Definition 50 (Unitary map, self-adjoint map) In a dagger category, a morphism $f : A \to B$ is called *unitary* if it is an isomorphism and $f^{-1} = f^\dagger$. A morphism $f : A \to A$ is called *self-adjoint* or *hermitian* if $f = f^\dagger$.

A *dagger functor* between dagger categories is a functor that satisfies $F(f^\dagger) = (Ff)^\dagger$ for all f.

Graphical Language

The graphical language of dagger categories extends that of categories. The adjoint of a morphism variable $f : A \to B$ is represented diagrammatically as follows:

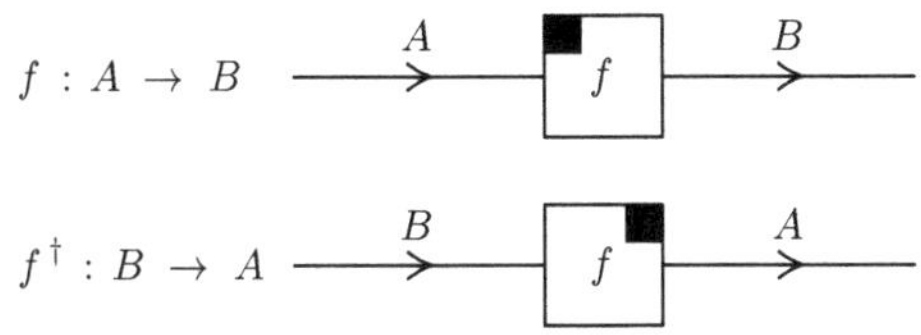

More generally, the adjoint of any diagram is its mirror image. Note that the mirror image of a box is visually distinguishable because we have marked the upper left corner of each box representing a morphism variable. Also note that, while we have taken the mirror image of each box, we have reversed the location, but not the direction, of the wires. Contrast this with (4.5).

Theorem 24 (Coherence for dagger categories) *A well-formed equation between two morphism terms in the language of dagger categories follows from the axioms of dagger categories if and only if it holds in the graphical language up to isomorphism of diagrams.*

Proof This is a consequence of coherence for categories, from Theorem 1. As usual, soundness is easy to check. For completeness, notice that any morphism term t of dagger categories can be transformed, via the axioms $(g \circ f)^\dagger = f^\dagger \circ g^\dagger$, $\mathrm{id}^\dagger = \mathrm{id}$, and $f^{\dagger\dagger} = f$, into an equivalent term t' with the property that $\dagger$ is only applied to morphism variables in t'. Such a term can be regarded as a term in the language of categories, over the extended set of morphism variables $\{f, f^\dagger, \ldots\}$. Now if t and s are two terms that have isomorphic diagrams, then by soundness, t' and s' have isomorphic diagrams. By Theorem 1, t' and s' are provably equal from the axioms of categories. Therefore t and s are provably equal from the axioms of dagger categories. □

We now consider "dagger notions" for the various monoidal categories from Sections 4.3, 4.4 and 4.5.

4.7.1 Dagger Monoidal Categories

Definition 51 A *dagger monoidal category* is a monoidal category that is a dagger category, such that the dagger structure is compatible with the monoidal structure in the following sense:

(a) $(f \otimes g)^\dagger = f^\dagger \otimes g^\dagger$, for all f, g;
(b) the canonical isomorphisms of the monoidal structure, $\alpha_{A,B,C} : (A \otimes B) \otimes C \to A \otimes (B \otimes C)$, $\lambda_A : I \otimes A \to A$, and $\rho_A : A \otimes I \to A$, are unitary.

Graphical Language

The graphical language of dagger monoidal categories is like the graphical language of monoidal categories, with the adjoint of a diagram given by its mirror image. For example,

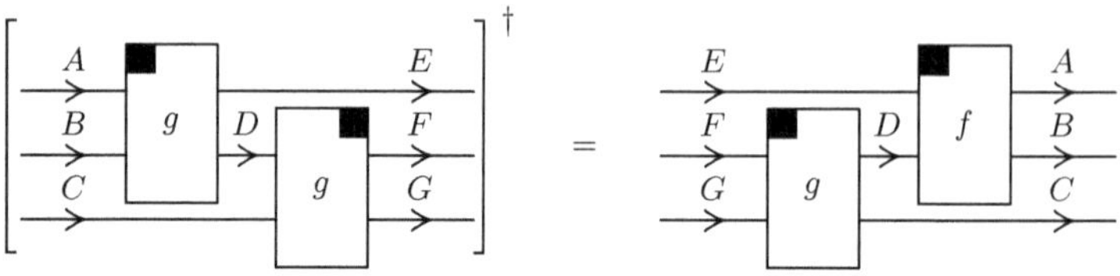

Theorem 25 (Coherence for planar dagger monoidal categories) *A well-formed equation between morphism terms in the language of dagger monoidal categories follows from the axioms of dagger monoidal categories if and only if it holds, up to planar isotopy, in the graphical language.*

Proof This is a consequence of coherence for planar monoidal categories, from Theorem 3. The proof is analogous to that of Theorem 24. Note that the axioms of dagger monoidal categories are precisely what is needed to ensure that all occurrences of † can be removed from a morphism term, except where applied directly to a morphism variable. □

4.7.2 Other Progressive Dagger Monoidal Notions

We can now "daggerize" the other progressive monoidal notions from Sect. 4.3:

Definition 52
- *A dagger monoidal category is* spacial *if it is spacial as a monoidal category.*
- *A* dagger braided monoidal category *is a dagger monoidal category with a unitary braiding* $c_{A,B} : A \otimes B \to B \otimes A$.
- *A* dagger balanced monoidal category *is a dagger braided monoidal category with a unitary twist* $\theta_A : A \to A$.
- *A* dagger symmetric monoidal category *[33] is a dagger braided monoidal category such that the unitary braiding is a symmetry.*

Graphical Languages

In each case, the graphical language extends the corresponding language from Sect. 4.3, with the dagger of a diagram taken to be its mirror image. Each notion has a coherence theorem, proved by the same method as Theorems 24 and 25. The requirements that the braiding and twist are unitary ensures that the dagger can be removed from the corresponding terms. The respective caveats from Sect. 4.3 also apply to the dagger cases.

Example 9 The category **Hilb** of Hilbert spaces is dagger symmetric monoidal, with the usual tensor product and symmetry.

4.7.3 Dagger Pivotal Categories

In defining dagger variants of the notions of Sect. 4.4, we find that the notion of a dagger autonomous category and a dagger pivotal category coincide. This is because the presence of a dagger structure on an autonomous category already induces a canonical isomorphism $A \cong A^{**}$, which automatically satisfies the pivotal axioms under mild assumptions.

To be more precise, consider a dagger monoidal category that is also right autonomous (as a monoidal category). Because $\eta_A : I \to A^* \otimes A$ has an adjoint $\eta_A^\dagger : A^* \otimes A \to I$, we can define a family of isomorphisms

$$i_A = A \xrightarrow{\cong} I \otimes A \xrightarrow{\eta_{A^*} \otimes \mathrm{id}_A} A^{**} \otimes A^* \otimes A \xrightarrow{\mathrm{id}_{A^{**}} \otimes \eta_A^\dagger} A^{**} \otimes I \xrightarrow{\cong} A^{**}.$$

We can represent this schematically as follows (but bearing in mind that we do not yet have a formal graphical language to work with):

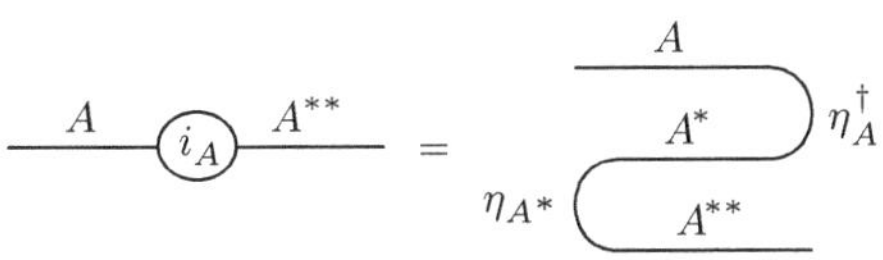

(4.1)

Lemma 7 *The following are equivalent in a right autonomous, dagger monoidal category:*

- *the family of isomorphisms $i_A : A \to A^{**}$, as defined above, determines a pivotal structure;*
- *for all A, B, the canonical isomorphisms $(A \otimes B)^* \cong B^* \otimes A^*$ and $I^* \cong I$ (determined by the right autonomous structure) are unitary, and for all $f : A \to B$, the equation $f^{*\dagger} = f^{\dagger *}$ holds.*

Proof By a direct calculation from the definitions, one can check three separate and independent facts:

- For any given $f : A \to B$, the diagram

$$\begin{array}{ccc} A & \xrightarrow{i_A} & A^{**} \\ {\scriptstyle f}\downarrow & & \downarrow{\scriptstyle f^{**}} \\ B & \xrightarrow{i_B} & B^{**} \end{array}$$

  commutes if and only if $f^{*\dagger} = f^{\dagger *}$. In particular, the family i_A is a natural transformation if and only if this condition holds for all f.
- The diagram from (4.3),

$$\begin{array}{ccc} & A \otimes B & \\ {\scriptstyle i_A \otimes i_B}\swarrow & & \searrow{\scriptstyle i_{A\otimes B}} \\ A^{**} \otimes B^{**} & \xrightarrow{\cong} & (A \otimes B)^{**} \end{array}$$

  commutes if and only if the canonical isomorphism $(A \otimes B)^* \cong B^* \otimes A^*$ is unitary.
- The morphism $i_I : I \to I^{**}$ is equal to the canonical isomorphism (from the right autonomous structure) if and only if the canonical isomorphism $I \to I^*$ is unitary.

Since the three conditions are the defining conditions for a pivotal structure, the lemma follows. □

Lemma 8 *Under the equivalent conditions of Lemma 7, the following hold:*

(a) i_A *is unitary.*

(b) $i_A = A \xrightarrow{\cong} A \otimes I \xrightarrow{id_A \otimes \epsilon^\dagger_{A^*}} A \otimes A^* \otimes A^{**} \xrightarrow{\epsilon_A \otimes id_{A^{**}}} I \otimes A^{**} \xrightarrow{\cong} A^{**}$:

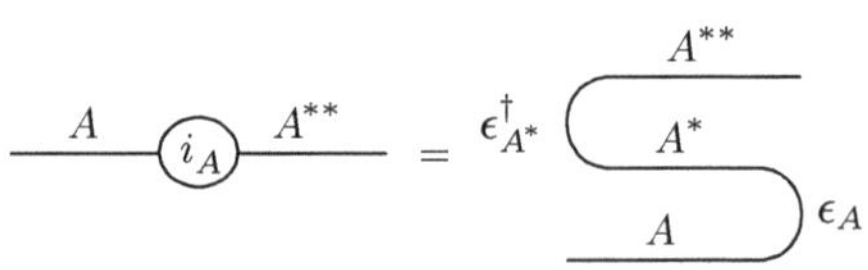

(c) $\eta^\dagger_A = \epsilon_{A^*} \circ (id_{A^*} \otimes i_A)$:

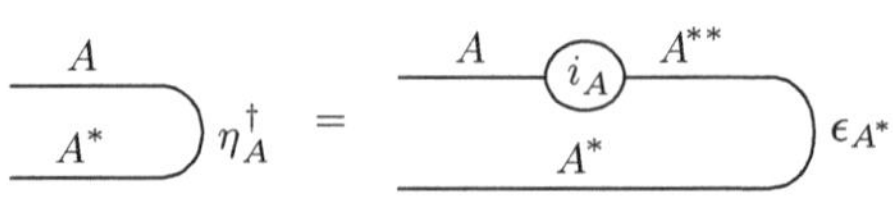

(d) $\epsilon_A^\dagger = (i_A^{-1} \otimes id_{A^*}) \circ \eta_{A^*}$:

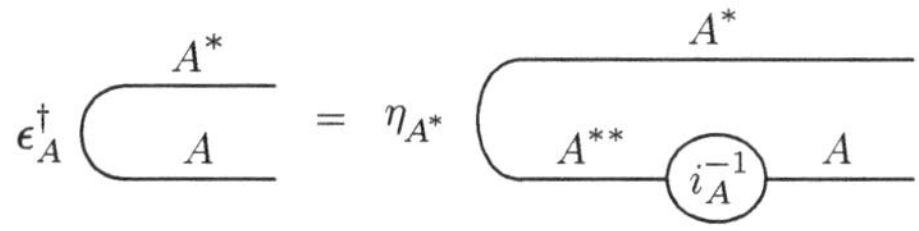

Proof To prove (a), first consider

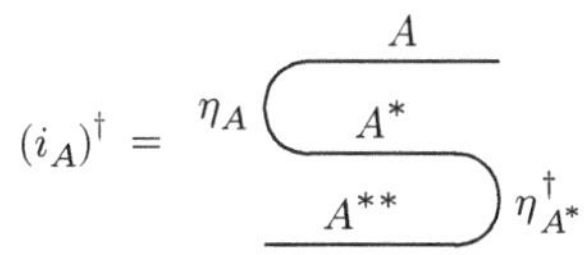

By definition of adjoint mates, we have

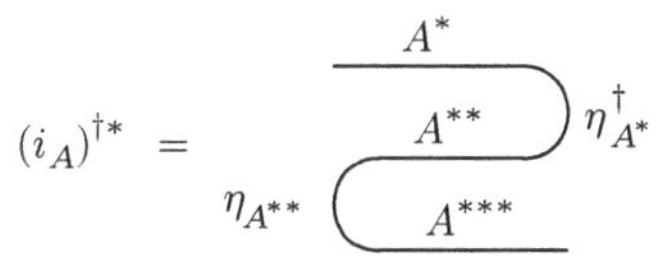

But this is just the definition of i_{A^*}, therefore $(i_A)^{\dagger *} = i_{A^*}$. By definition, i_A is unitary iff $(i_A)^\dagger = i_A^{-1}$, iff $(i_A)^{\dagger *} = (i_A^{-1})^*$, iff $i_{A^*} = (i_A^{-1})^* = (i_A^*)^{-1}$. Since i is a monoidal natural transformation, this holds by Saavedra Rivano's Lemma (Lemma 1).

To prove (b), note that the right-hand side is the inverse of $(i_A)^\dagger$. Therefore, (b) is equivalent to (a).

Finally, equations (c) and (d) are restatements of the definition of i_A from (4.1). □

Remark 28 The equivalence between (a) and (b) in Lemma 8 holds only if i_A is defined as in (4.1). It does not hold for an arbitrary pivotal structure on a right autonomous dagger monoidal category.

Armed with these results, we finally state the two equivalent definitions of a dagger pivotal category:

Definition 53 A *dagger pivotal category* is defined in one of the following equivalent ways:

1. as a dagger monoidal, right autonomous category such that the natural isomorphisms $(A \otimes B)^* \cong B^* \otimes A^*$ and $I^* \cong I$ (from the right autonomous structure) are unitary, and such that $f^{*\dagger} = f^{\dagger *}$ holds for all morphisms f; or
2. as a pivotal, dagger monoidal category satisfying the condition in Lemma 8(c) (or equivalently, (d)).

The first form of this definition is much easier to check in practice. The second form is more suitable for the proof of the coherence theorem below.

Remark 29 In a dagger pivotal category, the operation $(-)^*$ arises from an adjunction (in the categorical sense) of *objects*, with associated unit, counit, and adjoint mates. On the other hand, the operation $(-)^\dagger$ arises from an adjunction (in the linear algebra sense) of *morphisms*. The two concepts should not be confused with each other.

Graphical Language

The graphical language of dagger pivotal categories is like that of pivotal categories, where the adjoint of a diagram is given, as usual, by its mirror image. For example:

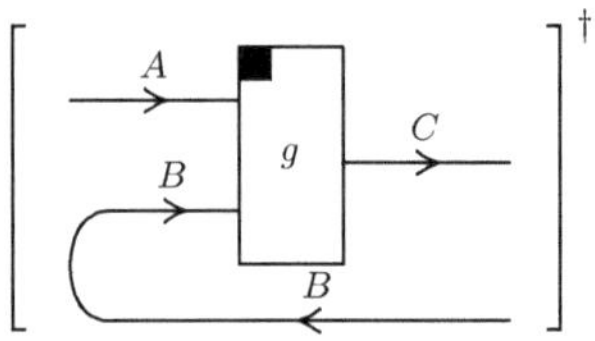

Note that in the graphical language, adjoint mates $f^* : B^* \to A^*$ are represented by rotation and adjoints $f^\dagger : B \to A$ by mirror image. Therefore, each morphism variable $f : A \to B$ induces four kinds of boxes:

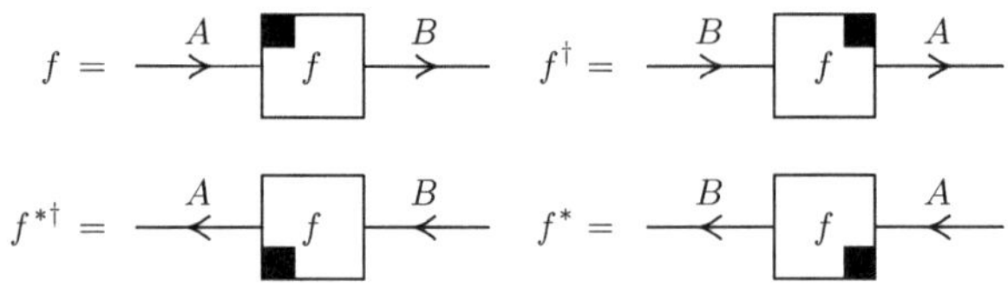

Also note that, unlike the informal notation used above, the graphical language does not explicitly display the isomorphism $i_A : A \to A^{**}$, and it does not explicitly distinguish $\eta_A : I \to A^* \otimes A$ from $\epsilon^\dagger_{A^*} : I \to A^* \otimes A^{**}$. This is justified by the following coherence theorem.

Theorem 26 (Coherence for dagger pivotal categories) *A well-formed equation between morphisms in the language of dagger pivotal categories follows from the axioms of dagger pivotal categories if and only if it holds in the graphical language up to planar isotopy, including rotation of boxes (by multiples of 180 degrees).*

Proof This follows from coherence of pivotal categories (Theorem 10), by the same argument used in the proof of Theorem 25. The equations from Lemma 8(c) and (d), and the fact that i_A is unitary, can be used to replace $\eta^\dagger_A$, $\epsilon^\dagger_A$, and $i^\dagger_A$ by equivalent terms not containing †. □

4.7.4 Other Dagger Pivotal Notions

It is possible to define dagger variants of the remaining pivotal notions from Sect. 4.4:

Definition 54 A dagger pivotal category is *spherical* (respectively *spacial*) if it is spherical (respectively spacial) as a pivotal category.

Definition 55 A *dagger braided pivotal category* is a dagger pivotal category with a unitary braiding $c_{A,B} : A \otimes B \to B \otimes A$.

Remark 30 Like any braided pivotal category, a dagger braided pivotal category is balanced by Lemma 4. However, in general the resulting twist $\theta_A : A \to A$ is not unitary. In fact, θ_A is unitary in this situation if and only if $\theta_{A^*} = (\theta_A)^*$, i.e., if and only if the category is tortile.

Definition 56 A *dagger tortile category* is defined in one of the following equivalent ways:

1. as a dagger braided pivotal category in which the canonical twist θ_A, defined as in Lemma 4, is unitary;
2. as a tortile, dagger monoidal category such that the braiding is unitary, and such that ϵ_A and η_A satisfy the (equivalent) conditions of Lemma 8(c) and (d); or
3. as a dagger balanced monoidal category that is right autonomous and satisfies

$\theta_A =$ 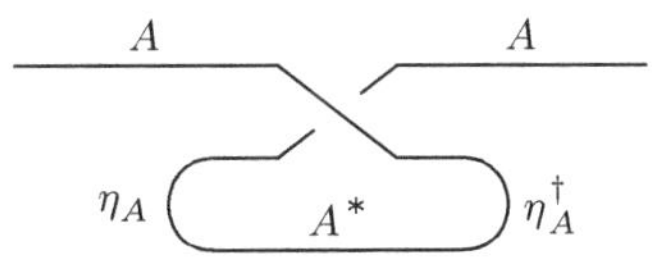

(4.2)

The first form of this definition emphasizes the relationship to dagger pivotal categories. The second form is easiest to check if a category is already known to be tortile. Finally, the third form takes ϵ_A, η_A, $c_{A,B}$ and θ_A as primitive operations and does not mention the pivotal structure i_A at all. The pivotal structure, in this case, is definable from (4.7) or (4.1), with the condition (4.2) ensuring that the two definitions coincide.

Definition 57 [1, 33] A *dagger compact closed category* is a dagger tortile category such that $\theta_A = \mathrm{id}_A$ for all A. Equivalently, it is a dagger symmetric monoidal category that is right autonomous and satisfies

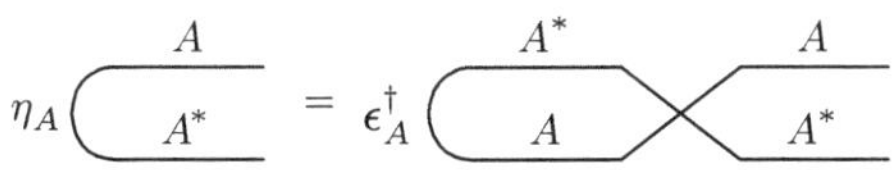

(4.3)

The equivalence of the two definition is immediate from the third form of the definition of dagger tortile categories. Note that (4.2) is equivalent to (4.3) in the symmetric case. Further, these conditions are equivalent to the condition in Lemma 8(d).

Example 10 The category **FdHilb** of finite dimensional Hilbert spaces is dagger compact closed, with A^* the usual dual space of linear functions from A to I, and with $f^\dagger$ the usual linear algebra adjoint.

Graphical Languages

Each of the notions defined in this section (except the spherical notion) has a graphical language, extending the corresponding graphical language from Sect. 4.4, with the dagger of a diagram taken to be its mirror image. Each notion has a coherence theorem, proved by the same method as Theorems 24 and 25. As expected, equivalence of diagrams is up to isomorphism (for spacial dagger pivotal categories); up to regular isotopy (for dagger braided pivotal categories); up to framed 3-dimensional isotopy (for dagger tortile categories); and up to isomorphism (for dagger compact closed categories).

4.7.5 Dagger Traced Categories

There is no difficulty in defining dagger variants of each of the traced notions of Sect. 4.5. A (left or right) trace on a dagger monoidal category is called a *dagger trace* if it satisfies

$$(\mathrm{Tr}\, f)^\dagger = \mathrm{Tr}(f^\dagger) \tag{4.4}$$

For example: a *dagger right traced category* is a right traced dagger monoidal category satisfying (4.4). A balanced traced category is *dagger balanced traced* if it is dagger balanced and satisfies (4.4). And similarly for the other notions. The representation theorems of Sect. 4.5 extend to these dagger variants:

Theorem 27 (Representation of dagger braided/balanced/symmetric traced categories) *Every dagger braided [balanced, symmetric] traced category can be fully and faithfully embedded in a dagger braided pivotal [dagger tortile, dagger compact closed] category, via a dagger braided [balanced, symmetric] traced functor.* □

The proof, in each case, is by Joyal, Street, and Verity's Int-construction [24], which respects the dagger structure.

Graphical Languages

The graphical language of each class of traced categories extends to the corresponding dagger traced categories, in a way suggested by Eq. (4.4). As usual, the dagger of a diagram is its mirror image, thus for example

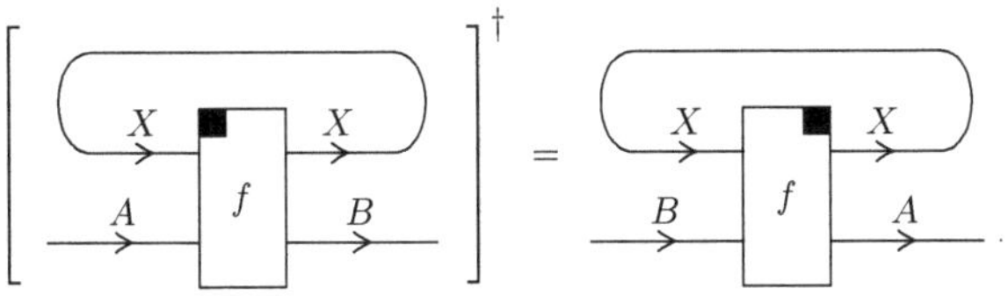

The coherence theorems of Sect. 4.5 extend to this setting.

4.7.6 Dagger Biproducts

In a dagger category, if $A \oplus B$ is a categorical product (with projections $\pi_1 : A \oplus B \to A$ and $\pi_2 : A \oplus B \to B$), then it is automatically a coproduct (with injections $\pi_1^\dagger : A \to A \oplus B$ and $\pi_2^\dagger : B \to A \oplus B$). It therefore makes sense to define a notion of *dagger biproduct*.

Definition 58 A *dagger biproduct category* is a biproduct category carrying a dagger structure, such that $\pi_i^\dagger = \iota_i : A_i \to A_1 \oplus A_2$ for $i = 1, 2$.

As in Sect. 4.6.3, we can equivalently define a dagger biproduct category as a dagger symmetric monoidal category, together with natural families of morphisms

$$\Delta_A : A \to A \otimes A, \qquad \Diamond_A : A \to I, \qquad \nabla_A : A \otimes A \to A, \qquad \Box_A : I \to A,$$

such that $\Delta_A^\dagger = \nabla_A$ and $\Diamond_A^\dagger = \Box_A$, satisfying the axioms in Table 4.7.

Graphical Language

The graphical language of dagger biproduct categories is like that of biproduct categories, where the dagger of a diagram is taken to be its mirror image. For example,

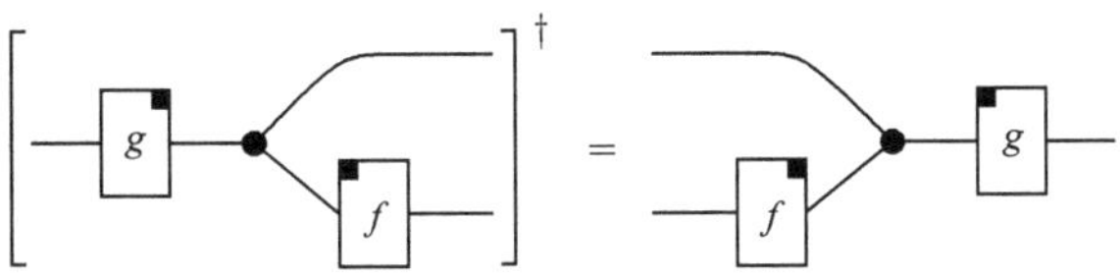

Theorem 28 (Coherence for dagger biproduct categories) *A well-formed equation between morphism terms in the language of dagger biproduct categories follows from the axioms of dagger biproduct categories if and only if it holds in the graphical language, up to isomorphism of diagrams, the diagrammatic manipulations shown in Table 4.8, and their duals.*

Proof By reduction to biproduct categories, as in the proofs of Theorems 24 and 25. The axioms $\Delta_A^\dagger = \nabla_A$ and $\Diamond_A^\dagger = \Box_A$ allow † to be removed from anywhere but a morphism variable. □

Finally, there is an obvious notion of *dagger traced biproduct category* (which is really a dagger traced dagger biproduct category), with graphical language derived from traced biproduct categories.

4.8 Bicategories

A bicategory [6] is a generalization of a monoidal category. In addition to objects $A, B, \ldots$ and morphisms $f, g, \ldots$, one now also considers *0-cells* $\alpha, \beta, \ldots$, which we can visualize as *colors*. For example, consider the following diagram. It is a standard diagram for monoidal categories, except that the areas between the wires have been colored.

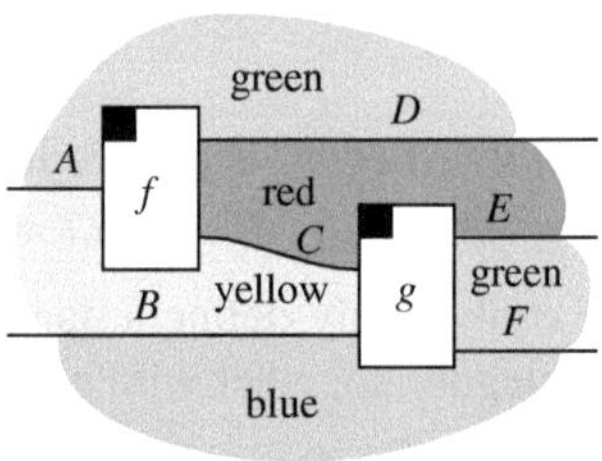

As usual, we have objects A, B, C, D, E, F and morphisms $f : A \to C \otimes D$ and $g : B \otimes C \to F \otimes E$. But now there are also 0-cells called green, red, yellow, and blue. In such diagrams, each object has a *source*, which is the 0-cell just above it, and a *target*, which is the 0-cell just below it. For example, we have A : green $\to$ yellow, B : yellow $\to$ blue, and so on. It is now clear that, to be consistently colored, such diagrams have to satisfy some coloring constraints. The constraints are:

- The tensor $B \otimes A$ of two objects may only be formed if the target of A is equal to the source of B. In symbols, for any 0-cells α, β, γ, if $A : \alpha \to \beta$ and $B : \beta \to \gamma$, then $B \otimes A : \alpha \to \gamma$.
- If $f : A \to B$ is a morphism, then A and B must have a common source and a common target. In symbols, if $f : A \to B$ and $A : \alpha \to \beta$, then $B : \alpha \to \beta$.
- One also requires a unit object $I_\alpha : \alpha \to \alpha$ for every color α.

As an illustration of the second property, consider $f : A \to C \otimes D$ in the above example, where A : green $\to$ yellow and $C \otimes D$: green $\to$ yellow. Subject to the above coloring constraints, a bicategory is then required to satisfy exactly the same axioms as a monoidal category. Notice, for example, that if $f : A \to B$ and $g : B \to C$ and f, g are well-colored, then so is $g \circ f : A \to C$. Also, the identity maps $\mathrm{id}_A : A \to A$, the associativity map $\alpha_{A,B,C} : (A \otimes B) \otimes C \xrightarrow{\cong} A \otimes (B \otimes C)$, and the other structural maps are well-colored. In particular, a monoidal category is the same thing as a one-object bicategory.

To give a detailed account of bicategories and their graphical languages is beyond the scope of this paper. We have already discussed over 30 different flavors of monoidal categories, and the reader can well imagine how many possible variations of bicategories there are, with 2-, 3-, and 4-dimensional graphical languages, once one considers bicategorical versions of braids, twists, adjoints, and traces. There are even more variations if one considers tricategories and beyond. We refer the reader to [6] for the definition and basic properties of bicategories, and to [41], [2, Sect. 7] for a taste of their graphical languages.

4.9 Beyond a Single Tensor Product

All the categorical notions that we have considered in this paper have just a single tensor product, which we represented as juxtaposition in the graphical languages. For notions of categories with more than one tensor product, the graphical languages get much more complicated. The details are beyond the scope of this paper, so we just outline the basics and give some references.

Examples of categories with more than one tensor are linearly distributive categories [13] and *-autonomous categories [4]. Both of these notions are models of multiplicative linear logic [17]. These categories have two tensors, often called "tensor" and "par", and written

$$A \otimes B \qquad \text{and} \qquad A \parr B.$$

The two tensors are related by some morphisms, such as $A \otimes (B \parr C) \to (A \otimes B) \parr C$, while other similar morphisms, such as $(A \otimes B) \parr C \to A \otimes (B \parr C)$, are not present.

To make a graphical language for more than one tensor product, one must label the wires by morphism terms, rather than morphism variables. One must also introduce special tensor and par nodes as shown here:

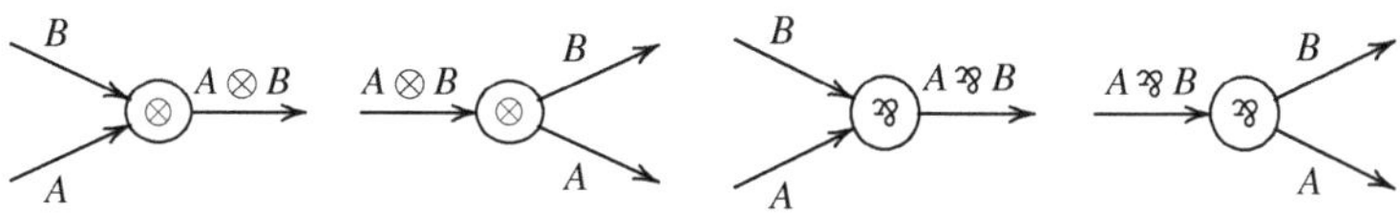

along with similar nodes for the units. Equivalence of diagrams must be taken up to axiomatic manipulations, such as the following, which is called *cut elimination* in logic:

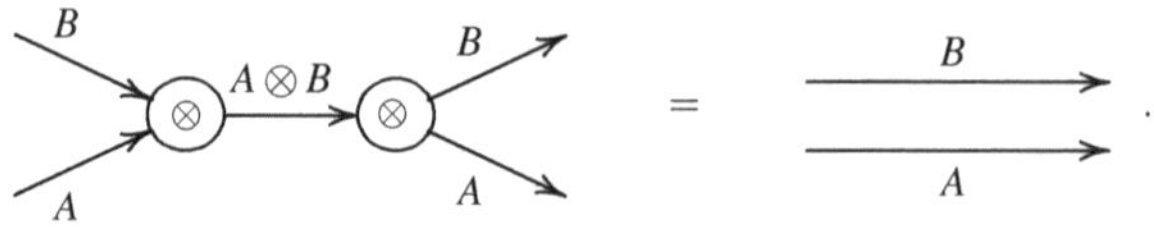

Finally, one must state a *correctness criterion*, to explain why certain diagrams, such as the left one following, are well-formed, while others, such as the right one, are not well-formed.

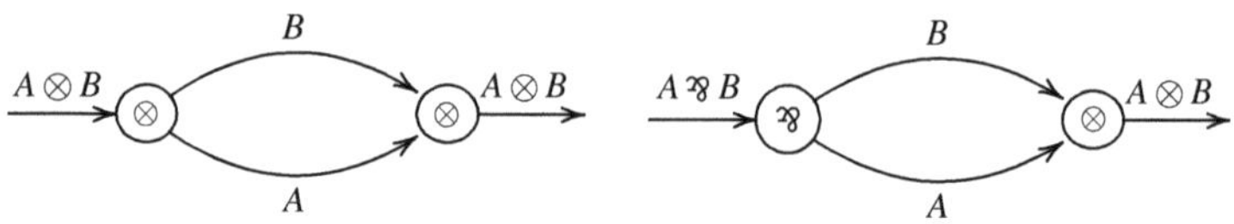

The resulting theory is called the theory of *proof nets*, and was first given by Girard for unit-free multiplicative linear logic [17]. It was later extended to include the tensor units by Blute et al. [8].

4.10 Summary

Table 4.10 summarizes the graphical languages from Sects. 4.2, 4.3, 4.4, 4.5 and 4.6. The name of each class of categories is shown along with a typical diagram or equation. The arrows indicate forgetful functors. We have omitted spherical categories, because they do not possess a graphical language modulo a natural notion of isotopy.

The letter d indicates the dimension of the diagrams, and the letter i indicates the dimension of the ambient space for isotopy. If $i > d$, then isotopy coincides with isomorphism of diagrams. Special cases are "3f" for framed diagrams and framed isotopy in 3 dimensions; "2+" for two-dimensional diagram with crossings (i.e., isotopy is taken on 2-dimensional projections, rather than on 3-dimensional diagrams); "reg" for regular isotopy; and "rot" to indicate that isotopy includes rotation of boxes. Finally, "eqn" indicates that equivalence of diagrams is taken modulo equational axioms.

The letter c indicates the status of a coherence theorem. This is usually a reference to a proof of the theorem, or "conj" if the result is conjectured. A checkmark "√" indicates a result that is folklore or whose proof is trivial. "int" indicates that the coherence theorem follows from a version of Joyal, Street, and Verity's Int-construction, and the corresponding coherence theorem for pivotal categories. An asterisk "∗" indicates that the result has only been proved for simple signatures.

Dagger variants can be defined of all of the notions shown in Table 4.10, except the planar autonomous and braided autonomous notions. Finally, bicategories require their own (presumably much larger) table and are not included here.

Table 4.10 Summary of monoidal notions and their graphical languages

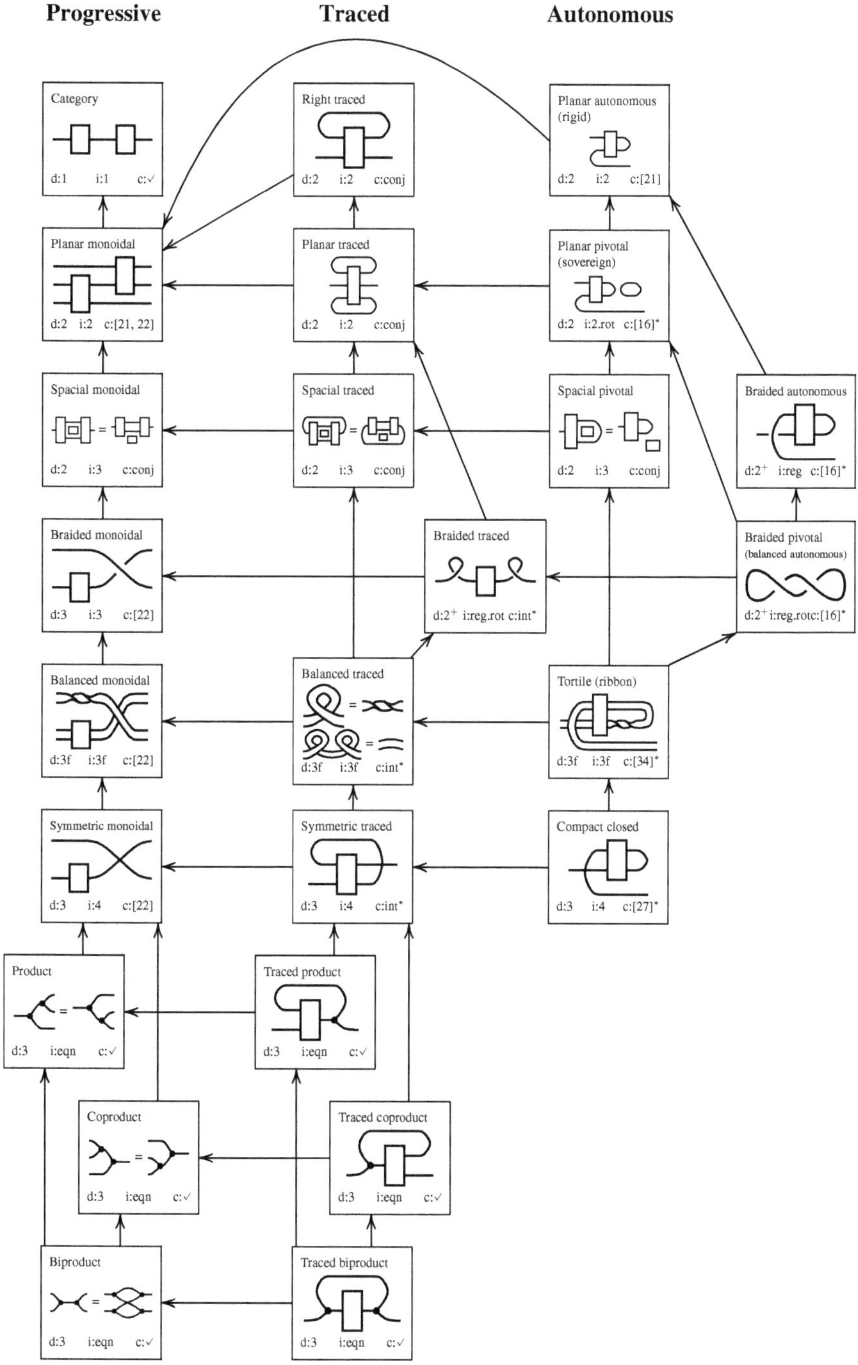

References

1. Abramsky, S., Coecke, B.: A categorical semantics of quantum protocols. In: Proceedings of the 19th Annual IEEE Symposium on Logic in Computer Science, LICS 2004, pp. 415–425. IEEE Computer Society Press, Washington, DC (2004)
2. Baez, J.C., Dolan, J.: Higher-dimensional algebra and topological quantum field theory. J. Math. Phys. **36**(11), 6073–6105 (1995)
3. Bainbridge, E.S.: Feedback and generalized logic. Inf. Control **31**, 75–96 (1976)
4. Barr, M.: *-Autonomous Categories. Lectures Notes in Mathematics 752. Springer, New York (1979)
5. Barrett, J.W., Westbury, B.W.: Spherical categories. Adv. Math. **143**, 357–375 (1999)
6. Bénabou, J.: Introduction to bicategories, part I. In: Reports of the Midwest Category Seminar, Lecture Notes in Mathematics 47, pp. 1–77. Springer, New York (1967)
7. Bloom, S.L., Esik, Z.: Axiomatizing schemes and their behaviors. J. Comput. Syst. Sci. **31**, 375–393 (1985)
8. Blute, R.F., Cockett, J.R.B., Seely, R.A.G., Trimble, T.H.: Natural deduction and coherence for weakly distributive categories. J. Pure Appl. Algebra **113**(3), 229–296 (1996)
9. Căzănescu, V.-E.: On context-free trees. Theor. Comput. Sci. **41**, 33–50 (1985)
10. Căzănescu, V.-E., Stefănescu, G.: Towards a new algebraic foundation of flowchart scheme theory. Fundamenta Informaticae **13**, 171–210 (1990). Also appeared as: INCREST Preprint Series in Mathematics 43, Bucharest (1987)
11. Căzănescu, V.-E., Stefănescu, G.: Feedback, iteration and repetition. In: Păun, G. (ed.) Mathematical Aspects of Natural and Formal Languages, pp. 43–62. World Scientific, Singapore (1995). Also appeared as: INCREST Preprint Series in Mathematics 42, Bucharest (1988)
12. Căzănescu. V.-E., Ungureanu, C.: Again on advice on structuring compilers and proving them correct. Preprint Series in Mathematics 75, INCREST, Bucharest (1982)
13. Cockett, J.R.B., Seely, R.A.G.: Weakly distributive categories. Pure Appl. Algebra **114**(2), 133–173 (1997)
14. Esik, Z.: Identities in iterative and rational algebraic theories. Comput. Linguistics Comput. Lang. **XIV**, 183–207 (1980)
15. Freyd, P.J., Yetter, D.N.: Braided compact closed categories with applications to low dimensional topology. Adv. Math. **77**, 156–182 (1989)
16. Freyd, P.J., Yetter, D.N.: Coherence theorems via knot theory. J. Pure Appl. Algebra **78**, 49–76 (1992)
17. Girard, J.-Y.: Linear logic. Theor. Comput. Sci. **50**, 1–102 (1987)
18. Hasegawa, M.: Models of Sharing Graphs: A Categorical Semantics of let and letrec. PhD thesis, Department of Computer Science, University of Edinburgh (July 1997)
19. Joyal, A., Street, R.: The geometry of tensor calculus II. Unpublished draft, available from Ross Street's website
20. Joyal, A., Street, R.: Braided monoidal categories. Mathematics Report 860081, Macquarie University (November 1986)
21. Joyal, A., Street, R.: Planar diagrams and tensor algebra. Unpublished manuscript, available from Ross Street's website (September 1988)
22. Joyal, A., Street, R.: The geometry of tensor calculus I. Adv. Math. **88**(1), 55–112 (1991)
23. Joyal, A., Street, R.: Braided tensor categories. Adv. Math. **102**, 20–78 (1993)
24. Joyal, A., Street, R., Verity, D.: Traced monoidal categories. Math. Proc. Camb. Philoso. Soc. **119**, 447–468 (1996)
25. Kauffman, L.H.: An invariant of regular isotopy. Trans. Am. Math. Soc. **318**(2), 417–471 (1990)
26. Kelly, G.M.: An abstract approach to coherence. In: Mac Lane, S. (ed.) Coherence in Categories, Lecture Notes in Mathematics 281, pp. 106–147. Springer, New York (1972)
27. Kelly, G.M., Laplaza, M.L.: Coherence for compact closed categories. J. Pure Appl. Algebra **19**, 193–213 (1980)

28. Mac Lane, S.: Natural associativity and commutativity. Rice Uni. Stud. **49**, 28–46 (1963)
29. Mac Lane, S.: Categories for the Working Mathematician. Graduate Texts in Mathematics 5. Springer, New York (1971)
30. Penrose, R.: Applications of negative dimensional tensors. In: Welsh, D.J.A. (ed.) Combinatorial Mathematics and Its Applications, pp. 221–244. Academic Press, New York (1971)
31. Reidemeister, K.: Knotentheorie. Springer, Berlin (1932); Chelsea, New York (1948). English translation: Knot Theory, BCS Associates (1983)
32. Saavedra Rivano, N.: Categories Tanakiennes. Lecture Notes in Mathematics 265. Springer, New York (1972)
33. Selinger, P.: Dagger compact closed categories and completely positive maps. In: Proceedings of the 3rd International Workshop on Quantum Programming Languages, Electronic Notes in Theoretical Computer Science 170, pp. 139–163. Elsevier Science, Amsterdam (2007)
34. Shum, M.C.: Tortile tensor categories. J. Pure Appl. Algebra **93**, 57–110 (1994)
35. Stefănescu, G.: An algebraic theory of flowchart schemes. In: Proceedings of the 11th Colloquium on Trees in Algebra and Programming, CAAP'86, Lecture Notes in Computer Science 214, pp. 60–73. Springer, New York (1986)
36. Stefănescu, G.: On flowchart theories, part I. the deterministic case. J. Comput. Syst. Sci. **35**(2):163–191 (1987)
37. Stefănescu, G.: On flowchart theories: Part II. The nondeterministic case. TCS **52**, 307–340 (1987)
38. Stefănescu, G.: Feedback theories (a calculus for isomorphism classes of flowchart schemes). Revue Roumaine de Mathématiques Pures et Appliquées **35**, 73–79 (1990). Also appeared as: INCREST Preprint Series in Mathematics 24, Bucharest (1986)
39. Stefănescu, G.: Algebra of flownomials. Technical Report TUM-I9437, Technische Univerität München (1994)
40. Stefănescu, G.: Network Algebra. Springer, New York (2000)
41. Street, R.: Low-dimensional topology and higher-order categories. In: Proceedings of the International Category Theory Conference (CT'95), 1995. Available from http://www.mta.ca/_cat-dist/ct95.html
42. Turaev, V.G.: Quantum Invariants of Knots and 3-Manifolds. Studies in Mathematics 18. Walter De Gruyter & Co., Berlin (1994)
43. Yetter, D.N.: Framed tangles and a theorem of Deligne on braided deformations of tannakian categories. In: Gerstenhaber, M., Stasheff, J.D. (eds.) Deformation Theory and Quantum Groups with Applications to Mathematical Physics, Contemporary Mathematics 134, pp. 325–349. Americal Mathematical Society, Providence (1992)

Chapter 5
Geometry of Interaction and the Dynamics of Proof Reduction: A Tutorial

E. Haghverdi and P. Scott

Abstract Girard's Geometry of Interaction (GoI) is a program that aims at giving mathematical models of algorithms independently of any extant languages. In the context of proof theory, where one views algorithms as proofs and computation as cut-elimination, this program translates to providing a mathematical modelling of the dynamics of cut-elimination. The kind of logics we deal with, such as Girard's linear logic, are resource sensitive and have their proof-theory intimately related to various monoidal (tensor) categories. The GoI interpretation of dynamics aims to develop an algebraic/geometric theory of invariants for information flow in networks of proofs, via feedback.

We shall give an introduction to the categorical approach to GoI, including background material on proof theory, categorical logic, traced and partially traced monoidal $*$-categories, and orthogonalities.

5.1 Introduction

In the 1930s, Gerhard Gentzen developed a profound approach to Hilbert's proof theory, in which formal laws for deriving logical entailments $\Gamma \vdash \Delta$ (i.e. premisses Γ entail conclusions Δ) were carefully systematized, breaking the laws of logic into three groups: (i) the Axiom and Cut-Rule, (ii) Structural Rules, and (iii) Logical Rules. Gentzen's work revealed the hidden symmetries in logical syntax, and

E. Haghverdi (✉)
School of Informatics, Indiana University, Bloomington, IN, USA
e-mail: ehaghver@indiana.edu

P. Scott (✉)
Department of Mathematics and Statistics, University of Ottawa, Ottawa ON, Canada
e-mail: phil@mathstat.uottawa.ca

P. Scott Research partially supported by a Discovery Grant from NSERC, Canada.

Haghverdi, E., Scott, P.: *Geometry of Interaction and the Dynamics of Proof Reduction: A Tutorial*. Lect. Notes Phys. **813**, 357–417 (2011)
DOI 10.1007/978-3-642-12821-9_5

his remarkable Cut-Elimination Theorem, one of the deepest in logic, has also had considerable significance for theoretical computer science.

In these lectures we shall give the background, both logical and categorical, to a remarkable new approach to Gentzen's work, stemming from J-Y Girard's introduction of Linear Logic in 1987 [Gi87]. Linear Logic, a radical analysis of the Gentzen rules of traditional logic, is based upon studying the use of resources in these rules, e.g. in duplicating and eliminating premisses and conclusions in a logical inference. We may think of proofs as *dynamical systems*, with inputs and outputs being the hypotheses and the conclusions respectively, and we think of the rules involved in transforming, i.e. in rewriting, proof trees (in Gentzen's Cut-Elimination Algorithm) as *interaction* between these dynamical systems. We are looking for mathematical invariants for the dynamics of these systems.

Girard's *Geometry of Interaction* (GoI) project began in the late 1980s [Gi89, Gi89a]. The first paper on GoI was set in an operator algebraic context: proofs were interpreted as operators on the Hilbert space of square summable sequences. The GoI interpretation of cut-elimination was given by a finite sum, which was finite due to nilpotency of the summands. This already pointed to the usefulness of the GoI view of logic: one has a degree of nilpotency that measures the complexity of cut-elimination (=computation). This also inspired a different line of work in GoI research, the so called path-semantics with relationships to lambda calculus, a fundamental model of computation [DR95].

One might ask: why is this important? The answer lies in realizing that one way to model computation is precisely as an instance of Gentzen's algorithm. We search for mathematical models of this dynamical process of cut-elimination, expecting that such an analysis will shed deep light on the very nature of computation and its complexity. Indeed, there are connections of the whole project with complexity, as we mention in Remark 5.5.2 in these notes.

The early work on understanding the categorical framework of GoI was begun in lectures of Abramsky and of Hyland in the early 1990s. This brought the notion of abstract trace (in the sense of Joyal, Street, and Verity [JSV96]) into the picture. Work by Hyland, by Abramsky [Abr96] and later by us [AHS02, HS04a] has emphasized the role of abstract traces in modelling cut-elimination in GoI. Our categorical modelling of GoI has recently led us to the use of $*$-categories (see Sect. 5.8), already familiar to theoretical physicists in the work by Doplicher, Roberts and others. This approach to GoI offers a potential connection to the literature in several areas of interest in mathematics and physics, for example to knot theory, where trace appears under the name *braid closure* (cf. [Abr07]). The most recent work by Girard [Gi08], makes use of type II_1 von Neumann algebras to offer a new interpretation of GoI, although the categorical meaning is totally open. It is our strong hope and belief that the categorical and logical structures outlined in these notes will be conducive to non-trivial and productive connections with applications to physics.

5.2 From Monoidal Categories to *-Autonomy

5.2.1 Monoidal Categories

Monoidal (tensor) categories are a fundamental mathematical structure arising in many areas of mathematics, theoretical computer science and physics, and increasingly in mathematical logic. The subject is a vast one, so we will just include definitions and examples relevant to these lectures. For general background, the reader is referred to standard category theory texts [Bor93, Mac98]. For general surveys of monoidal categories in relation to categorical and linear logics, see the articles [Sc00, BS04, Mel07] and further references given below.

Definition 5.2.1 A *monoidal (or tensor) category* $(\mathcal{C}, \otimes, I, \alpha, \ell, r)$ is a category $\mathcal{C}$, with functor $\otimes : \mathcal{C} \times \mathcal{C} \to \mathcal{C}$, unit object $I \in ob(\mathcal{C})$, and specified isomorphisms (natural in A, B, C): $\alpha_{ABC} : (A \otimes B) \otimes C \stackrel{\cong}{\longrightarrow} A \otimes (B \otimes C)$, $\ell_A : I \otimes A \stackrel{\cong}{\longrightarrow} A$, $r_A : A \otimes I \stackrel{\cong}{\longrightarrow} A$ satisfying the following equations (in diagrammatic form): $\ell_I = r_I : I \otimes I \to I$, as well as:

$$\begin{array}{ccc} (A\otimes I)\otimes C & \xrightarrow{\alpha} & A\otimes(I\otimes C) \\ \downarrow{\scriptstyle r_A\otimes id_C} & & \downarrow{\scriptstyle id_A\otimes \ell_C} \\ A\otimes C & = & A\otimes C \end{array} \qquad \begin{array}{ccccc} A(B(CD)) & \xleftarrow{\alpha} & (AB)(CD) & \xleftarrow{\alpha} & ((AB)C)D \\ \downarrow{\scriptstyle id_A\otimes\alpha} & & & & \downarrow{\scriptstyle \alpha\otimes id_D} \\ A((BC)D) & & \xleftarrow{\qquad\alpha\qquad} & & (A(BC))D \end{array}$$

where we omit $\otimes$'s and subscripts in the second diagram for typographical reasons. This latter diagram is known as the Mac Lane pentagon. It expresses an equality between the two a priori different natural isomorphisms between $((A \otimes B) \otimes C) \otimes D$ and $A \otimes (B \otimes (C \otimes D))$.

Monoidal structure is not generally unique nor canonical: there may be several (nonisomorphic) tensor structures on the same category. An interesting special case is when the isos α, ℓ, r are all identity morphisms. In that case, we say the monoidal category is *strict*.

Definition 5.2.2 A *strict monoidal category* is a category $\mathcal{C}$ with a functor $\otimes : \mathcal{C} \times \mathcal{C} \to \mathcal{C}$ and $I \in ob(\mathcal{C})$ satisfying the following equations:

- $(A \otimes B) \otimes C = A \otimes (B \otimes C)$.
- $A \otimes I = A = I \otimes A$.
- $(f \otimes g) \otimes h = f \otimes (g \otimes h)$ for any arrows f, g, h.
- $f \otimes id_I = f = id_I \otimes f$, for any arrow $f : A \to B$.

Many concrete examples of strict monoidal categories arise in knot theory, quantum groups and related areas (e.g. [KRT97]). More generally, the Mac Lane Coherence Theorem [Mac98] states that every monoidal category is equivalent to a strict one. This essentially says that in an arbitrary monoidal category $\mathcal{C}$, every "formal" diagram of arrows (from a source object to a target object) which is built from instances of the maps α, ℓ, r under the monoidal category operations automatically commutes. Thus, without loss of generality (up to equivalence) we can assume our monoidal categories are strict. Notice in a strict monoidal category, the objects form a monoid (= semigroup with unit) under $\otimes$.

From now on we write $(\mathcal{C}, \otimes, I)$ for monoidal categories, omitting the remaining structure maps α, ℓ, r when it is clear. We introduce some standard graphical notation for arrows in Fig. 5.1.

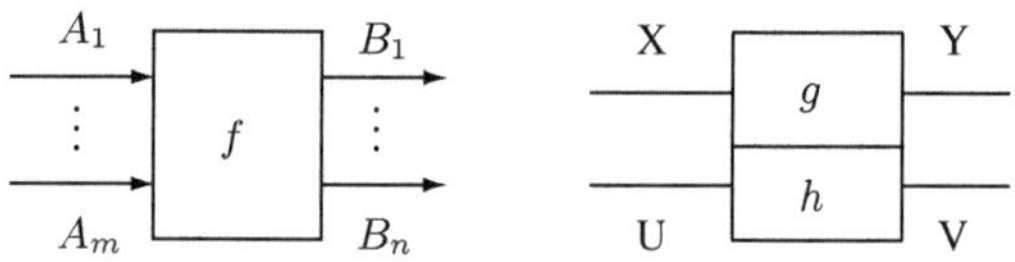

$$A_1 \otimes \cdots \otimes A_m \xrightarrow{f} B_1 \otimes \cdots \otimes B_n \qquad X \otimes U \xrightarrow{g\otimes h} Y \otimes V$$

Fig. 5.1 Pictorial representation of morphisms

In any monoidal category $(\mathcal{C}, \otimes, I)$, we can define the *monoid of scalars* to be $\mathcal{C}(I, I)$. For example, in the monoidal category $(\mathbf{Vec}, \otimes, I)$ of $\mathbf{k}$-vector spaces and linear maps, with the usual notion of algebraic tensor product, and $I = \mathbf{k}$ (the base field) observe $\mathbf{Vec}(I, I) \cong I$. The following result is from [KL80] (see also [Abr05]).

Proposition 5.2.3 (Kelly-Laplaza) *In any monoidal category, the scalars form a commutative monoid.*

There are many additional structures one may add to this basic definition. We shall introduce below such notions as *symmetric, closed, $*$-autonomous*, and *traced* structure, which are key to modelling proofs in linear logic.
Suppose first that there is a natural isomorphism $s_{AB} : A \otimes B \to B \otimes A$ (called a *braiding*) making the following three diagrams commutative.

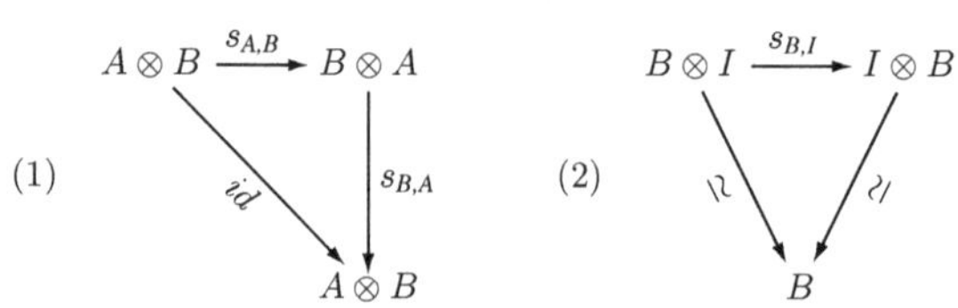

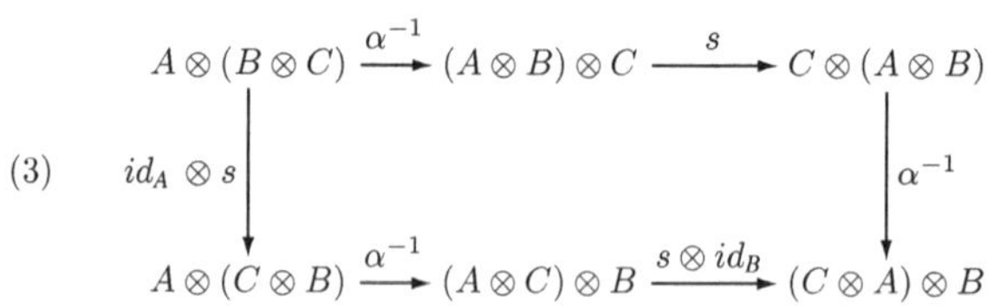

where in (3) we have omitted subscripts for typographical reasons. We say $\mathcal{C}$ is *symmetric* if Diagrams (1), (2), and (3) commute. Notice in a symmetric monoidal category, $s_{AB} = (s_{BA})^{-1}$.

More generally, a *braided* monoidal category is a monoidal category satisfying the commutativity of diagrams (2), (3), and (3'), where (3') is like (3) but replacing α^{-1} by α, $id \otimes s$ by $s \otimes id$ and appropriately relabelling the nodes. Such categories arise in knot theory and physics [JS91, JS93, KRT97] as well as in recent semantical studies in Quantum Computing [AbCo04, Abr05].

Let us give some examples that will be useful later.

Examples 5.2.4 (Symmetric Monoidal Categories)

1. Any cartesian category (=finite products), with $\otimes = \times$.
2. Any co-cartesian category (= finite coproducts), with $\otimes = +$
3. $\mathbf{Rel}_{\times}$. This is the category **Rel** whose objects are sets and whose arrows are binary relations. Recall the composition of two arrows is their relational product: given $A \xrightarrow{R} B \xrightarrow{S} C$, define $A \xrightarrow{S \circ R} C$ to be the relation defined by

$$a(S{\circ}R)c \text{ iff } \exists_{b \in B}\, aRb \wedge bSc\ .$$

The identity morphism $A \xrightarrow{id_A} A$ is simply the diagonal relation $\Delta_A = \{(a, a) \mid a \in A\}$. The functor $\otimes : \mathbf{Rel} \times \mathbf{Rel} \to \mathbf{Rel}$ is defined as follows. On objects, $\otimes = \times$, the cartesian product of sets; on arrows, $A \otimes B \xrightarrow{R \otimes S} C \otimes D$ is the relation given by: $(a, b) R \otimes S (c, d)$ iff aRc & bSd. The tensor unit $I = \{*\}$, any one element set.
4. $\mathbf{Rel}_{+}$. This is again the category **Rel**, except $\otimes = +$ (disjoint union), where disjoint union in **Set** is given by: $X + Y = X \times \{1\} \cup Y \times \{2\}$. On arrows, $A \otimes B \xrightarrow{R \otimes S} C \otimes D$ is the relation given by:

   $(x, i) R \otimes S (y, j)$ iff $[(i = j = 1 \text{ and } xRy) \text{ or } (i = j = 2 \text{ and } xSy)]$
   Here the tensor unit $I = \emptyset$.
5. Two important monoidal subcategories of $\mathbf{Rel}_{+}$ are:

   (i) **Pfn**: Sets and partial functions. Here the morphisms between sets are relations which are *functional*, i.e. binary relations $A \xrightarrow{R} B$ satisfying: $\forall_{x \in A} \forall_{y, y' \in B}\, [xRy \wedge xRy' \to y = y']$.
   (ii) **PInj**: Sets and partial injective functions. This is the subcategory of **Pfn** consisting of those partial functions which are also injective on their domains: $\forall_{x, x' \in A} \forall_{y \in B}\, [xRy \wedge x'Ry \to x = x']$.

6. $\mathbf{Vec}_{fd}$ and **Vec**: (finite dimensional) vector spaces over **k**, where **k** is a field. Here $V \otimes W$ is taken to be the usual tensor product, and $I = \mathbf{k}$.
7. The categories $\mathbf{Hilb}_{\oplus}$, $(\mathbf{Hilb}_{\oplus})_{fd}$ of Hilbert spaces (resp. finite dimensional Hilbert spaces) and bounded linear maps with the direct sum $\oplus$ as tensor. Similarly, we may consider the categories **Ban** and $\mathbf{Ban}_{fd}$ of Banach spaces (resp. finite dimensional Banach spaces) and bounded linear maps. Important

subcategories of the above include **cBan** and **cHilb**, where the maps are (nonexpansive) contractions, i.e. linear maps L satisfying $\|L(x)\| \leq \|x\|$.

8. $\mathbf{Hilb}_{\otimes}$ is the category of Hilbert spaces and bounded linear maps, with the tensor being the usual Hilbert space tensor product. There are also a variety of tensor products on Banach spaces, but we shall not require that theory.

5.2.2 Closed Structure

In order to deal with internal function spaces, we introduce the notion of *closedness*, as an adjoint functor to $\otimes$:

Definition 5.2.5 A *symmetric monoidal closed category* (smcc) $\mathcal{C}$ is a symmetric monoidal category such that for all $A \in \mathcal{C}$, the functor $- \otimes A : \mathcal{C} \to \mathcal{C}$ has a right adjoint $A \multimap -$, i.e. there is an isomorphism, natural in B, C, satisfying

$$\mathcal{C}(C \otimes A, B) \cong \mathcal{C}(C, A \multimap B) \tag{5.1}$$

We say $A \multimap B$ is the "linear exponential" or "linear function space". In particular, the isomorphism (5.1) induces *evaluation* and *coevaluation* maps $(A \multimap B) \otimes A \to B$ and $C \to (A \multimap (C \otimes A))$, satisfying the adjoint equations.

Examples 5.2.6

1. Any ccc, with $A \otimes B = A \times B$ and $A \multimap B = A \Rightarrow B$.
2. A poset $\mathcal{P} = (P, \leq)$ is an smcc iff there are operations $\otimes, \multimap : P^2 \to P$, $1 \in P$ satisfying:

   (i) $(P, \otimes, 1)$ is a commutative monoid.
   (ii) $\otimes, \multimap$ are functorial in the posetal sense: i.e. $x \leq x'$, $y \leq y'$ implies $x \otimes y \leq x' \otimes y'$ and $x' \multimap y \leq x \multimap y'$
   (iii) (Closedness) $x \otimes y \leq z$ iff $x \leq y \multimap z$.

3. *Girard's Phase Semantics*: This is a posetal smcc, in the sense of Example 2 above. Let $M = (M, \cdot, e)$ be a commutative monoid. Consider the poset $\mathcal{P}(M)$, the powerset of M. We view $\mathcal{P}(M)$ as a poset ordered by inclusion. For $X, Y \in \mathcal{P}(M)$, define

$$X \otimes Y = XY =_{def} \{x{\cdot}y \mid x \in X, y \in Y\}$$
$$X \multimap Y = \{z \in M \mid z{\cdot}X \subseteq Y\} \text{ and } I = \{e\}$$

4. **Vec**, where $V \otimes W$ is the usual algebraic tensor product and $V \multimap W = Lin(V, W)$. More generally, consider $\mathcal{R}$-Modules over a commutative ring $\mathcal{R}$, with the standard algebraic notions of $V \otimes_{\mathcal{R}} W$ and $V \multimap W = Hom(V, W)$.
5. **MOD**(G). This example extends groups acting on sets to groups acting linearly on vector spaces. Let G be a group and V a vector space. A *representation of*

G *on* V is a group homomorphism $\rho : G \to Aut(V)$; equivalently, it is a left G-action $G \times V \stackrel{\cdot}{\longrightarrow} V$ (satisfying the same equations as a G-set) such that $v \mapsto g{\cdot}v$ is a linear automorphism, for each $g \in G$. The pair (ρ, V) is called a G-*module* or G-*space*. $\mathbf{MOD}(G)$ has as objects the G-modules and as morphisms the linear maps commuting with the G-actions. Define the smcc structure of $\mathbf{MOD}(G)$ as follows:

$$\begin{aligned} V \otimes W &= \text{the usual tensor product, with action determined by} \\ & \quad g{\cdot}(v \otimes w) = g{\cdot}v \otimes g{\cdot}w \\ V \multimap W &= Lin(V, W), \text{ with action } (g{\cdot}f)(v) = g{\cdot}f(g^{-1}{\cdot}v)\,, \\ & \quad \textit{the contragredient action.} \end{aligned}$$

5.2.3 Monoidal Categories with Duality

For the purposes of studying linear logic, as well as general duality theories, we need to consider monoidal categories equipped with a notion of involutive negation (or "duals"). A general categorical theory of such dualities, including many traditional mathematical duality theories, was developed by M. Barr [Barr79] in the mid 1970's, some ten years before linear logic.

Definition 5.2.7 ([Barr79]) A $*$-*autonomous category* $(\mathcal{C}, \otimes, I, \multimap, \bot)$ is an smcc with a distinguished *dualizing* object $\bot$, such that (letting $A^* = A \multimap \bot$), the canonical map $\mu_A : A \to A^{**}$ is an iso, for all A (i.e. "all objects are reflexive").

Facts about $*$-autonomous categories $\mathcal{C}$:

- The operation $(-)^*$ induces a contravariant dualizing functor $\mathcal{C}^{op} \stackrel{(\,)^*}{\longrightarrow} \mathcal{C}$ such that $\mathcal{C}(A, B) \cong \mathcal{C}(B^*, A^*)$ which is a natural iso and which satisfies all natural coherence equations.
- $\mathcal{C}$ is closed under duality of categorical constructions: e.g. $\mathcal{C}$ has products iff it has coproducts, $\mathcal{C}$ is complete iff it is co-complete, etc.
- $(A \multimap B)^* \cong A \otimes B^*$ and $I \cong \bot^*$ Also $A \multimap B \cong B^* \multimap A^*$.
- We may define $A \parr B = (A^* \otimes B^*)^*$, a kind of "de Morgan dual" of $\otimes$. In linear logic, this is the connective "par", a kind of "parallel disjunction". In general, $\otimes \neq \parr$, and (in general) there is not even a $\mathcal{C}$-morphism $A \otimes B \to A \parr B$.
- As we shall see below, categorical models of multiplicative, additive linear logic will be $*$-autonomous categories with products (hence coproducts).

The first two examples are from Example 5.2.4 above.

Example 5.2.8 $\mathbf{Rel}_\times$. The category of relations $\mathbf{Rel}_\times$ is probably the simplest $*$-autonomous category. For sets A, B, $A \otimes B = A \multimap B = A \times B$. Let the dualizing object $\bot = \{*\}$, any one-element set. As for the dualizing functor $(-)^*$, on objects define $A^* = A$. On arrows, given a relation $R : A \to B$, we define $R^* = R^{op} : B \to A$ to be the *opposite* relation (so that bR^*a iff aRb, for any $a \in A, b \in B$). Notice: $(A \otimes B)^* = A^* \otimes B^* = A \times B$.

Example 5.2.9 **Vec** is symmetric monoidal closed, where $I = \mathbf{k}$ (the base field), $V \otimes W$ is the usual algebraic tensor product and $V \multimap W = Lin(V, W)$. Note that in the category **Vec** of vector spaces over the field **k**, V satisfies $V \cong V^{**}$(via the canonical map μ_V) iff V is finite dimensional (see [Ger85], p. 68). Hence, $\mathbf{Vec}_{fd}$ is $*$-autonomous, where $I = \mathbf{k}$ (the base field) and $V^* = V \multimap I$ is the usual dual space.

Unfortunately, the above two examples $\mathbf{Rel}_\times$ and $\mathbf{Vec}_{fd}$ are "degenerate" $*$-autonomous categories (from the viewpoint of linear logic), since $\parr \cong \otimes$. That is, $(A \otimes B)^* \cong A^* \otimes B^*$. We shall mention these below, as examples of *compact* categories. Indeed, from the viewpoint of linear logic, it is quite hard to find nice examples of nondegenerate $*$-autonomous categories. One of the motivations that led to [Barr79] was that such categories arise quite naturally in various topological duality theories. The following discussion is a quick summary, primarily based on work of M. Barr (e.g. [Barr79]) and the treatment in Blute [Bl96], based on a topology originally due to Lefschetz ([Lef]). See also [BS96]. Let **TVec** denote the category whose objects are vector spaces equipped with linear topologies, and whose morphisms are linear continuous maps.

Barr showed that **TVec** is a symmetric monoidal closed category, when $V \multimap W$ is defined to be the vector space of linear continuous maps, topologized with the topology of pointwise convergence. (It is shown in [Barr79] that the forgetful functor $\mathbf{TVec} \to \mathbf{Vec}$ is tensor-preserving). Let V^* denote $V \multimap \mathbf{k}$. Lefschetz proved that the canonical embedding $V \to V^{**}$ is always a bijection, but need not be an isomorphism. Now we just cut down to so-called *reflexive* spaces: those for which the embedding $V \to V^{**}$ is actually an isomorphism:

Theorem 5.2.10 (Barr) ***RTVec****, the full subcategory of reflexive objects in* ***TVec****, is a complete, cocomplete $*$-autonomous category, with $I^* = I = \mathbf{k}$ the dualizing object. Moreover, in* ***RTVec****, $\otimes$ and $\parr$ are not isomorphic.*

More generally, other classes of $*$-autonomous categories arise by taking a continuous linear analog of G-sets, namely categories of group representations, using the category **RTVec**.

Definition 5.2.11 Let G be a group. A *continuous G-module* is a linear action of G on a space V in **TVec**, such that for all $g \in G$, the induced map $g\cdot(\ ) : V \to V$ is continuous. Let $\mathbf{TMOD}(G)$ denote the category of continuous G-modules and continuous equivariant maps. Let $\mathbf{RTMOD}(G)$ denote the full subcategory of reflexive objects.

Theorem 5.2.12 *The category* $\mathbf{TMOD}(G)$ *is symmetric monoidal closed. The category* $\mathbf{RTMOD}(G)$ *is $*$-autonomous, and a reflective subcategory of* $\mathbf{TMOD}(G)$ *via the functor* $(\ \ )^{**}$*. Furthermore the forgetful functor* $|\ \ | : \mathbf{RTMOD}(G) \to \mathbf{RTVec}$ *preserves the $*$-autonomous structure.*

Following [Bl96], still more general classes of $*$-autonomous categories arise analogously using the category $\mathbf{RTMOD}(\mathsf{H})$, the reflective subcategory of linearly topologized H -modules, for a cocommutative Hopf algebra H .

The next notion is much more familiar mathematically, although logically it corresponds to a rather degenerate case of linear logic: the case where $\otimes = ⅋$:

Definition 5.2.13 A *compact closed category* [KL80] is a symmetric monoidal category such that for each object A there exists a dual object A^*, and canonical morphisms:

$$\nu : I \to A \otimes A^*$$
$$\psi : A^* \otimes A \to I$$

such that evident equations hold. In the case of a strict monoidal category, these equations reduce to the usual adjunction triangles.

*Remark 5.2.14 (From Compactness to *-Autonomy)* For constructing new models of multiplicative linear logic, there is a general categorical construction *Double Glueing* which can be used to turn compact closed categories into nontrivial *-autonomous ones, essentially by breaking the isomorphism between $\otimes$ and ⅋ . This is described in detail in [HylSc03]. Indeed, double gluing can be iterated, to obtain interesting categories, e.g. see [HaSc07].

We have already remarked above that the categories $\mathbf{Rel}_\times$ and $\mathbf{Vec}_{fd}$ are compact closed. In the study of quantum computing, Abramsky and Coecke [AbCo04] have shown the utility of *strongly compact closed* categories, those with additional structure abstracting the theory of inner product spaces.

Lemma 5.2.15

- *Compact closed categories are $*$-autonomous, with the tensor unit as dualizing object.*
- *Recall $A ⅋ B = (A^* \otimes B^*)^*$. In any $*$-autonomous category in which the tensor unit is the dualizing object, there is a canonical morphism* *$A \otimes B \to A ⅋ B$ given by: $\mu_A \otimes \mu_B : A \otimes B \to A^{**} \otimes B^{**} \cong (A^* \otimes B^*)^*$. In a compact closed category, this morphism is an isomorphism.*

What are monoidal functors between monoidal categories? Here there can be several notions. Let us pick an important one:

Definition 5.2.16 A *monoidal functor* between monoidal categories is a 3-tuple (F, m_I, m) where $F : \mathcal{C} \to \mathcal{D}$ is a functor, together with a morphism $m_I : I \longrightarrow F(I)$ and a natural transformation $m_{UV} : F(U) \otimes F(V) \longrightarrow F(U \otimes V)$ satisfying some coherence diagrams (which we omit). F is *strict* if m_I, m_{UV} are identities. A monoidal functor is *symmetric* if m commutes with the symmetries: $m_{B,A} s_{FA,FB} = F(s_{A,B}) m_{A,B}$, for all A, B.

Finally, we need an appropriate notion of natural transformation for monoidal functors.

Definition 5.2.17 A natural transformation between monoidal functors $\alpha : F \to G$ is *monoidal* if it is compatible with both m_I and m_{UV}, for all U, V, in the sense that

the following equations hold:
(i) $\alpha_I \circ m_I = m_I$ and
(ii) $m_{UV} \circ (\alpha_U \otimes \alpha_V) = \alpha_{U\otimes V} \circ m_{UV}$.

Remark 5.2.18 (Alternative Treatments of $*$-autonomy) There are alternative definitions of $*$-autonomous categories, some based on attempts to axiomatize a fully faithful dualizing functor $(-)^* : \mathcal{C}^{op} \to \mathcal{C}$. This leads to thorny problems concerning what are the appropriate categorical coherence equations to impose. Recent work of Robin Houston [Hou07] has shown that this is subtle and is inadequately addressed in the literature, so we omit discussing it.

A more radical alternative categorical treatment of the various layers of linear logic (and thus of $*$-autonomous categories) arose in work of Cockett and Seely and coworkers [CS97, BCST96, BCS00]. Their idea is to first consider *linearly (or weakly) distributive categories*: monoidal categories with two monoidal structures tensor ($\otimes$) and cotensor ($⅋$), together with various coherence and (weak) distributive laws relating them. This corresponds to a kind of multiplicative linear logic of just conjunction/disjunction, without any negation or duality relating the two tensors. On top of this structure, one can impose an involutive negation $(-)^*$ (which will satisfy a De Morgan duality between tensor and cotensor); in addition, one may adjoin products $\times$ (and thus, coproducts $+$) for the additive structure. Finally, one may impose the further exponential structure of linear logic (see below). The authors in the above papers also take extra care in handling the logical units (for tensor and cotensor) and the various categorical coherence problems these require.

5.3 Linear Logic and Categorical Proof Theory

5.3.1 Gentzen's Proof Theory

Gentzen's approach to Hilbert's proof theory [GLT, Mel07], especially his *sequent calculi* and his fundamental theorem on Cut-Elimination, have had a profound influence not only in logic, but recently in category theory and computer science as well. The connections of Gentzen proof theory with categorical logic (and linear logic) are discussed in various survey papers and books [LS86, Sc00, BS04, Mel07]. Let us just introduce some basic terminology.

A *sequent* for a logical language $\mathcal{L}$ is an expression

$$A_1, A_2, \cdots, A_m \vdash B_1, B_2, \cdots, B_n \tag{5.2}$$

where $A_1, A_2, \cdots, A_m$ and $B_1, B_2, \cdots, B_n$ are finite lists (possibly empty) of formulas of $\mathcal{L}$. Sequents are denoted $\Gamma \vdash \Delta$, where Γ and Δ are lists of formulas. We think of sequent (5.2) as a formal *entailment* relationship between the premisses Γ and (potential) conclusions Δ.

Traditional logicians would give the semantical meaning (of the truth) of the sequent (5.2) as: *the conjunction of the* A_i *entails the disjunction of the* B_j. More generally, following Lambek and Lawvere, category theorists interpret *proofs* of such sequents (modulo equivalence of proofs) as *arrows* in appropriate (freely generated) monoidal categories. For logics $\mathcal{L}$ similar to Girard's linear logic [Gi87], we interpret a proof π of sequent (5.2) in a *-autonomous category $(\mathcal{C}, \otimes, I, \multimap, \perp)$ with a "cotensor" $\parr$ (see Definition 5.2.7 above) as an arrow of the following form

$$[\![A_1]\!] \otimes [\![A_2]\!] \otimes \cdots \otimes [\![A_m]\!] \stackrel{[\![\pi]\!]}{\longrightarrow} [\![B_1]\!] \parr [\![B_2]\!] \parr \cdots \parr [\![B_n]\!] \quad (5.3)$$

Here $[\![-]\!] : \mathcal{L} \to \mathcal{C}$ is an interpretation function of formulas and proofs (of the logic $\mathcal{L}$) into the objects and arrows of $\mathcal{C}$. We interpret formulas A_i as objects $[\![A_i]\!] \in \mathcal{C}$ by induction, starting with an arbitrary interpretation of the atoms (as objects of $\mathcal{C}$).

Remark 5.3.1 (Notation) We abuse notation for $[\![\pi]\!]$ above and omit writing $[\![-]\!]$ on formulas when it is clear; thus we write the arrow (5.3) above as

$$A_1 \otimes A_2 \otimes \cdots \otimes A_m \stackrel{[\![\pi]\!]}{\longrightarrow} B_1 \parr B_2 \parr \cdots \parr B_n \quad (5.4)$$

as an interpretation of sequent (5.2) above in category $\mathcal{C}$.

Gentzen's approach to proof theory gives rules for generating *formal proofs* of sequents. These formal proofs are trees generated by certain rules (called *rules of inference*) for building new sequents from old sequents, starting from initially given sequents called *axioms*. Thus, a (formal) proof of $\Gamma \vdash \Delta$ is a tree with root labelled by $\Gamma \vdash \Delta$ and in which every node is labelled by a rule of inference and in which the leaves are labelled by instances of axioms.

Lambek [L89] pointed out that Gentzen's sequent calculus was analogous to Bourbaki's method of bilinear maps. For example, given lists $\Gamma = A_1 \cdots A_m$ and $\Delta = B_1 B_2 \cdots B_n$ of $R - R$ bimodules of a given ring R, there is a natural isomorphism

$$Mult(\Gamma A B \Delta, C) \cong Mult(\Gamma A \otimes B \Delta, C) \quad (5.5)$$

between $m+n+2$-linear and $m+n+1$-linear maps. Bourbaki derived many aspects of tensor products just from this universal property. Such a formal bijection is at the heart of Linear Logic, whose rules we now present briefly.

Gentzen's Rules

Gentzen's rules analyze the deep structure and implicit symmetries hidden in logical syntax. Gentzen broke down the manipulations of logic into two classes of rules applied to sequents: *structural rules* and *logical rules* (including Axiom and Cut rules.) All rules come in pairs (left/right) applying to the left (resp. right) side of a sequent.

Gentzen's Structural Rules (Left/Right)

$$\text{Permutation} \quad \frac{\Gamma \vdash \Delta}{\sigma(\Gamma) \vdash \Delta} \quad \frac{\Gamma \vdash \Delta}{\Gamma \vdash \tau(\Delta)} \quad \sigma, \tau \text{ permutations.}$$

$$\text{Contraction} \quad \frac{\Gamma, A, A \vdash \Delta}{\Gamma, A \vdash \Delta} \quad \frac{\Gamma \vdash \Delta, B, B}{\Gamma \vdash \Delta, B}$$

$$\text{Weakening} \quad \frac{\Gamma \vdash \Delta}{\Gamma, A \vdash \Delta} \quad \frac{\Gamma \vdash \Delta}{\Gamma \vdash \Delta, B}$$

Permutation says that if Γ entails Δ (i.e. if the sequent $\Gamma \vdash \Delta$ holds), then we can permute arbitrarily the order of the lists of premisses and conclusions and still have a valid inference. *Contraction* says (for the Left rule) that if Γ together with two copies of premise A entail Δ, then we can still infer Δ from Γ but using only one copy of A; dually for contraction on the right. *Weakening* (on the left) says if Γ entails Δ, then adding extra premisses to Γ still entails Δ, and dually for the right hand rule (see [GSS, Abr93, L89]).

In linear logic we do *not* allow such uncontrolled contraction and weakening rules; rather, formulas which can be contracted or weakened are marked with $!A$ (for left rules) and $?A$ for right rules. We shall mention more on this below.

By controlling (and making explicit) these traditional structural rules, logic takes on a completely different character.

Definition 5.3.2 (Linear Logic) *Formulas* of the theory **LL** (linear logic) are generated from atoms and their negations $p, p^{\perp}, q, q^{\perp}, \cdots$, constants $I, \perp, \mathbf{1}, \mathbf{0}$ using the binary connectives $\otimes, ⅋, \times, +$ and unary operations $!, ?$. Negation is extended by de Morgan duality to all expressions as follows: $p^{\perp\perp} = p$, $(A \otimes B)^{\perp} = A^{\perp} ⅋ B^{\perp}$ and dually, as well as $(A \times B)^{\perp} = A^{\perp} + B^{\perp}$ and $(!A)^{\perp} = ?(A^{\perp})$ and dually. Finally, $A \multimap B$ is defined to be $A^{\perp} ⅋ B$ (this connective is redundant, but useful for understanding the categorical semantics of linear logic).

We think of $A \otimes B$ and $A \times B$ as kinds of conjunctions, $A ⅋ B$ and $A+B$ as kinds of disjunctions, and $A^{\perp}$ as (linear) negation. As suggested by the notation, such logics will be interpretable in $*$-autonomous categories with additional structure (see below).

The Rules of Linear Logic are in Fig. 5.2. Previously equivalent notions (of traditional classical logic) now split into subtle variants based on resource allocation. For example, the rules for *Multiplicative* connectives simply concatenate their input hypotheses Γ and Γ', whereas the rules for *Additive* connectives merge two input hypotheses Γ into one. The situation is analogous for conclusions Δ and Δ'. The names of the rules suggest their categorical meaning.

The Exponential rules are the rules for the connectives ! and ? (e.g. contraction and weakening on the left side of sequents). Using the rules of negation, one can obtain the dual laws (e.g. contraction and weakening on the right side) by using the dual ? connective.

The logical connectives in linear logic can represent linguistic distinctions related to resource use which are simply impossible to formulate in traditional logic (see [Gi89, Abr93]). For example, we think of a linear entailment $A_1, \cdots, A_m \vdash B$

Structural	*Perm*	$\dfrac{\Gamma \vdash \Delta}{\sigma(\Gamma) \vdash \tau(\Delta)}$	σ, τ permutations.
Axiom & Cut	*Axiom*	$A \vdash A$	
	Cut	$\dfrac{\Gamma \vdash A, \Delta \quad \Gamma', A \vdash \Delta'}{\Gamma, \Gamma' \vdash \Delta, \Delta'}$	
Negation		$\dfrac{\Gamma \vdash A, \Delta}{\Gamma, A^{\perp} \vdash \Delta}$	$\dfrac{\Gamma, A \vdash \Delta}{\Gamma \vdash A^{\perp}, \Delta}$
Multiplicatives	*Tensor*	$\dfrac{\Gamma, A, B \vdash \Delta}{\Gamma, A \otimes B \vdash \Delta}$	$\dfrac{\Gamma \vdash A, \Delta \quad \Gamma' \vdash B, \Delta'}{\Gamma, \Gamma' \vdash A \otimes B, \Delta, \Delta'}$
	Par	$\dfrac{\Gamma, A \vdash \Delta \quad \Gamma', B \vdash \Delta'}{\Gamma, \Gamma', A \invamp B \vdash \Delta, \Delta'}$	$\dfrac{\Gamma \vdash A, B, \Delta}{\Gamma \vdash A \invamp B, \Delta}$
	Units	$\dfrac{\Gamma \vdash \Delta}{\Gamma, I \vdash \Delta}$	$\vdash I$
		$\perp \vdash$	$\dfrac{\Gamma \vdash \Delta}{\Gamma \vdash \perp, \Delta}$
	Implication	$\dfrac{\Gamma \vdash A, \Delta \quad \Gamma', B \vdash \Delta'}{\Gamma, \Gamma', A \multimap B \vdash \Delta, \Delta'}$	$\dfrac{\Gamma, A \vdash B, \Delta}{\Gamma \vdash A \multimap B, \Delta}$
Additives	*Product*	$\dfrac{\Gamma, A \vdash \Delta}{\Gamma, A \times B \vdash \Delta} \quad \dfrac{\Gamma, B \vdash \Delta}{\Gamma, A \times B \vdash \Delta}$	$\dfrac{\Gamma \vdash A, \Delta \quad \Gamma \vdash B, \Delta}{\Gamma \vdash A \times B, \Delta}$
	Coproduct	$\dfrac{\Gamma, A \vdash \Delta \quad \Gamma, B \vdash \Delta}{\Gamma, A + B \vdash \Delta}$	$\dfrac{\Gamma \vdash A, \Delta}{\Gamma \vdash A + B, \Delta} \quad \dfrac{\Gamma \vdash B, \Delta}{\Gamma \vdash A + B, \Delta}$
	Units	$\Gamma, \mathbf{0} \vdash \Delta$	$\Gamma \vdash \mathbf{1}, \Delta$
Exponentials	*Weakening*	$\dfrac{\Gamma \vdash \Delta}{\Gamma, !A \vdash \Delta}$	*Contraction* $\dfrac{\Gamma, !A, !A \vdash \Delta}{\Gamma, !A \vdash \Delta}$
	Storage	$\dfrac{!\Gamma \vdash A}{!\Gamma \vdash !A}$	*Dereliction* $\dfrac{\Gamma, A \vdash \Delta}{\Gamma, !A \vdash \Delta}$

Fig. 5.2 Rules for classical propositional LL

as an *action*–a kind of process that in a single step consumes the inputs A_i and produces output B. Think of a chemical reaction, in which B is produced in 1-step from the reactants A_i. For example, this permits representing in a natural manner the step-by-step behaviour of various abstract machines, certain models of concurrency like Petri Nets, etc. Thus, linear logic permits us to describe the instantaneous state of a system, and its step-wise evolution, intrinsically within the logic itself (e.g. with no need for explicit time parameters, etc.)

We should note that linear logic is not about simply removing Gentzen's structural rules, but rather modulating their use. The particular connective $!A$, which indicates that contraction and weakening may be applied to formula A, yields the *Exponential* connectives in Fig. 5.2. From a resource viewpoint, an hypothesis $!A$ is one which can be reused arbitrarily. It is roughly like an infinite tensor power $\otimes^{\omega} A$, and more generally (for physicists) something like an exterior algebra or Fock-space like construction.

Moreover, this connective permits decomposing intuitionistic implication "$\Rightarrow$" (categorically, the cartesian closed function space) into more basic notions:

$$A \Rightarrow B = (!A) \multimap B$$

Remark 5.3.3 (1-sided Sequents & Theories) Observe that in classical linear logic **LL**, two-sided sequents can be replaced by one-sided sequents, since $\Gamma \vdash \Delta$ is equivalent to $\vdash \Gamma^{\perp}, \Delta$, with $\Gamma^{\perp}$ the list $A_1^{\perp}, \cdots, A_n^{\perp}$, where Γ is $A_1, \cdots, A_n$. This permits halving the number of rules, and we shall use this notation frequently, see Fig. 5.3. Finally, we end with the following standard terminology of subtheories of **LL** in Fig. 5.2. *The literature usually presents the theories below using 1-sided sequents.*

MLL: *multiplicative linear logic* is built from the atoms and multiplicative units $\{I, \perp\}$ using the connectives $\{\otimes, \parr, (\ )^{\perp}\}$. The rules include the structural (permutation), axioms, cut, negation and the multiplicative rules. This theory corresponds semantically to $*$-autonomous categories (in which we interpret linear negation by $(\ )^*$).

Structural	*Perm*	$\dfrac{\vdash \Gamma}{\vdash \tau(\Gamma)}$	τ a permutation.	
Axiom & Cut	*Axiom*	$\vdash A^{\perp}, A$		
	Cut	$\dfrac{\vdash A, \Gamma \quad \vdash A^{\perp}, \Gamma'}{\vdash \Gamma, \Gamma'}$		
Multiplicatives	*Tensor*	$\dfrac{\vdash A, \Gamma \quad \vdash B, \Gamma'}{\vdash A \otimes B, \Gamma, \Gamma'}$		
	Par	$\dfrac{\vdash A, B, \Gamma}{\vdash A \parr B, \Gamma}$		
	Units	$\vdash I$	$\dfrac{\vdash \Gamma}{\vdash \perp, \Gamma}$	
Additives	*Product*	$\dfrac{\vdash A, \Gamma \quad \vdash B, \Gamma}{\vdash A \times B, \Gamma}$		
	Coproduct	$\dfrac{\vdash A, \Gamma}{\vdash A + B, \Gamma} \quad \dfrac{\vdash B, \Gamma}{\vdash A + B, \Gamma}$		
	Units	$\vdash \mathbf{1}, \Gamma$		
Exponentials	*Weakening*	$\dfrac{\vdash \Gamma}{\vdash ?A, \Gamma}$	*Contraction*	$\dfrac{\vdash ?A, ?A, \Gamma}{\vdash ?A, \Gamma}$
	Storage	$\dfrac{\vdash ?\Gamma, A}{\vdash ?\Gamma, !A}$	*Dereliction*	$\dfrac{\vdash A, \Gamma}{\vdash ?A, \Gamma}$

Fig. 5.3 1-Sided rules for classical propositional **LL**

MALL: *multiplicative additive linear logic* is built from the atoms and the units $\{I, \bot, \mathbf{0}, \mathbf{1}\}$ using the connectives $\{\otimes, \parr, (\ )^{\bot}, \times, +\}$. The rules include the **MLL** rules together with the additive rules. This theory corresponds semantically to $*$-autonomous categories with products (hence coproducts).

MELL: *multiplicative exponential linear logic* is built from those formulas of **LL** that do not use any of the additive structure: that is, formulas built from the atoms and multiplicative units $\{I, \bot\}$ using only the connectives $\{\otimes, \parr, (\ )^{\bot}, !, ?\}$. The rules include the structural (permutation), axioms, cut, negation and the multiplicative and exponential rules.

CATEGORICAL PROOF THEORY

One of the basic ideas of categorical logic and categorical proof theory is that (the proof theory of) various logics generate interesting classes of *free categories*: free cartesian, cartesian closed, monoidal, monoidal closed, *-autonomous, toposes, etc. The intuition is:

- Formulas of a logic should be the objects of a category.
- Proofs (or, rather, equivalence classes of proofs) should be the morphisms.

The subject began in the work of Lawvere and of Lambek in the 1960s and is discussed in detail in [LS86] (cf. also the expository treatment in [BS04]). One of the early applications of Lambek was to apply these methods to solve coherence problems for various monoidal categories.

5.3.2 Categorical Models of Linear Logic

We are interested in finding the categories appropriate to modelling linear logic proofs (just as cartesian closed categories modelled intuitionistic $\wedge, \Rightarrow, \top$ proofs). The basic equations we certainly *must* postulate arise from the operational semantics–that is *cut-elimination of proofs*. If we have a proof-rewriting

$$\begin{array}{c} \pi \\ \vdots \\ A_1, \cdots, A_m \vdash B_1, \cdots, B_n \end{array} \quad \rightsquigarrow \quad \begin{array}{c} \pi' \\ \vdots \\ A_1, \cdots, A_m \vdash B_1, \cdots, B_n \end{array}$$

then the categorical interpretation $[\![-]\!]$ of these proofs (as arrows in an appropriate category as in (5.4) above) should be to give *equal* arrows $[\![\pi]\!] = [\![\pi']\!]$: $A_1 \otimes \cdots \otimes A_m \rightarrow B_1 \parr \cdots \parr B_n$.

In the case of sequent calculi, this rewriting is generated by the rules of Gentzen's Cut-Elimination algorithm [GLT]. However, there are sometimes natural categorical equations (e.g. the universal property of cartesian products) which are not decided by traditional proof theoretic rewriting, and need to be postulated separately (otherwise, conjunction only gives a "weak product" [LS86]). Precisely which equations to add, to make a mathematically natural and beautiful structure, is an important question. The problem is further compounded in linear logic (at the level of the exponentials) where the equations and coherences are more subtle, with more variations possible.

The first attempted categorical semantics of **LL** is in Seely's paper [See89] which is still a good resource (although some fine details have turned out to require modification). Since that time, considerable effort by many researchers has led to major clarifications and quite different axiomatizations. An excellent survey of the current state-of-the-art is in Melliès [Mel07]. In the case of Multiplicative-Additive classical

Arrow-generating Rules	Equations
$A \xrightarrow{id} A$ $\qquad \dfrac{A \xrightarrow{f} B \quad B \xrightarrow{g} C}{A \xrightarrow{gf} C}$	equations of a category
$\dfrac{A \xrightarrow{f} B \quad A' \xrightarrow{g} B'}{A \otimes A' \xrightarrow{f \otimes g} B \otimes B'}$	$\otimes$ is a functor : $ff' \otimes gg' = (f \otimes g)(f' \otimes g')$ $id \otimes id = id$
$(A \otimes B) \otimes C \xrightarrow{\alpha} A \otimes (B \otimes C)$ $A \otimes B \xrightarrow{s} B \otimes A$ $I \otimes A \xrightarrow{\ell} A$	α, s, ℓ are natural isos equations for symmetric monoidal structure
$\dfrac{A \otimes B \xrightarrow{f} C}{A \xrightarrow{f^*} (B \multimap C)}$ $\multimap$R $(A \multimap B) \otimes A \xrightarrow{ev} B$	equations for monoidal closedness (this gives smcc's)
$\dfrac{C \xrightarrow{f} A \quad C \xrightarrow{g} B}{C \xrightarrow{\langle f,g \rangle} A \times B}$ $A \times B \xrightarrow{\pi_1} A \quad A \times B \xrightarrow{\pi_2} B$ $A \xrightarrow{!_A} \top$	cartesian products (this gives smcc's + products)
$\dfrac{A \xrightarrow{f} B}{B^* \xrightarrow{f^*} A^*}$	$(-)^*$ is a contravariant functor
$A^* \longrightarrow (A \multimap \bot)$ $(A \multimap \bot) \longrightarrow A^*$	these are natural isos
$(A \multimap B) \to (B^* \multimap A^*)$	natural strength iso
$A \to ((A \multimap \bot) \multimap \bot)$	natural iso

Fig. 5.4 *-Autonomous categories equationally

linear logic **MALL**, there is little controversy: the syntax should generate a free $*$-autonomous category with products (and thus coproducts). In more detail, in Fig. 5.4 we present the categorical structure of (free) $*$-autonomous categories considered as symmetric monoidal closed categories (smcc's) with dualizing objects $\bot$, as in the discussion above. We may think of the arrows $A \xrightarrow{f} B$ as proofs of very simple sequents $A \vdash B$ (where premisses and conclusions are lists of length 1). For example, the identity map $A \xrightarrow{id} A$ corresponds to the axiom $A \vdash A$. The remaining laws of linear logic follow from the arrow-generating rules. The associated equations guarantee that: (i) we get all the axioms of $*$-autonomous categories with products, but also (ii) these are the equations between proofs we must postulate to get a nice categorical structure (this is relevant to our next section on Cut-Elimination).

At this point we could also add coproducts, denoted $+$, and their associated equations, dual to products. But once we have the equations of $*$-autonomous categories (at the bottom of Fig. 5.4) we get coproducts for free, essentially by De Morgan duality. Finally we add any necessary coherence equations, as in Barr's monograph [Barr79].

5.3.3 Adding Exponentials: Full Linear Logic

By far the most subtle question is how to model the linear modality !. We begin with seven basic derivation forms, arising from the rules of linear logic and then postulate equations which arise directly from the categorical viewpoint.

Exercise 5.3.4 Prove the laws in Fig. 5.5, using the 2-sided rules in Fig. 5.2. Let us give two examples. As mentioned above, we think of $A \xrightarrow{f} B$ as a proof f of the sequent $A \vdash B$.

Functoriality	$\dfrac{A \xrightarrow{f} B}{!A \xrightarrow{!f} !B}$	
Monoidalness	$I \xrightarrow{m_I} !I$	$!A \otimes !B \xrightarrow{m_{AB}} !(A \otimes B)$
Products	$I \xrightarrow{n_I} !\top$	$!A \otimes !B \xrightarrow{n_{AB}} !(A \times B)$
Dereliction	$!A \xrightarrow{\varepsilon_A} A$	
Weakening	$!A \xrightarrow{\varepsilon'_A} I$	
Contraction	$!A \xrightarrow{\delta'_A} !A \otimes !A$	
Digging (Storage)	$!A \xrightarrow{\delta_A} !!A$	

Fig. 5.5 Basic exponential laws

- Functoriality: $\dfrac{\dfrac{\begin{array}{c} f \\ \vdots \\ A \vdash B \end{array}}{!A \vdash B}\,\text{Derel.}}{!A \vdash !B}\,\text{Storage}$, where f is the given proof of $A \vdash B$.
- Contraction: Applying functoriality to the axiom $A \vdash A$, we get a proof $\pi : !A \vdash !A$. Now use π twice in the following proof tree:

$$\dfrac{\dfrac{\begin{array}{c}\pi \\ \vdots \\ !A \vdash !A\end{array} \quad \begin{array}{c}\pi \\ \vdots \\ !A \vdash !A\end{array}}{!A, !A \vdash !A \otimes !A}\,\otimes\text{R}}{!A \vdash !A \otimes !A}\,\text{Contr}$$

So, what is a model of full linear logic? The state-of-the-art is described in work of Hyland-Schalk [HylSc03] and especially Melliès [Mel07]. Here is one class of structure that is popular to impose: let $\mathcal{C}$ be a model of **MALL** proofs, i.e. a $*$-autonomous category with products (and hence coproducts). We add:

- $(!, m_I, m_{AB}) : \mathcal{C} \to \mathcal{C}$ is a monoidal endofunctor
- $!A \xrightarrow{\varepsilon_A} A$ and $!A \xrightarrow{\delta_A} !!A$ are monoidal natural transformations.
- $(!, \delta, \varepsilon)$ is a monoidal comonad.
- n_I, n_{AB} are isomorphisms, natural in A, B.
- The associated adjunction structure $\langle F, U, \eta, \varepsilon \rangle$ between the co-Kleisli category of ! and $\mathcal{C}$ is monoidal.
- Various coherence equations [BCS96, Mel07].

However, for the purposes of Geometry of Interaction, we shall not need all this elaborate structure of the exponentials and the associated properties of cocommutative comonoids, etc. Indeed, beyond the basic derivations in Fig. 5.5, one merely needs the exponential structure associated to a *Linear Combinatory Algebra* [AHS02], as we shall see.

5.3.4 Cut Elimination: Gentzen's Operational Semantics of Proofs

Let us briefly discuss the Cut-Elimination theorem in proof theory. For more details, the reader may examine the works of Girard (e.g. [GLT, Gi87]) or the survey of Melliès [Mel07] or the textbook [TrSc]. Recall the Cut-Rule, which is a kind of generalized composition law:

$$\dfrac{\Gamma \vdash \Delta, A \quad \Gamma', A \vdash \Delta'}{\Gamma, \Gamma' \vdash \Delta, \Delta'}\,\text{Cut}$$

A fundamental theorem of logic is the following result of Gentzen:

Cut-elimination (Gentzen's Haupsatz, 1934): If π is a proof of $\Gamma \vdash \Delta$, then there is a proof π' of $\Gamma \vdash \Delta$ which does not use the cut rule.

It is the basis of Proof Theory, at the very foundations of Hilbert's approach to logic, and has applications in a wide range of areas of both logic and theoretical computer science.

For usual sequent calculus, Gentzen gave a Non-Deterministic algorithm $\pi \rightsquigarrow \pi'$ (the cut-elimination procedure) for transforming proofs π into proofs π'. The details of the rewriting steps (for each proof rule, Left and Right) become rather intricate. Here is an example of one rewriting step, with respect to the Contraction-Right Rule:

$$\dfrac{\dfrac{\begin{array}{c}\vdots\\ \Gamma \vdash B, B\end{array}}{\Gamma \vdash B}\,Contr \qquad \begin{array}{c}\vdots\\ B \vdash \Delta\end{array}}{\Gamma \vdash \Delta}\,\text{Cut} \quad \rightsquigarrow \quad \begin{array}{c}\dfrac{\dfrac{\begin{array}{c}\vdots\\ \Gamma \vdash B, B\end{array} \quad \begin{array}{c}\vdots\\ B \vdash \Delta\end{array}}{\Gamma \vdash B, \Delta}\,\text{Cut} \qquad \begin{array}{c}\vdots\\ B \vdash \Delta\end{array}}{\Gamma \vdash \Delta, \Delta}\,\text{Cut}\\ \vdots\; Contrs\&Perms\\ \Gamma \vdash \Delta\end{array}$$

Notice that this is slightly strange: starting from the root of the tree and going upwards, the subproof of $B \vdash \Delta$ in the original left proof is now duplicated in the right proof *higher up in the tree*. So moving from the LHS proof to the RHS proof, we have replaced a single Cut (on B) by two cuts on B higher up in the tree (beyond the contraction); at the same time we have postponed the contractions until later, lower down in the proof. But since the duplicated proof of $B \vdash \Delta$ may be arbitrarily complex (millions of lines long) it is not obvious that the rewriting above has "simplified" anything. The point is that Gentzen, with respect to subtle complexity measures, is able to show that there is a measure which decreases, thus the process terminates. This is explained in more details in [GLT, GSS, TrSc]. Thus, to every proof we obtain a "cut-free" proof, i.e. its *normal form*. One sometimes calls the process *proof normalization*.

For the systems of linear logic we deal with in this paper, the rewriting/cut-elimination process yields *unique* normal forms, that is the cut-free form of a proof is independent of the order of applying the rewriting steps. This is proved by a Church-Rosser (or Diamond Lemma) type of argument [LS86, GLT]. In GoI, we shall obtain analogs of this property (e.g., see Lemma 5.6.8).

5.4 Traced Monoidal Categories

The theory of traces has had a fundamental impact within diverse areas of mathematics, from functional analysis and noncommutative geometry to topology and knot theory. More recently, abstract traces have arisen in logic and theoretical computer science. For example, in the 1980s and 1990s it was realized there was a need for algebraic structures modelling cyclic operations. parametrized fixedpoints and feedback in such areas as: flowchart schemes, dataflow, network algebra, and more recently in quantum computing and biological modelling.

Traced monoidal categories were introduced by Joyal, Street, and Verity [JSV96]. These categories and their variants have turned out to be key ingredients in discussing the above phenomena. As quoted by [JSV96],

> This paper introduces axioms for an abstract trace on a monoidal category. This trace can be interpreted in various contexts where it could alternatively be called contraction, feedback, Markov trace or braid closure...

There have been various extensions of traces and partial traces: we discuss more of this in Sect. 5.7 as well as in Remark 5.4.4 below.

Definition 5.4.1 A *traced symmetric monoidal category* is a symmetric monoidal category $(\mathcal{C}, \otimes, I, s)$ with a family of functions $\mathrm{Tr}^U_{X,Y} : \mathcal{C}(X \otimes U, Y \otimes U) \longrightarrow \mathcal{C}(X, Y)$ pictured in Fig. 5.6, called a *trace*, subject to the following axioms:

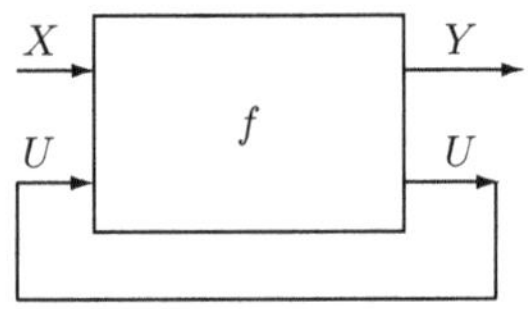

Fig. 5.6 The trace $\mathrm{Tr}^U_{X,Y}(f)$

1. **Natural** in X, $\mathrm{Tr}^U_{X,Y}(f)g = \mathrm{Tr}^U_{X',Y}(f(g \otimes 1_U))$, where $f : X \otimes U \longrightarrow Y \otimes U$, $g : X' \longrightarrow X$,
2. **Natural** in Y, $g\mathrm{Tr}^U_{X,Y}(f) = \mathrm{Tr}^U_{X,Y'}((g \otimes 1_U)f)$, where $f : X \otimes U \longrightarrow Y \otimes U$, $g : Y \longrightarrow Y'$,
3. **Dinatural** in U, $\mathrm{Tr}^U_{X,Y}((1_Y \otimes g)f) = \mathrm{Tr}^{U'}_{X,Y}(f(1_X \otimes g))$, where $f : X \otimes U \longrightarrow Y \otimes U'$, $g : U' \longrightarrow U$,
4. **Vanishing (I, II)**, $\mathrm{Tr}^I_{X,Y}(f) = f$ and $\mathrm{Tr}^{U \otimes V}_{X,Y}(g) = \mathrm{Tr}^U_{X,Y}(\mathrm{Tr}^V_{X \otimes U, Y \otimes U}(g))$, for $f : X \otimes I \longrightarrow Y \otimes I$ and $g : X \otimes U \otimes V \longrightarrow Y \otimes U \otimes V$.
5. **Superposing**,
$$g \otimes \mathrm{Tr}^U_{X,Y}(f) = \mathrm{Tr}^U_{W \otimes X, Z \otimes Y}(g \otimes f)$$
for $f : X \otimes U \longrightarrow Y \otimes U$ and $g : W \longrightarrow Z$.
6. **Yanking**, $\mathrm{Tr}^U_{U,U}(s_{U,U}) = 1_U$.

Given $f : X \otimes U \to Y \otimes U$, we think of $\mathrm{Tr}^U_{X,Y}(f)$ as "feedback along U", as in Fig. 5.6. Similarly, the axioms of traced monoidal categories have suitable geometrical representation, given in Appendix 1 (cf. also [JSV96, AHS02, Has08]).

Observe that if $X = Y = I$, up to isomorphism we have $\mathrm{Tr}^U_{X,Y}(f) : \mathcal{C}(U, U) \to \mathcal{C}(I, I)$ is a *scalar-valued* trace (cf. Proposition 5.2.3).

Exercise 5.4.2 (Generalized Yanking) Let $\mathcal{C}$ be a traced symmetric monoidal category, with arrows $f : X \to Y$ and $g : Y \to Z$. Then $g \circ f = \mathtt{Tr}^{Y}_{X,Z}(s_{Y,Z} \circ (f \otimes g))$. Geometrically, stare at the diagram in Fig. 5.7, and do a "string-pulling" argument (For an algebraic proof, see Proposition 2.4 in [AHS02])

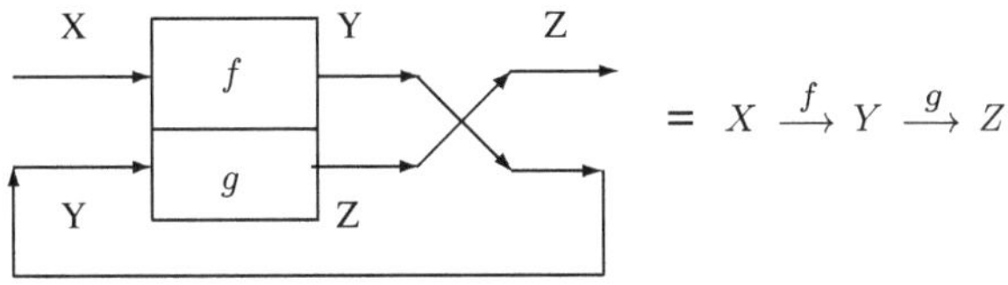

Fig. 5.7 Generalized yanking

Note that this exercise actually says that composition $g \circ f$ in a traced monoidal category is definable from tensor and symmetries. More generally, in [AHS02] we have the following normal-form theorem for arrows in traced symmetric monoidal categories:

Theorem 5.4.3 *Let $\mathcal{C}$ be a traced symmetric monoidal category, and T a collection of arrows in $\mathcal{C}$. Then any expression E built from arrows in T using tensor product, composition, and trace can be represented as $Tr(\pi F \tau)$ where F consists of tensor products of arrows in T and π, τ are permutations (built from symmetry and identity maps).*

Let us remark that for logicians, the discussion above prefigures the Execution Formula (see Eq. (5.10) and Fig. 5.9(b) below), since it illustrates the reduction of general composition ("cut") to a global trace applied to primitive compositions of permutations and tensoring.

Generalized yanking is also often used in some axiomatizations for partial traces [ABP99, Pl03] although for our purposes it is equivalent to yanking [AHS02].

Remark 5.4.4 (Some traces literature) In computer science, there has been a long tradition of studying theories related to traces and partial traces in the analysis of feedback, fixed points, iteration theories, and related notions in network algebra and flowcharts. Detailed and fundamental categorical work by Manes and Arbib [MA86], Bloom and Esik [BE93], and Stefanescu [Ste00] have greatly influenced our development here. We should mention very interesting work on circuits and feedback categories in a series of papers by Katis, Sabadini, and Walters (e.g. [KSW02]). They also introduce an interesting notion of partial trace, an important topic we introduce (for purposes of GoI) in Sect. 5.7 below. We should also mention work of P. Hines [Hi97, Hi03] both on analyzing GoI and studies of abstract machines. Finally, a survey of recent results on traced monoidal categories is in Hasegawa [Has08].

5.4.1 Wave vs. Particle Style Traces

Many examples of traces can be divided into two styles [Abr96, AHS02]: *Product Style* and *Sum Style*, or more evocatively (following Abramsky) "wave style" and "particle style". These refer, respectively, to whether the monoidal tensor $\otimes$ is given by a cartesian product versus whether it is given by a disjoint union. As explained in [Abr96, AHS02], product-style traces may be thought of as passing information in a "global information wave" while sum-style traces can be modelled by streams of particles or tokens flowing around a network (cf. [AHS02, Hag00, Hi97]). We shall now illustrate both styles of trace.

Examples 5.4.5 (Product Style Traces)

1. The category $\mathbf{Rel}_\times$ is traced. Let $R : X \times U \longrightarrow Y \times U$ be a morphism in $\mathbf{Rel}_\times$. Then $\mathtt{Tr}^U_{X,Y}(R) : X \longrightarrow Y$ is defined by: $\mathtt{Tr}^U_{X,Y}(R)(x, y) = \exists u.R(x, u, y, u)$.
2. The category $\mathbf{Vec}_{fd}$ is traced. Let $f : V \otimes U \longrightarrow W \otimes U$ be a linear map, where U, V, W are finite dimensional vector spaces with bases $\{u_i\}, \{v_j\}, \{w_k\}$. We define $\mathtt{Tr}^U_{V,W}(f) : V \longrightarrow W$ by:

$$\mathtt{Tr}^U_{V,W}(f)(v_i) = \sum_{j,k} a^{kj}_{ij} w_k \qquad \text{where } f(v_i \otimes u_j) = \sum_{k,m} a^{km}_{ij} w_k \otimes u_m.$$

   This reduces to the usual trace of $f : U \longrightarrow U$ when V and W are one dimensional.
3. Note that both $\mathbf{Rel}_\times$ and $\mathbf{Vec}_{fd}$ are compact closed categories. More generally [JSV96], every compact closed category has a *unique* canonical trace given by:

$$\mathtt{Tr}^U_{A,B}(f) = A \cong A \otimes I \stackrel{id\otimes v}{\longrightarrow} A \otimes U \otimes U^* \stackrel{f\otimes id}{\longrightarrow} B \otimes U \otimes U^* \stackrel{id\otimes\psi\circ s}{\longrightarrow} B \otimes I \cong B.$$

   Uniqueness of this trace is shown in [Has08].
4. *Coherent Logic and ∃-Doctrines.* A slight generalization of Example (1) is to consider any theory in multisorted coherent logic, that is the fragment $\{\exists, \wedge\}$ of ordinary logic (here it doesn't matter if one picks intuitionist or classical logic) [KR77]. The objects are Sorts (assumed closed under $\times$), denoted X, Y, Z, etc. Morphisms are (equivalence classes of) formulas, thought-of as relations between sorts: $R(x, y) : X \to Y$, modulo provable equivalence. Composition is defined like relational composition: $X \stackrel{R(x,y)}{\longrightarrow} Y \stackrel{S(y,z)}{\longrightarrow} Z = X \stackrel{T(x,z)}{\longrightarrow} Z$, where

$$T(x, z) = \exists y.R(x, y) \wedge S(y, z)$$

   This is a well-defined operation, using laws of coherent logic. Omitting pairing symbols, given $R(x, u, y, u') : X \times U \longrightarrow Y \times U$, define

$$\mathtt{Tr}^U_{X,Y}(x, y) : X \to Y = \exists u R(x, u, y, u)$$

The same calculations used in Example (1) can be mimicked in Coherent Logic to show that this yields a trace. The close connections of Coherent Logic with Regular Categories provides a large stock of examples of these styles of trace. Indeed, still more generally, the calculations are true for Lawvere's Existential Doctrines with $\exists_\pi$-quantifiers along projections, Frobenius Reciprocity, Beck-Chevalley, and in which equality is definable by $\exists_\Delta$, existential quantification along a diagonal [Law69, Law70].

5. The category ω-$\mathbf{CPO}_\perp$ consists of objects of ω-$\mathbf{CPO}$ with a least element $\perp$, and maps of ω-$\mathbf{CPO}$ that do not necessarily preserve $\perp$. Here $\otimes = \times$, $I = \{\perp\}$. The (dinatural) family of least-fixed-point combinators $\mathsf{Y}_U : U^U \to U$ induces a trace, given as follows (using informal lambda calculus notation): for any $f : X \times U \to Y \times U$, $\mathtt{Tr}^U_{X,Y}(f)(x) = f_1(x, \mathsf{Y}_U(\lambda u. f_2(x,u)))$, where $f_1 = \pi_1 \circ f : X \times U \to Y$, $f_2 = \pi_2 \circ f : X \times U \to U$ and $\mathsf{Y}_U(\lambda u. f_2(x,u)) =$ the least element u' of U such that $f_2(x,u') = u'$.
6. (cf. Katis, Sabadini,Walters [KSW02]) Take any (Lawvere) equational theory, for example the theory of rings. Define a category whose objects are of the form R^n for a fixed ring R, where $n \in \mathbb{N}$. Define $Hom(R^n, R^m)$ = m-tuples of polynomials in n indeterminates, with composition being substitution. For example, the identity map $R^n \xrightarrow{id} R^n$ is given by the list of n polynomials $p_1, \cdots, p_n$, where $p_i(x_1, \cdots, x_n) = x_i$. Here $\otimes$ is cartesian product.
A morphism $\vec{f}, \vec{g} \in Hom(R^n \times R^p, R^m \times R^p)$ is a list of $m+p$ polynomials in $n+p$ unknowns. We can write it as a system of polynomial equations:

$$
\begin{aligned}
y_1 &= f_1(\vec{x}, \vec{u}) \\
&\vdots \\
y_m &= f_m(\vec{x}, \vec{u}) \\
u'_1 &= g_1(\vec{x}, \vec{u}) \\
&\vdots \\
u'_p &= g_p(\vec{x}, \vec{u})
\end{aligned}
$$

The operation of trace or feedback is the formal identification of the variables u'_i on the LHS of the equations with the u_i on the RHS. Of course, to know this setting is consistent (yielding a nontrivial category) we should provide models in which there exist nontrivial solutions of such simultaneous feedback equations. These are discussed, for example, in [KSW02] above. This example admits many generalizations: for example, to general Lawvere theories, in which morphisms are represented by (equivalence classes of) terms with free variables, modulo provable equality in the theory.

Unfortunately, the above examples do not really illustrate the notion of feedback as data flow: the movement of tokens through a network. This latter view, emphasized in work of Abramsky and later Haghverdi and Hines (cf. [Abr96,

AHS02, Hag00, Hi97]), is illustrated by examples based on sum-style monoidal structure. They are related to dataflow interpretations of graphical networks. We illustrate this view with categories connected to **Rel**.

Examples 5.4.6 (Sum-style Traces)

1. $\mathbf{Rel}_+$, the category **Rel** with $\otimes = +$, disjoint union. Suppose $X + U \xrightarrow{R} Y + U$ is a relation. The coproduct injections induce four restricted relations : $R_{UU}, R_{UY}, R_{XY}, R_{XU}$ (for example, $R_{XY} \subseteq X \times Y$ is such that $R_{XY}(x, y) = R(in_1^{X,U}(x), in_1^{Y,U}(y))$. Let R^* be the reflexive, transitive closure of the relation R. A trace can be defined as follows:

$$\mathtt{Tr}^U_{X,Y}(R) = R_{XY} \cup \bigcup_{n\geq 0} R_{UY} \circ R^n_{UU} \circ R_{XU}$$
$$= R_{XY} \cup R_{UY} \circ R^*_{UU} \circ R_{XU}. \tag{5.6}$$

2. The categories **Pfn** and **PInj** of sets and partial functions (resp. sets and partial injective functions), as monoidal subcategories of $\mathbf{Rel}_+$. The tensor product is given by the disjoint union of sets, where we identify $A+B = A\times\{1\}\cup B\times\{2\}$ (note that this is not a coproduct in **PInj**, although it is a coproduct in **Pfn**). There are the obvious injections $in_1^{A,B} : A \to A + B$ and $in_2^{A,B} : B \to A + B$ as well as "quasiprojections" $\rho_1 : A + B \longrightarrow A$ given by $\rho_1((a, 1)) = a$ (where $\rho_1((b, 2))$ is undefined) and similarly for $\rho_2 : A + B \longrightarrow B$.
   Given a morphism $f : X + U \longrightarrow Y + U$, we may consider its four "components" $f_{XY} : X \to Y$, $f_{XU} : X \to U$, $f_{UX} : U \to X$, and $f_{UU} : U \to U$ obtained by pre- and post-composing with injections and quasiprojections: for example, $f_{XY} = X \xrightarrow{in_1} X + U \xrightarrow{f} Y + U \xrightarrow{\rho_1} Y$, (See Fig. 5.8).

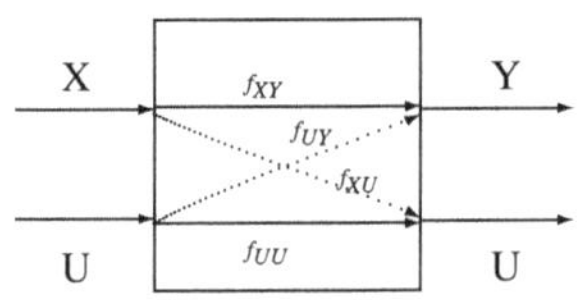

Fig. 5.8 Components of $f : X + U \to Y + U$

Both **Pfn** and **Pinj** are traced, the trace being given by the following iterative formula

$$\mathtt{Tr}^U_{X,Y}(f) = f_{XY} + \sum_{n\in\omega} f_{UY} f^n_{UU} f_{XU}, \tag{5.7}$$

which we interpret as follows:
For the category **Pfn** (respectively **PInj**), a family $\{h_i\}_{i\in I} : X \longrightarrow Y$ is said to be *summable* if the h_i's have pairwise disjoint domains (respectively, have pairwise disjoint domains and codomains). In either case, we define the sum of the family to be:

$$\left(\sum_{i\in I} h_i\right)(x) = \begin{cases} h_j(x), & \text{if } x \in \mathtt{Dom}(h_j) \text{ for some } j \in I; \\ \text{undefined}, & \text{else.} \end{cases}$$

From a dataflow view, particles enter through X, travel around a loop on U some number n of times, then exit through Y. Numerous other examples of such "particle-style" traces are studied in [AHS02, Hag00]. We shall now introduce a general theory of such traces, based upon Haghverdi's *Unique Decomposition Categories*.

5.4.2 Unique Decomposition Categories and Particle-Style Traces

How do we make sense of sums such as in Eq. (5.7) above? Haghverdi [Hag00, Hag00a] introduced symmetric monoidal categories whose homsets come equipped with (technically, are *enriched in*) an abstract summability structure, called a Σ-monoid. Sigma monoids, and their variants, permit forming certain infinite sums of maps, in a manner compatible with the monoidal category structure. Haghverdi's work is a generalization of the work of Manes and Arbib [MA86] who introduced *partially additive categories* in programming language semantics. These categories form a useful general framework for speaking of while-loops, and axiomatizing Elgot's work on feedback and iteration, as well as fixed-point semantics.

Recently, Hines and Scott have investigated the the work of Haghverdi and Manes-Arbib in more general Σ-structures with certain partially defined traces (cf Sect. 5.7 below), aimed at a general theory of "quantum while-loops" in quantum computing.

In what follows, we give a basic framework for Σ structures sufficient for our purposes.

Definition 5.4.7 A Σ*-monoid* consists of a pair (M, Σ) where M is a nonempty set and Σ is a partial operation on the countable families in M (we say that $\{x_i\}_{i\in I}$ is *summable* if $\sum_{i\in I} x_i$ is defined), subject to the following axioms:

1. *Partition-Associativity Axiom.* If $\{x_i\}_{i\in I}$ is a countable family and if $\{I_j\}_{j\in J}$ is a (countable) partition of I, then $\{x_i\}_{i\in I}$ is summable if and only if $\{x_i\}_{i\in I_j}$ is summable for every $j \in J$ and $\sum_{i\in I_j} x_i$ is summable for $j \in J$. In that case, $\sum_{i\in I} x_i = \sum_{j\in J}(\sum_{i\in I_j} x_i)$.
2. *Unary Sum Axiom.* Any family $\{x_i\}_{i\in I}$ in which I is a singleton is summable and $\sum_{i\in I} x_i = x_j$ if $I = \{j\}$.

A morphism of Σ monoids is a function that preserves sums of countably-indexed summable families: i.e. if $\{x_i\}_{i\in I}$ is summable, then so is $\{f(x_i)\}_{i\in I}$ and $f(\sum_{i\in I} x_i) = \sum_{i\in I} f(x_i)$. Σ-monoids form a symmetric monoidal closed category $\Sigma\mathbf{Mon}$.

A $\Sigma\mathbf{Mon}$-category $\mathcal{C}$ is a category enriched in $\Sigma\mathbf{Mon}$; i.e. its homsets are enriched with a partial infinitary sum, compatible with composition. Such categories have non-empty homsets, e.g. they have zero morphisms $0_{XY} : X \longrightarrow Y = \sum_{i\in\emptyset} f_i$ for $f_i \in \mathcal{C}(X, Y)$. For details see [MA86, Hag00].

Definition 5.4.8 A *unique decomposition category* (UDC) $\mathcal{C}$ is a symmetric monoidal Σ**Mon**-category which satisfies the following axiom:

(A) For all $j \in I$ there are morphisms called *quasi injections*: $\iota_j : X_j \longrightarrow \otimes_I X_i$, and *quasi projections*: $\rho_j : \otimes_I X_i \longrightarrow X_j$, such that

(i) $\rho_k \iota_j = 1_{X_j}$ if $j = k$ and $0_{X_j X_k}$ otherwise.
(ii) $\sum_{i\in I} \iota_i \rho_i = 1_{\otimes_I X_i}$.

Proposition 5.4.9 (Finite Matrix Representation) *Given* $f : \otimes_J X_j \longrightarrow \otimes_I Y_i$ *in a UDC with* $|I| = m$ *and* $|J| = n$*, there exists a unique family* $\{f_{ij}\}_{i\in I, j\in J} : X_j \longrightarrow Y_i$ *with* $f = \sum_{i\in I, j\in J} \iota_i f_{ij} \rho_j$*, namely,* $f_{ij} = \rho_i f \iota_j$.

Thus every morphism $f : \otimes_J X_j \longrightarrow \otimes_I Y_i$ in a UDC can be represented by a matrix; for example f above (with $|I| = m$ and $|J| = n$) is represented by the $m \times n$ matrix $[f_{ij}]$. Composition of morphisms in a UDC then corresponds to matrix multiplication.

Proposition 5.4.10 (Standard Trace Formula) *Let* $\mathcal{C}$ *be a unique decomposition category such that for every* X, Y, U *and* $f : X \otimes U \longrightarrow Y \otimes U$*, the sum* $f_{11} + \sum_{n=0}^{\infty} f_{12} f_{22}^n f_{21}$ *exists, where* f_{ij} *are the components*[1] *of* f*. Then,* $\mathcal{C}$ *is traced and* $Tr_{X,Y}^U(f) = f_{11} + \sum_{n=0}^{\infty} f_{12} f_{22}^n f_{21}$.

The trace formula above is called the standard trace, and a UDC with such a trace is called a *traced UDC with standard trace*. Note that a UDC can be traced with a trace different from the standard one. In this paper all traced UDCs are the ones with the standard trace.

We now present some more examples. For further details, see [AHS02, Hag00].

Examples 5.4.11 (Traced UDC's)

1. All the categories in Example 5.4.6 above. In **Rel**$_+$, *all* countable families are summable, and $\sum_{i\in I} R_i = \cup_i R_i$. In the case of **Pfn** and **PInj**, summability of a family of morphisms $\{f_i\}_{i\in I}$ is as given above in the Examples. In this case, the two trace Formulas (5.6) and (5.7) exactly correspond to the standard trace formula in Proposition 5.4.10 above.
2. **SRel**, the category of stochastic relations. Here the objects are measurable spaces $(X, \mathcal{F}_X)$ and maps $f : (X, \mathcal{F}_X) \rightarrow (Y, \mathcal{F}_Y)$ are stochastic kernels, i.e. $f : X \times \mathcal{F}_Y \rightarrow [0, 1]$ which are bounded measurable in the first variable and subprobability measures in the second. Composition $g \circ f(x, C) = \int_Y g(-, C) df(x, -)$ where $f(x, \_)$ is the measure for integration. This category has finite and countable coproducts (which form the tensor). A family $\{f_i\}_{i\in I}$ is summable iff $\sum_{i\in I} f_i(x, Y) \leq 1$ for all $x \in X$.

---

[1] Here $f_{11} : X \rightarrow Y$, $f_{12} : U \rightarrow X$, $f_{21} : X \rightarrow U$, $f_{22} : U \rightarrow U$.

3. **Hilb**$_2$. Consider the category **cHilb** of Hilbert spaces and linear contractions (norm ≤ 1). Barr [Barr92] defined a contravariant faithful functor ℓ_2 : $\mathbf{PInj}^{op} \longrightarrow \mathbf{cHilb}$ by: for a set X, $\ell_2(X)$ is the set of all complex valued functions a on X for which the (unordered) sum $\sum_{x \in X} |a(x)|^2$ is finite. $\ell_2(X)$ is a Hilbert space with norm given by $||a|| = (\sum_{x \in X} |a(x)|^2)^{1/2}$ and inner product given by $< a, b >= \sum_{x \in X} a(x)\overline{b(x)}$ for $a, b \in \ell_2(X)$. Given a partial injection $f : X \to Y$ in **PInj**, then $l_2(f) : \ell_2(Y) \to \ell_2(X)$ is defined by

$$\ell_2(f)(b)(x) = \begin{cases} b(f(x)) & x \in Dom(f) \\ 0 & \text{otherwise.} \end{cases}$$

This gives a correspondence between partial injective functions and partial isometries on Hilbert spaces (see also [Gi95a, Abr96].) Let $\mathbf{Hilb}_2 = \ell_2[\mathbf{PInj}]$. Its objects are $\ell_2(X)$ for a set X and morphisms $u : \ell_2(X) \longrightarrow \ell_2(Y)$ are of the form $\ell_2(f)$ for some partial injective function $Y \xrightarrow{f} X$. Hence, $\mathbf{Hilb}_2$ is a nonfull subcategory of **Hilb**. It forms a traced UDC with respect to the induced ℓ_2 structure, as follows:

- $\ell_2(X) \oplus \ell_2(Y) \cong \ell_2(X \uplus Y)$ is a tensor product in $\mathbf{Hilb}_2$ (but is a biproduct in **Hilb**) with unit $\ell_2(\emptyset)$.
- Quasi injections and projections = their ℓ_2 images from **PInj**.
- Define: A $\mathbf{Hilb}_2$ family $\{\ell_2(f_i)\}$ is *summable* if
  - $\{f_i\}$ is summable in **PInj**
  - In that case, $\sum_i \ell_2(f_i) =_{def} \ell_2(\sum_i f_i)$.
- $\mathbf{Hilb}_2$ is traced. Given

$$u : \ell_2(X) \oplus \ell_2(U) \longrightarrow \ell_2(Y) \oplus \ell_2(U)$$
$$\mathtt{Tr}(u) =_{def} \ell_2(Tr^U_{Y,X}(f))$$

  where $u = \ell_2(f)$ with $f : Y \uplus U \longrightarrow X \uplus U \in \mathbf{PInj}$.
- Since **PInj** is self-dual, $\ell_2 : \mathbf{PInj} \to \mathbf{Hilb}_2$ is an equivalence of categories. Here is a chart giving some explicit equivalences:

$\mathbf{PInj}(X, Y)$	$\mathbf{Hilb}(\ell_2(Y), \ell_2(X))$
f	$\ell_2(f)$
Partial injective function	Partial isometry
Total	Isometry
Total and surjective	Unitary
$X = Y$ and f is identity on $Dom(f)$	Projection

- Many (although not all) of the above examples of traced UDC's are special cases of the Partially Additive Categories of Manes and Arbib [MA86]. Those also form traced UDC's with standard trace formula.

5.4.3 The Int Construction

Starting with a symmetric traced monoidal category $\mathcal{C}$, we now describe a compact closed category **Int**($\mathcal{C}$) given in [JSV96] (which is isomorphic to the category $\mathcal{G}(\mathcal{C})$ in [Abr96]). We follow the treatment in [Abr96], and actually give the construction for $\mathcal{G}(\mathcal{C})$; for simplicity, we call both these categories the **Int** construction. The reason for the name is in Exercise 5.4.13 below.

Definition 5.4.12 (The Int Construction) Given a traced monoidal category $\mathcal{C}$ we define a compact closed category **Int**($\mathcal{C}$) $\cong \mathcal{G}(\mathcal{C})$ as follows:

- Objects: Pairs of objects (A^+, A^-) where A^+ and A^- are objects of $\mathcal{C}$.
- Arrows: An arrow $f : (A^+, A^-) \longrightarrow (B^+, B^-)$ in **Int**($\mathcal{C}$) is an arrow $f : A^+ \otimes B^- \longrightarrow A^- \otimes B^+$ in $\mathcal{C}$.
- Identity: $1_{(A^+,A^-)} = s_{A^+,A^-}$, the symmetry or "twist" map.
- Composition: Arrows $f : (A^+, A^-) \longrightarrow (B^+, B^-)$ and $g : (B^+, B^-) \longrightarrow (C^+, C^-)$ have composite $g \circ f : (A^+, A^-) \longrightarrow (C^+, C^-)$ given by:

$$g \circ f = \mathrm{Tr}^{B^- \otimes B^+}_{A^+ \otimes C^-, A^- \otimes C^+}(\beta(f \otimes g)\alpha)$$

where $\alpha = (1_{A^+} \otimes 1_{B^-} \otimes s_{C^-,B^+})(1_{A^+} \otimes s_{C^-,B^-} \otimes 1_{B^+})$ and $\beta = (1_{A^-} \otimes 1_{C^+} \otimes s_{B^+,B^-})(1_{A^-} \otimes s_{B^+,C^+} \otimes 1_{B^-})(1_{A^-} \otimes 1_{B^+} \otimes s_{B^-,C^+})$. Pictorially, $g \circ f$ is given by symmetric feedback:

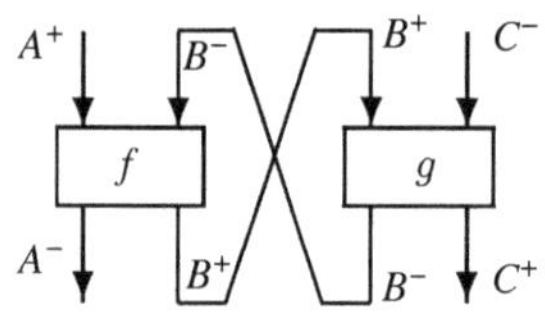

- Tensor: $(A^+, A^-) \otimes (B^+, B^-) = (A^+ \otimes B^+, A^- \otimes B^-)$ and for $(A^+, A^-) \longrightarrow (B^+, B^-)$ and $g : (C^+, C^-) \longrightarrow (D^+, D^-)$, $f \otimes g = (1_{A^-} \otimes s_{B^+,C^-} \otimes 1_{D^+})(f \otimes g)(1_{A^+} \otimes s_{C^+,B^-} \otimes 1_{D^-})$
- Unit: (I, I).
- Duality: The dual of (A^+, A^-) is given by $(A^+, A^-)^* = (A^-, A^+)$ where the unit $\eta : (I, I) \longrightarrow (A^+, A^-) \otimes (A^+, A^-)^* =_{def} s_{A^-,A^+}$ and the counit map $\epsilon : (A^+, A^-)^* \otimes (A^+, A^-) \longrightarrow (I, I) =_{def} s_{A^-,A^+}$.
- Internal Homs: As usual, $(A^+, A^-) \multimap (B^+, B^-) = (A^+, A^-)^* \otimes (B^+, B^-) = (A^- \otimes B^+, A^+ \otimes B^-)$.

Following Abramsky [Abr96], we interpret the objects of **Int**($\mathcal{C}$) in a game-theoretic manner: A^+ is the type of "moves by Player (the System)" and A^- is the type of "moves by Opponent (the Environment)". The composition of morphisms in **Int**($\mathcal{C}$) is connected to Girard's execution formula (see below). In [Abr96] it is pointed out that $\mathcal{G}$(**PInj**) captures the essence of the original Girard GoI interpretation in [Gi89a] (we discuss this in more detail below), while $\mathcal{G}(\omega\text{-CPO}_\perp)$ is the model of GoI in [AJ94a].

Exercise 5.4.13 (Why Int?) The **Int** construction above is analogous to (in fact, it yields) the construction of the integers $\mathbb{Z}$ from the natural numbers $\mathbb{N}$. Indeed (using the notation above): put an equivalence relation on $\mathbb{N} \times \mathbb{N}$ by defining: $(A^+, A^-) \sim (B^+, B^-)$ iff $A^+ + B^- = A^- + B^+$ in $\mathbb{N}$. Prove this yields $\mathbb{Z}$. Harder question: show how this is a special case of the **Int** construction.

Translating the work of [JSV96] in our setting we obtain that **Int**$(\mathcal{C})$ is a kind of "free compact closure" of $\mathcal{C}$ at the bicategorical level (for which the reader is referred to [JSV96]):

Proposition 5.4.14 *Let $\mathcal{C}$ be a traced symmetric monoidal category*

1. ***Int**$(\mathcal{C})$ defined above is a compact closed category. Moreover,* $F_\mathcal{C} : \mathcal{C} \longrightarrow$ ***Int**$(\mathcal{C})$ defined by $F_\mathcal{C}(A) = (A, I)$ and $F_\mathcal{C}(f) = f$ is a full and faithful embedding.*
2. *The inclusion[2] of 2-categories **CompCl** $\hookrightarrow$ **TraMon** of compact closed categories into traced monoidal ones has a left biadjoint with unit having component at $\mathcal{C}$ given by $F_\mathcal{C}$.*

We remark that [Has08] shows (in the general setting of [JSV96]) that a traced monoidal category $\mathcal{C}$ is closed iff the canonical inclusion $\mathcal{C} \hookrightarrow$ **Int**$(\mathcal{C})$ has a right adjoint. Finally, we should remark that the **Int** construction has seen other applications in recent categorical studies of the semantics of quantum computing, arising from the fundamental paper [AbCo04].

5.5 What is the Geometry of Interaction?

5.5.1 Dynamical Invariants for Cut-Elimination

Recall the earlier discussion of Cut-Elimination and the rewriting theory of proofs. We begin with some general questions:

- How do we mathematically model the dynamics of cut-elimination (i.e. the movement of information in the rewriting of the proof trees)?
- Are there dynamical (mathematical) invariants φ for proof normalization, that is: if π rewrites to π', then $\varphi(\pi) = \varphi(\pi')$?
- In what sense is cut-elimination related to recent theories of abstract algorithms?

Recall that in categorical proof theory, for any logic $\mathcal{L}$, we may interpret proofs of sequents $\Gamma \vdash \Delta$ as arrows (in an appropriate structured category $\mathcal{C}$) as in (5.4)

[2] Recent work of Hasegawa and Katsumata [HK09] has shown that the notion of 2-cell in [JSV96] must be changed to *invertible* monoidal natural transformation.

above. This gives an interpretation function (call a *denotation*) $[\![-]\!]_d : \mathcal{L} \to \mathcal{C}$ which satisfies: for any rewriting step $\rightsquigarrow$ in the cut-elimination process, if $\pi \rightsquigarrow \pi'$ then $[\![\pi]\!]_d = [\![\pi']\!]_d$. Such functions $[\![\pi]\!]_d$ lead to a rather bland notion of "invariant" for cut-elimination. Indeed, "$\rightsquigarrow$" implies simply denotational equality, the equations one must impose to give the appropriate algebraic structure of the category of proofs (depending on the logic): e.g. cartesian, cartesian closed, monoidal closed, etc. We search for more meaningful invariants, with deeper connections to the dynamics.

Girard's Geometry of Interaction (GoI) program was the first attempt to model, in a mathematically sophisticated way, the dynamics of cut-elimination, and in particular to find an invariant (the Execution Formula) with more subtle features. The first proposal appeared in [Gi89], followed by an important series of papers [Gi89a, Gi88, Gi95a] written in the language of operator algebras. His recent work [Gi07, Gi08] has moved towards the framework of von Neumann algebras. However, it became clear early on, from lectures of Abramsky [AJ94a, Abr96] and also Hyland in the early 1990s that more simple conceptual machinery, now understood to be based on traced monoidal categories, suffices to understand many of the fundamental algebraic and geometric ideas underlying early GoI. This was explored by us in a series of papers [AHS02, HS04a, HS04b, HS05a, Hag06]. In what follows we shall explore some algebraic aspects of Girard's early GoI 1, and the notion of information flow. We leave it an open question how to connect this up with Girard's more recent ideas based on von Neumann algebras [Gi07, Gi08].

5.5.2 Girard's GoI 1 Framework: An Overview

The basic idea of [Gi89a] is to consider proofs as certain matrix operators on a C^*-algebra $\mathcal{B}(\mathsf{H})$ of bounded linear operators on a Hilbert space H. We shall look at proofs π of 1-sided sequents in **LL**, say $\pi : \vdash \Gamma$, where Γ is a list of formulas. A key notion in Girard's work was to keep track of all the cut formulas used in a proof. These *general proofs* have the form $\pi : \vdash [\Delta], \Gamma$ where Δ is a list of all the Cut formulas generated from applying the Cut Rule, as follows:

$$\frac{\vdash [\Delta], \Gamma, A \quad \vdash [\Delta'], A^{\perp}, \Gamma'}{\vdash [\Delta, \Delta', A, A^{\perp}]\ \Gamma,\ \Gamma'}\ \text{Cut}$$

We think of this Cut Rule as taking the cut formulas $A, A^{\perp}$ (in that order) and putting them on a stack: the (ordered) list $[\Delta, \Delta', A, A^{\perp}]$.

Thus, in a general proof $\pi : \vdash [\Delta], \Gamma$, we have that Δ is an even length list of cut formulas, say $\Delta = A_1, A_1^{\perp}, \cdots, A_m, A_m^{\perp}$. In general suppose $|\Delta| = 2m$ and $|\Gamma| = n$, so that $\vdash [\Delta], \Gamma$ has $n + 2m$ formulas. Let us informally describe the GoI ingredients.

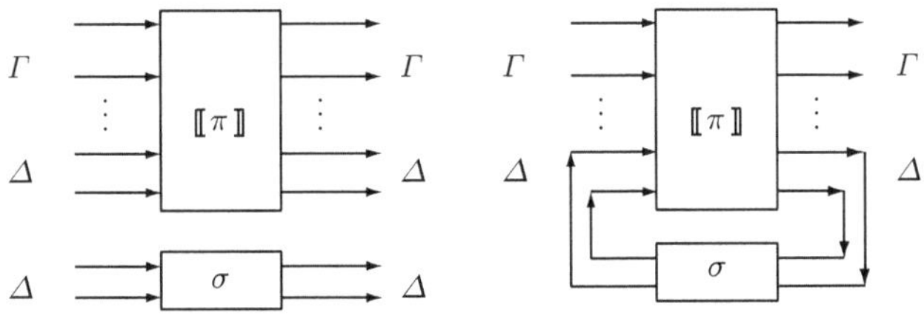

$(\llbracket \pi \rrbracket, \sigma)$ $\qquad$ $Ex(\llbracket \pi \rrbracket, \sigma) = \mathtt{Tr}^{\otimes\Delta}_{\otimes\Gamma,\otimes\Gamma}((1_{\otimes\Gamma} \otimes \sigma)\llbracket \pi \rrbracket)$

(a) $\qquad$ (b)

Fig. 5.9 Proofs of $\vdash [\Delta], \Gamma$ as I/O boxes

A key aspect of Girard's interpretation is to consider a *Dynamic Interpretation* $\llbracket - \rrbracket$ of proofs. A proof $\pi : \vdash [\Delta], \Gamma$ will be modelled by a pair of I/O (input-output) boxes (Fig. 5.9(a)), in which σ represents the set of cuts Δ. Cut-elimination will be modelled by a diagram involving the feedback on σ (Fig. 5.9(b).)

Formulas in sequents are interpreted (uniformly) using a special object U in the category $\mathcal{C}$. In Girard's GoI 1, $\mathcal{C}$ = **Hilb**, the category of Hilbert spaces and bounded linear maps and $U = \ell_2(\mathbb{N}) = \ell^2$, the Hilbert space of square summable sequences. Indeed, the interpretation actually occurs in **Hilb**$_2$ (Example 5.4.11 (3)). We know that **Hilb**$_2$ is equivalent to **PInj** under the ℓ_2 functor; it follows that the GoI 1 interpretation below may equally well be thought-of as occurring in **PInj**, with $U = \mathbb{N}$.

In the GoI interpretation of logic, formulas are interpreted as *types* via a notion of *orthogonality*, $(\ )^{\perp}$, on certain hom-sets. Such notions of orthogonality are needed both to define types (as sets equal to their biorthogonal) as well as to give convergence-like properties of the Execution Formula. Below we introduce such notions concretely in Definition 5.6.3, and more abstractly (following [HylSc03]) in Definition 5.7.9.

Proofs on the other hand are interpreted as morphisms in **Int**($\mathcal{C}$). Suppose we have a proof π of a sequent $\vdash [\Delta], \Gamma$, with $|\Gamma| = n$ and $|\Delta| = 2m$. This is interpreted as a morphism $\llbracket \pi \rrbracket$ in **Int**($\mathcal{C}$) from (U^n, U^{2m}) to itself, where U^k is a shorthand for the k-fold tensor product of U with itself: equivalently, as a map $U^{n+2m} \xrightarrow{\llbracket \pi \rrbracket} U^{n+2m}$ in $\mathcal{C}$. Notice that all formulas Γ and Δ occur twice (i.e. as both inputs and outputs to $\llbracket \pi \rrbracket$) in Fig. 5.9(a).

Remark 5.5.1 (GoI Notation) For ease of computing the GoI interpretation of proofs π (using their graphical representations as in Fig. 5.9 above), we often label the inputs and outputs by the I/O formulas themselves (e.g. Γ, Δ in Fig. 5.9), rather than the object U (which uniformly interprets all formulas).

The interpretation of proofs is completed by defining the morphism $\sigma := s^{\otimes m}$ representing Δ, where s is the symmetry (i.e. the identity map in **Int**($\mathcal{C}$)). The precise sense in which we interpret formulas and proofs will be described in Sect. 5.6 below.

To recap, we will interpret proofs-with-cuts $\pi : \vdash [\Delta], \Gamma$ as pairs $([\![\pi]\!], \sigma)$ such that:

- $[\![\pi]\!] : U^{n+2m} \longrightarrow U^{n+2m}$ is defined inductively on proofs, and
- $\sigma : U^{2m} \longrightarrow U^{2m} = s^{\otimes m}$ (the m-fold tensor product of the symmetry morphism $s_{U,U}$ with itself) represents the cuts Δ.

Here, $|\Delta| = 2m$ and $|\Gamma| = n$. If $\Delta = \emptyset$, π is cut-free and $\sigma = 0$ will be a zero morphism. (This will always exist, since our categories will be Σ-monoid enriched). We note that in Girard's model $\mathbf{Hilb}_2$ and our $*$-category approach in Proposition 5.8.7 below, $([\![\pi]\!], \sigma)$ are partial symmetries).

As we are working in a traced UDC, we can use the matricial representation of arrows (see Proposition 5.4.9) to write $[\![\pi]\!]$ as a block matrix:

$$[\![\pi]\!] = \left(\begin{array}{c|c} \pi_{11} & \pi_{12} \\ \hline \pi_{21} & \pi_{22} \end{array}\right)$$

The dynamics of proofs (cut-elimination) will be interpreted using the Execution Formula defined in formula (5.8) below. This is illustrated in Fig. 5.9(b). In any traced UDC, this can be represented as a sum, as in (5.9) below.

Execution/Trace Formula

$$EX([\![\pi]\!], \sigma) =_{def} \mathrm{Tr}^{\otimes\Delta}_{\otimes\Gamma,\otimes\Gamma}((1_{\otimes\Gamma} \otimes \sigma)[\![\pi]\!]) \tag{5.8}$$

$$= \pi_{11} + \sum_{n\geq 0} \pi_{12}(\sigma\pi_{22})^n(\sigma\pi_{21}) \tag{5.9}$$

Note that the underlying category $\mathcal{C}$ is a traced UDC and more generally in Sect. 5.8, a traced category, where $[\![\otimes\Delta]\!] = U^{2m}$, $[\![\otimes\Gamma]\!] = U^n$. Thus $EX([\![\pi]\!], \sigma) : U^n \rightarrow U^n$ exists as a $\mathcal{C}$-morphism.

The essential mathematical ingredients at work in GoI were understood to consist of a traced symmetric monoidal category, a traced endofunctor and the special object U we alluded to above, called a *reflexive* object. Such structures are called *GoI Situations*, see below for detailed definitions. In [HS04a], we showed that Girard's GoI 1 can be modeled categorically, using GoI Situations where the underlying category $\mathcal{C}$ is a traced UDC. In particular, we proved that the original operator algebraic framework in [Gi89a] is captured by the GoI Situation on the category $\mathbf{Hilb}_2$, see Proposition 5.6.13 below.

Remark 5.5.2 (GoI, Path-Based Computing, Complexity) Important approaches to GoI arose in work of V. Danos and L. Regnier and coworkers [Dan90, DR95, Lau01, MR91]. This work analyzes information flow in β-reduction of untyped lambda calculus, using paths in proof-nets. The GoI execution formula may be analyzed as a certain kind of sum-of-paths formula, breaking down β-reduction to local reversible asynchronous steps. The authors give detailed and profound analyses of the kinds of paths and information flow this viewpoint represents, together with fundamental algebraic models for computation. This leads to important connections with previous work in the geometry of β-reduction; in particular, to relations of GoI with optimal reduction [GAL92].

GoI has also had some connections with complexity [GSS, BP01] in particular with evaluation strategies and rates of growth of numerical measures assigned to proofs in bounded logics. In the case of Traced UDC-style models of GoI and GoI Situations as studied here, more recently Schöpp [Sch07] used this machinery to study fragments of bounded (affine) linear logic suitable for studying logarithmic space.

In a different direction, in [AHS02], a general analysis of algebraic models of GoI is carried out. There it is shown how to use GoI Situations to obtain models of the $\{!, \multimap\}$ fragment of linear logic, presented in terms of *linear combinatary algebras*. These are certain combinatory algebras $(\mathcal{A}, \cdot)$ equipped with a map $! : \mathcal{A} \to \mathcal{A}$ and constants $B, C, I, K, W, D, \delta, F$ satisfying the combinatory identities for a Hilbert-style axiomatization of $\{!, \multimap\}$. The method is sketched as follows.

Let $\mathcal{C}$ be a traced smc, with an endofunctor $T : \mathcal{C} \to \mathcal{C}$ and an object (called a *reflexive object*) $U \in \mathcal{C}$ with retractions $U \otimes U \lhd U$, $I \lhd U$, and $TU \lhd U$. Then if T satisfies some reasonable axioms and setting $V = (U, U)$ and $I = (I, I)$, it is shown in [AHS02] how the homset $\mathbf{Int}(\mathcal{C})(I, V) = \mathcal{C}(U, U)$ naturally inherits the structure of a linear combinatory algebra. For example, in the case of $\mathcal{C} = \mathbf{PInj}$, $\mathbb{N}$ is such a reflexive object, with endofunctor $T(-) = \mathbb{N} \times (-)$. This example underlies the original Girard GoI constructions. The model in [AJ94a] likewise arises from $\mathbf{Int}(\mathbf{CPO}_\perp)$. Moreover, Girard's original operator-theoretic models (in the category of Hilbert spaces), as well as Danos-Regnier's *small model* [DR95] are also captured in the above framework using some additional functorial structure (see [Hag00]).

We should mention Abramsky's paper [Abr07] which, while discussing Temperley-Lieb Algebra in knot theory, develops a version of *planar* GoI. In a different vein, Fuhrman and Pym [FP07] develop a categorical framework for obtaining models for classical logic using a GoI/Int construction applied to certain extensions of symmetric linearly distributive categories, along the lines of the work of Blue-Cockett-Seely.

5.6 GoI Interpretation of MELL

The Geometry of Interaction interprets an underlying logical system at three levels: *formulas, proofs* and *cut-elimination.* We shall carry out this interpretation for **MELL** without units in the following sections. There are two fundamental ingredients in a GoI interpretation: (i) A GoI Situation containing the underlying traced UDC, and (ii) A notion of orthogonality. We begin by defining these ingredients. We shall discuss generalizations and extensions of these notions in later sections.

Definition 5.6.1 A *GoI Situation* is a triple $(\mathcal{C}, T, U)$ where:

1. $\mathcal{C}$ is a traced symmetric monoidal category
2. $T : \mathcal{C} \longrightarrow \mathcal{C}$ is a traced symmetric monoidal functor with the following *monoidal* retractions (i.e. the retraction pairs are monoidal natural transformations):

   a) $TT \lhd T\ (e, e')$ (Comultiplication)
   b) $Id \lhd T\ (d, d')$ (Dereliction)
   c) $T \otimes T \lhd T\ (c, c')$ (Contraction)
   d) $\mathcal{K}_I \lhd T\ (w, w')$ (Weakening). Here $\mathcal{K}_I$ is the constant I functor.

3. U is an object of $\mathcal{C}$, called *a reflexive object*, with retractions:
   (a) $U \otimes U \lhd U\ (j, k)$, (b) $I \lhd U$, and (c) $TU \lhd U\ (u, v)$.

Here $TT \lhd T\ (e, e')$ means that there are monoidal natural transformations $e_X : TTX \longrightarrow TX$ and $e'_X : TX \longrightarrow TTX$ such that $e'e = 1_{TT}$. We say that TT *is a retract of* T. Similarly for the other items.

Before we proceed, let's consider some examples of GoI Situations $(\mathcal{C}, T, U)$. For comparison of our notation with the notation of Girard and his students, see Appendix 2.

Examples 5.6.2

1. (**PInj**, $\mathbb{N} \times -, \mathbb{N}$). Here $\mathbb{N}$ is the set of natural numbers. The functor $T = \mathbb{N} \times -$, is defined as $TX = \mathbb{N} \times X$ and for a morphism $f : X \longrightarrow Y, Tf = 1_{\mathbb{N}} \times f$. We shall refer the reader to [AHS02] for details on this and the following examples. However, we include a few definitions for illustration. For example, consider the cases for $U \otimes U \lhd U\ (j, k)$, Comultiplication and Contraction:

   - $\mathbb{N} \uplus \mathbb{N} \lhd \mathbb{N}\ (j, k)$ is defined by $j : \mathbb{N} \uplus \mathbb{N} \longrightarrow \mathbb{N}$, $j(1, n) = 2n$, $j(2, n) = 2n+1$ and $k : \mathbb{N} \longrightarrow \mathbb{N} \uplus \mathbb{N}$,

   $$k(n) = \begin{cases} (1, n/2), & \text{if } n \text{ even;} \\ (2, (n-1)/2), & \text{if } n \text{ odd}. \end{cases}$$

   Clearly $kj = 1_{\mathbb{N} \uplus \mathbb{N}}$.

   - (Comultiplication) $\mathbb{N} \times (\mathbb{N} \times X) \xrightarrow{e_X} \mathbb{N} \times X$ and $\mathbb{N} \times X \xrightarrow{e'_X} \mathbb{N} \times (\mathbb{N} \times X)$
   $\mathbb{N} \times (\mathbb{N} \times X) \xrightarrow{e_X} \mathbb{N} \times X$ is defined by, $e_X(n_1, (n_2, x)) = (\langle n_1, n_2 \rangle, x)$. Given $f : X \longrightarrow Y$, $(1_{\mathbb{N}} \times f)e_X((n_1, (n_2, x)) = (\langle n_1, n_2 \rangle, f(x)) = e_Y(1_{\mathbb{N}} \times (1_{\mathbb{N}} \times f)(n_1, (n_2, x))$ for all $n_1, n_2 \in \mathbb{N}$ and $x \in X$ proving the naturality of e_X.
   $e'_X(n, x) = (n_1, (n_2, x))$ where $\langle n_1, n_2 \rangle = n$.
   $e'_X e_X(n_1, (n_2, x)) = e'_X(\langle n_1, n_2 \rangle, x) = (n_1, (n_2, x))$ for all $n_1, n_2 \in \mathbb{N}$ and $x \in X$.
   - (Contraction) $(\mathbb{N} \times X) \uplus (\mathbb{N} \times X) \xrightarrow{c_X} \mathbb{N} \times X$ and $\mathbb{N} \times X \xrightarrow{c'_X} (\mathbb{N} \times X) \uplus (\mathbb{N} \times X)$.

   $$c_X = \begin{cases} (1, (n, x)) \mapsto (2n, x) \\ (2, (n, x)) \mapsto (2n+1, x) \end{cases}$$

Given $f : X \longrightarrow Y$, $(1_{\mathbb{N}} \times f)c_X(1, (n, x)) = (2n, f(x)) = c_Y(1_{\mathbb{N}} \times f \uplus 1_{\mathbb{N}} \times f)(1, (n, x))$ for all $n \in \mathbb{N}$ and $x \in X$. Similarly $(1_{\mathbb{N}} \times f)c_X(2, (n, x)) = (2n + 1, f(x)) = c_Y(1_{\mathbb{N}} \times f \uplus 1_{\mathbb{N}} \times f)(2, (n, x))$ for all $n \in \mathbb{N}$ and $x \in X$, proving the naturality of c_X.

$$c'_X(n, x) = \begin{cases} (1, (n/2, x)), & \text{if } n \text{ is even;} \\ (2, ((n-1)/2, x), & \text{if } n \text{ is odd.} \end{cases}$$

Finally, $c'_X c_X(1, (n, x)) = c'_X(2n, x) = (1, (n, x))$ and $c'_X c_X(2, (n, x)) = c'_X(2n + 1, x) = (2, (n, x))$.

2. $(\mathbf{Pfn}, \mathbb{N} \times -, \mathbb{N})$.
3. $(\mathbf{Rel}_+, \mathbb{N} \times -, \mathbb{N})$.
4. $(\mathbf{SRel}, T, \mathbb{N}^\infty)$. Here $T : \mathbf{SRel} \longrightarrow \mathbf{SRel}$ is defined as $T(X, \mathcal{F}_X) = (\mathbb{N} \times X, \mathcal{F}_{\mathbb{N}\times X})$ where $\mathcal{F}_{\mathbb{N}\times X}$ is the σ-field on $X \uplus X \uplus X \cdots$ (ω copies). For a given $f : (X, \mathcal{F}_X) \longrightarrow (Y, \mathcal{F}_Y)$, $Tf((n, x), \biguplus_{i\in\omega} B_i) = f(x, B_n)$.

Note that throughout this section we shall be working with GoI Situations where the underlying category $\mathcal{C}$ is a traced UDC. Formulas and proofs will be interpreted in the endomorphism monoid of the reflexive object, i.e. $\mathcal{C}(U, U)$.

Definition 5.6.3 (Orthogonality and Types) Let f, g be morphisms in $\mathcal{C}(U, U)$. We say that f is *nilpotent* if $f^k = 0$ for some $k \geq 1$. We say that f is *orthogonal* to g, denoted $f \perp g$ if gf is nilpotent. Orthogonality is a symmetric relation and it makes sense because 0_{UU} exists. Also, $0 \perp f$ for all $f \in \mathcal{C}(U, U)$.

Given a subset X of $\mathcal{C}(U, U)$, we define

$$X^\perp = \{f \in \mathcal{C}(U, U) | \forall g(g \in X \Rightarrow f \perp g)\}$$

A *type* is any subset X of $\mathcal{C}(U, U)$ such that $X = X^{\perp\perp}$. Note that types are inhabited, since 0_{UU} belongs to every type.

5.6.1 GoI Interpretation of Formulas

Formulas are interpreted by *types* as defined above, by induction.

Definition 5.6.4 Consider a GoI situation $(\mathcal{C}, T, U)$ as above with j_1, j_2, k_1, k_2 components of j and k respectively. Let A be an **MELL** formula. We define *the GoI interpretation of* A, denoted θA, inductively as follows:

1. If $A \equiv \alpha$ that is A is an atom, then $\theta A = X$ an arbitrary type.
2. If $A \equiv \alpha^\perp$, $\theta A = X^\perp$, where $\theta\alpha = X$ is given by assumption.
3. If $A \equiv B \otimes C, \theta A = Y^{\perp\perp}$, where $Y = \{j_1 a k_1 + j_2 b k_2 | a \in \theta B, b \in \theta C\}$.
4. If $A \equiv B \mathbin{⅋} C$, $\theta A = Y^\perp$, where $Y = \{j_1 a k_1 + j_2 b k_2 | a \in (\theta B)^\perp, b \in (\theta C)^\perp\}$.

5. If $A \equiv\ !B$, $\theta A = Y^{\perp\perp}$, where $Y = \{uT(a)v | a \in \theta B\}$.
6. If $A \equiv ?B$, $\theta A = Y^{\perp}$, where $Y = \{uT(a)v | a \in (\theta B)^{\perp}\}$.

An easy consequence of the definition is $(\theta A)^{\perp} = \theta A^{\perp}$ for any formula A.

5.6.2 GoI Interpretation of Proofs

In this section we formally define the GoI interpretation for proofs of **MELL** without the units in a GoI situation. Proofs are interpreted in the homset $\mathcal{C}(U, U)$ of endomorphisms of U. In what follows, we urge the reader to re-examine the overview remarks of Sect. 5.5.2 and refer back to the feedback diagrams in Fig. 5.9.

Convention: All identity morphisms are on tensor copies of U; however we adopt the convention of writing 1_{Γ} instead of 1_{U^n} with $|\Gamma| = n$, where U^n denotes the n-fold tensor product of U with itself. The retraction pairs are fixed once and for all.

Every **MELL** sequent will be of the form $\vdash [\Delta], \Gamma$ where Γ is a sequence of formulas and Δ is a sequence of cut formulas that have already been made in the proof of $\vdash \Gamma$ (e.g. $A, A^{\perp}, B, B^{\perp}$). This is used to keep track of the cuts that are already made in the proof of $\vdash \Gamma$. Suppose that Γ consists of n and Δ consists of $2m$ formulas. Then a proof π of $\vdash [\Delta], \Gamma$ is represented by a morphism $[\![\pi]\!] \in \mathcal{C}(U^{n+2m}, U^{n+2m})$. Recall that this corresponds to a morphism from U to itself, using the retraction morphisms $U \otimes U \lhd U\ (j, k)$. However, it is much more convenient to work in $\mathcal{C}(U^{n+2m}, U^{n+2m})$ (matrices on $\mathcal{C}(U, U)$). Define the morphism $\sigma : U^{2m} \longrightarrow U^{2m}$, as $\sigma = s \otimes \cdots \otimes s$ (m-copies) where s is the symmetry morphism, the 2×2 antidiagonal matrix $[a_{ij}]$, where $a_{12} = a_{21} = 1$; $a_{11} = a_{22} = 0$. Here σ represents the cuts in the proof of $\vdash \Gamma$, i.e. it models Δ. If Δ is empty (that is for a cut-free proof), we define $\sigma : I \longrightarrow I$ to be the zero morphism 0_{II}. Note that $U^0 = I$ where I is the unit of the tensor in the category $\mathcal{C}$.

Given block matrices A, B, by $A \otimes B$ we mean the block matrix with A and B on the main diagonal (the rest zeros). Thus σ above is the $2m \times 2m$ block matrix with the 2×2 matrix s along the main diagonal.

Definition 5.6.5 (The GoI Interpretation) Let π be a proof of $\vdash [\Delta], \Gamma$. We define *the GoI interpretation of* π, denoted by $[\![\pi]\!]$, by induction on the length of the proof as follows. We illustrate two key cases (Cut and Contraction) geometrically below. The other cases have a similar geometric form and are left as exercises.[3]

[3] The GoI interpretation of proofs involves manipulation and rearrangment of the interface wires of a proof box. GoI situations, with their reflexive object U and monoidal retracts, give the essential mechanism for modelling the "permuting, splitting, merging, and manipulating" of wires underlying the GoI interpretation of proofs. This is illustrated here for the Cut and Contraction Rules.

1. π is an *axiom* $\vdash A, A^{\perp}$, then $m = 0, n = 2$ and $[\![\pi]\!] = s = \begin{bmatrix} 0 & 1 \\ 1 & 0 \end{bmatrix}$.
2. π is obtained using the *cut* rule on π' and π'' that is

$$\dfrac{\begin{array}{c}\pi' \\ \vdots \\ \vdash [\Delta'], \Gamma', A\end{array} \quad \begin{array}{c}\pi'' \\ \vdots \\ \vdash [\Delta''], A^{\perp}, \Gamma''\end{array}}{\vdash [\Delta', \Delta'', A, A^{\perp}], \Gamma', \Gamma''}\ \text{cut}$$

Then we define $[\![\pi]\!]$ as follows: $[\![\pi]\!] = \tau^{-1}\left([\![\pi']\!] \otimes [\![\pi'']\!]\right)\tau$ where τ and τ^{-1} are the permutations of the interface (the identity on Γ', Δ') indicated below:

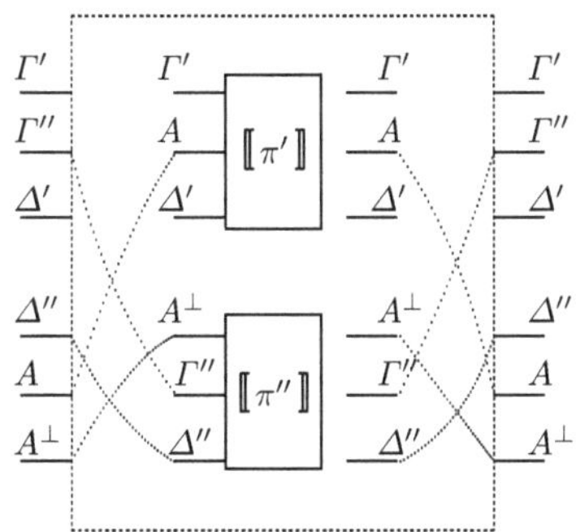

3. π is obtained using the *exchange* rule on the formulas A_i and A_{i+1} in Γ'. That is π is of the form

$$\dfrac{\begin{array}{c}\pi' \\ \vdots \\ \vdash [\Delta], \Gamma'\end{array}}{\vdash [\Delta], \Gamma}\ \text{exchange}$$

where in Γ' we have A_i, A_{i+1}. Then, $[\![\pi]\!]$ is obtained from $[\![\pi']\!]$ by interchanging the rows i and $i+1$. So suppose that $\Gamma' = \Gamma_1', A_i, A_{i+1}, \Gamma_2'$, then $\Gamma = \Gamma_1', A_{i+1}, A_i, \Gamma_2'$ and $[\![\pi]\!] = \tau^{-1}[\![\pi']\!]\tau$, where $\tau = 1_{\Gamma_1'} \otimes s \otimes 1_{\Gamma_2' \otimes \Delta}$.
4. π is obtained using an application of the *par* rule, that is π is of the form:

$$\dfrac{\begin{array}{c}\pi' \\ \vdots \\ \vdash [\Delta], \Gamma', A, B\end{array}}{\vdash [\Delta], \Gamma', A \parr B}\ \parr$$

Then $[\![\pi]\!] = g[\![\pi']\!]f$, where $f = 1_{\Gamma'} \otimes k \otimes 1_{\Delta}$ and $g = 1_{\Gamma'} \otimes j \otimes 1_{\Delta}$, recalling that $U \otimes U \lhd U\ (j, k)$.

5. π is obtained using an application of the *times* rule, that is π has the form

$$\dfrac{\begin{array}{cc} \pi' & \pi'' \\ \vdots & \vdots \\ \vdash [\Delta'], \Gamma', A & \vdash [\Delta''], \Gamma'', B \end{array}}{\vdash [\Delta', \Delta''], \Gamma', \Gamma'', A \otimes B} \text{ times}$$

Then (similarly to the Cut rule), define $[\![\pi]\!] = g\tau^{-1}\left([\![\pi']\!] \otimes [\![\pi'']\!]\right)\tau f$, where τ is a permutation, $f = 1_{\Gamma' \otimes \Gamma''} \otimes k \otimes 1_{\Delta' \otimes \Delta''}$ and $g = 1_{\Gamma' \otimes \Gamma''} \otimes j \otimes 1_{\Delta' \otimes \Delta''}$.

6. π is obtained from π' by an *of course* rule; that is π has the form :

$$\dfrac{\begin{array}{c} \pi' \\ \vdots \\ \vdash [\Delta], ?\Gamma', A \end{array}}{\vdash [\Delta], ?\Gamma', !A} \text{ of course}$$

Then $[\![\pi]\!] = ((ue_U)^{\otimes n} \otimes u \otimes u^{\otimes 2m})\varphi^{-1}T((v^{\otimes n} \otimes 1_A \otimes 1_\Delta)[\![\pi']\!](u^{\otimes n} \otimes 1_A \otimes 1_\Delta))\varphi((e'_U v)^{\otimes n} \otimes v \otimes v^{\otimes 2m})$, where $TT \lhd T\ (e, e')$, $|\Gamma'| = n$, $|\Delta| = 2m$, and $\varphi : (T^2U)^{\otimes n} \otimes TU \otimes (TU)^{\otimes 2m} \longrightarrow T((TU)^{\otimes n} \otimes U \otimes U^{\otimes 2m})$ is the canonical isomorphism.

7. π is obtained using the *contraction* rule on π', that is

$$\dfrac{\begin{array}{c} \pi' \\ \vdots \\ \vdash [\Delta], \Gamma', ?A, ?A \end{array}}{\vdash [\Delta], \Gamma', ?A} \text{ contraction}$$

Then we define $[\![\pi]\!]$ as follows, , where $T \otimes T \lhd T\ (c, c')$:

$$[\![\pi]\!] = (1_{\Gamma'} \otimes (u(c_U v \otimes v)) \otimes 1_\Delta)[\![\pi']\!](1_{\Gamma'} \otimes (u \otimes u)c'_U v \otimes 1_\Delta)$$

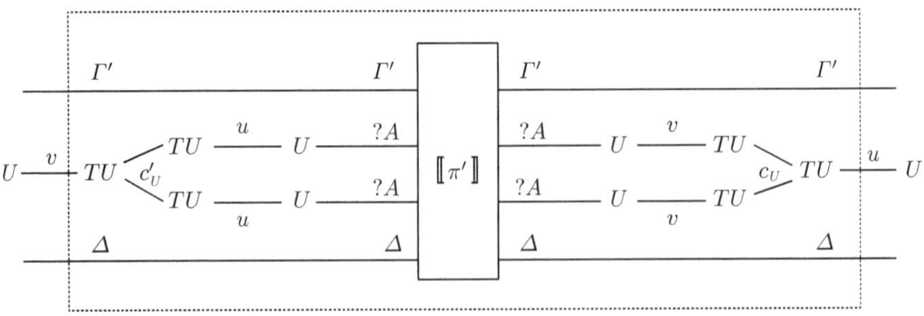

8. π is obtained from π' by the *dereliction* rule, that is π is of the form:

$$\dfrac{\begin{array}{c}\pi' \\ \vdots \\ \vdash [\Delta], \Gamma', A\end{array}}{\vdash [\Delta], \Gamma', ?A}\ \text{dereliction}$$

Then $[\![\pi]\!] = (1_{\Gamma'} \otimes ud_U \otimes 1_\Delta)[\![\pi']\!](1_{\Gamma'} \otimes d'_U v \otimes 1_\Delta)$ where $Id \lhd T\,(d, d')$.

9. π is obtained from π' by the *weakening* rule, that is π is of the form:

$$\dfrac{\begin{array}{c}\pi' \\ \vdots \\ \vdash [\Delta], \Gamma'\end{array}}{\vdash [\Delta], \Gamma', ?A}\ \text{weakening}$$

Then $[\![\pi]\!] = (1_{\Gamma'} \otimes uw_U \otimes 1_\Delta)[\![\pi']\!](1_{\Gamma'} \otimes w'_U v \otimes 1_\Delta)$, where $\mathcal{K}_I \lhd T\,(w, w')$.

This finishes the GoI interpretation for **MELL**.

Example 5.6.6 Let π be the following proof (of cut applied to the axioms). Categorically it corresponds to $id \circ id$.

$$\dfrac{\vdash A^\perp, A \qquad \vdash A^\perp, A}{\vdash [A, A^\perp], A^\perp, A}\ \text{cut}$$

Then the GoI semantics of this proof (see the *Cut* rule above) is given by conjugation with a permutation matrix τ:

$$[\![\pi]\!] = \begin{bmatrix} 1\,0\,0\,0 \\ 0\,0\,0\,1 \\ 0\,1\,0\,0 \\ 0\,0\,1\,0 \end{bmatrix} \left[\begin{array}{cc|cc} 0 & 1 & 0 & 0 \\ 1 & 0 & 0 & 0 \\ \hline 0 & 0 & 0 & 1 \\ 0 & 0 & 1 & 0 \end{array}\right] \begin{bmatrix} 1\,0\,0\,0 \\ 0\,0\,1\,0 \\ 0\,0\,0\,1 \\ 0\,1\,0\,0 \end{bmatrix} = \begin{bmatrix} \mathbf{0} & Id_2 \\ Id_2 & \mathbf{0} \end{bmatrix}$$

where Id_2 is the 2×2 identity matrix, $\mathbf{0}$ is the 2×2 zero matrix and the middle matrix is $s \otimes s$.

5.6.3 GoI Interpretation of Cut-Elimination

Dynamics is at the heart of the GoI interpretation as compared to denotational semantics and it is hidden in the cut-elimination process. The mathematical model of cut-elimination is given by the *execution formula* defined as follows:

$$EX([\![\pi]\!], \sigma) = \mathrm{Tr}^{U^{2m}}_{U^n, U^n}((1_{U^n} \otimes \sigma)[\![\pi]\!]) \tag{5.10}$$

where π is a proof of the sequent $\vdash [\Delta], \Gamma$. Pictorially this can be represented as in Fig. 5.9(b) in Sect. 5.5.2 above.

Note that $EX([\![\pi]\!], \sigma)$ is a morphism from $U^n \longrightarrow U^n$ and it always makes sense since the trace of any morphism in $\mathcal{C}(U^{2m+n}, U^{2m+n})$ is defined. Since we are working with a traced UDC with the standard trace, we can rewrite the execution Formula (5.10) in a more familiar form:

$$EX([\![\pi]\!], \sigma) = \pi_{11} + \sum_{n \geq 0} \pi_{12}(\sigma\pi_{22})^n(\sigma\pi_{21})$$

where $[\![\pi]\!] = \begin{bmatrix} \pi_{11} & \pi_{12} \\ \pi_{21} & \pi_{22} \end{bmatrix}$. Note that the execution formula defined in this categorical framework *always* makes sense; that is, we do not need a convergence criterion.

The intention here is to prove that the result of this execution formula is what corresponds to the cut-free proof obtained from π using Gentzen's cut-elimination procedure. We will also show that for any proof π of **MELL** the execution formula is a finite sum, which corresponds to termination of computation as opposed to divergence.

Example 5.6.7 Consider the proof π in Example 5.6.6 above. Recall also that $\sigma = s$ in this case ($m = 1$). Then

$$EX([\![\pi]\!], \sigma) = \mathrm{Tr}^{U^2}_{U^2,U^2}\left(\begin{bmatrix} 1&0&0&0 \\ 0&1&0&0 \\ 0&0&0&1 \\ 0&0&1&0 \end{bmatrix}\begin{bmatrix} 0&0&1&0 \\ 0&0&0&1 \\ 1&0&0&0 \\ 0&1&0&0 \end{bmatrix}\right)$$

$$= \begin{bmatrix} 0&0 \\ 0&0 \end{bmatrix} + \sum_{n \geq 0}\begin{bmatrix} 1&0 \\ 0&1 \end{bmatrix}\begin{bmatrix} 0&0 \\ 0&0 \end{bmatrix}^n\begin{bmatrix} 0&1 \\ 1&0 \end{bmatrix} = \begin{bmatrix} 0&1 \\ 1&0 \end{bmatrix} = [\![\vdash A^{\perp}, A]\!].$$

Note that in this case we have obtained the GoI interpretation of the cut-free proof obtained by applying Gentzen's Hauptsatz to the proof π. (Categorically, this just says $id \circ id = id$ in **Int**($\mathcal{C}$), where the composition is obtained dynamically by running the Execution formula). This is generalized in Theorem 5.6.12 below.

5.6.4 Soundness of the GoI Interpretation: Running the Execution Formula

In order to ensure that the definition above yields a semantics, we need to prove the soundness of the GoI interpretation. In other words, we have to show that if a proof π is reduced (via cut-elimination) to its cut-free form π', then $EX([\![\pi]\!], \sigma)$ is a finite sum and $EX([\![\pi]\!], \sigma) = [\![\pi']\!]$. Intuitively this says that if one thinks of cut-elimination as computation then $[\![\pi]\!]$ can be thought of as an algorithm. The computation takes place as follows: if we run $EX([\![\pi]\!], \sigma)$, it terminates after finitely many steps (cf. finite sum) and yields a *datum* (cf. cut-free proof). This intuition will be made precise in this section through the definition of *type* and the main theorems (see Theorems 5.8.12, 5.8.14). The next result is the analog of the Church-Rosser (or Diamond) property in our setting.

Lemma 5.6.8 (Associativity of cut) *Let π be a proof of $\vdash [\Gamma, \Delta], \Lambda$ and σ and τ be the morphisms representing the cut-formulas in Γ and Δ respectively. Then*

$$EX([\![\pi]\!], \sigma \otimes \tau) = EX(EX([\![\pi]\!], \tau), \sigma)$$

Proof Follows from naturality and vanishing II properties of trace. □

Definition 5.6.9 Let $\Gamma = A_1, \cdots, A_n$. A *datum* of type $\theta\Gamma$ is a morphism $M : U^n \longrightarrow U^n$ such that for any $\beta_1 \in \theta(A_1^\perp), \cdots, \beta_n \in \theta(A_n^\perp)$, $(\beta_1 \otimes \cdots \otimes \beta_n)M$ is nilpotent. An *algorithm* of type $\theta\Gamma$ is a morphism $M : U^{n+2m} \longrightarrow U^{n+2m}$ for some integer m such that for $\sigma : U^{2m} \longrightarrow U^{2m}$ defined in the usual way, $EX(M, \sigma) = \mathrm{Tr}^{U^{2m}}_{U^n,U^n}((1 \otimes \sigma)M)$ is a finite sum and a datum of type $\theta\Gamma$.

Lemma 5.6.10 *Let $M : U^n \longrightarrow U^n$ and $a : U \longrightarrow U$. Define $CUT(a, M) = (a \otimes 1_{U^{n-1}})M : U^n \longrightarrow U^n$. Note that the matrix representation of $CUT(a, M)$ is the matrix obtained from M by multiplying its first row by a. Then $M = [m_{ij}]$ is a datum of type $\theta(A, \Gamma)$ iff for any $a \in \theta A^\perp$, am_{11} is nilpotent and the morphism $ex(CUT(a, M)) = Tr^A(s^{-1}_{\Gamma,A} CUT(a, M)s_{\Gamma,A})$ is in $\theta(\Gamma)$. Here $s_{\Gamma,A}$ is the symmetry morphism from $\Gamma \otimes A$ to $A \otimes \Gamma$.*

Theorem 5.6.11 (Proofs as Algorithms) *Let Γ be a sequent, and π be a proof of Γ. Then $[\![\pi]\!]$ is an algorithm of type $\theta\Gamma$*

Theorem 5.6.12 (Ex is an invariant) *Let π be a proof of a sequent $\vdash [\Delta], \Gamma$ in **MELL**. Then*

(i) $EX([\![\pi]\!], \sigma)$ is a finite sum.
(ii) If π reduces to π' by any sequence of cut-eliminations and $?A$ does not occur in Γ for any formula A, then $EX([\![\pi]\!], \sigma) = EX([\![\pi']\!], \tau)$. So $EX([\![\pi]\!], \sigma)$ is an invariant of reduction.
(iii) In particular, if π' is any cut-free proof obtained from π by cut-elimination, then $EX([\![\pi]\!], \sigma) = [\![\pi']\!]$.

In [HS04a] we show that we obtain the same execution formula as Girard. Note that in Girard's original execution formula $[\![\pi]\!]$ and σ are both $2m + n$ by $2m + n$ matrices. To connect up with our previous notation, let $\tilde{\sigma} = s \otimes \cdots \otimes s$ (m-times.)

Proposition 5.6.13 (Original Execution Formula) *Let π be a proof of $\vdash [\Delta], \Gamma$. Then in Girard's model **Hilb**$_2$,*

$$((1 - \sigma^2)\sum_{n=0}^{\infty}[\![\pi]\!](\sigma[\![\pi]\!])^n(1 - \sigma^2))_{n\times n} = Tr^{U^{2m}}_{U^n,U^n}((1 \otimes \tilde{\sigma})[\![\pi]\!])$$

where $(A)_{n\times n}$ *is the submatrix of* A *consisting of the first* n *rows and the first* n *columns.*

In the next two sections we discuss further generalizations of the notions of trace and orthogonality. These notions play crucial roles in GoI interpretations.

5.7 Partial Trace and Abstract Orthogonality

In this section we look at partial traces. The idea of generalizing the abstract trace of [JSV96] to the partial setting is not new. For example, partial traces were already studied in work of Abramsky, Blute, and Panangaden [ABP99], in unpublished lecture notes of Gordon Plotkin [Pl03], work of Blute, Cockett, and Seely [BCS00] (see Remark 5.7.2), and others. The guiding example in [ABP99] is the relationship between trace class operators on a Hilbert space and Hilbert-Schmidt operators. This allows the authors to establish a close correspondence between trace and nuclear ideals in a tensor ∗-category. Plotkin's work develops a theory of Conway ideals on biproduct categories, and an associated categorical trace theory. Unfortunately none of these extant theories is appropriate for Girard's GoI. So we present an axiomatization for partial traces suitable for our purposes.

Recall, following Joyal, Street, and Verity [JSV96], a (parametric) trace in a symmetric monoidal category $(\mathcal{C}, \otimes, I, s)$ is a family of maps

$$\mathtt{Tr}^U_{X,Y} : \mathcal{C}(X \otimes U, Y \otimes U) \longrightarrow \mathcal{C}(X, Y),$$

satisfying various well-known naturality equations. A *partial* (parametric) trace requires instead that each $\mathtt{Tr}^U_{X,Y}$ be a partial map (with domain denoted $\mathbb{T}^U_{X,Y}$) and satisfy various closure conditions.

Definition 5.7.1 (Trace Class) Let $(\mathcal{C}, \otimes, I, s)$ be a symmetric monoidal category. A *(parametric) trace class* in $\mathcal{C}$ is a choice of a family of subsets, for each object U of $\mathcal{C}$, of the form

$$\mathbb{T}^U_{X,Y} \subseteq \mathcal{C}(X \otimes U, Y \otimes U) \text{ for all objects } X, Y \text{ of } \mathcal{C}$$

together with a family of functions, called a *(parametric) partial trace*, of the form

$$\mathtt{Tr}^U_{X,Y} : \mathbb{T}^U_{X,Y} \longrightarrow \mathcal{C}(X, Y)$$

subject to the following axioms. Here the parameters are X and Y and a morphism $f \in \mathbb{T}^U_{X,Y}$, by abuse of terminology, is said to be *trace class*.

- **Naturality** in X and Y: For any $f \in \mathbb{T}^U_{X,Y}$ and $g : X' \longrightarrow X$ and $h : Y \longrightarrow Y'$,

$$(h \otimes 1_U) f (g \otimes 1_U) \in \mathbb{T}^U_{X',Y'},$$

and $$\mathtt{Tr}^U_{X',Y'}((h \otimes 1_U) f (g \otimes 1_U)) = h\, \mathtt{Tr}^U_{X,Y}(f)\, g.$$

- **Dinaturality** in U: For any $f : X \otimes U \longrightarrow Y \otimes U'$, $g : U' \longrightarrow U$,

$$(1_Y \otimes g)f \in \mathbb{T}^U_{X,Y} \text{ iff } f(1_X \otimes g) \in \mathbb{T}^{U'}_{X,Y},$$
and
$$\mathtt{Tr}^U_{X,Y}((1_Y \otimes g)f) = \mathtt{Tr}^{U'}_{X,Y}(f(1_X \otimes g)).$$

- **Vanishing I**: $\mathbb{T}^I_{X,Y} = \mathcal{C}(X \otimes I, Y \otimes I)$, and for $f \in \mathbb{T}^I_{X,Y}$

$$\mathtt{Tr}^I_{X,Y}(f) = \rho_Y f \rho_X^{-1}.$$

Here $\rho_A : A \otimes I \longrightarrow A$ is the right unit isomorphism of the monoidal category.
- **Vanishing II**: For any $g : X \otimes U \otimes V \longrightarrow Y \otimes U \otimes V$, if $g \in \mathbb{T}^V_{X\otimes U, Y\otimes U}$, then

$$g \in \mathbb{T}^{U\otimes V}_{X,Y} \text{ iff } \mathtt{Tr}^V_{X\otimes U, Y\otimes U}(g) \in \mathbb{T}^U_{X,Y},$$
and in the latter case
$$\mathtt{Tr}^{U\otimes V}_{X,Y}(g) = \mathtt{Tr}^U_{X,Y}(\mathtt{Tr}^V_{X\otimes U, Y\otimes U}(g)).$$

- **Superposing**: For any $f \in \mathbb{T}^U_{X,Y}$ and $g : W \longrightarrow Z$,

$$g \otimes f \in \mathbb{T}^U_{W\otimes X, Z\otimes Y},$$
and
$$\mathtt{Tr}^U_{W\otimes X, Z\otimes Y}(g \otimes f) = g \otimes \mathtt{Tr}^U_{X,Y}(f).$$

- **Yanking**: $s_{UU} \in \mathbb{T}^U_{U,U}$, and $\mathtt{Tr}^U_{U,U}(s_{U,U}) = 1_U$.

A symmetric monoidal category $(\mathcal{C}, \otimes, I, s)$ with such a trace class is called a *partially traced category*, or a *category with a trace class*.

If we let X and Y be I (the unit of the tensor), we get a family of operations $\mathtt{Tr}^U_{I,I} : \mathbb{T}^U_{I,I} \longrightarrow \mathcal{C}(I, I)$ defining what we call a *non-parametric (scalar-valued) trace*.

Remark 5.7.2 An early definition of a partial parametric trace is due to Abramsky, Blute and Panangaden in [ABP99]. Our definition is different but related to theirs. First, we have used the Yanking axiom in Joyal, Street and Verity [JSV96], whereas in [ABP99] they use a conditional version of the so-called "generalized yanking"; that is, for $f : X \longrightarrow U$ and $g : U \longrightarrow Y$, $\mathtt{Tr}^U_{X,Y}(s_{U,Y}(f \otimes g)) = gf$ *whenever* $s_{U,Y}(f \otimes g)$ is trace class. In our theory s_{UU} is traceable for all U; on the other hand, many examples in [ABP99] do not have this property. More importantly, we do not require one of the ideal axioms in [ABP99]. Namely, we do **not** ask that for $f \in \mathbb{T}^U_{X,Y}$ and any $h : U \longrightarrow U$, $(1_Y \otimes h)f$ and $f(1_X \otimes h)$ be in $\mathbb{T}^U_{X,Y}$. Indeed in the next section we prove that the categories $(\mathbf{Vec}_{fd}, \oplus)$ of finite dimensional vector spaces, and $(\mathbf{CMet}, \times)$ of complete metric spaces are partially traced. It can be shown that in both categories the above ideal axiom and Vanishing II of [ABP99] fail and hence they are not traced in the sense of [ABP99].

In [BCS00], Blute, Cockett, and Seely develop an interesting and detailed theory of trace (and fixpoint) combinators in a linearly distributive category, including an appropriate version of the *Int* construction of [JSV96] in that setting. The authors take a local view of the trace combinator: rather than assuming that a trace is available at every object, they consider the effect of particular objects having a trace (partiality of trace), as well as restricting to "compatible classes" of trace operators (which guarantees that an object may have at most one trace structure.)

One is obliged to say that there are many different approaches to partial categorical traces and ideals; ours is geared to the details of Girard's GoI. We believe our traceability conditions are most naturally formulated as we did above, as properties of morphisms rather than objects, but this may be a matter of taste.

5.7.1 Examples of Partial Traces

(a) Finite Dimensional Vector Spaces

The category $\mathbf{Vec}_{fd}$ of finite dimensional vector spaces and linear transformations is a symmetric monoidal, indeed an additive, category (see [Mac98]), with monoidal product taken to be $\oplus$, the direct sum (biproduct). Hence, given $f : \oplus_I X_i \longrightarrow \oplus_J Y_j$ with $|I| = n$ and $|J| = m$, we can write f as an $m \times n$ matrix $f = [f_{ij}]$ of its components, where $f_{ij} : X_j \longrightarrow Y_i$ (notice the switch in the indices i and j).

We give a trace class structure on the category $(\mathbf{Vec}_{fd}, \oplus, \mathbf{0})$ as follows. We shall say an $f : X \oplus U \longrightarrow Y \oplus U$ is *trace class* iff $(I - f_{22})$ is invertible, where I is the identity matrix, and I and f_{22} have size $dim(U)$. In that case, we write

$$\mathtt{Tr}^U_{X,Y}(f) = f_{11} + f_{12}(I - f_{22})^{-1} f_{21} \tag{5.11}$$

This definition is motivated by a generalization of the fact that for a matrix A, $(I - A)^{-1} = \sum_i A^i$, whenever the infinite sum converges. Clearly this sum converges when the matrix norm of A is strictly less than 1, or when A is nilpotent, but in both cases the general idea is the desire to have $(I - A)$ invertible. If the infinite sum for $(I - f_{22})^{-1}$ exists, the above formula for $\mathtt{Tr}^U_{X,Y}(f)$ becomes the usual "particle-style" trace in [Abr96, AHS02, HS04a]. One advantage of formula (5.11) is that it does not *a priori* assume the convergence of the sum, nor even that $(I - f_{22})^{-1}$ be computable by iterative methods.

Proposition 5.7.3 $(\mathrm{Vec}_{fd}\oplus, \mathbf{0})$ *is partially traced, with trace class as above.*

The proof is sketched in [HS05a]. Proposition 5.7.3 uses the following standard facts from linear algebra:

Lemma 5.7.4 *Let* $M = \begin{bmatrix} A & B \\ C & D \end{bmatrix}$ *be a partitioned matrix with blocks A $(m \times m)$, B $(m \times n)$, C $(n \times m)$ and D $(n \times n)$. If D is invertible, then M is invertible iff $A - BD^{-1}C$ (the Schur Complement of D) is invertible.*

Lemma 5.7.5 *Given A $(m \times n)$ and B $(n \times m)$, $(I_m - AB)$ is invertible iff $(I_n - BA)$ is invertible. Moreover $(I_m - AB)^{-1}A = A(I_n - BA)^{-1}$.*

Proof (Proposition 5.7.3) We shall verify a couple of axioms.

- **Naturality** in X and Y: Suppose $f \in \mathbb{T}^U_{X,Y}$ and $g : X' \longrightarrow X$ and $h : Y \longrightarrow Y'$, $(h \oplus 1_U)f(g \oplus 1_U)$ can be represented by its matrix $\begin{bmatrix} hf_{11}g & hf_{12} \\ f_{21}g & f_{22} \end{bmatrix}$ whose component from U to itself is f_{22} and hence $(h \oplus 1_U)f(g \oplus 1_U) \in \mathbb{T}^U_{X',Y'}$ and it is easy to see that $h\mathtt{Tr}^U_{X,Y}(f)g = \mathtt{Tr}^U_{X',Y'}((h \oplus 1_U)f(g \oplus 1_U))$.
- **Dinaturality** in U: Let $f : X \oplus U \longrightarrow Y \oplus U'$, $g : U' \longrightarrow U$. $(1_Y \oplus g)f \in \mathbb{T}^U_{X,Y}$ iff $I - gf_{22}$ is invertible iff $I - f_{22}g$ is invertible by Lemma 5.7.5 and thus iff $f(1_X \oplus g) \in \mathbb{T}^{U'}_{X,Y}$.

$$\begin{aligned} \mathtt{Tr}^U_{X,Y}((1_Y \oplus g)f) &= f_{11} + f_{12}(I - gf_{22})^{-1} g f_{21} \\ &= f_{11} + f_{12} g (I - f_{22} g)^{-1} f_{21} \quad \text{by Lemma 5.7.5.} \\ &= \mathtt{Tr}^{U'}_{X,Y}(f(1_X \oplus g)). \end{aligned}$$

□

As discussed in Remark 5.7.2, the category $(\mathbf{Vec}_{fd}, \oplus)$ is *not* partially traced in the sense of ABP.

(b) Metric Spaces

Consider the category **CMet** of complete metric spaces with non-expansive maps, that is $f : (M, d_M) \longrightarrow (N, d_N)$ such that $d_N(f(x), f(y)) \le d_M(x, y)$, for all $x, y \in M$. Note that the tempting collection of complete metric spaces and contractions $(d_N(f(x), f(y)) < d_M(x, y))$ is not a category: there are no identity morphisms! **CMet** has products, namely given (M, d_M) and (N, d_N) we define $(M \times N, d_{M \times N})$ with $d_{M\times N}((m, n), (m', n')) = max\{d_M(m, m'), d_N(n, n')\}$.

We define the trace class structure on **CMet**(where $\otimes = \times$) as follows. We say that a morphism $f : X \times U \longrightarrow Y \times U$ is in $\mathbb{T}^U_{X,Y}$ iff for every $x \in X$ the induced map $\pi_2 \lambda u. f(x, u) : U \longrightarrow U$ has a unique fixed point; in other words, iff for every $x \in X$, there is a unique u, and a y, such that $f(x, u) = (y, u)$. Note that in this case y is necessarily unique. Also, note that contractions have unique fixed points, by the Banach fixed point theorem.

Suppose $f \in \mathbb{T}^U_{X,Y}$. We define $\mathtt{Tr}^U_{X,Y}(f) : X \longrightarrow Y$ by $\mathtt{Tr}^U_{X,Y}(f)(x) = y$, where $f(x, u) = (y, u)$ for the unique u. Equivalently, $\mathtt{Tr}^U_{X,Y}(f)(x) = \pi_1 f(x, u)$ where u is the unique fixed point of $\pi_2 \lambda t. f(x, t)$.

Proposition 5.7.6 (CMet, $\times$, $\{*\}$) *is a partially traced category with trace class as above.*

Lemma 5.7.7 *Let A and B be sets, $f : A \longrightarrow B$ and $g : B \longrightarrow A$. Then, gf has a unique fixed point if and only if fg does. Moreover, let $a \in A$ be the unique fixed point of $gf : A \longrightarrow A$ and $b \in B$ be the unique fixed point of $fg : B \longrightarrow B$. Then $f(a) = b$ and $g(b) = a$.*

Proof (Proposition 5.7.6) We shall verify the dinaturality axiom. For $f : X \times U \longrightarrow Y \times U$ and $x \in X$, we will use f_x to denote the map $\lambda u.f(x,u) : U \longrightarrow Y \times U$.

Dinaturality in U: Let $f : X \times U \longrightarrow Y \times U'$, $g : U' \longrightarrow U$. Note that for any $x \in X$, $\pi_2((1_Y \times g)f)_x = g(\pi_2 f_x)$ and $\pi_2(f(1_X \times g))_x = (\pi_2 f_x)g$ and $g(\pi_2 f_x)$ has a unique fixed point iff $(\pi_2 f_x)g$ has a unique fixed point, by Lemma 5.7.7. Thus $(1_Y \times g)f \in \mathbb{T}^U_{X,Y}$ iff $f(1_X \times g) \in \mathbb{T}^{U'}_{X,Y}$.

$$
\begin{aligned}
\mathtt{Tr}^U_{X,Y}((1_Y \times g)f)(x) &= \pi_1(1 \times g)f(x,u) \quad u \text{ is the unique} \\
&\quad \text{fixed point of } g(\pi_2 f_x) \\
&= \pi_1 f(x,u) \\
&= \pi_1 f(x, g(u')) \text{ by Lemma 5.7.7} \\
&\quad \text{where } u' \text{ is the unqiue fixed point of } (\pi_2 f_x)g \\
&= \mathtt{Tr}^{U'}_{X,Y}(f(1_X \times g))(x).
\end{aligned}
$$

□

Proposition 5.7.6 remains valid for the category (**Set**, $\times$) of sets and mappings. The latter then becomes a partially traced category with the same definition for trace class morphisms as in **CMet**. However, this fails for the category (**Rel**, $\times$), of sets and relations, as Lemma 5.7.7 is no longer valid: consider the sets $A = \{a\}$, $B = \{b, b'\}$, and let $f = \{(a,b), (a,b')\}$ and $g = \{(b,a), (b',a)\}$.

(c) Total Traces

Of course, all (totally-defined) traces in the usual definition of a traced monoidal category yield a trace class, namely the entire homset is the domain of Tr.

Remark 5.7.8 (A Non-Example) Consider the structure (**CMet**, $\times$). Defining the trace class morphisms as those f such that $\pi_2\lambda u.f(x,u) : U \longrightarrow U$ is a contraction, for every $x \in X$, does not yield a partially traced category: all axioms are true except for dinaturality and Vanishing II.

For details and motivation on the orthogonality relation we refer the interested reader to [HS05a]. See also the important work by Hyland and Schalk in [HylSc03] for the general definition of orthogonality relations in a symmetric monoidal closed category and its connections to models of linear logic.

Definition 5.7.9 Let $\mathcal{C}$ be a traced symmetric monoidal category. A *(strong) orthogonality relation* on $\mathcal{C}$ is a family of relations $\perp_{UV}$ between maps $u : V \longrightarrow U$ and $x : U \longrightarrow V$, denoted $V \xrightarrow{u} U \perp_{UV} U \xrightarrow{x} V$, subject to the following axioms:

(i) *Isomorphism* : Let $f : U \otimes V' \longrightarrow V \otimes U'$ and $\hat{f} : U' \otimes V \longrightarrow V' \otimes U$ be such that $\mathtt{Tr}^{V'}(\mathtt{Tr}^{U'}((1 \otimes 1 \otimes s_{U',V'})\alpha^{-1}(f \otimes \hat{f})\alpha)) = s_{U,V}$ and $\mathtt{Tr}^{V}(\mathtt{Tr}^{U}((1 \otimes 1 \otimes s_{U,V})\alpha^{-1}(\hat{f} \otimes f)\alpha)) = s_{U',V'}$. Here $\alpha = (1 \otimes 1 \otimes s)(1 \otimes s \otimes 1)$ with s at appropriate types. Note that this simply means that $f : (U, V) \longrightarrow$

(U', V') and $\hat{f} : (U', V') \longrightarrow (U, V)$ are inverses of each other in $\mathcal{G}(\mathcal{C})$ (the compact closure of $\mathcal{C}$, [Hag00, AHS02].)
Then, for all $u : V \longrightarrow U$ and $x : U \longrightarrow V$,

$$u \perp_{UV} x \;\text{ iff }\; \mathtt{Tr}^{U}_{V',U'}(s_{U,U'}(u \otimes 1_{U'}) f s_{V',U}) \perp_{U'V'} \mathtt{Tr}^{V}_{U',V'}((1_{V'} \otimes x)\hat{f});$$

that is, orthogonality is invariant under isomorphism. This is so because the expressions above correspond to composition of u and f, and x and $\hat{f}$ in the compact closed category $\mathcal{G}(\mathcal{C})$.

(ii) *Precise Tensor:* For all $u : V \longrightarrow U$, $v : V' \longrightarrow U'$ and $h : U \otimes U' \longrightarrow V \otimes V'$,

$$(u \otimes v) \perp_{U \otimes U', V \otimes V'} h,$$
$$\text{iff}$$
$$v \perp_{U'V'} \mathtt{Tr}^{U}_{U',V'}(s_{U,V'}(u \otimes 1_{V'}) h s_{U',U}) \quad \text{and} \quad u \perp_{UV} \mathtt{Tr}^{U'}_{U,V}((1 \otimes v)h)$$

(iii) *Identity* : For all $u : V \longrightarrow U$ and $x : U \longrightarrow V$, $u \perp_{UV} x$ implies $1_I \perp_{II} \mathtt{Tr}^{V}_{I,I}(xu)$.

(iv) *Symmetry* : For all $u : V \longrightarrow U$ and $x : U \longrightarrow V$, $u \perp_{UV} x$ iff $x \perp_{VU} u$.

Example 5.7.10 (Orthogonality as trace class) Let $(\mathcal{C}, \otimes, I, \mathtt{Tr})$ be a partially traced category where $\otimes$ is the monoidal product with unit I, and $\mathtt{Tr}$ is the partial trace operator as in above. Let A and B be objects of $\mathcal{C}$. For $f : A \longrightarrow B$ and $g : B \longrightarrow A$, we can define an orthogonality relation by declaring $f \perp_{BA} g$ iff $gf \in \mathbb{T}^{A}_{I,I}$. The axioms can be checked easily and we shall not include the verification here. It turns out that this is a variation of the notion of *Focussed orthogonality* of Hyland and Schalk [HylSc03].

Hence, from our previous discussion on traces, we obtain the following examples:

- $\mathbf{Vec}_{fd}$. For $A \in \vec{fd}$, $f, g \in End(A)$, define $f \perp g$ iff $I - gf$ is invertible. Here I is the identity matrix of size $dim(A)$.
- **CMet**. Let $M \in$ **CMet**. For $f, g \in End(M)$, define $f \perp g$ iff gf has a unique fixed point.

5.8 Typed GoI for MELL in *-Categories

The GoI interpretation we presented in Sect. 5.6 was carried out using a GoI Situation with the underlying category a traced UDC, and using an orthogonality relation, defined based on nilpotency. Moreover, formulas and proofs were interpreted based on a single reflexive object U. It is possible to extend this framework vastly beyond these limits, indeed it is possible to give a multi-object (typed) GoI (MGOI) interpretation for **MELL** using a GoI Situation with additional structure and a compatible abstract orthogonality relation. We shall briefly highlight what is involved without getting into details. Interested readers can refer to [Hag06] and [HS09].

For the purposes of this general version we shall need an additional structure on a monoidal category, namely that of contravariant functor $(\,)^*$. In the following we shall recall the definition of monoidal $*$-categories from [ABP99]. Nevertheless, note that our definition is different from that in [ABP99], as we do not require a conjugation functor, and we demand stronger conditions on the functor $(\,)^*$. Categories such as these with further structure on the homsets (W^*-categories) were first introduced in [GLR85]. The idea there was to generalize the notions and machinery of von Neumann algebras to a categorical setting. Later, similar categories (C^*-categories) were defined in [DopR89] and studied in depth. The motivation in this work was to present a new duality theory for compact groups, itself motivated by the work in early seventies on superselection structure in quantum field theory. Both [GLR85] and [DopR89] are excellent sources for examples of $*$-categories we define here.

Definition 5.8.1 A *symmetric monoidal $*$-category* $\mathcal{C}$ is a symmetric monoidal category with a strict symmetric monoidal functor $(\,)^* : \mathcal{C}^{op} \longrightarrow \mathcal{C}$ which is strictly involutive and the identity on objects. Note that this in particular implies that $(f \otimes g)^* = f^* \otimes g^*$, and $s^*_{A,B} = s_{B,A}$ where $s_{A,B}$ is the symmetry morphism.

We say that a morphism $f : A \longrightarrow A$ is *Hermitian* if $f^* = f$. A morphism $f : A \longrightarrow B$ is called a *partial isometry* if $f^* f f^* = f^*$ or equivalently, if $f f^* f = f$. A morphism $f : A \longrightarrow A$ is called a *partial symmetry* if it is Hermitian and a partial isometry. That is, if $f^* = f$ and $f^3 = f$. Note that there is no underlying Hilbert space structure on the homsets of $\mathcal{C}$; the terminology here is borrowed from operator algebras to account for the similar properties of such morphisms, which can be expressed in the more general setting of $*$-categories.

An obvious example is the category $\mathbf{Hilb}_{\otimes}$ of Hilbert spaces and bounded linear maps with tensor product of Hilbert spaces as the monoidal product. Given $f : H \longrightarrow K$, $f^* : K \longrightarrow H$ is given by the adjoint of f, defined uniquely by $\langle f(x), y\rangle = \langle x, f^*(y)\rangle$. It is not hard to see that all the required properties are satisfied. Note that the category $\mathbf{Hilb}_{\oplus}$ of Hilbert spaces and bounded linear maps but with direct sum as the monoidal product is a $*$-category too, with the same definition for the $(\,)^*$ functor.

Another example is the category $\mathbf{Rel}_{\times}$ of sets and relations with the cartesian product of sets as the monoidal product. Given $f : X \longrightarrow Y$, $f^* = \overline{f}$ where $\overline{f}$ is the converse relation. Again, note that the category $\mathbf{Rel}_{\oplus}$ of sets and relations with monoidal product, the disjoint union (categorical biproduct) is a monoidal $*$-category too, with the same definition for the $(\,)^*$ functor.

Yet another example that shows up frequently in the context of GoI is the category $\mathbf{PInj}_{\uplus}$ of sets and partial injective maps, with disjoint union as the monoidal product. Given $f : X \longrightarrow Y$, $f^* = f^{-1}$.

Other examples include $\mathbf{Hilb}_{fd}$ of finite dimensional Hilbert spaces and bounded linear maps, $\mathbf{URep}(G)$, finite representations of a compact group G, etc. For more details, examples and the ways that such categories show up in logic and computer science, see [ABP99].

Definition 5.8.2 A *GoI category* is a triple $(\mathcal{C}, T, \perp)$ where $\mathcal{C}$ is a partially traced $*$-category as in Section 5.7, $T = (T, \psi, \psi_I) : \mathcal{C} \longrightarrow \mathcal{C}$ is a traced symmetric monoidal functor, that is if $f \in \mathbb{T}^U_{X,Y}$, then $\psi^{-1}_{Y,U} T(f) \psi_{X,U} \in \mathbb{T}^{TU}_{TX,TY}$ and $\mathrm{Tr}^{TU}_{TX,TY}(\psi^{-1}_{Y,U} T(f) \psi_{X,U}) = T(\mathrm{Tr}^U_{X,Y}(f))$. Here $\perp$ is an orthogonality relation on $\mathcal{C}$ as in the above. Furthermore, we require that,

- The following natural retractions exist:

  (i) $\mathcal{K}_I \lhd T\ (w, w^*)$, $\quad \mathcal{K}_I$ denotes the constant I functor.
  (ii) $Id \lhd T\ (d, d^*)$
  (iii) $T^2 \lhd T\ (e, e^*)$
  (iv) $T \otimes T \lhd T\ (c, c^*)$

- The orthogonality relation must be *GoI compatible*, that is, it must satisfy the following additional axioms:

  (c1) For all $f : V \longrightarrow U$, $g : U \longrightarrow V$,
  $$f \perp_{U,V} g \ \text{ implies } \ d_U f d^*_V \perp_{TU,TV} Tg.$$
  (c2) For all $f : U \longrightarrow U$ and $g : I \longrightarrow I$,
  $$w_U g w^*_U \perp_{TU,TU} Tf.$$
  (c3) For all $f : TV \otimes TV \longrightarrow TU \otimes TU$ and $g : U \longrightarrow V$,
  $$f \perp_{TU\otimes TU, TV\otimes TV} \ Tg \otimes Tg \ \text{ implies } \ c_U f c^*_V \perp_{TU,TV} Tg.$$

- The functor T commutes with $(\ )^*$, that is $(T(f))^* = T(f^*)$. Moreover, $\psi^* = \psi^{-1}$ and $\psi^*_I = \psi^{-1}_I$.

Proposition 5.8.3 *Suppose $\mathcal{C}$ is a partially traced $*$-category that is in addition equipped with an endofunctor T and monoidal retractions as in Definition 5.8.2. Then, the orthogonality relation $\perp$ defined as in Example 5.7.10 is GoI compatible.*

Proof We shall verify the compatibility axioms of Definition 5.8.2.
(c1) $\mathrm{Tr}^{TV}(T(g) d_U f d^*_V) = \mathrm{Tr}^{TV}(d_V g f d^*_V) = \mathrm{Tr}^V(gf)$.
(c2) $\mathrm{Tr}^{TU}(T(f) w_U g w^*_U) = \mathrm{Tr}^{TU}(w_U g w^*_U) = \mathrm{Tr}^I(g)$.
Recall that $\mathbb{T}^I_{I,I} = \mathcal{C}(I, I)$.
(c3) $\mathrm{Tr}^{TV}(T(g) c_U f c^*_V) = \mathrm{Tr}^{TV}(c_V (Tg \otimes Tg) f c^*_V) = \mathrm{Tr}^{TV\otimes TV}((Tg \otimes Tg) f)$.
□

GoI categories are the main mathematical structures in our semantic interpretation in the following section. Here are a few examples of GoI categories.

Examples 5.8.4 (a) ($\mathbf{PInj}_{\uplus}, T, \perp$). This is a GoI situation (see Examples 5.6.2). We define, $f \perp g$ iff gf is nilpotent. It can be easily checked that this definition satisfies the axioms for an orthogonality relation.

Let us verify the compatibility axioms:

- For $f : V \longrightarrow U$ and $g : U \longrightarrow V$, suppose gf is nilpotent, say $(gf)^n = 0$, then $(T(g)d_U f d_V^*)^n = (d_V gf d_V^*)^n$ by naturality of d_U, but as $d_V^* d_V = 1_V$ we have $(d_V gf d_V^*)^n = d_V (gf)^n d_V^* = 0$.
- As $I = \emptyset$ and $w_I = 0$, we have that $T(f) w_U g w_U^*$ is nilpotent.
- For $f : TV \otimes TV \longrightarrow TU \otimes TU$ and $g : U \longrightarrow V$, suppose $(Tg \otimes Tg)f$ is nilpotent, say $((Tg \otimes Tg)f)^n = 0$, Then $(T(g)c_U f c_V^*)^n = (c_V (Tg \otimes Tg) f c_V^*)^n$, by naturality of c_V, but as $c_V^* c_V = 1_{TV \otimes TV}$ we have $(c_V (Tg \otimes Tg) f c_V^*)^n = c_V ((Tg \otimes Tg) f)^n c_V^* = 0$.

Finally, for any $f : X \longrightarrow Y$, $(Tf)^* = T(f^*)$.

(b) ($\mathbf{Hilb}_{\oplus}, T, \perp$), where **Hilb** is the category of Hilbert spaces and bounded linear maps. The monoidal product is the direct sum of Hilbert spaces. By the above, $\mathbf{Hilb}_{\oplus}$ is a partially traced $*$-category. Define:

$T(H) = \ell^2 \otimes H$ where ℓ^2 is the space of square summable sequences.

We define $f \perp g$ iff $(1 - gf)$ is an invertible linear transformation. Compatibility follows from Proposition 5.8.3, because for $f : H \longrightarrow K$, $g : K \longrightarrow H$, $f \perp g$ iff $gf \in \mathbb{T}^H$. Finally, as $\mathbf{Hilb}_{\otimes}$ is also a $*$-category with f^* the adjoint of f, we have that for any $f : H \longrightarrow K$, $(Tf)^* = T(f^*)$.

(c) ($\mathbf{Rel}_{\oplus}, T, \perp$) is a GoI-category with the same definitions for T and $\perp$ as in the case of **PInj**. Note that disjoint union, denoted $\oplus$, is in fact the categorical biproduct in **Rel**.

Multiobject Geometry of Interaction (MGoI) was introduced in [HS05a] and was used to interpret **MLL** without units. It was later extended to exponentials in [Hag06]. The main idea in [HS05a] was to keep the types of the formulas that were defined by a denotational semantics map during the GoI interpretation. For the multiplicative case this also implies that the MGoI interpretation becomes "localized" to different endomorphism monoids, rather than the endomorphisms of a fixed reflexive object U as in usual GoI (described previously above). Now there is no need for a reflexive object U and this makes the interpretation of **MLL** possible in categories like finite dimensional vector spaces.

On the other hand, in the case of exponentials, we soon observe that infinity forces itself into the framework: it is no longer possible to carry out the MGoI interpretation in finite dimensions. This transition to infinity occurs, for example, when we are forced to admit a retraction $TTA \lhd TA$ for any object A in the relevant category. Note that, although in this way reflexive objects reappear, they are not used to collapse types as in the GoI interpretation using a single object U.

5.8.1 MGoI Interpretation of Formulas

Given a GoI category $(\mathcal{C}, T, \perp)$, let A be an object of $\mathcal{C}$ and let $f, g \in End(A)$. We say that f is *orthogonal to* g, denoted $f \perp g$, if $(f, g) \in \perp$. Also given $X \subseteq End(A)$ we define

$$X^{\perp} = \{f \in End(A) \mid \forall g \in X, f \perp g\}.$$

We can define an operator on the objects of $\mathcal{C}$ as follows: given an object A, we look at the subsets of $End(A)$ which equal their bi-orthogonal: $\mathcal{T}(A) = \{X \subseteq End(A) \mid X^{\perp\perp} = X\}$.

We wish to define the MGoI interpretation of formulas. First we define an interpretation map $[\![-]\!]$ on the formulas of **MELL** as follows. Given the value of $[\![-]\!]$ on the atomic propositions as objects of $\mathcal{C}$, we extend it to all formulas by:

- $[\![A^{\perp}]\!] = [\![A]\!]$
- $[\![A \parr B]\!] = [\![A \otimes B]\!] = [\![A]\!] \otimes [\![B]\!]$.
- $[\![!A]\!] = [\![?A]\!] = T[\![A]\!]$.

The MGoI-interpretation for formulas is then defined as follows.

- $\theta(\alpha) \in \mathcal{T}([\![\alpha]\!])$, where α is an atomic formula.
- $\theta(\alpha^{\perp}) = \theta(\alpha)^{\perp}$, where α is an atomic formula.
- $\theta(A \otimes B) = \{a \otimes b \mid a \in \theta(A), b \in \theta(B)\}^{\perp\perp}$
- $\theta(A \parr B) = \{a \otimes b \mid a \in \theta(A)^{\perp}, b \in \theta(B)^{\perp}\}^{\perp}$
- $\theta(!A) = \{Ta \mid a \in \theta(A)\}^{\perp\perp}$
- $\theta(?A) = \{Ta \mid a \in \theta(A^{\perp})\}^{\perp}$

Easy consequences of the definition are: (i) for any formula A, $(\theta A)^{\perp} = \theta A^{\perp}$, (ii) $\theta(A) \subseteq End([\![A]\!])$, and (iii) $\theta(A)^{\perp\perp} = \theta(A)$.

5.8.2 MGoI Interpretation of Proofs

In this section we define the MGoI interpretation for proofs of **MELL** without units. All references from now on refer to this MGoI interpretation unless stated otherwise.

As before, every **MELL** sequent will be of the form $\vdash [\Delta], \Gamma$ where Γ is a sequence of formulas and Δ is a sequence of cut formulas that have already been made in the proof of $\vdash \Gamma$. As before, this device is used to keep track of the cuts in a proof of $\vdash \Gamma$. As mentioned earlier, in MGoI proofs are interpreted locally in endomorphism monoids. A proof π of $\vdash [\Delta], \Gamma$ is represented by a morphism $[\![\pi]\!] \in End(\otimes [\![\Gamma]\!] \otimes [\![\overline{\Delta}]\!])$. Here, if $\Gamma = A_1, \cdots, A_n$, $\otimes [\![\Gamma]\!]$ stands for $[\![A_1]\!] \otimes \cdots \otimes [\![A_n]\!]$, and with $\Delta = B_1, B_1^{\perp}, \cdots B_m, B_m^{\perp}$, $[\![\overline{\Delta}]\!] = T^k([\![B_1]\!] \otimes \cdots \otimes [\![B_m]\!]^{\perp})$, for some non-negative integer k, with T^0 being the identity functor. We drop the double brackets wherever there is no danger of confusion. We also define $\sigma = s \otimes \cdots \otimes s$ (m-copies) where s is the symmetry map at different types (omitted for convenience), and $|\Delta| = 2m$. The morphism σ represents the cuts in the proof of $\vdash \Gamma$, i.e. it models Δ. In the case where Δ is empty (that is for a cut-free proof), we define $\sigma : I \longrightarrow I$ to be 1_I where I is the unit of the monoidal product in $\mathcal{C}$.

Definition 5.8.5 (The MGoI Interpretation) Let π be a proof of $\vdash [\Delta], \Gamma$. We define *the MGoI interpretation of* π, denoted by $[\![\pi]\!]$, by induction on the length of the proof as follows. As in ordinary GoI above, the reader is encouraged to draw the diagrams representing the interpretation of each rule.

1. π is an *axiom* $\vdash A, A^{\perp}$, $[\![\pi]\!] := s_{V,V}$ where $[\![A]\!] = [\![A^{\perp}]\!] = V$.
2. π is obtained using the *cut* rule on π' and π'' that is,

$$\dfrac{\begin{array}{c}\pi' \\ \vdots \\ \vdash [\Delta'], \Gamma', A\end{array} \qquad \begin{array}{c}\pi'' \\ \vdots \\ \vdash [\Delta''], A^{\perp}, \Gamma''\end{array}}{\vdash [\Delta', \Delta'', A, A^{\perp}], \Gamma', \Gamma''}\ cut$$

   Define $[\![\pi]\!] = \tau^{-1}\left([\![\pi']\!] \otimes [\![\pi'']\!]\right)\tau$, where τ is the permutation $\Gamma' \otimes \Gamma'' \otimes \overline{\Delta'} \otimes \overline{\Delta''} \otimes A \otimes A^{\perp} \stackrel{\tau}{\longrightarrow} \Gamma' \otimes A \otimes \overline{\Delta'} \otimes A^{\perp} \otimes \Gamma'' \otimes \overline{\Delta''}$. (double brackets and $\otimes$ are dropped for the sake of readability).
3. π is obtained using the *exchange* rule on the formulas A_i and A_{i+1} in Γ'. That is π is of the form

$$\dfrac{\begin{array}{c}\pi' \\ \vdots \\ \vdash [\Delta], \Gamma'\end{array}}{\vdash [\Delta], \Gamma}\ exchange$$

   where $\Gamma' = \Gamma_1', A_i, A_{i+1}, \Gamma_2'$ and $\Gamma = \Gamma_1', A_{i+1}, A_i, \Gamma_2'$. Then, $[\![\pi]\!]$ is obtained from $[\![\pi']\!]$ by interchanging the rows i and $i+1$. So, $[\![\pi]\!] = \tau^{-1}[\![\pi']\!]\tau$, where $\tau = 1_{\Gamma_1'} \otimes s \otimes 1_{\Gamma_2' \otimes \overline{\Delta}}$.
4. π is obtained using an application of the *par* rule, that is π is of the form:

$$\dfrac{\begin{array}{c}\pi' \\ \vdots \\ \vdash [\Delta], \Gamma', A, B\end{array}}{\vdash [\Delta], \Gamma', A \parr B}\ \parr \quad . \text{ Then } [\![\pi]\!] = [\![\pi']\!].$$

5. π is obtained using an application of the *times* rule, that is π is of the form:

$$\dfrac{\begin{array}{c}\pi' \\ \vdots \\ \vdash [\Delta'], \Gamma', A\end{array} \qquad \begin{array}{c}\pi'' \\ \vdots \\ \vdash [\Delta''], \Gamma'', B\end{array}}{\vdash [\Delta', \Delta''], \Gamma', \Gamma'', A \otimes B}\ \otimes$$

Then $[\![\pi]\!] = \tau^{-1}\left([\![\pi']\!] \otimes [\![\pi'']\!]\right)\tau$, where τ is the permutation $\Gamma' \otimes \Gamma'' \otimes A \otimes B \otimes \overline{\Delta'} \otimes \overline{\Delta''} \xrightarrow{\tau} \Gamma' \otimes A \otimes \overline{\Delta'} \otimes \Gamma'' \otimes B \otimes \overline{\Delta''}$.

6. π is obtained from π' by an *of course* rule, that is π has the form :

$$\dfrac{\begin{array}{c}\pi' \\ \vdots \\ \vdash [\Delta], ?\Gamma', A\end{array}}{\vdash [\Delta], ?\Gamma', !A}\ \text{of course}$$

   Then $[\![\pi]\!] = (e_{\Gamma'} \otimes 1_{TA} \otimes 1_{\overline{\Delta}})\,\varphi^{-1}T([\![\pi']\!])\varphi(e^*_{\Gamma'} \otimes 1_{TA} \otimes 1_{\overline{\Delta}})$, where $TT \lhd T\,(e, e^*)$, with $\Gamma' = A_1, \cdots, A_n$, $e_{\Gamma'} = e_{A_1} \otimes \cdots \otimes e_{A_n}$, similarly for e^*, and φ is the canonical isomorphism $T^2(\Gamma') \otimes TA \otimes T(\overline{\Delta}) \longrightarrow T(T(\Gamma') \otimes A \otimes \overline{\Delta})$ defined using the isomorphism $\psi_{X,Y} : TX \times TY \longrightarrow T(X \otimes Y)$. With $\Gamma' = A_1, \cdots, A_n$, $T(\Gamma')$ is a short hand for $TA_1 \otimes \cdots \otimes TA_n$, similarly for $T(\overline{\Delta})$.

7. π is obtained from π' by the *dereliction* rule, that is, π is of the form :

$$\dfrac{\begin{array}{c}\pi' \\ \vdots \\ \vdash [\Delta], \Gamma', A\end{array}}{\vdash [\Delta], \Gamma', ?A}\ \text{dereliction}$$

   Then $[\![\pi]\!] = (1_{\Gamma'} \otimes d_A \otimes 1_{\overline{\Delta}})[\![\pi']\!](1_{\Gamma'} \otimes d^*_A \otimes 1_{\overline{\Delta}})$ where $Id \lhd T\,(d, d^*)$.

8. π is obtained from π' by the *weakening* rule, that is, π is of the form:

$$\dfrac{\begin{array}{c}\pi' \\ \vdots \\ \vdash [\Delta], \Gamma'\end{array}}{\vdash [\Delta], \Gamma', ?A}\ \text{weakening}$$

   Then $[\![\pi]\!] = (1_{\Gamma'} \otimes w_A \otimes 1_{\overline{\Delta}})[\![\pi']\!](1_{\Gamma'} \otimes w^*_A \otimes 1_{\overline{\Delta}})$, where $\mathcal{K}_I \lhd T\,(w, w^*)$.

9. π is obtained from π' by the *contraction* rule, that is, π is of the form :

$$\dfrac{\begin{array}{c}\pi' \\ \vdots \\ \vdash [\Delta], \Gamma', ?A, ?A\end{array}}{\vdash [\Delta], \Gamma', ?A}\ \text{contraction}$$

   Then $[\![\pi]\!] = (1_{\Gamma'} \otimes c_A \otimes 1_{\overline{\Delta}})[\![\pi']\!](1_{\Gamma'} \otimes c^*_A \otimes 1_{\overline{\Delta}})$, where $T \otimes T \lhd T\,(c, c^*)$.

Examples 5.8.6 (a) Let π be the following proof:

$$\dfrac{\vdash A, A^{\perp} \qquad \vdash A, A^{\perp}}{\vdash [A^{\perp}, A], A, A^{\perp}}\ \mathit{cut}$$

Then the MGoI interpretation of this proof is given by $[\![\pi]\!] = \tau^{-1}(s \otimes s)\tau = s_{V\otimes V, V\otimes V}$ where $\tau = (1 \otimes 1 \otimes s)(1 \otimes s \otimes 1)$ and $[\![A]\!] = [\![A^{\perp}]\!] = V$.

(b) Now consider the following proof

$$\cfrac{\cfrac{\cfrac{\vdash A, A^{\perp}}{\vdash A, ?A^{\perp}}}{\vdash !A, ?A^{\perp}} \quad \vdash B, B^{\perp}}{\vdash !A \otimes B, ?A^{\perp} \parr B^{\perp}}$$

Given $[\![A]\!] = V$ and $[\![B]\!] = W$, we have $[\![\pi]\!] = (1 \otimes s \otimes 1)(1 \otimes e \otimes 1 \otimes 1)$ $(\psi^{-1}T(h)\psi \otimes s)(1 \otimes e^{*} \otimes 1 \otimes 1)(1 \otimes s \otimes 1)$ where $h = (1 \otimes d_V)s(1 \otimes d_V^{*})$.

Proposition 5.8.7 *Let π be an* **MELL** *proof of $\vdash [\Delta], \Gamma$. Then $[\![\pi]\!]$ is a partial symmetry.*

Proof The proof follows by induction on the length of the proofs, noting that the functor $(\ )^{*}$ is a strict symmetric monoidal functor, $T(f)^{*} = T(f^{*})$, $\psi^{*} = \psi^{-1}$, and $\psi_I^{*} = \psi_I^{-1}$. □

5.8.3 Interpretation of Cut-Elimination

As we saw previously, the mathematical model of cut-elimination is given by the *execution formula* as in (5.8), defined as follows:

$$EX([\![\pi]\!], \sigma) = \mathrm{Tr}^{\otimes \Delta}_{\otimes \Gamma, \otimes \Gamma}((1 \otimes \sigma)[\![\pi]\!])$$

where π is a proof of the sequent $\vdash [\Delta], \Gamma$, and $\sigma = s^{\otimes m}$ models Δ, where $|\Delta| = 2m$. Note that $EX([\![\pi]\!], \sigma)$ is a morphism from $\otimes\Gamma \longrightarrow \otimes\Gamma$, when it exists. We shall prove below (see Theorem 5.8.12) that the execution formula always exists for any **MELL** proof π. Informally, this means $EX(f, \sigma)$ "converges" whenever $f = [\![\pi]\!]$, for some proof π in **MELL**.

Example 5.8.8 Consider the proof π in Example 5.6.6 above. Recall also that $\sigma = s$ in this case ($m = 1$). Then

$$EX([\![\pi]\!], \sigma) = \mathrm{Tr}((1 \otimes s_{V,V})s_{V\otimes V, V\otimes V}) = s_{V,V}$$

5.8.4 Soundness of the Interpretation and Modelling Computation

In this section we discuss the soundness of the MGoI interpretation. We show that if a proof π is reduced (via cut-elimination) to another proof π', then $EX\,([\![\pi]\!], \sigma) = EX\left([\![\pi']\!], \tau\right)$; that is, $EX\,([\![\pi]\!], \sigma)$ is an invariant of reduction. In particular, if π' is cut-free (i.e. a normal form) we have $EX\,([\![\pi]\!], \sigma) = EX\left([\![\pi']\!], 1_I\right) = [\![\pi']\!]$.

Intuitively this says that if one thinks of cut-elimination as computation then $[\![\pi]\!]$ can be thought of as an algorithm. The computation takes place as follows: if

$EX(\llbracket \pi \rrbracket, \sigma)$ exists then it yields a datum (cf. cut-free proof). This intuition will be made precise below (Theorems 5.8.12 and 5.8.14).

We shall not give the proof of the soundness here, but will mention the main lemmas used in this proof.

Lemma 5.8.9 (Associativity of cut) *Let π be a proof of $\vdash [\Gamma, \Delta], \Lambda$ and σ and τ be the morphisms representing the cut-formulas in Γ and Δ respectively. Then*

$$EX(\llbracket \pi \rrbracket, \sigma \otimes \tau) = EX(EX(\llbracket \pi \rrbracket, \tau), \sigma) = EX(EX((1 \otimes s)\llbracket \pi \rrbracket(1 \otimes s), \sigma), \tau),$$

whenever all traces exist.

Definition 5.8.10 Let $\Gamma = A_1, \cdots, A_n$ and $V_i = \llbracket A_i \rrbracket$.

- A *datum of type $\theta\Gamma$* is a morphism $M : \otimes_i V_i \longrightarrow \otimes_i V_i$ such that for any $a_i \in \theta(A_i^{\perp})$, $\otimes_i a_i \perp M$ and

  $$M{\bullet}a_1 := \mathrm{Tr}^{V_1}(s^{-1}_{\otimes_{i\neq 1} V_i, V_1}(a_1 \otimes 1_{V_2} \otimes \cdots \otimes 1_{V_n}) M s_{\otimes_{i\neq 1} V_i, V_1})$$

  and

  $$M\hat{\bullet}(a_2 \otimes \cdots \otimes a_n) := \mathrm{Tr}^{V_2 \otimes \cdots \otimes V_n}((1 \otimes a_2 \otimes \cdots \otimes a_n) M)$$

  both exist.
- An *algorithm of type $\theta\Gamma$* is a morphism $M : \otimes_i V_i \otimes \llbracket \Delta \rrbracket \longrightarrow \otimes_i V_i \otimes \llbracket \Delta \rrbracket$ for some $\Delta = B_1, B_2, \cdots, B_{2m}$ with m a nonnegative integer and $B_{i+1} = B_i^{\perp}$ for $i = 1, 3, \cdots, 2m-1$, such that if $\sigma : \otimes_{i=1}^{2m} \llbracket B_i \rrbracket \longrightarrow \otimes_{i=1}^{2m} \llbracket B_i \rrbracket$ is $\otimes_{i=1,odd}^{2m-1} s_{\llbracket B_i \rrbracket, \llbracket B_{i+1} \rrbracket}$, $EX(M, \sigma)$ exists and is a datum of type $\theta\Gamma$. (Here σ is defined to be 1_I for $m = 0$, that is when Δ is empty.)

Lemma 5.8.11 *Let $\widetilde{\Gamma} = A_2, \cdots, A_n$ and $\Gamma = A_1, \widetilde{\Gamma}$. Let $V_i = \llbracket A_i \rrbracket$, and $M : \otimes_i V_i \longrightarrow \otimes_i V_i$, for $i = 1, \cdots, n$. Then, M is a datum of type $\theta(\Gamma)$ iff for all $a_i \in \theta(A_i^{\perp})$, $M{\bullet}a_1$ and $M\hat{\bullet}(a_2 \otimes \cdots \otimes a_n)$ (defined as above) exist and are in $\theta(\widetilde{\Gamma})$, and $\theta(A_1)$, respectively.*

Theorem 5.8.12 (Proofs as algorithms) *Let π be an* **MELL** *proof of a sequent $\vdash [\Delta], \Gamma$. Then $\llbracket \pi \rrbracket$ is an algorithm of type $\theta\Gamma$.*

Corollary 5.8.13 (Existence of Dynamics) *Let π be an* **MELL** *proof of a sequent $\vdash [\Delta], \Gamma$. Then $EX(\llbracket \pi \rrbracket, \sigma)$ exists.*

Theorem 5.8.14 (EX is an invariant) *Let π be an MELL proof of a sequent $\vdash [\Delta], \Gamma$ such that $?A$ does not occur in Γ for any formula A. Then,*

- *If π reduces to π' by any sequence of cut-elimination steps, then $EX(\llbracket \pi \rrbracket, \sigma) = EX\left(\llbracket \pi' \rrbracket, \tau\right)$. So $EX(\llbracket \pi \rrbracket, \sigma)$ is an invariant of reduction.*
- *In particular, if π' is any cut-free proof obtained from π by cut-elimination, then $EX(\llbracket \pi \rrbracket, \sigma) = EX\left(\llbracket \pi' \rrbracket, 1_I\right) = \llbracket \pi' \rrbracket$.*

5.9 Concluding Remarks

We have mentioned several open questions in the tutorial, and the reader will be able to find many interesting questions in following up the literature in the Bibliography. Still, a few questions seem particularly apt.

(i) The GoI interpretation does not seem to deal well with units in **LL**. Thus, one should formulate GoI taking into account **-autonomous categories without units*. One such study is in R. Houston's thesis [Hou07].
(ii) The question of how to take into account the additives of **LL** in GoI and the associated categorical analysis of [Gi95a], both along the style here, as well as in the style of [AHS02], is still open.
(iii) Finding examples of our GoI situations in von Neumann algebras, and categorically analyzing Girard's recent notions of GoI [Gi07, Gi08] is a challenge, and presumably would need to accommodate categorical versions of Polarized Linear Logics, as in [HaSc07].

Appendix 1: Graphical Representation of The Trace Axioms

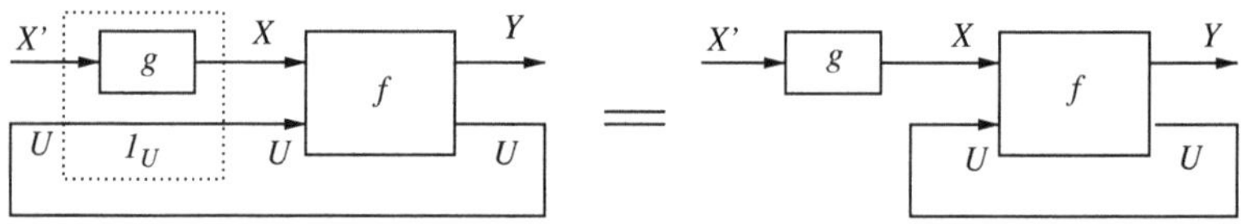

Naturality in X

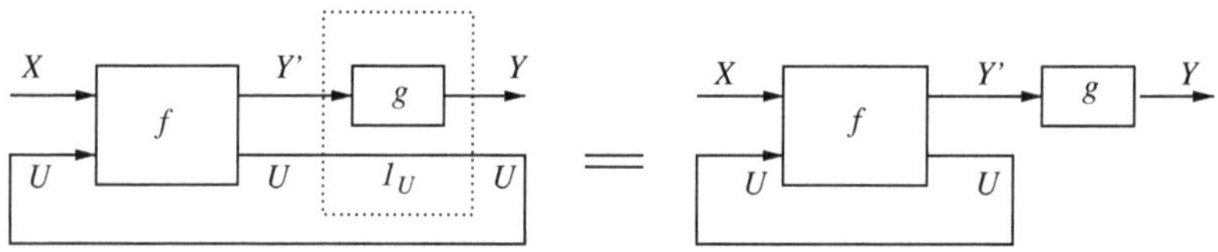

Naturality in Y

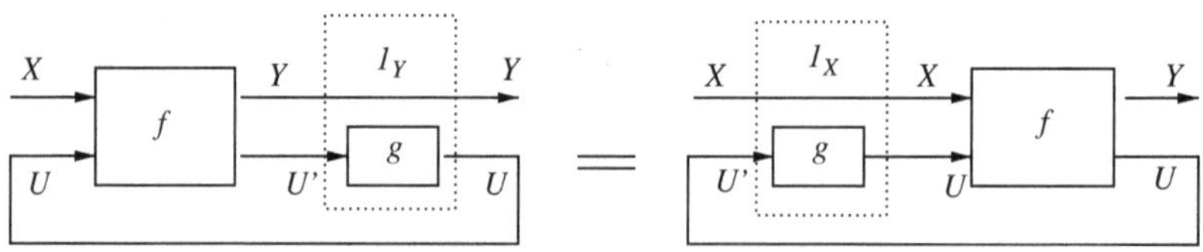

Dinaturality in U

Vanishing I

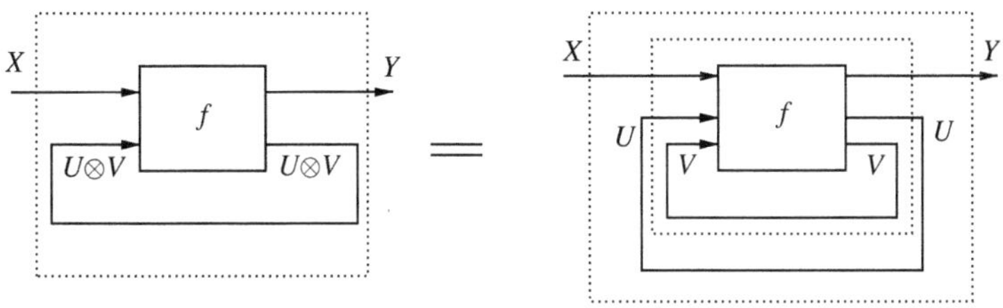

Vanishing II

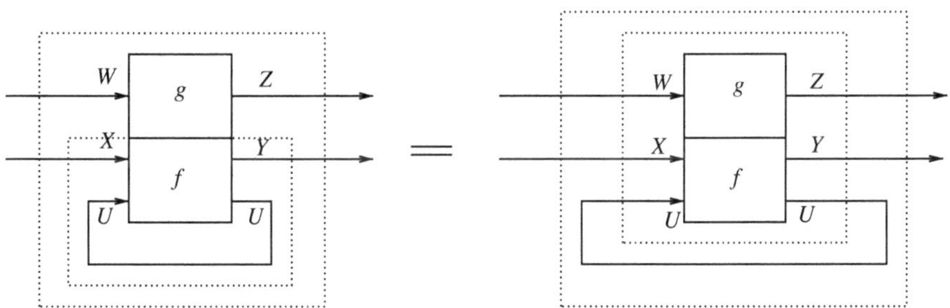

Superposing

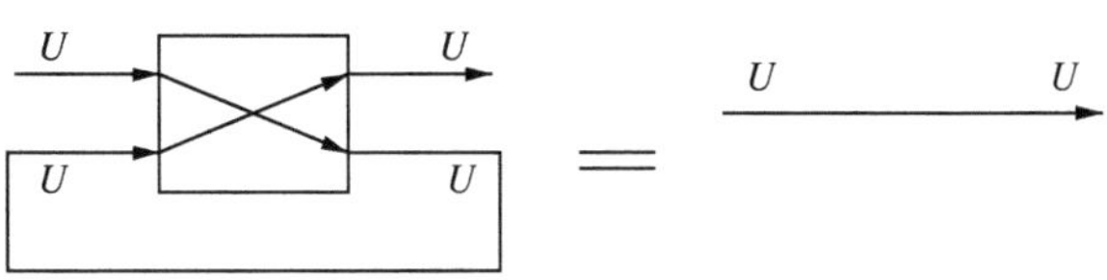

Yanking

Appendix 2: Comparing GoI Notation

Girard	*This Paper*
$1 \otimes a$	$uT(a)v$
p, p^*	j_1, k_1
q, q^*	j_2, k_2
$(1 \otimes r), (1 \otimes r^*)$	uc_1v, uc'_1v
$(1 \otimes s), (1 \otimes s^*)$	uc_2v, uc'_2v
t, t^*	$ue_U(Tv)v, u(Tu)e'_Uv$
d, d^*	ud_U, d'_Uv

References

Abr93. Abramsky, S.: Computational interpretations of linear logic. Theor. Comput. Sci. **111**, 3–57 (1993)

Abr96. Abramsky, S.: Retracing some paths in process algebra. In: CONCUR 96, Springer LNCS 1119, pp. 1–17 (1996)

Abr05. Abramsky, S.: Abstract scalars, loops, and free traced and strongly compact closed categories. In: CALCO 2005, vol. 3629, pp. 1–31. Springer Lecture Notes in Computer Science (2005)

Abr07. Abramsky, S.: Temperley-Lieb algebra: From knot theory to logic and computation via quantum mechanics. In: Chen, G., Kauffman, L., Lomonaco, S. (eds.) Mathematics of Quantum Computing and Technology, pp. 515–558. Taylor and Francis, Abington (2007)

ABP99. Abramsky, S., Blute, R., Panangaden, P.: Nuclear and trace ideals in tensored *-categories. J. Pure Appl. Algebra **143**, 3–47 (1999)

AbCo04. Abramsky, S., Coecke, B.: A categorical semantics of quantum protocols. In: Proceedings of the 19th Annual IEEE Symposium on Logic in Computer Science (LICS), pp. 415–425. IEEE Computer Science Press (2004)

AHS02. Abramsky, S., Haghverdi, E., Scott, P.J.: Geometry of Interaction and Linear Combinatory Algebras. MSCS, vol. 12(5), pp. 625–665, CUP (2002)

AJ94a. Abramsky, S., Jagadeesan, R.: New foundations for the geometry of interaction. Inf. Comput. **111**(1), 53–119 (1994)

Bail95. Baillot, P. (1995), Abramsky-Jagadeesan-Malacaria strategies and the geometry of interaction, mémoire de DEA, p. 7. Universite Paris, Paris (1995)

BP01. Baillot, P., Pedicini, M.: Elementary complexity and geometry of interaction. Fundamenta Informaticae **45**(1–2), 1–34 (2001)

Barr79. Barr, M.: *-Autonomous Categories. Springer Lecture Notes in Mathematics 752 (1979)

Barr92. Barr, M.: Algebraically Compact Functors. JPAA **82**, 211–231 (1992)

BE93. Bloom, S.L., Esik, Z.: Iteration theories: Equational logic of iterative processes. EATCS Monographs on Theoretical Computer Science. Springer, New York (1993)

Bl96. Blute, R.: Hopf algebras and linear logic. Math. Struct. Comput. Sci. **6**, 189–212 (1996)

BCST96. Blute, R., Cockett, J.R.B., Seely, R.A.G., Trimble, T.: Natural deduction and coherence for weakly distributive categories. J. Pure Appl. Algebra **13**, 229–296 (1996)

BCS96. Blute, R., Cockett, J.R.B., Seely, R.A.G.: ! and ?: Storage as tensorial strength. Math. Struct. Comput. Sci. **6**, 313–351 (1996)

BCS00. Blute, R., Cockett, J.R.B., Seely, R.A.G.: Feedback for linearly distributive categories: Traces and fixpoints, Bill (Lawvere) Fest. J. Pure Appl. Algebra **154**, 27–69 (2000)

BS96. Blute, R., Scott, P.: Linear Lauchli semantics. Ann. Pure Appl. Logic **77**, 101–142 (1996)

BS04. Blute, R., Scott, P.: Category Theory for Linear Logicians, in Linear Logic in Computer Science, pp. 3–64. Cambridge University Press, Cambridge (2004)

Bor93. Borceux, F.: Handbook of Categorical Algebra. Cambridge University Press, Cambridge (1993)

CS97. Cockett, J.R.B., Seely, R.A.G.: Weakly distributive categories. J. Pure Appl. Algebra **114**, 133–173 (1997)

Dan90. Danos, V. La logique linéaire appliquée à l'étude de divers processus de normalisation et principalement du *-calcul. PhD thesis, p. VII. Université Paris, Paris (1990)

DR95. Danos, V., Regnier, L.: Proof-nets and the Hilbert space. In: Advances in Linear Logic, London Math. Soc. Notes, 222, CUP, pp. 307–328 (1995)

DopR89. Doplicher, S., Roberts, J.E.: A new duality for compact groups. Invent. Math. **98**, 157–218 (1989)

FP07. Fuhrman, C., Pym, D.: On categorical models of classical logic and the geometry of interaction. Mathematical Structures in Computer Science, pp. 957–1027. Cambridge (2007)

Ger85. Geroch, R.: Mathematical Physics. University of Chicago Press, Chicago (1985)

GLR85. Ghez, P., Lima, R., Roberts, J.E.: W*–categories. Pacific J. Math. **120**, 79–109 (1985)

Gi87. Girard, J.-Y.: Linear logic. Theor. Comput. Sci. **50**(1), 1–102 (1987)

Gi88. Girard, J.-Y.: Geometry of interaction II: Deadlock-free algorithms. In: Proceedings of COLOG'88, LNCS 417, pp. 76–93. Springer, New York (1988)

Gi89. Girard, J-Y.: Towards a geometry of interaction. In: Gray, J.W., Scedrov, A. (eds.) Categories in Computer Science and Logic. Contemp. Math, 92, pp. 69–108. AMS, (1989)

Gi89a. Girard, J.-Y.: Geometry of interaction I: Interpretation of system F. In: Proceedings of the Logic Colloquium 88, pp. 221–260. North Holland (1989a)

Gi95a. Girard, J.-Y.: Geometry of interaction III: Accommodating the additives. In: Advances in Linear Logic, LNS 222, CUP, pp. 329–389 (1995)

Gi07. Girard, J.-Y.: Le Point Aveugle I, II, Hermann Editeurs, Paris, 567 + pp (2007)

Gi08. Girard, J.-Y.: Geometry of Interaction V: logic in the hyperfinite factor, manuscript (2008)

GLT. Girard, J.-Y., Lafont, Y., Taylor, P.: Proofs and Types, Cambridge Tracts in Theoretical Computer Science 7 (1989)

GSS. Girard, J.-Y., Scedrov, A., Scott, P.J.: Bounded linear logic. Theor. Comput. Sci. **97**, 1–66 (1992)

GAL92. Gonthier, G., Abadi, M., Lévy, J.-J.: The geometry of optimal lambda reduction. In: Proceedings of Logic in Computer Science, vol. 9, pp. 15–26 (1992)

Hag00. Haghverdi, E.: A Categorical Approach to Linear Logic, Geometry of Proofs and Full Completeness, PhD Thesis, University of Ottawa, Canada (2000)

Hag00a. Haghverdi, E.: Unique Decomposition Categories, Geometry of Interaction and combinatory logic. Math. Struct. Comput. Sci. **10**, 205–231 (2000)

Hag06. Haghverdi, E.: Typed GoI for Exponentials. In: Bugliesi, M. et al. (eds.) Proceedings of ICALP 2006, Part II, LNCS 4052, pp. 384–395. Springer, New York (2006)

HS04a. Haghverdi, E., Scott, P.J.: A categorical model for the Geometry of Interaction, Theoretical Computer Science Volume 350, Issues 2–3, Feb 2006, pp. 252–274. (Preliminary Version in: Automata, Languages, Programming (ICALP 2004), Springer LNCS 3142, pp. 708–720)

HS04b. Haghverdi, E., Scott, P.J.: From Geometry of Interaction to Denotational Semantics. Proceedings of CTCS2004. In ENTCS, vol. 122, pp. 67–87. Elsevier (2004)

HS05a. Haghverdi, E., Scott, P.J.: Towards a Typed Geometry of Interaction, CSL2005 (Computer Science Logic), Luke Ong, Ed. SLNCS 3634, pp. 216–231 (2005)

HS09. Haghverdi, E., Scott, P.J.: Towards a Typed Geometry of Interaction, Full version of [HS05a], in preparation

HaSc07. Hamano, M., Scott, P.: A categorical semantics for polarized MALL. Ann. Pure Appl. Logic **145**, 276–313 (2007)

Has97. Hasegawa, M.: Recursion from Cyclic Sharing: Traced Monoidal Categories and Models of Cyclic Lambda Calculus. Springer LNCS 1210, pp. 196–213 (1997)

Has08. Hasegawa, M.: On traced monoidal closed categories. Math. Struct. Comput. Sci. **19**(2), 217–244 (2009)

HK09. Hasegawa, M., Katsumata, S.: A note on the biadjunction between 2- categories of traced monoidal categories and tortile monoidal categories, Mathematical Proceedings of the Cambridge Philosophical Society, to appear (2009)

Hi97. Hines, P.: The Algebra of Self-Similarity and its Applications. Thesis. University of Wales (1997)

Hi03. Hines, P.: A categorical framework for finite state machines. Math. Struct. Comput. Sci. **13**, 451–480 (2003)

Hou07. Houston, R.: Modelling Linear Logic without Units, PhD Thesis, Dept. of Computer Science, Manchester University (2007)

HylSc03. Hyland, M., Schalk, A.: Glueing and orthogonality for models of linear logic. Theor. Comput. Sci. **294**, 183–231 (2003)

JS91. Joyal, A., Street, R.: The geometry of tensor calculus I. Adv. Math. **88**, 55–112 (1991)

JS93. Joyal, A., Street, R.: Braided tensor categories. Adv. Math. **102** (1), 20–79 (1993)

JSV96. Joyal, A., Street, R., Verity, D.: Traced monoidal categories. Math. Proc. Camb. Phil. Soc. **119**, 447–468 (1996)

KRT97. Kassel, C., Rosso, M., Turaev, V.: Quantum Groups and Knot Invariants. Soc. Mathématique de France (1997)

KSW02. Katis, P., Sabadini, N., Walters, R.F.C.: Feedback, trace and fixed-point semantics. Theor. Inf. Appl. **36**, 181–194 (2002)

KL80. Kelly, G.M., Laplaza, M.: Coherence for compact closed categories. J. Pure Appl. Algebra **19**, 193–213 (1980)

KR77. Kock, A., Reyes, G.: Doctrines in categorical logic. In: Barwise, J. (ed.) Handbook of Mathematical Logic. North-Holland, Amsterdam (1977)

LS86. Lambek, J., Scott, P.J.: Introduction to Higher Order Categorical Logic, Cambridge Studies in Advanced Mathematics 7. Cambridge University Press, Cambridge (1986)

L89. Lambek, J.: Multicategories revisited. Contemp. Math. **92**, 217–239 (1987)

Lau01. Laurent, O.: A Token Machine for Full Geometry of Interaction. In TLCA '01, SLNCS 2044, pp. 283–297 (2001),

Law69. Lawvere, F.W.: Adjointness in foundations. Dialectica **23**, 281–296 (1969)

Law70. Lawvere, F.W.: Equality in hyperdoctrines and comprehension schema as an adjoint functor, Applications of Category Theory, Proceedings of A.M.S. Symposia on Pure Math XVII, AMS, Providence, RI (1970)

Lef. Lefschetz, S.: Algebraic Topology, Am. Math. Soc. Colloquium Publications (1942)

Mac98. Mac Lane, S.: Categories for the Working Mathematician, 2nd edn. Springer, New York (1998)

MR91. Malacaria, P., Regnier, L.: Some Results on the Interpretation of *-calculus in Operator Algebras. Proceedings of Logic in Computer Science (LICS), pp. 63–72. IEEE Press (1991)

MA86. Manes, E., Arbib, M.: Algebraic Approaches to Program Semantics. Springer, New York (1986)

Mel07. Melliès, P.-A.: Categorical semantics of linear logic: A survey, 132pp. (in preparation). See website http://www.pps.jussieu.fr/mellies/

Pl03. Plotkin, G.: Trace Ideals, MFPS 2003 invited lecture, Montreal (unpublished)

Reg92. Regnier, L.: Lambda-calcul et Réseaux, PhD Thesis, VII. Université Paris, Paris (1992)

Sch07. Schöpp, U.: Stratified Bounded Affine Logic for Logarithmic Space. Proceedings of Logic in Computer Science (LICS), pp. 411–420. IEEE (2007)

Sc00. Scott, P.: Some Aspects of Categories in Computer Science. In: Hazewinkel, M. (ed.) Handbook of Algebra, vol. 2, pp. 3–77. Elsevier, Amsterdam (2000)

See89. Seely, R.A.G.: Linear logic, *-autonomous categories and cofree coalgebras. Contemporary Mathematics, vol. 92. American Mathematical Society (1989)

Ste00. Stefanescu, G.: Network Algebra, Springer, New York (2000)

TrSc. Troelstra, A.S., Schwichtenberg, H.: Basic Proof Theory. Cambridge University Press, Cambridge (1996)

Part III
More Example Applications

Chapter 6
Dagger Categories and Formal Distributions

R. Blute and P. Panangaden

Abstract A *nuclear ideal* is an ideal contained in an ambient monoidal dagger category which has all of the structure of a compact closed category, except that it lacks identities. Intuitively, the identities are too "singular" to live in the ideal. Typical examples include the ideal of Hilbert-Schmidt maps contained in the category of Hilbert spaces, or the ideal of test functions contained in the category DRel of tame distributions on Euclidean space.

In this paper, we construct a category of *tame formal distributions* with coefficients in an associative algebra. We show that there is a formal analogue of the nuclear ideal constructed in DRel, and hence there is a partial trace operation on the category. By taking formal distributions with coefficients in the dual of a cocommutative Hopf algebra, we obtain a categorical generalization of the Borcherds' notion of *elementary vertex group*. Furthermore, when considering the algebra of symmetric endomorphisms of an object in such a category, we obtain a vertex group in Borcherds' sense. The nuclear ideal structure induces a partial trace operator on such vertex groups.

6.1 Introduction

The Abramsky-Coecke notion of abstract quantum mechanics [2] is a proposal to abstract quantum theory away from the usual category of (possibly finite-dimensional) Hilbert spaces and determine the underlying structures which should be taken as primitive. Unlike more traditional quantum logic, which is based on lattice theory, the Abramsky-Coecke approach is explicitly categorical in nature. The authors argue that the minimal necessary structure for interpreting quantum theory

R. Blute (✉)
Department of Mathematics, University of Ottawa, Ottawa, ON, Canada
e-mail: rblute@uottawa.ca

P. Panangaden (✉)
School of Computer Science, McGill University, Montreal, QC, Canada
e-mail: prakash@cs.mcgill.ca

Blute, R., Panangaden, P.: *Dagger Categories and Formal Distributions*. Lect. Notes Phys. **813**, 421–436 (2011)
DOI 10.1007/978-3-642-12821-9_6

is that of a *(monoidal) dagger category*, i.e. a category with a strict involution which is the identity on objects.

They show that this framework provides a rich semantics for quantum computing and quantum information theory. For example, the Born rule emerges naturally from their axiomatization and one can express the correctness of various protocols, such as teleportation [7], as the commutativity of certain diagrams.

In subsequent work [3], the authors provide a diagrammatic language which simultaneously gives the free such category and provides a graphical language for reasoning about quantum systems.

Since their initial papers, the subject of abstract quantum theory and dagger categories has become quite active, and has developed important results. We mention in particular the work of Selinger [25]. Aside from developing another graphical language, the author considers the construction of *completely positive maps* in a general dagger category. CPMs are used, for example in the axiomatic description of quantum operations as described in [10]. (We also note that we use Selinger's notation and terminology throughout.)

Also important is the work of Coecke and Pavlovic [11], where they show that monoidal dagger categories even provide a framework for considering the existence of classical objects in a quantum universe. This is the subject of enormous research in quantum physics, see for example [15]. Furthermore, the description of classical structure in this setting is extremely elegant. A classical object in such a category is one with a compatible coalgebra structure. The comultiplication then models copying, and the counit models deleting, the two operations that define classical objects. Thus traditional algebraic/categorical structures are brought into consideration. See also [12] for further work in this direction.

Finally we mention Abramsky's paper [5]. Aside from summarizing much of the previous work discussed above, the author stresses the importance of *abstract scalars*. In any monoidal category, the scalars are the endomorphisms of the tensor unit. Traditionally, since quantum mechanics was carried out in the category of Hilbert spaces, the scalars were the complex numbers, this being the base field. But an abstract approach allows for considering other possibilities for scalars and Abramsky emphasizes the importance of being able to consider dagger categories with other choices for scalars. One of the interesting properties of the construction in this paper is that we will consider formal distributions with coefficients in an arbitrary commutative, associative algebra A, and the elements of A will act as our scalars.

Monoidal dagger categories were considered by Abramsky, Blute and Panangaden [1] under the guise of *tensored $*$-categories*, (using teminology of Doplicher and Roberts [13]). We were interested in various extensions and elaborations of the category Rel of sets and binary relations. In particular, we were interested in developing a category whose objects are "continuously varying relations". So objects would be open subsets of Euclidean space, and morphisms would be continuous functions $\alpha\colon X \times Y \to \mathsf{C}$ (where C is the field of complex numbers.) Similarly, we wished to replace the usual relational composition:

$$a(\mathcal{R};\mathcal{S})c \text{ if and only if } \exists b \text{ such that } a\mathcal{R}b \text{ and } b\mathcal{S}c$$

with the following "continuous analogue":

$$(\alpha;\beta)(x,z) = \int_Y \alpha(x,y)\beta(y,z)dy$$

This idea led to the construction of the category **DRel** described below, and in [1]. Basically the objects of this category are open subsets of Euclidean space, and morphisms are certain well-behaved *distributions*. Distributions were introduced by Schwartz [24] to capture in a mathematically rigorous fashion the Dirac delta "function", which satisfied the relation

$$(\alpha;\delta)(x,y') = \int_Y \alpha(x,y)\delta(y,y')dy = \alpha(x,y')$$

and its symmetric variant. In fact, no such function exists [6], though physicists made frequent use of such a δ. As a recent example, quantum fields are today frequently modelled as *operator-valued distributions* in the Wightman axiomatization [18]. Schwartz axiomatizes the above δ as a *generalized function* or *function with singularities*. The distributions described in [1] are well-behaved in the sense that, when viewed as generalized functions, they have only mild singularities.

In this paper, we introduce a "formal" analogue of the **DRel** construction. Formal distributions, i.e. formal power series in both x and x^{-1}, have played a fundamental role in algebraic and axiomatic approaches to quantum field theory. See, for example, [18, 22]. Indeed, they are the basis for the axiomatization of the notion of vertex algebra [18] and the notion of *locality* [18, 20], both of which figure in the present work. In this paper, we consider formal distributions with coefficients in a commutative algebra. We show that there is a formal notion of tameness inspired by the construction of **DRel**.

Previous work on formal distributions has focused on algebras of distributions. See for example the works [18, 19]. However, in this paper, we wish to build a category of such distributions. In keeping with the passage from untyped to typed λ-calculus, we obtain a category by considering typed distributions. Atomic types are first assigned to the variables, and then a type for the distribution is inferred from these atomic types. One thus obtains a monoidal category, which we call **ARel**. We will see that the resulting category is a monoidal dagger category.

We also demonstrate that this category has a *nuclear ideal*, in the sense of [1]. In that paper, the authors observed that one of the key aspects of the category of sets and relations, the most elementary example of a monoidal dagger category, is that one has "transfer of variables" i.e. one can use the closed structure and the involution to move variables from "input" to "output". The category of Hilbert spaces does not allow such transfer of variables arbitrarily. Instead, one has a large class of morphisms which can be transposed in this fashion. These are the *Hilbert-Schmidt maps*. The notion of nuclear ideal captures the idea of "partially defined transpose". This idea was suggested by the definition of a nuclear morphism between Banach spaces, due to Grothendieck [16], and subsequent work of Higgs and Rowe [17]. Higgs and Rowe axiomatized the notion of nuclearity for a symmetric monoidal closed category, and is appropriate for the analysis of nuclearity for Banach spaces. The concept of nuclearity in analysis can be viewed as describing when one can think of linear maps as matrices. In the case of a compact closed dagger category such as Rel, all morphisms are nuclear, while in the category of Hilbert spaces, the nuclear morphisms are precisely the Hilbert-Schmidt maps [21].

In the category DRel discussed above, the Schwartz kernel theorem provides an inclusion of the space of test functions into the space of tame distributions, and such distributions form a nuclear ideal. Thus, another way of viewing the axioms of the definition of nuclear ideal is as an axiomitization of categories of (possibly) singular functions, containing a class of nonsingular functions. We show here that a formal analogue of this construction holds in our category ARel of formal distributions.

Another goal of this paper is to relate the notions arising in this paper and the *vertex groups* of Borcherds [9, 26]. Both can be viewed as axiomatizing the notion of singular map. In the former case, we have a category of singular maps, containing an ideal of nonsingular maps. In Borcherds' work, singular maps are viewed as an algebra over an algebra of nonsingular maps defined on some "group", (in fact, a Hopf algebra.). We show that when one considers the category of tame formal distributions with coefficients in the dual of a cocommutative Hopf algebra, one obtains examples of *vertex categories*, i.e. "many-object vertex groups".

The notion of (monoidal) dagger category has appeared in a number of guises. They appeared as *tensored ∗-categories* in the work of Doplicher and Roberts [13, 14]. Their work involved considering categories of unitary representations of compact groups, one of the most significant examples of a monoidal dagger category. They considered such categories intheir analysis of superselection sectors, and proved a fundamental represnetation theorem. Any compact closed monoidal dagger category with certain normed structure (making it a C^*-category), is equivalent tothe category of representations of a compact group. Given the use of monoidal dagger categories and formal distributions in several axiomatizations of quantum field theory, it is our hope that the structures in this paper will be of use in extending the Abramsky-Coecke framework to include QFT.

6.2 Dagger Categories and Nuclear Ideals

We here review the crucial definitions of *monoidal dagger category* and *nuclear ideal*. See [2, 25, 1] for more details, such as the appropriate coherence conditions.

Definition 6.2.1 A category $\mathcal{C}$ is a †-*category* if it is equipped with a functor $(-)^\dagger\colon \mathcal{C}^{op} \to \mathcal{C}$, which is strictly involutive and the identity on objects. We will also assume our †-categories are equipped with a *conjugate functor* $\overline{(\ )}\colon \mathcal{C} \to \mathcal{C}$. A †-category is †-*monoidal* if it is symmetric monoidal, $(f \otimes g)^\dagger = f^\dagger \otimes g^\dagger$, and the conjugate functor has natural isomorphisms $\overline{\overline{A}} \cong A$, $\overline{A \otimes B} \cong \overline{A} \otimes \overline{B}$, and $\overline{I} \cong I$. (We will generally take these to be equalities.) These must satisfy evident equations, see [25].

Definition 6.2.2 Let $\mathcal{C}$ be a monoidal †-category. A *nuclear ideal* for $\mathcal{C}$ consists of the following structure:

- For all objects $A, B \in \mathcal{C}$, a subset $\mathcal{N}(A, B) \subseteq Hom(A, B)$. We will refer to the union of these subsets as $\mathcal{N}(\mathcal{C})$ or $\mathcal{N}$. We will refer to the elements of $\mathcal{N}$ as *nuclear maps*. The class $\mathcal{N}$ must be closed under composition with arbitrary $\mathcal{C}$-morphisms, closed under $\otimes$, closed under $()^{\dagger}$, and the conjugate functor.
- A bijection $\theta : \mathcal{N}(A, B) \to Hom(I, \overline{A} \otimes B)$. The bijection θ must be natural and preserve the †-monoidal structure in an evident sense, see [1].

Examples

- The category Rel of sets and relations is a monoidal dagger category for which the entire category forms a nuclear ideal. Indeed any compact monoidal dagger category has this property.
- The category **Hilb** of Hilbert spaces and bounded linear maps maps is a well-known monoidal dagger category , which, in fact, led to the axiomatization [13]. Then the *Hilbert-Schmidt maps* form a nuclear ideal [1]. (This is one of the only examples where the conjugate functor is not merely the identity. Here it is the conjugate Hilbert space.)
- The category DRel of tame distributions on Euclidean space is a monoidal dagger category. The ideal of test functions (viewed as distributions) is a nuclear ideal. See [1] or the next section.
- We will define a subcategory of Rel called the category of *locally finite relations*. Let $R : A \to B$ be a binary relation and $a \in A$. Then let $R_a = \{b \in B | aRb\}$. Define R_b similarly for $b \in B$. Then we say that a relation is *locally finite* if, for all $a \in A, b \in B$, R_a, R_b are finite sets. Then it is straightforward to verify that we have a monoidal dagger category which is no longer compact closed. It is also easy to verify that the finite relations form a nuclear ideal.

6.3 Distributions as Relations

In this section, we review the construction of the category of tame distributions, denoted DRel [1]. We assume familiarity with basic notions from distribution theory. Suitable references are [24, 27, 6].

The idea was to build a category where composition is given by the formula:

$$\varphi(x, y); \psi(y, z) = \int \varphi(x, y)\psi(y, z)dy.$$

The intuition that guided our original work was that integration should generalize the existential quantification that appears in the definition of relational composition. The proper framework for constructing such a category is the theory of *distributions*. Recall that if Ω denotes a nonempty open subset of R^n, then $\mathcal{D}\Omega$ denotes the smooth (complex-valued) functions of compact support on Ω. We will refer to the elements of $\mathcal{D}\Omega$ as *test functions*. $\mathcal{D}\Omega$ is given the structure of a topological vector space. This structure is described for example in [6, 27]. Then we define a *distribution* on Ω to be a continuous, linear (complex-valued) functional on $\mathcal{D}\Omega$. Let $\mathcal{D}'(\Omega)$ denote the space of all distributions on Ω, equipped with the topology of pointwise convergence. We have a canonical inclusion

$$\iota\colon \mathcal{D}X \hookrightarrow \mathcal{D}'(X)$$

given as follows:

$$\phi(x) \mapsto [\psi(x) \in \mathcal{D}X \mapsto \int \phi(x)\psi(x)dx]$$

There is a canonical inclusion of $\mathcal{D}X \otimes \mathcal{D}Y$ into $\mathcal{D}X \times Y$ given by:

$$\varphi \otimes \psi \mapsto [(x, y) \mapsto \varphi(x)\psi(y)]$$

Proposition 6.3.1 *The space $\mathcal{D}X \otimes \mathcal{D}Y$ is sequentially dense in $\mathcal{D}X \times Y$.*

The construction of DRel makes essential use of the *Schwartz kernel theorem*, which gives conditions under which maps from $\mathcal{D}X$ to $\mathcal{D}'(Y)$ can be realized as distributions on $X \times Y$. We need the following notations to state the theorem. If f is a distribution on $X \times Y$ and $\phi \in \mathcal{D}X$ then $f_*(\phi)$ will be the function from $\mathcal{D}Y$ to the base field given by $\psi \in \mathcal{D}Y \mapsto f(\phi \otimes \psi)$ and $f^*(\psi)$ is given by the evident "transpose" formula. W The Schwartz kernel theorem states:

Theorem 6.3.2 *Let X and Y be two open subsets of R^n and R^m.*

1. *Let f be a distribution on $X \times Y$. For all functions $\phi \in \mathcal{D}X$ the linear map $f_*(\phi)$ is a distribution on Y. Furthermore, the map $\phi \mapsto f_*(\phi)$ from $\mathcal{D}X$ to $\mathcal{D}'(Y)$ is continuous.*
2. *Let f_* be a continuous linear map from $\mathcal{D}X$ to $\mathcal{D}'(Y)$. Then there exists a unique distribution on $X \times Y$ such that for $\phi \in \mathcal{D}X$ and $\psi \in \mathcal{D}Y$ the following holds:*

$$f(\phi \otimes \psi) = f_*(\phi)(\psi)$$

Evidently, by symmetry, the same result applies for f^*.

Definition 6.3.3 A *tame distribution* on $X \times Y$ is a distribution f on $X \times Y$ such that each of f^* and f_* factor continuously through the appropriate ι, where ι is the inclusion of the space of test functions into the space of distributions. Explicitly, there exist continuous linear maps

$$f_L\colon \mathcal{D}X \to \mathcal{D}Y$$
$$f_R\colon \mathcal{D}Y \to \mathcal{D}X$$

such that for every $\phi \in \mathcal{D}X$ and $\psi \in \mathcal{D}Y$, we have:

$$f_*(\phi)(\psi) = f^*(\psi)(\phi) = f(\phi \otimes \psi) = \int f_L(\phi)\psi dy = \int \phi f_R(\psi)dx$$

Intuitively, tame distributions are allowed to be mildly singular, in that composing with a test function "tames" the singularity.

6.3.1 Examples

- Let X be an open subset of R^n. The trace distribution on $X \times X$ is given by $Tr(\eta) = \int \eta(x, x)dx$ where $\eta(x, x') \in \mathcal{D}X \times X$. From this definition it follows that $Tr_*(\phi)(\psi) = Tr^*(\psi)(\phi) = Tr(\phi \otimes \psi) = \int \phi(x)\psi(x)dx$. Thus we clearly have $Tr_L(\phi) = Tr_R(\phi) = \phi$, which shows that δ is tame. This tame distribution will act as the identity in our category.
- Suppose that T is a regular distribution on $X \times Y$ with a test function $\beta(x, y)$ as its kernel, that is to say:

$$T(\alpha(x, y)) = \int_{X \times Y} \beta(x, y)\alpha(x, y)$$

Then T is tame with its associated functions being given by:

$$T_L(\phi) = \int_X \beta(x, y)\phi(x)$$
$$T_R(\psi) = \int_Y \beta(x, y)\psi(y)$$

We write $\mathcal{T}X, Y$ for the tame distributions on $X \times Y$.

Given tame distributions we can define the following operation which will serve as composition. Suppose that $f \in \mathcal{T}X, Y, g \in \mathcal{T}Y, Z$. We define $f; g \in \mathcal{T}X, Z$ as follows. Given that f is tame, we have a continuous function $f_L : \mathcal{D}X \to \mathcal{D}Y$. Applying the first part of the Schwartz kernel theorem to g, we obtain a morphism $g_* : \mathcal{D}Y \to \mathcal{D}'(Z)$. Composition gives a continuous map $\mathcal{D}X \to \mathcal{D}'(Z)$. By the second part of the kernel theorem, we obtain a distribution on $X \times Z$.

Definition 6.3.4 The category DRel has as objects open subsets on R^n, and, as morphisms, tame distributions. Composition is as described above.

Theorem 6.3.5 DRel *is a monoidal dagger category.*

The tensor product is given as follows. Given objects X and Y we define $X \otimes Y$ as the cartesian product space $X \times Y$. Given morphisms in DRel $f : X \to Y$ and $g : X' \to Y'$ we can define $f \otimes g : X \otimes X' \to Y \otimes Y'$ as follows. We first define $f \otimes g$ as a distribution on

$$\mathcal{D}X \otimes \mathcal{D}X' \otimes \mathcal{D}Y \otimes \mathcal{D}Y'$$

by the formula

$$(f \otimes g)(\phi(x) \otimes \phi'(x') \otimes \psi(y) \otimes \psi'(y')) = f(\phi \otimes \psi)g(\phi' \otimes \psi').$$

It is routine to verify that this is tame. We extend $f \otimes g$ to all of

$$\mathcal{D}X \times X' \times Y \times Y'$$

as above. The one-point space, written $I = \{*\}$, is the unit for the tensor (with measure $\mu(\{*\}) = 1$).

Finally the $*$-structure is the identity on objects. On morphisms, the only thing that changes is the role of f_L and f_R. The conjugate functor is taken to be the identity.

Theorem 6.3.6 *The sets $\mathcal{N}(Y, Z)$ form a nuclear ideal for* DRel.

One can also show:

Theorem 6.3.7 *The canonical nuclear ideal in* DRel *is traced.*

There is a more succinct description of the trace operator in DRel. Since $h = gf$ is nuclear, it has a kernel, $\alpha(x, x')$. Recall from Theorem 6.3.6 that the formula for α is given by:

$$\alpha(x, x') = f_R(\beta_g(y, x')) = \int_Y \beta_f(x, y)\beta_g(y, x')$$

Hence we may conclude that:

$$tr_A(h) = \int_X \alpha(x, x)$$

6.4 Categories of Formal Distributions

We now review the basic theory of formal distributions. Much of this theory was developped by Kac. Suitable references are [18, 19]. In the following, A will always denote a commutative, associative, unital algebra over some field k.

An expression of the form $\alpha(z) = \Sigma_{n \in Z} \alpha_n z^n$, where Z is the set of integers, $\alpha_n \in A$ and z is a variable, is called a *formal distribution with coefficients in A*. Similarly, one can speak of formal distributions in several variables. The set of formal distributions in a fixed set of variables forms an infinite dimensional vector space, denoted $A[[z, z^{-1}, w, w^{-1}, \ldots]]$.

The space of distributions has a great deal of structure, much of which is analogous to Schwartz's original theory of distributions. The key to defining such structure is the *residue* operation, defined by $Res_z(\alpha(z)) = \alpha_{-1} \in A$, i.e. the residue of α is the coefficient of z^{-1}. Similarly, if $\alpha(z, w) \in A[[z, z^{-1}, w, w^{-1}]]$, we can define $Res_z(\alpha(z, w)) \in A[[w, w^{-1}]]$.

We now observe that the space of Laurent polynomials $A[z, z^{-1}]$ can be viewed as the test functions for these formal distributions, with the evaluation $A[[z, z^{-1}]] \times A[z, z^{-1}] \to A$ being defined by

$$< \alpha(z), f(z) >= Res_z f(z)\alpha(z)$$

There is a formal analogue of the injection $\mathcal{D}Y \to \mathcal{D}'(Y)$ which is given simply by the inclusion $A[y, y^{-1}] \subseteq A[[y, y^{-1}]]$, and similarly in the multivariable case. There is a corresponding version of the Shwartz kernel theorem as well.

The *formal Dirac delta* is given by the distribution:

$$\delta(z, w) = z^{-1} \sum_{n \in Z} \left(\frac{z}{w}\right)^n$$

We have the fundamental property that for all $f(z) \in A[z, z^{-1}]$

$$< \delta(z, w), f(z) >= f(w)$$

Note that, in this equation, we are multiplying two distributions. In general, this cannot be done even formally, due to the possibility of infinite coefficients. We must have a notion of "tameness" to perform such multiplications. We will see that the Dirac delta is indeed tame.

One can also reiterate the process of taking residues. If α is a distribution, and $x_1, x_2, \ldots, x_n$ are among its variables, then we define

$$Res_{x_1, x_2, \ldots, x_n} \alpha = Res_{x_1}(Res_{x_2}(\ldots Res_{x_n} \alpha)) \ldots)$$

One can readily check that this is well-defined and independent of the order in which the residues are taken.

We also note that the space of formal distributions allows formal differentiation, i.e. we have operators:

$$\partial = \partial_z : A[[z, z^{-1}, w, w^{-1}, \ldots]] \to A[[z, z^{-1}, w, w^{-1}, \ldots]]$$

and that these satisfy equations analogous to those for differentiation of distributions, e.g.

$$Res_z \partial\alpha(z)\beta(z) = -Res_z \alpha(z)\partial\beta(z)$$

This is a formal analogue of integration by parts. Consult [18] for these and other results, such as the representation of distributions in terms of derivatives of deltas.

6.4.1 Tameness for Formal Distributions

We will now define a category which will be the formal analogue of DRel, and this category will have much of the same structure. We assume throughout the remainder of this section that $\mathcal{A}$ is a fixed associative unital algebra over a field k.

We assume the existence of an infinite set of *atomic types.* These will be type variables denoted A_1, A_2, B, ... Then the set of all types consists of all words of the form $A_1 \otimes A_2 \ldots \otimes A_n$. We refer to n as the *length* of the word. We also assume the existence of a unique word of length 0, denoted I. I is the tensor unit, and acts as the identity in the monoid of words. (Thus we will be working in a *strict* monoidal category). We also assume that we have an infinite stock of variables for each atomic type. These will be denoted $x\colon A$, but we will generally not write the type, if there is no danger of confusion.

Now we can talk about *typed* distributions. A formal distribution of type $A_1 \otimes A_2 \ldots \otimes A_m$ is an element of $\mathcal{A}[[x_1, x_1^{-1}, \ldots, x_m, x_m^{-1}]]$, where x_i is of type A_i. We say that a formal distribution $\alpha(x_1, x_2, \ldots, x_m, y_1, y_2, \ldots, y_n)$ of type $A_1 \otimes A_2 \ldots A_m \otimes B_1 \otimes \ldots B_n$ is *tame with respect to the type splitting* $A_1 \otimes A_2 \ldots A_m || B_1 \otimes \ldots B_n$ if, for all $f \in \mathcal{A}[x_1, x_1^{-1}, \ldots, x_m, x_m^{-1}]$,

$$Res_{x_1,x_2,\ldots,x_m}(f\alpha) \in \mathcal{A}\left[y_1, y_1^{-1}, \ldots, y_n, y_n^{-1}\right]$$

and dually for all $g \in \mathcal{A}[y_1, y_1^{-1}, \ldots, y_n, y_n^{-1}]$.

In other words, a tame distribution takes Laurent polynomials to Laurent polynomials. This is the obvious analogue of the notion of tameness used in [1], given that in the formal setting we are using Laurent polynomials as test functions.

Remark 6.4.1 We note that we consider two distributions (of the same type) equivalent if they are identical up to α-conversion, i.e. up to change of variable name (within the same type).

We are now ready to define the category ARel.

Definition 6.4.2 The category ARel is defined as follows. Objects are types. A morphism
$\alpha\colon A_1 \otimes A_2 \otimes \ldots \otimes A_n \to B_1 \otimes B_2 \ldots B_m$ is (the equivalence class of) a distribution of type $A_1 \otimes A_2 \otimes \ldots \otimes A_n \otimes B_1 \otimes B_2 \ldots B_m$ which is tame with respect to the type splitting $A_1 \otimes A_2 \otimes \ldots \otimes A_n || B_1 \otimes B_2 \ldots B_m$ Composition is defined as follows. Suppose that $\alpha\colon A_1 \otimes A_2 \otimes \ldots \otimes A_n \to B_1 \otimes B_2 \ldots B_m$ and that $\beta\colon B_1 \otimes B_2 \otimes \ldots \otimes B_m \to C_1 \otimes C_2 \ldots \otimes C_p$. Then we have

$$\beta\alpha(x_1, x_2, \ldots, x_n, y_1, \ldots, y_p) =$$
$$Res_{z_1,z_2,\ldots,z_m}[\alpha(x_1, \ldots, x_n, z_1, \ldots, z_m)\beta(z_1, \ldots, z_m, y_1, \ldots, y_p)]$$

Note that one must always be careful to use distinct variables in the two distributions being composed.

The identity is defined as:

$$id: A_1 \otimes A_2 \otimes \ldots \otimes A_n \to A_1 \otimes A_2 \otimes \ldots \otimes A_n = \Pi_{i=1}^n \delta_{A_i}$$

Also note that we set $Hom(I, I) = \mathcal{A}$, and more generally $Hom(I, A)$ is the space of Laurent polynomials on A. The justification for this is as in [1].

Theorem 6.4.3 ARel *is a category.*

Proof There are a number of things to check here, most are more or less straightforward. One must check that δ is tame, and that the product of δ's does indeed act as identity. One must check that the composite of two tame distributions is again tame, and finally associativity of composition follows from the observation that $\alpha Res_z\beta = Res_z\alpha\beta$, when z is not among α's variables. □

Theorem 6.4.4 ARel *is a monoidal* †*–category.*

Proof The tensor on objects is obvious. On morphisms, the tensor is given by multiplication. Again, when multiplying two distributions together, one must always make sure that the two distributions use distinct variables. The conjugate functor is taken to be the identity, and the †-functor reverses the order of variables. The necessary equations are straightforward to verify. □

Finally, we may state the following result which is also straightforward.

Theorem 6.4.5 *The Laurent polynomials form a nuclear ideal for* ARel.

Proof The bijection $\theta : \mathcal{N}(A, B) \to Hom(I, \overline{A} \otimes B)$ is the obvious injection of the test functions into the corresponding space of distributions. The necessary equations are all evident. □

6.4.2 Locality for Formal Distributions

We now review one of the crucial topics in formal distribution theory, the notion of *locality* of a formal distribution. This notion has been emphasized heavily by Kac [18–20]. These are a fundamental class of distributions which were inspired by the notion of locality in quantum field theory.

Definition 6.4.6 A formal distribution $\alpha(x, y)$ is *local* if there exists a positive integer N such that $(x - y)^N \alpha(x, y) = 0$.

The formal Dirac delta is local, as $(x - y)\delta(x, y) = 0$. Similarly, any derivative of the delta is local. We here collect some basic identities on derivatives of delta which are useful in proving such results.

Lemma 6.4.7 ([18], p. 16)

- $\delta(x, y) = \delta(y, x)$
- $\partial_x^j \delta(x, y) = (-\partial_y)^j \delta(x, y)$
- $(x - y)^{j+1} \partial_x^j \delta(x, y) = 0$

Now with the above formulas, one may characterize completely the local formal distributions:

Theorem 6.4.8 (Kac [18], p.18) *The local distributions are precisely those of the form:*

$$\alpha(x, y) = \sum_{j \in Z^+} c^j(y)\partial_y^{(j)}\delta(x, y)$$

where the above sum is finite and $c^j(y) = \mathcal{A}[[y, y^{-1}]]$. *The series* $c^j(y)$ *can be calculated by the formula:*

$$c^j(y) = Res_x \alpha(x, y)(x - y)^j$$

It is now straightforward to verify that the tame local distributions form a †-subcategory. The only thing remaining to verify is the following:

Lemma 6.4.9 *Suppose that* $\alpha(x, y)$ *and* $\beta(y, z)$ *are tame local distributions. Then* $Res_y[\alpha(x, y)\beta(y, z)]$ *is local as well. (In particular, it is well-defined.)*

Proof This follows from the above characterization of local distributions, and repeated application of the "integration by parts" formula. □

Now we define a category Loc-ARel, whose objects are atomic formal types, and morphisms are local distributions. Loc-ARel has an evident †-category structure.

6.4.3 Monoidal Structure for Loc-ARel

We now describe a tensor structure for the category Loc-ARel. This first requires defining an n-ary version of locality:

Definition 6.4.10 We suppose that

$$\alpha : A_1 \otimes A_2 \otimes \ldots A_n \to B_1 \otimes B_2 \otimes \ldots B_n$$

is a tame distribution, and that the corresponding variables are $x_1, x_2, \ldots, x_n, y_1, y_2, \ldots, y_n$. Then we say that α is *local* if there is a permutation σ of the set $\{1, 2, \ldots, n\}$ such that for all $i \in \{1, 2, \ldots, n\}$, there exists a natural number N_i such that:

$$(x_i - y_{\sigma(i)})^{N_i} \alpha = 0$$

Lemma 6.4.11 Loc-ARel *is a monoidal †-subcategory of* ARel.

Note however that there is no longer a nuclear ideal. However a slight modification of the notion of locality does yield a subcategory with a nuclear ideal. We say that a tame distribution $\alpha\colon A \to B$ is *stable* if it is of the form $\alpha = \alpha_1 + \alpha_2$, where α_1 is (tame) local, and α_2 is a Laurent polynomial. Thus the stable distributions only fail slightly to be local. It is straightforward to verify that we indeed have a category.

Lemma 6.4.12 *Let $\alpha(x, y)$ and $\beta(y, z)$ be stable distributions. Then*

$$Res_y[\alpha(x, y)\beta(y, z)]$$

is as well.

Proof One simply notes that the composition of two tame local distributions is tame and local, the composition of two nuclear morphisms is nuclear, and the compostion of a tame local distribution and a nuclear distribution is nuclear. The result now follows from the bilinearity of composition. □

So we define a category S-ARel whose objects are formal types and morphisms are stable distributions. It is evidently a monoidal †-subcategory of ARel. S-ARel is essentially the smallest extension of Loc-ARel for which there is a nuclear ideal.

Theorem 6.4.13 *The Laurent polynomials form a nuclear ideal in* S-ARel.

6.5 Vertex Groups and Categories

In this section, we review Borcherds' notion of an *elementary vertex group* [9], and then give a minor generalization of this notion, that being the notion of a *vertex category*, i.e. a many-object vertex group. We demonstrate that the category ARel of the previous section gives an example of a vertex category, whenever $\mathcal{A}$ is taken to be the dual of a cocommutative Hopf algebra H. We show further that when considering the algebra determined by the endomorphisms of an object of a vertex category, one obtains a vertex group in the Borcherds sense. We first review some basic facts about duals of Hopf algebras.

Before getting into the technical details of vertex groups, we recall some facts about duals of Hopf algebras. See [23] for details.

First recall that if H is a Hopf algebra, then H^*, the linear dual of H, is generally not a Hopf algebra, unless H is finite-dimensional. However, we have:

Lemma 6.5.1 *The dual of the comultiplication $\Delta\colon H \to H \otimes H$ induces an algebra structure on H^*, when composed with the canonical inclusion $H^* \otimes H^* \to (H \otimes H)^*$. If H is cocommutative, then H^* is a commutative algebra. Thus, if $f, g \in H^*$ and $h \in H$, then*

$$(fg)(h) = \sum_h f(h_1)g(h_2)$$

using the usual Sweedler notation, i.e.

$$\Delta(h) = \sum_h h_1 \otimes h_2$$

We will also make use of the fact that H^* has a canonical structure as a two-sided H-module via the formulas:

$$(hf)(h') = f(h'h) \qquad (fh)(h') = f(hh')$$

Remark 6.5.2 Finally we note that the existence of an involutive antipode gives a second possible monoidal †-structure on **ARel**. If $\alpha(x, y) = \sum \alpha_{ij} x^i y^j$ is a morphism from A to B, then define

$$\overline{\alpha} = \sum S^*(\alpha_{ij}) x^i y^j$$

and

$$\alpha^\dagger = \sum S^*(\alpha_{ij}) y^j x^i$$

In this section, we will always mean this monoidal †-structure.

The following definition is due to Borcherds [9]. It has been studied and elaborated on extensively by Snydal [26]. For examples, see either of these two references.

Definition 6.5.3 Let H be a cocommutative Hopf algebra over a field k. A vertex group *on* H *consists of a* k*-vector space* K, the *ring of singular functions* on H, with the following additional structure:

- K is an associative, unital algebra over the algebra H^*.
- K is a two-sided H-module. Further, the unit map $\eta\colon H^* \to K$ is a map of 2-sided H-modules.
- The product map on K, $\mu\colon K \otimes K \to K$ is equivariant under the left and right actions of H.
- There is a morphism $S_K : K \to K$ such that $S_K \circ \eta = \eta \circ S^*$.
  We further require that S_K be an antialgebra map, and that $S_K^2 = id$. If the algebra K is also commutative, then we say that we have a *commutative vertex group*.

Borcherds and Snydal only consider the commutative case, but the present work yields several natural noncommutative examples.

We now provide a categorical generalization of the previous definition by introducing the notion of a *vertex category*. This is the correct generalization in that a one-object vertex category is indeed a vertex group.

Definition 6.5.4 *Let* H *be a cocommutative Hopf algebra. An* H-vertex category *consists of a* †*-category* $\mathcal{C}$ *such that:*

- For all objects A, B in $\mathcal{C}$, we have that $Hom(A, B)$ is an H^*- module, a 2-sided H-module, and composition is H^*-bilinear.
- Composition also satisfies the following H*-invariance* property: If $f\colon A \to B$ and $g\colon B \to C$, then we have ($\triangleright$ and $\triangleleft$ denote the actions of H.)

$$h \triangleright (gf) = \Sigma_h (h_1 \triangleright g)(h_2 \triangleright f)$$
$$(gf) \triangleleft h = \Sigma_h (g \triangleleft h_1)(f \triangleleft h_2)$$

- We must also have the following *antipode condition.* First note that there is a canonical morphism $\eta\colon H^* \to Hom(A, A)$ which takes $f \in H^*$ to $f \triangleright id$. We require that η be a map of H-modules and that the following diagram commutes.

$$\begin{array}{ccc} H^* & \xrightarrow{\eta} & Hom(A,A) \\ {\scriptstyle S^*}\downarrow & & \downarrow{\scriptstyle (\,)^\dagger} \\ H^* & \xrightarrow{\eta} & Hom(A,A) \end{array}$$

The following results are all straightforward. All actions are defined by acting on coefficients.

Theorem 6.5.5

- *A one-object vertex category is a vertex group, with S_K being given by the dagger operation on Homsets.*
- *When A is the dual of a cocommutative Hopf algebra, then* ARel *is a vertex category.*
- *In any vertex category* C*, if $C \in$* C*, then $Hom(C, C)$ is a vertex group.*

6.6 Conclusion

The primary goal of the theory of formal distributions is to develop a more purely algebraic version of the Schwartz theory of distributions. Then the issue becomes the extent to which the original theory lifts to the algebraic setting. This is for example one of the goals of the monograph [18]. One is particularly interested in the many applications of distribution theory in quantum physics. In this paper, we have shown that the structure of the category DRel lifts to this formal setting in a straightforward way. Thus one is able to view these formal distributions as generalized relations, as discussed in [1]. We hope to explore this idea in the future.

Along the same lines, we have introduced the notion of a *vertex category* or multiobject vertex algebra. Connecting this idea with the original work of Borcherds [9] and Snydal [26] is also work we intend to explore.

References

1. Abramsky, S., Blute, R., Panangaden, P.: Nuclear and trace ideals in tensored *-categories. J. Pure Appl. Algebra **143**, 3–47(1999)
2. Abramsky, S., Coecke, B.: A categorical semantics of quantum protocols. In: Proceedings of the 19th Annual IEEE Symposium on Logic in Computer Science: LICS 2004, IEEE Computer Society, pp. 415–425 (2004)

3. Abramsky, S., Coecke, B.: Abstract physical traces. Theory Appl. Categories **14**, 111–124 (2005)
4. Abramsky, S., Jagadeesan, R.: New foundations for the geometry of interaction. Inf. Comput. **111**(1), 53–119 (1994)
5. Abramsky, S.: Abstract scalars, loops, and free traced and strongly compact closed categories. In: Proceedings of CALCO 2005, Springer Lecture Notes in Computer Science, vol. 3629, pp. 1–31 (2005)
6. M.A. Al-Gwaiz: Theory of Distributions. Dekker Pure and Applied Mathematics. Dekker, New York (1992)
7. Bennett, C., Brassard, G., Crpeau, C., Jozsa, R., Peres, A., Wootters, W.: Teleporting an unknown quantum state via dual classical and Einstein-Podolsky-Rosen channels. Phys. Rev. Lett. **70**, 1895–1899 (1993)
8. Blute, R.F., Cockett, J.R.B., Seely, R.A.G.: Feedback for linearly distributive categories: Traces and fixpoints. J. Pure Appl. Algebra **154**, 27–69 (2000)
9. Borcherds, R.: Vertex algebras. Preprint, q-alg/9706008. Also appeared in Topological field theory, primitive forms and related topics (Kyoto, 1996) Progress in Mathematics, vol. 60. Birkhäuser Boston, (1998)
10. Chuang, I., Nielsen, M., Quantum Computation and Quantum Information. Cambridge University Press, Cambridge (2000)
11. Coecke, B., Pavlovic, D.: Quantum measurements without sums. To appear in: Chen, G., Kauffman, L., Lomonaco, S. (eds.) The Mathematics of Quantum Computation and Technology, Taylor and Francis, Hoboken (2006)
12. Coecke, B., Paquette, E., Pavlovic, D.: POVMs and Naimark's theorem without sums, preprint (2006)
13. Doplicher, S., Roberts, J.: A new duality theory for compact groups. Inventiones Math. **98**, 157–218 (1989)
14. Ghez, P., Lima, R., Roberts, J.: w*-categories. Pacific J. Math. **120**, 79–109 (1985)
15. Giulini, D., Joos, E., Kiefer, C., Kupsch, J., Stamatescu, I.-O., Zeh, H.D.: Decoherence and the Appearance of a Classical World in Quantum Theory. Springer, New York (1996)
16. Grothendieck, A.: Produits Tensoriels Topologiques et Espaces Nucleaires. AMS Memoirs. American Mathematical Society (1955)
17. Higgs, D.A., Rowe, K.: Nuclearity in the category of complete semilattices. J. Pure Appl. Algebra **57**, 67–78 (1989)
18. Kac, V.: Vertex algebras for beginners. University Lecture Series, American Mathematical Society (1997)
19. Kac, V.: Formal distributions algebras and conformal algebras. In: Proceeding of te 12th International Congress of Mathematical Physics, pp. 80–97 (1997). q-alg/9709027
20. Kac, V.: The idea of locality, preprint q-alg/9709008
21. Kadison, R., Ringrose, J.: Fundamentals of the Theory of Operator Algebras. Academic Press, New York (1983)
22. Lian, B., Zuckerman, G.: Commutative quantum operator algebras. J. Pure Appl. Algebra **100**, 117–141 (1995)
23. Majid, S.: Foundations of Quantum Group Theory. Cambridge University Press, Cambridge (1995)
24. Schwartz, L.: Théorie des Distributions. Hermann, Paris (1957)
25. Selinger, P.: Dagger compact closed categories and completely positive maps. In: Proceedings of the 3rd International Workshop on Quantum Programming Languages (QPL 2005), Chicago. ENTCS 170, pp. 139–163 (2007)
26. Snydal, C.: Equivalence of Borcherds G-vertex algebras and axiomatic vertex algebras, preprint math.QA/9904104
27. Treves, F.: Topological Vector Spaces, Distributions and Kernels. Pure and Applied Mathematics, vol. 25. Academic Press, New York (1967)

Chapter 7
Proof Nets as Formal Feynman Diagrams

R. Blute and P. Panangaden

Abstract The introduction of linear logic and its associated proof theory has revolutionized many semantical investigations, for example, the search for fully-abstract models of PCF and the analysis of optimal reduction strategies for lambda calculi. In the present paper we show how proof nets, a graph-theoretic syntax for linear logic proofs, can be interpreted as operators in a simple calculus.

This calculus was inspired by Feynman diagrams in quantum field theory and is accordingly called the ϕ-calculus. The ingredients are formal integrals, formal power series, a derivative-like construct and analogues of the Dirac delta function.

Many of the manipulations of proof nets can be understood as manipulations of formulas reminiscent of a beginning calculus course. In particular, the "box" construct behaves like an exponential and the nesting of boxes phenomenon is the analogue of an exponentiated derivative formula. We show that the equations for the multiplicative-exponential fragment of linear logic hold.

7.1 Introduction

Girard's geometry of interaction programme [Gir89a, Gir89b, Gir95a] gave shape to the idea that computation is a branch of dynamical systems. The point is to give a mathematical theory of the dynamics of computation and not just a static description of the results as in denotational semantics.

The key intuition is that a proof net, a graphical representation of a proof, is decorated with operators at the nodes which direct the flow of information through the net. Now the process of normalization is not just described by a syntactic rewriting of the net, as is usually done in proof theory, but by the action of these operators. The operators are interpreted as linear operators on a suitable

R. Blute (✉)
Department of Mathematics, University of Ottawa, Ottawa, UN, Canada
e-mail: rblute@uottawa.ca

P. Panangaden (✉)
School of Computer Science, McGill University, Montreal, QC, Canada
e-mail: prakash@cs.mcgill.ca

Blute, R., Panangaden, P.: *Proof Nets as Formal Feynman Diagrams*. Lect. Notes Phys. **813**, 437–466 (2011)
DOI 10.1007/978-3-642-12821-9_7

Hilbert space. In this framework normalizability corresponds to nilpotence of a suitable operator. Given the correspondence between proof nets and the λ-calculus, a significant shift has occurred. One now has a *local, asynchronous* algorithm for β-reduction [DR93, ADLR94]. Abramsky and Jagadeesan [AJ94b] presented geometry of interaction using dataflow nets using fixed point theory instead of the apparatus of Hilbert spaces and operators. However, the information flow paradigm is clear in both presentations of the geometry of interaction.

In the present paper we begin an investigation into the notion of information flow. Our starting point is the notion of Feynman diagram in Quantum Field Theory [Fey49b, Fey49a, Fey62, IZ80]. These are graphical structures which can be seen as visualizations of interactions between elementary particles. The particles travel along the edges of the graph and interact at the vertices. Associated with these graphs are integrals whose values are related to the observable scattering processes. This intuitive picture can be justified from formal quantum field theory [Dys49]. Mathematically quantum field theory is about operators acting on Hilbert spaces, which describe the flow of particles. One can seek a formal analogy then with the framework of quantum field theory and the normalization process as described by the geometry of interaction.

We have, however, not yet reached a full understanding of the geometry of interaction. We have, instead, made a correspondence between proof nets and terms in a formal calculus, the ϕ-calculus, which closely mimics some of the ideas of quantum field theory. In particular we have imitated some of the techniques, called "functional methods" in the quantum field theory literature [IZ80], and shown how to represent the exponential types in linear logic as an exponential power series. The manipulations of boxes in linear logic amounts to certain simple exponential identities. Thus we have more than a pictorial correspondence; we have formal integrals whose evaluation corresponds to normalization.

This work was originally presented at the Newton Institute *Semantics of Computation* Seminar in December 1995. The publication of this edition provided an excellent opportunity to revive the work. We thank Bob Coecke for giving us the opportunity to do so.

7.2 Functional Integrals in Quantum Field Theory

Before we describe the ϕ-calculus in detail we will sketch the theory of functional integrals as they are used in quantum field theory. This section should be skipped by physicists. This section is very sketchy, but, it is hoped, it will provide an overview of the method of functional integrals and, more importantly, it will give a context for the ϕ-calculus to be introduced in the next section. It has been our experience that computer scientists, categorists and logicians, who have typically never heard of functional integrals tend to view the ϕ-calculus as an ad-hoc formalism "engineered" to capture the combinatorics of proof nets. In fact, almost everything that we introduce has an echo in quantum field theory.

There are numerous sources for functional integrals. The idea originated in Feynman's doctoral dissertation [Bro05] published in 1942, now available as a book. The basic idea is simple. Usually in nonrelativistic quantum mechanics one associates a wave function $\psi(\boldsymbol{x}, t)$ which obeys a partial differential equation govening its time evolution, the Schrödinger equation. The physical interpretation is that the probability density of finding the particle at location $\boldsymbol{x}$ at time t is given by $|\psi(\boldsymbol{x}, t)|^2$. The wave function describes how the particle is "smeared out" over space; it is called a *probability amplitude function.*

In the path integral approach, instead of associating a wave function with a particle one looks at all possible trajectories of the particle – whether dynamically possible or not according to classical mechanics – and associates a probability amplitude with each trajectory. Then one sums over all paths to obtain the overall probability amplitude function. This requires making sense of the "sum over all paths." It is well known that the naive integration theory cannot be used, since there are no non-trivial translation-invariant measures on infinite-dimensional spaces, like the space of all paths. However, Feynman made skillful use of approximation arguments and showed how one could calculate many quantities of interest in quantum mechanics [FH65]. Furthermore, this way of thinking inspired his later work on quantum electrodynamics [Fey49b, Fey49a]. Since then the theory of path integrals has been placed on a firm mathematical footing [GJ81, Sch81, Sim05].[1]

The functional integral is the extension of the path intgeral to infinite-dimensional systems. Moving to infinite dimensional systems raises the mathematical stakes considerably and led to much controversy about whether this is actually well-defined. In the past two decades a rigourous theory, due to Cartier and DeWitt-Morette has appeared [CDM95] but not everyone accepts that this formalises what physicists actually do when they make field-theoretic calculations.

What physicists do is to use a set of rules that make intuitive sense and which are guided by analogy with the ordinary calculus. Some of the ingredients of this formal calculus are entirely rigourous, for example, the variational derivative [GJ81]; but the existence of the integrals remains troublesome.

In classical field theory one has a function, say ϕ defined on the spacetime M. This function may be real or complex valued, and in addition, it may be vector or tensor or spinor valued. Let us consider for simplicity a real-valued function; this is called a *scalar* field. Thw field obeys a dynamical equation, for example the scalar field may obey an equation like $\Box\phi + m^2\phi = 0$ where $\Box$ is the four-dimensional laplacian.

In quantum field theory, the field is replaced by an operator acting on a Hilbert space of states. The quantum field is required to obey certain algebraic properties that capture aspects of causality, positivity of energy and relativistic invariance. The Hilbert space is usually required to have a special structure to accomodate the possibility of multiple particles. There is a distinguished state called the vacuum and one can vary the number of particles present by applying what are called *creation* and *annihilation* operators. There is a close relation between these operators and

[1] Actually a completely rigourous theory of path integration, due to Wiener, existed in the 1920s. It was, however, for statistical mechanics and worked with a gaussian measure rather than the kind of measure that Feynman needed.

the field operator: the field operator is required to be a sum of a creation and an annihilation operator.

The main idea of the functional integral approach is that the fundamental quantities of interest are transition amplitudes between states. These are usually states of a quantum field theory and are often given in terms of how many particles of each type and momentum are present in the field: this description of the states of a quantum field is called the *Fock* representation. The most important quantity is the vacuum-to-vacuum transition amplitude, written $\langle 0, - \mid 0, + \rangle$, where $\langle 0, - \mid$ represents the vacuum at early times and $\mid 0, + \rangle$ is the late vacuum. The idea of the functional approach is this can be obtained by summing a certain quantity—the action—over all field configurations interpolating between the initial and the final field.

This can be written as

$$W = \int [d\phi] \exp [- \int d^3\mathbf{x} \left(\frac{1}{2} \phi \mathcal{D} \phi \right)]$$

where $\mathcal{D}$ is some differential operator coming from the classical free field theory. The $[d\phi]$ is supposed to be the measure over all field configurations. Though we do not define it, we can do formal manipulations of this functional. In order to extract interesting results, we want not just the vacuum-to-vacuum trnasition probabilities but the expectation values for products of field operators, e.g. $\langle 0 \mid \phi(x)\phi(y) \mid 0 \rangle$ and other such combinations. In order to do this we add a "probe" to the field which couples to the field and which can be varied. This is fictitious, of course, and will be set to zero at the end. The probe (usually called a *current*) is typically written as $J(x)$. The form we now get for W is

$$W[J] = \int [d\phi] \exp [- \int d^3\mathbf{x} \left(\frac{1}{2} \phi \mathcal{D} \phi \right) - J\phi].$$

Note that W is now a functional of J.

Before we can do any calculations we need to rewrite the term. Consider, for the moment ordinary many-variable calculus. Suppose we write (,) for the inner product on $\mathbb{R}^n$ we can write a form

$$Q(x) = \frac{1}{2}(x, Ax) + (b, x).$$

Now recall the gaussian integral in many variables is just

$$\int d^n x \exp -\frac{1}{2}(x, Ax) = [det(A)^{-\frac{1}{2}}]$$

where A is an ordinary $n \times n$ matrix with positive eigenvalues and we have ignored factors involving 2π. To deal with Q we complete the square by setting $y = A^{-1}b$ and get

$$Q(y) + \frac{1}{2}(x - y, A(x - y)) = Q(x).$$

Using this and changing variables appropriately, we get

$$\int d^n x \exp - Q(x) = [detA]^{-\frac{1}{2}} \exp - Q(y).$$

Now in the functional case, the determinant of an operator does not make sense naively, we will just ignore it here. In actual practice these divergent determinants are made finite by a process called regularization and dealt with. It should be noted that there is a fascinating mathematical theory of these determinants that we will not pursue here.

Returning to our functional expression we get

$$W[J] = \exp \frac{1}{2} \int d^4 x d^y J(x) \mathcal{D}^{-1} J(y).$$

How are we to make sense of the inverse of a differential operator? It is well-known in mathematics and physics as the Green's function.[2] It is well-defined as a distribution. In quantum field theory, the Green's function with appropriately chosen boundary conditions is called the Feynman propagator.

In order to obtain interesting quantities we "differentiate" $W[J]$ with respect to the function $J(x)$. This is called the variational derivative and is well defined, see, for example, [GJ81]. Roughly speaking, one should think of this as a directional derivative in function space. The definition given in [GJ81] (page 202) is

$$(D_\psi(A))(\phi) = \lim_{\epsilon \to 0} [A(\phi + \epsilon\psi) - A(\phi)]/\epsilon.$$

This is the derivative of the functional A of ϕ in the direction of the function ϕ. Now we can consider the special case where ψ is the Dirac delta "function" $\delta(x)$ and write is using the common notation as

$$\frac{\delta}{\delta\phi(x)} A(\phi) \equiv D_{\delta(x)} A(\phi).$$

A very useful formula is

$$\frac{\delta}{\delta\phi(x)} \phi(y) = \delta(x, y).$$

Please note that some of the deltas are part of the variational derivative and some are Dirac distributions: unfortunate, but this is the common convention.

---

[2] The grammatically correct way to name this is a "Green function", but it is too late to change common practice.

How do we use these variational derivatives? If we consider the form of $W[J]$ before we rewrote it in terms of propagators we see that a variational derivative with respect to J brings down a factor of ϕ and thus, if we do this twice, for example, gives us $\langle 0 \mid \phi(x_1)\phi(x_2) \mid 0\rangle$. If we look at the trnasformed version of W the same derivatives tell us how to compte this quantity in terms of propagators. So in the end one gets explicit rules for calculating quantities of interest.

For a given field theory, the set of rules are called the *Feynman rules* and give explicit calculational prescriptions. The functional integral formalism is used to derive these rules. The rules can also be obtained more rigourously from the Hamiltonian form of the field theory as was shown by Dyson [Dys49]. All this can be made much more rigourous. The point is to show some of the calculational devices that physicists use. It is not be expected that after reading this section one will be able to calculate scattering cross-sections in quantum electrodynamics. The point is to the see that the ϕ-calculus is closely based on the formalisms commonly in use in quantum field theory.

7.3 Linear Realizability Algebra

This section is a summary of the theory of linear realizability algebras as developed by Abramsky [Abr91]. The presentation here closely follows that of Abramsky and Jagadeesan [AJ94a]. The basic idea is to take proofs in linear logic in sequent form and to interpret them as processes. The first step is to introduce *locations*; which one can think of as places through which information flows in or out of a proof. Of course the diagrammatic form of proof nets carry all this information without, in Girard's phrase "the bureaucracy of syntax". However, to make contact with an algebraic notation we have to reintroduce the locations to indicate how things are connected. In process terms these will correspond to ports or channels or names as in the π-calculus [Mil89]. In our formal calculus, locations will be introduced with roughly the same status. The set of *locations* $\mathcal{L}$ is ranged over by $x, y, z, \ldots$.

The next important idea is that of *located sequents*, of the form

$$\vdash x_1 : A_1, \ldots, x_k : A_k$$

where the x_i are distinct locations, and the A_i are formulas of $\mathtt{CLL}_2$. These sequents are to understood as *unordered*, *i.e.* as functions from $\{x_1, \ldots, x_k\}$- the sort of the sequent- to the set of $\mathtt{CLL}_2$ formulae.

A syntax of terms (Fig. 7.3) is introduced , which will be used as realizers for sequent proofs in $\mathtt{CLL}_2$. The symbols P, Q, R are used to range over these terms, and write $\mathtt{FN}(P)$ for the set of names occurring freely in P—its *sort*. With each term-forming operation one gives a linearity constraint on how it can be applied, and specifies its sort. In the very last case, the so-called "of course" modality, we have imposed a restriction that if a location is introduced by an "of course" we will require that all the other variables have been previously introduced by either derelictions, weakenings or contractions. We are interested in terms that arise from proof nets so we think of our terms as being typed; this is a major difference between our LRAs and those introduced by Abramsky and Jagadeesan [AJ94b].

Proof Rule	**Operation**	**Constraint**	**Sort**
Axiom	$I_{x,y}$		$\{x,y\}$
Cut	$P \cdot_x Q$	$\mathtt{FN}(P) \cap \mathtt{FN}(Q) = \{x\}$	$\mathtt{FN}(P) \cup \mathtt{FN}(Q) \setminus \{x\}$
Unit	U_x		$\{x\}$
Perp	$\perp_x (P)$	$x \notin \mathtt{FN}(P)$	$\mathtt{FN}(P) \cup \{x\}$
Times	$\otimes_z^{x,y}(P,Q)$	$x \in \mathtt{FN}(P), y \in \mathtt{FN}(Q)$ $\mathtt{FN}(P) \cap \mathtt{FN}(Q) = \emptyset$ $z \notin \mathtt{FN}(P) \cup \mathtt{FN}(Q)$	$\mathtt{FN}(P) \cup \mathtt{FN}(Q) \setminus \{x,y\} \cup \{z\}$
Par	$\invamp_z^{x,y}(P)$	$x, y \in \mathtt{FN}(P)$ $x \neq y$ $z \notin \mathtt{FN}(P)$	$\mathtt{FN}(P) \setminus \{x,y\} \cup \{z\}$
Plus Left	$L_z^x(P)$	$x \in \mathtt{FN}(P), z \notin \mathtt{FN}(P)$	$\mathtt{FN}(P) \setminus \{x\} \cup \{z\}$
Plus Right	$R_z^x(P)$	$x \in \mathtt{FN}(P), z \notin \mathtt{FN}(P)$	$\mathtt{FN}(P) \setminus \{x\} \cup \{z\}$
With	$\&_z^{x,y}(P)$	$x \in \mathtt{FN}(P), y \in \mathtt{FN}(Q)$ $\mathtt{FN}(P) \setminus \{x\} = \mathtt{FN}(Q) \setminus \{y\}$	$\mathtt{FN}(P) \setminus \{x\} \cup \{z\}$
Dereliction	$D_z^x(P)$	$x \in \mathtt{FN}(P), z \notin \mathtt{FN}(P)$	$\mathtt{FN}(P) \setminus \{x\} \cup \{z\}$
Weakening	$W_z(P)$	$z \notin \mathtt{FN}(P)$	$\mathtt{FN}(P) \cup \{z\}$
Contraction	$C_z^{x,y}(P)$	$x, y \in \mathtt{FN}(P)$ $x \neq y$ $z \notin \mathtt{FN}(P)$	$\mathtt{FN}(P) \setminus \{x,y\} \cup \{z\}$
Of course	$!_z^x(P)$	$x \in \mathtt{FN}(P), z \notin \mathtt{FN}(P)$ $\forall u \in FN(P) \setminus \{x\}.u$ is introduced by dereliction, weakening or contraction.	$\mathtt{FN}(P) \setminus \{x\} \cup \{z\}$

Fig. 7.1 Syntax: linear realizability algebra

There is an evident notion of *renaming* $P[x/y]$ and of α-conversion $P \equiv_\alpha Q$.

Terms are assigned to sequent proofs in CLL_2 as in Fig. 7.3.

The rewrite rules for terms, corresponding to cut-elimination of sequent proofs, can now be given. This is factored into two parts, in the style of [BB90]: a structural congruence $\equiv$ and a reduction relation $\to$.

The structural congruence is the least congruence $\equiv$ on terms such that:

(SC1) $P \equiv_x Q \Rightarrow P \equiv Q$
(SC2) $P \cdot_x Q \equiv Q \cdot_x P$
(SC3) $\omega(P_1, \ldots, P_k) \equiv \omega(P_1, \ldots, P_i \cdot_x Q, \ldots, P_k)$, if $x \in \mathtt{FN}(P_i)$.

The reductions are as follows:

(R1) $P \cdot_x I_{x,y} \to P[y/x]$.
(R3) $\invamp_z^{x,y}(P) \cdot_z \otimes_z^{x,y}(Q, R) \to P \cdot_x Q \cdot_y R$.
(R4) $L_z^x(P) \cdot_z \&_z^{x,y}(Q, R) \to P \cdot_x Q$.
(R5) $R_z^x(P) \cdot_z \&_z^{x,y}(Q, R) \to P \cdot_x R$.
(R6) $D_z^x(P) \cdot_z !_z^x(Q) \to P \cdot_x Q$.

Identity Group	$\overline{I_{x,y} \vdash x : A^{\perp}, y : A}$	$\dfrac{P \vdash \Gamma', x : A \quad Q \vdash \Gamma'', x : A^{\perp}}{P \cdot_x Q \vdash \Gamma', \Gamma''}$
Multiplicative Units	$\overline{U_x \vdash x : I}$	$\dfrac{P \vdash \Gamma}{\perp_x \vdash x : \perp, \Gamma}$
Multiplicatives	$\dfrac{P \vdash \Gamma', x : A \quad Q \vdash \Gamma'', y : B}{\otimes_z^{x,y}(P, Q) \vdash \Gamma', \Gamma'', z : A \otimes B}$	$\dfrac{P \vdash \Gamma', x : A, y : B}{⅋_z^{x,y}(P) \vdash \Gamma', z : A ⅋ B}$
Additives	$\dfrac{P \vdash \Gamma, x : A}{L_z^x(P) \vdash \Gamma, z : A \oplus B}$ $\dfrac{P \vdash \Gamma, x : B}{R_z^x(P) \vdash \Gamma, z : A \oplus B}$	$\dfrac{P \vdash \Gamma, x : A \quad Q \vdash \Gamma, y : B}{\&(P, Q) \vdash \Gamma, z : A \& B}$
Exponentials	$\dfrac{P \vdash \Gamma, x : A}{D_z^x(P) \vdash \Gamma, z : ?A}$ $\dfrac{P \vdash \Gamma}{W_z(P) \vdash \Gamma, z : ?A}$ $\dfrac{P \vdash \Gamma, x : ?A, y : ?A}{C_z^{x,y}(P) \vdash \Gamma, z : ?A}$	$\dfrac{P \vdash ?\Gamma, x : A}{!_z^x(P) \vdash ?\Gamma, z : !A}$

Fig. 7.2 Realizability semantics

(R7) $W_z(P) \cdot_z !_z^x(Q) \rightarrow W_{\boldsymbol{x}}(P)$, where $\mathtt{FN}(Q) \setminus \{x\} = \boldsymbol{x}$.

(R8) $C_z^{z',z''}(P) \cdot_z !_z^x(Q) \rightarrow C_{\boldsymbol{x}}^{\boldsymbol{x}',\boldsymbol{x}''}(P \cdot_{z'} !_{z'}^x(Q[\boldsymbol{x}'/\boldsymbol{x}]) \cdot_{z''} !_{z''}^x(Q[\boldsymbol{x}''/\boldsymbol{x}]))$, where $\mathtt{FN}(Q) \setminus \{x\} = \boldsymbol{x}$.

(R9) $!_z^x(P) \cdot_u !_u^v(Q) \rightarrow !_z^x(P \cdot_u !_u^v(Q))$, if $u \in \mathtt{FN}(P)$.

We are using the same numbering as in [AJ94b] and have left out **R2**, which talks about units. We write bold face to stand for sequences of variables.

These reductions can be applied in any context.

$$\frac{P \rightarrow Q}{C[P] \rightarrow C[Q]}$$

and are performed modulo structural congruence.

$$\frac{P' \equiv P \quad P \rightarrow Q \quad Q' \equiv Q}{P \rightarrow Q}$$

The basic theorem is that this algebra models cut elimination is classical linear logic. The precise statement is

Proposition 7.3.1 (Abramsky) *Let Π be a sequent proof of $\vdash \Gamma$ in $\mathtt{CLL}_2$ with corresponding realizing term P. If $P \rightarrow Q$, with Q cut-free (i.e. no occurrences of $\cdot_\alpha$), then Π reduces under cut-elimination to a cut-free sequent proof Π' with corresponding realizing term Q.*

In order to verify that one correctly models the process of cut-elimination in linear logic it suffices to verify the LRA equations **R1** through **R9**. In fact we will also check the following equation:

$$\&_z^{x,y}(P, Q)\cdot_u R \rightarrow \&_z^{x,y}(P\cdot_u R, Q\cdot_u R)\ [u \in \ \mathtt{FN}(P) \cap \ \mathtt{FN}(Q)]$$

This is the commutative with-reduction and is not satisfied in the extant examples of LRA.

7.4 The ϕ-Calculus

In this section we spell out the rules of our formal calculus. Briefly the ingredients are

Locations: which play the same role as the locations in the located sequents of LRA.
Basic terms: which, for the multiplicative fragment, play the role of LRA terms.
Operators: Which act on basic terms and which play the role of terms in the full LRA.

In order to define the basic terms we need locations, formal distributions, formal integrals and simple rules obeyed by these. In order to define operators we need to introduce a formal analogue of the variational derivative. This variational derivative construct is very closely modelled on the derivation of Feynman diagrams from a generating functional.

We assume that we are modelling a typed linear realizability algebra with given propositional atoms. We first formalize *locations*. We assume further that axiom links are only introduced for basic propositional atoms. We use the phrases "basic types" and "basic propositional atoms" interchangeably.

Definition 1 We assume that there are countably many distinct symbols, called **locations** for each basic type. We assume that there are the following operations on locations: if x and y are locations of types A and B respectively, then $\langle x, y\rangle$ and $[x, y]$ are locations of type $A \otimes B$ and $A \invamp B$ respectively. We use the usual sequent notation $x : A, y : B \vdash \langle x, y\rangle : A \otimes B$ and $x : A, y : B \vdash [x, y] : A \invamp B$ to express this.

Now we define *expressions*.

Definition 2 The collection of **expressions** is given by the following inductive definition. We also define, at the same time, the notion of the **sort** of an expression, which is the set of free locations, and their types, that appear in the expression.

1. Any real number r is an expression of sort $\emptyset$.
2. Given any two distinct locations, $x : A$ and $y : A^{\perp}$ $\delta(x, y)$ is an expression of sort $\{x : A, y : A^{\perp}\}$.
3. Given any two expressions P and Q, PQ and $P + Q$ are expressions of sort $\mathsf{S}(P) \cup \mathsf{S}(Q)$.
4. Given any expression P and any location $x : A$ in P, the expression $\int P dx$ is an expression of sort $\mathsf{S}(P) \setminus \{x : A\}$.

The expressions above look like the familiar expressions that one manipulates in calculus. The sorts describe the free locations that occur in expressions. The integral symbol is the only binding operator and is purely formal. Indeed any suitable notation for a binder will do, a more neutral one might be something like $Tr(e, x)$, which is more suggestive of a trace operation.

The equations obeyed by these expressions mirror the familiar rules of calculus. The only exotic ingredients are that the δ behaves like a Dirac delta "function". We will actually present a rewrite system rather than an equational system but one can think of these as equations.

We use the familiar notation $P(\ldots, y/x, \ldots)$ to mean the expression obtained by replacing all free occurrences of x by y with appropriate renaming of bound variables as needed to avoid capture; x and y must be of the same type of course. We now define equations that the terms obey.

Definition 3

1. $\delta(x, y) = \delta(y, x)$
2. $\int(\int P dx)dy = \int(\int P dy)dx$
3. $(P + Q) + R = P + (Q + R)$
4. $(P + Q) = (Q + P)$
5. $P \cdot (Q \cdot R) = (P \cdot Q) \cdot R$
6. $P \cdot Q = Q \cdot P$
7. $P \cdot (Q_1 + Q_2) = P \cdot Q_1 + P \cdot Q_2$
8. $P + 0 = P$.
9. $P \cdot 1 = P$.
10. $P \cdot 0 = 0$.
11. $\int P(\ldots, x, \ldots)\delta(x, y)dx = P(\ldots, y/x, \ldots)$.
12. $\delta([x, y], \langle u, v\rangle) = \delta(x, u)\delta(y, v)$.
13. If $P = P'$ *then* $PQ = P'Q$.
14. If $P = P'$ *then* $P + Q = P' + Q$.
15. If $P = P'$ *then* $\int P dx = \int P' dx$.

These equations are very straightforward and can be viewed as basic properties of functions and integration or about matrices and matrix multiplication. The only point is that with ordinary functions one cannot obtain anything with the behaviour of the δ-function; these are, however, easy to model with distributions or measures.

In order to model the exponentials and additives we need a rather more elaborate calculus. We introduce operators which are inspired by the use of generating functionals for Feynman diagrams in quantum field theory [IZ80, Ram81]. The two ingredients are formal power series and variational derivatives. In order to model pure linear logic the formal power series that arise as power-series expansions of exponentials are the only ones that are needed. We introduce a formal analogue of the variational derivative operator, commonly used in both classical and quantum field theories [Ram81]. For us the variational derivative plays the role of a mechanism that extracts a term from an exponential.

As before we have locations and expressions. We first introduce a new expression constructor.

Definition 4 If x is a location of type A then $\alpha_A(x)$ is an expression of sort $\{x : A\}$.

The point of α is to provide a "probe", which can be detected as needed. For each type and location there is a different α. We will usually not indicate the type subscript on the αs unless they are necessary. The last ingredient that we need in the world of expressions is an expression that plays the role of a "discarder", used, of course, for weakening.

Definition 5 If x is a location of type $\{x : A\}$, where A is a multiplicative type, then $W_A(x)$ is an expression of sort $\{x : A\}$. This satisfies the equations

1. $W_{A\otimes B}(\langle x, y\rangle) = W_A(x)W_B(y)$,
2. $W_{A\parr B}([x, y]) = W_A(x)W_B(y)$.

We can think of $W(x)$ intuitively as "grounding" in the sense of electrical circuits. In effect it provides a socket into which x is plugged but which is in turn connected to nothing else. So it is as if x were "grounded". If such a $W(x)$ is connected to a wire, $\int W(x)\delta(x, y)dx$ the result will be the same as grounding y. The other two equations express the fact that a complex W can be decomposed into simpler ones. We will introduce another decomposition rule for W after we have described the variational derivatives and a corresponding operator for weakening.

We introduce syntax for operators; these will be defined as maps from expressions to expressions.

Definition 6 Operators are given by the following inductive definition.

1. If M is any expression $\hat{M}$ is an operator of the same sort as M.
2. If $x : A$ is a location then $([.]|_{\alpha(x)=0})$ is an operator of sort $x : A$.
3. If $x : A$ is a location then $\frac{\delta}{\delta\alpha(x)}$ is an operator of sort $x : A$.
4. If P and Q are operators then so are $P + Q$ and $P \circ Q$ their sort is the union of the individual sorts.
5. If P is an operator then so is $\int P dx$; its sort is $\mathsf{S}(P) \setminus \{x\}$.

An operator of sort S acts on an expression of sort S' if $S \cap S'$ is not empty.

Operators map expressions to expressions. An important difference between the algebra of expressions and that of operators is that the (commutative) multiplication of expressions has been replaced by the (non-commutative) composition of operators.

The meaning of the operators above is given as follows. We use the meta-variables M, N for expressions and P, Q for operators. We begin with the definition of $\hat{M}$.

Definition 7 $\hat{M}(N) = M \cdot N$.

The notion of composition of operators is the standard one

Definition 8 $[P \circ Q](M) = P(Q(M))$.

Clearly we have $\hat{M} \circ \hat{N} = \hat{MN}$; thus we have an extension of the algebra of expressions. We will write 1 and 0 rather than, for example, $\hat{1}$ to denote the operators. The resulting ambiguity will rarely cause serious confusion.

The next set of rules define the operator $([.]|_{\alpha(x)=0})$. Intuitively this is the operation of "setting $\alpha(x)$ to 0" in an algebraic expression.

Definition 9 If M is an expression then the operator $([.]|_{\alpha(x)=0})$ acts as follows:

1. If $\alpha(x)$ does not appear in M then $([M]|_{\alpha(x)=0}) = M$.
2. $([MN]|_{\alpha(x)=0}) = (([M]|_{\alpha(x)=0}))(([N]|_{\alpha(x)=0}))$.
3. $([M+N]|_{\alpha(x)=0}) = (([M]|_{\alpha(x)=0})) + (([N]|_{\alpha(x)=0}))$.
4. $([M\alpha(x)]|_{\alpha(x)=0}) = 0$.

The rules for the variational derivative formalize what one would expect from a derivative, most notably the Leibniz rule, rule 5 below.

Definition 10 If M is an expression and x is a location we have the following equations:

1. If x and y are distinct locations then $\frac{\delta}{\delta\alpha(x)}\alpha(y) = 0$.
2. If $\alpha(x)$ does not occur in the expression M then $\frac{\delta}{\delta\alpha(x)} M = 0$.
3. $\frac{\delta}{\delta\alpha(x)}\alpha(x) = 1$.
4. $\frac{\delta}{\delta\alpha(x)}(M+N) = (\frac{\delta}{\delta\alpha(x)}M) + (\frac{\delta}{\delta\alpha(x)}N)$.
5. $\frac{\delta}{\delta\alpha(x)}(MN) = (M\frac{\delta}{\delta\alpha(x)}N) + ((\frac{\delta}{\delta\alpha(x)}M)N)$.

Clause 1 is, of course, a special case of clause 2 but is added for emphasis. Intuitively the variational derivative should be viewed as "probing for the presence of α". The reason that we use these variational derivatives rather than ordinary derivatives is that the dependence on the location is crucial for our purposes. Typically we use the combination of $([.]|_{\alpha(x)=0})$ and $\frac{\delta}{\delta\alpha(x)}$ so that we are "inserting probes", "testing for their presence" and then "removing them". The power of the formalism comes from the interaction between variational derivatives and exponentials.

The remaining rules defining the algebra of operators are given in the next definition.

Definition 11 Operators obey the following equations:

1. $\hat{0}(M) = 0$.
2. $\hat{1}(M) = M$.

3. $(P + Q)(M) = P(M) + Q(M)$.
4. $[\int P dx](M) = \int P(M)dx$, where x is not free in M.

One can prove the following easy lemma by structural induction on expressions.

Lemma 7.4.1 *If x and y are distinct locations* $\frac{\delta}{\delta\alpha(x)} \circ \frac{\delta}{\delta\alpha(y)}(M) = \frac{\delta}{\delta\alpha(y)} \circ \frac{\delta}{\delta\alpha(x)}(M)$.

This if of course the familiar notion of commuting.

Definition 12 Two operators P and Q which satisfy $P \circ Q = Q \circ P$ are said to **commute**.

It will turn out that commuting operators satisfy some nice properties that will be important in what follows.

The main mathematical gadget that we need is the notion of formal power series. For our purposes we will only need exponential power series but we give fairly general definitions.

Definition 13 If $(M_i|i \in I)$ is an indexed family of expressions then $\Sigma_I M_i$ is an expression. If $(P_i|i \in I)$ is an indexed family of operators then $\Sigma_I P_i$ is an operator. If I is a finite set, the result is the same as the ordinary sum; if I is infinite, the result is called a **formal power series**.

One may question the use of the word "power" in "power series" since there is nothing in the definition that says that we are working with powers of a single entity. Nevertheless we use this suggestive term since the series we are interested really are formal power series.

The meaning of a power series of operators is given in the evident way.

Definition 14 If $\Sigma_{i\in I} P_i$ is a formal power series of operators and M is an expression then $(\Sigma_{i\in I} P_i)(M) = \Sigma_{i\in I}(P_i(M))$.

The key power series that we use is the exponential. We first give a preliminary account and then a more refined account.

Definition 15 If M is an expression then the **exponential series** is

$$\Sigma_{k\geq 0} M^k/k!$$

and is written $\exp(M)$; here M^k means the k-fold product of M with itself.

What we are not making precise at the moment is the meaning of M^n.

A number of properties follow immediately from the preceding definition.

Lemma 7.4.2 *If the expression M contains no occurrence of $\alpha(x)$ then:*

1. $\frac{\delta}{\delta\alpha(x)}(MN) = M\frac{\delta}{\delta\alpha(x)}(N)$;
2. $(([.]|_{\alpha(x)=0}) \circ \frac{\delta}{\delta\alpha(x)})\exp(M\alpha(x)) = M$;
3. $(([.]|_{\alpha(x)=0}) \circ \frac{\delta}{\delta\alpha(x)} \circ \ldots n \ldots \circ \frac{\delta}{\delta\alpha(x)})\exp(M\alpha(x)) = M^n$.

The combination $\frac{\delta}{\delta\alpha(x)} \circ \ldots n \ldots \circ \frac{\delta}{\delta\alpha(x)}$ is often written $\frac{\delta^n}{\delta\alpha(x)^n}$.

The following facts about exponentials recall the usual elementary ideas about the exponential function from an introductory calculus course.

Lemma 7.4.3 *Suppose that M is an expression, the following equations hold.*

1. $\frac{\delta}{\delta\alpha(x)} \exp(M\alpha(x)) = M \cdot \exp(M\alpha(x))$.
2. $([\exp(M)]|_{\alpha(x)=0}) = \exp(([M]|_{\alpha(x)=0}))$.
3. $\exp(0) = 1$.

In fact the above definition of exponentials overlooks a subtlety which makes a difference as soon as we exponentiate operators. The factors of the form $1/(n!)$ are not just numerical factors, they indicate symmetrization. This is the key ingredient needed to model contraction in linear logic. We introduce a new syntactic primitive for symmetrization and give its rules.

Definition 16 If k is a positive integer and $x_1, \ldots, x_k$ and x are distinct locations then $\Delta^k(x_1, \ldots, x_k; x)$ is an expression, x is called the **principal** location. It has the following behaviour; all differently-named locations are assumed distinct.

1. $\int \Delta^{(k)}(x_1, \ldots, x_k; x) M(x, y_1, \ldots, y_l)\, dx =$
   $\int \prod_{i=1}^{k} M[x_i/x, y_1^i/y_1, \ldots, y_l^i/y_l] \prod_{j=1}^{l} \Delta^{(k)}(y_j^1, \ldots, y_j^k; y_j)\, dy_1^1 \ldots dy_l^k$.
2. $\int \Delta^{(k)}(x_1, \ldots, x_k; x) \Delta^{(m+1)}(x, x_{k+1}, \ldots, x_{k+m}; y)\, dx$
   $= \Delta^{(k+m)}(x_1, \ldots, x_k, x_{k+1}, \ldots, x_{k+m}; y)$.
3. $\int \Delta^{(k)}(u_1, \ldots, u_k; x) \Delta^{(k)}(u_1, \ldots, u_k; y)\, du_1 \ldots du_k = \delta(x, y)$.
4. $\int \Delta^{(k)}(x_1, \ldots, x_k; x) \Delta^{(m)}(y_1, \ldots, y_m; x) dx =$
   $\Sigma_{\iota \in inj(\{1,\ldots,k\},\{1,\ldots,m\})} \delta(x_1, y_{\iota(1)}) \ldots \delta(x_k, y_{\iota(k)})$,
   where $k \leq m$ and if S and T are sets, $inj(S, T)$ means injections from S to T.

The idea is that the Δ operators cause several previously distinct locations to be identified. The ordinary δ allows renaming of one location of another but the Δ allows, in effect, several locations to be renamed to the same one. The first rule says that when a location in an expression is connected to a symmetrizer we can multiply the expression k-fold using fresh locations. These locations are then symmetrized on the output. In other words this tells you how to push expressions through symmetrizers. We often write just Δ for Δ^2 and of course we never use Δ^1 since it is the ordinary δ. The important idea is the the Δs cause *symmetrization* of the locations being identified. The last rule in the definition of Δ is what makes it a symmetrizer. To see this more clearly, note the following special case which arises when $k = m$.

$$\int \Delta^{(k)}(x_1, \ldots, x_k; x) \Delta^{(k)}(y_1, \ldots, y_k; x) dx$$

$$= \Sigma_{\sigma \in perm\{1,\ldots,k\}} \delta(x_1, y_{\sigma(1)}) \ldots \delta(x_k, y_{\sigma(end)}).$$

Before we continue we emphasize that the rule for the δ expression applies to operators as well. We formalize this as the following lemma.

Lemma 7.4.4 *If $P(x', \mathbf{y})$ is an operator then $\int \delta(x, x')P dx' = P[x/x']$ in the sense that for any expression M, with x not free in M, $[\int \delta(x, x')P dx'](M) = P[x/x'](M[x/x'])$.*

Proof We give a brief sketch. First note that we can prove, by structural induction on M, for any expression M, that

$$\int \delta(x, x')\frac{\delta}{\delta\alpha(x')}(M)dx' = \frac{\delta}{\delta\alpha(x)}(M[x/x'])$$

which justifies the operator equation $\int \delta(x, x')\frac{\delta}{\delta\alpha(x')}dx' = \frac{\delta}{\delta\alpha(x)}$. Similarly one can show that $\int \delta(x, x')([M]|_{\alpha(x')=0}) = ([M[x/x']]|_{\alpha(x)=0})$. Now structural induction on P establishes the result.

We record a useful but obvious fact about derivatives of operators.

Lemma 7.4.5 *If $\alpha(x)$ does not occur in P then $\frac{\delta}{\delta\alpha(x)}(PQ) = P\frac{\delta}{\delta\alpha(x)}(Q)$.*

We are now ready to define the exponential of an operator.

Definition 17 Let $Q(x_1, \ldots, x_m)$ be an operator with its sort included in $\{x_1, \ldots, x_m\}$. The **exponential** of Q is the following power series, where we have used juxtaposition to indicate composition of the Qs:

$$\begin{aligned}
\exp(Q) = 1 + Q(x_1, \ldots, x_m) + \\
(1/2)\int Q(x_1', \ldots, x_m')Q(x_1'', \ldots, x_m'')\Delta(x_1', x_1''; x_1)\ldots\Delta(x_m', x_m''; x_m) \\
dx_1' \ldots dx_m' dx_1'' \ldots dx_m'' \\
+ \ldots \\
+ (1/(k!))\int Q(x_1^{(1)}, \ldots, x_m^{(k)})\ldots Q(x_1^{(k)}, \ldots, x_m^{(k)}) \\
\Delta^k(x_1^{(1)}, \ldots, x_1^{(k)}; x_1)\ldots\Delta^k(x_m^{(1)}, \ldots, x_m^{(k)}; x_m) \\
dx_1^{(1)} \ldots dx_m^{(1)} \ldots dx_1^{(k)} \ldots dx_m^{(k)} + \ldots
\end{aligned}$$

The above series is just the usual one for the exponential. What we have done is introduce the Δ operators rather than just writing Q^k for the k-fold composition of Q with itself. This makes precise the intuitive notation Q^k and interprets it as k-fold symmetrization of k distinct copies of Q. However, if we wish to speak informally, we can just forget about the Δ operators and pretend that we are working with the familiar notion of exponential power series of a single variable.

We have an important lemma describing how the Δs interacts with differentiating and evaluating at 0.

Lemma 7.4.6 *If M is an expression then*

$$\left[\int \Delta^k(x_1,\ldots,x_k;x)\left(([.]|_{\alpha(x_1)=0})\circ\frac{\delta}{\delta\alpha(x_1)}\right)\circ\ldots\right.$$
$$\left.\circ\left(([.]|_{\alpha(x_k)=0})\circ\frac{\delta}{\delta\alpha(x_k)}\right)dx_1\ldots dx_k\right](M)$$
$$=(1/(k!))\left(\int\Delta^k(x_1,\ldots,x_k;x)([.]|_{\alpha(x)=0})\left(\frac{\delta}{\delta\alpha(x_1)}\circ\ldots\circ\frac{\delta}{\delta\alpha(x_k)}\right)dx_1\ldots dx_k)(M).$$

Thus, we can remove the M and assert the evident operator equation.

Proof Note that the only terms that can survive the effect of the operator on the LHS are those of the form $M'\alpha(x_1)\ldots\alpha(x_k)$. If any $\alpha(x_j)$ occurred to a higher power it would be killed by the operator $([.]|_{\alpha(x_j)=0})$ that acts after the variational derivative. If any of the αs do not occur the term would be killed by the variational derivative. Now note that the $([.]|_{\alpha(x_i)=0})$ and the $\frac{\delta}{\delta\alpha(x_j)}$ commute if x_i and x_j are distinct. Thus we can move the $([.]|_{\alpha(x_j)=0})$ operators to the left. Now if we look at the kinds of terms that survive and compute the derivatives we get the result desired.

Intuitively we can think of the symmetrizers as "multiplexers". This is of course part of the definition of symmetrizer for expressions. The following lemma makes this precise. The proof is notationally tedious but is a routine structural induction, which we omit. We use the phrase *exponential-free* to mean an operator constructed out of ordinary expressions, and variational derivatives and setting α to 0, no exponentials (or other power-series) occur.

Lemma 7.4.7 *If* $P(x,y_1,\ldots,y_m))$ *is an exponential-free operator then*
$$\int\Delta^{(k)}(x_1,\ldots,x_k;x)P\ dx\ =$$
$$\int[\prod_{j=1}^m\Delta^{(k)}(y_j^1,\ldots,y_j^k;y_j)]\prod_{i=1}^k[P(x_i/x,y_1^i,\ldots,y_m^i)]$$
$$dx_1\ldots dx_k\ dy_1^1\ldots\ dy_m^k.$$

What the lemma says is that when we have a k-fold symmetrized location in an operator with m other locations we can make k copies of the operator with fresh copies of all the locations. In the y_j^l, the subscript (running from 1 to m) says which location was copied and the superscript (running from 1 to k) which of the k copies it is. The k fresh copies of the locations are each symmetrized to give the original locations; this explains the product of all the $\Delta^{(k)}$ terms.

We close this section by completing the definition of the weakening construct W by giving the decomposition rules for W.

Definition 18 (Decomposition Rules for W)

1. $W_\phi(u)([.]|_{\alpha(x)=0})\circ\frac{\delta}{\delta\alpha(x)}=W_{?\phi}(x)$
2. $\int\Delta^{(k)}(x_1,\ldots,x_k;x)W_\phi(x_1)\ldots W_\phi(x_k)\,dx_1\ldots dx_k=W_\phi(x)$

Term	Sort
r, a rational number	$\emptyset$
$\delta(x,y)$	$\{x : A, y : A^{\perp}\}$
$\Delta^{(k)}(x_1,\ldots,x_k;x)$	$\{x : A^{\perp}, x_1,\ldots,x_k : A\}$
$\alpha_A(x)$	$\{x : A\}$
$W_A(x)$	$\{x : A\}$
$P+Q, PQ$	$\mathsf{S}(P)\cup\mathsf{S}(P)$
$\int Pdx$	$\mathsf{S}(P)\setminus\{x\}$
$\sum_{i\in I} P_i$	$\cup_{i\in I}\mathsf{S}(P_i)$

Fig. 7.3 Syntax of expressions in the ϕ-calculus

Operator	Sort
M, an expression	$\mathsf{S}(M)$
$([\cdot]\vert_{\alpha(x)=0})$	$\{x : A\}$
$\frac{\delta}{\delta\alpha(x)}$	$\{x : A\}$
$P+Q, P\circ Q$	$\mathsf{S}(P)\cup\mathsf{S}(P)$
$\int Pdx$	$\mathsf{S}(P)\setminus\{x\}$
$\sum_{i\in I} P_i$	$\cup_{i\in I}\mathsf{S}(P_i)$

Fig. 7.4 Syntax of operators in the ϕ-calculus

7.5 Exponential Identities for Operators

Much of the combinatorial complexity of proof nets can be concealed within the formulas for derivatives of exponentials. In this section we collect these formulas in one place for easy of reference. There are, of course, infinitely many identities that one could write down, we will write only the ones that arise in the interpretation of linear logic.

We typically have to assume that various operators commute. If the operators do not commute one gets very complicated expressions, such as the Campbell-Baker formula, occurring in the study of Lie algebras and in quantum mechanics, for products of exponentials of operators. In the case of linear logic the linearity conditions will lead to operators that commute.

Proposition 7.5.1 *If the operator P contains no occurrence of $\alpha(x)$ then:*

1. $(([.]|_{\alpha(x)=0}) \circ \frac{\delta}{\delta\alpha(x)}) \exp(P\alpha(x)) = P$;
2.

$$\int (([.]|_{\alpha(x)=0}) \circ \frac{\delta}{\delta\alpha(x)} \ldots n \ldots \circ \frac{\delta}{\delta\alpha(x)}) \exp(P\alpha(x))dx$$

$$= \int \Delta^n(x^1,\ldots,x^n;x) \prod_{k=1}^{m} [\Delta^n(y_k^1,\ldots,y_k^n;y_k)]$$

$$\prod_{j=i}^{n}[P(y_1^j/y_1,\ldots,y_k^j/y_k,\ldots,y_m^j/y_m)]$$

$$dy_1^1\ldots dy_1^n\ldots dy_2^1\ldots dy_2^n\ldots\ldots dy_m^1\ldots dy_m^n,$$

where the RHS is essentially P^n.

In this formula the locations in P are $\{x, y_1, \ldots, y_m\}$ and there are n copies of P with fresh copies of each of these locations and there are $m+1$ symmetrizers, one to identify each of the n copies of the $m+1$ locations in P. Subscripts range (from 1 to m) over the locations in P and superscripts range (from 1 to n) over the different copies of the locations.

Proof The first part is just a special case of the second. One can just expand the power series and carry out the indicated composition term by term on a given expression using Lemma 7.4.6. There are three types of terms in the power-series expansion of the exponential: (1) terms of power less than n, (2) the term of power n and (3) terms of power greater than n. Terms of type (1) vanish when the variational derivative is carried out, terms of type (3) will have $\alpha(x)$ still present after the differentiation is done and will vanish when we carry out $([.]|_{\alpha(x)=0})$. The entire contribution comes from terms of type (2). From the definition of the exponential we get that the order n term is

$$(1/(n!))\cdot\int P(x^1/x, y_1^1/y^1,\ldots,y_1^m/y^m)\ldots P(x^n,y_1^n,\ldots,x_m^n)\alpha(x^1)\ldots\alpha(x^n)$$
$$\Delta^n(x^1,\ldots,x^n;x)\Delta^n(y_1^1,\ldots,y_1^n;x_1)\ldots\Delta^n(y_m^1,\ldots,y_m^n;y_m)$$
$$dy_1^1\ldots dy_m^1\ldots dy_1^n\ldots dy_m^n$$

where the free locations in P are $\{x, y^1, \ldots, y^m\}$. The variational derivatives are all of the form $\frac{\delta}{\delta\alpha(x)}$ and there are n of them. They can be written as

$$\int \Delta^n(u_1,\ldots,u_n;x)\prod_{j=1}^{n}\frac{\delta}{\delta\alpha(u_j)}du_1\ldots du_n.$$

Now when we do the x integral the only terms involving x are Δs. Thus using the definition of symmetrization 16 part (3) we get the sum over $n!$ combinations of the form $\prod_{i=1}^n \delta(u_i, x^{\sigma(i)})$ where σ is a permutation of $\{1,\ldots,n\}$; the sum is over all the permutations. Note however that the rest of the expression is completely symmetric with respect to any permutation of $\{x^1,\ldots,x^n\}$; thus we can replace this sum over all permutations with $n!$ times any one term, say $\prod_{i=1}^n \delta(u_i, x^i)$. Now doing the u_i integral with these δs replaces the $\frac{\delta}{\delta\alpha(u_i)}$ with $\frac{\delta}{\delta\alpha(x^i)}$. The variational derivatives are now exactly matched with the αs so we can carry out the indicated differentiations by just deleting $([x]|_{\alpha(.)=0})$, and all the $\frac{\delta}{\delta\alpha(.)}$ and all the αs. The $n!$ from the sum over permutations cancels the $1/(n!)$ in the expansion of the exponential giving the required result.

This proof is necessary to do once, to show that the symmetrizers interact in the right way to make sense of the $n!$ factors as permutations. Henceforth we will not give the same level of detail, instead we will use the abbreviation $\frac{\delta}{\delta\alpha(x)}^n$ and revert to the formal notation only after the intermediate steps are completed.

The following easy, but important, proposition can now be established. It is essentially the "nesting of boxes" formula.

Proposition 7.5.2 *Suppose that P and Q are operators not containing $\alpha(x)$ and suppose that they commute, then* $\exp(P \circ \frac{\delta}{\delta\alpha(x)}) \circ \exp(Q\alpha(x))$ *is the same as* $\exp(P \circ \frac{\delta}{\delta\alpha(x)} \circ \exp(Q\alpha(x)))$.

Proof We will outline the basic calculation which basically just uses Lemma 7.4.2 and Proposition 7.5.1. We will suppress the Δs in the following in order to keep the notation more readable. We expand the LHS in a formal power series to get

$$\Sigma_{k=0}^{\infty}(1/k!)P^k \frac{\delta^k}{\delta\alpha(x)^k} \circ \exp(Q\alpha(x)).$$

Because P, Q, $\frac{\delta}{\delta\alpha(x)}$ all commute we can Proposition 7.5.1 for the variational derivatives of exponentials and rearrange the order of terms to get

$$\Sigma_{k=0}^{\infty}(1/k!)P^k \circ Q^k \circ \exp(Q\alpha(x)).$$

Using the exponential formula to sum this series we get

$$\exp(P \circ Q) \circ \exp(Q\alpha(x)).$$

But this is exactly what the RHS expands to if we use the exponential formula.

The last proposition is a general version of promotion, we do not really need it but it shows the effect of multiply stacked exponentials.

Lemma 7.5.3 *If the operators P and Q have no locations in common then*

$$\int \Delta(x', x''; x)P(x', \ldots)\frac{\delta}{\delta\alpha(x'')} \exp(\alpha(x)Q(x, \boldsymbol{u})dxdx'dx''$$
$$= \int Q(x''/x, \boldsymbol{u}''/\boldsymbol{u})\Delta(\boldsymbol{u}, \boldsymbol{u}'; \boldsymbol{u}'')P(x', \ldots) \exp(\alpha(x')Q(x'/x, \boldsymbol{u}'/\boldsymbol{u})dx'dx''.$$

The proof is omitted, it is a routine calculation done by expanding each side and comparing terms. It allows us to finesse calculating the effect of multiply stacked exponentials of operators.

Finally we need the following lemma when proving that contraction works properly in conjunction with exponentiation. We suppress the renaming and symmetrizations to make the formula more readable, note that it is just a special case of a multiplexing formula of the kind defined in Lemma 7.4.7, with an exponentiated operator and $k = 2$.

Lemma 7.5.4

$$\int \Delta(y, z; u) \exp(Q(u, x_1, \ldots, x_k)) \; du =$$

$$\int \exp(Q(y/u, x'_1, \ldots, x'_k)) \exp(Q(z/u, x''_1, \ldots, x''_k)) \left[\prod_{j=1}^{k} \Delta(x'_j, x''_j; x_j)\right] dx'_1 \ldots dx'_k \, dx''_1 \ldots dx''_k.$$

Proof We proceed by induction on the exponential nesting depth. The base case is just the multiplexing formula proved in Lemma 7.4.7. In this proof we start from the right-hand side. Now we note that if we have two operators, A and B, *which commute with each other* then we have

$$e^{A+B} = e^A e^B.$$

This can easily be verified by the usual calculation which, of course, uses commutativity crucially. Now on the rhs of the equation we have the exponentials of two operators which commute because they have no variables in common. Using the formula then we get

$$\int \exp(Q(y/u, x'_1, \ldots, x'_k) + Q(z/u, x''_1, \ldots, x''_k)) \left[\prod_{j=1}^{k} \Delta(x'_j, x''_j; x_j)\right] dx'_1 \ldots dx'_k \, dx''_1 \ldots dx''_k.$$

The Δs symmetrize all the locations, it makes no difference if the power series is first expanded and then symmetrized or vice-verse thus all the Δs can be promoted to the exponential. Now using the inductive hypothesis we get the result.

Before we close this section we remark that if A and B do not commute we get a more complicated formula called the Campbell-Baker-Hausdorff formula. This formula does not arise in linear logic because of all the linearity constraints which ensure that operators do commute.

7.6 Interpreting Proof Nets

We now interpret terms in the linear realizability algebra as terms of the ϕ-calculus and show that the equations of the algebra are valid.
The translation is shown in Fig. 7.6. In the next section we show some example calculations. In this section we prove that this interpretation is sound.

The intuition behind the translation is as follows. The axiom link is just the identity which is modelled by the Dirac delta; in short we use a trivial propagator. The cut is modelled as an interaction, which means that we identify the common point

Proof Rule	LRA Term		Φ-Calculus
Axiom	$[\![I_{xy}]\!]$	=	$\delta(x,y)$
Cut	$[\![P \cdot_x Q]\!]$	=	$\int [\![P]\!][\![Q]\!]dx$
Tensor	$[\![\otimes_z^{x,y}(P,Q)]\!]$	=	$\int [\![P]\!][\![Q]\!]\delta(z,\langle x,y\rangle)dxdy$
Par	$[\![\parr_z^{x,y}(P)]\!]$	=	$\int [\![P]\!]\delta(z,[x,y])dxdy$
Dereliction	$[\![D_z^x(P)]\!]$	=	$[\![P[z/x]]\!]([.]\|_{\alpha(z)=0}) \circ \frac{\delta}{\delta\alpha(z)}$
Weakening	$[\![W_z(P)]\!]$	=	$[\![P]\!] \circ W(z)([.]\|_{\alpha(z)=0})\frac{\delta}{\delta\alpha(z)}$
Contraction	$[\![C_z^{x,y}]\!]$	=	$\int [\![P]\!]\Delta(x,y;z)dxdy$
Exponentiation	$[\![!_y^x(P)]\!]$	=	$\exp([\![P[y/x]]\!]\alpha_A(y))$

In the last line A is the type of the location x.

Fig. 7.5 Translation of LRA terms to the Φ-calculus

(the interaction is local) and we integrate over the possible interactions. The par and tensor links are constructing composite objects. They are modelled by using pairing of locations. The promotion corresponds to an exponentiation and dereliction is a variational derivative which probes for the presence of the α in an exponential. Weakening is like a dereliction, except that there is a W to perform discarding. Finally contraction is effected by a symmetrizer; we think of it like multiplexing.

We proceed to the formal soundness argument.

Theorem 7.6.1 *The interpretation of linear realizability algebra terms in the ϕ-calculus obeys the equations **R1,R3,R6,R7,R8,R9**.*

Proof The proof of **R1** is immediate from the definition of the Dirac delta function. For **R3** we calculate as follows.

$$[\![\otimes_z^{x,y}(P,Q)\cdot_z \parr_z^{u,v}(M)]\!]$$

$$= \int \delta(z,\langle x,y\rangle)\delta(z,[u,v])\ PQM\ dz\ dxyuv \qquad \text{by definition}$$

$$= \int \delta(\langle x,y\rangle,[u,v])\ PQM\ dxyuv \qquad \text{using } \delta \text{ to do the } z \text{ integral}$$

$$= \int \delta(x,u)\delta(y,v)\ PQM\ duvxy \qquad \text{decomposition of } \delta$$

$$= \int P[u/x]MQ[v/y]dudv \qquad \text{doing the } x, y \text{ integrals}$$

$$=[\![P[u/x]\cdot_u M\cdot_v Q[v/y]]\!] \qquad \text{by definition.}$$

For equation **R6** we calculate as follows.

$$[\![D_z^x(P)\cdot_z !_z^y(Q)]\!]$$

$$= \int P(x,\ldots)\delta(x,z)([.]|_{\alpha(z)=0}) \circ \frac{\delta}{\delta\alpha(z)}\delta(y,z)\exp(\alpha(z)Q)\ dzdxdy$$

by definition

$$= \int P(x, \ldots)\delta(x, z)([.]|_{\alpha(z)=0}) \circ \frac{\delta}{\delta\alpha(z)} \exp(\alpha(z) Q[z/y]) \, dxdz$$

using $\delta(y, z)$ to do the z integral

$$= \int P(x, \ldots)([.]|_{\alpha(x)=0}) \circ \frac{\delta}{\delta\alpha(x)} \exp(\alpha(x) Q[x/y]) \, dx$$

using Lemma 7.4.4 and $\delta(x, z)$

$$= \int P Q[x/y] \, dx$$

using Lemma 7.5.1

$$= [\![P \cdot_x Q[x/y]]\!]$$

by definition.

For **R7** we have to show

$$[\![W_z(P) \cdot_z !_z^y(Q)]\!] = [\![W_{\boldsymbol{u}}(P)]\!]$$

where $\boldsymbol{u} = \{u_1, \ldots, u_k\}$ is the set of free locations in Q other than y and $W_{\boldsymbol{u}}$ is shorthand for $W_{u_1} \ldots W_{u_k}$. The linearity constraints ensure that $\boldsymbol{u} \cap \mathsf{S}(P) = \emptyset$. If we use the translation and use the simple exponential identity Proposition 7.5.1, part 1, we get the formula below, which does not mention P,

$$\int W(z)([.]|_{\alpha(z)=0}) \circ \frac{\delta}{\delta\alpha(z)} \exp(\alpha(z) Q[z/y]) \, dz = \int W(z) Q[z/y] \, dz.$$

In fact P has nothing to do with this rule so we will ignore it in the rest of the discussion of this case. Now in order to complete the argument we must prove the following lemma:

Lemma 7.6.2

$$\int W_\phi(z) Q(z, \boldsymbol{u}) \, dz = \prod_i W(u_i)([.]|_{\alpha(u_i)=0}) \circ \frac{\delta}{\delta\alpha(u_i)}.$$

We have explicitly shown the formula ϕ that is being weakened on the left-hand side but not the subformulas of ϕ which are associated with the Ws on the right-hand side. We have implicitly used the decomposition formula for W that says

$$W_B(z)([.]|_{\alpha(z)=0}) \circ \frac{\delta}{\delta\alpha(z)} = W_{?B}(z).$$

Proof We begin with an induction on the complexity of the weakened formula ϕ. In the base case we assume that ϕ is atomic, say A. Now we do an induction on the structure of Q, which means that we look at the last rule used in the construction of the proof Q. Since we have an atomic formula being cut the last rule used in the construction of Q can only be a dereliction, or a weakening or a contraction. The dereliction case corresponds to Q being of the form $Q'([.]|_{\alpha(u)=0}) \circ \frac{\delta}{\delta\alpha(u)}$, where u is one of the locations in $\boldsymbol{u}$. Thus we have $\int W_A(z)Q'(z,u,\ldots)([.]|_{\alpha(u)=0}) \circ \frac{\delta}{\delta\alpha(u)}\, dz$. Now since Q' is a smaller proof by structural induction we have the result for all the locations other than u and we explicitly have the operator for u. The weakening case is exactly the same. For contraction we have the term $\int W_A(z)\Delta(z_1,z_2;z)Q'(z_1,z_2,\ldots)\, dzdz_1dz_2$. Using the second decomposition rule for weakening in definition 18 we get $\int W_A(z_1)W_A(z_2)Q'(z_1,z_2,\ldots)dz_1dz_2$ and now by the structural induction hypothesis we get the result. This completes the base case in the outer structural induction. The rest of the proof is essentially a use of the decomposition rules for W. We give one case. Suppose that ϕ is of the form $?\,\psi$. Then we have on the lhs $\int W_\psi(z)([\circ]|_{\alpha(z)=0})\frac{\delta}{\delta\alpha(z)}Q(z,\boldsymbol{u})\, dz$, where we have used the decomposition formula 18, part 1. But by the typing of proof nets Q itself must be $\exp(\alpha_\psi Q')$. Now using the usual calculation, from Formula 7.5.1, part 1, gives the result.

The proofs of **R8** and **R9** follow from the exponential identities. For **R8** we can use Lemma 7.5.4 directly. While for **R9** we can directly use the Lemma 7.5.2.

In this proof the most work went into analyzing weakening, the other rules really follow very easily from the basic framework. The reason for this is that weakening destroys a complex formula but the rest of the framework is local. Thus we have to decompose a W into its elementary pieces in order to get the components annihilated in atomic pieces.

7.7 Example Calculations

In this section we carry out some basic calculations that illustrate how the manipulations of proof nets are mimicked by the algebra of our operators. In the first two examples we will just use the formal terms needed for the multiplicatives and thereafter we use operators and illustrate them on examples using exponentials. It should be clear, after reading these examples, that carrying out the calculation with half a dozen contractions (the largest that we have tried by hand) is no more difficult than the examples below, even the bookkeeping with the locations is not very tedious. We do not explicitly give an example involving nested boxes because this would be very close to what is already shown in the proof of the last section.

A Basic Example with Cut

We reproduce the example from the section on multiplicatives. The simplest possible example involves an axiom link cut with another axiom link shown in Fig. 7.6.

The LRA term is $I_{x,u}\cdot_{u,v} I_{v,y}$. The expression in the ϕ-calculus is

$$\int \delta(x,u)\delta(u,v)\delta(v,y)dudv.$$

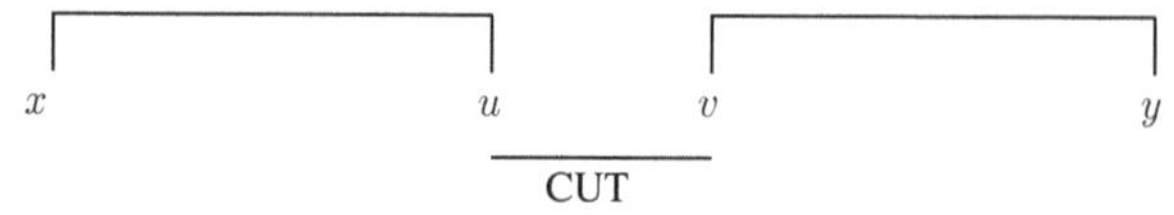

Fig. 7.6 Two axiom links CUT together

Carrying out the v integration and getting rid of the δ we get $\int \delta(x,u)\delta(u,y)du$. Using the convolution property of δ we get $\delta(x,y)$ which corresponds to the axiom link $I_{x,y}$.

Tensor and Par

Consider the proof net constructed as follows. We start with two axiom links, one for A and one for B. We form a single net by tensoring together the $A^{\perp}$ and the $B^{\perp}$. Now consider a second proof net constructed in the same way. With the second such net we introduce a par link connecting the A and the B. Now we cut the first net with the second net in the evident way, shown in Fig. 7.7.

The ϕ-calculus term, with locations introduced as appropriate is

$$\begin{aligned}\int \quad & \left[\int \delta_A(x,y)\delta_B(u,v)\delta(z,\langle y,v\rangle)dvdy\right] \\ & \left[\int \delta_A(p,q)\delta_B(r,s)\delta(t,\langle q,s\rangle)\delta(w,[p,r])dpdqdrds\right] \\ & \qquad\qquad \delta(w,z)dw\,dz.\end{aligned}$$

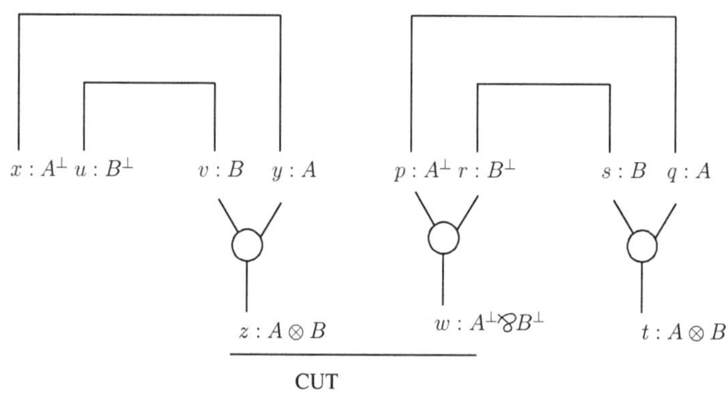

Fig. 7.7 Cutting a ⊗ link with a ⅋ link

We first do the w integral and eliminate the term $\delta(w,z)$. This will cause z to replace w. Now we do the z integral and eliminate the term $\delta(z,\langle y,v\rangle)$. This will yield the term $\delta(\langle y,v\rangle,[p,r])$, which can be decomposed into $\delta(y,p)\delta(v,r)$. The full term is now

$$\int \delta_A(x,y)\delta_B(u,v)\delta_A(p,q)\delta_B(r,s)\delta(y,p)\delta(v,r)\delta(t,\langle q,s\rangle)dydvdpdrdsdq.$$

This is the ϕ-calculus term that arises by translating the result of the first step of the cut-elimination process. Note that it has two cuts on the simpler formulas A and B. Now, as in the previous example we can perform the integrations over y and v using the formula for the δ and then we can perform the integrations over p and r using the convolution formula. The result is

$$\int \delta_A(x, q)\delta_B(u, s)\delta(t, \langle q, s\rangle)dqds,$$

which is indeed the form of the ϕ-calculus term that results from the cut-free proof.

A Basic Exponential Example

We consider the simplest possible cut involving exponential types. Consider the an axiom link for $A, A^{\perp}$. We can perform dereliction on the $A^{\perp}$. Now take another copy of this net and exponentiate on A. Finally cut the $?A^{\perp}$ with the $!A$. The proof net is shown in Fig. 7.8.

The result of translating this into the ϕ-calculus is (after some obvious simplifications)

$$\int \delta(x, y)([.]|_{\alpha(y)=0})\frac{\delta}{\delta\alpha(y)} \exp[\alpha(u)\delta(u, v)([.]|_{\alpha(v)=0})\frac{\delta}{\delta\alpha(v)}]\delta(y, u)dy\, du.$$

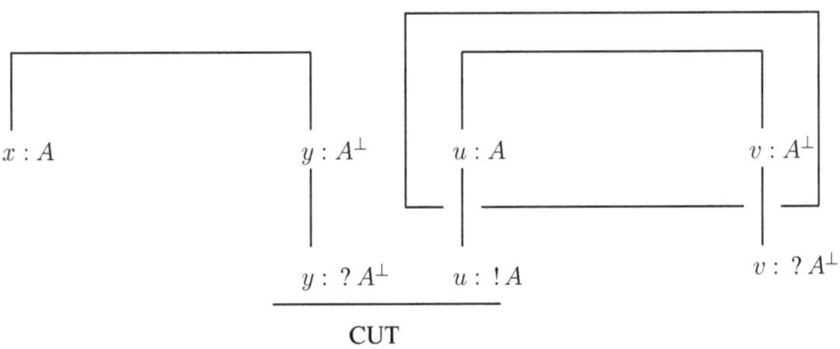

Fig. 7.8 The simplest possible example with exponentials

We can perform the u integration and eliminate the term $\delta(y, u)$. Then we can take the variational derivative of the exponential term which will yield

$$\int \delta(x, y)\delta(y, v)([.]|_{\alpha(v)=0})\frac{\delta}{\delta\alpha(v)}dy.$$

Now the last integral can be done with the convolution property of δ and we get

$$\delta(x, v)([.]|_{\alpha(v)=0})\frac{\delta}{\delta\alpha(v)}$$

which is what we expect from the cut-free proof.

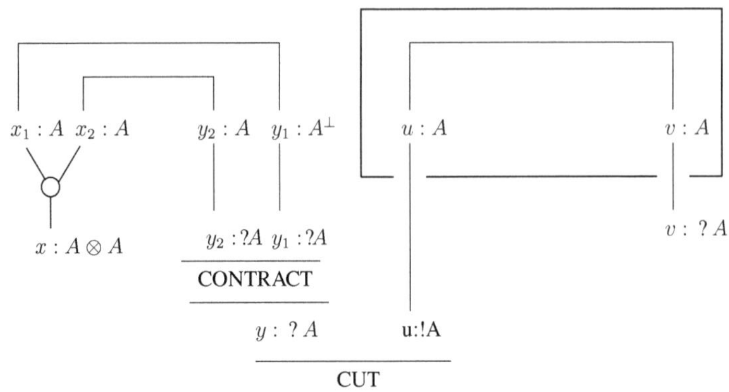

Fig. 7.9 An example with contraction

An Example with Contraction

We take a pair of axiom links for A, $A^\perp$ and derelict each one on the $A^\perp$ formula. We then combine them into a single net by tensoring the two A formulas. The two derelicted formulas are combined by contraction. Finally we take the basic exponentiated net, as in the last example and cut it with the proof net just constructed in the evident way. The resulting ϕ-calculus term is:

$$\int \delta(x_1, y_1)\delta(x_2, y_2)\delta(x, \langle x_1, x_2\rangle)$$
$$\frac{\delta}{\delta\alpha(y_1)}\frac{\delta}{\delta\alpha(y_2)}\Delta(y; y_1, y_2)\delta(y, u)$$
$$\exp(\alpha(u)\delta(u, v)\frac{\delta}{\delta\alpha(v)})dx_1dx_2dy_1dy_2dydu.$$

where we have written $\frac{\delta}{\delta\alpha(v)}$ rather than $([\frac{\delta}{\delta\alpha(v)}(\cdot)]|_{\alpha(v)=0})$ to avoid cluttering up the notation. We first get rid of the integration created by the cut so that u is replaced by y in the exponential. Next we extract the quadratic term from the power-series expansion of the exponential. All the other terms will vanish after taking derivatives. The relevant part of the exponential series is the term

$$(1/2)\Delta(y; y', y'')\alpha(y')\alpha(y'')\delta(y', v')\delta(y'', v'')\Delta(v; v', v'')\frac{\delta}{\delta\alpha(v')}\frac{\delta}{\delta\alpha(v'')}.$$

Now when we carry out the y integral the term $\Delta(y; y_1, y_2)\Delta(y; y', y'')$ becomes $2[\delta(y_1, y')\delta(y_2, y'')]$. The factor of 2 from the symmetrization cancels the factor of $1/2$ from the power-series expansion. Now we can carry out the y_1 and y_2 integrations to get

$$\int \delta(x_1, y')\delta(x_2, y'')\delta(x, \langle x_1, x_2\rangle)$$
$$\frac{\delta}{\delta\alpha(y')}\frac{\delta}{\delta\alpha(y'')}\alpha(y')\alpha(y'')\delta(y', v')\delta(y'', v'')\Delta(v; v', v'')\frac{\delta}{\delta\alpha(v')}\frac{\delta}{\delta\alpha(v'')}$$
$$dv'dv''dy'dy''dx_1dx_2.$$

Now we can do the derivatives and the y', y'' integrals to get

$$\int \delta(x_1, v')\delta(x_2, v'')\delta(x, \langle x_1, x_2\rangle)\Delta(v; v', v'')\frac{\delta}{\delta\alpha(v')}\frac{\delta}{\delta\alpha(v'')}dv'dv''dx_1dx_2.$$

this is what we expect after cut elimination. Notice how the argument to the exponentiation has become duplicated and has picked up a contraction on its other variables.

An Example with Contraction and Weakening

We consider a minor variation of the last example. Instead of using the tensor to obtain a pair of derelictions that need to be contracted we could have obtained a $?\,(A)^{\perp}$ by weakening. The ϕ-calculus term would then be

$$\begin{aligned}&\int \delta(x, y_1)W(y_2)\frac{\delta}{\delta\alpha(y_1)}\frac{\delta}{\delta\alpha(y_2)}\Delta(y; y_1, y_2)\\ &\qquad \delta(y, u)\exp(\alpha(u)\delta(u, v)\frac{\delta}{\delta\alpha(v)})dy_1dy_2dydu.\end{aligned}$$

We can reproduce the calculations as before to get

$$\begin{aligned}&\int \delta(x, y')W(y'')\delta(y', v')\delta(y'', v'')\\ &\qquad \Delta(v; v', v'')\frac{\delta}{\delta\alpha(v')}\frac{\delta}{\delta\alpha(v'')}dv'dv''dy'dy''.\end{aligned}$$

Carrying out the, by now routine, simplifications, we get
$\int \delta(x, v')W(v'')\Delta(v; v', v'')\frac{\delta}{\delta\alpha(v')}\frac{\delta}{\delta\alpha(v'')}dv'dv''$.
In this example note how the original weakening at the location y'' has turned into a weakening at the location v''.

7.8 Conclusions

We feel that the most interesting feature of this work is that the subtle combinatorics of proof nets is captured by the elementary rules of the ϕ-calculus. More specifically, the formal devices of a variational derivative, formal power series, symmetrizers and integrals. The fact that the equations of a linear realizability algebra are obeyed for our fragment shows that the basic normalization behaviour of proof nets is captured.

But the main caveats are as follows. We have to posit quite a lot of rules to make weakening behave correctly. This reflects the idea that we are using up resources piece by piece, whereas weakening causes a "large" type to appear all at once. Thus, in the reductions, we have to decompose this before throwing it away. We have not addressed additives in the present paper. It turns out that the same kind of variational derivative formalism works. There are some interesting features, we model additives with superposition rather than choice and as a result one can push a cut inside a "with box."

Originally, we had sought to model the exponential type using the so called Fock space construction of quantum field theory [Blu93]; this led to our present investigations. Fock space—also known as the symmetric tensor algebra—can be viewed as the space of analytic functions on a Banach space and, in a formal sense can be viewed as an exponential. Our original work [RBS93] fell short of modelling linear logic. Girard [Gir95b] later succeeded in modelling linear logic using analytic functions on what he called coherent Banach spaces. A key idea in that work is that the exponentials correspond to the Fock-space construction.

The connection with quantum field theory may be mere analogy but the use of formal power series and variational derivatives is more than that. The technical result of this paper is that the combinatorics of proof nets (at least for multiplicative-exponential linear logic) have been captured by the mathematical structures that we have introduced. Furthermore, these structures have an independent mathematical existence that has nothing to do with proof nets or linear logic or quantum field theory. They form the basis therefore of a research program to investigate several topics that have recently been based on linear logic. Foremost among these are the spectacular results of Abramsky, Jagadeesan and Malacaria [AJM00], and Hyland and Ong [HO00] and Nickau [Nic94] which have led to semantically-presented fully-abstract models of PCF [Mil77, Plo77]. These models are based on the intuitions of games and the flow of information between the players of the games. The variational derivative, as we have used it, seems to embody the same ideas. It is used to query a term for the presence of an exponential.

Since this work was first presented in 1995, there have been some interesting developments. A categorical view of quantum computation[AC04], and indeed of quantum mechanics, has taken hold and been vigourously pursued [Sel07].

Important work, in terms of the relevance to the present work, is found in the investigations of Marcelo Fiore et al. where similar formal differential structure is discussed in the context of bicategories, see, for example, recent papers and slides available on Fiore's web page [Fio06, FGHW07, Fio07]. There are close connections between the structure that he finds and the creation and annihilation operators of quantum field theory which act on Fock space. There has also been independent work by Jamie Vickary, so far unpublished, which develops a theory of creation and annihilation operators on Fock space in the context of categorical quantum mechanics [AC04].

Also clearly relevant is the work of Ehrhard and Regnier [ER03, ER06] in the notion of *differential* λ*-calculus* and *differential linear logic*. These papers provide an extension of the usual notions of λ-calculus and linear logic to include a differential combinator, and explore the syntactic consequences. The possible relationships to the present work are striking. Ehrhard and Regnier's work was subsequently categorified in [BCS06].

Finally it is possible to construct a mathematical model for the ϕ-calculus. The manipulations that we have done with variational derivatives and exponentials are very close to the calculations that one does to derive Feynman diagrams from the generating functional of a quantum field theory [Ram81]. A precise calculus for these functional derivatives viewed as operators and for propagators appears in the treatment of functional integration in "Quantum Physics" by Glimm and Jaffe [GJ81].

Acknowledgments We would like to thank Samson Abramsky, Martin Hyland, Radha Jagadeesa and Mikhail Gromov for interesting discussions.

References

Abr91. Abramsky, S.: Computational interpretations of linear logic. Lecture at Category Theory Meeting in Montreal (1991)

AC04. Abramsky, S., Coecke, B.: A categorical semantics of quantum protocols. In: Proceedings of the 19th Annual IEEE Symposium on Logic in Computer Science: LICS 2004, pp. 415–425. IEEE Computer Society (2004)

ADLR94. Asperti, A., Danos, V., Laneve, C., Regnier, L.: Paths in the *-calculus. In: Proceedings of the 9th Annual IEEE Symposium On Logic In Computer Science, Paris, France, July 1994

AJ94a. Abramsky, S., Jagadeesan, R.: Games and full completeness for multiplicative linear logic. Symbolic J. Logic **59**(2), 543–574 (1994)

AJ94b. Abramsky, S., Jagadeesan, R.: New foundations for the geometry of interaction. Inf. Comput. **111**(1), 53–119 (May 1994)

AJM00. Abramsky, S., Jagadeesan, R., Malacaria, P.: Full abstraction for PCF. Inf. Comput. **163**, 409–470 (2000)

BB90. Berry, G., Boudol, G.: The chemical abstract machine. In: Proceedings of the 17th Annual ACM Symposium on Principles of Programming Languages, pp. 81–94. ACM (1990)

BCS06. Blute, R., Cockett, R., Seely, R.: Differential categories. Math. Struct. Comput. Sci. **16**, 1049–1083 (2006)

Blu93. Blute, R.: Linear logic, coherence and dinaturality. Theor. Comput. Sci. **115**, 3–41 (1993)

Bro05. Brown, L.M. (ed.): Feynman's thesis. World Scientific, 2005. Book version of R.P. Feynman's PhD thesis from Princeton University (1942)

CDM95. Cartier, P., DeWitt-Morette, C.: A new perspective on functional integration. J. Math. Phys. **36**, 2237–2312 (1995)

DR93. Danos, V., Regnier, L.: Local and asynchronous *-reduction. In: Proceedings of the 8th IEEE Symposium on Logic in Computer Science, Montréal. IEEE Press, Canada, June 1993

Dys49. Dyson, F.J.: The radiation theories of Tomanaga, Schwinger and Feynman. Phys. Rev. **75**, 486–502 (1949)

ER03. Ehrhard, T., Regnier, L.: The differential lambda-calculus. Theor. Comput. Sci. **309**, 1–41 (2003)

ER06. Ehrhard, T., Regnier, L.: Differential interaction nets. Theor. Comput. Sci. **364**, 166–195 (2006)

Fey49a. Feynman, R.P.: The space-time approach to quantum electrodynamics. Phys. Rev. **76**, 769–789 (1949)

Fey49b. Feynman, R.P.: The theory of positrons. Phys. Rev. **76**, 749–759 (1949)

Fey62. Feynman, R.P.: Quantum Electrodynamics. Lecture Note and Preprint Series. Benjamin, W.A. Inc. (1962)

FGHW07. Fiore, M.P., Gambino, N., Hyland, M., Winskel, G.: The cartesian closed bicategory of generalised species of structures. J. London Math. Soc. **77**, 203–220 (2007)

FH65. Feynman, R.P., Hibbs, A.R.: Quantum mechanics and path integrals. McGraw Hill, New York (1965)

Fio06. Fiore, M.P.: Adjoints and fock space in the context of profunctors. Talk given at the Cats, Kets and Cloisters Workshop, Oxford University, Oxford. (July 2006)

Fio07. Fiore, M.P.: Differential structure in models of multiplicative biadditive intuitionistic linear logic. In: Typed Lambda Calculi and Applications, Number 4583 in Lecture Notes in Computer Science, pp. 163–177 (2007)

Gir89a. Girard, J.-Y.: Geometry of interaction I: Interpretation of system F. In: Ferro, R. et. al. (eds.) Proceedings Logic Colloquium 88, pp. 221–260. North-Holland, Amsterdam (1989)

Gir89b. Girard, J.-Y.: Geometry of interaction II: Deadlock free algorithms. In: Martin-Lof, P., Mints, G. (eds.) Proceedings of COLOG 88, Number 417 in Lecture Notes In Computer Science, pp. 76–93. Springer, New York 1989

Gir95a. Girard, J.-Y.: Geometry of interaction III: Accomodating the additives. In: Girard, J.-Y., Lafont, Y., Regnier, L. (eds.) Advances in Linear Logic, Number 222 in London Mathematics Society Lecture Note Series, pp. 329–389. Cambridge University Press, Cambridge (1995)

Gir95b. Girard, J.-Y.: Linear logic: Its syntax and semantics. In: Girard, J.-Y., Lafont, Y., Regnier, L. (eds.) Advances in Linear logic, Number 222 in London Mathematics Society Lecture Note Series, pp. 1–42. Cambridge University Press, Cambridge (1995)

GJ81. Glimm, J., Jaffe, A.: Quantum Physics: A Functional Integral Point of View. Springer, New York (1981)

HO00. Hyland, J.M.E., Ong, C.-H.L.: On full abstraction for pcf: I. models, observables and the full abstraction problem, II. dialogue games and innocent strategies, III. a fully abstract and universal game model. Inf. Comput. **163**, 285–408 (2000)

IZ80. Itzykson, C., Zuber, J.-B.: Quantum Field Theory. McGraw-Hill, New York (1980)

Mil77. Milner, R.: Fully abstract models of typed lambda-calculi. Theor. Comput. Sci. **4**(1), 1–23 (1977)

Mil89. Milner, R.: Communication and Concurrency. Prentice-Hall, Upper Saddle River (1989)

Nic94. Nickau, H.: Hereditarily sequential functionals. In: Nerode, A., Matiyasevich, Yu.V. (eds.) In: Proceedings of the Symposium on Logical Foundations of Computer Science: Logic at St. Petersburg, volume 813 of Lecture Notes in Computer Science, pp. 253–264. Springer, New York (1994)

Plo77. Plotkin, G.D.: LCF considered as a programming language. Theor. Comput. Sci. **5**(3), 223–256 (1977)

Ram81. Ramond, P.: Field Theory, A Modern Primer. Frontiers in Physics. Benjamin-Cummings (1981)

RBS93. Panangaden, P., Blute, R., Seely, R.: Holomorphic functions as a model of exponential types in linear logic. In: Brookes, S. et al. (eds.) Proceedings of the 9th International Conference on Mathematical Foundations of Programming Semantics, volume 802 of Lecture Notes In Computer Science. Springer, New York (1993)

Sch81. Schulman, L.S.: Techniques and applications of path integrals. Wiley, New York (1981)

Sel07. Selinger, P.: Dagger compact closed categories and completely positive maps. In: Proceedings of the 3rd International Workshop on Quantum Programming Languages (QPL 2005), Number 170 in ENTCS, pp. 139–163 (2007)

Sim05. Simon, B.: Functional Integration and Quantum Physics. AMS Chelsea, 2nd edn. AMS Chelsea Publishing, Providence (2005)

Chapter 8
Compact Monoidal Categories from Linguistics to Physics

J. Lambek

Abstract This is largely an expository paper, revisiting some ideas about *compact* 2-categories, in which each 1-cell has both a left and a right adjoint. In the special case with only one 0-cell (where the 1-cells are usually called "objects") we obtain a *compact strictly monoidal category*. Assuming furthermore that the 2-cells describe a partial order, we obtain a compact partially ordered monoid, which has been called a *pregroup*. Indeed, a pregroup in which the left and right adjoints coincide is just a partially ordered group (= pogroup).

A brief exposition of recent joint work with Preller and Lambek "Mathematical Structures in Computer Science", 17, (2007) will be given here, investigating *free* compact strictly monoidal categories, which may be said to describe computations in pregroups. Free pregroups lend themselves to the study of grammar in natural languages such as English. While one would not expect to find a connection between linguistics and physics, applications of (free) compact symmetric monoidal categories to physics have been made by Coecke "The Logic of Entanglement" (2004), Abramsky and Coecke "Proceedings of 19th IEEE Conference on Logic in Computer Science", pp. 415–425 (2004), Abramsky and Duncan "Mathematical Structures in Computer Science", 16, 469–489 (2006), Selinger "Electronic Notes in Theoretical Computer Science", 170, 139–163 (2007).

Compact symmetric monoidal categories had already been studied by Kelly and Laplaza "Journal of Pure and Applied Algebra", 19, 193–213 (1980), who called them "compact closed" and by Barr "Lecture Notes in Mathematics", 752 (1979), "Journal of Pure and Applied Algebra", 111, 1–20 (1996), "Theoretical Computer Science", 139, 115–130 (1995), who called them "compact star-autonomous". I had intended to show that Feynman diagrams for quantum electro-dynamics (QED) could be described by certain compact Barr-autonomous categories, but was disappointed to find that these reduced to a rather degenerate case, that of partially

J. Lambek (✉)
Department of Mathematics and Statistics, McGill University, Montreal, QC, Canada
e-mail: lambek@math.mcgill.ca

Lambek, J.: *Compact Monoidal Categories from Linguistics to Physics*. Lect. Notes Phys. **813**, 467–487 (2011)
DOI 10.1007/978-3-642-12821-9_8

ordered groups (= pogroups). Still, I will reluctantly present an extension of this idea from QED to the Standard Model. Finally, I will briefly review the transition from 2-categories to the bicategories of Bénabou "Lecture Notes in Mathematics" 47, 1–77 (1967), using methods of Bourbaki "Algebre multilineaire" (1948) and Gentzen (see Kleene "Introduction to metamathematics" (1952)), which may also be of interest in physics.

8.1 Compact 2-Categories and Pregroups

A *2-category* has 0-cells, 1-cells and 2-cells. A typical 2-category (Cat) is that of all small categories with

- 0-cells = small categories,
- 1-cells = functors,
- 2-cells = natural transformations.

We recall that 2-cells have, in addition to the *vertical composition* (represented by a small circle)

$$\frac{t : F \to G \quad u : G \to H}{u \circ t : F \to H}$$

also a *horizontal composition* (represented by juxtaposition)

$$\frac{s : H \to K \quad t : F \to G}{st : HF \to KG}$$

defined by the diagonal of the commutative square:

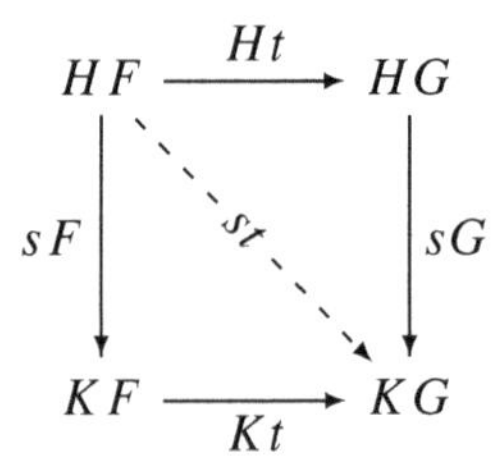

The equations in a 2-category are described formally in Mac Lane's "Categories for the working mathematician" [23], but should be familiar from Cat.

The notion of adjoint functor is known in Cat, but exactly the same definition works in any 2-category. Thus $(F, U, \eta, \varepsilon)$ defines an *adjoint pair* if

$$\eta : 1 \to FU \qquad \text{and} \qquad \varepsilon : UF \to 1$$

are 2-cells such that the following *triangular equations* hold:

$$F \xrightarrow{\eta F} FUF \xrightarrow{F\varepsilon} F = F \xrightarrow{1_F} F \tag{8.1}$$

$$U \xrightarrow{U\eta} UFU \xrightarrow{\varepsilon U} U = U \xrightarrow{1_U} U \tag{8.2}$$

Special cases of 2-categories are
strictly monoidal categories: with only one 0-cell;
partially ordered categories: with only one 2-cell from any 1-cell to another;
partially ordered monoids: both of the above.

A 2-category is said to be *compact* if every 1-cell G has both a *left adjoint* G^ℓ and a *right adjoint* G^r. We describe the two adjoint pairs thus:

$$(G, G^r, \eta_G, \varepsilon_G), \qquad (G^\ell, G, \eta_{G^\ell}, \varepsilon_{G^\ell}).$$

Of special interest are compact partially ordered monoids, which I have called *pregroups*. In any pregroup we have

$$GG^r \to 1 \to G^r G, \qquad G^\ell G \to 1 \to GG^\ell$$

and the triangular equations hold automatically, since the arrow denotes a partial order. If $G^\ell = G^r$ for all 1-cells G, the pregroup is just a *partially ordered group*, more precisely a partially ordered monoid in which each element has an inverse.

Pregroups that are not partially ordered groups are not easy to come by. My favourite example is the monoid of all unbounded order-preserving mappings $\mathbb{Z} \to \mathbb{Z}$, with multiplication and order defined as follows:

$$(fg)(z) = f(g(z)), \qquad f \to g \Leftrightarrow f(z) \leq g(z)$$

for all $z \in \mathbb{Z}$. Adjoints are defined thus:

$$g^\ell(x) = \inf\{y \in \mathbb{Z} | x \leq g(y)\}, \qquad g^r(x) = \sup\{y \in \mathbb{Z} | g(y) \leq x\}.$$

To see that this is not a group, consider $g(x) = 2x$, then

$$g^r(x) = [x/2], \qquad g^\ell(x) = [(x+1)/2].$$

The following equations hold for the elements of any pregroup, or even for the arrows (1-cells) in any compact partially ordered category:

$$1^\ell = 1 = 1^r,\ a^{r\ell} = a = a^{\ell r},\ (ab)^\ell = b^\ell a^\ell,\ (ab)^r = b^r a^r.$$

Moreover, adjoints are unique and

$$a \to b \Rightarrow b^{\ell} \to a^{\ell} \Rightarrow a^{\ell\ell} \to b^{\ell\ell} \Rightarrow \cdots ,$$
$$a \to b \Rightarrow b^{r} \to a^{r} \Rightarrow a^{rr} \to b^{rr} \Rightarrow \cdots .$$

8.2 Pregroups for Grammar

Pregroups freely generated by a partially ordered set have recently found an application to the grammar of natural languages. To illustrate this with a tiny fragment of English grammar, consider the poset of *basic types*:

$\mathbf{q}_1$ = yes-or-no question in present tense,
$\mathbf{q}$ = yes-or-no question in any tense,
$\overline{\mathbf{q}}$ = question (including Wh-question),
$\mathbf{i}$ = infinitive of intransitive verb,
π_3 = third person singular subject,
π_2 = second person singular or any plural subject,
$\mathbf{o}$ = direct object,
$\mathbf{p}$ = plural noun phrase,

with *basic arrows* (partial orders)

$$\mathbf{q}_1 \to \mathbf{q} \to \overline{\mathbf{q}},\ \mathbf{p} \to \pi_2,\ \mathbf{p} \to \mathbf{o}.$$

Here are three sample questions with their associated *types* (elements of the free pregroup):

does he go with her ?

$$(\mathbf{q}_1\ \mathbf{i}^{\ell}\ \pi_3^{\ell})\pi_3\ \mathbf{i}\ (\mathbf{i}^{r}\ \mathbf{i}\ \mathbf{o}^{\ell})\mathbf{o} \ \to\ \mathbf{q}_1$$

whom does he go with — ?

$$(\overline{\mathbf{q}}\ \mathbf{o}^{\ell\ell}\ \mathbf{q}^{\ell})(\mathbf{q}_1\ \mathbf{i}^{\ell}\ \pi_3^{\ell})\pi_3\ \mathbf{i}\ (\mathbf{i}^{r}\ \mathbf{i}\ \mathbf{o}^{\ell}) \ \to\ \overline{\mathbf{q}}$$

with whom does he go ?

$$(\overline{\mathbf{q}}\ \mathbf{o}^{\ell\ell\ell}\ \overline{\mathbf{q}}^{\ell})(\overline{\mathbf{q}}\ \mathbf{o}^{\ell\ell}\ \mathbf{q}^{\ell})(\mathbf{q}_1\ \mathbf{i}^{\ell}\ \pi_3^{\ell})\pi_3\ \mathbf{i} \ \to\ \overline{\mathbf{q}}$$

where

$$\mathbf{q}^{\ell}\mathbf{q}_1 \to \mathbf{q}^{\ell}\mathbf{q} \to 1 .$$

The underlinks in a similar enterprise were first introduced by Zellig Harris [12]. They may be viewed as degenerate proofnets. The dash at the end of the second question represents what Chomsky calls a *trace*, inserted here to facilitate comparison with mainstream linguistics.

The reader may wonder why the above calculations involve only *contractions*

$$a^{\ell} a \to 1 \qquad \text{and} \qquad a a^{r} \to 1$$

and no *expansions*

$$1 \to a a^{\ell} \qquad \text{or} \qquad 1 \to a^{r} a\,.$$

The reason is the following:

Switching Lemma. Without loss of generality, one may assume that, in any calculation in a freely generated pregroup, all contractions precede all expansions.

This implies, of course, that, when the right hand side is a *simple type* (obtained from a basic type by adjunctions), no expansions are needed. The proof of the lemma [19] depends on the triangular equations.

Note that, already in the first sample question above, the contraction $\mathbf{i}^{\ell}\mathbf{i} \to 1$ was postponed in order to ensure that the question does not end after *go*. Here the postponement was obligatory, but often it is optional, allowing different interpretations. For example, consider

old men and women

$$(\mathbf{p}\,\mathbf{p}^{\ell})\,\mathbf{p}\,(\mathbf{p}^{r}\,\mathbf{p}\,\mathbf{p}^{\ell})\,\mathbf{p} \quad \to \mathbf{p}$$

versus

old men and women

$$(\mathbf{p}\,\mathbf{p}^{\ell})\,\mathbf{p}\,(\mathbf{p}^{r}\,\mathbf{p}\,\mathbf{p}^{\ell})\,\mathbf{p} \quad \to \mathbf{p}$$

In the first noun phrase only the men are described as being old, in the second both men and women are.

This suggests that we should think of the arrow not just as a partial order, but as a *derivation*. In other words, we should replace the pregroup by a compact strictly monoidal category, or even by a compact 2-category.

8.3 Free Compact 2-Categories

Free compact 2-categories were studied by Preller and Lambek [24]. To convey our main ideas, let me sketch briefly here how to construct the compact 2-category with one 0-cell freely generated by a given *basic category*.

- *basic* 1-cells are objects of the basic category;
- *simple* 1-cells have the form $A^{(z)}$, where A is a basic 1-cell and $z \in \mathbb{Z}$;
- 1-*cells* are strings of simple ones, the empty string to be denoted by 1;
- *composition* of 1-cells is concatenation of strings;
- *adjoints* of 1-cells are formed by reversing the order and decreasing the superscript by one unit for left adjoints, increasing it by 1 for right adjoints, but the empty string is its own left and right adjoint.

A description of 2-cells will be given presently. For this we need to introduce

(3.1) *simple arrows* of the form $f^{(z)} : A^{(z)} \to B^{(z)}$, where

- either z is even and $f : A \to B$ is basic
- or z is odd and $f : B \to A$ is basic,

(3.2) *contractions* $A^{(z)}A^{(z+1)} \to 1$ and *expansions* $1 \to A^{(z+1)}A^{(z)}$.

For example, if $f^0 = f : A \to B$ is an arrow in the basic category, we obtain $f^r : B^r \to A^r$ as follows:

$$B^r \to A^r A B^r \to A^r B A^r \to A^r.$$

This assumes that we have already introduced contractions $\varepsilon_B : BB^r \to 1$ and expansions $\eta_A : 1 \to A^r A$ when A and B are basic 1-cells. We may then also define

$$\varepsilon_{(A^r)} = (\eta_A)^r, \qquad \eta_{(B^r)} = (\varepsilon_B)^r.$$

Repeating this process, we obtain $f^{rr} : A^{rr} \to B^{rr}$ as well as $\varepsilon_{(B^{rr})}$ and $\eta_{(A^{rr})}$, etc. This will account for positive z, but negative z may be treated similarly.

The triangular equations for basic 1-cells must be postulated. But then we can infer them also for simple 1-cells, provided we postulate that adjunction acts contravariently on both horizontal and vertical composition. For example,

$$\begin{aligned}\varepsilon_{(A^r)}A^r \circ A^r\eta_{(A^r)} &= (\eta_A)^r A^r \circ A^r(\varepsilon_A)^r \\ &= (A\eta_A)^r \circ (\varepsilon_A A)^r \\ &= (\varepsilon_A A \circ A\eta_A)^r \\ &= (1_A)^r \\ &= 1_{(A^r)}.\end{aligned}$$

2-cells from one 1-cell to another are obtained by performing a sequence of "deductions" with the help of simple arrows, contractions and expansions, as follows:

$$\Gamma A^{(z)}\Delta \to \Gamma B^{(z)}\Delta, \quad \Gamma A^{(z)}A^{(z+1)}\Delta \to \Gamma\Delta, \quad \Gamma\Delta \to \Gamma A^{(z+1)}A^{(z)}\Delta,$$

where Γ and Δ are strings of 1-cells. However, these 2-cells are subject to the triangular equations discussed earlier. To obtain a canonical representation of 2-cells, it will be convenient to introduce generalized contractions and expansions, which already absorb certain simple arrows.

(3.3) *Generalized contractions* have the form ε_f, where $f : A \to B$ is a simple arrow and ε_f is the diagonal of the commutative square

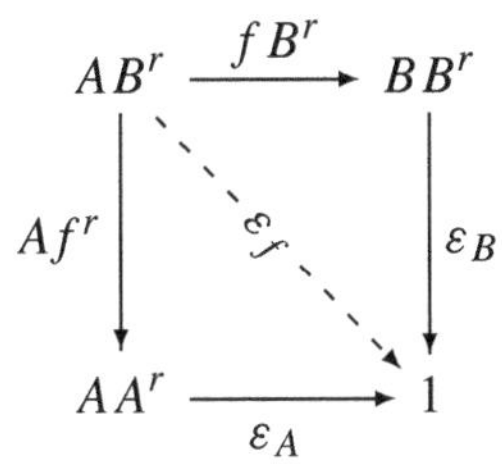

and *generalized expansions* have the form η_g, where $g : C \to B$ is a simple arrow and η_g is the diagonal of the commutative square

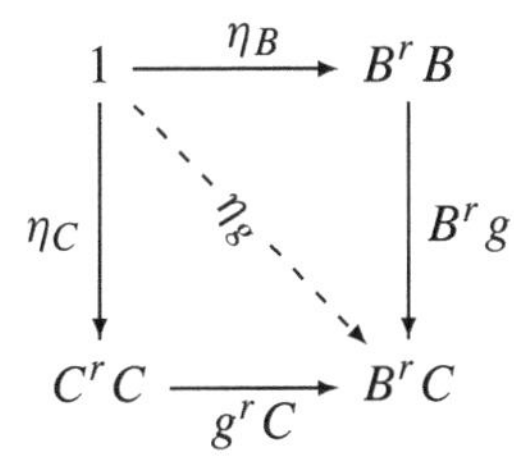

The Switching Lemma mentioned earlier for free pregroups can be sharpened to hold also for free compact 2-categories with one 0-cell.

Categorical Switching Lemma

Without loss of generality one may assume that a 2-cell consists of generalized contractions followed by simple arrows followed by generalized expansions.

Here is an indication of a crucial step in the proof: Suppose a generalized expansion immediately precedes a generalized contraction, as in

$$A \xrightarrow{A\eta_g} AB^rC \xrightarrow{\varepsilon_f C} C$$

where $f : A \to B$ and $g : B \to C$ are simple errors, then the compound arrow

$$(\varepsilon_f C) \circ (A\eta_g) = g \circ f$$

may be replaced by the simple arrow $g \circ f : A \to C$.

To see this look at the following commutative diagram:

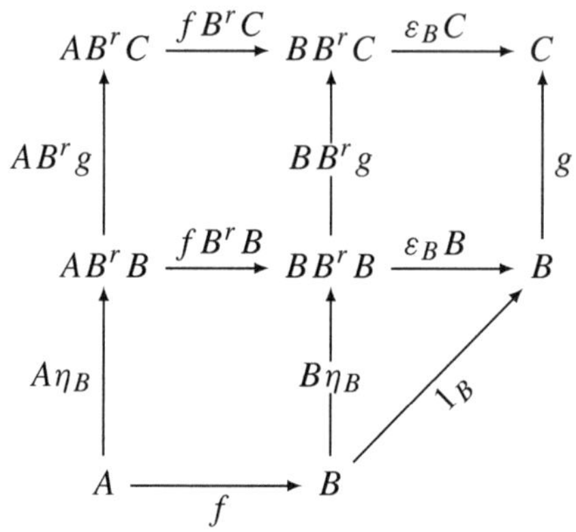

and note that the compound arrow on the top is $\varepsilon_f C$ and the compound arrow on the left is A_{η_g}.

To check the commutativity of the above squares, pretend you are in Cat, then apply the naturality of f, ε_B and f again.

We have thus proved the generalized triangle equality

$$(\varepsilon_f C) \circ (A\eta_g) = g \circ f$$

and can show similarly that

$$(A^r \varepsilon_g) \circ (\eta_f C^r) = f^r \circ g^r.$$

We may then represent 2-cells by geometric diagrams called *transition systems* by Preller and Lambek [24]. For example, given simple arrows

$$f : A \to F,\ g : C \to D,\ \ell : B \to E,\ i : G \to H,\ j : I \to J,$$

we obtain a 2-cell

$$ABCD^rE^r \to FG^rHI^rJ$$

as a vertical composition as follows:

$$(FG^rH\eta_j) \circ (F\eta_i) \circ (f\varepsilon_h) \circ (AB\varepsilon_g E^r),$$

which is represented horizontally thus:

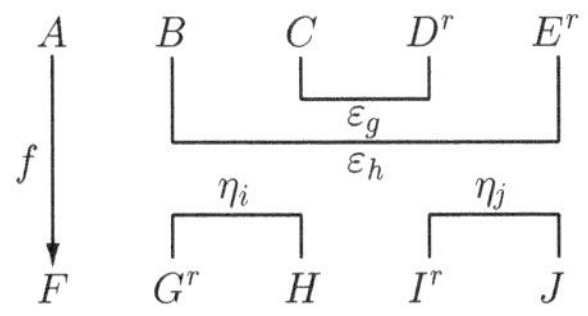

When describing a transition system between two 1-cells Γ and Δ, we must ensure that any simple 1-cell of Γ or Δ is at an endpoint of exactly one simple arrow, underlink or overlink, and that these don't cross.

The Switching Lemma ensures that the composition of two transition systems is again a transition system, by a process we called "combing", but which others have called "yanking". For example:

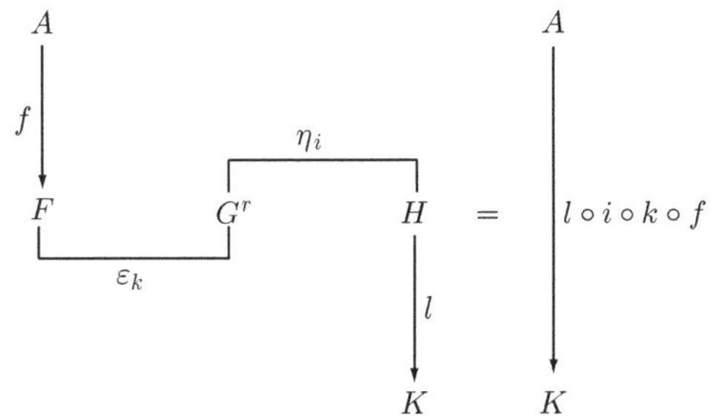

For more details see [24], where it is also shown that the free compact 2-category thus constructed has the expected universal property.

Buszkowski [6] had shown that the original Switching Lemma for free pregroups is essentially a cut-elimination theorem for compact bilinear logic. Our categorical version shows that the composition of 2-cells in free compact 2-categories (with one 0-cell) can be performed without mentioning vertical composition, except that of basic arrows, from which other simple arrows are easily constructed. I believe that this is the true rôle of cut-elimination also in other categorical contexts. (The restriction that there is only one 0-cell was made for expository purposes and may of course be removed.)

8.4 In Search of a Compact Feynman Category

From now on let us assume that we are in a compact 2-category with one 0-cell, also known as a compact strictly monoidal category. Let $U = F^r$ be the right adjoint of F, hence $F = U^\ell$ the left adjoint of U. The triangular Eq. (8.1) may be represented geometrically as an equation between diagrams:

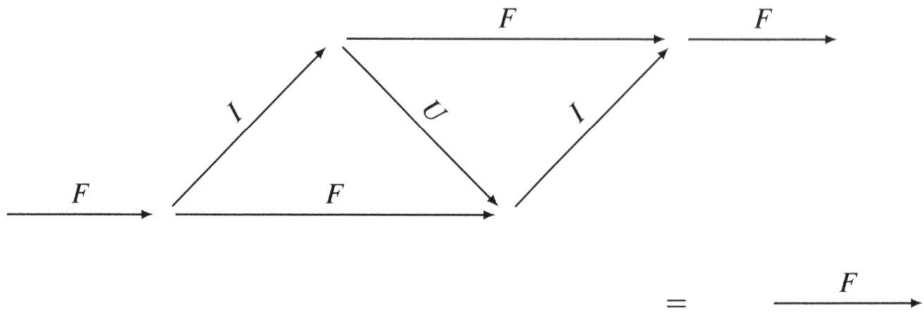

It is tempting to give a physical interpretation to this in quantum electro-dynamics (QED):

- $F = e^- =$ electron,
- $U = e^+ =$ positron,
- $I = \gamma =$ photon.

It does not seem profitable to distinguish between right and left adjoints here, so we will assume from now on that $G^\ell = G^r$ for any 1-cell G.

The equation $G^\ell = G^r$ will hold automatically if the 2-category is *symmetric*, that is, if composition of 1-cells is commutative, as we shall assume from now on. This requires, in particular, that any two composable 1-cells must have the same source and target, as is ensured by our assumption that there is only one 0-cell. Our compact 2-category thus becomes what Kelly and Laplaza [15] call a *compact closed category* (closure being a consequence of compactness, as defined here) and what Barr calls a *compact $*$-autonomous* category (the star being the common symbol for the superscripts ℓ and r), although here the tensor product is assumed to be associative on the nose. The second triangular Eq. (8.2) is now a consequence of the first (8.1).

Diagrams such as (4.1) were introduced by Feynman as an aid to calculating probabilities. For example, the probability of what happens at any vertex of (4.1) is given by the (idealized) charge of the electron.

The equal sign in (4.1) must be taken with a grain of salt. What actually happens is that the electron goes from point x to point y in space-time in many different ways. Each of the ways has associated with it a certain complex number, its *amplitude*, depending on the energy-momentum 4-vector. These amplitudes must be added up and the square of the absolute value of their sum is interpreted as the probability for an electron to go from x to y. Hence the equal sign really holds between equivalence classes of alternative motions.

The easiest way to ensure the equality in (4.1) is to let the arrow stand for a partial order. Then we would also predict

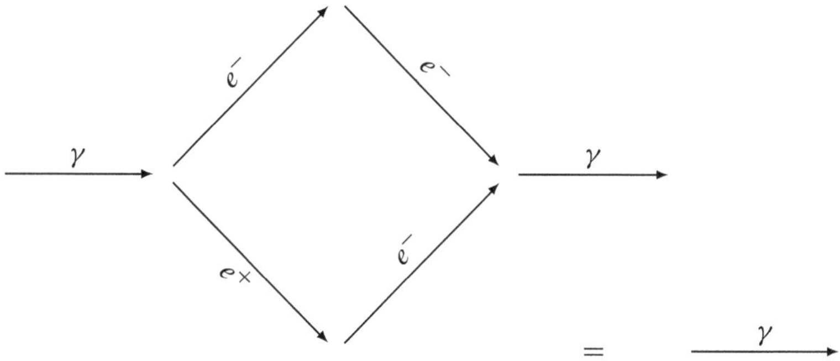

in line with what physicists call "vacuum polarization". Disappointingly, this will imply that our compact 2-category degenerates into a partially ordered group, in which adjoints are just inverses.

8.5 A Pogroup for QED

I had hoped to describe an interesting freely generated compact monoidal category for application to quantum electro-dynamics. But, after all the i-s were dotted and all the t-s were crossed, I realized that all I had was a partially ordered group. I will now describe a provisional version of this pogroup, provisional because I have not taken into account the spin of the electron and the polarization of light. These had also been downplayed, if not ignored, by Feynman [9], whose beautiful exposition I am relying on.

We take as 1-cells all finite multisets (to be explained presently) of *fundamental* 1-cells. These are pairs (x, a), where $x = (x_0, x_1, x_2, x_3) \in \mathbb{R}^4$ is a point in space-time and $a = (a_0, a_1, a_2, a_3) \in \{-1, 0, +1\}^4$ represents a fundamental particle. For expository purposes we will write $(x, a) = x^a$, thus suggesting that $x^0 = 1$ does not depend on x.

A *multiset* is a string, liable to arbitrary permutations of its elements. The composition of 1-cells, usually called "tensor product" in monoidal categories, is obtained by combining two multisets into one. The empty multiset is the unity element 1. (Conceivably, these multisets should be replaced by sets, but this should only be done after the spin of the electron has been taken into account.)

2-cells, that is arrows between multisets, are "made up" from the following:

- motions: $x^a \to y^a$,
- contractions: $x^a y^b \to x^{a+b}$,
- expansions: $x^{a+b} \to x^a x^b$.

The last two are subject to the condition that

$$a_i = 0 \quad \text{or} \quad b_i = 0 \quad \text{or} \quad a_i + b_i = 0$$

for all $i \in \{0, 1, 2, 3\}$. We recall that the arrow represents a *partial* order (not a pre-order) so that $\leftrightarrow$ means equality.

We will leave the complete interpretation of the quadruple a until later. For the moment let us only mention that

- $e = (1, -1, -1, -1)$ represents the electron e^-,
- $-e = (-1, 1, 1, 1)$ represents the positron e^+,
- $0 = (0, 0, 0, 0)$ represents the photon γ.

Hence the contraction

$$x^e x^0 \to x^e$$

describes an electron at x absorbing a photon. If the arrow is reversed, the expansion describes emission of a photon. The contraction

$$x^e x^{-e} \to x^0 = 1$$

describes the annihilation of an electron-positron pair. If the arrow is reversed, the expansion describes pair creation. This should suffice for *QED*.

What is meant by saying that 2-cells are "made up" from motions, contractions and expansions? Without giving a tedious formal definition, let me illustrate this by a calculation:

$$u^e \to v^e \to v^0 v^e \to w^0 v^e \to w^e w^{-e} v^e \to w^e x^{-e} x^e \to w^e x^0 \to y^e y^0 \to y^e \to z^e$$

which is furthermore illustrated by the Feynman diagram, the lefthand side of (4.1):

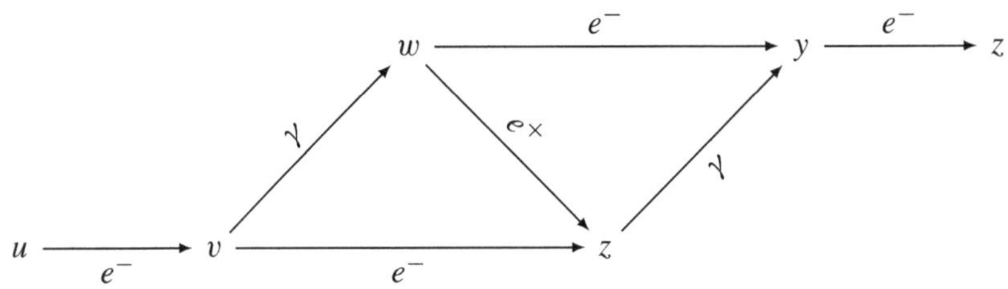

As another illustration, consider two motions $x^a \to y^a$ and $u^a \to v^a$. They generate a 2-cell $x^a u^a \to y^a v^a$. But, since $y^a v^a = v^a y^a$, this 2-cell is also generated by the motions $x^a \to v^a$ and $u^a \to y^a$. According to our interpretation, this should imply that the amplitudes of the two processes are to be added before calculating the probability of the transition $x^a u^a \to y^a v^a$. This is indeed the case, as Feynman pointed out.

Although I had expected to find an interesting compact closed category, all we ended up with was a partially ordered group with $x^0 = y^0 = 1$ and inverse $(x^a)^{-1} = x^{-a}$. It is not the free pogroup generated by the x^a, since we have additional equalities $x^a x^b = x^{a+b}$, when one of a_i, b_i or $a_i + b_i$ is 0 for each i.

We might have obtained a more interesting Feynman 2-category (with one 0-cell), had we not assumed that the 2-cells describe a partial order, but that all "ways" of going from one point in space-time to another count as 2-cells. However, the resulting strictly monoidal category would not be compact and would not be relevant for the present discussion. I have not investigated what happens if we assume that the symmetry is not exact or if it is replaced by braiding, as in [13].

Already the ancient philosopher Parmenides believed that the flow of time is an illusion, not shared by the gods. It is therefore of some interest to show formally that $x^a \to y^a$ implies (and is implied by) $y^{-a} \to x^{-a}$, meaning that any particle may be viewed as the corresponding anti-particle moving backwards in time. Assuming $x^a \to y^a$, we calculate

$$y^{-a} \to y^{-a}y^0 \to y^{-a}x^0 \to y^{-a}x^a x^{-a} \to y^{-a}y^a x^{-a} \to y^0 x^{-a} \to x^0 x^{-a} \to x^{-a}.$$

To avoid an overabundance of 2-cells, we will not allow $x^a \to y^b$ unless $a = b$ and we will postulate

$$x^a = y^a \text{ if and only if } x = y \text{ or } a = 0. \tag{8.3}$$

The above treatment ignores the Pauli exclusion principle, which asserts that two identical electrons (with the same spin direction) cannot occupy the same position in space-time. We could have overcome this objection had we replaced "multisets" by "sets" in our definition of 1-cells. But this would not do either, since two identical photons or two electrons with opposite spin can be at the same place.

8.6 From QED to the Standard Model

Had we only been interested in QED and weak interactions, we could have taken a to be a pair (a_0, a_1) with a_1 representing the electric charge, if the charge of the electron is taken as -1. The minus sign here results from an arbitrary choice by Benjamin Franklin as to what constitutes positive versus negative charge. We have chosen $a = (a_0, a_1, a_2, a_3)$ to account also for strong interactions, with

$$a_1 + a_2 + a_3 = 3 \times \text{ electric charge.}$$

Other "colourless" particles in which $a_1 = a_2 = a_3$ are the following:

- neutrino $\nu = (1, 0, 0, 0)$,
- anti-neutrino $\overline{\nu} = (-1, 0, 0, 0)$,

and the weak vector bosons

- $W^+ = (0, 1, 1, 1)$,
- $W^- = (0, -1, -1, -1)$,

and

- $Z^0 = (0, 0, 0, 0)$

unfortunately sharing the same quadruple with the photon.

To account for the strong forces, one requires some new fermions, called "quarks", and some new bosons called "gluons", for which a_1, a_2 and a_3 are no longer equal. Thus we have the

- (red) up-quark $u = (1, 0, 1, 1)$

and the

- (red) down-quark $d = (1, -1, 0, 0)$

with two "colour" variants, depending on the position of the 0 and the -1 respectively, as well as the corresponding anti-particles $-u$ and $-d$. There are six gluons to allow for changes of colour, e.g. $(0, 1, -1, 0)$ permits

$$(1, 0, 1, 1) + (0, 1, -1, 0) \to (1, 1, 0, 1),$$

combining with a red up-quark to yield, say, a blue one. Allegedly, there are also two so-called "diagonal" gluons, which have not been described here. Altogether, our quadruples represent 25 known fundamental particles and anti-particles: 4 leptons, 12 quarks, 3 weak vector bosons (not distinguishing Z^0 from γ) and 6 gluons.

Let us illustrate this with just one Feynman diagram:

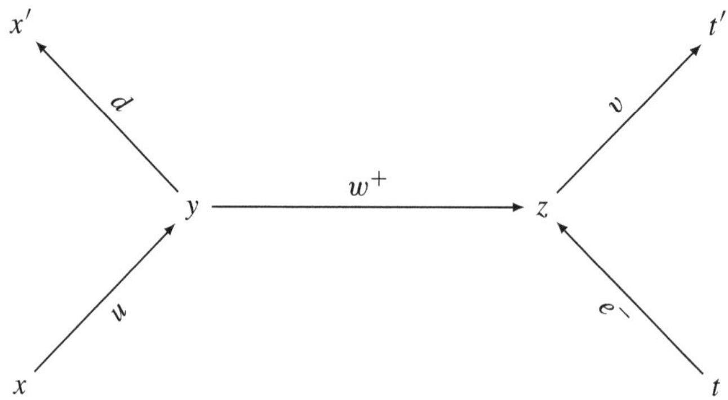

Showing how an up-quark decomposes into a down-quark of the same colour and a positive weak vector boson, which then combines with an electron to form a neutrino. We calculate

$$x^{(1,0,1,1)} t^{(1,-1,-1,-1)} \to y^{(1,0,1,1)} t^{(1,-1,-1,-1)} \to x'^{(1,-1,0,0)} z^{(0,1,1,1)} t^{(1,-1,-1,-1)} \\ \to x'^{(1,-1,0,0)} t'^{(1,0,0,0)}.$$

We have used a_0 to represent the *fermion number*:

- $a_0 = 1$ for fermions,
- $a_0 = -1$ for anti-fermions,
- $a_0 = 0$ for bosons.

Actually, only the number of leptons and the number of quarks are preserved separately in known physical interactions. Having adopted the fermion number instead, we allow in principle that quarks can be transformed into leptons with the help of some not yet discovered bosons, e.g.

$$(1, -1, 0, 0) + (0, 1, 0, 0) \to (1, 0, 0, 0).$$

As Feynman points out, this might predict the instability of the proton, which has not yet been verified experimentally.

Our representation of fundamental fermions was inspired by the more concrete representation proposed by Harari [11] and Shupe [26], but that of bosons departs from theirs. Here is a rather "odd" observation, depending on Benjamin Franklin's arbitrary choice: out of a possible $3^4 = 81$ quadruples with components $-1, 0$ and $+1$, the number of $+1$s and the number of -1s in the quadruples occurring above are both odd or zero. This would still be the case if we admitted six additional bosons, variants of the hypothetical $(0, 1, 0, 0)$ mentioned above, making a total of 31. However, we have not accounted for the diagonal gluons and the conjectured graviton and Higgs particle. If our "odd" observation is taken seriously, there would still be six other potential elementary particles, represented by variants of $(1, -1, 1, 1)$, bringing the total up to 37.

I am indebted to Derek Wise for bringing to my attention a recent article by Bilson-Thompson [3], which also offers an abstract development of the Harari-Shupe model. Rather than invoking a fermion number, he represents a fermion by a braided triple of "helons", namely twists of a ribbon through $\pm 2\pi$ or 0, and he distinguishes fermions from their anti-particles by associating them with braids and anti-braids respectively, thus bringing in the braid group B_3. His ideas are further developed in a joint article with Markopoulou and Smolin [4]. If braiding is not used to distinguish electrons from positrons, could it be invoked to distinguish spin-up from spin-down?

8.7 From 2-Categories to Bicategories

Bicategories were introduced by Jean Bénabou [2]. They are like 2-categories, except that composition of 1-cells, usually called "tensor product", is associated only up to coherent isomorphism. Bicategories with a single 0-cell are better known as monoidal categories. Symmetric monoidal categories, albeit with an additional operation "dagger", play a rôle in the categorical approach to quantum mechanics by Abramsky, Coecke and Selinger [1, 25]. I would like to take a closer look at bicategories, if only to remind people that the usual coherence and other properties need not be postulated, but can be proved if the right definition is adopted. I will follow [18, 22].

A typical bicategory is that of *bimodules*:

- 0-cells = rings $R, S, \cdots$,
- 1-cells = bimodules ${}_RA_S,\ {}_SB_T,\ \cdots$,
- 2-cells = bimodule homomorphisms (= linear mappings).

Composition of 1-cells is the usual tensor product

$$({}_RA_S,\ {}_SB_T)\ \mapsto\ {}_R(A\otimes B)_T.$$

Its many properties can all be deduced from Bourbaki's [5] definition, which prescribes a bilinear mapping $m_{AB} : AB \to A\otimes B$ with the universal property: given any bilinear mapping $f : AB \to C$, there is a unique linear mapping $f^{\S} : A\otimes B \to C$ such that $f^{\S} m_{AB} = f$.

Given elements $a\in A$ and $b\in B$ and abbreviating

$$m_{AB}ab = (a, b),$$

we may write the above equation as

$$f^{\S}(a, b) = fab.$$

From this the usual properties of the tensor product are easily deduced.

For example, if $f : A \to A'$ and $g : B \to B'$, we may define

$$f\otimes g : A\otimes B \to A'\otimes B'$$

by putting $f\otimes g = h^{\S}$, where $h : AB \to A'\otimes B'$ is given by

$$hab = (fa, gb),$$

hence

$$(f\otimes g)(a, b) = (fa, gb).$$

To show that $\otimes$ is a bifunctor, we require e.g. that

$$(f'\otimes g')(f\otimes g) = f'f\otimes g'g,$$

when $f' : A' \to A''$ and $g' : B' \to B''$. We prove this by calculating

$$\begin{aligned}(f'\otimes g')(f\otimes g)(a, b) &= (f'\otimes g')(fa, gb)\\ &= (f'fa,\ g'gb)\\ &= (f'f\otimes g'g)(a, b).\end{aligned}$$

The associative arrow

$$\alpha_{ABC} : (A \otimes B) \otimes C \to A \otimes (B \otimes C)$$

is defined by the equation

$$\alpha_{ABC}((a, b), c) = (a, (b, c)).$$

It is easily checked that α is a natural transformation. Similarly one defines

$$\alpha^{-1}_{ABC} : A \otimes (B \otimes C) \to (A \otimes B) \otimes C$$

and checks that $\alpha\alpha^{-1} = 1$ and $\alpha^{-1}\alpha = 1$ (omitting subscripts). Mac Lane's famous *pentagonal coherence* condition asserts the commutativity of the following diagram:

$$\begin{array}{ccccc}
((A \otimes B) \otimes C) \otimes D & \xrightarrow{\alpha} & (A \otimes B) \otimes (C \otimes D) & \xrightarrow{\alpha} & A \otimes (B \otimes (C \otimes D)) \\
\downarrow{\scriptstyle \alpha \otimes D} & & & & \uparrow{\scriptstyle A \otimes \alpha} \\
(A \otimes (B \otimes C)) \otimes D & & \xrightarrow[\alpha]{\qquad\qquad\qquad} & & A \otimes ((B \otimes C) \otimes D)
\end{array}$$

This is proved by pointing out that there is a unique arrow

$$f : (((A \otimes B) \otimes C) \otimes D) \to A \otimes (B \otimes (C \otimes D))$$

such that

$$f(((a, b), c), d) = (a, (b, (c, d))).$$

The identity 1-cell $I_S : S \to S$ for the tensor product is of course the bimodule ${}_S S_S$ obtained from the ring S.

Passing from the concrete bicategory of bimodules to arbitrary bicategories, we need to treat multilinear maps abstractly. This was done with the help of multicategories (see e.g. [18]), called "operads" by some people.

A *multicategory*, as viewed most recently [22], is essentially a 2-category, except that 1-cells are freely generated from *basic* 1-cells, and 2-cells are restricted to intuitionistic Gentzen sequents (see e.g. Kleene [16]), which are composed by *cuts*:

- Given basic 1-cells $R \overset{A}{\leftarrow} S,\ S \overset{B}{\leftarrow} T,\ T \overset{C}{\leftarrow} U, \ldots$, we form (compound) 1-cells $R \overset{AB}{\leftarrow} T,\ R \overset{ABC}{\leftarrow} U$, etc.
- We must also admit the empty 1-cells $R \overset{\emptyset_R}{\leftarrow} R,\ S \overset{\emptyset_S}{\leftarrow} S$, etc.
- The only 2-cells we retain are of the form $\Gamma \to G$, where Γ is any 1-cell and G is a basic one. A *cut* has the form

$$\frac{f : \Lambda \to A \; g : \Gamma A \Delta \to B}{g\langle f\rangle : \Gamma \Lambda \Delta \to B}$$

where

$$g\langle f\rangle = g \circ \Gamma f \Delta.$$

The equations holding in a multicategory are all inherited from those of the encompassing 2-category, even though we have discarded all 2-cells except those whose targets are basic 1-cells.

A tensor product of 1-cells can be introduced by a 2-cell

$$\mathbf{m}_{AB} : AB \to A \otimes B$$

together with a rule

$$\frac{f : \Gamma AB\Delta \to C}{f^{\S} : \Gamma(A \otimes B)\Delta \to C},$$

which associates to any $f : \Gamma AB\Delta \to C$ a *unique* $f^{\S} : \Gamma(A \otimes B)\Delta \to C$ such that

$$f^{\S}\langle \mathbf{m}_{AB}\rangle = f.$$

The uniqueness may also be expressed equationally by saying that, for any $g : \Gamma(A \otimes B)\Delta \to C$,

$$(g\langle \mathbf{m}_{AB}\rangle)^{\S} = g.$$

With any 0-cell R there is associated an identity 1-cell I_R, introduced by the 2-cell

$$\mathbf{i}_R : \emptyset_R \to I_R$$

and a rule

$$\frac{f : \Gamma\Delta \to C}{f^{\#} : \Gamma I_R \Delta \to C},$$

which associates to any $f : \Gamma\Delta \to C$ a unique $f^{\#} : \Gamma I_R \Delta \to C$ such that

$$f^{\#}\langle \mathbf{i}_R\rangle = f.$$

The uniqueness amounts to the equation

$$(g\langle \mathbf{i}_R\rangle)^{\#} = g$$

for any $g : \Gamma I_R \Delta \to C$.

The arguments we employed for bimodules carry over to any multicategory, provided we replace elements $a \in A$ by *indeterminate* arrows $a : \Lambda \to A$, or better by *variables* of type A. This can be done formally by invoking the *internal language* of a multicategory, see [18] for details of this approach.

8.8 Other Operations in Bilinear Logic

It may be of interest to point out that other operations occurring in bilinear (= non-commutative linear) logic can be introduced in the same way (see e.g. [19]). For example, the operation "over" whose dual operation "under" is represented by a lollipop by Girard (see e.g. Troelstra [27]), is introduced as follows:

$$\mathbf{e}_{DA} : (D/A)A \to D,$$

the rule

$$\frac{f : \Gamma A \to D}{f^* : \Gamma \to D/A},$$

which associates to every 1-cell $f : \Gamma A \to D$ a unique 1-cell $f^* : \Gamma \to D/A$ such that

$$\mathbf{e}_{DA}\langle f^* \rangle = f.$$

The uniqueness can be expressed by the equation

$$(\mathbf{e}_{DA}\langle g \rangle)^* = g$$

for any $g : \Gamma \to D/A$.

The logical conjunction (= categorical direct product) can be introduced by two 1-cells

$$\mathbf{p}_{AB} : A \wedge B \to A, \qquad \mathbf{q}_{AB} : A \wedge B \to B$$

and the rule

$$\frac{f : \Lambda \to A \; g : \Lambda \to B}{\langle f, g \rangle : \Lambda \to A \wedge B},$$

which associates to any pair of 1-cells $f : \Lambda \to A$ and $g : \Lambda \to B$ a unique 1-cell $\langle f, g \rangle : \Lambda \to A \wedge B$ such that

$$\mathbf{p}_{AB}\langle f, g \rangle = f, \qquad \mathbf{q}_{AB}\langle f, g \rangle = g,$$

The uniqueness can be expressed by the equation

$$\langle \mathbf{p}_{AB}, \mathbf{q}_{AB} \rangle = 1_{A \wedge B}.$$

For a discussion of these and other operations see e.g. [19]. It was there assumed that there is only one 0-cell, but the arguments carry over to the general case.

Adjoints of 1-cells can be defined in any bicategory. Thus $(F, U, \eta, \varepsilon)$ is an *adjoint pair* if

$$F : R \to S,\ U : S \to R,\ \eta : I_S \to F \otimes U,\ \varepsilon : U \otimes F \to I_R$$

such that

$$U \xrightarrow{\sim} U \otimes I_S \xrightarrow{1_U \otimes \eta} U \otimes (F \otimes U) \xrightarrow{\alpha^{-1}} (U \otimes F) \otimes U \xrightarrow{\varepsilon} I_R \otimes U \xrightarrow{\sim} U = U \xrightarrow{1_U} U$$

and a similar equation holds for F.

A bicategory is said to be *compact* if every 1-cell has both a left and a right adjoint. To exhibit a concrete example of a compact bicategory, I find myself turning to the exercises in [17]. It is shown there that a right module A_S has a left adjoint ${}_SA^{\ell}$ if and only if it is finitely generated and projective. The left module ${}_SA^{\ell}$ is again finitely generated and may be identified with S_S/A_S. If S is a division ring, all one-sided S-modules are automatically projective. Similar considerations apply to left modules ${}_RA$ and their right adjoints $A^r_R = {}_RA \backslash {}_RR$. We infer that the following concrete bicategory is compact:

- 0-cells = division rings,
- 1-cells = bimodules finitely generated on both sides,
- 2-cells = bimodule homomorphisms.

A compact monoidal category of possible interest in Physics is the category of all $\mathbf{H} - \mathbf{H}$-bimodules finitely generated on both sides, where $\mathbf{H}$ is the division ring of quaternions. A special object of this monoidal category is the ring of all 4×4 real matrices, which is known to be isomorphic to $\mathbf{H} \otimes \mathbf{H}^{\mathrm{op}}$.

8.9 Postscript

After writing this paper, I became aware of the article by Joyal and Street [19]. They carried out something resembling what I have been trying to do in Sect. 8.3, but for monoidal categories which are not strictly monoidal. They also constructed free monoidal, symmetric monoidal and braided monoidal categories. They had in mind an (as yet unpublished) application to Feynman diagrams. Their work involves many technical details and definitions, which I admit not having had the patience to absorb.

I wish to thank Anne Preller for her careful reading of the manuscript.

References

1. Abramsky, S., Coecke, B.: A categorical semantics of quantum protocols, In: Proceedings of 19th IEEE conference on Logic in Computer Science, pp. 415–425. IEEE Press Washington, DC (2004)
2. Bénabou, J.: Introduction to bicategories. Lect. Notes Math. **47**, 1–77 (1967)
3. Bilson-Thompson, S.O.: A topological model of composite preons, Preprint 2006. arXiv:hep-ph/0503213
4. Bilson-Thompson, S.O., Markopoulou, F., Smolin, L.: Quantum Gravity and the Standard Model, Preprint 2006. arXiv:hep-th/0603022
5. Bourbaki, N.: Algèbre multilinéaire**,** Hermann, Paris (1948)
6. Buszkowski, W.: Cut elimination for the Lambek calculus of adjoints. In: Abrusci, V.M. et al. (eds.) Papers in formal linguistics and logic. Bulzoni, Rome (2002)
7. Doşen, K.: Cut elimination in categories, Kluwer, Dordrecht (1999)
8. Doşen, K., Petrić, Z.: Proof-net categories, Project 1650. Math. Institute, Belgrade (2005)
9. Feynman, R.P.: QED: The Strange Theory of Light and Matter. Princeton University Press, Princeton (1985)
10. Glass, A.: Partially ordered groups, World Scientific, Singapore (1998)
11. Harari, H.: Phys. Lett. **86B**, 83–86 (1979)
12. Harris, Z.: A cyclic cancellation automaton for sentence well-formedness. Int. Comput. Centre Bull. **5**, 69–94 (1966)
13. Joyal, A., Street, R.: Braided monoidal categories, Macquarrie University Report No. 800081 (1986)
14. Joyal, A., Street, R.: The geometry of tensor calculus I. Adv. Math. **88**, 55–112 (1991)
15. Kelly, G.M., Laplaza, M.L.: Coherence for compact closed categories. J. Pure Appl. Algebra **19**, 193–213 (1980)
16. Kleene, S.C.: Introduction to metamathematics, Van Nostrand (1952)
17. Lambek, J.: Lectures on Rings and Modules, pp. 1976. Waltham, Mass (1986)
18. Lambek, J.: Multicategories revisited. Contem. Math. **92**, 217–239 (1989)
19. Lambek, J.: Logic without structural rules. (Another look at cut elimination). In: Doşen, K. et al, (eds.) Substructural Logic, Studies in Logic and Computation 2, pp. 129–206. Oxford Science Publications, Oxford (1993)
20. Lambek, J.: Type grammar revisited. In: Lecomte, A. et al. (eds.) Logical Aspects of Computational Linguistics, Springer LNAI **1582**, 1–27 (1999)
21. Lambek, J.: Four-vector representation of fundamental particles. Int. J. Theor. Phys. **39** (2253)–2258 (2000)
22. Lambek, J.: Bicategories in algebra and linguistics. In: Ehrhard, T. et al. (eds.) Linear Logic in Computer Science. LondonMathematical Society Lecture Note Series 316, pp. 325–245. Cambridge University Press, Cambridge (2004)
23. Mac Lane S.: Categories for the working mathematician, Springer, New York (1971)
24. Preller, A., Lambek, J.: Free compact 2-categories. Math. Struct. Comput. Sci. **17**, 309–340 (2007)
25. Selinger, P.: Dagger compact closed categories and completely positive maps. Electron. Notes Theor. Comput. Sci. **170**, 139–163 (2007)
26. Shupe, M.A.: Phys. Lett. **86B**, 87–92 (1979)
27. Troelstra, A.S.: Lectures on linear logic, CSLI Lecture Notes 29, Stanford (1992)

Part IV
Informatic Geometry

Chapter 9
Domain Theory and Measurement

K. Martin

Abstract Lecture notes on domain theory and measurement, driven by applications to physics, computer science and information theory, with a hint of provocation.

9.1 Introduction

9.1.1 History

I loved everything about being a graduate student except going to class, doing homework, taking exams, fulfilling requirements associated with earning a degree and being severely underpaid. What I especially loved was being able to do mathematics. One semester I signed up for a course on domain theory. I went to the first lecture and heard about dcpo's and the fixed point theorem: every Scott continuous map $f : D \to D$ on a dcpo D with a least element $\perp$ has a least fixed point given by

$$\text{fix}(f) := \bigsqcup_{n \geq 0} f^n(\perp)$$

I thought it was neat, so I skipped the rest of my classes that day and immediately went home to try it out and see how it worked. I wrote down an example of a function on the interval domain $D = \mathbf{I}\mathbb{R}$, this one: for a continuous $f : \mathbb{R} \to \mathbb{R}$ on the real line, define

$$\text{split}_f : C(f) \to C(f)$$
$$\text{split}_f[a, b] = \begin{cases} \text{left}[a, b] & \text{if left}[a, b] \in C(f); \\ \text{right}[a, b] & \text{otherwise,} \end{cases}$$

K. Martin (✉)
Naval Research Laboratory, Washington, DC, USA
e-mail: keye.martin@nrl.navy.mil

Martin, K.: *Domain Theory and Measurement*. Lect. Notes Phys. **813**, 491–591 (2011)
DOI 10.1007/978-3-642-12821-9_9

where $C(f)$ is the subset of $\mathbf{IR}$ where f changes sign,

$$C(f) = \{[a,b] : f(a) \cdot f(b) \leq 0\},$$

and left$[a,b] = [a,(a+b)/2]$ and right$[a,b] = [(a+b)/2,b]$. If we begin from any interval $[a,b] \in C(f)$ on which f changes sign, then

$$\bigsqcup_{n \geq 0} \text{split}_f^n[a,b]$$

is a fixed point of split_f, just like the fixed point theorem says it should be. Because $\text{fix}(\text{split}_f) = \{x \in \mathbf{IR} : \text{split}_f(x) = x\} = \{[r] : f(r) = 0\}$, iterating split_f is a scheme for calculating a solution of the equation $f(x) = 0$.

The problem was: split_f was *not* Scott continuous, so the fixed point theorem could not be used to explain its behavior on $\mathbf{IR}$. And there was an especially easy way to see it: plenty of functions f have more than one zero on a given interval x – but if split_f is Scott continuous, its least fixed point on $\uparrow x$ is unique (being maximal), implying that f has only one zero on x. So then the question became: why did this function behave as though it were continuous? I set about to find an answer. In the process, I became so interested in domain theory that I never went back to class again.

What I learned was that there was a reason that this function behaved as though it were continuous: it wasn't, but its *measure* was. Its measure was an important thing: the length function $\mu[a,b] = b-a$ was different from other functions. I later learned that all recursive functions could be modeled in this way and that the *measurement* μ was intertwined with the structure of the domain itself. It provided a measure of information content and one could use this idea to measure the 'rate' at which a process on a domain manipulated information and to do more things than should be mentioned in an introduction.

I was never really able to finish telling the story in my doctoral thesis the way I thought it should have been told. Nevertheless, at two hundred pages, I decided to stop typing and go to sunny England, where the theory advanced, with the same structure found in computation (domains and measurements) also being found in quantum mechanics and general relativity. Later, it was realized that the same structure was also present in information theory: there was a domain of binary channels, for instance, with capacity as a measurement. In all these cases, there are neat applications and new perspectives offered on ideas we previously misled ourselves into believing we understood. The interaction between these areas is what makes the study of domains and measurements very exciting.

That brings us to now. This is a "tutorial" on domain theory and measurement. It is about what we believe we know today. It's also about what we believe we don't know today. There are also new results and ideas here never published before.

9.1.2 Overview

In Sect. 9.2, the basic elements of domains and measurements are introduced where our goal is to explain partiality and content as concepts and to explain how one models them with domains and measurements in practice so that new problems can be solved. In essence, the goal is to teach "the method" of finding domains and measurements in nature. A dozen or so basic examples are given. In Sect. 9.3, we give an important example of what one does with domains and measurements: applies fixed point theorems. Applications include numerical methods and fractals. In Sect. 9.4, we give more advanced examples of partiality and content: the domain of real analytic mappings, the domain of finite probability distributions, the domain of quantum mixed states and from general relativity, the domain of spacetime intervals. Applications of these domains are to the computation of real analytic mappings, the maximum entropy state in statistical mechanics, classical and quantum communication and to the reconstruction of spacetime from a countable set, including its geometry.

In Sect. 9.5, we discuss the informatic derivative: when an operator on a domain iterates to a fixed point, its informatic derivative measures the rate at which the iterates converge. Applications are to numerical analysis, to the computation of the Holevo capacity in quantum information theory and to the complexity of list algorithms. The informatic derivative applies to both continuous and *discrete* data. In Sect. 9.6, we discuss additional models of "process" that have proven themselves useful in the measurement formalism: the renee equation, trajectories, vectors. Applications include showing that each order on a domain gives rise to a natural notion of computability, such as the primitive and partial recursive functions; the analysis of algorithms as trajectories on domains, such as Grover's algorithm for quantum searching, whose complexity is the amount of time it takes a trajectory to reach its maximum in the information order; an analysis of how noise affects communication with qubits; the derivation of lower bounds on the complexity of algorithms such as sorting and searching and the fixed point theorem: entropy is the least fixed point of the copying operator that is above algorithmic complexity.

In Sect. 9.7, we give a brief overview of where things currently stand in the study of domains and measurements, and try to persuade domain theorists in search of a decent job to send us an email.

9.1.3 To the Student

A *student* is any person young enough at heart to be open to new ideas. This paper is written for students. We have tried to strike a balance between *philosophy, mathematics* and *applicability*. Philosophy: what is the big picture? Mathematics: how do we learn about the big picture? Applicability: what can the big picture teach us about the world we live in? Philosophy is good because *thinking* is good. Mathematics is good because knowing *what* you are thinking about is good. Applicability is good because knowing *why* you are thinking what you are thinking about is good. It is pretty rare that a set of ideas starts off with all three of these in equal measure. Sometimes there is only philosophy, sometimes only math, sometimes only a question. But as a set of ideas evolves, one hopes to see the appearance of all three.

9.2 The Basic Elements

Most newcomers to domain theory stop reading when they see domains presented as a seemingly endless list of axioms satisfied by partial orders. If this is your first time reading about domain theory, perhaps you should consider a different approach. Try first reading Sect. 9.2.1 to understand the ideas intuitively. Then go to Sect. 9.2.2, but ignore the technical definitions and just look at the dozen or so examples given instead. After those examples, have a look at Sect. 9.4, where there are more involved examples. Then ask yourself a question: given the intuitions on partiality and information content combined with the numerous instances of the idea that you have seen, how would *you* formally capture those ideas?

If you find a formal mathematical definition of domain and measurement that captures all of the examples, compare it to the formalizations given in Sects. 9.2.2 and 9.2.3. If your formalization differs, it might be time to stop reading these notes and to pursue your own direction. If it is the same, then you will understand the basic definitions of domain theory and measurement in a way few people do. And if you are unable to come up with a formalization that captures the basic examples, then you will better appreciate definitions like "continuous dcpo" and "measurement"—you will see them for what they are: a significant step toward a mathematical definition of "information".

Major references: [1, 15].

9.2.1 Intuition

A *domain* $(D, \sqsubseteq)$ is a set of objects D together with a partial order $\sqsubseteq$ that has certain intrinsic notions of completeness and approximation defined by the order. The order $\sqsubseteq$ is thought of as an *information order*. Intuitively, $x \sqsubseteq y$ means "x contains information about y" or that "x carries information about y." We might also say y is at least as informative as x—though this is really just mathematical uptightness that obscures the essence of the idea: when talking to one's friends, people always just say that $x \sqsubseteq y$ means y is *more informative* than x. Elements that compare in the information order are *comparable* and the thing to remember about comparable elements is that *one of them carries information about the other.*

The *completeness* in a domain refers to the fact that certain results generated by processes have "limits". For instance, if a process generates a sequence (x_n) of elements that *increase* with respect to the information order, $x_n \sqsubseteq x_{n+1}$ for all n, then it should 'go somewhere' i.e.

$$x_1 \sqsubseteq x_2 \sqsubseteq \ldots \Longrightarrow \bigsqcup_{n \in \mathbb{N}} x_n \in D$$

The element $\bigsqcup_{n \geq \mathbb{N}} x_n$ is not only above each x_n in the information order, it is the "best" such object. Intuitively, if the process generating (x_n) is an algorithm repeatedly producing iterates x_n, then $\bigsqcup_{n \geq \mathbb{N}} x_n$ is the *final answer.*

The notion of *approximation* is a special case of the information order. If x approximates y, we write $x \ll y$. What it means intuitively is that x carries *essential* information about y. But what does "essential" mean? One view of essential is that any process that produces a sequence (x_n) of values with $\bigsqcup x_n = y$ must satisfy $x \sqsubseteq x_n$ for all but a finite number of the x_n. That is, we cannot compute y without first computing in *finite time* an object that x carries information about. Thus, x can also be thought of as a *finite approximation* of y. Put yet another way, $x \ll y$ means that all informatic paths to y must pass through x.

An *ideal* (or total) object x in a domain D is one that we can only get to using a process that constructs a sequence of finite approximations. For example, a *maximal element* $x \in D$ is an object that cannot be improved upon i.e.

$$(\forall y \in D)\, x \sqsubseteq y \Rightarrow x = y.$$

Each maximal element is an example of an ideal element. Any object that is not ideal (or total) is called *partial.* Let us give several intuitive examples of ideal and partial objects.

A compact interval $[a, b]$ of the real line provides a *partial* description of a real number; a one point interval $[x, x]$ is total. The uniform probability distribution $\perp = (1/n, \ldots, 1/n)$ provides incomplete information on the expected outcome of an experiment, while the finite probability distribution $(1, 0, \ldots, 0)$ predicts the outcome with certainty. The polynomial $1 + x$ is a finite approximation of the analytic mapping e^x. A pure state $|\psi\rangle\langle\psi|$ in quantum mechanics is *total*; a mixed state like $\perp = I/n$ is partial. An infinite set of natural numbers is total while a finite subset of it provides a finite approximation.

A *measurement* $\mu : D \to [0, \infty)$ is a function on a domain D that to each informative object $x \in D$ assigns a number μx that measures the *amount of partiality* in x. The amount of partiality, or *uncertainty*, in an object is also called its *information content.* For instance, we would expect uncertainty to decrease as we move up in the information order,

$$x \sqsubseteq y \Rightarrow \mu x \geq \mu y.$$

If a process calculates $x = \bigsqcup x_n$, we would expect

$$\mu\left(\bigsqcup_{n \in \mathbb{N}} x_n\right) = \lim_{n \to \infty} \mu x_n.$$

If x and y are comparable and $\mu x = \mu y$, then this means that one carries information about the other and that they have the same information content, so we would expect $x = y$. In particular, if $\mu x = 0$, so that x is an object with no uncertainty, then we would expect that x cannot be improved upon. That is, we would expect x to be maximal in the information order.

9.2.2 Domains

In this section, we give several basic examples of domains, including the formal definition of a continuous dcpo. At no point in this section will we define "domain," though we will quite frequently make statements like "such and such is an example of a domain." There is a good reason for our vagueness, but at this point in time, we intend to remain vague about it.

The intrinsic notion of completeness in a domain is at least partially captured by the fact that it forms a dcpo:

Definition 1 Let $(P, \sqsubseteq)$ be a partially ordered set or *poset*. A nonempty subset $S \subseteq P$ is *directed* if $(\forall x, y \in S)(\exists z \in S)\, x, y \sqsubseteq z$. The *supremum* $\bigsqcup S$ of $S \subseteq P$ is the least of its upper bounds when it exists. A *dcpo* is a poset in which every directed set has a supremum.

One way to formalize the intrinsic notion of approximation possessed by a domain is *continuity*:

Definition 2 Let $(D, \sqsubseteq)$ be a dcpo. For elements $x, y \in D$, we write $x \ll y$ iff for every directed subset S with $y \sqsubseteq \bigsqcup S$, we have $x \sqsubseteq s$, for some $s \in S$. We set

- $\twoheaddownarrow x := \{y \in D : y \ll x\}$ and $\twoheaduparrow x := \{y \in D : x \ll y\}$
- $\downarrow x := \{y \in D : y \sqsubseteq x\}$ and $\uparrow x := \{y \in D : x \sqsubseteq y\}$

A set $B \subseteq D$ is a *basis* when $B \cap \twoheaddownarrow x$ is directed with supremum x for each $x \in D$. A dcpo is *continuous* when it has a basis and *ω-continuous* when it has a countable basis.

Remark Any continuous dcpo is an example of a *domain*.

Example 1 The collection of compact intervals of the real line

$$\mathbf{I}\mathbb{R} = \{[a, b] : a, b \in \mathbb{R} \;\&\; a \leq b\}$$

ordered under reverse inclusion

$$[a, b] \sqsubseteq [c, d] \Leftrightarrow [c, d] \subseteq [a, b]$$

is an ω-continuous dcpo:

- For directed $S \subseteq \mathbf{I}\mathbb{R}$, $\bigsqcup S = \bigcap S$,
- $I \ll J \Leftrightarrow J \subseteq \text{int}(I)$, and
- $\{[p, q] : p, q \in \mathbb{Q} \;\&\; p \leq q\}$ is a countable basis for $\mathbf{I}\mathbb{R}$.

The domain $\mathbf{I}\mathbb{R}$ is called the *interval domain*. If we replace $\mathbb{R}$ by $[0, 1]$, then we obtain the *interval domain* $\mathbf{I}[0, 1]$ over the unit interval.

A binary channel has two inputs ("0" and "1") and two outputs ("0" and "1"). An input is sent through the channel to a receiver. Because of noise in the channel, what arrives may not necessarily be what the sender intended. The effect of noise on input data is modelled by a noise matrix u. If data is sent through the channel according to the distribution x, then the output is distributed as $y = x \cdot u$. The noise matrix u is given by

$$u = \begin{pmatrix} a & \bar{a} \\ b & \bar{b} \end{pmatrix}$$

where $a = P(0|0)$ is the probability of receiving 0 when 0 is sent and $b = P(0|1)$ is the probability of receiving 0 when 1 is sent and $\bar{x} := 1 - x$ for $x \in [0, 1]$. Thus, the noise matrix of a binary channel can be represented by a point (a, b) in the unit square $[0, 1]^2$ and all points in the unit square represent the noise matrix of some binary channel.

Example 2 Binary channels. The set of nonnegative noise matrices

$$\mathbb{N} = \left\{ (a, b) = \begin{pmatrix} a & \bar{a} \\ b & \bar{b} \end{pmatrix} : a \geq b \;\&\; a, b \in [0, 1] \right\}$$

is in bijective correspondence with $\mathbf{I}[0, 1]$ via $(a, b) \mapsto [b, a]$. With the order it inherits from $\mathbf{I}[0, 1]$, $\mathbb{N}$ is called the *domain of binary channels*.

Example 3 Let X be a locally compact Hausdorff space. Its *upper space*

$$\mathbf{U}X = \{\emptyset \neq K \subseteq X : K \text{ is compact}\}$$

ordered under reverse inclusion

$$A \sqsubseteq B \Leftrightarrow B \subseteq A$$

is a continuous dcpo:

- For directed $S \subseteq \mathbf{U}X$, $\bigsqcup S = \bigcap S$, and
- $A \ll B \Leftrightarrow B \subseteq \text{int}(A)$.

Example 4 Given a metric space (X, d), the *formal ball model* [6]

$$\mathbf{B}X = X \times [0, \infty)$$

is a poset when ordered via

$$(x, r) \sqsubseteq (y, s) \Leftrightarrow d(x, y) \leq r - s.$$

The approximation relation is characterized by

$$(x, r) \ll (y, s) \Leftrightarrow d(x, y) < r - s.$$

The poset $\mathbf{B}X$ is continuous. However, $\mathbf{B}X$ is a dcpo iff the metric d is complete. In addition, $\mathbf{B}X$ has a countable basis iff X is a separable metric space.

Definition 3 An element x of a poset is *compact* if $x \ll x$. A poset is *algebraic* if its compact elements form a basis; it is *ω-algebraic* if it has a countable basis of compact elements.

Example 5 The powerset of the naturals

$$\mathcal{P}\omega = \{x : x \subseteq \omega\}$$

ordered by inclusion $x \sqsubseteq y \Leftrightarrow x \subseteq y$ is an ω-algebraic dcpo:

- For directed set $S \subseteq \mathcal{P}\omega$, $\bigsqcup S = \bigcup S$,
- $x \ll y \Leftrightarrow x \sqsubseteq y$ & x is finite, and
- $\{x \in \mathcal{P}\omega : x \text{ is finite}\}$ is a countable basis for $\mathcal{P}\omega$.

Example 6 Binary strings. The collection of functions

$$\Sigma^\infty = \{\, s \mid s : \{1, \ldots, n\} \to \{0, 1\}, 0 \le n \le \infty \,\}$$

ordered by extension

$$s \sqsubseteq t \Leftrightarrow |s| \le |t| \;\&\; (\,\forall\, 1 \le i \le |s|\,)\; s(i) = t(i),$$

where $|s|$ is the cardinality of $\mathrm{dom}(s)$, is an ω-algebraic dcpo:

- For directed $S \subseteq \Sigma^\infty$, $\bigsqcup S = \bigcup S$,
- $s \ll t \Leftrightarrow s \sqsubseteq t$ & $|s| < \infty$,
- $\{s \in \Sigma^\infty : |s| < \infty\}$ is a countable basis for Σ^∞,
- The least element $\bot$ is the unique s with $|s| = 0$.

A *partial function* (or partial map) on a set X is a function $f : A \to X$ where $A \subseteq X$. We write $\mathrm{dom}(f) = A$ and denote partial maps as $f : X \rightharpoonup X$. They are equivalent to functions of the form $f : X \to X \cup \{\bot\}$. The next domain is of central importance in recursion theory:

Example 7 The set of partial mappings on the naturals

$$[\mathbb{N} \rightharpoonup \mathbb{N}] = \{\, f \mid f : \mathbb{N} \rightharpoonup \mathbb{N} \text{ is a partial map}\}$$

ordered by extension

$$f \sqsubseteq g \Leftrightarrow \mathrm{dom}(f) \subseteq \mathrm{dom}(g) \;\&\; f = g \text{ on } \mathrm{dom}(f)$$

is an ω-algebraic dcpo:

- For directed set $S \subseteq [\mathbb{N} \rightharpoonup \mathbb{N}]$, $\bigsqcup S = \bigcup S$,
- $f \ll g \Leftrightarrow f \sqsubseteq g$ & $\mathrm{dom}(f)$ is finite, and
- $\{f \in [\mathbb{N} \rightharpoonup \mathbb{N}] : \mathrm{dom}(f) \text{ finite}\}$ is a countable basis for $[\mathbb{N} \rightharpoonup \mathbb{N}]$.

Algebraic domains may seem "discrete" in some sense, or at least more discrete than domains that are continuous but not algebraic, such as $\mathbf{I}\mathbb{R}$. However, the reader should not go around believing that the continuous and the discrete are irreversibly divided—they are not. In domain theory and measurement, it is often possible to take a unified view of the two. To partially illustrate this point, let us now consider a *continuous extension* of the finite powerset $\mathcal{P}\{1, \ldots, n\}$ to the set of finite probability distributions

$$\Delta^n := \left\{x \in [0, 1]^n : \sum x_i = 1\right\}.$$

Each set $A \in \mathcal{P}\{1, \ldots, n\}$ has a characteristic map $\chi_A : \{1, \ldots, n\} \to \{0, 1\}$ defined by

$$\chi_A(i) := \begin{cases} 1 & \text{if } i \in A; \\ 0 & \text{otherwise.} \end{cases}$$

for which we have

$$A \supseteq B \iff \chi_A \geq \chi_B$$

where $\geq$ is the pointwise order on functions of type $\{1, \ldots, n\} \to \{0, 1\}$ and $1 \geq 0$. But each $A \in \mathcal{P}\{1, \ldots, n\} \setminus \{\emptyset\}$ corresponds to a canonical $x \in \Delta^n$ given by

$$x_i := \begin{cases} x^+ & \text{if } i \in A; \\ 0 & \text{otherwise,} \end{cases}$$

where x^+ refers to the largest probability in x. Thus, we can think of *any* $x \in \Delta^n$ as having a characteristic function $\chi_x : \{1, \ldots, n\} \to [0, 1]$ given by

$$\chi_x(i) := \begin{cases} 1 & \text{if } x_i = x^+; \\ x_i & \text{otherwise.} \end{cases}$$

Example 8 The set of *classical states*

$$\Delta^n := \left\{x \in [0, 1]^n : \sum x_i = 1\right\}$$

is a continuous dcpo in its *implicative order* [23]

$$x \sqsubseteq y \equiv \chi_x \geq \chi_y.$$

The implicative order can also be characterized as

$$x \sqsubseteq y \equiv (\forall i)\; x_i < y_i \Rightarrow x_i = x^+$$

where again x^+ refers to the largest probability in x. Thus, only a maximum probability is allowed to increase as we move up in the information order on Δ^n. If the maximum probability refers to a solution of a problem, then moving up in this order ensures that we are getting closer to the answer.

Example 9 The set of *decreasing* classical states

$$\Lambda^n := \{x \in \Delta^n : (\forall 1 \leq i < n)\; x_i \geq x_{i+1}\}$$

with the *majorization relation* $\leq$ given by

$$x \leq y \equiv (\forall k < n) \sum_{i=1}^{k} x_i \leq \sum_{i=1}^{k} y_i$$

is a continuous dcpo $(\Lambda^n, \leq)$. If the implicative $\sqsubseteq$ order is restricted to Λ^n, then we have $(\Lambda^n, \sqsubseteq) \subseteq (\Lambda^n, \leq)$, and this inclusion is strict.

A *list* over S is a function $x : \{1, \ldots, n\} \to S$ for $n \geq 0$ and the set of all such x is denoted $[S]$. The *length* of a list x is $|\text{dom}\, x|$. A list x can be written as $[x(1), \ldots, x(n)]$, where the *empty list* (the list of length 0) is written $[\,]$. We can also write lists as $a :: x$, where $a \in S$ is the *first element* of the list $a :: x$ and $x \in [S]$ is the *rest* of the list $a :: x$. For example, the list $[1, 2, 3]$ can be written $1 :: [2, 3]$.

A set $K \subseteq \mathbb{N}$ is *convex* if $a, b \in K \;\&\; a \leq x \leq b \Rightarrow x \in K$. Given a finite convex set $K \subseteq \mathbb{N}$, the map $\text{scale}(K) : \{1, \ldots, |K|\} \to K$ given by

$$\text{scale}(K)(i) = \min\; K + i - 1$$

relabels the elements of K so that they begin with one.

Example 10 The domain of finite lists. The set of finite lists $[S]$ with $\sqsubseteq$ given by reverse convex containment

$$x \sqsubseteq y \equiv (\exists \text{ convex } K \subseteq \{1, \ldots, \text{length}(y)\})\; y \circ \text{scale}(K) = x.$$

is an algebraic dcpo in which all elements are compact. If $x \sqsubseteq y$, we say that y is a *sublist* of x.

For instance, if $L = [1, 2, 3, 4, 5, 6]$, then $[1, 2, 3], [4, 5, 6], [3, 4, 5], [2, 3, 4], [3, 4], [5]$ and $[\,]$ are all sublists of L, while $[1, 4, 5, 6], [1, 3]$ and $[2, 4]$ are *not* sublists of L. The set $[S]$ is also called *the free monoid* over S.

Example 11 Products of domains. If D and E are dcpo's then

$$D \times E := \{(d, e) : d \in D \;\&\; e \in E\}$$

is a dcpo in the pointwise order

$$(x_1, y_1) \sqsubseteq (x_2, y_2) \equiv x_1 \sqsubseteq x_2 \;\&\; y_1 \sqsubseteq y_2.$$

If D and E are both continuous, then so is $D \times E$, where

$$(x_1, y_1) \ll (x_2, y_2) \equiv x_1 \ll x_2 \;\&\; y_1 \ll y_2.$$

Having discussed the information order, let us turn now to the question of *information content.*

9.2.3 Measurement

From Sect. 9.2.1, a measurement $\mu : D \to [0, \infty)$ should satisfy:

1. For all $x, y \in D$, $x \sqsubseteq y \Rightarrow \mu x \geq \mu y$, and
2. If (x_n) is an increasing sequence in D, then

$$\mu\left(\bigsqcup_{n \geq 1} x_n\right) = \lim_{n \to \infty} \mu x_n.$$

On all the domains that we will work with, a mapping will have these two properties exactly when it is *Scott continuous.*

Definition 4 For a subset $X \subseteq D$ of a dcpo D, define

$$\uparrow X := \bigcup_{x \in X} \uparrow x \;\;\&\;\; \downarrow X := \bigcup_{x \in X} \downarrow x$$

A subset $U \subseteq D$ of a dcpo D is *Scott open* when it is an *upper set* $U = \uparrow U$ that is *inaccessible by directed suprema*:

$$\bigsqcup S \in U \Rightarrow S \cap U \neq \emptyset$$

for all directed $S \subseteq D$.

The Scott open sets on a dcpo form a topology. A subset $C \subseteq D$ is *Scott closed* when it is a *lower set* $C = \downarrow C$ that contains the supremum of every directed set it contains. Of particular importance for us is that the Scott topology on a continuous dcpo has the collection $\{\Uparrow x : x \in D\}$ as a basis. That is, it is a topology determined by approximation.

Example 12 A basic Scott open set in $\mathbf{I}[0, 1]$ is

$$\Uparrow [a, b] = \{x \in \mathbf{I}[0, 1] : x \subseteq \text{int}([a, b])\}.$$

In the domain of binary channels $\mathbb{N} = \{(a, b) : a \geq b \ \& \ a, b \in [0, 1]\}$, drawn with a on the x-axis and b on the y-axis, such a set forms a right triangle whose hypotenuse lies along the diagonal, but whose other two sides are removed.

Definition 5 A function $f : D \to E$ between dcpo's is *Scott continuous* if the inverse image of a Scott open set in E is Scott open in D.

Scott continuity can be characterized order theoretically [1]:

Theorem 1 *A function $f : D \to E$ is Scott continuous iff f is monotone,*

$$(\forall x, y \in D)\, x \sqsubseteq y \Rightarrow f(x) \sqsubseteq f(y),$$

and preserves directed suprema:

$$f\left(\bigsqcup S\right) = \bigsqcup f(S),$$

for all directed $S \subseteq D$.

Thus, on the *very* reasonable assumption that the domains which arise in practice always allow us to replace directed sets with increasing sequences, a measurement should at least be a Scott continuous function $\mu : D \to [0, \infty)^*$, where $[0, \infty)^*$ is the domain of nonnegative reals in its dual order:

$$x \sqsubseteq y \equiv y \leq x$$

and $\leq$ refers to the usual way of ordering real numbers. But there is more to the story of content than just continuity.

Imagine that we would like to compute an ideal element $x \in D$ in some domain D that cannot be computed exactly. How accurately would we like to calculate it? We wish to calculate it to within an accuracy of $a \ll x$. Now we proceed to calculate, using some process to determine a sequence of values $x_1, \ldots, x_n \ldots$ each x_i containing information about x, that is, $x_i \sqsubseteq x$. When do we stop? We stop when some *measure* of information content μ says that we are 'close enough' to the answer x. How does μ say this?

It tells us that *if* x_n contains information about x and *if* x and x_n are close enough in information content, *then* we have succeeded in calculating x to within the desired accuracy. That is, we have found an x_n such that $a \ll x_n$. In symbols,

$$(\exists \varepsilon > 0)(\forall n)(x_n \sqsubseteq x \ \& \ |\mu x - \mu x_n| < \varepsilon \Rightarrow a \ll x_n)$$

Now the thing to realize is that other computations may take *other* paths (x_n) to x and that we may also be interested in *other* levels of accuracy a. Since we want μ to guarantee accuracy for these processes too, we want μ to satisfy

$$(\forall a \ll x)(\exists \varepsilon > 0)(\forall y \in D)(y \sqsubseteq x \;\&\; |\mu x - \mu y| < \varepsilon \Rightarrow a \ll y)$$

If μ can provide this for the element $x \in D$, then μ must be measuring the *information content* of x. If the last statement holds, then it also holds when we can quantify over *all* Scott open sets U since sets of the form $\twoheaduparrow a$ are a basis for the Scott topology at x. For a dcpo D, we arrive at the following:

Definition 6 A Scott continuous $\mu : D \to [0, \infty)^*$ is said to *measure the content* of $x \in D$ if for all Scott open sets $U \subseteq D$,

$$x \in U \Rightarrow (\exists \varepsilon > 0)\, x \in \mu_\varepsilon(x) \subseteq U$$

where

$$\mu_\varepsilon(x) := \{y \in D : y \sqsubseteq x \;\&\; |\mu x - \mu y| < \varepsilon\}$$

are called the *ε-approximations* of x.

We often refer to μ as simply "measuring" $x \in D$ or as measuring $X \subseteq D$ when it measures each element of X. Minimally, a *measurement* should measure the content of its kernel:

Definition 7 A *measurement* $\mu : D \to [0, \infty)^*$ is a Scott continuous map that measures the content of $\ker(\mu) := \{x \in D : \mu x = 0\}$.

The order on a domain D defines a clear sense in which one object has "more information" than another: a *qualitative* view of information content. The definition of measurement attempts to identify those monotone mappings μ which offer a *quantitative* measure of information content in the sense specified by the order. The essential point in the definition of measurement is that μ measure content in a manner that is consistent with the particular view offered by the order. There are plenty of monotone mappings that are not measurements—and while some of them may measure information content in *some other sense*, each sense must first be specified by a different information order. The definition of measurement is then a minimal test that a function μ must pass if we are to regard it as providing a measure of information content.

Lemma 1 *Let $\mu : D \to [0, \infty)^*$ be a measurement.*

(i) *If $x \in \ker(\mu)$, then $x \in \max(D) = \{x \in D : \uparrow x = \{x\}\}$.*
(ii) *If μ measures the content of $y \in D$, then*

$$(\forall x \in D)\, x \sqsubseteq y \;\&\; \mu x = \mu y \Rightarrow x = y.$$

These results say (i) elements with no uncertainty are maximal in the information order and (ii) comparable elements with the same information content are equal. The converse of (i) is not true and there are many important cases (see Sect. 9.3.4 for instance) where the applicability of measurement is greatly heightened by the fact that $\ker \mu$ need not consist of *all* maximal elements.

Example 13 Canonical measurements

(i) $(\mathbf{I}\mathbb{R}, \mu)$ the interval domain with the length measurement $\mu[a, b] = b - a$.
(ii) $(\mathcal{P}\omega, |\cdot|)$ the powerset of the naturals with $|\cdot| : \mathcal{P}\omega \to [0, \infty)^*$ given by

$$|x| = 1 - \sum_{n \in x} \frac{1}{2^{n+1}}.$$

(iii) $([\mathbb{N} \rightharpoonup \mathbb{N}], \mu)$ the partial functions on the naturals with

$$\mu f = |\mathrm{dom}(f)|$$

where $|\cdot|$ is the previous measurement on $\mathcal{P}\omega$.
(iv) $(\Sigma^\infty, 1/2^{|\cdot|})$ the binary strings where $|\cdot| : \Sigma^\infty \to [0, \infty]$ is the length of a string.
(v) $(\mathbf{U}X, \mathrm{diam})$ the upper space of a locally compact metric space (X, d) with

$$\mathrm{diam}\, K = \sup\{d(x, y) : x, y \in K\}.$$

(vi) $(\mathbf{B}X, \pi)$ the formal ball model of a complete metric space (X, d) with

$$\pi(x, r) = r$$

(vii) (Δ^n, μ) the classical states in their implicative order with $\mu x = 1 - x^+$. Shannon entropy

$$H(x) = -\sum_{i=1}^{n} x_i \log_2(x_i)$$

is also a measurement on Δ^n.
(viii) $(\mathbb{N}, c)$ the nonnegative binary channels with capacity from information theory (Shannon)

$$c(a, b) = \log_2 \left(2^{\frac{\bar{a}H(b) - \bar{b}H(a)}{a-b}} + 2^{\frac{bH(a) - aH(b)}{a-b}} \right)$$

where $c(a, a) := 0$ and $H(x) = -x \log_2(x) - (1-x)\log_2(1-x)$ is the binary entropy.

(ix) ($[S]$, length) lists with length as a measurement
(x) Products: if (D, μ) and (E, λ), are domains with measurements, then $D \times E$ is a domain with $\max\{\mu, \lambda\}$ and $\mu + \lambda$ as measurements.[1]

In each case, we have $\ker \mu = \max(D)$.

We will see other examples in Sect. 9.4, including the domains of analytic mappings, quantum states and spacetime intervals. The reader who is impatient to find out what one does with a measurement can skip ahead to any of the other sections as long as they promise to eventually return. The reader interested in understanding the ideas should continue reading.

The view of information content taken in the study of measurement is that of a structural relationship between two classes of objects which, generally speaking, arises when one class may be viewed as a simplification of the other. The process by which a member of one class is simplified and thereby "reduced" to an element of the other is what we mean by "the measurement process" in domain theory [16]. One of the classes may well be a subset of real numbers, but the 'structural relationship' underlying content should not be forgotten. Here is the definition of measurement in this more general case:

Definition 8 A Scott continuous map $\mu : D \to E$ between dcpo's is said to *measure the content of* $x \in D$ if

$$x \in U \Rightarrow (\exists \varepsilon \in \sigma_E)\, x \in \mu_\varepsilon(x) \subseteq U,$$

whenever $U \in \sigma_D$ is Scott open and

$$\mu_\varepsilon(x) := \mu^{-1}(\varepsilon) \cap \downarrow\! x$$

are the elements ε close to x in content. The map μ *measures* X if it measures the content of each $x \in X$.

Definition 9 A *measurement* is a Scott continuous map $\mu : D \to E$ between dcpo's that measures $\ker \mu := \{x \in D : \mu x \in \max(E)\}$.

In the case $E = [0, \infty)^*$, the new definition of "measures the content of x" is equivalent to the one given earlier, so we reserve the right to denote the set $\mu_{[0,\varepsilon)}(x)$ by $\mu_\varepsilon(x)$ in contrast to how we first defined $\mu_\varepsilon(x)$, though we will always be clear about how we are using this notation. In addition, Lemma 1 remains valid in this more general case; the 'reflective' nature of measurement is covered in more detail in [15]. In addition, with the more abstract formulation of measurement, it becomes clear that measurements compose. That, for example, is why it was easy to measure the domain $[\mathbb{N} \rightharpoonup \mathbb{N}]$ of partial functions in the last example.

---

[1] In *principle*, it is possible to measure the dcpo of Scott continuous maps $[D \to E]$. In practice, though, the question is how to do so *simply*. See [21, 43] for more.

9.2.4 Distance, Content and Topology

Why should there be any relation between *topology* and *information content*? To answer this, we have to remember that we are not just talking about any topology, but rather, the Scott topology, which as we have seen is the topology of approximation. Second, we have to recall the subtle relation between information content and the desire to obtain accurate approximations of ideal elements discussed in the last section.

As it turns out, one way to think of a measurement is essentially as being the informatic analogue of 'metric' for domain theory. There are several senses in which this is true. Let us consider one by returning to the elements ε close to $x \in D$, abbreviated to

$$\mu_\varepsilon(x) := \mu_{[0,\varepsilon)}(x) = \{y \in D : y \sqsubseteq x \ \&\ \mu y < \varepsilon\},$$

for $\varepsilon > 0$.

Theorem 2 *Let D be a continuous dcpo. If $\mu : D \to [0,\infty)^*$ measures $X \subseteq D$, then*

$$\{\uparrow\mu_\varepsilon(x) \cap X : x \in X, \varepsilon > 0\}$$

is a basis for the relative Scott topology on X.

Thus, in the presence of a measurement, we can understand the Scott topology as being derived from ε-approximations of points, similar to the way the topology of a metric space is specified.

To further develop the analogy between metric and measurement hinted at in the last result, suppose that a continuous dcpo D has the property that for any $x, y \in D$ there is $z \in D$ with $z \sqsubseteq x, y$. Notice that in Example 13, domains (i)–(ix) all have this property. If we encounter a continuous dcpo that does not have this property, we can always adjoin a bottom element $\bot$, and scale the measurement so that $\mu\bot = 1$. See chapter five of [15] for more. Then we can define $d : D^2 \to [0,\infty)^*$ given by

$$d(x, y) = \inf\{\mu z : z \ll x, y\} = \inf\{\mu z : z \sqsubseteq x, y\}$$

Because μ is monotone, d is Scott continuous. Because μ is Scott continuous, we have $d(x, y) = \mu x$ when $x \sqsubseteq y$. The distance function d associated to μ is sometimes denoted $d(\mu)$.

Definition 10 For a monotone map $\mu : D \to [0,\infty)^*$ on a continuous dcpo D with $d = d(\mu)$ defined,

$$B_\varepsilon(x) := \{y \in D : d(x, y) < \varepsilon\}$$

for all $x \in D$, $\varepsilon > 0$.

Happily, distance and content are related as follows.

Theorem 3 *If $\mu : D \to [0, \infty)^*$ is Scott continuous on a continuous dcpo D with $d = d(\mu)$ defined, then*

$$B_\varepsilon(x) = \uparrow\mu_\varepsilon(x),$$

for each $x \in D$ and $\varepsilon > 0$. Consequently,

$$\{B_\varepsilon(x) \cap X : x \in X, \varepsilon > 0\}$$

is a basis for the relative Scott topology on X whenever μ measures X.

Example 14 For $(\mathbf{I}\mathbb{R}, \mu)$,

$$d([a], [b]) = |a - b|,$$

for all $a, b \in \mathbb{R}$. Because d is the Euclidean metric on $\mathbb{R}$, we can conclude that $\max(\mathbf{I}\mathbb{R})$ in its relative Scott topology is homeomorphic to $\mathbb{R}$.

The last example is also true for $\mathbf{I}[0, 1]$. But now something interesting happens, because Theorem 3 says that *any measurement* on $\mathbf{I}[0, 1]$ induces the Euclidean topology on its kernel. Recalling that the *capacity* $c : \mathbb{N} \to [0, 1]^*$ of a binary channel

$$c(a, b) = \log_2\left(2^{\frac{\bar{a}H(b)-\bar{b}H(a)}{a-b}} + 2^{\frac{bH(a)-aH(b)}{a-b}}\right)$$

is a measurement on the domain of binary channels $\mathbb{N} \simeq \mathbf{I}[0, 1]$ from Example 13(viii), its associated distance function on $\ker(c) = \max(\mathbb{N})$ is

$$\rho([a], [b]) = c(a, b) = c(b, a)$$

Then, just like Euclidean distance, capacity $c : [0, 1]^2 \to [0, 1]$ also has the following three properties:

(i) $c(a, b) = c(b, a)$,
(ii) $c(a, b) = 0$ iff $a = b$,
(iii) The sets $\{y \in [0, 1] : c(x, y) < \varepsilon\}$ for $\varepsilon > 0$ form a basis for the Euclidean topology on $[0, 1]$.

Capacity does not satisfy the triangle inequality, so it is not a priori obvious that the sets in (iii) form a basis for any topology, let alone the Euclidean topology. Let us state this another way: the topology of certain spaces can be derived from a notion of distance that is defined in terms of the amount of information that can be transmitted between two points [29].

Now we consider another sense in which measurements are the domain theoretic counterpart to metrics.

Definition 11 A measurement $\mu : D \to [0,\infty)^*$ on a continuous dcpo D satisfies *the triangle inequality* if for all consistent pairs $x, y \in D$, there is an element $z \sqsubseteq x, y$ such that $\mu z \leq \mu x + \mu y$.

When a measurement satisfies the triangle inequality, its corresponding notion of distance is a metric on the set of elements with measure zero.

Theorem 4 *Let (D,μ) be a domain with a measurement satisfying the triangle inequality. Then $d(\mu) : \ker(\mu) \times \ker(\mu) \to [0,\infty)$ is a metric which yields the relative Scott topology on* $\ker(\mu)$.

For instance, many of the measurements in Example 13 satisfy the triangle inequality, including (i)–(v) and (xi). More generally, the class of *Lebesgue measurements*, discussed in Sect. 9.3.4, allows one to conclude that $\ker(\mu)$ is metrizable. In fact, in most cases, we can construct a metric from μ, though the construction is more involved. One such case is when there is an element $z \sqsubseteq x, y$ with $\mu z \leq 2 \cdot \max\{\mu x, \mu y\}$, see chapter five of [15] for more on this.

The relation between measurement and topology does not end with the observation that they are like metrics. It turns out that measuring a domain is *equivalent* to being able to generate a certain topology.

Definition 12 The μ *topology* on a continuous dcpo D has

$$\{\twoheaduparrow a \cap \downarrow x : a, x \in D\}$$

as a basis.

Unexpectedly, the μ topology is *always* zero dimensional and Hausdorff.

Theorem 5 *Let D be a continuous dcpo. A Scott continuous $\mu : D \to [0,\infty)^*$ measures D iff $\{\mu_\varepsilon(x) : x \in D \ \& \ \varepsilon > 0\}$ is a basis for the μ topology.*

In the above result, $\mu_\varepsilon(x) = \mu_{[0,\varepsilon)}$ is defined as it was earlier in this section. We pause for a moment now to look at a few of the things one does with domains and measurements.

9.3 Fixed Points

A *least element* in a dcpo D is an element $\perp$ such that $\perp \sqsubseteq x$ for all $x \in D$. The first theorem I ever heard about in domain theory is:

Theorem 6 *Let D be a dcpo with a least element $\perp$. If $f : D \to D$ is Scott continuous, it has a least fixed point given by*

$$\mathrm{fix}(f) := \bigsqcup_{n \geq 0} f^n(\perp)$$

A useful corollary is that f has a least fixed point on $\uparrow x$ if $x \sqsubseteq f(x)$.

Exercise: Prove that split_f from the introduction is *not* Scott continuous by showing that it is *not* monotone. (*Hint*: Cheat, by reading this section).

Major references: [15]

9.3.1 Fixed Points of Nonmonotonic Mappings

Ordinarily, this discussion would be deferred to the section "forms of process evolution" but we include it here so that the reader gets some quick examples of what one does with measurement.

Definition 13 A *splitting* on a dcpo D is a function $s : D \to D$ with $x \sqsubseteq s(x)$ for all $x \in D$.

Theorem 7 *Let D be a dcpo with a measurement μ that measures D. If $I \subseteq D$ is closed under directed suprema and $s : I \to I$ is a splitting whose measure*

$$\mu \circ s : I \to [0, \infty)^*$$

is Scott continuous, then

$$(\forall x \in I) \bigsqcup_{n \geq 0} s^n(x) \text{ is a fixed point of } s.$$

Moreover, the set of fixed points $\text{fix}(s) = \{x \in I : s(x) = x\}$ *is a dcpo.*

In applications, a slightly weaker formulation can be useful: if for every increasing sequence (x_n) in I we have

$$\mu s\left(\bigsqcup x_n\right) = \lim_{n \to \infty} \mu s(x_n),$$

then

$$\bigsqcup_{n \geq 0} s^n(x) \in \text{fix}(s),$$

for every $x \in I$. In addition, $\text{fix}(s) = I \cap \ker(\mu)$ iff $\mu s(x) < \mu x$ for all $x \in I$ with $\mu x > 0$. The point being: we do not need to check that $\mu \circ s$ is monotone in order to establish the existence of fixed points.

Example 15 Let $f : \mathbb{R} \to \mathbb{R}$ be a continuous map on the real line. Denote by $C(f)$ the subset of $\mathbf{I}\mathbb{R}$ where f changes sign, that is,

$$C(f) = \{[a, b] : f(a) \cdot f(b) \leq 0\}.$$

The continuity of f ensures that this set is closed under directed suprema, and the mapping

$$\text{split}_f : C(f) \to C(f)$$

given by

$$\text{split}_f[a, b] = \begin{cases} \text{left}[a, b] & \text{if left}[a, b] \in C(f); \\ \text{right}[a, b] & \text{otherwise,} \end{cases}$$

is a splitting where left$[a, b] = [a, (a+b)/2]$ and right$[a, b] = [(a+b)/2, b]$. The measure of this mapping

$$\mu\,\text{split}_f[a, b] = \frac{\mu[a, b]}{2}$$

is Scott continuous, so Theorem 7 implies that

$$\bigsqcup_{n \geq 0} \text{split}_f^n[a, b] \in \text{fix}(\text{split}_f).$$

However, fix(split$_f$) $= \{[r] : f(r) = 0\}$, which means that iterating split$_f$ is a scheme for calculating a solution of the equation $f(x) = 0$. This numerical technique is called *the bisection method.*

Proposition 1 *For a continuous selfmap $f : \mathbb{R} \to \mathbb{R}$ which has at least one zero, the following are equivalent:*

(i) *The map* split$_f$ *is monotone.*
(ii) *The map f has a unique zero r and*

$$C(f) = \{[a, r] : a \leq r\} \cup \{[r, b] : r \leq b\}.$$

That is, if split$_f$ is monotone, then in order to calculate the solution r of $f(x) = 0$ using the bisection method, we must first *know* the solution r.

Example 16 A function $f : [a, b] \to \mathbb{R}$ is *unimodal* if it has a maximum value assumed at a unique point $x^* \in [a, b]$ such that

(i) f is strictly increasing on $[a, x^*]$, and
(ii) f is strictly decreasing on $[x^*, b]$.

Unimodal functions have the important property that

$$x_1 < x_2 \Rightarrow \begin{cases} x_1 \leq x^* \leq b & \text{if } f(x_1) < f(x_2), \\ a \leq x^* \leq x_2 & \text{otherwise.} \end{cases}$$

This observation leads to an algorithm for computing x^*. For a unimodal map $f : [a, b] \to \mathbb{R}$ with maximizer $x^\star \in [a, b]$ and a constant $1/2 < r < 1$, define a dcpo by

$$I_{x^*} = \{\bar{x} \in \mathbf{I}\mathbb{R} : [a, b] \sqsubseteq \bar{x} \sqsubseteq [x^*]\},$$

and a splitting by

$$\max{}_f : I_{x^*} \to I_{x^*}$$
$$\max{}_f[a, b] = \begin{cases} [l(a, b), b] & \text{if } f(l(a, b)) < f(r(a, b)); \\ [a, r(a, b)] & \text{otherwise,} \end{cases}$$

where $l(a, b) = (b-a)(1-r)+a$ and $r(a, b) = (b-a)r+a$. The measure of $\max_f$ is Scott continuous since $\mu\ \max_f(\bar{x}) = r \cdot \mu(\bar{x})$, for all $\bar{x} \in I_{x^*}$. By Theorem 7,

$$\bigsqcup_{n \geq 0} \max{}_f^n(\bar{x}) \in \text{fix}(\max{}_f),$$

for any $\bar{x} \in I_{x^*}$. However, any fixed point of $\max_f$ has measure zero, and the only element of I_{x^*} with measure zero is $[x^*]$. Thus, $\bigsqcup \max_f^n[a, b] = [x^*]$, which means that iterating $\max_f$ yields a method for calculating x^*. This technique is called the *r-section search.*

Finally, observe that $\max_f$ is not monotone. Let $-1 < \alpha < 1$ and $f(x) = 1 - x^2$. The function f is unimodal on any compact interval. Since $\max_f[-1, 1] = [-1, 2r - 1]$, we see that

$$\begin{aligned} \max{}_f[-1, 1] \sqsubseteq \max{}_f[\alpha, 1] &\Rightarrow 1 \leq 2r - 1 \text{ or } r(\alpha, 1) \leq 2r - 1 \\ &\Rightarrow 1 \leq r \text{ or } \alpha + 1 \leq r(\alpha + 1) \\ &\Rightarrow r \geq 1, \end{aligned}$$

which contradicts $r < 1$. Thus, for no value of r is the algorithm monotone.

The previous examples make it clear that there are natural and important examples of processes on domains that are fundamentally nonmonotonic but which nevertheless have fixed points whose existence can be easily established by measurement based results. Moreover, the previous fixed point theorem is a strict generalization of the usual fixed point theorem in domain theory:

Example 17 If $f : D \to D$ is a Scott continuous map on a dcpo D with a measurement μ that measures D, then we consider its restriction to the set of points where it improves

$$I(f) = \{x \in D : x \sqsubseteq f(x)\}.$$

This yields a splitting $f : I(f) \to I(f)$ on a dcpo with continuous measure. By Theorem 7,

$$(\forall x \in I(f)) \bigsqcup_{n\geq 0} f^n(x) \text{ is a fixed point of } f.$$

For instance, if D is ω-continuous with basis $\{b_n : n \in \mathbb{N}\}$, then

$$\mu x = |\{n : b_n \ll x\}|$$

defines such a measurement. Notice, however, that with this construction we normally have $\ker \mu = \emptyset$.

9.3.2 Numerical Methods

Numerical methods provide an interesting application of domains and measurements. The two examples in the last section, the bisection and the golden section search, really only scratch the surface of what is possible in this regard. So in this section, we take a closer look.

9.3.2.1 A Topological Question: What the **** are We Computing?

By Theorem 2, a measurement μ allows one to derive the Scott topology on $\ker(\mu)$. This fundamental fact ensures that what *appears* to be computation *actually is* computation.

Example 18 Recall the bisection method $\text{split}_f : C(f) \to C(f)$ from Example 15. By Theorem 7,

$$\bigsqcup_{n\geq 0} \text{split}_f^n(x) \in \text{fix}(\text{split}_f),$$

for all $x \in C(f)$. But $\text{fix}(\text{split}_f) = \{[r] : f(r) = 0\}$, which means that iterating split_f is a scheme for calculating a zero of f. Right?

Well, almost. Let's take a closer look at things. In the zero finding problem, the desired result is a *number* that approximates the zero r, not an interval. In practice, we calculate a small enough interval x, and then choose a point within it as an approximation of r. The true reason that split_f is an algorithm for computing r is that if we begin with any $x \in C(f)$, and then choose any sequence $x_n \in \text{split}_f^n(x)$, we *always* have

$$|x_n - r| \leq \mu \, \text{split}_f^n(x) \leq \frac{\mu x}{2^n},$$

and hence $x_n \to r$ in the *usual topology* on the real line.

Then what we need to know is that computation on a domain actually corresponds to computation in reality. For the splittings of Prop. 7, the following result confirms exactly this.

Proposition 2 *Let D be a continuous dcpo with a map μ that measures D, $I \subseteq D$ a set closed under suprema of increasing sequences and $s : I \to I$ a splitting with $\mu s \leq c \cdot \mu$ for a constant $0 \leq c < 1$. Then for all $x \in I$, if $x_n \in \uparrow s^n(x) \cap \ker \mu$, we have*

$$x_n \to \bigsqcup_{n \geq 0} s^n(x) \in \text{fix}(s) \subseteq \ker \mu,$$

in the relative Scott topology on $\ker \mu$.

For instance, in the case of the interval domain $\mathbf{I}\mathbb{R}$, we have

$$\ker \mu = \max(\mathbf{I}\mathbb{R}) = \{[x] : x \in \mathbb{R}\} \simeq \mathbb{R}$$

where the homemorphism is between the relative Scott topology on $\ker \mu$ and the usual topology on the real line. Thus, to say that $\bigsqcup_{n \geq 0} \text{split}_f^n(x)$ computes a zero of f means exactly the same thing as it does in numerical analysis.

9.3.2.2 Numerical Methods and the Information Order

Some numerical methods manipulate information in a manner that is fundamentally different than a *bracketing method*, such as the bisection, or a *one point method*, like Newton's method. Each way of manipulating information corresponds to a different information order.

Example 19 Let $D = [0, 1]$ be the unit interval in its usual order. Then

$$\mathcal{P}_C(D) = \{[a, b] : a, b \in [0, 1] \ \& \ a \leq b\}$$

is called *the convex powerdomain* over D and its order is given by

$$[a, b] \sqsubseteq [x, y] \Leftrightarrow a \leq x \ \& \ b \leq y.$$

We can measure this object by

$$\mu[a, b] = (1 - a) + (1 - b).$$

Note that $\ker \mu = \max(\mathcal{P}_C(D)) = \{[1]\}$.

The measurement above has a natural explanation [19].

9.3.2.3 One Point Methods

A one point method amounts to iterating a continuous $f : [a, b] \to [a, b]$ until we reach a fixed point, so it should come as no surprise that we can model them domain theoretically with a copy of $[a, b] \simeq [0, 1]$ in its usual order. However, there is another way. We can exploit the fact that

$$[0,1] \simeq \{[x] : x \in [0,1]\} \subseteq \mathcal{P}_C[0,1].$$

This subset we name the *total reals* and for this reason we refer to the other elements of $\mathcal{P}_C[0,1]$ as *partial reals*.

Example 20 Let f be concave increasing on $[a,b]$ with $f(a) < 0$ and $f(b) > 0$. Then consider the partial function $I_f : \mathcal{P}_C[a,r] \rightharpoonup \mathcal{P}_C[a,r]$ given by

$$[x] \mapsto [x - f(x)/f'(x)],$$

which is defined only on the subset of total reals. By Theorem 7,

$$\bigsqcup_{n\geq 0} I_f^n[x] \in \text{fix}(I_f) = \{[r]\},$$

and so *Newton's method* converges for any initial guess $x \in \text{dom}(I_f)$.

One of the standard reasons for avoiding Newton's method is that it requires the calculation of a derivative. A common method for overcoming this difficulty is to approximate the derivative by calculating a difference quotient using *two values*which simultaneously also serve to approximate the zero r. The most famous of the *interpolation methods*, as they are called, is probably the secant method.

9.3.2.4 An Analysis of the Secant Method

One point methods are nothing more than iterating a function on some part of the real line, so domain theory is not *necessary* for describing them. However, with multi-point or *interpolation methods*, i.e., those which use more than one point to determine the next approximation in an iterative scheme, we arrive at our first example where pursuit of the uniformity ideal mandates a domain theoretic approach.

Example 21 The secant method. If we have a real valued function f, the following scheme is very useful for zero finding: choose *two* initial guesses x_0 and x_1 and then proceed inductively according to

$$x_{n+1} = x_n - f(x_n) \cdot \frac{x_n - x_{n-1}}{f(x_n) - f(x_{n-1})}$$

for $n \geq 1$. The hope is that this sequence converges to a zero of f.

At each iteration of this algorithm, instead of one value, as with Newton's method, there are two values to be used in calculating the next approximation. We visualize it as a sequence of intervals:

$$[x_0, x_1] \to [x_1, x_2] \to [x_2, x_3] \to \cdots$$

The arrow indicates that we are moving up in the information order. These intervals are almost never nested. Happily, though, they often form an increasing sequence in the domain $\mathcal{P}_C[a,b]$ of partial reals.

If we have a function f, its derivative $df[x] = f'(x)$ can be extended from the total reals to the set of *all* partial reals $\mathcal{P}_C[a,b]$ by

$$df[x,y] = \frac{f(y)-f(x)}{y-x} \text{ if } y > x.$$

And just like that, we can model the secant method.

Theorem 8 *Let f be concave and increasing on $[a,b]$ with a zero $r \in (a,b)$. Then iterating the splitting* $\sec_f : \mathcal{P}_C[a,r] \to \mathcal{P}_C[a,r]$ *given by*

$$\sec_f[x,y] = \left[y, y - \frac{f(y)}{df[x,y]}\right]$$

is an algorithm for calcuating r. That is,

$$\bigsqcup_{n\geq 0} \sec_f^n(x) = [r],$$

for any $x \in \mathcal{P}_C[a,r]$.

For a total real $[x]$, we have $\sec_f[x] = [x, x - f(x)/f'(x)]$, which says that the secant method arises as the extension of a reversible formulation of Newton's method from the set of total reals to the set of *all* partial reals.

An interesting consequence here is that if we are able to compute the value of f' at just one $x \in [a,r)$, then the problem of generating two initial guesses for the secant method is eliminated: given such an $[x]$, we are then assured that we have enough information to calculate the partial real $\sec_f[x]$, and from there, Theorem 8 ensures that the iterates $\sec_f^n[x]$ converge to $[r]$.

So we have seen enough to find it plausible that the one point methods, the bracketing methods and the interpolation methods all have natural domain theoretic models and that the question of their correctness amounts *in all cases* to proving that some operator has a fixed point. Notice that numerical analysis only uses the fixed point approach for one point methods. This provides a nice uniform approach. But to really be able to believe in it, we need domain theory to teach us something new and significant about zero finding—perhaps something that someone other than a domain theorist would care about.

9.3.2.5 A New Method for Zero Finding

The zero finding problem really is one of the great problems in the history of mathematics: given a real valued function f on an interval $[a,b]$, find a *zero* of f, that is, a number x such that $f(x) = 0$. Evariste Galois proved that one must resort to algorithms in solving this problem by showing that polynomials of degree five and higher have no solution by radicals, i.e., their zeroes are not in general expressible by a formula.

If one assumes nothing about f except continuity, then there are many senses in which *the bisection method* is the optimal algorithm for zero finding ([3, 12]). However, for a class of Lipschitz mappings [4], the bisection method is no longer optimal. But Lipschitz mappings have derivatives almost everywhere [38]. In addition, the optimal algorithm makes use of the Lipschitz constant [4], which is a bound on its derivative. Another case in which bisection is not optimal is the class of convex mappings [8]. But there again, one finds that convex mappings are differentiable everywhere except on a countable set [38]. These two examples raise the following question: If we have a nontrivial class $\mathcal{C}$ of functions and a zero finding algorithm which is better than the bisection for the members of $\mathcal{C}$, must the functions in $\mathcal{C}$ possess some amount of differentiability? In short, is differentiability in some form necessary in order to beat the bisection method?

We are going to prove that the answer to this question is no. For the class of *Hölder continuous* mappings, which contains all of the well-known examples of nowhere differentiable functions, including those arising in the analysis of Brownian motion and fractals [7], we use domain theory and measurement to design and analyze a new algorithm for zero finding which is better than the bisection method at every iteration.

We will design the method for Hölder continuous functions which have a simple zero on a compact interval $[a, b]$.

Definition 14 A map $f : [a, b] \to \mathbb{R}$ is *Hölder continuous* if there are positive constants $c > 0$ and $\alpha > 0$ such that

$$|f(x) - f(y)| \leq c \cdot |x - y|^{\alpha}$$

for all $x, y \in [a, b]$.

Example 22 Weierstrass's function. The function introduced in 1872 by Weierstrass,

$$f(x) = \sum_{n=0}^{\infty} a^n \cos(b^n \pi x),$$

is nowhere differentiable for $0 < a < 1$ and b an odd integer with $ab > 1 + 3\pi/2$. It is Hölder continuous [46] with $\alpha = \log(1/a)/\log b$.

Definition 15 A function $f : [a, b] \to \mathbb{R}$ has a *simple zero* $r \in [a, b]$ if

$$\operatorname{sgn} f(x) = \operatorname{sgn}(x - r)$$

for $x \in [a, b]$, where $\operatorname{sgn}(x) = x/|x|$ for $x \neq 0$, and $\operatorname{sgn}(0) = 0$. Write

$$\Box f := \{x \in \mathbf{I}\mathbb{R} : [a, b] \sqsubseteq x \sqsubseteq [r]\}$$

for the set of intervals where f changes sign.

Then a function has a simple zero r if it is positive to the right of r and negative to the left of r. We will make use of the following operators on $\mathbf{I}\mathbb{R}$:

Definition 16

- $l : \mathbf{I}\mathbb{R} \to \mathbb{R} :: [a, b] \mapsto a$
- $m : \mathbf{I}\mathbb{R} \to \mathbb{R} :: [a, b] \mapsto (a + b)/2$
- $r : \mathbf{I}\mathbb{R} \to \mathbb{R} :: [a, b] \mapsto b$

These are abbreviated $l_x := l(x)$, $r_x := r(x)$ and $m_x := m(x)$.

For instance, if $f : [a, b] \to \mathbb{R}$ has a simple zero r on $[a, b]$, then the bisection $\text{split}_f : \Box f \to \Box f$ can be written compactly as

$$\text{split}_f(x) = \begin{cases} [l_x, m_x] & \text{if } f(m_x) > 0; \\ [m_x, r_x] & \text{otherwise.} \end{cases}$$

This formulation of the bisection will help us understand its relation to the new method:

Theorem 9 *Let $f : [a, b] \to \mathbb{R}$ be a Hölder continuous map with a simple zero r. Then iterating the splitting $s_f : \Box f \to \Box f$ given by*

$$s_f(x) = \begin{cases} \left[l_x, m_x - (f(m_x)/c)^{1/\alpha}\right] & \text{if } f(m_x) > 0; \\ \\ \left[m_x + (|f(m_x)|/c)^{1/\alpha}, r_x\right] & \text{otherwise;} \end{cases}$$

is an algorithm for computing r. That is,

$$\bigsqcup_{n \geq 0} s_f^n(x) = [r],$$

for all $x \in \Box f$. Thus, for all $x \in \Box f$, if $x_n \in s_f^n(x)$ for each n, then $x_n \to r$.

The method also easily extends to the case where $|fx - fy| \leq c \cdot g(|x - y|)$, for a left invertible $g : [0, \infty) \to [0, \infty)$ satisfying $g(0) = 0$.

9.3.2.6 A Comparison with the Bisection

If s_1 and s_2 are two algorithms, then a natural intuition stemming from domain theory is to say that s_2 is a better algorithm than s_1 if

$$s_1 \sqsubseteq s_2 \equiv (\forall x)\, s_1(x) \sqsubseteq s_2(x).$$

However, in the analysis of numerical techniques, one should not expect to be able to make absolute statements such as "Algorithm 1 is better than Algorithm 2 always and there is nothing more to be said." For instance, sometimes the bisection method is better than Newton's method, if the derivatives of a function are difficult (or impossible) to calculate, while an advantage of Newton's method is its quadratic convergence when close enough to the root. Aside from the fact that our method requires one to determine the constants α and c – which is not necessarily a simple matter – we can in a lot of cases say that s_f is simply better than the bisection:

Proposition 3 *Let $f : [a, b] \to \mathbb{R}$ be a Hölder continuous map with a simple zero r. Then* $\text{split}_f \sqsubseteq s_f$ *and for any $x \in \Box f$,*

$$\mu s_f(x) = \mu\, \text{split}_f(x) - \left(\frac{|f(m_x)|}{c}\right)^{1/\alpha} \leq \mu\, \text{split}_f(x),$$

with equality only in the unlikely event that $m_x = r$.

For instance, if r is a computable irrational and we begin with an input x having rational endpoints, then s_f is a strict improvement over the bisection.

Corollary 1 *Let $f : [a, b] \to \mathbb{R}$ be a Hölder continuous map with a simple irrational zero r. Then for any $x \in \Box f$ with rational endpoints $l_x, r_x \in \mathbb{Q}$,*

$$\mu s_f^n(x) < \mu\, \text{split}_f^n(x),$$

for all iterations $n \geq 1$.

And in general we can see the same is true anytime the input interval does not contain r as its midpoint: once s_f gains an advantage over the bisection, it keeps this advantage forever. While the qualitative statement $\text{split}_f \sqsubseteq s_f$ is certainly a strong one for numerical methods, when taken on its own, it leaves something to be desired: how are we to know the inputs where they are equal? Even if we know that $\text{split}_f \sqsubseteq s_f$ and $\text{split}_f \neq s_f$, they may only differ on a single input, which doesn't say very much.

But when we incorporate the quantitative as well, then the clarity of what we are saying improves greatly: split_f and s_f are equal iff their measures are iff one can magically choose an input whose midpoint is the zero (which amounts to guessing the answer). This provides a simple and clear example of the "extra something" that measurement adds to the standard order theoretic setting and illustrates how precise an analysis is possible when the qualitative and quantitative are united.

To summarize, domain theory and measurement provides a language for expressing zero finding algorithms which renders the verification process systematic and *uniform:* it enables us to turn the question of correctness into one about fixed points for *all* zero finding methods, whereas this is only normally achieved in numerical analysis for one point schemes like Newton's method. And because it also produced something new, it is okay to believe in it now if you want to.

9.3.3 Unique Fixed Points

So measurement can be used to generalize the Scott fixed point theorem so as to include important nonmonotonic processes. But it can also improve upon it for monotone maps as well, by giving a technique that guarantees *unique* fixed points.

Definition 17 Let D be a continuous dcpo with a measurement μ. A monotone map $f : D \to D$ is a *contraction* if there is a constant $c < 1$ with

$$\mu f(x) \leq c \cdot \mu x$$

for all $x \in D$.

Theorem 10 *Let D be a continuous dcpo with a measurement μ such that*

$$(\forall\, x, y \in \ker \mu\,)(\,\exists\, z \in D\,)\; z \sqsubseteq x, y.$$

If $f : D \to D$ is a contraction and there is a point $x \in D$ with $x \sqsubseteq f(x)$, then

$$x^{\star} = \bigsqcup_{n \geq 0} f^n(x) \in \max(D)$$

is the unique fixed point of f on D. Furthermore, $x^{\star}$ is an attractor in two different senses:

(i) *For all $x \in \ker \mu$, $f^n(x) \to x^{\star}$ in the Scott topology on $\ker \mu$, and*
(ii) *For all $x \sqsubseteq x^{\star}$, $\bigsqcup_{n \geq 0} f^n(x) = x^{\star}$, and this supremum is a limit in the Scott topology on D.*

When a domain has a least element, the last result is easier to state.

Corollary 2 *Let D be a domain with least element $\bot$ and measurement μ. If $f : D \to D$ is a contraction, then*

$$x^{\star} = \bigsqcup_{n \geq 0} f^n(\bot) \in \max D$$

is the unique fixed point of f on D. In addition, the other conclusions of Theorem 10 hold as well.

Example 23 Let $f : X \to X$ be a contraction on a complete metric space X with Lipschitz constant $c < 1$. The mapping $f : X \to X$ extends to a monotone map on the formal ball model $\bar{f} : \mathbf{B}X \to \mathbf{B}X$ given by

$$\bar{f}(x, r) = (fx, c \cdot r),$$

which satisfies

$$\pi \bar{f}(x,r) = c \cdot \pi(x,r),$$

where $\pi : \mathbf{B}X \to [0,\infty)^*$ is the standard measurement on $\mathbf{B}X$, $\pi(x,r) = r$. Now choose r so that $(x,r) \sqsubseteq \bar{f}(x,r)$. By Theorem 10, $\bar{f}$ has a unique attractor which implies that f does also because $X \simeq \ker\pi$.

We can also use the upper space (**U**X, diam) to prove the Banach contraction theorem for compact metric spaces by applying the technique of the last example. In [17], a domain theoretic result is given which generalizes the Banach contraction theorem. Next up: probably the most overused example of a Scott continuous map in domain theory. Here is something new about it:

Example 24 Consider the well-known functional

$$\phi : [\mathbb{N} \rightharpoonup \mathbb{N}] \to [\mathbb{N} \rightharpoonup \mathbb{N}]$$
$$\phi(f)(k) = \begin{cases} 1 & \text{if } k = 0, \\ kf(k-1) & \text{if } k \geq 1 \;\&\; k-1 \in \operatorname{dom} f. \end{cases}$$

which is easily seen to be monotone. Applying $\mu : [\mathbb{N} \rightharpoonup \mathbb{N}] \to [0,\infty)^*$, we compute

$$\begin{aligned}
\mu\phi(f) &= |\operatorname{dom}(\phi(f))| \\
&= 1 - \sum_{k \in \operatorname{dom}(\phi(f))} \frac{1}{2^{k+1}} \\
&= 1 - \left(\frac{1}{2^{0+1}} + \sum_{k-1 \in \operatorname{dom}(f)} \frac{1}{2^{k+1}} \right) \\
&= 1 - \left(\frac{1}{2} + \sum_{k \in \operatorname{dom}(f)} \frac{1}{2^{k+2}} \right) \\
&= \frac{1}{2} \left(1 - \sum_{k \in \operatorname{dom}(f)} \frac{1}{2^{k+1}} \right) \\
&= \frac{\mu f}{2}
\end{aligned}$$

which means ϕ is a contraction on the domain $[\mathbb{N} \rightharpoonup \mathbb{N}]$. By the contraction principle,

$$\bigsqcup_{n \in \mathbb{N}} \phi^n(\bot) = \mathrm{fac}$$

is the unique fixed point of ϕ on $[\mathbb{N} \rightharpoonup \mathbb{N}]$, where $\bot$ is the function defined nowhere.

9.3.4 Fractals

We now consider certain nontrivial examples of contractions and some of their fixed points: *fractals*. By induction, a continuous map $\mu : D \to [0,\infty)^*$ is a measurement iff for all *finite* $F \subseteq \ker\mu$ and all open sets $U \subseteq D$,

$$F \subseteq U \Rightarrow (\exists\,\varepsilon > 0)(\forall x \in F)\ \mu_\varepsilon(x) \subseteq U.$$

If we require this to hold, not only for finite sets F, but for *all* compact sets K, we have exactly a *Lebesgue measurement.*

Definition 18 A *Lebesgue measurement* $\mu : D \to [0,\infty)^*$ is a continuous map such that for all compact sets $K \subseteq \ker\mu$ and all open sets $U \subseteq D$,

$$K \subseteq U \Rightarrow (\exists\,\varepsilon > 0)(\forall x \in K)\ \mu_\varepsilon(x) \subseteq U.$$

Not all measurements are Lebesgue (Example 5.3.2 of [15]). The existence of a Lebesgue measurement on a domain implies an important relationship between the Scott topology and the *Vietoris topology*:

Definition 19 The *Vietoris hyperspace* of a Hausdorff space X is the set of all nonempty compact subsets $\mathcal{P}_{com}(X)$ with the *Vietoris topology:* it has a basis given by all sets of the form

$$\sigma(U_1,\cdots,U_n) := \{K \in \mathcal{P}_{com}(X) : K \subseteq \bigcup_{i=1}^{n} U_i \text{ and } K \cap U_i \neq \emptyset,\ 1 \le i \le n\},$$

where U_i is a nonempty open subset of X, for each $1 \le i \le n$.

Given a finite number of contractions on a domain (D,μ) with a Lebesgue measurement μ, their union is modelled by a contraction on the *convex powerdomain* which then has a unique fixed point and yields the following result from [24]:

Theorem 11 *Let D be a continuous dcpo such that*

$$(\forall x, y \in D)(\exists z \in D)\ z \sqsubseteq x, y.$$

If $f : D \to D$ and $g : D \to D$ are contractions for which

$$(\exists x \in D)\ x \sqsubseteq f(x)\ \&\ x \sqsubseteq g(x),$$

then there is a unique $K \in \mathcal{P}_{com}(\ker\mu)$ such that $f(K) \cup g(K) = K$. In addition, it is an attractor:

$$(\forall C \in \mathcal{P}_{com}(\ker\mu))\ (f \cup g)^n(C) \to K,$$

in the Vietoris topology on $\mathcal{P}_{com}(\ker\mu)$.

In order to apply these results, we need a simple and clear way to recognize Lebesgue measurements. Let $f : [0,\infty)^2 \to [0,\infty)$ be a function such that $f(x_n, y_n) \to 0$ whenever $x_n, y_n \to 0$.

Theorem 12 *If $\mu : D \to [0,\infty)^*$ is a measurement such that for all pairs $x, y \in D$ with an upper bound,*

$$(\exists\, z \sqsubseteq x, y)\ \mu z \leq f(\mu x, \mu y),$$

then μ is a Lebesgue measurement.

The value of this result is that it identifies a condition satisfied by many of the Lebesgue measurements encountered in practice. For instance, just consider the number of examples covered by $f(s,t) = 2 \cdot \max\{s,t\}$.

Example 25 Lebesgue measurements.

(i) The domain of streams $(\Sigma^\infty, 1/2^{|\cdot|})$.
(ii) The powerset of the naturals $(\mathcal{P}\omega, |\cdot|)$.
(iii) The domain of partial maps $([\mathbb{N} \rightharpoonup \mathbb{N}], |\mathrm{dom}|)$.
(iv) The interval domain $(\mathbf{I}\mathbb{R}, \mu)$.
(v) The upper space $(\mathbf{U}X, \mathrm{diam})$ of a locally compact metric space (X, d).
(vi) The formal ball model $(\mathbf{B}X, \pi)$ of a complete metric space (X, d).

In fact, $f(s,t) = s + t$ applies to (i)–(v), the *triangle inequality*.

We are now going to apply Theorem 11 to obtain the classical result of [11] for hyperbolic iterated function systems on complete metric spaces.

Definition 20 An *iterated function system* (IFS) on a space X is a nonempty finite collection of continuous selfmaps on X. We write an IFS as $(X; f_1, \ldots, f_n)$.

Definition 21 An IFS $(X; f_1, \ldots, f_n)$ is *hyperbolic* if X is a complete metric space and f_i is a contraction for all $1 \leq i \leq n$.

Definition 22 Let (X, d) be a metric space. The *Hausdorff metric* on $\mathcal{P}_{com}(X)$ is

$$d_H(A, B) = \max\{\sup_{a \in A} d(a, B), \sup_{b \in B} d(b, A), \}$$

for $A, B \in \mathcal{P}_{com}(X)$.

Hyperbolic iterated function systems are used to model fractals: Given a fractal image, one searches for a hyperbolic IFS which models it. But what does it mean to model an *image*? The answer is given by Hutchinson's fundamental result [11].

Theorem 13 (Hutchinson) *If $(X; f_1, \ldots, f_n)$ is a hyperbolic IFS on a complete metric space X, then there is a unique nonempty compact subset $K \subseteq X$ such that*

$$K = \bigcup_{i=1}^{n} f_i(K).$$

Moreover, for any nonempty compact set $C \subseteq X$, $\left(\bigcup_{i=1}^{n} f_i\right)^k (C) \to K$ *in the Hausdorff metric* d_H *as* $k \to \infty$.

At this stage, we can see that what will be most difficult in proving such a result is the convergence in the Hausdorff metric. Luckily, this topology is *independent* of the metric d on X.

Theorem 14 *Let* (X, d) *be a metric space. Then the topology induced by the Hausdorff metric* d_H *on* $\mathcal{P}_{com}(X)$ *is the Vietoris topology on* $\mathcal{P}_{com}(X)$.

In [6], the formal ball model $\mathbf{B}X$ is used to give a domain theoretic proof of the existence and uniqueness of the set K in Theorem 13 for any complete metric space (X, d). What is missing from that discussion is the important issue that K is also an attractor with respect to the Hausdorff metric d_H.

Example 26 If we have two contractions $f, g : X \to X$ on a complete metric space X, they have Scott continuous extensions

$$\bar{f}, \bar{g} : \mathbf{B}X \to \mathbf{B}X$$

which are contractions on $\mathbf{B}X$ with respect to $\pi(x, r) = r$. But π is a Lebesgue measurement on a domain which has the property that for all $(x, r), (y, s) \in \mathbf{B}X$, there is an element $z = (x, r + s + d(x, y)) \in \mathbf{B}X$ with $z \sqsubseteq (x, r), (y, s)$. In addition, for any $x \in X$, choosing r so that

$$r \geq \frac{d(x, fx)}{1 - c_f} \text{ and } r \geq \frac{d(x, gx)}{1 - c_g},$$

where $c_f, c_g < 1$ are the Lipschitz constants for f and g, respectively, gives a point $(x, r) \sqsubseteq \bar{f}(x, r), \bar{g}(x, r)$. By Theorem 11,

$$(\exists! K \in \mathcal{P}_{com}(\ker \pi))\ \bar{f}(K) \cup \bar{g}(K) = K.$$

However, because $\ker \pi \simeq X$ and the mappings $\bar{f}, \bar{g}$ extend f and g, it is clear that

$$(\exists! K \in \mathcal{P}_{com}(X))\ f(K) \cup g(K) = K$$

Finally, by Theorems 11 and 14, K is an attractor for $f \cup g$ on $\mathcal{P}_{com}(X)$.

If a space may be realized as the kernel of a Lebesgue measurement on a continuous *dcpo* D, then Theorem 11 implies that Hutchinson's result holds for any finite family of contractions which extend to D. *Necessarily*, two questions arise:

- Which spaces arise as the kernel of a Lebesgue measurement?
- When does a domain admit a Lebesgue measurement?

The answer to the first question is that a space is completely metrizable iff it is the kernel of a Lebesgue measurement on a continuous dcpo, and metrizable iff it is the kernel of a Lebesgue measurement on a continuous poset. The answer to the second question, for an ω continuous dcpo D, is that the set of maximal elements $\max(D)$ is regular iff it is metrizable iff it is the kernel of a Lebesgue measurement on D.

All of this is explained in more detail in [24]. Such results also have interesting implications for general relativity [27].

9.4 Instances of Partiality

We now consider four examples of domains whose descriptions are nontrivial. Our first example is from analysis: the domain of real analytic mappings. The basic idea is to be able to write things like

$$1 \sqsubseteq 1+x \sqsubseteq 1+x+\frac{x^2}{2!} \sqsubseteq \ldots \sqsubseteq \bigsqcup_{n\geq 0}\left(1+\ldots+\frac{x^n}{n!}\right) = e^x$$

Here the polynomials $(1+\ldots+x^n/n!)$ in a Taylor expansion are *partial*, while the analytic map e^x is *total*. On this domain, we see that analytic mappings arise as fixed points of monotone operators which provide schemes for how to compute them as a limit of 'finite approximations' (polynomials).

Our second example concerns finite probability distributions or *classical states*: $e_1 = (1,0,\ldots,0)$, $e_2 = (0,1,0,\ldots,0)$, $\ldots$, $e_n = (0,0,\ldots,1)$ are *total*, while all others are *partial*; in particular, the least informative distribution is $\perp = (1/n,\ldots,1/n)$, which we expect to be a least element in the "domain" of classical states. On this domain, the maximum entropy state of statistical mechanics arises as the least fixed point of a Scott continuous operator that gives a scheme for calculating it.

Our third example is the quantum analogue of the second: the domain of quantum states. In it, pure states $|\psi\rangle\langle\psi|$ are *total*, while all others (the mixed states) are *partial*; in particular, its least element is the completely mixed state $\perp = I/n$. On this domain, unital quantum channels will be seen to have the same domain theoretic properties as binary symmetric channels from classical information theory: they are Scott continuous and have a Scott closed set of fixed points. Later, after we have studied the informatic derivative, we will see that this domain enables us to calculate the Holevo capacity of a unital qubit channel. The domain of quantum states can also be used to recover classical and quantum logic in a unified manner.

Our fourth example is from general relativity: the domain of spacetime intervals. In it, single events $[x,x]$ are *total*, while nontrivial intervals $[p,q]$ are *partial*, in a manner completely analogous to the interval domain $\mathbf{I}\mathbb{R}$. In fact, these two domains have the exact same formal structure, as we will see. The domain of spacetime intervals is used to explain how spacetime, including its geometry, can be reconstructed in a purely order theoretic manner beginning from only a countable dense set. This result may be of interest to those concerned with the causal set approach to quantum gravity.

References: The results in this section are from the following sources: Section 3.1 is from some of the author's unpublished notes (1998), Sect. 3.2 is from [5, 31], Sect.3.3 is from [5, 30] and Sect. 3.4 is from [26, 32, 27].

9.4.1 Analytic Mappings

Real analytic mappings will be represented as infinite lists of rational numbers.

Definition 23 A *list* over $\mathbb{Q}$ is a function $x : \{0 \ldots, n\} \to \mathbb{Q}$ for $n \in \mathbb{N} \cup \{\infty\}$. The *length* of a list x is $|\text{dom}(x)|$. $\mathbb{Q}^\infty$ is the set of both finite and infinite lists over $\mathbb{Q}$.

A finite list x is usually written as a vector $[x_0, \ldots, x_n]$, where $x_i = x(i)$. The empty list has been excluded above because there is no empty polynomial.

Definition 24 The prefix order $\sqsubseteq$ on $\mathbb{Q}^\infty$ is given by

$$x \sqsubseteq y \equiv \text{dom}(x) \subseteq \text{dom}(y) \text{ and } (\forall i \in \text{dom}(x))\, x_i = y_i.$$

In this way, $\mathbb{Q}^\infty$ is an ω-algebraic Scott domain whose compact elements are exactly $\mathbb{Q}_{fin}$.

Definition 25 The *norm* of a power series

$$f(x) = \sum_{n=0}^{\infty} a_n\, x^n$$

is

$$\| f \|(x) = \sum_{n=0}^{\infty} |a_n\, x^n|$$

provided that this sum exists.

To say that a power series has a norm on $[a, b]$ means exactly that it converges absolutely on $[a, b]$.

Lemma 2 *If a function* f *is defined by an absolutely convergent power series on* $[a, b]$*, then* f *and* $\| f \|$ *are both continuous on* $[a, b]$.

Recall that $C[a, b]$ denotes the space of continuous real value function defined on $[a, b]$.

Definition 26 The *degree* of a list p is $|p| :=$ (length p) $- 1$, with the understanding that the degree of an infinite list is ∞. For a list of rationals $a \in \mathbb{Q}^\infty$, we set

$$(\sigma a)x = \sum_{n=0}^{|a|} a_n\, x^n,$$

whenever such a sum exists for all $x \in [a, b]$. This gives a partial mapping $\sigma : \mathbb{Q}^\infty \to C[a, b]$. A list p is *analytic* on $[a, b]$ if $\|(\sigma p)r\| < \infty$ where $r = \max\{|a|, |b|\}$.

Observe that σ is defined for *any* analytic list.

Corollary 3 *For every analytic $p \in \mathbb{Q}^\infty$, $\sigma p \in C[a, b]$ & $\|\sigma p\| \in C[a, b]$.*

For $f, g \in C[a, b]$, the *uniform metric* is

$$d(f, g) = \sup\{|f(x) - g(x)| : x \in [a, b]\}$$

With these preliminaries out the way, we can now order analytic mappings:

Definition 27 The set

$$\mathbb{P}^\infty[a, b] := \{(p, r) : p \text{ analytic}, r \in [0, \infty)^*\}$$

is ordered by

$$(p, r) \sqsubseteq (q, s) \equiv p \sqsubseteq q \text{ and } d(\|\sigma p\|, \|\sigma q\|) \leq r - s.$$

Theorem 15 *$\mathbb{P}^\infty[a, b]$ is an ω-continuous dcpo with a countable basis given by*

$$\{(p, r) : p \text{ finite}, r \in \mathbb{Q} \,\&\, r \geq 0\}.$$

Its approximation relation is

$$(p, r) \ll (q, s) \Leftrightarrow p \text{ finite} \,\&\, d(||\sigma p||, ||\sigma q||) < r - s$$

and its natural measurement $\mu : \mathbb{P}^\infty[a, b] \to [0, \infty)^*$ given by

$$\mu(p, r) = r + \frac{1}{2^{|p|}}$$

measures all of $\mathbb{P}^\infty[a, b]$, has $\ker \mu = \max(\mathbb{P}^\infty[a, b])$ and satisifies the triangle inequality: for all pairs $x, y \in \mathbb{P}^\infty[a, b]$ with an upper bound, there is $z \sqsubseteq x, y$ with $\mu z \leq \mu x + \mu y$.

We adopt the convention of writing

$$\mathcal{P}_{fin}[a, b] = \{(p, r) \in \mathbb{P}^\infty[a, b] : p \text{ finite}, r \geq 0\}$$

Proposition 4 *If $f : \mathcal{P}_{fin}[a, b] \to E$ is a monotone map into a dcpo such that*

$$f(p, r) = \bigsqcup f(p, r + 1/n)$$

for p finite, then f may be extended uniquely to a Scott continuous map on all of $\mathbb{P}^\infty[a, b]$.

A mapping f of the type discussed in the previous result is said to be *invariant on polynomials*.

Example 27 The unary operation addition by 1

$$(p, r) \mapsto ([a_0 + 1, \dots, a_n], r)$$

is monotone and invariant on polynomials, so it extends uniquely to $\mathbb{P}^\infty[a, b]$.

At times we may blur the distinction between polynomials and lists of rational numbers, that is, we will treat them as one and the same for the purpose of illustrating various points about mappings on $\mathbb{P}^\infty[a, b]$ and the *functions* they act on. Now for a nontrivial example.

Example 28 Let $[a, b]$ be an interval containing 0 and define

$$I : \mathcal{P}_{fin}[a, b] \to \mathbb{P}^\infty[a, b]$$
$$I(p, r) = (\int_0^x p(t)\, dt, m \cdot r)$$

where $m = \max\{|a|, |b|\}$ and $\int_0^x p(t)\ dt$ is the list operation taking $p = [a_0, \dots, a_n]$ to $[0, a_0, \dots, a_n/(n + 1)]$. This mapping is monotone and invariant on polynomials so it has a unique Scott continuous extension to all of $\mathbb{P}^\infty[a, b]$, which we denote by $\int_0^x \cdot$.

From a *symbolic* definition of integral for polynomials, the one we normally program when implementing the polynomial data type, we systematically obtain a definition of integral for analytic mappings.

Example 29 The exponential map. Consider the operator

$$\exp : \mathbb{P}^\infty[-c, c] \to \mathbb{P}^\infty[-c, c]$$
$$\exp(p, r) = 1 + \int_0^x (p, r)$$

for $0 < c < 1$, the Scott continuous map $\int_0^x$ composed with the unary Scott continuous operator that adds 1. Since $(1, r) \sqsubseteq \exp(1, r)$ for $r \geq 1$, Scott continuity gives a fixed point

$$\text{fix}(\exp) = \bigsqcup_{n \geq 0} \exp^n(1, 1)$$

that is easily seen to be the exponential map e^x. This fixed point is *unique*.

First, because exp is a contraction with respect to the natural measurement μ with $\mu(\exp) \leq \max\{c, 1/2\} \cdot \mu$, any other fixed point $\exp(p, r) = (p, r)$ yields $\mu(p, r) = 0$. But any fixed point $(p, 0)$ must also have $1 \sqsubseteq p$. Let $r := d(1, \|\sigma p\|) + 1 \geq 1$.

Then since $(1, r) \sqsubseteq (p, 0)$, we have a lower bound for both $(p, 0)$ and fix(exp), which gives $(p, 0) = \text{fix}(\exp)$ since exp is a contraction.

Notice that $\exp^n(1, 1)$ is the nth-degree Taylor approximation of the maximal element e^x. In addition, the smaller the interval $[-c, c]$, the smaller that c is, the quicker that exp converges to e^x. Thus, using the domain of analytic mappings we see that *fewer terms of the Taylor series* are required to approximate e^x on the interval $[-c/2, c/2]$ than on the interval $[-c, c]$.

In a similar way one can realize the sine and cosine functions as unique fixed points of Scott continuous mappings.

Example 30 The trigonometric functions. The operator for the sine is

$$\phi(p, r) = x - \int_0^x \int_0^y (p, r).$$

The operator for the cosine is

$$\phi(p, r) = 1 - \int_0^x \int_0^y (p, r).$$

The iteration for the first begins with the polynomial x, and the second begins with the polynomial 1.

9.4.2 Classical States

Definition 28 Let $n \geq 2$. The *classical states* are

$$\Delta^n := \left\{ x \in [0, 1]^n : \sum_{i=1}^{n} x_i = 1 \right\}.$$

A classical state $x \in \Delta^n$ is *pure* when $x_i = 1$ for some $i \in \{1, \ldots, n\}$; we denote such a state by e_i.

Pure states $\{e_i\}_i$ are the actual states a system can be in, while general mixed states x and y are epistemic entities. Imagine that one of n different outcomes is possible. If our knowledge of the outcome is $x \in \Delta^n$, and then by some means we determine that outcome i is not possible, our knowledge improves to

$$p_i(x) = \frac{1}{1 - x_i}(x_1, \ldots, \hat{x}_i, \ldots, x_{n+1}) \in \Delta^n,$$

where $p_i(x)$ is obtained by first removing x_i from x and then renormalizing. The partial mappings which result, $p_i : \Delta^{n+1} \rightharpoonup \Delta^n$ with $\text{dom}(p_i) = \Delta^{n+1} \setminus \{e_i\}$, are called the *Bayesian projections* and lead one to the following relation on classical states.

Definition 29 For $x, y \in \Delta^{n+1}$,

$$x \sqsubseteq y \equiv (\forall i)(x, y \in \mathrm{dom}(p_i) \Rightarrow p_i(x) \sqsubseteq p_i(y)).$$

For $x, y \in \Delta^2$,

$$x \sqsubseteq y \equiv (y_1 \leq x_1 \leq 1/2) \text{ or } (1/2 \leq x_1 \leq y_1).$$

The relation $\sqsubseteq$ on Δ^n is called the *Bayesian order.*

As we can see, the definition of Δ^{n+1} from Δ^n is natural. The order on Δ^2, is derived from the graph of entropy $H(x) = -x \log_2(x) - (1-x)\log_2(1-x)$ as follows:

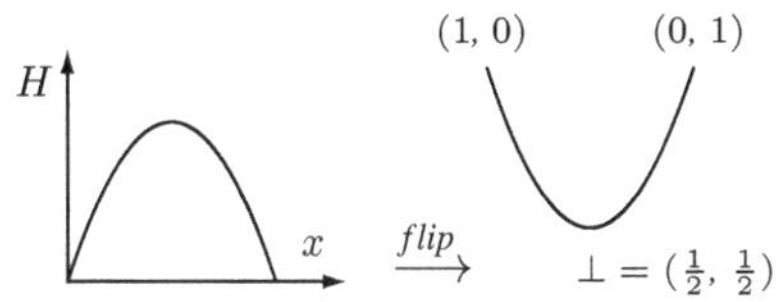

It is canonical in the following sense:

Theorem 16 *There is a unique partial order on Δ^2 that satisfies the mixing law*

$$x \sqsubseteq y \text{ and } p \in [0, 1] \Rightarrow x \sqsubseteq (1-p)x + py \sqsubseteq y$$

and has $\perp := (1/2, 1/2)$ as a least element. It is the Bayesian order on classical two states.

The Bayesian order was discovered in [5] where the following is proven:

Theorem 17 *$(\Delta^n, \sqsubseteq)$ is a dcpo with least element $\perp := (1/n, \ldots, 1/n)$ and $\max(\Delta^n) = \{e_i : 1 \leq i \leq n\}$. It has Shannon entropy*

$$\mu x = -\sum_{i=1}^{n} x_i \log x_i$$

as a measurement of type $\Delta^n \to [0, \infty)^*$.

A more subtle example of a measurement on Δ^n in its Bayesian order is the retraction $r : \Delta^n \to \Lambda^n$ which rearranges the probabilities in a classical state into descending order.

The Bayesian order has a more direct description: the *symmetric formulation.* Let $S(n)$ denote the group of permutations on $\{1, \ldots, n\}$ and

$$\Lambda^n := \{x \in \Delta^n : (\forall i < n)\, x_i \geq x_{i+1}\}$$

denote the collection of *monotone* decreasing classical states. It can then be shown [5] that for $x, y \in \Delta^n$, we have $x \sqsubseteq y$ iff there is a permutation $\sigma \in S(n)$ such that $x \cdot \sigma, y \cdot \sigma \in \Lambda^n$ and

$$(x \cdot \sigma)_i (y \cdot \sigma)_{i+1} \leq (x \cdot \sigma)_{i+1} (y \cdot \sigma)_i$$

for all i with $1 \leq i < n$. Thus, $(\Delta^n, \sqsubseteq)$ can be thought of as $n!$ many copies of the domain $(\Lambda^n, \sqsubseteq)$ identified along their common boundaries, where $(\Lambda^n, \sqsubseteq)$ is

$$x \sqsubseteq y \equiv (\forall i < n)\, x_i y_{i+1} \leq x_{i+1} y_i.$$

It should be remarked though that the problems of ordering Λ^n and Δ^n are very different, with the latter being far more challenging, especially if one also wants to consider quantum mixed states. Let us now consider an important application of the Bayesian order to give a method for calculating the maximum entropy state of statistical mechanics.

9.4.2.1 The Maximum Entropy Principle

The possible outcomes of an event are $a_1, \ldots, a_n$. It is repeated many times and an average value of E is observed. What is the probability p_i of a_i? The *maximum entropy principle* provides an approach to solve this problem: because Shannon entropy has a maximum value on the set

$$\left\{ p \in \Delta^n : \sum_{i=1}^{n} p_i \cdot a_i = E \right\}$$

that is assumed at exactly one point, one possibility is to use this state as the probability distribution that models our observed data. Beautiful—but how do we calculate this distribution?

Define

$$f(x) = \frac{\sum_{i=1}^{n} a_i e^{x a_i}}{\sum_{i=1}^{n} e^{x a_i}} - E, \quad I_f(x) = x - \frac{f(x)}{(a_n - a_1)^2}$$

for any $x \in \mathbb{R}$. Define $\lambda : \Delta^n \to \mathbb{R} \cup \{\pm\infty\}$ by

$$\lambda(x) = \begin{cases} \dfrac{\log\left(\frac{\text{sort}(x)_1}{\text{sort}(x)_2}\right)}{a_n - a_{n-1}} & \text{if } I_f(0) > 0; \\[2ex] \dfrac{\log\left(\frac{\text{sort}(x)_1}{\text{sort}(x)_2}\right)}{a_1 - a_2} & \text{otherwise.} \end{cases}$$

with the understanding for pure states that $\lambda x = \infty$ in the first case and $\lambda x = -\infty$ in the other. The map sort puts states into *decreasing* order.

Theorem 18 *Let $a_1 < E < a_n$. The map*

$$\phi : \Delta^n \to \Delta^n$$

given by

$$\phi(x) = (e^{I_f(\lambda x)a_1}, \ldots, e^{I_f(\lambda x)a_n}) \cdot \frac{1}{Z(x)}$$

$$Z(x) = \sum_{i=1}^{n} e^{I_f(\lambda x)a_i}$$

is Scott continuous in the Bayesian order. Its least fixed point is the maximum entropy state.

The maximum entropy principle has been successfully applied to perform image reconstruction from noisy data, probabilistic link extraction from intelligence data, natural language processing, stock price volatility, thermodynamics.

9.4.3 Quantum States

Let $\mathcal{H}^n$ denote an n-dimensional complex Hilbert space with specified inner product $\langle\cdot|\cdot\rangle$.

Definition 30 A *quantum state* is a density operator $\rho : \mathcal{H}^n \to \mathcal{H}^n$, i.e., a self-adjoint, positive, linear operator with $\mathrm{tr}(\rho) = 1$. The quantum states on $\mathcal{H}^n$ are denoted Ω^n.

Definition 31 A quantum state ρ on $\mathcal{H}^n$ is *pure* if

$$\mathrm{spec}(\rho) \subseteq \{0, 1\}.$$

The set of pure states is denoted Σ^n. They are in bijective correspondence with the one dimensional subspaces of $\mathcal{H}^n$.

Classical states are distributions on the set of pure states $\max(\Delta^n)$. By Gleason's theorem, an analogous result holds for quantum states: Density operators encode distributions on the set of pure states Σ^n.

Definition 32 A *quantum observable* is a self-adjoint linear operator $e : \mathcal{H}^n \to \mathcal{H}^n$.

An observable of a physical system is anything about it that we can measure. For example, *energy* is an observable. Observables in quantum mechanics are represented mathematically by self-adjoint operators.

If we have the operator e representing the energy observable of a system (for instance), then its set of eigenvalues $\mathrm{spec}(e)$, called the *spectrum* of e, consists of the actual energy values a system may assume. If our knowledge about the state of the system is represented by density operator ρ, then quantum mechanics predicts the probability that a measurement of observable e yields the value $\lambda \in \mathrm{spec}(e)$. It is

$$\mathrm{pr}(\rho \to e_\lambda) := \mathrm{tr}(p_e^\lambda \cdot \rho),$$

where p_e^λ is the projection corresponding to eigenvalue λ and e_λ is its associated eigenspace in the *spectral representation* of e.

Definition 33 Let e be an observable on $\mathcal{H}^n$ with $\mathrm{spec}(e) = \{1, \ldots, n\}$. For a quantum state ρ on Ω^n,

$$\mathrm{spec}(\rho|e) := (\mathrm{pr}(\rho \to e_1), \ldots, \mathrm{pr}(\rho \to e_n)) \in \Delta^n.$$

We assume that all observables e have $\mathrm{spec}(e) = \{1, \ldots, n\}$. For our purposes it is enough to assume $|\mathrm{spec}(e)| = n$; the set $\{1, \ldots, n\}$ is chosen for the sake of aesthetics. Intuitively, then, e is an experiment on a system which yields one of n different outcomes; if our a priori knowledge about the state of the system is ρ, then our knowledge about what the result of experiment e *will be* is $\mathrm{spec}(\rho|e)$. Thus, $\mathrm{spec}(\rho|e)$ determines our ability to *predict* the result of the experiment e.

Let us point out that $\mathrm{spec}(\rho) = \mathrm{Im}(\mathrm{spec}(\rho|e))$ and $\mathrm{spec}(\sigma) = \mathrm{Im}(\mathrm{spec}(\sigma|e))$ are equivalent to $[\rho, e] = 0$ and $[\sigma, e] = 0$, where $[a, b] = ab - ba$ is the commutator of operators.

Definition 34 Let $n \geq 2$. For quantum states $\rho, \sigma \in \Omega^n$, we have $\rho \sqsubseteq \sigma$ iff there is an observable $e : \mathcal{H}^n \to \mathcal{H}^n$ such that $[\rho, e] = [\sigma, e] = 0$ and $\mathrm{spec}(\rho|e) \sqsubseteq \mathrm{spec}(\sigma|e)$ in Δ^n.

This is called the *spectral order* on quantum states.

Theorem 19 *$(\Omega^n, \sqsubseteq)$ is a dcpo with maximal elements $\max(\Omega^n) = \Sigma^n$ and least element $\bot = I/n$, where I is the identity matrix. It has von Neumann entropy*

$$\sigma\rho = -\mathrm{tr}(\rho \log \rho)$$

as a measurement of type $\Omega^n \to [0, \infty)^*$.

Another natural measurement on Ω^n is the map $q : \Omega^n \to \Lambda^n$ which assigns to a quantum state its spectrum rearranged into descending order. It can be thought of as an important link between classical and quantum information theory.

There is one case where the spectral order can be described in an elementary manner.

Example 31 The 2×2 density operators Ω^2 can be represented as points on the unit ball in $\mathbb{R}^3$:

$$\Omega^2 \simeq \{(x, y, z) \in \mathbb{R}^3 : x^2 + y^2 + z^2 \leq 1\}.$$

For example, the origin $(0, 0, 0)$ corresponds to the completely mixed state $I/2$, while the points on the surface of the sphere describe the pure states. The order on Ω^2 then amounts to the following: $x \sqsubseteq y$ iff the line from the origin $\perp$ to y passes through x.

Let us now consider an application of Ω^2 to the study of communication.

9.4.3.1 Classical and Quantum Communication

The classical channels $f : \Delta^2 \to \Delta^2$ which increase entropy $(H(f(x)) \geq H(x))$ are exactly those f with $f(\perp) = \perp$. They are the *strict* mappings of domain theory, which are also known as *binary symmetric channels* in information theory. Similarly, the entropy increasing qubit channels are exactly those channels[2] $\varepsilon : \Omega^2 \to \Omega^2$ for which $\varepsilon(\perp) = \perp$. These are called *unital* in quantum information theory.

Definition 35 A qubit channel $\varepsilon : \Omega^2 \to \Omega^2$ is *unital* if $\varepsilon(\perp) = \perp$.

Theorem 20

- *A classical channel $f : \Delta^2 \to \Delta^2$ is binary symmetric iff it is Scott continuous and its set of fixed points is Scott closed.*
- *A quantum channel $f : \Omega^2 \to \Omega^2$ is unital if and only if it is Scott continuous and its set of fixed points is Scott closed.*

In fact, this last result hints at how to establish the uniqueness of Ω^2, in a manner completely similar to the corresponding result for Δ^2:

Theorem 21 *There is a unique partial order on Ω^2 with the following three properties:*

(i) *It has least element $\perp = I/2$,*
(ii) *It satisfies the mixing law: if $r \sqsubseteq s$, then $r \sqsubseteq tr+(1-t)s \sqsubseteq s$, for all $t \in [0, 1]$,*
(iii) *Every unital channel $f : \Omega^2 \to \Omega^2$ is Scott continuous and has a Scott closed set of fixed points.*

It is the spectral order, and gives Ω^2 the structure of a Scott domain.

Finally, let us turn to one last application of the spectral order.

9.4.3.2 Classical and Quantum Logic

The logics of Birkhoff and von Neumann consist of the propositions one can make about a physical system. Each proposition takes the form "The value of observable e is contained in $E \subseteq \text{spec}(e)$." For classical systems, the logic is $\mathcal{P}\{1, \ldots, n\}$, while for quantum systems it is $\mathbb{L}^n$, the lattice of (closed) subspaces of $\mathcal{H}^n$. In each case,

[2] Quantum channels are completely positive and convex linear, see [35] for more.

implication of propositions is captured by inclusion, and a fundamental distinction between classical and quantum—that there are pairs of quantum observables whose exact values cannot be simultaneously measured at a single moment in time—finds lattice theoretic expression: $\mathcal{P}\{1,\ldots,n\}$ is distributive; $\mathbb{L}^n$ is not.

Remarkably, the classical and quantum logics can be *derived* from the Bayesian and spectral orders using the *same* order theoretic technique.

Definition 36 An element x of a dcpo D is *irreducible* when

$$\bigwedge(\uparrow x \cap \max(D)) = x$$

The set of irreducible elements in D is written $\mathrm{Ir}(D)$.

The order dual of a poset $(D, \sqsubseteq_D)$ is written D^*; its order is $x \sqsubseteq y \Leftrightarrow y \sqsubseteq_D x$.

Theorem 22 *For $n \geq 2$, the classical lattices arise as*

$$\mathrm{Ir}(\Delta^n)^* \simeq \mathcal{P}\{1,\ldots,n\} \setminus \{\emptyset\},$$

and the quantum lattices arise as

$$\mathrm{Ir}(\Omega^n)^* \simeq \mathbb{L}^n \setminus \{0\}.$$

9.4.4 Spacetime Intervals

General relativity is Einstein's theory of gravity in which gravity is understood not in terms of mysterious "universal" forces but rather as part of the geometry of spacetime. It is profoundly beautiful and beautifully profound from both the physical and mathematical viewpoints and it teaches us clear lessons about the universe in which we live that are easily explainable. For example, it offers a wonderful explanation of gravity: if an apple falls from a tree, the path it takes is not determined by the Newtonian ideal of an "invisible force" but instead by the curvature of the space in which the apple resides: gravity is the curvature of spacetime. In addition, the presence of matter in spacetime causes it to "bend" and Einstein even gives us an equation that relates the curvature of spacetime to the matter present within it. *However.*

Since everything attracts everything else, a gravitating mass of sufficient size will eventually collapse. In 1965, Penrose [36] showed that any such collapse eventually leads to a singularity where the mathematical description of spacetime as a continuum breaks down. This leads to the need to reformulate gravity. It is hoped that the elusive quantum theory of gravity will resolve this problem.

Since the first singularity theorems [36, 10], causality has played a key role in understanding spacetime structure. The analysis of causal structure relies heavily on techniques of differential topology [37]. For the past decade Sorkin and others [42] have pursued a program for quantization of gravity based on causal structure. In this approach the causal relation is regarded as the fundamental ingredient and the topology and geometry are secondary.

In this section, we will see that the causal structure of spacetime is captured by a *domain* and learn the surprising connection between measurement, the Newtonian concept of time, and the geometry of spacetime.

Definition 37 A continuous poset $(P, \leq)$ is *bicontinuous* if

- For all $x, y \in P$, $x \ll y$ iff for all filtered $S \subseteq P$ with an infimum,

$$\bigwedge S \leq x \Rightarrow (\exists s \in S)\, s \leq y,$$

  and
- For each $x \in P$, the set $\mathord{\uparrow\!\!\uparrow} x$ is filtered with infimum x.

We tend to prefer the notation $\leq$ for the order on a poset that is known to be bicontinuous. For x, y in a poset $(P, \leq)$,

$$x < y \equiv x \leq y \;\&\; x \neq y.$$

In general, $<$ and $\ll$ are completely different ideas.

Example 32 $(\mathbb{R}, \leq)$, $(\mathbb{Q}, \leq)$ are bicontinuous.

Definition 38 The *interval topology* on a continuous poset P exists when sets of the form

$$(a, b) = \{x \in P : a \ll x \ll b\} \;\;\&\;\; \mathord{\uparrow\!\!\uparrow} x = \{y \in P : x \ll y\}$$

form a basis for a topology on P.

Notice that on a *bicontinuous poset*, the interval topology exists and has

$$(a, b) := \{x \in P : a \ll x \ll b\}$$

as a basis.

A *manifold* $\mathcal{M}$ is a locally Euclidean Hausdorff space that is connected and has a countable basis. Such spaces are paracompact. A *Lorentz metric* on a manifold is a symmetric, nondegenerate tensor field of type $(0, 2)$ whose signature is $(- + + +)$.

Definition 39 A *spacetime* is a real four-dimensional[3] smooth manifold $\mathcal{M}$ with a Lorentz metric g_{ab}.

Let $(\mathcal{M}, g_{ab})$ be a time-orientable spacetime. Let $\Pi^+_{\leq}$ denote the future directed causal curves, and $\Pi^+_{\ll}$ denote the future directed time-like curves.

---

[3] The results in the present paper work for any dimension $n \geq 2$ [26].

Definition 40 For $p \in \mathcal{M}$,

$$I^+(p) := \{q \in \mathcal{M} : (\exists \pi \in \Pi^+_{\ll})\, \pi(0) = p, \pi(1) = q\}$$

and

$$J^+(p) := \{q \in \mathcal{M} : (\exists \pi \in \Pi^+_{\leq})\, \pi(0) = p, \pi(1) = q\}$$

Similarly, we define $I^-(p)$ and $J^-(p)$.

We write the relation J^+ as

$$p \leq q \equiv q \in J^+(p).$$

The "Alexandroff topology" on a spacetime has $\{I^+(p) \cap I^-(q) : p, q \in \mathcal{M}\}$ as a basis; a spacetime $\mathcal{M}$ is strongly causal iff its Alexandroff topology is Hausdorff iff its Alexandroff topology is the manifold topology. Penrose has called *globally hyperbolic* spacetimes "the physically reasonable spacetimes [44]."

Definition 41 A spacetime $\mathcal{M}$ is *globally hyperbolic* if it is strongly causal and if $\uparrow a \cap \downarrow b$ is compact in the manifold topology, for all $a, b \in \mathcal{M}$.

Theorem 23 *If $\mathcal{M}$ is globally hyperbolic, then $(\mathcal{M}, \leq)$ is a bicontinuous poset with $\ll = I^+$ whose interval topology is the manifold topology.*

This result motivates the following definition:

Definition 42 A poset $(X, \leq)$ is *globally hyperbolic* if it is bicontinuous and each interval $[a, b] = \{x : a \leq x \leq b\}$ is compact in the interval topology.

Globally hyperbolic posets have rich enough structure that we can deduce many properties of spacetime from them *without* appealing to differentiable structure or geometry, such as the compactness of the space of causal curves [27]. We can also deduce new aspects of spacetime. Globally hyperbolic posets are very much like the real line. In fact, a well-known domain theoretic construction pertaining to the real line extends in perfect form to the globally hyperbolic posets:

Theorem 24 *The closed intervals of a globally hyperbolic poset X*

$$\mathbf{I}X := \{[a, b] : a \leq b \;\&\; a, b \in X\}$$

ordered by reverse inclusion

$$[a, b] \sqsubseteq [c, d] \equiv [c, d] \subseteq [a, b]$$

form a continuous domain with

$$[a, b] \ll [c, d] \equiv a \ll c \;\&\; d \ll b.$$

The poset X has a countable basis iff $\mathbf{I}X$ *is* ω*-continuous. Finally,*

$$\max(\mathbf{I}X) \simeq X$$

where the set of maximal elements has the relative Scott topology from $\mathbf{I}X$.

In fact, more is true: in [26] it is shown that the category of globally hyperbolic posets is naturally isomorphic to the category of *interval domains*. This observation – that spacetime has a canonical domain theoretic model – teaches us something new: from only a countable set of events and the causality relation, one can reconstruct spacetime in a purely order theoretic manner. Explaining this requires domain theory.

9.4.4.1 Reconstruction of the Spacetime Manifold

An *abstract basis* is a set $(C, \ll)$ with a *transitive* relation that is *interpolative* from the – *direction*:

$$F \ll x \Rightarrow (\exists y \in C)\, F \ll y \ll x,$$

for all finite subsets $F \subseteq C$ and all $x \in F$. Suppose, though, that it is also interpolative from the + *direction*:

$$x \ll F \Rightarrow (\exists y \in C)\, x \ll y \ll F.$$

Then we can define a new abstract basis of *intervals*

$$\mathrm{int}(C) = \{(a, b) : a \ll b\} = \ll \subseteq C^2$$

whose relation is

$$(a, b) \ll (c, d) \equiv a \ll c \,\&\, d \ll b.$$

Let $\mathbf{I}C$ denote the ideal completion of the abstract basis $\mathrm{int}(C)$.

Theorem 25 *Let C be a countable dense subset of a globally hyperbolic spacetime $\mathcal{M}$ and $\ll = I^+$ be timelike causality. Then*

$$\max(\mathbf{I}C) \simeq \mathcal{M}$$

where the set of maximal elements have the Scott topology.

Theorem 25 is very different from results like "Let $\mathcal{M}$ be a certain spacetime with relation $\leq$. Then the interval topology is the manifold topology." Here we identify, in abstract terms, a process by which a countable set with a causality relation determines a space. The process is entirely order theoretic in nature, spacetime is not required to understand or execute it (i.e., if we put $C = \mathbb{Q}$ and $\ll = <$, then $\max(\mathbf{I}C) \simeq \mathbb{R}$). In this sense, our understanding of the relation between causality and the topology of spacetime is now explainable independently of geometry. Ideally, one would now like to know what constraints on C in general imply that $\max(\mathbf{I}C)$ is a manifold.

9.4.4.2 Time and Measurement

A global time function $t : \mathcal{M} \to \mathbb{R}$ on a globally hyperbolic spacetime $\mathcal{M}$ is a continuous function such that $x < y \Rightarrow t(x) < t(y)$ and $t^{-1}(r) = \Sigma$ is a Cauchy surface for $\mathcal{M}$, for each $r \in \mathbb{R}$.

Theorem 26 *For any global time function $t : \mathcal{M} \to \mathbb{R}$ on a globally hyperbolic spacetime, the function $\Delta t : \mathcal{M} \to [0, \infty)^*$ given by $\Delta t[a, b] = t(b) - t(a)$ measures all of $\mathbf{I}(\mathcal{M})$. It is a measurement with* $\ker(\Delta t) = \max(\mathbf{I}(\mathcal{M}))$.

Let $d : \mathbf{I}(\mathcal{M}) \to [0, \infty)^*$ denote the Lorentz distance on a globally hyperbolic spacetime

$$d[a, b] = \sup_{\pi_{ab}} \operatorname{len}(\pi_{ab})$$

where the sup is taken over all causal curves that join a to b.

A function between continuous posets is *interval continuous* when each poset has an interval topology and the inverse image of an interval open set is interval open. By the bicontinuity of $\mathcal{M}$, the interval topology on $\mathbf{I}(\mathcal{M})$ exists, so we can consider interval continuity for functions $\mathbf{I}(\mathcal{M}) \to [0, \infty)^*$.

Theorem 27 *The Lorentz distance $d : \mathbf{I}(\mathcal{M}) \to [0, \infty)^*$ has the following properties:*

(i) *It is monotone: $x \leq y \Rightarrow d(x) \geq d(y)$,*
(ii) *It preserves the way below relation: $x \ll y \Rightarrow d(x) > d(y)$,*
(iii) *It is interval continuous and hence, by (i), Scott continuous.*

It does not measure $\mathbf{I}(\mathcal{M})$ at any point of $\ker(d)$.

That the Lorentz distance is not a measurement has all to do with relativity: it is a direct consequence of the fact that a clock travelling at the speed of light records no time as having elapsed i.e. the set of null intervals is equal to

$$\ker(d) \setminus \max(\mathbf{I}(\mathcal{M})) \neq \emptyset$$

but measurements μ always satisfy $\ker(\mu) \subseteq \max(D)$ (Lemma 1).

In fact, no interval continuous function $\mu : \mathbf{I}(\mathcal{M}) \to [0, \infty)^*$ can be a measurement: by interval continuity, $\mu x = 0$ for any x with $\Uparrow x = \emptyset$. Then just like the

Lorentz distance, an interval continuous μ will also assign 0 to "null intervals." In this way, we see that interval continuity captures an essential aspect of the Lorentz distance: interval continuous functions do not distinguish between single events and null intervals. In addition, since Δt is a measurement, it cannot be interval continuous. This provides a surprising *topological* distinction between the Newtonian and relativistic concepts of time: d is interval continuous, Δt is not. Put another way, Δt can be used to reconstruct the *topology* of spacetime (Theorem 2), while d is used to reconstruct its *geometry*.

9.4.4.3 Reconstruction of Spacetime Geometry

Specifically, if in addition to $\mathrm{int}(C)$ we also begin with a countable collection of numbers l_{ab} chosen for each $(a, b) \in \mathrm{int}(C)$ in such a way that the map

$$\mathrm{int}(C) \to [0, \infty)^* :: (a, b) \mapsto l_{ab}$$

is monotone, then in the process of reconstructing spacetime, we can also construct the Scott continuous function $d : \mathbf{I}C \to [0, \infty)^*$ given by

$$d(x) = \inf\{l_{ab} : (a, b) \ll x\}.$$

In the event that the countable number of l_{ab} chosen are the Lorentz distances $l_{ab} = d[a, b]$, then the function d constructed above yields the Lorentz distance for any spacetime interval, the reason being that both are Scott continuous and are equal on a basis of the domain.

Thus, from a countable dense set of events and a countable set of distances, we can reconstruct the spacetime manifold together with its geometry in a purely order theoretic manner.

9.5 The Informatic Derivative

Major references: [15, 20, 30]

9.5.1 In a Single Measurement

Recall the seemingly innocent definition of the μ *topology* from Sect. 9.2.4:

Definition 43 The μ *topology* on a continuous dcpo D has as a basis all sets of the form $\Uparrow x \cap \downarrow y$ where $x, y \in D$. It is denoted μ_D.

This also turns out to be the topology one needs to define rates of change on a domain. This comes as something of a surprise since the μ topology is *always zero-dimensional and Hausdorff.*

Definition 44 Let D be a continuous dcpo with a map $\mu : D \to [0, \infty)^*$ that measures $X \subseteq D$. If $f : D \to D$ is a function and $p \in X$ is not a compact element of D, then

$$df_\mu(p) := \lim_{x \to p} \frac{\mu f(x) - \mu f(p)}{\mu x - \mu p}$$

is called *the informatic derivative* of f at p with respect to μ, provided that it exists. The limit above is taken with respect to the μ topology.

If the limit above exists, then it is unique, since the μ topology is Hausdorff, and we are taking a limit at a point that is not *isolated*: $\{p\}$ is μ open iff p is compact. Notice too the importance of strict monotonicity of μ in Lemma 1: without it, we could not define the derivative. The definition of informatic derivative has a simple extension to functions $f : D \to E$ between domains with measurements (D, μ) and (E, λ) [15].

Our first example comes from calculus and provided the first relationship between domain theory and the differential calculus [15].

Theorem 28 *Let $f : \mathbb{R} \to \mathbb{R}$ be a continuous map on the real line with $p \in \mathbb{R}$. If $f'(p)$ exists, then*

$$d\bar{f}_\mu[p] = |f'(p)|$$

where $\bar{f}(x) = f(x)$ is the canonical extension of f to $\mathbf{I}\mathbb{R}$ and $\mu[a, b] = b - a$.

In particular, any iterative process with a classical derivative has an informatic derivative, and from the complexity viewpoint, they are equal. In fact, it can be shown that $d\bar{f}_\mu$ exists and is continuous iff f has a continuous first derivative i.e. the informatic derivative is equivalent to the classical derivative for C^1 functions. However, in general, informatic differentiability of $\bar{f}$ is strictly more general than classical differentiability [25].

9.5.2 The Derivative at a Fixed Point

It often happens that partial maps on spaces have fixed points which are *unknown*. For example, the polynomial $p : \mathbb{R} \to \mathbb{R}$ given by $p(x) = x^3 + x - 1$ has a zero on $[0, 1]$ because $p(0) \cdot p(1) < 0$. Consequently, $f(x) = x - p(x)$ has a fixed point on $[0, 1]$, even though we are not sure of what it is.

Because a partial map $f : X \rightharpoonup X$ on a space X may have an *unknown* fixed point p, methods for calculating it are important. A minimal requirement is usually that p be an *attractor*: that there exist an open set $U \subseteq X$ such that for all $x \in U$, $f^n(x) \to p$. This provides a simple scheme for approximating p: simply calculate the iterates $f^n(x)$ beginning with any $x \in U$.

In Sects. 9.3 and 9.4 we saw many examples of numerical methods and monotone maps (when restricted to $I(f) = \{x : x \sqsubseteq f(x)\}$) which give rise to partial splittings that converge to fixed points.

Lemma 3 *Let $s : D \rightharpoonup D$ be a partial splitting which maps into* $\mathrm{dom}(s)$. *If $s(p) = p$ and $ds_\mu(p)$ exists, then $ds_\mu(p) \le 1$.*

So we consider partial maps f with fixed points p such that $df_\mu(p) \le 1$. The identity map $1 : D \to D$ has $d(1)_\mu(p) = 1$ at any element which is not compact, meaning that a map whose derivative is unity need not have an attractive point. However, if $df_\mu(p) < 1$, then we can say something: for monotone maps with fixed points in the kernel, we have an attractor in the Scott topology.

Theorem 29 *Let $f : (D, \mu) \to (D, \mu)$ be a monotone mapping with $f(\ker\mu) \subseteq \ker\mu$. If $df_\mu(p) < 1$ at a fixed point $f(p) = p \in \ker\mu$, then there is an approximation $a \ll p$ such that*

(i) *For all $x \in D$, if $a \sqsubseteq x \sqsubseteq p$, then*

$$\bigsqcup_{n \ge 0} f^n(x) = p,$$

and this is a limit in the μ topology on D.

(ii) *The unique fixed point of f on $\uparrow a$ is p.*

(iii) *For all $x \in \ker\mu \cap \uparrow a$, $f^n(x) \to p$ in the Scott topology on* $\ker\mu$.

In [15], it is shown that (i) is equivalent to f being μ *continuous* at p, so we can take (i) as a definition of μ continuity at a fixed point. The bisection method split_f is not necessarily μ continuous at a fixed point if the corresponding zero of f is not isolated.

Corollary 4 *Let $f : \mathbb{R} \to \mathbb{R}$ be a continuous map on the real line with a fixed point $f(p) = p$. If $d\bar{f}_\mu[p] < 1$, then there is an $\varepsilon > 0$ such that*

$$(\forall x \in (p - \varepsilon, p + \varepsilon))\ f^n(x) \to p.$$

In particular, this holds if f is differentiable at p and $|f'(p)| < 1$.

The last corollary applies to continuous maps on the real line that have informatic derivatives but do not have classical derivatives [25]. As an application of Theorem 29, we will prove the correctness of Newton's Method *without* using Taylor's Theorem.

Example 33 Let $f : [a, b] \to \mathbb{R}$ be a continuous function with a zero $r \in (a, b)$. If f' is nonzero and continuous on $[a, b]$ and $f''(r)$ exists, we consider the continuous map $I_f : [a, b] \to \mathbb{R}$, given by

$$I_f(x) = x - \frac{f(x)}{f'(x)}.$$

It is easy to see that $I_f(r) = r$. By extending I_f to the real line in any way whatsoever, we appeal to Theorem 28 and obtain

$$\frac{d\bar{I}_f}{d\mu}[r] = 0.$$

By Corollary 4, we see that there is an $\varepsilon > 0$ such that $I_f(x) \to r$ for all $x \in (r - \varepsilon, r + \varepsilon)$.

But what is achieved by avoiding Taylor's theorem? To prove the correctness of Newton's method using Taylor's theorem, we must assume that f'' exists on an *open interval* containing the zero r. The proof we gave in Example 33 assumes only that $f''(r)$ exists. This gives one definite advantage to using Theorem 29 in place of Taylor's theorem: we can prove that Newton's method works on a larger class of functions.

Of course, once we know that an iterative process works *correctly*, the next question inevitably concerns the *rate* at which it works. In classical numerical analysis, the efficiency of an iterative algorithm is determined by calculating its *order of convergence*.

Definition 45 Let (x_n) be a sequence of reals with $x_n \to p$. If

$$0 < \lim_{n\to\infty} \frac{|x_{n+1} - p|}{|x_n - p|^\alpha} = r < \infty,$$

for some $\alpha \geq 1$, then α is called the *order of convergence* of the sequence. If $\alpha = 1$ then r is called the *rate of convergence* of (x_n).

In this definition, the sequence (x_n) is generated by a numerical algorithm designed to calculate p. The larger that α is, the quicker the convergence of (x_n) to p, the better the algorithm.

If $\alpha = 1$, the algorithm is said to converge *linearly*. For $\alpha = 2$, the convergence is *quadratic*. Two linearly convergent algorithms may be compared based on their rates of convergence.

Notice that orders of convergence are calculated using the uncertainty $|x_n - p|$. To extend the idea to the setting of domains with measurements, we consider sequences (x_n) which converge to their suprema p in the μ topology on D, and replace $|x_n - p|$ with $|\mu x_n - \mu p|$.

Definition 46 Let D be a dcpo and let μ measure $X \subseteq D$. If (x_n) is a sequence in D which converges to its supremum $p \in X$ in the μ topology and

$$0 < \lim_{n\to\infty} \frac{\mu x_{n+1} - \mu p}{(\mu x_n - \mu p)^\alpha} = r < \infty,$$

for some $\alpha \geq 1$, then α is called the *order of convergence* of the sequence. If $\alpha = 1$ then r is called the *rate of convergence* of (x_n).

An *increasing* sequence (x_n) converges to its supremum p in the μ topology. We begin with linear processes: the informatic derivative enables the systematic computation of rates of convergence.

Lemma 4 *Let $s : (D, \mu) \rightharpoonup (D, \mu)$ be a partial map which maps into* dom(s) *and has a fixed point $p = \bigsqcup s^n x$ in the μ topology. If $ds_\mu(p)$ exists, then*

$$\lim_{n\to\infty} \frac{\mu s^{n+1}(x) - \mu p}{\mu s^n(x) - \mu p} = \frac{ds}{d\mu}(p),$$

provided $\mu s^n(x) - \mu p > 0$ for all $n \geq 0$.

Thus, to find the rate at which a linear algorithm s converges to a fixed point p, we find its derivative at p. But why is this a measure of efficiency?

Proposition 5 *Let $s : (D, \mu) \rightharpoonup (D, \mu)$ be a partial map which maps into* dom(s). *If s is μ continuous at a fixed point p and $0 < ds_\mu(p) < 1$, then for all $0 < \varepsilon < 1 - ds_\mu(p)$, there is an $a \ll p$ such that for all $x \in$* dom(s),

$$a \sqsubseteq x \sqsubseteq p \text{ and } n \geq \frac{\log(\varepsilon/(\mu x - \mu p))}{\log\left(ds_\mu(p) + \varepsilon\right)} \Rightarrow s^n x \sqsubseteq p \text{ and } |\mu s^n x - \mu p| < \varepsilon,$$

provided $x \neq p$ and $n \geq 1$.

Proposition 5 gives an upper bound on the number of iterations a linear process must do before it achieves ε accuracy. In order that this estimate hold, the input x must be sufficiently close. However, even in the presence of this mathematical annoyance, we can still use it to understand why rate of convergence is a measure of efficiency.

Suppose we have two linear processes s, t which have a common fixed point p and that $0 < ds_\mu(p) < dt_\mu(p) < 1$. Let $\varepsilon > 0$. Imagine we have different inputs for s and t which both have measure λ and that $\lambda - \mu p > \varepsilon$. (If $\lambda - \mu p \leq \varepsilon$, each process is already ε close.) Then

$$\frac{\log(\varepsilon/(\lambda - \mu p))}{\log\left(ds_\mu(p) + \varepsilon\right)} < \frac{\log(\varepsilon/(\lambda - \mu p))}{\log\left(dt_\mu(p) + \varepsilon\right)},$$

that is, the number of iterations which ensure t is ε close to p also guarantee that s is ε close to p. However, it may be that s can achieve ε accuracy with fewer iterations than t. Roughly speaking, s is a better algorithm than t for calculating p.

The estimate on the number of iterations in Proposition 5 is useful because of its generality. However, we often encounter linear processes which satisfy $\mu s(x) - \mu s(p) \leq (ds_\mu(p))(\mu x - \mu p)$ for $x \sqsubseteq p$. In this case, we use the estimate

$$n \geq \frac{\log(\varepsilon/(\mu x - \mu p))}{\log ds_\mu(p)}.$$

One question which springs to mind is: How can we know the values of μp and $ds_\mu(p)$ when p itself is unknown? Though we cannot always calculate these quantities independent of p, the estimates given for the number of iterations are still useful for comparing processes, as we saw above. On the other hand, in the case of Newton's Method, we actually know *a priori* that $\mu p = ds_\mu(p) = 0$. We can also calculate these quantities independent of p for the bisection method, the golden section search and for contraction mappings on complete metric spaces.

Example 34 For a continuous map $f : \mathbb{R} \to \mathbb{R}$, the bisection method is captured by the partial splitting

$$\text{split}_f : \mathbf{I}\mathbb{R} \rightharpoonup \mathbf{I}\mathbb{R}$$

and the data

- $\text{dom}(\text{split}_f) = C(f) = \{[a, b] \in \mathbf{I}\mathbb{R} : f(a) \cdot f(b) \leq 0\}$
- $\text{fix}(\text{split}_f) = \{[r] : f(r) = 0\}$
- $d(\text{split}_f)_\mu[r] = 1/2$ for all $[r] \in \text{fix}(\text{split}_f)$

If r is an isolated zero of f, then split_f is μ continuous at the associated fixed point $[r]$. By the remarks following Proposition 5, if r is an isolated zero of f and $x \in C(f)$ is a sufficiently small input around r, then

$$\text{split}_f^n x \text{ for } n \geq \frac{\log(\varepsilon/\mu x)}{\log(1/2)}$$

is an ε-approximation of r.

The estimate for the number of iterations given in the last example can fail without μ continuity. If we take $f(x) = x \cdot \sin(1/x)$ for $x \neq 0$ and $f(0) = 0$, then there are arbitrarily small intervals $\bar{x} \in C(f)$ with $\bar{x} \sqsubseteq [0]$, but for which $\text{split}_f \bar{x} \not\sqsubseteq [0]$. Beginning with *any* one of these intervals as input, and then doing $n \geq \log(\varepsilon/\mu x)/\log(1/2)$ iterations of split_f, leaves an interval of length $< \varepsilon$. The problem is that we are now on track to calculate a *different* zero $[r]$, rather than the one we *intended* to calculate, $[0]$.

The point is this: an estimate for the number of iterations is of little use if we do not know what we are calculating. This is why zeroes are normally assumed isolated in numerical analysis, as in Newton's method, where we assume $f'(r) \neq 0$. Thus, we expect iterative numerical methods to be μ continuous at fixed points when realized as partial maps on domains.

Example 35 The Golden Section Search. In Example 16, given a function $f : \mathbb{R} \to \mathbb{R}$ and a constant $1/2 < r < 1$, we defined the splitting

$$\max_f : \mathbf{IR} \to \mathbf{IR}$$

$$\max_f[a,b] = \begin{cases} [l(a,b), b] & \text{if } f(l(a,b)) < f(r(a,b)), \\ [a, r(a,b)] & \text{otherwise.} \end{cases}$$

where $l(a,b) = (b-a)(1-r) + a$ and $r(a,b) = (b-a)r + a$.

If f is unimodal on $[a,b]$ and its unique maximizer is $x^* \in \text{int}[a,b]$, then $\max_f$ is μ continuous at $[x^*]$ because

$$[a,b] \ll \bar{x} \sqsubseteq [x^*] \Rightarrow \max_f \bar{x} \sqsubseteq [x^*],$$

which was shown in Example 16, and because it has a derivative at $[x^*]$, given by

$$\frac{d(\max_f)}{d\mu}[x^*] = r.$$

Thus, if f is unimodal on $[a,b]$ and $x^* \in \text{int}[a,b]$, then

$$\max_f^n[a,b] \text{ for } n \geq \frac{\log(\varepsilon/(b-a))}{\log(r)}$$

is an ε-approximation of x^*.

Example 36 Contraction maps. If $f : X \to X$ is a contraction on a complete metric space (X, d) with constant $0 < c < 1$, its extension to the formal ball model

$$\bar{f} : \mathbf{B}X \to \mathbf{B}X, \quad \bar{f}(x,r) = (fx, c \cdot r)$$

has derivative $d\bar{f}_\pi(p) = c$, for all $p \in \mathbf{B}X$. The map $\bar{f}$ is Scott continuous and hence μ continuous at all points. If we take any $x \in X$ and $r \geq d(x, fx)/(1-c)$, then

$$\bar{f}^n(x,r) \text{ for } n \geq \frac{\log(r/\varepsilon)}{\log c}$$

is an ε-approximation of the unique attractor of f.

The presence of informatic linearity in the last three examples enables us to use the estimate mentioned after Proposition 5. The next example is more interesting.

Example 37 The Regula Falsi Method. For a function $f : [a,b] \to \mathbb{R}$ such that

(i) $f(a) < 0$ and $f(b) > 0$,
(ii) $f'(x) > 0$ for all $x \in [a,b]$, and
(iii) $f''(x) \geq 0$ for all $x \in [a,b]$,

we define the partial mapping

$$r_f : \mathbf{I}\mathbb{R} \rightharpoonup \mathbf{I}\mathbb{R}$$
$$r_f[x,b] = \left[b - f(b)\left(\frac{b-x}{f(b)-f(x)}\right), b\right]$$

whose domain is

$$\mathrm{dom}(r_f) = \{[x,b] : a \le x \le r\}$$

where $r \in (a,b)$ is the unique zero of f on $[a,b]$.

The map r_f is a Scott continuous splitting which maps the dcpo $\mathrm{dom}(r_f)$ into itself. For if $a \le x \le y \le r$, we have the string of inequalities

$$a \le x \le b - f(b)\left(\frac{b-x}{f(b)-f(x)}\right) \le b - f(b)\left(\frac{b-y}{f(b)-f(y)}\right) \le r,$$

where the second follows from $f(x) \le 0$, and the last two follow from

$$\frac{f(b)-f(x)}{b-x} \le \frac{f(b)-f(y)}{b-y} \le \frac{f(b)-f(r)}{b-r},$$

which is a consequence of the fact that f' is nondecreasing. This proves that r_f is a monotone splitting which takes $\mathrm{dom}(r_f)$ into itself. Finally, r_f is Scott continuous because its measure is Scott continuous.

By Proposition 7, if $\bar{x} \in \mathrm{dom}(r_f)$, then

$$\bigsqcup_{n\ge 0} r_f^n(\bar{x}) \in \mathrm{fix}(r_f),$$

but it is easy to see that $\mathrm{fix}(r_f) = \{[r,b]\}$. Thus, iterating r_f is an algorithm for approximating r, called the *Regula Falsi method*. But how efficient is it?

To answer this question, we calculate the informatic derivative of r_f at the fixed point $[r,b]$ as follows:

$$\begin{aligned}
\frac{dr_f}{d\mu}[r,b] &= \lim_{\bar{x}\to[r,b]} \frac{\mu r_f(\bar{x}) - \mu r_f[r,b]}{\mu\bar{x} - \mu[r,b]} \\
&= \lim_{x\to r^-} \frac{f(b)(r-x) + f(x)(b-r)}{(f(b)-f(x))(r-x)} \\
&= \lim_{x\to r^-} \left[\frac{f(b)}{f(b)-f(x)} + \frac{f(x)-f(r)}{r-x}\cdot\frac{b-r}{f(b)-f(x)}\right] \\
&= \frac{f(b)}{f(b)-f(r)} + (-1)f'(r)\cdot\frac{b-r}{f(b)-f(r)} \\
&= 1 - \frac{f'(r)(b-r)}{f(b)}.
\end{aligned}$$

By monotonicity of r_f, this derivative is nonnegative, and hence a number in the interval $[0, 1)$. In fact, we can see that

$$d(r_f)_\mu[r, b] \to 0 \text{ as } b \to r$$

so the efficiency of this algorithm is determined by the closeness of b to r. Notice that it does *not* depend on a.

Once we have the derivatives of two different algorithms which solve the same problem, we can compare them to understand their respective strengths and weaknesses.

Example 38 The Bisection versus Regula Falsi. If $f : \mathbb{R} \to \mathbb{R}$ is a continuous map and $[a, b]$ is an interval such that $f(a) < 0$ and $f(b) > 0$, $f' > 0$ on $[a, b]$ and $f'' \geq 0$ on $[a, b]$, then

$$\bigsqcup_{n\geq 0} \text{split}_f^n[a, b] = [r] \quad \text{and} \quad \bigsqcup_{n\geq 0} r_f^n[a, b] = [r, b]$$

are both schemes for calculating the unique zero r of f on $[a, b]$. But which one is better? We consider two examples.

If $f(x) = x^2 - x - 1$ and $[a, b] = [1, 2]$, then $r = (1 + \sqrt{5})/2$. Thus,

$$d(\text{split}_f)[r] = \frac{1}{2} \text{ and } d(r_f)[r, b] = \frac{7 - 3\sqrt{5}}{2} \approx 0.145898,$$

which means that *eventually* $\mu r_f(x) - \mu[r, b] \approx 0.14(\mu x - \mu[r, b])$, as compared to $\mu\, \text{split}_f(x) - \mu[r] = 0.5(\mu x - \mu[r])$ for the bisection. In other words, *eventually* the Regula Falsi method reduces the uncertainty in an interval by about 86%, while for the bisection uncertainty is always reduced by 50%. This suggests that r_f is preferable in this case. Six iterations of each gives

$$\text{split}_f^6[1, 2] = [1.59375, 1.625] \text{ and } r_f^6[1, 2] \approx [1.618025, 2].$$

The approximation of r offered by the bisection is the midpoint of $\text{split}_f^6[1, 2]$, 1.609375, while the approximation given by the Regula Falsi method is the left endpoint of $r_f^6[1, 2]$, around 1.618025. Thus, the Regula-Falsi method is accurate to four decimal places, while the bisection is only accurate to one. This supports the intuition offered by the informatic derivatives calculated above: r_f converges faster than split_f in this case.

If $f(x) = x^6 - x - 1$ and $[a, b] = [1, 2]$, then $r \approx 1.13472$. The informatic derivatives in this case are

$$d(\text{split}_f)[r] = \frac{1}{2} \text{ and } d(r_f)[r, b] \approx 0.85407,$$

which suggests that now it is split_f which converges faster. If we do sixteen iterations of each, we find that

$$\text{split}_f^{16}[1,2] \approx [1.134719, 1.134735] \text{ and } r_f^{16}[1,2] \approx [1.121308, 2].$$

Thus, the bisection gives the approximation $r \approx 1.13472$, while the Regula Falsi method is only accurate to one decimal place. In fact, it is only after 68 iterations that the Regula Falsi method can duplicate what the bisection achieves in 16:

$$r_f^{68}[1,2] \approx [1.13472, 2].$$

The intuition imparted by informatic derivative is also correct in this instance.

Example 39 The secant method. Recall from Theorem 8, the *secant method* $\sec_f : \mathcal{P}_C[a,r] \to \mathcal{P}_C[a,r]$ given by

$$\sec_f[x,y] = \left[y, y - \frac{f(y)}{df[x,y]}\right]$$

yields an algorithm for calculating r with $f(r) = 0$ given by

$$\bigsqcup_{n \geq 0} \sec_f^n(x) = [r],$$

for any $x \in \mathcal{P}_C[a,r]$. For the secant method $\sec_f$, we have $d(\sec_f)[r] = 0$.

Let $\bar{x} = [x,y] \sqsubseteq [r]$ with $\mu\bar{x} > 0$. Then by the mean value theorem and the triangle inequality,

$$0 \leq \frac{\mu \sec_f(\bar{x})}{\mu\bar{x}} \leq \frac{2(r-y)}{r-x+r-y} + \frac{|f(y)|}{f'(c)(r-x+r-y)},$$

where $c \in \bar{x}$. But since $r-x+r-y \geq 2(r-y)$ and $r-x+r-y \geq r-y$, the expression on the right is bounded by

$$\frac{2(r-y)}{2(r-y)} + \frac{|f(y)|}{f'(c)(r-y)} = 1 - \frac{|f(y)-f(r)|}{f'(c)(y-r)}.$$

As $\bar{x} \to [r]$ in the μ topology, we have $x, y \to r$ and $c \to r$. Hence,

$$0 \leq \lim_{\bar{x} \to [r]} \frac{\mu \sec_f(\bar{x})}{\mu\bar{x}} \leq \lim_{c,y \to r} \left(1 - \frac{|f(y)-f(r)|}{f'(c)(y-r)}\right) = 1 - 1 = 0,$$

proving the claim.

Thus, the convergence of the secant method is *superlinear*, in agreement with numerical analysis. This is an interesting example. The function $\sec_f$ does *not* correspond to iterating a classical real valued function, and the informatic derivative is *not* a classical derivative: the formula in Example 21 takes *two* real numbers as input, but returns only *one* as output.

Thus, to prove that a numerical method works correctly, we show it iterates to a fixed point. To go along with this uniform approach to the problem of correctness, we now have a uniform method for calculating rates of convergence of linear processes: simply take the informatic derivative of a map on a domain at a fixed point. This extends what is done is numerical analysis, enabling a unified treatment not previously possible. For instance, the secant method, the golden section search and the bisection method are iterative processes which have no classical descriptions as differentiable functions on the real line. Nevertheless, we have seen that they may be naturally described as mappings on domains which possess informatic derivatives.

9.5.3 Rates of Change in the Communication Process

A classical binary channel $f : \Delta^2 \to \Delta^2$ takes an input distribution to an output distribution. In a similar way, a qubit channel is a function of the form $\varepsilon : \Omega^2 \to \Omega^2$ that is convex linear and completely positive [35]. For our purposes, there is no need to get lost in too many details of the Hilbert space formulation: qubit channels can be represented as linear selfmaps on the unit ball in Euclidean three space as follows.

There is a 1–1 correspondence between density operators on a two dimensional state space and points on the unit ball $\mathbb{B}^3 = \{x \in \mathbb{R}^3 : |x| \leq 1\}$: each density operator $\rho : \mathcal{H}^2 \to \mathcal{H}^2$ can be written uniquely as

$$\rho = \frac{1}{2}\begin{pmatrix} 1 + r_z & r_x - ir_y \\ r_x + ir_y & 1 - r_z \end{pmatrix}$$

where $r = (r_x, r_y, r_z) \in \mathbb{R}^3$ satisfies $|r| = \sqrt{r_x^2 + r_y^2 + r_z^2} \leq 1$. The vector $r \in \mathbb{B}^3$ is called the *Bloch vector* associated to ρ. Bloch vectors have a number of aesthetically pleasing properties.

If ρ and σ are density operators with respective Bloch vectors r and s, then (i) the eigenvalues of ρ are $(1 \pm |r|)/2$, (ii) the von Neumann entropy of ρ is $S\rho = H((1+|r|)/2) = H((1-|r|)/2)$, where $H : [0, 1] \to [0, 1]$ is the base two Shannon entropy, (iii) if ρ and σ are pure states and $r + s = 0$, then ρ and σ are orthogonal, and thus form a basis for the state space; conversely, the Bloch vectors associated to a pair of orthogonal pure states form antipodal points on the sphere, (iv) the Bloch vector for a convex sum of mixed states is the convex sum of the Bloch vectors, (v) the Bloch vector for the completely mixed state $I/2$ is $0 = (0, 0, 0)$.

Because of the correspondence between Ω^2 and $\mathbb{B}^3$, let us now regard these two as equal.

A standard way of measuring the capacity of a quantum channel in quantum information is the Holevo capacity; it is sometimes called the product state capacity since input states are not allowed to be entangled across two or more uses of the channel.

Definition 47 For a quantum channel f, the *Holevo capacity* is given by

$$\mathrm{C}(f) = \sup_{\{x_i,\rho_i\}} \left[S\left(f\left(\sum_i x_i\rho_i\right)\right) - \sum_i x_i \cdot S(f(\rho_i)) \right]$$

where the supremum is taken over all ensembles $\{x_i, \rho_i\}$ of possible input states ρ_i to the channel.

The possible input states ρ_i to the channel are in general mixed and the x_i are probabilities with $\sum_i x_i = 1$. If f is the Bloch representation of a qubit channel, the Holevo capacity of f is given by

$$\mathrm{C}(f) = \sup_{\{x_i,r_i\}} \left[H\left(\frac{1+|f(\sum_i x_i r_i)|}{2}\right) - \sum_i x_i \cdot H\left(\frac{1+|f(r_i)|}{2}\right) \right]$$

where r_i are Bloch vectors for density operators in an ensemble, and we recall that eigenvalues of a density operator with Bloch vector r are $(1 \pm |r|)/2$.

Recall that the classical channels $f : \Delta^2 \to \Delta^2$ which increase entropy $(H(f(x)) \geq H(x))$ are exactly those f with $f(\bot) = \bot$. They are the *strict* mappings of domain theory, which are also known as *binary symmetric channels* in information theory. Similarly, the entropy increasing qubit channels are exactly those f for which $f(\bot) = \bot$. These are called *unital* in quantum information theory.

Theorem 30 *Let $\mu(x) = 1 - |x|$ denote the standard measurement on Ω^2. For any unital channel f and any $p \in \Omega^2$ different from $\bot$,*

$$df_\mu(p) = \frac{|f(p)|}{|p|}$$

Thus, the Holevo capacity of f is determined by the largest value of its informatic derivative. Explicitly,

$$\mathrm{C}(f) = 1 - H\left(\frac{1}{2} + \frac{1}{2} \sup_{x \in \ker(\mu)} df_\mu(x)\right)$$

Then $\mathrm{C}(f) = 1$ for *any* rotation f since $df_\mu = 1$. Notice that $df_\mu \equiv 1$ iff f is a rotation. For each $p \in [0, 1]$, the unique channel $f \sqsubseteq 1$ with $df_\mu = p$ is the depolarization channel $f = d_p = p \cdot I$, so that $\mathrm{C}(d_p) = 1 - H((1+p)/2)$. In fact, there is an isomorphism from binary symmetric channels onto the depolarization channels. The only unital qubit channel with capacity zero is 0 itself.

Example 40 The two Pauli channel in Bloch form is

$$\varepsilon(r) = pr + \left(\frac{1-p}{2}\right) s_x(r) + \left(\frac{1-p}{2}\right) s_y(r)$$

where s_x and s_y are the Bloch representations of the unitary channels derived from the Pauli spin operators σ_x and σ_y. This simplifies to

$$\varepsilon(r_x, r_y, r_z) = (pr_x, pr_y, -(1-p)r_z)$$

The matrix associated to ε is diagonal, so the diagonal element (eigenvalue) that has largest magnitude also yields the largest value of its informatic derivative. The capacity of the two Pauli channel is then

$$1 - H\left(\frac{1+\max\{p, 1-p\}}{2}\right)$$

where $p \in [0, 1]$.

The set of unital channels $\mathcal{U}$ is compact hence closed and thus forms a dcpo as a subset of the domain $[\Omega^2 \to \Omega^2]$.

Corollary 5 *The Holevo capacity* $\mathrm{C} : \mathcal{U} \to [0, 1]$ *is Scott continuous.*

Thus, the ability of a unital qubit channel to transmit information is determined by the largest value of its informatic derivative.

9.5.4 The Derivative at a Compact Element: A Discrete Derivative

If one looks closely at the definition of the informatic derivative above, it has a computationally restrictive aspect: the requirement that p not be isolated in the μ topology. This is equivalent to saying that p must not be a *compact* element of D. From the mathematical viewpoint, one does not object to this: mathematics offers us no way of obtaining unique "limits" at isolated points of topological spaces. Nevertheless, computationally, it is easy to write down simple examples of mappings on domains which *should have* derivatives, but are excluded simply because they work only with compact elements.

For instance, on the domain of lists $[S]$, the map rest: $[S] \to [S]$ which removes the first element from a nonempty list and sends the empty list to itself, satisfies

$$\mu\,\mathrm{rest}(x) = \mu(x) - 1$$

for $x \neq [\,]$, where μ is the length measurement. Thus, we ought to be able to say that $d(\mathrm{rest})_\mu(x) = 1$ for $x \neq [\,]$.

We now consider an extension of the definition of informatic derivative which applies at compact elements as long as they are not minimal. One of the benefits of this extension is that we are finally able to understand the sense in which the asymptotic notions of complexity used in numerical analysis (rates of convergence) are the same as those used in the analysis of "discrete" algorithms (for example, list processing). Another is the identification of an idea which allows us to systematically calculate both of these complexity notions in a uniform manner: informatic rates of change apply in both the continuous and discrete realms.

9.5.4.1 The Informatic Derivative at a Compact Element

Defining the informatic derivative of a selfmap on a domain D really only depends on our ability to define it for functions of the form $f : D \to \mathbb{R}$. If we set

$$df_{\mu}(p) = \lim_{x \to p} \frac{f(x) - f(p)}{\mu x - \mu p}$$

then for $f : D \to D$, we can set $df_{\mu}(p) = d(\mu f)_{\mu}(p)$, obtaining the usual definition of the informatic derivative. Of course, the problem is that this is only works when p is not compact i.e. when

$$p \notin K(D) := \{x \in D : x \ll x\}$$

These are precisely the points that are not isolated in the μ topology. The reason we must work with points which are not isolated is that there must be enough *nontrivial* μ open sets around p so that we can take a limit in the formal sense of topology – without enough nontrivial open sets, a limit may not be unique.

However, any point $p \notin \min(D) := \{x \in D : \downarrow x = \{x\}\}$ can be approximated from below using the nontrivial μ open subsets of D which are contained in $\downarrow p$ and which themselves contain p and at least one other element:

$$\text{approx}_{\mu}(p) = \{V \in \mu_D : p \in V \subseteq \downarrow p \text{ and } V \neq \{p\}\}.$$

Thus, the existence of approximations is not the problem – the problem is that we need a concept more applicable than 'limit'.

Definition 48 Let $f : D \to \mathbb{R}$ be a function and $p \in D$. We set

$$d^+ f_{\mu}(p) := \sup\{c : (\exists V \in \text{approx}_{\mu}(p))(\forall x \in V)\, f(x) - f(p) \geq c \cdot (\mu x - \mu p)\}$$

and

$$d^- f_{\mu}(p) := \inf\{c : (\exists V \in \text{approx}_{\mu}(p))(\forall x \in V)\, f(x) - f(p) \leq c \cdot (\mu x - \mu p)\},$$

provided p is not a *minimal element* of D, i.e., $p \notin \min(D)$.

The existence of the informatic derivative of a real-valued function in the usual case is expressible entirely in terms of $d^+ f_{\mu}$ and $d^- f_{\mu}$ as follows:

Theorem 31 *Let $f : D \to \mathbb{R}$ be a function with $p \in D \setminus K(D)$. Then $df_{\mu}(p)$ exists iff $d^+ f_{\mu}(p)$ exists, $d^- f_{\mu}(p)$ exists and $d^- f_{\mu}(p) \leq d^+ f_{\mu}(p)$. In either case, we have $df_{\mu}(p) = d^+ f_{\mu}(p) = d^- f_{\mu}(p)$.*

The previous theorem justifies the following definition.

Definition 49 Let $f : D \to \mathbb{R}$ be a function on a continuous dcpo D with a measurement μ which measures D at $p \in D \setminus \min(D)$. If $d^- f_\mu(p)$ exists, $d^+ f_\mu(p)$ exists and $d^- f_\mu(p) \leq d^+ f_\mu(p)$, then we define

$$df_\mu(p) := d^+ f_\mu(p)$$

and call this number the *informatic derivative* of f at p.

By Theorem 31, the new definition and the old definition agree in the continuous case ($p \notin K(D)$). We now turn our attention to the discrete case ($p \in K(D)$).

Theorem 32 *Let $f : D \to \mathbb{R}$ be a function on an algebraic dcpo D with a measurement μ that measures D at $p \in K(D) \setminus \min(D)$. Then the following are equivalent:*

(i) *The derivative $df_\mu(p)$ exists.*
(ii) *The supremum*

$$\sup \left\{ \frac{f(x) - f(p)}{\mu x - \mu p} : x \in K(D) \cap \downarrow p, x \neq p \right\}$$

exists and the infimum

$$\inf \left\{ \frac{f(x) - f(p)}{\mu x - \mu p} : x \in K(D) \cap \downarrow p, x \neq p \right\}$$

exists.

In either case, the value of $d^+ f_\mu(p)$ is the supremum in (ii), while the value of $d^- f_\mu(p)$ is the infimum in (ii).

Finally, the definition of derivative for selfmaps on a domain D.

Definition 50 Let $f : D \to D$ be a function on a domain (D, μ) with a map μ that measures D at $p \in D \setminus \min(D)$. If $d(\mu f)_\mu(p)$ exists, then we write

$$df_\mu(p) := d(\mu f)_\mu(p)$$

and call this number the *informatic derivative* of f at p with respect to μ. We also set $d^* f_\mu(p) := d^*(\mu f)_\mu(p)$ for $* \in \{+, -\}$.

It is easy to extend this definition for a map $f : (D, \mu) \to (E, \lambda)$, as was done for the original formulation of the derivative in the continuous case [15], but in the present paper there are no applications warranting such an abstraction.

Example 41 Derivatives of list operations.

(i) The map first : $[S] \to [S]$, $\text{first}(a :: x) = [a]$, $\text{first}[\,] = [\,]$. Using Theorem 32,

$$d(\text{first})_\mu(x) = d^+(\text{first})_\mu(x) = d^-(\text{first})_\mu(x) = 0,$$

for all $x \neq [\,]$. At $x = [\,]$, $d(\text{first})_\mu(x) = d^+(\text{first})_\mu(x) = 1 \geq 0 = d^-(\text{first})_\mu(x)$.

(ii) The map rest : $[S] \to [S]$, $\text{rest}(a :: x) = x$, $\text{rest}[\,] = [\,]$. Using Theorem 32,

$$d(\text{rest})_\mu(x) = d^+(\text{rest})_\mu(x) = d^-(\text{rest})_\mu(x) = 1,$$

for all $x \neq [\,]$. At $x = [\,]$, $d(\text{rest})_\mu(x) = d^+(\text{rest})_\mu(x) = 1 \geq 0 = d^-(\text{rest})_\mu(x)$.

There is something worth pointing out before we focus on the derivative in the discrete case. The definition of $df_\mu(p)$ splits into two cases, the continuous ($p \notin K(D)$) and the discrete ($p \in K(D)$). From this bifurcation appears a remarkable duality: In the continuous case the inequality $df^+_\mu(p) \leq df^-_\mu(p)$ always holds, but $df^-_\mu(p) \leq df^+_\mu(p)$ may not; in the discrete case the opposite is true, $df^-_\mu(p) \leq df^+_\mu(p)$ always holds, but $df^+_\mu(p) \leq df^-_\mu(p)$ may not.

The results of this section allow for only one interpretation of this phenomenon: In the continuous case, the derivative is determined by *local* properties of the function; in the discrete case, the derivative is determined by *global* properties of the function.

9.5.4.2 Measuring the Length of an Orbit

Throughout this section, we assume that (D, μ) is an algebraic dcpo whose compact elements $K(D)$ form a lower set $K(D) = \downarrow K(D)$. Some important examples of this are $\mathbb{N}^*$, $\mathbb{N}^\infty = \mathbb{N} \cup \{\infty\}$, $[S]$, $\mathcal{P}\omega$, Σ^∞, and $[\mathbb{N} \rightharpoonup \mathbb{N}]$. Computationally, this is not much of an assumption.

Theorem 33 (The Mean Value Theorem) *Let $f : D \to D$ be a function on (D, μ) such that $df_\mu(p)$ exists at a compact element p. Then*

$$(\mu x - \mu p) \cdot d^- f_\mu(p) \leq \mu f(x) - \mu f(p) \leq d^+ f_\mu(p) \cdot (\mu x - \mu p),$$

for all $x \sqsubseteq p$.

If a splitting r has a compact fixed point p reachable by iteration $\bigsqcup r^n(x) = p$, then the derivative of r at p can be used to provide a precise measure of the number of iterations required to get to p from an input of x. Later we will see that such quantities can play an integral role in determining the complexity of certain algorithms.

Definition 51 Let $r : D \to D$ be a splitting. An *orbit* is a sequence of iterates $(r^n x)$. An orbit is *compact* if

$$\bigsqcup_{n \geq 0} r^n(x) \in K(D).$$

The *length* of a compact orbit $(r^n x)$ is

$$|(r^n x)| := \inf\{n \geq 0 : r^{n+1}(x) = r^n(x)\}.$$

A compact orbit is *nontrivial* when $|(r^n x)| > 0$; otherwise it is a *fixed point*.

In this new language, we can say that we are interested in determining the length of nontrivial compact orbits of splittings. If $(r^n x)$ is a compact orbit, then $r^l(x)$ is a fixed point of r where $l = |(r^n x)|$. For this reason, we say that the orbit $(r^n x)$ *ends* at $p = r^l(x)$.

Lemma 5 *If a splitting $r : D \to D$ has a nontrivial compact orbit which ends at $p \in K(D)$, and $dr_\mu(p)$ exists, then $0 \leq dr_\mu(p) \leq 1$.*

Theorem 34 *Let r be a splitting with a nontrivial compact orbit $(r^n x)$ that ends at p. If $dr_\mu(p) = 0$, then $r(x) = p$. If $0 < dr_\mu(p) < 1$, then*

$$n \geq \left\lceil \frac{\log((\mu x - \mu p)/\varepsilon)}{\log(1/dr_\mu(p))} \right\rceil + 1 \Rightarrow |\mu r^n(x) - \mu p| < \varepsilon,$$

for any $\varepsilon > 0$.

By the compactness of p, there is a choice of $\varepsilon > 0$ which will ensure that $|\mu r^n(x) - \mu p| < \varepsilon \Rightarrow r^n(x) = p$, but at this level of generality we cannot give a precise description of it. It depends on μ. For lists, the value is $\varepsilon = 1$.

Example 42 Let r be a splitting on $[S]$ with $0 < dr_\mu(p) < 1$ at any fixed point p. Then for any x, there is some $k \geq 0$ such that $r^k(x) = p$ is a fixed point. By the last result, doing

$$n > \left\lceil \frac{\log(\mu x - \mu p)}{\log(1/dr_\mu(p))} \right\rceil$$

iterations implies that $r^n(x) = p$.

Let's consider an important example of this type.

Example 43 Contractive list operations. For a positive integer $x > 0$, define

$$m(x) = \begin{cases} x/2 & \text{if } x \text{ even;} \\ (x+1)/2 & \text{if } x \text{ odd.} \end{cases}$$

Consider the splittings

$$\text{left}(x) = [x(1), \cdots, x(m(\mu x) - 1)]$$
$$\text{right}(x) = [x(m(\mu x) + 1), \cdots, x(\mu x)]$$

each of which takes lists of length one or less to the empty list [ ]. Each has a derivative at its unique fixed point [ ] as follows.

First, since both of these maps are splittings and $p = [\,]$ has measure $\mu p = 0$, each has a derivative at p—it is simply a matter of determining d^+ at [ ] in each case. For this, if $x \neq [\,]$, then

$$\frac{\mu\,\text{left}(x)}{\mu x} \leq \frac{(\mu x/2) - (1/2)}{\mu x} = \frac{1}{2} \cdot \left(1 - \frac{1}{\mu x}\right) \leq \frac{1}{2}$$
$$\frac{\mu\,\text{right}(x)}{\mu x} \leq \frac{\mu x/2}{\mu x} = \frac{1}{2}$$

which means $d(\text{left})_\mu[\,] = d(\text{right})_\mu[\,] = 1/2$.

Notice that the case of 'left' is much more interesting than the case of "right." In the former, the value of the derivative is never attained by any of the quotients $\mu\,\text{left}/\mu$—it is determined by a 'limit' process which extracts global information about the mapping left.

Already we notice a relationship to processes in numerical analysis: the case $dr_\mu(p) = 0$ is an extreme form of *superlinear* convergence (extreme since in one iteration the computation finishes), while the case $0 < dr_\mu(p) < 1$ behaves just like ordinary *linear* convergence. However, unlike numerical analysis, we can actually say something about the case $dr_\mu(p) = 1$.

To do this is nontrivial, and in what follows, we seek only to illustrate the value of the informatic derivative in the discrete case by showing that the precise number of iterations required to calculate a fixed point p by iteration of a map r can be determined when $dr_\mu(p) = 1$—a case in which classical derivatives are notorious for yielding no information.

A compact element p that is not minimal has a natural set of *predecessors*, these are formally defined as the set of maximal elements in the dcpo $\downarrow p \setminus \{p\}$:

$$\text{pred}(p) = \max(\downarrow p \setminus \{p\}).$$

To see that this makes sense, notice that $\downarrow p \setminus \{p\}$ is nonempty since p is not minimal, and is closed in the μ topology, as the intersection of μ closed sets. But a μ closed set is closed under directed suprema, and so must have at least one maximal element.

Theorem 35 *Let $r : D \to D$ be a splitting on (D, μ) with a compact fixed point $p = r(p)$ such that*

$$(\forall x)\, x \sqsubseteq p \Rightarrow \bigsqcup_{n \geq 0} r^n(x) = p.$$

If $d^+r_\mu(x) = 1$ for all $x \sqsubseteq p$ and $d^-r_\mu(x) = 1$ for all $x \sqsubseteq p$ with $x \neq p$, then for all $x \sqsubseteq p$ with $x \neq p$, there is $q \in \mathrm{pred}(p)$ such that

$$r^n(x) = p \Leftrightarrow n = \frac{\mu x - \mu p}{\mu q - \mu p}.$$

It is interesting to notice in the last result that if $d^-r_\mu(p) = 1$, then we *must* have $r(x) = x$ for all $x \sqsubseteq p$. Of course, our hypotheses on r rule this out since the fixed point p must be an attractor on $\downarrow p$.

Example 44 In Example 41, we saw that the map rest : $[S] \to [S]$ is an example of the sort hypothesized in Theorem 35 with $p = [\,]$. The predecessors of p are the one element lists

$$\mathrm{pred}(p) = \{[x] : x \in S\}.$$

Thus, the last theorem says that

$$\mathrm{rest}^n(x) = [\,] \Leftrightarrow n = \mu x,$$

for any $x \neq [\,]$.

9.5.4.3 Complexity

We briefly consider how the informatic derivative offers a new perspective on the complexity of algorithms.

Example 45 Linear search. To search a list x for a key k consider

$$\mathrm{search} : [S] \times S \to \{\bot, \top\}$$

given by

$$\begin{aligned} \mathrm{search}([\,], k) &= \bot \\ \mathrm{search}(x, k) &= \top && \text{if first } x = k, \\ \mathrm{search}(x, k) &= \mathrm{search}(\mathrm{rest}\, x, k) && \text{otherwise.} \end{aligned}$$

Let $D = [S] \times S^\flat$ – the product of $[S]$ with the set S ordered flatly. We measure this domain as $\mu(x, k) = \mu x$. Let $r : D \to D$ be the splitting $r(x, k) = (\mathrm{rest}\, x, k)$.

On input (x, k) in the *worst case*, the number of comparisons n done by this algorithm is the same as the number of iterations needed to compute $r^n(x, k) = ([\,], k)$. Since $d^+r_\mu(x) = 1$ for all x and $d^-r_\mu(x) = 1$ for all $x \neq ([\,], k)$, Theorem 35 applies to give

$$r^n(x, k) = ([\,], k) \Leftrightarrow n = \mu(x, k) = \mu x,$$

which helps us understand how the complexity of a discrete algorithm can be determined by the derivative of a splitting which models its iteration mechanism.

Example 46 Binary search. To search a sorted list x for a key k, we use

$$\mathrm{bin} : [S] \times S \to \{\bot, \top\}$$

given by

$$\begin{array}{lll} \mathrm{bin}([\,], k) & = \bot & \\ \mathrm{bin}(x, k) & = \top & \text{if } \mathrm{mid}\, x = k, \\ \mathrm{bin}(x, k) & = \mathrm{bin}(\mathrm{left}\, x, k) & \text{if } \mathrm{mid}\, x > k, \\ \mathrm{bin}(x, k) & = \mathrm{bin}(\mathrm{right}\, x, k) & \text{otherwise.} \end{array}$$

where $\mathrm{mid}\, x := x(m(\mu x))$. Again $D = [S] \times S^\flat$ and $\mu(x, k) = \mu x$. This time we consider the splitting $r : D \to D$ by

$$r(x, k) = \begin{cases} (\mathrm{left}\, x, k) & \text{if } \mathrm{mid}\, x > k; \\ (\mathrm{right}\, x, k) & \text{otherwise.} \end{cases}$$

On input (x, k) in the *worst case*, the number of comparisons n must satisfy $r^n(x, k) = ([\,], k)$. In this case, we have $dr_\mu([\,], k) = 1/2$, so by Theorem 34,

$$n \leq \left\lceil \frac{\log(\mu x)}{\log(2)} \right\rceil + 1 = \lceil \log_2(\mu x) \rceil + 1,$$

since we know that the expression on the right is a number m that satisfies $r^m(x, k) = ([\,], k)$ but that n is the *least* of all such natural numbers because it was produced by the algorithm bin.

To summarize these simple examples: we have two different algorithms which solve the same problem recursively by iterating splittings r and s, respectively, on a domain (D, μ) in an effort to compute a fixed point p. If $dr_\mu(p) < ds_\mu(p)$, then the algorithm using r is faster than the one which uses s. In the case of linear search we have $ds_\mu(p) = 1$, while for binary search we have $dr_\mu(p) = 1/2$. As we have already seen, this is identical to the way one compares zero finding methods in numerical analysis – by comparing the derivatives of mappings at fixed points.

9.5.4.4 Thoughts on the Discrete Derivative

Theorem 31 is crucial in that it characterizes differentiability *independent* of its continuous component. Taking only this result as motivation for the definition of derivative leaves a few distinct possibilities. For instance, if we had called the derivative the interval $[d^- f_\mu(p), d^+ f_\mu(p)]$, we might notice more clearly the tendency of continuous information to collapse at a point. Another possibility is to say that the derivative is $d^- f_\mu(p)$. The author chose $d^+ f_\mu$ because it makes the most sense from an applied perspective. As an illustration, consider the intuitions we have about it: algorithms r with $dr_\mu(p) = 0$ belong to $O(1)$, those with $0 < dr_\mu(p) < 1$ belong to $O(\log n)$, while $dr_\mu(p) = 1$ indicates a process is in $O(n)$.

At first glance, an extension of the informatic derivative to the case of discrete data (compact elements) seems like an absurd idea. To begin, we have to confront the issue of essentially defining unique limits at isolated points. But even if we assume we have this, we need the new formulation to extend the previous, which means spotting a relationship between limits in the continuous realm versus finite sequences of discrete objects. But the truth is that all of this only sounds difficult because of what we are *taught*: that the continuous and discrete are "fundamentally different" and that one of the crucial distinctions between the two is the sensibility of the limit concept for continuous objects, as compared to the discrete case where 'limit' has no meaning. From this, we conclude that math students should spend less time attending lectures and more time coming up with new ideas.

The existence of a derivative in the discrete case means much more than it does in the continuous case. Most results on discrete derivatives do not hold in the continuous case. Just consider a quick example: let $r : D \to D$ be any continuous map with $p = r(p) \in K(D)$ and $dr_\mu(p) = 0$. If $x \sqsubseteq p$, then $r(x) = p$. Now compare this to the continuous case (like calculus on the real line), where one can only conclude that there is an $a \ll p$ such that $r^n(x) \to p$ for all x with $a \ll x \sqsubseteq p$. Again, this sharp contrast is due to the fact that discrete derivatives make use of *global* information, while continuous derivatives use only *local* information. Nevertheless, each is an instance of a common theme.

9.6 Forms of Process Evolution

9.6.1 Intuition

The idea in the measurement formalism is to analyze *processes*: a process is a thing that evolves in a space of informatic objects. The space of informatic objects is formally described by a domain with a measurement. By contrast, the measurement formalism allows for considerable flexibility in formalizing the notion of process. We have already seen one such notion of process: a function $f : D \to D$ that on input x produces iterates $(f^n(x))$ which converge to a fixed point $\bigsqcup f^n(x)$. This discrete form of evolution has various generalizations within the measurement formalism. We consider a few more of them in this section. The renee equation, which is a discrete extension of iteration, can be used to define recursion on general domains while still maintaining a first order view of evolution. The *trajectory* leads one to daydream about a *kinematics* of computation: for instance, the complexity of an algorithm is the amount of time it takes its trajectory in informatic space to achieve its order theoretic maximum. Lastly, we consider a third notion of process, one grounded on a 'thermodynamical' view of evolution, very different from the first two. The basic idea is this: before a process evolves, there are several possible states it may evolve to; when it finishes evolving, we gain information, but the acquisition of information is not free – how much does it cost?

Major references: [15, 22, 28, 29]

9.6.2 The Renee Equation

The renee equation is a model of recursion. After introducing this equation, we discuss two major results. The first is that every renee equation has a *unique* solution. The second is that the partial and primitive recursive functions on the naturals may be captured by taking closure under renee equations on the domains $\mathbb{N}^\infty$ and $\mathbb{N}^*$ – the naturals in their usual and opposite orders, respectively. This suggests that the information order on a domain determines a natural notion of computability, and that the renee equation yields a systematic method for determining this notion of computability. We will also see that one renee equation describing an algorithm leads to another which captures its complexity. This provides a qualitative and quantitative *first order* view of computation, one very much in line with actual program development.

9.6.2.1 Unique Solvability of the Equation

Recall the definition of the μ topology from Sect. 9.2.4.

Definition 52 Let $(X, +)$ be a Hausdorff space with a binary operation that is associative. If (x_n) is a sequence in X, then its *infinite sum* is

$$\sum_{n\geq 1} x_n := \lim_{n\to\infty} (x_1 + \cdots + x_n)$$

provided that the limit of the partial sums on the right exists.

Definition 53 Let $+ : D^2 \to D$ be a binary operation on a continuous dcpo. A point $x \in D$ is *idle* if there is a μ open set $\sigma(x)$ around x such that

(i) $(\sigma(x), +)$ is a semigroup, and
(ii) If (x_n) is any sequence in $\sigma(x)$ which converges to x in the μ topology, then

$$\sum_{n\geq 1} x_n \text{ exists and } \lim_{n\to\infty} \sum_{k\geq n} x_k = \lim_{n\to\infty} x_n.$$

The operation $+$ is said to be *idle* at x.

An idle point is one where the "unwinding" of a recursive definition stops. For example, $0 \in \mathbb{N}$, or the empty list.

Definition 54 Let D be a continuous dcpo. A μ continuous operation $+ : D^2 \to D$ is *iterative* if it has at least one idle point.

Here are a few simple examples of iterative operations.

Example 47 Data types.

(i) $([S], \cdot)$ concatenation of lists. The idle points are $\{[\,]\}$.
(ii) $(\mathbb{N}^*, +)$ addition of natural numbers. The idle points are $\{0\}$.

(iii) $(\mathbb{N}^*, \times)$ multiplication of natural numbers. The idle points are $\{0, 1\}$.
(iv) $(\{\bot, \top\}, \vee)$ Boolean 'or.' The idle points are $\{\bot, \top\}$.
(v) $(\{\bot, \top\}, \wedge)$ Boolean 'and.' The idle points are $\{\bot, \top\}$.

The μ topology on each domain above is discrete. The *fixed points* of a function $f : P \to P$ are $\text{fix}(f) = \{x \in P : f(x) = x\}$.

Definition 55 A splitting $r : D \to D$ on a dcpo D is *inductive* if for all $x \in D$, $\bigsqcup r^n x \in \text{fix}(r)$.

Definition 56 Let D be a dcpo and $(E, +)$ be a domain with an iterative operation. A function $\delta : D \to E$ *varies* with an inductive map $r : D \to D$ provided that

(i) For all $x \in \text{fix}(r)$, $\delta(x)$ is idle in E, and
(ii) For all $x \in D$, $\delta(r^n x) \to \delta(\bigsqcup r^n x)$ in the μ topology on E.

The function δ interprets the recursive part r of an algorithm in the domain $(E, +)$. A fixed point of r is mapped to an idle point in E: A point where recursion stops.

Definition 57 Let D be a dcpo and $(E, +)$ be a domain with an iterative operation. A *renee equation* on $D \to E$ is one of the form

$$\varphi = \delta + \varphi \circ r$$

where $\delta : D \to E$ varies with an inductive map $r : D \to D$.

Theorem 36 (Canonicity) *The renee equation*

$$\varphi = \delta + \varphi \circ r$$

has a unique solution which varies with r and agrees with δ on $\text{fix}(r)$.

Please stop and read the last theorem again. Thank you. The importance of φ varying with r is that it enables a verification principle [15]. Here are a few basic instances of the renee equation.

Example 48 The factorial function

$$\text{fac} : \mathbb{N} \to \mathbb{N}$$

is given by

$$\begin{aligned} &\text{fac } 0 = 1 \\ &\text{fac } n = n \times \text{ fac}(n - 1). \end{aligned}$$

Let $D = \mathbb{N}^*$ and $E = (\mathbb{N}^*, \times)$. Define $\delta : D \to E$ by

$$\delta(n) = \begin{cases} 1 & \text{if } n = 0, \\ n & \text{otherwise.} \end{cases}$$

and pred : $D \to D$ by $\text{pred}(n) = n - 1$, if $n > 0$, and $\text{pred}(0) = 0$. The unique solution of

$$\varphi = \delta \times \varphi \circ \text{ pred}$$

which satisfies $\varphi(0) = 1$ is the factorial function.

Example 49 The length of a list

$$\text{len} : [S] \to \mathbb{N}$$

is given by

$$\begin{aligned} \text{len } [\,] &= 0 \\ \text{len } a :: x &= 1 + \text{len } x. \end{aligned}$$

Let $D = [S]$ and $E = (\mathbb{N}^*, +)$. Define $\delta : D \to E$ by

$$\delta(x) = \begin{cases} 0 & \text{if } x = [\,], \\ 1 & \text{otherwise.} \end{cases}$$

and rest : $D \to D$ by $\text{rest}(a :: x) = x$ and $\text{rest}([\,]) = [\,]$. The unique solution of

$$\varphi = \delta + \varphi \circ \text{ rest}$$

which satisfies $\varphi([\,]) = 0$ is the length function.

Example 50 The merging of two sorted lists of integers

$$\text{merge} : [\text{int}] \times [\text{int}] \to [\text{int}]$$

is given by the following ML code

```
fun merge( [ ], ys )          = ys : int list
  | merge( xs, [ ] )          = xs
  | merge( x::xs, y::ys ) = if x <= y then
                                 x :: merge( xs, y :: ys )
                               else
                                 y :: merge( x :: xs, ys );
```

Let $D = [\text{int}] \times [\text{int}]$ and $E = ([\text{int}], \cdot)$. Define $\delta : D \to E$ by

$$\begin{aligned} \delta(x, [\,]) &= x \\ \delta([\,], y) &= y \\ \delta(x, y) &= [\min(\text{first } x, \text{first } y)], \text{ otherwise.} \end{aligned}$$

and $\pi : D \to D$ by

$$\begin{aligned} \pi(x, [\,]) &= ([\,], [\,]) \\ \pi([\,], y) &= ([\,], [\,]) \\ \pi(x, y) &= (\text{rest } x, y), \text{ if first } x \le \text{first } y; \\ \pi(x, y) &= (x, \text{rest } y), \text{ otherwise.} \end{aligned}$$

The unique solution of

$$\varphi = \delta \cdot \varphi \circ \pi$$

satisfying $\varphi([\,], [\,]) = [\,]$ is merge.

The last example is interesting because solving the equation yields a new iterative operation on [int]. We shall make use of this fact in the next example to solve an equation for sorting. In this way, we see that algorithms can be built up by solving sequences of renee equations.

Example 51 The prototypical bubblesort of a list of integers

$$\text{sort} : [\text{int}] \to [\text{int}]$$

is given by

$$\begin{aligned} \text{sort } [\,] &= [\,] \\ \text{sort } x &= \text{merge}(\,[\text{first } x],\ \text{sort rest } x\,) \end{aligned}$$

Let $D = [\text{int}]$ and $E = ([\text{int}], +)$ where

$$\begin{aligned} + : [\text{int}]^2 &\to [\text{int}] \\ (x, y) &\mapsto \text{merge}(x, y) \end{aligned}$$

is the merge operation of Example 50. Define $\delta : D \to E$ by

$$\delta(x) = \begin{cases} [\,] & \text{if } x = [\,] \\ [\text{first } x] & \text{otherwise} \end{cases}$$

and let rest : [int] $\to$ [int] be the usual splitting. The unique solution of

$$\varphi = \delta + \varphi \circ \text{rest}$$

satisfying $\varphi[\,] = [\,]$ is sort.

9.6.2.2 Computability from the Information Order

In this section, we will see that the primitive and partial recursive functions can both be captured using the renee equation: each arises as a canonical notion of computability derivable from a given information order.

Definition 58 Let $\mathbb{N}_\perp$ denote the set $\mathbb{N} \cup \{\perp\}$, where $\perp$ is an element that does not belong to $\mathbb{N}$.

For instance, one could take $\perp = \{\mathbb{N}\}$, should the need arise.

Definition 59 A *partial function* on the naturals is a function

$$f : \mathbb{N}^n \to \mathbb{N}_\perp,$$

where $n \geq 1$. We say that f is *undefined* at x exactly when $f(x) = \perp$.

Thinking of f as an algorithm, $f(x) = \perp$ means that the program f crashed when we sent it input x.

Definition 60 The *composition* of a partial map $f : \mathbb{N}^n \to \mathbb{N}_\perp$ with partial mappings $g_i : \mathbb{N}^k \to \mathbb{N}_\perp$, $1 \leq i \leq n$, is the partial map

$$f(g_1, \cdots, g_n) : \mathbb{N}^k \to \mathbb{N}_\perp$$

$$f(g_1, \cdots, g_n)(x) = \begin{cases} f(g_1(x), \cdots, g_n(x)) & \text{if } (\forall i)\ g_i(x) \neq \perp; \\ \perp & \text{otherwise.} \end{cases}$$

That is, if in the process of trying to run the program f, the computation of one of its inputs fails, then the entire computation fails.

Definition 61 A partial map $f : \mathbb{N}^{n+1} \to \mathbb{N}_\perp$ is defined by *primitive recursion* from $g : \mathbb{N}^n \to \mathbb{N}_\perp$ and $h : \mathbb{N}^{n+2} \to \mathbb{N}_\perp$ if

$$f(\bar{x}, y) = \begin{cases} g(\bar{x}) & \text{if } y = 0; \\ h(\bar{x}, y - 1, f(\bar{x}, y - 1)) & \text{otherwise.} \end{cases}$$

where we have written $\bar{x} \in \mathbb{N}^n$.

Computationally, primitive recursion is a counting loop.

Definition 62 The class of *primitive recursive functions* on the naturals is the smallest collection of functions $f : \mathbb{N}^n \to \mathbb{N}$ which contains the zero function, the successor, the projections, and is closed under composition and primitive recursion.

The analogue of a "while" loop is provided by minimization.

Definition 63 The *minimization* of a partial function $f : \mathbb{N}^{n+1} \to \mathbb{N}_\perp$ is the partial function

$$\mu f : \mathbb{N}^n \to \mathbb{N}_\perp$$

$$\mu f(x) = \min\{y \in \mathbb{N} : (\forall z < y)\ f(x, z) \neq \perp \ \&\ f(x, y) = 0\}$$

with the convention that $\mu f(x) = \perp$ if no such y exists.

Definition 64 The class of *partial recursive functions* on the naturals is the smallest collection of partial maps $f : \mathbb{N}^n \to \mathbb{N}_\perp$ which contains the zero function, the successor, the projections, and is closed under composition, primitive recursion, and minimization.

Let D be a domain which as a set satisfies $\mathbb{N} \subseteq D \subseteq \mathbb{N} \cup \{\infty\}$.

Definition 65 The sequence of domains $(D^n)_{n\geq 1}$ is given inductively by

$$\begin{aligned} D^1 &= D, \\ D^{n+1} &= D^n \times D^1,\ n > 0. \end{aligned}$$

We extend a few simple initial functions to D.

Definition 66 The initial functions.

(i) Addition of naturals $+ : D^2 \to D$ given by

$$(x, y) \mapsto \begin{cases} x + y & \text{if } x, y \in \mathbb{N}; \\ \infty & \text{otherwise.} \end{cases}$$

(ii) Multiplication of naturals $\times : D^2 \to D$ given by

$$(x, y) \mapsto \begin{cases} x \times y & \text{if } x, y \in \mathbb{N}; \\ \infty & \text{otherwise.} \end{cases}$$

(iii) The predicate $\leq: D^2 \to D$ given by

$$(x, y) \mapsto \begin{cases} x \leq y & \text{if } x, y \in \mathbb{N}; \\ \infty & \text{otherwise.} \end{cases}$$

(iv) The projections $\pi_i^n : D^n \to D$, for $n \geq 1$ and $1 \leq i \leq n$, given by

$$(x_1, \cdots, x_n) \mapsto \begin{cases} x_i & \text{if } (x_1, \cdots, x_n) \in \mathbb{N}^n; \\ \infty & \text{otherwise.} \end{cases}$$

A map $r : D^n \to D^n$ may be written in terms of its *coordinates* $r_i : D^n \to D$, for $1 \leq i \leq n$, as $r = (r_1, \cdots, r_n)$.

Definition 67 Let $\mathcal{C}(D)$ be the smallest class of functions $f : D^n \to D$ with the following properties:

(i) $\mathcal{C}(D)$ contains $+$, $\times$, $\leq$, and π_i^n, for $n \geq 1$ and $1 \leq i \leq n$,
(ii) $\mathcal{C}(D)$ is closed under substitution: If $f : D^n \to D$ is in $\mathcal{C}(D)$ and $g_i : D^k \to D$ is in $\mathcal{C}(D)$, for $1 \leq i \leq n$, then

$$f(g_1, \cdots, g_n) : D^k \to D \text{ is in } \mathcal{C}(D),$$

and

(iii) $\mathcal{C}(D)$ is closed under iteration: If $\delta : D^n \to D$ and $+ : D^2 \to D$ are in $\mathcal{C}(D)$, and $r : D^n \to D^n$ is a map whose coordinates are in $\mathcal{C}(D)$, then

$$\varphi = \delta + \varphi \circ r \in \mathcal{C}(D)$$

whenever this is a renee equation on $D^n \to D$.

$\mathcal{C}(D)$ contains maps of type $D^n \to D$. To obtain functions on the naturals, we simply restrict them to $\mathbb{N}^n$. In general, we obtain partial maps on the naturals, depending on whether or not D contains ∞.

Definition 68 The restriction of a mapping $f : D^n \to D$ to $\mathbb{N}^n$ is

$$|f| : \mathbb{N}^n \to \mathbb{N}_\perp$$
$$|f|(x) = \begin{cases} f(x) & \text{if } f(x) \in \mathbb{N}; \\ \perp & \text{otherwise.} \end{cases}$$

Let $\mathbb{N}^\infty$ denote the domain of naturals in their usual order with ∞ as a top element, $\mathbb{N}^*$ denote the domain of naturals in their dual order and $\mathbb{N}^\flat$ denote the domain of naturals ordered flatly: $x \sqsubseteq y \equiv x = y$.

The information order on a domain determines a notion of computability *because* it determines our ability to iterate.

Theorem 37

(i) $|\mathcal{C}(\mathbb{N}^\infty)|$ *is the class of partial recursive functions.*
(ii) $|\mathcal{C}(\mathbb{N}^*)|$ *is the class of primitive recursive functions.*
(iii) $|\mathcal{C}(\mathbb{N}^\flat)|$ *is the smallest class of functions containing the initial functions which is closed under substitution.*

9.6.2.3 A Renee Equation for Algorithmic Complexity

One renee equation describing an algorithm leads to another describing its complexity. If we have an algorithm $\varphi = \delta + \varphi \circ r$, then in order to calculate $\varphi(x)$, we must calculate $\delta(x)$, $r(x)$, $\varphi(rx)$ and $\delta(x) + \varphi(rx)$. Thus, the *cost* $c_\varphi(x)$ of calculating $\varphi(x)$ is the sum of the four costs associated with computing $\delta(x)$, $r(x)$, $\varphi(rx)$ and $\delta(x) + \varphi(rx)$. In symbols,

$$c_\varphi(x) = c_\delta(x) + c_r(x) + c_\varphi(rx) + c_+(\delta(x), \varphi(rx)).$$

When the functions (c_δ, c_+, c_r) actually describe the complexity of an algorithm, the equation above can be solved uniquely for c_φ.

Proposition 6 *Let* $\varphi = \delta + \varphi \circ r$ *be a renee equation on* $D \to E$. *If* $c_\delta : D \to \mathbb{N}^*$, $c_r : D \to \mathbb{N}^*$ *and* $c_+ : E^2 \to \mathbb{N}^*$ *are functions such that for all* $x \in D$,

$$\lim_{n\to\infty} c_\delta(r^n x) = \lim_{n\to\infty} c_r(r^n x) = \lim_{n\to\infty} c_+(\delta(r^n x), \varphi r(r^n x)) = 0,$$

then

$$c_\varphi = c_\delta + c_r + c_+(\delta, \varphi \circ r) + c_\varphi(r)$$

is a renee equation on $D \to (\mathbb{N}^*, +)$.

Thus, one renee equation describing an algorithm leads to another describing its complexity. Let's briefly consider a quick example just to check that the ideas work the way they should, we calculate the complexity of a sorting algorithm.

Example 52 Recall the prototypical bubblesort of a list of integers

$$\text{sort} : [\text{int}] \to [\text{int}]$$

is given by

$$\begin{aligned} &\text{sort}\ [\,] = [\,] \\ &\text{sort}\ x \ \ = \text{merge}([\text{first}\ x],\ \text{sort rest}\ x) \end{aligned}$$

Let $D = [\text{int}]$ and $E = ([\text{int}], +)$ where

$$+ : [\ \text{int}]^2 \to [\text{int}]$$

$$(x, y) \mapsto \text{merge}(x, y)$$

is the merge operation mentioned previously. Define $\delta : D \to E$ by

$$\delta(x) = \begin{cases} [\,] & \text{if } x = [\,] \\ [\text{first}\ x] & \text{otherwise} \end{cases}$$

and let $r : D \to D$ be the splitting $rx = \text{rest}\,x$. The unique solution of

$$\varphi = \delta + \varphi \circ r$$

satisfying $\varphi[\,] = [\,]$ is sort.

For the worst case analysis of $\text{sort} = \delta + \text{sort} \circ r$ the number of comparisons performed by r and δ on input x is zero. Hence,

$$c_r(x) = c_\delta(x) = 0,$$

while the cost of merging two lists x and y can be as great as $\mu x + \mu y$, so

$$c_+(x, y) = \mu x + \mu y.$$

By Proposition 6, we have a renee equation

$$c_{\text{sort}} = c_{+}(\delta, \text{sort} \circ r) + c_{\text{sort}}(r)$$

which *should* measure the complexity of bubblesort. But does it? By Theorem 36,

$$c_{\text{sort}}[\,] = 0,$$

while for any other list x, we have

$$\begin{aligned} c_{\text{sort}}(x) &= c_{+}(\delta(x), \text{sort}(rx)) + c_{\text{sort}}(rx) \\ &= \mu\,\delta(x) + \mu\,\text{sort}(rx) + c_{\text{sort}}(rx) \\ &= 1 + (\mu x - 1) + c_{\text{sort}}(rx) \\ &= \mu x + c_{\text{sort}}(rx). \end{aligned}$$

However, the function $f(x) = [\mu x(\mu x + 1)]/2$ varies with r, agrees with δ on $\text{fix}(r)$, and satisfies the equation above, so by the uniqueness in Theorem 36, we have

$$c_{\text{sort}}(x) = \frac{\mu x(\mu x + 1)}{2},$$

for all x.

One can go further with these ideas. In [18], the renee equation and measurement combine to provide a practical formal model of what a classical "search" method $\varphi = \delta + \varphi \circ r$ is. A particular highlight of the approach is that it does not force one to distinguish between discrete notions of searching, such as linear and binary searching of lists, and continuous notions of searching, such as zero finding methods like the bisection. The complexity c_φ of such methods is then shown to be determined by the number of iterations it takes r to get "close enough" to a fixed point. Thus, c_φ can also be calculated using the informatic derivative at a compact element.

9.6.3 Trajectories

Iterating an operator $f : D \to D$ yields a sequence $x, f(x), f^2(x), \ldots, f^n(x)$. Each $f^n(x)$ can be thought of as occuring at time n. It is natural to then wonder if an element $f^t(x)$ exists where $t \in [0, \infty)$. We would then have a trajectory $x : [0, \infty) \to D$ which describes the effect that f has had on x after t units of time. We could then take derivatives of x with respect to time and use them to learn things about a process. For instance, maybe the complexity of a process would amount to the point in time t when $x(s) \sqsubseteq x(t)$ for all s i.e. the "absolute maximum" of x. Maybe we could graph trajectories on the $t-\sqsubseteq$ axis to learn things about processes that we didn't know before. Maybe we should try this.

9.6.3.1 Kinematics

Proofs for this section can be found in [22].

Definition 69 A *variable* on a dcpo is a measurement $v : D \to [0, \infty)^*$ such that for all $x, y \in D$, we have $x \sqsubseteq y \;\&\; vx = vy \Rightarrow x = y$.

Definition 70 A *curve* on a domain D is a function $x : \mathrm{dom}(x) \to D$ where $\mathrm{dom}(x)$ is a nontrivial interval of the real line.

Each curve x determines a value of v at time t, which is the number $vx(t)$.

Definition 71 For a curve x and variable v on a dcpo,

$$\dot{x}_v(t) := \lim_{s \to t} \frac{vx(s) - vx(t)}{s - t}.$$

We then define

$$\ddot{x}_v := \frac{d\dot{x}_v}{dt}$$

and so on for higher order.

Because $\dot{x}_v : [0, \infty) \to \mathbb{R}$ is an ordinary function, higher order derivatives are calculated as usual – its the first derivative that requires theory.

Proposition 7 *Let x be a curve with $\dot{x}_v$ defined on (a, b).*

(i) *x is monotone increasing on $[a, b]$ iff $\dot{x}_v \le 0$ on (a, b) and $x[a, b]$ is a chain.*
(ii) *x is monotone decreasing on $[a, b]$ iff $\dot{x}_v \ge 0$ on (a, b) and $x[a, b]$ is a chain.*
(iii) *x is constant on $[a, b]$ iff $\dot{x}_v = 0$ on (a, b) and $x[a, b]$ is a chain.*

Notice that the sign of $\dot{x}_v$ is an indicator of how *uncertainty* behaves: If $\dot{x}_v \le 0$, then uncertainty is decreasing, so we are moving up in the order.

Definition 72 A curve x has a *relative maximum* at an interior point $t \in \mathrm{dom}(x)$ if there is an open set U_t containing t such that $x(s) \sqsubseteq x(t)$ for all $s \in U_t$. *Relative minimum* is defined dually, and these two give rise to *relative extremum*.

Notice that a *qualitative* relative maximum is a point in time where the *quantitative* uncertainty is a local minimum.

Lemma 6 *If a curve x has a relative extremum at interior point $t \in \mathrm{dom}(x)$, then for all variables v, either $\dot{x}_v(t) = 0$, or it does not exist.*

A nice illustration of why the qualitative idea $\sqsubseteq$ is important: if a curve has a derivative with respect to *just one* variable v, then its set of extreme points is contained in the set $\{t : \dot{x}_v(t) = 0\}$. This is quite valuable: we are free to choose the variable which makes the calculation as simple as possible.

Once we have the extreme points there is also a systematic way in the informatic setting to determine which (if any) are maxima or minima: the second derivative test, whose formalization requires one to acknowledge the qualitative structure on which it is implicitly founded.

Definition 73 A curve x is a *trajectory* if for all $t \in \mathrm{dom}(x)$ there is an open set U_t containing t such that

$$x(s) \sqsubseteq x(t) \text{ or } x(t) \sqsubseteq x(s)$$

for all $s \in U_t$.

Thus, a *trajectory* is a curve x with underlying qualitative structure; it is called C^2_v when $\ddot{x}_v$ is continuous, with respect to variable v.

Proposition 8 *Let x be a C^2_v trajectory. If $\dot{x}_v(t) = 0$ and $\ddot{x}_v(t) \neq 0$ for some interior point $t \in \mathrm{dom}(x)$, then x has a relative extremum at t.*

(i) *If $\ddot{x}_v(t) > 0$, then $x(t)$ is a relative maximum.*
(ii) *If $\ddot{x}_v(t) < 0$, then $x(t)$ is a relative minimum.*

In this work we will be mostly concerned with the strongest form of extrema on domains:

Definition 74 A curve x has an *absolute maximum* at $t \in \mathrm{dom}(x)$ if

$$x(s) \sqsubseteq x(t)$$

for all $s \in \mathrm{dom}(x)$. Absolute minimum is defined similarly.

Here is a simple but surprisingly useful way of establishing the existence of absolute extrema.

Proposition 9 *Let v be a variable on D and $x : [a, b] \to D$ a curve whose image is a chain. If $vx : [a, b] \to \mathbb{R}$ is Euclidean continuous, then*

(i) *The map x is continuous from the Euclidean to the Scott topology, and*
(ii) *The map x assumes an absolute maximum and an absolute minimum on $[a, b]$. In particular, its absolute maximum is*

$$x(t^*) = \bigsqcup_{t \in [a,b]} x(t)$$

for some $t^* \in [a, b]$, with a similar expression for the absolute minimum.

A valuable property of absolute maxima: If $x(t^*)$ is an absolute maximum, then for all variables υ,

$$\upsilon x(t^*) = \inf\{\upsilon x(t) : t \in \mathrm{dom}(x)\}.$$

That is, an absolute maximum is a point on a curve which simultaneously minimizes *all* variables.

9.6.3.2 Linear Searching

Suppose a list has $n > 0$ elements. Linear search begins with the first element in the list and proceeds to the next and so on until the key is located. At time t (after t comparisons), all elements with indices from 1 to t have been searched. Thus, a trajectory representing the information we have gained is $x(t) = t$ for $t \in [0, n]$. The natural space of informatic objects is $D = [0, n]$ whose natural measure of uncertainty is $\upsilon x \coloneqq n - x$.

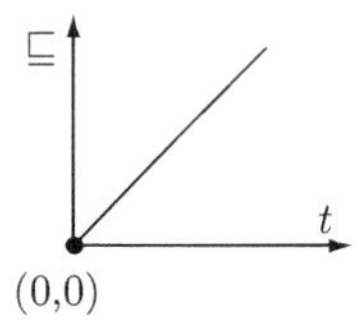

Next is a better example—one where the kinematics of computation will help us visualize a computation.

9.6.3.3 Binary Searching

This algorithm causes a trajectory on $(\mathbf{I}\mathbb{R}, \upsilon)$ with $\upsilon[a, b] = b - a$. For a continuous $f : \mathbb{R} \to \mathbb{R}$, let $\mathrm{split}_f : \mathbf{I}\mathbb{R} \to \mathbf{I}\mathbb{R}$ be the bisection method on the interval domain defined by

$$\mathrm{split}_f[a, b] := \begin{cases} \mathrm{left}[a, b] & \text{if } f(a) \cdot f((a + b)/2) \leq 0; \\ \mathrm{right}[a, b] & \text{otherwise.} \end{cases}$$

A given $x \in \mathbf{I}\mathbb{R}$ leads to a *trajectory* $x : [0, \infty) \to \mathbf{I}\mathbb{R}$ defined on natural numbers by

$$x(n) = \mathrm{split}_f^n(x)$$

and then extended to all intermediate times $n < t < n + 1$ by declaring $x(t)$ to be the *unique* element satisfying

$$x(n) \sqsubseteq x(t) \sqsubseteq x(n + 1) \text{ and } \mu x(t) = \frac{\upsilon x}{2^t}.$$

By definition, the trajectory of binary search is also increasing. But graphing it is more subtle. It looks like this:

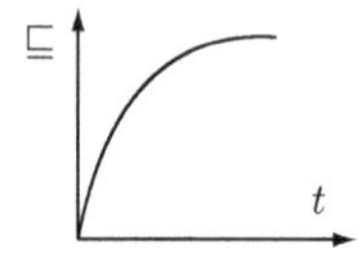

But why? Using the *kinematics of computation*, since $vx(t) = e^{-(\ln 2)t} \cdot vx(0)$, we have

$$\dot{x}_v(t) = (-\ln 2)vx(t) < 0$$

reflecting the fact that $x : [0, \infty) \to \mathbb{R}$ is *increasing*. In addition, $\ddot{x}_v(t) > 0$, so the graph is concave *down*. Notice that as $t \to \infty$, the trajectory should tend toward the answer as its velocity tends to zero.

Trajectories of classical search algorithms tend to increase with time. All of the curves basically look the same, so what's the point? It makes the dream of a "kinematics of computation" seem out of reach. But then, what is the point in dreaming of things that are *within* reach? Those aren't dreams, they're just things you plan to do.

9.6.3.4 Quantum Searching

Grover's algorithm [9] for searching is the only known quantum algorithm whose complexity is *provably better* than its classical counterpart. It searches a list L of length n (a power of two) for an element k known to occur in L precisely m times with $n > m \geq 1$. The register begins in the pure state

$$|\psi\rangle = \frac{1}{\sqrt{n}} \sum_{i=1}^{n} |i\rangle$$

and after j iterations of the Grover operator G

$$G^j|\psi\rangle = \frac{\sin(2j\theta + \theta)}{\sqrt{m}} \sum_{L(i)=k} |i\rangle + \frac{\cos(2j\theta + \theta)}{\sqrt{n-m}} \sum_{L(i)\neq k} |i\rangle$$

where $\sin^2\theta = m/n$. The probability that a measurement yields i after j iterations is

$$\sin^2(2j\theta + \theta)/m \text{ if } L(i) = k$$

and

$$\cos^2(2j\theta + \theta)/(n-m) \text{ if } L(i) \neq k.$$

To get the answer, we measure the state of the register in the basis $\{|i\rangle : 1 \leq i \leq n\}$; if we perform this measurement after j iterations of G, when the state of the register is $G^j|\psi\rangle$, our knowledge about the result is represented by the vector

$$x(j) = \left(\frac{\sin^2(2j\theta+\theta)}{m}, \ldots, \frac{\sin^2(2j\theta+\theta)}{m}, \right.$$
$$\left. \frac{\cos^2(2j\theta+\theta)}{n-m}, \ldots, \frac{\cos^2(2j\theta+\theta)}{n-m} \right)$$

The crucial step now is to *imagine* t iterations,

$$x(t) = \left(\frac{\sin^2(2t\theta+\theta)}{m}, \ldots, \frac{\sin^2(2t\theta+\theta)}{m}, \right.$$
$$\left. \frac{\cos^2(2t\theta+\theta)}{n-m}, \ldots, \frac{\cos^2(2t\theta+\theta)}{n-m} \right)$$

Thus, x is a curve on the domain Δ^n of classical states in its *implicative order* (Sect. 9.2.2)

$$x \sqsubseteq y \equiv (\forall i)\; x_i < y_i \Rightarrow x_i = x^+$$

where x^+ refers to the largest probability in x. Thus, only a maximum probability is allowed to increase as we move up in the information order on Δ^n. If the maximum probability refers to a solution of the search problem, then moving up in this order ensures that we are getting closer to the answer.

We will now use this trajectory to analyze Grover's algorithm using the kinematics of computation. Here are some crucial things our analysis will yield:

(a) The complexity of the algorithm,
(b) A qualitative property the algorithm possesses called *antimonotonicity*. Without knowledge of this aspect, an experimental implementation would almost certainly fail (for reasons that will be clear later).
(c) An explanation of the algorithm as being an attempt to calculate a *classical proposition*.

Precisely now, the classical state $x(t)$ is a vector of probabilities that do not increase for $t \in \text{dom}(x) = [a,b]$, $a = 0$ and $b = \pi/2\theta - 1$. The image of $x : [a,b] \to \Lambda^n$ is a chain in the implicative order, which is simplest to see by noting that it has the form

$$x = (f, \ldots, f, g, \ldots, g)$$

so that $g(s) \geq g(t) \Rightarrow x(s) \sqsubseteq x(t)$; otherwise, $x(t) \sqsubseteq x(s)$. We can now determine the exact nature of the motion represented by x using kinematics. Because $x : [a, b] \to D$ is a curve on a domain D whose image is a chain and whose time derivative $\dot{x}_v(t)$ exists with respect to a variable v on Δ^n, we know that

(i) The curve x has an absolute maximum on $[a, b]$: There is $t^* \in [a, b]$ such that

$$x(t^*) = \bigsqcup_{t \in [a,b]} x(t),$$

and

(ii) Either $t^* = a$, $t^* = b$ or $\dot{x}_v(t^*) = 0$.

Part of the power of this simple approach is that we are free to choose any v we like. To illustrate, a tempting choice might be entropy $v = H$, but then solving $\dot{x}_v = 0$ means solving the equation

$$-m\dot{f}(1 + \log f) - (n - m)\dot{g}(1 + \log g) = 0$$

and we also have to determine the points where $\dot{x}_v$ is undefined, the set $\{t : g(t) = 0\}$. However, if we use

$$v = 1 - \sqrt{x^+},$$

we only have to solve a single elementary equation

$$\cos(2t\theta + \theta) = 0$$

for t, allowing us to conclude that the maximum must occur at $t = a$, $t = b$, or at points in

$$\{t : \dot{x}_v(t) = 0\} = \{b/2\}.$$

The absolute maximum of x is

$$x(b/2) = (1/m, \ldots, 1/m, 0, \ldots, 0)$$

because for the other points we find a minimum of

$$x(a) = x(b) = \bot = (1/n, \ldots, 1/n).$$

The value of knowing the absolute maximum is that it allows us to calculate the complexity of the algorithm: it is $O(b/2)$, the amount of time required to move to a state from which the likelihood of obtaining a correct result by measurement is maximized. This gives $O(\sqrt{n/m})$ using $\theta \geq \sin\theta \geq \sqrt{m/n}$ and then $b/2 \leq (\pi/4)\sqrt{n/m} - 1/2$.

From $\dot{x}_v(t) \leq 0$ on $[a, b/2]$ and $\dot{x}_v(t) \geq 0$ on $[b/2, b]$, we can also graph x:

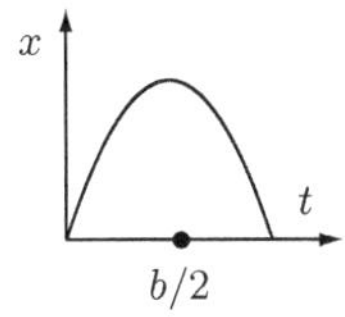

This is the "antimonotonicity" of Grover's algorithm: if $j = b/2$ iterations will solve the problem accurately, $2j$ iterations will mostly unsolve it! This means that our usual way of reasoning about iterative procedures like numerical methods, as in "we must do at least j iterations," no longer applies. We must say "do exactly j iterations; no more, no less." As is now clear, precise estimates like these have to be obtained before experimental realization is possible.

Finally, as explained in more detail in [23], we can view Grover's algorithm as an attempt to calculate as closely as possible the classical proposition

$$x(b/2) = (1/m, \ldots, 1/m, 0, \ldots, 0) \in \mathrm{Ir}(\Delta^n) = \left\{x : \bigwedge \uparrow x \cap \max(\Delta^n) = x\right\}.$$

It does so by generating *approximations*

$$x(t) \ll x(b/2)$$

for all $t \neq b/2$.

9.6.3.5 Amplitude Damping

Let $\mathcal{H}$ be the state space for a two dimensional quantum system. Two parties communicate with each other as follows. First, they agree up front on a fixed basis of $\mathcal{H}$, say $\{|\psi\rangle, |\phi\rangle\}$, which can be expressed in some basis $\{|0\rangle, |1\rangle\}$ as

$$|\psi\rangle = a|0\rangle + b|1\rangle \;\; \& \;\; |\phi\rangle = c|0\rangle + d|1\rangle$$

where the amplitudes a, b, c, d are all complex. The state $|\psi\rangle$ is taken to mean "0", while the state $|\phi\rangle$ is taken to mean "1". The first party, the sender, attempts to send one of these two qubits $|*\rangle \in \{|\psi\rangle, |\phi\rangle\}$ to the second party, the receiver. The second party receives *some* qubit and performs a measurement in the agreed upon basis. The result of this measurement is one of the qubits $\{|\psi\rangle, |\phi\rangle\}$, which is then interpreted as meaning either a "0" or a "1".

We say *some qubit* because as $|*\rangle$ travels, it suffers an unwanted interaction with its environment, whose effect on density operators can be described as

$$\varepsilon(\rho) = E_0 \rho E_0^\dagger + E_1 \rho E_1^\dagger$$

where the operation elements are given by

$$E_0 = \begin{pmatrix} 1 & 0 \\ 0 & \sqrt{1-\lambda} \end{pmatrix} \ \& \ E_1 = \begin{pmatrix} 0 & \sqrt{\lambda} \\ 0 & 0 \end{pmatrix}$$

This effect is known as *amplitude damping* and the parameter $\lambda \in [0, 1]$ can be thought of as the probability of losing a photon. Thus, the receiver does not necessarily acquire the qubit $|*\rangle$, but instead receives some degradation of it, describable by the density operator $\varepsilon(|*\rangle\langle*|)$.

The probability that '0' is received when '0' is sent is

$$\alpha = P(0|0) = -2|a|^4 p(\lambda) + |a|^2(\lambda + 2p(\lambda)) + 1 - \lambda$$

while the probability that '0' is received when '1' is sent is

$$\beta = P(0|1) = 2|a|^4 p(\lambda) + |a|^2(\lambda - 2p(\lambda))$$

where $p(\lambda) = -1 + \lambda + \sqrt{1-\lambda} \geq 0$. Thus, each choice of basis defines a classical binary channel (α, β). Notice that the probabilities α and β only depend on $|a|^2$ because $|c|^2 = |a|^2$ and $|b|^2 = |d|^2 = 1 - |a|^2$ by the orthogonality of $|\psi\rangle$ and $|\phi\rangle$, and because the initial expressions for α and β turn out to only depend on modulus squared terms. Because the basis is fixed, $|a|^2 \in [0, 1]$ is a constant and we obtain a function $x : [0, 1] \to \mathbb{N}$ of λ given by

$$x(\lambda) = (\alpha(\lambda), \beta(\lambda))$$

where we recall that $\mathbb{N} \simeq \mathbf{I}[0, 1]$ is the domain of binary channels. Its domain theoretic nature was first established in [29]:

Proposition 10 *The trajectory* $x : [0, 1] \to \mathbb{N}$ *is Scott continuous.*

One valuable aspect of x being Scott continuous is that we can now make precise the connection between quantum information's intuitive use of the word "noise" and information theory's precise account of it: the quantity $\mathrm{C}(x(\lambda))$ decreases as λ increases i.e. the amount of information that the two parties can communicate decreases as the the probability of losing a photon increases. In the extreme cases,

$$x(0) = (1, 0) \ \& \ x(1) = (|a|^2, |a|^2)$$

yielding respective capacities of 1 and 0. There is a more fundamental idea at work in this example and in many others like it: we have learned about capacity by only examining how the probabilities in the noise matrix change, and this more than justifies the domain theoretic approach. Imagine what would happen if we actually tried to calculate $\mathrm{C}(x(\lambda))$ explicitly: we would have to substitute $\alpha(\lambda) = -2|a|^4 p(\lambda) + |a|^2(\lambda + 2p(\lambda)) + 1 - \lambda$ for a and $\beta(\lambda) = 2|a|^4 p(\lambda) + |a|^2(\lambda - 2p(\lambda))$ for b into theformula

$$C(a,b) = \log_2\left(2^{\frac{\bar{a}H(b)-\bar{b}H(a)}{a-b}} + 2^{\frac{bH(a)-aH(b)}{a-b}}\right)$$

and then seek to show that the resulting quantity decreases as λ increases.

9.6.3.6 Decoherence Over Time

One interesting aspect of amplitude damping is that it is *not unital*. Any unital qubit channel will lead to a trajectory defined on some nontrivial interval since all classical channels derived from them are binary symmetric and the binary symmetric channels form a chain in $\mathbb{N}$. An interesting example in this last regard is phase damping as a function of time, whose effect on the pure state $|\psi\rangle = \alpha|0\rangle + \beta|1\rangle$ with density operator

$$\rho = |\psi\rangle\langle\psi| = \begin{pmatrix} |\alpha|^2 & \alpha\beta^* \\ \alpha^*\beta & |\beta|^2 \end{pmatrix}$$

after t units of time is

$$\rho(t) = \begin{pmatrix} |\alpha|^2 & \alpha\beta^* e^{-t/t_d} \\ e^{-t/t_d}\alpha^*\beta & |\beta|^2 \end{pmatrix}$$

where t_d is a constant known as the *decoherence time*. If the qubit decoheres for t units of time, then a "0" may no longer be a "0" and a "1" may no longer be a "1". Specifically, the probability that a "0" is still a "0" is

$$P(0|0) = |a|^4 + 2e^{-t/t_d}|a|^2|b|^2 + |b|^4$$

while the probability that a '1' changes into a '0' is

$$P(0|1) = 1 - P(0|0).$$

This gives rise to a binary symmetric channel.

9.6.4 Vectors

We can think of domains as a qualitative way of reasoning about informative objects, and measurement as a way of determining the amount of information in an object. But neither set of ideas attempts to *directly* answer the question "What is information?" In this section, we offer one possible answer to this question which has pragmatic value and is of interest to computer science.

To begin, we assume that the words "complexity" and "information" are just that—words. We start from a clean slate, forgetting the various connotations these words have in the sciences, and simply begin talking about them intuitively. We might say:

- The complexity of a *secret* is the amount of work required to *guess* it.
- The complexity of a *problem* is the amount of work required to *solve* it.
- The complexity of a *rocket* is the amount of work required to *escape gravity*.
- The complexity of a *probabilistic state* is the amount of work required to *resolve* it.

In all cases, there is a task we want to accomplish, and a way of measuring the work done by a process that *actually achieves* the task; such a process belongs to a prespecified class of processes which themselves are the stuff that science is meant to discover, study and understand. Then there are two points not to miss about complexity:

(i) It is relative to a prespecified class of processes,
(ii) The use of the word "required" necessitates the *minimization* of quantities like work over the class of processes.

Complexity is *process dependent*. Now, what is information in such a setting?

Information, in seeming stark contrast to complexity, is *process independent*. Here is what we mean: *information is complexity relative to the class of all conceivable processes*. For instance, suppose we wish to measure the complexity of an object x with respect to several different classes $P_1, \ldots, P_n$ of processes. Then the complexity of x varies with the notion of process: It will have complexities $c_1(x), \ldots, c_n(x)$, where c_i is calculated with respect to the class P_i. However, because information is complexity relative to the class of *all* conceivable processes, the information in an object like x will *not* vary. That is what we mean when we say information is process independent: it is an element present in *all* notions of complexity. So we expect

$$\text{complexity} \geq \text{information}$$

if only in terms of the mathematics implied by the discussion above. For example, this might allow us to *prove* that the amount of work you expect to do in solving a problem always exceeds the a priori uncertainty (information) you have about its solution: the less you know about the solution, the more work you should expect to do. An inequality like the one above could be valuable.

To test these ideas, we study the complexity of classical states relative to a class of processes. A class of processes will be derived from a domain (D, μ) with a measurement μ that supports a new notion called *orthogonality*. Write $c_D(x)$ for the complexity of a classical state x relative to (D, μ). Then we will see that

$$\inf_{D \in \Sigma} c_D = \sigma \tag{9.1}$$

where σ is Shannon entropy and Σ is the class of domains (D, μ). This equation provides a setting where it is clear that information in the sense of the discussion above is σ, and that the class of all conceivable processes is Σ. By (9.1), our intuitive development of 'complexity' turns out to be capable of deriving lower bounds on the complexity of algorithms such as sorting and searching. Another limit also exists,

$$\bigcap_{D\in\Sigma} \leq_D = \leq \tag{9.2}$$

where $\leq_D$ is a relation on classical states which means $x \leq_D y$ iff for all processes p on (D, μ), it takes more work for p to resolve x than y. This is *qualitative complexity*, and the value of the intersection above $\leq$ just happens to be the *majorization relation* from Sect. 9.2.2. Muirhead [34] discovered majorization in 1903, and in the last 100 years his relation has found impressive applications in areas such as economics, computer science, physics and pure mathematics [2, 14]. We will see that the complexity c_D is determined by its value on this subset.

The limits (9.1) and (9.2) comprise what we call *the universal limit*, because it is taken over the class of *all* domains. The pair $(\sigma, \leq)$ can also be derived on a *fixed* domain (D, μ) provided one has the ability to *copy* processes. The mathematics of copying necessitates the addition of algebraic structure $\otimes$ to domains (D, μ) already supporting orthogonality. It is from this setting, which identifies the essential mathematical structure required to execute classical information theory [41] over the class of semantic domains, that the *fixed point theorem* springs forth: as with recursive programs, the semantics of *information* can also be specified by a least fixed point:

$$\mathrm{fix}(\Phi) = \bigsqcup_{n\geq 0} \Phi^n(\perp) = \sigma$$

where Φ is the copying operator and $\perp$ is the complexity c_D, i.e., the least fixed point of domain theory connects complexity in computer science to entropy in physics. We thus learn that one can use domains to define the complexity of objects in such a way that information becomes a concept derived from complexity in a precise and systematic manner: as a least fixed point.

9.6.4.1 Processes

To study processes which may result in one of several different outcomes, we have to know what "different" means. This is what *orthogonality* does: It provides an order theoretic definition of "distinct."

Definition 75 A pair of elements $x, y \in D$ are *orthogonal* if $\mu(\uparrow x \cap \uparrow y) \subseteq \{0\}$. This is written $x \perp y$.

The word "domain" in this section means a continuous dcpo D with a least element $\perp$ and a map μ that measures all of D. By replacing μ with $\mu/\mu\perp$ if necessary, we can and will assume that $\mu\perp = 1$. Finally, we will assume that

$$\mu(\bigwedge F) \geq \sum_{x \in F} \mu x$$

for each finite set $F \subseteq D$ of pairwise orthogonal elements.

Example 53

(i) $\mathbf{I}[0, 1]$ with the length measurement μ is a domain.
(ii) Let $p \in \Delta^n$ be a classical state with all $p_k > 0$ and Σ^∞ the strings over the alphabet $\Sigma = \{0, \ldots, n-1\}$. Define $\mu : \Sigma^\infty \to [0, \infty)^*$ by $\mu\bot = 1$ and $\mu i = p_{i+1}$ for $i \in \Sigma$, and then extend it homomorphically by

$$\mu(s \cdot t) = \mu s \cdot \mu t$$

where the inner dot is concatenation of finite strings. The unique Scott continuous extension, which we call μ, yields a *domain* (D, μ).

An immediate corollary is the case $p = (1/2, 1/2) \in \Delta^2$ and $\Sigma = \{0, 1\} = 2$, the binary strings with the usual measurement: $(2^\infty, 1/2^{|\cdot|})$ is a domain. This is the basis for the study of binary codes. The fact that it is a domain *implies* the vital *Kraft inequality* of classical information theory.

Theorem 38 (Kraft) *We can find a finite antichain of Σ^∞ which has finite word lengths $a_1, a_2, \ldots, a_n$ iff*

$$\sum_{i=1}^{n} \frac{1}{|\Sigma|^{a_i}} \leq 1.$$

Finite antichains of finite words are sometimes also called *instantaneous codes*. The inequality in Kraft's result can be derived as follows:

Example 54 The Kraft inequality. We apply the last example with

$$p = (1/|\Sigma|, \ldots, 1/|\Sigma|) \in \Delta^{|\Sigma|}.$$

A finite subset of $\Sigma^{<\infty}$ is pairwise orthogonal iff it is an antichain. Thus,

$$\mu(\bigwedge F) \geq \sum_{x \in F} \mu x.$$

In particular, $1 = \mu\bot \geq \mu(\bigwedge F)$, using the monotonicity of μ. Notice that the bound we derive on the sum of the measures is more precise than the one given in the Kraft inequality. We call μ the *standard measurement* and assume *it* when writing (Σ^∞, μ), unless otherwise specified.

Finally, the order theoretic structure of (D, μ) gives rise to a notion of *process*: a set of outcomes which are (a) different and (b) achievable in finite time.

Definition 76 A *process* on (D, μ) is a function $p : \{1, \ldots, n\} \to D$ such that $p_i \perp p_j$ for $i \neq j$ and $\mu p > 0$. $P^n(D)$ denotes the set of all such processes.

9.6.4.2 Complexity (Quantitative)

There is a natural function $-\log \mu : P^n(D) \to (0, \infty)^n$ which takes a process $p \in P^n(D)$ to the positive vector

$$-\log \mu p = (-\log \mu p_1, \ldots, -\log \mu p_n).$$

By considering processes on the domain of binary strings $(2^\infty, \mu)$, it is clear that the expected work done by an algorithm which takes one of n different computational paths $p : \{1, \ldots, n\} \to D$ is $\langle -\log \mu p | x\rangle$. Thus, the complexity of a state $c : \Delta^n \to [0, \infty)^*$ is

$$c(x) := \inf\{\langle -\log \mu p | x\rangle : p \in P^n(D)\}.$$

The function sort$^+$ reorders the components of a vector so that they increase; its dual sort$^-$ reorders them so that they decrease.

Proposition 11 *For all* $x \in \Delta^n$,

$$c(x) = \inf\{\langle \mathrm{sort}^+(-\log \mu p) | \mathrm{sort}^-(x)\rangle : p \in P^n(D)\}.$$

In particular, the function c is symmetric.

So we can restrict our attention to monotone decreasing states Λ^n.

Definition 77 The *expectation* of $p \in P^n(D)$ is $\langle p\rangle : \Lambda^n \to [0, \infty)^*$ given by

$$\langle p\rangle x = \langle \mathrm{sort}^+(-\log \mu p) | x\rangle.$$

If the outcomes of process p are distributed as $x \in \Lambda^n$, then the work we expect p will do when taking one such computational path is $\langle p\rangle x$. And finally:

Definition 78 The *complexity* of a state $h : \Lambda^n \to [0, \infty)^*$ is

$$h(x) = \inf\{\langle p\rangle x : p \in P^n(D)\}.$$

Thus, the relation of h to c is that $c(x) = h(\mathrm{sort}^-(x))$ for all $x \in \Delta^n$. The *Shannon entropy* $\sigma : \Delta^n \to [0, \infty)$

$$\sigma x := -\sum_{i=1}^{n} x_i \log x_i$$

can also be viewed as a map on Λ^n, and as a map on *all*monotone states. Its type will be clear from the context.

Proposition 12 *If (D, μ) is a domain, then the complexity $h_D : (\Lambda^n, \leq) \to [0, \infty)^*$ is Scott continuous and $h_D \geq \sigma$ where σ is entropy and $\leq$ is majorization.*

We have now *proven* the following: the amount of work we expect to do when solving a problem exceeds our a priori uncertainty about the solution. That is, the less you know about the solution, the more work you should expect to do:

Example 55 Lower bounds on algorithmic complexity. Consider the problem of sorting lists of n objects by comparisons. Any algorithm which achieves this has a binary decision tree. For example, for lists with three elements, a_1, a_2, a_3, it is

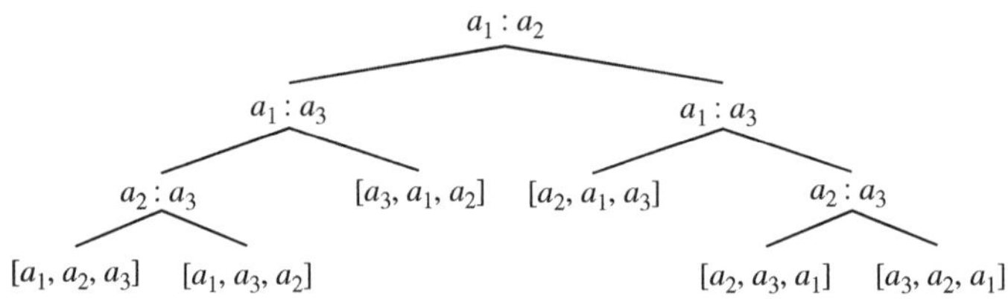

where a move left corresponds to a decision $\leq$, while a move right corresponds to a decision $>$. The leaves of this tree, which are labelled with lists representing potential outcomes of the algorithm, form an antichain of $n!$-many finite words in 2^∞ using the correspondence $\leq \mapsto 0$ and $> \mapsto 1$. This defines a process $p : \{1, \ldots, n!\} \to 2^\infty$. If our knowledge about the answer is $x \in \Lambda^{n!}$, then

$$\begin{aligned} \text{avg. comparisons} &= \langle -\log \mu p | x \rangle \\ &\geq \langle p \rangle(\mathrm{sort}^{-} x) \\ &\geq h(\mathrm{sort}^{-} x) \\ &\geq \sigma x. \end{aligned}$$

Assuming complete uncertainty about the answer, $x = \bot$, we get

$$\text{avg. comparisons} \geq \sigma \bot = \log n! \approx n \log n.$$

In addition, we can derive an entirely *objective conclusion:* In the *worst case*, we must do at least

$$\max(-\log \mu p) \geq \langle p \rangle \bot \geq \sigma \bot \approx n \log n$$

comparisons. Thus, sorting by comparisons is in general at least $O(n \log n)$. A similar analysis shows that searching by comparison is at least $O(\log n)$.

We have used *domain theoretic structure* as the basis for a new approach to counting the number of leaves in a binary tree. Just as different domains can give rise to different notions of computability (Sect. 9.6.2), different domains can also give rise to different complexity classes, for the simple reason that changing the order changes the notion of process. An example of this is $(L, \mu) \subseteq (2^\infty, \mu)$ which models linear search (Example 57).

9.6.4.3 Complexity (Qualitative)

Each domain (D, μ), because it implicitly defines a notion of process, provides an intuitive notion of what it means for one classical state to be more complex than another: x is more complex than y iff for all processes $p \in P^n(D)$, the work that p does in resolving x exceeds the work it does in resolving y. This is *qualitative complexity*.

Definition 79 For $x, y \in \Lambda^n$, the relation $\leq_D$ is

$$x \leq_D y \equiv (\forall p \in P^n(D))\, \langle p \rangle x \geq \langle p \rangle y.$$

Only one thing is clear about $\leq_D$: The qualitative analogue of Proposition 12.

Lemma 7 *For each domain* (D, μ), $\leq \subseteq \leq_D$.

The calculation of $\leq_D$ requires knowing more about the structure of D. We consider domains whose orders allow for the simultaneous description of *orthogonality* and *composition*. In the simplest of terms: These domains allow us to say what *different* outcomes are, and they allow us to form *composite* outcomes from pairs of outcomes.

Definition 80 A domain (D, μ) is *symbolic* when it has an associative operation $\otimes : D^2 \to D$ such that $\mu(x \otimes y) = \mu x \cdot \mu y$ and

$$x \perp u \text{ or } (x = u \;\&\; y \perp v) \Rightarrow x \otimes y \perp u \otimes v$$

for all $x, y, u, v \in D$.

Notice that $\otimes$ has a qualitative axiom and a quantitative axiom. One example of a symbolic domain is (Σ^∞, μ) for an alphabet Σ with $\otimes$ being concatenation.

Example 56 The $\otimes$ on $\mathbf{I}[0, 1]$ is

$$[a, b] \otimes [y_1, y_2] = [a + y_1 \cdot (b - a), a + y_2 \cdot (b - a)].$$

$(\mathbf{I}[0, 1], \otimes)$ is a monoid with $\perp \otimes x = x \otimes \perp = x$ and the measurement μ is a homomorphism! We can calculate zeroes of real-valued functions by repeatedly $\otimes$-ing left$(\perp) = [0, 1/2]$ and right$(\perp) = [1/2, 1]$, i.e., the bisection method.

We can $\otimes$ processes too: If $p : \{1, \ldots, n\} \to D$ and $q : \{1, \ldots, m\} \to D$ are processes, then $p \otimes q : \{1, \ldots, nm\} \to D$ is a process whose possible actions are $p_i \otimes q_j$, where p_i is any possible action of p, and q_j is any possible action of q. The exact indices assigned to these composite actions for our purposes is immaterial. We can characterize qualitative complexity on symbolic domains:

Theorem 39 *Let* $(D, \otimes, \mu)$ *be a symbolic domain. If there is a binary process* $p : \{1, 2\} \to D$, *then the relation* $\leq_D = \leq$.

9.6.4.4 The Universal Limit

We now see that $\leq$ and σ are two sides of the same coin: The former is a qualitative limit; the latter is a quantitative limit. Each is taken over the class of domains.

Theorem 40 *Let $\sigma : \Lambda^n \to [0, \infty)^*$ denote Shannon entropy and Σ denote the class of domains. Then*

$$\inf_{D \in \Sigma} h_D = \sigma$$

and

$$\bigcap_{D \in \Sigma} \leq_D = \leq$$

where the relation $\leq$ on Λ^n is majorization.

Corollary 6 *Shannon entropy $\sigma : (\Lambda^n, \leq) \to [0, \infty)^*$ is Scott continuous.*

By Theorem 40, the optimum value of $(h_D, \leq_D)$ is $(\sigma, \leq)$. But when does a domain have a value of $(h_D, \leq_D)$ that is close to $(\sigma, \leq)$? Though it is subtle, if we look at the case when $\leq_D$ achieves $\leq$ in the proof of Theorem 39, we see that a strongly contributing factor is the ability to *copy* processes—we made use of this idea when we formed the process $\bigotimes_{i=1}^n p$. We will now see that the ability to copy on a given domain *also guarantees* that h is close to σ.

9.6.4.5 Inequalities Relating Complexity to Entropy

We begin with some long overdue examples of complexity. It is convenient on a given domain (D, μ) to denote the complexity in dimension n by $h_n : \Lambda^n \to [0, \infty)$.

Example 57 Examples of h.

(i) On the lazy naturals $(L, \mu) \subseteq (2^\infty, \mu)$, where the L is for linear,

$$h_n(x) = x_1 + 2x_2 + \ldots + (n-1)x_{n-1} + (n-1)x_n$$

which is the average number of comparisons required to find an object among n using linear search.

(ii) On the domain of binary streams $(2^\infty, \mu)$,

$$h_2(x) \equiv 1$$
$$h_3(x) = x_1 + 2x_2 + 2x_3 = 2 - x_1$$
$$h_4(x) = \min\{2, x_1 + 2x_2 + 3x_3 + 3x_4\} = \min\{2, 3 - 2x_1 - x_2\}$$

In general, $h_n(x)$ is the average word length of an optimal code for transmitting n symbols distributed according to x.

(iii) On $(\mathbf{I}[0,1], \mu)$, $h_n(x) = -\sum_{i=1}^{n} x_i \log x_i$, Shannon entropy.

These examples do little to help us understand the relation of h to σ. What we need is some math. For each integer $k \geq 2$, let

$$c(k) := \inf\{\max(-\log \mu p) : p \in P^k(D)\}.$$

Intuitively, over the class $P^k(D)$ of algorithms with k outputs, $c(k)$ is the worst case complexity of the algorithm whose worst case complexity is *least.*

Theorem 41 *Let $(D, \otimes, \mu)$ be a symbolic domain with a process $p \in P^k(D)$. Then*

$$\sigma \leq h \leq \frac{c(k)}{\log k} \cdot (\log k + \sigma)$$

where h and σ can be taken in any dimension.

The mere existence of a process on a *symbolic* domain (D, μ) means not only that $\leq_D = \leq$ but also that h and σ are of the same order. Without the ability to 'copy' elements using $\otimes$, h and σ can be very different: Searching costs $O(n)$ on L, so h_L and σ are not of the same order. We need a slightly better estimate.

Definition 81 If $(D, \otimes, \mu)$ is a symbolic domain, then the integer

$$\inf\{k \geq 2 : c(k) = \log k\}$$

is called the *algebraic index* of (D, μ) when it exists.

By orthogonality, $c(k) \geq \log k$ always holds, so to calculate the algebraic index we need only prove $c(k) \leq \log k$. The value of the index for us is that:

Corollary 7 *If $(D, \otimes, \mu)$ is a symbolic domain with algebraic index $k \geq 2$, then*

$$\sigma \leq h \leq \log k + \sigma$$

where h and σ can be taken in any dimension.

There are results in [28] which explain why the algebraic index is a natural idea, but these use the *Gibbs map* and *partition function* from thermodynamics, which we do not have the space to discuss. But, it is simple to see that the algebraic index of $\mathbf{I}[0,1]$ is 2, the algebraic index of Σ^∞ is $|\Sigma|$ and in general, if there is a process $p \in P^n(D)$ on a symbolic domain with $(\mu p_1, \ldots, \mu p_n) = \bot \in \Lambda^n$ for some n, then D has an algebraic index $k \leq n$.

9.6.4.6 The Fixed Point Theorem

Let Λ be the set of *all* monotone decreasing states and let $\otimes : \Lambda \times \Lambda \to \Lambda$ be

$$x \otimes y := \mathrm{sort}^{-}(x_1 y, \ldots, x_n y).$$

That is, given $x \in \Lambda^n$ and $y \in \Lambda^m$, we multiply any x_i by any y_j and use these nm different products to build a vector in Λ^{nm}.

Definition 82 The copying operator $! : X \to X$ on a set X with a tensor $\otimes$ is

$$!x := x \otimes x$$

for all $x \in X$.

If $p \in P^n(D)$ is a process whose possible outputs are distributed as $x \in \Lambda^n$, then two independent copies of p considered together as a single process $!p$ will have outputs distributed according to $!x$. Now let $[\Lambda \to [0, \infty)^*]$ be the dcpo with the pointwise order $f \sqsubseteq g \equiv (\forall x)\, f(x) \geq g(x)$.

Theorem 42 *Let $(D, \otimes, \mu)$ be a symbolic domain whose algebraic index is $k \geq 2$. Then the least fixed point of the Scott continuous operator*

$$\Phi : [\Lambda \to [0, \infty)^*] \to [\Lambda \to [0, \infty)^*]$$
$$\Phi(f) = \frac{f!}{2}$$

on the set $\uparrow (h + \log k)$ *is*

$$\mathrm{fix}(\Phi) = \bigsqcup_{n \geq 0} \Phi^n(h + \log k) = \sigma,$$

where $h : \Lambda \to [0, \infty)$ *is the complexity on all states.*

This iterative process is very sensitive to where one begins. First, Φ has many fixed points above σ: Consider $c \cdot \sigma$ for $c < 1$. Thus, Φ cannot be a contraction on any subset containing $\uparrow h$. But Φ also has fixed points *below* σ: The map $f(x) = \log \dim(x) = \sigma\bot_{\dim(x)}$ is one such example. This proves that σ is genuinely a *least* fixed point.

The fixed point theorem can be used to derive Shannon's noiseless coding theorem [28]. In the proof of Theorem 42, we can regard Λ a continuous dcpo by viewing it as a disjoint union of domains. But we could just view it as a set. And if we do, the function space is still a dcpo, the theorem remains valid, and we obtain a new characterization of entropy:

Corollary 8 *Let $(D, \otimes, \mu)$ be a symbolic domain with algebraic index $k \geq 2$. Then there is a greatest function $f : \Lambda \to [0, \infty)$ which satisfies $h \geq f$ and $f(x \otimes x) \geq f(x) + f(x)$. It is Shannon entropy.*

The question then, "Does h approximate σ, or is it σ which approximates h" is capable of providing one with hours of entertainment. In closing, we should mention that Φ might also provide a systematic approach to defining information $\mathrm{fix}(\Phi)$ from complexity h in situations more general than symbolic domains.

9.6.4.7 The Quantum Case

The fixed point theorem also holds for quantum states where one replaces σ by von Neumann entropy, and $\otimes$ on domains by the algebraic tensor $\otimes$ of operators. (The domain theoretic $\otimes$ can also be mapped homomorphically onto the tensor of quantum states in such a way that domain theoretic orthogonality implies orthogonality in Hilbert space.) Several new connections emerge between computer science and quantum mechanics whose proofs combine new results with work dating as far back as Schrödinger [39] in 1936. The bridge that connects them is domain theory and measurement. One such result proves that reducing entanglement by a technique called local operations and classical communication is equivalent to simultaneously reducing the average case complexity of all binary trees, a major application of Theorem 39 that we could not include in this paper due to space limitations. These and related results are in [28].

9.7 Provocation

> *... and accordingly all experience hath shewn, that mankind are more disposed to suffer, while evils are sufferable, than to right themselves by abolishing the forms to which they are accustomed.*
>
> – Thomas Jefferson, The Declaration of Independence.

What is a Domain?

The "domains" of classical and quantum states are dcpo's with a definite notion of approximation, but they are *not* continuous. Their notion of approximation is

$$x \ll y \equiv (\forall \text{ directed } S)\; y = \bigsqcup S \Rightarrow (\exists s \in S)\, x \sqsubseteq s$$

On a *continuous dcpo*, the relation above is *equivalent* to the usual notion of approximation. In general, they are not equal, and the canonical examples are $(\Delta^n, \sqsubseteq)$ in the Bayesian order and $(\Omega^n, \sqsubseteq)$ in the spectral order. We forgot to mention this in Section 9.4 because we wanted to brainwash the reader, to convince them that the 'domain' illusion was real. Of course, in fairness to the author, we never said that we knew what a domain was exactly, just that they existed and that we would see lots of examples of them. Another possible example of a domain, the domain of *infinite* dimensional quantum states, is given in [33]. As a final example of something that is probably a domain, let us consider *the circle*.

I once had a prominent domain theorist tell me when I was a student that the circle could not be partially ordered in a natural way. I didn't believe it then and I don't believe it now. But now I have a reason:

The Circle

If we have two pure states $|\psi\rangle$ and $|\phi\rangle$ written in a basis $|i\rangle$ of n dimensional Hilbert space $\mathcal{H}^n$,

$$|\psi\rangle = \sum_{i=1}^{n} a_i |i\rangle \quad |\phi\rangle = \sum_{i=1}^{n} b_i |i\rangle$$

where the $a_i, b_i \in \mathbb{C}$ are complex, how can we order them so that (generally speaking) $|\psi\rangle \sqsubseteq |\phi\rangle$ means that the result of measuring the system in state ϕ is more predictable than the result of measuring the system in the state ψ? If we had an order $\sqsubseteq$ on classical probability distributions Δ^n, and another order $\sqsubseteq$ on phases $S^1 \cup \{0\}$, we could answer the question in what looks to be a natural way:

$$|\psi\rangle \sqsubseteq |\phi\rangle \equiv (|a_1|^2, \ldots, |a_n|^2) \sqsubseteq (|b_1|^2, \ldots, |b_n|^2) \;\&\; (\forall i)\, \text{phase}(a_i) \sqsubseteq \text{phase}(b_i).$$

Many orders on Δ^n are known. So the entire question is reduced to the ordering of phases.

It's Just a Phase

The phase of a complex number is either zero or a point on the circle, so the problem of ordering phases is really just the question of how to order the circle.

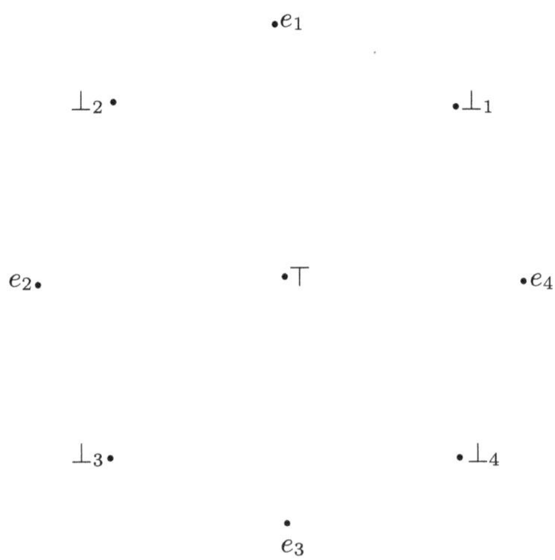

One way to order phases is to order the circle so that the arc from any $\perp_i$ to an adjacent e_j is isomorphic to $([0, 1], \leq)$, and that the center of the circle $\top = (0, 0)$ is above everything. Dynamically, if we start at e_4 and begin traversing the circle counterclockwise, then we move down until reaching $\perp_1$, at which point we begin moving up until e_1, down until $\perp_2$, up until e_2, down until $\perp_3$, up until e_3, down until $\perp_4$, and then up until returning to e_4. Notice that this is the kind of domain that Grover's algorithm, when viewed as acting on a two dimensional subpsace, seems to "move" in.

Another way to order phases is to use the discrete order: $x \sqsubseteq y$ iff $x = y$ or $y = (0, 0)$. This is very satisfying in that it does not leave one worried about the meaning of the order in the case where the classical distributions stay constant but the phases are allowed to vary.

Example 58 The reason that $\top = (0, 0)$ is above everything is so that relations like the following are satisfied:

$$\frac{1}{\sqrt{2}}(|0\rangle + |1\rangle) \sqsubseteq |0\rangle$$
$$\frac{1}{\sqrt{2}}(|0\rangle + |1\rangle) \sqsubseteq |1\rangle$$

What is a Measurement?

Though the more general formulation of approximation for domains like Ω^n is certainly meaningful, there are things that are missing. The definition of "domain" that we are looking for should allow one to do things like: prove sobriety of the Scott topology and give a satisfying definition of measurement. Yes, I realize that we defined measurement for a dcpo in Sect. 9.2.3, but I never said that we found the definition entirely convincing. The definition of measurement has more impact on a *continuous dcpo* as evidenced by results like Theorems 2 and 3.

Related to this question are two more pressing issues: (a) systematic methods for deriving higher order measurements from simpler measurements, and (b) techniques for proving that a given function is a measurement. For instance, to illustrate (a), if D is a domain and $F(D)$ is some higher order domain, like a powerdomain or an exponential object, can a measurement on D be used to *simply* construct one on $F(D)$? The question (b) is particularly urgent in physics and information theory: generally speaking, proving that functions like entropy and capacity are measurements is about as much fun as being a domain theorist in search of a decent job.[4]

References

1. Abramsky, S., Jung, A.: Domain theory. In: Abramsky, S., Gabbay, D.M., Maibaum, T.S.E. (eds.) Handbook of Logic in Computer Science III. Oxford University Press, Oxford (1994)
2. Alberti, P.M., Uhlmann, A.: Stochasticity and partial order: Doubly stochastic maps and unitary mixing. Dordrecht, Boston (1982)

[4] Any such domain theorist should send a CV and some recent papers to `keye.martin@nrl.navy.mil` immediately.

3. Brent, R.P.: Algorithms for minimization without derivatives. Prentice-Hall, New Jersey (1973)
4. Chernous'ko, F.L.: An optimal algorithm for finding the roots of an approximately computed function. Zh. vychisl. Mat. i mat. Fiz. **8**(4) 705–724 (1968), English translation
5. Coecke, B., Martin, K.: A partial order on classical and quantum states. Oxford University Computing Laboratory Research Report (August 2002)
6. Edalat, A., Heckmann, R.: A computational model for metric spaces. Theor. Comput. Sci. **193**, 53–73 (1998)
7. Falconer, K.: Fractal geometry. Wiley, Chichester (1990)
8. Gross, O., Johnson, S.M.: Sequential minimax search for a zero of a convex function. Math. Tables Other Aids Comput. **13**(65), 44–51 (1959)
9. Grover, L.K.: Quantum mechanics helps in searching for a needle in a haystack. Phys. Rev. Lett. **78**, 325 (1997)
10. Hawking, S.W., Ellis, G.F.R.: The large scale structure of space-time. Cambridge Monographs on Mathematical Physics. Cambridge University Press, Cambridge (1973)
11. Hutchinson, J.E.: Fractals and self-similarity. Indiana Univ. Math. J. **30**, 713–747 (1981)
12. Kowalski, M., Sikorski, K., Stenger, F.: Selected topics in approximation and computation. Oxford University Press, Oxford (1995)
13. Kraft, L.G.: A device for quantizing, grouping and coding amplitude modulated pulses. Thesis, M.S., Electrical Engineering Department, MIT (1949)
14. Marshall, A.W., Olkin, I.: Inequalities: Theory of majorization and its applications. Academic Press Inc., New York (1979)
15. Martin, K.: A foundation for computation. Ph.D. Thesis, Tulane University, Department of Mathematics (2000)
16. Martin, K.: The measurement process in domain theory. Lecture Notes in Computer Science, Vol. 1853. Springer, New York (2000)
17. Martin, K.: Unique fixed points in domain theory. Elect. Notes Theor. Comput. Sci. **45**, 258–268 (2001)
18. Martin, K.: A renee equation for algorithmic complexity. Lecture Notes in Computer Science, vol. 2215, pp. 201–218. Springer, New York (2001)
19. Martin, K.: Powerdomains and zero finding. Elect. Notes Theor. Comput. Sci. **59**(3), 173–184 (2001)
20. Martin, K.: The informatic derivative at a compact element. Lecture Notes in Computer Science, vol. 2303, pp. 103–148. Springer, New York (2002)
21. Martin, K.: B-sides. Oxford University Computing Lab, Research Report (2003)
22. Martin, K.: Epistemic motion in quantum searching. Oxford University Computing Laboratory, Research Report (2003)
23. Martin, K.: A continuous domain of classical states. Oxford University Computing Laboratory, Research Report (2003)
24. Martin, K.: Fractals and domain theory. Mathematical Structures in Computer Science, vol. 14(6), pp. 833–851. Cambridge University Press, Cambridge (2004)
25. Martin, K., Ouaknine, J.: Informatic vs. classical differentiation on the real line. Elect. Notes Theor. Comput. Sci. **73**, 133–140 (2004)
26. Martin, K., Panangaden, P.: A domain of spacetime intervals in general relativity. Commun. Math. Phys. **267**(3), 563–586 (November 2006)
27. Martin, K.: Compactness of the space of causal curves. J. Class. Quantum Grav. **23**, 1241–1251 (2006)
28. Martin, K.: Entropy as a fixed point. Theoretical Computer Science (2006)
29. Martin, K.: Topology in information theory in topology. Theoretical Computer Science. To appear (2007)
30. Martin, K.: A domain theoretic model of qubit channels. To appear (2008)
31. Martin, K.: The maximum entropy state. Logical Methods in Computer Science. To appear
32. Martin, K., Panangaden, P.: In preparation (2008)

33. Mashburn, J.: A spectral order for infinite dimensional quantum spaces. Elect. Notes Theor. Comput. Sci. **173**, 263–273 (2007)
34. Muirhead, R.F.: Some methods applicable to identities and inequalities of symmetric algebraic functions of n letters. Proc. Edinburgh Math. Soc. **21**, 144–157 (1903)
35. Nielsen, M., Chuang, I.: Quantum computation and quantum information. Cambridge University Press, Cambridge (2000)
36. Penrose, R.: Gravitational collapse and space-time singularities. Phys. Rev. Lett. **14**, 57–59 (1965)
37. Penrose, R.: Techniques of differential topology in relativity. Soc. Ind. Appl. Math. (1972)
38. Royden, H.L.: Real Analysis, 3rd edn. Macmillan Publishing, New York (1988)
39. Schrödinger, E.: Proc. Cambridge Philosophical Soc. **32**, 446 (1936)
40. Scott, D.: Outline of a mathematical theory of computation. Technical Monograph PRG-2, Oxford University Computing Laboratory (November 1970)
41. Shannon, C.E.: A mathematical theory of communication. Bell Syst. Tech. J. **27**, 379–423, 623–656 (1948)
42. Sorkin, R.: Spacetime and causal sets. In: D'Olivo, J. et al. (eds.) Relativity and Gravitation: Classical and Quantum. World Scientific, Singapore (1991)
43. Spreen, D. (2001) On some constructions in quantitative domain theory. Extended Abstract. http://www.informatik.uni-siegen.de/ spreen/
44. Wald, R.M.: General relativity. The University of Chicago Press, Chicago (1984)
45. Waszkiewicz, P.: Quantitative continuous domains. PhD Thesis, University of Birmingham (2002)
46. Yamaguti, M., Hata, M., Kigami, J.: Mathematics of fractals. Translations of Mathematical Monographs, vol. 167. American Math Society (1997)

Chapter 10
A Partial Order on Classical and Quantum States

B. Coecke and K. Martin

Abstract We introduce a partial order on classical and quantum mixed states which reveals that these *sets* are actually *domains*: Directed complete partially ordered sets with an intrinsic notion of approximation. The operational significance of the partial orders involved conclusively establishes that physical information has a natural domain theoretic structure. For example, the set of maximal elements in the domain of quantum states is precisely the set of *pure states*, while the completely mixed ensemble I/n is the order theoretic least element $\perp$.

In the same way that the order on a domain provides a rigorous qualitative definition of information, a special type of mapping on a domain called a *measurement* provides a formal account of the intuitive notion "information content." Not only is physical information domain theoretic, but so too is physical entropy: Shannon entropy is a measurement on the domain of classical states; von Neumann entropy is a measurement on the domain of quantum states.

These results yield a foundation from which one can (a) reason qualitatively about probability, (b) derive the lattices of Birkhoff and von Neumann in a unified manner, suggesting that domains may provide a formalism for the logic of partial belief, and (c) develop new techniques for studying phenomena like noise and entanglement. Along the way, new lines of investigation open up within various subdisciplines of physics, mathematics and theoretical computer science.

10.1 Introduction

One of the great lessons of the differential and integral calculus is that we can conquer the infinite, and in particular, the continuous, by means of the discrete. An infinite sum may be understood as a limit of finite sums, the area beneath a curve as

---

B. Coecke (✉)
OUCL, University of Oxford, Oxford, UK
e-mail: Bob.Coecke@comlab.ox.ac.uk

K. Martin (✉)
Naval Research Laboratory, Washington, DC, USA
e-mail: keye.martin@nrl.navy.mil

Coecke, B., Martin, K.: *A Partial Order on Classical and Quantum States*. Lect. Notes Phys. **813**, 593–683 (2011)
DOI 10.1007/978-3-642-12821-9_10

the limit of areas of approximating rectangles, the line tangent to a curve at a point is the limit of the secant lines joining points nearby.

The philosophy espoused is unambiguous: The *ideal* can be realized as a *limit* of the *partial;* the abstract, as a limit of the concrete; the continuous, a limit of the discrete, and so on. And this powerful ideology, as it arises in the context of recursive functionals, is part of what the axioms of domain theory are intended to capture. But even in Scott's prelude to the subject, it is difficult to keep the imagination from wandering beyond the confines of computation [16]:

> Maybe it would be better to talk about *information*; thus, $x \sqsubseteq y$ means that x and y want to approximate the same entity, but y gives more information about it. This means we have to allow *incomplete* entities, like x, containing only *partial* information.

In its purest interpretation, domain theory is a branch of mathematics which offers an exclusively qualitative account of information: A proposal for how we might find information structured in a universe where all things arise as a limit of the partial.

Physics of course is also the study of information. But with one caveat: In physics, the term "information" normally manages to escape rigorous mathematical definition, and in those cases where it does not, its formulations tend toward the purely *quantitative*. But what is self-evident is that only qualitative accounts of physical phenomena are capable of imparting "structural laws of general validity." From A. Einstein,

> I do not believe in micro- and macro-laws, but only in (structural) laws of general validity.

Now this is not to say that physics ought to be done in laboratories without numbers, simply that our *understanding* of physical reality should be mathematically expressible in such a way that the laws of nature are clearly delineated from the conventions of man.

Thus, at least on the surface, there is a good match between what domain theory offers, and what physics needs: Domain theory can provide the structures of reality, physics in turn can explicate the reality of the structures. A research program in this direction begins with the demonstration contained herein that the density operator formulation of quantum mechanics is an instance of domain theory: Its partial elements are the mixed states, its total or idealized elements are the pure states.

The route to this discovery passes through *the measurement formalism*, a theory [8] which allows for the quantitative expression $\mu : D \to [0, \infty)^*$ of the qualitative notion captured by a domain $(D, \sqsubseteq)$. In doing so, it yields an indispensable methodology for uncovering the structural aspects of information which often enough seem to appear in purely quantitative disguise. Such is the case with classical and quantum information, for instance, which are normally formulated in terms of Shannon and von Neumann entropy.

Our method of transport is a philosophy still advocated in certain studies on the foundations of physics [3, 9, 13]: Every formal idea should represent a meaningful physical notion, and each successive mathematical development ought to have a clear counterpart in physical reality. To illustrate, the partial order on classical states is defined inductively in terms of Bayesian state update, which corresponds to the process by which an observer looks for an object and updates his knowledge according to what he finds. Similarly, the partial order on quantum states relies on the physical notion of a measurement process.

On our way, this vehicle escorts classical and quantum probability into a genuine formal realization of the Bayesian ideal, elegantly captured by F. P. Ramsey [14]: "Probability is the logic of partial belief."

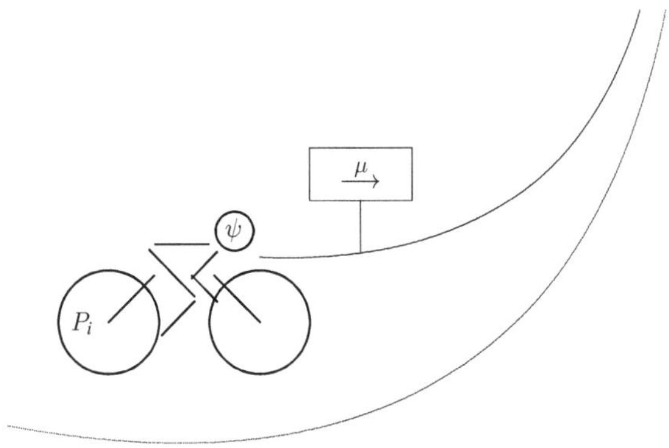

Concretely, we introduce the domain of classical states, which has Shannon entropy as a measurement. The partial order on classical states extends to yield a domain of quantum states with von Neumann entropy as a measurement. As already mentioned, the operational significance of the partial orders involved unquestionably demonstrates that physical information has a natural domain theoretic structure. By recognizing this structure, the present work achieves unity across various subdisciplines of physics and information theory. For example, the Birkhoff-von Neumann contrast, between classical and quantum, which arises in the logical aspect, is in perfect harmony with Shannon and von Neumann entropy, which arises in more "pragmatic" pursuits. All of these are part of a single, and it would appear, more complete, picture of physical reality.

10.2 Classical States

The information an observer has a priori about the result of an event in which one of n different outcomes is possible can be described by a function $x : \{1, \ldots, n\} \to [0, 1]$ that assigns a probability x_i indicating the degree to which outcome i is likely. These are called classical states.

Definition 1 The *classical n-states* are

$$\Delta^n := \{x \in [0, 1]^n : \sum_{i=1}^{n} x_i = 1\},$$

where $x = (x_1, \ldots, x_n)$ and $n \geq 1$.

In this section we will introduce a natural partial order on classical states that is probably best referred to as the *Bayesian order*. Before doing so, here is a brief indication of how this order was discovered and our original motivation for studying it.

In contrast to a classical n-state, a *quantum n-state* is a self-adjoint, positive, trace one, linear operator $\rho : \mathcal{H}^n \to \mathcal{H}^n$ on a n dimensional complex Hilbert space $\mathcal{H}^n$. In particular, ρ is an $n \times n$ matrix of complex numbers whose n eigenvalues $\lambda_i \geq 0$ for $1 \leq i \leq n$ add up to one. Thus, to each quantum state ρ we can associate a classical state $\text{spec}(\rho) = (\lambda_1, \ldots, \lambda_n)$.

Thus, *if* we have a partial order $\sqsubseteq$ on Δ^n, we might be able to use the connection between quantum and classical given above to derive a natural candidate for a partial order on quantum states as follows:

$$\rho \sqsubseteq \sigma \Leftrightarrow \text{spec}(\rho) \sqsubseteq \text{spec}(\sigma) \text{ and (insert magic here).}$$

And then the questions start: (i) Can we really order matrices by ordering their eigenvalues? (ii) How exactly do we form the *list* $(\lambda_1, \ldots, \lambda_n)$, when in actuality the eigenvalues $\text{spec}(\rho)$ of ρ only form a *set*? (iii) How do we order classical states?

The first two questions will be answered in the next section, but the short answers are: (i) Yes, *if* we have the right order on Δ^n, and (ii) quantum measurement. Let us then get on with the answering of (iii).

10.2.1 Two States and the Parabola

Begin by imagining $n + 1$ boxes

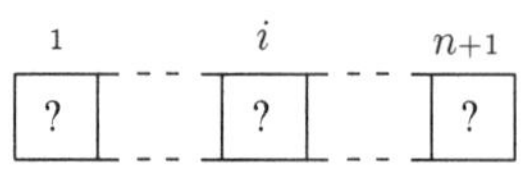

In one of these boxes, there lies a tenured position in the land of free expression. There are two observers searching frantically for its location. The knowledge an observer has about its location is a classical state $x \in \Delta^{n+1}$, formed by assigning a probability x_i which indicates the likelihood that the tenured position is located in box i:

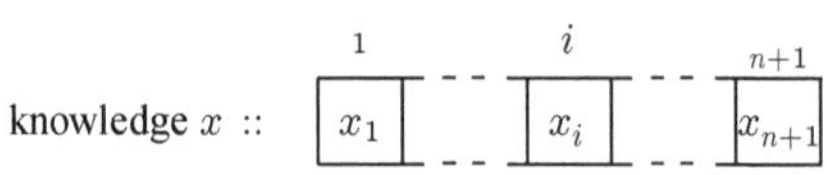

For example, if the observer is frustrated beyond belief because he has no earthly idea which box contains the tenured position, then his knowledge would be the *completely mixed state*

$$\perp = (1/(n + 1), \ldots, 1/(n + 1)),$$

indicative of the fact that he regards all boxes as equally likely:

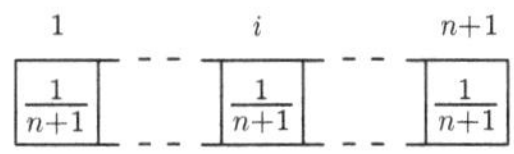

On the other hand, if the observer knows the tenured position is located in box i, then his knowledge would be the *pure state*

$$e_i = (0, \dots, 1, \dots, 0),$$

where the one occurs at index i:

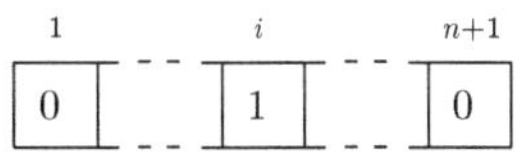

In general, the actual location is always represented by a pure state. This much is independent of all observers.

Let k be the actual location of the tenured position, $x \in \Delta^{n+1}$ represent the knowledge of the first observer and $y \in \Delta^{n+1}$ the knowledge of the second observer:

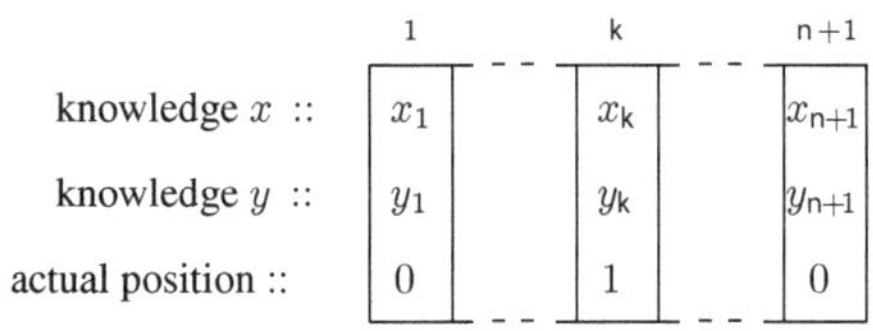

In the interest of holding the reader's attention, let $x \neq y$. If $\sqsubseteq$ is a partial order on Δ^{n+1} that expresses what it means for one state to be more informative than another, and in this order we have $x \sqsubseteq y$, then observer one knows less about the location of the tenured position than observer two.

But now suppose that each observer looks into box i only to discover that it does not contain the tenured position. Then $x_i < 1$ and $y_i < 1$. In addition, the knowledge of the first observer changes to

$$p_i(x) = \frac{1}{1 - x_i}(x_1, \dots, \widehat{x_i}, \dots, x_{n+1}) \in \Delta^n,$$

while the state of the second observer's knowledge updates to $p_i(y)$:

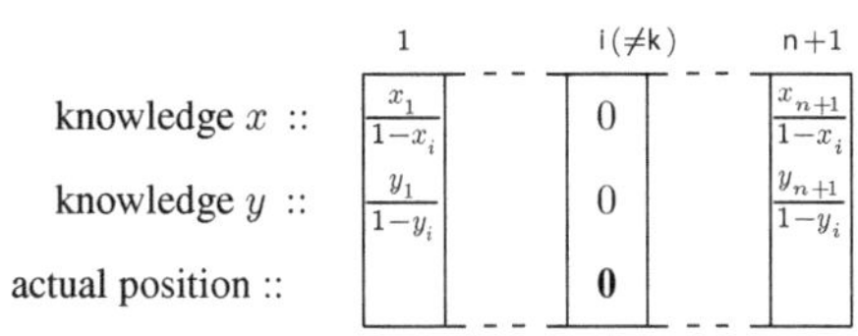

Because the second observer knew more than the first before observation, and because they have both increased their knowledge by the same amount (they both now additionally know that it is not in box i), we must conclude that the second *still* knows more than the first. That is,

$$p_i(x) \sqsubseteq p_i(y)$$

whenever $i \neq k$. But this reasoning should apply in *all* situations, i.e., it should not depend on the actual location of the tenured position: We should allow for the reality that k could be any of the values in $\{1, \ldots, n+1\}$. Thus, we arrive at a potential definition of $(\Delta^{n+1}, \sqsubseteq)$ in terms of $(\Delta^n, \sqsubseteq)$:

$$x \sqsubseteq y \Leftrightarrow (\forall i)(x_i, y_i < 1 \Rightarrow p_i(x) \sqsubseteq p_i(y)).$$

This leaves just one question: How do we order Δ^2? The answer appears when we imagine that the order $\sqsubseteq$ on Δ^n is known, and then use it to *formally* express some of the well-known *intuitions* used in physics when reasoning about classical states as information:

- The *completely mixed state* should be the *least element* of $(\Delta^n, \sqsubseteq)$,

$$(\forall x) \perp \sqsubseteq x,$$

- The set of *pure states* should be the set of *maximal elements*,

$$\max(\Delta^n) = \{e_i : 1 \leq i \leq n\},$$

- The observer's a priori uncertainty, *Shannon entropy*

$$\mu x = -\sum_{i=1}^{n} x_i \log x_i,$$

  should be a *measurement* in the sense of domain theory [8]. In particular, as states become more informative, uncertainty should decrease:

$$x \sqsubseteq y \Rightarrow \mu x \geq \mu y,$$

  i.e., as a map from the poset Δ^n to the poset $[0, \infty)^*$ of nonnegative reals in their opposite order, it should be *monotone*.
- **The mixing law** should be respected by $\sqsubseteq$:

$$x \sqsubseteq y \text{ and } p \in [0,1] \Rightarrow x \sqsubseteq (1-p)x + py \sqsubseteq y.$$

  The state $(1-p)x + py$ is a mixture of x and y whose composition consists of $(1-p)$ percent x and p percent y. Thus, the mixing law says that if y is more informative than x, then any mixture of the two is more informative than x, but less informative than y.

  This leaves only one way of ordering Δ^2.

Definition 2 For $x, y \in \Delta^2$, we order classical two states by

$$(x_1, x_2) \sqsubseteq (y_1, y_2) \Leftrightarrow (y_1 \leq x_1 \leq 1/2) \text{ or } (1/2 \leq x_1 \leq y_1).$$

We will prove the uniqueness of this order after explaining its derivation. For the latter, look at the graph of Shannon entropy μ on two states:

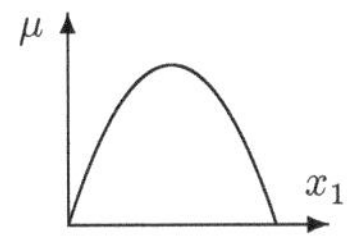

Remembering that the order $\sqsubseteq$ on Δ^2 should be defined so that Shannon entropy $\mu : \Delta^2 \to [0, \infty)^*$ is a measurement in the sense of domain theory (and hence *monotone*), a natural candidate for $\sqsubseteq$ appears when we flip the parabola upside down:

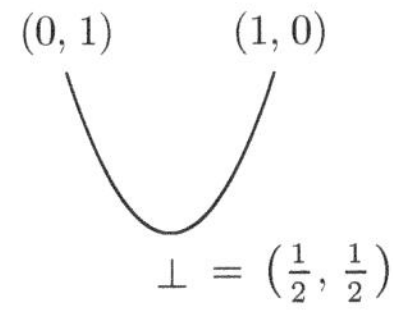

The order suggested by this picture is simply a copy of $[0, 1/2]^*$ and $[1/2, 1]$ joined at $1/2$,

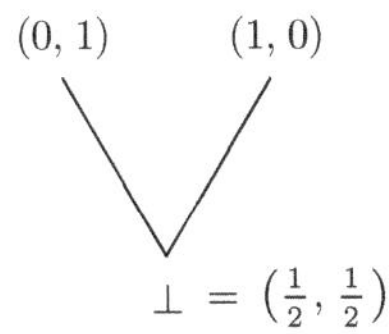

which is exactly how we defined the order on Δ^2 (Definition 2).

For its uniqueness, first realize that there are at least two reasonable interpretations of the mixing law: (i) (Informatically) When two comparable states are mixed, a loss of information is experienced from one point of view that is simultaneously a gain of information from the other, (ii) (Geometrically) The line connecting two comparable states moves up in the order.

Lemma 1 *A partial order $\sqsubseteq$ on Δ^n respects the mixing law iff the map $f : [0, 1] \to \Delta^n$ given by $f(t) = (1 - t)x + ty$ is monotone for each pair of comparable states $x \sqsubseteq y$.*

Proof The monotonicity of f implies the mixing law. For the converse, let $s < t$. By the mixing law, $x \sqsubseteq f(t) \sqsubseteq y$, so applying the mixing law again to $x \sqsubseteq f(t)$, gives

$$x \sqsubseteq \left(1 - \frac{s}{t}\right)x + \frac{s}{t} \cdot f(t) \sqsubseteq f(t),$$

which finishes the proof since $f(s) = \left(1 - \frac{s}{t}\right)x + \frac{s}{t} f(t)$.

Now the uniqueness of $(\Delta^2, \sqsubseteq)$ is transparent.

Theorem 1 *There is a unique partial order on Δ^2 which satisfies the mixing law and has $\bot = (1/2, 1/2)$. It is the order on classical two states.*

Proof Let $\leq$ be *any* partial order on Δ^2 which respects the mixing law and has least element $\bot = (1/2, 1/2)$.

Because $\bot \leq e_1 = (1, 0)$, Lemma 1 implies that the straight line path f_1 from $f_1(0) = \bot$ to $f_1(1) = e_1$ is monotone. Similarly, the line f_2 from $f_2(0) = \bot$ to $f_2(1) = e_2 = (0, 1)$ is monotone. Thus, $\sqsubseteq \subseteq \leq$.

To prove $\leq \subseteq \sqsubseteq$, suppose $x \leq y$. First, we must have either $x_1, y_1 \leq 1/2$ or $1/2 \geq x_1, y_1$: Otherwise, the line f from $f(0) = x$ to $f(1) = y$ passes through $\bot$, and since f is monotone by the mixing law, we have $x = \bot \sqsubseteq y$. But this means that either $x \sqsubseteq y$ or $y \sqsubseteq x$. In the first case, the proof is done. In the latter, we must have $y \leq x$, which by the antisymmetry of $\leq$, gives $x = y$, and hence $x \sqsubseteq y$. □

Thus far, we have not defined terms like "domain" and "measurement." At this stage, there is no need to. Let us simply point out that Δ^2 is a domain with Shannon entropy μ as a measurement such that

$$\ker \mu = \max(\Delta^2) = \{e_1, e_2\}.$$

The precise definitions of these terms will become apparent as we proceed.

10.2.2 A Partial Order on Classical States

When an observer looks in box i and discovers that the object of his desire is not there, the classical state x representing his knowledge of its location collapses to one $p_i(x)$ in a lower dimension as follows.

Definition 3 Let $n \geq 2$. The projection which collapses the ith outcome is the partial map $p_i : \Delta^{n+1} \rightharpoonup \Delta^n$ given by

$$p_i(x) = \frac{1}{1 - x_i}(x_1, \ldots, \widehat{x_i}, \ldots, x_{n+1})$$

for $1 \leq i \leq n + 1$ and $0 \leq x_i < 1$. It is defined on $\mathrm{dom}(p_i) = \Delta^{n+1} \setminus \{e_i\}$.

In the way of needless (but fun) geometric illustration, consider the case of the *triangle* Δ^3. If $x = (x_1, x_2, x_3)$, then although $p_i(x)$ is technically a member of Δ^2, we can still picture its effect on x as follows:

Recalling the definition of $(\Delta^2, \sqsubseteq)$ from the last section, we can now completely specify the order on classical states.

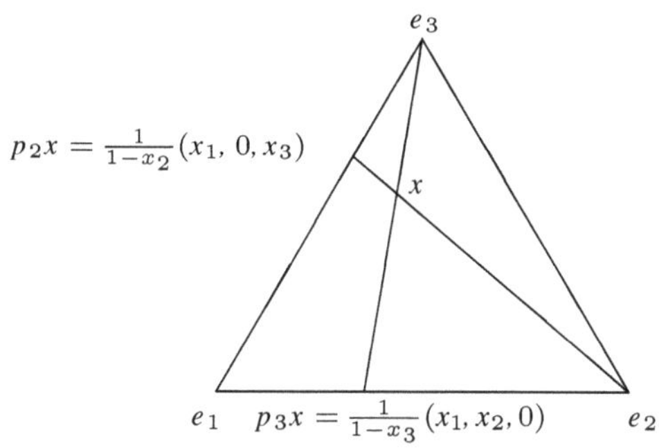

Definition 4 Let $n \geq 2$. For $x, y \in \Delta^{n+1}$, we define

$$x \sqsubseteq y \Leftrightarrow (\forall i)(x, y \in \mathrm{dom}(p_i) \Rightarrow p_i(x) \sqsubseteq p_i(y)),$$

where i ranges over the set $\{1, \ldots, n+1\}$.

To be perfectly clear, notice that $x, y \in \mathrm{dom}(p_i)$ iff $x_i, y_i < 1$. The following operators on classical states will prove indispensable in what follows.

Definition 5 Let $n \geq 2$. For $x \in \Delta^n$, we set

$$x^+ := \max_{1 \leq i \leq n} x_i \text{ and } x^- := \min_{1 \leq i \leq n} x_i.$$

We have $x^- \in [0, 1/n]$ and $x^+ \in [1/n, 1]$.

For example, a state x is pure iff $x^+ = 1$, while $\perp$ is the unique classical state x with $x^+ = x^-$.

Lemma 2 *Let $x, y \in \Delta^n$ for $n \geq 2$. Then*

(i) *If $x \sqsubseteq y$ with $x_i = 1$, then $y_i = 1$.*
(ii) *If $x \sqsubseteq \perp$, then $x = \perp$.*

Proof (i) Assume for $n \geq 2$. For $n+1$, suppose $x_i = 1$. First, we claim there is some $k \neq i$ with $y_k = 0$. If not, then because $n + 1 \geq 3$, there is some $k \neq i$ such that

$$0 < y_k < \sum_{k \neq i} y_k = 1 - y_i \leq 1,$$

as the sum above involves at least two positive numbers. Because $k \neq i$, $x_k = 0$, so the inductive hypothesis applied to $p_k(x) \sqsubseteq p_k(y)$ gives

$$\frac{y_i}{1-y_k} = \frac{x_i}{1-x_k} = 1 \implies y_k = 1 - y_i,$$

which contradicts $y_k < 1 - y_i$. Thus, there is $k \neq i$ with $x_k = y_k = 0$, and the inductive hypothesis applied to $p_k(x) \sqsubseteq p_k(y)$ yields

$$y_i = \frac{y_i}{1 - y_k} = \frac{x_i}{1 - x_k} = 1,$$

finishing the proof.

(ii) We know $x^+ < 1$, since otherwise by (i) we would have $\perp^+ = 1$. Now the proof is a trivial induction: Since $x \sqsubseteq \perp$, we have $p_i(x) \sqsubseteq p_i(\perp) = \perp_n$, and by the inductive hypothesis, $p_i(x) = \perp_n$ for all $i \in \{1, \ldots, n+1\}$. The only possibility is $x = \perp$.

Lemma 3 *For classical n-states* $x \sqsubseteq y$, *either* $x = \perp$, $y = \perp$ *or there is an index* $k \in \{1, \ldots, n\}$ *such that* $x_k \leq y_k$, $x_k > x^-$ *and* $y_k > y^-$.

Proof The result is true for $n = 2$. Assume it for n. To prove it for $n+1$, we start with $x, y \neq \perp$. Immediately, we have $x \neq y$ (otherwise, $x \neq \perp \Rightarrow x_k = y_k = x^+ > x^-$), and by virtue of Lemma 2(i), $x^+ < 1$.

Now let i be an index with $x_i \geq y_i$. Throughout, $y_i < 1$, since otherwise $x = y$. Then either (1) $p_i(x) \neq \perp_n$ or (2) $p_i(x) = \perp_n$.

In case (1), we cannot have $p_i(y) = \perp_n$ (Lemma 2(ii)), so the inductive hypothesis applies, yielding an index of $p_i(x)$ and $p_i(y)$ which we can relabel as an index k of x and y with

$$\frac{x^-}{1 - x_i} \leq p_i(x)^- < \frac{x_k}{1 - x_i} \leq \frac{y_k}{1 - y_i} > p_i(y)^- \geq \frac{y^-}{1 - y_i}.$$

Then $x_k > x^-$ and $y_k > y^-$. In addition, since $x_i \geq y_i$,

$$x_k \leq \frac{1 - x_i}{1 - y_i} \cdot y_k \leq 1 \cdot y_k,$$

which finishes the proof in case (1).

In case (2), $p_i(x) = \perp$. It helps to picture x as the $(n+1)$-state

$$\left(\frac{1 - x_i}{n}, \ldots, x_i, \ldots, \frac{1 - x_i}{n}\right),$$

though our proof does not depend on this informal remark. Because $x \neq \perp$, $x_i \neq (1 - x_i)/n$, so either $(1 - x_i)/n > x_i$ or $x_i > (1 - x_i)/n$.

The case $(1 - x_i)/n > x_i$ is simple: Since $x_i \geq y_i$, there must exist $k \neq i$ with $x_k \leq y_k$, or else we could derive $x_i < y_i$. Then we have

$$y_k \geq x_k = \frac{1 - x_i}{n} > x_i \geq y_i,$$

which also makes it clear that $x_k > x^- = x_i$ and $y_k > y_i \geq y^-$.

For the last case, we have $x_i > (1 - x_i)/n$. First we eliminate the possibility $y^+ = 1$. If $y^+ = 1$, then there is an index j with $y_j = 0$. Delicately, we can take $j \neq i$ because $n + 1 \geq 3$. Then $p_j(x) \neq \bot_n$ and $p_j(y) \neq \bot_n$, so the inductive hypothesis applies to yield an index k

$$\frac{x^-}{1 - x_j} \leq p_j(x)^- < \frac{x_k}{1 - x_j} \leq \frac{y_k}{1 - y_j} > p_j(y)^- \geq \frac{y^-}{1 - y_j}.$$

But $x_k > x^-$ implies that $k = i$. In addition, we have known from the start that $y_i < 1$, which means $y_i = 0$ because $y^+ = 1$. But then $0 = y_i = y_k \geq x_k = x_i = 0$, which contradicts $x_i > (1 - x_i)/n \geq 0$.

To finish case (2), we have $x^+, y^+ < 1$ and $x_i = x^+ > (1 - x_i)/n$. What we will prove is that $x_i > x^-$, $y_i > y^-$ and $x_i \leq y_i$. The first of these is clear. For the other two, let k be *any* index different from i. Then $p_k(x) \neq \bot_n$, which means $p_k(y) \neq \bot_n$ since $p_k(x) \sqsubseteq p_k(y)$. By the inductive hypothesis, there is an index j such that

$$\frac{x^-}{1 - x_k} \leq p_k(x)^- < \frac{x_j}{1 - x_k} \leq \frac{y_j}{1 - y_k} > p_k(y)^- \geq \frac{y^-}{1 - y_k}.$$

Again, $x_j > x^-$ implies $j = i$. Hence, $y_i > y^-$. But this also gives us $x_i(1 - y_k) \leq y_i(1 - x_k)$, for *all* $k \neq i$, which enables

$$x_i \sum_{k \neq i} (1 - y_k) \leq y_i \sum_{k \neq i} (1 - x_k) \implies \frac{x_i}{y_i} \leq \frac{n - 1 + x_i}{n - 1 + y_i},$$

ending in $(n - 1)x_i + x_i y_i \leq (n - 1)y_i + x_i y_i$. □

Lemma 4 *If $x \sqsubseteq y$ in Δ^n for $n \geq 2$, then there is an index $i \in \{1, \ldots, n\}$ such that $x_i = x^- \geq y^- = y_i$.*

Proof If $x = \bot$ the claim is trivial; thus, $y \neq \bot$, by Lemma 2(ii). By Lemma 3, there is an index $k \in \{1, \ldots, n\}$ such that $x_k \leq y_k$, $x_k > x^-$ and $y_k > y^-$.

If $x_k = 1 \leq y_k$, then $x = y$ and the proof is done. If $y_k = 1$, then let i be an index where $x_i = x^-$. We cannot have $i = k$ since $x_k > x^-$. Thus, $x_i = x^- \geq y_i = 0$.

Assume for n. For $n + 1$, $p_k(x) \sqsubseteq p_k(y)$, and the inductive hypothesis applies to yield an index i of x and y with

$$p_k(x)^- = \frac{x_i}{1 - x_k} \geq \frac{y_i}{1 - y_k} = p_k(y)^-.$$

Because $x_k \leq y_k$, $x_i \geq y_i$. But since $x_k > x^-$ and $y_k > y^-$, we have $x_i = x^-$ and $y_i = y^-$.

Now that we understand the behavior of minima, the nature of the maxima is immediate (and fundamental).

Proposition 1 *Let $x, y \in \Delta^n$ and e_i be the pure states in Δ^n.*

(i) *If $x \sqsubseteq y$, then there is an index i such that $x_i = x^+ \leq y^+ = y_i$.*
(ii) *For any i, $x_i = x^+$ if and only if $x \sqsubseteq e_i$.*
(iii) *If $x \sqsubseteq y$ and $x^+ = y^+$, then $x = y$.*

Proof All of these statements are proved by induction. The arguments below all assume that the respective claims are true for n and give the argument for the $n+1$ case. That they are true for $n = 2$ is clear.

(i) By Lemma 4, there is an index k with $x_k = x^- \geq y^- = y_k$, so we apply the inductive hypothesis to $p_k(x) \sqsubseteq p_k(y)$ to obtain an index i such that

$$p_k(x)^+ = \frac{x_i}{1 - x^-} \leq \frac{y_i}{1 - y^-} = p_k(y)^+.$$

Since $x_i \geq x^- = x_k$ and $x_i \geq x_j$ for all $j \neq k$ using $p_k(x)^+ = x_i/(1 - x^-)$, we have $x_i = x^+$. Similarly, $y_i = y^+$. That $x_i \sqsubseteq y_i$ now follows from

$$x_i \leq \frac{1 - x^-}{1 - y^-} \cdot y_i \leq 1 \cdot y_i$$

since $x^- \geq y^-$.

(ii) Let i be an index where $x_i = x^+$ and $e_i \in \Delta^{n+1}$ be the associated pure state whose value at index i is one. To prove that $x \sqsubseteq e_i$, we must show that $p_k(x) \sqsubseteq p_k(e_i)$ for all $k \neq i$. Fix an arbitrary $k \neq i$.

First, let j be the index of $p_k(x)$ corresponding to index i in x. This index exists because $k \neq i$. The value of $p_k(x)$ at index j is

$$p_k(x)^+ = \frac{x_i}{1 - x_k}.$$

Second, $p_k(e_i)$ is a pure state in Δ^n whose value at index j is one. By the inductive hypothesis, $p_k(x) \sqsubseteq p_k(e_i)$, for all $k \neq i$, which means $x \sqsubseteq e_i$.

For the converse, suppose $x \sqsubseteq y := e_i$. By (i), there is an index k such that $x_k = x^+$ and $y_k = y^+$. But y is pure, so we must have $k = i$, which means $x_i = x_k = x^+$.

(iii) Starting with $x \sqsubseteq y$ and $x^+ = y^+$, we use Lemma 4 to project away the minima $x_k = x^- \geq y^- = y_k$, obtaining $p_k(x) \sqsubseteq p_k(y)$. Applying (i) to this pair yields an index i with

$$p_k(x)^+ = \frac{x_i}{1 - x^-} \leq \frac{y_i}{1 - y^-} = p_k(y)^+.$$

As in the proof of (i), $x_i = x^+$ and $y_i = y^+$. But since $x_i = y_i > 0$ and $p_k(x)^+ \leq p_k(y)^+$, we have $x^- \leq y^-$, which gives $x^- = y^-$. This means $p_k(x)^+ = p_k(y)^+$, and since $p_k(x) \sqsubseteq p_k(y)$, the inductive hypothesis applies, leaving $p_k(x) = p_k(y)$. Because we also have $x_k = y_k$, the states x and y are equal. □

Proposition 1(ii) shows that an outcome with maximum probability in a classical state has a certain *qualitative* character to it. In general, it is the *only* outcome we can say this about.

Theorem 2 *Δ^n is a partially ordered set for each $n \geq 2$. Its maximal elements are the pure states,*

$$\max(\Delta^n) = \{x \in \Delta^n : x^+ = 1\},$$

and its least element is the completely mixed state $\bot := (1/n, \ldots, 1/n)$.

Proof The proof is by induction. It is true for $n = 2$. Assume the result for n. Then for $n + 1$, the reflexivity and transitivity are clear.

For antisymmetry, let $x \sqsubseteq y$ and $y \sqsubseteq x$. By Proposition 1(i), we have $x^+ \leq y^+$ and $y^+ \leq x^+$. By Proposition 1(iii), $x = y$.

That the least element is $\bot$ follows from $p_i(\bot) = \bot_n$ for all i. For its maximal elements, first suppose $x^+ = 1$ and that $x \sqsubseteq y$. By Proposition 1(i), there is an index i with $x_i = x^+ = 1 \leq y^+ = y_i$, so $y_i = 1$, which means $x = y$. Hence, $x \in \max(\Delta^{n+1})$.

Conversely, if $x \in \max(\Delta^{n+1})$, then $x \sqsubseteq e_i$ by Proposition 1(ii), where e_i is the pure state corresponding to $x_i = x^+$. By the maximality of x, $x = e_i$, which means $x^+ = 1$. □

The next result displays some fundamental properties of the order on classical states—the crucial *degeneration* lemma.

Lemma 5 (Degeneration) *If $x \sqsubseteq y$ in Δ^n, then*

$$(x_i = 0 \Rightarrow y_i = 0) \;\&\; (y_i = y_j > 0 \Rightarrow x_i = x_j)$$

for all $1 \leq i, j \leq n$.

Proof Both of these are proved by induction. For $n = 2$ they are easily seen to be true; we give the arguments for $n + 1$ assuming n.

For $(x_i = 0 \Rightarrow y_i = 0)$, we can assume $x^+, y^+ < 1$: If $x^+ = 1$, then $x = y$ since x is maximal; If $y^+ = 1$, then either $y_i = 0$, which finishes the proof, or $y_i = 1$, in which case Proposition 1(i) gives $y_i = y^+ = 1 \geq x^+ = x_i > 0$, contradicting $x_i = 0$. Thus, since $x^+, y^+ < 1$, any $k \neq i$ yields $p_k(x) \sqsubseteq p_k(y)$, and since $x_i/(1 - x_k) = 0$, the inductive hypothesis gives $y_i/(1 - y_k) = 0$ hence $y_i = 0$.

For the other claim, suppose $y_i = y_j > 0$ with $i \neq j$. Then $y^+ < 1$. In addition, $x^+ < 1$ or else $x = y$ and we are done. Then because $n + 1 \geq 3$, there is $k \in \{1, \ldots, n+1\} \setminus \{i, j\}$. For any such index, we have $p_k(x) \sqsubseteq p_k(y)$, so the inductive hypothesis gives $x_i/(1 - x_k) = x_j/(1 - x_k)$, i.e., $x_i = x_j$. □

The standard projections $\pi_k : \Delta^n \to [0, 1]$ are $\pi_k(x) = x_k$ for $1 \leq k \leq n$. Lemma 4 extends to increasing sequences as follows.

Lemma 6 *If* (x_i) *is an increasing sequence in* Δ^n, *then*

(i) *There is an index* k *with* $\pi_k(x_i) = x_i^-$ *for all* i.
(ii) *There is an index* k *with* $\pi_k(x_i) = x_i^+$ *for all* i.

Proof (i) Before starting, a crucial consequence of Lemma 5 for the present argument is that

$$\{k : y_k = y^-\} \subseteq \{k : x_k = x^-\}$$

provided that $x \sqsubseteq y$ and $y^- > 0$. Thus, *any* increasing sequence (x_i) with $x_i^- > 0$ leads to a decreasing sequence of nonempty finite sets. The intersection of such a sequence must be nonempty, and any member k in this intersection will satisfy $\pi_k(x_i) = x_i^-$ for all i.

Thus, for our sequence $(x_i)_{i\geq 1}$, we may assume that there is a *least* integer $m \geq 1$ with $x_m^- = 0$. First, the proof is finished if we find k with $\pi_k(x_i) = x_i^-$ for all $i \leq m$, since then we have $\pi_k(x_m) = 0$ and hence $\pi_k(x_i) = 0$ for all $i \geq m$, by Lemma 5, which means $\pi_k(x_i) = x_i^-$ for all $i \geq 1$.

The case $m = 1$ is trivial. If $m > 1$, then for the subsequence $(x_i)_{i<m}$, we have $x_i^- > 0$ for $i < m$, by the choice of m, so our opening remarks give $\pi_k(x_i) = x_i^-$, for all $i < m$, where k is *any* index in $\{k : \pi_k(x_{m-1}) = x_{m-1}^-\}$. By Lemma 4, there is k with $\pi_k(x_{m-1}) = x_{m-1}^-$ and $\pi_k(x_m) = x_m^-$. This value of k gives $\pi_k(x_i) = x_i^-$ for all $i \leq m$.

(ii) We simplify modify the proof of Proposition 1(i) using (i).

Now we take our first step toward proving that Δ^n is a domain.

Definition 6 A subset S of a poset is *directed* if it is nonempty and

$$(\forall x, y \in S)(\exists z \in S)\, x, y \sqsubseteq z\,.$$

A directed-complete partial order, or *dcpo*, is a poset in which every directed subset has a supremum.

A familiar example of a directed set is an *increasing sequence*: A sequence (x_i) such that $x_i \sqsubseteq x_{i+1}$ for all i. Joyfully, on classical states, one can always replace directed sets with increasing sequences, so we never have to think about the former.

Proposition 2 *The classical states* Δ^n *are a dcpo. In more detail,*

(i) *If* (x_i) *is an increasing sequence, then*

$$\bigsqcup_{i\geq 1} x_i = (\lim_{i\to\infty} \pi_1(x_i), \ldots, \lim_{i\to\infty} \pi_n(x_i)).$$

(ii) *Every directed subset of* Δ^n *contains an increasing sequence with the same supremum.*

Proof We first prove (i) by induction. It is true for $n = 2$. Assume for n. Given an increasing sequence (x_i), Lemma 6 yields an index k such that $\pi_k(x_i) = x_i^-$ for all i. The sequence $(p_k(x_i))$ is increasing in Δ^n, so by the inductive hypothesis, we know that

$$\lim_{i\to\infty}\left(\frac{\pi_j(x_i)}{1-x_i^-}\right)$$

exists for all $j \neq k$. The sequence (x_i^-) is decreasing and contained in $[0, 1/(n+1)]$, so it has a limit $s_k = \lim \pi_k(x_i) < 1$, which means $(1 - x_i^-)$ has a limit that is not zero. Thus,

$$s_j := \lim_{i\to\infty} \pi_j(x_i) = \lim_{i\to\infty}\left(\frac{\pi_j(x_i)}{1-x_i^-}\right)\cdot \lim_{i\to\infty}(1-x_i^-)$$

exists for $j \neq k$. Notice that

$$\sum_{j=1}^{n+1} s_j = \sum_{j=1}^{n+1} \lim_{i\to\infty} \pi_j(x_i) = \lim_{i\to\infty} \sum_{j=1}^{n+1} \pi_j(x_i) = 1,$$

which means that $s = (s_1, \ldots, s_{n+1})$ is a classical state. We claim that s is the supremum of (x_i).

To avoid needless complication, we can assume $x_i^+ < 1$, since otherwise (x_i) has finitely many distinct elements, and then the claim is obvious. To prove that $x_i \sqsubseteq s$ for all i, we must show

$$(\forall i)(\forall j)\, s_j < 1 \Rightarrow p_j(x_i) \sqsubseteq p_j(s).$$

Fix an index j with $s_j < 1$. Then the sequence $(p_j(x_i))_{i\geq 1}$ is increasing in Δ^n, so by the inductive hypothesis, it has a supremum

$$\bigsqcup_{i\geq 1} p_j(x_i) = \left(\lim_{i\to\infty}\frac{\pi_1(x_i)}{1-\pi_j(x_i)}, \ldots, \lim_{i\to\infty}\frac{\widehat{\pi_j(x_i)}}{1-\pi_j(x_i)}, \ldots, \lim_{i\to\infty}\frac{\pi_{n+1}(x_i)}{1-\pi_j(x_i)}\right)$$

which is equal to $p_j(s)$ since $s_j = \lim_{i\to\infty} \pi_j(x_i) < 1$. Hence, $p_j(x_i) \sqsubseteq p_j(s)$ for all i and j with $s_j < 1$, which means $x_i \sqsubseteq s$ for all i.

To prove that s is the supremum of (x_i), let u be any upper bound of (x_i). We must show that $s \sqsubseteq u$, i.e.,

$$(\forall j)\, s_j < 1 \,\&\, u_j < 1 \Rightarrow p_j(s) \sqsubseteq p_j(u).$$

Let j be any index with $s_j < 1$ and $u_j < 1$. Then since $p_j(x_i) \sqsubseteq p_j(u)$ for all i, we have $p_j(s) = \bigsqcup_{i\geq 1} p_j(x_i) \sqsubseteq p_j(u)$, using the inductive hypothesis and that $s_j < 1$. Thus, $s \sqsubseteq u$, which proves $s = \bigsqcup x_i$.

The directed completeness of Δ^n and (ii) now follow from a theorem in [8] provided there is a strictly monotone map $f : \Delta^n \to [0, \infty)^*$ which preserves suprema of increasing sequences. To see that $f(x) = 1 - x^+$ is one such map, if (x_i) is an increasing sequence, Lemma 6(ii) yields an index k with $\pi_k(x_i) = x_i^+$ for all i, so

$$\left(\bigsqcup_{i\geq 1} x_i\right)^+ = \lim_{i\to\infty} \pi_k(x_i) = \lim_{i\to\infty} x_i^+,$$

which makes it clear that f preserves suprema of increasing sequences. That f is strictly monotone follows from Proposition 1(iii). □

Definition 7 A map $f : D \to E$ between dcpo's is *Scott continuous* if it is *monotone*

$$x \sqsubseteq y \Rightarrow f(x) \sqsubseteq f(y)$$

and it *preserves directed suprema*:

$$f\left(\bigsqcup S\right) = \bigsqcup f(S)$$

for any directed set $S \subseteq D$.

Corollary 1 *A monotone map $f : \Delta^n \to E$ into a dcpo E is Scott continuous iff for each increasing sequence (x_i) in Δ^n, $f\left(\bigsqcup x_i\right) = \bigsqcup f(x_i)$.*

Proof If $S \subseteq \Delta^n$ is directed, then $\bigsqcup f(S) \sqsubseteq f(\bigsqcup S)$ by monotonicity. For the other direction, Proposition 12(ii) gives an increasing sequence (x_i) in S with $\bigsqcup S = \bigsqcup x_i$, enabling

$$f\left(\bigsqcup S\right) = f\left(\bigsqcup x_i\right) = \bigsqcup f(x_i) \sqsubseteq \bigsqcup f(S),$$

confirming that f preserves suprema of *all* directed sets provided it does so for increasing sequences. □

For instance, the map

$$\Delta^n \to [0, 1] :: x \mapsto x^+$$

is Scott continuous, while $x \mapsto 1 - x^+$ is Scott continuous as a map $\Delta^n \to [0, 1]^*$. An amusing example of a Scott continuous map that is not Euclidean continuous is the natural retraction from Δ^2 onto

$$\partial\Delta^2 = \max(\Delta^2)\,.$$

Generally speaking, the *entropy* of an event with probability p is $-\log p$. If forced to choose a single probability representative of an entire classical state x,

x^+ would be the most sensible choice, because of its qualitative significance in Proposition 1. Thus, one might say that $s(x) = -\log x^+$ measures the entropy of a classical state.

Corollary 2 *The map* $s : \Delta^n \to [0, \infty)^*$ *given by* $s(x) = -\log x^+$ *is Scott continuous. It has the following properties:*

(i) *For all* $x, y \in \Delta^n$, *if* $x \sqsubseteq y$ *and* $s(x) = s(y)$, *then* $x = y$.
(ii) *For all* $x \in \Delta^n$, *we have* $s(x) = 0$ *iff* $x \in \max(\Delta^n)$.
(iii) *For all* $x \in \Delta^n$, *we have* $s(x) = \log n$ *iff* $x = \perp$.

Proof The map is well-defined because $x^+ \in [1/n, 1]$. That s is strictly monotone follows from Proposition 1. (ii) and (iii) follow from combinations of direct calculation and applications of (i).

By the last result, a monotone map $f : D \to \Delta^n$ from a dcpo D is Scott continuous iff $s \circ f$ is Scott continuous. We will take a closer look at entropy later on.

10.2.3 Symmetries for Classical States

We now establish the fundamental role played by the symmetric group

$$S(n) = \{\sigma | \sigma : \{1, \dots, n\} \simeq \{1, \dots, n\}\}$$

of bijections on the set $\{1, \dots, n\}$. These we also refer to as *permutations* or *symmetries*. The composition of $x \in \Delta^n$ and $\sigma \in S(n)$ is written $x \cdot \sigma$.

Definition 8 A state $x \in \Delta^n$ is *monotone* if $x_i \geq x_{i+1}$ for all $i < n$.

A classical state $x \in \Delta^n$ can be completely described by a monotone state $x \cdot \sigma$ and a symmetry σ^{-1}. The order on Δ^n has an analogous representation.

Lemma 7 *For states* $x, y \in \Delta^2$, *we have* $x \sqsubseteq y$ *iff there is a permutation* σ *of* $\{1, 2\}$ *such that* $x \cdot \sigma = (x^+, x^-)$, $y \cdot \sigma = (y^+, y^-)$ *and* $x^+ y^- \leq x^- y^+$.

Theorem 3 *For* $x, y \in \Delta^n$, *we have* $x \sqsubseteq y$ *iff there is a permutation* σ *of* $\{1, \dots, n\}$ *such that* $x \cdot \sigma$ *and* $y \cdot \sigma$ *are monotone and*

$$(x \cdot \sigma)_i (y \cdot \sigma)_{i+1} \leq (x \cdot \sigma)_{i+1} (y \cdot \sigma)_i$$

for all i *with* $1 \leq i < n$.

Proof By the last lemma, the claim is true for $n = 2$. Assume the result for $n \geq 2$. For the case $n + 1$, we prove both implications separately.

First suppose $x \sqsubseteq y$. Let k be an index with $x_k = x^- \geq y^- = y_k$. By the inductive hypothesis applied to $p_k(x) \sqsubseteq p_k(y)$, there is a permutation σ of

$\{1, \ldots, n\}$ such that $p_k(x) \cdot \sigma$ and $p_k(y) \cdot \sigma$ are monotone. Now compose σ with the natural bijection that maps indices of $p_k(x)$ and $p_k(y)$ to indices of x and y, and since there is no harm in doing so, call the resulting bijection $\sigma : \{1, \ldots, n\} \to \{1, \ldots, n+1\} \setminus \{k\}$.

We extend σ to a permutation of $\{1, \ldots, n+1\}$ by setting $\sigma(n+1) := k$. It is then clear that $x \cdot \sigma$ and $y \cdot \sigma$ are monotone and that

$$(x \cdot \sigma)_i (y \cdot \sigma)_{i+1} \leq (x \cdot \sigma)_{i+1} (y \cdot \sigma)_i$$

for all $1 \leq i < n$. To finish this direction, we need to prove

$$(x \cdot \sigma)_n (y \cdot \sigma)_{n+1} \leq (x \cdot \sigma)_{n+1} (y \cdot \sigma)_n.$$

Because $(x \cdot \sigma)_{n+1} = x_k = x^- \geq y^- = y_k = (y \cdot \sigma)_{n+1}$, we need only consider the case that $(y \cdot \sigma)_n < (x \cdot \sigma)_n$.

First, $x^+ < 1$, since $(x \cdot \sigma)_n > 0 \Rightarrow (x \cdot \sigma)_1 = x^+ < 1$. Next, $y^+ < 1$, since otherwise $(y \cdot \sigma)_{n+1} = 0$, in which case the inequality is trivial. Now let j be an index with $x_j = x^+ \leq y^+ = y_j$. By the inductive hypothesis applied to $p_j(x) \sqsubseteq p_j(y)$, we obtain a permutation π of $\{1, \ldots, n\}$. Similar to the case of σ, we regard π as a bijection

$$\{2, \ldots, n+1\} \to \{1, \ldots, n+1\} \setminus \{j\}$$

and then extend it to a permutation of $\{1, \ldots, n+1\}$ by setting $\pi(1) := j$. Again $x \cdot \pi$ and $y \cdot \pi$ are monotone, and in this case we have

$$(x \cdot \pi)_i (y \cdot \pi)_{i+1} \leq (x \cdot \pi)_{i+1} (y \cdot \pi)_i$$

for $2 \leq i < n+1$. Because $x \cdot \pi = x \cdot \sigma$ and $y \cdot \pi = y \cdot \sigma$, setting $i = n$ yields the desired inequality, finishing this direction.

For the other, let σ be a permutation of $\{1, \ldots, n+1\}$ with $x \cdot \sigma$, $y \cdot \sigma$ monotone and $(x \cdot \sigma)_i (y \cdot \sigma)_{i+1} \leq (x \cdot \sigma)_{i+1} (y \cdot \sigma)_i$ for all i with $1 \leq i < n+1$. First notice that slightly more is true:

$$(*) \quad (x \cdot \sigma)_i (y \cdot \sigma)_j \leq (x \cdot \sigma)_j (y \cdot \sigma)_i$$

for $1 \leq i \leq j \leq n+1$.

In the cases $(x \cdot \sigma)_i = 0$ and $(y \cdot \sigma)_j = 0$, $(*)$ is clear; if $(x \cdot \sigma)_i > 0$ and $(y \cdot \sigma)_j > 0$, then $(y \cdot \sigma)_k > 0$ for $k \leq j$ by monotonicity, which means $(x \cdot \sigma)_k > 0$ for all $i \leq k \leq j$ as well, since for $k > i$ we have

$$(x \cdot \sigma)_k \geq \frac{(x \cdot \sigma)_{k-1} (y \cdot \sigma)_k}{(y \cdot \sigma)_{k-1}} > 0$$

assuming $(x \cdot \sigma)_{k-1} > 0$. Without division by zero to worry about, $(*)$ is now clear.

To prove $x \sqsubseteq y$ we must show $p_k(x) \sqsubseteq p_k(y)$ for all k with $x_k < 1$, $y_k < 1$. To this end, fix one such k. We restrict σ to a bijection

$$\{1, \ldots, n+1\} \setminus \sigma^{-1}(k) \to \{1, \ldots, n+1\} \setminus \{k\}$$

which then yields a permutation σ_k of $\{1, \ldots, n\}$ such that $p_k(x) \cdot \sigma_k$ and $p_k(y) \cdot \sigma_k$ are monotone. By (*), we have

$$(p_k(x) \cdot \sigma_k)_i (p_k(y) \cdot \sigma_k)_{i+1} \leq (p_k(x) \cdot \sigma_k)_{i+1} (p_k(y) \cdot \sigma_k)_i$$

for all $1 \leq i < n$. By the inductive hypothesis, $p_k(x) \sqsubseteq p_k(y)$, finishing the proof. □

The explicit nature of the representation using symmetries can be advantageous in establishing certain properties of the order.

Lemma 8 *The map $x \mapsto x \cdot \sigma$ is an order isomorphism of Δ^n for each $\sigma \in S(n)$.*

Proof Let $f(x) = x \cdot \sigma$. To see that f is monotone, if $x \sqsubseteq y$, then there is $\nu \in S(n)$ with $x \cdot \nu$ and $y \cdot \nu$ monotone satisfying the inequalities of Theorem 3. But the same is true of $x \cdot \sigma$ and $y \cdot \sigma$ if we apply the permutation $\sigma^{-1} \cdot \nu$ to each. Thus, $f(x) = x \cdot \sigma \sqsubseteq y \cdot \sigma = f(y)$.

The same argument shows $g(x) = x \cdot \sigma^{-1}$ is monotone. Because f and g are also inverse to one another, each is an order isomorphism. □

It is now time to take a more in depth look at the order on classical states. To keep things simple initially, we start on the outside and work our way inward. The *boundary* of Δ^{n+1},

$$\partial \Delta^{n+1} = \bigcup_{1 \leq i \leq n+1} \ker \pi_i,$$

can be understood *geometrically* as $n+1$ copies of Δ^n identified at certain points. The same result holds *order theoretically*. That is, the dcpo $\partial \Delta^{n+1}$ is order isomorphic to $n+1$ copies of the dcpo Δ^n identified along their common boundaries.

Proposition 3 *For $n \geq 1$, we have an order isomorphism*

$$\Delta^n \simeq \{x \in \Delta^{n+1} : \pi_i(x) = 0\},$$

for any of the standard projections $\pi_i : \Delta^{n+1} \to [0, 1]$ with $1 \leq i \leq n+1$.

Proof First, $i_{n+1} : \Delta^n \to \Delta^{n+1} :: x \mapsto (x, 0)$ is an order embedding. It is order reflecting:

$$i_{n+1}(x) \sqsubseteq i_{n+1}(y) \implies x = p_{n+1}(i_{n+1}(x)) \sqsubseteq p_{n+1}(i_{n+1}(y)) = y.$$

For its monotonicity, let $x \sqsubseteq y$. By Theorem 3, there is $\sigma \in S(n)$ with $x \cdot \sigma$ and $y \cdot \sigma$ monotone such that the usual inequalities hold.

Now extend σ to a permutation in $S(n+1)$ by setting $\sigma(n+1) = n+1$. Because the value of the state $i_{n+1}(x)$ at index $n+1$ is zero, $i_{n+1}(x) \cdot \sigma$ and $i_{n+1}(y) \cdot \sigma$ are monotone and satisfy the inequalities of Theorem 3. Thus, $i_{n+1}(x) \sqsubseteq i_{n+1}(y)$.

The other n maps, i_k for $1 \leq k \leq n$, which produce an $n+1$ state having value zero at index k, all arise as the composition of isomorphisms (derived from right multiplication by a symmetry) followed by i_{n+1}. □

Thus, the boundary of the triangle Δ^3 is a dcpo made of three copies of Δ^2:

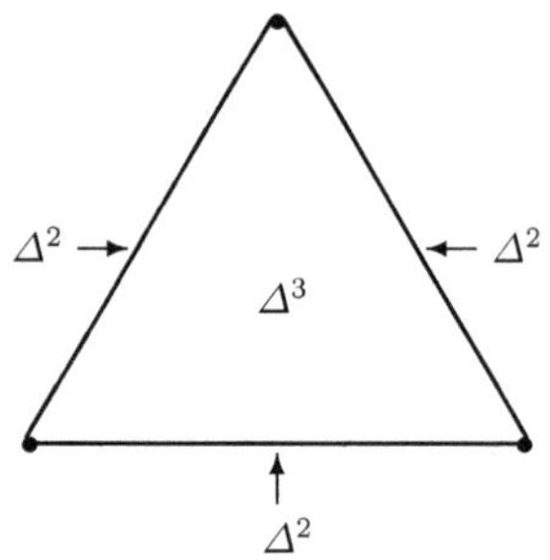

To get an idea of what the order is like on $\mathrm{int}(\Delta^n)$, we need to look a little closer. First, some long overdue notation.

Definition 9 The monotone classical states are denoted

$$\Lambda^n := \{x \in \Delta^n : (\forall i < n)\, x_i \geq x_{i+1}\}.$$

For $\sigma \in S(n)$,

$$\Delta^n_\sigma := \{x \in \Delta^n : x \cdot \sigma \in \Lambda^n\}.$$

Notice that $\Delta^n_1 = \Lambda^n$.

As we have already seen, the order on monotone states can be characterized purely algebraically. For the sake of emphasis:

Lemma 9 *For $x, y \in \Lambda^n$, $x \sqsubseteq y$ iff $(\forall\, 1 \leq i < n)\, x_i y_{i+1} \leq y_i x_{i+1}$.*

Just as was the case with its boundary, there is also a natural way of dividing Δ^n itself into regions: For each $n \geq 1$,

$$\Delta^n := \bigcup_{\sigma \in S(n)} \Delta^n_\sigma.$$

And just as before, these regions are identical (order-theoretically).

Proposition 4 *Let $n \geq 2$.*

(i) *For each $\sigma \in S(n)$, Δ^n_σ is closed under directed suprema in Δ^n.*
(ii) *For an increasing sequence (x_i) in Δ^n, there is $\sigma \in S(n)$ with $x_i \in \Delta^n_\sigma$ for all i.*

(iii) *The natural map*

$$r : \Delta^n \to \Lambda^n$$

is a Scott continuous retraction whose restriction to Δ^n_σ *is an order isomorphism* $\Delta^n_\sigma \simeq \Lambda^n$ *for each* $\sigma \in S(n)$.

Proof (i) Since every directed set contains an increasing sequence with the same supremum, we only have to prove this result for increasing sequences (x_i) in Δ^n_σ. By Lemma 8 and the formula for suprema,

$$\left(\bigsqcup x_i\right) \cdot \sigma = \bigsqcup (x_i \cdot \sigma) = (\lim_{i\to\infty} \pi_1(x_i \cdot \sigma), \ldots, \lim_{i\to\infty} \pi_n(x_i \cdot \sigma)).$$

But the state on the far right is monotone because all the $x_i \cdot \sigma$ are. This proves that $\bigsqcup x_i \in \Delta^n$ also belongs to Δ^n_σ.

(ii) This is a straightforward induction using Lemma 6(i).

(iii) For $\sigma \in S(n)$, we denote the order isomorphism in Lemma 8 by $r_\sigma(x) = x \cdot \sigma$. Set $r(x) = r_\sigma(x)$ for $x \in \Delta^n_\sigma$. This map is well defined and its restriction to Δ^n_σ is an order isomorphism: $\Delta^n_\sigma = r_\sigma^{-1}(\Lambda^n) \simeq \Lambda^n$.

It is monotone: If $x \sqsubseteq y$, then by Theorem 3, there is $\sigma \in S(n)$ with $x, y \in \Delta^n_\sigma$ which gives $r(x) = r_\sigma(x) \sqsubseteq r_\sigma(y) = r(y)$.

It is Scott continuous: If $\mu : \Delta^n \to [0, \infty)^*$ is strictly monotone, Scott continuous and $\mu r = \mu$, then r itself is Scott continuous, since it is monotone and has continuous measure μr. Let $\mu x = -\log x^+$ (Corollary 2).

Finally, $r|_{\Lambda^n} = 1_{\Lambda^n}$, which proves that r is a retraction.

Thus, we can think of Δ^n as being $n!$-many copies of the retract Λ^n identified along their common boundaries. For instance, Δ^3 splits into six different regions, all order isomorphic to Λ^3:

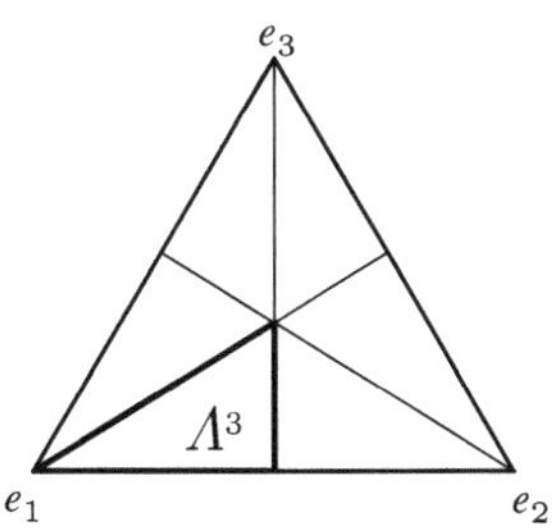

This, combined with an elementary analysis of Λ^3, allows us to determine the upper sets of $(\Delta^3, \sqsubseteq)$ shown in Fig. 10.1.

We now have our first example of an *intuition* about classical states that has been *formally justified*. Consider a closed cylinder of volume V partitioned into smaller volumes V_i as follows:

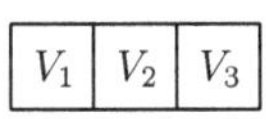

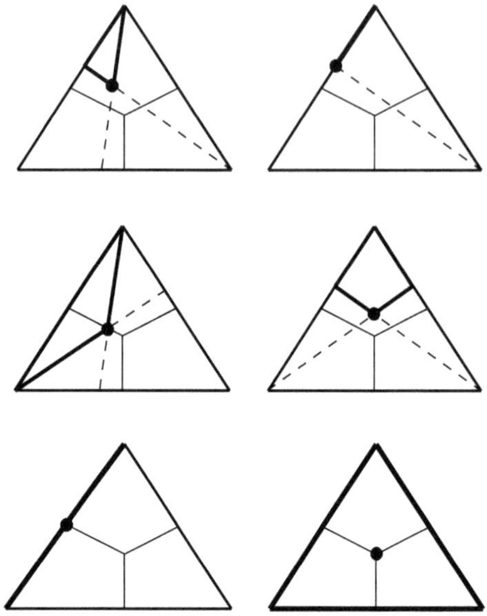

Fig. 10.1 Pictures of $\uparrow x$ for $x \in \Delta^3$

The cylinder is known a priori to contain a single molecule. With no other information available to us, our knowledge of the molecule's location is (p_1, p_2, p_3) where $p_i = V_i/V$. Or is it? Well, it is if we *assume* that the volumes are labelled from left to right as 1,2,3. But if they have been labelled in the reverse order, as

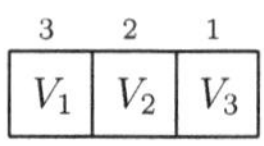

then our knowledge is (p_3, p_2, p_1).

Naturally, we *intuitively* understand that in the grand scheme of things it makes no difference how we label things—as long as all statements made about the experiment are made with respect to the same choice of labels, we will not encounter any trouble: What is physically true for one choice of labels is also true for any other. But that's where the magic is! We have *derived* this simple truth: For each $\sigma \in S(n)$, the map $x \mapsto x \cdot \sigma$ is an order isomorphism.

In short, there is a definite physical reason why Δ^n is divided into *different* regions Δ^n_σ all of which are "identical" ($\Delta^n_\sigma \simeq \Delta^n_\nu$). For the very same reason, measures of information content in such experiments tend to be *symmetric*.

Definition 10 A function $f : \Delta^n \to E$ is *symmetric* if for all $\sigma \in S(n)$, we have $f(x \cdot \sigma) = f(x)$.

Lemma 10 *Let E be a dcpo. Then*

(i) *Every function $f : \Lambda^n \to E$ determines a unique symmetric extension $\bar{f} : \Delta^n \to E$ given by $\bar{f} = f \circ r$ where r is the natural retraction.*

(ii) *Monotonicity, strict monotonicity and Scott continuity are inherited by $\bar{f}$ whenever they are possessed by f.*

Proof (i) For the uniqueness of $\bar{f}$, if $g : \Delta^n \to E$ is another symmetric extension of f, then for any $x \in \Delta^n_\sigma$, we can write

$$g(x) = g(x \cdot \sigma) = f(x \cdot \sigma) = \bar{f}(x \cdot \sigma) = \bar{f}(x)$$

using that g is symmetric, followed by the fact that $g = f$ on Λ^n, and then the fact that $\bar{f}$ is a symmetric extension of f.

(ii) Each property is preserved by composition and satisfied by r.

Example 1 Canonical symmetric functions on Δ^n

(i) The maps $\Delta^n \to [0, 1] :: x \mapsto x^+$ and $\Delta^n \to [0, 1]^* :: x \mapsto 1 - x^+$.
(ii) Entropy $s(x) = -\log x^+$.
(iii) The natural retraction $r : \Delta^n \to \Lambda^n$.
(iv) Shannon entropy

$$\mu x = -\sum_{i=1}^{n} x_i \log x_i.$$

As the last result illustrates, the retraction $r : \Delta^n \to \Lambda^n$ provides us with a general approach for solving problems involving classical states: First solve it for Λ^n, and then for Δ^n in general.

10.2.4 Approximation of Classical States

A decent understanding of approximation can provide insight about the nature of *partiality*. Partiality, as we will see in the next section, is imperative for a meaningful discussion on entropy.

Definition 11 Let D be a dcpo. For $x, y \in D$, we write $x \ll y$ iff for all directed sets $S \subseteq D$,

$$y = \bigsqcup S \Rightarrow (\exists s \in S)\, x \sqsubseteq s.$$

The *approximations* of $x \in D$ are

$$\mathord{\downarrow\!\!\downarrow} x := \{y \in D : y \ll x\},$$

and D is called *exact* if $\mathord{\downarrow\!\!\downarrow} x$ is directed with supremum x for all $x \in D$.

A *continuous* dcpo is exact, and in that case, the "way below" relation and our notion of approximation above are equivalent. In addition, the two notions also coincide on maximal elements.

Lemma 11 *Let D be a dcpo. For each $x \in D$, the set $\mathord{\downarrow}x$ is directed with supremum x iff it contains a directed set with supremum x.*

The ability to approximate classical states is provided by the mixing law.

Proposition 5 (The mixing law) *If $x \sqsubseteq y$ in Δ^n, then*

$$x \sqsubseteq (1-p)x + py \sqsubseteq y$$

for all $p \in [0,1]$.

Proof Let z denote the classical state $(1-p)x + py$. Because $x \sqsubseteq y$, there is a symmetry σ with $x \cdot \sigma$, $y \cdot \sigma$ monotone. First,

$$\begin{aligned}(z\cdot\sigma)_i &= (1-p)(x\cdot\sigma)_i + p(y\cdot\sigma)_i \\ &\geq (1-p)(x\cdot\sigma)_{i+1} + p(y\cdot\sigma)_{i+1} \\ &= (z\cdot\sigma)_{i+1},\end{aligned}$$

for $1 \leq i < n$, which means $z \cdot \sigma$ is monotone. Thus, $x \sqsubseteq z$ follows from

$$\begin{aligned}(x\cdot\sigma)_i(z\cdot\sigma)_{i+1} &\leq (x\cdot\sigma)_{i+1}(z\cdot\sigma)_i \\ &\Leftrightarrow \\ p(x\cdot\sigma)_i(y\cdot\sigma)_{i+1} &\leq p(x\cdot\sigma)_{i+1}(y\cdot\sigma)_i,\end{aligned}$$

while $z \sqsubseteq y$ follows similarly. □

A *path* from x to y in a space X is a continuous map

$$p : [0,1] \to X$$

with $p(0) = x$ and $p(1) = y$. A *segment* of a path p is $p[a,b]$ for $b > a$. Any monotone path into Δ^n with its Euclidean topology is Scott continuous. For instance, by the mixing law (Lemma 1), the *straight line path* from x to y,

$$\pi_{xy}(t) = (1-t)x + ty$$

is Scott continuous iff $x \sqsubseteq y$.

Lemma 12 *Let $x \sqsubseteq y$ with $x \in \Delta^n$ and $y \in \Lambda^n$. Then*

(i) *If $y_i > 0$ for all i, then $x \in \Lambda^n$.*
(ii) *If $x \ll y$, then $x \in \Lambda^n$.*

Proof (i) The proof is by induction. For the $n+1$ case, Lemma 4 gives an index i with $x_i = x^- \geq y_i = y^-$, while the monotonicity of y yields $y_{n+1} = y^- = y_i > 0$. By degeneration (Lemma 5), $x_{n+1} = x_i = x^- > 0$.

Now we can apply the inductive hypothesis to $p_{n+1}(x) \sqsubseteq p_{n+1}(y)$, since $p_{n+1}(y) \in \Lambda^n$ and all its values are positive, to deduce that $p_{n+1}(x) \in \Lambda^n$. But since $x_{n+1} = x^-$, we have $x \in \Lambda^{n+1}$.

(ii) We apply (i). By the Scott continuity of $\pi_{\perp y}$,

$$y = \bigsqcup_{t<1} \pi_{\perp y}(t),$$

and since $x \ll y$, we have $x \sqsubseteq \pi_{\perp y}(t)$ for some $t < 1$. Because $y \in \Lambda^n$, $\pi_{\perp y}(t) \in \Lambda^n$, and $\pi_{\perp y}(t)_i > 0$ for all i since $t < 1$. By (i), $x \in \Lambda^n$. □

Proposition 6 *Let* $r : \Delta^n \to \Lambda^n$ *be the natural retraction.*

(i) *If* $x, y \in \Delta^n_\sigma$ *and* $x \sqsubseteq y$, *then* $\pi_{xy}(t) \in \Delta^n_\sigma$ *for all* $t \in [0, 1]$.
(ii) *For* $x, y \in \Delta^n$, *we have* $x \ll y$ *iff*

$$(\forall \sigma \in S(n))(y \in \Delta^n_\sigma \Rightarrow x \in \Delta^n_\sigma) \text{ and } (r(x) \ll r(y) \text{ in } \Lambda^n).$$

Proof (i) This was shown in the proof of the mixing law. (ii) First recall that right multiplication by $\sigma \in S(n)$, $r_\sigma(x) = x \cdot \sigma$, is an order isomorphism of Δ^n. If $x \ll y$, then $x \sqsubseteq y$, which means $x, y \in \Delta^n_\sigma$ for some $\sigma \in S(n)$. Because r_σ is an order isomorphism,

$$x \ll y \Rightarrow r_\sigma(x) \ll r_\sigma(y) \text{ in } \Delta^n.$$

But $r(x) = r_\sigma(x)$ and $r(y) = r_\sigma(y)$, which means $r(x) \ll r(y)$ in Δ^n. However, $r(x), r(y) \in \Lambda^n$ *and in addition* Λ^n is closed under directed suprema in Δ^n by Prop. 4(i), which means that the supremum in Λ^n of a directed set $S \subseteq \Lambda^n$ is *equal* to the supremum it has as a subset of Δ^n. Thus, $r(x) \ll r(y)$ in Λ^n.

To finish this direction, suppose $y \in \Delta^n_\sigma$. Then $x \cdot \sigma = r_\sigma(x) \ll r_\sigma(y) = y \cdot \sigma$ in Δ^n, since r_σ is an order isomorphism. But $y \cdot \sigma$ is monotone, so Lemma 12 implies that $x \cdot \sigma$ is too, i.e., $x \in \Delta^n_\sigma$.

For the other direction, if we choose any $\sigma \in S(n)$ with $y \in \Delta^n_\sigma$, then $x \in \Delta^n_\sigma$. By assumption, we have $r_\sigma(x) = r(x) \ll r(y) = r_\sigma(y)$ in Λ^n. If we show that $r_\sigma(x) \ll r_\sigma(y)$ in Δ^n, then because r_σ is an order isomorphism, we may conclude $x \ll y$ in Δ^n.

Let (y_i) be an increasing sequence in Δ^n with $r_\sigma(y) = \bigsqcup y_i$. By Proposition 4, there is $\nu \in S(n)$ with $y_i \in \Delta^n_\nu$ for all i, and hence $y \cdot \sigma \in \Delta^n_\nu$. Then because $y \in \Delta^n_{\sigma\cdot\nu}$, we have $x \in \Delta^n_{\sigma\cdot\nu}$ by assumption, so the following relation involves only states in Λ^n:

$$x \cdot (\sigma \cdot \nu) = x \cdot \sigma \ll y \cdot \sigma = y \cdot (\sigma \cdot \nu) = \bigsqcup (y_i \cdot \nu).$$

Because $x \cdot \sigma \ll y \cdot \sigma$ in Λ^n, we must have $x \cdot (\sigma \cdot \nu) \sqsubseteq y_i \cdot \nu$ for some i, i.e., $r_\nu(r_\sigma(x)) \sqsubseteq r_\nu(y_i)$ which gives $r_\sigma(x) \sqsubseteq y_i$. Then $r_\sigma(x) \ll r_\sigma(y)$ in Δ^n, and now the proof is finished. □

Theorem 4 *The classical states Δ^n are exact.*

(i) *For every $x \in \Delta^n$, $\pi_{\perp x}(t) \ll x$ for all $t < 1$.*
(ii) *The approximation relation $\ll$ is interpolative: If $x \ll y$ in Δ^n, then there is $z \in \Delta^n$ with $x \ll z \ll y$.*

Proof The exactness of Δ^n follows from (i), the Scott continuity of $\pi_{\perp x}$, and Lemma 11. To prove (i), we first show that $\pi_{\perp x}(t) \ll x$ in Λ^n for any $x \in \Lambda^n$ and $t < 1$. Notice that $\pi_{\perp x}(t) \in \Lambda^n$ for all $t \in [0,1]$ by Proposition 6(i). Let $x = \bigsqcup y_k \in \Lambda^n$ for an increasing sequence $(y_k)_{k\geq 1}$ in Λ^n.

For $i < n$ fixed, we will show that there is an integer k_i such that

$$\left(\frac{1-t}{n} + tx_i\right)\pi_{i+1}(y_k) \leq \pi_i(y_k)\left(\frac{1-t}{n} + tx_{i+1}\right)$$

for all $k \geq k_i$. If $x_i = 0$, then $x_{i+1} = 0$ by the monotonicity of x, and then we can take $k_i = 1$, by the monotonicity of each y_k. Thus, we can assume $x_i > 0$.

If $x_{i+1} = 0$, then we can write

$$\left(\frac{1-t}{n} + tx_i\right)\lim_{k\to\infty}\pi_{i+1}(y_k) = 0 < \delta < x_i\left(\frac{1-t}{n}\right) = \lim_{k\to\infty}\pi_i(y_k)\left(\frac{1-t}{n}\right),$$

where $\delta > 0$ is some constant, and we use $x = \bigsqcup y_k$. This makes it clear that such a k_i exists in this case. Thus, we can also assume $x_{i+1} > 0$.

If $x_i = x_{i+1} > 0$, then because $y_k \sqsubseteq x$, degeneration (Lemma 5) gives $\pi_i(y_k) = \pi_{i+1}(y_k) > 0$ for each k. In this case, we can again take $k_i = 1$. Thus, we assume $x_i > x_{i+1} > 0$. By the degeneration lemma, this also implies $\pi_i(y_k) > 0$ and $\pi_{i+1}(y_k) > 0$ for all k. But then we get

$$\frac{(1-t)/n + tx_i}{(1-t)/n + tx_{i+1}} < \frac{x_i}{x_{i+1}} = \lim_{k\to\infty}\frac{\pi_i(y_k)}{\pi_{i+1}(y_k)},$$

using $x_i > x_{i+1} > 0$, $t < 1$ and $\bigsqcup y_k = x$. Thus, in this case there is also a large enough k_i such that the desired inequality holds for all $k \geq k_i$.

Then $\pi_{\perp x}(t) \sqsubseteq y_k$ where $k \geq \max\{k_i : 1 \leq i < n\}$, which proves $\pi_{\perp x}(t) \ll x$ in Λ^n for all $t < 1$. To finish the proof, let x be *any* classical state and $r : \Delta^n \to \Lambda^n$ the natural retract. We know

- $x \in \Delta^n_\sigma \Rightarrow \pi_{\perp x}(t) \in \Delta^n_\sigma$ for all $t \in [0,1]$, and
- $r(\pi_{\perp x}(t)) = \pi_{\perp r(x)}(t) \ll r(x)$ in Λ^n, for all $t < 1$,

where the first follows from Proposition 6(i), and the second from what we proved above. By Prop. 6(ii), these two give $\pi_{\perp x}(t) \ll x$ for all $t < 1$.

(ii) First, for any $x \in \Delta^n$, we have $\pi_{\perp x}(s) \ll \pi_{\perp x}(t)$ whenever $s < t$. This easily follows from (i): For $p := \pi_{\perp x}(t)$ we have

$$\pi_{\perp x}(s) = \pi_{\perp p}(s/t) \ll p = \pi_{\perp x}(t).$$

If $x \ll y$, then $x \sqsubseteq \pi_{\perp y}(t) \ll y$ for some $t < 1$. Thus,

$$x \sqsubseteq \pi_{\perp y}(t) \ll \pi_{\perp y}((t+1)/2) \ll y,$$

so taking $z := \pi_{\perp y}((t+1)/2)$ finishes the proof. □

The last result demonstrates the existence of a natural approximative structure on classical states: The dcpo Δ^n can rightfully be called a *domain*. As we said at the start, domains normally have *partial* elements, and *total* or ideal elements. We now explain the relationship between the qualitative notion of approximation $\ll$ and the natural intuitive notions of "partiality" and "totality" for classical states.

Intuitively, a classical state x is *partial* iff it offers no certainty about any outcome iff $(\forall i)\, 0 < x_i < 1$ iff $(\forall i)\, x_i > 0$. One may object that $x = (1/2, 1/2, 0) \in \Delta^3$ seems partial but is excluded from the above. However, only "some" of x is partial, the element $p_3(x) = \perp \in \Delta^2$. As a state in Δ^3, though, x is not genuinely partial because it imparts certainty about the third outcome.

On the other hand, if we *assume* that the order theoretic structure of Δ^n has captured our *intuitive* understanding of classical states, we easily arrive at an alternative formulation of partiality: An object is partial when it approximates something. The latter of course is *purely qualitative* and provides exactly what one hopes for: A formalization of intuition.

Lemma 13 (Partiality) *For each $x \in \Delta^n$, the set $\Uparrow\! x$ is nonempty iff $x_i > 0$ for all i.*

Proof If $x \ll y$, there there is $t < 1$ with $x \sqsubseteq \pi_{\perp y}(t)$. Because $t < 1$, $\pi_{\perp y}(t)_i > 0$, so degeneration (Lemma 5) gives $x_i > 0$. For the other direction, let $x_i > 0$ for all i.

Intuitively, because x is in the interior of Δ^n, the line segment from $\perp$ to x can be extended *nontrivially* to a point y on the boundary of Δ^n, for which we then have $x \ll y$. Formally now, we can assume $x \neq \perp$. Then

$$0 < x^- < 1/n \Rightarrow \lambda := \frac{1}{1 - nx^-} > 1.$$

Let y be the classical state defined pointwise by

$$y_i = \frac{1}{n} \cdot (1 - \lambda) + \lambda \cdot x_i$$

for each $1 \le i \le n$. To see that y is in fact a classical state, notice that

$$0 = \frac{1}{n} \cdot (1 - \lambda) + \lambda x^- \le y_i \le \sum_{i=1}^{n} y_i = 1.$$

Since $0 \le 1/\lambda < 1$, $\pi_{\perp y}(1/\lambda) = x \ll y$, which proves $\Uparrow\! x \neq \emptyset$. □

The "opposite" of partiality is totality: A classical state is *total* when it imparts certainty about *all* of its outcomes. Thus, the total or *ideal* classical states are exactly the pure states e_i, which we have already characterized qualitatively as being precisely $\max(\Delta^n)$. But the approximation relation can offer additional insight about the sense in which pure states are total.

To understand the connection between the two, let's begin by thinking about $x \ll y$, which we could say means that

- All paths (y_i) to y must *qualitatively* exceed x after some finite stage,

which can be read as

- All paths to y essentially begin with x,

and finally

- A process (y_i) can only end up in state $y = \bigsqcup y_i$ provided that it has the information represented by x: x is *necessary* for having y, i.e., the only way to know y is to first know x.

In each version of $\ll$ above, some reference to a process is made (a path is assumed to be generated by some process), providing us with a crucial distinction between $\ll$ and $\sqsubseteq$: $x \ll y$ is a statement about *processes*, $x \sqsubseteq y$ is a statement about information. The difference between these two becomes clear by considering states x, y, z with $x \ll y \sqsubseteq z$ but *not* $x \ll z$.

Example 2 Let $\perp \neq x \ll y := (1/2, 1/2, 0) \sqsubseteq z := e_1$. Then $x \not\ll z$. Here are two equivalent perspectives:

(i) In terms of knowledge: We are not *required* to know that an object is not in box 3 before we can know that it is in box 1.
(ii) In terms of processes:From an initial state of $\perp$, one way to conclude the object is in box 1 is to begin by ruling out box 3 as a possibility, and then look in one of the others—but this does not describe *all* ways. We could just look in box 1.

Thus, $\sqsubseteq$ makes statements about *potential* evolutions of state; $\ll$ is concerned with what we must know in order to obtain information using the process of observation.

This example suggests that $\ll$ is capable of expressing a characteristic of totality: The only time we expect the implication

$$(\forall y, z)\; y \in \mathord{\uparrow\!\!\!\uparrow} x \text{ and } y \sqsubseteq z \Rightarrow z \in \mathord{\uparrow\!\!\!\uparrow} x$$

to hold nontrivially is when x is a state from which a *unique* outcome is likely, i.e., x approximates a unique pure state. When $\mathord{\uparrow\!\!\!\uparrow} x$ satisfies the implication above, it is called an *upper set*.

Proposition 7 (Approximation of pure states) *Let* $n \geq 2$

(i) *For all* $x \in \Delta^n$, $x \ll e_i$ *iff* $x = \pi_{\perp e_i}(t)$ *for some* $t < 1$.
(ii) *For all* $x \in \Delta^n$, $\twoheaduparrow x$ *is an upper set iff it is empty, all of* Δ^n, *or contains a unique pure state.*

Proof (i) For $n = 2$ this is clear. Assume $n \geq 3$. Because $x \ll e_i$, there is $s < 1$ with $x \sqsubseteq \pi_{\perp e_i}(s)$. By degeneration, $(\exists a > 0)(\forall k \neq i)(x_k = a)$, which now makes the claim obvious.

(ii) For ($\Rightarrow$), every nonempty upper set contains at least one maximal element. By (i), either $x = \perp$, or $\twoheaduparrow x$ contains a unique pure state.

For the other direction, we need to prove that $\twoheaduparrow x$ is an upper set when it contains a unique pure state e_i. Suppose $x \ll y \sqsubseteq z$. First, because $x \ll e_i$, it is routine to show that $x_i = x^+$ and $x_k = x^-$ for all $k \neq i$. Because $\twoheaduparrow x$ contains a unique pure state, $x \neq \perp$, which means $x^+ > x^- > 0$. To apply Proposition 6(ii), we first show $z \in \Delta^n_\sigma \Rightarrow x \in \Delta^n_\sigma$.

Let $z \cdot \sigma$ be monotone. Because $x \sqsubseteq z$, $x \cdot \sigma \sqsubseteq z \cdot \sigma$, which means there is an index k with $(x \cdot \sigma)_k = x^+ \leq (z \cdot \sigma)_k = z^+$. By the monotonicity of $z \cdot \sigma$ and degeneration,

$$(z \cdot \sigma)_k = (z \cdot \sigma)_1 = z^+ \geq x^+ > 0 \Longrightarrow (x \cdot \sigma)_k = (x \cdot \sigma)_1 = x^+ > 0,$$

which means $x \cdot \sigma$ is monotone, since the only other value it assumes is x^-.

To finish, we need to show $r(x) \ll r(z)$ in Λ^n. First, $r(z)_2 > 0$, since otherwise $r(z) = e_1$, for which we already know $r(x) \ll r(e_i) = e_1 = r(z)$. By degeneration, this also means $r(y)_2 > 0$. Because $r(x) \ll r(y)$ in Λ^n, there is $t < 1$ with $r(x) \sqsubseteq \pi_{\perp r(y)}(t)$. Thus,

$$\frac{r(x)_1}{r(x)_2} \leq \frac{(1/n)(1-t) + tr(y)_1}{(1/n)(1-t) + tr(y)_2} < \frac{r(y)_1}{r(y)_2} \leq \frac{r(z)_1}{r(z)_2},$$

where the strict inequality follows from $r(y)_1 > r(y)_2 > 0$ (which is a consequence of degeneration using $r(x)_1 > r(x)_2 > 0$ and $r(x) \sqsubseteq r(y)$). Because $r(x)_i / r(x)_{i+1} = 1$ for $1 < i < n$, it is clear that $r(x) \ll r(z)$ in Λ^n. □

An approximation a of a pure state x defines a region $\twoheaduparrow a$ of Δ^n known in domain theory as a *Scott open set.*

Definition 12 A subset U of a dcpo D is *Scott open* if

- U is an *upper set*: $(\forall x \in U)(\forall y \in D)\, x \sqsubseteq y \Rightarrow y \in U$, and
- U is *inaccessible by directed suprema*: For any directed set $S \subseteq D$,

$$\bigsqcup S \in U \Rightarrow S \cap U \neq \emptyset\,.$$

The collection of all Scott open subsets of D is σ_D.

Notice that a map $f : D \to E$ between dcpo's is Scott continuous in the sense defined earlier iff $f^{-1}(U)$ is Scott open in D whenever U is Scott open in E.

Lemma 14 *For all* $x \in \Delta^n$, $\Uparrow x$ *is an upper set iff it is Scott open.*

Proof If $z = \bigsqcup S \in \Uparrow x$, then by interpolation (Theorem 4), there is $y \in \Delta^n$ with $x \ll y \ll z$. Thus, by $y \ll z$, there is $s \in S$ with $y \sqsubseteq s$, and since $\Uparrow x$ is an upper set, $s \in \Uparrow x$. Interestingly, one can also show that $\Uparrow x$ is Scott open iff $\uparrow(\Uparrow x)$ is Scott open.

The relation between approximation, partiality and purity can now be summarized as follows:

(i) The partial elements are those $x \in \Delta^n$ with $\Uparrow x \neq \emptyset$.
(ii) For a partial element $x \in \Delta^n$, $\Uparrow x$ is Scott open iff $x = \pi_{\perp e_i}(t)$ for some i and some $t < 1$ iff ($x = \perp$ or x approximates a unique pure state).

Thus, the "totality" of a pure state x is largely explained by the fact that $\Uparrow a$ is Scott open whenever $a \ll x$. To complete the picture,

Lemma 15 *A subset* $U \subseteq \Delta^n$ *is Scott open iff*

- *Any monotone path from* $x \in U$ *to a pure state lies in* U*, and*
- *The line from* $\perp$ *to* $x \in U$ *has a segment contained in* U*,*

and for pure states x, there is an equivalence between "approximation of x" and "Scott open set containing x": Given any $a \ll x$, the set $\Uparrow a$ is Scott open, while given any Scott open U with $x \in U$, we can (by exactness) find an approximation $a \in U$ of x with $x \in \Uparrow a \subseteq U$.

Approximation can also describe things of a more concrete nature. Because of its close connection to the mixing law, which is especially evident in the case of pure states (Proposition 7(i)), we can sometimes reinterpret mixing as approximation. This, for instance, can be useful when one seeks to explain the sense in which certain forms of noise work "against" the state σ of a system.

Example 3 The depolarization channel The map $d_p : \Delta^n \to \Delta^n$ by

$$d_p(\sigma) = p\perp + (1 - p)\sigma$$

describes the process by which a state $\sigma \in \Delta^n$ is *depolarized* with probability $p > 0$ (has all bias and hence all information removed from it) and is otherwise unaltered. But notice:

$$d_p(\sigma) = \pi_{\perp\sigma}(1 - p),$$

which means $d_p(\sigma) \ll \sigma$ for $p > 0$. In particular, the effect of depolarization on a state is *qualitative*.

To say that the effect of noise is qualitative essentially means that while the state of the system has suffered, it has not been "degraded beyond recognition." This is not always the case: Some forms of noise are more destructive than others and the order on classical states can at times capture this.

Example 4 Classical bit flipping A state $\sigma \in \Delta^2$ suffering the effect of a magnetic field is "flipped" with probability p and otherwise left alone

$$f_p(\sigma) = p\sigma^* + (1-p)\sigma,$$

where $*$ is the *involution* $(x, y)^* = (y, x)$. In this case, we have

$$(\forall \sigma . f_p(\sigma) \ll \sigma) \Leftrightarrow 0 < p \leq 1/2,$$

i.e., the effect of the noise is qualitative iff the field is weak enough.

Those familiar with classical information theory may know what we call classical bit flipping by another name, the *binary symmetric channel*. In this important example, a bit (a "0" or a "1") is transmitted correctly through a channel with probability $1 - p$ and reversed with probability p:

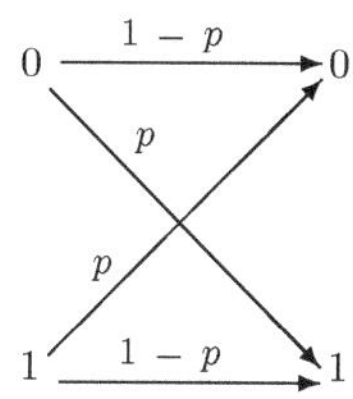

Given that information is sent through the binary symmetric channel, we want to determine the information that is actually received. The information sent is modelled by $\sigma = (x, y) \in \Delta^2$, where x is the probability that 0 is sent and y is the probability that 1 is sent. The effect that the channel has on information passing through it (σ) is captured by its *channel matrix*

$$\begin{bmatrix} 1-p & p \\ p & 1-p \end{bmatrix}$$

To determine the information received when (x, y) is sent, we calculate a distribution for the output using the channel matrix as follows:

$$\begin{bmatrix} 1-p & p \\ p & 1-p \end{bmatrix} \cdot \begin{bmatrix} x \\ y \end{bmatrix} = \begin{bmatrix} (1-p)x + py \\ px + (1-p)y \end{bmatrix}$$

All of this is implicit in the operator $f_p(\sigma) = p\sigma^* + (1-p)\sigma$ of Example 4: The distribution for the output is $f_p(\sigma)$, the 0 bit is $e_1 = (1, 0)$, the 1 bit is $e_2 = (0, 1)$, and reversing σ means applying the involution $*$ to obtain σ^*.

10.2.5 Entropy, Content and Partiality

We have already seen how the use of $\sqsubseteq$ on Δ^n enables a precise formulation of what it means to say that a classical state is "information." One of the advantages in taking this approach to *defining* information is that the structure of a domain can then be used to define the notion "information content," i.e., we can say what it *means* to measure the *content* of information.

The idea introduced in [8] is this: Assuming that information is formally specified as a domain, measuring content *means* measuring partiality, i.e., the amount of partiality in an object. The importance of this conceptually is that partiality, as we have already seen, is intimately connected to the order theoretic structure of a domain.

To slightly motivate the formal definition we are about to see, suppose that $\mu : \Delta^n \to [0, \infty)$ is a measure of content on classical states. Then μx is the amount of uncertainty (or partiality) in x. As we move up in the order $\sqsubseteq$ on Δ^n, states become more informative, so uncertainty *decreases*:

$$x \sqsubseteq y \Rightarrow \mu x \geq \mu y.$$

That is, as a map from Δ^n to $[0, \infty)^*$, μ is monotone. If μ is defined in terms of the usual formulae from physics (arithmetic, logarithms, other elementary functions), then it is continuous in the sense of analysis, and hence *Scott continuous* from Δ^n to $[0, \infty)^*$.

The essence of the distinction between content and a random continuous map on a domain is subtle. Consider a pure state $x \in \max(\Delta^n)$ and one of its approximations $a \ll x$, so that a is information any process must have before it can evolve to x. Then we also expect $a \ll y$ *provided* that

(i) y is a state from which it is possible to evolve to x, and
(ii) y is "close enough" to x in content.

The first translates as "$y \sqsubseteq x$"; the second translates as "$|\mu x - \mu y| < \varepsilon$," on the *assumption* that μ measures information content. Putting everything together now, *if* μ is a measure of content, *then* we expect that

$$x \in \mathord{\uparrow\!\!\uparrow} a \Rightarrow (\exists \varepsilon > 0)(y \sqsubseteq x \ \&\ |\mu x - \mu y| < \varepsilon \Rightarrow y \in \mathord{\uparrow\!\!\uparrow} a).$$

Because x is pure, we can replace $\mathord{\uparrow\!\!\uparrow} a$ with a Scott open set $U \subseteq \Delta^n$, as we saw in the last section.

Definition 13 A Scott continuous map $\mu : D \to [0, \infty)^*$ on a dcpo is said to *measure the content of* $x \in D$ if

$$x \in U \Rightarrow (\exists \varepsilon > 0)\, x \in \mu_\varepsilon(x) \subseteq U,$$

whenever $U \in \sigma_D$ is Scott open and

$$\mu_\varepsilon(x) := \{y \in D : y \sqsubseteq x \;\&\; |\mu x - \mu y| < \varepsilon\}$$

are the elements ε close to x in content. The map μ *measures* X if it measures the content of each $x \in X$.

In order for a map μ to be regarded "a measure of content," it must *minimally* be capable of distinguishing those elements which *it claims* are maximally informative. That is, μ must measure all of the objects which it regards as possessing no uncertainty $\ker \mu := \{x : \mu x = 0\}$.

Definition 14 A *measurement* is a Scott continuous map $\mu : D \to [0, \infty)^*$ on a dcpo that measures $\ker \mu := \{x \in D : \mu x = 0\}$.

The measurement formalism [8] teaches that the ability to measure content is indicative of a purely structural relationship that exists between two classes of informative objects. Neither class need consist of numbers. This relationship is formally expressed by a map $\mu : D \to E$ whose general nature is to *reflect* properties of simpler objects E onto more complex objects D.

The motivation for the idea stems from the empirical fact that it is often easier to reason about D in terms of E rather than deal with D directly [8]. Hence the reflective nature of μ: It confirms that we *actually can* learn about $x \in D$ by studying the properties of its simplification $\mu x \in E$.

Definition 15 A Scott continuous map $\mu : D \to E$ between dcpo's is said to *measure the content of* $x \in D$ if

$$x \in U \Rightarrow (\exists \varepsilon \in \sigma_E)\, x \in \mu_\varepsilon(x) \subseteq U,$$

whenever $U \in \sigma_D$ is Scott open and

$$\mu_\varepsilon(x) := \mu^{-1}(\varepsilon) \cap \downarrow\! x$$

are the elements ε close to x in content. The map μ *measures* X if it measures the content of each $x \in X$.

Definition 16 A *measurement* is a Scott continuous map $\mu : D \to E$ between dcpo's that measures $\ker \mu := \{x \in D : \mu x \in \max(E)\}$.

These definitions are easily seen to be equivalent to the quantitative formulations we saw earlier by setting $E = [0, \infty)^*$. To establish the reflective nature of content, we use the following relationship between the order $\sqsubseteq$ on a dcpo D and its Scott open sets σ_D:

$$x \sqsubseteq y \Leftrightarrow (\forall U \in \sigma_D)(x \in U \Rightarrow y \in U).$$

Proposition 8 *Let* $\mu : D \to E$ *be a measurement and* x *an object that it measures.*

(i) *If* $\mu x \in \max(E)$*, then* $x \in \max(D)$.

(ii) *If* $\mu x = \perp$, *then* $x = \perp$, *provided* $\perp \in D$ *exists.*
(iii) *If* $y \sqsubseteq x$ *and* $\mu x = \mu y$, *then* $x = y$.
(iv) *If* $x_n \sqsubseteq x$ *and* (μx_n) *is directed with supremum* μx, *then* $\bigsqcup x_n = x$.

In addition, the composition of measurements is again a measurement.

Proof The proofs here are essentially taken verbatim from [8], where other properties of content can be found.

(i) Let $x \in U$. Then $y \in \mu_\varepsilon(x) \subseteq U$, for some $\varepsilon \in \sigma_E$. Since U was arbitrary, $x \sqsubseteq y$. By antisymmetry, $x = y$.
(ii) If $x \sqsubseteq y$, then $\mu x = \mu y$, since $\mu x \in \max(E)$, which gives $y \in \ker \mu$. Since μ is a measurement, it measures y, so $x = y$ by (iii).
(iii) First, $\mu(\perp) \sqsubseteq \mu x = \perp$, so $\mu(\perp) = \perp = \mu x$. Since $\perp \sqsubseteq x$, we can apply (i) to obtain $x = \perp$.
(iv) Let $x_n \sqsubseteq u$ for all n. If $x \in U$, then $x \in \mu_\varepsilon(x) \subseteq U$, which means

$$\mu x = \bigsqcup \mu x_n \in \varepsilon,$$

and so $\mu x_n \in \varepsilon$ for some n, which gives $x_n \in U$ and hence $u \in U$. Since U was arbitrary, $x \sqsubseteq u$. Thus, $\bigsqcup x_n = x$.

Finally, if we have measurements $D \xrightarrow{\mu} E \xrightarrow{\lambda} F$, then $\lambda\mu$ measures $\ker \lambda\mu$ as follows. First, if $x \in \ker \lambda\mu$ and $x \in U \in \sigma_D$, then $x \in \ker \mu$ so there is $\varepsilon \in \sigma_E$ with $x \in \mu_\varepsilon(x) \subseteq U$. Then, since $\mu x \in \varepsilon$ and $\mu x \in \ker \lambda$, there is $\delta \in \sigma_F$ with $\mu x \in \lambda_\delta(\mu x) \subseteq \varepsilon$. We have

$$x \in (\lambda\mu)^{-1}(\delta) \cap \downarrow\! x \subseteq \mu_\varepsilon(x) \subseteq U,$$

which finishes the proof. □

With the benefit of the abstract formulation of content, let us take a second look at uncertainty ($E = [0, \infty)^*$). By Proposition 8(i), we know that

$$\mu x = 0 \Rightarrow x \in \max(D),$$

for any measurement $\mu : D \to [0, \infty)^*$. That is, quantitative certainty implies qualitative certainty. As a case in point, if $D = \Delta^n$, then, as we will see shortly, Shannon entropy $\mu : D \to [0, \infty)^*$ given by

$$\mu x = -\sum_{i=1}^{n} x_i \log x_i$$

is a measurement. Thus, any classical state x with entropy $\mu x = 0$ is *pure*. But now we have an explanation for why such properties hold:

(i) In the sense of the measurement formalism, μ is a measure of content between the domains Δ^n and $[0, \infty)^*$, and
(ii) Measures of content between domains *always* reflect maximality.

The same is true of the von Neumann entropy on quantum states (that we will see later). But the moral of the last result is what is most important: *Subject to moderate hypotheses, information behaves in the same manner as its content.*

Proposition 9 *The natural retraction $r : \Delta^n \to \Lambda^n$ is a measurement.*

Proof To start, notice that $\ker r = \max(\Delta^n)$. Let $U \subseteq \Delta^n$ be a Scott open set that contains the pure state x. By exactness, there is $\bot \neq a \ll x$ with $a \in U$. By Prop. 6(ii), $r(a) \ll r(x)$ in Λ^n. Because x is pure, $\varepsilon := \Uparrow r(a)$ is a Scott open subset of Λ^n (a corollary of Theorem 4 and Proposition 7). We claim $x \in r_\varepsilon(x) \subseteq \Uparrow a \subseteq U$ as follows.

First, $x \in r_\varepsilon(x)$ by $r(a) \ll r(x)$. Then, if $y \in r_\varepsilon(x)$, we have $r(a) \ll r(y)$ in Λ^n and $y \sqsubseteq x$. To prove that $a \ll y$ in Δ^n and finish the proof, we must show $y \in \Delta^n_\sigma \Rightarrow a \in \Delta^n_\sigma$.

For this subtle point, a takes its maximum at a unique index, because $a \neq \bot$ and it approximates a pure state (Proposition 7(i)). Then $r(a)$ does as well. Since $r(a) \sqsubseteq r(y)$, degeneration implies the same is true of $r(y)$ and hence of y. Thus, because y takes its maximum at a unique index, and because $y \sqsubseteq x \in \max(\Delta^n)$, we have $y \in \Delta^n_\sigma \Rightarrow x \in \Delta^n_\sigma$, while $a \ll x$ then implies $a \in \Delta^n_\sigma$. □

We have made intuitive use of this fact numerous times: Whenever we prove a statement about classical states by first proving it for monotone states, we are *implicitly* appealing to the fact that $r(x)$ provides a decent measure of the content of x.

Example 5 The standard variable $v : \Delta^n \to [0, \infty)^*$ given by

$$v(x) = 1 - x^+$$

is a measurement with $\ker v = \max(\Delta^n)$. To prove as much, we need only show that its restriction to Λ^n, $\lambda := v|_{\Lambda^n}$, is a measurement, since then $v = \lambda \circ r$ must be another.

To this end, let $U \subseteq \Lambda^n$ be a Scott open set and $x \in \ker \lambda$. Because U is Scott open, there is $t < 1$ with $a := \pi_{\bot x}(t) \in U$. We then have

$$x \in \lambda_\varepsilon(x) \subseteq \uparrow a \subseteq U,$$

where

$$\varepsilon := \frac{1}{2} \cdot \frac{a_2}{a_1 + a_2} > 0.$$

Example 6 The entropy $s : \Delta^n \to [0, \infty)^*$ given by

$$s(x) = -\log x^+$$

is a measurement with $\ker s = \max(\Delta^n)$. First, $s(x) \geq \upsilon(x)$, using the classic inequality $\log t \leq t - 1$ for $t > 0$. Thus,

$$x \in s_\varepsilon(x) \subseteq \upsilon_\varepsilon(x),$$

for any pure state x and $\varepsilon > 0$. Because υ is a measurement, so is s.

Now for Shannon entropy.

Lemma 16 *Let $x \sqsubseteq y$ be monotone classical states in Δ^n. Then there is $k \in \{1, \ldots, n\}$ such that*

(i) $(\forall i < k)\, x_i \leq y_i$, *and*
(ii) $(\forall i \geq k)\, x_i \geq y_i$.

Proof First, since $x \sqsubseteq y$, we have by induction that $x_i y_{i+j} \leq y_i x_{i+j}$, for each $j \in \{0, \ldots, n-i\}$. Thus, if $x_i \geq y_i$, then $x_{i+j} \geq y_{i+j}$ for each $j \in \{0, \ldots, n-i\}$. Now let k be the least integer $1 \leq k \leq n$ with $x_k \geq y_k$. Notice that such a k exists since $x_n \geq y_n$. This finishes the proof.

The *relative Shannon entropy* of y given x is

$$\mu(y\|x) := \sum_{i=1}^{n} y_i \log(y_i/x_i)$$

where $x, y \in \Delta^n$. This quantity is always nonnegative and is zero iff $x = y$.

Theorem 5 *Let $\mu : \Delta^n \to [0, \infty)^*$ be the Shannon entropy on classical states*

$$\mu x = -\sum_{i=1}^{n} x_i \log x_i$$

where the logarithm is natural. Then μ is a measurement. In addition,

(i) *For all $x, y \in \Delta^n$, if $x \sqsubseteq y$ and $\mu(x) = \mu(y)$, then $x = y$.*
(ii) *For all $x \in \Delta^n$, we have $\mu(x) = 0$ iff $x \in \max(\Delta^n)$.*
(iii) *For all $x \in \Delta^n$, we have $\mu(x) = \log n$ iff $x = \bot$.*

Proof Because μ is symmetric, its Scott continuity follows if we show that its restriction to the dcpo Λ^n is Scott continuous. First we prove its monotonicity into $[0, \infty)^*$.

Let $x \sqsubseteq y$ be monotone classical states. By Lemma 16, there is an integer $k \in \{1, \ldots, n\}$ such that $x_i \leq y_i$ for $i < k$ and $x_i \geq y_i$ for $i \geq k$. Then

$$\sum_{i=1}^{n}(y_i - x_i)\log x_i = \sum_{i<k}(y_i - x_i)\log(x_i/x_k) + \sum_{i>k}(y_i - x_i)\log(x_i/x_k) \geq 0.$$

Notice that if $x_k = 0$ then the sum of the $i > k$ vanishes, while the sum of the $i < k$ blows up, but is nevertheless nonnegative. From the nonnegativity of this sum, we have

$$\mu x \geq -\sum_{i=1}^{n} y_i \log x_i \geq \mu y,$$

where the second inequality follows from $\mu(y\|x) \geq 0$. This proves that μ is monotone into $[0, \infty)^*$.

If in addition to $x \sqsubseteq y$ we also have $\mu x = \mu y$, then the inequality above immediately gives $\mu(y\|x) = 0$, which implies $x = y$. This establishes that μ is strictly monotone. For its Scott continuity, if (x_i) is increasing, then

$$\begin{aligned}\mu\left(\bigsqcup x_i\right) &= \mu(\lim_{i\to\infty} \pi_1(x_i), \ldots, \lim_{i\to\infty} \pi_n(x_i)) \\ &= \lim_{i\to\infty} \mu(\pi_1(x_i), \ldots, \pi_n(x_i)) \\ &= \lim_{i\to\infty} \mu x_i,\end{aligned}$$

where the first equality uses Proposition 12 and the second uses the continuity of μ with respect to the Euclidean topology. By Lemma 1, μ is Scott continuous. Finally, μ is a measurement: For $x \in \Delta^n$, we have

$$\mu x \geq -x^+ \log x^+ \geq \frac{1}{n} \cdot v(x),$$

using $\log t \leq t-1$ for $t > 0$ and $x^+ \geq 1/n$, where v is the variable from Example 5. Since v is a measurement, $(1/n)\cdot v$ is a measurement, which means that μ is as well. □

It is important to realize that the minimal account of content given here is more substantial than it may seem: There are natural mappings which do not measure content.

Example 7 Numbers are not enough. For $n \geq 3$, consider

$$f : \Delta^n \to [0, \infty)^* :: x \mapsto x^-.$$

It is Scott continuous, symmetric and assumes its order theoretic minimum at $\perp$. Furthermore, even though $f(x) = 0$ for all $x \in \max(\Delta^n)$, f *does not* measure the content of a single pure state.

For instance, suppose f measured the content of $e_1 \in \Delta^3$. Then given any open $U \subseteq \Delta^n$ with $e_1 \in U$, there would exist $\varepsilon > 0$ with $e_1 \in f_\varepsilon(e_1) \subseteq U$. Then $(1/2, 1/2, 0) \in U$. But because this applies to *any* open set U, we now have a proof that $e_1 \sqsubseteq (1/2, 1/2, 0)$.

More intuitively: Many states are assigned maximal measure by f which are not pure. For instance $f(x, y, 0) = 0$ on Δ^3, even though the only time $(x, y, 0)$ is pure is when $x = 1$ or $y = 1$.

Here is a summary.

Example 8 Canonical measures of content on Δ^n

(i) The maps $\Delta^n \to [0, 1] :: x \mapsto x^+$ and $\Delta^n \to [0, 1]^* :: x \mapsto 1 - x^+$.
(ii) Entropy $s(x) = -\log x^+$.
(iii) The natural retraction $r : \Delta^n \to \Lambda^n$.
(iv) Shannon entropy

$$\mu x = -\sum_{i=1}^{n} x_i \log x_i .$$

10.3 Quantum States

We now pursue the idea which motivated our study of the Bayesian order on classical states: The *spectral order* on quantum states. Later we will see that the spectral order can be characterized in a manner completely analogous to the order on classical states:

- The *inductive formulation*, in terms of quantum projections, and
- The *symmetric formulation*, in terms of unitary transformations.

These two accounts of the quantum order, when restricted to a class of states exhibiting classical behavior, are equivalent to the inductive and symmetric characterizations of the Bayesian order on classical states studied in the last section.

10.3.1 Essentials

An n-dimensional complex Hilbert space $\mathcal{H}^n$ is an n-dimensional vector space over $\mathbb{C}$ with specified inner product $\langle \cdot \mid \cdot \rangle$.

Definition 17 A *base* of $\mathcal{H}^n$ is a sequence $(\psi_i)_{i=1}^n$ of *unit vectors,*

$$\langle \psi_i \mid \psi_i \rangle = 1,$$

which are *mutually orthogonal*:

$$i \neq j \Rightarrow \langle \psi_i \mid \psi_j \rangle = 0.$$

We write $x \perp y$ to express the orthogonality of two vectors $x, y \in \mathcal{H}^n$, and as is customary, extend this notation to subspaces of $\mathcal{H}^n$ as follows:

$$\Psi \perp \Phi \;\Leftrightarrow\; \forall \psi \in \Psi \setminus \{o\}, \forall \phi \in \Phi \setminus \{o\} : \psi \perp \phi$$

where o is the zero of $\mathcal{H}^n$.

Definition 18 A linear operator $\rho : \mathcal{H}^n \to \mathcal{H}^n$ is *self-adjoint* if

$$\langle \phi \mid \rho\psi \rangle = \langle \rho\phi \mid \psi \rangle,$$

for all $\phi, \psi \in \mathcal{H}^n$, *positive* when

$$\langle \psi \mid \rho\psi \rangle \geq 0$$

for all $\psi \in \mathcal{H}^n$, and *idempotent* when $\rho^2 := \rho \circ \rho = \rho$.

The *spectral theorem* of von Neumann [12], roughly speaking, states that each self-adjoint operator on a Hilbert space decomposes into a sum of simple operators called *projections.*

Definition 19 A *projection* or *projector* is a self-adjoint, linear, idempotent operator. The set of projections is denoted $\mathbb{P}^n$. A projection $P \in \mathbb{P}^n$ is fully characterized by its subspace of fixed points $\mathsf{fix}(P) \subseteq \mathcal{H}^n$.

All we need here is the finite dimensional case of the spectral theorem.

Theorem 6 *A self-adjoint linear operator $\rho : \mathcal{H}^n \to \mathcal{H}^n$ decomposes uniquely into a linear combination of mutually orthogonal projections*

$$\rho = \sum_{\lambda \in \mathsf{spec}(\rho)} \lambda \cdot P_\rho^\lambda \quad \text{with} \quad \sum_{\lambda \in \mathsf{spec}(\rho)} P_\rho^\lambda = I$$

whose images span $\mathcal{H}^n$. The set $\mathsf{spec}(\rho) \subseteq \mathbb{R}$ is called the spectrum of ρ.

We write the fact that the images of the projections span $\mathcal{H}^n$ as

$$\mathsf{span}\left(\bigcup_{\lambda \in \mathsf{spec}(\rho)} \mathsf{fix}(P_\rho^\lambda) \right) = \mathcal{H}^n\,, \tag{10.1}$$

where by idempotence we have $\mathsf{fix}(P) = \mathsf{Im}(P) = P(\mathcal{H}^n)$.

Definition 20 The *trace* of a linear operator ρ on $\mathcal{H}^n$ is

$$\mathsf{tr}(\rho) := \sum_i \langle \psi_i \mid \rho\psi_i \rangle,$$

where $\{\psi_i\}$ is any base of $\mathcal{H}^n$. If A is *any* matrix representation of ρ, then $\mathsf{tr}(\rho) = \sum A_{ii}$ is the sum of the elements on the diagonal of A.

The standard kinematical account of a quantum system includes both a description of the *states* a system can take, and of its *observables*, i.e., the *measurements* that can be performed on the system.

Definition 21 A *density operator* ρ on $\mathcal{H}^n$ is a self-adjoint, positive, linear operator with $\mathsf{tr}(\rho) = 1$. A *quantum n-state* is a density operator. The class of quantum n-states is denoted Ω^n.

Definition 22 A quantum state ρ is *pure* if $\mathsf{spec}(\rho) \subseteq \{0, 1\}$. The set of pure states is written Σ^n.

A classical state is a distribution on the set of pure states $\max(\Delta^n)$. Similarly, Gleason's theorem [5] establishes that density operators encode precisely the measures on the closed subspaces of $\mathcal{H}^n$, i.e., density operators are distributions on the set of pure states.

Definition 23 A *quantum n-measurement* is a self-adjoint linear operator $e : \mathcal{H}^n \to \mathcal{H}^n$.

For instance, if e is the energy observable, then its spectrum $\mathsf{spec}(e)$ contains the actual energy values a system can assume. According to quantum mechanics, if the density operator of a system is ρ, then a measurement of the observable e yields $\lambda \in \mathsf{spec}(e)$ as the result with probability

$$\mathsf{prob}_e^\lambda(\rho) := \mathsf{tr}(P_e^\lambda \cdot \rho).$$

Now what we want to do is rewrite all of this in a form more amenable to the task at hand.

Definition 24 $\mathbb{L}^n$ is the set of closed subspaces of $\mathcal{H}^n$.

By the spectral theorem, we can write a self-adjoint operator e as

$$e\psi = \sum_{\lambda \in \mathsf{spec}(e)} (\lambda \cdot P_e^\lambda)\psi.$$

By mutual orthogonality, $e\psi = \lambda\psi \Leftrightarrow P_e^\lambda\psi = \psi$, so the eigenspaces

$$e_\lambda := \{\psi \in \mathcal{H}^n \mid e\psi = \lambda\psi\} = \mathsf{fix}(P_e^\lambda)$$

give rise to a labeled collection of mutually orthogonal subspaces

$$\mathcal{D}_e := \{e_\lambda \mid \lambda \in \mathsf{spec}(e)\}$$

which span $\mathcal{H}^n$.

Definition 25 A *decomposition* of $\mathcal{H}^n$ is a family of mutually orthogonal subspaces of $\mathcal{H}^n$ of dimension at least one which span $\mathcal{H}^n$. The decompositions of $\mathcal{H}^n$ are denoted $\mathbb{D}^n$.

We will also refer to the union $\bigcup \mathcal{D}$ of a decomposition $\mathcal{D}$ as being the decomposition itself since the first characterizes the latter.

Definition 26 A *spectral decomposition* of $\mathcal{H}^n$ is an injective function $f : X \to \mathbb{L}^n$ defined on a nonempty set $X \subseteq \mathbb{R}$ with $f(X) \in \mathbb{D}^n$. The domain of f is written $\mathsf{spec}(f) = X$ and called the *spectrum* of f.

Equivalently, a spectral decomposition is a partial injection $f : \mathbb{R} \rightharpoonup \mathbb{L}^n$ with $\mathsf{Im}(f) \in \mathbb{D}^n$ and $\mathsf{spec}(f) := \mathsf{dom}(f)$.

Lemma 17 *There is a one to one correspondence between self-adjoint operators on $\mathcal{H}^n$ and spectral decompositions of $\mathcal{H}^n$.*

Thus, we frequently use the operator and decomposition language interchangeably. For example, here is an alternate formulation of quantum states:

Definition 27 A *density operator* is a spectral decomposition r with

$$\sum_{\lambda \in \mathsf{spec}(r)} \lambda \cdot \mathsf{dim}(r_\lambda) = 1$$

and $\mathsf{spec}(r) \subseteq [0, \infty)$.

In particular, a *pure state* $r \in \Sigma^n$ is a decomposition $r : \{0, 1\} \to \mathbb{L}^n$ with

$$\sum_{\lambda \in \{0,1\}} \lambda \cdot \mathsf{dim}(r_\lambda) = 1.$$

From this equation we see that the subspace r_1 is one-dimensional. In fact, r_1 serves to characterize r, since r_0 must then be a certain $n-1$ dimensional subspace known as the *orthocomplement* of r_1,

$$r_0 = r_1^{\perp} := \{\psi \in \mathcal{H}^n \mid \psi \perp r_1\}.$$

We have proven the following.

Lemma 18 *The pure states on $\mathcal{H}^n$ are in bijective correspondence with the one dimensional subspaces of $\mathcal{H}^n$.*

So much for states. For observables, we will consider only those e on $\mathcal{H}^n$ with the maximum number of distinguishable outcomes n. By simple renaming then, we can take $\mathsf{spec}(e) = \{1, \ldots, n\}$. This convention highlights the role played by measurements: They are *labelings*, i.e., to each outcome $1 \leq i \leq n$, a measurement assigns those states e_i of the system for which observable e has value i with certainty (probability one).

Definition 28 A *labeling* is a spectral decomposition

$$e : \{1, \ldots, n\} \to \mathbb{L}^n.$$

Notice that non-degeneration of $\mathsf{spec}(e)$ implies that the decomposition $\mathcal{D}_e$ consists only of one-dimensional subspaces (pure states), i.e., $\mathcal{D}_e$ cannot be refined any further:

Definition 29 A decomposition $\mathcal{D}$ is a *refinement* of decomposition $\mathcal{D}'$ iff

$$\bigcup \mathcal{D} \subseteq \bigcup \mathcal{D}' .$$

Finally, the probabilities. Recall that the probability of obtaining outcome i in a measurement of observable e on a system with density operator r is given by

$$\mathsf{prob}_e^i(r) := \mathsf{tr}(P_e^i \cdot r) .$$

For a state r and a labeling e, $\langle r|e_i\rangle$ denotes the ith diagonal element of the matrix representation of r when expressed in a base B in which all P_e^i diagonalize, and thus, by the spectral decomposition theorem, in which e itself diagonalizes. Writing $P_e^i \cdot r$ in base B then yields

$$\begin{pmatrix} 0 & & & & & & \\ & \ddots & & & 0 & & \\ & & 0 & & & & \\ & & & 1 & & & \\ & & & & 0 & & \\ & 0 & & & & \ddots & \\ & & & & & & 0 \end{pmatrix} \cdot \begin{pmatrix} \langle r|e_1\rangle & & & & \\ & \ddots & & ? & \\ & & \langle r|e_i\rangle & & \\ & ? & & \ddots & \\ & & & & \langle r|e_n\rangle \end{pmatrix} = \begin{pmatrix} 0 & & & ? & & & \\ & \ddots & & \vdots & & 0 & \\ & & 0 & ? & & & \\ ? & \cdots & ? & \langle r|e_i\rangle ? & \cdots & ? & \\ & & & ? & 0 & & \\ & 0 & & \vdots & & \ddots & \\ & & & ? & & & 0 \end{pmatrix}$$

and thus

$$\mathsf{tr}(P_e^i \cdot r) = \langle r|e_i\rangle ,$$

from which we conclude

$$\mathsf{prob}_e^i(r) = \langle r|e_i\rangle . \tag{10.2}$$

In particular, $\langle r|e_i\rangle$ does *not* depend on B, but only on r and e_i, since it is equal to $\mathsf{tr}(P_e^i \cdot r)$. Further, since $\mathsf{tr}(r) = 1$,

$$\sum_{i=1}^{n} \langle r|e_i\rangle = 1,$$

which simply says that a measurement for observable e yields some outcome $i \in \mathsf{spec}(e)$ with probability one.

Definition 30 For a state r and labeling e, we define

$$\mathsf{spec}(r|e) := (\langle r|e_1\rangle, \ldots, \langle r|e_n\rangle) \in \Delta^n .$$

Notice that $\mathsf{spec}(r|e)$ is a *list*, while $\mathsf{spec}(r)$ is a *set*.

In general, $\mathsf{spec}(r|e)$ may not consist of eigenvalues of r, i.e., elements of $\mathsf{spec}(r)$. However, if r also diagonalizes in base B, then its diagonal consists of eigenvalues of r. And this is the case we are most interested in.

Definition 31 A state r *admits* a labeling e if

$$\mathsf{Im}(\mathsf{spec}(r|e)) = \mathsf{spec}(r).$$

Any state admits at least $n!$ labelings, corresponding with different permutations of $\mathsf{spec}(r|e)$. More generally, the following result tells us exactly when a labeling yields the spectrum of a state.

Proposition 10 *The following are equivalent for state r and labeling e:*

- *r admits labeling e.*
- *$\mathcal{D}_e$ is a refinement of $\mathcal{D}_r$.*
- *r diagonalizes in a base B in which e is diagonal.*
- *r and e commute, that is, $[r, e] = r \cdot e - e \cdot r = 0$.*

The following are equivalent for states r and s:

- *They admit a joint labeling e.*
- *They admit joint refinement $\mathcal{D}$.*
- *They diagonalize in a common base B.*
- *They commute, that is, $[r, s] = 0$.*

Additionally, states r and s admit labeling e iff

$$[r, s] = [r, e] = [s, e] = 0\,.$$

In particular we have in all the above cases that

$$B \subset \bigcup \mathcal{D}_e \subseteq \bigcup \mathcal{D} \subseteq \bigcup \mathcal{D}_r \cap \bigcup \mathcal{D}_s \subseteq \bigcup \mathcal{D}_r$$

whenever one of the inclusions applies.

Proof Given a base B, for all $\psi \in B$ we have $\psi \in \bigcup \mathcal{D}_e$ iff all $\psi \in B$ are eigenvectors of e iff e diagonalizes in the base B. Thus, any self-adjoint operator e diagonalizes in a base B iff $B \subseteq \bigcup \mathcal{D}_e$.

We already showed above that r admits labeling e when there exists a base B in which both r and e diagnalize, that is, whenever B is included both in $\bigcup \mathcal{D}_r$ and $\bigcup \mathcal{D}_e$ and as such

$$B \subseteq \bigcup \{\mathsf{span}(\psi) \mid \psi \in B\} \subseteq \bigcup \mathcal{D}_r$$

where since e is non-degenerated we have

$$\{\mathsf{span}(\psi) \mid \psi \in B\} = \mathcal{D}_e$$

and thus $\bigcup \mathcal{D}_e \subseteq \bigcup \mathcal{D}_r$. Two states r and s then admit a joint labeling e whenever

$$\bigcup \mathcal{D}_e \subseteq \bigcup \mathcal{D}_r \cap \bigcup \mathcal{D}_s \, .$$

The converses of these derivations is obvious. Whenever r and s admit diagonalization in a common base B, when representing them in B commutation reduces to commutation of reals. For the other results with respect to commutation, in particular the fact that self-adjoint operators diagonalize in a common base if they commute, we refer to relevant literature.

Lemma 19 *Let* r *be a state and* e *be a labeling with* $[r, e] = 0$. *Then*

$$\langle r|e_i\rangle = \lambda \Leftrightarrow \psi_i \in r_\lambda \Leftrightarrow e_i \subseteq r_\lambda \tag{10.3}$$

and

$$\mathsf{dim}(r_\lambda) = \mathsf{card}(\{1 \leq i \leq n \mid \langle r|e_i\rangle = \lambda\}) \, . \tag{10.4}$$

In particular, $\mathsf{dim}(r_\lambda)$ *does not depend on the choice of* e*, so neither do the multiplicities of eigenvalues.*

Finally, the following result is indispensable and we will appeal to it time and time again (often implicitly).

Lemma 20 (Definability) *For any labeling* e *and classical state* x*, there is a unique quantum state* $r \in \Omega^n$ *with* $[r, e] = 0$ *and* $\mathsf{spec}(r|e) = x$.

Although the notions decomposition, refinement and labeling as well as the representation of states and measurements as maps that label subspaces in terms of spectra are not standard in orthodox quantum theory [12], they prove to be useful in our setting since they highlight degeneration of spectra, a fundamental ingredient in the ordering of both classical and quantum states.

10.3.2 A Partial Order on Quantum States

Here is the *spectral order* on quantum states Ω^n.

Definition 32 For states $r, s \in \Omega^n$, we write $r \sqsubseteq s$ iff there exists a labeling e such that

- e is admitted both by r and s,
- $\mathsf{spec}(r|e) \sqsubseteq \mathsf{spec}(s|e)$ in Δ^n.

Though the order on quantum states only requires that *there exist* a single joint labeling, it nevertheless applies to *all* labels shared by r and s. This is like the way that $x \sqsubseteq y$ for classical states implies $x \cdot \sigma \sqsubseteq y \cdot \sigma$, for any $\sigma \in S(n)$ with $x, y \in \Delta^n_\sigma$.

Proposition 11 *If $r \sqsubseteq s$ in Ω^n, then $\mathsf{spec}(r|e) \sqsubseteq \mathsf{spec}(s|e)$ in Δ^n, for any labeling e with $[r, e] = [s, e] = 0$.*

Proof We prove the equivalent statement that

$$\mathsf{spec}(r|e) \sqsubseteq \mathsf{spec}(s|e) \Leftrightarrow \mathsf{spec}(r|e') \sqsubseteq \mathsf{spec}(s|e') \tag{10.5}$$

whenever $[r, e] = [s, e] = [r, e'] = [s, e'] = 0$. Since

$$\begin{aligned} \bigcup \mathcal{D}_r \cap \bigcup \mathcal{D}_s &= \bigcup \{r_\lambda \mid \lambda \in \mathsf{spec}(r)\} \cap \bigcup \{s_{\lambda'} \mid \lambda' \in \mathsf{spec}(s)\} \\ &= \bigcup \{r_\lambda \cap s_{\lambda'} \mid \lambda \in \mathsf{spec}(r)\,, \lambda' \in \mathsf{spec}(s)\}\,, \end{aligned}$$

and, since whenever $[r, e] = [s, e] = 0$ we have

$$\bigcup \mathcal{D}_e \subseteq \bigcup \mathcal{D}_r \cap \bigcup \mathcal{D}_s$$

by Proposition 10, it follows that

$$\bigcup \mathcal{D}_e \subseteq \bigcup \{r_\lambda \cap s_{\lambda'} \mid \lambda \in \mathsf{spec}(r)\,, \lambda' \in \mathsf{spec}(s)\}\,,$$

where, since $r_\lambda \perp r_{\lambda'}$ and $s_\lambda \perp s_{\lambda'}$ for $\lambda \neq \lambda'$, the subspaces $r_\lambda \cap s_{\lambda'}$ are mutually orthogonal for non-coincideng labels (λ, λ') and thus mutually exclusive. Since their union includes $\bigcup \mathcal{D}_e$ they span $\mathcal{H}^n$, so they constitute a decomposition

$$\mathcal{D}_{r,s} := \{r_\lambda \cap s_{\lambda'} \mid \lambda \in \mathsf{spec}(r)\,, \lambda' \in \mathsf{spec}(s)\}$$

with $\mathcal{D}_e$ as a refinement.

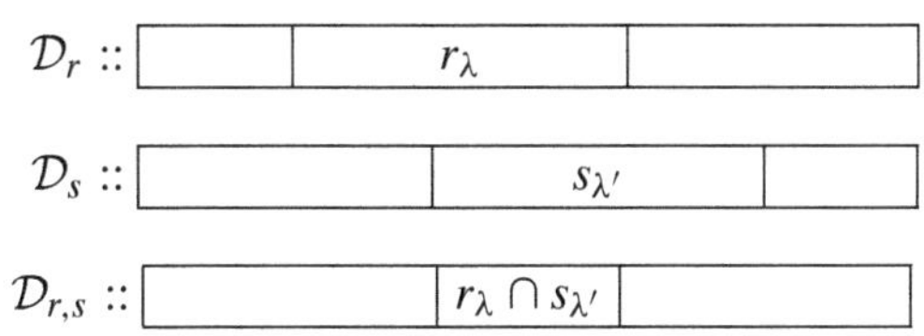

Since $\mathcal{D}_e$ is a refinement of $\mathcal{D}_{r,s}$ it also follows that

$$\mathsf{dim}(r_\lambda \cap s_{\lambda'}) = \mathsf{card}\left(\{i \in \{1, \ldots, n\} \mid e_i \subseteq r_\lambda \cap s_{\lambda'}\}\right)$$

where the quantity on the left does not depend on e. Since,

$$\begin{aligned} e_i \subseteq r_\lambda \cap s_{\lambda'} &\Leftrightarrow e_i \subseteq r_\lambda, e_i \subseteq s_{\lambda'} \\ &\Leftrightarrow \lambda = \langle r|e_i\rangle, \lambda' = \langle s|e_i\rangle \\ &\Leftrightarrow (\langle r|e_i\rangle, \langle s|e_i\rangle) = (\lambda, \lambda') \end{aligned}$$

for $(\lambda, \lambda') \in \mathsf{spec}(r) \times \mathsf{spec}(s)$, we have

$$\mathsf{card}\Big(\big\{i \in \{1, \ldots, n\} \mid (\langle r|e_i\rangle, \langle s|e_i\rangle) = (\lambda, \lambda')\big\}\Big) = \mathsf{dim}(r_\lambda \cap s_{\lambda'})\,.$$

Thus, writing

$$\mathsf{spec}(r, s|e) = \Big((\langle r|e_1\rangle, \langle s|e_1\rangle)\,,\ \ldots\ ,(\langle r|e_n\rangle,\ \langle s|e_n\rangle)\Big)$$

it follows that the list $\mathsf{spec}(r, s|e)$ contains a fixed collection of elements

$$\Big(\ldots, \underbrace{(\lambda, \lambda')\,,\ \ldots\ , (\lambda, \lambda')}_{\mathsf{dim}(r_\lambda \cap s_{\lambda'})}, \ldots\Big),$$

where all $(\lambda, \lambda') \in \mathsf{spec}(r) \times \mathsf{spec}(s)$, independent on the choice of e except for the order of the elements in this list, that is, given e and e' such that $[r, e] = [s, e] = [r, e'] = [s, e'] = 0$ we have

$$\mathsf{spec}(r, s|e') = \mathsf{spec}(r, s|e) \cdot \sigma$$

for some permutation $\sigma : \{1, \ldots, n\} \to \{1, \ldots, n\}$ and thus

$$\mathsf{spec}(r|e') = \mathsf{spec}(r|e) \cdot \sigma \ \text{ and } \ \mathsf{spec}(s|e') = \mathsf{spec}(s|e) \cdot \sigma\,.$$

But these are classical states, so

$$\mathsf{spec}(r|e) \sqsubseteq \mathsf{spec}(s|e) \Leftrightarrow \mathsf{spec}(r|e) \cdot \sigma \sqsubseteq \mathsf{spec}(s|e) \cdot \sigma,$$

from which implication (10.5) follows. □

The last result uses one of the two fundamental properties possessed by the Bayesian order on Δ^n: It is *symmetric*, i.e., the map

$$\Delta^n \to \Delta^n :: x \mapsto x \cdot \sigma$$

is an order isomorphism, for any $\sigma \in S(n)$. This *label independence* of the Bayesian order is a simple case of a more general notion satisfied by the spectral order which we will study in the section on symmetries. To hint at the connection: The equation

$$\mathsf{spec}(r|e) \cdot \sigma = \mathsf{spec}(r|e \cdot \sigma)$$

indicates that permuting the classical state $\mathsf{spec}(r|e)$ is the same as permuting the subspaces $(e_i)_{i=1}^{i=n}$ of the labeling e.

The second crucial property of the Bayesian order on Δ^n is that it is *degenerative*:

$$x \sqsubseteq y \Rightarrow (y_i = y_j > 0 \Rightarrow x_i = x_j > 0).$$

Here is the quantum version of the *degeneration lemma* for classical states.

Lemma 21 *If* $r \sqsubseteq s$ *in* Ω^n *then*

$$r_0 \subseteq s_0 \tag{10.6}$$

and

$$\bigcup s_{>0} \subseteq \bigcup r_{>0}\,, \tag{10.7}$$

where

$$r_{>0} := \mathcal{D}_r \setminus \{r_0\} \qquad \text{and} \qquad s_{>0} := \mathcal{D}_s \setminus \{s_0\}\,.$$

Proof Since $r \sqsubseteq s$ they admit a labeling e such that $\mathsf{spec}(r|e) \sqsubseteq \mathsf{spec}(s|e)$ and thus by degeneration for classical states (Lemma 5), we have

$$\{1 \le i \le n \mid \langle r|e_i\rangle = 0\} \subseteq \{1 \le i \le n \mid \langle s|e_i\rangle = 0\}$$

so eq. (10.6) follows. Analogously, for

$$\langle s|e_i\rangle \in \mathsf{spec}_0(s) := \mathsf{spec}(s) \setminus \{0\}$$

classical degeneration again yields

$$\{1 \le j \le n \mid \langle s|e_j\rangle = \langle s|e_i\rangle\} \subseteq \{1 \le j \le n \mid \langle r|e_j\rangle = \langle r|e_i\rangle\}$$

so eq. (10.7) follows. □

Recall that an increasing sequence (x_i) of classical states must be confined to some region Δ^n_σ. Here is the analogous result for the spectral order.

Lemma 22 *Let* $(r_i)_{i\ge1}$ *be a sequence such that for all* $i \ge 1$ *we have that* $r_i \sqsubseteq r_{i+1}$. *Then there exists a joint refinement* $\mathcal{D}_{(r_i)}$ *of* $(\mathcal{D}_{r_i})_{i\ge1}$ *and thus the states* $(r_i)_{i\ge1}$ *admit joint labeling.*

Proof We agree that the first index for states refers to the sequence index and that the second refers to eigenvalues. First note that by Lemma 21, since $r_i \sqsubseteq r_{i+1}$ we have

$$r_{i,0} \subseteq r_{i+1,0} \tag{10.8}$$

$$\bigcup r_{i+1,>0} \subseteq \bigcup r_{i,>0}\,. \tag{10.9}$$

We now proceed by induction.

As base case we take r_1 as its own refinement. Note that the spectrum of a state r decomposes in a zero and a non-zero part to which we refer as $\mathsf{spec}_0(r)$. Let $\mathcal{D}_i$ be the constructed joint refinement for $(r_1, \ldots, r_i)$. Set

$$\mathcal{D}_{i+1} = (\mathcal{D} \cup \mathcal{E} \cup \mathcal{F}) \setminus \{o\}\,,$$

where

$$\begin{aligned} \mathcal{D} &= \{a \cap r_{i,0} \mid a \in \mathcal{D}_i\} \\ \mathcal{E} &= \{a \cap r_{i+1,0} \mid a \in r_{i,>0}\} \\ \mathcal{F} &= r_{i+1,>0}\,. \end{aligned}$$

Graphically, in terms of decompositions of $\mathcal{H}^n$ in subspaces,

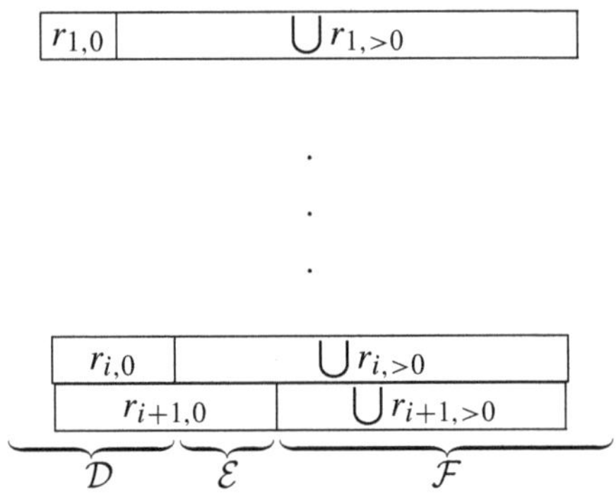

We now prove that $\mathcal{D}_{i+1}$ is a joint refinement for $(r_1, \ldots, r_{i+1})$. Since $r_i \sqsubseteq r_{i+1}$ they admit a joint refinement $\mathcal{G}$ so we have by Proposition 10 that

$$\begin{aligned} \bigcup \mathcal{G} &\subseteq \bigcup \mathcal{D}_{r_i} \cap \bigcup \mathcal{D}_{r_{i+1}} \\ &= \left(r_{i,0} \cup \bigcup r_{i,>0}\right) \cap \left(r_{i+1,0} \cup \bigcup r_{i+1,>0}\right) \\ &= \left(r_{i,0} \cap r_{i+1,0}\right) \cup \left(\bigcup r_{i,>0} \cap r_{i+1,0}\right) \cup \left(\bigcup r_{i,>0} \cap \bigcup r_{i+1,>0}\right) \\ &= r_{i,0} \cup \left(\bigcup r_{i,>0} \cap r_{i+1,0}\right) \cup \bigcup r_{i+1,>0} \end{aligned}$$

by Eqs. (10.8) and (10.9) and since

$$r_{i,0} \cap \bigcup r_{i+1,>0} = \emptyset\,.$$

We moreover have

$$\begin{aligned}\mathsf{span}\left(\bigcup r_{i,>0} \cap r_{i+1,0}\right) &= \mathsf{span}\left(\bigcup \{a \cap r_{i+1,0} \mid a \in r_{i,>0}\}\right)\\ &= \mathsf{span}\left(\bigcup \mathcal{E}\right)\end{aligned}$$

Since $\mathcal{D}_i$ is a refinement for $\mathcal{D}_{r_i}$ it also follows that

$$\begin{aligned}\mathsf{span}(r_{i,0}) &= \mathsf{span}\left(\bigcup \{a \cap r_{i,0} \mid a \in \mathcal{D}_i\}\right)\\ &= \mathsf{span}\left(\bigcup \mathcal{D}\right).\end{aligned}$$

Thus,

$$\begin{aligned}\mathcal{H}^n &= \mathsf{span}(\bigcup \mathcal{G})\\ &= \mathsf{span}\left(\bigcup \mathcal{D}_{r_i} \cap \bigcup \mathcal{D}_{r_{i+1}}\right)\\ &= \mathsf{span}\left(\bigcup \mathcal{D} \cup \bigcup \mathcal{E} \cup \bigcup \mathcal{F}\right)\\ &= \mathsf{span}\left(\bigcup \mathcal{D}_{i+1}\right).\end{aligned}$$

The elements in $\mathcal{D}$, $\mathcal{E}$ and $\mathcal{F}$ are mutually orthogonal since $\mathcal{D}_i$, $r_{i,>0}$ and $r_{i+1,>0}$ consist of mutually orthogonal elements. Moreover, the sets $\bigcup \mathcal{D}$, $\bigcup \mathcal{E}$ and $\bigcup \mathcal{F}$ are themselves mutually orthogonal since

- $\bigcup \mathcal{F} = \bigcup r_{i+1,>0} \perp r_{i+1,0} \supseteq \bigcup \mathcal{E}$,
- $\bigcup \mathcal{D} \subseteq r_{i,0} \perp \bigcup r_{i,>0} \supseteq \bigcup \mathcal{E}$, and,
- $\bigcup \mathcal{D} \subseteq r_{i,0} \perp \bigcup r_{n,>0} \supseteq \bigcup r_{i+1,>0} = \bigcup \mathcal{F}$,

where the last inclusion follows from Eq. (10.9). Thus, $\mathcal{D}_{i+1}$ is a decomposition of $\mathcal{H}^n$. Since

- $\bigcup \mathcal{F} = \bigcup r_{i+1,>0}$,
- $\bigcup \mathcal{E} = r_{i+1,0}$, and,
- $\bigcup \mathcal{D} \subseteq r_{i,0} \subseteq r_{i+1,0}$,

by eq. (10.8), it follows that

$$\bigcup \mathcal{D}_{i+1} \subseteq r_{i+1,0} \cup \bigcup r_{i+1,>0} = \bigcup \mathcal{D}_{r_{i+1}}$$

so $\mathcal{D}_{i+1}$ is a refinement of $\mathcal{D}_{r_{i+1}}$. Since

- $\bigcup \mathcal{E} = \bigcup r_{i+1,>0} \subseteq \bigcup r_{i,>0} \subseteq \bigcup \mathcal{D}_n$, by Eq. (10.9) and the inductive assumption,
- $\bigcup \mathcal{F} \subseteq \bigcup r_{i,>0} \subseteq \bigcup \mathcal{D}_i$, and,
- $\bigcup \mathcal{D} \subseteq \bigcup \mathcal{D}_i$,

it follows that $\mathcal{D}_{i+1}$ is a refinement of $\mathcal{D}_i$ and thus of all $\mathcal{D}_{r_j}$ for $1 \leq j \leq i$.

Finally, consider an infinite sequence $(r_i)_{i\geq 1}$ such that for all $i \geq 1$ we have that $r_i \sqsubseteq r_{i+1}$ and let $(\mathcal{D}_i)_{i\geq 1}$ be the corresponding series of refinements, each member being the above constructed common refinement of $(\mathcal{D}_{r_1}, \ldots, \mathcal{D}_{r_i})$. Note that $(\bigcup \mathcal{D}_i)_{i\geq 1}$ is decreasing with respect to intersection. Then, since $\mathcal{H}^n$ is n-dimensional, there can only be n distinct decompositions contained in $(\mathcal{D}_i)_{i\geq 1}$, that is $n-1$ non-trivial refinements steps $\mathcal{D}_i \mapsto \mathcal{D}_{i+1}$. Thus,

$$\bigcup \mathcal{D}_{(r_i)} := \bigcap_{i\geq 1} \bigcup \mathcal{D}_i$$

is equal to the intersection of a finitely many decreasing sets and thus must be equal to its smallest member, which is a common refinement for $(\mathcal{D}_{r_i})_{i\geq 1}$. □

Theorem 7 *Ω^n is a partially ordered set for each $n \geq 2$. Its maximal elements are the pure states,*

$$\max(\Omega^n) = \Sigma^n,$$

while its least element is the completely mixed state

$$\bot := \begin{pmatrix} 1/n & & 0 \\ & \ddots & \\ 0 & & 1/n \end{pmatrix}.$$

Proof For reflexivity, consider any labeling e admitted by state r. Then, due to reflexivity in Δ^n (Theorem 2), reflexivity in Ω^n follows.

For anti-symmetry assume that $r \sqsubseteq s$ and $s \sqsubseteq r$. By Lemma 22 there exists a joint labeling e and thus by definition 32 we have

$$\mathsf{spec}(r|e) \sqsubseteq \mathsf{spec}(s|e) \text{ and } \mathsf{spec}(s|e) \sqsubseteq \mathsf{spec}(r|e).$$

Due to anti-symmetry in Δ^n we obtain $\mathsf{spec}(r|e) = \mathsf{spec}(s|e)$ so $r = s$ by Lemma 20.

For transitivity assume that $r \sqsubseteq s$ and $s \sqsubseteq t$. By Lemma 22 there exists a joint labeling e and thus we have

$$\mathsf{spec}(r|e) \sqsubseteq \mathsf{spec}(s|e) \text{ and } \mathsf{spec}(s|e) \sqsubseteq \mathsf{spec}(t|e).$$

Thus, due to transitivity in Δ^n we obtain $\mathsf{spec}(r|e) \sqsubseteq \mathsf{spec}(t|e)$ and thus by Definition 32 we have $r \sqsubseteq t$.

Since $\mathsf{spec}(r) = \{0, 1\}$ for any $r \in \Sigma^n$, when $s \in \Omega^n$ satisfies $r \sqsubseteq s$ it follows for any labeling e admitted by r and s that we have

$$\mathsf{spec}(r|e) = (1, 0, \ldots, 0) \cdot \sigma \sqsubseteq \mathsf{spec}(s|e)$$

in Δ^n for some permutation σ, so $\mathsf{spec}(r|e) = \mathsf{spec}(s|e)$ since $\mathsf{spec}(r|e) \in \max(\Delta^n)$, and thus $r = s$.

Conversely, for any state $r \in \Omega^n$ expressed in a base $B \in \mathcal{D}_e$ in which it diagonalizes, we have

$$\mathsf{spec}(r|e) \sqsubseteq (1, 0, \ldots, 0) \cdot \sigma$$

in Δ^n for some permutation σ, so r has a pure state above it, and thus the pure quantum states are the only maximal elements of Ω^n.

Since $\mathsf{spec}(\perp) = \{1/n\}$ we have $\mathcal{D}_\perp = \mathcal{H}^n$ and thus $\perp$ admits any labeling. Given $r \in \Omega^n$ and labeling e admitted by e we then have

$$\mathsf{spec}(\perp|e) = (1/n, \ldots, 1/n) \sqsubseteq \mathsf{spec}(r|e)\,,$$

so $\perp \sqsubseteq r$ and thus $\perp$ is the least element of Ω^n. □

Examining the proofs given so far reveals that the technique used in defining the spectral order serves to distinguish an interesting class of partial orders on classical states for which the Bayesian order is the canonical member.

Corollary 3 *If $\sqsubseteq$ is a symmetric and degenerative partial order on Δ^n, then the relation in Definition 32 is a partial order on Ω^n. Moreover,*

- $\max(\Omega^n) = \Sigma^n$ *whenever* $\max(\Delta^n) = \{e_i : 1 \le i \le n\}$, *and*
- *The completely mixed state is the bottom of Ω^n whenever $(1/n, \ldots, 1/n)$ is the bottom of Δ^n.*

By Lemma 20, we can define a quantum state r by specifying two pieces of information: (i) a labeling e which it admits, that is $[r, e] = 0$, and (ii) a classical state x for which $\mathsf{spec}(r|e) := x$. We use this idea in what follows.

Proposition 12 *The quantum states Ω^n are a dcpo. In more detail,*

(i) *If $(r_i)_{i\ge1}$ is an increasing sequence, then its supremum $\bigsqcup_{i\ge1} r_i$ exists and is implicitly defined by*

$$\mathsf{spec}\Big(\bigsqcup_{i\ge1} r_i \;\Big|\; e\Big) = \Big(\lim_{i\to\infty} \langle r_i|e_1\rangle, \ldots, \lim_{i\to\infty} \langle r_i|e_n\rangle\Big) \tag{10.10}$$

for some and thus any joint labeling e of $(r_i)_{i\ge1}$.

(ii) *Every directed subset of Ω^n contains an increasing sequence with the same supremum.*

Proof By Lemma 22 there exists a joint labeling e for $(r_i)_{i\ge1}$ and thus by Definition 32 it follows that $(\mathsf{spec}(r_i|e))_{i\ge1}$ is an increasing sequence in Δ^n. Then by Proposition 12 we know that the pointwise limit

$$\lim_{i\to\infty} \mathsf{spec}(r_i|e) := \left(\lim_{i\to\infty}\langle r_i|e_1\rangle, \ldots, \lim_{i\to\infty}\langle r_i|e_n\rangle\right)$$

exists. We define a state r implicitly via

$$\mathsf{spec}(r|e) = \lim_{i\to\infty} \mathsf{spec}(r_i|e)\,.$$

We first show that this state r is independent on the choice of e. Since

$$\begin{aligned}\bigcup \mathcal{D}_{(r_i)} &= \bigcap_i \bigcup \mathcal{D}_{r_i} = \bigcap_i \bigcup \{r_{i,\lambda_i} | \lambda_i \in \mathsf{spec}(r_i)\} \\ &= \bigcup \left\{ \bigcap_i r_{i,\lambda_i} \;\middle|\; \forall i : \lambda_i \in \mathsf{spec}(r_i) \right\},\end{aligned}$$

where we leave the proof of the first equality to the reader (straightforward verification via the inductive definition of $\mathcal{D}_{(r_i)}$), so

$$\mathcal{D}_{(r_i)} = \left\{ \bigcap_i r_{i,\lambda_i} \;\middle|\; \forall i : \lambda_i \in \mathsf{spec}(r_i) \right\}.$$

Note that by $\bigcup \mathcal{D}_{(r_i)} = \bigcap_i \bigcup \mathcal{D}_{r_i}$ it also follows that for any joint labeling e of $(r_i)_{i\geq 1}$, since

$$\forall i \geq 1 : \bigcup \mathcal{D}_e \subseteq \bigcup \mathcal{D}_{r_i} \Rightarrow \bigcup \mathcal{D}_e \subseteq \bigcap_i \bigcup \mathcal{D}_{r_i}\,,$$

we have $\bigcup \mathcal{D}_e \subseteq \bigcup \mathcal{D}_{(r_i)}$, that is, $\bigcup \mathcal{D}_{(r_i)}$ contains all joint labelings of $(r_i)_{i\geq 1}$ (and only those, so it is maximal with respect to this property).

If e is a joint labeling of $(r_i)_{i\geq 1}$ and $\bigcap_i r_{i,\lambda_i} \neq \emptyset$, where

$$(\lambda_i)_{i\leq 1} \in \prod_{i\leq 1} \mathsf{spec}(r_i)\,,$$

then there exists some $e_j \in \mathcal{D}_e$ such that $e_j \subseteq \bigcap_i r_{i,\lambda_i}$ for which we have

$$\begin{aligned} e_j \subseteq \bigcap_i r_{i,\lambda_i} &\Leftrightarrow \forall i : e_j \subseteq r_{i,\lambda_i} \\ &\Leftrightarrow \forall i : \langle r_i|e_j\rangle = \lambda_i \\ &\Leftrightarrow \big(\langle r_i|e_j\rangle\big)_{i\geq 1} = (\lambda_i)_{i\geq 1}\,. \end{aligned}$$

Since $\lim_{i\to\infty}\langle r_i|e_j\rangle$ exists, $\lim_{i\to\infty}\lambda_i$ exists and is equal to it. However, $(\lambda_i)_{i\geq 1}$ does not depend on any labeling so neither does its limit $\lim_{i\to\infty}\lambda_i$. can define r now as follows without any reference to a labeling:

$$\mathsf{spec}(r) := \left\{ \lim_{i\to\infty} \lambda_i \;\middle|\; (\lambda_i)_{i\leq 1} \in \prod_{i\leq 1} \mathsf{spec}(r_i)\,,\ \bigcap_i r_{i,\lambda_i} \neq \emptyset \right\},$$

$$r : \mathsf{spec}(r) \to \mathbb{L}^n :: \lim_{i\to\infty} \lambda_i \mapsto \bigcap_i r_{i,\lambda_i}\,.$$

Next we prove that r is an upper bound of $(r_i)_{i\geq 1}$. By Proposition 12 we have

$$\begin{aligned}\bigsqcup_{i\geq 1} \mathsf{spec}(r_i|e) &= \bigsqcup_{i\geq 1}\Big(\langle r_i|e_1\rangle, \ldots, \langle r_i|e_n\rangle\Big)\\ &= \lim_{i\to\infty} \mathsf{spec}(r_i|e)\\ &= \mathsf{spec}(r|e)\,.\end{aligned}$$

so for all $i \geq 1$ we have $\mathsf{spec}(r_i|e) \sqsubseteq \mathsf{spec}(r|e)$ and thus by definition of the order on quantum states it follows that $r_i \sqsubseteq r$ for all $i \geq 1$.

We now show that r is the least upper bound of $(r_i)_{i\geq 1}$. Let s be any upper bound of the sequence $(r_i)_{i\geq 1}$, i.e., for all $i \geq 1$, $r_i \sqsubseteq s$. We now prove that $r \sqsubseteq s$. By the proof of Lemma 22 we know that there exists a finite subsequence of $(r_i)_{i\geq 1}$ (of which we can assume that it has n members) which yields the same common refinement of $(r_i)_{i\geq 1}$ for the given construction—since there are only $n-1$ refinement steps possible. Denote this finite subsequence by $(r_{i_j})_{j=1}^{j=n}$. Then, since

$$r_{i_1} \sqsubseteq \ldots \sqsubseteq r_{i_n} \sqsubseteq s$$

they admit a common refinement and thus a common labeling e, which is also a common labeling for the whole sequence (r_i), and which we can assume to be the one by means of which we defined r since the definition of r does not depend on the choice of labeling. In this labeling we then have for each $i \geq 1$ that

$$\mathsf{spec}(r|e) = \bigsqcup_{i\geq 1} \mathsf{spec}(r_i|e) \sqsubseteq \mathsf{spec}(s|e)$$

in Δ^n since for all $i \geq 1$ we have $\mathsf{spec}(r_i|e) \sqsubseteq \mathsf{spec}(s|e)$. Thus $r \sqsubseteq s$ by definition of the order on quantum states.

We conclude $r = \bigsqcup_{i\geq 1} r_i$ from which Eq. (10.10) then follows.

(ii) The map $\Omega^n \to [0,1] :: r \mapsto \max(\mathsf{spec}(r))$ preserves suprema of increasing sequences and is strictly monotone. □

Thus, we can think of $\mathsf{spec}(\cdot|\cdot)$ as being Scott continuous in its first argument: For any observable e,

$$\mathsf{spec}\left(\bigsqcup r_i|e\right) = \bigsqcup \mathsf{spec}(r_i|e)$$

whenever (r_i) is an increasing sequence in Ω^n.

10.3.3 Symmetries for Quantum States

We introduce symmetries, the quantum analogue of permutations for classical states.

Definition 33 A *unitary transformation* is a surjective linear operator $U : \mathcal{H}^n \to \mathcal{H}^n$ which preserves angles:

$$\langle U\phi \mid U\psi\rangle = \langle \phi \mid \psi\rangle ,$$

for all $\psi, \phi \in \mathcal{H}^n$. U is called a *quantum n-symmetry.*

In particular, the inverse U^{-1} of a unitary operator U is unitary.

Lemma 23 *Let U be a quantum symmetry on $\mathcal{H}^n$. For any labeling e,*

$$U \cdot e : \{1, \ldots, n\} \to \mathbb{L}^n :: i \mapsto \{U(\psi) \in \mathcal{H}^n \mid \psi \in e_i\}$$

is a labeling, while for any state r,

$$U \cdot r : \mathsf{spec}(r) \to \mathbb{L}^n :: \lambda \mapsto \{U(\psi) \in \mathcal{H}^n \mid \psi \in r_\lambda\}$$

is a state with $\mathsf{spec}(U \cdot r) = \mathsf{spec}(r)$.

In both cases only the action of U on subspaces comes into play. Thus, two unitary operators U and U' related by $U = re^{i\theta} \cdot U'$ with $r > 0$ and $\theta \in [0, 2\pi)$ should be thought of as equivalent. The linearity of the maps, in conjuction with the coincidence of the action of U and U' on subspaces does force them to essentially be the same [4], e.g. $\mathsf{span}(e^{i\theta}\psi) = \mathsf{span}(\psi)$, though for $\psi \neq \phi$ and both nonzero we find $\mathsf{span}(\phi + e^{i\theta}\psi) \neq \mathsf{span}(\phi + \psi)$ for $\theta \neq 0$. Thus, a *quantum n-symmetry* should be conceived of as a class of unitary operators on $\mathcal{H}^n$ with equivalent action on subspaces. We will freely represent such a class by one of its representatives.

Lemma 24 *For a state r and a labeling e with $[r, e] = 0$,*

$$\langle r|(U \cdot e)_i\rangle = \langle U^{-1} \cdot r|e_i\rangle . \tag{10.11}$$

Proof First note that

$$(U \cdot e)_i = \{U(\psi) \in \mathcal{H}^n \mid \psi \in e_i\} = \{\psi \in \mathcal{H}^n \mid U^{-1}(\psi) \in e_i\}$$

and

$$(U^{-1} \cdot r)_\lambda = \{U^{-1}(\psi) \in \mathcal{H}^n \mid \psi \in r_\lambda\} = \{\psi \in \mathcal{H}^n \mid U(\psi) \in r_\lambda\} .$$

Next, following Eq. (10.3) we have

$$\begin{aligned}\langle r|(U\cdot e)_i\rangle=\lambda &\Leftrightarrow (U\cdot e)_i\subseteq r_\lambda\\ &\Leftrightarrow \forall\psi\in\mathcal{H}^n: U^{-1}(\psi)\in e_i\Rightarrow \psi\in r_\lambda\\ &\Leftrightarrow \forall\psi\in\mathcal{H}^n: \psi\in e_i\Rightarrow U(\psi)\in r_\lambda\\ &\Leftrightarrow e_i\subseteq (U^{-1}\cdot r)_\lambda\\ &\Leftrightarrow \langle U^{-1}\cdot r|e_i\rangle=\lambda,\end{aligned}$$

which completes the proof. □

Now we give a symmetric characterization of the spectral order analogous to the symmetric characterization of the Bayesian order on classical states. Equation (10.11) leads us to the following dual formulations, which we call the *active* and *passive* (cfr. active and passive transformations in classical mechanics are those acting on the system and the reference frame, respectively). We assume for both theorems that *a labeling e has been fixed in advance.*

Theorem 8 (Active) *For $r, s\in\Omega^n$, we have $r\sqsubseteq s$ iff there exists a quantum symmetry $U:\mathcal{H}^n\to\mathcal{H}^n$ such that*

- $\mathsf{spec}(U\cdot r|e)$ *and* $\mathsf{spec}(U\cdot s|e)$ *are monotone*
- $[r,e]=[s,e]=0$

and

$$\langle U\cdot r|e_i\rangle\langle U\cdot s|e_{i+1}\rangle\le\langle U\cdot r|e_{i+1}\rangle\langle U\cdot s|e_i\rangle$$

for all i with $1\le i<n$.

Theorem 9 (Passive) *For $r, s\in\Omega^n$, we have $r\sqsubseteq s$ iff there exists a quantum symmetry $U:\mathcal{H}^n\to\mathcal{H}^n$ such that*

- $\mathsf{spec}(r|\,U\cdot e)$ *and* $\mathsf{spec}(s|\,U\cdot e)$ *are monotone*
- $[r,U\cdot e]=[s,U\cdot e]=0$

and

$$\langle r|(U\cdot e)_i\rangle\langle s|(U\cdot e)_{i+1}\rangle\le\langle r|(U\cdot e)_{i+1}\rangle\langle s|(U\cdot e)_i\rangle$$

for all i with $1\le i<n$.

Proof Any labeling e' can be obtained from a given one e as $U\cdot e$ for some unitary transformation U. Indeed, in terms of linear operators this correspondence translates as $e'=U\circ e\circ U^{-1}$ so $e\cdot\psi=i\,\psi$ iff $e'\cdot U(\psi)=i\,U(\psi)$, that is, $\psi\in e_i\Leftrightarrow U(\psi)\in e'_i$ yielding the definition of $U\cdot e$ in terms of labelings. The result then straightforwardly follows from Theorem 3 and Lemma 24. □

The following is now merely an observation.

Proposition 13 *The map $(U\cdot -):\Omega^n\to\Omega^n$ is an order isomorphism for any quantum symmetry $U:\mathcal{H}^n\to\mathcal{H}^n$.*

Theorem 8 is the quantum counterpart of Theorem 3 for classical states: The action on states $(U \cdot -) : \Omega^n \to \Omega^n$ in terms of a unitary transformation U corresponds to the action on states $(- \cdot \sigma) : \Delta^n \to \Delta^n$ in terms of a permutation σ. But what is the classical analogue of the passive formulation of the spectral order?

Definition 34 A *classical labeling* is an injective function

$$e : \{1, \ldots, n\} \to \max(\Delta^n).$$

The *standard labeling* is 1 defined by $1(i) = e_i$.

Like the quantum case, we can write a classical state x from the point of view of a classical labeling e as

$$\mathsf{spec}(x|e) := (\langle x|e_1\rangle, \ldots, \langle x|e_n\rangle),$$

where $\langle\cdot|\cdot\rangle$ is the standard inner product on $\mathbb{R}^n$. For $e = 1$, $\mathsf{spec}(x|1) = x$. Notice too that $\langle e_i|e_j\rangle = 0$ for $i \neq j$, so the image of a classical labeling e is by definition a mutually orthogonal collection of pure states.

A classical labeling e induces a permutation $1^{-1} \circ e \in S(n)$. Thus, a *classical labeling* is merely a way of *rearranging* a *fixed* set of n orthogonal pure states $\max(\Delta^n)$. By contrast, a quantum labeling corresponds to selecting n orthogonal pure states from an infinite set of potential pure states *and* arranging the n pure states chosen.

Because classical labelings and symmetries are essentially the same, Theorem 3 is the passive formulation of the Bayesian order when we fix the standard classical label 1 as our reference frame. All other classical labels e can be written as $e = 1 \circ \sigma$ for some $\sigma \in S(n)$, analogous to the quantum case. To summarize:

	Classically	Quantum
Labeling	Permutation e	Self-adjoint operator e with spectrum $\{1, \ldots, \mathsf{n}\}$
Symmetry	Permutation σ	Unitary transformation U

The equivalence of "symmetry" and "labeling" for classical states suggests the following analogy: Symmetries are to classical states as labelings are to quantum states. Though this is not entirely conceptually satisfying, it is a useful mathematical view of things. To illustrate, notice the strong resemblance between the following characterization of the spectral order, in terms of labels, and the symmetric characterization of the Bayesian order.

Theorem 10 *For $r, s \in \Omega^n$, we have $r \sqsubseteq s$ iff there is a quantum labeling e such that*

- $\mathsf{spec}(r|e)$ *and* $\mathsf{spec}(s|e)$ *are monotone*
- $[r, e] = [s, e] = 0$

and

$$\langle r|e_i\rangle\langle s|e_{i+1}\rangle \leq \langle r|e_{i+1}\rangle\langle s|e_i\rangle$$

for all i with $1 \leq i < n$.

Compared to Theorem 8 and Theorem 9, in this result it is the *act* of labeling itself that transforms a state into a monotone classical state. In the classical case, it is obviously a permuation (classical label) which converts a state to monotone form.

As a second example, first recall that the symmetric group $S(n)$ divides Δ^n into order isomorphic regions,

$$\Delta^n := \bigcup_{\sigma\in S(n)} \Delta^n_\sigma,$$

where $\Delta^n_\sigma \simeq \Lambda^n$. Similarly, quantum states are divided into order isomorphic regions by the class of measurement operators:

$$\Omega^n := \bigcup_e \Omega^n|e,$$

where $\Omega^n|e := \{r \in \Omega^n : [r,e] = 0\}$, i.e., the set of quantum states admitted by measurement e. Here is the quantum version of Proposition 4.

Proposition 14 *Let* $n \geq 2$. *Then*

(i) *For each labeling e,* $\Omega^n|e$ *is closed under directed suprema.*
(ii) *For an increasing sequence* (r_i), *there is a labeling e with* $r_i \in \Omega^n|e$ *for all i.*
(iii) *The natural map*

$$q : \Omega^n \to \Lambda^n$$

is Scott continuous, strictly monotone and restricts to a retraction

$$r_e : \Omega^n|e \simeq \Delta^n \to \Lambda^n$$

for each e.

Proof The precise definition of q is as follows: For $s \in \Omega|e$, we define $q(s) := r(\mathsf{spec}(s|e))$, where $r : \Delta^n \to \Lambda^n$ is the natural retraction. □

In particular, the Bayesian order on classical states is an instance of the spectral order on quantum states, which is realized whenever we specify a labeling e. It may interest the reader to know that both authors claim that $q : \Omega^n \to \Lambda^n$ *cannot* be factored into a composition of monotone maps

$$\Omega^n \xrightarrow{?} \Delta^n \xrightarrow{r} \Lambda^n,$$

where $? : \Omega^n \to \Delta^n$ denotes a monotone map that probably doesn't exist.

10.3.4 Approximation of Quantum States

Like classical states, the ability to approximate quantum states order theoretically is a consequence of the mixing law.

Proposition 15 *If $r \sqsubseteq s$ in Ω^n, then*

$$r \sqsubseteq (1-p)r + ps \sqsubseteq s$$

for all $p \in [0,1]$.

Proof First, $(1-p)r + ps$ is a density operator. Because $r \sqsubseteq s$, there is a labeling e with $[r,e] = [s,e] = 0$. Then

$$\begin{aligned}[(1-p)r + ps, e] &= ((1-p)r + ps)e - e((1-p)r + ps) \\ &= (1-p)[r,e] + p[s,e] \\ &= 0.\end{aligned}$$

Next,

$$\mathsf{spec}((1-p)r + ps|e) = (1-p)\mathsf{spec}(r|e) + p \cdot \mathsf{spec}(s|e),$$

because $(1-p)r + ps$, r and s are *diagonal* when written in the base e. The result now follows from the mixing law for classical states. □

Like the classical case, the mixing law is equivalent to saying that the path $\pi_{rs} : [0,1] \to \Omega^n$ from r to s given by

$$\pi_{rs}(t) = (1-t)r + ts$$

is Scott continuous iff $r \sqsubseteq s$.

Lemma 25 *If $r \sqsubseteq s$ in Ω^n and $\mathsf{spec}(s) \subseteq (0,\infty)$, then*

$$[s,e] = 0 \Rightarrow [r,e] = 0,$$

for any labeling e.

Proof First recall that $[s,e] = 0$ means that $\mathcal{D}_e$ is a refinement of $\mathcal{D}_s$. But the spectrum of s is *positive*, so Lemma 21 implies that $\mathcal{D}_s$ is a refinement of $\mathcal{D}_r$. Thus, $\mathcal{D}_e$ is a refinement of $\mathcal{D}_r$, which means $[r,e] = 0$.

The last result is the quantum analogue of Lemma 12(i) for classical states. The next few results further demonstrate the parallel between Δ^n_σ for classical states and $\Omega^n|e$ for quantum states.

Proposition 16 *Let $n \geq 2$.*

(i) *If $r, s \in \Omega^n|e$, then $\pi_{rs}(t) \in \Omega^n|e$ for all $t \in [0,1]$.*

(ii) *For* $r, s \in \Omega^n$, *we have* $r \ll s$ *iff for any labeling* e, *if* $s \in \Omega^n|e$, *then* $r \in \Omega^n|e$ *and* $\mathsf{spec}(r|e) \ll \mathsf{spec}(s|e)$ *in* Δ^n.

Proof (i) This was established in the proof of the mixing law.

(ii) ($\Rightarrow$) Let $r \ll s$. Then for some $t < 1$, $r \sqsubseteq \pi_{\perp s}(t)$. If $[s, e] = 0$, then by (i), $[\pi_{\perp s}(t), e] = 0$ for all t, since we always have $[\perp, e] = 0$. However, because $t < 1$, the spectrum of $\pi_{\perp s}(t)$ is *positive*, which is clear since

$$\mathsf{spec}(\pi_{\perp s}(t)|e) = (1 - t)\perp + t \cdot \mathsf{spec}(s|e).$$

By Lemma 25, $[r, e] = 0$. The other part is obvious.

(ii)($\Leftarrow$) Suppose $s = \bigsqcup s_i$ for an increasing sequence (s_i). Then there is a labeling e with $[s_i, e] = 0$ for all i and $[s, e] = 0$. By assumption, $[r, e] = 0$, and since

$$\mathsf{spec}(r|e) \ll \mathsf{spec}(s|e) = \bigsqcup_{i \geq 1} \mathsf{spec}(s_i|e) \text{ in } \Delta^n,$$

we have $\mathsf{spec}(r|e) \sqsubseteq \mathsf{spec}(s_i|e)$ for some i, and hence $r \sqsubseteq s_i$. Thus, $r \ll s$. □

Ω^n is a domain: A dcpo with an intrinsic notion of approximation.

Theorem 11 *The quantum states* Ω^n *are exact. In addition,*

(i) *For all* $r \in \Omega^n$, $\pi_{\perp r}(t) \ll r$ *for all* $t < 1$.
(ii) *The approximation relation* $\ll$ *is interpolative: If* $r \ll s$ *in* Ω^n, *then there is* $q \in \Omega^n$ *with* $r \ll q \ll s$.

Proof (i) By Prop. 16(i), $[\pi_{\perp r}(t), e] = 0$ whenever $[r, e] = 0$. Since

$$\mathsf{spec}(\pi_{\perp r}(t)|e) = (1 - t)\perp + t \cdot \mathsf{spec}(r|e) \ll \mathsf{spec}(r|e) \text{ in } \Delta^n,$$

Proposition 16 (ii) gives $\pi_{\perp r}(t) \ll r$ for all $t < 1$.

The map $\pi_{\perp r}$ is Scott continuous, so r is the supremum of an increasing sequence of approximations. This implies that $\twoheaddownarrow r$ is directed with supremum r, proving the exactness of Ω^n.

(ii) Mimic the argument for classical states to show $\pi_{\perp r}(t_1) \ll \pi_{\perp r}(t_2)$ whenever $t_1 < t_2$. □

The notion of partiality derivable from $\ll$ on Ω^n is worth taking a brief look at. As before, we call $r \in \Omega^n$ partial iff $\twoheaduparrow r \neq \emptyset$.

Lemma 26 (Partiality) *For* $r \in \Omega^n$, *the set* $\twoheaduparrow r \neq \emptyset$ *iff* $\mathsf{spec}(\mathsf{r}) \subseteq (0, \infty)$.

Proof The only direction which requires proof is ($\Leftarrow$). Let e be a labeling with $[r, e] = 0$. Let $x := \mathsf{spec}(r|e) \in \Delta^n$. From the proof of Lemma 13 for classical states, there is $y \in \Delta^n$ such that $\pi_{\perp y}(t) = x$ for some $t < 1$.

Let $s \in \Omega^n$ with $[s, e] = 0$ and $\mathsf{spec}(s|e) = y$. First, $\pi_{\perp s}(t) = r$, since $\pi_{\perp s}(t), r \in \Omega^n|e$ and

$$\mathsf{spec}(\pi_{\perp s}(t)|e) = \pi_{\perp y}(t) = x = \mathsf{spec}(r|e).$$

Because $t < 1$, $r = \pi_{\perp s}(t) \ll s$ in Ω^n. □

Thus, a quantum state which is partial cannot be pure. In addition, by exactness, all quantum states r arise as the supremum of an increasing sequence

$$(\pi_{\perp r}(1 - 1/n))_{n \geq 1}$$

of partial states which approximate r.

Lemma 27 (Approximation of pure states) *Let $n \geq 2$ and $\psi \in \max(\Omega^n)$ be a pure state. For all $r \in \Omega^n$, $r \ll \psi \Leftrightarrow r = \pi_{\perp\psi}(t)$ for some $t < 1$.*

Proof Let $r \ll \psi$. Let e be any labeling with $[\psi, e] = 0$. Then $[r, e] = 0$ and

$$x := \mathsf{spec}(r|e) \ll y := \mathsf{spec}(\psi|e) \in \max(\Delta^n)\,.$$

Thus, by Prop. 7,

$$(\exists t < 1)\, x = \pi_{\perp y}(t).$$

But $\pi_{\perp\psi}(t) \in \Omega^n|e$ and $\mathsf{spec}(\pi_{\perp\psi}(t)|e) = \pi_{\perp y}(t) = x = \mathsf{spec}(r|e)$, so $r = \pi_{\perp\psi}(t)$, since each is diagonal in e and their spectra are equal. □

Thus, the order theoretic approximations of pure states ψ are precisely the mixtures of ψ with the completely mixed ensemble $\perp$.

Example 9 The depolarization channel $d_p : \Omega^n \to \Omega^n$ describes the process by which the density operator of a system has all bias removed from it with probability p

$$d_p(r) = p \cdot I/n + (1 - p)r.$$

It can be rewritten as

$$d_p(r) = p\perp + (1 - p)r,$$

very similar to the classical case we considered earlier. Just as in the classical case, we also have $d_p(r) \ll r$ for $p > 0$.

10.3.5 Entropy

The word "measurement" is used in domain theory and in quantum mechanics. They are related as follows: Domain theoretically, to measure the content of an object x, we must do something to x that will convert the information it represents into a simpler form μx that can be *understood*. Physically, as it turns out, the content of a quantum state r can be measured by selecting an appropriate quantum measurement e that converts r into a monotone classical state $\mathsf{spec}(r|e)$. (We might think of e as a way of extracting classical information from r.) This defines the map

$$q : \Omega^n \to \Lambda^n,$$

completely analogous to $r : \Delta^n \to \Lambda^n$ for classical states, which is a measurement in the sense of domain theory.

Proposition 17 *The map $q : \Omega^n \to \Lambda^n$ is a measurement.*

Proof First, q is Scott continuous, strictly monotone, and preserves and reflects maximal elements. To show that it measures $\ker q = \max(\Omega^n)$, let $\psi \in \Omega^n$ and $U \subseteq \Omega^n$ be Scott open with $\psi \in U$.

Then there is $0 < t < 1$ with $a := \pi_{\perp\psi}(t) \in U$. Because $a \ll \psi$ in Ω^n, $q(a) \ll q(\psi)$ in Λ^n, which means $\varepsilon := \twoheaduparrow q(a)$ is a Scott open subset of Λ^n. We claim that $\psi \in q_\varepsilon(\psi) \subseteq \uparrow a \subseteq U$. That $\psi \in q_\varepsilon(\psi)$ is clear.

Now let $s \in q_\varepsilon(\psi)$. Then there is a labeling e with $[s, e] = [\psi, e] = 0$. Because $a \ll \psi$, we must have $[a, e] = 0$. But we also know

$$\perp \neq r(\mathsf{spec}(a|e)) = q(a) \ll q(s) = r(\mathsf{spec}(s|e)),$$

where $r : \Delta^n \to \Lambda^n$ is the natural retraction. Now the proof that r is a measurement gives

$$\mathsf{spec}(a|e) \ll \mathsf{spec}(s|e) \text{ in } \Delta^n,$$

which implies that $a \sqsubseteq s$, and thus $s \in U$, finishing the proof. □

In particular, we can measure the content of a quantum state with a classical state.

Example 10 The content of a density operator ρ can also be measured with its largest eigenvalue,

$$\rho \mapsto \max(\mathsf{spec}(\rho)).$$

This is a measurement into $[0, 1]$ since it factors as $q(\rho)^+$. Similarly,

$$\rho \mapsto 1 - q(\rho)^+$$

and

$$\rho \mapsto -\log q(\rho)^+$$

are measurements into $[0, \infty)^*$.

The measures of content in the last example are the quantum versions of the maps $x \mapsto x^+$, $x \mapsto 1 - x^+$ and $x \mapsto -\log x^+$ on classical states. The extension of Shannon entropy to quantum states is called *von Neumann entropy*.

Theorem 12 *Let $\sigma : \Omega^n \to [0, \infty)^*$ be the von Neumann entropy on quantum states*

$$\sigma(r) = -\mathrm{tr}(r \cdot \log r)$$

where the logarithm is natural. Then σ is a measurement in the sense of domain theory. In addition,

(i) *For all $r, s \in \Omega^n$, if $r \sqsubseteq s$ and $\sigma(r) = \sigma(s)$, then $r = s$.*
(ii) *For all $r \in \Omega^n$, we have $\sigma(r) = 0$ iff $r \in \max(\Omega^n) = \Sigma^n$.*
(iii) *For all $r \in \Omega^n$, we have $\sigma(r) = \log n$ iff $r = \perp$.*

Proof The von Neumann entropy σ factors as

$$\sigma = \mu \circ q$$

where $q : \Omega^n \to \Lambda^n$ assigns to a quantum state its *monotone* spectrum, and $\mu : \Lambda^n \to [0, \infty)^*$ is Shannon entropy. Since q and μ have all the properties mentioned in this result, so does σ. □

By now it is clear that quantum information is more intricate than classical information, if for no other reason than the superficial observation that a density operator is "more complicated" than a classical state. What we now want is a precise formulation of the intuitive idea that there is more information in the quantum than in the classical.

One hint is provided by Proposition 14: We can associate each classical state to a quantum state in such a way that *information is conserved*:

$$\begin{gathered}\text{conservation of information}\\ =\\ (\text{qualitative conservation}) + (\text{quantitative conservation})\\ =\\ (\text{order embedding}) + (\text{preservation of entropy}).\end{gathered}$$

And this is what we now prove: While each classical state can be associated to a quantum state in such a way that information is conserved, the converse is *never true*.

Theorem 13 *Let $n \geq 2$. Then*

- *There is an order embedding $\phi : \Delta^n \to \Omega^n$ such that $\sigma \circ \phi = \mu$.*
- *For $m \geq 2$, there is no order embedding $\phi : \Omega^n \to \Delta^m$ with $\mu \circ \phi = \sigma$.*

Proof For the first, Proposition 14 gives an embedding which preserves entropy. For the second, if there is an embedding of Ω^n into Δ^m which preserves entropy, it yields an injection of $\max(\Omega^n)$ into $\max(\Delta^m)$, which is impossible since the first of these sets is infinite, while the latter is finite.

The reader may be interested to know that the authors both claim that the above result holds independent of entropic considerations.

10.4 Synthesis

We now obtain a unified perspective on classical and quantum which leads to a methodology applicable in any setting where one has (i) a notion of state and (ii) a notion of state update as the result of observation.

10.4.1 Classical Projections

We first turn back to classical states to show that they admit a more general class of projectors and that the inductive definition of the Bayesian order extends to this larger class. First note that a classical projection

$$p_i : \Delta^{n+1} \rightharpoonup \Delta^n$$

is undefined in the singleton

$$\mathsf{fix}(p_i)^{\perp} := \{x \in \Delta^{n+1} \mid x_i = 1\}$$

and has as "fixed points"

$$\mathsf{fix}(p_i) := \{x \in \Delta^{n+1} \mid x_i = 0\}\,.$$

Any projection p_i moreover has a complementary projector, namely

$$p_i^{\perp} : \Delta^{n+1} \rightharpoonup \Delta^1 : x \mapsto (1)$$

which is undefined in

$$\mathsf{fix}(p_i)^{\perp} := \{x \in \Delta^{n+1} \mid x_i = 0\}\,.$$

This projector expresses the update that the observer experiences when he looks in box i and the object of his desire is actually there. Equivalently, this corresponds to looking in all boxes except box i and not finding the object. The condition

$$x \sqsubseteq y \Rightarrow p_i^{\perp}(x) \sqsubseteq p_i^{\perp}(y)$$

is however trivially satisfied whenever $p_i^\perp$ is defined both in x and y. As such, one could have included it in Definition 2.7 providing an interpretation "whatever outcome we obtain when looking in box i, the corresponding collapse of knowledge preserves the partial order".

We define general projectors, encoding knowledge update when looking in several boxes at once. Let $n \geq 2$ and $1 \leq k \leq n$. The map which collapses all $i_1, \ldots, i_k^{th}$ outcomes is

$$p_{i_1,\ldots,i_k} : \Delta^n \rightharpoonup \Delta^{n-k}$$
$$p_{i_1,\ldots,i_k}(x) = \frac{1}{1 - \sum_j x_{i_j}}(x_1, \ldots, \widehat{x_{i_1}}, \ldots, \widehat{x_{i_j}}, \ldots, \widehat{x_{i_k}}, \ldots, x_n)$$

for $1 \leq i_1, \ldots, i_k \leq n$ and $0 \leq x_{i_1}, \ldots, x_{i_j}, \ldots, x_{i_k} < 1$. The projector corresponding to "looking in all boxes except" is

$$p^\perp_{i_1,\ldots,i_k} : \Delta^n \rightharpoonup \Delta^{n-k}$$
$$p^\perp_{i_1,\ldots,i_k}(x) = \frac{1}{\sum_j x_{i_j}}(\widehat{x_1}, \ldots, x_{i_1}, \ldots, x_{i_j}, \ldots, x_{i_k}, \ldots, \widehat{x_n})$$

for $1 \leq i_1, \ldots, i_k \leq n$.

The set of projectors as defined constitute a Boolen algebra isomorphic to the powerset $\mathcal{P}(\{1, \ldots, n\})$ when we adjoin the empty map

$$p_{1,\ldots,n} : \Delta^n \rightharpoonup \emptyset$$

and the identity

$$p : \Delta^n \to \Delta^n .$$

In particular we have

$$p^\perp_{1,\ldots\widehat{i_1},\ldots,\widehat{i_j},\ldots,\widehat{i_k},\ldots,k} = p_{i_1,\ldots,i_k} ,$$

that is, projections inherit orthogonality from the complementation of the Boolean algebra $\mathcal{P}(\{1, \ldots, n\})$.

Proposition 18 *Let $x, y \in \Delta^{n+1}$. Then*

$$x \sqsubseteq y \Leftrightarrow (\forall \{i_1, \ldots, i_k\} \subseteq \{1, \ldots, n\})\ p_{i_1,\ldots,i_k}(x) \sqsubseteq p_{i_1,\ldots,i_k}(y).$$

Proof Given $p_i : \Delta^n \rightharpoonup \Delta^{n-1}$ define $\tilde{p}_i : \Delta^n \rightharpoonup \Delta^n$ via

$$\tilde{p}_i : \Delta^n \overset{p_i}{\rightharpoonup} \Delta^{n-1} \overset{\iota_i}{\to} \Delta^n$$

where

$$\begin{array}{ll} \pi_j(\iota_i(x)) = x_j & \text{for } j < i \\ \pi_j(\iota_i(x)) = 0 & \text{for } j = i \\ \pi_j(\iota_i(x)) = x_{j+1} & \text{for } j > i\,. \end{array}$$

Analogously we introduce the map

$$\tilde{p}_{i_1,\ldots,i_k} = \iota_{i_1,\ldots,i_k} \cdot p_{i_1,\ldots,i_k} : \Delta^n \rightharpoonup \Delta^n$$

where

$$\begin{array}{ll} \pi_j(\iota_i(x)) = x_{j+l-1} & \text{for } i_{l-1} < j < i_l \\ \pi_j(\iota_i(x)) = 0 & \text{for } j = i_l \\ \pi_j(\iota_i(x)) = x_{j+l} & \text{for } i_l < j < i_{l+1} \end{array}$$

when assuming that $i_1, \ldots, i_k$ is monotone and formally setting $i_0 = 0$ and $i_{k+1} = n + 1$. We then have

$$\tilde{p}_{i_1,\ldots,i_k} = \tilde{p}_{i_1} \cdot \ldots \cdot \tilde{p}_{i_k}$$

from which the result follows by induction. □

Clearly, we rely on the following.

Proposition 19 *Projections commute with respect to composition.*

In particular, the Boolean algebra of projections is defined from concatenated action of projections

$$\tilde{p} \leq \tilde{q} \Leftrightarrow \tilde{p} \cdot \tilde{q} = \tilde{p}$$

where $\tilde{p}$ and $\tilde{q}$ are defined as in the proof of Proposition 18.

10.4.2 Quantum Projections

We now show that the projective structure of classical states and its corresponding inductive definition of the Bayesian order are preserved by the natural embeddings of Δ^n into Ω^n. In particular, the classical projections become instances of Hilbert space projectors.

The Hilbert space projectors $\mathbb{P}^n$ also constitute an orthocomplemented lattice for the partial order

$$P \leq Q \Leftrightarrow P \cdot Q = P\,.$$

This lattice is however no longer distributive, e.g. [2, 3], and as such is not a Boolean algebra. Related to this, commutativity for projections as we have in Proposition 19 is not valid anymore for Hilbert space projectors.

Analogous to the introduction of $\tilde{p}$ given p in order to be able to compose projections, we now will have to do the converse for Hilbert space projectors in order to state an inductive definition of the partial ordering ofthe quantum states.

Any projector $P \in \mathbb{P}^n$ can be equivalently represented as a partial surjective map

$$P : \mathcal{H}^n \rightharpoonup \mathsf{fix}(P)$$

which is undefined in $\mathsf{fix}(P)^{\perp}$. When an isomorphism

$$h : \mathsf{fix}(P) \to \mathcal{H}^k$$

is specified, with $0 \leq k = \mathsf{dim}(\mathsf{fix}(P)) \leq n$, we can define

$$P^{\downarrow} : \mathcal{H}^n \stackrel{P}{\rightharpoonup} \mathsf{fix}(P) \stackrel{h}{\to} \mathcal{H}^k$$

of which the codomain does not depend on P anymore.

Note that such a map $P^{\downarrow}$ as well as P itself is fully characterized by its kernel, thus these maps are in bijective correspondence with the subspaces $\mathbb{L}^n$ via

$$\mathbb{P}^n \to \mathbb{L}^n : P \mapsto \mathsf{fix}(P)$$

and also with $\{0, 1\}$-labeled decompositions, or equivalently, two element ordered decompositions, via

$$P \mapsto \left(\mathsf{fix}(P), \mathsf{fix}(P)^{\perp}\right).$$

Abstracting over the $\{0, 1\}$-labeling we set

$$\mathcal{D}_P := \left\{\mathsf{fix}(P), \mathsf{fix}(P)^{\perp}\right\}.$$

Proposition 20 *The following are equivalent for state r and projector P:*

- *They admit joint labeling e.*
- *$\mathcal{D}_r$ and $\mathcal{D}_P$ admit a joint refinement $\mathcal{D}$.*
- *They diagonalize in a common base B.*
- $[r, P] = 0$.

The following are equivalent for states r and s and projector P:

- *They admit joint labeling e, i.e., $\mathcal{D}_r$, $\mathcal{D}_s$ and $\mathcal{D}_P$ admit a joint refinement $\mathcal{D}$, i.e., they diagonalize in a common base B.*
- *They pairwise admit joint labeling, i.e., $\mathcal{D}_r$, $\mathcal{D}_s$ and $\mathcal{D}_P$ pairwise admit joint refinement, i.e., they pairwise diagonalize in a common base.*
- $[r, P] = [s, P] = [r, s] = 0$.

Proof Equivalence of the first four conditions follows from Proposition 10 since P is a state up to normaliztion, that is,

$$\frac{1}{\dim\left(\mathrm{fix}(P)\right)} P \in \Omega^n .$$

Given a joint refinement $\mathcal{D}$ for r, s and P any labeling e such that $\bigcup \mathcal{D}_e = \bigcup \mathcal{D}$ is a joint labeling, and a base $B \in \bigcup \mathcal{D}_e$ yields joint diagonalization. At its turn, given a base B in which r, s and P diagonalize then

$$\bigcup \{\mathrm{span}(\psi) \mid \psi \in B\}$$

is a joint refinement.

Whenever we have a joint refinement, a joint labeling or a joint base for s, t and P then we have pairwise existence of one too. For the converse statement we provide a proof. We are going to prove a more general statement however, namely, that whenever we have a set of decompositions $\{\mathcal{D}_i \mid i \in I\}$, for technical simplicity envisioned as being finite, and such that for all $i, j \in I$ we have that $\mathcal{D}_i$ and $\mathcal{D}_j$ admits a joint refinement, then $\{\mathcal{D}_i \mid i \in I\}$ as a whole admits one. (This fact is implied by well-known results in the study of quantum structures [2, 3, 7, 13], though the terminology there is different from ours. For the sake of a self-contained discussion, we provide a complete proof.)

We call $a, b \in \mathbb{L}^n$ compatible, denoted $a \leftrightarrow b$, iff $\{a, a^\perp\}$ and $\{b, b^\perp\}$ admit joint refinement—in lattice terms this means that they generate a subalgebra of $\mathbb{L}^n$ which is Boolean [3]. Then we have that

$$\mathrm{span}\left(a \cap b\,,\ a \cap b^\perp\right) = a\,. \tag{10.12}$$

Indeed, existence of a joint refinement for $\{a, a^\perp\}$ and $\{b, b^\perp\}$ implies

$$\begin{aligned}
\mathcal{H}^n &= \mathrm{span}\left((a \cup a^\perp) \cap (b \cup b^\perp)\right) \\
&= \mathrm{span}\left((a \cap (b \cup b^\perp)) \cup (a^\perp \cap (b \cup b^\perp))\right) \\
&= \mathrm{span}\left(\mathrm{span}\left(a \cap (b \cup b^\perp)\right)\,,\ \mathrm{span}\left(a^\perp \cap (b \cup b^\perp)\right)\right)
\end{aligned}$$

and since

$$\mathrm{span}\left(a \cap (b \cup b^\perp)\right) \subseteq a \quad \text{and} \quad \mathrm{span}\left(a^\perp \cap (b \cup b^\perp)\right) \subseteq a^\perp$$

are subspaces of $\mathcal{H}^n$ this forces Eq. (10.12).

The fact that each $\mathcal{D}_i$ is a decomposition, implying mutual orthogonality of its members, and that we have pairwise existence of a joint refinement for all decompositions in $\{\mathcal{D}_i \mid i \in I\}$, implies that

$$\forall a, b \in \bigcup\{\mathcal{D}_i \mid i \in I\} : a \leftrightarrow b \,.$$

We will now construct a joint refinement inductively, that is, we build a series (d_i) containing all elements of $\bigcup\{\mathcal{D}_i \mid i \in I\}$ and construct a joint refinement $\mathcal{E}_{k+1}$for $(d_1, \ldots, d_{k+1})$ given a joint refinement $\mathcal{E}_k$ for $(d_1, \ldots, d_k)$, taking as base case $\mathcal{E}_1 := \{d_1, d_1^\perp\}$. Set

$$\mathcal{E}_{k+1} := \{a \cap d_{k+1}, a \cap d_{k+1}^\perp \mid a \in \mathcal{E}_k\} \setminus \{o\} \,.$$

It clearly follows that $\bigcup \mathcal{E}_{k+1} \subseteq \bigcup \mathcal{E}_k$ so we obtain a decreasing sequence.

We then also have that

- $\mathsf{span}(\mathcal{E}_1) = \mathcal{H}^n$, and,
- $\mathsf{span}(\mathcal{E}_{k+1}) = \mathsf{span}\left(\{a \cap d_{k+1}, a \cap d_{k+1}^\perp \mid a \in \mathcal{E}_k\}\right) = \mathsf{span}\left(\mathcal{E}_k\right)$

due to Eq. (10.12), what proves that the inductive procedure preserves spanning $\mathcal{H}^n$. Mutual orthogonality of the elements in $\mathcal{E}_{k+1}$ also follows construction.

It remains to be proven that

$$\bigcap_{j \in I} \bigcup \mathcal{E}_j \subseteq \bigcup \mathcal{D}_i$$

for all $i \in I$. Let $\#(d_i)$ be the length of (d_i). From the construction it follows that the elements of $\mathcal{E}_{\#(d_i)}$ are of the form $a_1 \cap \ldots \cap a_{\#(d_i)}$ where $a_j \in \{d_j, d_j^\perp\}$. For every $\mathcal{D}_i$ there is a subsequence of elements a_{i_j} such that $d_{i_j} \in \mathcal{D}_i$. Let $a_{i_1} \cap \ldots \cap a_{i_{\#\mathcal{D}_i}}$ be the corresponding subterm. We claim that the only non-empty such terms are those for which there is exactly one $1 \leq j \leq \#\mathcal{D}_i$ such that $a_{i_j} = d_{i_j}$ and for all others $k \neq j$ we have $a_{i_j} = d_{i_k}^\perp$. That there is at most one follows from the fact that the elements in $\mathcal{D}_i$ are mutually orthogonal. That there is necessarily one follows from the fact that we otherwise have $d_{i_1}^\perp \cap \ldots \cap d_{i_{\#\mathcal{D}_i}}^\perp$ as this subterm what implies that any vector contained in it should be orthogonal to all $d_{i_j} \in \mathcal{D}_i$, what is impossible since $\mathcal{D}_i$ spans $\mathcal{H}^n$. So every subterm $a_{i_1} \cap \ldots \cap a_{i_{\#\mathcal{D}_i}}$ and thus also every term $a_1 \cap \ldots \cap a_{\#(d_i)}$ contains $d_{i_j} \in \mathcal{D}_i$ and thus

$$a_1 \cap \ldots \cap a_{\#(d_i)} \subseteq \bigcup \mathcal{D}_i$$

so

$$\bigcap_{j \in I} \bigcup \mathcal{E}_j = \bigcup \mathcal{E}_{\#(d_i)} \subseteq \bigcup \mathcal{D}_i$$

for all $i \in I$ what completes the proof. □

It is well-known that projectors on $\mathcal{H}^n$ induce maps on Ω^n in terms of Luders' rule [11], that is

$$P[-] : \Omega^n \rightharpoonup \Omega^n : r \mapsto \frac{P \cdot r \cdot P}{\mathrm{tr}(P \cdot r)}$$

for $\mathrm{tr}(P \cdot r) > 0$. Note that $P : \Omega^n \rightharpoonup \Omega^n$ is still idempotent since

$$P \cdot (P \cdot r \cdot P) \cdot P = P \cdot r \cdot P\,,$$

so we can set

$$\mathrm{fix}(P[-]) := \{P \cdot r \cdot P \mid r \in \Omega^n\}\,.$$

Given an isomorphism

$$g : \mathrm{fix}(P[-]) \to \Omega^k,$$

with $0 \leq k = \mathrm{dim}(\mathrm{fix}(P)) \leq n$, and possibly induced by an isomorphism h on the underlying Hilbert spaces, we can define a map

$$\Omega^n \stackrel{P[-]}{\rightharpoonup} \mathrm{fix}(P[-]) \stackrel{g}{\to} \Omega^k\,. \tag{10.13}$$

This will be of our view of projectors in this section, except for an additional extension of the kernel to those density matrices that do not commute with P.

- By a projector $P^{\downarrow}[-] : \Omega^n \rightharpoonup \Omega^k$ we refer to the partial map induced by $P \in \mathbb{P}^n$ which has as kernel those states $x \in \Omega^n$ which are such that either
  - $\mathrm{tr}(P \cdot x) = 0$
  - $[x, P] \neq 0$

  and with the images defined by Eq. (10.13).

When writing down $P^{\downarrow}[-]$ we as such assume that an isomorphism h, or equivalently, g has been specified.

We introduce some dialectics analogous to that of labelings.

- A state r *admits* a projector $P^{\downarrow}[-]$ iff $P^{\downarrow}[\,r\,]$ is defined.

Then by Proposition 20 P and r admit joint labeling.

Let $\mathbb{I}(k, n)$ be the collection of monotone maps of the form

$$\iota : \{1, \ldots, k\} \to \{1, \ldots, n\}$$

for $0 \leq k \leq n$, where the monotonicity is with respect to the usual order on natural numbers, and let

$$\mathbb{I}^n = \bigcup\{\mathbb{I}(k, n) \mid 0 \leq k \leq n\}\,.$$

Let

$$\iota^* : \{1, \dots, n\} \rightharpoonup \{1, \dots, k\}$$

be the partial inverse for any given ι and let $\mathbb{P}^n|e$ be the projectors $P \in \mathbb{P}^n$ that admit a given labeling e, that is,

$$\mathbb{P}^n|e := \left\{ P \in \mathbb{P}^n \,\middle|\, \bigcup \mathcal{D}_e \subseteq \bigcup \mathcal{D}_P \right\} .$$

Lemma 28 *Given a labeling e of $\mathcal{H}^n$ then $\mathbb{P}^n|e \cong \mathbb{I}^n$.*

Proof Given $I \subseteq \{1, \dots, n\}$ define $\iota \in \mathbb{I}^n$ such that I is its range. It then follows that

$$\mathcal{P}(\{1, \dots, n\}) \cong \mathbb{I}^n$$

via $I \mapsto \iota$ due to monotonicity of ι.

We moreover have that $P \in \mathbb{P}^n|e$ iff $\bigcup \mathcal{D}_e \subseteq \bigcup \{\mathsf{fix}(P), \mathsf{fix}(P)^\perp\}$ iff there exists $I_P \subseteq \{1, \dots, n\}$ such that $e_i \in \mathsf{fix}(P) \Leftrightarrow i \in I_P$, and thus we have $\mathbb{P}^n|e \cong \mathcal{P}(\{1, \dots, n\})$ via $P \mapsto I_P$.

In Proposition 18, we characterized the Bayesian order in terms of projections. Here is the formulation for the spectral order in terms of quantum projections.

Theorem 14 *Let $n \geq 2$. For $r, s \in \Omega^n$, we have*

$$r \sqsubseteq s \Leftrightarrow P^\downarrow[\,r\,] \sqsubseteq P^\downarrow[\,s\,] \tag{10.14}$$

- *for all projectors $P^\downarrow[-]$ admitting both r and s, and,*
- *provided there are enough projectors admitting both r and s,*

where we adopt the base cases

- $\Omega^0 := \emptyset$;
- $\Omega^1 := \{(1)\}$ *with* $(1) \sqsubseteq (1)$;
- *For $r, s \in \Omega^2$ we have $r \sqsubseteq s$ iff there exist $p, q \in [0, 1]$ with $p \leq q$ and a pure state $t \in \Sigma^n$ such that*

$$r = (1 - p)\bot + pt \qquad s = (1 - q)\bot + qt ,$$

with

$$\bot := \begin{pmatrix} \frac{1}{2} & 0 \\ 0 & \frac{1}{2} \end{pmatrix} .$$

Proof Any self-adjoint operator on $\mathcal{H}^2$ either has a non-degenerated spectrum or is $\bot$. Excluding the latter case, given $r \in \Omega^2$ there exists a unique labeling e (up to permutation of the labels) such that r admits e. This labeling is obtained by setting

$e_1 = r_+$ and $e_2 = r_+^\perp$, where r_+ are the eigenvectors for eigenvalue $\mathsf{max}(\mathsf{spec}(r))$. In view of Definition 32, it then follows that if e is admitted both by r and s and $\mathsf{spec}(r|e) \sqsubseteq \mathsf{spec}(s|e)$ then

$$\langle r|e_1\rangle = \mathsf{max}(\mathsf{spec}(r)) \leq \mathsf{max}(\mathsf{spec}(s)) = \langle s|e_1\rangle$$

with $s_+ = r_+$ is necessary and sufficient for $r \sqsubseteq s$. Defining $p, q \in [0, 1]$ by

$$p = 2\langle r|e_1\rangle - 1 \quad \text{and} \quad q = 2\langle s|e_1\rangle - 1\,,$$

and the pure state $t \in \Omega^n$ such that $t_1 = e_1$ we obtain

$$(1-p)\perp + pt = (1-p)\begin{pmatrix} \frac{1}{2} & 0 \\ 0 & \frac{1}{2} \end{pmatrix} + p\begin{pmatrix} 1 & 0 \\ 0 & 0 \end{pmatrix} = \begin{pmatrix} \langle r|e_1\rangle & 0 \\ 0 & 1-\langle r|e_1\rangle \end{pmatrix} = r$$
$$\text{and} \quad (1-q)\perp + qt = \begin{pmatrix} \langle s|e_1\rangle & 0 \\ 0 & 1-\langle s|e_1\rangle \end{pmatrix} = s$$

what encodes $s_+ = r_+$ and $\langle r|e_1\rangle \leq \langle s|e_1\rangle$ provided that $p \leq q$.

Let $r \sqsubseteq s$ according to Definition 32. We need to prove that r and s admit enough projectors and that they satisfy Eq. (10.14) with respect to those admitted. So let us first define what we mean by enough projectors.

We mean by this having at least enough to consitute a family of mutually orthogonal projectors $\{P_e^i \mid 1 \leq i \leq n\}$ that support the spectral decomposition of a labeling

$$e = \sum_i i\, P_e^i\,.$$

One verifies that this is equivalent to saying that there exists a family of mutually orthogonal projectors $\{P_i \mid 1 \leq i \leq n\}$ such that $\bigcap_i \bigcup \mathcal{D}_{P_i}$ is the decomposition of some labeling e. It is moreover not restrictive to assume that for all i we have $\mathsf{dim}(\mathsf{fix}(P_i)) = n - 1$.

Since $r \sqsubseteq s$ they admit a joint labeling e that admits projectors $\mathbb{P}^n|e$, among which we have those defined by $\mathsf{fix}(P_i) = e_i$. Then $\mathcal{D}_e = \bigcap_i \bigcup \mathcal{D}_{P_i}$ so r and s admit enough projectors.

We now show that $r \sqsubseteq s$ implies $P^\downarrow[\,r\,] \sqsubseteq P^\downarrow[\,s\,]$ provided $P^\downarrow[-]$ is admitted both by r and s. By Proposition 20 we have, since there exists a pairwise common refinement for r, s and P, that they all together admit a joint labeling e for which we then moreover have $\mathsf{spec}(r|e) \sqsubseteq \mathsf{spec}(s|e) \in \Delta^n$. Let

$$I_P := \{1 \leq i \leq n \mid e_i \in \mathsf{fix}(P)\}$$

and set $I_P^\perp := \{0, \ldots, 1\} \setminus I_P$. By Proposition 18 we then have that

$$p_{I_P^\perp}(\mathsf{spec}(r|e)) \sqsubseteq p_{I_P^\perp}(\mathsf{spec}(s|e))\,.$$

Let $\iota \in \mathbb{I}(k, n)$ with $k = \mathsf{dim}(\mathsf{fix}(P))$ such that it range coincides with I_P. Next, given any isomorphism $h : \mathsf{fix}(P) \to \mathcal{H}^k$, and thus also a corresponding one on states

$$g : \mathsf{fix}(P[-]) \to \Omega^k$$

choose a labeling (e'_i) of $\mathcal{H}^k$ such that $h(e_{\iota(i)}) = e'_i$—we slightly abusively refer to a base vector by the subspace of the labeling in which it is contained. We obtain commutation of the following maps

$$\begin{array}{ccc} \{1,\ldots,n\} & \xrightarrow{\;e\;} & \mathbb{L}^n \\ {\scriptstyle \iota}\uparrow & & \downarrow{\scriptstyle \tilde{h}} \\ \{1,\ldots,k\} & \xrightarrow{\;e'\;} & \mathbb{L}^k \end{array}$$

where $\tilde{h} : \mathbb{L}^n \rightharpoonup \mathbb{L}^k$ is here the partial surjective map arising when $h : \mathsf{fix}(P) \to \mathcal{H}^k$ is applied pointwisely to those one-dimensional $a \in \mathbb{L}^n$ which are such that $a \subseteq \mathsf{fix}(P)$. Indeed, we have

$$i \overset{\iota}{\mapsto} \iota(i) \overset{e}{\mapsto} e_{\iota(i)} \overset{\tilde{h}}{\mapsto} e'_i \,.$$

Since due to $\bigcup \mathcal{D}_e \subseteq \bigcup \mathcal{D}_P$ we have

$$\begin{aligned} P^{\downarrow}(r) &= g\left(\frac{1}{\mathsf{tr}(P \cdot r)} P \cdot r \cdot P\right) \\ &= \frac{1}{\sum_{i \in I_P} \langle r | e_i \rangle} \begin{pmatrix} \langle r | e_{\iota(1)} \rangle & & 0 \\ & \ddots & \\ 0 & & \langle r | e_{\iota(k)} \rangle \end{pmatrix} \quad \text{in } (e'_i) \end{aligned}$$

It then follows that

$$\begin{aligned} \pi_j\left(p_{I_P^{\perp}}(\mathsf{spec}(r|e))\right) &= \frac{1}{\sum_{i \in I_P} \langle r | e_i \rangle} \langle r | e_{\iota(j)} \rangle \\ &= \left\langle P^{\downarrow}(r) \,\middle|\, e'_j \right\rangle \\ &= \pi_j\left(\mathsf{spec}(P^{\downarrow}(r)|e')\right) \end{aligned}$$

so

$$p_{I_P^{\perp}}(\mathsf{spec}(r|e)) = \mathsf{spec}(P^{\downarrow}(r)|e')$$

and thus

$$\mathsf{spec}(P^{\downarrow}(r)|e') \sqsubseteq \mathsf{spec}(P^{\downarrow}(s)|e').$$

We then conclude $P^{\downarrow}(r) \sqsubseteq P^{\downarrow}(s)$.

Conversely, assume that there exists mutually orthogonal projectors $\{P_i \mid 1 \leq i \leq n\}$ such that $\bigcap_i \bigcup \mathcal{D}_{P_i}$ is the decomposition of some labeling e and for which we have $P^{\downarrow}[\,r\,]_i \sqsubseteq P^{\downarrow}[\,s\,]_i$. Since r and s are admitted by all P_i we have by the proof of Proposition 20 that there exists a joint labeling for r and s and all P_i, which as such can only be e itself due to $\mathcal{D}_e = \bigcap_i \bigcup \mathcal{D}_{P_i}$. By constructing the isomorphisms

$$h_e : \Omega^n|e \to \Delta^n \text{ and } h_{e'} : \Omega^{n-1}|e' \to \Delta^{n-1}$$

for each projector P_i such that we have commutation of

$$\begin{array}{ccc} \Omega^n|e & \xrightarrow{h_e} & \Delta^n \\ {\scriptstyle P_i^{\downarrow}}\downarrow & & \downarrow{\scriptstyle p_i} \\ \Omega^{n-1}|e' & \xrightarrow{h_{e'}} & \Delta^{n-1} \end{array}$$

taking into account the isomorphisms $g_i : \mathsf{fix}(P_i[-]) \to \Omega^k$ that are different for each P_i, we can embed the quantum case for projectors P_i and states r and s that admit a fixed labeling e in the classical one with projectors p_i. Then $r \sqsubseteq s$ follows from $\mathsf{spec}(r|e) \sqsubseteq \mathsf{spec}(s|e) \in \Delta^n$ which itself results from the inductive definition for classical states. □

Denote by $\mathbb{P}^n_{\bullet}$ projectors on $n-1$ dimensional subspaces of $\mathcal{H}^n$. We have the following analogy between classical and quantum states:

	Classical	Quantum
States	Δ^n	Ω^n
Pure states	$\max(\Delta^n)$	Σ^n
Primitive projections	$\{p_i \mid 0 \leq i \leq n\}$	$\mathbb{P}^n_{\bullet}$
General projections	$\{p_{i_1,\ldots,i_k} \mid 0 \leq i_1,\ldots,i_k \leq n\}$	$\mathbb{P}^n$

10.4.3 The Lattices of Birkhoff and Von Neumann

In the spectral order, quantum states are ordered by requiring of a labeling e that it *commute* with the states r and s under consideration. In the corresponding inductive formulation, a condition of commutation with states is imposed on projections. Knowing the structural importance of non-commutativity of observables in quantum mechanics, it may surprise the reader to learn that the lattices of Birkhoff and von Neumann [2], the powerset $\mathcal{P}\{1,\ldots,n\}$ and the collection $\mathbb{L}^n$ of subspaces of $\mathcal{H}^n$

ordered by inclusion, can be recovered from Δ^n and Ω^n in a *purely order theoretic manner*.

Recall here that the fundamentally different nature of quantum versus classical observables can also be explained in order theoretic terms, roughly, by the distributivity of $\mathcal{P}\{1,\ldots,n\}$ versus the non-distributivity of $\mathbb{L}^n$ [3, 7, 13]. The relation between observables and these lattices is as follows.

A (real-valued) observable of a classical system, with pure states $\Sigma^n_{\mathrm{cl}} \cong \{1,\ldots,n\}$, is a map $a : \Sigma^n_{\mathrm{cl}} \to \mathbb{R}$ with range $\mathsf{spec}(a)$ that assigns to each pure state the value of that observable. As a consequence, any proposition about the system of the form

"The value of classical observable a is contained in $E \subseteq \mathsf{spec}(a)$"

encodes as the set $a^{-1}(E)$. By considering all observables, that is, all such maps a, we obtain $\mathcal{P}\{1,\ldots,n\} \cong \mathcal{P}(\Sigma^n_{\mathrm{cl}})$ as the algebra of propositions about the system. Inclusion of sets encodes implication of propositions.

Envision the pure states Σ^n_{qm} of a quantum system as the one-dimensional subspaces of $\mathcal{H}^n$, that is, a pure state is the set of fixed points r_1 of a density operator r with $\mathsf{spec}(r) \subseteq \{0,1\}$. For an observable on a quantum system, i.e., a self-adjoint operator A with spectrum $\mathsf{spec}(A)$, any proposition about the system of the form

"The value of quantum observable A is contained in $E \subseteq \mathsf{spec}(A)$"

encodes as the fixed points of the projector P^E_A, since the states included in it will yield an outcome in E in a measurement of the observable A (the probability of all other outcomes is zero). This set of fixed points is a subspace of $\mathcal{H}^n$. By considering all observables, that is, all self-adjoint operators A, we obtain $\mathbb{L}^n \subset \mathcal{P}(\Sigma^n_{\mathrm{qm}})$ as the algebra of propositions about the system. And once again, inclusion of subspaces encodes implication of propositions.

And so now, to briefly state things, in this section we establish the following: Though the domain of quantum states is grounded in commutativity, it is nevertheless a genuine *quantum structure* in the sense of present day theoretical physics.

Definition 35 Let D be a dcpo. An element $x \in D$ is *irreducible* if

$$\bigwedge(\uparrow x \cap \max(D)) = x.$$

The set of irreducible elements in D is $\mathrm{Ir}(D)$.

Our first result establishes that the irreducible states in Δ^n have an unmistakable operational significance: They are precisely the states one derives by applying all possible combinations of projections p_i to the initial state $\perp \in \Delta^n$.

Lemma 29 *For $x \in \Delta^n$, the following are equivalent:*

(i) *The state x is irreducible.*

(ii) *For all* $i \in \{1, \ldots, n\}$, *either* $x_i = x^+$ *or* $x_i = 0$.
(iii) *There is a nonempty subset* $X \subseteq \max(\Delta^n)$ *with* $x = \bigwedge X$.

Proof Recall from Proposition 1(ii) that for any classical state x and any index i, $x \sqsubseteq e_i \Leftrightarrow x_i = x^+$. This fact is used implicitly in what follows.

(i) $\Rightarrow$ (ii): Let x be irreducible. Let $y \in \Delta^n$ be the classical state with $y_i = y^+ \Leftrightarrow e_i \in \uparrow x \cap \max(\Delta^n)$ and $y_i = 0$ otherwise. Then y is a lower bound for $\uparrow x \cap \max(\Delta^n)$, so $y \sqsubseteq x$. We claim that $y = x$. If $y^+ < x^+$, then

$$\sum_{i=1}^{n} x_i \geq \sum_{x_i = x^+} x_i > \sum_{y_i = y^+} y_i = 1,$$

which contradicts the fact that x is a classical state. Then $y^+ \geq x^+$. But since $y \sqsubseteq x$, we know $y^+ \leq x^+$. Thus, $x^+ = y^+$, which gives $x = y$. This proves (ii).

(ii) $\Rightarrow$ (i): The proof is by induction. It is true for $n = 2$. Assume it for Δ^n, and let $x \in \Delta^{n+1}$ be a state of the desired form. We can assume x is not pure, since otherwise x is clearly irreducible.

Let $y \in \Delta^{n+1}$ be a lower bound for the set $\uparrow x \cap \Delta^{n+1}$. Because x is not pure, y cannot be pure either. Let i be any index with $1 \leq i \leq n+1$. Then $p_i(x)$ has the form mentioned in (ii), so it is irreducible by the inductive hypothesis. The state $p_i(y)$ is a lower bound of $\uparrow p_i(x) \cap \max(\Delta^n)$, so the irreduciblity of $p_i(x)$ gives $p_i(y) \sqsubseteq p_i(x)$. Then $y \sqsubseteq x$. This puts $x \in \mathrm{Ir}(\Delta^{n+1})$.

(iii) $\Rightarrow$ (i): Let y be the state with $y_i = y^+$ iff $e_i \in X$ and $y_i = 0$ otherwise. By (i)=(ii), y is irreducible, while $X = \uparrow y \cap \max(\Delta^n)$ gives

$$x = \bigwedge X = \bigwedge(\uparrow y \cap \max(\Delta^n)) = y,$$

which shows that x is irreducible. □

(i) $\Rightarrow$ (iii): Obvious.

Now we prove that $\mathcal{P}\{1, \ldots, n\}$ is recoverable from the irreducible elements of Δ^n. Specifically, $\mathrm{Ir}(D)$ in the order it inherits from D is order isomorphic to a subset of $\mathcal{P}(\max(D))$ ordered by *reverse inclusion*, so we must consider $\mathrm{Ir}(D)$ in its dual order, $\mathrm{Ir}(D)^*$. Second, since every $x \in D$ has a maximal element above it, the empty set is not represented in $\mathrm{Ir}(D)^*$, so we adjoin a least element 0 to obtain the poset

$$\mathrm{Ir}(D)^*_\perp := \mathrm{Ir}(D)^* \cup \{0\}.$$

Proposition 21 *For any* $n \geq 2$,

$$\mathrm{Ir}(\Delta^n)^*_\perp \simeq \mathcal{P}\{1, \ldots, n\}.$$

Proof Let $e : \{1, \ldots, n\} \to \max(\Delta^n)$ be the natural bijection that takes an outcome i to its associated pure state $e(i) = e_i$. An order isomorphism

$$\varphi : \mathrm{Ir}(\Delta^n)^* \to \mathcal{P}\{1, \cdots, n\} \setminus \{\emptyset\}$$

is then given by

$$\varphi(x) = e^{-1}\big(\uparrow x \cap \max(\Delta^n)\big).$$

First, φ is surjective: Given $\emptyset \neq X \in \mathcal{P}\{1, \ldots, n\}$,

$$\varphi\Big(\bigwedge e(X)\Big) = X,$$

using Lemma 29(iii). Next, it is an order embedding: For $x, y \in \mathrm{Ir}(\Delta^n)^*$,

$$\begin{aligned} x \sqsubseteq y &\Leftrightarrow \uparrow x \cap \max(\Delta^n) \subseteq \uparrow y \cap \max(\Delta^n) \\ &\Leftrightarrow e^{-1}\big(\uparrow x \cap \max(\Delta^n)\big) \subseteq e^{-1}\big(\uparrow y \cap \max(\Delta^n)\big) \\ &\Leftrightarrow \varphi(x) \subseteq \varphi(y). \end{aligned}$$

Now we simply extend φ to an order isomorphism from $\mathrm{Ir}(\Delta^n)^*_\perp$ to $\mathcal{P}\{1, \cdots, n\}$ by setting $\varphi(0) = \emptyset$, and the proof is finished. □

We turn now to the analogous result for quantum states. To stress the analogy with classical states we denote pure states now as $\max(\Omega^n)$ rather than as Σ^n.

First we prove the analogue of Proposition 1(ii) for quantum states. Denote by r^+ the subspace of eigenvectors for the largest eigenvalue, that is $r^+ = r_\lambda$ for $\lambda = \max\big(\mathsf{spec}(r)\big)$.

Lemma 30 *For $r \in \Omega^n$ and $t \in \max(\Omega^n)$, we have*

$$r \sqsubseteq t \Leftrightarrow t_1 \subseteq r^+$$

and thus

$$r^+ = \bigcup \{t_1 \mid t \in \uparrow r \cap \max(\Omega^n)\}.$$

Proof Let $t \in \max(\Omega^n)$ be such that $t_1 \subseteq r^+$. Define a labeling e that satisfies

- $e_1 = t_1$,
- $e_2, \ldots, e_{\mathsf{dim}(r^+)} \in r^+ \cap t_1^\perp$, and,
- $e_{\mathsf{dim}(r^+)+1}, \ldots, e_n \in \bigcup \mathcal{D}_r \cap (r^+)^\perp$.

By Proposition 1(ii) we then have $\mathsf{spec}(r|e) \sqsubseteq \mathsf{spec}(t|e)$ and thus $r \sqsubseteq t$.

Conversely, let $t \in \max(\Omega^n)$ be such that $r \sqsubseteq t$. Then there exists labeling e such that $[r, e] = [t, e] = 0$, that is $\bigcup \mathcal{D}_e \subseteq \bigcup \mathcal{D}_r \cap \bigcup \mathcal{D}_t$, which implies since $\mathcal{D}_t = \{t_1^\perp, t_1\}$ with t_1 one-dimensional that there exists $i \in \{1, \ldots, n\}$ such that $t_1 = e_i$, say $i = 1$. Then

$$\mathsf{spec}(r|e) \sqsubseteq \mathsf{spec}(t|e) = (1, 0, \ldots, 0)$$

in Δ^n. By Prop. 1(ii) it then follows that $\langle r|e_i\rangle = \mathsf{spec}(r|e)^+$ so $e_i \subseteq r^+$ and thus $t_1 \subseteq r^+$. □

A quantum state is irreducible iff its spectrum can be viewed as an irreducible classical state.

Lemma 31 *For $r \in \Omega^n$, the following are equivalent:*

(i) *The state r is irreducible.*
(ii) *There is a labeling e with $[r, e] = 0$ and $\mathsf{spec}(r|e) \in \mathrm{Ir}(\Delta^n)$.*
(iii) *Either there exists $\lambda \in (0, 1]$ such that $\mathsf{spec}(r) = \{0, \lambda\}$ or $r = \perp$.*

In either case, $\mathsf{spec}(r|e) \in \mathrm{Ir}(\Delta^n)$ for any labeling e with $[r, e] = 0$.

Proof By Lemma 29, (ii) ⇔ (iii) is obvious. The rest of the proof essentially relies on the analogous result for classical states.

(i) ⇒ (iii) Let r be irreducible and e be a labeling with $[r, e] = 0$. Let

$$X = \{\mathsf{spec}(t|e) : t \in \uparrow r \cap \max(\Omega^n) \cap (\Omega^n|e)\}.$$

By Lemma 29(iii), the infimum of X is an irreducible classical state, and we use this to implicitly define a quantum state $s \in \Omega^n$ by

$$\mathsf{spec}(s|e) := \bigwedge X \in \mathrm{Ir}(\Delta^n).$$

By the definition of $\mathsf{spec}(s|e)$, we immediately have $\mathsf{spec}(r|e) \sqsubseteq \mathsf{spec}(s|e)$, which implies $r \sqsubseteq s$ in Ω^n.

We claim that $r = s$. To prove this, we need only show that

$$\uparrow r \cap \max(\Omega^n) \subseteq \uparrow s \cap \max(\Omega^n),$$

for then we have

$$\begin{aligned} r \sqsubseteq s &\Rightarrow \uparrow r \cap \max(\Omega^n) = \uparrow s \cap \max(\Omega^n) \\ &\Rightarrow r = \bigwedge \uparrow r \cap \max(\Omega^n) = \bigwedge \uparrow s \cap \max(\Omega^n) \\ &\Rightarrow s \sqsubseteq r \end{aligned}$$

using the irreducibility of r.

Let $t \in \uparrow r \cap \max(\Omega^n)$. By Lemma 30, $t_1 \subseteq r^+$, and since $r^+ \subseteq s^+$, $t_1 \subseteq s^+$, which again by Lemma 30 gives $t \in \uparrow s \cap \max(\Omega^n)$. But why do we have $r^+ \subseteq s^+$?

This is the crucial part of the argument: Since $[r, e] = 0$, $\bigcup \mathcal{D}_e \subseteq \bigcup \mathcal{D}_r$, so $\mathcal{D}_e$ contains a subset S of cardinality $\mathsf{dim}(r^+)$ whose union is contained in r^+. Each element of S is a one dimensional subspace of $\mathcal{H}^n$, so the usual bijection allows us to treat S as a collection of pure states.

For each pure state $t \in S$, we have $[t, e] = 0$, since $t_1 \in \mathcal{D}_e$, and $t \in \uparrow r \cap \max(\Omega^n)$, using Lemma 30 and $t_1 \subseteq r^+$. Then by the definition of s, $s \sqsubseteq t$, while Lemma 30 gives $t_1 \subseteq s^+$. But then, because s^+ is a subspace, we clearly have

$$r^+ = \mathsf{span}(\{t_1 : t \in S\}) \subseteq s^+,$$

which proves $r = s$. Thus, $\mathsf{spec}(r|e) = \mathsf{spec}(s|e)$ is irreducible in Δ^n.

(iii) $\Rightarrow$ (i) Let $r = \perp$. Then $\uparrow \perp \cap \max(\Omega^n) = \max(\Omega^n)$. If $s \sqsubseteq \max(\Omega^n)$ (pointwisely) then by Lemma 30 it follows that

$$\mathcal{H}^n = \{t_1 \mid t \in \max(\Omega^n)\} \subseteq s^+ .$$

Thus $s = \perp$ so $s \sqsubseteq r$ and as such since trivially $\perp \sqsubseteq \max(\Omega^n)$ (pointwisely) we conclude $\perp = \bigwedge(\uparrow \perp \cap \max(\Omega^n))$.

Let $\mathsf{spec}(r) = \{0, \lambda\}$. First, $r \sqsubseteq \uparrow r \cap \max(\Omega^n)$ (pointwisely) is again trivial. Second, let $s \sqsubseteq \uparrow r \cap \max(\Omega^n)$ (pointwisely). Then,

$$\uparrow r \cap \max(\Omega^n) \subseteq \uparrow s \cap \max(\Omega^n)$$

so it follows by Lemma 30 that

$$\begin{aligned} r^+ &= \bigcup \{t_1 \mid t \in \uparrow r \cap \max(\Omega^n)\} \\ &\subseteq \bigcup \{t_1 \mid t \in \uparrow s \cap \max(\Omega^n)\} \\ &= s^+ . \end{aligned}$$

Now define a labeling e that satisfies

- $e_1, \ldots, e_k \subseteq r^+$ for $k := \mathsf{dim}(r^+)$,
- $e_{k+1}, \ldots, e_{k+l} \subseteq (r^+)^\perp \cap s^+$ for $l := \mathsf{dim}(s^+) - \mathsf{dim}(r^+)$, and,
- $e_{k+l+1}, \ldots, e_n \subseteq (s^+)^\perp$ where $1 - l - k = 1 - \mathsf{dim}(s^+)$.

We have $[r, e] = [s, e] = 0$, while Lemma 29 gives $\mathsf{spec}(s|e) \sqsubseteq \mathsf{spec}(r|e)$ since $\mathsf{spec}(r|e)$ is irreducible in Δ^n. Thus, $s \sqsubseteq r$. □

Theorem 15 *For any* $n \geq 2$,

$$\mathrm{Ir}(\Omega^n)^*_\perp \simeq \mathbb{L}^n.$$

Proof An order isomorphism $\varphi : \mathrm{Ir}(\Omega^n)^* \to \mathbb{L}^n \setminus \{0\}$ is given by

$$\varphi(r) = r^+.$$

For its surjectivity, given any $A \in \mathbb{L}^n \setminus \{0\}$, define an irreducible quantum state $r : \{0, \lambda\} \to \mathbb{L}^n$ by $r_\lambda = A$ and $r_0 = A^\perp$, where $\lambda = 1/\mathsf{dim}(A) > 0$. Then $\varphi(r) = A$. The fact that it is an order isomorphism follows straightforwardly from quantum degeneration (Lemma 21).

The particular nature of this proof, which essentially relies on how we recover $\mathcal{P}\{1, \ldots, n\}$ from Δ^n, exhibits how much of the structure of Ω^n is already present in the partial order on Δ^n.

To summarize, we are able to recover $\mathbb{L}^n$, the basic quantum structure from which all other are derivable, from the domain of quantum states in a purely order theoretic manner. Here is an analogy worth remembering: Ω^n is to $\mathbb{L}^n$ as density operators are to pure states. More to the point, in view of the fact that

- The canonical order theoretic structure corresponding to quantum mechanics in terms of only pure states is $\mathbb{L}^n$,

we are tempted to claim that

- The canonical order theoretic structure corresponding to quantum mechanics in terms of density operators is Ω^n.

In short, because the density operator formulation offers a more complete picture than simply working with pure states, the domain Ω^n offers a more complete picture than the lattice $\mathbb{L}^n$.

Finally, let us add one last twist to the story: Not only does this more complete picture emerge as the result of *commutative considerations*, but any natural approach to ordering states which allows non-commutativity seems destined to fail!

Fact 1 *If we define $r \sqsubseteq s$ for $r, s \in \Omega^n$ by either*

(i) *"there exists a labeling e such that* $\mathsf{spec}(r|e) \sqsubseteq \mathsf{spec}(s|e)$ *in Δ^n," or*
(ii) *"for all labelings e we have* $\mathsf{spec}(r|e) \sqsubseteq \mathsf{spec}(s|e)$ *in Δ^n,"*

where e does not necessarily commute with r and s, then in both cases, the relation $\sqsubseteq$ is not an information order.

Justification We will only provide explicit proofs of the following partial statements for the case of $n = 2$ (arguments in higher dimensions are essentially of the same nature):

- In case (i), all states (including bottom) are above all pure states.
- In case (ii), no state (including bottom) is strictly below a pure state.

Let r be a pure state with $\psi \in r_1$ and let $\psi^\perp \in r_0$. Then there exists a labeling e such that $\psi + \psi^\perp \in e_1$ and $\psi - \psi^\perp \in e_2$. For this labeling e we have $\langle r|e_1\rangle = \langle r|e_2\rangle = 1/2$, that is, $\mathsf{spec}(r|e) = \perp$ in Δ^2. Thus, for all $s \in \Omega^2$ we have $\mathsf{spec}(r|e) \sqsubseteq \mathsf{spec}(s|e)$ in Δ^2. In case (i) this implies $r \sqsubseteq s$ in Ω^2. In case (ii) this implies that we cannot have $s \sqsubset r$ in Ω^2. □

Thus, the commutativity implied by the existence of a joint labeling in the spectral order seems unavoidable if one wants to obtain a non-trivial partial order. This can be physically explained as follows: Quantum mechanics bears as one of its most fundamental principles that the maximal knowledge an observer can have about a system at a single point in time amounts to knowing the values of a class of observables that constitute a maximally commuting family; any knowledge beyond this is forbidden. Thus, on the assumption that a partial order on quantum states should make statements about knowledge we possess about a system, commutativity at some level is probably unavoidable.

10.5 Applications

We consider some basic applications of classical and quantum states.

10.5.1 A Calculus for Noise

One of the basic ideas in the measurement formalism [8] is that one can differentiate functions $f : D \to E$ between collections of informative objects with respect to underlying notions of content. Speaking abstractly, it offers a definition of "informatic rate of change," i.e., the *rate* at which (the content of) the output of a process changes with respect to (the content of) its input.

As we have seen, the domains of classical and quantum states have many natural notions of content, so in principle we ought to be able to study informatic rates of change in these settings as a means of improving our understanding about the behavior of various phenomena.

One such example arises easily in the study of noise: By modelling the effect of noise as a selfmap on classical or quantum states, we can apply the informatic derivative with respect to a preferred notion of content μ to gain a precise measure of the effect a given form of noise f has on a given state σ. For ease of exposition, we illustrate the idea on Δ^2. Here are some natural candidates for μ:

- $\mu x = 1 - x^+$
- $\mu x = 2x^+x^-$
- $\mu x = -x^+ \log x^+ - x^- \log x^-$ (Shannon entropy)

We'll use the first since it is the simplest.

Definition 36 A *noise operator* is a function $f : \Delta^2 \to \Delta^2$ such that $f\sigma \sqsubseteq \sigma$.

The intuition in this definition is that noise *qualitatively* increases uncertainty. Now, suppose a system is in state σ when it suffers an unwanted interaction with its environment, which changes its state to $f\sigma$. How can we measure the effect of the noise on the state of the system?

First, we write down a "grammar" which allows for the description of noise: A simple class of noise operators $\mathbb{N}$ is

- $\perp, 1 \in \mathbb{N}$
- $f, g \in \mathbb{N} \Rightarrow f \circ g \in \mathbb{N}$
- $f, g \in \mathbb{N} \Rightarrow pf + (1-p)g \in \mathbb{N}$ for $p \in [0, 1]$,
- $f, g \in \mathbb{N}$ & $f \sqsubseteq g \Rightarrow pf^* + (1-p)g \in \mathbb{N}$ for $p \in [0, 1/2]$,

where $*$ is the involution $(x, y)^* = (y, x)$. It is straightforward to check that the class of noise operators on Δ^2 are closed under the operations mentioned above.

Now the effect that channel $f \in \mathbb{N}$ has on state σ can be systematically calculated as follows:

Theorem 16 *If $f, g \in \mathbb{N}$, then*

- $d(\perp)_\mu(\sigma) = 0$,
- $d(1)_\mu(\sigma) = 1$,
- $d(f \circ g)_\mu(\sigma) = df_\mu(g\sigma) \cdot dg_\mu(\sigma)$,
- $d(pf + (1-p)g)_\mu(\sigma) = pdf_\mu(\sigma) + (1-p)dg_\mu(\sigma)$,
- $d(pf^* + (1-p)g)_\mu(\sigma) = (1-p)dg_\mu(\sigma) - pdf_\mu(\sigma)$,

for any $\sigma \neq \perp$.

This theorem allows us to verify *inductively* that $df_\mu(\sigma)$ is a measure of *reliability*. For instance, if $df_\mu(\sigma) = 0$, then the noise f has had a very strong effect on σ (as a channel, f is *unreliable* for the transmission of σ), while if $df_\mu(\sigma) = 1$, we intuitively expect $f(\sigma) = \sigma$, i.e., f is completely reliable.

Lemma 32 *If f is a noise operator and $f\sigma = \sigma$, then either $df_\mu(\sigma) \geq 1$ or it does not exist.*

We now have a fun and *systematic* approach to an interesting problem: Determining the states that a particular type of noise does not affect.

Example 11 Consider the depolarization of a classical state,

$$f\sigma = p\perp + (1-p)\sigma.$$

For $\sigma \neq \perp$, we have

$$df_\mu = pd(\perp)_\mu + (1-p)d(1)_\mu = 1 - p,$$

so the only unaffected state is $\perp$ for $p > 0$.

Example 12 The effect of a magnetic field on data stored on a disk is

$$f\sigma = p\sigma^* + (1-p)\sigma.$$

For $\sigma \neq \perp$, we have

$$df_\mu = -pd(1)_\mu + (1-p)d(1)_\mu = 1 - 2p.$$

Thus, if you are a state, it is better to be depolarized than flipped.

In quantum mechanics, the study of noise and how to beat it is called *decoherence*. In the quantum case, some neat measures of content arise, corresponding to the classical ones:

- $\mu x = 1 - x^+ \implies \mu\rho = 1 - \mathrm{spec}(\rho)^+$,
- $\mu x = 2x^+x^- \implies \mu\rho = 1 - \mathrm{tr}(\rho^2)$,
- $\mu x = -x^+ \log x^+ - x^- \log x^- \implies \mu\rho = -\mathrm{tr}(\rho \log \rho)$.

For consistency, we use the first one here as well.

Example 13 Depolarization of quantum states is

$$f(r) = t \cdot \frac{I}{n} + (1-t)r = t\bot + (1-t)r.$$

Once again, $df_\mu(r) = 1 - t$. But the reason is physical. For instance, in the two dimensional case we have

$$f(r) = (1-t)r + \frac{t}{3}(\sigma_x r \sigma_x + \sigma_y r \sigma_y + \sigma_z r \sigma_z)$$

It affects the entire state in a uniform way. Quantum bit/phase flipping, by contrast, only affects "part" of r. Things get more interesting then.

10.5.2 The Axioms of Domain Theory

This work led to the introduction of a new class of domains, the *exact domains*. We will show in this section that exact domains offer a new perspective on the more traditional, *continuous* domains [1]. With the benefit of this new point of view, it then becomes possible to ask certain foundational questions that some domain theorists may find intriguing.

Recall that in the study of approximation on classical states, we learned that $x \ll y$ is a statement which implicitly carries a specific *context*. In order to conclude $x \ll z$ when $y \sqsubseteq z$, we need to know that the statement $y \sqsubseteq z$ is being made in the same context as $x \ll y$. Aside from the case when x approximates a pure state (Prop. 7), there is another way of ensuring this: If *all* entities involved (x, y, z) can be regarded as necessary for a *single* state ($\Uparrow z \neq \emptyset$).

Proposition 22 (Context) *For all $x, y, z \in \Delta^n$, if $x \ll y \sqsubseteq z$ and $\Uparrow z \neq \emptyset$, then $x \ll z$.*

Proof First we prove that if $x, y, z \in \Lambda^n$ with $x \ll y \sqsubseteq z$ in Λ^n and $z_i > 0$ for all i, then $x \ll z$ in Λ^n. Let $z = \bigsqcup w_k$ where $(w_k)_{k\geq 1}$ is increasing in Λ^n. Now we proceed just as in the proof of Theorem 4. For $x_i = x_{i+1} > 0$, we can take $k_i = 1$, since the monotonicity of w_k implies

$$\frac{x_i}{x_{i+1}} = 1 \leq \frac{\pi_i(w_k)}{\pi_{i+1}(w_k)}$$

for all k, while in the case of $x_i > x_{i+1} > 0$, degeneration (Lemma 5) gives $y_i > y_{i+1} > 0$, which accounts for the strict inequality in

$$\frac{x_i}{x_{i+1}} < \frac{y_i}{y_{i+1}} \leq \frac{z_i}{z_{i+1}} = \lim_{k\to\infty} \frac{\pi_i(w_k)}{\pi_{i+1}(w_k)}.$$

The definition of limit again makes it clear that the required k_i exists.

More generally, if $x \ll y \sqsubseteq z \ll w$ in Δ^n, we use Proposition 6(ii) to prove $x \ll z$. First, $z \in \Delta^n_\sigma \Rightarrow x \in \Delta^n_\sigma$ follows from Lemma 12(i) since $z_i > 0$ for all i using Lemma 13 and $z \ll w$. And second, since Proposition 6(ii) gives $r(x) \ll r(y) \sqsubseteq r(z)$ in Λ^n, our opening argument now applies leaving $r(x) \ll r(z)$ in Λ^n.

The value of this observation is that it provides a theoretical explanation for why the approximation relation on Δ^n is interpolative:

Lemma 33 *If D is an exact dcpo such that for all* $x, y, z \in D$,

$$x \ll y \sqsubseteq z \Rightarrow x \ll z,$$

whenever $\uparrow\!\!\uparrow z \neq \emptyset$, *then* $\ll$ *is interpolative. Moreover, a dcpo is continuous iff it is exact and* $x \ll y \sqsubseteq z \Rightarrow x \ll z$ *for all* $x, y, z \in D$.

Proof The proof given in [1] applies unchanged. □

From this we can see that exact domains require *precision* when reasoning about approximation. By contrast, the single most important aspect of approximation on a continuous domain is *not* that it is interpolative [1], but rather that it is *context independent*. The present work seems to provide sufficient impetus for investigating domains beyond the continuous variety.

10.5.3 Qualitative Measures of Entanglement

Quantum entanglement is the essential feature in quantum communication schemes and quantum cryptographic protocols that distinguishes them from their classical counterparts. For the particular dialectics used here we refer to the standard literature on the matter.

We illustrate by means of a series of examples how the results of this paper can be applied to the study of entanglement. A full development on the matter is in preparation.

Example 14 Measures of entanglement of bipartite quantum systems. Let $\mathcal{H}^n$ be a n-dimensional complex Hilbert space. According to Schmidt's biorthogonal decomposition theorem [15], any bipartite state $\Psi \in \mathcal{H}^n \otimes \mathcal{H}^n$ can be rewritten as

$$\Psi = \sum_i c_i \psi_i \otimes \phi_i$$

with (ψ_i) and (ϕ_i) orthonormal bases of $\mathcal{H}^n$ and (c_i) positive real coifficients which are as a set uniquely defined. In particular we have $\sum_i c_i^2 = 1$ due to normalization of Ψ, so every $\Psi \in \mathcal{H}^n \otimes \mathcal{H}^n$ defines a unique classical state $c := (c_i^2)$. We can then qualitatively measure entanglement using the dcpo Λ^n as

$$\mathsf{Ent} : \mathcal{H}^n \otimes \mathcal{H}^n \to \Lambda^n : \Psi \mapsto r(c)$$

where r is the usual retraction on classical states.

Moreover, every measure of content

$$\mu : \Lambda^n \to [0, 1]^*$$

gives rise to a quantitative measure of entanglement

$$\mu \cdot \mathsf{Ent} : \mathcal{H}^n \otimes \mathcal{H}^n \to [0, 1]^* .$$

When taking as μ Shannon entropy we find the usual quantitative measure of entanglement for bipartite quantum systems.

The maximal element of Λ^n then encodes the non-entangled states, that is, the pure tensors $\psi \otimes \phi$. The minimal element of Λ^n then encodes the maximally entangled state, that is

$$\sum_i \frac{1}{\sqrt{n}} \psi_i \otimes \phi_i \in \mathcal{H}^n \otimes \mathcal{H}^n$$

which does not depend on the choice of bases.

Since $\mu : \Lambda^2 \to [0, 1]^*$ is a duality, using Λ^2 rather than in $[0, 1]^*$ doesn't teach us much for the case $n = 2$, that is, for a pair of qubits. For qutrits however, $n = 3$, we capture essential qualitative differences by valuating in Λ^3.

Consider for example the state

$$\mathsf{S} := \frac{1}{\sqrt{2}}(\psi_1 \otimes \phi_1 + \psi_2 \otimes \phi_2) \in \mathcal{H}^3 \otimes \mathcal{H}^3,$$

that is,

$$\mathsf{S} = \frac{1}{\sqrt{2}}(|00\rangle + |11\rangle).$$

The state S is entangled but this entanglement has essentially a qubit nature, that is, we can express the state by only using a subbase of $\mathcal{H}^3$ that contains two vectors. In particular, the entanglement coincides with that of the EPR or singlet state so it is maximal as qubit entanglement.

On the other hand, the states

$$\mathsf{T}_q := q(\psi_1 \otimes \phi_1) + \frac{1-q}{2}(\psi_2 \otimes \phi_2 + \psi_3 \otimes \phi_3) \in \mathcal{H}^3 \otimes \mathcal{H}^3$$

for $1/3 < q < 1$, that is,

$$\mathsf{T}_q := q(|00\rangle) + \frac{1-q}{2}(|11\rangle + |22\rangle)\,,$$

exhibits genuine qutrit entanglement.

Unfortunately, for q ranging in $(\frac{1}{3}, 1)$ Shannon entropy ranges in $(0, 1)$ so some T_q have entropy higher than S and some have entropy less than S. The valuation $\mu \cdot \mathsf{Ent}$ as such doesn't capture the qualitative feature that distinquishes between maximal qubit-type entanglement and essentially qutrit type entanglement.

However, Λ^3 does. Indeed, consider

$$\mathsf{Ent}(\mathsf{S}) = r\left(\frac{1}{2}, \frac{1}{2}, 0\right) \quad \text{and} \quad \mathsf{Ent}(\mathsf{T}_q) = r\left(q, \frac{1-q}{2}, \frac{1-q}{2}\right),$$

that is,

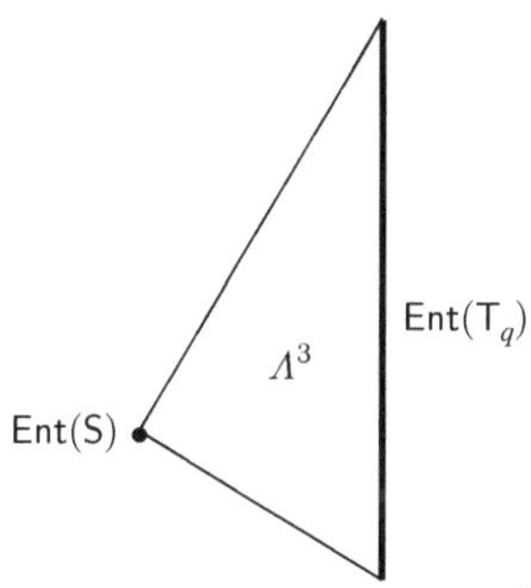

in graphical terms. Since there is no value for $q \in (\frac{1}{3}, 1)$ for which we have that $(1/2, 1/2, 0)$ and $(q, (1-q)/2, (1-q)/2)$ compare in Λ^3, it follows for all T_q with $q \in (\frac{1}{3}, 1)$ that

$$\mathsf{Ent}(\mathsf{S}) \not\sqsubseteq \mathsf{Ent}(\mathsf{T}_q) \quad \text{and} \quad \mathsf{Ent}(\mathsf{T}_q) \not\sqsubseteq \mathsf{Ent}(\mathsf{S})\,.$$

The states $\Psi \in \mathcal{H}^3 \otimes \mathcal{H}^3$ for which we have $\mathsf{Ent}(\Psi) \sqsubseteq \mathsf{Ent}(\mathsf{S})$ are those which are such that

$$\mathsf{Ent}(\Psi) = r\left(q, \frac{1-q}{2}, \frac{1-q}{2}\right)$$

for $0 \leq q \leq 1/3$, that is, that are convex combinations of S and the maximally entangled state in $\mathcal{H}^3 \otimes \mathcal{H}^3$,

$$\frac{1}{\sqrt{3}}(|00\rangle + |11\rangle + |22\rangle)\,,$$

for which we set

$$\top := \mathsf{Ent}\left(\frac{1}{\sqrt{3}}(|00\rangle + |11\rangle + |22\rangle)\right) .$$

Graphically,

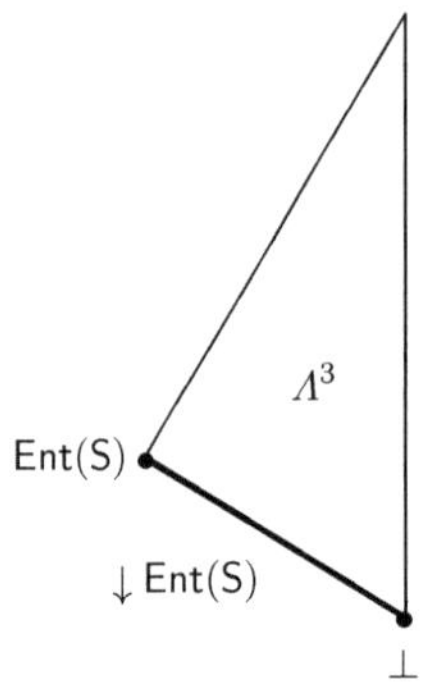

where $\downarrow \mathsf{Ent}(\mathsf{S})$ is the lower set of $\mathsf{Ent}(\mathsf{S})$ in Λ^3.

The states $\Psi \in \mathcal{H}^3 \otimes \mathcal{H}^3$ for which we have $\mathsf{Ent}(\mathsf{S}) \sqsubseteq \mathsf{Ent}(\Psi)$ are those which are such that

$$\mathsf{Ent}(\Psi) = r\,(q, 1-q, 0)$$

for $0 \le q \le 1/2$, that is, convex combinations of S and the minimally entangled state in $\mathcal{H}^3 \otimes \mathcal{H}^3$ (the pure tensor $|00\rangle$), for which we set

$$\top := \mathsf{Ent}(|00\rangle) .$$

Graphically,

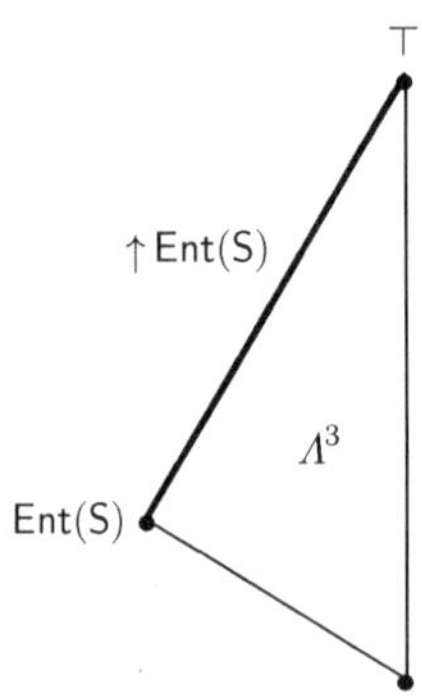

where $\uparrow \mathsf{Ent}(\mathsf{S})$ is the upper set of $\mathsf{Ent}(\mathsf{S})$ in Λ^3.

We can now refine our qualitative representation of entanglement for bipartite states using the order on quantum states.

Example 15 Qualitative entanglement of bipartite quantum systems In Example 14, the quantitative valuation $\mu \cdot \mathsf{Ent}$ with μ Shannon entropy, that is, the usual valuation attributed to a bipartite quantum system in order to measure entanglement, can equivalently be defined as the von Neumann entropy of one of the quantum states $\rho_1(\Psi)$ or $\rho_2(\Psi)$ for $\Psi \in \mathcal{H}^n \otimes \mathcal{H}^n$ that arise by tracing over the other system.

Explicitly, for $\Psi = \sum_i c_i \psi_i \otimes \phi_i$ we obtain

$$\rho_1(\Psi) := \mathsf{tr}_2(\Psi) = \begin{pmatrix} c_1^2 & & 0 \\ & \ddots & \\ 0 & & c_n^2 \end{pmatrix} \text{ in } (\psi_i)$$

$$\rho_2(\Psi) := \mathsf{tr}_1(\Psi) = \begin{pmatrix} c_1^2 & & 0 \\ & \ddots & \\ 0 & & c_n^2 \end{pmatrix} \text{ in } (\phi_i)\,.$$

Since the diagonals coincide, von Neumann entropy coincides and in either case gives the same value.

This implies that we can refine the valuation of entanglement Ent in Example 14 as

$$\mathsf{Ent}^{\Omega} : \mathcal{H}^n \otimes \mathcal{H}^n \to \Omega^n \times \Omega^n : \Psi \mapsto (\rho_1(\Psi), \rho_2(\Psi))$$

where $\Omega^n \times \Omega^n$ is ordered pointwisely, that is,

$$(r_1, r_2) \sqsubseteq (s_1, s_2) \;\Leftrightarrow\; r_1 \sqsubseteq r_2 \text{ and } s_1 \sqsubseteq s_2\,.$$

On $\Omega^n \times \Omega^n$ we can then define as a measure of content

$$\mu_{1,2} : \Omega^n \times \Omega^n \to [0,1]^* : (r_1, r_2) \mapsto \frac{\mu(r_1) + \mu(r_2)}{2}$$

where μ is von Neumann entropy. This results in a quantitative measure of entanglement on $\mathcal{H}^n \otimes \mathcal{H}^n$ that exactly coincides with the usual one. Indeed,

$$\mu_{1,2}\left(\mathsf{Ent}^{\Omega}(\Psi)\right) = \frac{\mu_1(\rho_1(\Psi)) + \mu_2(\rho_2(\Psi))}{2} = \mu(\rho_1(\Psi)) = \mu(\rho_2(\Psi))\,.$$

Note here in particular that Ent^{Ω} "almost" turns the states in $\mathcal{H}^n \otimes \mathcal{H}^n$ into a domain by setting

$$\Psi \sqsubseteq \Phi \;\Leftrightarrow\; \mathsf{Ent}^{\Omega}(\Psi) \sqsubseteq \mathsf{Ent}^{\Omega}(\Phi)\,.$$

We obtain a preorder that has pure tensors as maximal elements and that has $\perp$ as a minimum.

We however lose some anti-symmetry in this passage. In particular, when considering the Schmidt base, the order loses track of relative phases between base vectors. Indeed,

$$\mathsf{Ent}^{\Omega}(\psi_1 \otimes \phi_1 + \psi_2 \otimes \phi_2) = \mathsf{Ent}^{\Omega}(\psi_1 \otimes \phi_1 + i\psi_2 \otimes \phi_2)$$

although

$$\mathsf{ray}(\psi_1 \otimes \phi_1 + \psi_2 \otimes \phi_2) \neq \mathsf{ray}(\psi_1 \otimes \phi_1 + i\psi_2 \otimes \phi_2)$$

so these vectors do not encode the same state.

However, this can be fixed by taking into account their phases in defining the order. We will provide the details in a future paper.

Pure tensors avoid this since

$$\psi \otimes (i\phi) = (i\psi) \otimes \phi = i(\psi \otimes \phi)$$

for which we have

$$\mathsf{ray}(i\psi \otimes \phi) = \mathsf{ray}(\psi \otimes \phi)\ .$$

The maximally entangled states do not depend on the bases at all.

The essential difference between the qualitative valuations Ent^{Ω} and Ent is the fact that Ent^{Ω} takes into account the identity of pure tensors above.

Example 16 Qualitative entanglement of multipartite quantum systems In Example 14 we measured entanglement of bipartite quantum systems using unicity of the coefficients in the Schmidt biorthogonal decomposition. There however does not exist a similar construction for arbitrary multipartite sytems, that is, there is no Schmidt-type decomposition theorem for arbitrary $\mathcal{H}^n \otimes \ldots \otimes \mathcal{H}^n$.

In particular, up to now there was not even a satisfactory notion of maximal entanglement e.g. see [10]. Indeed, when considering three partite qubit states, for the Greenberger-Horn-Zeilinger state [6]

$$\mathsf{GHZ} := \frac{1}{\sqrt{2}}(|000\rangle + |111\rangle)$$

and the W-state

$$\mathsf{W} := \frac{1}{\sqrt{3}}(|100\rangle + |010\rangle + |001\rangle)$$

there are conflicting arguments about which one is maximally entangled. The general favourite is however GHZ in particular in view of its maximal violation of certain type of inequalities (e.g. Bell's) that are characteristic for entanglement.

The solution of this conflict lies in specification of a context with respect to which one measures entanglement, in the sense of Example 15.

We propose here a qualitative measure for multipartite entanglement that favours GHZ as the maximally entangled state, allong the lines of the valuation in Example 15 for bipartite entanglement.

Define

$$\mathsf{Ent}^{\Omega} : \mathcal{H}^n \otimes \ldots \otimes \mathcal{H}^n \to \Omega^n \times \ldots \times \Omega^n : \Psi \mapsto \Big(\rho_1(\Psi), \ldots, \rho_m(\Psi)\Big)$$

where $\rho_i(\Psi)$ arises by tracing over all systems except the ith. We can do this for example by considering the Schmidt decomposition for $\mathcal{H}^n \otimes (\mathcal{H}^n \otimes \ldots \otimes \mathcal{H}^n)$ where the single Hilbert space encodes the ith system.

We then obtain for the above examples that

$$\mathsf{Ent}^{\Omega}(\mathsf{GHZ}) = \left(\begin{pmatrix} 1/2 & 0 \\ 0 & 1/2 \end{pmatrix}, \begin{pmatrix} 1/2 & 0 \\ 0 & 1/2 \end{pmatrix}, \begin{pmatrix} 1/2 & 0 \\ 0 & 1/2 \end{pmatrix}\right)$$

since we have

$$\mathsf{GHZ} = \frac{1}{\sqrt{2}}(|0\rangle|00\rangle + |1\rangle|11\rangle)$$

with respect to the 1st component and

$$\mathsf{Ent}^{\Omega}(\mathsf{W}) = \left(\begin{pmatrix} 2/3 & 0 \\ 0 & 1/3 \end{pmatrix}, \begin{pmatrix} 2/3 & 0 \\ 0 & 1/3 \end{pmatrix}, \begin{pmatrix} 2/3 & 0 \\ 0 & 1/3 \end{pmatrix}\right)$$

since for example

$$\mathsf{W} := \frac{\sqrt{2}}{\sqrt{3}}|0\rangle\left(\frac{1}{\sqrt{2}}(|10\rangle + |01\rangle)\right) + \frac{1}{\sqrt{3}}|1\rangle|00\rangle$$

and as such it follows that

$$\mathsf{Ent}^{\Omega}(\mathsf{GHZ}) \sqsubset \mathsf{Ent}^{\Omega}(\mathsf{W})\,.$$

Depicting only the part of Ω^2 containing the relevant pure states $|0\rangle$ and $|1\rangle$ here, that is, a copy of Δ^2, this represents graphically as $\mathsf{Ent}^{\Omega}(\mathsf{GHZ})$

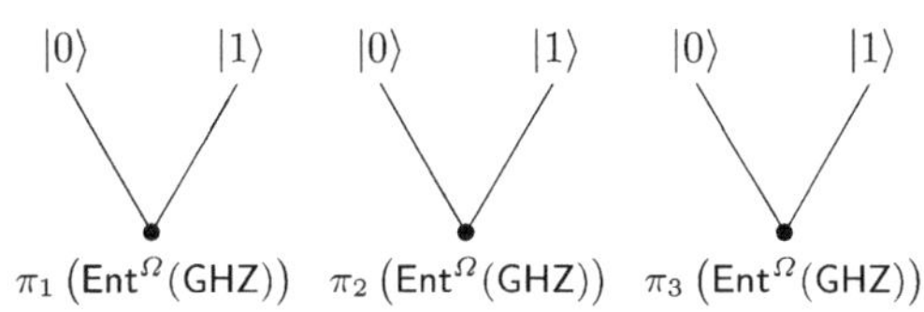

versus $\mathsf{Ent}^{\Omega}(\mathsf{W})$

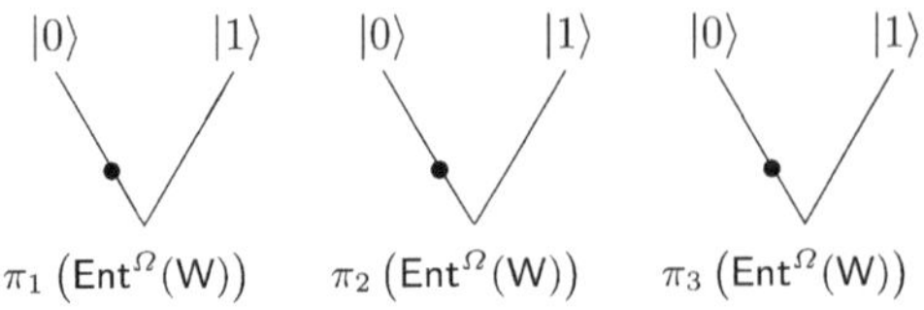

where the maps π_1, π_2 and π_3 represent the components of Ent^{Ω}.

We can define a quantitative measure of entanglement on $\mathcal{H}^n \otimes \ldots \otimes \mathcal{H}^n$ via composition of Ent^{Ω} and

$$\mu_{1,\ldots,m} : \Omega^n \times \ldots \times \Omega^n \to [0,1]^* : (r_1, \ldots, r_n) \mapsto \frac{1}{m} \sum_i \mu(r_i)$$

where μ is again von Neumann entropy. We obtain as such the desired values on pure tensors and the maximally entangled state. In particular do we obtain

$$\mu_{1,2,3}\left(\mathsf{Ent}^{\Omega}(\mathsf{GHZ})\right) = 1 \,.$$

When one prefers to abstract over the identity of the pure tensors above, it is clear that all the above still holds by substituting Λ^n for Ω^n, that is, Ent for Ent^{Ω}.

Acknowledgments To Samson Abramsky and Prakash Panangaden – who supported the authors when ideas combining semantics and physics were far less popular than they are today. This paper appears in its original, *unpublished* form in tribute to a memorable time and place.

References

1. Abramsky, S., Jung, A.: Domain theory. In: Abramsky, S., Gabbay, D.M., Maibaum, T.S.E. (eds.) Handbook of Logic in Computer Science, vol. III. Oxford University Press, Oxford (1994)
2. Birkhoff, G., von Neumann, J.: The logic of quantum mechanics. Ann. Math. **37**, 823–843 (1936)
3. Coecke, B., Moore, D.J., Wilce, A. (eds.): Current Research in Operational Quantum logic: Algebras, Categories, Languages. Kluwer Academic Publishers, Dordrecht (2000)
4. Faure, Cl.-A., Frölicher, A.: Modern projective Geometry. Kluwer Academic Publishers, Dordrecht (2000)
5. Gleason, A.M.: Measures on the closed subspaces of a Hilbert space. J. Math. Mech. **6**, 885–893 (1957)
6. Greenberger, D.M., Horn, M.A., Shimony, A., Zeilinger, A.: Bell's theorem without inequalities. Am. J. Phys. **58**, 1131 (1990)
7. Kalmbach, G.: Orthomodular Lattices. Academic Press, New York (1983)
8. Martin, K. A foundation for computation. Ph.D. Thesis, Department of Mathematics, Tulane University (2000)
9. Mackey, G.M.: The Mathematical Foundations of Quantum Mechanics. W. A. Benjamin, New York (1963)
10. van Loock, P., Braunstein, S.L.: Multipartite entanglement, 2002. arXiv:quantph/0205068

11. Lüders, G.: Über die Zustandsänderung durch den Messprozess (German). Annalen der Physik **8**, 322–328 (1951)
12. von Neumann, J.: Mathematische Grundlagen der Quantenmechanik. Springer, New York (1932); Translation, Mathematical Foundations of Quantum mechanics. Princeton University Press, Princeton (1955)
13. Piron, C.: Foundations of quantum physics. W. A. Benjamin, Reading (1976)
14. Ramsey, F.P.: Truth and probability, 1926. In: Braithwaite, R.B. (ed.) The Foundations of Mathematics and Other Logical Essays, Ch. 7. Routledge, London (1931)
15. Schmidt, E.: Math. Ann. **63**, 433 (1907)
16. Scott, D.: Outline of a mathematical theory of computation. TechnicalMonograph PRG-2. (November 1970)

Part V
Spatio-Temporal Geometry

Chapter 11
Domain Theory and General Relativity

K. Martin and P. Panangaden

Abstract We discuss the current state of investigations into the domain theoretic structure of spacetime, including recent developments which explain the connection between measurement, the Newtonian concept of time and the Lorentz distance.

11.1 Introduction

Domains [AJ94, GKK] are special types of posets that have played an important role in theoretical computer science since the late 1960s when they were discovered by Dana Scott [Sco70] for the purpose of providing a semantics for the lambda calculus. They are partially ordered sets that carry intrinsic (order theoretic) notions of completeness and approximation. The basic intuition is that the order relation captures the idea of approximation qualitatively. There is an abstract notion of finite piece of information, or of finite approximation, which plays a key role in the analysis of computation.

These posets have a number of topologies defined on them: the Scott topology and the interval topology, in particular. The Scott topology is particularly important in that continuity with respect to this topology captures some of the information processing aspects of computability. In particular, a Scott continuous function has the following property: a finite piece of information about the output requires only a finite piece of information about the input. While this does not completely reduce Turing computability to topology it captures a very crucial information processing aspect of computable functions.

K. Martin (✉)
Naval Research Laboratory, Washington, DC, USA
e-mail: keye.martin@nrl.navy.mil

P. Panangaden (✉)
School of Computer Science, McGill University, Montreal, QC, Canada
e-mail: prakash@cs.mcgill.ca

Martin, K., Panangaden, P.: *Domain Theory and General Relativity*. Lect. Notes Phys. **813**, 687–703 (2011)
DOI 10.1007/978-3-642-12821-9_11

General relativity is Einstein's theory of gravity in which gravity is understood not in terms of mysterious "universal" forces but rather as part of the geometry of spacetime. It is profoundly beautiful and beautifully profound from both the physical and mathematical viewpoints and it teaches us clear lessons about the universe in which we live that are easily explainable. For example, it offers a wonderful explanation of gravity: if an apple falls from a tree, the path it takes is not determined by the Newtonian ideal of an "invisible force" but instead by the curvature of the space in which the apple resides: gravity is the curvature of spacetime. In addition, the presence of matter in spacetime causes it to "bend" and Einstein even gives us an equation that relates the curvature of spacetime to the matter present within it.

The study of spacetime structure from an abstract viewpoint—i.e., not from the viewpoint of solving differential equations—was initiated by Penrose [Pen65] in a dramatic paper in which he showed a fundamental inconsistency of gravity. It was known since Chandrasekhar [Cha31] that since everything attracts everything else a gravitating mass of sufficient size will eventually collapse. What Penrose showed was that any such collapse eventually leads to a singularity where the mathematical description of spacetime as a continuum breaks down. This leads to the need to reformulate gravity. It is hoped that the elusive quantum theory of gravity will resolve this problem.

Since the first singularity theorems [Pen65, HE73] causality has played a key role in understanding spacetime structure. The analysis of causal structure relies heavily on techniques of differential topology [Pen72]. For the past decade Sorkin and others [Sor91] have pursued a program for quantization of gravity based on causal structure. In this approach the causal relation is regarded as the fundamental ingredient and the topology and geometry are secondary.

In a paper that appeared in 2006 [KP], we prove that the causality relation is much more than a relation—it turns a globally hyperbolic spacetime into what is known as a *bicontinuous poset*. The order on a bicontinuous poset allows one to define an intrinsic topology called *the interval topology*. On a globally hyperbolic spacetime, the interval topology is the manifold topology. Theorems that reconstruct the spacetime topology have been known [Pen72] and Malament [Mal77] has shown that the class of time-like curves determines the causal structure. We establish these results as well though in a purely order theoretic fashion: there is no need to know what "smooth curve" means.

Our more abstract stance also teaches us something *new*: a globally hyperbolic spacetime *itself* can be reconstructed in a purely order theoretic manner, beginning from only a countable dense set of events and the causality relation. The ultimate reason for this is that the category of globally hyperbolic posets, which contains the globally hyperbolic spacetimes, is *equivalent* to a very special category of posets called *interval domains*. This provides a profound connection between domain theory, first introduced for the purposes of assigning semantics to programming languages, and general relativity, a theory meant to explain gravity. Even from a purely mathematical perspective this equivalence is surprising, since globally hyperbolic spacetimes are usually not order theoretically complete, but interval domains always are.

Measurements were introduced by Martin in [Mar00a]. One thing they provide is a way of incorporating quantitative information into domain theory. More recently we have also shown how the *geometry* of spacetime can be reconstructed order theoretically. The reason is that the Lorentz distance defines a Scott continuous function on the domain of spacetime intervals. What is even more interesting, though, is that our setting provides a way to *topologically* distinguish between Newtonian and relativistic notions of time. Every global time function defines a measurement on the domain of spacetime intervals, in particular, it is Scott continuous. The Lorentz distance is not only Scott continuous, but satisfies a stronger property, that it is interval continuous. An interval continuous function must assign zero to any element which approximates nothing. Thus, no interval continuous function on the domain of spacetime intervals can ever be a measurement and the reason for this has entirely to do with relativity: a clock moving at the speed of light records no time as having elapsed, so an interval continuous function is incapable of distinguishing between a single event and a null interval.

11.2 Domains, Continuous Posets and Topology

A *poset* is a partially ordered set, i.e., a set together with a reflexive, antisymmetric and transitive relation.

Definition 1 Let $(P, \sqsubseteq)$ be a partially ordered set. A nonempty subset $S \subseteq P$ is *directed* if $(\forall x, y \in S)(\exists z \in S)\, x, y \sqsubseteq z$. The *supremum* of $S \subseteq P$ is the least of all its upper bounds provided it exists. This is written $\bigsqcup S$.

These ideas have duals that will be important to us: a nonempty $S \subseteq P$ is *filtered* if $(\forall x, y \in S)(\exists z \in S)\, z \sqsubseteq x, y$. The *infimum* $\bigwedge S$ of $S \subseteq P$ is the greatest of all its lower bounds provided it exists.

Definition 2 For a subset X of a poset P, set

$$\uparrow\! X := \{y \in P : (\exists x \in X)\, x \sqsubseteq y\} \;\;\&\;\; \downarrow\! X := \{y \in P : (\exists x \in X)\, y \sqsubseteq x\}.$$

We write $\uparrow x = \uparrow \{x\}$ and $\downarrow x = \downarrow \{x\}$ for elements $x \in X$.

A partial order allows for the derivation of several intrinsically defined topologies. Here is our first example.

Definition 3 A subset U of a poset P is *Scott open* if

(i) U is an upper set: $x \in U \;\&\; x \sqsubseteq y \Rightarrow y \in U$, and
(ii) U is inaccessible by directed suprema: For every directed $S \subseteq P$ with a supremum,

$$\bigsqcup S \in U \Rightarrow S \cap U \neq \emptyset.$$

The collection of all Scott open sets on P is called the *Scott topology*.

Posets can have a variety of completeness properties. The following completeness condition has turned out to be particularly useful in applications.

Definition 4 A *dcpo* is a poset in which every directed subset has a supremum. The *least element* in a poset, when it exists, is the unique element $\bot$ with $\bot \sqsubseteq x$ for all x.

The set of *maximal elements* in a dcpo D is

$$\max(D) := \{x \in D : \uparrow x = \{x\}\}.$$

Each element in a dcpo has a maximal element above it.

Definition 5 For elements x, y of a poset, write $x \ll y$ iff for all directed sets S with a supremum,

$$y \sqsubseteq \bigsqcup S \Rightarrow (\exists s \in S)\, x \sqsubseteq s.$$

We set $\twoheaddownarrow x = \{a \in D : a \ll x\}$ and $\twoheaduparrow x = \{a \in D : x \ll a\}$.

For the symbol "$\ll$," read "approximates."

Definition 6 A *basis* for a poset D is a subset B such that $B \cap \twoheaddownarrow x$ contains a directed set with supremum x for all $x \in D$. A poset is *continuous* if it has a basis. A poset is *ω-continuous* if it has a countable basis.

Continuous posets have an important property, they are *interpolative*.

Proposition 1 *If $x \ll y$ in a continuous poset P, then there is $z \in P$ with $x \ll z \ll y$.*

This enables a clear description of the Scott topology,

Theorem 1 *The collection $\{\twoheaduparrow x : x \in D\}$ is a basis for the Scott topology on a continuous poset.*

Definition 7 A *continuous dcpo* is a continuous poset which is also a dcpo. A *domain* is a continuous dcpo.

The next example is due to Scott[Sco70] and worth keeping in mind when we consider the analogous construction for globally hyperbolic spacetimes.

Example 1 The collection of compact intervals of the real line

$$\mathbf{I}\mathbb{R} = \{[a, b] : a, b \in \mathbb{R} \;\&\; a \leq b\}$$

ordered under reverse inclusion

$$[a, b] \sqsubseteq [c, d] \Leftrightarrow [c, d] \subseteq [a, b]$$

is an ω-continuous dcpo:

- For directed $S \subseteq \mathbf{I}\mathbb{R}$, $\bigsqcup S = \bigcap S$,
- $I \ll J \Leftrightarrow J \subseteq \text{int}(I)$, and
- $\{[p, q] : p, q \in \mathbb{Q} \;\&\; p \leq q\}$ is a countable basis for $\mathbf{I}\mathbb{R}$.

The domain $\mathbf{I}\mathbb{R}$ is called the *interval domain.*

We also have $\max(\mathbf{I}\mathbb{R}) \simeq \mathbb{R}$ in the Scott topology. Approximation can help explain why:

Example 2 A basic Scott open set in $\mathbf{I}\mathbb{R}$ is

$$\twoheaduparrow[a, b] = \{x \in \mathbf{I}\mathbb{R} : x \subseteq (a, b)\}.$$

One of the interesting things about $\mathbf{I}\mathbb{R}$ is that it is a domain that is derived from an underlying poset with an abundance of order theoretic structure. Part of this structure is that the real line is *bicontinuous*, a fundamental notion in the present work:

Definition 8 A continuous poset $(P, \leq)$ is *bicontinuous* if

- For all $x, y \in P$, $x \ll y$ iff for all filtered $S \subseteq P$ with an infimum,

$$\bigwedge S \leq x \Rightarrow (\exists s \in S)\, s \leq y,$$

  and
- For each $x \in P$, the set $\twoheaduparrow x$ is filtered with infimum x.

Example 3 $\mathbb{R}$, $\mathbb{Q}$ are bicontinuous.

Definition 9 On a bicontinuous poset P, sets of the form

$$(a, b) := \{x \in P : a \ll x \ll b\}$$

form a basis for a topology called *the interval topology.*

The proof uses interpolation and bicontinuity. In contrast to a domain, a bicontinuous poset P has $\twoheaduparrow x \neq \emptyset$ for each x, so it is rarely a dcpo. We tend to prefer the notation $\leq$ for the order on a poset that is known to be bicontinuous. Otherwise, we use the notation $\sqsubseteq$.

Definition 10 For x, y in a poset $(P, \leq)$,

$$x < y \equiv x \leq y \;\&\; x \neq y.$$

In general, $<$ and $\ll$ are completely different ideas.

11.3 The Causal Structure of Spacetime

A *manifold* $\mathcal{M}$ is a locally Euclidean Hausdorff space that is connected and has a countable basis. Such spaces are paracompact. A *Lorentz metric* on a manifold is a symmetric, nondegenerate tensor field of type (0, 2) whose signature is $(- + + +)$.

Definition 11 A *spacetime* is a real four-dimensional[1] smooth manifold $\mathcal{M}$ with a Lorentz metric g_{ab}.

Let $(\mathcal{M}, g_{ab})$ be a time-orientable spacetime. Let $\Pi^+_\leq$ denote the future directed causal curves, and $\Pi^+_\ll$ denote the future directed time-like curves.

Definition 12 For $p \in \mathcal{M}$,

$$I^+(p) := \{q \in \mathcal{M} : (\exists \pi \in \Pi^+_\ll)\, \pi(0) = p, \pi(1) = q\}$$

and

$$J^+(p) := \{q \in \mathcal{M} : (\exists \pi \in \Pi^+_\leq)\, \pi(0) = p, \pi(1) = q\}$$

Similarly, we define $I^-(p)$ and $J^-(p)$.

We write the relation J^+ as

$$p \leq q \equiv q \in J^+(p).$$

The following properties from [HE73] are very useful:

Proposition 2 *Let $p, q, r \in \mathcal{M}$. Then*

(i) *The sets $I^+(p)$ and $I^-(p)$ are open.*
(ii) $p \leq q$ *and* $r \in I^+(q) \Rightarrow r \in I^+(p)$
(iii) $q \in I^+(p)$ *and* $q \leq r \Rightarrow r \in I^+(p)$
(iv) $\mathrm{Cl}(I^+(p)) = \mathrm{Cl}(J^+(p))$ *and* $\mathrm{Cl}(I^-(p)) = \mathrm{Cl}(J^-(p))$.

We always assume the chronology conditions that ensure $(\mathcal{M}, \leq)$ is a partially ordered set. We also assume *strong causality* which can be characterized as follows [Pen72]:

Theorem 2 *A spacetime $\mathcal{M}$ is strongly causal iff its Alexandroff topology is Hausdorff iff its Alexandroff topology is the manifold topology.*

The Alexandroff topology on a spacetime has $\{I^+(p) \cap I^-(q) : p, q \in \mathcal{M}\}$ as a basis [Pen72].[2]

11.4 Global Hyperbolicity

Penrose has called *globally hyperbolic* spacetimes "the physically reasonable spacetimes [Wal84]."

---

[1] The results in the present paper work for any dimension $n \geq 2$ [J93].

[2] This terminology is common among relativists but order theorists use the phrase "Alexandrov topology" to mean something else: the topology generated by the upper sets.

Definition 13 A spacetime $\mathcal{M}$ is *globally hyperbolic* if it is strongly causal and if $\uparrow a \cap \downarrow b$ is compact in the manifold topology, for all $a, b \in \mathcal{M}$.

Theorem 3 ([KP]) *If $\mathcal{M}$ is globally hyperbolic, then $(\mathcal{M}, \leq)$ is a bicontinuous poset with $\ll = I^+$ whose interval topology is the manifold topology.*

This result motivates the following definition:

Definition 14 A poset $(X, \leq)$ is *globally hyperbolic* if it is bicontinuous and each interval $[a, b] = \{x : a \leq x \leq b\}$ is compact in the interval topology.

Globally hyperbolic posets have rich enough structure that we can deduce many properties of spacetime from them *without* appealing to differentiable structure or geometry. Here is one such example:

Definition 15 Let $(X, \leq)$ be a globally hyperbolic poset. A subset $\pi \subseteq X$ is a *causal curve* if it is compact, connected and linearly ordered. We define

$$\pi(0) := \bot \text{ and } \pi(1) := \top$$

where $\bot$ and $\top$ are the least and greatest elements of π. For $P, Q \subseteq X$,

$$C(P, Q) := \{\pi : \pi \text{ causal curve}, \pi(0) \in P, \pi(1) \in Q\}$$

and call this *the space of causal curves* between P and Q.

This definition is motivated by the fact that a subset of a globally hyperbolic spacetime $\mathcal{M}$ is the image of a causal curve iff it is the image of a continuous monotone increasing $\pi : [0, 1] \to \mathcal{M}$ iff it is a compact connected linearly ordered subset of $(\mathcal{M}, \leq)$.

Theorem 4 ([Mar06]) *If $(X, \leq)$ is a separable globally hyperbolic poset, then the space of causal curves $C(P, Q)$ is compact in the Vietoris topology and hence in the upper topology.*

This result plays an important role in the proofs of certain singularity theorems [Wal84], in establishing the existence of maximum length geodesics [HE73], and in the proof of certain positive mass theorems [Pen93]. Moreover, while events in spacetime are maximal elements of $\mathbf{I}\mathcal{M}$, causal curves are maximal elements in a higher order domain $\mathrm{C}(\mathbf{I}\mathcal{M})$, called the *convex powerdomain* of $\mathbf{I}\mathcal{M}$. This is discussed in more detail in [Mar06].

We can also deduce new aspects of spacetime. Globally hyperbolic posets are very much like the real line. In fact, a well-known domain theoretic construction pertaining to the real line extends in perfect form to the globally hyperbolic posets:

Theorem 5 ([KP]) *The closed intervals of a globally hyperbolic poset X*

$$\mathbf{I}X := \{[a, b] : a \leq b \;\&\; a, b \in X\}$$

ordered by reverse inclusion

$$[a, b] \sqsubseteq [c, d] \equiv [c, d] \subseteq [a, b]$$

form a continuous domain with

$$[a, b] \ll [c, d] \equiv a \ll c \,\&\, d \ll b.$$

The poset X has a countable basis iff $\mathbf{I}X$ *is ω-continuous. Finally,*

$$\max(\mathbf{I}X) \simeq X$$

where the set of maximal elements has the relative Scott topology from $\mathbf{I}X$.

This observation—that spacetime has a canonical domain theoretic model—teaches us something new: from only a countable set of events and the causality relation, one can reconstruct spacetime in a purely order theoretic manner. Explaining this requires domain theory.

11.5 Spacetime from a Discrete Causal Set

An *abstract basis* is a set $(C, \ll)$ with a *transitive* relation that is *interpolative* from the—*direction*:

$$F \ll x \Rightarrow (\exists y \in C)\, F \ll y \ll x,$$

for all finite subsets $F \subseteq C$ and all $x \in F$. Suppose, though, that it is also interpolative from the $+$ *direction*:

$$x \ll F \Rightarrow (\exists y \in C)\, x \ll y \ll F.$$

Then we can define a new abstract basis of *intervals*

$$\text{int}(C) = \{(a, b) : a \ll b\} = \ll \subseteq C^2$$

whose relation is

$$(a, b) \ll (c, d) \equiv a \ll c \,\&\, d \ll b.$$

Let $\mathbf{I}C$ denote the ideal completion of the abstract basis $\text{int}(C)$.

Theorem 6 ([KP]) *Let C be a countable dense subset of a globally hyperbolic spacetime $\mathcal{M}$ and $\ll = I^+$ be timelike causality. Then*

$$\max(\mathbf{I}C) \simeq \mathcal{M}$$

where the set of maximal elements have the Scott topology.

In "ordering the order" I^+, taking its completion, and then the set of maximal elements, we recover spacetime by reasoning only about the causal relationships between a countable dense set of events. One objection to this might be that we begin from a *dense* set C, and then order theoretically recover the space $\mathcal{M}$—but *dense* is a topological idea so we need to know the topology of $\mathcal{M}$ before we can recover it! But the denseness of C can be expressed in purely causal terms:

$$C \text{ dense} \equiv (\forall x, y \in \mathcal{M})(\exists z \in C)\, x \ll z \ll y.$$

Now the objection might be that we still have to reference $\mathcal{M}$. We too would like to not reference $\mathcal{M}$ at all. However, some global property needs to be assumed, either directly or indirectly, in order to reconstruct $\mathcal{M}$.

Theorem 6 is very different from results like "Let $\mathcal{M}$ be a certain spacetime with relation $\leq$. Then the interval topology is the manifold topology." Here we identify, in abstract terms, a process by which a countable set with a causality relation determines a space. The process is entirely order theoretic in nature, spacetime is not required to understand or execute it (i.e., if we put $C = \mathbb{Q}$ and $\ll = <$, then $\max(\mathbf{I}C) \simeq \mathbb{R}$). In this sense, our understanding of the relation between causality and the topology of spacetime is now explainable independently of geometry.

Ideally, one would now like to know what constraints on C in general imply that $\max(\mathbf{I}C)$ is a manifold.

11.6 Spacetime as a Domain

The category of globally hyperbolic posets is naturally isomorphic to a special category of domains called interval domains.

Definition 16 An *interval poset* is a poset D that has two functions left : $D \to \max(D)$ and right : $D \to \max(D)$ such that

(i) Each $x \in D$ is an "interval" with left(x) and right(x) as endpoints:

$$(\forall x \in D)\, x = \text{left}(x) \sqcap \text{right}(x),$$

(ii) The union of two intervals with a common endpoint is another interval: For all $x, y \in D$, if right$(x) = $ left(y), then

$$\text{left}(x \sqcap y) = \text{left}(x) \quad \& \quad \text{right}(x \sqcap y) = \text{right}(y),$$

(iii) Each point $p \in \uparrow\!x \cap \max(D)$ of an interval $x \in D$ determines two subintervals, left$(x) \sqcap p$ and $p \sqcap$ right(x), with endpoints:

$$\text{left}(\text{left}(x) \sqcap p) = \text{left}(x) \quad \& \quad \text{right}(\text{left}(x) \sqcap p) = p$$
$$\text{left}(p \sqcap \text{right}(x)) = p \quad \& \quad \text{right}(p \sqcap \text{right}(x)) = \text{right}(x)$$

Notice that a nonempty interval poset D has $\max(D) \neq \emptyset$ by definition. With interval posets, we only assume that infima indicated in the definition exist; in particular, we do not assume the existence of all binary infima.

Definition 17 For an interval poset (D, left, right), the relation $\leq$ on $\max(D)$ is

$$a \leq b \equiv (\exists x \in D)\, a = \text{left}(x) \;\&\; b = \text{right}(x)$$

for $a, b \in \max(D)$.

The axioms of interval posets imply that $(\max(D), \leq)$ is a poset.

Definition 18 An *interval domain* is an interval poset (D, left, right) where D is a continuous dcpo such that

(i) If $p \in \Uparrow\! x \cap \max(D)$, then

$$\Uparrow\!(\text{left}(x) \sqcap p) \neq \emptyset \quad \& \quad \Uparrow\!(p \sqcap \text{right}(x)) \neq \emptyset.$$

(ii) For all $x \in D$, the following are equivalent:

(a) $\Uparrow\! x \neq \emptyset$
(b) $(\forall y \in [\,\text{left}(x), \cdot\,])(\, y \sqsubseteq x \;\Rightarrow\; y \ll \text{right}(y) \text{ in } [\,\cdot, \text{right}(y)\,]\,)$
(c) $(\forall y \in [\cdot, \text{right}(x)])(\, y \sqsubseteq x \;\Rightarrow\; y \ll \text{left}(y) \;\text{ in } [\,\text{left}(y), \cdot\,]\,)$

(iii) Invariance of endpoints under suprema:

(a) For all directed $S \subseteq [p, \cdot]$

$$\text{left}(\bigsqcup S) = p \;\;\&\;\; \text{right}(\bigsqcup S) = \text{right}(\bigsqcup T)$$

for any directed $T \subseteq [q, \cdot]$ with $\text{right}(T) = \text{right}(S)$.
(b) For all directed $S \subseteq [\cdot, q]$

$$\text{left}(\bigsqcup S) = \text{left}(\bigsqcup T) \;\;\&\;\; \text{right}(\bigsqcup S) = q$$

for any directed $T \subseteq [\cdot, p]$ with $\text{left}(T) = \text{left}(S)$.

(iv) Intervals are compact: For all $x \in D$, $\uparrow x \cap \max(D)$ is Scott compact.

Interval domains are interval posets whose axioms also take into account the completeness and approximation present in a domain: (i) says if a point p belongs to the interior of an interval $x \in D$, the subintervals $\text{left}(x) \sqcap p$ and $p \sqcap \text{right}(x)$ both have nonempty interior; (ii) says an interval has nonempty interior iff all intervals that contain it have nonempty interior locally; (iii) explains the behavior of endpoints when taking suprema.

For a globally hyperbolic $(X, \leq)$, we define:

$$\text{left} : \mathbf{I}X \to \mathbf{I}X :: [a, b] \mapsto [a]$$

and

$$\text{right} : \mathbf{I}X \to \mathbf{I}X :: [a, b] \mapsto [b].$$

Lemma 1 *If* $(X, \leq)$ *is a globally hyperbolic poset, then* $(\mathbf{I}X$, *left, right*$)$ *is an interval domain.*

In essence, this is the only example.

Lemma 2 *If* $(D$, *left, right*$)$ *is an interval domain, then* $(\max(D), \leq)$ *is a globally hyperbolic poset.*

The equivalence between globally hyperbolic posets and interval domains is as follows:

Definition 19 The category $\underline{\mathrm{IN}}$ of interval domains and commutative maps is given by

- **objects** Interval domains $(D, \text{left}, \text{right})$.
- **arrows** Scott continuous $f : D \to E$ that commute with left and right, i.e., such that both

$$\begin{array}{ccc} D & \xrightarrow{\text{left}_D} & D \\ {\scriptstyle f}\downarrow & & \downarrow{\scriptstyle f} \\ E & \xrightarrow[\text{left}_E]{} & E \end{array} \qquad \text{and} \qquad \begin{array}{ccc} D & \xrightarrow{\text{right}_D} & D \\ {\scriptstyle f}\downarrow & & \downarrow{\scriptstyle f} \\ E & \xrightarrow[\text{right}_E]{} & E \end{array}$$

  commute.
- **identity** $1 : D \to D$.
- **composition** $f \circ g$.

Definition 20 The category $\underline{\mathrm{G}}$ is given by

- **objects** Globally hyperbolic posets $(X, \leq)$.
- **arrows** Continuous in the interval topology, monotone.
- **identity** $1 : X \to X$.
- **composition** $f \circ g$.

It is routine to verify that $\underline{\mathrm{IN}}$ and $\underline{\mathrm{G}}$ are categories.

Proposition 3 *The correspondence* $\mathbf{I} : \underline{\mathrm{G}} \to \underline{\mathrm{IN}}$ *given by*

$$(X, \leq) \mapsto (\mathbf{I}X, \mathit{left}, \mathit{right})$$
$$(f : X \to Y) \mapsto (\bar{f} : \mathbf{I}X \to \mathbf{I}Y)$$

is a functor between categories.

Proposition 4 *The correspondence* $\max : \underline{\mathrm{IN}} \to \underline{\mathrm{G}}$ *given by*

$$(D, \mathit{left}, \mathit{right}) \mapsto (\max(D), \leq)$$
$$(f : D \to E) \mapsto (f|_{\max(D)} : \max(D) \to \max(E))$$

is a functor between categories.

Before the statement of the main theorem in this section, we recall the definition of a natural isomorphism.

Definition 21 A *natural transformation* $\eta : F \to G$ between functors $F : \mathcal{C} \to \mathcal{D}$ and $G : \mathcal{C} \to \mathcal{D}$ is a collection of arrows $(\eta_X : F(X) \to G(X))_{X \in \mathcal{C}}$ such that for any arrow $f : A \to B$ in $\mathcal{C}$,

$$\begin{array}{ccc} F(A) & \xrightarrow{\eta_A} & G(A) \\ {\scriptstyle F(f)}\downarrow & & \downarrow{\scriptstyle G(f)} \\ F(B) & \xrightarrow[\eta_B]{} & G(B) \end{array}$$

commutes. If each η_X is an isomorphism, η is a *natural isomorphism*.

Categories $\mathcal{C}$ and $\mathcal{D}$ are *equivalent* when there are functors $F : \mathcal{C} \to \mathcal{D}$ and $G : \mathcal{D} \to \mathcal{C}$ and natural isomorphisms $\eta : 1_{\mathcal{C}} \to GF$ and $\mu : 1_{\mathcal{D}} \to FG$.

Theorem 7 ([KP]) *The category of globally hyperbolic posets is equivalent to the category of interval domains.*

This result suggests that questions about spacetime can be converted to domain theoretic form, where we can use domain theory to answer them, and then translate the answers back to the language of physics (and vice-versa). Notice too that the category of interval posets and commutative maps is equivalent to the category of posets and monotone maps.

It also shows that causality between events is equivalent to an order on *regions* of spacetime. Most importantly, we have shown that globally hyperbolic spacetime with causality is equivalent to a structure $\mathbf{I}X$ whose origins are "discrete." This is the formal explanation for why spacetime can be reconstructed from a countable dense set of events in a purely order theoretic manner.

11.7 Time and Measurement

A *domain* is a partially ordered set with intrinsic notions of completeness and approximation defined by the order. A *measurement* is a function μ that to each informative object x assigns a number μx which measures the information content of the object x. Let us now define the latter term precisely before discussing it further.

A function $f : D \to E$ between domains is *Scott continuous* if the inverse image of a Scott open set in E is Scott open in D. This is equivalent [AJ94] to saying that f is *monotone*,

$$(\forall x, y \in D)\, x \sqsubseteq y \Rightarrow f(x) \sqsubseteq f(y),$$

and that it *preserves directed suprema*:

$$f\left(\bigsqcup S\right) = \bigsqcup f(S),$$

for all directed $S \subseteq D$. In particular, for the domain $[0, \infty)^*$ of nonnegative reals in their opposite order, a Scott continuous function $\mu : D \to [0, \infty)^*$ will satisfy

1. For all $x, y \in D$, $x \sqsubseteq y \Rightarrow \mu x \geq \mu y$, and
2. If (x_n) is an increasing sequence in D, then

$$\mu\left(\bigsqcup_{n\geq 1} x_n\right) = \lim_{n\to\infty} \mu x_n.$$

This is the case of Scott continuity that we are most interested in presently:

Definition 22 A Scott continuous $\mu : D \to [0, \infty)^*$ is said to *measure the content* of $x \in D$ if for all Scott open sets $U \subseteq D$,

$$x \in U \Rightarrow (\exists \varepsilon > 0)\, x \in \mu_\varepsilon(x) \subseteq U$$

where

$$\mu_\varepsilon(x) := \{y \in D : y \sqsubseteq x \;\&\; |\mu x - \mu y| < \varepsilon\}$$

are called the *ε-approximations* of x.

We often refer to μ as simply "measuring" $x \in D$ or as measuring $X \subseteq D$ when it measures each element of X. The last definition, as well as the next, easily extend to maps μ that take values in an arbitrary domain E.

Definition 23 A *measurement* $\mu : D \to [0, \infty)^*$ is a Scott continuous map that measures the content of $\ker(\mu) := \{x \in D : \mu x = 0\}$.

The order on a domain D defines a clear sense in which one object has "more information" than another: a *qualitative* view of information content. The definition of measurement attempts to identify those monotone mappings μ which offer a *quantitative* measure of information content in the sense specified by the order. The essential point in the definition of measurement is that μ measure content in a manner that is consistent with the particular view offered by the order. There are plenty of monotone mappings that are not measurements—and while some of them may measure information content in *some other sense*, each sense must first be specified by a different information order. The definition of measurement is then a minimal test that a function μ must pass if we are to regard it as providing a measure of information content.

We now consider a few properties that measures of information content have which arbitrary monotone mappings in general need not have: qualities that make them 'different' from maps that are simply monotone. Other such properties may be found in [Mar00a].

Theorem 8 ([Mar00a]) *Let $\mu : D \to [0, \infty)^*$ be a measurement.*

(i) *If $x \in \ker(\mu)$, then $x \in \max(D) = \{x \in D : \uparrow x = \{x\}\}$.*
(ii) *If μ measures the content of $y \in D$, then*

$$(\forall x \in D)\ x \sqsubseteq y \ \&\ \mu x = \mu y \Rightarrow x = y.$$

(iii) *If μ measures $X \subseteq D$, then*

$$\{\uparrow \mu_\varepsilon(x) \cap X : x \in X, \varepsilon > 0\}$$

is a basis for the Scott topology on X.

A global time function $t : \mathcal{M} \to \mathbb{R}$ on a globally hyperbolic spacetime $\mathcal{M}$ is a continuous function such that $x < y \Rightarrow t(x) < t(y)$ and $t^{-1}(r) = \Sigma$ is a Cauchy surface for $\mathcal{M}$, for each $r \in \mathbb{R}$.

Theorem 9 *For any global time function $t : \mathcal{M} \to \mathbb{R}$ on a globally hyperbolic spacetime, the function $\Delta t : \mathcal{M} \to [0, \infty)^*$ given by $\Delta t[a, b] = t(b) - t(a)$ measures all of $\mathbf{I}(\mathcal{M})$. It is a measurement with* $\ker(\Delta t) = \max(\mathbf{I}(\mathcal{M}))$.

Let $d : \mathbf{I}(\mathcal{M}) \to [0, \infty)^*$ denote the Lorentz distance on a globally hyperbolic spacetime

$$d[a, b] = \sup_{\pi_{ab}} \mathrm{len}(\pi_{ab})$$

where the sup is taken over all causal curves that join a to b.

Definition 24 The *interval topology* on a continuous poset P exists when sets of the form

$$(a, b) = \{x \in P : a \ll x \ll b\} \quad \& \quad \Uparrow x = \{y \in P : x \ll y\}$$

form a basis for a topology on P.

For bicontinuous posets, this definition of interval topology is equivalent to the definition considered earlier. A function between continuous posets is *interval continuous* when each poset has an interval topology and the inverse image of an interval open set is interval open. By the bicontinuity of $\mathcal{M}$, the interval topology on $\mathbf{I}(\mathcal{M})$ exists, so we can consider interval continuity for functions $\mathbf{I}(\mathcal{M}) \to [0, \infty)^*$.

Theorem 10 *The Lorentz distance* $d\colon \mathbf{I}(\mathcal{M}) \to [0, \infty)^*$ *has the following properties:*

(i) *It is monotone:* $x \leq y \Rightarrow d(x) \geq d(y)$,
(ii) *It preserves the way below relation:* $x \ll y \Rightarrow d(x) > d(y)$,
(iii) *It is interval continuous and hence Scott continuous.*

It does not measure $\mathbf{I}(\mathcal{M})$ *at any point of* $\ker(d)$.

That the Lorentz distance is not a measurement is a direct consequence of the fact that a clock travelling at the speed of light records no time as having elapsed i.e. the set of null intervals is equal to

$$\ker(d) \setminus \max(\mathbf{I}(\mathcal{M})) \neq \emptyset$$

but measurements always have the property that $\mu x = 0$ implies $x \in \max(D)$ (Theorem 8).

In fact, no interval continuous function $\mu : \mathbf{I}(\mathcal{M}) \to [0, \infty)^*$ can be a measurement: by interval continuity, $\mu x = 0$ for any x with $\Uparrow x = \emptyset$. Just like the Lorentz distance, an interval continuous μ will also assign 0 to "null intervals." In this way, we see that interval continuity captures an essential aspect of the Lorentz distance. In addition, since Δt is a measurement, it cannot be interval continuous. This provides a surprising *topological* distinction between the Newtonian and relativistic concepts of time: d is interval continuous, Δt is not. Put another way, Δt can be used to reconstruct the *topology* of spacetime (Theorem 8(iii)), while d is used to reconstruct its *geometry*.

11.8 Spacetime Geometry from a Discrete Causal Set

Let us return now to the reconstruction of spacetime (Sect. 11.5) from a countable dense set $(C, \ll)$. Specifically, we take the rounded ideal completion $\mathbf{I}(C)$of the abstract basis of *intervals*

$$\text{int}(C) = \{(a, b) : a \ll b\} = \ll \subseteq C^2$$

whose relation is

$$(a, b) \ll (c, d) \equiv a \ll c \;\&\; d \ll b.$$

We are then able to recover spacetime as

$$\max(\mathbf{I}C) \simeq \mathcal{M}$$

where the set of maximal elements have the Scott topology. Let us now suppose that in addition to $\mathrm{int}(C)$ that we also begin with a countable collection of numbers l_{ab} chosen for each $(a, b) \in \mathrm{int}(C)$ in such a way that the map

$$\mathrm{int}(C) \to [0, \infty)^* :: (a, b) \mapsto l_{ab}$$

is monotone. Then in the process of reconstructing spacetime, we can also construct the Scott continuous function $d : \mathbf{I}C \to [0, \infty)^*$ given by

$$d(x) = \inf\{l_{ab} : (a, b) \ll x\}.$$

In the event that the countable number of l_{ab} chosen are the Lorentz distances $l_{ab} = d[a, b]$, then the function d constructed above yields the Lorentz distance for any spacetime interval, the reason being that both are Scott continuous and are equal on a basis of the domain.

Thus, from a countable dense set of events and a countable set of distances, we can reconstruct the spacetime manifold together with its geometry in a purely order theoretic manner.

11.9 Conclusions

We have seen the following ideas in this paper:

1. how to reconstruct the spacetime topology from the causal structure using purely order-theoretic ideas,
2. an abstract order-theoretic definition of global hyperbolicity,
3. that one can reconstruct spacetime, meaning its topology and *geometry*, from a countable dense subset,
4. an equivalence of categories between the category of interval domains and the category of globally hyperbolic posets.
5. a topological distinction between Newtonian and relativistic notions of time.

References

AJ94. Abramsky, S., Jung, A.: Domain theory. In Maibaum, T.S.E., Abramsky, S., Gabbay, D. M. (eds.), Handbook of Logic in Computer Science, vol. III. Oxford University Press, Oxford (1994)

Beem. Beem, J.K., Ehrlich, P.E., Easley, K.L.: Global Lorentzian Geometry. CRC Press, Boca Raton (1996)

Cha31. Chandrasekhar, S.: The maximum mass of ideal white dwarfs. Astrophys. J. **74**, 81–82 (1931)

Faj00. Fajstrup, L.: Loops, ditopology and deadlocks: Geometry and concurrency. Math. Struct. Comput. Sci. **10**(4), 459–480 (2000)

GFR98. Goubault, E., Fajstrup, L., Raussen, M.: Algebraic topology and concurrency. Department of Mathematical Sciences RR-98-2008, Aalborg University, 1998. Presented at MFCS 1998 London

GKK. Gierz, G., Hoffman, K.H., Keimel, K., Lawson, J.D., Mislove, M., Scott, D.S.: Continuous lattices and domains. Number 93 in Encyclopedia of Mathematics and its Applications. Cambridge University Press, Cambridge (2003)

HE73. Hawking, S.W., Ellis, G.F.R. The large scale structure of space-time. Cambridge Monographs on Mathematical Physics. Cambridge University Press, Cambridge (1973)

J93. Joshi, P.S.: Global aspects in gravitation and cosmology. International Series of Monographs on Physics 87. Oxford Science Publications, Oxford (1993)

Mal77. Malement, D.: The class of continuous timelike curves determines the topology of spacetime. J. Math. Phys. **18**(7), 1399–1404 (1977)

Mar00a. Martin, K.: A foundation for computation. PhD thesis, Department of Mathematics, Tulane University (2000)

Mar00b. Martin K.: The measurement process in domain theory. In: Proceedings of the 27th International Colloquium on Automata, Languages and Programming (ICALP'00), Number 1853 in Lecture Notes In Computer Science, pp. 116–126. Springer, New York (2000)

Mar06. Martin, K. Compactness of the space of causal curves. J. Class. Quant. Grav. **23**, 1241–1251 (2006)

KP. Martin, K., Panangaden, P.: A domain of spacetime intervals in general relativity. Commun. Math. Phys. **267**(3), 563–586 (November 2006)

Pen65. Penrose, R.: Gravitational collapse and space-time singularities. Phys. Rev. Lett. **14**, 57–59 (1965)

Pen72. Penrose, R.: Techniques of differential topology in relativity. Society for Industrial and Applied Mathematics (1972)

Pen93. Penrose, R., Sorkin, R.D., Woolgar, E.: A positive mass theorem based on the focusing and retardation of null geodesics. Submitted to Commun. Math. Phys., arXiv:grqc/9301015 (1993)

Sco70. Scott, D.: Outline of a mathematical theory of computation. Technical Monograph PRG-2, Oxford University Computing Laboratory (1970)

Sor91. Sorkin, R.: Spacetime and causal sets. In: D'Olivo J. et. al. (eds.) Relativity and Gravitation: Classical and Quantum. World Scientific, Singapore (1991)

SW96. Sorkin, R.D., Woolgar, E.: A causal order for spacetimes with c 0 Lorentzian metrics: Proof of compactness of the space of causal curves. Class. Quant. Grav. **13**, 1971–1994 (1996)

Wal84. Wald, R.M.: General Relativity. The University of Chicago Press, Chicago (1984)

Chapter 12
Process, Distinction, Groupoids and Clifford Algebras: an Alternative View of the Quantum Formalism

B.J. Hiley

Abstract In this paper we start from a basic notion of process, which we structure into two groupoids, one orthogonal and one symplectic. By introducing additional structure, we convert these groupoids into orthogonal and symplectic Clifford algebras respectively. We show how the orthogonal Clifford algebra, which include the Schrödinger, Pauli and Dirac formalisms, describe the *classical* light-cone structure of space-time, as well as providing a basis for the description of quantum phenomena. By constructing an orthogonal Clifford bundle with a Dirac connection, we make contact with quantum mechanics through the Bohm formalism which emerges quite naturally from the connection, showing that it is a structural feature of the mathematics. We then generalise the approach to include the symplectic Clifford algebra, which leads us to a non-commutative geometry with projections onto shadow manifolds. These shadow manifolds are none other than examples of the phase space constructed by Bohm. We also argue that this provides us with a mathematical structure that fits the implicate-explicate order proposed by Bohm.

12.1 The Algebra of Process

Traditionally basic theories of quantum phenomena are described in terms of the dynamical properties of particles-in-interaction, or more basically, fields-in-interaction built on an a priori given manifold. Special relativity demands this manifold is a Minkowski space-time, while general relativity demands a more general manifold with a metric carrying the properties of the gravitational field. In this paper we explore the possibility of starting from a primitive notion of *process*, in which flux, activity or movement is taken as basic and from which physical phenomena in general and quantum phenomena in particular emerge. Indeed we expect the space-time manifold itself to arise from this basic process.

B.J. Hiley (✉)
Birkbeck College, University of London, London, UK
e-mail: b.hiley@bbk.ac.uk

Hiley, B.J.: *Process, Distinction, Groupoids and Clifford Algebras: an Alternative View of the Quantum Formalism*. Lect. Notes Phys. **813**, 705–752 (2011)
DOI 10.1007/978-3-642-12821-9_12

The problem with the traditional view is that it is mechanistic and reductionist in spirit and is in stark contrast to the perception and deep insights of Bohr [1], who has already argued that it is the notion of wholeness that is essential to our understanding of quantum processes, a notion that he felt could only be handled mathematically through the abstract quantum algorithm, together with the principle of complementarity. We will argue that ideas based on a fundamental notion of process offers an alternative, more coherent view which will also provide an ontology.

The hope of finding a better understanding of Nature through a process philosophy is not new. Already Whitehead [2] carries this analysis much further and proposes that reality is essentially an *organism* in which the whole determines the properties of the parts rather than the parts determining the whole. Less well known is the work of Bohm [3] who carries these ideas much further in general terms, as well as attempting to articulate them in a mathematical form [4, 5]. Apart from the well known conceptual difficulties presented by the usual approach to quantum phenomena, a strong motivation and support for these ideas comes from the entangled state phenomenon known as quantum non-locality. Here the properties of groups of individual systems cannot be derived from *a priori* given individual properties, but inherit properties derived from the whole system, supporting Whitehead's use of the term "ogranic".

How then are we to start to build such a theory and develop it into a sound mathematical structure? I believe that Grassmann [6] and Hamilton [7] had already begun to show us how this might be achieved in terms of basic elements that form an algebra. Indeed Grassmann was motivated by a notion of *becoming*, a notion that eventually led him to what we now call a Grassmann algebra. This structure plays an important role in physics today. However the full possibilities of Grassmann's ideas have been lost because the original motivation has been forgotten. With this loss, the exploitation of this potentially rich structure has been stifled [8–10].

Grassmann began his discussion by drawing a distinction between *form* and *magnitude*. Today's physics is all about magnitude, about measured values in a given form, but for Grassmann it was about exploring and developing new forms. The notion of form is very broad and more commonly occurs in 'thought' or 'thought form' as Grassmann put it [11]. Now thought is about becoming and becoming yields a continuous form. Essentially we could ask how one thought becomes another? Is the new thought independent of the old or is there some essential dependence? The answer to the first question is clearly "no", because the old thought contains the potentiality of the new thought, while the new thought contains a trace of the old. Hence again as Grassmann puts it "the continuous form is a twofold act of placement and conjunction, the two are united in a single act and thus proceed together as an indivisible unit". Why not a similar idea for the unfolding of material processes? If we succeed we will have the possibility of removing the difference between material processes and thought.

Let us try to formalise Grassmann's ideas and regard T_1 and T_2 as the opposite distinguishable poles of an indivisible process. Notice however we have made a distinction in the overall process as Kauffman [13] would put it. Making a distinction does not deny the implicit indivisibility of the process, it simply notes the

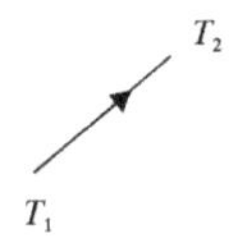

Fig. 12.1 One-simplex

differences. Therefore let us write the mathematical expression for this process as $[T_1 T_2]$, the brace emphasising its indivisibility. We can represent this brace in the form of a diagram (see Fig. 12.1).

The arrow emphasises that T_1 and T_2, although distinguishable, cannot be separated.

When applied to space, we will write $[P_1 P_2]$, where each distinguishable region of space will be denoted by P_i. Grassmann called this brace an *extensive*. Thus we will call quantities like $[A\ B]$ *extensions*. They denote the activity of one region transforming into another. For more complex structures, we can generalise these basic processes to those shown in Fig. 12.2.

In this way we have a field of extensives which can be related and organised into a multiplex of relations of process, activity, movement.[1] The sum total of all such relations constitutes what Bohm and I have called the *holomovement* [15]. Thus for Grassmann, space was a particular realisation of the general notion of process.

Let us now consider the meaning of $[P\ P]$ where there is only one region. This implies that region P is continually transforming into itself. It is a dynamic entity essentially remaining the same, yet continually transforming into itself. It is an invariant feature in the total flux. Mathematically it is an idempotent as $[P\ P] \bullet [P\ P] = [P\ P]$. We will see that idempotents play a central role in our approach.

Notice we are working with quite general notions and anything that remains invariant in a dynamical process can be treated as an idempotent. Thus an atom is, on one level, an entity, not of "rock-like" existence, but a process continually transforming into itself. But on another level it is composed of sub-processes, electrons,

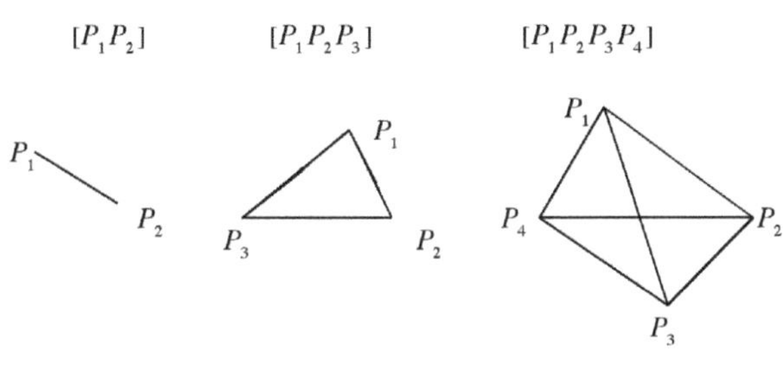

Fig. 12.2 One, two and three-simplexes

[1] The term "movement" is being used, not to describe movements of objects but in a more general sense, implying more subtle orders of change, development and evolution of every kind [14].

nucleons, quarks and so on, each of which can be represented by idempotents. A group of molecules that remain stable can also be treated as a relative idempotent in a more complex structure of process. Clearly this can be extended to viruses, animals, plants right up to human beings and beyond.

At any level the idempotent can be thought of as a "point" which has structure, but what are the consequences if it is taken to be a point in some geometry? Here the notion of a point becomes a dynamic entity and not the changeless entity of classical geometry. The geometry is dynamic. It is just what we want, for example, in general relativity, showing that there is no separation between the "geometry" and the "matter" it contains. If we consider space to be made up of "points" then space itself cannot be a static receptacle for matter. It is a dynamic, flowing structure of process or activity. This is the holomovement. In this sense it has much in common with the quantum vacuum from which quasi-invariant structures emerge and into which quasi-invariant structures decay.

Let us now introduce some mathematical structure. Let us assume these processes form an structure over the real field[2] in the following sense:

- multiplication by a real scalar denotes the strength of the process.
- the process is assumed to be oriented. Thus $[P_1P_2] = -[P_2P_1]$.
- next we introduce an inner multiplication defined by

$$[P_1P_2] \bullet [P_2P_3] = \pm[P_1P_3] \tag{12.1}$$

  This can be regarded as the *order of succession*. The choice of $+$ or $-$ will turn out ultimately to do with what is conventionally called the metric of the space constructed from the processes. We will develop this idea as we go along.
- to complete the notion of an algebra we need to define a notion of addition of two processes to produce a new process. A mechanical analogy of this, although inadequate in many ways, is the motion that arises when two harmonic oscillations at right angles are combined. It is well-known that these produce an elliptical motion when the phases are adjusted appropriately. This addition process can be regarded as an expression of the *order of co-existence* completing our Leibnizian view of process. We will clarify what addition means for the physics as we go along.

We can show that the product is associative and that the mathematical structure we have defined is a Brandt groupoid [16], the details of which will be discussed in Sect. (12.3.1).

We now turn our attention to showing how the symmetries of space-time are carried by an algebra and, indeed, to show how the points are ordered into an overall structure. For the sake of simplicity we will consider how we may order a discrete structure process [17].

---

[2] We assume the real field for convenience and leave open the possibility for a more fundamental structure.

12.2 Some Specific Algebras of Process

12.2.1 Quaternions

Now that we have a definition of a product to denote succession, we can sharpen up what we mean by a "point". Recall a point is something that continually turns into itself, so that $[P_0P_0] \bullet [P_0P_0] = [P_0P_0]$. Thus $[P_0P_0]$ represents the point P_0. Thus the "point" is an *idempotent* as we have discussed above. It is interesting to note that this general notion of an *element of existence* was first introduced by Eddington [18].

We are working in a domain of continuous process or activity. To make a distinction we must have a form that is stable, and what makes a better stable process than that process that turns continually into itself. Mathematically it is the idempotent, ϵ, that continually turns into itself, $\epsilon\epsilon = \epsilon$ hence $\epsilon\epsilon \ldots \epsilon = \epsilon$.

Now let us move on to consider two independent processes $[P_0P_1]$ and $[P_0P_2]$. Suppose we want to turn $[P_0P_1]$ into $[P_0P_2]$. How is this done? Introduce a process $[P_1P_2]$, then using the product rule we get

$$[P_0P_1] \bullet [P_1P_2] = [P_0P_2]$$

Here we are using the $+$ in the product. In Sect. 12.2.2 will show when it is appropriate to use the minus. We can also show

$$\begin{aligned}
[P_0P_2] \bullet [P_1P_2] &= -[P_0P_1] \\
[P_0P_1] \bullet [P_0P_2] &= -[P_1P_2] \\
[P_0P_1] \bullet [P_0P_1] &= -[P_0P_0] = -[P_1P_1] = -1 \\
[P_0P_2] \bullet [P_0P_2] &= -[P_0P_0] = -[P_2P_2] = -1
\end{aligned}$$

Note we have put $[P_0P_0] = [P_1P_1] = [P_2P_2] = 1$ because in the algebra, the point behaves like the identity. Thus there exists three two-sided units in our algebra. The multiplication table now closes and can be written as

1	$[P_0P_1]$	$[P_0P_2]$	$[P_1P_2]$
$[P_0P_1]$	-1	$-[P_1P_2]$	$[P_0P_2]$
$[P_0P_2]$	$[P_1P_2]$	-1	$-[P_0P_1]$
$[P_1P_2]$	$-[P_0P_2]$	$[P_0P_1]$	-1

This structure is isomorphic to the quaternion algebra $\mathcal{C}_{0,2}$. To show this let us make the following identification:- $[P_0P_1] = i$, $[P_0P_2] = j$ and $[P_1P_2] = k$ where the $\{i, j, k\}$ are the quaternions. The multiplication table above then becomes

1	i	j	k
i	-1	$-k$	j
j	k	-1	$-i$
k	$-j$	i	-1

which should be immediately be recognised as the quaternion algebra.

Rather than use the traditional notation for the quaternions, let us use the notation that is now standard in Clifford algebra texts such as Lounesto [19]. Thus we will write $[P_0P_1] = e_1$, $[P_0P_2] = e_2$ and $[P_1P_2] = e_{21}$. Here $e_{21} = e_2e_1$. We use this notation from now on.

Having established that the essentials of quaternion Clifford algebra lies at the heart of our process algebra, and we know that the quaternions generate rotations when applied to a vector space. To link with these ideas, let us first notice that $\mathcal{C}_{0,2}$ is normal and simple so that every automorphism is inner. Thus all automorphisms are inner. Thus we can write

$$A' = g(\theta)Ag^{-1}(\theta) \tag{12.2}$$

where $g(\theta)$ is an element of the Clifford group and θ is some parameter which turns out to be an angle. A typical element of this group is given by

$$g(\theta) = \cos(\theta/2) + e_{12}\sin(\theta/2) \tag{12.3}$$

In the simple structure we are considering here, all elements of the group are of this form.

Suppose we choose a specific rotation with $\theta = \pi/2$, then we find

$$g(\pi/2) = (1 + e_{12})/\sqrt{2}; \qquad g^{-1}(\pi/2) = (1 - e_{12})/\sqrt{2}$$

With these two expressions we can show exactly why the rules introduced above actually work. Take for example $A = e_1$, and put $\theta = \pi/2$, then

$$g(\pi/2)e_1g^{-1}(\pi/2) = e_1g^{-2}(\pi/2) = -e_1e_1e_2 = e_2.$$

Translating this back into the original notation this gives

$$[P_0P_1] \bullet [P_1P_2] = [P_0P_2]$$

which is where we started.

Clearly what we have constructed is a method of rotating through right angles in a plane. However if we allow θ to take continuous values $0 \geq \theta \geq 2\pi$, we can create a continuum of points on a circle, since

$$e_1' = g(\theta)e_1g^{-1}(\theta) = \cos^2(\theta/2) - \sin^2(\theta/2)e_1 + 2\cos(\theta/2)\sin(\theta/2)e_2$$

or

$$e_1' = \cos(\theta)e_1 + \sin(\theta)e_2$$

Of course this relation looks very familiar but notice we are approaching things from a different point of view. We can formalise this by introducing a mapping from our algebra of process, our Clifford algebra, $\mathcal{C}_{0,2}$ to a Euclidean vector space V, viz:

$$\eta : \mathcal{C} \rightarrow V \qquad \text{such that} \quad e_1 = \hat{e}_1 \text{ and } e_2 = \hat{e}_2$$

where $\{\hat{e}_1, \hat{e}_2\}$ is a pair of othonormal basis vectors in V satisfying $\hat{e}_i^2 = 1$; $(\hat{e}_1.\hat{e}_2) = 0$. The inner automorphism in $\mathcal{C}_{0,2}$ then induce rotations in V. Thus we have reversed the traditional argument of starting from a given vector space and constructing an algebra. We claim that it is also possible and, for our purposes, profitable to construct a vector space from the group of movements, or processes. This is a more primitive example of the Gel'fand construction [12]. We suggest that this possibility would be worth exploring when investigating so called quantum space-times.

So far we have confined our attention to a very simple case where we only have two degrees of freedom. In Sect. 12.2.3 we will show how the above rules can be generalised and applied to three basic movements $[P_0P_1]$, $[P_0P_2]$ and $[P_0P_3]$ to produce the Pauli Clifford, which in turn produces a 3-dimensional vector space, $R_{3,0}$. The processes here are all of a polar kind.

Let us now generalise our approach by including processes of a temporal kind, so that we are led to Lorentz-type Clifford groups. In the next sub-section we show exactly how this can be done, again by simply considering two basic processes $[P_0P]$ and $[P_0T]$, the latter being a temporal process. This may come as a surprise as it is usual to think of "succession" as a succession in time. However here we use "succession" in a more general way, implying a general notion of order as in a "succession of points".

12.2.2 Lorentz Group

Let us illustrate how to generalise our approach by considering one movement in space and one in time. We will show that this results in the Clifford algebra $\mathcal{C}_{1,1}$ which will enable us to construct the mini-Minkowski space $R_{1,1}$.

In this case the basic processes are $[P_0P]$ and $[P_0T]$. We will follow Kauffman [13] and call $[P_0P]$ a *polar* extensive while $[P_0T]$ will be called a *temporal* extensive. We now extend the product to

$$[P_0P] \bullet [P_0P] = -[P_0P] \bullet [PP_0] = -[P_0P_0] = -1$$
$$[P_0T] \bullet [P_0T] = -[P_0T] \bullet [TP_0] = +[P_0P_0] = +1$$

Here we have used the relations $[P_0P_0] = 1$; $[P\ P] = 1$ while $[T\ T] = -1$. The difference in sign between these two terms will ultimately emerge in the signature of the metric g_{ij} when constructed. Thus we have $[P\ P] = g_{PP}$ and $[T\ T] = g_{TT}$.

We then get the multiplication table

1	$[P_0T]$	$[P_0P]$	$[P\ T]$
$[P_0T]$	1	$[P\ T]$	$[P_0P]$
$[P_0P]$	$-[P\ T]$	-1	$[P_0T]$
$[P\ T]$	$-[P_0P]$	$-[P_0T]$	1

If we identify $[P_0T] = e_0$, $[P_0P] = e_1$ and $[P\,T] = e_{10}$, we see this multiplication table is isomorphic to the multiplication table of the Clifford algebra $\mathcal{C}_{1,1}$.

In this algebra it is of interest to examine the sum, $[P_0T] + [P_0P]$, and the difference $[P_0T] - [P_0P]$. Here

$$\Big[[P_0T] \pm [P_0P]\Big] \bullet [P\,T] = -[P_0T] \bullet [T\,P] \pm [P_0P] \bullet [P\,T] = \Big[[P_0T] \pm [P_0P]\Big]$$

In other words $\Big[[P_0T] \pm [P_0P]\Big]$ transforms into itself. Now writing this in a more familiar form, we have

$$\Big[[P_0T] \pm [P_0P]\Big] = e_0 \pm e_1 \quad \xrightarrow{\eta} \quad t \pm x$$

which we should immediately recognise as the light cone co-ordinates used by Kauffman [13]. In our approach $\Big[[P_0T] + [P_0P]\Big] = \eta$ corresponds to a movement along one light ray, while $\Big[[P_0T] - [P_0P]\Big] = \xi$ corresponds to a movement along a perpendicular light ray as shown in Fig. 12.3.

Since we have a movement in time and a movement in space, how then do we envisage the notion of velocity? In other words can we give a meaning to dividing one vector by another? Suppose we first form the reciprocal $1/[P_0T]$. Then clearly

$$[P_0T]^{-1} = \frac{1}{[P_0T]} \times \frac{[P_0T]}{[P_0T]} = \frac{[P_0T]}{[P_0T]} \times \frac{1}{[P_0T]} = [P_0T]$$

But this leaves us with a problem. Do we write for the velocity

$$[P_0P] \bullet [P_0T]^{-1} = -[P\,T] \qquad \text{or} \qquad [P_0T]^{-1} \bullet [P_0P] = [P\,T]?$$

The conventional approach in the Dirac theory identifies the positive direction of the velocity with $\alpha_1 = \gamma_{01}$. In our simple Clifford algebra $\mathcal{C}_{1,1}$, the equivalent to α_1 is $e_{01} = [P_0P] \bullet [P_0T]^{-1}$. We then identify e_{01} with the positive direction of the velocity.[3] With this identification we can write an element of the Clifford group as

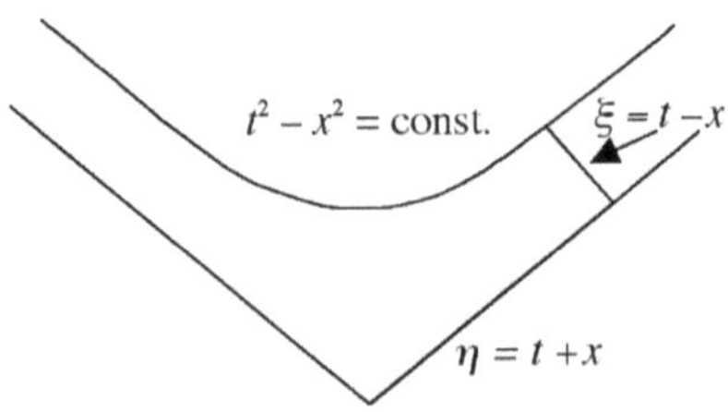

Fig. 12.3 Light cone coordinates

[3] Note this immediately gives us an explanation as to *why* we identify α_{0i} with the velocity in the Dirac theory.

$$g^{\pm}(\lambda) = \cosh(\lambda/2) \pm e_{01} \sinh(\lambda/2)$$

where λ, the rapidity is given by $\tanh(\lambda) = v/c$. In terms of these transformations it is easily shown that

$$e_0' + e_1' = g(e_0 + e_1)g^{-1} = k^{-1}(e_0 + e_1)$$

and

$$e_0' - e_1' = g(e_0 - e_1)g^{-1} = k(e_0 - e_1)$$

where $k = \sqrt{\frac{1+v}{1-v}}$. Since we can write

$$e_0 \xrightarrow{\eta} t \qquad \text{and} \qquad e_1 \xrightarrow{\eta} x$$

we see that

$$t' + x' = k^{-1}(t + x) \qquad \text{and} \qquad t' - x' = k(t - x)$$

This is just the defining expression for the k-calculus from which Kauffman [13] extracts all the transformations required for special relativity. This result also holds when $\mathcal{C}_{1,1}$ is generalised to $\mathcal{C}_{1,3}$. Thus we provide an alternative but complementary approach to the Kauffman approach. We will discuss Kauffman's approach to special relativity a little further later in the paper.

12.2.3 The Pauli Clifford

Let us now move onto generating the Pauli Clifford. As we have already remarked we need three basic polar extensives $[P_0P_1], [P_0P_2], [P_0P_3]$. We require these to describe three independent movements. We also require three movements, one to take us from $[P_0P_1]$ to $[P_0P_2]$, another to take form $[P_0P_1]$ to $[P_0P_3]$ and a third to take us from $[P_0P_2]$ to $[P_0P_3]$. Call these movements $[P_1P_2], [P_1P_3]$ and $[P_2P_3]$.

If we are to construct the Pauli algebra we need to assume the following relations,

$$[P_0P_0] = 1; \quad [P_1P_1] = [P_2P_2] = [P_3P_3] = -1$$

We should notice the change in sign in this case. This is to take account of the different metric we are choosing and because of that we need to ensure $[P_0P_1] \bullet [P_0P_1] = [P_0P_2] \bullet [P_0P_2] = [P_0P_3] \bullet [P_0P_3] = 1$. At this stage the notation is looking a bit clumsy so it will be simplified by writing the six basic movements as $[a], [b], [c], [ab], [ac], [bc]$.

We now use the order of succession to establish

$$[aa] = [bb] = [cc] = 1$$
$$[ab] \bullet [bc] = [ac]$$
$$[ac] \bullet [cb] = [ab]$$
$$[ba] \bullet [ac] = [bc]$$

where the rule for the product is self evident. There exist in the algebra, three two-sided units, $[aa]$, $[bb]$, and $[cc]$. For simplicity we will replace these elements by the unit element 1. This can be justified by the following results $[aa] \bullet [ab] = [ab]$; $[ba] \bullet [aa] = [ba]$; $[bb] \bullet [ba] = [ba]$, etc.

Now

$$[ab] \bullet [ab] = -[ab] \bullet [ba] = [aa] = -1$$
$$[ac] \bullet [ac] = -[ac] \bullet [ca] = [cc] = -1$$
$$[bc] \bullet [bc] = -[bc] \bullet [cb] = [bb] = -1$$

There is the possibility of forming $[abc]$. This gives

$$[abc] \bullet [abc] = -[abc] \bullet [acb] = [abc] \bullet [cab] = -[ab] \bullet [ab] = -1$$
$$[abc] \bullet [cb] = -[ab] \bullet [b] = [a], \qquad \text{etc.}$$

In the last expression we need to know that $[P_0 P_i] \bullet [P_j P_k] = [P_j P_k] \bullet [P_0 P_i] \quad \forall\ i \neq j \neq k$. Thus the algebra closes on itself. If we now change the notation to bring it in line with standard Clifford variables we have

$$[P_0 P_i] = e_i; \quad [P_i P_j] = e_{ji}; \quad [abc] = -e_{123}$$

then it is straight forward to show that the algebra is isomorphic to the Clifford algebra $R_{3,0}$ which we was called the Pauli-Clifford algebra in Frescura and Hiley (1980).

The significance of this algebra is that it carries the rotational symmetries of a three-dimensional Euclidean space. The movements $[ab]$, $[ac]$ and $[bc]$ generate the Lie algebra of SO(3), the group of rotations in three-space, but because we have constructed a Clifford algebra, the rotations are carried by inner automorphisms of the type shown in Eq. (12.2). A typical form of $g(\theta)$ is shown in Eq. (12.3). In the three dimensional case we use the bivectors $[ab] = e_{21}$, $[ac] = e_{31}$ and $[bc] = e_{32}$ to define three infinitesimal rotations about the three independent axes.

We can extend the method we have outlined above to generate the Dirac Clifford algebra and the conformal Clifford algebra which contains the twistor originally introduced by Penrose [20]. We will not discuss these structures in this paper.

In summary then we have shown, or rather sketched out how to generate the Clifford algebras from a primitive notion of process or activity. A more rigorous mathematical approach to these topics will be found in Griffor [24]. The algebras we have discussed in this paper so far only use orthogonal groups and so we have, in effect, developed a *directional calculus*. What is missing from our discussion so far is the translation symmetry. The generalisation to include translations will be discussed later

12.3 Connections with Other Mathematical Approaches

12.3.1 Formal Mathematical Structure

We have already indicated that we have identified the structure we are exploring as a Brandt groupoid [16] but we don't have to return to this early work to get a formal understanding of this structure. Ronnie Brown [25] has an excellent review of this structure which we will briefly explain here.

A groupoid G is a small category in which every morphism is an isomorphism. This category comprises a set of morphisms, together with a set of objects or points, $Ob(G)$. We also have a pair of functions $s, t : G \rightarrow Ob(G)$, together with a function $i : Ob(G) \rightarrow G$ such that $si = ti = 1$. The functions s, t are called the source and target maps respectively.

The elements of process that we have been using can easily be connected with this language. Consider a general movement a_j. We can identify sa_j as P_0 and ta_j as P_j, then

$$[P_0 P_j] \qquad \Rightarrow \qquad a_j : sa_j \rightarrow ta_j$$

If we have a pair of movements a and b then we can compose these movements if $ta = sb$. Thus

$$[sa \xrightarrow{a} ta][sb \xrightarrow{b} tb] = [sa \xrightarrow{ab} tb] \qquad \text{iff} \quad ta = sb.$$

It is not difficult to show that this product is associative. Thus the mathematical structure we considered in the last section can be formally identified with a groupoid.

It should be noted that our approach is also related to the approach of Abramski and Coecke [26] who apply category theory to quantum mechanics, again attempting to build a process view of quantum phenomena. It would be interesting to compare the two approaches in detail but such a comparison will not be attempted in this paper.

12.3.2 Comparison with Kauffman's Calculus of Distinctions

In the earlier sections of this paper, we have referred to the work of Kauffman [13, 23, 30] which provides us with a calculus of distinctions, or as it is sometimes called, the iterant algebra. Kauffman also introduces a binary symbol $[A, B]$ which he motivates in the following way. He wants to start with a basic idea from which all further discussions follow. Start with an undifferentiated two-dimensional canvas. Then make a distinction by defining a boundary, dividing the canvas into two regions marking the "inside" A and the "outside" B symbolizing this distinction by $[A, B]$.

How is this distinction related to the basic movement that we have used earlier? To bring this out simply, we can also write this as $\overline{\text{A}\,|} = B$. Here we use $\overline{\quad\;|}$ to denote Spencer Brown's "cross" indicating we must cross the boundary. Kauffman [13] uses this analogy to argue that not only do we make a distinction but we must act out the distinction by actually crossing the boundary. Crossing the boundary enables us to discuss change, to discuss movement, to discuss process.

Now in order to structure these movements, Kauffman introduces a multiplication rule

$$[A, B] * [C, D] = [AC, BD] \tag{12.4}$$

Furthermore he assumes that $AC = CA$ and $BD = DB$, i.e. they are commutative products. In this case, suppose we take $B = C$ then

$$[A, B] * [B, C] = [AB, BC] = B[A, C]$$

Then if we put $B = \pm 1$ we obtain the same product rule that we introduced earlier in this paper. This means the common element is replaced by ± 1 in the product. Thus we see that our algebra is a special case of that introduced by Kauffman.

The sum that Kauffman introduces is also a generalisation of the sum we introduce. He defines

$$[A, B] + [C, D] = [A + C, B + D]$$

In Kauffman [23] the k-calculus is set up by directly considering $[t + x]$ and $[t - x]$ but it is difficult to understand how "marks" A and B become co-ordinates. In Sect. 12.2.1 above we have a clearer way of showing how the co-ordinates arise since we have a well defined mapping, η, taking us from the Clifford algebra, $\mathcal{C}_{n,m}$, to an underlying vector space $V_{n,m}$. Once again notice the order, $\mathcal{C}_{n,m}$ is primary, $V_{n,m}$ is derived.

12.3.3 Some Deeper Relations Between Our Approach and that of the Calculus of Distinction

Let us go deeper into the relation between the general structure of Clifford algebras and its relationship to the Kauffman calculus [13]. However we first want to point out that Schönberg [28] and Fernandes [29] have shown us how to build any orthogonal Clifford algebra from a pair of dual Grassmann algebras whose generators satisfy the relationship

$$[a_i, a_j^\dagger]_+ = g_{ij} \qquad [a_i, a_j]_+ = 0 = [a_i^\dagger, a_j^\dagger]_+ \tag{12.5}$$

These will be recognised as *vector* fermionic "annihilation" and "creation" operators.[4] Notice these are vector operators and not the spinor operators used in particle physics. Using these we find

$$\sigma_x = a + a^\dagger \qquad \text{and} \qquad \sigma_z = a - a^\dagger \tag{12.6}$$

Now Kauffman [13] introduces an operation p that destroys the inside but leaves the outside alone. This can be written as

$$p * [A, B] = [0, B]. \tag{12.7}$$

One way to do this is to let $p = [0, 1]$ and use the product $*$ defined in Eq. (12.4). Then clearly Eq. (12.7) will follow. Kauffman develops these ideas further and shows

$$\sigma_x * [A, B] = [B, A] \qquad \text{and} \qquad \sigma_z * [A, B] = [A, -B] \tag{12.8}$$

Now let us ask what our operators a and $a^\dagger$ produce when we operate on $[A, B]$. It is straightforward to show

$$a * [A, B] = [B, 0] \qquad \text{and} \qquad a^\dagger * [A, B] = [0, A] \tag{12.9}$$

Thus we see that here the annihilation operator a destroys the inside and puts the outside inside. On the other hand the creation operator $a^\dagger$ destroys the outside and puts the inside outside!

We can actually carry this further and ask what action the algebraic spinors (minimal left ideals) have on $[A, B]$. It is not difficult to show that in this case we have two algebraic spinors given by

$$\psi_{L_1} = aa^\dagger + a^\dagger \qquad \text{and} \qquad \psi_{L_2} = a^\dagger a + a. \tag{12.10}$$

Now let us ask what happens when we use these spinors to produce a change in $[A, B]$. What is this change? Again the answer is easy because

$$\psi_{L_1} * [A, B] = (aa^\dagger + a^\dagger) * [A, B] = [A, A] \tag{12.11}$$

$$\psi_{L_2} * [A, B] = (a^\dagger a + a) * [A, B] = [B, B] \tag{12.12}$$

[4] This suggests that these operators could be used to describe the creation and annihilations of extensions.

This means that one type of spinor, ψ_{L_1}, destroys the outside and puts a copy of the inside outside, while the other spinor, ψ_{L_2}, destroys the inside and puts a copy of the outside inside! In other words these operators remove the original distinction, but in different ways. More importantly from the point of view that we are exploring here is that we see how the algebraic spinor itself is active in producing a specific change in the overall process and not merely a vector in an abstract Hilbert space.

Physicists treat these two spinors, Ψ_{L_1} and Ψ_{L_2} as equivalent and map them onto an external spinor space, the Hilbert space of ordinary spinors ψ. In other words this single spinor, ψ, is an equivalence class of the algebraic spinor when it is projected onto a Hilbert space (see Bratteli and Robinson [31]). The projection means that we have lost the possibility of exploiting the additional dynamical structure offered by the algebraic spinors.

Notice that these spinors are themselves part of the algebra. The whole thrust of our argument is that we must exploit the properties of the algebra and not confine ourselves to an external Hilbert space. When we do this we can continue with our idea that the elements of our algebra describe activity or process even in the quantum domain.

Let us take this whole discussion a little further. Consider the following relationship

$$a * [A, 0] = [0, 0] \tag{12.13}$$

This should be compared with the physicist's definition of a vacuum state,

$$a|0\rangle = a \Downarrow = 0 \tag{12.14}$$

where we have introduced the notation used by Finkelstein [32]. Thus Eq. (12.14) shows that $[A, 0]$ acts like the vacuum state in physics.

Furthermore $a^{\dagger} * [0, B] = [0, 0]$ should be compared with $a^{\dagger} \Uparrow = 0$. Finkelstein calls, $\Uparrow$, the plenum. Thus we see that $[0, B]$ acts like the plenum. Thus we see that here the vacuum state is not empty. Internally it has content but externally it is empty. For the plenum, it is the other way round, so unlike Parmenides we can have "movement" from outside to inside! We can take this a bit further by recalling that we can write the projector onto the vacuum as $V = |0\rangle\langle 0|$, then we have $aV = 0$. If the projector onto the plenum is $P = |\infty\rangle\langle\infty|$, we find $a^{\dagger} P = 0$. It is interesting to note that relations like these are central to the approach of Schönberg [28].

12.3.4 Extensions and the Incident Algebra of Raptis and Zapatrin

All of the above discussion suggests that we could perhaps write $[A, B]$ as $|A\rangle\langle B|$. It should be noticed that this is only a symbolic change and the bra/ket should not be identified with vectors in an external vector space. Nevertheless by making this identification we can bring out the relationship of our work to that of Zapatrin [33] and of Raptis and Zapatrin [27] who developed an approach through what they called the incident algebra. In this structure the product rule is written in the form

$$|A\rangle\langle B|\cdot|C\rangle\langle D| = |A\rangle\langle B|C\rangle\langle D| = \delta_{BC}|A\rangle\langle D|$$

Again this multiplication rule is essentially rule (12.1), the order of succession above. But there is a major difference. When $B \neq C$ the product equals zero, whereas we leave it undefined at this level. So tempting as it seems we must not *identify* their $|A\rangle\langle B|$ with our use of the same symbols. Nevertheless a good notation can take us to places where we did not expect to go as we will see!

12.4 Some Radical New Ideas

12.4.1 Intersection of the Past with the Future

We now want to look more deeply into the structure based on relationships like $[P_1 P_2]$ by, as it were, "getting inside" the connection between P_1 and P_2. Remember we are focusing on process or movement and we are symbolising the notion of *becoming* by $[P_1 P_2]$. Ultimately we want to think of these relationships as an ordered structure defining what we have previously called "pre-space" (see Bohm [3] and Hiley [34]). In other words, these relationships are not to be thought of as occurring in space-time, but rather the properties of space-time are to be abstracted from this pre-space. We have already suggested how we can achieve this through the mapping $\eta : \mathcal{C} \rightarrow V$ but much more work has to be done. Let us go more slowly.

Conventionally physical processes are always assumed to unfold in space-time, and furthermore, time evolution is always assumed to be from point to point. In other words, physics always tries to talk about time-development *at an instant*. Any change always involves the limiting process $\lim_{\Delta t\rightarrow 0}(\Delta x)/(\Delta t)$.

However before taking the limit we were taking points in the past (x_1, t_1) and relating them to a points in the future (x_2, t_2), that is we are relating what *was* to what *will be*. But we try to hide the significance of that by going to the limit $t_2 - t_1 \rightarrow 0$ when we interpret the change to take place at an instant, t. Yet curiously the instant is a set of measure zero sandwiched between the infinity of that which has passed and the infinity of that which is not yet. This is fine for evolution of point-like entities but not for the evolution of structures.

At this point I wish to recall Feynman's classic paper where he sets out his thinking that led to his "sum over paths" approach [35]. There he starts by dividing space-time into two regions R' and R''. R' consists of a region of space occupied by the wave function before time t', while R'' is the region occupied by the wave function after time t'', i.e $t' < t''$. Then he suggested that we should regard the wave function in region R' as contain information coming from the "past", while the conjugate wave function in the region R'' represents information coming from the "future".[5] The possible "present" is then the intersection between the two, which

[5] This is essentially the same idea that led to the notion of the anti-particle "going backward in time", but at this stage we are not considering anti-matter.

is simply represented by the transition probability amplitude $\langle\psi(R'')|\psi(R')\rangle$.[6] But what I want to discuss here is $|\psi(R')\rangle\langle\psi(R'')|$. This is where all the action is!

Before taking up this point I would like to call attention to a similar notion introduced by Stuart Kauffman [37] in his discussion of biological evolution. Here the discussion is about the evolution of *structure*. He talks about the evolution of biological structures from their present form into the "adjacent possible". This means that only certain forms can develop out of the past. Thus not only does the future form contain a trace of the past, but it is also constrained by what is "immediately" possible, its *potentialities*. So any development is governed by the *tension between the persistence of the past, and an anticipation of the future*.

What I would now like to do is to build this notion into a dynamics. Somehow we have to relate the past to the possible futures, not in a completely deterministic way, but in a way that constrains the possible future development. My basic notion is that physical processes involve an extended structure in both space and time. I have elsewhere called this structure a "moment" [38]. In spatial terms, it is fundamentally non-local in space, however it is also "non-local" in time. I see this more in terms of a-local concepts, with locality yet to be defined. It is a kind of 'extension in time' or a 'duron', an idea that has a resonance with what Grassmann was trying to do as we outlined earlier. This extension in time may seem too extreme but remember quantum theory must accommodate the energy-time uncertainty principle. This implies that a process with a given sharp energy cannot be described as unfolding at an instant of time except, perhaps, in some approximation. The notion of a moment captures the essential ambiguity implicit in quantum theory.

The idea of process that we introduced in Sect. 12.2 is naturally suited to describing this notion of a moment. All we need to do is to regard the two aspects of $[P_1 P_2]$ as functions of two times so that we can write this bracket in the form $[A(t_1), B(t_2)]$ and show that, as Feynman [35] actually demonstrates for the Schrödinger case, we capture the usual equations of motion in the limit $t_1 \to t_2$.

12.4.2 A Change in Notation

In Sect. 12.3.3 we have already shown that if write $|0\rangle\langle 0|$ for $[0, 0]$ we have the possibility of a ready made link between our notation and the bra-ket notation of Dirac. In fact Dirac in his work on spinors [39] has already suggested that we can factorise an element of the Clifford algebra into a pair $|A\rangle$ and $\langle B|$. He shows that this ket is, in fact, a spinor, while the bra is a dual spinor.

Let us therefore follow Dirac and replace $[P_1, P_2]$ by $|P_1\rangle\langle P_2|$.[7] In Sect. 12.2.1 we introduced the Clifford group without any justification in terms of the ideas we were developing. Rather we called on our prior knowledge of the structure of

[6] While I was preparing this article, Aharonov reminded me of the similarity of these ideas to his on pre- and post-selection [36].

[7] $|P_1\rangle$ and $\langle P_2|$ are not to be taken as elements in a Hilbert space and its dual.

Clifford algebras to introduce the idea. However we can justify this choice once we have assumed that the basic process can be factorised. Now we can argue that under a rotation, we have the transformations

$$|P'\rangle = R|P\rangle \qquad \text{and} \qquad \langle P'| = \langle P|R^{-1}$$

This gives us the added advantage of showing immediately that $\langle P|P\rangle$ is rotationally invariant.

This also helps us the understand Kauffman's [23, 40] notation when he writes the Lorentz transformation in the form

$$\mathcal{L}[A, B] = [k^{-1}A, kB]$$

where $k = \sqrt{\frac{1+v}{1-v}}$ is a function of the relative velocity. First we consider a boost in the x-direction, $g(v)$, and write

$$\mathcal{L}[A, B] = g(v)|A\rangle\langle B|g^{-1}(v) \tag{12.15}$$

which does not in general reduce to Kauffman's expression. However if we form

$$e_0' + e_1' = g(v)\Big(|P_0\rangle(\langle T| + \langle P|)\Big)g^{-1}(v) = k^{-1}\Big(|P_0\rangle(\langle T| + \langle P|)\Big) = k^{-1}(e_0 + e_1)$$

and similarly

$$e_0' - e_1' = g(v)\Big(|P_0\rangle(\langle T| - \langle P|)\Big)g^{-1}(v) = k\Big(|P_0\rangle(\langle T| - \langle P|)\Big) = k(e_0 - e_1)$$

We arrive at Kauffman's result provided we write $A = |P_0\rangle\langle T| + |P_0\rangle\langle P|$ and $B = |P_0\rangle\langle T| - |P_0\rangle\langle P|$.

The traditional way to proceed is to assume that $|P\rangle$ is a vector in an external Hilbert space. As Frescura and Hiley [21] have shown there is, in fact, no need to go outside the algebra. We can identify the ket as an element, Ψ_L, of a left ideal, $\mathcal{I}_L$, in the algebra. One way to generate such an element is to choose a suitable primitive idempotent, ϵ. Then we multiply from the left by any element of the algebra to form the element Ψ_L giving

$$\Psi_L = A\epsilon$$

where the general element A can be written in the form $A = a_0 + \sum a_i e_i + \sum a_{ij} e_{ij} + \cdots + a_n e_{12\ldots n}$, when the Clifford algebra is generated by $\{1, e_1 \ldots e_n\}$.[8]

[8] Physicists should not be put off by the notation because the es are exactly the γs used in standard Dirac theory We have changed notation simply to bring out the fact that we do not need to go to a matrix representation, although that possibility is always open to us should we wish to take advantage of it.

There is a simplification with which we will avail ourselves and that is that, in the Clifford algebras in which we are interested, we can generate an element of the left ideal by choosing A to comprise only the even elements of the algebra. In this case we can write

$$A = \psi_L = a_0 + \sum a_{ij}e_{ij} + \sum a_{ijkn}e_{ijkn} + \dots.$$

The dual element corresponding to $\langle B|$ is an element of the minimal right ideal generated from ϵ and is given by

$$\Psi_R = \epsilon B$$

We can also obtain Ψ_R from the element of the minimal left ideal by *conjugation*. Conjugation is defined as an anti-involution on $\mathcal{A}$ induced by the orthogonal involution -1_X, X being the vector space induced by the Clifford algebra. In this case $\Psi_R = \widetilde{\Psi}_L$. Thus in our new notation, we have

$$[A, B] \rightarrow \Psi_L \Psi_R = \psi_L \epsilon \psi_R = \hat{\rho}$$

We will call this element the *Clifford density element*. $\hat{\rho}$ is a key element in our Clifford algebra approach to quantum theory [41–43]. As we showed in those papers, we can reproduce all the results of quantum mechanics without the need to introduce Hilbert space and the wave function. Furthermore as we will show later, we are led directly to the Bohm model showing that this model is merely a re-formulation of quantum theory in which the complex numbers are not necessary. Before embarking on this discussion we will continue to explore the structure of space-time offered by our methods.

12.4.3 The Lorentz Transformations from Light Rays and a Clock

In this sub-section we sketch how Kauffman's use of the k-calculus, introduced originally by Page [45] and popularised by Bondi [46], emerges from the results obtained in Sect. 12.2.2 and in the previous sub-section. We have shown that the movements along light rays lead directly to the light cone co-ordinates. We did this by noting that the light cone movements $e_0 \pm e_1$ transform through

$$e_0' \pm e_1' = k^{\mp}(v)(e_0 \pm e_1)$$

where $k(v) = \sqrt{\frac{1+v}{1-v}}$. We then use the projection η to obtain the light cone co-ordinates so that

$$t' \pm x' = k^{\mp}(v)(t \pm x)$$

The basic idea of the k-calculus is to explore the geometry of space armed only with a radar gun and a clock. The gun is fired from the origin of the co-ordinate system at time t_1 so that this event has co-ordinates $(t_1, 0)$. The radar signal returns to the origin at time t_2 after being reflected off an object at the point (t, x). Assuming the speed of light is unity ($c = 1$) both ways, we see that

$$t_2 - t_1 = 2x \qquad \text{and} \qquad t_2 + t_1 = 2t.$$

In terms of the Kauffman's iterant calculus, we can write

$$[t_2, t_1] = [t + x, t - x]$$

But we have already shown in Eq. (12.16) how the RHS of this equation transforms under a Lorentz transformation. Thus we have

$$[t_2', t_1'] = [t' + x', t' - x'] = [k^{-1}(v)(t + x), k(v)(t - x)] = [k^{-1}(v)t_2, k(v)t_1]$$

This means that

$$t_1' = kt_1(v) \qquad \text{and} \qquad t_2 = k(v)t_2' \tag{12.16}$$

These are just the starting relations of the k-calculus. Its assumption is that if observer A sends, at time t_1, a signal to observer A' moving at a constant relative speed v, then it is received at A', at time t_1', where $t_1' = k(v)t_1$. Because of the principle of relativity if A' sends at time t_2' a signal to A then it will be received at time t_2, where $t_2 = k(v)t_2'$. This is just the result obtained in Eq. (12.16).

Thus it is possible to use the principle of Galilean relativity, the constancy of the speed of light and like-light movements to abstract Minkowski space-time. Light-like movements are, of course, simply light signals. For more details of this approach see Kauffman [23, 40].

12.4.4 The Light Cone Geometry

The results of the previous subsection indicates that the basic movements can be factorised into what seem to be a dual pair of spinors, namely, elements of a minimal left and right ideals. We now want to show that these elements can be given a meaning in terms of the light cone structures. This means that the properties of the light cones are implicit in the Clifford algebra itself.

Let us see how this works. Consider the Pauli Clifford algebra, $C_{3,0}$ and let us take the idempotent to be $(1 + e_3)/2$. Then we can form the element

$$\Psi_L(\mathbf{r}, t) = \psi_L(\mathbf{r}, t)(1 + e_3)/2 \tag{12.17}$$

where

$$\psi_L = g_0(\mathbf{r}, t) + g_1(\mathbf{r}, t)e_{23} + g_2(\mathbf{r}, t)e_{31} + g_3(\mathbf{r}, t)e_{12}$$

Let us now define a vector $\mathcal{V}$ in the Clifford algebra through the relation $\mathcal{V} = \Psi_L \Psi_R = \psi_L(1 + e_3)\psi_R$ where $\psi_R = \widetilde{\psi}_L$. Here $\sim$ denotes the Clifford conjugate. Then

$$\mathcal{V} = v_0 1 + v_1 e_1 + v_2 e_2 + v_3 e_3$$

where

$$\begin{aligned} v_0 &= g_0^2 + g_1^2 + g_2^2 + g_3^2 \qquad & v_1 &= 2(g_1 g_3 - g_0 g_2) \\ v_2 &= 2(g_0 g_1 + g_2 g_3) \qquad & v_3 &= g_0^2 - g_1^2 - g_2^2 + g_3^2. \end{aligned} \tag{12.18}$$

This will be immediately recognised as a Hopf map [47], which is not surprising since we are essentially dealing with light spheres as we will show in Sect. 12.4.6.

Furthermore Eq. (12.18) shows that $v_0^2 = v_1^2 + v_2^2 + v_3^2$. Thus if we now map $\phi : 1 \to e_0$ with $e_0^2 = -1$ and impose the condition $e_0 e_i + e_i e_0 = -2\delta_{0i}$ we have lifted the vector into the larger Clifford algebra $\mathcal{C}_{3,1}$. However since now $-v_0^2 + v_1^2 + v_2^2 + v_3^2 = 0$, we have constructed a null vector in this larger Clifford algebra. If we now project this Clifford vector into a vector in $V_{3,1}$ by $\eta : \mathcal{C}_{3,1} \to V_{3,1}$ we have constructed a light ray $\mathbf{v}(\mathbf{r}, t)$ in $V_{3,1}$. If we fix the vector at the origin of the co-ordinate system in $V_{3,1}$, then vary $\mathbf{r}, t$ we generate a light cone in $V_{3,1}$. Thus light-like movements in the Clifford algebra can be used to generate a light cone structure on the vector space $V_{3,1}$ which can be taken to be our space-time. This means that we have used elements of the minimal left ideal and its conjugate to generate the light cone.

This result is not that surprising because our elements of $\mathcal{I}_L$ are simply spinors in another guise. Normally we use matrices to represent spinors. Thus in more familiar form

$$|\Psi\rangle \to \begin{pmatrix} \psi_1 \\ \psi_2 \end{pmatrix} \tag{12.19}$$

Then we write

$$|\Psi\rangle\langle\Psi| = \begin{pmatrix} |\psi_1|^2 & \psi_1\psi_2^* \\ \psi_2\psi_1^* & |\psi_2|^2 \end{pmatrix}$$

If we now compare this with the matrix that Penrose introduces to describe a light ray,

$$X = \begin{pmatrix} t + z & x - iy \\ x + iy & t - z \end{pmatrix} \tag{12.20}$$

we find

$$t = |\psi_1|^2 + |\psi_2^*|^2 \quad x = \psi_1\psi_2^* + \psi_1^*\psi_2 \quad y = i(\psi_1\psi_2^* - \psi_1^*\psi_2) \quad z = |\psi_1|^2 - |\psi_2^*|^2$$

These relations are, in fact, identical to the expression in (12.18) provided first we map $(v_0, v_1, v_2, v_3) \rightarrow (t, x, y, z)$ and then use the relations between (g_0, g_1, g_2, g_3) and (ψ_1, ψ_2) presented in Hiley and Callaghan [42]. The mathematical reason for this to work runs as follows. We can map the equivalence class of minimal left ideals onto a Hilbert space $\rho : \Psi_L \rightarrow |\psi\rangle$. In this way we can represent the elements of the left ideal as matrices. Then $\rho : \Psi_L\Psi_R \rightarrow |\psi\rangle\langle\psi|$, the expression for the conventional density matrix.

12.4.5 The Dirac Clifford and $Sl(2\mathbb{C})$

In the previous sub-section although we started from the Pauli Clifford, $\mathcal{C}_{3,0}$, we lifted our structure into the Dirac Clifford $\mathcal{C}_{3,1}$ and found that we have generated light cones. Now it is well known that $Sl(2\mathbb{C})$ is the double cover of $O_+^\uparrow(1,3)$ [47]. The irreducible representations of $Sl(2\mathbb{C})$ are two-dmensional whereas the Dirac spinor is four-dimensional. The Dirc spinor, while being an irreducible representation of the Clifford algebra, is not an irreducible representation of the spin group $Sl(2\mathbb{C})$.

We can express this more formally by considering a Clifford bundle over an n-dimensional orientable manifold M. Call the set of sections of this bundle $\Delta(M)$. Then the Dirac spinor $\Psi_L \in \Delta(M)$ is irreducible. The irreducible representations of the spin group are classified by the eigenvalues of $e_{n+1} = ie_1 \ldots e_n$. In the Dirac case $(e_{n+1})^2 = 1$, so that the eigenvalues of e_{n+1} are ± 1. $\Delta(M)$ is then split into two eigenspaces

$$\Delta(M) = \Delta^+(M) \oplus \Delta^-(M) \tag{12.21}$$

We can now introduce two projection operators, $P^\pm$, defined by

$$P^\pm = (1 \pm e_{n+1})$$

so that

$$P^+\Psi_L = \begin{pmatrix} \Psi_L^+ \\ 0 \end{pmatrix} \in \Delta^+(M), \qquad P^-\Psi_L = \begin{pmatrix} 0 \\ \Psi_L^- \end{pmatrix} \in \Delta^-(M).$$

In this representation, a general element of the Lorentz group takes the form

$$L = \begin{pmatrix} \Lambda & 0 \\ 0 & (\Lambda^\dagger)^{-1} \end{pmatrix} \qquad L^{-1} = \begin{pmatrix} \Lambda^{-1} & 0 \\ 0 & \Lambda^\dagger \end{pmatrix}$$

Here Λ is the 2 by 2 matrix

$$\Lambda = \begin{pmatrix} \alpha & \beta \\ \gamma & \delta \end{pmatrix}$$

where $\alpha, \beta, \gamma, \delta \in \mathbb{C}$ and $\alpha\delta - \beta\gamma = 1$. Also

$$(\Lambda^{\dagger})^{-1} = C\Lambda^{*}C^{-1} = \begin{pmatrix} \delta^{*} & -\gamma^{*} \\ -\beta^{*} & \alpha^{*} \end{pmatrix}$$

and

$$C = \begin{pmatrix} 0 & 1 \\ -1 & 0 \end{pmatrix} = -C^{-1}$$

In this way we recognise that Λ is the irreducible representation of $Sl(2\mathbb{C})$. This group has a conjugate irreducible representation, $\Lambda^{*} \in \overline{Sl(2\mathbb{C})}$. These together with the respective dual irreducible representations $(\tilde{\Lambda})^{-1}$ and $(\Lambda^{\dagger})^{-1}$ provide all the necessary information to completely describe the role of the Dirac spinor. Here $\sim$ is the transpose of the matrix representation.

Indeed the spinor defined in Eq. (12.17) corresponds to $P^{+}\Psi_L$ and clearly transforms as

$$\Psi_L^{+\prime} = \Lambda \Psi_L^{+} \qquad \text{while} \qquad \Psi_L^{-\prime} = (\Lambda^{\dagger})^{-1}\Psi_L^{-}$$

On the other hand we also have $\Psi_R^{+\prime} = \Psi_R^{+}\Lambda^{\dagger}$, where $\Psi_R^{+\prime} = \widetilde{\Psi_L^{+\prime}}$. Here $\sim$ is the Clifford conjugate.

If we now form $\Psi_L^{+}\Psi_R^{+} = X(x)$ where X is given by Eq. (12.20). Then under a Lorentz transformation $X(x') = \Lambda X(x)\Lambda$, or more transparently

$$X(Lx) = \Lambda(L)X\Lambda(L)^{\dagger}$$

We can form a dual matrix $\overline{X}(x)$ which is constructed from $\psi_L(1 - e_3)\psi_R$. This means that in the matrix representation used above

$$\overline{X} = \begin{pmatrix} t - z & -x + iy \\ -x - iy & t + z \end{pmatrix}$$

This gives, under a Lorentz transformation

$$\overline{X}(Lx) = (\Lambda^{\dagger})^{-1}\overline{X}(x)\Lambda^{-1}$$

It may be of some interest to note the following results:

$$\overline{X}(x)X(y) + \overline{X}(y)X(x) = 2\mathbf{x}.\mathbf{y}$$
$$\overline{X}(x)X(y) - \overline{X}(y)X(x) = 2X(x^0\mathbf{y} - y^0\mathbf{x} + \mathbf{x} \wedge \mathbf{y})$$
$$detX(x) = \overline{X}(x)X(x) = \mathbf{x}^2$$
$$\overline{X}(x)X(y) - X(y)\overline{X}(x) = 2X(\mathbf{x} \wedge \mathbf{y}).$$

Finally let us return to the matrix representation of the Dirac algebra so that we can fit our approach into the more conventional one. We write

$$\Psi = \begin{pmatrix} \lambda \\ \rho \end{pmatrix}$$

where λ and ρ are column spinors of dimension 2. In fact

$$\Psi_L^+ \to \lambda \qquad\qquad \Psi_L^- \to \rho.$$

The two spinors λ and ρ correspond to the chiral Weyl spinors which are used to describe the left- and right-handed neutrino states. But notice that in our approach we have nowhere introduced quantum mechanics. The Clifford algebra is simply a way to describe the geometry of space-time. What is remarkable is the way physicists are introduced to the Dirac formalism. It appears that we are forced into the Clifford algebra by quantum mechanics, but this is manifestly not the case. Rather we should introduce the algebra from classical geometry and then show that quantum mechanics exploits the structure of this geometry. In this way quantum mechanics is not all that strange or novel.

Formally these ideas can be put into the language of fibre bundles. The left-handed Weyl spinor is a section of the spin bundle $(W, \pi, M, \mathbb{C}^2, Sl(2\mathbb{C}))$ while the right-handed Weyl spinor is a section of $(\overline{W}.\pi, M, \mathbb{C}, \overline{Sl(2\mathbb{C})})$. In this language the Dirac spinor is a section of $(D, \pi, M, \mathbb{C}, Sl(2\mathbb{C}) \oplus \overline{Sl(2\mathbb{C})})$.

12.4.6 Past and Future Light Cones

Consider a null ray defined by

$$\mathbf{r}^2 - c^2t^2 = 0$$

Then for a fixed t, the light cone intersects a space-like hypersurface in a sphere with raduis $r = \pm ct$. This means that if t is positive, the sphere is on the future light cone, while if t is negative, the sphere is on the past light cone. However the points on a sphere can be characterised by a stereographic projection onto points on a plane with coordinates $\lambda = (\xi, \eta)$ [48]. This we write as a matrix

$$\psi = \begin{pmatrix} \xi \\ \eta \end{pmatrix} \tag{12.22}$$

Under a Lorentz transformation we have

$$\psi' = \Lambda_1(L)\psi \tag{12.23}$$

Now we have seen that

$$x_0 = |\xi|^2 + |\eta|^2, \qquad x_1 = \xi\bar{\eta} + \bar{\xi}\eta \qquad x_2 = i(\xi\bar{\eta} - \bar{\xi}\eta) \qquad x_3 = |\xi|^2 - |\eta|^2$$

In the dotted and undotted notation of Penrose and Rindler [44] these results follow from

$$x_\mu = \sigma_\mu^{A\dot{A}} \psi_A \bar{\psi}_{\dot{A}}$$

where $\sigma_\mu^{A\dot{A}}$ are the spin frames

$$\sqrt{2}\sigma_0^{A\dot{A}} = \begin{pmatrix} 1 & 0 \\ 0 & 1 \end{pmatrix} \quad \sqrt{2}\sigma_1^{A\dot{A}} = \begin{pmatrix} 0 & 1 \\ 1 & 0 \end{pmatrix} \quad \sqrt{2}\sigma_2^{A\dot{A}} = \begin{pmatrix} 0 & i \\ -i & 0 \end{pmatrix} \quad \sqrt{2}\sigma_3^{A\dot{A}} = \begin{pmatrix} 1 & 0 \\ 0 & -1 \end{pmatrix}$$

Here we identify Eq. (12.22) with ψ_A via $\psi_1 = \xi$ and $\psi_2 = \eta$.

Let us look at the infinitesimal version of the transformation (12.23) by writing

$$\Lambda_1(L) = 1 + \epsilon \begin{pmatrix} \alpha & \beta \\ \gamma & \delta \end{pmatrix}$$

then

$$\begin{aligned} \Delta\xi &= \epsilon(\alpha\xi + \beta\eta) \qquad & \Delta\eta &= \epsilon(\gamma\xi + \delta\eta) \\ \Delta\bar{\xi} &= \epsilon(\bar{\alpha}\bar{\xi} + \bar{\beta}\bar{\eta}) \qquad & \Delta\bar{\eta} &= \epsilon(\bar{\gamma}\bar{\xi} + \bar{\delta}\bar{\eta}) \end{aligned}$$

Thus x_μ transforms as

$$x'_\mu = 1 + \omega_\mu^\nu x_\nu$$

with

$$\omega_\mu^\nu = \begin{pmatrix} 0 & -a_4 & -a_5 & -a_6 \\ -a_4 & 0 & a_3 & -a_2 \\ -a_5 & -a_3 & 0 & a_1 \\ -a_6 & a_2 & -a_1 & 0 \end{pmatrix} \tag{12.24}$$

where

$$\begin{aligned} 2a_1 &= i(-\beta+\bar{\beta}-\gamma+\bar{\gamma}) & 2a_2 &= (\beta+\bar{\beta}-\gamma-\bar{\gamma}) \\ 2a_3 &= i(-\alpha+\bar{\alpha}+\delta-\bar{\delta}) & 2a_4 &= (-\beta-\bar{\beta}-\gamma-\bar{\gamma}) \\ 2a_5 &= i(-\beta+\bar{\beta}+\gamma-\bar{\gamma}) & 2a_6 &= (-\alpha-\bar{\alpha}+\delta+\bar{\delta}). \end{aligned} \tag{12.25}$$

Now let us examine the past light cone. Let us again use the stereographic projection only this time using coordinates $\rho = (\sigma, \tau)$ which we can again write this in matrix form

$$\phi = \begin{pmatrix} \sigma \\ \tau \end{pmatrix}. \tag{12.26}$$

Under a Lorentz transformation we have

$$\phi' = \Lambda_2(L)\phi.$$

Then since the infinitesimal Lorentz transformation can be written as

$$\Lambda_2(L) = 1 + \epsilon \begin{pmatrix} A & B \\ \Gamma & \Delta \end{pmatrix},$$

we find

$$\begin{aligned} \Delta\sigma &= \epsilon(A\sigma + B\tau) & \Delta\tau &= \epsilon(\Gamma\sigma + \Delta\tau) \\ \Delta\bar{\sigma} &= \epsilon(\bar{A}\bar{\sigma} + \bar{B}\bar{\tau}) & \Delta\bar{\tau} &= \epsilon(\bar{\Gamma}\bar{\sigma} + \bar{\Delta}\bar{\tau}). \end{aligned}$$

Let us see what happens to a vector that can be written in the form

$$y_0 = -(|\sigma|^2+|\tau|^2), \qquad y_1 = \sigma\bar{\tau}+\bar{\sigma}\tau \qquad y_2 = i(\sigma\bar{\tau}-\bar{\sigma}\tau) \qquad y_3 = |\sigma|^2-|\tau|^2$$

This becomes in the dotted-undotted notation

$$y^\mu = \sigma^\mu_{\dot{A}A}\psi^A\overline{\psi}^{\dot{A}}$$

where we can identify $\overline{\psi}^{\dot{A}}$ with ϕ in Eq. (12.26) through the relations $\overline{\psi}^{\dot{1}} = \sigma$ and $\overline{\psi}^{\dot{2}} = \tau$. Writing

$$y'^\mu = 1 + \omega^\mu_\nu y^\nu$$

and comparing this with the expression (12.24), we find

$$\begin{aligned} 2a_1 &= i(-B+\bar{B}-\Gamma+\bar{\Gamma}) \quad 2a_2 = (B+\bar{B}-\Gamma\bar{\Gamma}) \quad 2a_3 = i(-A+\bar{A}+\Delta-\bar{\Delta}) \\ 2a_4 &= (B+\bar{B}+\Gamma+\bar{\Gamma}) \quad 2a_5 = i(B+\bar{B}-\Gamma+\bar{\Gamma}) \quad 2a_6 = (A+\bar{A}-\Delta-\bar{\Delta}) \end{aligned}$$

Comparing this expression with the expressions given in (12.25) we find

$$A = \bar{\delta} \qquad B = -\bar{\gamma} \qquad \Gamma = -\bar{\beta} \qquad \Delta = \bar{\alpha}$$

This means that the expression for Λ_2 becomes

$$\Lambda_2 = 1 + \epsilon \begin{pmatrix} \bar{\delta} & -\bar{\gamma} \\ -\bar{\beta} & \bar{\alpha} \end{pmatrix}$$

Thus

$$\Lambda_2(L) = (\Lambda^*)^{-1}$$

This means that $\Lambda_1 \in Sl(2\mathbb{C})$ and $\Lambda_2 \in \overline{Sl(2\mathbb{C})}$. With these results we can then argue that the spinor ψ_A can be used to describe the future light cone, while $\overline{\psi}^{\dot{A}}$ describes the past light cone.

Since in the matrix representation the Dirac spinor can be written as

$$\Psi = \begin{pmatrix} \psi_A \\ \overline{\psi}^{\dot{A}} \end{pmatrix}$$

we see the Dirac spinor contains information about the future and past light cones.

If we return to the Clifford algebra itself rather than a matrix representation, an element of the minimal left ideal of the Dirac Clifford Ψ_L contains the information necessary to describe both the future and past light cones.

It is interesting to note that we are able to show how ψ_A and $\overline{\psi}^{\dot{A}}$ are related to stereographic projections of a point on the light sphere emerging from the origin O. We illustrate this relationship in Fig. 12.4. The light ray from O passes through the light cone at the point P. If we project this point from the North pole, N, to the point Q_N we find the spinor co-ordinates of this point is $\overline{\psi}^{\dot{A}}$. If we project P from the South pole, S, to the point Q_S^d we find the spinor co-ordinates of this point is ψ_A. Under a Lorentz transformation Q_N and Q_S^d transform together. For more details on stereographic projections see Frescura and Hiley [48].

We can discuss all of this in terms of elements of the minimal ideals, which begins to explain why the Hopf transformation in Eq. (12.18) appears. To do this we

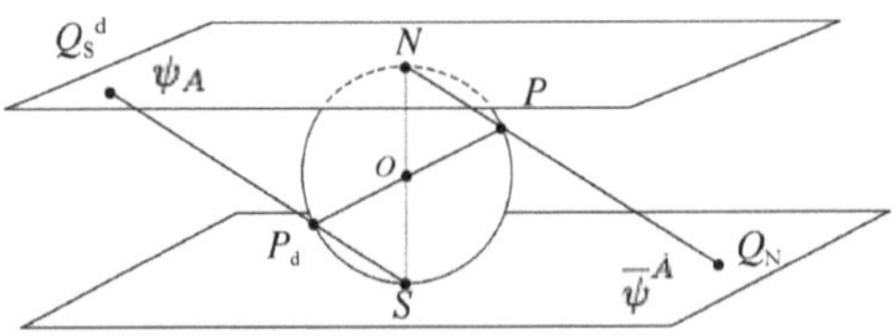

Fig. 12.4 Spinors describing stereographic projections

first note that Ψ_L contains elements from a pair of minimal left ideals of the covering group, one corresponds to $Sl(2\mathbb{C})$ itself, the other corresponds to $\overline{Sl(2\mathbb{C})}$ as we have discussed above. In fact if we introduce the projection operators $(1 \pm ie_5)/2$ we find that ψ_A has components

$$(1 - ie_5)(1 - e_{03})/4 \qquad \text{and} \qquad (1 - ie_5)(e_{13} + e_{01})/4.$$

The other element that we are interested in is $\bar{\psi}^{\dot{A}}$ which has components

$$(1 + ie_5)(e_0 - e_3)/4 \qquad \text{and} \qquad (1 + ie_5)(e_2 + e_{023})/4.$$

These projections give rise to the left- and right-handed helicity or Weyl states [49]. The notation can become more physically meaningful if we write

$$\Psi = \begin{pmatrix} \psi_\lambda \\ \psi_\rho \end{pmatrix},$$

where the suffixes λ and ρ denote the left- and right-handed Weyl spinors respectively.

It should be noted that so far all this discussion has been about *classical* space-time. We have done absolutely nothing to suggest that this mathematics has anything to do with quantum theory. Let us continue in this vein by going to the conformal Clifford $C_{2,4}$ where we will find the twistor of Penrose [20] appearing.

12.5 The Conformal Clifford $C_{2,4}$

12.5.1 The Twistor and Light-Cone Structures

In order to explain how the twistor plays a role in relating different light cones at different points in space-time, we will not start from the structure of elements of the minimum left ideals but from a specific matrix representation. This will help us to see how this bigger structure is related to the Dirac spinors discussed in Sect. 12.4.6.

We begin by considering a six-dimensional hypersphere given by the equation

$$(\beta_A \xi^A)^2 = \Omega^2 \qquad \text{where} \qquad A = (0, 1, 2, 3, 4, 5)$$

Here we have departed from our standard notation by using βs instead of es for the generators of the Clifford algebra. Thus we write $\{\beta_A\}$ for the generators of the conformal Clifford algebra $C_{2,4}$. We choose to write the metric tensor in the form $g_{AB} = (+ - - - - +)$ so that $[\beta_A, \beta_B] = g_{AB}$. The spinors of the six-space, ξ^A, are then defined through the relationship

$$(\beta_A \xi^A)\Psi = 0 \tag{12.27}$$

Now we will choose a specific representation of the generators such that

$$\beta_\mu = 1 \otimes \gamma_\mu \qquad \beta_4 = \sigma_1 \otimes \gamma_5 \qquad \beta_5 = \sigma_1\sigma_3 \otimes \gamma_5$$

$$\beta_i^2 = -1; \qquad \beta_0^2 = 1; \qquad \beta_4^2 = -1; \qquad \beta_5^2 = 1.$$

Here the γ_μ are the generators of the Dirac Clifford with $\mu = \{0, 1, 2, 3\}$. The σs are the generators of the Pauli Clifford algebra. By writing things in this way, it becomes clear how these algebras are related to each other. Equation (12.27) becomes

$$\begin{aligned}(\xi^4 + \xi^5)\psi_{\lambda_2} &= -i(\xi^0 - \sigma_i\xi^i)\psi_{\rho_1}\\ (\xi^4 - \xi^5)\psi_{\lambda_2} &= -i(\xi^0 - \sigma_i\xi^i)\psi_{\rho_2}\\ (\xi^4 + \xi^5)\psi_{\rho_2} &= i(\xi^0 + \sigma_i\xi^i)\psi_{\lambda_1}\\ (\xi^4 - \xi^5)\psi_{\rho_1} &= i(\xi^0 + \sigma_i\xi^i)\psi_{\lambda_2}\end{aligned} \quad \text{where} \quad \Psi = \begin{pmatrix}\psi_{\lambda_1}\\ \psi_{\lambda_2}\\ \psi_{\rho_1}\\ \psi_{\rho_2}\end{pmatrix} \tag{12.28}$$

Here Ψ is a bi-twistor which has eight components. The subscripts λ and σ again signify the left- and right-handed Weyl spinors, only now we have two pairs of Weyl spinors. Thus the bi-twistor contains of two Dirac four-spinors

$$\begin{pmatrix}\psi_{\lambda_1}\\ 0\\ \psi_{\rho_1}\\ 0\end{pmatrix} \quad \text{and} \quad \begin{pmatrix}0\\ \psi_{\lambda_2}\\ 0\\ \psi_{\rho_2}\end{pmatrix} \tag{12.29}$$

This means that we can describe two light cones within a single bi-twistor. We will see that it is this feature of the Penrose bi-twistor that will enable us to relate light cones at different points.

We have shown elsewhere [61] that the bi-twistor also contains a pair of Penrose twistors which can be written in the form

$$\begin{pmatrix}\psi_{\lambda_1}\\ 0\\ 0\\ \psi_{\rho_2}\end{pmatrix} \quad \text{and} \quad \begin{pmatrix}0\\ \psi_{\lambda_2}\\ \psi_{\rho_1}\\ 0\end{pmatrix}$$

As we will see the doubling will give a symmetrical relation between the pair of light cones described in Eq. (12.29).

Thus we see that there is a deep relationship between the Dirac spinors and the Penrose twistors which is sometimes overlooked in the component notation normally used to discuss twistors [20]. Furthermore since the Clifford algebra is a geometric algebra, this relationship must be geometric in nature and we emphasise, once again, that this has nothing per se to do with quantum mechanics. In other words the conformal spinor can be regarded as a combination of various sub-spinors and we need to find out the exact meaning of these spinors.

To do this let us go on to consider the relation between the conformal structure and Minkowski space. We will see that this is a projective relationship, which entails breaking the symmetry by fixing the light cone at infinity. To bring this feature out, we need to know how the co-ordinates x^μ of a point in Minkowski space-time are related to the projective co-ordinates ξ^A in the six-dimensional hyperspace. The projective relationship we need is

$$x^i = \frac{\xi^i}{\xi^4 + \xi^5}; \qquad \text{and} \qquad x^0 = \frac{\xi^0}{\xi^4 + \xi^5}. \tag{12.30}$$

This means that at the origin of our co-ordinate system in Minkowski space, we have $x^0 = x^i = \xi^0 = \xi^i = 0$. We need one more relationship to completely specify the origin in terms of the ξ^A. This relationship is

$$\left(\xi^4\right)^2 - \left(\xi^5\right)^2 = 0 = \left(\xi^4 - \xi^5\right)\left(\xi^4 + \xi^5\right) \tag{12.31}$$

Thus if $\left(\xi^4 + \xi^5\right) \neq 0$ then $\left(\xi^4 - \xi^5\right) = 0$. In terms of the projective co-ordinates, the origin is specified by

$$x^0 = x^i = \xi^0 = \xi^i = \left(\xi^4 - \xi^5\right) = 0$$

To see the implications for the spinors discussed above, lets us return to Eq. (12.28) where the spinors $(\psi_\lambda, \psi_\rho)$ are expressed in terms of ξ^A. If we examine the first and third equation in (12.28) we see immediately that $\psi_{\lambda_2} = \psi_{\rho_2} = 0$. This means that the bi-twistor at the origin is identified by

$$\Psi = \begin{pmatrix} \psi_{\lambda_1} \\ 0 \\ \psi_{\rho_1} \\ 0 \end{pmatrix}$$

Thus this bi-twistor represents light rays on the light cone at the origin (Fig. 12.5).

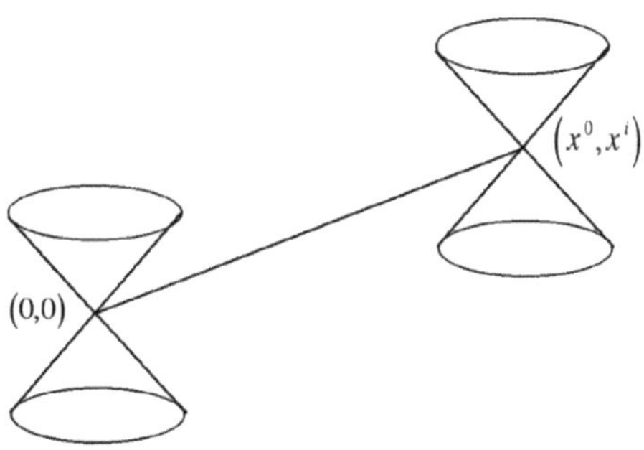

Fig. 12.5 Relative location of light cones

To see the meaning of the second term in Eq. (12.29) we must first recall the expression for the translation operator in the conformal group. This can be shown to be [61]

$$U(\Delta x) = 1 - \Delta x^{\mu} P_{\mu} \quad \text{where} \quad P_{\mu} = \frac{1}{2}\beta_{\mu}(\beta_4 - \beta_5)$$

Here Δx is the displacement produced.

Applying this to the bi-twistor we find

$$U(\Delta x)\begin{pmatrix} \psi_{\lambda_1} \\ \psi_{\lambda_2} \\ \psi_{\rho_1} \\ \psi_{\rho_2} \end{pmatrix} = \begin{pmatrix} \psi_{\lambda_1} \\ -i(\Delta x^0 - \sigma\,\Delta\mathbf{x})\psi_{\rho_1} \\ \psi_{\rho_1} \\ i(\Delta x^0 + \sigma\,\Delta\mathbf{x})\psi_{\lambda_1} \end{pmatrix} \tag{12.32}$$

which means that we have generated a new light ray on a new light cone at (x^0, x^i) given by

$$\psi_{\lambda_2} = -i(x^0 - \sigma_i x^i)\psi_{\rho_1} \quad \text{and} \quad \psi_{\rho_2} = -i(x^0 + \sigma_i x^i)\psi_{\lambda_1}$$

Thus our bi-twistor describes light rays on two light cones, one at the origin of the co-ordinates while the other is at some other point (x^0, x^i) as shown in Fig. 12.6.

Recalling that the origin has $\psi_{\lambda_2} = \psi_{\rho_2} = 0$, we get the pairing

$$\begin{pmatrix} \psi_{\lambda_1} \\ i(\Delta x^0 + \sigma.\Delta\mathbf{x})\psi_{\lambda_1} \end{pmatrix} \quad \text{and} \quad \begin{pmatrix} \psi_{\rho_1} \\ -i(\Delta x^0 - \sigma.\Delta\mathbf{x})\psi_{\rho_1} \end{pmatrix}$$

These are the two twistors that Penrose has introduced, one based on the forward light cone, the other on the past light cone.

All of these results can be obtained from the conformal Clifford using elements of the left ideals. However we will not discuss these here but will be presented elsewhere.

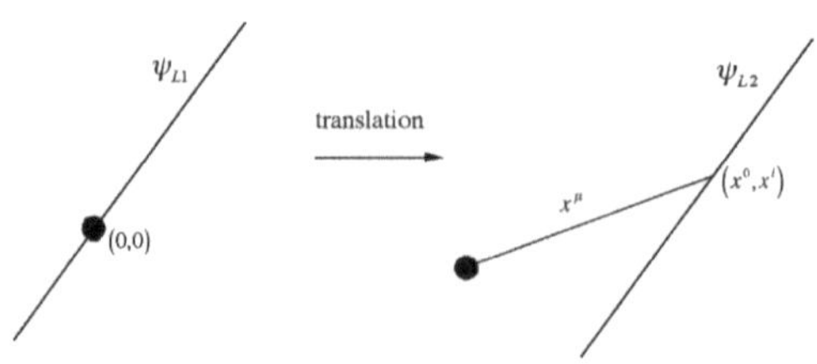

Fig. 12.6 Twistor translates light ray through the origin

12.5.2 Conclusions Drawn from the Orthogonal Clifford Algebras

Let us now summarise the position we have reached so far. We have shown that the orthogonal Clifford algebras carry a rich light-cone structure and that this structure can be abstracted from the algebra itself without the need to start from an a priori given vector space. This means that we do not have to start with a tensor structure on a vector space only to find later that we are forced to introduce the spinor structure, almost as an after thought, a feature that puzzled Eddington. He summed this up by famously remarking, "Something has slipped through the net" [62]. Rather we can start by taking process as basic and from the process itself, abstract a light-cone structure. Key to this approach is the interpretation of the structure of elements of the minimal left and right ideals of the algebra.

Recall that we started by considering an undivided whole in the form of the holomovement. We then followed Kauffman's [13] idea of making a mark or distinction in this total flux of process and then define a series of "extensons", which could be ordered into a groupoid. By adding structure, this groupoid can be developed into an algebra. We concentrated on the orthogonal symmetry that we see around us and showed how to construct a hierarchy of orthogonal Clifford algebras. We then show that from this algebra, we can abstract some of the usual properties of space-time by identify elements of the minimal left ideals with a light-cone structure. In fact we arrive at a structure of light-cones which ensures the underlying manifold we have constructed is Poincaré invariant.

In doing this we have identified the elements of a minimal left ideal with the spinors used in the usual approach. However we have a richer structure since we have sets of primitive idempotents from which we generate additional spinor structures, but in the usual approach these are all treated as equivalent and we therefore loose the possibility of exploiting this additional structure. One feature of this richer structure is that we are able to identify different distinctions with different idempotents, thus enabling us to obtain an insight to the implicate order introduced by Bohm [51] as we will explain later. There is a further advantage of our approach and this removes Eddington's worry because our resulting structure contains *both* spinors and tensors on an equal footing since both entities arise naturally within the algebra itself.

Let us look more carefully at how the distinctions are made and how each distinction "carves out" a specific space-time structure. In the Dirac algebra we make a distinction by picking out a specific time frame which is done by choosing a primitive idempotent $\epsilon = (1 + e_0)/2$. Using this idempotent we can then construct the light-cone structure based on this choice of idempotent.

Another observer can choose another distinction to produce a different time frame, namely, $\epsilon' = (1 + e_0')/2$. This will then produce a different light-cone structure. Of course these structures are equivalent since they are related through

$$\epsilon' = g\epsilon g^{-1}, \qquad \text{where} \qquad g \in G.$$

Here G is the Clifford group, which, in the case of the Dirac algebra, is simply the covering group of the Lorentz group.

The reaction to these ideas from the physicist brought up in the traditional way will probably be "Surely this is just the same as choosing a specific Lorentz frame of reference." Of course *mathematically* the two approaches are equivalent, but they are based on very different intuitive ideas. In the approach we are proposing here, the activity is going on in a structure in which there is no distinction between space and time. In fact, strictly there is no space or time. There is only a *pre-space-time*. Space-times are *constructed* by breaking the symmetry in the total process by choosing a specific idempotent is used to describe the process. In this special case a specific idempotent to define a specific time-frame. In other words the idempotents define equivalence classes of observers, namely Lorentz observers.

Perhaps Fig. 12.7 may help to bring the idea out more clearly. Speaking loosely we could say the activity is "going on" in the "space" E, whereas our description of what is going on is in the projections η_i, in what I have elsewhere [50] called "shadow manifolds".

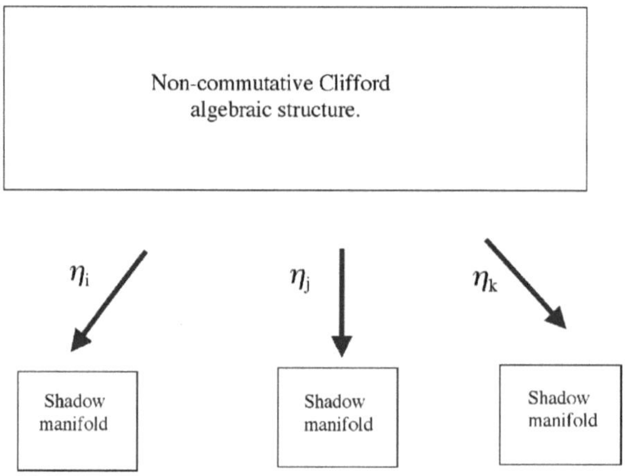

Fig. 12.7 The construction of shadow manifolds

In the language that David Bohm used [51], we can regard E as the holomovement or implicate order. The shadow manifolds are then the sets of explicate orders consistent with the implicate order. Once again we stress that the structure we have been discussing is purely classical and contains no feature of quantum mechanics even though we have a non-commutative algebra. To introduce quantum processes we must add more structure.

12.6 Connections in the Clifford Bundle

12.6.1 The Dirac Connection

In the previous section we have ended up with a mathematical structure which has remarkable similarities to a Clifford bundle. If we take just one of the shadow manifolds to be the base space then in fact we have a Clifford bundle. In this section we will develop our ideas further by exploiting known features of a Clifford bundle [52].

We will begin by considering a single shadow manifold and assume it to be a Riemannian manifold, M. We then construct a Clifford bundle $\mathcal{C}_{n,m}(M)$ with this manifold as the base space. Although not necessary, we will assume for simplicity the manifold to be flat. Let E be a bundle of left modules over $\mathcal{C}_{n,m}(M)$ so that at each point $x \in M$, the fibre is a left module over $\mathcal{C}_{n,m}(M)_x$.

In Sect. 12.5.1 we introduced a notion of space translation which led to the concept of the twistor. This looked after the geometric properties of the light cone structure. Now we want to consider particle dynamics so we introduce a first order differential operator $D : \Gamma(E) \to \Gamma(E)$, where $\Gamma(E)$ is the space of smooth cross-sections of E. This derivative is the algebraic equivalent of the directional derivative on the manifold. The derivative D defines a connection on M. If $\sigma \in \Gamma(E)$ then we define the connection through

$$D\sigma = \sum_{j=1}^{n} e_j \partial_{x_j}$$

where the $\{e_j\}$ are the generators of $\mathcal{C}_{n,m}$ satisfying the relation $\eta(e_j) = x_j$, x_j being the local co-ordinates of the point x on M. This connection is called the Dirac operator because it is easy to show that D^2 is the Dirac Laplacian ∇^2. Notice this definition can be used for *any* Clifford algebra and is not restricted to the relativistic algebra $\mathcal{C}_{1,3}$.

It is through this connection that we are able to handle the hierarchy of Clifford algebras, $\mathcal{C}_{0,1}, \mathcal{C}_{3,0}, \mathcal{C}_{1,3}$. We will call these algebras the Schrödinger, Pauli and Dirac Clifford algebras respectively. The reason for introducing the term "Schrödinger" is because ordinary non-relativistic quantum mechanics forms a trivial complex line bundle $E = \mathbb{R}^3 \otimes \mathbb{C}$. Here the wave function $\psi(x)$ is simply a section of E. This translates into our structure because $\mathcal{C}_{0,1} \cong \mathbb{C}$.

Notice as we remarked above, each algebra in the hierarchy has its own Dirac operator. These are

$D = e\sum_{i=1}^{3} \partial_{x_i}$	Schrödinger
$D = \sum_{i=1}^{3} \sigma_i \partial_{x_i}$	Pauli
$D = \sum_{\mu=1}^{3} \gamma_\mu \partial x^\mu$	Dirac

We should immediately recognise these as the momentum operators used in quantum mechanics. Notice the sum does not include the time derivative. The reason for this will become clear in the next sub-section.

Now it is also possible to view $\mathcal{C}_{n,m}(M)$ as a bundle of right modules which means we can exploit right multiplication, i.e. multiplication from the right. This means we also have the possibility of a "right" Dirac operator which we will denote as $\overleftarrow{D} = \sum_{j=1}^{n} \overleftarrow{\partial}_{x_j} e_j$. The 'left' Dirac operator introduced above will be denoted by $\overrightarrow{D}$.[9] Notice that $\overleftarrow{D}^2$ also produces the Dirac Laplacian.

12.6.2 Relevance to Quantum Mechanics

In the conventional approach it is the wave function that is singled out for special attention because it is assumed to contain all the information for describing the state of the system. We do not have a wave function. Instead we use elements of the minimal left and right ideals which we single out for special attention. However as we have pointed out earlier, the information contained in an element of the left ideal, say Ψ_L, is exactly the same as that contained in the conventional wave function.

We have already pointed out that the key element in our description is not Ψ_L alone but $\hat{\rho} = \Psi_L \Psi_R$. As we have also pointed out above $\hat{\rho}$ is equivalent to the density operator for a pure state ($\hat{\rho}^2 = \hat{\rho}$) in the conventional quantum theory. This means we must consider derivatives of the form $(\overrightarrow{D}\Psi_L)\Psi_R$ and $\Psi_L(\Psi_R \overleftarrow{D})$. Rather than treat these two derivatives separately we will consider expressions like

$$(\overrightarrow{D}\Psi_L)\Psi_R + \Psi_L(\Psi_R \overleftarrow{D}) \qquad \text{and} \qquad (\overrightarrow{D}\Psi_L)\Psi_R - \Psi_L(\Psi_R \overleftarrow{D})$$

which in turn suggests we also write the time derivatives as

$$(\partial_t \Psi_L)\Psi_R + \Psi(\partial_t \Psi_R) \qquad \text{and} \qquad (\partial_t \Psi_L)\Psi_R - \Psi(\partial_t \Psi_R)$$

Now the dynamics must include the Dirac derivatives, the external potentials, V, and the mass of the particle, so we will introduce two "types" of Hamiltonian, $\overrightarrow{H} = \overrightarrow{H}(\overrightarrow{D}, V, m)$ and $\overleftarrow{H} = \overleftarrow{H}(\overleftarrow{D}, V, m)$. Our defining dynamical equations now read

$$i[(\partial_t \Psi_L)\Psi_R + \Psi(\partial_t \Psi_R)] = i\partial_t \hat{\rho} = (\overrightarrow{H}\Psi_L)\Psi_R - \Psi_L(\Psi_R \overleftarrow{H}) \tag{12.33}$$

and

$$i[(\partial_t \Psi_L)\Psi_R - \Psi(\partial_t \Psi_R)] = (\overrightarrow{H}\Psi_L)\Psi_R + \Psi_L(\Psi_R \overleftarrow{H}) \tag{12.34}$$

The first of these Eq. (12.33) can be written in the more suggestive form

$$i\partial_t \hat{\rho} = [H, \hat{\rho}]_- \tag{12.35}$$

This equation has the form of Liouville's equation and can be shown to correspond exactly to the conservation of probability equation.

[9] It should also be noted that these derivatives are related to the sum and differences of the boundary, δ^*, and co-boundary, δ operators in the exterior calculus.

Equation (12.34) can be written in the form

$$i[(\partial_t \Psi_L)\Psi_R - \Psi(\partial_t \Psi_R)] = [H, \hat{\rho}]_+ \tag{12.36}$$

As far as I am aware this equation has not appeared in this form in the literature before it was introduced by Brown and Hiley [53]. They arrived at this equation using a different method and showed it was in fact the quantum Hamilton-Jacobi Equation that appears in Bohm's approach to quantum mechanics.

To bring out exactly what these equations mean in the approach we have adopted here, let us look at their form in the Schrödinger Clifford algebra $\mathcal{C}_{0,1}$. Here $\Psi_L = a(\mathbf{r}, t) + eb(\mathbf{r}, t)$ where e is the generator of the Clifford algebra $\mathcal{C}_{0,1}$ and $a, b \in \Re$. Similarly $\Psi_R = a(\mathbf{r}, t) - eb(\mathbf{r}, t)$. This means that $\Psi_L \Psi_R = a^2 + b^2$ which is just the probability density. Using $H = p^2/2m + V$ it is not difficult to show Eq. (12.35) becomes

$$\partial_t P + \nabla(P\nabla S/m) = 0$$

which is clearly the usual conservation of probability equation.

Equation (12.36) can be considerably simplified if we write $\Psi_L = R\exp[eS]$, then after some work we find the equation becomes the quantum Hamilton-Jacobi equation

$$\partial_t S + (\nabla S)^2/2m + Q + V = 0$$

where $Q = -\nabla^2 R/2mR$ is the usual expression for the quantum potential.

We have thus arrived at the two dynamical equations that form the basis of the Bohm approach to quantum mechanics. By generalizing this to the Pauli and Dirac Clifford algebras on can obtain a complete description of the quantum dynamical equations for the Pauli and Dirac particles. In each case we get a Louville type conservation of probability equation and in each case we have a quantum Hamilton-Jacobi equation. Thus we find the defining equations used in the Bohm approach appear in each of the three Clifford algebras considered in this paper. This means we have an approach to relativistic particle quantum mechanics that is completely analogous to the one used by Bohm for the non-relativistic Schrödinger equation. Thus the common misconception that the Bohm approach cannot be applied in the relativistic domain is just not correct.[10] The full details of our approach to the relativistic domain will be found in Hiley and Callaghan [43].

The method we have outlined in this paper improves considerably on the previous attempts to extend the Bohm approach to spin and to the relativistic domain. For example the Pauli equation has already been treated by Bohm, Schiller, and Tiomno [55] and Hestenes and Gurtler [56], while the Dirac theory has been only

[10] The results discussed here are for a fixed number of particles. To discuss creation and annihilation of particles we must go to a field theory as discussed in Bohm et al. [54].

been partially treated in terms of the Bohm approach by Bohm and Hiley [57] and by Hestenes [8]. Our method presented here gives a systematic approach arising directly from connections in the appropriate Clifford bundles.

12.7 Expectation Values

12.7.1 How to Find the Coefficients in $\hat{\rho}$

In the previous section we have shown how the state of the system is described by the Clifford density element, $\hat{\rho} = \Psi_L \Psi_R$, evolves in time. However we need to extract information about the properties of the quantum system from $\hat{\rho}$. To do this we need to examine the details of $\hat{\rho}$. In general this element takes the form

$$\hat{\rho} = \rho^s + \sum \rho^i e_i + \sum \rho^{ij} e_{ij} + \dots$$

Clearly the coefficients, $\rho^{\dots i \dots}$, characterise different aspects of the state of the system. How then do we abstract these coefficients from the general expression?

Notice first that any Clifford element, B, can also be written in the form

$$B = b^s + \sum b^i e_i + \sum b^{ij} e_{ij} + \dots$$

Secondly, all Clifford algebras contain a notion of trace, where the trace of the identity element is the dimension of the $\mathcal{I}_{L_m}$ and the trace of each e_A is zero except when $A = 0$. Then

$$tr(B) = b^s n$$

Thus the scalar coefficient corresponds to taking the trace. To find the other coefficients we note that if we multiply B by the appropriate basis element $e_{\dots ij \dots}$, we will always be left with a scalar term. This term will then be the trace. But of course this will just give the coefficient that stands in front of the particular $e_{\dots ij \dots}$ chosen. For example if we want to find b^k, we form $tr(Be_k)$ which then picks out b^k because

$$tr(Be_k) = nb^k$$

In this way we can systematically find all the coefficients of any element in the algebra.

To illustrate this in the particular case of the Dirac Clifford, we write

$$B = b^s + \sum b^i e_i + \sum b^{ij} e_{ij} + \sum b^{ijk} e_{ijk} + b^5 e_5$$

Then we have 5 trace-types

$$b^s = tr(B1)/4 \qquad b^i = tr(Be_i)/4 \qquad b^{ij} = -tr(Be_{ij})/4$$
$$b^{ijk} = -tr(Be_{ijk})/4 \qquad b^5 = tr(Be_5)/4$$

Thus we can find all the coefficients of any element of the Dirac Clifford.

12.7.2 How to Find Expectation Values

In standard quantum mechanics, expectation values of operators are found via:

$$\langle B\rangle = \langle\psi|B|\psi\rangle = tr(B\rho)$$

Here ρ is the standard density matrix. These mean values can be taken over directly into the algebraic approach by replacing the standard density matrix by the $\hat{\rho}$ as can easily be seen from the following formal correspondences:-

$$|\psi\rangle \rightarrow \Psi_L = \psi_L\epsilon \qquad\qquad \langle\psi| \rightarrow \Psi_R = \epsilon\psi_R$$

Then we can write

$$\langle B\rangle = tr(\epsilon\psi_R B\psi_L\epsilon)$$

Since the trace is invariant under cyclic permutations, we can arrive at the form

$$\langle B\rangle = tr(\psi_R B\psi_L\epsilon) = tr(B\psi_L\epsilon\psi_R) \tag{12.37}$$

But $\psi_L\epsilon\psi_R = \hat{\rho}$, so we have

$$\langle B\rangle = tr(B\hat{\rho})$$

Thus the mean value of any dynamical element, B, in the Clifford algebra becomes

$$\begin{aligned} tr(B\hat{\rho}) &= tr(b^s\hat{\rho} + \sum b^i e_i\hat{\rho} + \sum b^{ij} e_{ij}\hat{\rho} + \ldots) \\ &= b^s tr(\hat{\rho}) + \sum b^i tr(e_i\hat{\rho}) + \sum b^{ij} tr(e_{ij}\hat{\rho})/2\ldots \end{aligned}$$

This shows that the state of our system is specified by a set of bilinear invariants

$tr(1\hat{\rho}) \rightarrow$ scalar $\quad tr(e_j\hat{\rho}) \rightarrow$ vector $\quad tr(e_{ij}\hat{\rho}) \rightarrow$ bivector $\quad tr(\ldots) \rightarrow \ldots$

These bilinear invariants play an important role in the algebraic approach.

12.7.3 Some Specific Expectation Values

We now give a specific example of how we calculate expectation values in the Pauli Clifford algebra. Suppose we require the expectation value of a component of the spin of a Pauli particle. We identify the spin components with the elements $\{e_i\}$ of the algebra. Then, specifically for $i = j$, we find that the expectation value of the jth component of the spin is given by

$$\langle S_j \rangle = tr(e_j \hat{\rho}). \tag{12.38}$$

In order to evaluate $\hat{\rho} = \psi_L \epsilon \psi_R$, we need to chose a specific expression for ϵ. Since it is conventional to choose the z-direction as the direction of any applied homogeneous magnetic field, we choose for our idempotent $(1 + e_3)/2$. Thus

$$2\langle S_j \rangle = tr(e_j \psi_L (1 + e_3) \psi_R) = tr(e_j \psi_L e_3 \psi_R).$$

Let us define $\rho S = \psi_L e_3 \psi_R$ where Ψ_L is given by

$$\Psi_L = \psi_L \epsilon = (a + e_{123} d)(1 + e_3)/2 + (c + e_{123} b)(e_1 + e_{13})/2 \tag{12.39}$$

Here a, b, c and d are related to the components of the usual matrix representation (See Eq. 12.19)

$$\psi = \begin{pmatrix} \psi_1 \\ \psi_2 \end{pmatrix} \tag{12.40}$$

via

$$2a = \psi_1 + \psi_1^* \qquad 2c = \psi_2 + \psi_2^*$$
$$2e_{123} d = \psi_1 - \psi_1^* \qquad 2e_{123} b = \psi_2 - \psi_2^*. \tag{12.41}$$

Using Eqs. (12.39) and (12.41), it can be shown that S is the usual expression for the spin vector

$$\rho S = (\psi_1 \psi_2^* + \psi_2 \psi_1^*) e_1 + i(\psi_1 \psi_2^* - \psi_2 \psi_1^*) e_2 + (|\psi_1|^2| - |\psi_2|^2) e_3 \tag{12.42}$$

Thus we find

$$\langle S_j \rangle = tr(e_j \rho S) = (e_j \cdot \rho S)/2$$

Here we use the conventional notation that $(A \cdot B)/2$ is the scalar part of the Clifford product. This simply picks out the appropriate coefficient of e_j in Eq. (12.42). Thus for example, the mean value of the x-component of the spin is

$$\langle S_1 \rangle = (\psi_1 \psi_2^* + \psi_2 \psi_1^*)/2.$$

If the particle is in a state of spin "up" along the 3-axis, then we find

$$\langle S_1 \rangle = \langle S_2 \rangle = 0 \qquad \langle S_3 \rangle = 1/2$$

If, on the other hand spin is down along the 3-axis we find

$$\langle S_1 \rangle = \langle S_2 \rangle = 0 \qquad \langle S_3 \rangle = -1/2$$

Thus we see how the eigenvalues of a given physical element like spin are identified in the Clifford algebra approach without the need to bring in operators working in a Hilbert space. This whole process can be generalised to any element of any Clifford algebra. We will not discuss the details of this generalisation here.

12.7.4 Mean Values of Differential Elements

We will assume the elements of our $\mathcal{I}_{L_m}$ to be functions of the co-ordinates of the underlying manifold, therefore as we have seen above differentiation is possible. In order to continue making contact with standard quantum mechanics, we need to introduce elements like $E = B\partial/\partial t$ or $P = B\partial/\partial x_i$, where B is some general element of the Clifford algebra that has physical significance. Since we have to consider both differentiation from the left and the right, our expectation values take the form

$$2\langle P \rangle_{\pm} = tr\left[\epsilon\psi_R(P\psi_L)\epsilon\right] \pm tr\left[\epsilon\psi_L(\tilde{P}\psi_R)\epsilon\right] \tag{12.43}$$

Now let us write $P = B\partial$ where B is again any element of the Clifford algebra. Then

$$2\langle P \rangle_{\pm} = tr\left[B(\partial\psi_L)\epsilon\psi_R\right] \pm tr\left[\tilde{B}\psi_L\epsilon(\partial\psi_R)\right]. \tag{12.44}$$

To check the consistency of our approach to expectation values let us suppress the differential, ∂ in Eq. (12.43) and simply write $P = B$

$$2\langle P \rangle_{\pm} = tr\left[B\psi_L\epsilon\psi_R\right] \pm tr\left[\tilde{B}\psi_L\epsilon\psi_R\right]. \tag{12.45}$$

This expression reduces clearly reduces to Eq. (12.38) if $B = \tilde{B}$ which is what is desired for consistency.

12.7.5 Simple Example

We will now give an example to how this approach works in two simple cases. Let us find bilinear invariant for the energy in the Schrödinger case. We start from the Clifford energy element $e\partial_t$ and we evaluate the energy $\langle \hat{E} \rangle_+ \equiv \rho E$. Thus

$$2\rho E = e[(\partial_t \Psi_L)\Psi_R - \Psi_L(\partial_t \Psi_R)] \tag{12.46}$$

Here we have written $\epsilon = 1$ as this is the only idempotent in the Schrödinger Clifford algebra. In this algebra

$$\Psi_L = a + eb \qquad \text{while} \qquad \Psi_R = a - eb \tag{12.47}$$

where a and b are related to the conventional wave function via

$$2a = \psi + \psi^{\dagger} \qquad \text{and} \qquad 2b = \psi - \psi^{\dagger} \tag{12.48}$$

Substituting Eq. (12.47) into (12.46) gives

$$\rho E = (\partial_t a)b - a(\partial_t b).$$

and then using the relations between the real parameters and the standard wave functions given in Eq. (12.48), we find

$$2\rho E = e[(\partial_t \psi)\psi^* - (\partial_t \psi^*)\psi]$$

Finally if we write $\psi = R\exp[eS]$ we obtain

$$E = -\partial_t S. \tag{12.49}$$

Those familiar with the Bohm model will immediately recognize that this as the Bohm energy. Indeed if we work out the corresponding expression for the Clifford momentum element $\hat{P} = e\nabla$ we find it gives the Bohm momentum. To show this, substitute this expression into Eq. (12.45) and writing $\langle \hat{P} \rangle_+ = \rho \mathbf{P}$, we find $\mathbf{P} = \nabla S$. This will immediately be recognized as the Bohm momentum.

Thus it appears as if the Bohm model literally drops out of the Clifford algebra approach. Notice also that we are using the Clifford algebra over the reals, with the Clifford element e playing the role of i. We have discussed this consequence and all that follows from it in Hiley and Callaghan [42]. This method has been applied to Pauli Clifford in [42] and the Dirac Clifford in [43].

12.8 The Symplectic Group

So far we have concentrated on orthogonal properties and we have shown how the formulation using the notion of *process* leads to an orthogonal groupoid and then onto an orthogonal Clifford algebra. The rotations were handled using the Clifford group which operates on general elements of the Clifford algebra through inner automorphisms. This enabled us to construct the light cone structure of space-time.

We introduced the dynamical structure by exploiting the structure of the connections on a Clifford spin bundle. This actually breaks with the main aim of this work, namely to capture the underlying process completely in algebraic terms. Rather than capture the dynamical structure using connections we now introduce the symplectic groupoid. Here we are exploiting a suggestion of Connes [58] who has already pointed out that Heisenberg's earlier attempts to create quantum mechanics exploits this symplectic groupoid structure.

We will follow in the same spirit by considering how we go from a space-time structure to what is essentially a non-commutative phase space structure. To this end we now introduce a collection of qualitatively new movements that will double the algebraic structure we are exploring. Therefore we distinguish two types of movement. A space-time movement which we will now denote by $X[P_n P_m]$ and a new movement $P[P_n P_m]$ which will correspond to a movement in what we would conventionally be called the momentum space. However recall we start with no *a priori* given underlying manifold. It is our intention to generate any underlying manifolds from the algebra of process itself. Thus in this notation X and P simply label different qualities of movement. We then introduce the product $X[P_n P_k] \bullet P[P_k P_m] = XP[P_n P_m]$. We will assume that this product is not commutative and follow Heisenberg by assuming it satisfies the commutator

$$X[P_n P_k] \bullet P[P_k P_n] - P[P_n P_k] \bullet X[P_k P_n] = i = [X, P] \qquad (\hbar = 1)$$

In this way we arrive at a structure which has similarities with the structure used by Dirac [59] in his discussions of the quantum algebra. One of the technical problems of using the symplectic groupoid in the above form is that it is an infinite dimensional algebra and therefore it is difficult to see what is going on. We have found it much easier to illuminate the structure if we consider instead, the finite Weyl algebra, C_n^2 [60], where n is an integer $0 < n \leq \infty$. Then we have

$$X[P_n P_k] \bullet P[P_k P_n] = \omega P[P_n P_k] \bullet X[P_k P_n]$$

Here $\omega^n = 1$ so that ω is the nth root of unity. We can simplify the notation by writing $X[P_n P_k] = U$ and $P[P_k P_n] = V$ so that our algebra becomes

$$UV = \omega VU; \qquad U^n = 1 \qquad V^n = 1.$$

We will show that as $n \to \infty$, the discrete Weyl algebra becomes continuous and is essentially the Heisenberg algebra but with a small but significant difference, it includes the delta function as an idempotent. The discrete Weyl algebra allows us to explore what we will call a toy phase space.

This discrete Weyl algebra C_n^2 contains sets of idempotents, ϵ_j. We will choose a particular set given by

$$\epsilon_j = \frac{1}{n}\sum_k \omega^{-jk} R(0, k)$$

where

$$R(j,k) = \omega^{-jk/2} U^j V^k, \quad j,k = 0,1,\ldots,n-1$$

These idempotents satisfy the relation $\sum_k \epsilon_k \epsilon_j = \epsilon_j$. In keeping with the spirit we have been developing in this paper, we will regard the idempotents as "points" which we can map onto some underlying vector space.

If we now write $U = T = \exp\left[2\pi \delta x P/n\right]$ then we can generate a set of points using the inner automorphisms

$$\epsilon_{j+1} = T \epsilon_j T^{-1}$$

In this way we generate a set of points which we display in Fig. 12.8 below.

We can introduce an element of the algebra, $X = \delta x \sum_k \epsilon_k$ so that $X\epsilon_j = \delta x j \epsilon_j$. In other words the element X plays the role of a position operator enabling us to label the point with j. It should be noted that $\delta j \epsilon_j = \Psi_L$ is an element of a minimal left ideal.

We can also form a different set of idempotents

$$\epsilon'_j = \frac{1}{n} \sum_k \omega^{-jk} R(k,0)$$

This enables us to generate a set of other points using $V = T' = \exp\left[2\pi \delta p X/n\right]$ We can again introduce an element of the algebra $P = \delta p \sum_j j\epsilon'_j$ so that $P\epsilon'_j = \delta p j \epsilon'_j$. Thus P enables us to label the momentum points by, say, j_p. Again it should be noted that $\delta p j \epsilon'_j = \Phi_L$ is an element of another minimal left ideal (Fig. 12.9).

In other words we have generated a pair of complementary spaces, the position space and the momentum space. One important lesson we learn from this model is that it is not possible to generate both spaces simultaneously. This is because the two sets of idempotents are related by

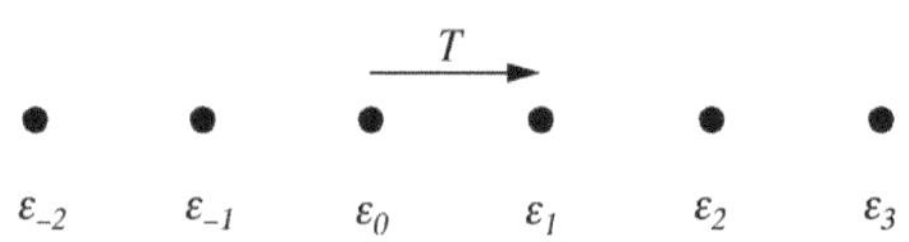

Fig. 12.8 Set of "space" points generated by T

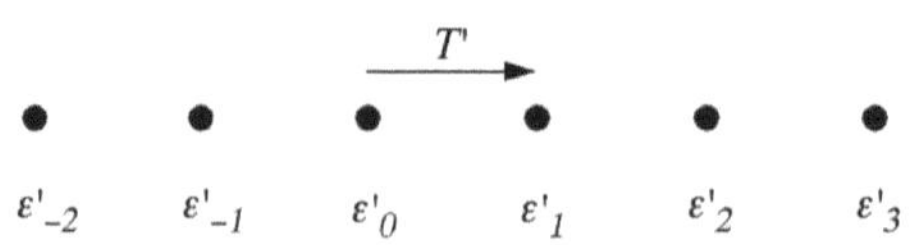

Fig. 12.9 Set of "momentum" points

$$\epsilon'_j = Z^{-1}\epsilon_j Z \qquad \text{with} \qquad Z = \frac{1}{\sqrt{n^3}}\sum_{ijk}\omega^{j(i-k)}R(j-i,k)$$

This transformation is, of course, a finite Fourier transformation. If this approach is correct, then it shows that the origins of the uncertainty principle is not that we disturb the phenomena with our "blunt" experimental instruments, but that the nature of physical processes themselves are such that it is structurally impossible to display position and momentum simultaneously.

In fact each finite Fourier transformation "explodes" every space point into every momentum point and *vice versa*. This means that the p-points are not "hidden", they are not manifest. We can regard them as *enfolded* (see Fig. 12.10). Essentially what we have is that the p-points are "hologramed" into the x-points and *vice versa*. In this way we have, as it were, an ontological principle of complementarity which is to be contrasted with Bohr's epistemic principle of complementarity. A more detailed discussion of the mathematics lying behind this approach will be found in Hiley and Monk [22]).

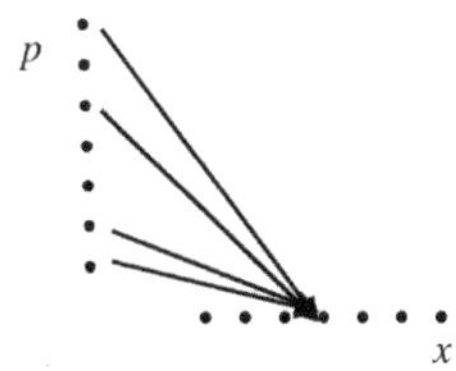

Fig. 12.10 Many p-points are enfolded into each x-point

12.8.1 The Continuum Limit

In this section we want to show how to proceed to the continuum limit essentially following the procedure outlined by Weyl [63]. To this end let us simplify the notation and write

$$\xi = \frac{2\pi\delta x}{n} \qquad \text{and} \qquad \eta = \frac{2\pi\delta p}{n}$$

Then we have

$$U = \exp[i\xi P] \qquad V = \exp[i\eta X] \qquad R(j,k) = \omega^{-jk/2}U^jV^k$$

with $\omega = \exp[i\xi\eta]$. In this case the idempotent becomes

$$\epsilon_0 = \frac{1}{n}\sum_k R(0,k) = \frac{1}{n}\sum_k V^k = \frac{1}{n}\sum_k \exp[ik\eta X]$$

Now $k\eta$ runs through all integer values, but as $\eta \propto 1/n$ and $n \to \infty$ $k\eta$ may assume all the real numbers from $-\infty$ to $+\infty$. In this case

$$\epsilon_0 = \frac{1}{n}\sum_k \exp[ik\eta X] \to \frac{1}{2\pi}\int d\beta \exp[i\beta q] = \delta(q)$$

If we now form

$$\begin{aligned}\epsilon_\xi = \exp[-i\xi P]\epsilon_0 \exp[i\xi P] &= \frac{1}{n}\sum_\beta \exp[i\xi P]\exp[i\beta X]\exp[i\xi P]\\ &= \frac{1}{n}\sum_\beta \exp[i\beta(X-\xi)]\\ \to \frac{1}{2\pi}\int d\beta \exp[i\beta(X-\xi)] = \delta(q-\xi)\end{aligned}$$

In this way we see the idempotents become delta functions and we have generated a continuum of points. We can now choose another set of points from ϵ'_0 using U to translate between points so that we can generate another set of points, the p-continuum. In this way we see that the lesson we learned in the discrete case generalises to the continuum.

12.8.2 The Schrödinger Representation

Let us take this a little further and write $\delta xj\epsilon_j = x_j$ so that $x_j \in \mathcal{I}_L$. Then

$$U^s : x_j \to x_{j-s} \qquad V^t : x_j \to \omega^{jt} x_j$$

Now let us consider the limit $n \to \infty$, then

$$s\xi \to \sigma \qquad t\eta \to \tau \qquad \omega^{kt} = \exp[i\xi kt\eta] = \exp[iq\tau]; \quad (k\xi = q)$$

Here k is an integer, but $\xi \propto 1/n$ which means that for n large, k runs from $-\infty$ to $+\infty$. This is because here k is considered as mod n, while $k\xi$ is mod $n\xi$, but $n\xi$ is a multiple of $2\pi/\eta$ so that $\pi/\eta \to \infty$ as $\eta \to 0$. In this case x_j becomes continuous variable X through the relation $x_j = \sqrt{\xi}X$. Then

$$U^s : X \to \exp[i\sigma P]X \to (X-\sigma) \qquad V^t : X \to \exp[i\tau X]X.$$

These relationships are in the symplectic Clifford algebra. It should immediately be recognised that this is one short step from the Schrödinger representation. In other words we have arrived at the usual continuum limit.

There is more work to be done to smoothly join the symplectic structure to the orthogonal Clifford algebra. Some of the mathematical details have been explored in Crumeyrolle [64] but space limits further discussion.

12.9 General Conclusions

In this article we have shown that by starting from a basic primitive notion of process, activity or movement, we have generated both orthogonal and symplectic Clifford algebras. We have shown that the orthogonal structure enables us to discuss the structure of space-time in terms of a light ray geometry. We found that this structure is entirely classical even though we found ourselves using the Pauli and Dirac Clifford algebras, normally assumed to be quantum in essence.

To introduce quantum aspects into our description, we used two approaches. The first exploited connections on a Clifford bundle. This led to two specific forms of connection, one operating from the left and the other from the right acting on what is the algebraic equivalent of the density "matrix". However our methods are independent of any specific matrix representation. The resulting dynamical equations were equivalent to the two dynamical equations exploited by the Bohm approach. This shows that the Bohm approach is not some peripheral structure in quantum mechanics but an essential part of standard algebraic quantum mechanics. This point is brought out clearly in terms of Clifford algebras.

We found the Schrödinger, Pauli and Dirac theories formed a hierarchy of Clifford algebras, each fitting naturally inside the other corresponding to the physical hierarchy, non-relativistic particle without spin, non-relativistic particle with spin and relativistic particle with spin. This provides a natural description for quantum phenomena, not in terms of representations in Hilbert space, although such a representation is available if desired, but in terms of Clifford algebras an appeal to a specific representations are not required.

The appearance of the defining equations of the Bohm approach provides a clear extension of the approach to the relativistic domain showing the criticism that the approach fails in the relativistic domain unfounded. The Dirac theory produces a "quantum potential" which can be shown to be the exact relativistic generalisation of the quantum potential found originally by de Broglie [65] and Bohm [66]. Space does not allow us to discuss these details in this paper but will be reported elsewhere in Hiley and Callaghan [42, 43].

In the final section we generalised the process structure by introducing a symplectic groupoid. We did this by using a discrete structure which in the limit approached the Heisenberg algebra. This algebra led to a general non-commutative structure in which it was not possible to create points in position and momentum simultaneously. This provides a clear ontological basis for what Bohr called the principle of complementarity.

The algebraic approach provides the details of the mathematical structure that lies behind Bohm's notion of the implicate order. The projections, the shadow manifolds, are examples of the explicate order. The simple original Bohm model of quantum mechanics, sometimes called "Bohmian mechanics", simply produces a phase space example of one of these shadow manifolds. Thus we have provided a more general mathematical framework within which it is possible to further explore the quantum world.

Acknowledgments I should like to thank in particular Bob Callaghan for his patience during the many discussions we had on various aspects of this subject. I would also like to thank Ernst Binz, Ray Brummelhuis, Bob Coecke, Maurice de Gosson and Clive Kilmister for their continual encouragement. Without Ray Brummelhuis' generous support this work would not have reached the light of day.

References

1. Bohr, N.: Atomic Physics and Human Knowledge. Science Editions, New York (1961)
2. Whitehead, A.N.: Process and Reality. Harper & Row, New York (1957)
3. Bohm, D.: Time, the Implicate Order and Pre-Space. In: Griffin D.R. (ed.) Physics and the Ultimate Significance of Time, pp. 172–208. SUNY Press, New York (1986)
4. Bohm, D.J., Davies, P.G., Hiley, B.J.: Algebraic QuantumMechanics and Pregeometry. In: Adenier, G., Krennikov, A.Yu., Nieuwenhuizen, Theo. M. (eds.) Quantum Theory: Reconsiderations of Foundations 3, Växjö, Sweden 2005, pp. 314–324, AIP, New York (2006)
5. Bohm, D., Hiley, B.J.: Generalization of the Twistor to Clifford Algebras as a basis for Geometry, Rev. Briasileira de Fisica, Volume Especial, Os 70 anos de Mario Schönberg (1984) 1–26
6. Grassmann, G.: A New Branch of Mathematics: The Ausdehnungslehre of 1844 and Other Works, translated by Kannenberg, L.C.. Open Court, Chicago (1995)
7. Hamilton, W.R.: Mathematical Papers, vol. 3. Algebra, Cambridge (1967)
8. Hestenes, D.: Spacetime physics with geometric algebra. Am. J. Phys. **71**, 691–704 (2003)
9. Doran, C., Lasenby, A.: Geometric algebra for physicists. Cambridge University Press, Cambridge (2003)
10. Rodrigues, Jr., W.A.: Algebraic and dirachestenes spinors and spinor fields. J. Math. Phys. **45**, 2908–2944 (2004)
11. Grassmann, G.: ibid. Sections 3 and 4
12. Demaret, J., Heller, M., Lambert, D.: Local and Global Properties of the world. Foundations Sci. **2**, 137–176 (1997)
13. Kauffman, L.H.: Space and time in computation and discrete physics. Int. J. General Syst. **27**, 249–273 (1998)
14. Bohm, D.: Fragmentation and Wholeness. The van Leer Jerusalem Foundation, Jerusalem (1976)
15. Bohm,D., Hiley, B.J., Stuart, A.E.G.: On a new mode of description in physics. Int. J. Theor. Phys. **3**, 171–183 (1970)
16. Brandt, H.: Uber eine Verallgemeinerung des Gruppenbegriffes. Math. Ann. **96**, 360–366 (1926)
17. Bohm, D.: Space, time, and the quantum theory understood in terms of discrete structural process. In: Proc. Int. Conf. on Elementary Particles, Kyoto, pp. 252–287 (1965)
18. Eddington, A.S.: The Philosophy of Physical Science. Ann Arbor Paperback, University of Michigan Press, Michigan (1958)
19. Lounesto, P.: Clifford Algebras and Spinors. Cambridge University Press, Cambridge (1997)
20. Penrose, R.: Twistor algebra. J. Math. Phys. **8**, 345–366 (1967)

21. Frescura, F.A.M., Hiley, B.J.: The implicate order, algebras, and the spinor. Found Phys. **10**, 7–31 (1980)
22. Hiley, B.J., Monk, N.: Quantum phase space and the discrete weyl algebra. Mod. Phys. Lett. **A8**, 3225–3233 (1993)
23. Kauffman, L.H.: Special relativity and a calculus of distinctions. In: Proceedings of the 9th International Meeting of ANPA, pp. 291–312, Cambridge, ANPA West, Palo Alto, Cal (1987)
24. Griffor, A.: From strings to clifford algebras. In: Bowden, K.G. (ed.) Implications: Scientific Aspects of AHPA 22, pp. 30–55 (2001)
25. Brown, R.: From groups to groupoids: A brief survey. London Math. Soc. **19**, 113–134 (1987)
26. Abramsky, S., Coecke, B.: A categorical semantics of quantum protocols. In: Proceedings of the 19th IEEE conference on Logic in Computer Science (LiCS'04). IEEE Computer Science Press (2004). Also quant-ph/0402130q-ph/0402130
27. Raptis, I., Zapatrin, R.R.: Algebraic description of spacetime foam. Classical Quantum Gravity **20**, 4187–4205 (2001). gr-qc/9904079
28. Schönberg, M.: Quantum kinematics and geometry. Nuovo Cimento **VI**, 356–380 (1957)
29. Fernandes, M.: Geometric Algebras and the Foundations of Quantum Theory. PhD Thesis, London University (1995)
30. Kauffman, K.L.: Sign and space. In: Religious Experience and Scientific Paradigms, Proceedings of the IAWSR Conf., Inst, Adv. Stud. World Religions, Stony Brook, pp. 118–164. New York (1982)
31. Bratteli, O., Robinson, D.W.: Operator Algebras and Quantum Statistical Mechanics I: C*- and W*-Algebras Symmetry Groups Decomposition of States. Springer, Berlin (1987)
32. Finkelstein, D.R.: Quantum Relativity: A Synthesis of the Ideas of Einstein and Heisenberg. Springer, Berlin (1996)
33. Zapatrin, R.R.: Incidence algebras of simplicial complexes. Pure Math. Appl. **11**, 105–118 (2001). eprint math.CO/0001065
34. Hiley, B.J.: Vacuum or Holomovement. In: Saunders, S., Brown, H.R. (eds.) The Philosophy of Vacuum, pp. 217–249. Clarendon Press, Oxford (1991)
35. Feynman, R.P.: Space-time approach to non-relativistic quantum mechanics. Rev. Mod. Phys. **20**, 367–387 (1948)
36. Aharonov, Y., Bergmann, P.G., Lebowitz, J.L.: Phys. Rev. **134**, B1410 (1964)
37. Kauffman, S.A.: Investigations. Oxford University Press, Oxford (2000)
38. Hiley, B.J.: In: Bowden, K.G. (ed.) Towards a dynamics of moments: The role of algebraic deformation and inequivalent vacuum states, Correlations. Proc. ANPA **23** 104–134 (2001)
39. Dirac, P.A.M.: Spinors in Hilbert Space. Plenum Press, New York (1974)
40. Kauffman, L.H.: Knots and Physics. World Scientific, Singapore (2001)
41. Hiley, B.J.: Some Notes on Quantum Mechanics using Clifford Algebras (Available in pre-print form.)
42. Hiley, B.J., Callaghan, R.: The Clifford Algebra approach to Quantum Mechanics A: The Schrödinger and Pauli Particles (Available in pre-print form.)
43. Hiley, B.J., Callaghan, R.: The Clifford Algebra approach to Quantum Mechanics B: The Dirac Particle (Available in pre-print form.)
44. Penrose, R., Rindler, W.: Spinors and Space-time, vol. 1. Cambridge University Press, Cambridge (1984)
45. Page, L.: A new relativity. Paper I. Fundamental principles and transformations between accelerated systems. Phys. Rev. **49**, 254–268 (1936)
46. Bondi, H.: Relativity and Common Sense. Dover, New York (1977)
47. Nakahara, M.: Geometry, Topology and Physics. Adam Hilger, Bristol (1990)
48. Frescura, F.A.M., Hiley, B.J.: Geometric interpretation of the Pauli spinor. Am. J Phys. **49**, 152–157 (1981)
49. Ryder, L.H.: Quantum Filed Theory. Cambridge University Press, Cambridge (1985)
50. Hiley, B.J.: Non-Commutative Geometry, the Bohm Interpretation and the Mind-Matter Relationship. In: Dubois, D.M. (ed.) Computing Anticipatory Systems: CASYS 2000–4th International Conference, pp. 77–88. AIP, New York (2001)

51. Bohm, D.: Wholeness and the Implicate Order. Routledge, London (1980)
52. Lawson, H.B., Michelsohn, M-L.: Spin Geometry. Princeton University Press, Princeton (1989)
53. Brown, M.R., Hiley, B.J.: Schrödinger revisited: an algebraic approach. quantph/0005026
54. Bohm, D., Hiley, B.J., Kaloyerou, P.N.: An Ontological basis for the quantum theory: II -A causal interpretation of quantum fields. Phys. Reports **144**, 349–375 (1987)
55. Bohm, D., Schiller, R., Tiomno, J.: A causal interpretation of the Pauli equation (A) and (B). Nuovo Cim. **1**, 48–66 and 67–91 (1955)
56. Hestenes, D., Gurtler, R.: Local observables in quantum theory. Am. J. Phys. **39**, 1028–1038 (1971)
57. Bohm, D., Hiley, B.J.: The Undivided Universe: An Ontological Interpretation of Quantum Theory. Routledge, London (1993)
58. Connes, A.: Non-commutative Geometry. Academic Press, San Diego (1990)
59. Dirac, P.A.M.: The fundamental equations of quantum mechanics. Proc. Roy. Soc. A **109**, 642–653 (1926)
60. Morris, A.O.: On a generalised clifford algebra. Quart. J. Math. **18**, 7–12 (1967)
61. Hiley, B.J.: Clifford Algebras. In preparation
62. Kilmister, C.W.: Eddington's Search for a Fundamental Theory. Cambridge University Press, Cambridge (1994)
63. Weyl, H.: The Theory of Groups and Quantum Mechanics. Dover Publication, London (1960)
64. Crumeyrolle, A.: Orthogonal and Symplectic Clifford Algebras: Spinor Structures. Kluwer, Dordrecht (1990)
65. de Broglie, L.: Non-linear Wave Mechanics: A Causal Interpretation. Elsevier, Amsterdam (1960)
66. Bohm, D.: A suggested interpretation of the quantum theory in terms of hidden variables, I. Phys. Rev. **85**, 166–179 (1952); and II, **85**, 180–193 (1952)

Chapter 13
"What is a Thing?": Topos Theory in the Foundations of Physics

A. Döring and C. Isham

From the range of the basic questions of metaphysics we shall here ask this one question: What is a thing? The question is quite old. What remains ever new about it is merely that it must be asked again and again. [40]

Martin Heidegger

Abstract The goal of this article is to summarise the first steps in developing a fundamentally new way of constructing theories of physics. The motivation comes from a desire to address certain deep issues that arise when contemplating quantum theories of space and time. In doing so we provide a new answer to Heidegger's timeless question "What is a thing?".

Our basic contention is that constructing a theory of physics is equivalent to finding a representation in a topos of a certain formal language that is attached to the system. Classical physics uses the topos of sets. Other theories involve a different topos. For the types of theory discussed in this article, a key goal is to represent any physical quantity A with an arrow $\breve{A}_\phi : \Sigma_\phi \to \mathcal{R}_\phi$ where Σ_ϕ and $\mathcal{R}_\phi$ are two special objects (the "state object" and "quantity-value object") in the appropriate topos, τ_ϕ.

We discuss two different types of language that can be attached to a system, S. The first, $\mathcal{PL}(S)$, is a propositional language; the second, $\mathcal{L}(S)$, is a higher-order, typed language. Both languages provide deductive systems with an intuitionistic logic. With the aid of $\mathcal{PL}(S)$ we expand and develop some of the earlier work[1] on topos theory and quantum physics. A key step is a process we term "daseinisation" by which a projection operator is mapped to a sub-object of the spectral presheaf $\underline{\Sigma}$—the topos quantum analogue of a classical state space. The topos concerned is

A. Döring (✉)
OUCL, University of Oxford, Oxford, UK
e-mail: doering@comlab.ox.ac.uk

C. Isham (✉)
The Blackett Laboratory, Imferial College,
e-mail: c.isham@imferial.ac.uk

[1] By CJI and collaborators.

Döring, A., Isham, C.: *"What is a Thing?": Topos Theory in the Foundations of Physics.* Lect. Notes Phys. **813**, 753–937 (2011)
DOI 10.1007/978-3-642-12821-9_13

$\mathbf{Sets}^{\mathcal{V}(\mathcal{H})^{\mathrm{op}}}$: the category of contravariant set-valued functors on the category (partially ordered set) $\mathcal{V}(\mathcal{H})$ of commutative sub-algebras of the algebra of bounded operators on the quantum Hilbert space $\mathcal{H}$.

There are two types of daseinisation, called "outer" and "inner": they involve approximating a projection operator by projectors that are, respectively, larger and smaller in the lattice of projectors on $\mathcal{H}$.

We then introduce the more sophisticated language $\mathcal{L}(S)$ and use it to study "truth objects" and "pseudo-states" in the topos. These objects play the role of states: a necessary development as the spectral presheaf has no global elements, and hence there are no microstates in the sense of classical physics.

One of the main mathematical achievements is finding a topos representation for self-adjoint operators. This involves showing that, for any bounded, self-adjoint operator $\hat{A}$, there is a corresponding arrow $\breve{\delta}^o(\hat{A}) : \underline{\Sigma} \to \underline{\mathbb{R}^{\succeq}}$ where $\underline{\mathbb{R}^{\succeq}}$ is the quantity-value object for this theory. The construction of $\breve{\delta}^o(\hat{A})$ is an extension of the daseinisation of projection operators.

The object $\underline{\mathbb{R}^{\succeq}}$ can serve as the quantity-value object if only outer daseinisation of self-adjoint operators is used in the construction of arrows $\breve{\delta}^o(\hat{A}) : \underline{\Sigma} \to \underline{\mathbb{R}^{\succeq}}$. If both inner and outer daseinisation are used, then a related presheaf $\underline{\mathbb{R}^{\leftrightarrow}}$ is the appropriate choice. Moreover, in order to enhance the applicability of the quantity-value object, one can consider a topos analogue of the Grothendieck extension of a monoid to a group, applied to $\underline{\mathbb{R}^{\succeq}}$ (resp. $\underline{\mathbb{R}^{\leftrightarrow}}$). The resulting object, $k(\underline{\mathbb{R}^{\succeq}})$ (resp. $k(\underline{\mathbb{R}^{\leftrightarrow}})$), is an abelian group-object in τ_ϕ.

Finally we turn to considering a *collection* of systems: in particular, we are interested in the relation between the topos representation of a composite system, and the representations of its constituents. Our approach to these matters is to construct a *category* of systems and to find coherent topos representations of the entire category.

This chapter is dedicated with respect to the memory of Professor Dr. Hans F. de Groote. 1944-2008.

13.1 Introduction

Many people who work in quantum gravity would agree that a deep change in our understanding of foundational issues will occur at some point along the path. However, opinions differ greatly on whether a radical revision is necessary at the very *beginning* of the process, or if it will emerge "along the way" from an existing, or future, research programme that is formulated using the current paradigms. For example, many (albeit not all) of the current generation of string theorists seem inclined to this view, as do a, perhaps smaller, fraction of those who work in loop quantum gravity.

In this article we take the iconoclastic view that a radical step is needed at the very outset. However, for anyone in this camp the problem is always knowing where to start. It is easy to talk about a "radical revision of current paradigms"—the phrase slips lightly off the tongue—but converting this pious hope into a concrete theoretical structure is a problem of the highest order.

For us, the starting point is quantum theory itself. More precisely, we believe that this theory needs to be radically revised, or even completely replaced, before a satisfactory theory of quantum gravity can be obtained.

In this context, a striking feature of the various current programmes for quantising gravity—including superstring theory and loop quantum gravity—is that, notwithstanding their disparate views on the nature of space and time, they almost all use more-or-less standard quantum theory. Although understandable from a pragmatic viewpoint (since all we have *is* more-or-less standard quantum theory) this situation is nevertheless questionable when viewed from a wider perspective.

For us, one of the most important issues is the use in the standard quantum formalism of critical mathematical ingredients that are taken for granted and yet which, we claim, implicitly assume certain properties of space and/or time. Such an *a priori* imposition of spatio-temporal concepts would be a major category[2] error if they turn out to be fundamentally incompatible with what is needed for a theory of quantum gravity.

A prime example is the use of the *continuum*[3] by which, in this context, is meant the real and/or complex numbers. These are a central ingredient in all the various mathematical frameworks in which quantum theory is commonly discussed. For example, this is clearly so with the use of (i) Hilbert spaces or C^*-algebras; (ii) geometric quantisation; (iii) probability functions on a non-distributive quantum logic; (iv) deformation quantisation; and (v) formal (i.e., mathematically ill-defined) path integrals and the like. The *a priori* imposition of such continuum concepts could be radically incompatible with a quantum-gravity formalism in which, say, space-time is fundamentally discrete: as, for example, in the causal-set programme.

As we shall argue later, this issue is closely connected with the question of what is meant by the "value" of a physical quantity. In so far as the concept is meaningful at all at the Planck scale, why should the value be a real number defined mathematically in the usual way?

Another significant reason for aspiring to change the quantum formalism is the peristalithic problem of deciding how a "quantum theory of cosmology" could be interpreted if one was lucky enough to find one. Most people who worry about foundational issues in quantum gravity would probably place the quantum-cosmology/closed-system problem at, or near, the top of their list of reasons for re-envisioning quantum theory. However, although we are deeply interested in such conceptual issues, the primary motivation for our research programme is not to find a new interpretation of quantum theory. Rather, our main goal is to find a novel structural framework within which new *types* of theories of physics can be constructed.

[2] The philosophy of Kant runs strongly in our veins.

[3] When used in this rather colloquial way, the word "continuum" suggests primarily the cardinality of the sets concerned, and, secondly, the topology that is conventionally placed on these sets.

However, having said that, in the context of quantum cosmology it is certainly true that the lack of any external "observer" of the universe "as a whole" renders inappropriate the standard Copenhagen interpretation with its instrumentalist use of counterfactual statements about what *would* happen *if* a certain measurement is performed. Indeed, the Copenhagen interpretation is inapplicable for *any*[4] system that is truly "closed" (or "self-contained") and for which, therefore, there is no "external" domain in which an observer can lurk. This problem has motivated much research over the years and continues to be of wide interest.

The philosophical questions that arise are profound, and look back to the birth of Western philosophy in ancient Greece, almost three thousand years ago. Of course, arguably, the longevity of these issues suggests that these questions are ill-posed in the first place, in which case the whole enterprise is a complete waste of time! This is probably the view of most, if not all, of our colleagues at Imperial College; but we beg to differ.[5]

When considering a closed system, the inadequacy of the conventional instrumentalist interpretation of quantum theory encourages the search for an interpretation that is more "realist" in some way. For over eighty years, this has been a recurring challenge for those concerned with the conceptual foundations of modern physics. In rising to this challenge we join our Greek ancestors in confronting once more the fundamental question[6]:

"What is a thing?"

Of course, as written, the question is itself questionable. For many philosophers, including Kant, would assert that the correct question is not "What is a thing?" but rather "What is a thing as it appears to *us*?". However, notwithstanding Kant's strictures, we seek the thing-in-itself, and, therefore, we persevere with Heidegger's form of the question.

Nevertheless, having said that, we can hardly ignore the last three thousand years of philosophy. In particular, we must defend ourselves against the charge of being "naïve realists".[7] At this point it become clear that theoretical physicists have a big advantage over professional philosophers. For we are permitted/required to study such issues in the context of specific mathematical frameworks for addressing the physical world; and one of the great fascinations of this process is the way in which various philosophical positions are implicit in the ensuing structures. For example,

[4] The existence of the long-range, and all penetrating, gravitational force means that, at a fundamental level, there is only *one* truly closed system, and that is the universe itself.

[5] Of course, it is also possible that our colleagues are right.

[6] "What is a thing?" is the title of one of the more comprehensible of Heidegger's works [40]. By this, we mean comprehensible to the authors of the present article. We cannot speak for our colleagues across the channel: from some of them we may need to distance ourselves.

[7] If we were professional philosophers this would be a terrible insult. :-)

the exact meaning of "realist" is infinitely debatable but, when used by a classical physicist, it invariably means the following:

1. The idea of "a property of the system" (for example, "the value of a physical quantity at a certain time") is meaningful, and mathematically representable in the theory.
2. Propositions about the system (typically asserting that the system has this or that property) are handled using Boolean logic. This requirement is compelling in so far as we humans are inclined to think in a Boolean way.
3. There is a space of "microstates" such that specifying a microstate[8] leads to unequivocal truth values for all propositions about the system: i.e., a state[9] encodes "the way things are". This is a natural way of ensuring that the first two conditions above are satisfied.

The standard interpretation of classical physics satisfies these requirements and provides the paradigmatic example of a realist philosophy in science. Heidegger's answer to his own question adopts a similar position [40]:

> A thing is always something that has such and such properties, always something that is constituted in such and such a way. This something is the bearer of the properties; the something, as it were, that underlies the qualities.

In quantum theory, the situation is very different. There, the existence of any such realist interpretation is foiled by the famous Kochen-Specker theorem [57]. This asserts that it is impossible to assign values to all physical quantities at once if this assignment is to satisfy the consistency condition that the value of a function of a physical quantity is that function of the value. For example, the value of 'energy squared' is the square of the value of energy.

Thus, from a conceptual perspective, the challenge is to find a quantum formalism that is "realist enough" to provide an acceptable alternative to the Copenhagen interpretation, with its instrumentally-construed intrinsic probabilities, whilst taking on board the implications of the Kochen-Specker theorem.

So, *in toto* what we seek is a formalism that is (i) free of *prima facie* prejudices about the nature of the values of physical quantities—in particular, there should be no fundamental use of the real or complex numbers; and (ii) "realist", in at least the minimal sense that propositions are meaningful, and are assigned 'truth values', not just instrumentalist probabilities of what would happen if appropriate measurements are made.

---

[8] In simple non-relativistic systems, the state is specified at any given moment of time. Relativistic systems (particularly quantum gravity!) require a more sophisticated understanding of "state", but the general idea is the same.

[9] We are a little slack in our use of language here and in what follows by frequently referring to a microstate as just a "state". The distinction only becomes important if one wants to introduce things like mixed states (in quantum theory), or macrostates (in classical physics) all of which are often just known as "states". Then one must talk about microstates (pure states) to distinguish them from the other type of state.

However, finding such a formalism is not easy: it is notoriously difficult to modify the mathematical framework of quantum theory without destroying the entire edifice. In particular, the Hilbert space structure is very rigid and cannot easily be changed; and the formal path-integral techniques do not fare much better.

To seek inspiration let us return briefly to the situation in classical physics. There, the concept of realism (as asserted in the three statements above) is encoded mathematically in the idea of a space of states, $\mathcal{S}$, where specifying a particular state (or "microstate"), $s \in \mathcal{S}$, determines entirely 'the way things are' for the system. In particular, this suggests that each physical quantity A should be associated with a real-valued function $\breve{A} : \mathcal{S} \to \mathbb{R}$ such that when the state of the system is s, the value of A is $\breve{A}(s)$. Of course, this is indeed precisely how the formalism of classical physics works.

In the spirit of general abstraction, one might wonder if this formalism can be generalised to a structure in which A is represented by an arrow $\breve{A} : \Sigma \to \mathcal{R}$ where Σ and $\mathcal{R}$ are objects in some category, τ, other than the category of sets, **Sets**? In such a theory, one would seek to represent propositions about the "values" (whatever that might mean) of physical quantities with sub-objects of Σ, just as in classical physics propositions are represented by subsets of the state space $\mathcal{S}$ (see Sect. 13.2.2 for more detail of this).

> Our central conceptual idea is that such a categorial structure constitutes a generalisation of the concept of 'realism' in which the 'values' of a physical quantity are coded in the arrow $\breve{A} : \Sigma \to \mathcal{R}$.

Clearly the propositions will play a key role in any such theory, and, presumably, the minimum required is that the associated sub-objects of Σ form some sort of "logic", just as the subsets of $\mathcal{S}$ form a Boolean algebra.

This rules out most categories since, generically, the sub-objects of an object do not have any logical structure. However, if the category τ is a 'topos' then the sub-objects of any object *do* have this property, and hence the current research programme.

Our suggestion, therefore, is to try to construct physical theories that are formulated in a topos other than **Sets**. This topos will depend on both the theory-type and the system. More precisely, if a theory-type (such as classical physics, or quantum physics) is applicable to a certain class of systems, then, for each system in this class, there is a topos in which the theory is to be formulated. For some theory-types the topos is system-independent: for example, classical physics always uses the topos of sets. For other theory-types, the topos varies from system to system: as we shall see, this is the case in quantum theory.

In somewhat more detail, any particular example of our suggested scheme will have the following ingredients:

1. There are two special objects in the topos τ_ϕ: the "state object"[10], Σ_ϕ, and the "quantity-value object", $\mathcal{R}_\phi$. Any physical quantity, A, is represented by an arrow $A_\phi : \Sigma_\phi \to \mathcal{R}_\phi$ in the topos. Whatever meaning can be ascribed to the concept of the 'value' of a physical quantity is encoded in (or derived from) this representation.
2. Propositions about a system are represented by sub-objects of the state object Σ_ϕ. These sub-objects form a Heyting algebra (as indeed do the sub-objects of any object in a topos): a distributive lattice that differs from a Boolean algebra only in that the *law of excluded middle* need not hold, i.e., $\alpha \vee \neg\alpha \preceq 1$. A Boolean algebra is a Heyting algebra with strict equality: $\alpha \vee \neg\alpha = 1$.
3. Generally speaking (and unlike in set theory), an object in a topos may not be determined by its "points". In particular, this may be so for the state object, in which case the concept of a microstate is not so useful.[11] Nevertheless, truth values can be assigned to propositions with the aid of a "truth object" (or "pseudo-state"). These truth values lie in another Heyting algebra.

Of course, it is not instantly obvious that quantum theory can be written in this way. However, as we shall see, there *is* a topos reformulation of quantum theory, and this has two immediate implications. The first is that we acquire a new type of "realist" interpretation of standard quantum theory. The second is that this new approach suggests ways of generalising quantum theory that make no fundamental reference to Hilbert spaces, path integrals, etc. In particular, there is no *prima facie* reason for introducing standard continuum quantities. As emphasised above, this is one of our main motivations for developing the topos approach. We shall say more about this later.

From a conceptual perspective, a central feature of our scheme is the "neo-realist" structure reflected mathematically in the three statements above. This neo-realism is the conceptual fruit of the fact that, from a categorial perspective, a physical theory expressed in a topos "looks" like *classical* physics expressed in the topos of sets.

The fact that (i) physical quantities are represented by arrows whose domain is the state object, Σ_ϕ; and (ii) propositions are represented by sub-objects of Σ_ϕ, suggests strongly that Σ_ϕ can be regarded as the topos-analogue of a classical state space. Indeed, for any classical system the topos is just the category **Sets** of small sets and functions, and the ideas above reduce to the familiar picture in which (i) there is a state space (set) $\mathcal{S}$; (ii) any physical quantity, A, is represented by a real-valued functions $\breve{A} : \mathcal{S} \to \mathbb{R}$; and (iii) propositions are represented by subsets of $\mathcal{S}$ with a logical structure given by the associated Boolean algebra.

Evidently the suggested mathematical structures could be used in two different ways. The first is that of the "conventional" theoretical physicist with little interest

[10] The meaning of the subscript "ϕ" is explained in the main text. It refers to a particular topos-representation of a formal language attached to the system.

[11] In quantum theory, the state object has no points/microstates at all. As we shall see, this statement is equivalent to the Kochen-Specker theorem.

in conceptual matters. For him/her, what we and our colleagues are developing is a new toolkit with which to construct novel types of theoretical models. Whether or not Nature has chosen such models remains to be seen, but, at the very least, the use of topoi certainly suggests new techniques.

For those physicists who are interested in conceptual issues, the topos framework gives a radically new way of thinking about the world. The neo-realism inherent in the formalism is described mathematically using the *internal* language that is associated with any topos. This describes how things look from "within" the topos: something that should be particularly useful in the context of quantum cosmology.[12]

On the other hand, the pragmatic theoretician with no interest in conceptual matters can use the "external" description of the topos in which the category of sets provides a metalanguage with which to formulate the theory. From a mathematical perspective, the interplay between the internal and external languages of a topos is one of the fascinations of the subject. However, much remains to be said about the significance of this interaction for real theories of physics.

This present article details the first steps in formulating one particular way of employing topos theory in physics. The scheme is very general but certainly not exclusive. Indeed, there are other potential uses of topos theory in physics and what is contained here is very much our own perspective on the subject. The paper is partly an amalgam of a series of four papers that we placed on the ArXiv server[13] in March 2007 [27–30]. However, we have added a significant amount of new material, and also made a few minor corrections (mainly typos).[14] We have also added some remarks about developments made by researchers other than ourselves since the ArXiv preprints were written. Of particular interest for our general programme is the work of Heunen et al. [42] which adds some further ingredients to the topoi-in-physics toolkit. Finally, we have included some background material from the earlier papers that formed the starting point for the current research programme [48–51].

We must emphasise again that this is *not* a review article about the general application of topos theory to physics; this would have made the article far too long. For example, Lawvere's seminal investigations of topoi were strongly motivated by his feeling that current paradigms in theoretical physics are inadequate in a deep way [61–63]. Then there is the fair amount of study of the use of synthetic differential geometry (SDG) in physics. The reader can find references to much of this on the, so-called, "Siberian toposes" web site.[15] There is also the work by Mallios and collaborators on "Abstract Differential Geometry" [69, 70, 76, 71]. The use of a form of intuitionistic logic for quantum theory has also been suggested by Coecke in [15]. He uses a construction discovered by Bruns and Lakser, the so-called injective

---

[12] In this context see the work of Markopoulou who considers a topos description of the universe as seen by different observers who live inside it [68].

[13] These are due to be published in *Journal of Mathematical Physics* in the Spring of 2008.

[14] Some of the more technical theorems have been placed in the Appendix with the hope that this makes the article a little easier to read.

[15] This is http://users.univer.omsk.su/~topoi/. See also Cecilia Flori's website that deals more generally with topos theory and physics: http://topos-physics.org/

hull of meet-semilattices (see also [82]) to embed a meet-semilattice of propositions into a Heyting algebra by introducing new joins to the meet-semilattice. There are no further obvious connections between this approach and the topos approach, but it would be interesting to compare both constructions with respect to the underlying geometric structures: both approaches formulate an intuitionistic form of quantum logic using Heyting algebras, and every complete Heyting algebra is a locale and hence a generalised topological space. Finally there is the work by Corbett et al. [3, 18] on "quantum" real numbers, or "qr-numbers". Of course, as always these days, Google will speedily reveal all that we have omitted.

But even less is this article a review of the use of category theory in general in physics. Indeed, there are many important topics that we do not mention at all. For example, Baez's advocation of n-categories [5, 6]; the important work on the use of symmetric monoidal categories with extra structure in quantum information theory and beyond [2, 86]; Takeuti's theory[16] of "quantum sets" [83]; and Crane's work on categorial models of space-time [19].

Finally, a word about the style in which this article is written. We spent much time pondering on this, as we did before writing the four ArXiv preprints. The intended audience is our colleagues who work in theoretical physics, especially those whose interests included foundational issues in quantum gravity and quantum theory. However, topos theory is not an easy branch of mathematics, and this poses the dilemma of how much background mathematics should be assumed of the reader, and how much should be explained as we go along.[17] We have approached this problem by including a short mathematical appendix on topos theory. However, reasons of space precluded a thorough treatment, and we hope that, fairly soon, someone will write an introductory review of topos theory in a style that is accessible to a typical theoretical-physicist reader.

This article is structured in the following way. We begin with a discussion of some of the conceptual background, in particular the role of the real numbers in conventional theoretical physics. Then in Sect. 13.3 we introduce the idea of attaching a propositional language, $\mathcal{PL}(S)$, to each physical system S. The intent is that each theory of S corresponds to a particular representation of $\mathcal{PL}(S)$. In particular, we show how classical physics satisfies this requirement in a very natural way.

Propositional languages have limited scope (they lack the quantifiers "$\forall$" and "$\exists$"), and in Sect. 13.4 we propose the use of a higher-order (typed) language $\mathcal{L}(S)$. Languages of this type are a central feature of topos theory and it is natural to consider the idea of representing $\mathcal{L}(S)$ in different topoi. Classical physics always takes place in the topos, **Sets**, of sets but our expectation is that other areas of physics will use different topoi.

This expectation is confirmed in Sect. 13.5 where we discuss in detail the representation of $\mathcal{PL}(S)$ for a quantum system (the representation of $\mathcal{L}(S)$ is discussed

---

[16] Takeuti's work is not exactly about category theory applied to quantum theory: it is more about the use of formal logic, but the spirit is similar. For a recent paper in this genre see [74].

[17] The references that we have found most helpful in our research are [65, 34, 59, 11, 66, 55].

in Sect. 13.8). The central idea is to represent propositions as sub-objects of the "spectral presheaf" $\underline{\Sigma}$ which belongs to the topos, $\mathbf{Sets}^{\mathcal{V}(\mathcal{H})^{\mathrm{op}}}$, of presheaves (set-valued, contravariant functors) on the category, $\mathcal{V}(\mathcal{H})$, of abelian sub-algebras of the algebra $B(\mathcal{H})$ of all bounded operators on $\mathcal{H}$. This representation employs the idea of "daseinisation" in which any given projection operator $\hat{P}$ is represented at each context/stage-of-truth V in $\mathcal{V}(\mathcal{H})$ by the "closest" projector to it in V. There are two variants of this: (i) "outer" daseinisation, in which $\hat{P}$ is approached from above (in the lattice of projectors in V); and (ii) "inner" daseinisation, in which $\hat{P}$ is approached from below.

The next key move is to discuss the "truth values" of propositions in a quantum theory. This requires the introduction of some analogue of the microstates of classical physics. We say "analogue" because the spectral presheaf $\underline{\Sigma}$—which is the quantum topos equivalent of a classical state space—has no global elements, and hence there are no microstates at all: this is equivalent to the Kochen-Specker theorem. The critical idea is that of a "truth object", or "pseudo-state" which, as we show in Sect. 13.6, is the closest one can get in quantum theory to a microstate.

In Sect. 13.7 we introduce the "de Groote" presheaves and the associated ideas that lead to the concept of daseinising an arbitrary bounded self-adjoint operator, not just a projector. Then, in Sect. 13.8, the spectral theorem is used to construct several possible models for the quantity-value presheaf in quantum physics. The simplest choice is $\underline{\mathbb{R}^{\succeq}}$, but this uses only outer daseinisation, and a more balanced choice is $\underline{\mathbb{R}^{\leftrightarrow}}$ which uses both inner and outer daseinisation. Another possibility is $k(\underline{\mathbb{R}^{\succeq}})$: the Grothendieck topos extension of the monoid object $\underline{\mathbb{R}^{\succeq}}$. A key result is the "non-commutative spectral theorem" which involves showing how each bounded, self-adjoint operator $\hat{A}$ can be represented by an arrow $\breve{A} : \underline{\Sigma} \to \underline{\mathbb{R}^{\leftrightarrow}}$.

In Sect. 13.10 we discuss the way in which unitary operators act on the quantum topos objects. Then, in Sects. 13.11, 13.12 and 13.13 we discuss the problem of handling "all" possible systems in a single coherent scheme. This involves introducing a category of systems which, it transpires, has a natural monoidal structure. We show in detail how this scheme works in the case of classical and quantum theory.

Finally, in Sect. 13.14 we discuss/speculate on some properties of the state object, quantity-value object, and truth objects that might be present in any topos representation of a physical system.

To facilitate reading this long article, some of the more technical material has been put in Appendix 1. In Appendix 2 there is a short introduction to some of the relevant parts of topos theory.

13.2 The Conceptual Background of Our Scheme

13.2.1 The Problem of Using Real Numbers A Priori

As mentioned in the Introduction, one of the main goals of our work is to find new tools with which to develop theories that are significant extensions of, or developments from, quantum theory but without being tied a priori to the use of the standard real or complex numbers.

In this context we note that real numbers arise in theories of physics in three different (but related) ways: (i) as the values of physical quantities; (ii) as the values of probabilities; and (iii) as a fundamental ingredient in models of space and time (especially in those based on differential geometry). All three are of direct concern vis-a-vis our worries about making unjustified, a priori assumptions in quantum theory. We shall now examine them in detail.

13.2.1.1 Why Are Physical Quantities Assumed to be Real-Valued?

One reason for assuming physical quantities are real-valued is undoubtedly grounded in the remark that, traditionally (i.e., in the pre-digital age), they are measured with rulers and pointers, or they are defined operationally in terms of such measurements. However, rulers and pointers are taken to be classical objects that exist in the physical space of classical physics, and this space is modelled using the reals. In this sense there is a direct link between the space in which physical quantities take their values (what we call the "quantity-value space") and the nature of physical space or space-time [45].

If conceded, this claim means the assumption that physical quantities are real-valued is problematic in any theory in which space, or space-time, is not modelled by a smooth manifold. Admittedly, if the theory employs a *background* space, or space-time—and if this background is a manifold—then the use of real-valued physical quantities *is* justified in so far as their value-space can be related to this background. Such a stance is particularly appropriate in situations where the background plays a central role in giving meaning to concepts like "observers" and "measuring devices", and thereby provides a basis for an instrumentalist interpretation of the theory.

But even here caution is needed since many theoretical physicists have claimed that the notion of a "space-time point in a manifold" is intrinsically flawed. One argument (due to Penrose) is based on the observation that any attempt to localise a "thing" is bound to fail beyond a certain point because of the quantum production of pairs of particles from the energy/momentum uncertainty caused by the spatial localisation. Another argument concerns the artificiality[18] of the use of real numbers as coordinates with which to identify a space-time point. There is also Einstein's famous "hole argument" in general relativity which asserts that the notion of a space-time point (in a manifold) has no physical meaning in a theory that is invariant under the group of space-time diffeomorphisms.

Another cautionary caveat concerning the invocation of a background is that this background structure may arise only in some "sector" of the theory; or it may exist only in some limiting, or approximate, sense. The associated instrumentalist

---

[18] The integers, and associated rationals, have a "natural" interpretation from a physical perspective since we can all count. On the other hand, the Cauchy-sequence and/or the Dedekind-cut definitions of the reals are distinctly un-intuitive from a physical perspective.

interpretation would then be similarly limited in scope. For this reason, if no other, a "realist" interpretation is more attractive than an instrumentalist one.

In fact, in such circumstances, the phrase "realist interpretation" does not really do justice to the situation since it tends to imply that there are other interpretations of the theory, particularly instrumentalism, with which the realist one can contend on a more-or-less equal footing. But, as we just argued, the instrumentalist interpretation may be severely limited as compared to the realist one. To flag this point, we will sometimes refer to a "realist formalism", rather than a "realist interpretation".[19]

13.2.1.2 Why are Probabilities Required to Lie in the Interval [0,1]?

The motivation for using the subset [0, 1] of the real numbers as the value space for probabilities comes from the relative-frequency interpretation of probability. Thus, in principle, an experiment is to be repeated a large number, N, times, and the probability associated with a particular result is defined to be the ratio N_i/N, where N_i is the number of experiments in which that result was obtained. The rational numbers N_i/N necessarily lie between 0 and 1, and if the limit $N \to \infty$ is taken—as is appropriate for a hypothetical "infinite ensemble"—real numbers in the closed interval [0, 1] are obtained.

The relative-frequency interpretation of probability is natural in instrumentalist theories of physics, but it is neither meaningful if there is no classical spatio-temporal background in which the necessary measurements could be made, nor if there is a background of a kind to which the relative-frequency interpretation cannot be adapted.

In the absence of a relative-frequency interpretation, the concept of "probability" must be understood in a different way. In the physical sciences, one of the most discussed approaches involves the concept of "potentiality", or "latency", as favoured by Heisenberg [41], Margenau [67], and Popper [75] (and, for good measure, Aristotle). In this case there is no compelling reason why the probability-value space should necessarily be a subset of the real numbers. The minimal requirement is that this value-space is an ordered set, so that one proposition can be said to be more or less probable than another. However, there is no *prima facie* reason why this set should be *totally* ordered: i.e., there may be pairs of propositions whose potentialities cannot be compared—something that seems eminently plausible in the context of non-commensurable quantities in quantum theory.

By invoking the idea of "potentiality", it becomes feasible to imagine a quantum-gravity theory with no spatio-temporal background but where probability is still a fundamental concept. However, it could also be that the concept of probability plays no fundamental role in such circumstances, and can be given a meaning *only* in the context of a sector, or limit, of the theory where a background does exist. This

[19] Of course, such discussions are unnecessary in classical physics since, there, if knowledge of the value of a physical quantity is gained by making a (ideal) measurement, the reason why we obtain the result that we do, is because the quantity *possessed* that value immediately before the measurement was made. In other words, "epistemology models ontology".

background could then support a limited instrumentalist interpretation which would include a (limited) relative-frequency understanding of probability.

In fact, most modern approaches to quantum gravity aspire to a formalism that is background independent [4, 17, 77, 78]. So, if a background space does arise, it will be in one of the restricted senses mentioned above. Indeed, it is often asserted that a proper theory of quantum gravity will not involve *any* direct spatio-temporal concepts, and that what we commonly call "space" and "time" will "emerge" from the formalism only in some appropriate limit [52]. In this case, any instrumentalist interpretation could only "emerge" in the same limit, as would the associated relative-frequency interpretation of probability.

In a theory of this type, there will be no *prima facie* link between the values of physical quantities and the nature of space or space-time, although, of course, this cannot be totally ruled out. In any event, part of the fundamental specification of the theory will involve deciding what the "quantity-value space" should be.

These considerations suggest that quantum theory must be radically changed if one wishes to accommodate situations where there is no background space/space-time manifold within which an instrumentalist interpretation can be formulated. In such a situation, some sort of "realist" formalism is essential.

These reflections also suggest that the quantity-value space employed in an instrumentalist realisation of a theory—or a "sector", or "limit", of the theory—need not be the same as the quantity-value space in a neo-realist formulation.

At first sight this may seem strange but, as is shown in Sect. 13.8, this is precisely what happens in the topos reformulation of standard quantum theory.

13.2.2 The Genesis of Topos Ideas in Physics

13.2.2.1 Why are Space and Time Modelled with Real Numbers?

Even setting aside the more exotic considerations of quantum gravity, one can still query the use of real numbers to model space and/or time. One might argue that (i) the use of (triples of) real numbers to model space is based on empirically-based reflections about the nature of "distances" between objects; and (ii) the use of real numbers to model time reflects our experience that "instants of time" appear to be totally ordered, and that intervals of time are always divisible.[20]

However, what does it really mean to say that two particles are separated by a distance of, for example, $\sqrt{2}$cms? From an empirical perspective, it would be impossible to make a measurement that could unequivocally reveal precisely that value from among the continuum of real numbers that lie around it. There will always be experimental errors of some sort: if nothing else, there are thermodynamical fluctuations in the measuring device; and, ultimately, uncertainties arising from quantum "fluctuations". Similar remarks apply to attempts to measure time.

---

[20] These remarks are expressed in the context of the Newtonian view of space and time, but it is easy enough to generalise them to special relativity.

Thus, from an operational perspective, the use of real numbers to label "points" in space and/or time is a theoretical abstraction that can never be realised in practice. But if the notion of a space/time/space-time "point" in a continuum is an abstraction, why do we use it? Of course it works well in theories used in normal physics, but at a fundamental level it must be seen as questionable.

These operational remarks say nothing about the structure of space (or time) "in itself", but, even assuming that this concept makes sense, which is debatable, the use of real numbers is still a metaphysical assumption with no fundamental justification.

Traditionally, we teach our students that measurements of physical quantities that are represented theoretically by real numbers give results that fall into "bins", construed as being subsets of the real line. This suggests that, from an operational perspective, it would be more appropriate to base mathematical models of space or time on a theory of "regions", rather than the real numbers themselves.

But then one asks "What is a region?", and if we answer "A subset of triples of real numbers for space, and a subset of real numbers for time", we are thrown back to the real numbers. One way of avoiding this circularity is to focus on relations between these "subsets" and see if they can be axiomatised in some way. The natural operations to perform on regions are (i) take intersections, or unions, of pairs of regions; and (ii) take the complement of a region. If the regions are modelled on Borel subsets of $\mathbb{R}$, then the intersections and unions could be extended to countable collections. If they are modelled on open sets, it would be arbitrary unions and finite intersections.

From a physical perspective, the use of open subsets as models of regions is attractive as it leaves a certain, arguably desirable, "fuzziness" at the edges, which is absent for closed sets. Thus, following this path, we would axiomatise that a mathematical model of space or time (or space-time) involves an algebra of entities called "regions", and with operations that are the analogue of unions and intersections for subsets of a set. This algebra would allow arbitrary "unions" and finite "intersections", and would distribute[21] over these operations. In effect, we are axiomatising that an appropriate mathematical model of space-time is an object in the category of locales.

However, a locale is the same thing as a complete Heyting algebra (for the definition see below), and, as we shall, Heyting algebras are inexorably linked with topos theory.

13.2.2.2 Another Possible Role for Heyting Algebras

The use of a Heyting algebra to model space/time/space-time is an attractive possibility, and was the origin of the interest in topos theory of one of us (CJI) some years ago. However, there is another motivation which is based more on logic, and the desire to construct a "neo-realist" interpretation of quantum theory.

---

[21] If the distributive law is dropped we could move towards the quantum-set ideas of [83]; or, perhaps, the ideas of non-commutative geometry instigated by Alain Connes [16].

To motivate topos theory as the source of neo-realism let us first consider classical physics, where everything is defined in the category, **Sets**, of sets and functions between sets. Then (i) any physical quantity, A, is represented by a real-valued function $\breve{A} : \mathcal{S} \to \mathbb{R}$, where $\mathcal{S}$ is the space of microstates; and (ii) a proposition of the form "$A \,\varepsilon\, \Delta$" (which asserts that the value of the physical quantity A lies in the subset Δ of the real line $\mathbb{R}$)[22] is represented by the subset[23] $\breve{A}^{-1}(\Delta) \subseteq \mathcal{S}$. In fact any proposition P about the system is represented by an associated subset, $\mathcal{S}_P$, of $\mathcal{S}$: namely, the set of states for which P is true. Conversely, every (Borel) subset of $\mathcal{S}$ represents a proposition.[24]

It is easy to see how the logical calculus of propositions arises in this picture. For let P and Q be propositions, represented by the subsets $\mathcal{S}_P$ and $\mathcal{S}_Q$ respectively, and consider the proposition "P and Q". This is true if, and only if, both P and Q are true, and hence the subset of states that represents this logical conjunction consists of those states that lie in both $\mathcal{S}_P$ and $\mathcal{S}_Q$—i.e., the set-theoretic intersection $\mathcal{S}_P \cap \mathcal{S}_Q$. Thus "$P$ and Q" is represented by $\mathcal{S}_P \cap \mathcal{S}_Q$. Similarly, the proposition "P or Q" is true if either P or Q (or both) are true, and hence this logical disjunction is represented by those states that lie in $\mathcal{S}_P$ plus those states that lie in $\mathcal{S}_Q$—i.e., the set-theoretic union $\mathcal{S}_P \cup \mathcal{S}_Q$. Finally, the logical negation "not P" is represented by all those points in $\mathcal{S}$ that do not lie in $\mathcal{S}_P$—i.e., the set-theoretic complement $\mathcal{S} \backslash \mathcal{S}_P$.

In this way, a fundamental relation is established between the logical calculus of propositions about a physical system, and the Boolean algebra of subsets of the state space. Thus the mathematical structure of classical physics is such that, *of necessity*, it reflects a "realist" philosophy, in the sense in which we are using the word.

One way to escape from the tyranny of Boolean algebras and classical realism is via topos theory. Broadly speaking, a topos is a category that behaves very much like the category of sets; in particular, the collection of sub-objects of an object forms a *Heyting algebra*, just as the collection of subsets of a set form a Boolean algebra. Our intention, therefore, is to explore the possibility of associating physical propositions with sub-objects of some object Σ (the analogue of a classical state space) in some topos.

A Heyting algebra, $\mathfrak{H}$, is a distributive lattice with a zero element, 0, and a unit element, 1, and with the property that to each pair $\alpha, \beta \in \mathfrak{H}$ there is an implication $\alpha \Rightarrow \beta$, characterised by

$$\gamma \preceq (\alpha \Rightarrow \beta) \text{ if and only if } \gamma \wedge \alpha \preceq \beta. \tag{13.1}$$

[22] In the rigorous theory of classical physics, the set $\mathcal{S}$ is a symplectic manifold, and Δ is a *Borel* subset of $\mathbb{R}$. Also, the function $\breve{A} : \mathcal{S} \to \mathbb{R}$ may be required to be measurable, or continuous, or smooth, depending on the quantity, A, under consideration. We will henceforth assume that $\Delta \subseteq \mathbb{R}$ is a Borel subset and all functions $\breve{A}$ are measurable.

[23] Throughout this article we will adopt the notation in which $A \subseteq B$ means that A is a subset of B that could equal B; while $A \subset B$ means that A is a *proper* subset of B; i.e., A does not equal B. Similar remarks apply to other pairs of ordering symbols like $\prec, \preceq$; or $\succ, \succeq$, etc.

[24] More precisely, every Borel subset of $\mathcal{S}$ represents *many* propositions about the values of physical quantities. Two propositions are said to be "physically equivalent" if they are represented by the same subset of $\mathcal{S}$.

The negation is defined as $\neg\alpha := (\alpha \Rightarrow 0)$ and has the property that the *law of excluded middle* need not hold, i.e., there may exist $\alpha \in \mathfrak{H}$, such that $\alpha \vee \neg\alpha \prec 1$ or, equivalently, there may exist $\alpha \in \mathfrak{H}$ such that $\neg\neg\alpha \succ \alpha$. This is the characteristic property of an intuitionistic logic.[25] A Boolean algebra is the special case of a Heyting algebra in which there is the strict equality: i.e., $\alpha \vee \neg\alpha = 1$ for all α. It is known from Stone's theorem [81] that each Boolean algebra is isomorphic to an algebra of (clopen, i.e., closed and open) subsets of a suitable (topological) space.

The elements of a Heyting algebra can be manipulated in a very similar way to those in a Boolean algebra. One of our claims is that, as far as theories of physics are concerned, Heyting logic is a viable[26] alternative to Boolean logic.

To give some idea of the difference between a Boolean algebra and a Heyting algebra, we note that the paradigmatic example of the former is the collection of all measurable subsets of a measure space X. Here, if $\alpha \subseteq X$ represents a proposition, the logical negation, $\neg\alpha$, is just the set-theoretic complement $X\backslash\alpha$.

On the other hand, the paradigmatic example of a Heyting algebra is the collection of all open sets in a topological space X. Here, if $\alpha \subseteq X$ is open, the logical negation $\neg\alpha$ is defined to be the *interior* of the set-theoretical complement $X\backslash\alpha$. Therefore, the difference between $\neg\alpha$ in the topological space X, and $\neg\alpha$ in the measurable space generated by the topology of X, is just the 'thin' boundary of the closed set $X\backslash\alpha$.

13.2.2.3 Our Main Contention About Topos Theory and Physics

We contend that, for a given theory-type (for example, classical physics, or quantum physics), each system S to which the theory is applicable is associated with a particular topos $\tau_\phi(S)$ within whose framework the theory, as applied to S, is to be formulated and interpreted. In this context, the "ϕ"-subscript is a label that changes as the theory-type changes. It signifies the representation of a system-language in the topos $\tau_\phi(S)$: we will come to this later.

The conceptual interpretation of this formalism is "neo-realist" in the following sense:

---

[25] Here, $\alpha \Rightarrow \beta$ is nothing but the category-theoretical exponential β^α and $\gamma \wedge \alpha$ is the product $\gamma \times \alpha$. The definition uses the adjunction between the exponential and the product, $\mathrm{Hom}(\gamma, \beta^\alpha) = \mathrm{Hom}(\gamma \times \alpha, \beta)$. A slightly easier, albeit "less categorical" definition is: a Heyting algebra, $\mathfrak{H}$, is a distributive lattice such that for any two elements $\alpha, \beta \in \mathfrak{H}$, the set $\{\gamma \in \mathfrak{H} \mid \gamma \wedge \alpha \preceq \beta\}$ has a maximal element, denoted by $(\alpha \Rightarrow \beta)$.

[26] The main difference between theorems proved using Heyting logic and those using Boolean logic is that proofs by contradiction cannot be used in the former. In particular, this means that one cannot prove that something exists by arguing that the assumption that it does not leads to contradiction; instead it is necessary to provide a *constructive* proof of the existence of the entity concerned. Arguably, this does not place any major restriction on building theories of physics. Indeed, over the years, various physicists (for example, Bryce DeWitt) have argued that constructive proofs should always be used in physics.

1. A physical quantity, A, is to be represented by an arrow $A_{\phi,S} : \Sigma_{\phi,S} \rightarrow \mathcal{R}_{\phi,S}$ where $\Sigma_{\phi,S}$ and $\mathcal{R}_{\phi,S}$ are two special objects in the topos $\tau_\phi(S)$. These are the analogues of, respectively, (i) the classical state space, $\mathcal{S}$; and (ii) the real numbers, $\mathbb{R}$, in which classical physical quantities take their values.
   In what follows, $\Sigma_{\phi,S}$ and $\mathcal{R}_{\phi,S}$ are called the "state object", and the "quantity-value object", respectively.
2. Propositions about the system S are represented by sub-objects of $\Sigma_{\phi,S}$. These sub-objects form a Heyting algebra.
3. Once the topos analogue of a state (a "truth object", or "pseudo-state") has been specified, these propositions are assigned truth values in the Heyting logic associated with the global elements of the sub-object classifier, $\Omega_{\tau_\phi(S)}$, in the topos $\tau_\phi(S)$.

Thus a theory expressed in this way *looks* very much like classical physics except that whereas classical physics always employs the topos of sets, other theories—including quantum theory and, we conjecture, quantum gravity—use a different topos.

One deep result in topos theory is that there is an *internal language* associated with each topos. In fact, not only does each topos generate an internal language, but, conversely, a language satisfying appropriate conditions generates a topos. Topoi constructed in this way are called "linguistic topoi", and every topos can be regarded as a linguistic topos. In many respects, this is one of the profoundest ways of understanding what a topos really "is".[27]

These results are exploited in Sect. 13.4 where we introduce the idea that, for any applicable theory-type, each physical system S is associated with a "local" language, $\mathcal{L}(S)$. The application of the theory-type to S is then involves finding a representation of $\mathcal{L}(S)$ in an appropriate topos; this is equivalent to finding a "translation" of $\mathcal{L}(S)$ into the internal language of that topos.

Closely related to the existence of this linguistic structure is the striking fact that a topos can be used as a *foundation* for mathematics itself, just as set theory is used in the foundations of "normal" (or "classical") mathematics. In this context, the key remark is that the internal language of a topos has a form that is similar in many ways to the formal language on which normal set theory is based. It is this internal, topos language that is used to *interpret* the theory in a "neo-realist" way.

The main difference with classical logic is that the logic of the topos language does not satisfy the law of excluded middle, and hence proofs by contradiction are not permitted. This has many intriguing consequences. For example, there are topoi in which there exist genuine *infinitesimals* that can be used to construct a rival to normal calculus. The possibility of such quantities stems from the fact that the normal proof that they do *not* exist is a proof by contradiction.

Thus each topos carries its own world of mathematics: a world which, generally speaking, is *not* the same as that of classical mathematics.

---

[27] This aspect of topos theory is discussed at length in the books by Bell [11], and Lambek and Scott [59].

Consequently, by postulating that, for a given theory-type, each physical system carries its own topos, we are also saying that to each physical system plus theory-type there is associated a framework for mathematics itself! Thus classical physics uses classical mathematics; and quantum theory uses "quantum mathematics"—the mathematics formulated in the topoi of quantum theory. To this we might add the conjecture: "Quantum gravity uses 'quantum gravity' mathematics"!

13.3 Propositional Languages and Theories of Physics

13.3.1 Two Opposing Interpretations of Propositions

Attempts to construct a naïve realist interpretation of quantum theory founder on the Kochen-Specker theorem. However, if, despite this theorem, some degree of realism is still sought, there are not that many options.

One approach is to focus on a particular, maximal commuting subset of physical quantities and declare by fiat that these are the ones that 'have' values; essentially, this is what is done in "modal" interpretations of quantum theory. However, this leaves open the question of why Nature should select this particular set, and the reasons proposed vary greatly from one scheme to another.

In our work, we take a completely different approach and try to formulate a scheme which takes into account *all* these different choices for commuting sets of physical quantities; in particular, equal ontological status is ascribed to all of them. This scheme is grounded in the topos-theoretic approach that was first proposed in [48–51]. This uses a technique whose first step is to construct a category, $\mathcal{C}$, the objects of which can be viewed as *contexts* in which the quantum theory can be displayed: in fact, they are just the commuting sub-algebras of operators in the theory. All this will be explained in more detail in Sect. 13.5.

In this earlier work, it was postulated that the logic for handling quantum propositions from this perspective is that associated with the topos of presheaves[28] (contravariant functors from $\mathcal{C}$ to **Sets**), $\mathbf{Sets}^{\mathcal{C}^{\mathrm{op}}}$. The idea is that a single presheaf will encode quantum propositions from the perspective of *all* contexts at once. However, in the original papers, the crucial "daseinisation" operation (see Sect. 13.5) was not known and, consequently, the discussion became rather convoluted in places. In addition, the generality and power of the underlying procedure was not fully appreciated by the authors.

For this reason, in the present article we return to the basic questions and reconsider them in the light of the overall topos structure that has now become clear.

We start by considering the way in which propositions arise, and are manipulated, in physics. For simplicity, we will concentrate on systems that are associated with "standard" physics. Then, to each such system S there is associated a set of physical

[28] In quantum theory, the category $\mathcal{C}$ is just a partially-ordered set, which simplifies many manipulations.

quantities—such as energy, momentum, position, angular momentum etc.[29]—all of which are real-valued. The associated propositions are of the form "$A \varepsilon \Delta$", where A is a physical quantity, and Δ is a Borel subset of $\mathbb{R}$.

From a conceptual perspective, the proposition "$A \varepsilon \Delta$" can be physics read in two, very different, ways:

(i) *The (naïve) realist interpretation:* "The physical quantity A *has* a value, and that value lies in Δ."
(ii) *The instrumentalist interpretation:* "*If* a measurement is made of A, the result will be found to lie in Δ."

The former is the familiar, "commonsense" understanding of propositions in both classical physics and daily life. The latter underpins the Copenhagen interpretation of quantum theory. Of course, the instrumentalist interpretation can also be applied to classical physics, but it does not lead to anything new. For, in classical physics, what is measured is what *is* the case: "Epistemology models ontology" (see footnote 19).

We will now study the role of propositions in physics more carefully, particularly in the context of "realist" interpretations.

13.3.2 The Propositional Language $\mathcal{PL}(S)$

13.3.2.1 Intuitionistic Logic and the Definition of $\mathcal{PL}(S)$

We are going to construct a formal language, $\mathcal{PL}(S)$, with which to express propositions about a physical system, S, and to make deductions concerning them. Our intention is to interpret these propositions in a "realist" way: an endeavour whose mathematical underpinning lies in constructing a representation of $\mathcal{PL}(S)$ in a Heyting algebra, $\mathfrak{H}$, that is part of the mathematical framework involved in the application of a particular theory-type to S.

In constructing $\mathcal{PL}(S)$ we suppose that we have first identified some set, $\mathcal{Q}(S)$, of *physical quantities*: this plays a fundamental role in our language. In addition, for *any* system S, we have the set, $P_B\mathbb{R}$ of (Borel) subsets of $\mathbb{R}$. We use the sets $\mathcal{Q}(S)$ and $P_B\mathbb{R}$ to construct the "primitive propositions" about the system S. These are of the form "$A \varepsilon \Delta$" where $A \in \mathcal{Q}(S)$ and $\Delta \in P_B\mathbb{R}$.

We denote the set of all such strings by $\mathcal{PL}(S)_0$. Note that what has been here called a "physical quantity" could better (but more clumsily) be termed the "name" of the physical quantity. For example, when we talk about the "energy" of a system, the word "energy" is the same, and functions in the same way in the formal language, irrespective of the details of the actual Hamiltonian of the system.

[29] This set does not have to contain "*all*" possible physical quantities: it suffices to concentrate on a subset that are deemed to be of particular interest. However, at some point, questions may arise about the "completeness" of the set.

The primitive propositions "$A \varepsilon \Delta$" are used to define "sentences". More precisely, a new set of symbols $\{\neg, \wedge, \vee, \Rightarrow\}$ is added to the language, and then a *sentence* is defined inductively by the following rules (see Chap. 6 in [34]):

1. Each primitive proposition "$A \varepsilon \Delta$" in $\mathcal{PL}(S)_0$ is a sentence.
2. If α is a sentence, then so is $\neg\alpha$.
3. If α and β are sentences, then so are $\alpha \wedge \beta$, $\alpha \vee \beta$, and $\alpha \Rightarrow \beta$.

The collection of all sentences, $\mathcal{PL}(S)$, is an elementary formal language that can be used to express and manipulate propositions about the system S. Note that, at this stage, the symbols $\neg$, $\wedge$, $\vee$, and $\Rightarrow$ have no explicit meaning, although of course the implicit intention is that they should stand for "not", "and", "or" and "implies", respectively. This implicit meaning becomes explicit when a representation of $\mathcal{PL}(S)$ is constructed as part of the application of a theory-type to S (see below). Note also that $\mathcal{PL}(S)$ is a *propositional* language only: it does not contain the quantifiers '$\forall$' or '$\exists$'. To include them requires a higher-order language. We shall return to this in our discussion of the language $\mathcal{L}(S)$.

The next step arises because $\mathcal{PL}(S)$ is not only a vehicle for expressing propositions about the system S: we also want to *reason* with it about the system. To achieve this, a series of axioms for a deductive logic must be added to $\mathcal{PL}(S)$. This could be either classical logic or intuitionistic logic, but we select the latter since it allows a larger class of representations/models, including representations in topoi in which the law of excluded middle fails.

The axioms for intuitionistic logic consist of a finite collection of sentences in $\mathcal{PL}(S)$ (for example, $\alpha \wedge \beta \Rightarrow \beta \wedge \alpha$), plus a single rule of inference, *modus ponens* (the "rule of detachment") which says that from α and $\alpha \Rightarrow \beta$ the sentence β may be derived.

Others axioms might be added to $\mathcal{PL}(S)$ to reflect the implicit meaning of the primitive proposition "$A \varepsilon \Delta$": i.e., (in a realist reading) "A has a value, and that value lies in $\Delta \subseteq \mathbb{R}$". For example, the sentence "$A \varepsilon \Delta_1 \wedge A \varepsilon \Delta_2$" ("$A$ belongs to Δ_1" *and* "A belongs to Δ_2") might seem to be equivalent to "A belongs to $\Delta_1 \cap \Delta_2$" i.e., "$A \varepsilon \Delta_1 \cap \Delta_2$". A similar remark applies to "$A \varepsilon \Delta_1 \vee A \varepsilon \Delta_2$".

Thus, along with the axioms of intuitionistic logic and detachment, we might be tempted to add the following axioms:

$$A \varepsilon \Delta_1 \wedge A \varepsilon \Delta_2 \Leftrightarrow A \varepsilon \Delta_1 \cap \Delta_2, \tag{13.2}$$

$$A \varepsilon \Delta_1 \vee A \varepsilon \Delta_2 \Leftrightarrow A \varepsilon \Delta_1 \cup \Delta_2. \tag{13.3}$$

These axioms are consistent with the intuitionistic logical structure of $\mathcal{PL}(S)$.

We shall see later the extent to which the axioms (13.2) and (13.3) are compatible with the topos representations of classical and quantum physics. However, the other obvious proposition to consider in this way—"It is *not* the case that A belongs to Δ"—is clearly problematical.

In classical logic, this proposition, "$\neg A \varepsilon \Delta$", is equivalent to "$A$ belongs to $\mathbb{R}\backslash\Delta$", where $\mathbb{R}\backslash\Delta$ denotes the set-theoretic complement of Δ in $\mathbb{R}$. This might suggest augmenting (13.2) and (13.3) with a third axiom

$$\neg A \,\varepsilon\, \Delta \Leftrightarrow A \,\varepsilon\, \mathbb{R} \backslash \Delta \tag{13.4}$$

However, applying "¬" to both sides of (13.4) gives

$$\neg\neg A \,\varepsilon\, \Delta \Leftrightarrow A \,\varepsilon\, \Delta \tag{13.5}$$

because of the set-theoretic result $\mathbb{R}\backslash(\mathbb{R}\backslash\Delta) = \Delta$. But in an intuitionistic logic we do not have $\alpha \Leftrightarrow \neg\neg\alpha$ but only $\alpha \Rightarrow \neg\neg\alpha$, and so (13.4) could be false in a Heyting-algebra representation of $\mathcal{PL}(S)$ that is not Boolean. Therefore, adding (13.4) as an axiom in $\mathcal{PL}(S)$ is not indicated if representations are to be sought in non-Boolean topoi.

13.3.2.2 Representations of $\mathcal{PL}(S)$

To use a language $\mathcal{PL}(S)$ "for real" for some specific physical system S one must first decide on the set $\mathcal{Q}(S)$ of physical quantities that are to be used in describing S. This language must then be represented in the concrete mathematical structure that arises when a theory-type (for example: classical physics, quantum physics, DI-physics,...) is applied to S. Such a representation, π, maps each primitive proposition, α, in $\mathcal{PL}(S)_0$ to an element, $\pi(\alpha)$, of some Heyting algebra (which could be Boolean), $\mathfrak{H}$, whose specification is part of the theory of S. For example, in classical mechanics, the propositions are represented in the Boolean algebra of all (Borel) subsets of the classical state space.

The representation of the primitive propositions can be extended recursively to all of $\mathcal{PL}(S)$ with the aid of the following rules [34]:

$$(a)\ \pi(\alpha \vee \beta) := \pi(\alpha) \vee \pi(\beta) \tag{13.6}$$
$$(b)\ \pi(\alpha \wedge \beta) := \pi(\alpha) \wedge \pi(\beta) \tag{13.7}$$
$$(c)\ \pi(\neg\alpha) := \neg\pi(\alpha) \tag{13.8}$$
$$(d)\ \pi(\alpha \Rightarrow \beta) := \pi(\alpha) \Rightarrow \pi(\beta) \tag{13.9}$$

Note that, on the left hand side of (13.6), (13.7), (13.8) and (13.9), the symbols $\{\neg, \wedge, \vee, \Rightarrow\}$ are elements of the language $\mathcal{PL}(S)$, whereas on the right hand side they denote the logical connectives in the Heyting algebra, $\mathfrak{H}$, in which the representation takes place.

This extension of π from $\mathcal{PL}(S)_0$ to $\mathcal{PL}(S)$ is consistent with the axioms for the intuitionistic, propositional logic of the language $\mathcal{PL}(S)$. More precisely, these axioms become tautologies: i.e., they are all represented by the maximum element, 1, in the Heyting algebra. By construction, the map $\pi : \mathcal{PL}(S) \to \mathfrak{H}$ is then a representation of $\mathcal{PL}(S)$ in the Heyting algebra $\mathfrak{H}$. A logician would say that $\pi : \mathcal{PL}(S) \to \mathfrak{H}$ is an $\mathfrak{H}$-*valuation*, or $\mathfrak{H}$-*model*, of the language $\mathcal{PL}(S)$.

Note that different systems, S, can have the same language. For example, consider a point-particle moving in one dimension, with a Hamiltonian function $H(x, p) = \frac{p^2}{2m} + V(x)$ and state space $T^*\mathbb{R}$. Different potentials V correspond to different systems (in the sense in which we are using the word "system"), but the physical quantities for these systems—or, more precisely, the "names" of these quantities, for example, "energy", "position", "momentum"—are the same for them all. Consequently, the language $\mathcal{PL}(S)$ is independent of V. However, the *representation* of, say, the proposition "$E \varepsilon \Delta$" (where 'E' is the energy), with a specific subset of the state space *will* depend on the details of the Hamiltonian.

Clearly, a major consideration in using the language $\mathcal{PL}(S)$ is choosing the Heyting algebra in which the representation is to take place. A fundamental result in topos theory is that the set of all sub-objects of any object in a topos is a Heyting algebra, and these are the Heyting algebras with which we will be concerned.

Of course, beyond the language, $\mathcal{S}$, and its representation π, lies the question of whether or not a proposition is "true". This requires the concept of a "state" which, when specified, yields "truth values" for the primitive propositions in $\mathcal{PL}(S)$. These can then be extended recursively to the rest of $\mathcal{PL}(S)$. In classical physics, the possible truth values are just "true" or "false". However, as we shall see, the situation in topos theory is more complex.

13.3.2.3 Using Geometric Logic

The inductive definition of $\mathcal{PL}(S)$ given above means that sentences can involve only a *finite* number of primitive propositions, and therefore only a finite number of disjunctions ("$\vee$") or conjunctions ("$\wedge$"). An interesting variant of this structure is the, so-called, "propositional geometric logic". This is characterised by modifying the language and logical axioms so that:

1. There are arbitrary disjunctions, including the empty disjunction ("0").
2. There are finite conjunctions, including the empty conjunction ("1")
3. Conjunction distributes over arbitrary disjunctions; disjunction distributes over finite conjunctions.

This structure does *not* include negation, implication, or infinite conjunctions.

From a conceptual viewpoint, this set of rules is obtained by considering what it means to actually "affirm" the propositions in $\mathcal{PL}(S)$. A careful analysis of this concept is given by Vickers [84]; the idea itself goes back to work by Smyth [79] and Abramsky [1]. The conclusion is that the set of "affirmable" propositions should satisfy the rules above.

Clearly such a logic is tailor-made for seeking representations in the open sets of a topological space—the paradigmatic example of a Heyting algebra. The phrase "geometric logic" is normally applied to a *first-order* logic with the properties above, and we will return to this in our discussion of the typed language $\mathcal{L}(S)$. What we have here is just the propositional part of this logic.

The restriction to geometric logic would be easy to incorporate into our languages $\mathcal{PL}(S)$: for example, the axiom (13.3) (if added) could be extended to read[30]

$$\bigvee_{i \in I} (A \,\varepsilon\, \Delta_i) = A \,\varepsilon \bigcup_{i \in I} \Delta_i \tag{13.10}$$

for all index sets I.

The move to geometric logic is motivated by a conception of truth that is grounded in the actions of making real measurements [84, 85]. This resonates strongly with the logical positivism that seems still to lurk in the collective unconscious of the physics profession, and which, of course, was strongly affirmed by Bohr in his analysis of quantum theory. However, our drive towards "neo-realism" involves replacing the idea of observation/measurement with that of "the way things are", albeit in a more sophisticated interpretation than that of the ubiquitous cobbler-in-the-market. Consequently, the conceptual reasons for using "affirmative" logic are less compelling. This issue deserves further thought: at the moment we are open-minded about it.

The use of geometric logic becomes more interesting in the context of the typed language $\mathcal{L}(S)$, and we shall return to this in Sect. 13.4.2.

13.3.2.4 Introducing Time Dependence

In addition to describing "the way things are" there is also the question of how the-way-things-are changes in time. In the form presented above, the language $\mathcal{PL}(S)$ may seem geared towards a "canonical" perspective in so far as the propositions concerned are implicitly taken to be asserted at a particular moment of time. As such, $\mathcal{PL}(S)$ deals with the values of physical quantities at that time. In other words, the underlying spatio-temporal perspective seems thoroughly "Newtonian".

However, this is only partly true since the phrase "physical quantity" can have meanings other than the canonical one. For example, one could talk about the "time average of momentum", and call that a physical quantity. In this case, the propositions would be about *histories* of the system, not just "the way things are" at a particular moment in time.

In practice, the question of time dependence can be addressed in various ways. One is to attach a (external) time label, t, to the physical quantities, so that the primitive propositions become of the form "$A_t \,\varepsilon\, \Delta$". This can be interpreted in two ways. The first is to think of $\mathcal{Q}(S)$ as including the symbols A_t for all physical quantities A and all values of time $t \in \mathbb{R}$. The second is to keep $\mathcal{Q}(S)$ fixed, but instead let the language itself becomes time-dependent, so that we should write $\mathcal{PL}(S)_t$, $t \in \mathbb{R}$.

---

[30] Note that the bi-implication $\Leftrightarrow$ used in, for example, (13.2) and (13.3), is not available if there is no implication symbol. Thus we have assumed that we are now working with a logical structure in which "equality" is a meaningful concept; hence the introduction of "=" in (13.10).

In the former case, $\mathcal{PL}(S)$ would naturally include *history* propositions of the form

$$(A_{1t_1} \varepsilon \Delta_1) \wedge (A_{2t_2} \varepsilon \Delta_2) \wedge \cdots \wedge (A_{nt_n} \varepsilon \Delta_n) \tag{13.11}$$

and other obvious variants of this. Here we assume that $t_1 \leq t_2 \leq \cdots \leq t_n$.

The sequential proposition in (13.11) is to be interpreted (in a realist reading) as asserting that "'The physical quantity A_1 has a value that lies in Δ_1 at time t_1' *and* 'the physical quantity A_2 has a value that lies in Δ_2 at time t_2' *and* $\cdots$ *and* 'the physical quantity A_n has a value that lies in Δ_n at time t_n'". Clearly what we have here is a type of *temporal* logic. Thus this would be an appropriate structure with which to discuss the 'consistent histories' interpretation of quantum theory, particularly in the, so-called, HPO (history projection formalism) [43]. In that context, (13.11) represents a, so-called, 'homogeneous' history.

From a general conceptual perspective, one might prefer to have an internal time object, rather than adding external time labels in the language. Indeed, in our later discussion of the higher-order, typed language $\mathcal{L}(S)$ we will strive to eliminate external entities. However, in the present case, $\Delta \subseteq \mathbb{R}$ is already an "external" (to the language) entity, as indeed is $A \in \mathcal{Q}(S)$, so there seems no particular objection to adding a time label too.

In the second approach, where there is only one time label, the representation π will map "$A_t \varepsilon \Delta$" to a time-dependent element, $\pi(A_t \varepsilon \Delta)$, of the Heyting algebra, $\mathfrak{H}$; one could say that this is a type of "Heisenberg picture".

This suggests another option, which is to keep the language free of any time labels, but allow the *representation* to be time-dependent. In this case, $\pi_t(A \varepsilon \Delta)$ is a time-dependent member of $\mathfrak{H}$.[31]

A different approach is to ascribe time dependence to the "truth objects" in the theory: this corresponds to a type of Schrödinger picture. The concept of a truth object is discussed in detail in Sect. 13.6.

13.3.2.5 The Representation of $\mathcal{PL}(S)$ in Classical Physics

Let us now look at the representation of $\mathcal{PL}(S)$ that corresponds to classical physics. In this case, the topos involved is just the category, **Sets**, of sets and functions between sets.

We will denote by π_{cl} the representation of $\mathcal{PL}(S)$ that describes the classical, Hamiltonian mechanics of a system, S, whose state space is a symplectic (or Poisson) manifold $\mathcal{S}$. We denote by $\breve{A} : \mathcal{S} \to \mathbb{R}$ the real-valued function[32] on $\mathcal{S}$ that represents the physical quantity A.

---

[31] Perhaps we should also consider the possibility that the Heyting algebra is time dependent, in which case $\pi_t(A \varepsilon \Delta)$ is a member of $\mathfrak{H}_t$.

[32] As mentioned in footnote 22, $\breve{A}$ is required to be measurable, continuous, or smooth, depending on the type of physical quantity that A is. However, for the most part, these details of classical

Then the representation π_{cl} maps the primitive proposition "$A \,\varepsilon\, \Delta$" to the subset of $\mathcal{S}$ given by

$$\pi_{\mathrm{cl}}(A \,\varepsilon\, \Delta) := \{s \in \mathcal{S} \mid \breve{A}(s) \in \Delta\} \\ = \breve{A}^{-1}(\Delta). \tag{13.12}$$

This representation can be extended to all the sentences in $\mathcal{PL}(S)$ with the aid of (13.6), (13.7), (13.8) and (13.9). Note that, since Δ is a Borel subset of $\mathbb{R}$, $\breve{A}^{-1}(\Delta)$ is a Borel subset of the state space $\mathcal{S}$. Hence, in this case, $\mathfrak{H}$ is equal to the Boolean algebra of all Borel subsets of $\mathcal{S}$ that can be obtained as inverse images of the form $\breve{A}^{-1}(\Delta)$ where $A \in \mathcal{Q}(S)$ and $\Delta \in P_B\mathbb{R}$.

We note that, for all (Borel) subsets Δ_1, Δ_2 of $\mathbb{R}$ we have

$$\breve{A}^{-1}(\Delta_1) \cap \breve{A}^{-1}(\Delta_2) = \breve{A}^{-1}(\Delta_1 \cap \Delta_2) \tag{13.13}$$
$$\breve{A}^{-1}(\Delta_1) \cup \breve{A}^{-1}(\Delta_2) = \breve{A}^{-1}(\Delta_1 \cup \Delta_2) \tag{13.14}$$
$$\neg\breve{A}^{-1}(\Delta_1) = \breve{A}^{-1}(\mathbb{R}\backslash\Delta_1) \tag{13.15}$$

and hence, in classical physics, all three conditions (13.2), (13.3) and (13.4) that we discussed earlier can be added consistently to the language $\mathcal{PL}(S)$.

Consider now the assignment of truth values to the propositions in this theory. This involves the idea of a "microstate" which, in classical physics, is simply an element s of the state space $\mathcal{S}$. Each microstate s assigns to each primitive proposition "$A \,\varepsilon\, \Delta$", a truth value, $\nu\big(A \,\varepsilon\, \Delta; s\big)$, which lies in the set {false, true} (which we identify with {0, 1}) and is defined as

$$\nu\big(A \,\varepsilon\, \Delta; s\big) := \begin{cases} 1 & \text{if } \breve{A}(s) \in \Delta; \\ 0 & \text{otherwise} \end{cases} \tag{13.16}$$

for all $s \in \mathcal{S}$.

13.3.2.6 The Failure to Represent $\mathcal{PL}(S)$ in Standard Quantum Theory

The procedure above that works so easily for classical physics fails completely if one tries to apply it to standard quantum theory.

In quantum physics, a physical quantity A is represented by a self-adjoint operator $\hat{A}$ on a Hilbert space $\mathcal{H}$, and the proposition "$A \,\varepsilon\, \Delta$" is represented by the projection operator $\hat{E}[A \in \Delta]$ which projects onto the subset $\Delta \cap \mathrm{sp}(\hat{A})$ of the spectrum, $\mathrm{sp}(\hat{A})$, of $\hat{A}$; i.e.,

$$\pi(A \,\varepsilon\, \Delta) := \hat{E}[A \in \Delta]. \tag{13.17}$$

---

mechanics are not relevant to our discussions, and usually we will not characterise $\breve{A} : \mathcal{S} \to \mathbb{R}$ beyond just saying that it is a measurable function/map from $\mathcal{S}$ to $\mathbb{R}$.

Of course, the set of all projection operators, $\mathcal{P}(\mathcal{H})$, in $\mathcal{H}$ has a "logic" of its own—the "quantum logic"[33] of the Hilbert space $\mathcal{H}$—but this is incompatible with the intuitionistic logic of the language $\mathcal{PL}(S)$, and the representation (13.17).

Indeed, since the "logic" $\mathcal{P}(\mathcal{H})$ is non-distributive, there will exist non-commuting operators $\hat{A}$, $\hat{B}$, $\hat{C}$, and Borel subsets Δ_A, Δ_B, Δ_C of $\mathbb{R}$ such that[34]

$$\hat{E}[A \in \Delta_A] \wedge \left(\hat{E}[B \in \Delta_B] \vee \hat{E}[C \in \Delta_C]\right) \neq \left(\hat{E}[A \in \Delta_A] \wedge \hat{E}[B \in \Delta_B]\right) \vee \left(\hat{E}[A \in \Delta_A] \wedge \hat{E}[C \in \Delta_C]\right) \quad (13.18)$$

while, on the other hand, the logical bi-implication

$$\alpha \wedge (\beta \vee \gamma) \Leftrightarrow (\alpha \wedge \beta) \vee (\alpha \wedge \gamma) \quad (13.19)$$

can be deduced from the axioms of the language $\mathcal{PL}(S)$.

This failure of distributivity bars any naïve realist interpretation of quantum logic. If an instrumentalist interpretation is used instead, the spectral projectors $\hat{E}[A \in \Delta]$ now represent counterfactual propositions about what would happen *if* a measurement is made, not propositions about what is "actually the case". And, of course, when a state is specified, this does not yield actual truth values but only the Born-rule probabilities of getting certain results.

13.4 A Higher-Order, Typed Language for Physics

13.4.1 The Basics of the Language $\mathcal{L}(S)$

We want now to consider the possibility of representing the physical quantities of a system by arrows in a topos other than **Sets**.

The physical meaning of such an arrow is not clear, a priori. Nor is it even clear *what* it is that is being represented in this way. However, what *is* clear is that in such a situation it is not correct to assume that the quantity-value object is necessarily the real-number object in the topos (assuming that there is one). Rather, this object has to be determined for each topos, and is therefore an important part of the "representation".

A powerful technique for allowing the quantity-value object to be system-dependent is to add a symbol "$\mathcal{R}$" to the system language. Developing this line

---

[33] For an excellent survey of quantum logic see [20]. This includes a discussion of a first-order axiomatisation of quantum logic, and with an associated sequent calculus. It is interesting to compare our work with what the authors of this paper have done. We hope to return to this at some time in the future.

[34] There is a well-known example that uses three rays in $\mathbb{R}^2$, so this phenomenon is not particularly exotic.

of thinking suggests that a symbol "Σ", too, should be added to the language, as a linguistic precursor of the state object, as well as a set of symbols of the form "$A : \Sigma \to \mathcal{R}$", to be construed as "what it is" (namely a physical quantity) that is represented by arrows in a topos. Similarly, there should be a symbol "Ω", to act as the linguistic precursor to the sub-object classifier in the topos; in the topos **Sets**, this is just the set $\{0, 1\}$.

The clean way of doing all this is to construct a "local language" [11]. Our basic assumption is that such a language, $\mathcal{L}(S)$, can be associated with each system S. A physical theory of S then corresponds to a representation of $\mathcal{L}(S)$ in an appropriate topos.

13.4.1.1 The Symbols of $\mathcal{L}(S)$

We first consider the minimal set of symbols needed to handle elementary physics. For more sophisticated theories in physics it will be necessary to change, or enlarge, this set of "ground type" symbols.

The symbols for the local language, $\mathcal{L}(S)$, are defined recursively as follows:

1. (a) The basic *type symbols* are $1, \Omega, \Sigma, \mathcal{R}$. The last two, Σ and $\mathcal{R}$, are known as *ground type symbols*. They are the linguistic precursors of the state object, and quantity-value object, respectively.
   If $T_1, T_2, \ldots, T_n$, $n \geq 1$, are type symbols, then so is[35] $T_1 \times T_2 \times \cdots \times T_n$.
   (b) If T is a type symbol, then so is PT.
2. (a) For each type symbol, T, there is associated a countable set of *variables of type T*.
   (b) There is a special symbol $*$.
3. (a) To each pair (T_1, T_2) of type symbols there is associated a set, $F_{\mathcal{L}(S)}(T_1, T_2)$, of *function symbols*. Such a symbol, A, is said to have *signature* $T_1 \to T_2$; this is indicated by writing $A : T_1 \to T_2$.
   (b) Some of these sets of function symbols may be empty. However, in our case, particular importance is attached to the set, $F_{\mathcal{L}(S)}(\Sigma, \mathcal{R})$, of function symbols $A : \Sigma \to \mathcal{R}$, and we assume this set is non-empty.

The function symbols $A : \Sigma \to \mathcal{R}$ represent the "physical quantities" of the system, and hence $F_{\mathcal{L}(S)}(\Sigma, \mathcal{R})$ will depend on the system S. In fact, the only parts of the language that are system-dependent are these function symbols. The set $F_{\mathcal{L}(S)}(\Sigma, \mathcal{R})$ is the analogue of the set, $\mathcal{Q}(S)$, of physical quantities associated with the propositional language $\mathcal{PL}(S)$.

For example, if S_1 is a point particle moving in one dimension, the set of physical quantities could be chosen to be $F_{\mathcal{L}(S_1)}(\Sigma, \mathcal{R}) = \{x, p, H\}$ which represent the position, momentum, and energy of the system. On the other hand, if S_2 is a particle moving in three dimensions, we could have $F_{\mathcal{L}(S_2)}(\Sigma, \mathcal{R}) = \{x, y, z, p_x, p_y, p_z, H\}$ to allow for three-dimensional position

[35] By definition, if $n = 0$ then $T_1 \times T_2 \times \cdots \times T_n := 1$.

and momentum (with respect to some given Euclidean coordinate system). Or, we could decide to add angular momentum too, to give the set $F_{\mathcal{L}(S_2)}(\Sigma, \mathcal{R}) = \{x, y, z, p_x, p_y, p_z, J_x, J_y, J_z, H\}$. A still further extension would be to add the quantities $\underline{x} \cdot \underline{n}$ and $\underline{p} \cdot \underline{m}$ for all unit vectors $\underline{n}$ and $\underline{m}$; and so on.

Note that, as with the propositional language $\mathcal{PL}(S)$, the fact that a given system has a specific Hamiltonian[36]—expressed as a particular function of position and momentum coordinates—is not something that is to be coded into the language: instead, such system dependence arises in the choice of *representation* of the language. This means that many different systems can have the same local language.

Finally, it should be emphasised that this list of symbols is minimal and one will certainly want to add more. One obvious, general, example is a type symbol $\mathbb{N}$ that is to be interpreted as the linguistic analogue of the natural numbers. The language could then be augmented with the axioms of Peano arithmetic.

13.4.1.2 The Terms of $\mathcal{L}(S)$

The next step is to enumerate the "terms" in the language, together with their associated types [11, 59]:

1. (a) For each type symbol T, the variables of type T are terms of type T.
   (b) The symbol $*$ is a term of type 1.
   (c) A term of type Ω is called a *formula*; a formula with no free variables is called a *sentence*.
2. If A is function symbol with signature $T_1 \to T_2$, and t is a term of type T_1, then $A(t)$ is term of type T_2.
   In particular, if $A : \Sigma \to \mathcal{R}$ is a physical quantity, and t is a term of type Σ, then $A(t)$ is a term of type $\mathcal{R}$.
3. (a) If $t_1, t_2, \ldots, t_n$ are terms of type $T_1, T_2, \ldots, T_n$, then $\langle t_1, t_2, \ldots, t_n\rangle$ is a term of type $T_1 \times T_2 \times \cdots \times T_n$.
   (b) If t is a term of type $T_1 \times T_2 \times \cdots \times T_n$, and if $1 \leq i \leq n$, then $(t)_i$ is a term of type T_i.
4. (a) If ω is a term of type Ω, and $\tilde{x}$ is a variable of type T, then $\{\tilde{x} \mid \omega\}$ is a term of type PT.
   (b) If t_1, t_2 are terms of the same type, then "$t_1 = t_2$" is a term of type Ω.
   (c) If t_1, t_2 are terms of type T, PT respectively, then $t_1 \in t_2$ is a term of type Ω.

Note that the logical operations are not included in the set of symbols. Instead, they can all be defined using what is already given. For example,

---

[36] It must be emphasised once more that the use of a local language is *not* restricted to standard, canonical systems in which the concept of a "Hamiltonian" is meaningful. The scope of the linguistic ideas is *much* wider than that and the canonical systems are only an example. Indeed, our long-term interest is in the application of these ideas to quantum gravity where the local language is likely to be very different from that used here. However, we anticipate that the basic ideas will be the same.

(i) $true := (* = *)$; and (ii) if α and β are terms of type Ω, then[37] $\alpha \wedge \beta := \big(\langle\alpha, \beta\rangle = \langle\text{true}, \text{true}\rangle\big)$. Thus, in terms of the original set of symbols, we have

$$\alpha \wedge \beta := \big(\langle\alpha, \beta\rangle = \langle * = *, * = *\rangle\big) \tag{13.20}$$

and so on. For details, see Chap. 3 in [11].

13.4.1.3 Terms of Particular Interest to us

Let A be a physical quantity in the set $F_{\mathcal{L}(S)}(\Sigma, \mathcal{R})$, and therefore a function symbol of signature $\Sigma \to \mathcal{R}$. In addition, let $\tilde{\Delta}$ be a variable (and therefore a term) of type $P\mathcal{R}$; and let $\tilde{s}$ be a variable (and therefore a term) of type Σ. Then some terms of particular interest to us are the following:

1. $A(\tilde{s})$ is a term of type $\mathcal{R}$ with a free variable, $\tilde{s}$, of type Σ.
2. '$A(\tilde{s}) \in \tilde{\Delta}$' is a term of type Ω with free variables (i) $\tilde{s}$ of type Σ; and (ii) $\tilde{\Delta}$ of type $P\mathcal{R}$.
3. $\{\tilde{s} \mid A(\tilde{s}) \in \tilde{\Delta}\}$ is a term of type $P\Sigma$ with a free variable $\tilde{\Delta}$ of type $P\mathcal{R}$.

As we shall see, "$A(\tilde{s}) \in \tilde{\Delta}$" is an analogue of the primitive propositions "$A \varepsilon \Delta$" in the propositional language $\mathcal{PL}(S)$. However, there is a crucial difference. In $\mathcal{PL}(S)$, the "Δ" in "$A \varepsilon \Delta$" is a specific subset of the external (to the language) real line $\mathbb{R}$. On the other hand, in the local language $\mathcal{L}(S)$, the "$\tilde{\Delta}$" in "$A(\tilde{s}) \in \tilde{\Delta}$" is an *internal* variable within the language.

13.4.1.4 Adding Axioms to the Language

To make the language $\mathcal{L}(S)$ into a deductive system we need to add a set of appropriate axioms and rules of inference. The former are expressed using *sequents*: defined as expressions of the form $\Gamma : \alpha$ where α is a formula (a term of type Ω) and Γ is a set of such formula. The intention is that "$\Gamma : \alpha$" is to be read intuitively as "the collection of formula in Γ 'imply' α". If Γ is empty we just write $: \alpha$.

The basic axioms include things like "$\alpha : \alpha$" (tautology), and "$: \tilde{t} \in \{\tilde{t} \mid \alpha\} \Leftrightarrow \alpha$" (comprehension) where $\tilde{t}$ is a variable of type T. These axioms[38] and the rules of inference (sophisticated analogues of *modus ponens*) give rise to a deductive system using intuitionistic logic. For the details see [11, 59].

For applications in physics we could, and presumably should, add extra axioms (in the form of sequents). For example, perhaps the quantity-value object should

[37] The parentheses ( ) are not symbols in the language, they are just a way of grouping letters and sentences. The same remark applies to the inverted commas ‘’.

[38] The complete set is [11]:

Tautology: $\alpha = \alpha$

Unity : $\tilde{x}_1 = *$ where $\tilde{x}_1$ is a variable of type 1.

always be an abelian-group object, or at least a semi-group?[39] This can be coded into the language by adding the axioms for an abelian group structure for $\mathcal{R}$. This involves the following steps:

1. Add the following symbols:
   (a) A "unit" function symbol $0 : 1 \to \mathcal{R}$; this will be the linguistic analogue of the unit element in an abelian group.
   (b) An "addition" function symbol $+ : \mathcal{R} \times \mathcal{R} \to \mathcal{R}$.
   (c) An "inverse" function symbol $- : \mathcal{R} \to \mathcal{R}$
2. Then add axioms like '$: \forall \tilde{r}\big(+ \langle \tilde{r}, 0(*)\rangle = \tilde{r}\big)$' where $\tilde{r}$ is a variable of type $\mathcal{R}$, and so on.

For another example, consider a point particle moving in three dimensions, with the function symbols $F_{\mathcal{L}(S)}(\Sigma, \mathcal{R}) = \{x, y, z, p_x, p_y, p_z, J_x, J_y, J_z, H\}$. As $\mathcal{L}(S)$ stands, there is no way to specify, for example, that "$J_x = yp_z - zp_y$". Such relations can only be implemented in a *representation* of the language. However, if this relation is felt to be "universal" (i.e., if it is expected to hold in all physically relevant representations), then it could be added to the language with the use of extra axioms.

One of the delicate decisions that has to be made about $\mathcal{L}(S)$ is what extra axioms to add to the base language. Too few, and the language lacks content; too many, and representations of potential physical significance are excluded. This is one of the places in the formalism where a degree of physical insight is necessary!

13.4.2 Representing $\mathcal{L}(S)$ in a Topos

The construction of a theory of the system S involves choosing a representation[40]/model, ϕ, of the language $\mathcal{L}(S)$ in a topos τ_ϕ.[41] The choice of both topos and representation depend on the theory-type being used.

Equality: $x = y, \alpha(\tilde{z}/x) : \alpha(\tilde{z}/y)$. Here, $\alpha(\tilde{z}/x)$ is the term α with $\tilde{z}$ replaced by the term x for each free occurrence of the variable $\tilde{z}$. The terms x and y must be of the same type as $\tilde{z}$.

Products: $: (\langle x_1, \ldots, x_n\rangle)_i = x_i$

$: x = \langle (x)_1, \ldots, (x)_n\rangle$

Comprehension: $: \tilde{t} \in \{\tilde{t} \mid \alpha\} \Leftrightarrow \alpha$

[39] One could go even further and add the axioms for real numbers. However, the example of quantum theory suggests that this is inappropriate: in general, the quantity-value object will *not* be the real-number object [29].

[40] The word "interpretation" is often used in the mathematical literature, but we want to reserve that for use in discussions of interpretations of quantum theory, and the like.

[41] A more comprehensive notation is $\tau_\phi(S)$, which draws attention to the system S under discussion; similarly, the state object could be written as $\Sigma_{\phi,S}$, and so on. This extended notation is used in Sect. 13.11 where we are concerned with the relations between *different* systems, and then it is

For example, consider a system, S, that can be treated using both classical physics and quantum physics, such as a point particle moving in three dimensions. Then, for the application of the theory-type "classical physics", in a representation denoted σ, the topos τ_σ is **Sets**, and Σ is represented by the symplectic manifold $\Sigma_\sigma := T^*\mathbb{R}^3$; $\mathcal{R}$ is represented by the usual real numbers $\mathbb{R}$.

On the other hand, as we shall see in Sect. 13.5, for the application of the theory-type "quantum physics", τ_ϕ is the topos, $\mathbf{Sets}^{\mathcal{V}(\mathcal{H})^{\mathrm{op}}}$, of presheaves over the category[42] $\mathcal{V}(\mathcal{H})$, where $\mathcal{H} \simeq L^2(\mathbb{R}^3, d^3x)$ is the Hilbert space of the system S. In this case, Σ is represented by $\Sigma_\phi := \underline{\Sigma}$, where $\underline{\Sigma}$ is the spectral presheaf; this representation is discussed at length in Sect. 13.5. For both theory types, the *details* of, for example, the Hamiltonian, are coded in the representation.

We now list the τ_ϕ-representation of the most significant symbols and terms in our language, $\mathcal{L}(S)$ (we have picked out only the parts that are immediately relevant to our programme: for full details see [11, 59]).

1. (a) The ground type symbols Σ and $\mathcal{R}$ are represented by objects Σ_ϕ and $\mathcal{R}_\phi$ in τ_ϕ. These are identified physically as the state object and quantity-value object, respectively.
   (b) The symbol Ω, is represented by $\Omega_\phi := \Omega_{\tau_\phi}$, the sub-object classifier of the topos τ_ϕ.
   (c) The symbol 1, is represented by $1_\phi := 1_{\tau_\phi}$, the terminal object in τ_ϕ.
2. For each type symbol PT, we have $(PT)_\phi := PT_\phi$, the power object of the object T_ϕ in τ_ϕ.
   In particular, $(P\Sigma)_\phi = P\Sigma_\phi$ and $(P\mathcal{R})_\phi = P\mathcal{R}_\phi$.
3. Each function symbol $A : \Sigma \to \mathcal{R}$ in $F_{\mathcal{L}(S)}(\Sigma, \mathcal{R})$ (i.e., each physical quantity) is represented by an arrow $A_\phi : \Sigma_\phi \to \mathcal{R}_\phi$ in τ_ϕ.
   We will generally require the representation to be *faithful*: i.e., the map $A \mapsto A_\phi$ is one-to-one.
4. A term of type Ω of the form "$A(\tilde{s}) \in \tilde{\Delta}$" (which has free variables $\tilde{s}, \tilde{\Delta}$ of type Σ and $P\mathcal{R}$ respectively) is represented by an arrow $[\![A(\tilde{s}) \in \tilde{\Delta}]\!]_\phi : \Sigma_\phi \times P\mathcal{R}_\phi \to \Omega_{\tau_\phi}$. In detail, this arrow is

$$[\![A(\tilde{s}) \in \tilde{\Delta}]\!]_\phi = e_{\mathcal{R}_\phi} \circ \langle [\![A(\tilde{s})]\!]_\phi, [\![\tilde{\Delta}]\!]_\phi \rangle \tag{13.21}$$

   where $e_{\mathcal{R}_\phi} : \mathcal{R}_\phi \times P\mathcal{R}_\phi \to \Omega_{\tau_\phi}$ is the usual evaluation map; $[\![A(\tilde{s})]\!]_\phi : \Sigma_\phi \to \mathcal{R}_\phi$ is the arrow A_ϕ; and $[\![\tilde{\Delta}]\!]_\phi : P\mathcal{R}_\phi \to P\mathcal{R}_\phi$ is the identity.
   Thus $[\![A(\tilde{s}) \in \tilde{\Delta}]\!]_\phi$ is the chain of arrows:

$$\Sigma_\phi \times P\mathcal{R}_\phi \xrightarrow{A_\phi \times \mathrm{id}} \mathcal{R}_\phi \times P\mathcal{R}_\phi \xrightarrow{e_{\mathcal{R}_\phi}} \Omega_{\tau_\phi}. \tag{13.22}$$

essential to indicate which system is meant. However, in the present article, only one system at a time is being considered, and so the truncated notation is fine.

[42] We just mention here that the objects in $\mathcal{V}(\mathcal{H})$ are the unital, commutative von Neumann sub-algebras of the algebra, $B(\mathcal{H})$, of all bounded operators on $\mathcal{H}$. Arrows in $\mathcal{V}(\mathcal{H})$ are the inclusions of smaller into larger algebras. We will explain, and motivate, this later.

We see that the analogue of the "Δ" used in the $\mathcal{PL}(S)$-proposition "$A \varepsilon \Delta$" is played by sub-objects of $\mathcal{R}_\phi$ (i.e., global elements of $P\mathcal{R}_\phi$) in the domain of the arrow in (13.22). These objects are, of course, representation-dependent (i.e., they depend on ϕ).

5. A term of type $P\Sigma$ of the form $\{\tilde{s} \mid A(\tilde{s}) \in \tilde{\Delta}\}$ (which has a free variable $\tilde{\Delta}$ of type $P\mathcal{R}$) is represented by an arrow $[\![\{\tilde{s} \mid A(\tilde{s}) \in \tilde{\Delta}\}]\!]_\phi : P\mathcal{R}_\phi \to P\Sigma_\phi$. This arrow is the power transpose[43] of $[\![A(\tilde{s}) \in \tilde{\Delta}]\!]_\phi$:

$$[\![\{\tilde{s} \mid A(\tilde{s}) \in \tilde{\Delta}\}]\!]_\phi = \ulcorner [\![A(\tilde{s}) \in \tilde{\Delta}]\!]_\phi \urcorner \tag{13.23}$$

6. A term, ω, of type Ω with no free variables is represented by a global element $[\![\omega]\!]_\phi : 1_{\tau_\phi} \to \Omega_{\tau_\phi}$. These will typically act as "truth values" for propositions about the system.
7. Any axioms that have been added to the language are required to be represented by the arrow $true : 1_{\tau_\phi} \to \Omega_{\tau_\phi}$.

13.4.2.1 The Local Set Theory of a Topos

We should emphasise that the decision to focus on the particular type of language that we have, is not an arbitrary one. Indeed, there is a deep connection between such languages and topos theory.

In this context, we first note that to any local language, $\mathcal{L}$, there is associated a "local set theory". This involves defining an "$\mathcal{L}$-set" to be a term X of power type (so that expressions of the form $x \in X$ are meaningful) and with no free variables. Analogues of all the usual set operations can be defined on $\mathcal{L}$-sets. For example, if X, Y are $\mathcal{L}$-sets of type PT, one can define $X \cap Y := \{\tilde{x} \mid \tilde{x} \in X \wedge \tilde{x} \in Y\}$ where $\tilde{x}$ is a variable of type T.

Furthermore, each local language, $\mathcal{L}$, gives rise to an associated topos, $\mathcal{C}(\mathcal{L})$, whose objects are equivalence classes of $\mathcal{L}$-sets, where $X \equiv Y$ is defined to mean that the equation $X = Y$ (i.e., a term of type Ω with no free variables) can be proved using the sequent calculus of the language with its axioms. From this perspective, a representation of the system-language $\mathcal{L}(S)$ in a topos τ is equivalent to a *functor* from the topos $\mathcal{C}(\mathcal{L}(S))$ to τ.

13.4.2.2 Theory Construction as a Translation of Languages

Conversely, for each topos τ there is a local language, $\mathcal{L}(\tau)$, whose ground type symbols are the objects of τ, and whose function symbols are the arrows in τ. It then follows that a representation of a local language, $\mathcal{L}$, in τ is equivalent to a 'translation' of $\mathcal{L}$ in $\mathcal{L}(\tau)$.

---

[43] One of the basic properties of a topos is that there is a one-to-one correspondence between arrows $f : A \times B \to \Omega$ and arrows $\ulcorner f \urcorner : B \to PA := \Omega^A$. In general, $\ulcorner f \urcorner$ is called the *power transpose* of f. If $B \simeq 1$ then $\ulcorner f \urcorner$ is known as the *name* of the arrow $f : A \to \Omega$.

Thus constructing a theory of physics is equivalent to finding a suitable *translation* of the system language, $\mathcal{L}(S)$, to the language, $\mathcal{L}(\tau)$, of an appropriate topos τ.

As we will see later, the idea of translating one local language into another plays a central role in the discussion of composite systems and sub-systems.

In the case of spoken languages, one can translate from, say, (i) English to German; or (ii) from English to Greek, and then from Greek to German. However, no matter how good the translators, these two ways of going from English to German will generally not agree. This is partly because the translation process is not unique, but also because each language possesses certain intrinsic features that simply do not admit of translation.

There is an interesting analogous question for the representation of the local languages $\mathcal{L}(S)$. Namely, suppose $\phi_1 : \mathcal{L}(S) \to \mathcal{L}(\tau_{\phi_1})$ and $\phi_2 : \mathcal{L}(S) \to \mathcal{L}(\tau_{\phi_2})$ are two different topos theories of the same system S (these could be, say, classical physics and quantum physics). The question is if/when will there be a translation $\phi_{12} : \mathcal{L}(\tau_{\phi_1}) \to \mathcal{L}(\tau_{\phi_2})$ such that

$$\phi_2 = \phi_{12} \circ \phi_1 \tag{13.24}$$

In terms of the representation functors from the topos $\mathcal{C}(\mathcal{L}(S))$ to the topoi τ_{ϕ_1} and τ_{ϕ_2}, the question is if there exists an interpolating functor from τ_{ϕ_1} to τ_{ϕ_2}.

In Sect. 13.12.2, we will introduce a certain category, $\mathcal{M}(\mathbf{Sys})$, whose objects are topoi and whose arrows are geometric morphisms between topoi. It would be natural to require the arrow from τ_{ϕ_1} to τ_{ϕ_2} (if it exists) to be an arrow in this category.

It is at this point that "geometric logic" enters the scene (cf. Sect. 13.3.2). A formula in $\mathcal{L}(S)$ is said to be *positive* if it does not contain the symbols[44] $\Rightarrow$ or $\forall$. These conditions imply that $\neg$ is also absent. In fact, a positive formula uses only $\exists$, $\wedge$ and $\vee$. A disjunction can have an arbitrary index set, but a conjunction can have only a finite index set. A sentence of the form $\forall x(\alpha \Rightarrow \beta)$ is said to be a *geometric implication* if both α and β are positive. Then a *geometric logic* is one in which only geometric implications are present in the language.

The advantage of using just the geometric part of logic is that geometric implications are *preserved* under geometric morphisms. This makes it appropriate to ask for the existence of 'geometric translations' $\phi_{12} : \mathcal{L}(\tau_{\phi_1}) \to \mathcal{L}(\tau_{\phi_2})$, as in (13.24), since these will preserve the logical structure of the language $\mathcal{L}(S)$.

[44] Here, the formula $\alpha \Rightarrow \beta$ is defined as $\alpha \Rightarrow \beta := (\alpha \wedge \beta) = \alpha$; $\forall$ is defined as $\forall x \alpha := (\{x \mid \alpha\} = \{x \mid \text{true}\})$; where true $:= * = *$.

The notion of "toinvariance" introduced recently by Landsman [60] can be interpreted within our structures as asserting that the translations $\phi_{12} : \mathcal{L}(\tau_{\phi_1}) \to \mathcal{L}(\tau_{\phi_2})$ should always exist; or, at least, they should under appropriate conditions. Of course, the significance of this depends on how much information about the system is reflected in the language $\mathcal{L}(S)$ and how much in the individual representations.

For example, in the case of classical and quantum physics, one might go so far as to include information about the dynamics of the system within the local language $\mathcal{L}(S)$. If the topoi ϕ_1 and ϕ_2 are those for the classical and quantum physics of S respectively (so that ϕ_1 is **Sets** and ϕ_2 is $\mathbf{Sets}^{\mathcal{V}(\mathcal{H})^{\mathrm{op}}}$), then an interpolating translation $\phi_{12} : \mathcal{L}(\mathbf{Sets}) \to \mathcal{L}(\mathbf{Sets}^{\mathcal{V}(\mathcal{H})^{\mathrm{op}}})$ would be a nice realisation of Landsman's long-term goal of regarding quantisation as some type of functorial operation.

Of course, introducing dynamics raises interesting questions about the status of the concept of "time" (cf. the discussion in Sect. 13.3.2.4). In particular, is time to be identified as an object in representing topos, or is it an external parameter, like the "Δ" quantities in the propositional languages $\mathcal{PL}(S)$?

13.4.3 Classical Physics in the Local Language $\mathcal{L}(S)$

The quantum theory representation of $\mathcal{L}(S)$ is studied in Sect. 13.5. Here we will look at the concrete form of the expressions above for the example of classical physics. In this case, for all systems S, and all classical representations, σ, the topos τ_σ is **Sets**. This representation of $\mathcal{L}(S)$ has the following ingredients:

1. (a) The ground type symbol Σ is represented by a symplectic manifold, Σ_σ, that is the state space for the system S.
   (b) The ground type symbol $\mathcal{R}$ is represented by the real line, i.e., $\mathcal{R}_\sigma := \mathbb{R}$.
   (c) The type symbol $P\Sigma$ is represented by the set, $P\Sigma_\sigma$, of all[45] subsets of the state space Σ_σ.
   The type symbol $P\mathcal{R}$ is represented by the set, $P\mathbb{R}$, of all subsets of $\mathbb{R}$.
2. (a) The type symbol Ω, is represented by $\Omega_{\mathbf{Sets}} := \{0, 1\}$: the sub-object classifier in **Sets**.
   (b) The type symbol 1, is represented by the singleton set: i.e., $1_{\mathbf{Sets}} = \{*\}$, the terminal object in **Sets**.
3. Each function symbol $A : \Sigma \to \mathcal{R}$, and hence each physical quantity, is represented by a real-valued (measurable) function, $A_\sigma : \Sigma_\sigma \to \mathbb{R}$, on the state space Σ_σ.
4. The term "$A(\tilde{s}) \in \tilde{\Delta}$" of type Ω (where $\tilde{s}$ and $\tilde{\Delta}$ are free variables of type Σ and $P\mathcal{R}$ respectively) is represented by the function $[\![A(\tilde{s}) \in \tilde{\Delta}]\!]_\sigma : \Sigma_\sigma \times P\mathbb{R} \to \{0, 1\}$ that is defined by (cf. (13.22))

[45] To be precise, we really need to use the collection $P_B\Sigma_\sigma$ of all *Borel* subsets of Σ_σ. Likewise for the subsets of $\mathbb{R}$.

$$[\![A(\tilde{s}) \in \tilde{\Delta}]\!]_\sigma(s, \Delta) = \begin{cases} 1 & \text{if } A_\sigma(s) \in \Delta; \\ 0 & \text{otherwise} \end{cases} \tag{13.25}$$

for all $(s, \Delta) \in \Sigma_\sigma \times P\mathbb{R}$.

5. The term $\{\tilde{s} \mid A(\tilde{s}) \in \tilde{\Delta}\}$ of type $P\Sigma$ (where $\tilde{\Delta}$ is a free variable of type $P\mathcal{R}$) is represented by the function $[\![\{\tilde{s} \mid A(\tilde{s}) \in \tilde{\Delta}\}]\!]_\sigma : P\mathbb{R} \to P\Sigma_\sigma$ that is defined by

$$\begin{aligned} [\![\{\tilde{s} \mid A(\tilde{s}) \in \tilde{\Delta}\}]\!]_\sigma(\Delta) &:= \{s \in \Sigma_\phi \mid A_\sigma(s) \in \Delta\} \\ &= A_\sigma^{-1}(\Delta) \end{aligned} \tag{13.26}$$

for all $\Delta \in P\mathbb{R}$.

13.4.4 Adapting the Language $\mathcal{L}(S)$ to Other Types of Physical System

Our central contention in this work is that (i) each physical system, S, can be equipped with a local language, $\mathcal{L}(S)$; and (ii) constructing an explicit theory of S in a particular theory-type is equivalent to finding a representation of $\mathcal{L}(S)$ in a topos which may well be other than the topos of sets.

There are many situations in which the language is independent of the theory-type, and then, for a given system S, the different topos representations of $\mathcal{L}(S)$ correspond to the application of the different theory-types to the same system S. We gave an example earlier of a point particle moving in three dimensions: the classical physics representation is in the topos **Sets**, but the quantum-theory representation is in the presheaf topos $\mathbf{Sets}^{\mathcal{V}(L^2(\mathbb{R}^3, d^3x))^{op}}$.

However, there are other situations where the relationship between the language and its representations is more complicated than this. In particular, there is the critical question about what features of the theory should go into the language, and what into the representation. The first step in adding new features is to augment the set of ground type symbols. This is because these represent the entities that are going to be of generic interest (such as a state object or quantity-value object). In doing this, extra axioms may also be introduced to encode the properties that the new objects are expected to possess in all representations of physical interest.

For example, suppose we want to use our formalism to discuss space-time physics: where does the information about the space-time go? If the subject is classical field theory in a curved space-time, then the topos τ is **Sets**, and the space-time manifold is part of the *background* structure. This makes it natural to have the manifold assumed in the representation; i.e., the information about the space-time is in the representation.

Alternatively, one can add a new ground type symbol, "M", to the language, to serve as the linguistic progenitor of "space-time"; thus M would have the same theoretical status as the symbols Σ and $\mathcal{R}$. In this context, we recall the brief discussion in Sect. 13.2.2 about the use of the real numbers in modelling space and/or time, and the motivation this provides for representing space-time as an object in a topos, and whose sub-objects represent the fundamental "regions".

If "M" is added to the language, a function symbol $\psi : M \to \mathcal{R}$ is then the progenitor of a physical field. In a representation, ϕ, the object M_ϕ plays the role of 'space-time' in the topos τ_ϕ, and $\psi_\phi : M_\phi \to \mathcal{R}_\phi$ is the representation of the field.

Of course, the language $\mathcal{L}(S)$ says nothing about what sort of entity M_ϕ is, except in so far as such information is encoded in extra axioms. For example, if the subject is classical field theory, then $\tau_\phi =$ **Sets**, and M_ϕ would be a standard differentiable manifold. On the other hand, if the topos τ_ϕ admits "infinitesimals", then M_ϕ could be a manifold according to the language of synthetic differential geometry [58].

The same type of argument applies to the status of "time" in a canonical theory. In particular, it would be possible to add a ground type symbol, $\mathcal{T}$, so that, in any representation, ϕ, the object $\mathcal{T}_\phi$ in the topos τ_ϕ is the analogue of the "time-line" for that theory. For standard physics in **Sets** we have $\mathcal{T}_\phi = \mathbb{R}$, but the form of $\mathcal{T}_\phi$ in a more general topos, τ_ϕ, would be a rich subject for speculation.

The addition of a "time-type" symbol, $\mathcal{T}$, to the language $\mathcal{L}(S)$ is a prime example of a situation where one might want to add extra axioms. These could involve ordering properties, or algebraic properties like those of an abelian group, and so on. In any topos representation, these properties would then be realised as the corresponding type of object in τ_ϕ. Thus abelian group axioms mean that $\mathcal{T}_\phi$ is an abelian-group object in the topos τ_ϕ; total-ordering axioms for the time-type $\mathcal{T}$ mean that $\mathcal{T}_\phi$ is a totally-ordered object in τ_ϕ, and so on.

As an interesting extension of this idea, one could have a space-time ground type symbol M, but then add the axioms for a partial ordering. In that case, M_ϕ would be a poset-object in τ_ϕ, which could be interpreted physically as the τ_ϕ-analogue of a causal set [32].

13.5 Quantum Propositions as Sub-objects of the Spectral Presheaf

13.5.1 Some Background Remarks

13.5.1.1 The Kochen-Specker Theorem

The idea of representing quantum theory in a topos of presheaves stemmed originally [48] from a desire to acquire a new perspective on the Kochen-Specker theorem [57]. It will be helpful at this stage to review some of this older material.

A commonsense belief, and one apparently shared by Heidegger, is that at any given time any physical quantity *must* have a value even if we do not know what it is. In classical physics, this is not problematic since the underlying mathematical structure is geared precisely to realise it. Specifically, if $\mathcal{S}$ is the state space of some classical system, and if the physical quantity A is represented by a real-valued function $\breve{A} : \mathcal{S} \to \mathbb{R}$, then the value $V^s(A)$ of A in any state $s \in \mathcal{S}$ is simply

$$V^s(A) = \breve{A}(s). \tag{13.27}$$

Thus all physical quantities possess a value in any state. Furthermore, if $h : \mathbb{R} \to \mathbb{R}$ is a real-valued function, a new physical quantity $h(A)$ can be defined by requiring the associated function $\breve{h(A)}$ to be

$$\breve{h(A)}(s) := h(\breve{A}(s)) \tag{13.28}$$

for all $s \in \mathcal{S}$; i.e., $\breve{h(A)} := h \circ \breve{A} : \mathcal{S} \to \mathbb{R}$. Thus the physical quantity $h(A)$ is *defined* by saying that its value in any state s is the result of applying the function h to the value of A; hence, by definition, the values of the physical quantities $h(A)$ and A satisfy the "functional composition principle"

$$V^s(h(A)) = h(V^s(A)) \tag{13.29}$$

for all states $s \in \mathcal{S}$.

However, standard quantum theory precludes any such naïve realist interpretation of the relation between formalism and physical world. And this obstruction comes from the mathematical formalism itself, in the guise of the famous Kochen-Specker theorem which asserts the impossibility of assigning values to all physical quantities whilst, at the same time, preserving the functional relations between them [57].

In a quantum theory, a physical quantity A is represented by a self-adjoint operator $\hat{A}$ on the Hilbert space of the system, and the first thing one has to decide is whether to regard a valuation as a function of the physical quantities themselves, or on the operators that represent them. From a mathematical perspective, the latter strategy is preferable, and we shall therefore define a valuation to be a real-valued function V on the set of all bounded, self-adjoint operators, with the properties that: (i) the value $V(\hat{A})$ of the physical quantity A represented by the operator $\hat{A}$ belongs to the spectrum of $\hat{A}$ (the so-called "value rule"); and (ii) the functional composition principle (or FUNC for short) holds:

$$V(\hat{B}) = h(V(\hat{A})) \tag{13.30}$$

for any pair of self-adjoint operators $\hat{A}$, $\hat{B}$ such that $\hat{B} = h(\hat{A})$ for some real-valued function h. If they existed, such valuations could be used to embed the set of self-adjoint operators in the commutative ring of real-valued functions on an underlying space of microstates, thereby laying the foundations for a hidden-variable interpretation of quantum theory.

Several important results follow from the definition of a valuation. For example, if $\hat{A}_1$ and $\hat{A}_2$ commute, it follows from the spectral theorem that there exists an operator $\hat{C}$ and functions h_1 and h_2 such that $\hat{A}_1 = h_1(\hat{C})$ and $\hat{A}_2 = h_2(\hat{C})$. It then follows from FUNC that

$$V(\hat{A}_1 + \hat{A}_2) = V(\hat{A}_1) + V(\hat{A}_2) \tag{13.31}$$

and

$$V(\hat{A}_1\hat{A}_2) = V(\hat{A}_1)V(\hat{A}_2). \tag{13.32}$$

The defining Eq. (13.30) for a valuation makes sense whatever the nature of the spectrum $sp(\hat{A})$ of the operator $\hat{A}$. However, if $sp(\hat{A})$ contains a continuous part, one might doubt the physical meaning of assigning one of its elements as a value. To handle the more general case, we shall view a valuation as primarily giving *truth values* to *propositions* about the values of a physical quantity, rather than assigning a specific value to the quantity itself.

As in Sect. 13.3, the propositions concerned are of the type "$A \,\varepsilon\, \Delta$", which (in a realist reading) asserts that the value of the physical quantity A lies in the (Borel) subset Δ of the spectrum $sp(\hat{A})$ of the associated operator $\hat{A}$. This proposition is represented by the spectral projector $\hat{E}[A \in \Delta]$, which motivates studying the general mathematical problem of assigning truth values to projection operators.

If $\hat{P}$ is a projection operator, the identity $\hat{P} = \hat{P}^2$ implies that $V(\hat{P}) = V(\hat{P}^2) = (V(\hat{P}))^2$ (from (13.32)); and hence, necessarily, $V(\hat{P}) = 0$ or 1. Thus V defines a homomorphism from the Boolean algebra $\{\hat{0}, \hat{1}, \hat{P}, \neg\hat{P} \equiv (\hat{1} - \hat{P})\}$ to the "false(0)-true(1)" Boolean algebra $\{0, 1\}$. More generally, a valuation V induces a homomorphism $\chi^V : W \to \{0, 1\}$ where W is any Boolean sub-algebra of the lattice $\mathcal{P}(\mathcal{H})$ of projectors on $\mathcal{H}$. In particular,

$$\hat{\alpha} \preceq \hat{\beta} \text{ implies } \chi^V(\hat{\alpha}) \leq \chi^V(\hat{\beta}) \tag{13.33}$$

where "$\hat{\alpha} \preceq \hat{\beta}$" refers to the partial ordering in the lattice $\mathcal{P}(\mathcal{H})$, and "$\chi^V(\hat{\alpha}) \leq \chi^V(\hat{\beta})$" is the ordering in the Boolean algebra $\{0, 1\}$.

The Kochen-Specker theorem asserts that no global valuations exist if the dimension of the Hilbert space $\mathcal{H}$ is greater than two. The obstructions to the existence of such valuations typically arise when trying to assign a single value to an operator $\hat{C}$ that can be written as $\hat{C} = g(\hat{A})$ and as $\hat{C} = h(\hat{B})$ with $[\hat{A}, \hat{B}] \neq 0$.

The various interpretations of quantum theory that aspire to use "beables", rather than "observables", are all concerned in one way or another with addressing this issue. Inherent in such schemes is a type of "contextuality" in which a value given to a physical quantity C cannot be part of a global assignment of values but must, instead, depend on some *context* in which C is to be considered. In practice, contextuality is endemic in any attempt to ascribe properties to quantities in a quantum theory. For example, as emphasized by Bell [12], in the situation where $\hat{C} = g(\hat{A}) = h(\hat{B})$, if the value of C is construed counterfactually as referring to what would be obtained *if* a measurement of A or of B is made—and with the value of C then being *defined* by applying to the result of the measurement the relation $C = g(A)$, or $C = h(B)$—then one can claim that the actual value obtained depends on whether the value of C is determined by measuring A, or by measuring B.

In the programme to be discussed here, the idea of a contextual valuation will be developed in a different direction from that of the existing modal interpretations in which "reality" is ascribed to only some commutative subset of physical quantities. In particular, rather than accepting such a limited domain of beables we shall propose a theory of "generalised" valuations that are defined globally on *all* propositions about values of physical quantities. However, the price of global existence is that any given proposition may have only a *generalised* truth value. More precisely, (i) the truth value of a proposition "$A \varepsilon \Delta$" belongs to a logical structure that is larger than $\{0, 1\}$; and (ii) these target-logics, and truth values, are context dependent.

It is clear that the main task is to formulate mathematically the idea of a contextual, truth value in such a way that the assignment of generalised truth values is consistent with an appropriate analogue of the functional composition principle, FUNC.

13.5.1.2 The Introduction of Coarse-Graining

In the original paper [48], this task is tackled using a type of "coarse-graining" operation. The key idea is that, although in a given situation in quantum theory it may not be possible to declare a particular proposition "$A \varepsilon \Delta$" to be true (or false), nevertheless there may be (Borel) functions f such that the associated propositions "$f(A) \varepsilon f(\Delta)$" *can* be said to be true. This possibility arises for the following reason.

Let W_A denote the spectral algebra of the operator $\hat{A}$ that represents a physical quantity A. Thus W_A is the Boolean algebra of projectors $\hat{E}[A \in \Delta]$ that project onto the eigenspaces associated with the Borel subsets Δ of the spectrum $\mathrm{sp}(\hat{A})$ of $\hat{A}$; physically speaking, $\hat{E}[A \varepsilon \Delta]$ represents the proposition "$A \varepsilon \Delta$". It follows from the spectral theorem that, for all Borel subsets J of the spectrum of $f(\hat{A})$, the spectral projector $\hat{E}[f(A) \varepsilon J]$ for the operator $f(\hat{A})$ is equal to the spectral projector $\hat{E}[A \varepsilon f^{-1}(J)]$ for $\hat{A}$. In particular, if $f(\Delta)$ is a Borel subset of $\mathrm{sp}(f(\hat{A}))$ then, since $\Delta \subseteq f^{-1}(f(\Delta))$, we have $\hat{E}[A \varepsilon \Delta] \preceq \hat{E}[A \varepsilon f^{-1}(f(\Delta))]$; and hence

$$\hat{E}[A \varepsilon \Delta] \preceq \hat{E}[f(A) \varepsilon f(\Delta)]. \tag{13.34}$$

Physically, the inequality in (13.34) reflects that the proposition "$f(A) \varepsilon f(\Delta)$" is generally weaker than the proposition "$A \varepsilon \Delta$" in the sense that the latter implies the former, but not necessarily vice versa. For example, the proposition "$f(A) = f(a)$" is weaker than the original proposition "$A = a$" if the function f is many-to-one and such that more than one eigenvalue of $\hat{A}$ is mapped to the same eigenvalue of $f(\hat{A})$. In general, we shall say that "$f(A) \varepsilon f(\Delta)$" is a *coarse-graining* of "$A \varepsilon \Delta$".

Now, if the proposition "$A \varepsilon \Delta$" is evaluated as "true" then, from (13.33) and (13.34), it follows that the weaker proposition "$f(A) \varepsilon f(\Delta)$" is also evaluated as 'true'.

This remark provokes the following observation. There may be situations in which, although the proposition "$A \varepsilon \Delta$" cannot be said to be either true or false, the weaker proposition "$f(A) \varepsilon f(\Delta)$" can. In particular, if the latter *can* be given the value "true", then—by virtue of the remark above—it is natural to suppose that any further coarse-graining to give an operator $g(f(\hat{A}))$ will yield a proposition "$g(f(A)) \in g(f(\Delta))$" that will also be evaluated as "true". Note that there may be more than one possible choice for the 'initial' function f, each of which can then be further coarse-grained in this way. This multi-branched picture of coarse-graining is one of the main justifications for our invocation of the topos-theoretic idea of a presheaf.

It transpires that the key remark above is the statement:

If "$f(A) \varepsilon f(\Delta)$" is true, then so is "$g(f(A) \varepsilon g(f(\Delta)$" for any function $g : \mathbb{R} \to \mathbb{R}$.

This is key because the property thus asserted can be restated by saying that the collection of all functions f such that "$f(A) \varepsilon f(\Delta)$" is "true" is a *sieve*[46]; and sieves are closely associated with global elements of the sub-object classifier in a category of presheaves, which is the presheaf of sieves (on objects of the base category).

To clarify this we start by defining a category $\mathcal{O}$ whose objects are the bounded, self-adjoint operators on $\mathcal{H}$. For the sake of simplicity, we will assume for the moment that $\mathcal{O}$ consists only of the operators whose spectrum is discrete. Then we say that there is a "morphism" from $\hat{B}$ to $\hat{A}$ if there exists a Borel function (more precisely, an equivalence class of Borel functions) $f : \mathrm{sp}(\hat{A}) \to \mathbb{R}$ such that $\hat{B} = f(\hat{A})$, where $\mathrm{sp}(\hat{A})$ is the spectrum of $\hat{A}$. Any such function on $\mathrm{sp}(\hat{A})$ is unique (up to the equivalence relation), and hence there is at most one morphism between any two operators. If $\hat{B} = f(\hat{A})$, the corresponding morphism in the category $\mathcal{O}$ will be denoted $f_{\mathcal{O}} : \hat{B} \to \hat{A}$. It then becomes clear that the statement in the box above is equivalent to the statement that the collection of all functions f such that "$f(A) \varepsilon f(\Delta)$" is "true", is a sieve[47] on the object $\hat{A}$ in the category $\mathcal{O}$.

This motivates very strongly looking at the topos category, $\mathbf{Sets}^{\mathcal{O}^{\mathrm{op}}}$ of contravariant,[48] set-valued functors on $\mathcal{O}$. Then, bearing in mind our discussion of values of physical quantities, it is rather natural to construct the following object in this topos:

Definition 1 The *spectral presheaf* on $\mathcal{O}$ is the contravariant functor $\underline{\Sigma} : \mathcal{O} \to \mathbf{Sets}$ defined as follows:

1. On objects: $\underline{\Sigma}(\hat{A}) := \mathrm{sp}(\hat{A})$.
2. On morphisms: If $f_{\mathcal{O}} : \hat{B} \to \hat{A}$, so that $\hat{B} = f(\hat{A})$, then $\underline{\Sigma}(f_{\mathcal{O}}) : \mathrm{sp}(\hat{A}) \to \mathrm{sp}(\hat{B})$ is defined by $\underline{\Sigma}(f_{\mathcal{O}})(\lambda) := f(\lambda)$ for all $\lambda \in \mathrm{sp}(\hat{A})$.

---

[46] A *sieve on an object* C in a category $\mathcal{C}$ is a collection of arrows in $\mathcal{C}$ with codomain C such that the following condition holds: if $f : B \to C$ is in the sieve and $g : A \to B$ is any other arrow in $\mathcal{C}$, then the composite arrow $f \circ g : A \to C$ is also contained in the sieve.

[47] It is a matter of convention whether this is called a sieve or a co-sieve.

[48] Ab initio, we could just as well have looked at covariant functors, but with our definitions the contravariant ones are more natural.

Note that $\underline{\Sigma}(f_{\mathcal{O}})$ is well-defined since, if $\lambda \in \sigma(\hat{A})$, then $f(\lambda)$ is indeed an element of the spectrum of $\hat{B}$; indeed, for these discrete-spectrum operators we have $\mathrm{sp}(f(\hat{A})) = f(\mathrm{sp}(\hat{A}))$.

The key remark now is the following. If $\mathcal{C}$ is any category, a *global element*, of a contravariant functor $\underline{X} : \mathcal{C} \to \mathbf{Sets}$ is defined to be a function γ that assigns to each object A in the category $\mathcal{C}$ an element $\gamma_A \in \underline{X}(A)$ in such a way that if $f : B \to A$ then $\underline{X}(f)(\gamma_A) = \gamma_B$ (see Appendix 2 for more details).

In the case of the spectral functor $\underline{\Sigma}$, a global element is therefore a function γ that assigns to each (bounded, discrete spectrum) self-adjoint operator $\hat{A}$, a real number $\gamma_A \in \mathrm{sp}(\hat{A})$ such that if $\hat{B} = f(\hat{A})$ then $f(\gamma_A) = \gamma_B$. But this is precisely the condition FUNC in Eq. (13.30) for a valuation!

> Thus, the Kochen-Specker theorem is equivalent to the statement that, if $\dim \mathcal{H} > 2$, the spectral presheaf $\underline{\Sigma}$ has no global elements.

It was this observation that motivated the original suggestion by one of us (CJI) and his collaborators that quantum theory should be studied from the perspective of topos theory. However, as it stands, the discussion above works only for operators with a discrete spectrum. This is fine for finite-dimensional Hilbert spaces, but in an infinite-dimensional space operators can have continuous parts in their spectra, and then things get more complicated.

One powerful way of tackling this problem is to replace the category of operators with a category, $\mathcal{V}(\mathcal{H})$, whose objects are commutative von Neumann sub-algebras of the algebra $B(\mathcal{H})$ of all bounded operators on $\mathcal{H}$. There is a close link with the category $\mathcal{C}$ since each self-adjoint operator generates a commutative von Neumann algebra, but using $\mathcal{V}(\mathcal{H})$ rather than $\mathcal{C}$ solves all the problems associated with continuous spectra [50].

Of course, this particular motivation for introducing $\mathcal{V}(\mathcal{H})$ is purely mathematical, but there are also very good physics reasons for this step. As we have mentioned earlier, one approach to handling the implications of the Kochen-Specker theorem is to "reify" only a subset of physical variables, as is done in the various "modal interpretations". The topos-theoretic extension of this idea of "partial reification", first proposed in [48–51], is to build a structure in which all possible reifiable sets of physical variables are included on an equal footing. This involves constructing a category, $\mathcal{C}$, whose objects are collections of quantum observables that *can* be simultaneously reified because the corresponding self-adjoint operators *commute*. The application of this type of topos scheme to an actual modal interpretation is discussed in the recent paper by Nakayama [72].

From a physical perspective, the objects in the category $\mathcal{C}$ can be viewed as *contexts* (or "world-views", or "windows on reality", or "classical snapshots") from whose perspectives the quantum theory can be displayed. This is the physical motivation for using commutative von Neumann algebras.

In the normal, instrumentalist interpretation of quantum theory, a context is therefore a collection of physical variables that can be measured simultaneously.

The physical significance of this contextual logic is discussed at length in [48–51, 53] and [28, 29].

13.5.1.3 Alternatives to von Neumann Algebras

It should be remarked that $\mathcal{V}(\mathcal{H})$ is not the only possible choice for the category of concepts. Another possibility is to construct a category whose objects are the Boolean sub-algebras of the non-distributive lattice of projection operators on the Hilbert space; more generally we could consider the Boolean sub-algebras of *any* non-distributive lattice. This option was discussed in [48].

Yet another possibility is to consider the abelian C^*-subalgebras of the algebra $B(\mathcal{H})$ of all bounded operators on $\mathcal{H}$. More generally, one could consider the abelian sub-algebras of *any* C^*-algebra; this is the option adopted by Heunen et al. [42] in their interesting recent development of our scheme. One disadvantage of a C^*-algebra is that in general it does not contain enough projectors, and if one wants to include them, it is necessary to move to AW^*-algebras, which are the abstract analogue of the concrete von Neumann algebras that we employ. For each of these choices there is a corresponding spectral object, and these different spectral objects are closely related.

It is clear that a similar procedure could be followed for *any* algebraic quantity $\mathfrak{A}$ that has an "interesting" collection of commutative sub-algebras. We will return to this remark in Sect. 13.14.1.

13.5.2 From Projections to Global Elements of the Outer Presheaf

13.5.2.1 The Definition of $\delta(\hat{P})_V$

The fundamental thesis of our work is that in constructing theories of physics one should seek representations of a formal language in a topos that may be other than **Sets**. We want now to study this idea closely in the context of the "toposification" of standard quantum theory, with particular emphasis on a topos representation of propositions. Most "standard" quantum systems (for example, one-dimensional motion with a Hamiltonian $H = \frac{p^2}{2m} + V(x)$) are obtained by "quantising" a classical system, and consequently the formal language is the same as it is for the classical system. Our immediate goal is to represent physical propositions with sub-objects of the spectral presheaf $\underline{\Sigma}$.

In this Section we concentrate on the propositional language $\mathcal{PL}(S)$ introduced in Sect. 13.3.2. Thus a key task is to find the map $\pi_{\text{qt}} : \mathcal{PL}(S)_0 \to \text{Sub}(\underline{\Sigma})$, where the primitive propositions in $\mathcal{PL}(S)_0$ are of the form "$A \,\varepsilon\, \Delta$". As we shall see, this is where the critical concept of *daseinisation* arises: the procedure whereby a projector $\hat{P}$ is transformed to a sub-object, $\underline{\delta(\hat{P})}$, of the spectral presheaf, $\underline{\Sigma}$, in the topos $\mathbf{Sets}^{\mathcal{V}(\mathcal{H})^{\text{op}}}$ (the precise definition of $\underline{\Sigma}$ is given in Sect. 13.5.3).

In standard quantum theory, a physical quantity is represented by a self-adjoint operator $\hat{A}$ in the algebra, $B(\mathcal{H})$, of all bounded operators on $\mathcal{H}$. If $\Delta \subseteq \mathbb{R}$ is a

Borel subset, we know from the spectral theorem that the proposition "$A \varepsilon \Delta$" is represented by[49] the projection operator $\hat{E}[A \in \Delta]$ in $B(\mathcal{H})$. For typographical simplicity, for the rest of this Section, $\hat{E}[A \in \Delta]$ will be denoted by $\hat{P}$.

We are going to consider the projection operator $\hat{P}$ from the perspective of the "category of contexts"—a keystone of the topos approach to quantum theory. As we have remarked earlier, there are several possible choices for this category most of which are considered in detail in the original papers [48–51]. Here we have elected to use the category $\mathcal{V}(\mathcal{H})$ of unital, abelian sub-algebras of $B(\mathcal{H})$. This partially-ordered set has a category structure in which (i) the objects are the abelian sub-algebras of $B(\mathcal{H})$; and (ii) there is an arrow $i_{V'V} : V' \to V$, where $V', V \in \mathrm{Ob}(\mathcal{V}(\mathcal{H}))$,[50] if and only if $V' \subseteq V$. By definition, the trivial sub-algebra $V_0 = \mathbb{C}\hat{1}$ is not included in the objects of $\mathcal{V}(\mathcal{H})$. A context could also be called a "world-view", a "classical snap-shot", a "window on reality", or even a Weltanschauung[51]; mathematicians often refer to it as a "stage of truth".

The critical question is what can be said about the projector $\hat{P}$ "from the perspective" of a particular context $V \in \mathrm{Ob}(\mathcal{V}(\mathcal{H}))$? If $\hat{P}$ belongs to V then a "full" image of $\hat{P}$ is obtained from this view-point, and there is nothing more to say. However, suppose the abelian sub-algebra V does *not* contain $\hat{P}$: what then?

We need to "approximate" $\hat{P}$ from the perspective of V, and an important ingredient in our work is to define this as meaning the "smallest" projection operator, $\delta(\hat{P})_V$, in V that is greater than, or equal to, $\hat{P}$:

$$\delta(\hat{P})_V := \bigwedge \{\hat{\alpha} \in \mathcal{P}(V) \mid \hat{\alpha} \succeq \hat{P}\}. \tag{13.35}$$

where "$\succeq$" is the usual ordering of projection operators, and where $\mathcal{P}(V)$ denotes the set of all projection operators in V.[52]

To see what this means, let $\hat{P}$ and $\hat{Q}$ represent the propositions "$A \varepsilon \Delta$" and "$A \varepsilon \Delta'$" respectively with $\Delta \subseteq \Delta'$, so that $\hat{P} \preceq \hat{Q}$. Since we learn less about the value of A from the proposition "$A \varepsilon \Delta'$" than from "$A \varepsilon \Delta$", the former proposition is said to be *weaker*. Clearly, the weaker proposition "$A \varepsilon \Delta'$" is implied by the stronger proposition "$A \varepsilon \Delta$". The construction of $\delta(\hat{P})_V$ as the smallest projection in V greater than or equal to $\hat{P}$ thus gives the strongest proposition expressible in V that is implied by $\hat{P}$ (although, if $\hat{A} \notin V$, the projection $\delta(\hat{P})_V$ cannot

---

[49] Note, however, that the map from propositions to projections is not injective: two propositions "$A \varepsilon \Delta_1$" and "$B \varepsilon \Delta_2$" concerning two distinct physical quantities, A and B, can be represented by the same projector: i.e., $\hat{E}[A \in \Delta_1] = \hat{E}[B \in \Delta_2]$.

[50] We denote by $\mathrm{Ob}(\mathcal{C})$ the collection of all objects in the category $\mathcal{C}$.

[51] "Weltanschauung" is a splendid German word. "Welt" means world; "schauen" is a verb and means to look, to view; "anschauen" is to look at; and "-ung" at the end of a word can make a noun from a verb. So it's Welt-an-schau-ung.

[52] We will later call the mapping $\underline{\delta}_V : \mathcal{P}(\mathcal{H}) \to \mathcal{P}(V)$ the *outer* daseinisation of projections to V and often denote it as $\underline{\delta}^o_V$; compare formula (13.83) below, where inner daseinisation is introduced.

usually be interpreted as a proposition *about* A).[53] Note that if $\hat{P}$ belongs to V, then $\delta(\hat{P})_V = \hat{P}$. The mapping $\hat{P} \mapsto \delta(\hat{P})_V$ was originally introduced by de Groote in [37], who called it the 'V-support' of $\hat{P}$.

The key idea in this part of our scheme is that rather than thinking of a quantum proposition, "$A \varepsilon \Delta$", as being represented by the *single* projection operator $\hat{E}[A \in \Delta]$, instead we consider the entire *collection* $\{\delta(\hat{E}[A \in \Delta])_V \mid V \in \mathrm{Ob}(\mathcal{V}(\mathcal{H}))\}$ of projection operators, one for each context V. As we will see, the link with topos theory is that this collection of projectors is a global element of a certain presheaf.

This "certain" presheaf is in fact the "outer" presheaf, which is defined as follows:

Definition 2 The outer[54] presheaf $\underline{O}$ is defined over the category $\mathcal{V}(\mathcal{H})$ as follows [48, 50]:

(i) On objects $V \in \mathrm{Ob}(\mathcal{V}(\mathcal{H}))$: We have $\underline{O}_V := \mathcal{P}(V)$
(ii) On morphisms $i_{V'V} : V' \subseteq V$: The mapping $\underline{O}(i_{V'V}) : \underline{O}_V \to \underline{O}_{V'}$ is given by $\underline{O}(i_{V'V})(\hat{\alpha}) := \delta(\hat{\alpha})_{V'}$ for all $\hat{\alpha} \in \mathcal{P}(V)$.

With this definition, it is clear that, for each projection operator $\hat{P}$, the assignment $V \mapsto \delta(\hat{P})_V$ defines a global element, denoted $\delta(\hat{P})$, of the presheaf $\underline{O}$. Indeed, for each context V, we have the projector $\delta(\hat{P})_V \in \mathcal{P}(V) = \underline{O}_V$, and if $i_{V'V} : V' \subseteq V$, then

$$\delta\big(\delta(\hat{P})_V\big)_{V'} = \bigwedge \big\{\hat{Q} \in \mathcal{P}(V') \mid \hat{Q} \succeq \delta(\hat{P})_V\big\} = \delta(\hat{P})_{V'} \tag{13.36}$$

and so the elements $\delta(\hat{P})_V$, $V \in \mathrm{Ob}(\mathcal{V}(\mathcal{H}))$, are compatible with the structure of the outer presheaf. Thus we have a mapping

$$\begin{aligned} \delta : \mathcal{P}(\mathcal{H}) &\to \Gamma\underline{O} \\ \hat{P} &\mapsto \{\delta(\hat{P})_V \mid V \in \mathrm{Ob}(\mathcal{V}(\mathcal{H}))\} \end{aligned} \tag{13.37}$$

from the projectors in $\mathcal{P}(\mathcal{H})$ to the global elements, $\Gamma\underline{O}$, of the outer presheaf.[55]

---

[53] Note that the definition in (13.35) exploits the fact that the lattice $\mathcal{P}(V)$ of projection operators in V is complete. This is the main reason why we chose von Neumann sub-algebras rather than C^*-algebras: the former contain enough projections, and their projection lattices are complete.

[54] In the original papers by CJI and collaborators, this was called the "coarse-graining" presheaf, and was denoted $\underline{G}$. The reason for the change of nomenclature will become apparent later.

[55] Vis-a-vis our use of the language $\mathcal{L}(S)$ a little further on, we should emphasise that the outer presheaf has no linguistic precursor, and in this sense, it has no fundamental status in the theory. In fact, we could avoid the outer presheaf altogether and always work directly with the spectral presheaf, $\underline{\Sigma}$, which, of course, *does* have a linguistic precursor. However, it is technically convenient to introduce the outer presheaf as an intermediate tool.

13.5.2.2 Properties of the Mapping $\delta : \mathcal{P}(\mathcal{H}) \to \Gamma\underline{O}$

Let us now note some properties of the map $\delta : \mathcal{P}(\mathcal{H}) \to \Gamma\underline{O}$ that are relevant to our overall scheme.

1. For all contexts V, we have $\delta(\hat{0})_V = \hat{0}$.
   The null projector represents all propositions of the form "$A \varepsilon \Delta$" with the property that $\mathrm{sp}(\hat{A}) \cap \Delta = \emptyset$. These propositions are trivially false.
2. For all contexts V, we have $\delta(\hat{1})_V = \hat{1}$.
   The unit operator $\hat{1}$ represents all propositions of the form "$A \varepsilon \Delta$" with the property that $\mathrm{sp}(\hat{A}) \cap \Delta = \mathrm{sp}(\hat{A})$. These propositions are trivially true.
3. There exist global elements of $\underline{O}$ that are *not* of the form $\delta(\hat{P})$ for any projector $\hat{P}$. This phenomenon will be discussed later. However, if $\gamma \in \Gamma\underline{O}$ *is* of the form $\delta(\hat{P})$ for some $\hat{P}$, then

$$\hat{P} = \bigwedge_{V \in \mathrm{Ob}(\mathcal{V}(\mathcal{H}))} \delta(\hat{P})_V, \tag{13.38}$$

   because $\delta(\hat{P})_V \succeq \hat{P}$ for all $V \in \mathrm{Ob}(\mathcal{V}(\mathcal{H}))$, and $\delta(\hat{P})_V = \hat{P}$ for any V that contains $\hat{P}$.

The next result is important as it means that "nothing is lost" in mapping a projection operator $\hat{P}$ to its associated global element, $\delta(\hat{P})$, of the presheaf $\underline{O}$.

Theorem 1 *The map $\delta : \mathcal{P}(\mathcal{H}) \to \Gamma\underline{O}$ is injective.*

This simply follows from (13.38): if $\delta(\hat{P}) = \delta(\hat{Q})$ for two projections $\hat{P}$, $\hat{Q}$, then

$$\hat{P} = \bigwedge_{V \in \mathrm{Ob}(\mathcal{V}(\mathcal{H}))} \delta(\hat{P})_V = \bigwedge_{V \in \mathrm{Ob}(\mathcal{V}(\mathcal{H}))} \delta(\hat{Q})_V = \hat{Q}. \tag{13.39}$$

13.5.2.3 A Logical Structure for $\Gamma\underline{O}$?

We have seen that the quantities $\delta(\hat{P}) := \{\delta(\hat{P})_V \mid V \in \mathrm{Ob}(\mathcal{V}(\mathcal{H}))\}$, $\hat{P} \in \mathcal{P}(\mathcal{H})$, are elements of $\Gamma\underline{O}$, and if they are to represent quantum propositions, one might expect/hope that (i) these global elements of $\underline{O}$ form a Heyting algebra; and (ii) this algebra is related in some way to the Heyting algebra of sub-objects of $\underline{\Sigma}$. Let us see how far we can go in this direction.

Our first remark is that any two global elements γ_1, γ_2 of $\underline{O}$ can be compared at each stage V in the sense of logical implication. More precisely, let $\gamma_{1V} \in \mathcal{P}(V)$ denote the V'th 'component' of γ_1, and ditto for γ_{2V}. Then we have the following result:

Definition 3 A partial ordering on $\Gamma\underline{O}$ can be constructed in a "local" way (i.e., "local" with respect to the objects in the category $\mathcal{V}(\mathcal{H})$) by defining

$$\gamma_1 \succeq \gamma_2 \text{ if, and only if, } \forall V \in \text{Ob}(\mathcal{V}(\mathcal{H})),\ \gamma_{1V} \succeq \gamma_{2V} \tag{13.40}$$

where the ordering on the right hand side of (13.40) is the usual ordering in the lattice of projectors $\mathcal{P}(V)$.

It is trivial to check that (13.40) defines a partial ordering on $\Gamma\underline{O}$. Thus $\Gamma\underline{O}$ is a partially ordered set.

Note that if $\hat{P}$, $\hat{Q}$ are projection operators, then it follows from (13.40) that

$$\delta(\hat{P}) \succeq \delta(\hat{Q}) \text{ if and only if } \hat{P} \succeq \hat{Q} \tag{13.41}$$

since $\hat{P} \succeq \hat{Q}$ implies $\delta(\hat{P})_V \succeq \delta(\hat{Q})_V$ for all contexts V.[56] Thus the mapping $\delta : \mathcal{P}(\mathcal{H}) \to \Gamma\underline{O}$ respects the partial order.

The next thing is to see if a logical "$\vee$"-operation can be defined on $\Gamma\underline{O}$. Once again, we try a "local" definition:

Theorem 2 *A "$\vee$"-structure on $\Gamma\underline{O}$ can be defined locally by*

$$(\gamma_1 \vee \gamma_2)_V := \gamma_{1V} \vee \gamma_{2V} \tag{13.42}$$

for all $\gamma_1, \gamma_2 \in \Gamma\underline{O}$, and for all $V \in Ob(\mathcal{V}(\mathcal{H}))$.

Proof It is not instantly clear that (13.42) defines a global element of $\underline{O}$. However, a key result in this direction is the following:

Lemma 1 *For each context V, and for all $\hat{\alpha}, \hat{\beta} \in \mathcal{P}(V)$, we have*

$$\underline{O}(i_{V'V})(\hat{\alpha} \vee \hat{\beta}) = \underline{O}(i_{V'V})(\hat{\alpha}) \vee \underline{O}(i_{V'V})(\hat{\beta}) \tag{13.43}$$

for all contexts V' such that $V' \subseteq V$.

The proof is a straightforward consequence of the definition of the presheaf $\underline{O}$.

One immediate consequence is that (13.42) defines a global element[57] of $\underline{O}$. Hence the theorem is proved.

It is also straightforward to show that, for any pair of projectors $\hat{P}, \hat{Q} \in \mathcal{P}(\mathcal{H})$, we have $\delta(\hat{P} \vee \hat{Q})_V = \delta(\hat{P})_V \vee \delta(\hat{Q})_V$, for all contexts $V \in \text{Ob}(\mathcal{V}(\mathcal{H}))$. This means that, as elements of $\Gamma\underline{O}$,

---

[56] On the other hand, in general, $\hat{P} \succ \hat{Q}$ does not imply $\delta(\hat{P})_V \succ \delta(\hat{Q})_V$ but only $\delta(\hat{P})_V \succeq \delta(\hat{Q})_V$.

[57] The existence of the $\vee$-operation on $\Gamma\underline{O}$ can be extended to $\underline{O}$ itself. More precisely, there is an arrow $\vee : \underline{O} \times \underline{O} \to \underline{O}$ where $\underline{O} \times \underline{O}$ denotes the product presheaf over $\mathcal{V}(\mathcal{H})$, whose objects are $(\underline{O} \times \underline{O})_V := \underline{O}_V \times \underline{O}_V$. Then the arrow $\vee : \underline{O} \times \underline{O} \to \underline{O}$ is defined at any context V by $\vee_V(\hat{\alpha}, \hat{\beta}) := \hat{\alpha} \vee \hat{\beta}$ for all $\hat{\alpha}, \hat{\beta} \in \underline{O}_V$.

$$\delta(\hat{P} \vee \hat{Q}) = \delta(\hat{P}) \vee \delta(\hat{Q}). \tag{13.44}$$

Thus the mapping $\delta : \mathcal{P}(\mathcal{H}) \to \Gamma \underline{O}$ preserves the logical "$\vee$" operation.

However, there is no analogous equation for the logical "$\wedge$"-operation. The obvious local definition would be, for each context V,

$$(\gamma_1 \wedge \gamma_2)_V := \gamma_{1V} \wedge \gamma_{2V} \tag{13.45}$$

but this does not define a global element of $\underline{O}$ since, unlike (13.43), for the $\wedge$-operation we have only

$$\underline{O}(i_{V'V})(\hat{\alpha} \wedge \hat{\beta}) \preceq \underline{O}(i_{V'V})(\hat{\alpha}) \wedge \underline{O}(i_{V'V})(\hat{\beta}) \tag{13.46}$$

for all $V' \subseteq V$. As a consequence, for all V, we have only the inequality

$$\delta(\hat{P} \wedge \hat{Q})_V \preceq \delta(\hat{P})_V \wedge \delta(\hat{Q})_V \tag{13.47}$$

and hence

$$\delta(\hat{P} \wedge \hat{Q}) \preceq \delta(\hat{P}) \wedge \delta(\hat{Q}). \tag{13.48}$$

It is easy to find examples where the inequality is strict. For example, let $\hat{P} \neq \hat{0}, \hat{1}$ and $\hat{Q} = \hat{1} - \hat{P}$. Then $\hat{P} \wedge \hat{Q} = 0$ and hence $\delta_V(\hat{P} \wedge \hat{Q}) = \hat{0}$, while $\delta(\hat{P})_V \wedge \delta(\hat{Q})_V$ can be strictly larger than $\hat{0}$, since $\delta(\hat{P})_V \succeq \hat{P}$ and $\delta(\hat{Q})_V \succeq \hat{Q}$.

13.5.2.4 Hyper-Elements of $\underline{O}$

We have seen that the global elements of $\underline{O}$, i.e., the elements of $\Gamma \underline{O}$, can be equipped with a partial-ordering and a "$\vee$"-operation, but attempts to define a "$\wedge$"-operation in the same way fail because of the inequality in (13.47).

However, the form of (13.46) and (13.47) suggests the following procedure. Let us define a *hyper-element* of $\underline{O}$ to be an association, for each stage $V \in \mathrm{Ob}(\mathcal{V}(\mathcal{H}))$, of an element $\gamma_V \in \underline{O}_V$ with the property that

$$\gamma_{V'} \succeq \underline{O}(i_{V'V})(\gamma_V) \tag{13.49}$$

for all $V' \subseteq V$. Clearly every element of $\Gamma \underline{O}$ is a hyper-element, but not conversely.

Now, if γ_1 and γ_2 are hyper-elements, we can define the operations "$\vee$" and "$\wedge$" locally as:

$$(\gamma_1 \vee \gamma_2)_V := \gamma_{1V} \vee \gamma_{2V} \tag{13.50}$$

$$(\gamma_1 \wedge \gamma_2)_V := \gamma_{1V} \wedge \gamma_{2V} \tag{13.51}$$

Because of (13.46) we have, for all $V' \subseteq V$,

$$\underline{O}(i_{V'V})\big((\gamma_1 \wedge \gamma_2)_V\big) = \underline{O}(i_{V'V})\big(\gamma_{1V} \wedge \gamma_{2V}\big) \tag{13.52}$$

$$\preceq \underline{O}(i_{V'V})(\gamma_{1V}) \wedge \underline{O}(i_{V'V})(\gamma_{2V}) \tag{13.53}$$

$$\preceq \gamma_{1V'} \wedge \gamma_{2V'} \tag{13.54}$$

$$= (\gamma_1 \wedge \gamma_2)_{V'} \tag{13.55}$$

so that the hyper-element condition (13.49) is preserved.

The occurrence of a logical "$\vee$" and "$\wedge$" structure is encouraging, but it is not yet what we want. For one thing, there is no mention of a negation operation; and, anyway, this is not the expected algebra of sub-objects of a "state space" object. To proceed further we must study more carefully the sub-objects of the spectral presheaf.

13.5.3 Daseinisation: Heidegger Encounters Physics

13.5.3.1 From Global Elements of $\underline{O}$ to Sub-objects of $\underline{\Sigma}$

The spectral presheaf, $\underline{\Sigma}$, played a central role in the earlier discussions of quantum theory from a topos perspective [48–51]. Here is the formal definition.

Definition 4 The spectral presheaf, $\underline{\Sigma}$, is defined as the following functor from $\mathcal{V}(\mathcal{H})^{\mathrm{op}}$ to **Sets**:

1. On objects V: $\underline{\Sigma}_V$ is the Gel'fand spectrum of the unital, abelian sub-algebra V of $B(\mathcal{H})$; i.e., the set of all multiplicative linear functionals $\lambda : V \to \mathbb{C}$ such that $\langle \lambda, \hat{1} \rangle = 1$.
2. On morphisms $i_{V'V} : V' \subseteq V$: $\underline{\Sigma}(i_{V'V}) : \underline{\Sigma}_V \to \underline{\Sigma}_{V'}$ is defined by $\underline{\Sigma}(i_{V'V})(\lambda) := \lambda|_{V'}$; i.e., the restriction of the functional $\lambda : V \to \mathbb{C}$ to the sub-algebra $V' \subseteq V$.

One central result of spectral theory is that $\underline{\Sigma}_V$ has a topology that is compact and Hausdorff, and with respect to which the Gel'fand transforms[58] of the elements of V are continuous functions from $\underline{\Sigma}_V$ to $\mathbb{C}$. This will be important in what follows [56].

The spectral presheaf plays a fundamental role in our research programme as applied to quantum theory. For example, it was shown in the earlier work that the Kochen-Specker theorem [57] is equivalent to the statement that $\underline{\Sigma}$ has no global elements. However, $\underline{\Sigma}$ *does* have sub-objects, and these are central to our scheme:

Definition 5 A *sub-object* $\underline{S}$ of the spectral presheaf $\underline{\Sigma}$ is a functor $\underline{S} : \mathcal{V}(\mathcal{H})^{op} \to$ **Sets** such that

1. $\underline{S}_V$ is a subset of $\underline{\Sigma}_V$ for all V.
2. If $V' \subseteq V$, then $\underline{S}(i_{V'V}) : \underline{S}_V \to \underline{S}_{V'}$ is just the restriction $\lambda \mapsto \lambda|_{V'}$ (i.e., the same as for $\underline{\Sigma}$), applied to the elements $\lambda \in \underline{S}_V \subseteq \underline{\Sigma}_V$.

---

[58] If $\hat{A} \in V$, the Gel'fand transform, $\overline{A} : \underline{\Sigma}_V \to \mathbb{C}$, of $\hat{A}$ is defined by $\overline{A}(\lambda) := \langle \lambda, \hat{A} \rangle$ for all $\lambda \in \underline{\Sigma}_V$.

This definition of a sub-object is standard. However, for our purposes we need something slightly different, namely concept of a "clopen" sub-object. This is defined to be a sub-object $\underline{S}$ of $\underline{\Sigma}$ such that, for all V, the set $\underline{S}_V$ is a *clopen*[59] subset of the compact, Hausdorff space $\underline{\Sigma}_V$. We denote by $\mathrm{Sub}_{\mathrm{cl}}(\underline{\Sigma})$ the set of all clopen sub-objects of $\underline{\Sigma}$. We will show later (in the Appendix) that, like $\mathrm{Sub}(\underline{\Sigma})$, the set $\mathrm{Sub}_{\mathrm{cl}}(\underline{\Sigma})$ is a Heyting algebra. In Sect. 13.6.5 we show that there is an object $P_{\mathrm{cl}}\underline{\Sigma}$ whose global elements are precisely the clopen sub-objects of $\underline{\Sigma}$.

This interest in clopen sets is easy to explain. For, according to the Gel'fand spectral theory, a projection operator $\hat{\alpha} \in \mathcal{P}(V)$ corresponds to a unique clopen subset, $S_{\hat{\alpha}}$ of the Gel'fand spectrum, $\underline{\Sigma}_V$. Furthermore, the Gel'fand transform $\overline{\alpha} : \underline{\Sigma}_V \to \mathbb{C}$ of $\hat{\alpha}$ takes the values 0, 1 only, since the spectrum of a projection operator is just $\{0, 1\}$.

It follows that $\overline{\alpha}$ is the characteristic function of the subset, $S_{\hat{\alpha}}$, of $\underline{\Sigma}_V$, defined by

$$S_{\hat{\alpha}} := \{\lambda \in \underline{\Sigma}_V \mid \langle \lambda, \hat{\alpha} \rangle = 1\}. \tag{13.56}$$

The clopen nature of $S_{\hat{\alpha}}$ follows from the fact that, by the spectral theory, the function $\overline{\alpha} : \underline{\Sigma}_V \to \{0, 1\}$ is continuous.

In fact, there is a lattice isomorphism between the lattice $\mathcal{P}(V)$ of projectors in V and the lattice $\mathcal{CL}(\underline{\Sigma}_V)$ of clopen subsets of $\underline{\Sigma}_V$,[60] given by

$$\hat{\alpha} \mapsto S_{\hat{\alpha}} := \{\lambda \in \underline{\Sigma}_V \mid \langle \lambda, \hat{\alpha} \rangle = 1\}. \tag{13.57}$$

Conversely, given a clopen subset $S \in \mathcal{CL}(\underline{\Sigma}_V)$, we get the corresponding projection $\hat{\alpha}$ as the (inverse Gel'fand transform of the) characteristic function of S. Hence, each $S \in \mathcal{CL}(\underline{\Sigma}_V)$ is of the form $S = S_{\hat{\alpha}}$ for some $\hat{\alpha} \in \mathcal{P}(V)$.

Our claim is the following:

Theorem 3 *For each projection operator $\hat{P} \in \mathcal{P}(\mathcal{H})$, the collection*

$$\underline{\delta(\hat{P})} := \{S_{\delta(\hat{P})_V} \subseteq \underline{\Sigma}_V \mid V \in Ob(\mathcal{V}(\mathcal{H}))\} \tag{13.58}$$

forms a (clopen) sub-object of the spectral presheaf $\underline{\Sigma}$.

Proof To see this, let $\lambda \in S_{\delta(\hat{P})_V}$. Then if V' is some abelian sub-algebra of V, we have $\delta(\hat{P})_{V'} = \bigwedge\left\{\hat{\alpha} \in \mathcal{P}(V') \mid \hat{\alpha} \succeq \delta(\hat{P})_V\right\} \succeq \delta(\hat{P})_V$. Now let $\hat{\alpha} := \delta(\hat{P})_{V'} - \delta(\hat{P})_V$. Then $\langle \lambda, \delta(\hat{P})_{V'} \rangle = \langle \lambda, \delta(\hat{P})_V \rangle + \langle \lambda, \hat{\alpha} \rangle = 1$, since $\langle \lambda, \delta(\hat{P})_V \rangle = 1$ and $\langle \lambda, \hat{\alpha} \rangle \in \{0, 1\}$. This shows that

59 A "clopen" subset of a topological space is one that is both open and closed.

60 The lattice structure on $\mathcal{CL}(\underline{\Sigma}_V)$ is defined as follows: if $(U_i)_{i \in I}$ is an arbitrary family of clopen subsets of $\underline{\Sigma}_V$, then the *closure* $\overline{\bigcup_{i \in I} U_i}$ is the maximum. The closure is necessary since the union of infinitely many closed sets need not be closed. The *interior* $\mathrm{int} \bigcap_{i \in I} U_i$ is the minimum of the family. One must take the interior since $\bigcap_{i \in I} U_i$ is closed, but not necessarily open.

$$\{\lambda|_{V'} \mid \lambda \in S_{\delta(\hat{P})_V}\} \subseteq S_{\delta(\hat{P})_{V'}}. \tag{13.59}$$

However, the left hand side of (13.59) is the subset $\underline{O}(i_{V'V})(S_{\delta(\hat{P})_V}) \subseteq \underline{\Sigma}_{V'}$ of the outer-presheaf restriction of elements in $S_{\delta(\hat{P})_V}$ to $\underline{\Sigma}_{V'}$, and the restricted elements all lie in $S_{\delta(\hat{P})_{V'}}$. It follows that the collection of sets

$$\underline{\delta(\hat{P})} := \{S_{\delta(\hat{P})_V} \subseteq \underline{\Sigma}_V \mid V \in \mathrm{Ob}(\mathcal{V}(\mathcal{H}))\} \tag{13.60}$$

forms a (clopen) sub-object of the spectral presheaf $\underline{\Sigma}$.

By these means we have constructed a mapping

$$\begin{aligned} \delta : \mathcal{P}(\mathcal{H}) &\longrightarrow \mathrm{Sub}_{\mathrm{cl}}(\underline{\Sigma}) \\ \hat{P} &\mapsto \underline{\delta(\hat{P})} := \{S_{\delta(\hat{P})_V} \mid V \in \mathrm{Ob}(\mathcal{V}(\mathcal{H}))\} \end{aligned} \tag{13.61}$$

which sends projection operators on $\mathcal{H}$ to clopen sub-objects of $\underline{\Sigma}$. As a matter of notation, we will denote the clopen subset $S_{\delta(\hat{P})_V} \subseteq \underline{\Sigma}_V$ as $\underline{\delta(\hat{P})}_V$. The notation $\delta(\hat{P})_V$ refers to the element (i.e., projection operator) of $\underline{O}_V$ defined earlier.

13.5.3.2 The Definition of Daseinisation

As usual, the projection $\hat{P}$ is regarded as representing a proposition about the quantum system. Thus δ maps propositions about a quantum system to (clopen) sub-objects of the spectral presheaf. This is strikingly analogous to the situation in classical physics, in which propositions are represented by subsets of the classical state space.

Definition 6 The map δ in (13.61) is a fundamental part of our constructions. We call it the (outer) daseinisation of $\hat{P}$. We shall use the same word to refer to the operation in (13.37) that relates to the outer presheaf.

The expression "daseinisation" comes from the German word *Dasein*, which plays a central role in Heidegger's existential philosophy. Dasein translates to "existence" or, in the very literal sense often stressed by Heidegger, to being-there-in-the-world.[61] Thus daseinisation "brings-a-quantum-property-into-existence"[62] by hurling it into the collection of all possible classical snap-shots of the world provided by the category of contexts.

We will summarise here some useful properties of daseinisation.

[61] The hyphens are *very* important.

[62] The hyphens are *very* important.

1. The null projection $\hat{0}$ is mapped to the empty sub-object of $\underline{\Sigma}$:

$$\underline{\delta(\hat{0})} = \{\varnothing_V \mid V \in \mathrm{Ob}(\mathcal{V}(\mathcal{H}))\} \tag{13.62}$$

2. The identity projection $\hat{1}$ is mapped to the unit sub-object of $\underline{\Sigma}$:

$$\underline{\delta(\hat{1})} = \{\underline{\Sigma}_V \mid V \in \mathrm{Ob}(\mathcal{V}(\mathcal{H}))\} = \underline{\Sigma} \tag{13.63}$$

3. Since the daseinisation map $\delta : \mathcal{P}(\mathcal{H}) \to \Gamma \underline{O}$ is injective (see Sect. 13.5.2), and the mapping $\Gamma \underline{O} \to \Gamma(P_{\mathrm{cl}}\underline{\Sigma})$ is injective (because there is a monic arrow $\underline{O} \to P_{\mathrm{cl}}\underline{\Sigma}$ in $\mathbf{Sets}^{\mathcal{V}(\mathcal{H})^{\mathrm{op}}}$; see Sect. 13.6.5), it follows that the daseinisation map $\delta : \mathcal{P}(\mathcal{H}) \to \Gamma(P_{\mathrm{cl}}\underline{\Sigma}) \simeq \mathrm{Sub}_{\mathrm{cl}}(\underline{\Sigma})$ is also injective. Thus no information about the projector $\hat{P}$ is lost when it is daseinised to become $\underline{\delta(\hat{P})}$.

13.5.4 *The Heyting Algebra Structure on* $\mathbf{Sub}_{\mathrm{cl}}(\underline{\Sigma})$

The reason for daseinising projections is that the set, $\mathrm{Sub}(\underline{\Sigma})$, of sub-objects of the spectral presheaf forms a *Heyting algebra*. Thus the idea is to find a map $\pi_{\mathrm{qt}} : \mathcal{PL}(S)_0 \to \mathrm{Sub}(\underline{\Sigma})$ and then extend it to all of $\mathcal{PL}(S)$ using the simple recursion ideas discussed in Sect. 13.3.2.

In our case, the act of daseinisation gives a map from the projection operators to the clopen sub-objects of $\mathrm{Sub}(\underline{\Sigma})$, and therefore a map $\pi_{\mathrm{qt}} : \mathcal{PL}(S)_0 \to \mathrm{Sub}_{\mathrm{cl}}(\underline{\Sigma})$ can be defined by

$$\pi_{\mathrm{qt}}(A \,\varepsilon\, \Delta) := \underline{\delta\big(\hat{E}[A \in \Delta]\big)} \tag{13.64}$$

However, to extend this definition to $\mathcal{PL}(S)$, it is necessary to show that the set of clopen sub-objects, $\mathrm{Sub}_{\mathrm{cl}}(\underline{\Sigma})$, is a Heyting algebra. This is not completely obvious from the definition alone. However, it is true, and the proof is given in Theorem 15 in the Appendix.

In conclusion: daseinisation can be used to give a representation/model of the language $\mathcal{PL}(S)$ in the Heyting algebra $\mathrm{Sub}_{\mathrm{cl}}(\underline{\Sigma})$.[63]

13.5.5 *Daseinisation and the Operations of Quantum Logic*

It is interesting to ask to what extent the map $\delta : \mathcal{P}(\mathcal{H}) \to \mathrm{Sub}_{\mathrm{cl}}(\underline{\Sigma})$ respects the lattice structure on $\mathcal{P}(\mathcal{H})$. Of course, we know that it cannot be *completely* preserved since the quantum logic $\mathcal{P}(\mathcal{H})$ is non-distributive, whereas $\mathrm{Sub}_{\mathrm{cl}}(\underline{\Sigma})$ is a Heyting algebra, and hence distributive.

---

[63] Since the clopen sub-objects of $\underline{\Sigma}$ correspond bijectively to the hyper-elements of the outer presheaf $\underline{O}$, it is clear that the hyper-elements of $\underline{O}$ form a Heyting algebra, too.

We saw in Sect. 13.5.2 that, for the mapping $\delta : \mathcal{P}(\mathcal{H}) \to \Gamma \underline{O}$, we have

$$\delta(\hat{P} \vee \hat{Q})_V = \delta(\hat{P})_V \vee \delta(\hat{Q})_V, \tag{13.65}$$
$$\delta(\hat{P} \wedge \hat{Q})_V \preceq \delta(\hat{P})_V \wedge \delta(\hat{Q})_V \tag{13.66}$$

for all contexts V in $\mathrm{Ob}(\mathcal{V}(\mathcal{H}))$.

The clopen subset of $\underline{\Sigma}_V$ that corresponds to $\delta(\hat{P})_V \vee \delta(\hat{Q})_V$ is $S_{\delta(\hat{P})_V} \cup S_{\delta(\hat{Q})_V}$. This implies that the daseinisation map $\delta : \mathcal{P}(\mathcal{H}) \to \mathrm{Sub}_{\mathrm{cl}}(\underline{\Sigma})$ is a morphism of $\vee$-semi-lattices.

On the other hand, $\delta(\hat{P})_V \wedge \delta(\hat{Q})_V$ corresponds to the subset $S_{\delta(\hat{P})_V} \cap S_{\delta(\hat{Q})_V}$ of $\underline{\Sigma}_V$. Therefore, since $S_{\delta(\hat{P} \wedge \hat{Q})_V} \subseteq S_{\delta(\hat{P})_V} \cap S_{\delta(\hat{Q})_V}$, daseinisation is *not* a morphism of $\wedge$-semi-lattices. In summary, for all projectors $\hat{P}$, $\hat{Q}$ we have

$$\underline{\delta(\hat{P} \vee \hat{Q})} = \underline{\delta(\hat{P})} \vee \underline{\delta(\hat{Q})} \tag{13.67}$$
$$\underline{\delta(\hat{P} \wedge \hat{Q})} \preceq \underline{\delta(\hat{P})} \wedge \underline{\delta(\hat{Q})} \tag{13.68}$$

where the logical connectives on the left hand side lie in the quantum logic $\mathcal{P}(\mathcal{H})$, and those on the right hand side lie in the Heyting algebra $\mathrm{Sub}_{\mathrm{cl}}(\underline{\Sigma})$, as do the symbols "=" and "$\preceq$".

As remarked above, it is not surprising that (13.68) is not an equality. Indeed, the quantum logic $\mathcal{P}(\mathcal{H})$ is non-distributive, whereas the Heyting algebra $\mathrm{Sub}_{\mathrm{cl}}(\underline{\Sigma})$ *is* distributive, and so it would be impossible for both (13.67) and (13.68) to be equalities. The inequality in (13.68) is the price that must be paid for liberating the projection operators from the shackles of quantum logic and transporting them to the existential world of Heyting algebras.

13.5.5.1 The Status of the Possible Axiom "$A\,\varepsilon\,\Delta_1 \wedge A\,\varepsilon\,\Delta_2 \Leftrightarrow A\,\varepsilon\,\Delta_1 \cap \Delta_2$"

We have the representation in (13.64), $\pi_{\mathrm{qt}}(A\,\varepsilon\,\Delta) := \underline{\delta\big(\hat{E}[A \in \Delta]\big)}$, of the primitive propositions $A\,\varepsilon\,\Delta$, and, as explained in Sect. 13.3.2, this can be extended to compound sentences by making the obvious definitions:

$$(a)\;\; \pi_{\mathrm{qt}}(\alpha \vee \beta) := \pi_{\mathrm{qt}}(\alpha) \vee \pi_{\mathrm{qt}}(\beta) \tag{13.69}$$
$$(b)\;\; \pi_{\mathrm{qt}}(\alpha \wedge \beta) := \pi_{\mathrm{qt}}(\alpha) \wedge \pi_{\mathrm{qt}}\beta) \tag{13.70}$$
$$(c)\;\; \pi_{\mathrm{qt}}(\neg\alpha) := \neg\pi_{\mathrm{qt}}(\alpha) \tag{13.71}$$
$$(d)\;\; \pi_{\mathrm{qt}}(\alpha \Rightarrow \beta) := \pi_{\mathrm{qt}}(\alpha) \Rightarrow \pi_{\mathrm{qt}}(\beta) \tag{13.72}$$

As a result, we necessarily get a representation of the full language $\mathcal{PL}(S)$ in the Heyting algebra $\mathrm{Sub}_{\mathrm{cl}}(\underline{\Sigma})$. However, we then find that:

$$\pi_{\mathrm{qt}}(A\,\varepsilon\,\Delta_1 \wedge A\,\varepsilon\,\Delta_2) := \pi_{\mathrm{qt}}(A\,\varepsilon\,\Delta_1) \wedge \pi_{\mathrm{qt}}(A\,\varepsilon\,\Delta_2) \tag{13.73}$$
$$= \underline{\delta(\hat{E}[A \in \Delta_1])} \wedge \underline{\delta(\hat{E}[A \in \Delta_2])} \tag{13.74}$$

$$\succeq \underline{\delta(\hat{E}[A \in \Delta_1] \wedge \hat{E}[A \in \Delta_2])} \quad (13.75)$$

$$= \underline{\delta(\hat{E}[A \in \Delta_1 \cap \Delta_2)])} \quad (13.76)$$

$$= \pi_{\rm qt}(A \,\varepsilon\, \Delta_1 \cap \Delta_2) \quad (13.77)$$

where, (13.75) comes from (13.68), and in (13.76) we have used the property of spectral projectors that $\hat{E}[A \in \Delta_1] \wedge \hat{E}[A \in \Delta_2] = \hat{E}[A \in \Delta_1 \cap \Delta_2)]$. Thus, although by definition, $\pi_{\rm qt}(A \,\varepsilon\, \Delta_1 \wedge A \,\varepsilon\, \Delta_2) = \pi_{\rm qt}(A \,\varepsilon\, \Delta_1) \wedge \pi_{\rm qt}(A \,\varepsilon\, \Delta_2)$, we only have the inequality

$$\pi_{\rm qt}(A \,\varepsilon\, \Delta_1 \cap \Delta_2) \preceq \pi_{\rm qt}(A \,\varepsilon\, \Delta_1 \wedge A \,\varepsilon\, \Delta_2) \quad (13.78)$$

On the other hand, the same line of argument shows that

$$\pi_{\rm qt}(A \,\varepsilon\, \Delta_1 \vee A \,\varepsilon\, \Delta_2) = \pi_{\rm qt}(A \,\varepsilon\, \Delta_1 \cup \Delta_2) \quad (13.79)$$

Thus it would be consistent to add the axiom

$$A \,\varepsilon\, \Delta_1 \vee A \,\varepsilon\, \Delta_2 \Leftrightarrow A \,\varepsilon\, \Delta_1 \cup \Delta_2 \quad (13.80)$$

to the language $\mathcal{PL}(S)$, but not

$$A \,\varepsilon\, \Delta_1 \wedge A \,\varepsilon\, \Delta_2 \Leftrightarrow A \,\varepsilon\, \Delta_1 \cap \Delta_2 \quad (13.81)$$

Of, course, both axioms are consistent with the representation of $\mathcal{PL}(S)$ in classical physics.

It should be emphasised that there is nothing wrong with this result: indeed, as stated above, it is the necessary price to be paid for forcing a non-distributive algebra to have a "representation" in a Heyting algebra.

13.5.5.2 Inner Daseinisation and $\delta(\neg\hat{P})$

In the same spirit, one might ask about "$\neg(A \,\varepsilon\, \Delta)$". By definition, as in (13.9), we have $\pi_{\rm qt}(\neg(A \,\varepsilon\, \Delta)) := \neg\pi_{\rm qt}(A \,\varepsilon\, \Delta) = \neg\underline{\delta\big(\hat{E}[A \in \Delta]\big)}$. However, the question then is how, if at all, this is related to $\underline{\delta(\hat{E}[A \in \mathbb{R}/\Delta])} = \underline{\delta(\neg\hat{E}[A \in \Delta])}$, bearing in mind the axiom

$$\neg(A \,\varepsilon\, \Delta) \Leftrightarrow A \,\varepsilon\, \mathbb{R}\backslash\Delta \quad (13.82)$$

that can be added to the classical representation of $\mathcal{PL}(S)$. Thus something needs to be said about $\underline{\delta(\neg\hat{P})}$, where $\neg\hat{P} = \hat{1} - \hat{P}$ is the negation operation in the quantum logic $\mathcal{P}(\mathcal{H})$.

To proceed further, we need to introduce another operation:

Definition 7 The inner daseinisation, $\delta^i(\hat{P})_V$, of $\hat{P}$ to a context V is defined (for each $V \in \mathrm{Ob}(\mathcal{V}(\mathcal{H}))$) as

$$\delta^i(\hat{P})_V := \bigvee \{\hat{\beta} \in \mathcal{P}(V) \mid \hat{\beta} \preceq \hat{P}\}. \tag{13.83}$$

This should be contrasted with the definition of outer daseinisation in (13.35).

Thus $\delta^i(\hat{P})_V$ is the best approximation that can be made to $\hat{P}$ by taking the "largest" projector in V that implies $\hat{P}$.

As with the other daseinisation construction, this operation was first introduced by de Groote in [37] where he called it the *core* of the projection operator $\hat{P}$. We prefer to use the phrase "inner daseinisation", and then to refer to (13.35) as the "outer daseinisation" operation on $\hat{P}$. The existing notation $\delta(\hat{P})_V$ will be replaced with $\delta^o(\hat{P})_V$ if there is any danger of confusing the two daseinisation operations.

With the aid of inner daseinisation, a new presheaf, $\underline{I}$, can be constructed as an exact analogue of the outer presheaf, $\underline{O}$, defined in Sect. 13.5.2. Specifically:

Definition 8 The inner presheaf $\underline{I}$ is defined over the category $\mathcal{V}(\mathcal{H})$ as follows:

(i) On objects $V \in \mathrm{Ob}(\mathcal{V}(\mathcal{H}))$: We have $\underline{I}_V := \mathcal{P}(V)$
(ii) On morphisms $i_{V'V} : V' \subseteq V$: The mapping $\underline{I}(i_{V'V}) : \underline{I}_V \to \underline{I}_{V'}$ is given by $\underline{I}(i_{V'V})(\hat{\alpha}) := \delta^i(\hat{\alpha})_V$ for all $\hat{\alpha} \in \mathcal{P}(V)$.

It is easy to see that the collection $\{\delta^i(\hat{P})_V \mid V \in \mathrm{Ob}(\mathcal{V}(\mathcal{H}))\}$ of projection operators given by (13.83) is a global element of $\underline{I}$.

It is also straightforward to show that

$$\underline{O}(i_{V'V})(\neg\hat{\alpha}) = \neg\, \underline{I}(i_{V'V})(\hat{\alpha}) \tag{13.84}$$

for all projectors $\hat{\alpha}$ in V, and for all $V' \subseteq V$. It follows from (13.84) that

$$\delta^o(\neg\hat{P})_V = \hat{1} - \delta^i(\hat{P})_V \tag{13.85}$$

for all projectors $\hat{P}$ and all contexts V.

It is clear from (13.84) that the negation operation on projectors defines a map $\neg : \Gamma\underline{O} \to \Gamma\underline{I}$, $\gamma \mapsto \neg\gamma$; i.e., for all contexts V, we map $\gamma(V) \mapsto \neg\gamma(V) := \hat{1} - \gamma(V)$. Actually, one can go further than this and show that the presheaves $\underline{O}$ and $\underline{I}$ are isomorphic in the category $\mathbf{Sets}^{\mathcal{V}(\mathcal{H})^{\mathrm{op}}}$. This means that, in principle, we can always work with one presheaf only. However, for reasons of symmetry it is sometime useful to invoke both presheaves.

As with outer daseinisation, inner daseinisation can also be used to define a mapping from projection operators to sub-objects of the spectral presheaf. Specifically, if $\hat{P}$ is a projection, for each $V \in \mathrm{Ob}(\mathcal{V}(\mathcal{H}))$ define

$$T_{\delta^i(\hat{P})_V} := \{\lambda \in \underline{\Sigma}_V \mid \langle\lambda, \delta^i(\hat{P})_V\rangle = 0\}. \tag{13.86}$$

It is easy to see that these subsets form a clopen sub-object, $\underline{\delta^i(\hat{P})}$, of $\underline{\Sigma}$. It follows from (13.85) that $T_{\delta^i(\hat{P})_V} = S_{\delta^o(\neg\hat{P})_V}$.

13.5.5.3 Using Boolean Algebras as the Base Category

As we have mentioned several times already, the collection, $\mathcal{V}(\mathcal{H})$, of all commutative von Neumann sub-algebras of $B(\mathcal{H})$ is not the only possible choice for the base category over which to construct presheaves. In fact, if we are only interested in the propositional language $\mathcal{PL}(S)$, a somewhat simpler choice is the collection, $\mathcal{B}l(\mathcal{H})$, of all Boolean sub-algebras of the non-distributive lattice, $\mathcal{P}(\mathcal{H})$, of projection operators on $\mathcal{H}$. More abstractly, for any non-distributive lattice $\mathfrak{B}$, one could use the category of Boolean sub-algebras of $\mathfrak{B}$. This possibility was raised in the original paper [48] but has not been used much thereafter. However, it does have some interesting features.

The analogue of the (von Neumann algebra) spectral presheaf, $\underline{\Sigma}$, is the so-called *dual* presheaf, $\underline{D}$:

Definition 9 The dual presheaf on $\mathcal{B}l(\mathcal{H})$ is the contravariant functor $\underline{D} : \mathcal{B}l(\mathcal{H}) \to$ **Sets** defined as follows:

1. On objects in $\mathcal{B}l(\mathcal{H})$: $\underline{D}(B)$ is the *dual* of B; i.e., the set $\mathrm{Hom}(B, \{0, 1\})$ of all homomorphisms from the Boolean algebra B to the Boolean algebra $\{0, 1\}$.
2. On morphisms in $\mathcal{B}l(\mathcal{H})$: If $i_{B_2 B_1} : B_2 \subseteq B_1$ then $\underline{D}(i_{B_2 B_1}) : \underline{D}(B_1) \to \underline{D}(B_2)$ is defined by $\underline{D}(i_{B_2 B_1})(\chi) := \chi|_{B_2}$, where $\chi|_{B_2}$ denotes the restriction of $\chi \in \underline{D}(B_1)$ to the sub-algebra $B_2 \subseteq B_1$.

A global element of the functor $\underline{D} : \mathcal{B}l(\mathcal{H})^{\mathrm{op}} \to \mathrm{Set}$ is then a function γ that associates to each $B \in \mathrm{Ob}(\mathcal{B}l(\mathcal{H}))$ an element γ_B of the dual of B such that if $i_{B_2 B_1} : B_2 \to B_1$ then $\gamma_{B_1}|_{B_2} = \gamma_{B_2}$; thus, for all $\hat{\alpha} \in B_2$,

$$\gamma_{B_2}(\hat{\alpha}) = \gamma_{B_1}((i_{B_2 B_1}(\hat{\alpha})). \tag{13.87}$$

Since each projection operator, $\hat{\alpha}$ belongs to at least one Boolean algebra (for example, the algebra $\{\hat{0}, \hat{1}, \hat{\alpha}, \neg\hat{\alpha}\}$) it follows that a global element of the presheaf $\underline{D}$ associates to each projection operator $\hat{\alpha}$ a number $V(\hat{\alpha})$ which is either 0 or 1, and is such that, if $\hat{\alpha} \wedge \hat{\beta} = \hat{0}$, then $V(\hat{\alpha} \vee \hat{\beta}) = V(\hat{\alpha}) + V(\hat{\beta})$. These types of valuation are often used in the proofs of the Kochen-Specker theorem that focus on the construction of specific counter-examples. In fact, it is easy to see the following:

> The Kochen-Specker theorem is equivalent to the statement that, if $\dim \mathcal{H} > 2$, the dual presheaf $\underline{D} : \mathcal{B}l(\mathcal{H})^{\mathrm{op}} \to$ **Sets** has no global elements.

It is easy to apply the concept of "daseinisation" to the topos $\mathbf{Sets}^{\mathcal{B}l(\mathcal{H})^{\mathrm{op}}}$. In the case of von Neumann algebras, the outer daseinisation of a projection operator $\hat{P}$ was defined as (see (13.35))

$$\delta(\hat{P})_V := \bigwedge \left\{\hat{\alpha} \in \mathcal{P}(V) \mid \hat{\alpha} \succeq \hat{P}\right\} \tag{13.88}$$

where $\mathcal{P}(V)$ denotes the collection of all projection operators in the commutative von Neumann algebra V. In this form, $\delta(\hat{P})$ appears as a global element of the outer presheaf $\underline{O}$.

When using the base category, $\mathcal{B}l(\mathcal{H})$, of Boolean sub-algebras of $\mathcal{P}(\mathcal{H})$, we define

$$\delta(\hat{P})_B := \bigwedge \{\hat{\alpha} \in B \mid \hat{\alpha} \succeq \hat{P}\} \tag{13.89}$$

for each Boolean sub-algebra B of projection operators on $\mathcal{H}$.[64] Clearly, the (outer) daseinisation, $\delta(\hat{P})$, is now a global element of the obvious $B(\mathcal{H})$-analogue of the outer presheaf $\underline{O}$. There are parallel remarks for the inner daseinisation and inner presheaf. The existence of these daseinisation operations means that the propositional language $\mathcal{PL}(S)$ can be represented in the topos $\mathbf{Sets}^{\mathcal{B}l(\mathcal{H})^{\mathrm{op}}}$ in a way that is closely analogous to that used above for the topos $\mathbf{Sets}^{\mathcal{V}(\mathcal{H})^{\mathrm{op}}}$.

Note that (i) each Boolean algebra of projection operators B generates a commutative von Neumann algebra, B'', (the double commutant); and, conversely, (ii) to each von Neumann algebra V there is associated the complete Boolean algebra $\mathcal{P}(V)$ of the projection operators in V. This implies that the operation

$$\phi : \mathcal{B}l(\mathcal{H}) \to \mathcal{V}(\mathcal{H}) \tag{13.90}$$

$$B \mapsto B'' \tag{13.91}$$

defines a full and faithful functor between the categories $\mathcal{B}l(\mathcal{H})$ and $\mathcal{V}(\mathcal{H})$. This functor can be used to pull-back the spectral presheaf, $\underline{\Sigma}$, in $\mathbf{Sets}^{\mathcal{V}(\mathcal{H})^{\mathrm{op}}}$ to the object $\phi^*\underline{\Sigma} := \underline{\Sigma} \circ \phi$ in $\mathbf{Sets}^{\mathcal{B}l(\mathcal{H})^{\mathrm{op}}}$. This pull-back is closely related to the dual presheaf $\underline{D}$.

13.5.6 The Special Nature of Daseinised Projections

13.5.6.1 Daseinised Projections as Optimal Sub-objects

We have shown how daseinisation leads to an interpretation/model of the language $\mathcal{PL}(S)$ in the Heyting algebra $\mathrm{Sub}_{\mathrm{cl}}(\underline{\Sigma})$. In particular, any primitive proposition "$A \,\varepsilon\, \Delta$" is represented by the clopen sub-object $\underline{\delta(\hat{E}[A \in \Delta])}$.

We have seen that, in general, the "and", $\underline{\delta(\hat{P})} \wedge \underline{\delta(\hat{Q})}$, of the daseinisation of two projection operators $\hat{P}$ and $\hat{Q}$, is not itself of the form $\underline{\delta(\hat{R})}$ for any projector $\hat{R}$. The same applies to the negation $\neg\underline{\delta(\hat{P})}$.

This raises the question of whether the sub-objects of $\underline{\Sigma}$ that *are* of the form $\underline{\delta(\hat{P})}$ can be characterised in a simple way. Rather interestingly, the answer is "yes", as we will now see.

---

[64] To be precise, we assume that B is complete such that the infimum in (13.89) is well-defined and lies in B.

Let $V', V \in \mathrm{Ob}(\mathcal{V}(\mathcal{H}))$ be such that $V' \subseteq V$. As would be expected, there is a close connection between the restriction $\underline{O}(i_{V'V}) : \underline{O}_V \to \underline{O}_{V'}, \delta(\hat{P})_V \mapsto \delta(\hat{P})_{V'}$, of the outer presheaf, and the restriction $\underline{\Sigma}(i_{V'V}) : \underline{\Sigma}_V \to \underline{\Sigma}_{V'}, \lambda \mapsto \lambda|_{V'}$, of the spectral presheaf. Indeed, if $\hat{P} \in \mathcal{P}(\mathcal{H})$ is a projection operator, and $S_{\delta(\hat{P})_V} \subseteq \underline{\Sigma}_V$ is defined as in (13.56), we have the following result:

$$S_{\underline{O}(i_{V'V})(\delta(\hat{P})_V)} = \underline{\Sigma}(i_{V'V})(S_{\delta(\hat{P})_V}). \tag{13.92}$$

The proof is given in Theorem 16 in the Appendix.

This result shows that the sub-objects $\underline{\delta(\hat{P})} = \{S_{\delta(\hat{P})_V} \mid V \in \mathrm{Ob}(\mathcal{V}(\mathcal{H}))\}$ of $\underline{\Sigma}$ are of a very special kind. Namely, they are such that the restrictions

$$\underline{\Sigma}(i_{V'V}) : S_{\delta(\hat{P})_V} \to S_{\delta(\hat{P})_{V'}} \tag{13.93}$$

are *surjective* mapping of sets.

For an arbitrary sub-object $\underline{K}$ of $\underline{\Sigma}$, this will not be the case and $\underline{\Sigma}(i_{V'V})$ only maps $\underline{K}_V$ *into* $\underline{K}_{V'}$. Indeed, this is essentially the *definition* of a sub-object of a presheaf. Thus we see that the daseinised projections $\underline{\delta(\hat{P})} = \{S_{\delta(\hat{P})_V} \mid V \in \mathrm{Ob}(\mathcal{V}(\mathcal{H}))\}$ are optimal in the following sense. As we go "down the line" to smaller and smaller sub-algebras of a context V—for example, from V to $V' \subseteq V$, then to $V'' \subseteq V'$ etc.—then the subsets $S_{\delta(\hat{P})_{V'}}, S_{\delta(\hat{P})_{V''}}, \ldots$ are as small as they can be; i.e., $S_{\delta(\hat{P})_{V'}}$ is the *smallest* subset of $\underline{\Sigma}_{V'}$ such that $\underline{\Sigma}(i_{V'V})(S_{\delta(\hat{P})_V}) \subseteq S_{\delta(\hat{P})_{V'}}$, likewise $S_{\delta(\hat{P})_{V''}}$ is the smallest subset of $\underline{\Sigma}_{V''}$ such that $\underline{\Sigma}(i_{V''V'})(S_{\delta(\hat{P})_{V'}}) \subseteq S_{\delta(\hat{P})_{V''}}$, and so on.

It is also clear from this result that there are lots of sub-objects of $\underline{\Sigma}$ that are *not* of the form $\underline{\delta(\hat{P})}$ for any projector $\hat{P} \in \mathcal{P}(\mathcal{H})$.

These more general sub-objects of $\underline{\Sigma}$ show up explicitly in the representation of the more sophisticated language $\mathcal{L}(S)$. This will be discussed thoroughly in Sect. 13.8 when we analyse the representation, ϕ, of the language $\mathcal{L}(S)$ in the topos $\mathbf{Sets}^{\mathcal{V}(\mathcal{H})^{\mathrm{op}}}$. This involves constructing the quantity-value object $\mathcal{R}_\phi$ (to be denoted $\underline{\mathcal{R}}$), and then finding the representation of a function symbol $A : \Sigma \to \mathcal{R}$ in $\mathcal{L}(S)$, in the form of a specific arrow $\breve{A} : \underline{\Sigma} \to \underline{\mathcal{R}}$ in the topos. The generic sub-objects of $\underline{\Sigma}$ are then of the form $\breve{A}^{-1}(\underline{\Xi})$ for sub-objects $\underline{\Xi}$ of $\underline{\mathcal{R}}$. This is an illuminating way of studying the sub-objects of $\underline{\Sigma}$ that do not come from the propositional language $\mathcal{PL}(S)$.

13.6 Truth Values in Topos Physics

13.6.1 The Mathematical Proposition "$x \in K$"

So far we have concentrated on finding a Heyting-algebra representation of the propositions in quantum theory, but of course there is more to physics than that. We also want to know if/when a certain proposition is *true*: a question which, in physical theories, is normally answered by specifying a *(micro)state* of the system, or something that can play an analogous role.

In classical physics, the situation is straightforward (see Sect. 13.3.2). There, a proposition "$A\,\varepsilon\,\Delta$" is represented by the subset $\pi_{\mathrm{cl}}(A\,\varepsilon\,\Delta) := \breve{A}^{-1}(\Delta) \subseteq \mathcal{S}$ of the state space $\mathcal{S}$; and then, the proposition is true in a state s if and only if $s \in \breve{A}^{-1}(\Delta)$; i.e., if and only if the (micro)state s belongs to the subset, $\pi_{\mathrm{cl}}(A\,\varepsilon\,\Delta)$, of $\mathcal{S}$ that represents the proposition.

Thus, each state s assigns to any primitive proposition "$A\,\varepsilon\,\Delta$", a truth value, $\nu\big(A\,\varepsilon\,\Delta; s\big)$, which lies in the set {false, true} (which we identify with $\{0, 1\}$) and is defined as

$$\nu\big(A\,\varepsilon\,\Delta; s\big) := \begin{cases} 1 & \text{if } s \in \pi_{\mathrm{cl}}(A\,\varepsilon\,\Delta) = \breve{A}^{-1}(\Delta); \\ 0 & \text{otherwise.} \end{cases} \tag{13.94}$$

However, the situation in quantum theory is very different. There, the spectral presheaf $\underline{\Sigma}$—which is the analogue of the classical state space $\mathcal{S}$—has no global elements at all. Our expectation is that this will be true in any topos-based theory that goes "beyond quantum theory": i.e., $\Gamma\Sigma_\phi$ is empty; or, if Σ_ϕ does have global elements, there are not enough of them to determine Σ_ϕ as an object in the topos. In this circumstance, a new concept is required to replace the familiar idea of a "state of the system". As we shall see, this involves the concept of a "truth object", or "pseudo-state".

In physics, the propositions of interest are of the form "$A\,\varepsilon\,\Delta$", which refers to the value of a *physical* quantity. However, in constructing a theory of physics, such physical propositions must first be translated into *mathematical* propositions. The concept of "truth" is then studied in the context of the latter.

Let us start with set-theory based mathematics, where the most basic proposition is of the form "$x \in K$", where K is a subset of a set X, and x is an element of X. Then the truth value, denoted $\nu\big(\, x \in K \,\big)$, of the proposition "$x \in K$" is

$$\nu\big(\, x \in K \,\big) = \begin{cases} 1 & \text{if } x \text{ belongs to } K; \\ 0 & \text{otherwise.} \end{cases} \tag{13.95}$$

Thus the proposition "$x \in K$" is true if, and only if, x belongs to K. In other words, $x \mapsto \nu\big(\, x \in K \,\big)$ is the characteristic function of the subset K of X; cf. (13.512) in the Appendix.

This remark is the foundation of the assignment of truth values in classical physics. Specifically, if the state is $s \in \mathcal{S}$, the truth value, $\nu\big(A\,\varepsilon\,\Delta; s\big)$, of the *physical* proposition "$A\,\varepsilon\,\Delta$" is defined to be the truth value of the *mathematical* proposition "$\breve{A}(s) \in \Delta$"; or, equivalently, of the mathematical proposition "$s \in \breve{A}^{-1}(\Delta)$". Thus, using (13.95), we get, for all $s \in \mathcal{S}$,

$$\nu\big(A\,\varepsilon\,\Delta; s\big) := \begin{cases} 1 & \text{if } s \text{ belongs to } \breve{A}^{-1}(\Delta); \\ 0 & \text{otherwise.} \end{cases} \tag{13.96}$$

which reproduces (13.94).

We now consider the analogue of the above in a general topos τ. Let X be an object in τ, and let K be a sub-object of X. Then K is determined by a characteristic arrow $\chi_K : X \to \Omega_\tau$, where Ω_τ is the sub-object classifier; equivalently, we have an arrow $\ulcorner K \urcorner : 1_\tau \to PX$.

Now suppose that $x : 1_\tau \to X$ is a global element of X; i.e., $x \in \Gamma X := \mathrm{Hom}_\tau(1_\tau, X)$. Then the truth value of the mathematical proposition "$x \in K$" is defined to be

$$\nu\big(x \in K\big) := \chi_K \circ x \tag{13.97}$$

where $\chi_K \circ x : 1_\tau \to \Omega_\tau$. Thus $\nu\big(x \in K\big)$ is an element of $\Gamma\Omega_\tau$; i.e., it is a global element of the sub-object classifier Ω_τ.

The connection with the result (13.95) (in the topos **Sets**) can be seen by noting that, in (13.95), the characteristic function of the subset $K \subseteq X$ is the function $\chi_K : X \to \{0, 1\}$ such that $\chi_K(x) = 1$ if $x \in K$, and $\chi_K(x) = 0$ otherwise. It follows that (13.95) can be rewritten as

$$\nu\big(x \in K\big) = \chi_K(x) \tag{13.98}$$

$$= \chi_K \circ x \tag{13.99}$$

where in (13.99), x denotes the function $x : \{*\} \to X$ that is defined by $x(*) := x$. The link with (13.97) is clear when one remembers that, in the topos **Sets**, the terminal object, $1_{\mathbf{Sets}}$, is just the singleton set $\{*\}$.

In quantum theory, the topos is $\mathbf{Sets}^{\mathcal{V}(\mathcal{H})^{\mathrm{op}}}$, and so the objects are all presheaves. In particular, at each stage V, the sub-object classifier $\underline{\Omega}_V := \Omega_{\mathbf{Sets}^{\mathcal{V}(\mathcal{H})^{\mathrm{op}}}}(V)$ is the set of sieves on V (see e.g. [66]). In this case, if $\underline{K}$ is a sub-object of $\underline{X}$, and $x \in \Gamma\underline{X}$, the explicit form for (13.99) is the sieve

$$\nu\big(x \in \underline{K}\big)_V := \{V' \subseteq V \mid x_{V'} \in \underline{K}_{V'}\} \tag{13.100}$$

at each stage $V \in \mathrm{Ob}(\mathcal{V}(\mathcal{H}))$. In other words, at each stage/context V, the truth value of the mathematical proposition "$x \in \underline{K}$" is defined to be all those stages $V' \subseteq V$ "down the line" such that the "component", $x_{V'}$ of x at that stage *is* an element of the component, $\underline{K}_{V'} \subseteq \underline{X}_{V'}$, of $\underline{K}$ at that stage.

The definitions (13.97) and (13.100) play a central role in constructing truth values in our quantum topos scheme. However, as $\underline{\Sigma}$ has no global elements, these truth values cannot be derived from some expression $\nu\big(s \in \underline{K}\big)$ with $s : 1_{\mathbf{Sets}^{\mathcal{V}(\mathcal{H})^{\mathrm{op}}}} \to \underline{\Sigma}$. Therefore, we must proceed differently, as will become clear by the end of the following Section.

However, before we do so, let us make one final remark concerning (13.95). Namely, in normal set theory the proposition "$x \in K$" is true if, and only if,

$$\{x\} \subseteq K \tag{13.101}$$

i.e., if an only if the set $\{x\}$ is a subset of K. The transition from the proposition "$x \in K$" to the proposition "$\{x\} \subseteq K$" is seemingly trivial, but in a topos other than **Sets** it takes on a new significance. In particular, as we shall see shortly, although the spectral presheaf, $\underline{\Sigma}$, has no global elements, it does have certain 'minimal' sub-objects that are as 'close' as one can get to a global element, and then the topos analogue of (13.101) is very important.

13.6.2 Truth Objects

13.6.2.1 Linguistic Aspects of Truth Objects

To understand how "truth values" of physical propositions arise we return again to our earlier discussion of local languages. In this Section we will employ the local language $\mathcal{L}(S)$ rather than the propositional language, $\mathcal{PL}(S)$, that was used earlier in this article.

Thus, let $\mathcal{L}(S)$ be the local language for a system S. This is a typed language whose minimal set of ground type symbols is Σ and $\mathcal{R}$. In addition, there is a non-empty set, $F_{\mathcal{L}(S)}(\Sigma, \mathcal{R})$, of function symbols $A : \Sigma \to \mathcal{R}$ that correspond to the physical quantities of S.

Now consider a representation, ϕ, of $\mathcal{L}(S)$ in a topos τ_ϕ. As discussed earlier, the propositional aspects of the language $\mathcal{L}(S)$ are captured in the term "$A(\tilde{s}) \in \tilde{\Delta}$" of type Ω, where $\tilde{s}$ and $\tilde{\Delta}$ are variables of type Σ and $P\mathcal{R}$ respectively [27]. In a topos representation, ϕ, the representation, $[\![\, A(\tilde{s}) \in \tilde{\Delta} \,]\!]_\phi$, of the term "$A(\tilde{s}) \in \tilde{\Delta}$" is given by the chain of arrows[65] [11] (cf. (13.22))

$$\Sigma_\phi \times P\mathcal{R}_\phi \xrightarrow{A_\phi \times \mathrm{id}} \mathcal{R}_\phi \times P\mathcal{R}_\phi \xrightarrow{e_{\mathcal{R}_\phi}} \Omega_{\tau_\phi} \tag{13.102}$$

in the topos τ_ϕ. Then, if $\ulcorner \Xi \urcorner : 1_{\tau_\phi} \to P\mathcal{R}_\phi$ is the name of a sub-object, Ξ, of the quantity-value object $\mathcal{R}_\phi$, we get the chain

$$\Sigma_\phi \simeq \Sigma_\phi \times 1_{\tau_\phi} \xrightarrow{\mathrm{id} \times \ulcorner \Xi \urcorner} \Sigma_\phi \times P\mathcal{R}_\phi \xrightarrow{A_\phi \times \mathrm{id}} \mathcal{R}_\phi \times P\mathcal{R}_\phi \xrightarrow{e_{\mathcal{R}_\phi}} \Omega_{\tau_\phi}. \tag{13.103}$$

which is the characteristic arrow of the sub-object of Σ_ϕ that represents the physical proposition "$A \,\varepsilon\, \Xi$".

Equivalently, we can use the term, $\{\tilde{s} \mid A(\tilde{s}) \in \tilde{\Delta}\}$, which has a free variable $\tilde{\Delta}$ of type $P\mathcal{R}$ and is of type $P\Sigma$. This term is represented by the arrow $[\![\, \{\tilde{s} \mid A(\tilde{s}) \in \tilde{\Delta}\} \,]\!]_\phi : P\mathcal{R}_\phi \to P\Sigma_\phi$, which is the power transpose of $[\![\, A(\tilde{s}) \in \tilde{\Delta} \,]\!]_\phi$ (cf. (13.23)):

$$[\![\, \{\tilde{s} \mid A(\tilde{s}) \in \tilde{\Delta}\} \,]\!]_\phi = \ulcorner [\![\, A(\tilde{s}) \in \tilde{\Delta} \,]\!]_\phi \urcorner \tag{13.104}$$

[65] In (13.102), $e_{\mathcal{R}_\phi} : \mathcal{R}_\phi \times P\mathcal{R}_\phi \to \Omega_{\tau_\phi}$ is the evaluation arrow associated with the power object $P\mathcal{R}_\phi$.

The proposition "$A \,\varepsilon\, \Xi$" is then represented by the arrow $[\![\{\tilde{s} \mid A(\tilde{s}) \in \tilde{\Delta}\}]\!]_\phi \circ \ulcorner \Xi \urcorner : 1_{\tau_\phi} \to P\Sigma_\phi$; this is the name of the sub-object of Σ_ϕ that represents "$A \,\varepsilon\, \Xi$".

We note an important difference from the analogous situation for the language $\mathcal{PL}(S)$. In propositions of the type "$A \,\varepsilon\, \Delta$", the symbol "Δ" is a specific subset of $\mathbb{R}$ and is hence *external* to the language. In particular, it is *independent* of the representation of $\mathcal{PL}(S)$. However, in the case of $\mathcal{L}(S)$, the variable $\tilde{\Delta}$ is *internal* to the language, and the quantity Ξ in the proposition "$A \,\varepsilon\, \Xi$" is a sub-object of $\mathcal{R}_\phi$ in a *specific* topos representation, ϕ, of $\mathcal{L}(S)$.

So, this is how physical propositions are represented mathematically. But how are truth values to be assigned to these propositions? In the topos τ_ϕ a truth value is an element of the Heyting algebra $\Gamma\Omega_{\tau_\phi}$. Thus the challenge is to assign a global element of Ω_{τ_ϕ} to each proposition associated with the representation of the term $\{\tilde{s} \mid A(\tilde{s}) \in \tilde{\Delta}\}$ of type $P\Sigma$; (or, equivalently, the representation of the term "$A(\tilde{s}) \in \tilde{\Delta}$").

Let us first pose this question at a linguistic level. In a representation ϕ, an element of $\Gamma\Omega_{\tau_\phi}$ is associated with a representation of a term of type Ω with no free variables. Hence the question can be rephrased as asking how a term, t, in $\mathcal{L}(S)$ of type $P\Sigma$ can be "converted" into a term of type Ω? At this stage, we are happy to have free variables, in which case the desired term will be represented by an arrow in τ_ϕ whose co-domain is Ω_{τ_ϕ}, but whose domain is other than 1_{τ_ϕ}. This would be an intermediate stage to obtaining a global element of Ω_{τ_ϕ}.

In the context of the language $\mathcal{L}(S)$ there are three obvious ways of 'converting' the term t of type $P\Sigma$ to a term of type Ω:

1. Choose a term, s, of type Σ; then the term "$s \in t$" is of type Ω. We will call this the "microstate" option.
2. Choose a term, $\mathbb{T}$, of type $PP\Sigma$; then the term "$t \in \mathbb{T}$" is of type Ω. We shall refer to this as the "truth object" option.
3. Choose a term, $\mathfrak{w}$, of type $P\Sigma$; then the term "$\mathfrak{w} \subseteq t$" is of type Ω.[66] For reasons that will become clear later we shall refer to this as the "pseudo-state" option.

13.6.2.2 The Micro-State Option

In regard to the first option, the simplest example of a term of type Σ is a variable $\tilde{s_1}$ of type Σ. Then, the term "$\tilde{s_1} \in \{\tilde{s} \mid A(\tilde{s}) \in \tilde{\Delta}\}$" is of type Ω with the free variables $\tilde{s_1}$ and $\tilde{\Delta}$ of type Σ and $P\mathcal{R}$ respectively. However, the axiom of comprehension in $\mathcal{L}(S)$ says that

$$\tilde{s_1} \in \{\tilde{s} \mid A(\tilde{s}) \in \tilde{\Delta}\} \Leftrightarrow A(\tilde{s_1}) \in \tilde{\Delta} \tag{13.105}$$

and so we are back with the term "$A(\tilde{s}) \in \tilde{\Delta}$", which is of type Ω and with the free variable $\tilde{s}$ of type Σ.

---

[66] In general, if t and s are set-like terms (i.e., terms of power type, PX, say), then "$t \subseteq s$" is defined as the term '$\forall \tilde{x} \in t(\tilde{x} \in s)$'; here, $\tilde{x}$ is a variable of type X.

As stated above, the ϕ-representation, $[\![A(\tilde{s}) \in \tilde{\Delta}]\!]_\phi$, of "$A(\tilde{s}) \in \tilde{\Delta}$" is the chain of arrows in (13.102). Now, suppose the representation, ϕ, is such that there exist global elements, $s : 1_{\tau_\phi} \to \Sigma_\phi$, of Σ_ϕ. Then each such element can be regarded as a "(micro)state" of the system in that topos representation. Furthermore, let $\ulcorner \Xi \urcorner : 1_{\tau_\phi} \to P\mathcal{R}_\phi$ be the name of a sub-object, Ξ, of the quantity-value object $\mathcal{R}_\phi$. Then, by the basic property of the product $\Sigma_\phi \times P\mathcal{R}_\phi$, there is an arrow $\langle s, \ulcorner \Xi \urcorner \rangle : 1_{\tau_\phi} \to \Sigma_\phi \times P\mathcal{R}_\phi$. This can be combined with the arrow $[\![A(\tilde{s}) \in \tilde{\Delta}]\!]_\phi : \Sigma_\phi \times P\mathcal{R}_\phi \to \Omega_{\tau_\phi}$ to give the arrow

$$[\![A(\tilde{s}) \in \tilde{\Delta}]\!]_\phi \circ \langle s, \ulcorner \Xi \urcorner \rangle : 1_{\tau_\phi} \longrightarrow \Omega_{\tau_\phi} \tag{13.106}$$

This is the desired global element of Ω_{τ_ϕ}.

In other words, when the "state of the system" is $s \in \Gamma \Sigma_\phi$, the "truth value" of the proposition "$A \,\varepsilon\, \Xi$" is the global element of Ω_{τ_ϕ} given by the arrow $[\![A(\tilde{s}) \in \tilde{\Delta}]\!]_\phi \circ \langle s, \ulcorner \Xi \urcorner \rangle : 1_{\tau_\phi} \to \Omega_{\tau_\phi}$.

This is the procedure that is adopted in classical physics when a truth value is assigned to propositions by specifying a microstate, $s \in \Sigma_\sigma$, where Σ_σ is the classical state space in the representation σ of $\mathcal{L}(S)$. Specifically, for all $s \in \Sigma_\sigma$, the truth value of the proposition "$A \,\varepsilon\, \Delta$" as given by (13.106) is (c.f. (13.94))

$$\nu\big(A \,\varepsilon\, \Delta; s\big) = [\![A(\tilde{s}) \in \tilde{\Delta}]\!]_\sigma(s, \Delta) = \begin{cases} 1 & \text{if } A_\sigma(s) \in \Delta; \\ 0 & \text{otherwise.} \end{cases} \tag{13.107}$$

where $[\![A(\tilde{s}) \in \tilde{\Delta}]\!]_\sigma : \Sigma_\sigma \times P\mathbb{R} \to \Omega_{\tau_\sigma} \simeq \{0, 1\}$. Thus we recover the earlier result (13.96).

13.6.2.3 The Truth Object Option

By hindsight, we know that the option to use global elements of Σ_ϕ is not available in the quantum case. For there the state object, $\underline{\Sigma}$, is the spectral presheaf, and this has no global elements by virtue of the Kochen-Specker theorem. The absence of global elements of the state object Σ_ϕ could well be true in many other topos models of physics (particularly those that go "beyond quantum theory"), and therefore an alternative general strategy is needed to that employing microstates $\ulcorner s \urcorner : 1_{\tau_\phi} \to \Sigma_\phi$.

This takes us to the second possibility: namely, to introduce a term, $\mathbb{T}$, of type $PP\Sigma$, and then work with the term "$\{\tilde{s} \mid A(\tilde{s}) \in \tilde{\Delta}\} \in \mathbb{T}$", which is of type Ω, and has whatever free variables are contained in $\mathbb{T}$, plus the variable $\tilde{\Delta}$ of type $P\mathcal{R}$.

The simplest choice is to let the term of type $PP\Sigma$ be a variable, $\tilde{\mathbb{T}}$, of type $PP\Sigma$, in which case the term '$\{\tilde{s} \mid A(\tilde{s}) \in \tilde{\Delta}\} \in \tilde{\mathbb{T}}$' has variables $\tilde{\Delta}$ and $\tilde{\mathbb{T}}$ of type $P\mathcal{R}$ and $PP\Sigma$ respectively. Therefore, in a topos representation it is represented by an arrow $[\![\{\tilde{s} \mid A(\tilde{s}) \in \tilde{\Delta}\} \in \tilde{\mathbb{T}}]\!]_\phi : P\mathcal{R}_\phi \times P(P\Sigma_\phi) \to \Omega_{\tau_\phi}$. In detail (see [11]) we have that

$$[\![\{\tilde{s} \mid A(\tilde{s}) \in \tilde{\Delta}\} \in \tilde{\mathbb{T}}]\!]_\phi = e_{P\Sigma_\phi} \circ [\![\{\tilde{s} \mid A(\tilde{s}) \in \tilde{\Delta}\}]\!]_\phi \times [\![\tilde{\mathbb{T}}]\!]_\phi \tag{13.108}$$

where $e_{P\Sigma_\phi} : P\Sigma_\phi \times P(P\Sigma_\phi) \to \Omega_{\tau_\phi}$ is the usual evaluation arrow. In using this expression we need the ϕ-representatives:

$$[\![\{\tilde{s} \mid A(\tilde{s}) \in \tilde{\Delta}\}]\!]_\phi : P\mathcal{R}_\phi \to P\Sigma_\phi \tag{13.109}$$

$$[\![\tilde{\mathbb{T}}]\!]_\phi : P(P\Sigma_\phi) \xrightarrow{\text{id}} P(P\Sigma_\phi) \tag{13.110}$$

Finally, let $\langle \ulcorner \Xi \urcorner, \ulcorner \mathbb{T} \urcorner \rangle$ be a pair of global elements in $P\mathcal{R}_\phi$ and $P(P\Sigma_\phi)$ respectively, so that $\ulcorner \Xi \urcorner : 1_{\tau_\phi} \to P\mathcal{R}_\phi$ and $\ulcorner \mathbb{T} \urcorner : 1_{\tau_\phi} \to P(P\Sigma_\phi)$. Thus, $\ulcorner \mathbb{T} \urcorner$ is the name of a "truth object", $\mathbb{T}$, in τ_ϕ. Then, for the physical proposition "$A \,\varepsilon\, \Xi$", we have the truth value

$$\nu\big(A \,\varepsilon\, \Xi; \mathbb{T}\big) = [\![\{\tilde{s} \mid A(\tilde{s}) \in \tilde{\Delta}\} \in \tilde{\mathbb{T}}]\!]_\phi \circ \langle \ulcorner \Xi \urcorner, \ulcorner \mathbb{T} \urcorner \rangle : 1_{\tau_\phi} \to \Omega_{\tau_\phi} \tag{13.111}$$

where $\langle \ulcorner \Xi \urcorner, \ulcorner \mathbb{T} \urcorner \rangle : 1_{\tau_\phi} \to P\mathcal{R}_\phi \times P(P\Sigma_\phi)$.

A small generalisation:

Slightly more generally, if $\tilde{J}$ and $\tilde{\mathbb{T}}$ are variables of type $P\Sigma$ and $P(P\Sigma)$ respectively, the term of interest is '$\tilde{J} \in \tilde{\mathbb{T}}$'. In the representation, ϕ, of $\mathcal{L}(S)$, this term maps to an arrow $[\![\tilde{J} \in \tilde{\mathbb{T}}]\!]_\phi : P\Sigma_\phi \times P(P\Sigma_\phi) \to \Omega_{\tau_\phi}$. Here, $[\![\tilde{J} \in \tilde{\mathbb{T}}]\!]_\phi = e_{P\Sigma_\phi} \circ [\![\tilde{J}]\!]_\phi \times [\![\tilde{\mathbb{T}}]\!]_\phi$ where $[\![\tilde{J}]\!]_\phi : P\Sigma_\phi \overset{\text{id}}{\to} P\Sigma_\phi$ and $[\![\tilde{\mathbb{T}}]\!]_\phi : P(P\Sigma_\phi) \xrightarrow{\text{id}} P(P\Sigma_\phi)$. Let $\ulcorner J \urcorner, \ulcorner \mathbb{T} \urcorner$ be global elements of $P\Sigma_\phi$ and $P(P\Sigma_\phi)$ respectively, so that $\ulcorner J \urcorner : 1_{\tau_\phi} \to P\Sigma_\phi$ and $\ulcorner \mathbb{T} \urcorner : 1_{\tau_\phi} \to P(P\Sigma_\phi)$. Then the truth of the (mathematical) proposition "$J \in \mathbb{T}$" is

$$\begin{aligned} \nu\big(J \in \mathbb{T}\big) &= [\![\tilde{J} \in \tilde{\mathbb{T}}]\!]_\phi \circ \langle \ulcorner J \urcorner, \ulcorner \mathbb{T} \urcorner \rangle \\ &= e_{P\Sigma_\phi} \circ \langle \ulcorner J \urcorner, \ulcorner \mathbb{T} \urcorner \rangle : 1_{\tau_\phi} \to \Omega_{\tau_\phi} \end{aligned} \tag{13.112}$$

13.6.2.4 The Example of Classical Physics

If classical physics is studied this way, the general formalism simplifies, and the term "$\{\tilde{s} \mid A(\tilde{s}) \in \tilde{\Delta}\} \in \tilde{\mathbb{T}}$" is represented by the function $\nu\big(A \,\varepsilon\, \Delta; \mathbb{T}\big) := [\![\{\tilde{s} \mid A(\tilde{s}) \in \tilde{\Delta}\} \in \tilde{\mathbb{T}}]\!]_\sigma : P\mathbb{R} \times P(P\Sigma_\sigma) \to \Omega_{\mathbf{Sets}} \simeq \{0, 1\}$ defined by

$$\begin{aligned} \nu\big(A \,\varepsilon\, \Delta; \mathbb{T}\big) &:= [\![\{\tilde{s} \mid A(\tilde{s}) \in \tilde{\Delta}\} \in \tilde{\mathbb{T}}]\!]_\sigma(\Delta, \mathbb{T}) \\ &= \begin{cases} 1 & \text{if } \{s \in \Sigma_\sigma \mid A_\sigma(s) \in \Delta\} \in \mathbb{T}; \\ 0 & \text{otherwise} \end{cases} \end{aligned} \tag{13.113}$$

$$= \begin{cases} 1 & \text{if } A_\sigma^{-1}(\Delta) \in \mathbb{T}; \\ 0 & \text{otherwise} \end{cases} \tag{13.114}$$

for all $\mathbb{T} \in P(P\Sigma_\sigma)$. We can clearly see the sense in which the truth object $\mathbb{T}$ is playing the role of a state. Note that the result (13.114) of classical physics is a special case of (13.111).

To recover the usual truth values given in (13.107), an appropriate truth object, $\mathbb{T}^s$, must be associated with each microstate $s \in \Sigma_\sigma$. The correct choice is

$$\mathbb{T}^s := \{J \subseteq \Sigma_\sigma \mid s \in J\} \tag{13.115}$$

for each $s \in \Sigma_\sigma$. It is clear that $s \in A_\sigma^{-1}(\Delta)$ (or, equivalently, $A_\sigma(s) \in \Delta$) if, and only if, $A_\sigma^{-1}(\Delta) \in \mathbb{T}^s$. Hence (13.114) can be rewritten as

$$\nu\big(A\,\varepsilon\,\Delta; \mathbb{T}^s\big) = \begin{cases} 1 & \text{if } s \in A_\sigma^{-1}(\Delta); \\ 0 & \text{otherwise.} \end{cases} \tag{13.116}$$

which reproduces (13.107) once $\nu\big(A\,\varepsilon\,\Delta; s\big)$ is identified with $\nu\big(A\,\varepsilon\,\Delta; \mathbb{T}^s\big)$.

13.6.3 Truth Objects in Quantum Theory

13.6.3.1 Preliminary Remarks

We can now start to discuss the application of these ideas to quantum theory. In order to use (13.111) (or (13.112)) we need to construct concrete truth objects, $\underline{\mathbb{T}}$, in the topos $\tau_\phi := \mathbf{Sets}^{\mathcal{V}(\mathcal{H})^{\mathrm{op}}}$. Thus the presheaf $\underline{\mathbb{T}}$ is a sub-object of $P\underline{\Sigma}$; equivalently, $\ulcorner\underline{\mathbb{T}}\urcorner : 1_{\tau_\phi} \to P(P\underline{\Sigma})$.

However, we have to keep in mind the need to restrict to *clopen* sub-objects of $\underline{\Sigma}$. In particular, we must show that there is a well-defined presheaf $P_{\mathrm{cl}}\underline{\Sigma}$ such that

$$\mathrm{Sub}_{\mathrm{cl}}(\underline{\Sigma}) \simeq \Gamma(P_{\mathrm{cl}}\underline{\Sigma}) \tag{13.117}$$

We will prove this in Sect. 13.6.5. Given (13.117) and $\underline{J} \in \mathrm{Sub}_{\mathrm{cl}}(\underline{\Sigma})$, it is then clear that a truth object, $\underline{\mathbb{T}}$, actually has to be a sub-object of $P_{\mathrm{cl}}\underline{\Sigma}$ in order that the valuation $\nu\big(\underline{J} \in \underline{\mathbb{T}}\big)$ in (13.112) is meaningful.

This truth value, $\nu\big(\underline{J} \in \underline{\mathbb{T}}\big)$, is a global element of $\underline{\Omega}$, and in the topos of presheaves, $\mathbf{Sets}^{\mathcal{V}(\mathcal{H})^{\mathrm{op}}}$, we have (see (13.100))

$$\nu\big(\underline{J} \in \underline{\mathbb{T}}\big)_V := \{V' \subseteq V \mid \underline{J}_{V'} \in \underline{\mathbb{T}}_{V'}\} \tag{13.118}$$

for each context V.

There are various examples of the presheaf $\underline{J}$ that are of interest to us. In particular, let $\underline{J} = \underline{\delta(\hat{P})}$ for some projector $\hat{P}$. Then, using the propositional language $\mathcal{PL}(S)$ introduced earlier, the 'truth' of the proposition represented by $\hat{P}$ (for example, "$A\,\varepsilon\,\Delta$") is

$$\nu\big(\underline{\delta(\hat{P})} \in \underline{\mathbb{T}}\big)_V = \{V' \subseteq V \mid \underline{\delta(\hat{P})}_{V'} \in \underline{\mathbb{T}}_{V'}\} \tag{13.119}$$

for all stages V.

When using the local language $\mathcal{L}(S)$, an important class of examples of the sub-object $\underline{J}$ of $\underline{\Sigma}$ are of the form $A_{\phi}^{-1}(\underline{\Xi})$, for some sub-object $\underline{\Xi}$ of $\underline{\mathcal{R}}$. This will yield the truth value, $\nu\big(A \,\varepsilon\, \underline{\Xi}; \underline{\mathbb{T}}\big)$, in (13.111). However, to discuss this further requires the representation of function symbols $A : \Sigma \to \mathcal{R}$ in the topos $\mathbf{Sets}^{\mathcal{V}(\mathcal{H})^{\mathrm{op}}}$, and this is deferred until Sect. 13.7.

13.6.3.2 The Truth Objects $\underline{\mathbb{T}}^{|\psi\rangle}$

The definition of truth objects in quantum theory was studied in the original papers [48–51]. It was shown there that to each quantum state $|\psi\rangle \in \mathcal{H}$, there corresponds a truth object, $\underline{\mathbb{T}}^{|\psi\rangle}$, which was defined as the following sub-object of the outer presheaf, $\underline{O}$:

$$\begin{aligned}\underline{\mathbb{T}}^{|\psi\rangle}_V &:= \{\hat{\alpha} \in \underline{O}_V \mid \mathrm{Prob}(\hat{\alpha};\, |\psi\rangle) = 1\} \\ &= \{\hat{\alpha} \in \underline{O}_V \mid \langle\psi|\,\hat{\alpha}\,|\psi\rangle = 1\}\end{aligned} \tag{13.120}$$

for all stages $V \in \mathrm{Ob}(\mathcal{V}(\mathcal{H}))$. Here, $\mathrm{Prob}(\hat{\alpha};\, |\psi\rangle)$ is the usual expression for the probability that the proposition represented by the projector $\hat{\alpha}$ is true, given that the quantum state is the (normalised) vector $|\psi\rangle$.

It is easy to see that (13.120) defines a genuine sub-object $\underline{\mathbb{T}}^{|\psi\rangle} = \{\underline{\mathbb{T}}^{|\psi\rangle}_V \mid V \in \mathrm{Ob}(\mathcal{V}(\mathcal{H}))\}$ of $\underline{O}$. Indeed, if $\hat{\beta} \succeq \hat{\alpha}$, then $\langle\psi|\,\hat{\beta}\,|\psi\rangle \geq \langle\psi|\,\hat{\alpha}\,|\psi\rangle$, and therefore, if $V' \subseteq V$ and $\hat{\alpha} \in \underline{O}_V$, then $\langle\psi|\,\underline{O}(i_{V'V})(\hat{\alpha})\,|\psi\rangle \geq \langle\psi|\,\hat{\alpha}\,|\psi\rangle$. In particular, if $\langle\psi|\,\hat{\alpha}\,|\psi\rangle = 1$ then $\langle\psi|\,\underline{O}(i_{V'V})(\hat{\alpha})\,|\psi\rangle = 1$.

The next step is to define the presheaf $P_{\mathrm{cl}}\underline{\Sigma}$, and show that there is a monic arrow $\underline{O} \to P_{\mathrm{cl}}\underline{\Sigma}$, so that $\underline{O}$ is a sub-object of $P_{\mathrm{cl}}\underline{\Sigma}$. Then, since $\underline{\mathbb{T}}^{|\psi\rangle}$ is a sub-object of $\underline{O}$, and $\underline{O}$ is a sub-object of $P_{\mathrm{cl}}\underline{\Sigma}$, it follows that $\underline{\mathbb{T}}^{|\psi\rangle}$ is a sub-object of $P_{\mathrm{cl}}\underline{\Sigma}$, as required. The discussion of the construction of $P_{\mathrm{cl}}\underline{\Sigma}$ is deferred to Sect. 13.6.5 so as not to break the flow of the presentation.

With this definition of $\underline{\mathbb{T}}^{|\psi\rangle}$, the truth value, (13.119), for the propositional language $\mathcal{PL}(S)$ becomes

$$\nu\big(\underline{\delta(\hat{P})} \in \underline{\mathbb{T}}^{|\psi\rangle}\big)_V = \{V' \subseteq V \mid \langle\psi|\,\delta(\hat{P})_{V'}\,|\psi\rangle = 1\} \tag{13.121}$$

It is easy to see that the definition of a truth object in (13.120) can be extended to a mixed state with a density-matrix operator $\hat{\rho}$: simply replace the definition in (13.120) with

$$\begin{aligned}\underline{\mathbb{T}}^{\hat{\rho}}_V &:= \{\hat{\alpha} \in \underline{O}_V \mid \mathrm{Prob}(\hat{\alpha};\, \rho) = 1\} \\ &= \{\hat{\alpha} \in \underline{O}_V \mid \mathrm{tr}(\hat{\rho}\hat{\alpha}) = 1\}\end{aligned} \tag{13.122}$$

However there is an important difference between the truth object associated with a vector state, $|\psi\rangle$, and the one associated with a density matrix, ρ. In the vector case, it is easy to see that the mapping $|\psi\rangle \to \underline{\mathbb{T}}^{|\psi\rangle}$ is one-to-one (up to a phase factor on $|\psi\rangle$) so that, in principle, the state $|\psi\rangle$ can be *recovered* from $\underline{\mathbb{T}}^{|\psi\rangle}$ (up to a phase-factor). On the other hand, there are simple counterexamples which show that, in general, the density matrix, ρ *cannot* be recovered from $\underline{\mathbb{T}}^{\hat{\rho}}$.

In a sense, this should not surprise us. The analogue of a density matrix in classical physics is a probability measure μ defined on the classical state space $\mathcal{S}$. Individual microstates $s \in \mathcal{S}$ are in one-to-one correspondence with probability measures of the form μ_s defined by $\mu_s(J) = 1$ if $s \in J$, $\mu_s(J) = 0$ if $s \notin J$.

However, one of the main claims of our programme is that any theory can be made to "look like" classical physics in the appropriate topos. This suggests that, in the topos version of quantum theory, a density matrix should be represented by some sort of measure on the state object $\underline{\Sigma}$ in the topos τ_ϕ.

It was shown in [24] that this is indeed true: every state of the von Neumann algebra $B(\mathcal{H})$ gives a certain kind of probability measure on (the clopen sub-objects of) the spectral presheaf, and conversely, each such measure determines a unique state on $B(\mathcal{H})$. Moreover, this result holds more generally for all von Neumann algebras with no direct summand of type I_2.[67]

13.6.4 The Pseudo-State Option

13.6.4.1 Some Background Remarks

We turn now to the third way mentioned above whereby a term, t, of type $P\Sigma$ in $\mathcal{L}(S)$ can be "converted" to a term of type Ω. Namely, choose a term, $\mathfrak{w}$, of type $P\Sigma$ and then use "$\mathfrak{w} \subseteq t$". As we shall see, this idea is easy to implement in the case of quantum theory and leads to an alternative way of thinking about truth objects.

Let us start by considering once more the case of classical physics. There, for each microstate s in the symplectic state manifold Σ_σ, there is an associated truth object, $\mathbb{T}^s$, defined by $\mathbb{T}^s := \{J \subseteq \Sigma_\sigma \mid s \in J\}$, as in (13.115). It is clear that the state s can be uniquely recovered from the collection of sets $\mathbb{T}^s$ as

$$s = \bigcap \{J \subseteq \Sigma_\sigma \mid s \in J\} \tag{13.123}$$

Note that (13.123) implies that $\mathbb{T}^s$ is an *ultrafilter* of subsets of Σ_σ.[68] As we shall shortly see, there is an intriguing analogue of this property for the quantum truth objects.

---

[67] Some related results on the topos-internal representation of states of a C^*-algebra can be found in the recent work by Heunen et al. [42].

[68] Let $\mathbb{L}$ be a lattice with zero element 0. A subset $F \subset \mathbb{L}$ is a "filter base" if (i) $0 \notin F$ and (ii) for all $a, b \in F$, there is some $c \in F$ such that $c \leq a \wedge b$. A subset $D \subset \mathbb{L}$ is called a "(proper) dual ideal" or a "filter" if (i) $0 \notin D$, (ii) for all $a, b \in D$, $a \wedge b \in D$ and (iii) $a \in D$ and $b > a$ implies $b \in D$. A maximal dual ideal/filter F in a complemented, *distributive* lattice $\mathbb{L}$ is called an "ultrafilter". It has the property that for all $a \in \mathbb{L}$, either $a \in F$ or $a' \in F$, where a' is the complement of a.

The analogue of (13.123) in the case of quantum theory is rather interesting. Now, of course, there are no microstates, but we do have the truth objects defined in (13.120), one for each vector state $|\psi\rangle \in \mathcal{H}$. To proceed further we note that $\langle\psi|\,\hat{\alpha}\,|\psi\rangle = 1$ if and only if $|\psi\rangle\langle\psi| \preceq \hat{\alpha}$. Thus $\underline{\mathbb{T}}^{|\psi\rangle}$ can be rewritten as

$$\underline{\mathbb{T}}_V^{|\psi\rangle} := \{\hat{\alpha} \in \underline{O}_V \mid |\psi\rangle\langle\psi| \preceq \hat{\alpha}\} \tag{13.124}$$

for each stage V. Note that, as defined in (13.124), $\underline{\mathbb{T}}^{|\psi\rangle}$ is a sub-object of $\underline{O}$; i.e., it is defined in terms of projection operators. However, as will be shown in Section 13.6.5.2, there is a monic arrow $\underline{O} \to P_{\rm cl}\underline{\Sigma}$, and by using this arrow, $\underline{\mathbb{T}}^{|\psi\rangle}$ can be regarded as a sub-object of $P_{\rm cl}\underline{\Sigma}$; hence $\Gamma\underline{\mathbb{T}}^{|\psi\rangle}$ is a collection of clopen sub-objects of $\underline{\Sigma}$. In this form, the definition of $\underline{\mathbb{T}}^{|\psi\rangle}$ involves clopen subsets of the spectral sets $\underline{\Sigma}_V$, $V \in \mathrm{Ob}(\mathcal{V}(\mathcal{H}))$.

It is clear from (13.124) that, for each V, $\underline{\mathbb{T}}_V^{|\psi\rangle}$ is a *filter* of projection operators in $\underline{O}_V \simeq \mathcal{P}(V)$; equivalently, it is a filter of clopen sub-sets of $\underline{\Sigma}_V$.

These ordering properties are associated with the following observation. If $|\psi\rangle$ is any vector state, we can collect together all the projection operators that are 'larger' or equal to $|\psi\rangle\langle\psi|$ and define:

$$T^{|\psi\rangle} := \{\hat{\alpha} \in \mathcal{P}(\mathcal{H}) \mid |\psi\rangle\langle\psi| \preceq \hat{\alpha}\} \tag{13.125}$$

It is clear that, for all stages/contexts $V \in \mathrm{Ob}(\mathcal{V}(\mathcal{H}))$, we have

$$\underline{\mathbb{T}}_V^{|\psi\rangle} = T^{|\psi\rangle} \cap V \tag{13.126}$$

Thus the presheaf $\underline{\mathbb{T}}^{|\psi\rangle}$ is obtained by "localising" $T^{|\psi\rangle}$ at each context V.

The significance of this localisation property is that $T^{|\psi\rangle}$ is a *maximal* (proper) filter in the non-distributive lattice, $\mathcal{P}(\mathcal{H})$, of all projection operators on $\mathcal{H}$. Such maximal filters in the projection lattices of von Neumann algebras were extensively discussed by de Groote [38] who called them "quasi-points". In particular, $T^{|\psi\rangle}$ is a, so-called, "atomic" quasi-point in $\mathcal{P}(\mathcal{H})$. Every pure state $|\psi\rangle$ gives rise to an atomic quasi-point, $T^{|\psi\rangle}$, and vice versa. We will return to these entities in Sect. 13.8.4.

13.6.4.2 Using Pseudo-States in Lieu of Truth Objects

The equation (13.123) from classical physics suggests that, in the quantum case, we look at the set-valued function on $\mathrm{Ob}(\mathcal{V}(\mathcal{H}))$ defined by

$$V \mapsto \bigwedge\{\hat{\alpha} \in \underline{\mathbb{T}}_V^{|\psi\rangle}\} = \bigwedge\{\hat{\alpha} \in \underline{O}_V \mid |\psi\rangle\langle\psi| \preceq \hat{\alpha}\} \tag{13.127}$$

where we have used (13.124) as the definition of $\underline{\mathbb{T}}^{|\psi\rangle}$. It is easy to check that this is a global element of $\underline{O}$; in fact, the right hand side of (13.127) is nothing but the outer daseinisation $\delta(\,|\psi\rangle\langle\psi|\,)$ of the projection operator $|\psi\rangle\langle\psi|$! Evidently, the quantity

$$\mathfrak{w}^{|\psi\rangle} := \delta(\,|\psi\rangle\langle\psi|\,) = V \mapsto \bigwedge\{\hat{\alpha} \in \underline{O}_V \mid |\psi\rangle\langle\psi| \preceq \hat{\alpha}\} \tag{13.128}$$

is of considerable interest. We shall refer to it as a "pseudo-state" for reasons that appear below.

Note that $\mathfrak{w}^{|\psi\rangle}$ is defined by (13.128) as an element of $\Gamma\underline{O}$. However, because of the monic $\underline{O} \to P_{\rm cl}\underline{\Sigma}$ we can also regard $\mathfrak{w}^{|\psi\rangle}$ as an element of $\Gamma(P_{\rm cl}\underline{\Sigma}) \simeq \mathrm{Sub}_{\rm cl}(\underline{\Sigma})$. The corresponding (clopen) sub-object of $\underline{\Sigma}$ will be denoted $\underline{\mathfrak{w}}^{|\psi\rangle} := \underline{\delta(|\psi\rangle\langle\psi|)}$.

We know that the map $|\psi\rangle \mapsto \underline{\mathbb{T}}^{|\psi\rangle}$ is injective. What can be said about the map $|\psi\rangle \mapsto \underline{\mathfrak{w}}^{|\psi\rangle}$? In this context, we note that $\underline{\mathbb{T}}^{|\psi\rangle}$ is readily recoverable from $\mathfrak{w}^{|\psi\rangle} \in \Gamma\underline{O}$ as

$$\underline{\mathbb{T}}_V^{|\psi\rangle} = \{\hat{\alpha} \in \underline{O}_V \mid \hat{\alpha} \succeq \mathfrak{w}_V^{|\psi\rangle}\} \tag{13.129}$$

for all contexts V. From these relations it follows that is $|\psi\rangle \mapsto \underline{\mathfrak{w}}^{|\psi\rangle}$ is injective.

Note that (13.129) essentially follows from the fact that, for each V, the collection, $\underline{\mathbb{T}}_V^{|\psi\rangle}$ of projectors in $\underline{O}_V$ is an *upper* set (in fact, as remarked earlier, it is a filter). In this respect, the projectors/clopen subsets $\underline{\mathbb{T}}_V^{|\psi\rangle}$ behave like the filter of *clopen neighbourhoods* of a subset in a topological space. This remark translates globally to the relation of the collection, $\Gamma\underline{\mathbb{T}}^{|\psi\rangle}$, of sub-objects of $\underline{\Sigma}$ to the specific sub-object $\underline{\mathfrak{w}}^{|\psi\rangle}$.

It follows that there is a one-to-one correspondence between truth objects, $\underline{\mathbb{T}}^{|\psi\rangle}$, and pseudo-states, $\underline{\mathfrak{w}}^{|\psi\rangle}$. However, the former is (a representation of) a term of type $P(P\Sigma)$, whereas the latter is of type $P\Sigma$. So how is this reflected in the assignment of generalised truth values?

Note first that, from the definition of $\mathfrak{w}^{|\psi\rangle}$, it follows that if $\hat{\alpha} \in \underline{\mathbb{T}}_V^{|\psi\rangle}$ then $\hat{\alpha} \succeq \mathfrak{w}_V^{|\psi\rangle}$. On the other hand, from (13.129) we have that if $\hat{\alpha} \succeq \mathfrak{w}_V^{|\psi\rangle}$ then $\hat{\alpha} \in \underline{\mathbb{T}}_V^{|\psi\rangle}$. Thus we have the simple, but important, result:

$$\hat{\alpha} \in \underline{\mathbb{T}}_V^{|\psi\rangle} \text{ if, and only if } \hat{\alpha} \succeq \mathfrak{w}_V^{|\psi\rangle} \tag{13.130}$$

In particular, for any projector $\hat{P}$ we have $\delta(\hat{P})_V \in \underline{\mathbb{T}}_V^{|\psi\rangle}$ if, and only if $\delta(\hat{P})_V \succeq \mathfrak{w}_V^{|\psi\rangle}$.

In terms of sub-objects of $\underline{\Sigma}$, we have $\delta(\hat{P})_V \succeq \mathfrak{w}_V^{|\psi\rangle}$ if and only if $\underline{\delta(\hat{P})}_V \supseteq \underline{\mathfrak{w}}_V^{|\psi\rangle}$. Hence, (13.130) can be rewritten as

$$\delta(\hat{P})_V \in \underline{\mathbb{T}}_V^{|\psi\rangle} \text{ if, and only if } \underline{\delta(\hat{P})}_V \supseteq \underline{\mathfrak{w}}_V \tag{13.131}$$

and so (13.119) can be written as

$$\nu\big(\underline{\delta(\hat{P})} \in \underline{\mathbb{T}}\big)_V = \{V' \subseteq V \mid \underline{\delta(\hat{P})}_{V'} \supseteq \underline{\mathfrak{w}}_{V'}^{|\psi\rangle}\} \tag{13.132}$$

However, the right hand side of (13.132) is just the topos truth value, $\nu(\underline{\mathfrak{w}}^{|\psi\rangle} \subseteq \underline{\delta(\hat{P})})$. It follows that

$$\text{“}\underline{\delta(\hat{P})} \in \underline{\mathbb{T}}^{|\psi\rangle}\text{” is equivalent to “}\underline{\mathfrak{w}}^{|\psi\rangle} \subseteq \underline{\delta(\hat{P})}\text{”} \tag{13.133}$$

and hence we can use the generalised truth values $\nu\big(\underline{\delta(\hat{P})} \in \underline{\mathbb{T}}^{|\psi\rangle}\big)$ or $\nu\big(\underline{\mathfrak{w}}^{|\psi\rangle} \subseteq \underline{\delta(\hat{P})}\big)$ interchangeably.

Thus, if desired, a truth object in quantum theory can be regarded as a sub-object of $\underline{\Sigma}$, rather than a sub-object of $P\underline{\Sigma}$. In a sense, these sub-objects, $\underline{\mathfrak{w}}^{|\psi\rangle}$, of $\underline{\Sigma}$ are the "closest" we can get to global elements of $\underline{\Sigma}$. This is why we call them 'pseudo-states'. However, note that a pseudo-state is not a *minimal* element of the Heyting algebra $\mathrm{Sub_{cl}}(\underline{\Sigma})$ since there are clopen sub-objects $\underline{S}$ that include stalks that are empty sets, something that is not possible for a pseudo-state.[69]

13.6.4.3 Linguistic Implications

The result (13.133) is very suggestive for a more general development. In our existing treatment, in the formal language $\mathcal{L}(S)$ we have concentrated on propositions of the form "$\tilde{J} \in \tilde{\mathbb{T}}$" which, in a representation ϕ, maps to the arrow $[\![\tilde{J} \in \tilde{\mathbb{T}}]\!]_\phi : P\Sigma_\phi \times P(P\Sigma_\phi) \to \Omega_{\tau_\phi}$. Here $\tilde{J}$ and $\tilde{\mathbb{T}}$ are variables of type $P\Sigma$ and $P(P\Sigma)$ respectively.

> What is suggested by the discussion above is that we could equally focus on terms of the form "$\tilde{\mathfrak{w}} \subseteq \tilde{J}$", where both $\tilde{\mathfrak{w}}$ and $\tilde{J}$ are variables of type $P\Sigma$.

Note that, in general, the ϕ-representation of such a term is of the form

$$[\![\tilde{\mathfrak{w}} \subseteq \tilde{J}]\!]_\phi : P\Sigma_\phi \times P\Sigma_\phi \to \Omega_{\tau_\phi} \tag{13.134}$$

where the "first slot" on the right hand side of the pairing in (13.134) is a truth object (in pseudo-state form), and the second correspond to a proposition represented by a sub-object of Σ_ϕ.

However, this raises the rather obvious question "What *is* a pseudo-state?". More precisely, we would like to know a generic set of characteristic properties of those sub-objects of Σ_ϕ that can be regarded as "pseudo-states". A first step would be to answer this question in the case of quantum theory. In particular, are there any quantum pseudo-states that are *not* of the form $\underline{\mathfrak{w}}^{|\psi\rangle}$ for some vector $|\psi\rangle \in \mathcal{H}$?

In this context the localisation property expressed by (13.125) is rather suggestive. In the case that $\mathcal{H}$ has infinite dimension, de Groote has shown that there exist

---

[69] Note that the sub-objects $\underline{\mathfrak{w}}^{|\psi\rangle}$ do not have any global elements since any such would give a global element of $\underline{\Sigma}$ and, of course, there are none. Thus if one is seeking examples of presheaves with no global elements, the collection $\underline{\mathfrak{w}}^{|\psi\rangle}$, $|\psi\rangle \in \mathcal{H}$, afford many such.

quasi-points in $\mathcal{P}(\mathcal{H})$ that are not of the form $T^{|\psi\rangle}$ for some $|\psi\rangle \in \mathcal{H}$ [38].[70] If T is any such quasi-point, (13.125) suggests strongly that we define an associated presheaf, $\underline{T}$, by

$$\underline{T}_V = T \cap V \tag{13.135}$$

for all $V \in \mathrm{Ob}(\mathcal{V}(\mathcal{H}))$. This construction seems natural enough from a mathematical perspective, but we are not yet clear of the physical significance of the existence of such "quasi truth objects". The same applies to the associated "quasi pseudo-state", $\underline{\mathfrak{w}}^T$, defined by

$$\underline{\mathfrak{w}}^T_V := \bigwedge\{\hat{\alpha} \in \underline{T}_V\} = \bigwedge\{\hat{\alpha} \in T \cap V\} \tag{13.136}$$

13.6.4.4 Time-Dependence and the Truth Object

As emphasised at the end of Sect. 13.3.2, the question of time dependence depends on the theory-type being considered. The structure of the language $\mathcal{L}(S)$ that has been used so far is such that the time variable lies outside the language. In this situation, the time dependence of the system can be implemented in several ways.

For example, we can make the truth object time dependent, giving a family of truth objects, $t \mapsto \underline{\mathbb{T}}^t$, $t \in \mathbb{R}$. In the case of classical physics, with the truth objects $\mathbb{T}^s$, $s \in \Sigma_\sigma$, the time evolution comes from the time dependence, $t \mapsto s_t$, of the microstate in accordance with the classical equations of motion. This gives the family $t \mapsto \mathbb{T}^{s_t}$ of truth objects.

Something very similar happens in quantum theory, and we acquire a family, $t \mapsto \underline{\mathbb{T}}^{|\psi\rangle_t}$, of truth objects, where the states $|\psi\rangle_t$ satisfy the usual time-dependent Schrödinger equation. Thus both classical and quantum truth objects belong to a "Schrödinger picture" of time evolution. Of course, there is a pseudo-state analogue of this in which we get a one-parameter family, $t \mapsto \underline{\mathfrak{w}}^{|\psi\rangle_t}$, of clopen sub-objects of $\underline{\Sigma}$.

It is also possible to construct a "Heisenberg picture" where the truth object is constant but the physical quantities and associated propositions are time dependent. We will return to this in Sect. 13.10 when we discuss the use of unitary operators.

13.6.5 The Presheaf $P_{\mathrm{cl}}(\underline{\Sigma})$

13.6.5.1 The Definition of $P_{\mathrm{cl}}(\underline{\Sigma})$

We must now show that there really is a presheaf $P_{\mathrm{cl}}\underline{\Sigma}$.

The easiest way of defining $P_{\mathrm{cl}}\underline{\Sigma}$ is to start with the concrete expression for the normal power object $P\underline{\Sigma}$ [34]. First, if $\underline{F}$ is any presheaf over $\mathcal{V}(\mathcal{H})$, define the

[70] However, he has also shown that, in an appropriate topology, the set of all atomic quasi-points is *dense* in the set of all quasi-points. Of course, none of these intriguing structures arise in a finite-dimensional Hilbert space. So, in that sense, it is unlikely that they will play any fundamental role in explicating the topos representation of quantum theory.

restriction of $\underline{F}$ to V to be the functor $\underline{F}\downarrow V$ from the category[71] $\downarrow V$ to **Sets** that assigns to each $V_1 \subseteq V$, the set $\underline{F}_{V_1}$, and with the obvious induced presheaf maps.

Then, at each stage V, $P\underline{\Sigma}_V$ is the set of natural transformations from $\underline{\Sigma}\downarrow V$ to $\underline{\Omega}\downarrow V$. These are in one-to-one correspondence with families of maps $\sigma := \{\sigma_{V_1} : \underline{\Sigma}_{V_1} \to \underline{\Omega}_{V_1} \mid V_1 \subseteq V\}$, with the following commutative diagram for all $V_2 \subseteq V_1 \subseteq V$:[72]

$$\begin{array}{ccc} \underline{\Sigma}_{V_1} & \xrightarrow{\sigma_{V_1}} & \underline{\Omega}_{V_1} \\ {\scriptstyle \underline{\Sigma}(i_{V_2 V_1})}\downarrow & & \downarrow{\scriptstyle \underline{\Omega}(i_{V_2 V_1})} \\ \underline{\Sigma}_{V_2} & \xrightarrow[\sigma_{V_2}]{} & \underline{\Omega}_{V_2} \end{array}$$

The presheaf maps are defined by

$$P\underline{\Sigma}(i_{V_1 V}) : P\underline{\Sigma}_V \to P\underline{\Sigma}_{V_1} \tag{13.137}$$
$$\sigma \mapsto \{\sigma_{V_2} \mid V_2 \subseteq V_1\} \tag{13.138}$$

and the evaluation arrow $\mathrm{ev} : P\underline{\Sigma} \times \underline{\Sigma} \to \underline{\Omega}$, has the form, at each stage V:

$$\mathrm{ev}_V : P\underline{\Sigma}_V \times \underline{\Sigma}_V \to \underline{\Omega}_V \tag{13.139}$$
$$(\sigma, \lambda) \mapsto \sigma_V(\lambda) \tag{13.140}$$

Moreover, in general, given a map $\chi : \underline{\Sigma}_V \to \underline{\Omega}_V$, the subset of $\underline{\Sigma}_V$ associated with the corresponding sub-object is $\chi^{-1}(1)$, where 1 is the unit ("truth") in the Heyting algebra $\underline{\Omega}_V$.

This suggests strongly that an object, $P_{\mathrm{cl}}\underline{\Sigma}$, in $\mathbf{Sets}^{\mathcal{V}(\mathcal{H})^{\mathrm{op}}}$ can be defined using the same definition of $P\underline{\Sigma}$ as above, except that the family of maps $\sigma := \{\sigma_{V_1} : \underline{\Sigma}_{V_1} \to \underline{\Omega}_{V_1} \mid V_1 \subseteq V\}$ must be such that, for all $V_1 \subseteq V$, $\sigma_{V_1}^{-1}(1)$ is a *clopen* subset of the (extremely disconnected) Hausdorff space $\underline{\Sigma}_{V_1}$. It is straightforward to check that such a restriction is consistent, and that $\mathrm{Sub}_{\mathrm{cl}}(\underline{\Sigma}) \simeq \Gamma(P_{\mathrm{cl}}\underline{\Sigma})$ as required.

13.6.5.2 The Monic Arrow From $\underline{O}$ to $P_{\mathrm{cl}}(\underline{\Sigma})$

We define $\iota : \underline{O} \times \underline{\Sigma} \to \underline{\Omega}$, with the power transpose $\ulcorner\iota\urcorner : \underline{O} \to P_{\mathrm{cl}}\underline{\Sigma}$, as follows. First recall that in any topos, τ there is a bijection $\mathrm{Hom}_\tau(A, C^B) \simeq \mathrm{Hom}_\tau(A \times B, C)$, and hence, in particular, (using $P\underline{\Sigma} = \underline{\Omega}^{\underline{\Sigma}}$)

---

[71] The notation $\downarrow V$ means the partially-ordered set of all sub-algebras $V' \subseteq V$.

[72] Note that any sub-object, $\underline{J}$ of $\underline{\Sigma}$, gives rise to such a natural transformation from $\underline{\Sigma}\downarrow V$ to $\underline{\Omega}\downarrow V$ for all stages V. Namely, for all $V_1 \subseteq V$, $\sigma_{V_1} : \underline{\Sigma}_{V_1} \to \underline{\Omega}_{V_1}$ is defined to be the characteristic arrow $\chi_{\underline{J}_{V_1}} : \underline{\Sigma}_{V_1} \to \underline{\Omega}_{V_1}$ of the sub-object $\underline{J}$ of $\underline{\Sigma}$.

$$\mathrm{Hom}_{\mathbf{Sets}^{\mathcal{V}(\mathcal{H})^{\mathrm{op}}}}(\underline{O}, P\underline{\Sigma}) \simeq \mathrm{Hom}_{\mathbf{Sets}^{\mathcal{V}(\mathcal{H})^{\mathrm{op}}}}(\underline{O} \times \underline{\Sigma}, \underline{\Omega}). \tag{13.141}$$

Now let $\hat{\alpha} \in \mathcal{P}(V)$, and let $S_{\hat{\alpha}} := \{\lambda \in \underline{\Sigma}_V \mid \langle\lambda, \hat{\alpha}\rangle = 1\}$ be the clopen subset of $\underline{\Sigma}_V$ that corresponds to the projector $\hat{\alpha}$ via the spectral theorem; see (13.56). Then we define $\iota : \underline{O} \times \underline{\Sigma} \to \underline{\Omega}$ at stage V by

$$\iota_V(\hat{\alpha}, \lambda) := \{V' \subseteq V \mid \underline{\Sigma}(i_{V'V})(\lambda) \in S_{\underline{O}(i_{V'V})(\hat{\alpha})}\} \tag{13.142}$$

for all $(\hat{\alpha}, \lambda) \in \underline{O}_V \times \underline{\Sigma}_V$.

On the other hand, the basic result relating coarse-graining to subsets of $\underline{\Sigma}$ is

$$S_{\underline{O}(i_{V'V})(\delta(\hat{\alpha})_V)} = \underline{\Sigma}(i_{V'V})(S_{\delta(\hat{\alpha})_V}) \tag{13.143}$$

for all $V' \subseteq V$ and for all $\hat{\alpha} \in \underline{O}_V$. It follows that

$$\iota_V(\hat{\alpha}, \lambda) := \{V' \subseteq V \mid \underline{\Sigma}(i_{V'V})(\lambda) \in \underline{\Sigma}(i_{V'V})(S_{\hat{\alpha}})\} \tag{13.144}$$

for all $(\hat{\alpha}, \lambda) \in \underline{O}_V \times \underline{\Sigma}_V$. In this form is is clear that $\iota_V(\hat{\alpha}, \lambda)$ is indeed a *sieve* on V; i.e., an element of $\underline{\Omega}_V$.

The next step is to show that the collection of maps $\iota_V : \underline{O}_V \times \underline{\Sigma}_V \to \underline{\Omega}_V$ defined in (13.142) constitutes a natural transformation from the object $\underline{O} \times \underline{\Sigma}$ to the object $\underline{\Omega}$ in the topos $\mathbf{Sets}^{\mathcal{V}(\mathcal{H})^{\mathrm{op}}}$. This involves chasing around a few commutative squares, and we will spare the reader the ordeal. There is some subtlety, since we really want to deal with $\mathrm{Hom}_{\mathbf{Sets}^{\mathcal{V}(\mathcal{H})^{\mathrm{op}}}}(\underline{O}, P_{\mathrm{cl}}\underline{\Sigma})$, not $\mathrm{Hom}_{\mathbf{Sets}^{\mathcal{V}(\mathcal{H})^{\mathrm{op}}}}(\underline{O}, P\underline{\Sigma})$; but all works in the end.

To prove that $\ulcorner\iota\urcorner : \underline{O} \to P_{\mathrm{cl}}\underline{\Sigma}$ is monic, it suffices to show that the map $\ulcorner\iota\urcorner_V : \underline{O}_V \to P_{\mathrm{cl}}\underline{\Sigma}_V$ is injective at all stages V. This is a straightforward exercise and the details will not be given here.

The conclusion of this exercise is that, since $\ulcorner\iota\urcorner : \underline{O} \to P_{\mathrm{cl}}\underline{\Sigma}$ is monic, the truth sub-objects $\underline{\mathbb{T}}^{|\psi\rangle}$ of $\underline{O}$ can also be regarded as sub-objects of $P_{\mathrm{cl}}\underline{\Sigma}$, and hence the truth value assignment in (13.119) is well-defined.

Finally then, for any given quantum state $|\psi\rangle$ the basic proposition "$A \,\varepsilon\, \Delta$" can be assigned a generalised truth value $\nu(A \,\varepsilon\, \Delta;\, |\psi\rangle)$ in $\Gamma\underline{\Omega}$, where $\tau := \mathbf{Sets}^{\mathcal{V}(\mathcal{H})^{\mathrm{op}}}$ is the topos of presheaves over $\mathcal{V}(\mathcal{H})$. This is defined at each stage/context V as

$$\begin{aligned}\nu\big(A \,\varepsilon\, \Delta;\, |\psi\rangle\big)_V &:= \nu\big(\underline{\delta(\hat{E}[A \in \Delta])} \in \underline{\mathbb{T}}^{|\psi\rangle}\big)_V \\ &= \{V' \subseteq V \mid \underline{\delta\big(\hat{E}[A \in \Delta]\big)}_{V'} \in \underline{\mathbb{T}}^{|\psi\rangle}_{V'}\}\end{aligned} \tag{13.145}$$

13.6.6 *Yet Another Perspective on the K-S Theorem*

In classical physics, the pseudo-state $\mathfrak{w}^s \subseteq \mathcal{S}$ associated with the microstate $s \in \mathcal{S}$ is just $\mathfrak{w}^s := \{s\}$. This gives the diagram

$$
\begin{array}{ccc}
\mathfrak{w}^s & \longrightarrow & \mathcal{S} \\
\Big\downarrow & & \Big\downarrow \pi \\
\{*\} & \xrightarrow[\ulcorner \mathfrak{w}^s \urcorner]{} & P\mathcal{S}
\end{array}
\tag{13.146}
$$

where $\ulcorner \mathfrak{w}^s \urcorner (*) := \{s\}$ and π is the canonical map

$$
\begin{aligned}
\pi : \mathcal{S} &\longrightarrow P\mathcal{S} \\
s &\mapsto \{s\}
\end{aligned}
\tag{13.147}
$$

The singleton $\{*\}$ is the terminal object in the category, **Sets**, of sets, and the subset embedding $\mathfrak{w}^s \to \mathcal{S}$ in (13.146) is the categorical pull-back by π of the monic $\ulcorner \mathfrak{w}^s \urcorner : \{*\} \to P\mathcal{S}$.

In the quantum case, the analogue of the diagram (13.146) is

$$
\begin{array}{ccc}
\underline{\mathfrak{w}}^{s\,|\psi\rangle} & \longrightarrow & \underline{\Sigma} \\
\Big\downarrow & & \Big\downarrow \pi \\
\underline{1} & \xrightarrow[\ulcorner \underline{\mathfrak{w}}^{s\,|\psi\rangle} \urcorner]{} & P\underline{\Sigma}
\end{array}
\tag{13.148}
$$

where the arrow $\pi : \underline{\Sigma} \to P\underline{\Sigma}$ has yet to be defined. To proceed further, let us first return to the set-theory map

$$
\begin{aligned}
X &\to PX \\
x &\mapsto \{x\}
\end{aligned}
\tag{13.149}
$$

where X is any set.

We can think of (13.149) as the power transpose, $\ulcorner \beta \urcorner : X \to PX$, of the map $\beta : X \times X \to \{0, 1\}$ defined by

$$
\beta(x, y) := \begin{cases} 1 & \text{if } x = y; \\ 0 & \text{otherwise.} \end{cases}
\tag{13.150}
$$

In our topos case, the obvious definition for the arrow $\pi : \underline{\Sigma} \to P\underline{\Sigma}$ is the power transpose $\ulcorner \beta \urcorner : \underline{\Sigma} \to P\underline{\Sigma}$, of the arrow $\beta : \underline{\Sigma} \times \underline{\Sigma} \to \underline{\Omega}$, defined by

$$
\beta_V(\lambda_1, \lambda_2) := \{V' \subseteq V \mid \lambda_1|_{V'} = \lambda_2|_{V'}\}
\tag{13.151}
$$

for all stages V. Note that, in linguistic terms, the arrow defined in (13.151) is just the representation in the quantum topos $\mathbf{Sets}^{\mathcal{V}(\mathcal{H})^{\mathrm{op}}}$, of the term "$\tilde{\sigma}_1 = \tilde{\sigma}_2$", where $\tilde{\sigma}_1$ and $\tilde{\sigma}_2$ are terms of type Σ; i.e., $[\![\, \tilde{\sigma}_1 = \tilde{\sigma}_2 \,]\!] : \underline{\Sigma} \times \underline{\Sigma} \to \underline{\Omega}$.

With this definition of π, the diagram in (13.148) becomes meaningful: in particular the monic $\underline{\mathfrak{w}}^{|\psi\rangle} \hookrightarrow \underline{\Sigma}$ is the categorical pull-back by π of the monic $\ulcorner \underline{\mathfrak{w}}^{|\psi\rangle} \urcorner : \underline{1} \to P\underline{\Sigma}$.

There is, however, a significant difference between (13.148) and its classical analogue (13.146). In the latter case, the function $\ulcorner \mathfrak{w}^s \urcorner : \{*\} \to P\mathcal{S}$ can be "lifted" to a function $\ulcorner \mathfrak{w}^s \urcorner^{\uparrow} : \{*\} \to \mathcal{S}$ to give a commutative diagram: i.e., such that

$$\pi \circ \ulcorner \mathfrak{w}^s \urcorner^{\uparrow} = \ulcorner \mathfrak{w}^s \urcorner. \tag{13.152}$$

Indeed, simply define

$$\ulcorner \mathfrak{w}^s \urcorner^{\uparrow}(*) := s \tag{13.153}$$

However, in the quantum case there can be no "lift" $\ulcorner \underline{\mathfrak{w}}^{|\psi\rangle} \urcorner^{\uparrow} : \underline{1} \to \underline{\Sigma}$, as this would correspond to a global element of the spectral presheaf $\underline{\Sigma}$, and of course there are none. Thus, from this perspective, the Kochen-Specker theorem can be understood as asserting the existence of an *obstruction* to lifting the arrow $\ulcorner \underline{\mathfrak{w}}^{|\psi\rangle} \urcorner : \underline{1} \to P\underline{\Sigma}$.

Lifting problems of the type

$$\begin{array}{ccc} & & A \\ & & \downarrow{\scriptstyle \pi} \\ C & \xrightarrow[\phi]{} & B \end{array} \tag{13.154}$$

occur in many places in mathematics. A special, but very well-known, example of (13.154) arises when trying to construct cross-sections of a non-trivial principle fiber bundle $\pi : P \to M$. In diagrammatic terms we have

$$\begin{array}{ccc} & & P \\ & & \downarrow{\scriptstyle \pi} \\ M & \xrightarrow[\mathrm{id}]{} & M \end{array} \tag{13.155}$$

A cross-section of this bundle corresponds to a lifting of the map $\mathrm{id} : M \to M$.

The obstructions to lifting $\text{id} : M \to M$ through π can be studied in various ways. One technique is to decompose the bundle $\pi : P \to M$ into a series of interpolating fibrations $P \to P_1 \to P_2 \to \cdots M$ where each fibration $P_i \to P_{i+1}$ has the special property that the fiber is a particular Eilenberg-McLane space (this is known as a "Postnikov tower"). One then studies the sequential lifting of the function $\text{id} : M \to M$, i.e., first try to lift it through the fibration $P_1 \to M$; if that is successful try to lift it through $P_2 \to P_1$; and so on. Potential obstructions to performing these liftings appear as elements of the cohomology groups $H^k(M; \pi^{k-1}(F))$, $k = 1, 2, \ldots$, where F is the fiber of the bundle.

We have long felt that it should possible to describe the non-existence of global elements of $\underline{\Sigma}$ (i.e., the Kochen-Specker theorem) in some cohomological way, and the remark above suggests one possibility. Namely, perhaps there is some analogue of a "Postnikov factorisation" for the arrow $\underline{\pi} : \underline{\Sigma} \to P\underline{\Sigma}$ that could give a cohomological description of the obstructions to a global element of $\underline{\Sigma}$, i.e., to the lifting of a pseudo-state $\ulcorner \underline{\mathfrak{w}}^{|\psi\rangle} \urcorner : 1 \to P\underline{\Sigma}$ through the arrow $\underline{\pi} : \underline{\Sigma} \to \underline{P\Sigma}$ to give an arrow $\underline{1} \to \underline{\Sigma}$.

Related to this is the question of if there is a "pseudo-state object", $\underline{\mathbb{W}}$, with the defining property that $\Gamma\underline{\mathbb{W}}$ is equal to the set of all pseudo-states. Of course, to do this properly requires a definition of a pseudo-state that goes beyond the specific constructions of the objects $\underline{\mathfrak{w}}^{|\psi\rangle}$, $|\psi\rangle \in \mathcal{H}$. In particular, are there pseudo-states that are *not* of the form $\underline{\mathfrak{w}}^{|\psi\rangle}$?

If such an object, $\underline{\mathbb{W}}$, *can* be found then $\underline{\mathbb{W}}$ will be a sub-object of $P\underline{\Sigma}$, and in the diagram in (13.148) one could then look to replace $P\underline{\Sigma}$ with $\underline{\mathbb{W}}$.

13.7 The de Groote Presheaves of Physical Quantities

13.7.1 Background Remarks

Our task now is to consider the representation of the local language, $\mathcal{L}(S)$, in the case of quantum theory. We assume that the relevant topos is the same as that used for the propositional language $\mathcal{PL}(S)$, i.e., $\mathbf{Sets}^{\mathcal{V}(\mathcal{H})^{\text{op}}}$, but the emphasis is very different.

From a physics perspective, the key symbols in $\mathcal{L}(S)$ are (i) the ground type symbols, Σ and $\mathcal{R}$—the linguistic precursors of the state object and the quantity-value object respectively—and (ii) the function symbols $A : \Sigma \to \mathcal{R}$, which are the precursors of physical quantities. In the quantum-theory representation, ϕ, of $\mathcal{L}(S)$, the representation, Σ_ϕ, of Σ is defined to be the spectral presheaf $\underline{\Sigma}$ in the topos $\mathbf{Sets}^{\mathcal{V}(\mathcal{H})^{\text{op}}}$.

The critical question is to find the object, $\mathcal{R}_\phi$ (provisionally denoted as a presheaf $\underline{\mathcal{R}}$), in $\mathbf{Sets}^{\mathcal{V}(\mathcal{H})^{\text{op}}}$ that represents $\mathcal{R}$, and is hence the quantity-value object. One might anticipate that $\underline{\mathcal{R}}$ is just the real-number object in the topos $\mathbf{Sets}^{\mathcal{V}(\mathcal{H})^{\text{op}}}$, but that turns out to be quite wrong, and the right answer cannot just be guessed. In fact, the correct choice for $\underline{\mathcal{R}}$ is found indirectly by considering a related question:

namely, how to represent each function symbol $A : \Sigma \to \mathcal{R}$, with a concrete arrow $A_\phi : \Sigma_\phi \to \mathcal{R}_\phi$ in $\mathbf{Sets}^{\mathcal{V}(\mathcal{H})^{\mathrm{op}}}$, i.e., with a natural transformation $\breve{A} : \underline{\Sigma} \to \underline{\mathcal{R}}$ between the presheaves $\underline{\Sigma}$ and $\underline{\mathcal{R}}$.

Critical to this task are the daseinisation operations on projection operators that were defined earlier as (13.35) and (13.83), and which are repeated here for convenience:

Definition 10 If $\hat{P}$ is a projection operator, and $V \in \mathrm{Ob}(\mathcal{V}(\mathcal{H}))$ is any context/stage, we define:

1. The "outer daseinisation" operation is

$$\delta^o(\hat{P})_V := \bigwedge \{\hat{\alpha} \in \mathcal{P}(V) \mid \hat{P} \preceq \hat{\alpha}\}. \tag{13.156}$$

   where "$\preceq$" denotes the usual ordering of projection operators, and where $\mathcal{P}(V)$ is the set of all projection operators in V.
2. Similarly, the "inner daseinisation" operation is defined in the context V as

$$\delta^i(\hat{P})_V := \bigvee \{\hat{\beta} \in \mathcal{P}(V) \mid \hat{\beta} \preceq \hat{P}\}. \tag{13.157}$$

Thus $\delta^o(\hat{P})_V$ is the best approximation to $\hat{P}$ in V from "above", being the smallest projection in V that is larger than or equal to $\hat{P}$. Similarly, $\delta^i(\hat{P})_V$ is the best approximation to $\hat{P}$ from "below", being the largest projection in V that is smaller than or equal to $\hat{P}$.

In Sect. 13.6.5, we showed that the outer presheaf is a sub-object of the power object $P_{\mathrm{cl}}\underline{\Sigma}$ (in the category $\mathbf{Sets}^{\mathcal{V}(\mathcal{H})^{\mathrm{op}}}$), and hence that the global element $\delta^o(\hat{P})$ of $\underline{O}$ determines a (clopen) sub-object, $\underline{\delta^o(\hat{P})}$, of the spectral presheaf $\underline{\Sigma}$. By these means, the quantum logic of the lattice $\mathcal{P}(\mathcal{H})$ is mapped into the Heyting algebra of the set, $\mathrm{Sub}_{\mathrm{cl}}(\underline{\Sigma})$, of clopen sub-objects of $\underline{\Sigma}$.

Our task now is to perform the second stage of the programme: namely (i) identify the quantity-value presheaf, $\underline{\mathcal{R}}$; and (ii) show that any physical quantity can be represented by an arrow from $\underline{\Sigma}$ to $\underline{\mathcal{R}}$.

13.7.2 The Daseinisation of an Arbitrary Self-Adjoint Operator

13.7.2.1 Spectral Families and Spectral Order

We now want to extend the daseinisation operations from projections to arbitrary (bounded) self-adjoint operators. To this end, consider first a bounded, self-adjoint operator, $\hat{A}$, whose spectrum is purely discrete. Then the spectral theorem can be used to write $\hat{A} = \sum_{i=1}^{\infty} a_i \hat{P}_i$, where $a_1, a_2, \ldots$ are the eigenvalues of $\hat{A}$, and $\hat{P}_1, \hat{P}_2, \ldots$ are the spectral projection operators onto the corresponding eigenspaces.

A construction that comes immediately to mind is to use the daseinisation operation on projections to define

$$\delta^o(\hat{A})_V := \sum_{i=1}^{\infty} a_i\, \delta^o(\hat{P}_i)_V \tag{13.158}$$

for each stage V. However, this procedure is rather unnatural. For one thing, the projections, $\hat{P}_i, i = 1, 2, \ldots$ form a complete orthonormal set:

$$\sum_{i=1}^{\infty} \hat{P}_i = \hat{1}, \tag{13.159}$$

$$\hat{P}_i \hat{P}_j = \delta_{ij} \hat{P}_i, \tag{13.160}$$

whereas, in general, the collection of daseinised projections, $\delta^o(\hat{P}_i)_V, i = 1, 2, \ldots$ will not satisfy either of these conditions. In addition, it is hard to see how the expression $\delta^o(\hat{A})_V := \sum_{i=1}^{\infty} a_i\, \delta^o(\hat{P}_i)_V$ can be generalised to operators, $\hat{A}$, with a continuous spectrum.

The answer to this conundrum lies in the work of de Groote. He realised that although it is not useful to daseinise the spectral projections of an operator $\hat{A}$, it *is* possible to daseinise the *spectral family* of $\hat{A}$ [37, 39].

Spectral families.

We first recall that a *spectral family* is a family of projection operators $\hat{E}_\lambda$, $\lambda \in \mathbb{R}$, with the following properties:

1. If $\lambda_2 \leq \lambda_1$ then $\hat{E}_{\lambda_2} \preceq \hat{E}_{\lambda_1}$.
2. The net $\lambda \mapsto \hat{E}_\lambda$ of projection operators in the lattice $\mathcal{P}(\mathcal{H})$ is bounded above by $\hat{1}$, and below by $\hat{0}$. In fact,

$$\lim_{\lambda\to\infty} \hat{E}_\lambda = \hat{1}, \tag{13.161}$$

$$\lim_{\lambda\to-\infty} \hat{E}_\lambda = \hat{0}. \tag{13.162}$$

3. The map $\lambda \mapsto \hat{E}_\lambda$ is right-continuous:[73]

$$\bigwedge_{\epsilon\downarrow 0} \hat{E}_{\lambda+\epsilon} = \hat{E}_\lambda \tag{13.163}$$

   for all $\lambda \in \mathbb{R}$.

The spectral theorem asserts that for any self-adjoint operator $\hat{A}$, there exists a spectral family, $\lambda \mapsto \hat{E}^A_\lambda$, such that

$$\hat{A} = \int_{\mathbb{R}} \lambda\, d\hat{E}^A_\lambda \tag{13.164}$$

[73] It is a matter of convention whether one chooses right-continuous or left-continuous.

We are only concerned with bounded operators, and so the (weak Stieljes) integral in (13.164) is really over the bounded spectrum of $\hat{A}$ which, of course, is a compact subset of $\mathbb{R}$. Conversely, given a bounded spectral family $\{\hat{E}_\lambda\}_{\lambda\in\mathbb{R}}$,[74] there is a bounded self-adjoint operator $\hat{A}$ such that $\hat{A} = \int_{\mathbb{R}} \lambda \, d\hat{E}_\lambda$.

The spectral order.

A key element for our work is the so-called *spectral order* that was introduced in [73].[75] It is defined as follows. Let $\hat{A}$ and $\hat{B}$ be (bounded) self-adjoint operators with spectral families $\{\hat{E}^A_\lambda\}_{\lambda\in\mathbb{R}}$ and $\{\hat{E}^B_\lambda\}_{\lambda\in\mathbb{R}}$, respectively. Then define:

$$\hat{A} \preceq_s \hat{B} \text{ if and only if } \hat{E}^B_\lambda \preceq \hat{E}^A_\lambda \text{ for all } \lambda \in \mathbb{R}. \tag{13.165}$$

It is easy to see that (13.165) defines a genuine partial ordering on $B(\mathcal{H})_{\text{sa}}$ (the self-adjoint operators in $B(\mathcal{H})$). In fact, $B(\mathcal{H})_{\text{sa}}$ is a 'boundedly complete' lattice with respect to the spectral order, i.e., each bounded set S of self-adjoint operators has a minimum $\bigwedge S \in B(\mathcal{H})_{\text{sa}}$ and a maximum $\bigvee S \in B(\mathcal{H})_{\text{sa}}$ with respect to this order.

If $\hat{P}$, $\hat{Q}$ are projections, then

$$\hat{P} \preceq_s \hat{Q} \text{ if and only if } \hat{P} \preceq \hat{Q}, \tag{13.166}$$

so the spectral order coincides with the usual partial order on $\mathcal{P}(\mathcal{H})$. To ensure this, the "reverse" relation in (13.165) is necessary, since the spectral family of a projection $\hat{P}$ is given by

$$E^{\hat{P}}_\lambda = \begin{cases} \hat{0} & \text{if } \lambda < 0 \\ \hat{1} - \hat{P} & \text{if } 0 \leq \lambda < 1 \\ \hat{1} & \text{if } \lambda \geq 1. \end{cases} \tag{13.167}$$

If $\hat{A}$, $\hat{B}$ are self-adjoint operators such that (i) either $\hat{A}$ or $\hat{B}$ is a projection, or (ii) $[\hat{A}, \hat{B}] = \hat{0}$, then $\hat{A} \preceq_s \hat{B}$ if and only if $\hat{A} \preceq \hat{B}$. Here "$\preceq$" denotes the usual ordering on $B(\mathcal{H})_{\text{sa}}$.[76]

Moreover, if $\hat{A}$, $\hat{B}$ are arbitrary self-adjoint operators, then $\hat{A} \preceq_s \hat{B}$ implies $\hat{A} \preceq \hat{B}$, but not vice versa in general. Thus the spectral order is a partial order on $B(\mathcal{H})_{\text{sa}}$ that is coarser than the usual one.

13.7.2.2 Daseinisation of Self-Adjoint Operators

De Groote's crucial observation was the following. Let $\lambda \mapsto \hat{E}_\lambda$ be a spectral family in $\mathcal{P}(\mathcal{H})$ (or, equivalently, a self-adjoint operator $\hat{A}$). Then, for each stage V, the following maps:

---

[74] That is to say, there are $a, b \in \mathbb{R}$ such that $\hat{E}_\lambda = \hat{0}$ for all $\lambda \leq a$ and $\hat{E}_\lambda = \hat{1}$ for all $\lambda \geq b$.

[75] The spectral order was later reinvented by de Groote, see [36].

[76] The 'usual' ordering is $\hat{A} \preceq \hat{B}$ if $\langle\psi|\,\hat{A}\,|\psi\rangle \leq \langle\psi|\,\hat{B}\,|\psi\rangle$ for all vectors $|\psi\rangle \in \mathcal{H}$.

$$\lambda \mapsto \bigwedge_{\mu>\lambda} \delta^o(\hat{E}_\mu)_V \tag{13.168}$$

$$\lambda \mapsto \delta^i(\hat{E}_\lambda)_V \tag{13.169}$$

also define spectral families.[77] These spectral families lie in $\mathcal{P}(V)$ and hence, by the spectral theorem, define self-adjoint operators in V. This leads to the definition of the two daseinisations of an arbitrary self-adjoint operator:

Definition 11 Let $\hat{A}$ be an arbitrary self-adjoint operator. Then the *outer* and *inner* daseinisations of $\hat{A}$ are defined at each stage V as:

$$\delta^o(\hat{A})_V := \int_{\mathbb{R}} \lambda \, d\left(\delta^i_V(\hat{E}^A_\lambda)\right), \tag{13.170}$$

$$\delta^i(\hat{A})_V := \int_{\mathbb{R}} \lambda \, d\left(\bigwedge_{\mu>\lambda} \delta^o_V(\hat{E}^A_\mu)\right), \tag{13.171}$$

respectively.

Note that for all $\lambda \in \mathbb{R}$, and for all stages V, we have

$$\delta^i(\hat{E}_\lambda)_V \preceq \bigwedge_{\mu>\lambda} \delta^o(\hat{E}_\mu)_V \tag{13.172}$$

and hence, for all V,

$$\delta^i(\hat{A})_V \preceq_s \delta^o(\hat{A})_V. \tag{13.173}$$

This explains why the "i" and "o" superscripts in (13.170) and (13.171) are defined the way round that they are.

Both outer daseinisation (13.170) and inner daseinisation (13.171) can be used to "adapt" a self-adjoint operator $\hat{A}$ to contexts $V \in \mathrm{Ob}(\mathcal{V}(\mathcal{H}))$ that do not contain $\hat{A}$. (On the other hand, if $\hat{A} \in V$, then $\delta^o(\hat{A})_V = \delta^i(\hat{A})_V = \hat{A}$.)

13.7.2.3 Properties of Daseinisation

We will now list some useful properties of daseinisation.

1. It is clear that the outer, and inner, daseinisation operations can be extended to situations where the self-adjoint operator $\hat{A}$ does not belong to $B(\mathcal{H})_{\mathrm{sa}}$, or where V is not an *abelian* sub-algebra of $B(\mathcal{H})$. Specifically, let $\mathcal{N}$ be an arbitrary von

[77] The reason (13.168) and (13.169) have a different form is that $\lambda \mapsto \delta^i(\hat{E}_\lambda)_V$ is right continuous whereas $\lambda \mapsto \delta^o(\hat{E}_\lambda)_V$ is not. On the other hand, the family $\lambda \mapsto \bigwedge_{\mu>\lambda} \delta^o(\hat{E}_\mu)_V$ *is* right continuous.

Neumann algebra, and let $\mathcal{S} \subset \mathcal{N}$ be a proper von Neumann sub-algebra such that $\hat{1}_{\mathcal{N}} = \hat{1}_{\mathcal{S}} = \hat{1}$. Then outer and inner daseinisation can be defined as the mappings

$$\begin{aligned} \delta^o : \mathcal{N}_{\mathrm{sa}} &\to \mathcal{S}_{\mathrm{sa}} \\ \hat{A} &\mapsto \int_{\mathbb{R}} \lambda \, d\left(\delta^i_{\mathcal{S}}\left(\hat{E}^A_\lambda\right)\right), \end{aligned} \tag{13.174}$$

$$\begin{aligned} \delta^i : \mathcal{N}_{\mathrm{sa}} &\to \mathcal{S}_{\mathrm{sa}} \\ \hat{A} &\mapsto \int_{\mathbb{R}} \lambda \, d\left(\bigwedge_{\mu > \lambda} \delta^o_{\mathcal{S}}(\hat{E}^A_\mu)\right). \end{aligned} \tag{13.175}$$

A particular case is $\mathcal{N} = V$ and $\mathcal{S} = V'$ for two contexts V, V' such that $V' \subset V$. Hence, a self-adjoint operator can be restricted from one context to a sub-context.

For the moment, we will let $\mathcal{N}$ be an arbitrary von Neumann algebra, with $\mathcal{S} \subset \mathcal{N}$.

2. By construction,

$$\delta^o(\hat{A})_{\mathcal{S}} = \bigwedge\{\hat{B} \in \mathcal{S}_{\mathrm{sa}} \mid \hat{B} \succeq_s \hat{A}\}, \tag{13.176}$$

where the minimum is taken with respect to the spectral order; i.e., $\delta^o(\hat{A})_{\mathcal{S}}$ is the smallest self-adjoint operator in $\mathcal{S}$ that is spectrally larger than (or equal to) $\hat{A}$. This implies $\delta^o(\hat{A})_{\mathcal{S}} \succeq \hat{A}$ in the usual order. Likewise,

$$\delta^i(\hat{A})_{\mathcal{S}} = \bigvee\{\hat{B} \in \mathcal{S}_{\mathrm{sa}} \mid \hat{B} \preceq_s \hat{A}\}, \tag{13.177}$$

so $\delta^i(\hat{A})_{\mathcal{S}}$ is the largest self-adjoint operator in $\mathcal{S}$ spectrally smaller than (or equal to) $\hat{A}$, which implies $\delta^i(\hat{A})_{\mathcal{S}} \preceq \hat{A}$.

3. In general, neither $\delta^o(\hat{A})_{\mathcal{S}}$ nor $\delta^i(\hat{A})_{\mathcal{S}}$ can be written as Borel functions of the operator $\hat{A}$, since daseinisation changes the elements of the spectral family, while a function merely "shuffles them around".

4. Let $\hat{A} \in \mathcal{N}$ be self-adjoint. The spectrum, $\mathrm{sp}(\hat{A})$, consists of all $\lambda \in \mathbb{R}$ such that the spectral family $\{\hat{E}^A_\lambda\}_{\lambda\in\mathbb{R}}$ is non-constant on any neighbourhood of λ. By definition, outer daseinisation of $\hat{A}$ acts on the spectral family of $\hat{A}$ by sending $\hat{E}^A_\lambda$ to $\hat{E}^{\delta^o(\hat{A})_{\mathcal{S}}}_\lambda = \delta^i(\hat{E}^A_\lambda)_{\mathcal{S}}$. If $\{\hat{E}^A_\lambda\}_{\lambda\in\mathbb{R}}$ is constant on some neighbourhood of λ, then the spectral family $\{\hat{E}^{\delta^o(\hat{A})_{\mathcal{S}}}_\lambda\}_{\lambda\in\mathbb{R}}$ of $\delta^o(\hat{A})_{\mathcal{S}}$ is also constant on this neighbourhood. This shows that

$$\mathrm{sp}(\delta^o(\hat{A})_{\mathcal{S}}) \subseteq \mathrm{sp}(\hat{A}) \tag{13.178}$$

for all self-adjoint operators $\hat{A} \in \mathcal{N}_{\mathrm{sa}}$ and all von Neumann sub-algebras $\mathcal{S}$. Analogous arguments apply to inner daseinisation.

Heuristically, this result implies that the spectrum of the operator $\delta^o(\hat{A})_{\mathcal{S}}$ is more degenerate than that of $\hat{A}$; i.e., the effect of daseinisation is to 'collapse' eigenvalues.

5. Outer and inner daseinisation are both non-linear mappings, even on commuting operators. We will show this for projections explicitly. For example, let $\hat{Q} := \hat{1} - \hat{P}$. Then $\delta^o(\hat{Q} + \hat{P})_{\mathcal{S}} = \delta^o(\hat{1})_{\mathcal{S}} = \hat{1}$, while $\delta^o(\hat{1} - \hat{P})_{\mathcal{S}} \succ \hat{1} - \hat{P}$ and $\delta^o(\hat{P})_{\mathcal{S}} \succ \hat{P}$ in general, so $\delta^o(\hat{1} - \hat{P})_{\mathcal{S}} + \delta^o(\hat{P})_{\mathcal{S}}$ is the sum of two non-orthogonal projections in general (and hence not equal to $\hat{1}$). For inner daseinisation, we have $\delta^i(\hat{1} - \hat{P})_{\mathcal{S}} \prec \hat{1} - \hat{P}$ and $\delta^i(\hat{P})_{\mathcal{S}} \prec \hat{P}$ in general, so $\delta^i(\hat{1} - \hat{P})_{\mathcal{S}} + \delta^i(\hat{P})_{\mathcal{S}} \prec \hat{1} = \delta^i(\hat{1} - \hat{P} + \hat{P})_{\mathcal{S}}$ in general.

6. If $a \geq 0$, then $\delta^o(a\hat{A})_{\mathcal{S}} = a\delta^o(\hat{A})_{\mathcal{S}}$ and $\delta^i(a\hat{A})_{\mathcal{S}} = a\delta^i(\hat{A})_{\mathcal{S}}$. If $a < 0$, then $\delta^o(a\hat{A})_{\mathcal{S}} = a\delta^i(\hat{A})_{\mathcal{S}}$ and $\delta^i(a\hat{A})_{\mathcal{S}} = a\delta^o(\hat{A})_{\mathcal{S}}$. This is due the behaviour of spectral families under the mapping $\hat{A} \mapsto -\hat{A}$.

7. Let $\hat{A}$ be a self-adjoint operator, and let $\hat{E}[A \leq \lambda] = \hat{E}^A_\lambda$ be an element of the spectral family of $\hat{A}$. From (13.170) we get

$$\hat{E}[\delta^o_{\mathcal{S}}(A) \leq \lambda] = \delta^i_{\mathcal{S}}\big(\hat{E}[A \leq \lambda]\big) \tag{13.179}$$

and then

$$\begin{aligned} \hat{E}[\delta^o(\hat{A})_{\mathcal{S}} > \lambda] &= \hat{1} - \hat{E}[\delta^o(\hat{A})_{\mathcal{S}} \leq \lambda] && (13.180)\\ &= \hat{1} - \delta^i_{\mathcal{S}}\big(\hat{E}[A \leq \lambda]\big) && (13.181)\\ &= \delta^o_{\mathcal{S}}\big(\hat{1} - \hat{E}[A \leq \lambda]\big) && (13.182)\end{aligned}$$

where we have used the general result that, for any projection $\hat{P}$, we have $\hat{1} - \delta^i(\hat{P})_{\mathcal{S}} = \delta^o_{\mathcal{S}}(\hat{1} - \hat{P})$. Then, (13.182) gives

$$\hat{E}[\delta^o(\hat{A})_{\mathcal{S}} > \lambda] = \delta^o\big(\hat{E}[A > \lambda]\big)_{\mathcal{S}}. \tag{13.183}$$

13.7.2.4 The de Groote Presheaves

We know that $V \mapsto \delta^o(\hat{P})_V$ and $V \mapsto \delta^i(\hat{P})_V$ are global elements of the outer presheaf, $\underline{O}$, and inner presheaf, $\underline{I}$, respectively. Using the daseinisation operation for self-adjoint operators, it is straightforward to construct analogous presheaves for which $V \mapsto \delta^o(\hat{A})_V$ and $V \mapsto \delta^i(\hat{A})_V$ are global elements. One of these presheaves was briefly considered in [37]. We call these the "de Groote presheaves" in recognition of the importance of de Groote's work.

Definition 12 The outer de Groote presheaf, $\underline{\mathbb{O}}$, is defined as follows:

(i) On objects $V \in \mathrm{Ob}(\mathcal{V}(\mathcal{H}))$: We define $\underline{\mathbb{O}}_V := V_{\mathrm{sa}}$, the collection of self-adjoint members of V.
(ii) On morphisms $i_{V'V} : V' \subseteq V$: The mapping $\underline{\mathbb{O}}(i_{V'V}) : \underline{\mathbb{O}}_V \to \underline{\mathbb{O}}_{V'}$ is given by

$$\begin{aligned} \underline{\mathbb{O}}(i_{V'V})(\hat{A}) &:= \delta^o(\hat{A})_{V'} && (13.184)\\ &= \int_{\mathbb{R}} \lambda\, d\big(\delta^i(\hat{E}^A_\lambda)_{V'}\big) && (13.185)\end{aligned}$$

$$= \int_{\mathbb{R}} \lambda \, d\big(\underline{I}(i_{V'V})(\hat{E}^A_\lambda)\big) \tag{13.186}$$

for all $\hat{A} \in \underline{\mathbb{O}}_V$.

Here we used the fact that the restriction mapping $\underline{I}(i_{V'V})$ of the inner presheaf $\underline{I}$ is the inner daseinisation of projections $\delta^i : \mathcal{P}(V) \to \mathcal{P}(V')$.

Definition 13 The inner de Groote presheaf, $\underline{\mathrm{II}}$, is defined as follows:

(i) On objects $V \in \mathrm{Ob}(\mathcal{V}(\mathcal{H}))$: We define $\underline{\mathrm{II}}_V := V_{\mathrm{sa}}$, the collection of self-adjoint members of V.
(ii) On morphisms $i_{V'V} : V' \subseteq V$: The mapping $\underline{\mathrm{II}}(i_{V'V}) : \underline{\mathrm{II}}_V \to \underline{\mathrm{II}}_{V'}$ is given by

$$\underline{\mathrm{II}}(i_{V'V})(\hat{A}) := \delta^i(\hat{A})_{V'} \tag{13.187}$$

$$= \int_{\mathbb{R}} \lambda \, d\left(\bigwedge_{\mu>\lambda} (\delta^o(\hat{E}^A_\mu)_{V'}\right) \tag{13.188}$$

$$= \int_{\mathbb{R}} \lambda \, d\left(\bigwedge_{\mu>\lambda} (\underline{O}(i_{V'V})(\hat{E}^A_\mu)\right) \tag{13.189}$$

for all $\hat{A} \in \underline{\mathbb{O}}_V$ (where $\underline{O}(i_{V'V}) = \delta^o : \mathcal{P}(V) \to \mathcal{P}(V')$).

It is now clear that, by construction, $\delta^o(\hat{A}) := V \mapsto \delta^o(\hat{A})_V$ is a global element of $\underline{\mathbb{O}}$, and $\delta^i(\hat{A}) := V \mapsto \delta^i(\hat{A})_V$ is a global element of $\underline{\mathrm{II}}$.

De Groote found an example of an element of $\Gamma\underline{\mathbb{O}}$ that is *not* of the form $\delta^o(\hat{A})$ (as mentioned in [37]). The same example can be used to show that there are global elements of the outer presheaf $\underline{O}$ that are not of the form $\delta^o(\hat{P})$ for any projection $\hat{P} \in \mathcal{P}(\mathcal{H})$.

On the other hand, we have:

Theorem 4 *The mapping*

$$\delta^i : B(\mathcal{H})_{sa} \to \Gamma\underline{II} \tag{13.190}$$

$$\hat{A} \mapsto \delta^i(\hat{A}) \tag{13.191}$$

from self-adjoint operators in $B(\mathcal{H})$ to global sections of the outer de Groote presheaf is injective. Likewise,

$$\delta^o : B(\mathcal{H})_{sa} \to \Gamma\underline{\mathbb{O}} \tag{13.192}$$

$$\hat{A} \mapsto \delta^o(\hat{A}) \tag{13.193}$$

is injective.

Proof By construction, $\hat{A} \geq_s \delta^i(\hat{A})_V$ for all $V \in \mathrm{Ob}(\mathcal{V}(\mathcal{H}))$. Since $\hat{A}$ is contained in at least one context, we have

$$\hat{A} = \bigvee_{V \in \mathrm{Ob}(\mathcal{V}(\mathcal{H}))} \delta^i(\hat{A})_V, \tag{13.194}$$

where the maximum is taken with respect to the spectral order. If $\delta^i(\hat{A}) = \delta^i(\hat{B})$, then we have

$$\hat{A} = \bigvee_{V \in \mathrm{Ob}(\mathcal{V}(\mathcal{H}))} \delta^i(\hat{A})_V = \bigvee_{V \in \mathrm{Ob}(\mathcal{V}(\mathcal{H}))} \delta^i(\hat{B})_V = \hat{B}. \tag{13.195}$$

Analogously, $\hat{A} \leq_s \delta^o(\hat{A})_V$ for all $V \in \mathrm{Ob}(\mathcal{V}(\mathcal{H}))$, so

$$\hat{A} = \bigwedge_{V \in \mathrm{Ob}(\mathcal{V}(\mathcal{H}))} \delta^o(\hat{A})_V. \tag{13.196}$$

If $\delta^o(\hat{A}) = \delta^o(\hat{B})$, then we have

$$\hat{A} = \bigwedge_{V \in \mathrm{Ob}(\mathcal{V}(\mathcal{H}))} \delta^o(\hat{A})_V = \bigwedge_{V \in \mathrm{Ob}(\mathcal{V}(\mathcal{H}))} \delta^o(\hat{B})_V = \hat{B}. \tag{13.197}$$

The same argument also holds more generally for arbitrary von Neumann algebras, not just $B(\mathcal{H})$.

13.7.3 Daseinisation from Galois Connections

In this short subsection, we show how daseinisation of projections and self-adjoint operators relates to certain Galois connections: i.e., adjunctions between lattices of projections resp. self-adjoint operators. The observation that daseinisation can be understood in this way is due to F. William Lawvere[78] and Pedro Resende[79], whom we thank. A forthcoming paper [31] will give a more detailed account.

Let $V \in \mathcal{V}(\mathcal{H})$ be an abelian von Neumann sub-algebra of $B(\mathcal{H})$. Then there is an inclusion

$$\iota_{VB(\mathcal{H})} : \mathcal{P}(V) \longrightarrow \mathcal{P}(\mathcal{H}) \tag{13.198}$$

$$\hat{P} \longmapsto \hat{P} \tag{13.199}$$

of the projection lattice of V into the projection lattice of $B(\mathcal{H})$. Both lattices are complete, and the inclusion $\iota_{VB(\mathcal{H})}$ obviously preserves arbitrary meets of projec-

[78] Private communication.
[79] Private communication.

tions. However, the meet operation is the product in the poset $\mathcal{P}(V)$, viewed as a category, and hence $\iota_{VB(\mathcal{H})}$ is an order-preserving functor that preserves arbitrary products. Such a functor always has a left adjoint. In our case, we have the following result.

Theorem 5 *The left adjoint of $\iota_{VB(\mathcal{H})}$ is the outer daseinisation to V of projections in $\mathcal{P}(\mathcal{H})$:*

$$\delta^o_V : \mathcal{P}(\mathcal{H}) \longrightarrow \mathcal{P}(V) \tag{13.200}$$

$$\hat{Q} \longmapsto \delta^o(\hat{Q})_V, \tag{13.201}$$

Proof To prove this we must show that for $\hat{Q} \in \mathcal{P}(\mathcal{H})$ and $\hat{P} \in \mathcal{P}(V)$,

$$\hat{Q} \preceq \iota_{VB(\mathcal{H})}(\hat{P}) \text{ iff } \delta^o(\hat{Q})_V \sqsubseteq \hat{P}. \tag{13.202}$$

Here, we use the symbol $\sqsubseteq$ for the partial order on $\mathcal{P}(V)$ in order to distinguish it from the partial order on $\mathcal{P}(\mathcal{H})$. Assume that

$$\hat{Q} \preceq \iota_{VB(\mathcal{H})}(\hat{P}). \tag{13.203}$$

Then

$$\delta^o(\hat{Q})_V \sqsubseteq \delta^o_V(\iota_{VB(\mathcal{H})}(\hat{P})), \tag{13.204}$$

since $\delta^o_V : \mathcal{P}(\mathcal{H}) \to \mathcal{P}(V)$ is an order-preserving mapping. Obviously, we have $\delta^o_V(\iota_{VB(\mathcal{H})}(\hat{P})) = \hat{P}$, hence

$$\delta^o(\hat{Q})_V \sqsubseteq \hat{P}. \tag{13.205}$$

Conversely, assume that $\delta^o(\hat{Q})_V \sqsubseteq \hat{P}$. Then

$$\iota_{VB(\mathcal{H})}(\delta^o(\hat{Q})_V) \succeq \hat{Q}, \tag{13.206}$$

which simply means that $\delta^o(\hat{Q})_V \succeq \hat{Q}$ in $\mathcal{P}(\mathcal{H})$. Hence, since $\iota_{VB(\mathcal{H})} : \mathcal{P}(V) \to \mathcal{P}(\mathcal{H})$ is order-preserving, we obtain

$$\hat{Q} \preceq \iota_{VB(\mathcal{H})}(\delta^o(\hat{Q})_V) \preceq \iota_{VB(\mathcal{H})}(\hat{P}). \tag{13.207}$$

This proves that δ^o_V is indeed the left adjoint of $\iota_{VB(\mathcal{H})}$. Such an adjunction between posets is called a *Galois connection.*

The inclusion functor $\iota_{VB(\mathcal{H})} : \mathcal{P}(V) \to \mathcal{P}(\mathcal{H})$ also preserves arbitrary joins. Categorically, these are coproducts, and hence the functor $\iota_{VB(\mathcal{H})}$ also has a right adjoint. This is nothing but the inner daseinisation of projections to V:

$$\delta^i_V : \mathcal{P}(\mathcal{H}) \longrightarrow \mathcal{P}(V) \tag{13.208}$$
$$\hat{Q} \longmapsto \delta^i(\hat{Q})_V. \tag{13.209}$$

The proof that δ^i_V is indeed the right adjoint of $\iota_{VB(\mathcal{H})}$ is very similar to the proof that δ^o_V is the left adjoint, and we will not give the details here.

Interestingly, the results can be extended to all self-adjoint operators. The lattices involved are the lattice $(B(\mathcal{H}), \leq_s)$ of self-adjoint operators in $B(\mathcal{H})$, equipped with the *spectral* order, and the lattice $(V_{\mathrm{sa}}, \sqsubseteq_s)$ of self-adjoint operators in V, also with respect to the spectral order. On the sublattice of projections, the spectral and the usual order coincide (both for $\mathcal{P}(\mathcal{H})$ and $\mathcal{P}(V)$).

The obvious inclusion functor is

$$\iota_{V_{\mathrm{sa}}B(\mathcal{H})_{\mathrm{sa}}} : V_{\mathrm{sa}} \longrightarrow B(\mathcal{H})_{\mathrm{sa}} \tag{13.210}$$
$$\hat{A} \longmapsto \hat{A}. \tag{13.211}$$

However, since the lattices $(B(\mathcal{H}), \leq_s)$ and $(V_{\mathrm{sa}}, \sqsubseteq_s)$ are only boundedly complete, the existence of left and right adjoints of $\iota_{V_{\mathrm{sa}}B(\mathcal{H})_{\mathrm{sa}}}$ is not guaranteed automatically. Nevertheless, it is straightforward to show that the outer daseinisation functor

$$\delta^o_V : B(\mathcal{H})_{\mathrm{sa}} \longrightarrow V_{\mathrm{sa}} \tag{13.212}$$
$$\hat{A} \longmapsto \delta^o(\hat{A})_V \tag{13.213}$$

is left adjoint to $\iota_{V_{\mathrm{sa}}B(\mathcal{H})_{\mathrm{sa}}}$, and the inner daseinisation

$$\delta^i_V : B(\mathcal{H})_{\mathrm{sa}} \longrightarrow V_{\mathrm{sa}} \tag{13.214}$$
$$\hat{A} \longmapsto \delta^i(\hat{A})_V \tag{13.215}$$

is right adjoint to $\iota_{V_{\mathrm{sa}}B(\mathcal{H})_{\mathrm{sa}}}$.[80]

13.8 The Presheaves $\underline{\mathrm{sp}(\hat{A})^{\succeq}}$, $\underline{\mathbb{R}^{\succeq}}$ and $\underline{\mathbb{R}^{\leftrightarrow}}$

13.8.1 Background to the Quantity-Value Presheaf $\underline{\mathcal{R}}$

Our goal now is to construct a "quantity-value" presheaf $\underline{\mathcal{R}}$ with the property that inner and/or outer daseinisation of an self-adjoint operator $\hat{A}$ can be used to define an arrow, i.e., a natural transformation, from $\underline{\Sigma}$ to $\underline{\mathcal{R}}$.[81]

---

[80] Of course, taking a single context V and inner and outer daseinisation to this context is not sufficient in the application to quantum theory. One rather has to consider all contexts $V \in \mathcal{V}(\mathcal{H})$ in order to construct sub-objects of $\underline{\Sigma}$ (from outer daseinisation of projections), and natural transformations that represent physical quantities (from inner and outer daseinisation of self-adjoint operators, see following sections).

[81] In fact, we will define several closely related presheaves that can serve as a quantity-value object.

The arrow corresponding to a self-adjoint operator $\hat{A} \in B(\mathcal{H})$ is denoted for now by $\breve{A} : \underline{\Sigma} \to \underline{\mathcal{R}}$. At each stage V, we need a mapping

$$\breve{A}_V : \underline{\Sigma}_V \to \underline{\mathcal{R}}_V \tag{13.216}$$

$$\lambda \mapsto \breve{A}_V(\lambda), \tag{13.217}$$

and we make the basic assumption that this mapping is given by evaluation. More precisely, $\lambda \in \underline{\Sigma}_V$ is a spectral element[82] of V and hence can be evaluated on operators lying in V. And, while $\hat{A}$ will generally not lie in V, both the inner daseinisation $\delta^i(\hat{A})_V$ and the outer daseinisation $\delta^o(\hat{A})_V$ do.

Let us start by considering the operators $\delta^o(\hat{A})_V$, $V \in \mathrm{Ob}(\mathcal{V}(\mathcal{H}))$. Each of these is a self-adjoint operator in the commutative von Neumann algebra V, and hence, by the spectral theorem, can be represented by a function, (the Gel'fand transform[83]) $\overline{\delta^o(\hat{A})_V} : \underline{\Sigma}_V \to \mathrm{sp}(\delta^o(\hat{A})_V)$, with values in the spectrum $\mathrm{sp}(\delta^o(\hat{A})_V)$ of the self-adjoint operator $\delta^o(\hat{A})_V$. Since the spectrum of a self-adjoint operator is a subset of $\mathbb{R}$, we can also write $\overline{\delta^o(\hat{A})_V} : \underline{\Sigma}_V \to \mathbb{R}$. The question now is whether the collection of maps $\overline{\delta^o(\hat{A})_V} : \underline{\Sigma}_V \to \mathbb{R}$, $V \in \mathrm{Ob}(\mathcal{V}(\mathcal{H}))$, can be regarded as an arrow from $\underline{\Sigma}$ to some presheaf $\underline{\mathcal{R}}$.

To answer this we need to see how these operators behave as we go "down a chain" of sub-algebras $V' \subseteq V$. The first remark is that if $V' \subseteq V$ then $\delta^o(\hat{A})_{V'} \succeq \delta^o(\hat{A})_V$. When applied to the Gel'fand transforms, this leads to the equation

$$\overline{\delta^o(\hat{A})_{V'}}(\lambda|_{V'}) \geq \overline{\delta^o(\hat{A})_V}(\lambda) \tag{13.218}$$

for all $\lambda \in \underline{\Sigma}_V$, where $\lambda|_{V'}$ denotes the restriction of the spectral element $\lambda \in \underline{\Sigma}_V$ to the sub-algebra $V' \subseteq V$. However, the definition of the spectral presheaf is such that $\lambda|_{V'} = \underline{\Sigma}(i_{V'V})(\lambda)$, and hence (13.218) can be rewritten as

$$\overline{\delta^o(\hat{A})_{V'}}\big(\underline{\Sigma}(i_{V'V})(\lambda)\big) \geq \overline{\delta^o(\hat{A})_V}(\lambda) \tag{13.219}$$

for all $\lambda \in \underline{\Sigma}_V$.

It is a standard result that the Dedekind real number object, $\underline{\mathbb{R}}$, in a presheaf topos $\mathbf{Sets}^{\mathcal{C}^{op}}$ is the *constant* functor from $\mathcal{C}^{op}$ to $\mathbb{R}$ [66]. It follows that the family of Gel'fand transforms, $\overline{\delta^o(\hat{A})_V}$, $V \in \mathrm{Ob}(\mathcal{V}(\mathcal{H}))$, of the daseinised operators $\delta^o(\hat{A})_V$, $V \in \mathrm{Ob}(\mathcal{V}(\mathcal{H}))$, does *not* define an arrow from $\underline{\Sigma}$ to $\underline{\mathbb{R}}$, as this would require an equality in (13.219), which is not true. Thus the quantity-value presheaf, $\underline{\mathcal{R}}$, in the topos $\mathbf{Sets}^{\mathcal{V}(\mathcal{H})^{\mathrm{op}}}$ is *not* the real-number object $\underline{\mathbb{R}}$, although clearly $\underline{\mathcal{R}}$ has *something*

[82] A "spectral element", $\lambda \in \underline{\Sigma}_V$ of V, is a multiplicative, linear functional $\lambda : V \to \mathbb{C}$ with $\langle \lambda, \hat{1} \rangle = 1$, see also Definition 4.

[83] This use of the "overline" symbol for the Gel'fand transform should not be confused with our later use of the same symbol to indicate a co-presheaf.

to do with the real numbers. We must take into account the growth of these real numbers as we go from V to smaller sub-algebras V'. Similarly, if we consider inner daseinisation, we get a series of falling real numbers.

The presheaf, $\underline{\mathcal{R}}$, that we will choose, and which will be denoted by $\underline{\mathbb{R}^{\leftrightarrow}}$, incorporates both aspects (growing and falling real numbers).

13.8.2 Definition of the Presheaves $\underline{sp(\hat{A})^{\succeq}}$, $\underline{\mathbb{R}^{\succeq}}$ and $\underline{\mathbb{R}^{\leftrightarrow}}$

The inapplicability of the real-number object $\underline{\mathbb{R}}$ may seem strange at first,[84] but actually it is not that surprising. Because of the Kochen-Specker theorem, we do not expect to be able to assign (constant) real numbers as values of physical quantities, at least not globally. Instead, we draw on some recent results of M. Jackson [54], obtained as part of his extensive study of measure theory on a topos of presheaves. Here, we use a single construction in Jackson's thesis: the presheaf of "order-preserving functions" over a partially ordered set—in our case, $\mathcal{V}(\mathcal{H})$. In fact, we will need both order-reversing and order-preserving functions.

Definition 14 Let $(\mathcal{Q}, \preceq)$ and $(\mathcal{P}, \preceq)$ be partially ordered sets. A function

$$\mu : \mathcal{Q} \to \mathcal{P} \tag{13.220}$$

is order-preserving if $q_1 \preceq q_2$ implies $\mu(q_1) \preceq \mu(q_2)$ for all $q_1, q_2 \in \mathcal{Q}$. It is order-reversing if $q_1 \preceq q_2$ implies $\mu(q_1) \succeq \mu(q_2)$. We denote by $\mathcal{OP}(\mathcal{Q}, \mathcal{P})$ the set of order-preserving functions $\mu : \mathcal{Q} \to \mathcal{P}$, and by $\mathcal{OR}(\mathcal{Q}, \mathcal{P})$ the set of order-reversing functions.[85]

We note that if μ is order-preserving, then $-\mu$ is order-reversing, and vice versa.

Adapting Jackson's definitions slightly, if $\mathcal{P}$ is any partially-ordered set, we have the following.

Definition 15 The $\mathcal{P}$-valued presheaf, $\underline{\mathcal{P}^{\succeq}}$, of order-reversing functions over $\mathcal{V}(\mathcal{H})$ is defined as follows:

(i) On objects $V \in \mathrm{Ob}(\mathcal{V}(\mathcal{H}))$:

$$\underline{\mathcal{P}^{\succeq}}_V := \{\mu : \downarrow\! V \to \mathcal{P} \mid \mu \in \mathcal{OR}(\downarrow\! V, \mathcal{P})\} \tag{13.221}$$

where $\downarrow\! V \subset \mathrm{Ob}(\mathcal{V}(\mathcal{H}))$ is the set of all von Neumann sub-algebras of V.

(ii) On morphisms $i_{V'V} : V' \subseteq V$: The mapping $\underline{\mathcal{P}^{\succeq}}(i_{V'V}) : \underline{\mathcal{P}^{\succeq}}_V \to \underline{\mathcal{P}^{\succeq}}_{V'}$ is given by

---

84 Indeed, it puzzled us for a while!

85 Order-preserving functions often are called *monotone*, while order-reversing functions are called *antitone*.

$$\underline{\mathcal{P}^{\succeq}}(i_{V'V})(\mu) := \mu_{|_{V'}} \tag{13.222}$$

where $\mu_{|_{V'}}$ denotes the restriction of the function μ to $\downarrow V' \subseteq \downarrow V$.

Jackson uses order-preserving functions with $\mathcal{P} := [0, \infty)$ (the non-negative reals), with the usual order $\leq$.

Clearly, there is an analogous definition of the $\mathcal{P}$-valued presheaf, $\underline{\mathcal{P}^{\preceq}}$, of order-preserving functions from $\downarrow V$ to $\mathcal{P}$. It can be shown that $\underline{\mathcal{P}^{\succeq}}$ and $\underline{\mathcal{P}^{\preceq}}$ are isomorphic objects in $\mathbf{Sets}^{\mathcal{V}(\mathcal{H})^{\mathrm{op}}}$.

Let us first consider $\underline{\mathcal{P}^{\succeq}}$. For us, the key examples for the partially ordered set $\mathcal{P}$ are (i) $\mathbb{R}$, the real numbers with the usual order $\leq$, and (ii) $\mathrm{sp}(\hat{A}) \subset \mathbb{R}$, the spectrum of some bounded self-adjoint operator $\hat{A}$, with the order $\leq$ inherited from $\mathbb{R}$. Clearly, the associated presheaf $\underline{\mathrm{sp}(\hat{A})^{\succeq}}$ is a sub-object of the presheaf $\underline{\mathbb{R}^{\succeq}}$.

Now let $\hat{A} \in B(\mathcal{H})_{\mathrm{sa}}$, and let $V \in \mathrm{Ob}(\mathcal{V}(\mathcal{H}))$. Then to each $\lambda \in \underline{\Sigma}_V$ there is associated the function

$$\breve{\delta}^o(\hat{A})_V(\lambda) : \downarrow V \to \mathrm{sp}(\hat{A}), \tag{13.223}$$

given by

$$\left(\breve{\delta}^o(\hat{A})_V(\lambda)\right)(V') := \overline{\delta^o(\hat{A})_{V'}}(\underline{\Sigma}(i_{V'V})(\lambda)) \tag{13.224}$$

$$= \overline{\delta^o(\hat{A})_{V'}}(\lambda|_{V'}) \tag{13.225}$$

$$= \langle \lambda|_{V'}, \delta^o(\hat{A})_{V'} \rangle \tag{13.226}$$

$$= \langle \lambda, \delta^o(\hat{A})_{V'} \rangle \tag{13.227}$$

for all $V' \subseteq V$. We note that as V' becomes smaller, $\delta^o(\hat{A})_{V'}$ becomes larger (or stays the same) in the spectral order, and hence in the usual order on operators. Therefore, $\breve{\delta}^o(\hat{A})_V(\lambda) : \downarrow V \to \mathrm{sp}(\hat{A})$ is an *order-reversing function*, for each $\lambda \in \underline{\Sigma}_V$.

It is worth noting that daseinisation of $\hat{A}$, i.e., the approximation of the self-adjoint operator $\hat{A}$ in the *spectral* order, allows to define a function $\breve{\delta}^o(\hat{A})_V(\lambda)$ (for each $\lambda \in \underline{\Sigma}_V$) with values in the spectrum of $\hat{A}$, since we have $\mathrm{sp}(\delta^o(\hat{A})_V) \subseteq \mathrm{sp}(\hat{A})$, see (13.178). If we had chosen an approximation in the usual *linear* order on $B(\mathcal{H})_{\mathrm{sa}}$, then the approximated operators would not have a spectrum that is contained in $\mathrm{sp}(\hat{A})$ in general.

Let

$$\breve{\delta}^o(\hat{A})_V : \underline{\Sigma}_V \to \underline{\mathrm{sp}(\hat{A})^{\succeq}}_V \tag{13.228}$$

$$\lambda \mapsto \breve{\delta}^o(\hat{A})_V(\lambda) \tag{13.229}$$

denote the set of order-reversing functions from $\downarrow V$ to $\mathrm{sp}(\hat{A})$ obtained in this way. We then have the following, fundamental, result which can be regarded as a type of "non-commutative" spectral theorem in which each bounded, self-adjoint operator $\hat{A}$ is mapped to an arrow from $\underline{\Sigma}$ to $\underline{\mathbb{R}^{\succeq}}$:

Theorem 6 *The mappings* $\breve{\delta}^o(\hat{A})_V$, $V \in Ob(\mathcal{V}(\mathcal{H}))$, *are the components of a natural transformation/arrow* $\breve{\delta}^o(\hat{A}) : \underline{\Sigma} \to \underline{sp(\hat{A})^{\succeq}}$.

Proof We only have to prove that, whenever $V' \subset V$, the diagram

$$\begin{array}{ccc} \underline{\Sigma}_V & \xrightarrow{\breve{\delta}^o(\hat{A})_V} & \underline{\mathrm{sp}(\hat{A})^{\succeq}}_V \\ \downarrow & & \downarrow \\ \underline{\Sigma}_{V'} & \xrightarrow[\breve{\delta}^o(\hat{A})_{V'}]{} & \underline{\mathrm{sp}(\hat{A})^{\succeq}}_{V'} \end{array}$$

commutes. Here, the vertical arrows are the restrictions of the relevant presheaves from the stage V to $V' \subseteq V$.

In fact, the commutativity of the diagram follows directly from the definitions. For each $\lambda \in \underline{\Sigma}_V$, the composition of the upper arrow and the right vertical arrow gives

$$(\breve{\delta}^o(\hat{A})_V(\lambda))|_{V'} = \breve{\delta}^o(\hat{A})_{V'}(\lambda|_{V'}), \tag{13.230}$$

which is the same function that we get by first restricting λ from $\underline{\Sigma}_V$ to $\underline{\Sigma}_{V'}$ and then applying $\breve{\delta}^o(\hat{A})_{V'}$.

In this way, to each physical quantity $\hat{A}$ in quantum theory there is assigned a natural transformation $\breve{\delta}^o(\hat{A})$ from the state object $\underline{\Sigma}$ to the presheaf $\underline{\mathrm{sp}(\hat{A})^{\succeq}}$. Since $\underline{\mathrm{sp}(\hat{A})^{\succeq}}$ is a sub-object of $\underline{\mathbb{R}^{\succeq}}$ for each $\hat{A}$, $\breve{\delta}^o(\hat{A})$ can also be seen as a natural transformation/arrow from $\underline{\Sigma}$ to $\underline{\mathbb{R}^{\succeq}}$. Hence the presheaf $\underline{\mathbb{R}^{\succeq}}$ in the topos $\mathbf{Sets}^{\mathcal{V}(\mathcal{H})^{\mathrm{op}}}$ is one candidate for the quantity-value object of quantum theory. Note that it follows from Theorem 4 that the mapping

$$\theta : B(\mathcal{H})_{\mathrm{sa}} \to \mathrm{Hom}_{\mathbf{Sets}^{\mathcal{V}(\mathcal{H})^{\mathrm{op}}}}(\underline{\Sigma}, \underline{\mathbb{R}^{\succeq}}) \tag{13.231}$$

$$\hat{A} \mapsto \breve{\delta}^o(\hat{A}) \tag{13.232}$$

is injective.[86]

If S denotes our quantum system, then, on the level of the formal language $\mathcal{L}(S)$, we expect the mapping $A \to \hat{A}$ to be injective, where A is a function symbol of

[86] Interestingly, these results all carry over to an arbitrary von Neumann algebra $\mathcal{N} \subseteq B(\mathcal{H})$. In this way, the formalism is flexible enough to adapt to situations where we have symmetries (which can described mathematically by a von Neumann algebra $\mathcal{N}$ that has a non-trivial commutant) and super-selection rules (which corresponds to $\mathcal{N}$ having a non-trivial centre).

signature $\Sigma \to \mathcal{R}$. It follows that we have obtained a a faithful representation of these function symbols by arrows $\breve{\delta}^o(\hat{A}) : \underline{\Sigma} \to \underline{\mathbb{R}^{\succeq}}$ in the topos $\mathbf{Sets}^{\mathcal{V}(\mathcal{H})^{\mathrm{op}}}$.

Similarly, there is an order-preserving function

$$\breve{\delta}^i(\hat{A})_V(\lambda) : \downarrow V \to \mathrm{sp}(\hat{A}), \tag{13.233}$$

that is defined for all $V' \subseteq V$ by

$$\left(\breve{\delta}^i(\hat{A})_V(\lambda)\right)(V') = \overline{\delta^i(\hat{A})_{V'}}(\underline{\Sigma}(i_{V'V})(\lambda)) \tag{13.234}$$

$$= \langle \lambda, \delta^i(\hat{A})_{V'} \rangle. \tag{13.235}$$

Since $\delta^i(\hat{A})_{V'}$ becomes smaller (or stays the same) as V' gets smaller, $\breve{\delta}^i(\hat{A})_V(\lambda)$ indeed is an *order-preserving* function from $\downarrow V$ to $\mathrm{sp}(\hat{A})$ for each $\lambda \in \underline{\Sigma}_V$. Again, approximation in the spectral order (in this case from below) allows us to define a function with values in $\mathrm{sp}(\hat{A})$, which would not be possible when using the linear order.

Clearly, we can use the functions $\breve{\delta}^i(\hat{A})_V(\lambda)$, $\lambda \in \underline{\Sigma}_V$, to define a natural transformation $\breve{\delta}^i(\hat{A}) : \underline{\Sigma} \to \underline{\mathrm{sp}(\hat{A})^{\preceq}}$ from the spectral presheaf, $\underline{\Sigma}$, to the presheaf $\underline{\mathrm{sp}(\hat{A})^{\preceq}}$ of real-valued, order-preserving functions on $\downarrow V$ with values in $\mathrm{sp}(\hat{A})$. The components of $\breve{\delta}^i(\hat{A})$ are

$$\breve{\delta}^i(\hat{A})_V : \underline{\Sigma}_V \to \underline{\mathrm{sp}(\hat{A})^{\preceq}}_V \tag{13.236}$$

$$\lambda \mapsto \breve{\delta}^i(\hat{A})_V(\lambda). \tag{13.237}$$

Since $\underline{\mathrm{sp}(\hat{A})^{\preceq}}$ is a sub-presheaf of $\underline{\mathbb{R}^{\preceq}}$, the presheaf of real-valued, order-preserving functions, we also obtain a natural transformation from $\underline{\Sigma}$ to $\underline{\mathbb{R}^{\preceq}}$. It follows from Theorem 4 that the mapping from self-adjoint operators to natural transformations $\breve{\delta}^i(\hat{A}) : \underline{\Sigma} \to \underline{\mathbb{R}^{\preceq}}$ is injective.

The functions obtained from inner and outer daseinisation can be combined to give yet another presheaf, and one that will be particularly useful for the physical interpretation of these constructions. The general definition is the following.

Definition 16 Let $\mathcal{P}$ be a partially-ordered set. The $\mathcal{P}$-valued presheaf, $\underline{\mathcal{P}^{\leftrightarrow}}$, of order-preserving and order-reversing functions on $\mathcal{V}(\mathcal{H})$ is defined as follows:

(i) On objects $V \in Ob(\mathcal{V}(\mathcal{H}))$:

$$\underline{\mathcal{P}^{\leftrightarrow}}_V := \{(\mu, \nu) \mid \mu \in \mathcal{OP}(\downarrow V, \mathcal{P}), \nu \in \mathcal{OR}(\downarrow V, \mathcal{P}), \mu \leq \nu\} \tag{13.238}$$

where $\downarrow V \subset \mathrm{Ob}(\mathcal{V}(\mathcal{H}))$ is the set of all sub-algebras V' of V. Note that we introduce the condition $\mu \leq \nu$, i.e., for all $V' \in \downarrow V$ we demand $\mu(V') \leq \nu(V')$.

(ii) On morphisms $i_{V'V} : V' \subseteq V$:

$$\underline{\mathcal{P}^{\leftrightarrow}}(i_{V'V}) : \underline{\mathcal{P}^{\leftrightarrow}}_V \longrightarrow \underline{\mathcal{P}^{\leftrightarrow}}_{V'} \tag{13.239}$$
$$(\mu, \nu) \longmapsto (\mu|_{V'}, \nu|_{V'}), \tag{13.240}$$

where $\mu|_{V'}$ denotes the restriction of μ to $\downarrow V' \subseteq \downarrow V$, and analogously for $\nu|_{V'}$.

Note that since we have the condition $\mu \leq \nu$ in (i), the presheaf $\underline{\mathcal{P}^{\leftrightarrow}}$ is not simply the product of the presheaves $\underline{\mathcal{P}^{\succeq}}$ and $\underline{\mathcal{P}^{\preceq}}$.

As we will discuss shortly, the presheaf, $\underline{\mathbb{R}^{\leftrightarrow}}$, of order-preserving and order-reversing, real-valued functions is closely related to the "k-extension" of the presheaf $\underline{\mathbb{R}^{\succeq}}$ (see the Appendix for details of the k-extension procedure).

Now let

$$\breve{\delta}(\hat{A})_V := \left(\breve{\delta}^i(\hat{A})_V(\cdot), \breve{\delta}^o(\hat{A})_V(\cdot)\right) : \underline{\Sigma}_V \to \underline{\mathbb{R}^{\leftrightarrow}}_V \tag{13.241}$$

denote the set of all pairs of order-preserving and order-reversing functions from $\downarrow V$ to $\mathbb{R}$ that can be obtained from inner and outer daseinisation. It is easy to see that we have the following result:

Theorem 7 *The mappings $\breve{\delta}(\hat{A})_V$, $V \in Ob(\mathcal{V}(\mathcal{H}))$, are the components of a natural transformation $\breve{\delta}(\hat{A}) : \underline{\Sigma} \to \underline{\mathbb{R}^{\leftrightarrow}}$.*

Again from Theorem 4, the mapping from self-adjoint operators to natural transformations, $\hat{A} \to \breve{\delta}(\hat{A})$, is injective.

Since $\breve{\delta}^i(\hat{A})_V(\lambda) \leq \breve{\delta}^o(\hat{A})_V(\lambda)$ for all $\lambda \in \underline{\Sigma}_V$, we can interpret each pair $(\breve{\delta}^i(\hat{A})_V(\lambda), \breve{\delta}^o(\hat{A})_V(\lambda))$ of values as an *interval*, which gives a first hint at the physical interpretation.

13.8.3 Inner and Outer Daseinisation from Functions on Filters

There is a close relationship between inner and outer daseinisation, and certain functions on the filters in the projection lattice $\mathcal{P}(\mathcal{H})$ of $B(\mathcal{H})$. We give a summary of these results here: details can be found in de Groote's work [37, 39], the article [22], and a forthcoming paper [25]. This Subsection serves as a preparation for the physical interpretation of the arrows $\breve{\delta}(\hat{A}) : \underline{\Sigma} \to \underline{\mathbb{R}^{\leftrightarrow}}$.

13.8.3.1 Filter Bases, Filters and Ultrafilters

We first need some basic definitions. Let $\mathbb{L}$ be a lattice with zero element 0. A subset f of $\mathbb{L}$ is called a *filter base* if (i) $0 \neq f$ and (ii) for all $a, b \in f$, there is a $c \in f$ such that $c \leq a \wedge b$.

A subset F of a lattice $\mathbb{L}$ with zero element 0 is a (proper) *filter* (or *dual ideal*) if (i) $0 \notin F$; (ii) $a, b \in F$ implies $a \wedge b \in F$; and (iii) $a \in F$ and $b \geq a$ imply $b \in F$. In other words, a filter is an upper set in the lattice $\mathbb{L}$ that is closed under finite minima.

By Zorn's lemma, every filter is contained in a maximal filter. Obviously, such a maximal filter is also a maximal filter base.

Let $\mathbb{L}'$ be a sublattice of $\mathbb{L}$ (with common 0), and let F' be a filter in $\mathbb{L}'$. Then, seen as a subset of $\mathbb{L}$, F' is a filter base in $\mathbb{L}$. The smallest filter in $\mathbb{L}$ that contains F' is the *cone over F' in* $\mathbb{L}$:

$$\mathcal{C}_{\mathbb{L}}(F') := \{b \in \mathbb{L} \mid \exists a \in F' : a \leq b\}. \tag{13.242}$$

This is nothing but the upper set $\uparrow F'$ of F' in $\mathbb{L}$.

In our applications, $\mathbb{L}$ typically is the lattice $\mathcal{P}(\mathcal{H})$ of projections in $B(\mathcal{H})$, and $\mathbb{L}'$ is the lattice $\mathcal{P}(V)$ of projections in an abelian sub-algebra V.

If $\mathbb{L}$ is a *Boolean* lattice, i.e., if it is a distributive lattice with minimal element 0 and maximal element 1, and a complement (negation) $\neg : \mathbb{L} \to \mathbb{L}$ such that $a \vee \neg a = 1$ for all $a \in \mathbb{L}$, then we define an *ultrafilter* $\tilde{F}$ to be a maximal filter in $\mathbb{L}$. An ultrafilter $\tilde{F}$ is characterised by the following property: for all $a \in \mathbb{L}$, either $a \in \tilde{F}$ or $\neg a \in \tilde{F}$. This follows from the fact that for all $a \in \mathbb{L}$, we have $a \vee \neg a = 1$.

Specifically, let us assume that $\tilde{F}$ is an ultrafilter and $a \notin \tilde{F}$. This means that there is some $b \in \tilde{F}$ such that $b \wedge a = 0$. Using distributivity of the lattice $\mathbb{L}$, we get

$$b = b \wedge (a \vee \neg a) = (b \wedge a) \vee (b \wedge \neg a) = b \wedge \neg a, \tag{13.243}$$

so $b \leq \neg a$. Since $b \in \tilde{F}$ and $\tilde{F}$ is a filter, this implies $\neg a \in \tilde{F}$. Conversely, if $\neg a \notin \tilde{F}$, we obtain $a \in \tilde{F}$.

The projection lattice $\mathcal{P}(V)$ of an abelian von Neumann algebra V is a Boolean lattice. The maximal element is the identity operator $\widehat{1}$ and, as we saw earlier, the complement of a projection is given as $\neg\hat{\alpha} = \hat{1} - \hat{\alpha}$. Each ultrafilter $\tilde{F}$ in $\mathcal{P}(V)$ hence contains either $\hat{\alpha}$ or $\hat{1} - \hat{\alpha}$ for all $\hat{\alpha} \in \mathcal{P}(V)$.

13.8.3.2 Spectral Elements and Ultrafilters

Let $V \in \mathrm{Ob}(\mathcal{V}(\mathcal{H}))$, and let $\lambda \in \underline{\Sigma}_V$ be a spectral element of the von Neumann algebra V. This means that λ is a multiplicative functional on V. For all projections $\hat{\alpha} \in \mathcal{P}(V)$, we have therefore

$$\langle \lambda, \hat{\alpha} \rangle = \langle \lambda, \hat{\alpha}^2 \rangle = \langle \lambda, \hat{\alpha} \rangle \langle \lambda, \hat{\alpha} \rangle, \tag{13.244}$$

and so $\langle \lambda, \hat{\alpha} \rangle \in \{0, 1\}$. Moreover, $\langle \lambda, \hat{0} \rangle = 0$, $\langle \lambda, \hat{1} \rangle = 1$, and if $\langle \lambda, \hat{\alpha} \rangle = 0$, then $\langle \lambda, \hat{1} - \hat{\alpha} \rangle = 1$ (since $\langle \lambda, \hat{\alpha} \rangle + \langle \lambda, \hat{1} - \hat{\alpha} \rangle = \langle \lambda, \hat{1} \rangle$). Hence, for each $\hat{\alpha} \in \mathcal{P}(V)$ we have either $\langle \lambda, \hat{\alpha} \rangle = 1$ or $\langle \lambda, \hat{1} - \hat{\alpha} \rangle = 1$. This shows that the family

$$F_\lambda := \{\hat{\alpha} \in \mathcal{P}(V) \mid \langle \lambda, \hat{\alpha} \rangle = 1\} \tag{13.245}$$

of projections is an ultrafilter in $\mathcal{P}(V)$. Conversely, each $\lambda \in \underline{\Sigma}_V$ is uniquely determined by the set $\{\langle \lambda, \hat{\alpha} \rangle \mid \hat{\alpha} \in \mathcal{P}(V)\}$ and hence by an ultrafilter in $\mathcal{P}(V)$. This shows that there is a bijection between the set $\mathcal{Q}(V)$ of ultrafilters in $\mathcal{P}(V)$ and the Gel'fand spectrum $\underline{\Sigma}_V$.

13.8.3.3 Observable and Antonymous Functions

Let $\mathcal{N}$ be a von Neumann algebra, and let $\mathcal{F}(\mathcal{N})$ be the set of filters in the projection lattice $\mathcal{P}(\mathcal{N})$ of $\mathcal{N}$. De Groote has shown [39] that to each self-adjoint operator $\hat{A} \in \mathcal{N}$, there corresponds a, so-called, "observable function" $f_{\hat{A}} : \mathcal{F}(\mathcal{N}) \to \mathrm{sp}(\hat{A})$. If $\mathcal{N}$ is abelian, $\mathcal{N} = V$, then $f_{\hat{A}}|_{\mathcal{Q}(V)}$ is just the Gel'fand transform of $\hat{A}$. However, it is striking that $f_{\hat{A}}$ can be defined even if $\mathcal{N}$ is *non*-abelian; for us, the important example is $\mathcal{N} = B(\mathcal{H})$.

If $\{\hat{E}^A_\mu\}_{\mu\in\mathbb{R}}$ is the spectral family of $\hat{A}$, then $f_{\hat{A}}$ is defined as

$$\begin{aligned} f_{\hat{A}} : \mathcal{F}(\mathcal{N}) &\to \mathrm{sp}(\hat{A}) \\ F &\mapsto \inf\{\mu \in \mathbb{R} \mid \hat{E}^A_\mu \in F\}. \end{aligned} \tag{13.246}$$

Conversely, given a bounded function $f : \mathcal{F}(\mathcal{N}) \to \mathbb{R}$ with certain properties, one can find a unique self-adjoint operator $\hat{A} \in \mathcal{N}$ such that $f = f_{\hat{A}}$.

It can be shown that each observable function is completely determined by its restriction to the space of maximal filters [39]. Let $\mathcal{Q}(\mathcal{N})$ denote the space of maximal filters in $\mathcal{P}(\mathcal{N})$. The sets

$$\mathcal{Q}_{\hat{P}}(\mathcal{N}) := \{F \in \mathcal{Q}(\mathcal{N}) \mid \hat{P} \in F\}, \quad \hat{P} \in \mathcal{P}(\mathcal{N}), \tag{13.247}$$

form the base of a totally disconnected topology on $\mathcal{Q}(\mathcal{N})$. Following de Groote, this space is called the *Stone spectrum* of $\mathcal{N}$. If $\mathcal{N}$ is abelian, $\mathcal{N} = V$, then, upon the identification of maximal filters (which are ultrafilters) in $\mathcal{P}(V)$ and spectral elements in $\underline{\Sigma}_V$, the Stone spectrum $\mathcal{Q}(V)$ is the Gel'fand spectrum $\underline{\Sigma}_V$ of V.

This shows that for an arbitrary von Neumann algebra $\mathcal{N}$, the Stone spectrum $\mathcal{Q}(\mathcal{N})$ is a generalisation of the Gel'fand spectrum (the latter is only defined for abelian algebras). The observable function $f_{\hat{A}}$ is a generalisation of the Gel'fand transform of $\hat{A}$.

We want to show that the observable function $f_{\delta^o(\hat{V})_A}$ of the outer daseinisation of $\hat{A}$ to V can be expressed by the observable function $f_{\hat{A}}$ of $\hat{A}$ directly. Since this works for all $V \in \mathrm{Ob}(\mathcal{V}(\mathcal{H}))$, we obtain a nice encoding of all the functions $f_{\delta^o(\hat{A})_V}$ and hence of the self-adjoint operators $\delta^o(\hat{A})_V$. The result (already shown in [37]) is that, for all stages $V \in \mathrm{Ob}(\mathcal{V}(\mathcal{H}))$ and all filters F in $\mathcal{F}(V)$,

$$f_{\delta^o(\hat{A})_V}(F) = f_{\hat{A}}(\mathcal{C}_{B(\mathcal{H})}(F)). \tag{13.248}$$

We want to give an elementary proof of this. We need

Lemma 2 *Let $\mathcal{N}$ be a von Neumann algebra, $\mathcal{S}$ a von Neumann sub-algebra of $\mathcal{N}$, and let $\delta^i_{\mathcal{S}} : \mathcal{P}(\mathcal{N}) \to \mathcal{P}(\mathcal{S})$ be the inner daseinisation map on projections. Then, for all filters $F \in \mathcal{F}(\mathcal{S})$,*

$$(\delta^i_{\mathcal{S}})^{-1}(F) = \mathcal{C}_{\mathcal{N}}(F). \tag{13.249}$$

Proof If $\hat{Q} \in F \subset \mathcal{P}(\mathcal{S})$, then $(\delta^i_{\mathcal{S}})^{-1}(\hat{Q}) = \{\hat{P} \in \mathcal{P}(\mathcal{N}) \mid \delta^i(\hat{P})_{\mathcal{S}} = \hat{Q}\}$. Let $\hat{P} \in \mathcal{P}(\mathcal{N})$ be such that there is a $\hat{Q} \in F$ with $\hat{Q} \leq \hat{P}$, i.e., $\hat{P} \in \mathcal{C}_{\mathcal{N}}(F)$. Then $\delta^i(\hat{P})_{\mathcal{S}} \geq \hat{Q}$, which implies $\delta^i(\hat{P})_{\mathcal{S}} \in F$, since F is a filter in $\mathcal{P}(\mathcal{S})$. This shows that $\mathcal{C}_{\mathcal{N}}(F) \subseteq (\delta^i_{\mathcal{S}})^{-1}(F)$. Now let $\hat{P} \in \mathcal{P}(\mathcal{N})$ be such that there is no $\hat{Q} \in F$ with $\hat{Q} \leq \hat{P}$. Since $\delta^i(\hat{P})_{\mathcal{S}} \leq \hat{P}$, there also is no $\hat{Q} \in F$ with $\hat{Q} \leq \delta^i(\hat{P})_{\mathcal{S}}$, so $\hat{P} \notin (\delta^i_{\mathcal{S}})^{-1}(F)$. This shows that $(\delta^i_{\mathcal{S}})^{-1}(F) \subseteq \mathcal{C}_{\mathcal{N}}(F)$.

We now can prove

Theorem 8 *Let $\hat{A} \in \mathcal{N}_{sa}$. For all von Neumann sub-algebras $\mathcal{S} \subseteq \mathcal{N}$ and all filters $F \in \mathcal{F}(\mathcal{S})$, we have*

$$f_{\delta^o(\hat{A})_{\mathcal{S}}}(F) = f_{\hat{A}}(\mathcal{C}_{\mathcal{N}}(F)). \tag{13.250}$$

Proof We have

$$\begin{aligned}
f_{\delta^o(\hat{A})_{\mathcal{S}}(F)} &= \inf\{\lambda \in \mathbb{R} \mid \hat{E}^{\delta^o(\hat{A})_{\mathcal{S}}}_\lambda \in F\} \\
&= \inf\{\lambda \in \mathbb{R} \mid \delta^i(\hat{E}^A_\lambda)_{\mathcal{S}} \in F\} \\
&= \inf\{\lambda \in \mathbb{R} \mid \hat{E}^A_\lambda \in (\delta^i_{\mathcal{S}})^{-1}(F)\} \\
&= \inf\{\lambda \in \mathbb{R} \mid \hat{E}^A_\lambda \in \mathcal{C}_{\mathcal{N}}(F)\} \\
&= f_{\hat{A}}(\mathcal{C}_{\mathcal{N}}(F)).
\end{aligned}$$

The second equality is the definition of outer daseinisation (on the level of spectral projections, see (13.169)). In the penultimate step, we used Lemma 2.

This clearly implies (13.248). We saw above that to each $\lambda \in \underline{\Sigma}_V$ there corresponds a unique ultrafilter $F_\lambda \in \mathcal{Q}(V)$. Since $\delta^o(\hat{A})_V \in V_{\rm sa}$, the observable function $f_{\delta^o(\hat{A})_V}$ is the Gel'fand transform of $\delta^o(\hat{A})_V$, and so, upon identifying the ultrafilter F_λ with the spectral element λ, we have

$$f_{\delta^o(\hat{A})_V}(F_\lambda) = \overline{\delta^o(\hat{A})_V}(\lambda) = \langle \lambda, \delta^o(\hat{A})_V \rangle. \tag{13.251}$$

From (13.248) we have

$$\langle \lambda, \delta^o(\hat{A})_V \rangle = f_{\delta^o(\hat{A})_V}(F_\lambda) = f_{\hat{A}}(\mathcal{C}_{B(\mathcal{H})}(F_\lambda)) \tag{13.252}$$

for all $V \in \text{Ob}(\mathcal{V}(\mathcal{H}))$ and for all $\lambda \in \underline{\Sigma}_V$. In this sense, the observable function $f_{\hat{A}}$ encodes all the outer daseinisations $\delta^o(\hat{A})_V$, $V \in \text{Ob}(\mathcal{V}(\mathcal{H}))$, of $\hat{A}$.

There is also a function, $g_{\hat{A}}$, on the filters in $\mathcal{P}(\mathcal{H})$ that encodes all the inner daseinisations $\delta^i(\hat{A})_V$, $V \in \text{Ob}(\mathcal{V}(\mathcal{H}))$. This function is given for an arbitrary von Neumann algebra $\mathcal{N}$ by

$$g_{\hat{A}} : \mathcal{F}(\mathcal{N}) \to \mathrm{sp}(\hat{A}) \tag{13.253}$$

$$F \mapsto \sup\{\lambda \in \mathbb{R} \mid \hat{1} - \hat{E}^A_\lambda \in F\} \tag{13.254}$$

and is called the *antonymous function* of $\hat{A}$ [22]. If $\mathcal{N}$ is abelian, then $g_{\hat{A}}|_{\mathcal{Q}(V)}$ is the Gel'fand transform of $\hat{A}$ and coincides with $f_{\hat{A}}$ on the space $\mathcal{Q}(V)$ of maximal filters, i.e.,ultrafilters in $\mathcal{P}(V)$. As functions on $\mathcal{F}(V)$, $f_{\hat{A}}$ and $g_{\hat{A}}$ are different also in the abelian case. For an arbitrary von Neumann algebra $\mathcal{N}$, the antonymous function $g_{\hat{A}}$ is another generalisation of the Gel'fand transform of $\hat{A}$.

There is a close relationship between observable and antonymous functions [39, 22]: for all von Neumann algebras $\mathcal{N}$ and all self-adjoint operators $\hat{A} \in \mathcal{N}_{\mathrm{sa}}$, it holds that

$$-f_{\hat{A}} = g_{-\hat{A}}. \tag{13.255}$$

There is a relation analogous to (13.248) for antonymous functions: for all $V \in \mathrm{Ob}(\mathcal{V}(\mathcal{H}))$ and all filters F in $\mathcal{F}(V)$,

$$g_{\delta^i(\hat{A})_V}(F) = g_{\hat{A}}(\mathcal{C}_{B(\mathcal{H})}(F)). \tag{13.256}$$

This follows from

Theorem 9 *Let $\hat{A} \in \mathcal{N}_{sa}$. For all von Neumann sub-algebras $\mathcal{S} \subseteq \mathcal{N}$ and all filters $F \in \mathcal{F}(\mathcal{S})$, we have*

$$g_{\delta^i(\hat{A})_{\mathcal{S}}}(F) = g_{\hat{A}}(\mathcal{C}_{\mathcal{N}}(F)). \tag{13.257}$$

Proof We have

$$\begin{aligned}
g_{\delta^i(\hat{A})_{\mathcal{S}}}(F) &= \sup\{\lambda \in \mathbb{R} \mid \hat{1} - \hat{E}^{\delta^i(\hat{A})_{\mathcal{S}}}_\lambda \in F\} \\
&= \sup\{\lambda \in \mathbb{R} \mid \hat{1} - \bigwedge_{\mu>\lambda} \delta^o(\hat{E}^A_\mu)_{\mathcal{S}} \in F\} \\
&= \sup\{\lambda \in \mathbb{R} \mid \hat{1} - \delta^o(\hat{E}^A_\lambda)_{\mathcal{S}} \in F\} \\
&= \sup\{\lambda \in \mathbb{R} \mid \delta^i(\hat{1} - \hat{E}^A_\lambda)_{\mathcal{S}} \in F\} \\
&= \sup\{\lambda \in \mathbb{R} \mid \hat{1} - \hat{E}^A_\lambda \in (\delta^i_{\mathcal{S}})^{-1}(F)\} \\
&= \sup\{\lambda \in \mathbb{R} \mid \hat{1} - \hat{E}^A_\lambda \in \mathcal{C}_{\mathcal{N}}(F)\} \\
&= g_{\hat{A}}(\mathcal{C}_{\mathcal{N}}(F)),
\end{aligned}$$

where in the penultimate step we used Lemma 2.

Let $\lambda \in \underline{\Sigma}_V$, and let $F_\lambda \in \mathcal{Q}(V)$ be the corresponding ultrafilter. Since $\delta^i(\hat{A})_V \in V$, the antonymous function $g_{\delta^i(\hat{A})_V}$ is the Gel'fand transform of $\delta^i(\hat{A})_V$, and we have

$$g_{\delta^i(\hat{A})_V}(F_\lambda) = \overline{\delta^i(\hat{A})_V}(\lambda) = \langle \lambda, \delta^i(\hat{A})_V \rangle. \tag{13.258}$$

From (13.256), we get

$$\langle \lambda, \delta^i(\hat{A})_V \rangle = g_{\delta^i(\hat{A})_V}(F_\lambda) = g_{\hat{A}}(\mathcal{C}_{B(\mathcal{H})}(F_\lambda)) \tag{13.259}$$

for all $V \in \mathrm{Ob}(\mathcal{V}(\mathcal{H}))$ and all $\lambda \in \underline{\Sigma}_V$. Thus the antonymous function $g_{\hat{A}}$ encodes all the inner daseinisations $\delta^i(\hat{A})_V$, $V \in \mathrm{Ob}(\mathcal{V}(\mathcal{H}))$, of $\hat{A}$.

13.8.4 A Physical Interpretation of the Arrow $\breve{\delta}(\hat{A}) : \underline{\Sigma} \to \underline{\mathbb{R}^{\leftrightarrow}}$

Let $|\psi\rangle \in \mathcal{H}$ be a unit vector in the Hilbert space of the quantum system. The expectation value of a self-adjoint operator $\hat{A} \in B(\mathcal{H})$ in the state $|\psi\rangle$ is given by

$$\langle\psi|\,\hat{A}\,|\psi\rangle = \int_{-||\hat{A}||}^{||\hat{A}||} \lambda\, d\langle\psi|\,\hat{E}^A_\lambda\,|\psi\rangle. \tag{13.260}$$

In the discussion of truth objects in Sect. 13.6, we introduced the maximal filter $T^{|\psi\rangle}$ in $\mathcal{P}(\mathcal{H})$,[87] given by (cf. (13.125))

$$T^{|\psi\rangle} := \{\hat{\alpha} \in \mathcal{P}(\mathcal{H}) \mid \hat{\alpha} \succeq |\psi\rangle\langle\psi|\}, \tag{13.261}$$

where $|\psi\rangle\langle\psi|$ is the projection onto the one-dimensional subspace of $\mathcal{H}$ generated by $|\psi\rangle$. As shown in [22], the expectation value $\langle\psi|\,\hat{A}\,|\psi\rangle$ can be written as

$$\langle\psi|\,\hat{A}\,|\psi\rangle = \int_{g_{\hat{A}}(T^{|\psi\rangle})}^{f_{\hat{A}}(T^{|\psi\rangle})} \lambda\, d\langle\psi|\,\hat{E}^A_\lambda\,|\psi\rangle. \tag{13.262}$$

In an instrumentalist interpretation,[88] one would interpret $g_{\hat{A}}(T^{|\psi\rangle})$, resp. $f_{\hat{A}}(T^{|\psi\rangle})$, as the smallest, resp. largest, possible result of a measurement of the physical quantity A when the state is $|\psi\rangle$. If $|\psi\rangle$ is an eigenstate of $\hat{A}$, then $\langle\psi|\,\hat{A}\,|\psi\rangle$ is an eigenvalue of $\hat{A}$, and in this case, $\langle\psi|\,\hat{A}\,|\psi\rangle \in \mathrm{sp}(\hat{A})$; moreover,

$$\langle\psi|\,\hat{A}\,|\psi\rangle = g_{\hat{A}}(T^{|\psi\rangle}) = f_{\hat{A}}(T^{|\psi\rangle}). \tag{13.263}$$

If $|\psi\rangle$ is not an eigenstate of $\hat{A}$, then

$$g_{\hat{A}}(T^{|\psi\rangle}) < \langle\psi|\,\hat{A}\,|\psi\rangle < f_{\hat{A}}(T^{|\psi\rangle}). \tag{13.264}$$

---

[87] Since $\mathcal{P}(\mathcal{H})$ is not distributive, $T^{|\psi\rangle}$ is not an ultrafilter; i.e., there are projections $\hat{P} \in \mathcal{P}(\mathcal{H})$ such that neither $\hat{P} \in T^{|\psi\rangle}$ nor $\hat{1} - \hat{P} \in T^{|\psi\rangle}$.

[88] Which we avoid in general, of course!

For details, see [22].

Let V be an abelian sub-algebra of $B(\mathcal{H})$ such that $\underline{\Sigma}_V$ contains the spectral element, $\lambda^{|\psi\rangle}$, associated with $|\psi\rangle$.[89] The corresponding ultrafilter in $\mathcal{P}(V)$ consists of those projections $\hat{\alpha} \in \mathcal{P}(V)$ such that $\hat{\alpha} \succeq |\psi\rangle\langle\psi|$. This is just the evaluation, $\underline{\mathbb{T}}^{|\psi\rangle}_V$, at stage V of our truth object, $\underline{\mathbb{T}}^{|\psi\rangle}$; see (13.124).

Hence the cone $\mathcal{C}(\underline{\mathbb{T}}^{|\psi\rangle}_V) := \mathcal{C}_{B(\mathcal{H})}(\underline{\mathbb{T}}^{|\psi\rangle}_V)$ consists of all projections $\hat{R} \in \mathcal{P}(\mathcal{H})$ such that $\hat{R} \succeq |\psi\rangle\langle\psi|$; and so, for all stages V such that $|\psi\rangle\langle\psi| \in \mathcal{P}(V)$ we have

$$\mathcal{C}(\underline{\mathbb{T}}^{|\psi\rangle}_V) = T^{|\psi\rangle}. \tag{13.265}$$

This allows us to write the expectation value as

$$\langle\psi|\,\hat{A}\,|\psi\rangle = \int_{g_{\hat{A}}(\mathcal{C}(\underline{\mathbb{T}}^{|\psi\rangle}_V))}^{f_{\hat{A}}(\mathcal{C}(\underline{\mathbb{T}}^{|\psi\rangle}_V))} \lambda\, d\langle\psi|\,\hat{E}^A_\lambda\,|\psi\rangle \tag{13.266}$$

$$= \int_{g_{\delta^i(\hat{A})_V}(\underline{\mathbb{T}}^{|\psi\rangle}_V)}^{f_{\delta^o(\hat{A})_V}(\underline{\mathbb{T}}^{|\psi\rangle}_V)} \lambda\, d\langle\psi|\,\hat{E}^A_\lambda\,|\psi\rangle \tag{13.267}$$

for these stages V.

Equations (13.251) and (13.258) show that $f_{\delta^o(\hat{A})_V}(\underline{\mathbb{T}}^{|\psi\rangle}_V) = \langle\psi|\,\delta^o(\hat{A})_V\,|\psi\rangle$ and $g_{\delta^i(\hat{A})_V}(\underline{\mathbb{T}}^{|\psi\rangle}_V) = \langle\psi|\,\delta^i(\hat{A})_V\,|\psi\rangle$. In the language of instrumentalism, for stages V for which $\lambda^{|\psi\rangle} \in \underline{\Sigma}_V$, the value $\langle\psi|\,\delta^i(\hat{A})_V\,|\psi\rangle \in \mathrm{sp}(\hat{A})$ is the smallest possible measurement result for $\hat{A}$ in the quantum state $|\psi\rangle$; and $\langle\psi|\,\delta^o(\hat{A})_V\,|\psi\rangle \in \mathrm{sp}(\hat{A})$ is the largest possible result.

These results depend on the fact that we use (inner and outer) daseinisation, i.e., approximations in the spectral, not the linear order.

If $\lambda \in \underline{\Sigma}_V$ is not of the form $\lambda = \lambda^{|\psi\rangle}$, for some $|\psi\rangle \in \mathcal{H}$, then the cone $\mathcal{C}(F_\lambda)$ over the ultrafilter F_λ corresponding to λ cannot be identified with a vector in $\mathcal{H}$. Nevertheless, the quantity $\mathcal{C}(F_\lambda)$ is well-defined, and (13.248) and (13.256) hold. If we go from V to a sub-algebra $V' \subseteq V$, then $\delta^i(\hat{A})_{V'} \preceq \delta^i(\hat{A})_V$ and $\delta^o(\hat{A})_{V'} \succeq \delta^o(\hat{A})_V$, hence

$$\langle\lambda, \delta^i(\hat{A})_{V'}\rangle \leq \langle\lambda, \delta^i(\hat{A})_V\rangle, \tag{13.268}$$

$$\langle\lambda, \delta^o(\hat{A})_{V'}\rangle \geq \langle\lambda, \delta^o(\hat{A})_V\rangle \tag{13.269}$$

for all $\lambda \in \underline{\Sigma}_V$.

We can interpret the function

$$\breve{\delta}(\hat{A})_V : \underline{\Sigma}_V \to \underline{\mathbb{R}^{\leftrightarrow}}_V \tag{13.270}$$

---

[89] This is the element defined by $\lambda^{|\psi\rangle}(\hat{A}) := \langle\psi|\,\hat{A}\,|\psi\rangle$ for all $\hat{A} \in V$. It is characterised by the fact that $\lambda^{|\psi\rangle}(|\psi\rangle\langle\psi|) = 1$ and $\lambda^{|\psi\rangle}(\hat{Q}) = 0$ for all $\hat{Q} \in \mathcal{P}(V)$ such that $\hat{Q}\,|\psi\rangle\langle\psi| = \hat{0}$. We have $\lambda^{|\psi\rangle} \in \underline{\Sigma}_V$ if and only if $|\psi\rangle\langle\psi| \in \mathcal{P}(V)$.

$$\lambda \mapsto \breve{\delta}(\hat{A})_V(\lambda) = \left(\breve{\delta}^i(\hat{A})_V(\lambda), \breve{\delta}^o(\hat{A})_V(\lambda)\right) \tag{13.271}$$

as giving the "spread" or "range" of the physical quantity A at stages $V' \subseteq V$. Each element $\lambda \in \underline{\Sigma}_V$ gives its own "spread" $\breve{\delta}(\hat{A})_V(\lambda) : \downarrow V \to \mathrm{sp}(\hat{A}) \times \mathrm{sp}(\hat{A})$. The intuitive idea is that at stage V, given a point $\lambda \in \underline{\Sigma}_V$, the physical quantity A "spreads over" the subset of the spectrum, $\mathrm{sp}(\hat{A})$, of $\hat{A}$ given by the closed interval of $\mathrm{sp}(\hat{A}) \subset \mathbb{R}$ defined by

$$[\breve{\delta}^i(\hat{A})_V(\lambda)(V), \breve{\delta}^o(\hat{A})_V(\lambda)(V)] \cap \mathrm{sp}(\hat{A}) = [\langle\lambda, \delta^i(\hat{A})_V\rangle, \langle\lambda, \delta^o(\hat{A}))_V\rangle] \cap \mathrm{sp}(\hat{A}). \tag{13.272}$$

For a proper sub-algebra $V' \subset V$, the spreading is over the (potentially larger) subset

$$[\breve{\delta}^i(\hat{A})_V(\lambda)(V'), \breve{\delta}^o(\hat{A})_V(\lambda)(V')] \cap \mathrm{sp}(\hat{A}) = [\langle\lambda, \delta^i(\hat{A})_{V'}\rangle, \langle\lambda, \delta^o(\hat{A})_{V'}\rangle] \cap \mathrm{sp}(\hat{A}). \tag{13.273}$$

All this is local in the sense that these expressions are defined at a stage V and for sub-algebras, V', of V, where $\lambda \in \underline{\Sigma}_V$. No similar global construction or interpretation is possible, since the spectral presheaf $\underline{\Sigma}$ has no global elements, i.e., no points (while the *set* $\underline{\Sigma}_V$ does have points).

As we go down to smaller sub-algebras $V' \subseteq V$, the spread gets larger. This comes from the fact that $\hat{A}$ has to be adapted more and more as we go to smaller sub-algebras V'. More precisely, $\hat{A}$ is approximated from below by $\delta^i(\hat{A})_{V'} \in V'$ and from above by $\delta^o(\hat{A})_{V'} \in V'$. This approximation gets coarser as V' gets smaller, which basically means that V' contains less and less projections.

It should be remarked that $\breve{\delta}(\hat{A})$ does not assign actual values to the physical quantity A, but rather the possible *range* of such values; and these are independent of any state $|\psi\rangle$. This is analogous to the classical case where physical quantities are represented by real-valued functions on state space. The range of possible values is state-independent, but the actual value possessed by a physical quantity *does* depend on the state of the system.

13.8.4.1 The Quantity-Value Presheaf $\underline{\mathbb{R}^{\leftrightarrow}}$ and the Interval Domain

Heunen et al. observed [42] that the presheaf $\underline{\mathbb{R}^{\leftrightarrow}}$ is closely related to the *interval domain* in our topos. This object has mainly been considered in theoretical computer science [33] and can be used to systematically encode situations where real numbers are only known—or can only be defined—up to a certain degree of accuracy. Approximation processes can be well described using the mathematics of domain theory. Clearly, this has close relations to our physical situation, where the real numbers are spectral values of self-adjoint operators and coarse-graining (or rather the inverse process of fine-graining) can be understood as a process of approximation.

13.8.5 The Value of a Physical Quantity in a Quantum State

We now want to discuss how physical quantities, represented by natural transformations $\breve{\delta}(\hat{A})$, acquire "values" in a given quantum state. Of course, this is not as straightforward as in the classical case, since from the Kochen-Specker theorem, we know that physical quantities do not have real numbers as their values. As we saw, this is related to the fact that there are no microstates, i.e., the spectral presheaf has no global elements.

In classical physics, a physical quantity A is represented by a function $\breve{A} : \mathcal{S} \to \mathbb{R}$ from the state space $\mathcal{S}$ to the real numbers. A point $s \in \mathcal{S}$ is a microstate, and the physical quantity A has the value $\breve{A}(s)$ in this state.

We want to mimic this as closely as possible in the quantum case. In order to do so, we take a pseudo-state

$$\mathfrak{w}^{|\psi\rangle} := \delta(|\psi\rangle\langle\psi|) = V \mapsto \bigwedge\{\hat{\alpha} \in \underline{O}_V \mid |\psi\rangle\langle\psi| \preceq \hat{\alpha}\} \tag{13.274}$$

(see (13.128)) and consider it as a sub-object of $\underline{\Sigma}$. This means that at each stage $V \in \mathrm{Ob}(\mathcal{V}(\mathcal{H}))$, we consider the set

$$\underline{\mathfrak{w}}^{|\psi\rangle}_V := \{\lambda \in \underline{\Sigma}_V \mid \langle\lambda, \delta^o(|\psi\rangle\langle\psi|)_V\rangle = 1\} \subseteq \underline{\Sigma}_V. \tag{13.275}$$

Of course, the sub-object of $\underline{\Sigma}$ that we get simply is $\underline{\delta^o(|\psi\rangle\langle\psi|)}$. Sub-objects of this kind are as close to microstates as we can get, see the discussion in Sect. 13.6.3 and [23]. We can then form the composition

$$\underline{\mathfrak{w}}^{|\psi\rangle} \to \underline{\Sigma} \xrightarrow{\breve{\delta}(\hat{A})} \underline{\mathbb{R}^{\leftrightarrow}}, \tag{13.276}$$

which is also denoted by $\breve{\delta}(\hat{A})(\underline{\mathfrak{w}}^{|\psi\rangle})$. One can think of this arrow as being the "value" of the physical quantity A in the state described by $\underline{\mathfrak{w}}^{|\psi\rangle}$.

The first question is if we actually obtain a sub-object of $\underline{\mathbb{R}^{\leftrightarrow}}$ in this way. Let $V, V' \in \mathrm{Ob}(\mathcal{V}(\mathcal{H}))$, $V' \subseteq V$. We have to show that

$$\underline{\mathbb{R}^{\leftrightarrow}}(i_{V'V})(\breve{\delta}(\hat{A})_V(\underline{\mathfrak{w}}^{|\psi\rangle}_V)) \subseteq \breve{\delta}(\hat{A})_{V'}(\underline{\mathfrak{w}}^{|\psi\rangle}_{V'}). \tag{13.277}$$

Let $\lambda \in \underline{\mathfrak{w}}^{|\psi\rangle}_V$, then

$$\underline{\mathbb{R}^{\leftrightarrow}}(i_{V'V})(\breve{\delta}(\hat{A})_V(\lambda)) = (\breve{\delta}(\hat{A})_V(\lambda))|_{V'} = \breve{\delta}(\hat{A})_{V'}(\lambda|_{V'}). \tag{13.278}$$

By definition, we have

$$\underline{\mathfrak{w}}^{|\psi\rangle}_{V'} = \underline{\Sigma}(i_{V'V})(\underline{\mathfrak{w}}^{|\psi\rangle}_V) = \{\lambda|_{V'} \mid \lambda \in \underline{\mathfrak{w}}^{|\psi\rangle}_V\}, \tag{13.279}$$

that is, *every* $\lambda' \in \underline{\mathfrak{w}}^{|\psi\rangle}_{V'}$ is given as the restriction of some $\lambda \in \underline{\mathfrak{w}}^{|\psi\rangle}_{V}$. This implies that we even obtain the equality

$$\underline{\mathbb{R}^{\leftrightarrow}}(i_{V'V})(\breve{\delta}(\hat{A})_V(\underline{\mathfrak{w}}^{|\psi\rangle}_{V})) = \breve{\delta}(\hat{A})_{V'}(\underline{\mathfrak{w}}^{|\psi\rangle}_{V'}), \tag{13.280}$$

so $\breve{\delta}(\hat{A})(\underline{\mathfrak{w}}^{|\psi\rangle})$ is indeed a sub-object of $\underline{\mathbb{R}^{\leftrightarrow}}$.

13.8.5.1 Values as Pairs of Functions and Eigenvalues

At each stage $V \in \mathrm{Ob}(\mathcal{V}(\mathcal{H}))$, we have pairs of order-preserving and order-reversing functions $\breve{\delta}(\hat{A})(\lambda)$, one function for each $\lambda \in \underline{\mathfrak{w}}^{|\psi\rangle}_{V}$. If $|\psi\rangle$ is an eigenstate of $\hat{A}$ and V is an abelian sub-algebra that contains $\hat{A}$, then $\delta^i(\hat{A})_V = \delta^o(\hat{A})_V = \hat{A}$. Moreover, $\underline{\mathfrak{w}}^{|\psi\rangle}_{V}$ contains the single element $\lambda_{|\psi\rangle\langle\psi|} \in \underline{\Sigma}_V$, which is the pure state that assigns 1 to $|\psi\rangle\langle\psi|$ and 0 to all projections in $\mathcal{P}(V)$ orthogonal to $|\psi\rangle\langle\psi|$.

Evaluating $\breve{\delta}(\hat{A})(\underline{\mathfrak{w}}^{|\psi\rangle})$ at V hence gives a pair, consisting of an order-preserving function $\breve{\delta}^i(\hat{A})_V(\lambda_{|\psi\rangle\langle\psi|}) : \downarrow\!V \to \mathrm{sp}(\hat{A})$ and an order-reversing function $\breve{\delta}^o(\hat{A})_V(\lambda_{|\psi\rangle\langle\psi|}) : \downarrow\!V \to \mathrm{sp}(\hat{A})$:

$$\breve{\delta}(\hat{A})_V(\underline{\mathfrak{w}}^{|\psi\rangle}_{V}) = (\breve{\delta}^i(\hat{A})_V(\lambda_{|\psi\rangle\langle\psi|}), \breve{\delta}^o(\hat{A})_V(\lambda_{|\psi\rangle\langle\psi|})). \tag{13.281}$$

The value of both functions at stage V is $\overline{\hat{A}}(\lambda_{|\psi\rangle\langle\psi|}) = \langle\lambda_{|\psi\rangle\langle\psi|}, \hat{A}\rangle$, which is the eigenvalue of $\hat{A}$ in the state $|\psi\rangle$. In this sense, we get back the ordinary eigenvalue of $\hat{A}$ when the system is in the eigenstate ψ.

13.8.5.2 A Simple Example

We consider the value of the self-adjoint projection operator $|\psi\rangle\langle\psi|$, seen as (the representative of) a physical quantity, in the (pseudo-)state $\underline{\mathfrak{w}}^{|\psi\rangle}$. We remark that $\mathrm{sp}(|\psi\rangle\langle\psi|) = \{0, 1\}$. By definition,

$$\breve{\delta}(|\psi\rangle\langle\psi|)_V(\lambda) = (\breve{\delta}^i(|\psi\rangle\langle\psi|)_V(\lambda), \breve{\delta}^o(|\psi\rangle\langle\psi|)_V(\lambda)) \tag{13.282}$$

for all $V \in \mathrm{Ob}(\mathcal{V}(\mathcal{H}))$ and all $\lambda \in \underline{\Sigma}_V$. In particular, the function

$$\breve{\delta}^o(|\psi\rangle\langle\psi|)_V(\lambda) : \downarrow\!V \to \{0, 1\} \tag{13.283}$$

is given as (see (13.227), for all $V' \subseteq V$,

$$\breve{\delta}^o(|\psi\rangle\langle\psi|)_V(\lambda)(V') = \langle\lambda, \delta^o(|\psi\rangle\langle\psi|)_{V'}\rangle. \tag{13.284}$$

If $\lambda \in \underline{\mathfrak{w}}^{|\psi\rangle}_{V}$, then $\langle\lambda, \delta^o(|\psi\rangle\langle\psi|)_V\rangle = 1$, see (13.275). Hence, for all $\lambda \in \underline{\mathfrak{w}}^{|\psi\rangle}_{V}$, we obtain, for all $V' \subseteq V$,

$$\breve{\delta}^o(\,|\psi\rangle\langle\psi|\,)_V(\lambda)(V') = \langle\lambda, \delta^o(\,|\psi\rangle\langle\psi|\,)_{V'}\rangle = 1. \tag{13.285}$$

If we denote the constant function on $\downarrow V$ with value 1 as $1_{\downarrow V}$, then we can write

$$\breve{\delta}(\,|\psi\rangle\langle\psi|\,)_V(\lambda) = (\breve{\delta}^i(\,|\psi\rangle\langle\psi|\,)_V(\lambda), 1_{\downarrow V}) \tag{13.286}$$

for all V and all $\lambda \in \underline{\mathfrak{w}}^{|\psi\rangle}_V$. The constant function $1_{\downarrow V}$ trivially is an order-reversing function from $\downarrow V$ to $\mathrm{sp}(\,|\psi\rangle\langle\psi|\,)$. We now consider the function

$$\breve{\delta}^i(\,|\psi\rangle\langle\psi|\,)_V(\lambda) : \downarrow V \to \{0, 1\}. \tag{13.287}$$

It is given as (see (13.235), for all $V' \subseteq V$,

$$\breve{\delta}^i(\,|\psi\rangle\langle\psi|\,)_V(\lambda)(V') = \langle\lambda, \delta^i(\,|\psi\rangle\langle\psi|\,)_{V'}\rangle. \tag{13.288}$$

If $|\psi\rangle\langle\psi| \in \mathcal{P}(V')$, then, for all $\lambda \in \underline{\mathfrak{w}}^{|\psi\rangle}_{V'}$, we have $\langle\lambda, \delta^i(\,|\psi\rangle\langle\psi|\,)_{V'}\rangle = \langle\lambda, |\psi\rangle\langle\psi|\,\rangle = 1$. If $|\psi\rangle\langle\psi| \notin \mathcal{P}(V')$, then $\delta^i(\,|\psi\rangle\langle\psi|\,)_{V'} = \hat{0}$, since

$$\delta^i(\,|\psi\rangle\langle\psi|\,)_V \preceq |\psi\rangle\langle\psi| \tag{13.289}$$

and $|\psi\rangle\langle\psi|$ is a projection onto a one-dimensional subspace, so $\delta^i(\,|\psi\rangle\langle\psi|\,)_{V'}$ must project onto the zero-dimensional subspace.

Thus we get, for all V, for all $\lambda \in \underline{\mathfrak{w}}^{|\psi\rangle}_V$ and all $V' \subseteq V$:

$$\breve{\delta}^i(\,|\psi\rangle\langle\psi|\,)_V(\lambda)(V') = \begin{cases} 1 \text{ if } |\psi\rangle\langle\psi| \in V' \\ 0 \text{ if } |\psi\rangle\langle\psi| \notin V' \end{cases} \tag{13.290}$$

Summing up, we have completely described the "value" $\breve{\delta}(\,|\psi\rangle\langle\psi|\,)(\underline{\mathfrak{w}}^{|\psi\rangle})$ of the physical quantity described by $|\psi\rangle\langle\psi|$ in the pseudo-state given by $\underline{\mathfrak{w}}^{|\psi\rangle}$.

There is an immediate generalisation of one part of this result: Consider an arbitrary non-zero projection $\hat{P} \in \mathcal{V}(\mathcal{H})$,[90] the corresponding sub-object $\underline{\delta^o(\hat{P})}$ of $\underline{\Sigma}$ obtained from outer daseinisation, and the sub-object $\breve{\delta}(\hat{P})(\underline{\delta^o(\hat{P})})$ of $\underline{\mathbb{R}^{\leftrightarrow}}$. For all $V \in \mathrm{Ob}(\mathcal{V}(\mathcal{H}))$ and all $\lambda \in \underline{\delta^o(\hat{P})}_V$, a completely analogous argument to the one given above shows that for the order-reversing functions $\breve{\delta}^o(\hat{P})_V(\lambda) : \downarrow V \to \{0, 1\}$, we always obtain the constant function $1_{\downarrow V}$.

The behaviour of the order-preserving functions $\breve{\delta}^i(\hat{P})_V(\lambda) : \downarrow V \to \{0, 1\}$ is more complicated than in the case that $\hat{P}$ projects onto a one-dimensional subspace. In general, $\hat{P} \notin V'$ does not imply $\breve{\delta}^i(\hat{P})_{V'}(\lambda)(V') = 0$ for $\lambda \in \underline{\delta^o(\hat{P})}_V$, so the analogue of (13.290) does not hold in general.

---

[90] Of course, if $\hat{P}$ is not a projection onto a one-dimensional subspace, then it cannot be identified with a state.

13.8.6 Properties of $\underline{\mathbb{R}^{\leftrightarrow}}$

From the perspective of our overall programme, Theorem 7 is a key result and shows that $\underline{\mathbb{R}^{\leftrightarrow}}$ is a possible choice for the quantity-value object for quantum theory. To explore this further, we start by noting some elementary properties of the presheaf $\underline{\mathbb{R}^{\leftrightarrow}}$. Analogous arguments apply to the presheaves $\underline{\mathbb{R}^{\preceq}}$ and $\underline{\mathbb{R}^{\succeq}}$.

1. The presheaf $\underline{\mathbb{R}^{\leftrightarrow}}$ has global elements: namely, pairs of order-preserving and order-reversing functions on the partially-ordered set $\mathrm{Ob}(\mathcal{V}(\mathcal{H}))$ of objects in the category $\mathcal{V}(\mathcal{H})$; i.e., pairs of functions $(\mu, \nu) : \mathrm{Ob}(\mathcal{V}(\mathcal{H})) \to \mathbb{R}$ such that:

$$\forall V_1, V_2 \in \mathrm{Ob}(\mathcal{V}(\mathcal{H})), V_2 \subseteq V_1 : \mu(V_2) \leq \mu(V_1), \nu(V_2) \geq \nu(V_1). \quad (13.291)$$

2. (a) Elements of $\Gamma\underline{\mathbb{R}^{\leftrightarrow}}$ can be added: i.e., if $(\mu_1, \nu_1), (\mu_2, \nu_2) \in \Gamma\underline{\mathbb{R}^{\leftrightarrow}}$, define $(\mu_1, \nu_1) + (\mu_2, \nu_2)$ at each stage V by

$$((\mu_1, \nu_1) + (\mu_2, \nu_2))(V') := (\mu_1(V') + \mu_2(V'), \nu_1(V') + \nu_2(V')) \quad (13.292)$$

for all $V' \subseteq V$. Note that if $V_2 \subseteq V_1 \subseteq V$, then $\mu_1(V_2) \leq \mu_1(V_1)$ and $\mu_2(V_2) \leq \mu_2(V_1)$, and so $\mu_1(V_2) + \mu_2(V_2) \leq \mu_1(V_1) + \mu_2(V_1)$. Likewise, $\nu_1(V_2) + \nu_2(V_2) \geq \nu_1(V_1) + \nu_2(V_1)$. Thus the definition of $(\mu_1, \nu_1) + (\mu_2, \nu_2)$ in (13.292) makes sense. Obviously, addition is commutative and associative.

(b) However, it is *not* possible to define "$(\mu_1, \nu_1) - (\mu_2, \nu_2)$" in this way since the difference between two order-preserving functions may not be order-preserving, nor need the difference of two order-reversing functions be order-reversing. This problem is addressed in Sect. 13.9.

(c) A zero/unit element can be defined for the additive structure on $\Gamma\underline{\mathbb{R}^{\leftrightarrow}}$ as $0(V) := (0, 0)$ for all $V \in \mathrm{Ob}(\mathcal{V}(\mathcal{H}))$, where $(0, 0)$ denotes a pair of two copies of the function that is constantly 0 on $\mathrm{Ob}(\mathcal{V}(\mathcal{H}))$.
It follows from (a) and (c) that $\Gamma\underline{\mathbb{R}^{\leftrightarrow}}$ is a commutative monoid (i.e., a semi-group with a unit).

The commutative monoid structure for $\Gamma\underline{\mathbb{R}^{\leftrightarrow}}$ is a reflection of the stronger fact that $\underline{\mathbb{R}^{\leftrightarrow}}$ is a *commutative-monoid* object in the topos $\mathbf{Sets}^{\mathcal{V}(\mathcal{H})^{\mathrm{op}}}$. Specifically, there is an arrow

$$+ : \underline{\mathbb{R}^{\leftrightarrow}} \times \underline{\mathbb{R}^{\leftrightarrow}} \to \underline{\mathbb{R}^{\leftrightarrow}}, \quad (13.293)$$

$$+_V((\mu_1, \nu_1), (\mu_2, \nu_2)) := (\mu_1 + \mu_2, \nu_1 + \nu_2) \quad (13.294)$$

for all $(\mu_1, \nu_1), (\mu_2, \nu_2) \in \underline{\mathbb{R}^{\leftrightarrow}}_V$, and for all stages $V \in \mathrm{Ob}(\mathcal{V}(\mathcal{H}))$. Here, $(\mu_1 + \mu_2, \nu_1 + \nu_2)$ denotes the real-valued function on $\downarrow\!V$ defined by

$$(\mu_1 + \mu_2, \nu_1 + \nu_2)(V') := (\mu_1(V') + \mu_2(V'), \nu_1(V') + \nu_2(V')) \quad (13.295)$$

for all $V' \subseteq V$.

3. The real numbers, $\mathbb{R}$, form a ring, and so it is natural to see if a multiplicative structure can be put on $\Gamma\underline{\mathbb{R}^{\leftrightarrow}}$. The obvious "definition" would be, for all V,

$$(\mu_1, \nu_1)(\mu_2, \nu_2)(V) := (\mu_1(V)\mu_2(V), \nu_1(V)\nu_2(V)) \tag{13.296}$$

for $(\mu_1, \nu_1), (\mu_2, \nu_2) \in \Gamma\underline{\mathbb{R}^{\leftrightarrow}}$. However, this fails because the right hand side of (13.296) may not be a pair consisting of an order-preserving and an order-reversing function. This problem arises, for example, if $\nu_1(V)$ and $\nu_2(V)$ become negative: then, as V gets smaller, the product $\nu_1(V)\nu_2(V)$ gets larger and thus defines an *order-preserving* function.

13.8.7 The Representation of Propositions From Inverse Images

In Sect. 13.3.2, we introduced a simple propositional language, $\mathcal{PL}(S)$, for each system S, and discussed its representations for the case of classical physics. Then, in Sect. 13.5 we analysed the, far more complicated, quantum-theoretical representation of this language in the set of clopen subsets of the spectral presheaf, $\underline{\Sigma}$, in the topos $\mathbf{Sets}^{\mathcal{V}(\mathcal{H})^{\mathrm{op}}}$. This gives a representation of the primitive propositions "$A \varepsilon \Delta$" as sub-objects of $\underline{\Sigma}$:

$$\pi_{\mathrm{qt}}(A \,\varepsilon\, \Delta) := \underline{\delta^o\big(\hat{E}[A \in \Delta]\big)} \tag{13.297}$$

where "δ^o" is the (outer) daseinisation operation, and $\hat{E}[A \in \Delta]$ is the spectral projection corresponding to the subset $\Delta \cap \mathrm{sp}(\hat{A})$ of the spectrum, $\mathrm{sp}(\hat{A})$, of the self-adjoint operator $\hat{A}$.

We now want to remark briefly on the nature, and representation, of propositions using the "local" language $\mathcal{L}(S)$.

In any classical representation, σ, of $\mathcal{L}(S)$ in **Sets**, the representation, $\mathcal{R}_\sigma$, of the quantity-value symbol $\mathcal{R}$ is always just the real numbers $\mathbb{R}$. Therefore, it is simple to take a subset $\Delta \subseteq \mathbb{R}$ of $\mathbb{R}$, and construct the propositions "$A \varepsilon \Delta$". In fact, if $A_\sigma : \Sigma_\sigma \to \mathbb{R}$ is the representation of the function symbol A with signature $\Sigma \to \mathcal{R}$, then $A_\sigma^{-1}(\Delta)$ is a subset of the symplectic manifold Σ_σ (the representation of the ground type Σ). This subset, $A_\sigma^{-1}(\Delta) \subseteq \Sigma_\sigma$, represents the proposition "$A \varepsilon \Delta$" in the Boolean algebra of all (Borel) subsets of Σ_σ.

We should consider the analogue of these steps in the representation, ϕ, of the same language, $\mathcal{L}(S)$, in the topos $\tau_\phi := \mathbf{Sets}^{\mathcal{V}(\mathcal{H})^{\mathrm{op}}}$. In fact, the issues to be discussed apply to a representation in *any* topos.

We first note that if Ξ is a sub-object of $\mathcal{R}_\phi$, and if $A_\phi : \Sigma_\phi \to \mathcal{R}_\phi$, then there is an associated sub-object of Σ_ϕ, denoted $A_\phi^{-1}(\Xi)$. Specifically, if $\chi_\Xi : \mathcal{R}_\phi \to \Omega_{\tau_\phi}$ is the characteristic arrow of the sub-object Ξ, then $A_\phi^{-1}(\Xi)$ is defined to be the sub-object of Σ_ϕ whose characteristic arrow is $\chi_\Xi \circ A_\phi : \Sigma_\phi \to \Omega_{\tau_\phi}$. These sub-objects are analogues of the subsets, $A_\sigma^{-1}(\Delta)$, of the classical state space Σ_σ: as such, they can represent propositions. In this spirit, we could denote by "$A \varepsilon \Xi$"

the proposition which the sub-object $A_\phi^{-1}(\Xi)$ represents. Of course, the proposition "$A \varepsilon \Xi$" must be interpreted with respect to the quantity-value object and the state object of the topos under consideration.[91]

In the case of quantum theory, the arrows $A_\phi : \Sigma_\phi \to \mathcal{R}_\phi$ are of the form $\breve{\delta}(\hat{A}) : \underline{\Sigma} \to \underline{\mathbb{R}^{\leftrightarrow}}$ where $\mathcal{R}_\phi = \underline{\mathbb{R}^{\leftrightarrow}}$. It follows that the propositions in our $\mathcal{L}(S)$-theory are represented by the sub-objects $\breve{\delta}(\hat{A})^{-1}(\underline{\Xi})$ of $\underline{\Sigma}$, where $\underline{\Xi}$ is a sub-object of $\underline{\mathbb{R}^{\leftrightarrow}}$.

To interpret such propositions, note first that in the $\mathcal{PL}(S)$-propositions "$A \varepsilon \Delta$", the range "Δ" belongs to the world that is external to the language. Consequently, the *meaning* of Δ is given independently of $\mathcal{PL}(S)$. This "externally interpreted" Δ is then inserted into the quantum representation of $\mathcal{PL}(S)$ via the daseinisation of propositions discussed in Sect. 13.5.

However, the situation is very different for the $\mathcal{L}(S)$-propositions "$A \varepsilon \Xi$". Here, the quantity "Ξ" belongs to the particular topos τ_ϕ, and hence it is representation dependent. The implication is that the "meaning" of "$A \varepsilon \Xi$" can only be discussed from "within the topos" using the internal language that is associated with τ_ϕ, which, we recall, carries the translation of $\mathcal{L}(S)$ given by the topos-representation ϕ.

From a conceptual perspective, this situation is "relational", with the meanings of the various propositions being determined by their relations to each other as formulated in the internal language of the topos. Concomitantly, the meaning of "truth" cannot be understood using the *correspondence theory* (much favoured by instrumentalists) for there is nothing external to which a proposition can "correspond". Instead, what is needed is more like a *coherence* theory of truth in which a whole body of propositions is considered together [35]. This is a fascinating subject, but further discussion must be deferred to later work.

13.8.8 The Relation Between the Formal Languages $\mathcal{L}(S)$ and $\mathcal{PL}(S)$

In the propositional language $\mathcal{PL}(S)$, we have symbols "$A \varepsilon \Delta$" representing primitive propositions. In the quantum case, such a primitive proposition is represented by the outer daseinisation $\underline{\delta^o(\hat{P})}$ of the projection corresponding to the proposition. (The spectral theorem gives the link between propositions and projections.)

We now want to show that the sub-objects of $\underline{\Sigma}$ of the form $\underline{\delta^o(\hat{P})}$ for $\hat{P} \in \mathcal{P}(\mathcal{H})$ also are part of the language $\mathcal{L}(S)$. More precisely, we will show that $\underline{\delta^o(\hat{P})}$ can be obtained as the inverse image of a certain sub-object of $\underline{\mathbb{R}^{\leftrightarrow}}$.

The sub-object of $\underline{\mathbb{R}^{\leftrightarrow}}$ that we consider is $\breve{\delta}(\hat{P})(\underline{\delta^o(\hat{P})})$. We take the inverse image of this sub-object under the natural transformation $\breve{\delta}(\hat{P}) : \underline{\Sigma} \to \underline{\mathbb{R}^{\leftrightarrow}}$. This

[91] Compare Sect. 13.8.5, where it is shown for the quantum topos $\mathbf{Sets}^{\mathcal{V}(\mathcal{H})^{\mathrm{op}}}$ how physical quantities aquire values.

means that we assume that the language $\mathcal{L}(S)$ contains a function symbol $P : \Sigma \to \mathcal{R}$ that is represented by the natural transformation $\breve{\delta}(\hat{P})$.

One more remark: although it may look as if we put in from the start the sub-object $\underline{\delta^o(\hat{P})}$ that we want to construct, this is not the case: we can take the inverse of an *arbitrary* sub-object of $\underline{\mathbb{R}^{\leftrightarrow}}$, and we happen to choose $\breve{\delta}(\hat{P})(\underline{\delta^o(\hat{P})})$. Forming the inverse image of a sub-object of $\underline{\mathbb{R}^{\leftrightarrow}}$ under the natural transformation $\breve{\delta}(\hat{P})$ is analogous to taking the inverse image $f^{-1}\{r\}$ of some real value r under some real-valued function f. The real value r can be given as the value $r = f(x)$ of the function at some element x of its domain. This does not imply that $f^{-1}\{r\} = \{x\}$: the inverse image may contain more elements than just $\{x\}$. Likewise, we have to discuss whether the inverse image $\breve{\delta}(\hat{P})^{-1}(\breve{\delta}(\hat{P})(\underline{\delta^o(\hat{P})}))$ equals $\underline{\delta^o(\hat{P})}$ or is some larger sub-object of $\underline{\Sigma}$.

We start with the case that $\hat{P} = |\psi\rangle\langle\psi|$, i.e., $\hat{P}$ is the projection onto a one-dimensional subspace. We use the fact that $\underline{\delta^o(|\psi\rangle\langle\psi|)} = \underline{\mathfrak{w}}^{|\psi\rangle}$ (see Definition (13.128) and discussion thereafter).

Theorem 10 *The inverse image* $\breve{\delta}(|\psi\rangle\langle\psi|)^{-1}(\breve{\delta}(|\psi\rangle\langle\psi|)(\underline{\mathfrak{w}}^{|\psi\rangle}))$ *is* $\underline{\mathfrak{w}}^{|\psi\rangle}$.

Proof Let $\underline{S} := \breve{\delta}(|\psi\rangle\langle\psi|)^{-1}(\breve{\delta}(|\psi\rangle\langle\psi|)(\underline{\mathfrak{w}}^{|\psi\rangle}))$. In any case, we have

$$\underline{S} \supseteq \underline{\mathfrak{w}}^{|\psi\rangle}. \tag{13.298}$$

Let us assume that the inclusion is proper. Then there exists some $V \in \mathrm{Ob}(\mathcal{V}(\mathcal{H}))$ such that

$$\underline{S}(V) \supset \underline{\mathfrak{w}}_V^{|\psi\rangle} = \{\lambda \in \underline{\Sigma}_V \mid \langle\lambda, \delta^o(|\psi\rangle\langle\psi|)_V\rangle = 1\}, \tag{13.299}$$

which is equivalent to the existence of some $\lambda_0 \in \underline{S}_V$ such that

$$\langle\lambda_0, \delta^o(|\psi\rangle\langle\psi|)_V\rangle = 0. \tag{13.300}$$

By definition, we have $\underline{S}(V) = \{\lambda \in \underline{\Sigma}_V \mid \breve{\delta}(|\psi\rangle\langle\psi|)_V(\lambda) \in \breve{\delta}(|\psi\rangle\langle\psi|)_V(\underline{\mathfrak{w}}^{|\psi\rangle})\}$. For all $\tilde{\lambda} \in \underline{\mathfrak{w}}_V^{|\psi\rangle}$, it holds that

$$\langle\tilde{\lambda}, \delta^o(|\psi\rangle\langle\psi|)_V\rangle = 1, \tag{13.301}$$

see (13.285). This implies that for all $\lambda \in \underline{S}(V)$, we must have $\langle\lambda, \delta^o(|\psi\rangle\langle\psi|)_V\rangle = 1$, which contradicts (13.300). Hence there cannot be a proper inclusion $\underline{S} \supset \underline{\mathfrak{w}}^{|\psi\rangle}$, and we rather have equality

$$\underline{S}(V) = \underline{\mathfrak{w}}_V^{|\psi\rangle} \tag{13.302}$$

for all $V \in \mathrm{Ob}(\mathcal{V}(\mathcal{H}))$.

The proof is based on the fact that the order-reversing functions of the form $\breve{\delta}^o(|\psi\rangle\langle\psi|)_V(\lambda) : \downarrow\! V \to \{0, 1\}$, where $V \in \mathrm{Ob}(\mathcal{V}(\mathcal{H}))$ and $\lambda \in \underline{\mathfrak{w}}_V^{|\psi\rangle}$, are constant functions $1_{\downarrow V}$. The remark at the end of Sect. 13.8.5 shows that this holds more generally for arbitrary non-zero projections $\hat{P}$. Hence we obtain:

Corollary 1 *The inverse image* $\breve{\delta}(\hat{P})^{-1}(\breve{\delta}(\hat{P})(\underline{\delta^o(\hat{P})}))$ *is* $\underline{\delta^o(\hat{P})}$.

13.9 Extending the Quantity-Value Presheaf to an Abelian Group Object

13.9.1 Preliminary Remarks

We have shown how each self-adjoint operator, $\hat{A}$, on the Hilbert space $\mathcal{H}$ gives rise to an arrow $\breve{\delta}(\hat{A}) : \underline{\Sigma} \to \underline{\mathbb{R}^{\leftrightarrow}}$ in the topos $\mathbf{Sets}^{\mathcal{V}(\mathcal{H})^{\mathrm{op}}}$. Thus, in the topos representation, ϕ, of $\mathcal{L}(S)$ for the theory-type 'quantum theory', the arrow $\breve{\delta}(\hat{A}) : \underline{\Sigma} \to \underline{\mathbb{R}^{\leftrightarrow}}$ is one possible choice[92] for the representation, $A_\phi : \Sigma_\phi \to \mathcal{R}_\phi$, of the function symbol, $A : \Sigma \to \mathcal{R}$.

This implies that the quantity-value object, $\mathcal{R}_\phi$, is the presheaf, $\underline{\mathbb{R}^{\leftrightarrow}}$. However, although such an identification is possible, it does impose certain restrictions on the formalism. These stem from the fact that $\underline{\mathbb{R}^{\leftrightarrow}}$ is only a *monoid*-object in $\mathbf{Sets}^{\mathcal{V}(\mathcal{H})^{\mathrm{op}}}$, and $\Gamma\underline{\mathbb{R}^{\leftrightarrow}}$ is only a monoid, whereas the real numbers of standard physics are an abelian group; indeed, they are a commutative ring.

In standard classical physics, $\mathrm{Hom}_{\mathbf{Sets}}(\Sigma_\sigma, \mathbb{R})$ is the set of real-valued functions on the manifold Σ_σ; as such, it possesses the structure of a commutative ring. On the other hand, the set of arrows $\mathrm{Hom}_{\mathbf{Sets}^{\mathcal{V}(\mathcal{H})^{\mathrm{op}}}}(\underline{\Sigma}, \underline{\mathbb{R}^{\leftrightarrow}})$ has only the structure of an additive monoid. This additive structure is defined locally in the following way. Let $\alpha, \beta \in \mathrm{Hom}_{\mathbf{Sets}^{\mathcal{V}(\mathcal{H})^{\mathrm{op}}}}(\underline{\Sigma}, \underline{\mathbb{R}^{\leftrightarrow}})$. At each stage $V \in \mathrm{Ob}(\mathcal{V}(\mathcal{H}))$, α_V is a pair $(\mu_{1,V}, \nu_{1,V})$, consisting of a function $\mu_{1,V}$ from $\underline{\Sigma}_V$ to $\underline{\mathbb{R}^{\preceq}}_V$, and a function $\nu_{1,V}$ from $\underline{\Sigma}_V$ to $\underline{\mathbb{R}^{\succeq}}_V$. For each $\lambda \in \underline{\Sigma}_V$, one has an order-preserving function $\mu_{1,V}(\lambda) : \downarrow V \to \mathbb{R}$, and an order-reversing function $\nu_{1,V}(\lambda) : \downarrow V \to \mathbb{R}$. We use the notation $\alpha_V(\lambda) := (\mu_{1,V}(\lambda), \nu_{1,V}(\lambda))$.

Similarly, β is given at each stage V by a pair of functions $(\mu_{2,V}, \nu_{2,V})$, and for all $\lambda \in \underline{\Sigma}_V$, we write $\beta_V(\lambda) := (\mu_{2,V}(\lambda), \nu_{2,V}(\lambda))$

We define, for all $V \in \mathrm{Ob}(\mathcal{V}(\mathcal{H}))$, and all $\lambda \in \underline{\Sigma}_V$, (c.f. (13.295))

$$(\alpha + \beta)_V(\lambda) = ((\mu_{1,V}(\lambda), \nu_{1,V}(\lambda)) + (\mu_{2,V}(\lambda), \nu_{2,V}(\lambda))) \tag{13.303}$$

$$:= (\mu_{1,V}(\lambda) + \mu_{2,V}(\lambda), \nu_{1,V}(\lambda) + \nu_{2,V}(\lambda)) \tag{13.304}$$

$$= \alpha_V(\lambda) + \beta_V(\lambda), \tag{13.305}$$

It is clear that $(\alpha + \beta)_V(\lambda)$ is a pair consisting of an order-preserving and an order-reversing function for all V and all $\lambda \in \underline{\Sigma}_V$, so that $\alpha + \beta$ is well defined.[93]

---

[92] Another choice is to use the presheaf $\underline{\mathbb{R}^{\succeq}}$ as the quantity-value object, or the isomorphic presheaf $\underline{\mathbb{R}^{\preceq}}$.

[93] To avoid confusion we should emphasise that, in general, the sum $\breve{\delta}(\hat{A}) + \breve{\delta}(\hat{B})$ is *not* equal to $\breve{\delta}(\hat{A} + \hat{B})$.

Arguably, the fact that $\mathrm{Hom}_{\mathbf{Sets}^{\mathcal{V}(\mathcal{H})^{\mathrm{op}}}}(\underline{\Sigma}, \underline{\mathbb{R}^{\leftrightarrow}})$ is only a monoid[94] is a weakness in so far as we are trying to make quantum theory "look" as much like classical physics as possible. Of course, in more obscure applications such as Planck-length quantum gravity, the nature of the quantity-value object is very much open for debate. But when applied to regular physics, we might like our formalism to look more like classical physics than the monoid-only structure of $\mathrm{Hom}_{\mathbf{Sets}^{\mathcal{V}(\mathcal{H})^{\mathrm{op}}}}(\underline{\Sigma}, \underline{\mathbb{R}^{\leftrightarrow}})$.

The need for a subtraction, i.e., some sort of abelian group structure on $\underline{\mathbb{R}^{\leftrightarrow}}$, brings to mind the well-known Grothendieck k-construction that is much used in algebraic topology and other branches of pure mathematics. This gives a way of "extending" an abelian semi-group to become an abelian group, and this technique can be adapted to the present situation. The goal is to construct a "Grothendieck completion", $k(\underline{\mathbb{R}^{\leftrightarrow}})$, of $\underline{\mathbb{R}^{\leftrightarrow}}$ that is an abelian-group object in the topos $\mathbf{Sets}^{\mathcal{V}(\mathcal{H})^{\mathrm{op}}}$.[95]

Of course, we can apply the k-construction also to the presheaf $\underline{\mathbb{R}^{\succeq}}$ (or $\underline{\mathbb{R}^{\preceq}}$, if we like). This comes with an extra advantage: it is then possible to define the square of an arrow $\breve{\delta}^{o}(\hat{A}) : \underline{\Sigma} \to \underline{\mathbb{R}^{\succeq}}$, as is shown in the Appendix. Hence, given arrows $\breve{\delta}^{o}(\hat{A})$ and $\breve{\delta}^{o}(\hat{A^2})$, we can define an "intrinsic dispersion":[96]

$$\nabla(\hat{A}) := \breve{\delta}^{o}(\hat{A^2}) - \breve{\delta}^{o}(\hat{A})^2. \tag{13.306}$$

Since the whole k-construction is quite complicated and is not used in this article beyond the present section, we have decided to put all the relevant definitions into the Appendix where it can be read at leisure by anyone who is interested.

Interestingly, there is a close relation between $\underline{\mathbb{R}^{\leftrightarrow}}$ and $k(\underline{\mathbb{R}^{\succeq}})$, as shown in the next Subsection.

13.9.2 The Relation Between $\underline{\mathbb{R}^{\leftrightarrow}}$ and $k(\underline{\mathbb{R}^{\succeq}})$

In Sect. 13.8.2, we considered the presheaf $\underline{\mathbb{R}^{\leftrightarrow}}$ of order-preserving and order-reversing functions as a possible quantity-value object. The advantage of this presheaf is the symmetric utilisation of inner and outer daseinisation, and the associated physical interpretation of arrows from $\underline{\Sigma}$ to $\underline{\mathbb{R}^{\leftrightarrow}}$.

It transpires that $\underline{\mathbb{R}^{\leftrightarrow}}$ is closely related to $k(\underline{\mathbb{R}^{\succeq}})$. Namely, for each V, we can define an equivalence relation $\equiv$ on $\underline{\mathbb{R}^{\leftrightarrow}}_V$ by

---

94 An internal version of this result would show that the exponential object $\underline{\mathbb{R}^{\leftrightarrow}}^{\underline{\Sigma}}$ is a monoid object in the topos $\mathbf{Sets}^{\mathcal{V}(\mathcal{H})^{\mathrm{op}}}$. This could well be true, but we have not studied it in detail.

95 Ideally, we might like $k(\underline{\mathbb{R}^{\succeq}})$ or $k(\underline{\mathbb{R}^{\leftrightarrow}})$ to be a commutative-ring object, but this is not true.

96 The notation used here is potentially a little misleading. We have not given any meaning to "A^2" in the language $\mathcal{L}(S)$; i.e., in its current form, the language does not give meaning to the square of a function symbol. Therefore, when we write $\breve{\delta}^{o}(\hat{A^2})$ this must be understood as being the Gel'fand transform of the outer daseinisation of the operator $\hat{A}^2$.

$$(\mu_1, \nu_1) \equiv (\mu_2, \nu_2) \text{ iff } \mu_1 + \nu_1 = \mu_2 + \nu_2. \tag{13.307}$$

Then $\underline{\mathbb{R}^{\leftrightarrow}}/\equiv$ is isomorphic to $k(\underline{\mathbb{R}^{\succeq}})$ under the mapping

$$[\mu, \nu] \mapsto [\nu, -\mu] \in k(\underline{\mathbb{R}^{\succeq}})_V \tag{13.308}$$

for all V and all $[\mu, \nu] \in (\underline{\mathbb{R}^{\leftrightarrow}}/\equiv)_V$.[97]

However, there is a difference between the arrows that represent physical quantities. The arrow $[\breve{\delta}^o(\hat{A})] : \underline{\Sigma} \to k(\underline{\mathbb{R}^{\succeq}})$ is given by first sending $\hat{A} \in B(\mathcal{H})_{\text{sa}}$ to $\breve{\delta}^o(\hat{A})$ and then taking k-equivalence classes—a construction that only involves outer daseinisation. On the other hand, there is an arrow $[\breve{\delta}(\hat{A})] : \underline{\Sigma} \to \underline{\mathbb{R}^{\leftrightarrow}}/\equiv$, given by first sending $\hat{A}$ to $\breve{\delta}(\hat{A})$ and then taking the equivalence classes defined in (13.307). This involves both inner and outer daseinisation.

We can show that $[\breve{\delta}^o(\hat{A})]$ uniquely determines $\hat{A}$ as follows: Let

$$[\breve{\delta}^o(\hat{A})] : \underline{\Sigma} \to k(\underline{\mathbb{R}^{\succeq}}) \tag{13.309}$$

denote the natural transformation from the spectral presheaf to the abelian group-object $k(\underline{\mathbb{R}^{\succeq}})$, given by first sending $\hat{A}$ to $\breve{\delta}^o(\hat{A})$ and then taking the k-equivalence classes at each stage V. The monoid $\underline{\mathbb{R}^{\succeq}}$ is embedded into $k(\underline{\mathbb{R}^{\succeq}})$ by sending $\nu \in \underline{\mathbb{R}^{\succeq}}_V$ to $[\nu, 0] \in k(\underline{\mathbb{R}^{\succeq}})_V$ for all V, which implies that $\hat{A}$ is also uniquely determined by $[\breve{\delta}^o(\hat{A})]$.[98] We note that, currently, it is an open question if $[\breve{\delta}(\hat{A})]$ also fixes $\hat{A}$ uniquely.

We now have constructed several presheaves that are abelian group objects within $\mathbf{Sets}^{\mathcal{V}(\mathcal{H})^{\text{op}}}$, namely $k(\underline{\mathbb{R}^{\leftrightarrow}})$, $k(\underline{\mathbb{R}^{\succeq}})$ and $\underline{\mathbb{R}^{\leftrightarrow}}/\equiv$. The latter two are isomorphic presheaves, as we have shown. All three presheaves can serve as the quantity-value presheaf if one wants to have an abelian-group object for this purpose. Intuitively, if the quantity-value object is only an abelian-monoid object like $\underline{\mathbb{R}^{\leftrightarrow}}$, then the "values" can only be added, while in the case of an abelian-group object, they can be added and subtracted.

13.9.3 Algebraic Properties of the Potential Quantity-Value Presheaves

As matters stand, we have several possible choices for the quantity-value presheaf, which is the representation for quantum theory of the symbol $\mathcal{R}$ of the formal language $\mathcal{L}(S)$ that describes our physical system. In this sub-section, we want to

---

97 This identification also explains formula (13.307), which may look odd at first sight. Recall that $[\mu, \nu] \in (\underline{\mathbb{R}^{\leftrightarrow}}/\equiv)_V$ means that μ is order-preserving and ν is order-reversing.

98 In an analogous manner, one can show that the arrows $\breve{\delta}^i(\hat{A}) : \underline{\Sigma} \to \underline{\mathbb{R}^{\preceq}}$ and $[\breve{\delta}^i(\hat{A})] : \underline{\Sigma} \to k(\underline{\mathbb{R}^{\preceq}})$ uniquely determine $\hat{A}$, and that the arrow $\breve{\delta}(\hat{A}) : \underline{\Sigma} \to \underline{\mathbb{R}^{\leftrightarrow}}$ also uniquely determines $\hat{A}$.

compare the algebraic properties of these various presheaves. In particular, we will consider the presheaves $\underline{\mathbb{R}^{\succeq}}$, $k(\underline{\mathbb{R}^{\succeq}})$, $\underline{\mathbb{R}^{\leftrightarrow}}$ and $k(\underline{\mathbb{R}^{\leftrightarrow}})$.[99]

13.9.3.1 Global Elements

We first note that all these presheaves have global elements. For example, a global element of $\underline{\mathbb{R}^{\succeq}}$ is given by an order-reversing function $\nu : \mathcal{V}(\mathcal{H}) \to \mathbb{R}$. As remarked in the Appendix, we have $\Gamma k(\underline{\mathbb{R}^{\succeq}}) \simeq k(\Gamma\underline{\mathbb{R}^{\succeq}})$. Global elements of $\underline{\mathbb{R}^{\leftrightarrow}}$ are pairs (μ, ν) consisting of an order-preserving function $\mu : \mathcal{V}(\mathcal{H}) \to \mathbb{R}$ and an order-reversing function $\nu : \mathcal{V}(\mathcal{H}) \to \mathbb{R}$. Finally, it is easy to show that $\Gamma k(\underline{\mathbb{R}^{\leftrightarrow}}) \simeq k(\Gamma\underline{\mathbb{R}^{\leftrightarrow}})$.

13.9.3.2 The Real Number Object as a Sub-object

In a presheaf topos $\mathbf{Sets}^{\mathcal{C}^{op}}$, the Dedekind real number object $\underline{\mathbb{R}}$ is the constant functor from $\mathcal{C}^{op}$ to $\mathbb{R}$. The presheaf $\underline{\mathbb{R}}$ is an internal field object (see e.g. [55]).

The presheaf $\underline{\mathbb{R}^{\succeq}}$ contains the constant presheaf $\underline{\mathbb{R}}$ as a sub-object: let $V \in \mathrm{Ob}(\mathcal{V}(\mathcal{H}))$ and $r \in \underline{\mathbb{R}}_V \simeq \mathbb{R}$. Then the function $c_{r,V} :\downarrow V \to \mathbb{R}$ that has the constant value r is an element of $\underline{\mathbb{R}^{\succeq}}_V$ since it is an order-reversing function. Moreover, the global sections of $\underline{\mathbb{R}}$ are given by constant functions $r : \mathcal{V}(\mathcal{H}) \to \mathbb{R}$, and such functions are also global sections of $\underline{\mathbb{R}^{\succeq}}$.

The presheaf $\underline{\mathbb{R}^{\succeq}}$ can be seen as a sub-object of $k(\underline{\mathbb{R}^{\succeq}})$: let $V \in \mathcal{V}(\mathcal{H})$ and $\nu \in \underline{\mathbb{R}^{\succeq}}_V$, then $[\nu, 0] \in k(\underline{\mathbb{R}^{\succeq}})_V$. Thus the real number object $\underline{\mathbb{R}}$ is also a sub-object of $k(\underline{\mathbb{R}^{\succeq}})$.

A real number $r \in \underline{\mathbb{R}}_V$ defines the pair $(c_{r,V}, c_{r,V})$ consisting of two copies of the constant function $c_{r,V} :\downarrow V \to \mathbb{R}$. Since $c_{r,V}$ is both order-preserving and order-reversing, $(c_{r,V}, c_{r,V})$ is an element of $\underline{\mathbb{R}^{\leftrightarrow}}_V$ and hence $\underline{\mathbb{R}}$ is a sub-object of $\underline{\mathbb{R}^{\leftrightarrow}}$. Since $\underline{\mathbb{R}^{\leftrightarrow}}$ is a sub-object of $k(\underline{\mathbb{R}^{\leftrightarrow}})$, the latter presheaf also contains the real number object $\underline{\mathbb{R}}$ as a sub-object.

13.9.3.3 Multiplying with Real Numbers and Vector Space Structure

Let $c_r \in \Gamma\underline{\mathbb{R}}$ be the constant function on $\mathcal{V}(\mathcal{H})$ with value r. The global element c_r of $\underline{\mathbb{R}}$ defines locally, at each $V \in \mathcal{V}(\mathcal{H})$, a constant function $c_{r,V} : \downarrow V \to \mathbb{R}$. We want to consider if, and how, multiplication with these constant functions is defined in the various presheaves. We will call this "multiplying with a real number".

Let $\mu \in \underline{\mathbb{R}^{\succeq}}_V$. For all $V' \in\downarrow V$, we define the product

$$(c_{r,V}\mu)(V') := c_{r,V}(V')\mu(V') = r\mu(V'). \tag{13.310}$$

If $r \geq 0$, then $r\mu :\downarrow V \to \mathbb{R}$ is an order-reversing function again. However, if $r < 0$, then $r\mu$ is order-preserving and hence not an element of $\underline{\mathbb{R}^{\succeq}}_V$. This shows that for the presheaf $\underline{\mathbb{R}^{\succeq}}$ only multiplication by non-negative real numbers is well-defined.

---

[99] The presheaf $\underline{\mathbb{R}^{\preceq}}$ is isomorphic to $\underline{\mathbb{R}^{\succeq}}$ and hence will not be considered separately.

However, if we consider $k(\underline{\mathbb{R}^{\succeq}})$ then multiplication with an arbitrary real number is well-defined. For simplicity, we first consider $r = -1$, i.e., negation. Let $[\nu, \kappa] \in \underline{\mathbb{R}^{\succeq}}_V$, then, for all $V' \in \downarrow V$,

$$
\begin{aligned}
(c_{-1,V}[\nu,\kappa])(V') &:= c_{-1,V}(V')[\nu(V'),\kappa(V')] &(13.311)\\
&= -[\nu(V'),\kappa(V')] &(13.312)\\
&= [\kappa(V'),\nu(V')], &(13.313)
\end{aligned}
$$

so we have

$$
c_{-1,V}[\nu,\kappa] = -[\nu,\kappa] = [\kappa,\nu]. \tag{13.314}
$$

This multiplication with the real number -1 is, of course, defined in such a way that it fits in with the additive group structure on $k(\underline{\mathbb{R}^{\succeq}})$.

It follows that multiplying an element $[\nu, \kappa]$ of $k(\underline{\mathbb{R}^{\succeq}})_V$ with an arbitrary real number r can be defined as

$$
c_{r,V}[\nu,\kappa] := \begin{cases} [c_{r,V}\nu, c_{r,V}\kappa] = [r\nu, r\kappa] & \text{if } r \geq 0 \\ -[c_{-r,V}\nu, c_{-r,V}\kappa] = [-r\kappa, -r\nu] & \text{if } r < 0. \end{cases} \tag{13.315}
$$

Remark 1 In this way, the group object $k(\underline{\mathbb{R}^{\succeq}})$ in $\mathbf{Sets}^{\mathcal{V}(\mathcal{H})^{\mathrm{op}}}$ becomes a vector space object, with the field object $\underline{\mathbb{R}}$ as the scalars.

Interestingly, one can define multiplication with arbitrary real numbers also for $\underline{\mathbb{R}^{\leftrightarrow}}$, although this presheaf is not a group object in $\mathbf{Sets}^{\mathcal{V}(\mathcal{H})^{\mathrm{op}}}$. Let $(\mu, \nu) \in \underline{\mathbb{R}^{\leftrightarrow}}_V$, so that $\mu : \downarrow V \to \mathbb{R}$ is an order-preserving function and $\nu : \downarrow V \to \mathbb{R}$ is order-reversing. Let r be an arbitrary real number. We define

$$
c_{r,V}(\mu,\nu) := \begin{cases} (c_{r,V}\mu, c_{r,V}\nu) = (r\mu, r\nu) & \text{if } r \geq 0 \\ (c_{r,V}\nu, c_{r,V}\mu) = (r\nu, r\mu) & \text{if } r < 0. \end{cases} \tag{13.316}
$$

This is well-defined since if μ is order-preserving, then $-\mu$ is order-reversing, and if ν is order-reversing, then $-\nu$ is order-preserving. For $r = -1$, we obtain

$$
c_{-1,V}(\mu,\nu) = -(\mu,\nu) = (-\nu,-\mu). \tag{13.317}
$$

But this does *not* mean that $-(\mu, \nu)$ is an additive inverse of (μ, ν). Such inverses do not exist in $\underline{\mathbb{R}^{\leftrightarrow}}$, since it is not a group object. Rather, we get

$$
(\mu,\nu) + (-(\mu,\nu)) = (\mu,\nu) + (-\nu,-\mu) = (\mu-\nu, \nu-\mu). \tag{13.318}
$$

If, for all $V' \in \downarrow V$, we interpret the absolute value $|(\mu - \nu)(V')|$ as a measure of uncertainty as given by the pair (μ, ν) at stage V', then we see from (13.318) that adding (μ, ν) and $(-\nu, -\mu)$ gives a pair $(\mu - \nu, \nu - \mu) \in \underline{\mathbb{R}^{\leftrightarrow}}_V$ concentrated around $(0,0)$, but with an uncertainty twice as large (for all stages V'). We call $-(\mu,\nu) = (-\nu,-\mu)$ the *pseudo-inverse* of $(\mu,\nu) \in \underline{\mathbb{R}^{\leftrightarrow}}$.

More generally, we can define a second monoid structure (besides addition) on $\underline{\mathbb{R}^{\leftrightarrow}}$, called *pseudo-subtraction* and given by

$$(\mu_1,\nu_1) - (\mu_2,\nu_2) := (\mu_1,\nu_1) + (-\nu_2,-\mu_2) = (\mu_1 - \nu_2, \nu_1 - \mu_2). \quad (13.319)$$

This operation has a neutral element, namely $(c_{0,V}, c_{0,V})$, for all stages $V \in \mathcal{V}(\mathcal{H})$, which of course is also the neutral element for addition. In this sense, $\underline{\mathbb{R}^{\leftrightarrow}}$ is close to being a group object. Taking equivalence classes as described in (13.307) makes $\underline{\mathbb{R}^{\leftrightarrow}}$ into a group object, $\underline{\mathbb{R}^{\leftrightarrow}}/\equiv$, isomorphic to $k(\underline{\mathbb{R}^{\succeq}})$.

Since multiplication with arbitrary real numbers is well-defined, the presheaf $\underline{\mathbb{R}^{\leftrightarrow}}$ is "almost a vector space object" over $\underline{\mathbb{R}}$.

Elements of $k(\underline{\mathbb{R}^{\leftrightarrow}})_V$ are of the form $[(\mu_1,\nu_1),(\mu_2,\nu_2)]$. Multiplication with an arbitrary real number r is defined in the following way:

$$c_{r,V}[(\mu_1,\nu_1),(\mu_2,\nu_2)] := \begin{cases} [(r\mu_1, r\nu_1),(r\mu_2, r\nu_2)] & \text{if } r \geq 0 \\ [(-r\mu_2, -r\nu_2),(-r\mu_1, -r\nu_1)] & \text{if } r < 0. \end{cases} \quad (13.320)$$

The additive group structure on $k(\underline{\mathbb{R}^{\leftrightarrow}})$ implies

$$-[(\mu_1,\nu_1),(\mu_2,\nu_2)] = [(\mu_2,\nu_2),(\mu_1,\nu_1)], \quad (13.321)$$

so the multiplication with the real number -1 fits in with the group structure. On the other hand, this negation is completely different from the negation on $\underline{\mathbb{R}^{\leftrightarrow}}$ (where $-(\mu,\nu) = (-\nu,-\mu)$ for all $(\mu,\nu) \in \underline{\mathbb{R}^{\leftrightarrow}}_V$ and all $V \in \mathrm{Ob}(\mathcal{V}(\mathcal{H}))$).

Remark 2 The presheaf $k(\underline{\mathbb{R}^{\leftrightarrow}})$ is a vector space object in $\mathbf{Sets}^{\mathcal{V}(\mathcal{H})^{\mathrm{op}}}$, with $\underline{\mathbb{R}}$ as the scalars.

13.10 The Role of Unitary Operators

13.10.1 The Daseinisation of Unitary Operators

Unitary operators play an important role in the formulation of quantum theory, and we need to understand the analogue of this in our topos formalism.

Unitary operators arise in the context of both "covariance" and "invariance". In elementary quantum theory, the "covariance" aspect comes the fact that if we have made the associations

$$\text{Physical state} \mapsto \text{state vector } |\psi\rangle \in \mathcal{H}$$

$$\text{Physical observable } A \mapsto \text{self-adjoint operator } \hat{A} \text{ acting on } \mathcal{H}$$

then the same physical predictions will be obtained if the following associations are used instead

$$\text{Physical state} \mapsto \text{state vector } \hat{U}\,|\psi\rangle \in \mathcal{H} \tag{13.322}$$

$$\text{Physical observable } A \mapsto \text{self-adjoint operator } \hat{U}\hat{A}\hat{U}^{-1}\text{acting on } \mathcal{H}$$

for any unitary operator $\hat{U}$. Thus the mathematical representatives of physical quantities are defined only up to arbitrary transformations of the type above. In non-relativistic quantum theory, this leads to the canonical commutation relations; the angular-momentum commutator algebra; and the unitary time displacement operator. Similar considerations in relativistic quantum theory involve the Poincaré group.

The "invariance" aspect of unitary operators arises when the operator commutes with the Hamiltonian, giving rise to conserved quantities.

13.10.1.1 Daseinisation of Unitary Operators

As a side remark, we first consider the question if daseinisation can be applied to a unitary operator $\hat{U}$. The answer is clearly "yes", via the spectral representation:

$$\hat{U} = \int_{\mathbb{R}} e^{i\lambda} d\hat{E}^U_\lambda \tag{13.323}$$

where $\lambda \mapsto E^{\hat{U}}_\lambda$ is the spectral family for $\hat{U}$. Then, in analogy with (13.170) and (13.171) we have the following:

Definition 17 The outer daseinisation, $\delta^o(\hat{U})$, resp. the inner daseinisation, $\delta^i(\hat{U})$, of a unitary operator $\hat{U}$ are defined as follows:

$$\delta^o(\hat{U})_V := \int_{\mathbb{R}} e^{i\lambda}\, d\big(\delta^i_V(\hat{E}^U_\lambda)\big), \tag{13.324}$$

$$\delta^i(\hat{U})_V := \int_{\mathbb{R}} e^{i\lambda}\, d\Big(\bigwedge_{\mu>\lambda} \delta^o_V(\hat{E}^U_\mu)\Big), \tag{13.325}$$

at each stage V.

To interpret these entities[100] we need to introduce a new presheaf defined as follows.

Definition 18 The outer, unitary de Groote presheaf, $\underline{\mathbb{U}}$, is defined by:

(i) On objects $V \in \mathrm{Ob}(\mathcal{V}(\mathcal{H}))$: $\underline{\mathbb{U}}_V := V_{\mathrm{un}}$, the collection of unitary operators in V.

[100] It would be possible to "complexify" the presheaf $k(\underline{\mathbb{R}^{\succeq}})$ in order to represent unitary operators as arrows from $\underline{\Sigma}$ to $\mathbb{C}k(\underline{\mathbb{R}^{\succeq}})$. Similar remarks apply to the presheaf $\underline{\mathbb{R}^{\leftrightarrow}}$. However, there is no obvious physical use for this procedure.

(ii) On morphisms $i_{V'V} : V' \subseteq V$: The mapping $\underline{\mathrm{U}}(i_{V'V}) : \underline{\mathrm{U}}_V \to \underline{\mathrm{U}}_{V'}$ is given by

$$\underline{\mathrm{U}}(i_{V'V})(\hat{\alpha}) := \delta^o(\hat{\alpha})_{V'} \tag{13.326}$$

$$= \int_{\mathbb{R}} e^{i\lambda}\, d\big(\delta^i(\hat{E}^{\alpha}_{\lambda})_{V'}\big) \tag{13.327}$$

$$= \int_{\mathbb{R}} e^{i\lambda}\, d\big(\underline{I}(i_{V'V})(\hat{E}^{\alpha}_{\lambda})\big) \tag{13.328}$$

for all $\hat{\alpha} \in \underline{\mathrm{U}}_V$.

Clearly, (i) there is an analogous definition of an "inner", unitary de Groote presheaf; and (ii) the map $V \mapsto \delta^o(\hat{U})_V$ defines a global element of $\underline{\mathrm{U}}$.

This definition has the interesting consequence that, at each stage V,

$$\delta^o(e^{i\hat{A}})_V = e^{i\delta^o(\hat{A})_V} \tag{13.329}$$

A particular example of this construction is the one-parameter family of unitary operators, $t \mapsto e^{it\hat{H}}$, where $\hat{H}$ is the Hamiltonian of the system.

Of course, in our case everything commutes. Thus suppose $g \mapsto \hat{U}_g$ is a representation of a Lie group G on the Hilbert space $\mathcal{H}$. Then these operators can be daseinised to give the map $g \mapsto \delta^o(\hat{U}_g)$, but generally this is not a representation of G (or of its Lie algebra) since, at each stage V we have

$$\delta^o(\hat{U}_{g_1})_V \delta^o(\hat{U}_{g_2})_V = \delta^o(\hat{U}_{g_2})_V \delta^o(\hat{U}_{g_1})_V \tag{13.330}$$

for all $g_1, g_2 \in G$. Clearly, there is an analogous result for inner daseinisation.

13.10.2 *Unitary Operators and Arrows in* $\mathbf{Sets}^{\mathcal{V}(\mathcal{H})^{op}}$

13.10.2.1 The Definition of $\ell_{\hat{U}} : \mathrm{Ob}(\mathcal{V}(\mathcal{H})) \to \mathrm{Ob}(\mathcal{V}(\mathcal{H}))$

In classical physics, the analogue of unitary operators are "canonical transformations"; i.e., symplectic diffeomorphisms from the state space $\mathcal{S}$ to itself. This suggests that should try to associate arrows in $\mathbf{Sets}^{\mathcal{V}(\mathcal{H})^{\mathrm{op}}}$ with each unitary operator $\hat{U}$.

Thus we want to see if unitary operators can act on the objects in $\mathbf{Sets}^{\mathcal{V}(\mathcal{H})^{\mathrm{op}}}$. In fact, if $\mathcal{U}(\mathcal{H})$ denotes the group of all unitary operators in $\mathcal{H}$, we would like to find a *realisation* of $\mathcal{U}(\mathcal{H})$ in the topos $\mathbf{Sets}^{\mathcal{V}(\mathcal{H})^{\mathrm{op}}}$.

As a first step, if $\hat{U} \in \mathcal{U}(\mathcal{H})$ and $V \in \mathrm{Ob}(\mathcal{V}(\mathcal{H}))$ is an abelian von Neumann sub-algebra of $B(\mathcal{H})$, let us define

$$\ell_{\hat{U}}(V) := \{\hat{U}\hat{A}\hat{U}^{-1} \mid \hat{A} \in V\}. \tag{13.331}$$

It is clear that $\ell_{\hat{U}}(V)$ is a unital, abelian algebra of operators, and that self-adjoint operators are mapped into self-adjoint operators. Furthermore, the map $\hat{A} \mapsto \hat{U}\hat{A}\hat{U}^{-1}$ is continuous in the weak-operator topology, and hence, if $\{\hat{A}_i\}_{i\in I}$ is a weakly-convergent net of operators in V, then $\{\hat{U}\hat{A}_i\hat{U}^{-1}\}_{i\in I}$ is a weakly-convergent net of operators in $\ell_{\hat{U}}(V)$, and vice versa.

It follows that $\ell_{\hat{U}}(V)$ is an abelian von Neumann algebra (i.e., it is weakly closed), and hence $\ell_{\hat{U}}$ can be viewed as a map $\ell_{\hat{U}} : \mathrm{Ob}(\mathcal{V}(\mathcal{H})) \to \mathrm{Ob}(\mathcal{V}(\mathcal{H}))$. We note the following:

1. Clearly, for all $\hat{U}_1, \hat{U}_2 \in \mathcal{U}(\mathcal{H})$,

$$\ell_{\hat{U}_1} \circ \ell_{\hat{U}_2} = \ell_{\hat{U}_1\hat{U}_2} \tag{13.332}$$

   Thus $\hat{U} \mapsto \ell_{\hat{U}}$ is a realisation of the group $\mathcal{U}(\mathcal{H})$ as a group of transformations of $\mathrm{Ob}(\mathcal{V}(\mathcal{H}))$.
2. For all $U \in \mathcal{U}(\mathcal{H})$, V and $\ell_{\hat{U}}(V)$ are *isomorphic* sub-algebras of $B(\mathcal{H})$, and $\ell_{\hat{U}}^{-1} = \ell_{\hat{U}^{-1}}$.
3. If $V' \subseteq V$, then, for all $\hat{U} \in \mathcal{U}(\mathcal{H})$,

$$\ell_{\hat{U}}(V') \subseteq \ell_{\hat{U}}(V). \tag{13.333}$$

   Hence, each transformation $\ell_{\hat{U}}$ preserves the partial-ordering of the poset category $\mathcal{V}(\mathcal{H})$.
   From this it follows that each $\ell_{\hat{U}} : \mathrm{Ob}(\mathcal{V}(\mathcal{H})) \to \mathrm{Ob}(\mathcal{V}(\mathcal{H}))$ is a *functor* from the category $\mathcal{V}(\mathcal{H})$ to itself.
4. One consequence of the order-preserving property of $\ell_{\hat{U}}$ is as follows. Let S be a sieve of arrows on V, i.e., a collection of sub-algebras of V with the property that if $V' \in S$, then, for all $V'' \subseteq V'$ we have $V'' \in S$. Then

$$\ell_{\hat{U}}(S) := \{\ell_{\hat{U}}(V') \mid V' \in S\} \tag{13.334}$$

   is a sieve of arrows on $\ell_{\hat{U}}(V)$.[101]
5. It is easy to see that $\mathcal{U}(\mathcal{H})$ acts as a group of transformations on the set, $\mathrm{Sub}_{\mathrm{cl}}(\underline{\Sigma})$ of clopen subobjcts of $\underline{\Sigma}$. Namely, if $\underline{S} \in \mathrm{Sub}_{\mathrm{cl}}(\underline{\Sigma})$ we have an associated family, $\hat{S}_V$, of projection operators in V, and then we define, for all V,

$$(\rho_{\hat{U}}\hat{S})_V := \hat{U}\hat{S}_{\ell_{\hat{U}^{-1}}(V)}\hat{U}^{-1} \tag{13.335}$$

   where have used the fact that if $\hat{\alpha} \in \mathcal{PL}(W)$ then $\hat{U}\hat{\alpha}\hat{U}^{-1}$ belongs to $\mathcal{PL}(\ell_{\hat{U}}W)$. It is easy to see that (13.335) satisfies

---

[101] In the partially ordered set $\mathcal{V}(\mathcal{H})$, an arrow from V' to V can be identified with the sub-algebra $V' \subseteq V$, since there is exactly one arrow from V' to V.

$$\rho_{\hat{U}_2} \circ \rho_{\hat{U}_1} = \rho_{\hat{U}_2 \hat{U}_1} \tag{13.336}$$

for all $\hat{U}_1, \hat{U}_2 \in \mathcal{U}(\mathcal{H})$. Thus we do indeed obtain an action of the group $\mathcal{U}(\mathcal{H})$ on $\mathrm{Sub_{cl}}(\underline{\Sigma})$.

13.10.2.2 The Effect of $\ell_{\hat{U}}$ on Daseinisation

We recall that if $\hat{P}$ is any projection, then the (outer) daseinisation, $\delta^o(\hat{P})_V$, of $\hat{P}$ at stage V is (13.35)

$$\delta^o(\hat{P})_V := \bigwedge \{\hat{Q} \in \mathcal{P}(V) \mid \hat{Q} \succeq \hat{P}\} \tag{13.337}$$

where we have resorted once more to using the propositional language $\mathcal{PL}(S)$. Thus

$$\begin{aligned} \hat{U}\delta^o(\hat{P})_V\hat{U}^{-1} &= \hat{U} \bigwedge \{\hat{Q} \in \mathcal{P}(V) \mid \hat{Q} \succeq \hat{P}\}\hat{U}^{-1} \\ &= \bigwedge \{\hat{U}\hat{Q}\hat{U}^{-1} \in \mathcal{P}(\ell_{\hat{U}}(V)) \mid \hat{Q} \succeq \hat{P}\} \\ &= \bigwedge \{\hat{U}\hat{Q}\hat{U}^{-1} \in \mathcal{P}(\ell_{\hat{U}}(V)) \mid \hat{U}\hat{Q}\hat{U}^{-1} \succeq \hat{U}\hat{P}\hat{U}^{-1}\} \\ &= \delta^o(\hat{U}\hat{P}\hat{U}^{-1})_{\ell_{\hat{U}}(V)} \end{aligned} \tag{13.338}$$

where we used the fact that the map $\hat{Q} \mapsto \hat{U}\hat{Q}\hat{U}^{-1}$ is weakly continuous.

Thus we have the important result

$$\hat{U}\delta^o(\hat{P})_V\hat{U}^{-1} = \delta^o(\hat{U}\hat{P}\hat{U}^{-1})_{\ell_{\hat{U}}(V)} \tag{13.339}$$

for all unitary operators $\hat{U}$, and for all stages V. There is an analogous result for inner daseinisation. Note that (13.335) and (13.339) together imply that

$$(\rho_{\hat{U}}\delta^o(\hat{P}))_V = \delta^o(\hat{U}\hat{P}\hat{U}^{-1})_V \tag{13.340}$$

for all stages V. Thus the action of $\mathcal{U}(\mathcal{H})$ on $\hat{P}H$ marches in harmony with its action on $\mathrm{Sub_{cl}}(\underline{\Sigma})$ via daseinisation.

Note that Eq. (13.339) can be applied to the de Groote presheaf $\underline{\mathbb{O}}$ to give

$$\hat{U}\delta^o(\hat{A})_V\hat{U}^{-1} = \delta^o(\hat{U}\hat{A}\hat{U}^{-1})_{\ell_{\hat{U}}(V)} \tag{13.341}$$

for unitary operators $\hat{U}$, and all stages V.

13.10.2.3 The Covariance of Truth Values

We now turn to the crucial issue of how unitary operators act on truth values of physical propositions. We recall that the truth sub-object, $\mathbb{T}^{|\psi\rangle}$, of the outer presheaf, $\underline{O}$, is defined at each stage V by (cf. (13.83))

$$\underline{\mathbb{T}}^{|\psi\rangle}_V := \{\hat{\alpha} \in \underline{O}_V \mid \text{Prob}(\hat{\alpha};\ |\psi\rangle) = 1\}$$
$$= \{\hat{\alpha} \in \underline{O}_V \mid \langle\psi|\,\hat{\alpha}\,|\psi\rangle = 1\} \tag{13.342}$$

The neo-realist, physical interpretation of $\underline{\mathbb{T}}^{|\psi\rangle}$ is that the "truth" of the proposition represented by $\hat{P}$ is

$$\nu\big(\underline{\delta}^o(\hat{P}) \in \underline{\mathbb{T}}^{|\psi\rangle}\big)_V := \{V' \subseteq V \mid \underline{\delta}^o(\hat{P})_{V'} \in \underline{\mathbb{T}}^{|\psi\rangle}_{V'}\} \tag{13.343}$$
$$= \{V' \subseteq V \mid \langle\psi|\,\delta^o(\hat{P})_{V'}\,|\psi\rangle = 1\} \tag{13.344}$$

for all stages V. We then get

$$\ell_{\hat{U}}\ \big(\nu\big(\underline{\delta}^o(\hat{P}) \in \underline{\mathbb{T}}^{|\psi\rangle}\big)_V$$
$$= \ell_{\hat{U}}\{V' \subseteq V \mid \langle\psi|\,\delta^o(\hat{P})_{V'}\,|\psi\rangle = 1\} \tag{13.345}$$
$$= \{\ell_{\hat{U}}(V') \subseteq \ell_{\hat{U}}(V) \mid \langle\psi|\,\delta^o(\hat{P})_{V'}\,|\psi\rangle = 1\} \tag{13.346}$$
$$= \{\ell_{\hat{U}}(V') \subseteq \ell_{\hat{U}}(V) \mid \langle\psi|\,\hat{U}^{-1}\hat{U}\delta^o(\hat{P})_{V'}\hat{U}^{-1}\hat{U}\,|\psi\rangle = 1\} \tag{13.347}$$
$$= \{\ell_{\hat{U}}(V') \subseteq \ell_{\hat{U}}(V) \mid \langle\psi|\,\hat{U}^{-1}\delta^o(\hat{U}\hat{P}\hat{U}^{-1})_{\ell_{\hat{U}}(V)}\hat{U}\,|\psi\rangle = 1\} \tag{13.348}$$
$$= \nu\big(\underline{\delta}^o(\hat{U}\hat{P}\hat{U}^{-1}) \in \underline{\mathbb{T}}^{\hat{U}\,|\psi\rangle}\big)_{\ell_{\hat{U}}(V)}. \tag{13.349}$$

Thus we get the important result

$$\nu\big(\underline{\delta}^o(\hat{U}\hat{P}\hat{U}^{-1}) \in \underline{\mathbb{T}}^{\hat{U}\,|\psi\rangle}\big)_{\ell_{\hat{U}}(V)} = \ell_{\hat{U}}\big(\nu\big(\underline{\delta}^o(\hat{P}) \in \underline{\mathbb{T}}^{|\psi\rangle}\big)_V\big). \tag{13.350}$$

This can be viewed as the topos analogue of the statement in (13.322) about the invariance of the results of quantum theory under the transformations $|\psi\rangle \mapsto \hat{U}\,|\psi\rangle$, $\hat{A} \mapsto \hat{U}\hat{A}\hat{U}^{-1}$. Of course, there is a pseudo-state analogue of all these expressions involving the sub-objects $\underline{\mathfrak{w}}^{|\psi\rangle}$, $|\psi\rangle \in \mathcal{H}$.

13.10.2.4 The $\hat{U}$-Twisted Presheaf

It is important to realise that (13.350) is all that is needed from a physics perspective: the unitary group, $\mathcal{U}(\mathcal{H})$, acts on the set $\text{Sub}_{\text{cl}}(\underline{\Sigma})$ of clopen sub-objects of $\underline{\Sigma}$ in such a way that the physical results (truth values) transform in a suitably covariant way.

However, it is natural to see if this action can be derived from an internal arrow $\rho_{\hat{U}} : \underline{\Sigma} \to \underline{\Sigma}$ in the topos. To investigate this let us return once more to the definition (13.331) of the functor $\ell_{\hat{U}} : \mathcal{V}(\mathcal{H}) \to \mathcal{V}(\mathcal{H})$. As we shall see later, any such functor induces a "geometric morphism" from $\mathbf{Sets}^{\mathcal{V}(\mathcal{H})^{\text{op}}}$ to $\mathbf{Sets}^{\mathcal{V}(\mathcal{H})^{\text{op}}}$. The exact definition is not needed here: it suffices to remark that part of this geometric morphism is an arrow $\ell^*_{\hat{U}} : \mathbf{Sets}^{\mathcal{V}(\mathcal{H})^{\text{op}}} \to \mathbf{Sets}^{\mathcal{V}(\mathcal{H})^{\text{op}}}$ defined by

$$\underline{F} \mapsto \ell^*_{\hat{U}}\underline{F} := \underline{F} \circ \ell_{\hat{U}}. \tag{13.351}$$

Note that, if $\hat{U}_1, \hat{U}_2 \in \mathcal{U}(\mathcal{H})$ then, for all presheaves $\underline{F}$,

$$\begin{aligned} \ell^*_{\hat{U}_2}(\ell^*_{\hat{U}_1}\underline{F}) &= (\ell^*_{\hat{U}_1}\underline{F}) \circ \ell_{\hat{U}_2} = (\underline{F} \circ \ell_{\hat{U}_1}) \circ \ell_{\hat{U}_2} \\ &= \underline{F} \circ (\ell_{\hat{U}_1} \circ \ell_{\hat{U}_2}) = \underline{F} \circ \ell_{\hat{U}_1\hat{U}_2} \\ &= \ell^*_{\hat{U}_1\hat{U}_2}\underline{F}. \end{aligned} \tag{13.352}$$

Since this is true for all functors $\underline{F}$ in $\mathbf{Sets}^{\mathcal{V}(\mathcal{H})^{\mathrm{op}}}$, we deduce that

$$\ell^*_{\hat{U}_2} \circ \ell^*_{\hat{U}_1} = \ell^*_{\hat{U}_1\hat{U}_2} \tag{13.353}$$

and hence the map $\hat{U} \mapsto \ell^*_{\hat{U}}$ is an (anti-)representation of the group $\mathcal{U}(\mathcal{H})$ by arrows in the topos $\mathbf{Sets}^{\mathcal{V}(\mathcal{H})^{\mathrm{op}}}$.

Of particular interest to us are the presheaves $\ell^*_{\hat{U}}\underline{\Sigma}$ and $\ell^*_{\hat{U}}k(\underline{\mathbb{R}^{\succeq}})$. We denote them by $\underline{\Sigma}^{\hat{U}}$ and $k(\underline{\mathbb{R}^{\succeq}})^{\hat{U}}$ respectively and say that they are '$\hat{U}$-twisted'.

Theorem 11 *For each $\hat{U} \in \mathcal{U}(\mathcal{H})$, there is a natural isomorphism $\iota : \underline{\Sigma} \to \underline{\Sigma}^{\hat{U}}$ as given in the following diagram*

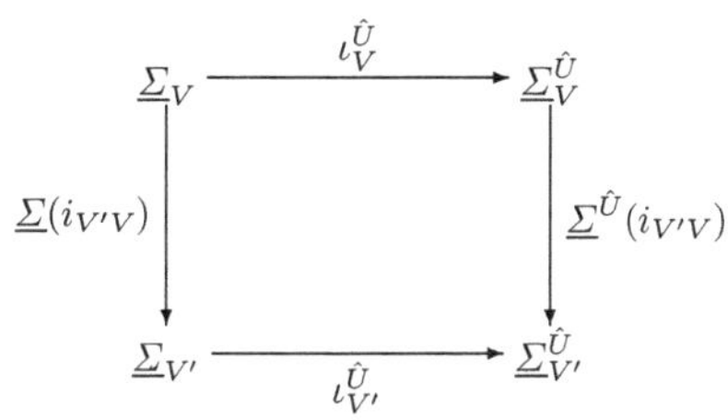

where, at each stage V,

$$(\iota^U_V(\lambda))(\hat{A}) := \langle\lambda, \hat{U}^{-1}\hat{A}\hat{U}\rangle \tag{13.354}$$

for all $\lambda \in \underline{\Sigma}_V$, and all $\hat{A} \in V_{sa}$.

The proof, which just involves chasing round the diagram above using the basic definitions, is not included here.

Even simpler is the following theorem:

Theorem 12 *For each $\hat{U} \in \mathcal{U}(\mathcal{H})$, there is a natural isomorphism $\kappa^{\hat{U}} : \underline{\mathbb{R}^{\succeq}} \to (\underline{\mathbb{R}^{\succeq}})^{\hat{U}}$ whose components $\kappa^{\hat{U}}_V : \underline{\mathbb{R}^{\succeq}}_V \to (\underline{\mathbb{R}^{\succeq}})^{\hat{U}}_V$ are given by*

$$\kappa^{\hat{U}}_V(\mu)(\ell_{\hat{U}}(V')) := \mu(V') \tag{13.355}$$

for all $V' \subseteq V$.

Here, we recall $\mu \in \underline{\mathbb{R}^{\succeq}}_V$ is a function $\mu :\downarrow V \rightarrow \mathbb{R}$ such that if $V_2 \subseteq V_1 \subseteq V$ then $\mu(V_2) \geq \mu(V_1)$, i.e., an order-reversing function. In (13.355) we have used the fact that there is a bijection between the sets $\downarrow \ell_{\hat{U}}(V)$ and $\downarrow V$.

Finally,

Theorem 13 *We have the following commutative diagram:*

$$\begin{array}{ccc} \underline{\Sigma} & \xrightarrow{\iota^{\hat{U}}} & \underline{\Sigma}^{\hat{U}} \\ {\scriptstyle \breve{\delta}(\hat{A})}\downarrow & & \downarrow{\scriptstyle \breve{\delta}(\hat{U}^{-1}\hat{A}\hat{U})} \\ \underline{\mathbb{R}^{\succeq}} & \xrightarrow[\kappa^{\hat{U}}]{} & \underline{\mathbb{R}^{\succeq}}^{\hat{U}}. \end{array}$$

13.10.3 Covariance Transformations for a General Topos

It is interesting to reflect on the analogue of the above constructions for a general topos. It soon becomes clear that, once again, we encounter the antithetical concepts of "internal" and "external".

For example, in the discussion above, the unitary operators and the group $\mathcal{U}(\mathcal{H})$ lie outside the topos $\mathbf{Sets}^{\mathcal{V}(\mathcal{H})^{\mathrm{op}}}$ and enter directly from the underlying, standard quantum formalism. As such, they are external to both the languages $\mathcal{PL}(S)$ and $\mathcal{L}(S)$. And, as we have seen, the direct action of the group is on the set, $\mathrm{Sub}_{\mathrm{cl}}(\underline{\Sigma})$ of sub-objects of $\underline{\Sigma}$, and this does not derive from an internal action of the form $\rho_{\hat{U}} : \underline{\Sigma} \rightarrow \underline{\Sigma}$ (or, indeed from an arrow $\rho_{\hat{U}} : P_{\mathrm{cl}}\underline{\Sigma} \rightarrow P_{\mathrm{cl}}\underline{\Sigma}$)

This, essentially external, action of $\mathcal{U}(\mathcal{H})$ is all that is needed from a physical perspective. In particular, it encodes all that is necessary for the notion of quantum covariance to be implemented in this topos scheme.

Time evolution is simply obtained by using the one-parameter family of pseudo-states, $t \mapsto \underline{\mathfrak{w}}^{|\psi\rangle_t}$, and it is easy to see how the notion of 'symmetry' (invariance under time evolutions) is incorporated. Specifically, we say that the quantum theory is symmetric under the action of a unitary operator $\hat{U}$ if

$$\rho_{\hat{U}}\underline{\mathfrak{w}}^{|\psi\rangle_t} = \underline{\mathfrak{w}}^{|\psi\rangle_t} \tag{13.356}$$

for all time t. It is easy to check that (13.356) is equivalent to

$$\hat{U}e^{-i\hat{H}t}\,\hat{U}^{-1} = e^{-i\hat{H}t} \tag{13.357}$$

for all t, where $\hat{H}$ is the hamiltonian. The Eq. (13.357), of course, is just the usual quantum theory definition of a system to be symmetric under the action of a unitary operator $\hat{U}$.

One might anticipate that notions of "covariance" and "symmetry" have applications well beyond those in classical physics and quantum physics. However, it is worth noting that many covariance transformations that arise in physics are related to external, fixed structures in the theory. This is true *par excellence* in the case of anything to do with space-time. In classical general relativity, there is a fixed space-time manifold whose diffeomorphism group plays a major role in the theory. When one comes to canonical quantum gravity only the spatial diffeomorphisms are manifestly present, this being just one feature of the infamous "problem of time" in canonical quantum gravity whereby time is not regarded as an external quantity but rather something that is defined/determined by the field content of the theory. If one could construct a theory in which physical space "emerged" from some more basic structure, then the spatial diffeomorphisms would also not be present at the basic level.

It seems likely that this will be a general feature of any topos theory of physics. That is, external covariance groups will be represented externally by bijective transformations of the set, $\mathrm{Sub}(\Sigma_{\phi,S})$, of sub-objects of the state object $\Sigma_{\phi,S}$, but not by arrows internal to the topos. However, there may also be an internal, "covariance" group object in the topos, which acts via arrows in the topos.

The notion of "symmetry" could also play a significant role in a general topos theory. However, this, of course, is closely related to the concept to time, and time development, which opens up a Pandora's box of possibilities in regard to internal realisations of these and related concepts. These issues are important, and await further development.

13.11 The Category of Systems

13.11.1 Background Remarks

We now return to the more general aspects of our theory, and study its application to a *collection* of systems, each one of which may be associated with a different topos. For example, if S_1, S_2 is a pair of systems, with associated topoi $\tau(S_1)$ and $\tau(S_2)$, and if S_1 is a sub-system of S_2, then we wish to consider how $\tau(S_1)$ is related to $\tau(S_2)$. Similarly, if a composite system is formed from a pair of systems S_1, S_2, what relations are there between the topos of the composite system and the topoi of the constituent parts?

Of course, in one sense, there is only one true "system", and that is the universe as a whole. Concomitantly, there is just one local language, and one topos. However, in practice, the science community divides the universe conceptually into portions that are sufficiently simple to be amenable to theoretical and/or empirical discussion. Of course, this division is not unique, but it must be such that the coupling between portions is weak enough that, to a good approximation, their theoretical models can be studied in isolation from each other. Such an essentially isolated[102] portion of the universe is called a "sub-system". By an abuse of language, sub-systems

[102] The ideal monad has no windows.

of the universe are usually called "systems" (so that the universe as a whole is one super-system), and then we can talk about "sub-systems" of these systems; or "composites" of them; or sub-systems of the composite systems, and so on.

In practice, references by physicists to systems and sub-systems[103] do not generally signify *actual* sub-systems of the real universe but rather idealisations of possible systems. This is what a physics lecturer means when he or she starts a lecture by saying "Consider a point particle moving in three dimensions.....".

To develop these ideas further we need mathematical control over the systems of interest, and their interrelations. To this end, we start by focussing on some collection, **Sys**, of physical systems to which a particular theory-type is deemed to be applicable. For example, we could consider a collection of systems that are to be discussed using the methodology of classical physics; or systems to be discussed using standard quantum theory; or whatever. For completeness, we require that every sub-system of a system in **Sys** is itself a member of **Sys**, as is every composite of members of **Sys**.

We shall assume that the systems in **Sys** are all associated with local languages of the type discussed earlier, and that they all have the *same* set of ground symbols which, for the purposes of the present discussion, we take to be just Σ and $\mathcal{R}$. It follows that the languages $\mathcal{L}(S)$, $S \in$ **Sys**, differ from each other only in the set of function symbols $F_{\mathcal{L}(S)}(\Sigma, \mathcal{R})$; i.e., the set of *physical quantities*.

As a simple example of the system-dependence of the set of function symbols let system S_1 be a point particle moving in one dimension, and let the set of physical quantities be $F_{\mathcal{L}(S_1)}(\Sigma, \mathcal{R}) = \{x, p, H\}$. In the language $\mathcal{L}(S_1)$, these function-symbols represent the position, momentum, and energy of the system respectively. On the other hand, if S_2 is a particle moving in three dimensions, then in the language $\mathcal{L}(S_2)$ we could have $F_{\mathcal{L}(S_2)}(\Sigma, \mathcal{R}) = \{x, y, z, p_x, p_y, p_z, H\}$ to allow for three-dimensional position and momentum. Or, we could decide to add angular momentum as well, to give the set $F_{\mathcal{L}(S_2)}(\Sigma, \mathcal{R}) = \{x, y, z, p_x, p_y, p_z, J_x, J_y, J_z, H\}$.

13.11.2 The Category Sys

13.11.2.1 The Arrows and Translations for the Disjoint Sum $S_1 \sqcup S_2$

The use of local languages is central to our overall topos scheme, and therefore we need to understand, in particular, (i) the relation between the languages $\mathcal{L}(S_1)$ and $\mathcal{L}(S_2)$ if S_1 is a sub-system of S_2; and (ii) the relation between $\mathcal{L}(S_1)$, $\mathcal{L}(S_2)$ and $\mathcal{L}(S_1 \diamond S_2)$, where $S_1 \diamond S_2$ denotes the composite of systems S_1 and S_2.

---

[103] The word "sub-system" does not only mean a collection of objects that is spatially localised. One could also consider sub-systems of field systems by focussing on a just a few modes of the fields as is done, for example, in the Robertson-Walker model for cosmology. Another possibility would be to use fields localised in some fixed space, or space-time region provided that this is consistent with the dynamics.

These discussions can be made more precise by regarding **Sys** as a category whose objects are the systems.[104] The arrows in **Sys** need to cover two basic types of relation: (i) that between S_1 and S_2 if S_1 is a "sub-system" of S_2; and (ii) that between a composite system, $S_1 \diamond S_2$, and its constituent systems, S_1 and S_2.

This may seem straightforward but, in fact, care is needed since although the idea of a "sub-system" seems intuitively clear, it is hard to give a physically acceptable definition that is universal. However, some insight into this idea can be gained by considering its meaning in classical physics. This is very relevant for the general scheme since one of our main goals is to make all theories "look" like classical physics in the appropriate topos.

To this end, let S_1 and S_2 be classical systems whose state spaces are the symplectic manifolds $\mathcal{S}_1$ and $\mathcal{S}_2$ respectively. If S_1 is deemed to be a sub-system of S_2, it is natural to require that $\mathcal{S}_1$ is a *sub-manifold* of $\mathcal{S}_2$, i.e., $\mathcal{S}_1 \subseteq \mathcal{S}_2$. However, this condition cannot be used as a *definition* of a "sub-system" since the converse may not be true: i.e., if $\mathcal{S}_1 \subseteq \mathcal{S}_2$, this does not necessarily mean that, from a physical perspective, S_1 could, or would, be said to be a sub-system of S_2.[105]

On the other hand, there are situations where being a sub-manifold clearly *does* imply being a physical sub-system. For example, suppose the state space $\mathcal{S}$ of a system S is a disconnected manifold with two components $\mathcal{S}_1$ and $\mathcal{S}_2$, so that $\mathcal{S}$ is the disjoint union, $\mathcal{S}_1 \coprod \mathcal{S}_2$, of the sub-manifolds $\mathcal{S}_1$ and $\mathcal{S}_2$. Then it seems physically appropriate to say that the system S itself is disconnected, and to write $S = S_1 \sqcup S_2$ where the symplectic manifolds that represent the sub-systems S_1 and S_2 are $\mathcal{S}_1$ and $\mathcal{S}_2$ respectively.

One reason why it is reasonable to call S_1 and S_2 "sub-systems" in this particular situation is that any continuous dynamical evolution of a state point in $\mathcal{S} \simeq \mathcal{S}_1 \sqcup \mathcal{S}_2$ will always lie in either one component or the other. This suggests that perhaps, in general, a necessary condition for a sub-manifold $\mathcal{S}_1 \subseteq \mathcal{S}_2$ to represent a physical sub-system is that the dynamics of the system S_2 must be such that $\mathcal{S}_1$ is mapped into itself under the dynamical evolution on $\mathcal{S}_2$; in other words, $\mathcal{S}_1$ is a *dynamically-invariant* sub-manifold of $\mathcal{S}_2$. This correlates with the idea mentioned earlier that sub-systems are weakly-coupled with each other.

However, such a dynamical restriction is not something that should be coded into the languages, $\mathcal{L}(S_1)$ and $\mathcal{L}(S_2)$: rather, the dynamics is to be associated with the *representation* of these languages in the appropriate topoi.

Still, this caveat does not apply to the disjoint sum $S_1 \sqcup S_2$ of two systems S_1, S_2, and we will assume that, in general, (i.e., not just in classical physics) it is legitimate to think of S_1 and S_2 as being sub-systems of $S_1 \sqcup S_2$; something that we indicate by defining arrows $i_1 : S_1 \to S_1 \sqcup S_2$, and $i_2 : S_2 \to S_1 \sqcup S_2$ in **Sys**.

---

[104] To control the size of **Sys** we assume that the collection of objects/systems is a *set* rather than a more general class.

[105] For example, consider the diagonal sub-manifold $\Delta(\mathcal{S}) \subset \mathcal{S} \times \mathcal{S}$ of the symplectic manifold $\mathcal{S} \times \mathcal{S}$ that represents the composite $S \diamond S$ of two copies of a system S. Evidently, the states in $\Delta(\mathcal{S})$ correspond to the situation in which both copies of S "march together". It is doubtful if this would be recognised physically as a sub-system.

To proceed further it is important to understand the connection between the putative arrows in the category **Sys**, and the "translations" of the associated languages. The first step is to consider what can be said about the relation between $\mathcal{L}(S_1 \sqcup S_2)$, and $\mathcal{L}(S_1)$ and $\mathcal{L}(S_2)$. All three languages share the same ground type symbols, and so what we are concerned with is the relation between the function symbols of signature $\Sigma \to \mathcal{R}$ in these languages.

By considering what is meant intuitively by the disjoint sum, it seems plausible that each physical quantity for the system $S_1 \sqcup S_2$ produces a physical quantity for S_1, and another one for S_2. Conversely, specifying a pair of physical quantities—one for S_1 and one for S_2—gives a physical quantity for $S_1 \sqcup S_2$. In other words,

$$F_{\mathcal{L}(S_1 \sqcup S_2)}(\Sigma, \mathcal{R}) \simeq F_{\mathcal{L}(S_1)}(\Sigma, \mathcal{R}) \times F_{\mathcal{L}(S_2)}(\Sigma, \mathcal{R}) \tag{13.358}$$

However, it is important not to be too dogmatic about statements of this type since in non-classical theories new possibilities can arise that are counter to intuition.

Associated with (13.358) are the maps $\mathcal{L}(i_1) : F_{\mathcal{L}(S_1 \sqcup S_2)}(\Sigma, \mathcal{R}) \to F_{\mathcal{L}(S_1)}(\Sigma, \mathcal{R})$ and $\mathcal{L}(i_2) : F_{\mathcal{L}(S_1 \sqcup S_2)}(\Sigma, \mathcal{R}) \to F_{\mathcal{L}(S_2)}(\Sigma, \mathcal{R})$, defined as the projection maps of the product. In the theory of local languages, these transformations are essentially *translations* [11] of $\mathcal{L}(S_1 \sqcup S_2)$ in $\mathcal{L}(S_1)$ and $\mathcal{L}(S_2)$ respectively; a situation that we denote $\mathcal{L}(i_1) : \mathcal{L}(S_1 \sqcup S_2) \to \mathcal{L}(S_1)$, and $\mathcal{L}(i_2) : \mathcal{L}(S_1 \sqcup S_2) \to \mathcal{L}(S_2)$.

To be more precise, these operations are translations if, taking $\mathcal{L}(i_1)$ as the explanatory example, the map $\mathcal{L}(i_1) : F_{\mathcal{L}(S_1 \sqcup S_2)}(\Sigma, \mathcal{R}) \to F_{\mathcal{L}(S_1)}(\Sigma, \mathcal{R})$ is supplemented with the following map from the ground symbols of $\mathcal{L}(S_1 \sqcup S_2)$ to those of $\mathcal{L}(S_1)$:

$$\mathcal{L}(i_1)(\Sigma) := \Sigma, \tag{13.359}$$

$$\mathcal{L}(i_1)(\mathcal{R}) := \mathcal{R}, \tag{13.360}$$

$$\mathcal{L}(i_1)(1) := 1, \tag{13.361}$$

$$\mathcal{L}(i_1)(\Omega) := \Omega. \tag{13.362}$$

Such a translation map is then extended to all type symbols using the definitions

$$\mathcal{L}(i_1)(T_1 \times T_2 \times \cdots \times T_n) = \mathcal{L}(i_1)(T_1) \times \mathcal{L}(i_1)(T_2) \times \cdots \times \mathcal{L}(i_1)(T_n), \tag{13.363}$$

$$\mathcal{L}(i_1)(PT) = P[\mathcal{L}(i_1)(T)] \tag{13.364}$$

for all finite n and all type symbols $T, T_1, T_2, \ldots, T_n$. This, in turn, can be extended inductively to all terms in the language. Thus, in our case, the translations act trivially on all the type symbols.

Arrows in **Sys** *are Translations*

Motivated by this argument we now turn everything around and, in general, *define* an arrow $j : S_1 \to S$ in the category **Sys** to mean that there is some *physically meaningful* way of transforming the physical quantities in S to physical quantities in S_1. If, for any pair of systems S_1, S there is more than one such transformation, then there will be more than one arrow from S_1 to S.

To make this more precise, let **Loc** denote the collection of all (small[106]) local languages. This is a category whose objects are the local languages, and whose arrows are translations between languages. Then our basic assumption is that the association $S \mapsto \mathcal{L}(S)$ is a covariant functor from **Sys** to $\mathbf{Loc}^{\mathrm{op}}$, which we denote as $\mathcal{L} : \mathbf{Sys} \to \mathbf{Loc}^{\mathrm{op}}$.

Note that the combination of a pair of arrows in **Sys** exists in so far as the associated translations can be combined.

13.11.2.2 The Arrows and Translations for the Composite System $S_1 \diamond S_2$

Let us now consider the composition $S_1 \diamond S_2$ of a pair of systems. In the case of classical physics, if $\mathcal{S}_1$ and $\mathcal{S}_2$ are the symplectic manifolds that represent the systems S_1 and S_2 respectively, then the manifold that represents the composite system is the cartesian product $\mathcal{S}_1 \times \mathcal{S}_2$. This is distinguished by the existence of the two projection functions $\mathrm{pr}_1 : \mathcal{S}_1 \times \mathcal{S}_2 \to \mathcal{S}_1$ and $\mathrm{pr}_2 : \mathcal{S}_1 \times \mathcal{S}_2 \to \mathcal{S}_2$.

It seems reasonable to impose the same type of structure on **Sys**: i.e., to require there to be arrows $p_1 : S_1 \diamond S_2 \to S_1$ and $p_2 : S_1 \diamond S_2 \to S_2$ in **Sys**. However, bearing in mind the definition above, these arrows p_1, p_2 exist if, and only if, there are corresponding translations $\mathcal{L}(p_1) : \mathcal{L}(S_1) \to \mathcal{L}(S_1 \diamond S_2)$, and $\mathcal{L}(p_2) : \mathcal{L}(S_2) \to \mathcal{L}(S_1 \diamond S_2)$. But there *are* such translations: for if A_1 is a physical quantity for system S_1, then $\mathcal{L}(p_1)(A_1)$ can be defined as that same physical quantity, but now regarded as pertaining to the combined system $S_1 \diamond S_2$; and analogously for system S_2.[107] We shall denote this translated quantity, $\mathcal{L}(p_1)(A_1)$, by $A_1 \diamond 1$.

Note that we do *not* postulate any simple relation between $F_{\mathcal{L}(S_1 \diamond S_2)}(\Sigma, \mathcal{R})$ and $F_{\mathcal{L}(S_1)}(\Sigma, \mathcal{R})$ and $F_{\mathcal{L}(S_2)}(\Sigma, \mathcal{R})$; i.e., there is no analogue of (13.358) for combinations of systems.

The definitions above of the basic arrows suggest that we might also want to impose the following conditions:

1. The arrows $i_1 : S_1 \to S_1 \sqcup S_2$, and $i_2 : S_2 \to S_1 \sqcup S_2$ are *monic* in **Sys**.
2. The arrows $p_1 : S_1 \diamond S_2 \to S_1$ and $p_2 : S_1 \diamond S_2 \to S_2$ are *epic* arrows in **Sys**.

However, we do *not* require that $S_1 \sqcup S_2$ and $S_1 \diamond S_2$ are the co-product and product, respectively, of S_1 and S_2 in the category **Sys**.

---

106 This means that the collection of symbols is a set, not a more general class.

107 For example, if A is the energy of particle 1, then we can talk about this energy in the combination of a pair of particles. Of course, in—for example—classical physics there is no reason why the energy of particle 1 should be *conserved* in the composite system, but that, dynamical, question is a different matter.

13.11.2.3 The Concept of "Isomorphic" Systems

We also need to decide what it means to say that two systems S_1 and S_2 are *isomorphic*, to be denoted $S_1 \simeq S_2$. As with the concept of sub-system, the notion of isomorphism is to some extent a matter of definition rather than obvious physical structure, albeit with the expectation that isomorphic systems in **Sys** will correspond to isomorphic local languages, and be represented by isomorphic mathematical objects in any concrete realisation of the axioms: for example, by isomorphic symplectic manifolds in classical physics.

To a considerable extent, the physical meaning of "isomorphism" depends on whether one is dealing with actual physical systems, or idealisations of them. For example, an electron confined in a box in Cambridge is presumably isomorphic to one confined in the same type of box in London, although they are not the same physical system. On the other hand, when a lecturer says "Consider an electron trapped in a box....", he/she is referring to an idealised system.

One could, perhaps, say that an idealised system is an *equivalence class* (under isomorphisms) of real systems, but even working only with idealisations does not entirely remove the need for the concept of isomorphism.

For example, in classical mechanics, consider the (idealised) system S of a point particle moving in a box, and let 1 denote the 'trivial system' that consists of just a single point with no internal or external degrees of freedom. Now consider the system $S \diamond 1$. In classical mechanics this is represented by the symplectic manifold $\mathcal{S} \times \{*\}$, where $\{*\}$ is a single point, regarded as a zero-dimensional manifold. However, $\mathcal{S} \times \{*\}$ is isomorphic to the manifold $\mathcal{S}$, and it is clear physically that the system $S \diamond 1$ is isomorphic to the system S. On the other hand, one cannot say that $S \diamond 1$ is literally *equal* to S, so the concept of "isomorphism" needs to be maintained.

One thing that *is* clear is that if $S_1 \simeq S_2$ then $F_{\mathcal{L}(S_1)}(\Sigma, \mathcal{R}) \simeq F_{\mathcal{L}(S_2)}(\Sigma, \mathcal{R})$, and if any other non-empty sets of function symbols are present, then they too must be isomorphic.

Note that when introducing a trivial system, 1, it necessary to specify its local language, $\mathcal{L}(1)$. The set of function symbols $F_{\mathcal{L}(1)}(\Sigma, \mathcal{R})$ is not completely empty since, in classical physics, one does have a preferred physical quantity, which is just the number 1. If one asks what is meant in general by the "number 1" the answer is not trivial since, in the reals $\mathbb{R}$, the number 1 is the multiplicative identity. It would be possible to add the existence of such a unit to the axioms for $\mathcal{R}$ but this would involve introducing a multiplicative structure and we do not know if there might be physically interesting topos representations that do not have this feature.

For the moment then, we will say that the trivial system has just a single physical quantity, which in classical physics translates to the number 1. More generally, for the language $\mathcal{L}(1)$ we specify that $F_{\mathcal{L}(1)}(\Sigma, \mathcal{R}) := \{I\}$, i.e., $F_{\mathcal{L}(1)}(\Sigma, \mathcal{R})$ has just a single element, I, say. Furthermore, we add the axiom

$$: \forall \tilde{s}_1 \forall \tilde{s}_2, I(\tilde{s}_1) = I(\tilde{s}_2), \tag{13.365}$$

where $\tilde{s}_1$ and $\tilde{s}_2$ are variables of type Σ. In fact, it seems natural to add such a trivial quantity to the language $\mathcal{L}(S)$ for *any* system S, and from now on we will assume that this has been done.

A related issue is that, in classical physics, if A is a physical quantity, then so is rA for any $r \in \mathbb{R}$. This is because the set of classical quantities $A_\sigma : \Sigma_\sigma \to \mathcal{R}_\sigma \simeq \mathbb{R}$ forms a *ring* whose structure derives from the ring structure of $\mathbb{R}$. It would be possible to add ring axioms for $\mathcal{R}$ to the language $\mathcal{L}(S)$, but this is too strong, not least because, as shown earlier, it fails in quantum theory. Clearly, the general question of axioms for $\mathcal{R}$ needs more thought: a task for later work.

If desired, an *empty* system, 0, can be added too, with $F_{\mathcal{L}(0)}(\Sigma, \mathcal{R}) := \emptyset$. This, so called, "pure language", $\mathcal{L}(0)$, is an initial object in the category **Loc**.

13.11.2.4 An Axiomatic Formulation of the Category Sys

Let us now summarise, and clarify, our list of axioms for a category **Sys**:

1. The collection **Sys** is a small category se objects are the systems of interest (or, if desired, morphism classes of such systems) and whose ows are defined as above. Thus the fundamental property of an arrow $j : S_1 \to S$ in **Sys** is that it induces, and is essentially *defined by*, a translation $\mathcal{L}(j) : \mathcal{L}(S) \to \mathcal{L}(S_1)$. Physically, this corresponds to the physical quantities for system S being 'pulled-back' to give physical quantities for system S_1.
   Arrows of particular interest are those associated with "sub-systems" and "composite systems", as discussed above.
2. The axioms for a category are satisfied because:
   (a) Physically, the ability to form composites of arrows follows from the concept of "pulling-back" physical quantities. From a mathematical perspective, if $j : S_1 \to S_2$ and $k : S_2 \to S_3$, then the translations give functions $\mathcal{L}(j) : F_{\mathcal{L}(S_2)}(\Sigma, \mathcal{R}) \to F_{\mathcal{L}(S_1)}(\Sigma, \mathcal{R})$ and $\mathcal{L}(k) : F_{\mathcal{L}(S_3)}(\Sigma, \mathcal{R}) \to F_{\mathcal{L}(S_2)}(\Sigma, \mathcal{R})$. Then clearly $\mathcal{L}(j) \circ \mathcal{L}(k) : F_{\mathcal{L}(S_3)}(\Sigma, \mathcal{R}) \to F_{\mathcal{L}(S_1)}(\Sigma, \mathcal{R})$, and this can thought of as the translation corresponding to the arrow $k \circ j : S_1 \to S_3$.
   The associativity of the law of arrow combination can be proved in a similar way.
   (b) We add by hand a special arrow $\mathrm{id}_S : S \to S$ which is defined to correspond to the translation $\mathcal{L}(\mathrm{id}_S)$ that is given by the identity map on $F_{\mathcal{L}(S)}(\Sigma, \mathcal{R})$. Clearly, $\mathrm{id}_S : S \to S$ acts an an identity morphism should.
3. For any pair of systems S_1, S_2, there is a *disjoint sum*, denoted $S_1 \sqcup S_2$. The disjoint sum has the following properties:
   (a) For all systems S_1, S_2, S_3 in **Sys**:

$$(S_1 \sqcup S_2) \sqcup S_3 \simeq S_1 \sqcup (S_2 \sqcup S_3). \tag{13.366}$$

(b) For all systems S_1, S_2 in **Sys**:

$$S_1 \sqcup S_2 \simeq S_2 \sqcup S_1. \tag{13.367}$$

(c) There are arrows in **Sys**:

$$i_1 : S_1 \to S_1 \sqcup S_2 \text{ and } i_2 : S_2 \to S_1 \sqcup S_2 \tag{13.368}$$

that are associated with translations in the sense discussed in Sect. 13.11.2. These are associated with the decomposition

$$F_{\mathcal{L}(S_1 \sqcup S_2)}(\Sigma, \mathcal{R}) \simeq F_{\mathcal{L}(S_1)}(\Sigma, \mathcal{R}) \times F_{\mathcal{L}(S_2)}(\Sigma, \mathcal{R}). \tag{13.369}$$

We assume that if S_1, S_2 belong to **Sys**, then **Sys** also contains $S_1 \sqcup S_2$.

4. For any given pair of systems S_1, S_2, there is a *composite* system in **Sys**, denoted[108] $S_1 \diamond S_2$, with the following properties:

(a) For all systems S_1, S_2, S_3 in **Sys**:

$$(S_1 \diamond S_2) \diamond S_3 \simeq S_1 \diamond (S_2 \diamond S_3). \tag{13.370}$$

(b) For all systems S_1, S_2 in **Sys**:

$$S_1 \diamond S_2 \simeq S_2 \diamond S_1. \tag{13.371}$$

(c) There are arrows in **Sys**:

$$p_1 : S_1 \diamond S_2 \to S_1 \text{ and } p_2 : S_1 \diamond S_2 \to S_2 \tag{13.372}$$

that are associated with translations in the sense discussed in Sect. 13.11.2.2.

We assume that if S_1, S_2 belong to **Sys**, then **Sys** also contains the composite system $S_1 \diamond S_2$.

5. It seems physically reasonable to add the axiom

$$(S_1 \sqcup S_2) \diamond S \simeq (S_1 \diamond S) \sqcup (S_2 \diamond S) \tag{13.373}$$

for all systems S_1, S_2, S. However, physical intuition can be a dangerous thing, and so, as with most of these axioms, we are not dogmatic, and feel free to change them as new insights emerge.

---

108 The product operation in a monoidal category is often written "⊗". However, a different symbol has been used here to avoid confusion with existing usages in physics of the tensor product sign "⊗".

6. There is a trivial system, 1, such that for all systems S, we have

$$S \diamond 1 \simeq S \simeq 1 \diamond S \tag{13.374}$$

7. It may be convenient to postulate an "empty system", 0, with the properties

$$S \diamond 0 \simeq 0 \diamond S \simeq 0 \tag{13.375}$$
$$S \sqcup 0 \simeq 0 \sqcup S \simeq S \tag{13.376}$$

   for all systems S.
   Within the meaning given to arrows in **Sys**, 0 is a *terminal object* in **Sys**. This is because the empty set of function symbols of signature $\Sigma \to \mathcal{R}$ is a subset of any other set of function symbols of this signature.

It might seem tempting to postulate that composition laws are well-behaved with respect to arrows. Namely, if $j : S_1 \to S_2$, then, for any S, there is an arrow $S_1 \diamond S \to S_2 \diamond S$ and an arrow $S_1 \sqcup S \to S_2 \sqcup S$.[109]

In the case of the disjoint sum, such an arrow can be easily constructed using (13.369). First split the function symbols in $F_{\mathcal{L}(S_1 \sqcup S)}(\Sigma, \mathcal{R})$ into $F_{\mathcal{L}(S_1)}(\Sigma, \mathcal{R}) \times F_{\mathcal{L}(S)}(\Sigma, \mathcal{R})$ and the function symbols in $F_{\mathcal{L}(S_2 \sqcup S)}(\Sigma, \mathcal{R})$ into $F_{\mathcal{L}(S_2)}(\Sigma, \mathcal{R}) \times F_{\mathcal{L}(S)}(\Sigma, \mathcal{R})$. Since there is an arrow $j : S_1 \to S_2$, there is a translation $\mathcal{L}(j) : \mathcal{L}(S_2) \to \mathcal{L}(S_1)$, given by a mapping $\mathcal{L}(j) : F_{\mathcal{L}(S_2)}(\Sigma, \mathcal{R}) \to F_{\mathcal{L}(S_1)}(\Sigma, \mathcal{R})$. Of course, then there is also a mapping $\mathcal{L}(j) \times \mathcal{L}(\mathrm{id}_S) : F_{\mathcal{L}(S_2)}(\Sigma, \mathcal{R}) \times F_{\mathcal{L}(S)}(\Sigma, \mathcal{R}) \to F_{\mathcal{L}(S_1)}(\Sigma, \mathcal{R}) \times F_{\mathcal{L}(S)}(\Sigma, \mathcal{R})$, i.e., a translation between $\mathcal{L}(S_2 \sqcup S)$ and $\mathcal{L}(S_1 \sqcup S)$. Since we assume that there is an arrow in **Sys** whenever there is a translation (in the opposite direction), there is indeed an arrow $S_1 \sqcup S \to S_2 \sqcup S$.

In the case of the composition, however, this would require a translation $\mathcal{L}(S_2 \diamond S) \to \mathcal{L}(S_1 \diamond S)$, and this cannot be done in general since we have no *prima facie* information about the set of function symbols $F_{\mathcal{L}(S_2 \diamond S)}(\Sigma, \mathcal{R})$. However, if we restrict the arrows in **Sys** to be those associated with sub-systems, combination of systems, and compositions of such arrows, then it is easy to see that the required translations exist (the proof of this makes essential use of (13.373)).

If we make this restriction of arrows, then the axioms (13.371), (13.374), (13.375), (13.376) and (13.377), mean that, essentially, **Sys** has the structure of a *symmetric monoidal*[110] category in which the monoidal product operation is "$\diamond$", and the left and right unit object is 1. There is also a monoidal structure associated with the disjoint sum "$\sqcup$", with 0 as the unit object.

---

[109] A more accurate way of capturing this idea is to say that the operation **Sys** $\times$ **Sys** $\to$ **Sys** in which

$$\langle S_1, S_2 \rangle \mapsto S_1 \diamond S_2 \tag{13.377}$$

is a *bi-functor* from **Sys** $\times$ **Sys** to **Sys**. Ditto for the operation in which $\langle S_1, S_2 \rangle \mapsto S_1 \sqcup S_2$.

[110] In the actual definition of a monoidal category the two isomorphisms in (13.374) are separated from each other, whereas we have identified them. Further more, these isomorphism are required to be natural. This seems a correct thing to require in our case, too.

We say "essentially" because in order to comply with all the axioms of a monoidal category, **Sys** must satisfy certain additional, so-called, "coherence" axioms. However, from a physical perspective these are very plausible statements about (i) how the unit object 1 intertwines with the $\diamond$-operation; how the null object intertwines with the $\sqcup$-operation; and (iii) certain properties of quadruple products (and disjoint sums) of systems.

A Simple Example of a Category **Sys**

It might be helpful at this point to give a simple example of a category **Sys**. To that end, let S denote a point particle that moves in three dimensions, and let us suppose that S has no sub-systems other than the trivial system 1. Then $S \diamond S$ is defined to be a pair of particles moving in three dimensions, and so on. Thus the objects in our category are $1, S, S \diamond S, \ldots, S \diamond S \diamond \cdots S \ldots$ where the "$\diamond$" operation is formed any finite number of times.

At this stage, the only arrows are those that are associated with the constituents of a composite system. However, we could contemplate adding to the systems the disjoint sum $S \sqcup (S \diamond S)$ which is a system that is either one particle or two particles (but, of course, not both at the same time). And, clearly, we could extend this to $S \sqcup (S \diamond S) \sqcup (S \diamond S \diamond S)$, and so on. Each of these disjoint sums comes with its own arrows, as explained above.

Note that this particular category of systems has the property that it can be treated using either classical physics or quantum theory.

13.11.3 Representations of Sys in Topoi

We assume that all the systems in **Sys** are to be treated with the same theory type. We also assume that systems in **Sys** with the *same* language are to be represented in the same topos. Then we define:[111]

Definition 19 A topos realisation of **Sys** is an association, ϕ, to each system S in **Sys**, of a triple $\phi(S) = \langle \rho_{\phi,S}, \mathcal{L}(S), \tau_\phi(S) \rangle$ where:

(i) $\tau_\phi(S)$ is the topos in which the theory-type applied to system S is to be realised.
(ii) $\mathcal{L}(S)$ is the local language in **Loc** that is associated with S. This is not dependent on the realisation ϕ.
(iii) $\rho_{\phi,S}$ is a representation of the local language $\mathcal{L}(S)$ in the topos $\tau_\phi(S)$. As a more descriptive piece of notation we write $\rho_{\phi,S} : \mathcal{L}(S) \rightsquigarrow \tau_\phi(S)$. The key part of this representation is the map

$$\rho_{\phi,S} : F_{\mathcal{L}(S)}\big(\Sigma, \mathcal{R}\big) \to \mathrm{Hom}_{\tau_\phi(S)}\big(\Sigma_{\phi,S}, \mathcal{R}_{\phi,S}\big) \tag{13.378}$$

[111] As emphasised already, the association $S \mapsto \mathcal{L}(S)$ is generally not one-to-one: i.e., many systems may share the same language. Thus, when we come discuss the representation of the language $\mathcal{L}(S)$ in a topos, the extra information about the system S is used in fixing the representation.

where $\Sigma_{\phi,S}$ and $\mathcal{R}_{\phi,S}$ are the state object and quantity-value object, respectively, of the representation ϕ in the topos $\tau_\phi(S)$. As a convenient piece of notation we write $A_{\phi,S} := \rho_{\phi,S}(A)$ for all $A \in F_{\mathcal{L}(S)}(\Sigma, \mathcal{R})$.

This definition is only partial; the possibility of extending it will be discussed shortly.

Now, if $j : S_1 \to S$ is an arrow in **Sys**, then there is a translation arrow $\mathcal{L}(j) : \mathcal{L}(S) \to \mathcal{L}(S_1)$. Thus we have the beginnings of a commutative diagram

$$\begin{array}{ccc} S_1 & \xrightarrow{\ \phi\ } & \langle \rho_{\phi,S_1}, \mathcal{L}(S_1), \tau_\phi(S_1)\rangle \\ {\scriptstyle j}\big\downarrow & & \big\uparrow {\scriptstyle ?\times\mathcal{L}(j)\times ?} \\ S & \xrightarrow[\ \phi\ ]{} & \langle \rho_{\phi,S}, \mathcal{L}(S), \tau_\phi(S)\rangle \end{array} \tag{13.379}$$

However, to be useful, the arrow on the right hand side of this diagram should refer to some relation between (i) the topoi $\tau_\phi(S_1)$ and $\tau_\phi(S)$; and (ii) the realisations $\rho_{\phi,S_1} : \mathcal{L}(S_1) \rightsquigarrow \tau_\phi(S_1)$ and $\rho_{\phi,S} : \mathcal{L}(S) \rightsquigarrow \tau_\phi(S)$: this is the significance of the two '?' symbols in the arrow written "$? \times \mathcal{L}(j) \times ?$".

Indeed, as things stand, Definition 19 says nothing about relations between the topoi representations of different systems in **Sys**. We are particularly interested in the situation where there are two different systems S_1 and S with an arrow $j : S_1 \to S$ in **Sys**.

We know that the arrow j is associated with a translation $\mathcal{L}(j) : \mathcal{L}(S) \to \mathcal{L}(S_1)$, and an attractive possibility, therefore, would be to seek, or postulate, a "covering" map $\phi(\mathcal{L}(j)) : \mathrm{Hom}_{\tau_\phi(S)}(\Sigma_{\phi,S}, \mathcal{R}_{\phi,S}) \to \mathrm{Hom}_{\tau_\phi(S_1)}(\Sigma_{\phi,S_1}, \mathcal{R}_{\phi,S_1})$ to be construed as a topos representation of the translation $\mathcal{L}(j) : \mathcal{L}(S) \to \mathcal{L}(S_1)$, and hence of the arrow $j : S_1 \to S$ in **Sys**.

This raises the questions of what properties these "translation representations" should possess in order to justify saying that they "cover" the translations. A minimal requirement is that if $k : S_2 \to S_1$ and $j : S_1 \to S$, then the map $\phi(\mathcal{L}(j \circ k)) : \mathrm{Hom}_{\tau_\phi(S)}(\Sigma_{\phi,S}, \mathcal{R}_{\phi,S}) \to \mathrm{Hom}_{\tau_\phi(S_2)}(\Sigma_{\phi,S_2}, \mathcal{R}_{\phi,S_2})$ factorises as

$$\phi(\mathcal{L}(j \circ k)) = \phi(\mathcal{L}(k)) \circ \phi(\mathcal{L}(j)). \tag{13.380}$$

We also require that

$$\phi(\mathcal{L}(\mathrm{id}_S)) = \mathrm{id} : \mathrm{Hom}_{\tau_\phi(S)}(\Sigma_{\phi,S}, \mathcal{R}_{\phi,S}) \to \mathrm{Hom}_{\tau_\phi(S)}(\Sigma_{\phi,S}, \mathcal{R}_{\phi,S}) \tag{13.381}$$

for all systems S.

The conditions (13.380) and (13.381) seem eminently plausible, and they are not particularly strong. A far more restrictive axiom would be to require the following diagram to commute:

$$\begin{array}{ccc} F_{\mathcal{L}(S)}\big(\Sigma,\mathcal{R}\big) & \xrightarrow{\rho_{\phi,S}} & \mathrm{Hom}_{\tau_\phi(S)}\big(\Sigma_{\phi,S},\mathcal{R}_{\phi,S}\big) \\ \Big\downarrow{\scriptstyle \mathcal{L}(j)} & & \Big\downarrow{\scriptstyle \phi(\mathcal{L}(j))} \\ F_{\mathcal{L}(S_1)}\big(\Sigma,\mathcal{R}\big) & \xrightarrow[\rho_{\phi,S_1}]{} & \mathrm{Hom}_{\tau_\phi(S_1)}\big(\Sigma_{\phi,S_1},\mathcal{R}_{\phi,S_1}\big) \end{array} \tag{13.382}$$

At first sight, this requirement seems very appealing. However, caution is needed when postulating "axioms" for a theoretical structure in physics. It is easy to get captivated by the underlying mathematics and to assume, erroneously, that what is mathematically elegant is necessarily true in the physical theory.

The translation $\phi(\mathcal{L}(j))$ maps an arrow from $\Sigma_{\phi,S}$ to $\mathcal{R}_{\phi,S}$ to an arrow from Σ_{ϕ,S_1} to $\mathcal{R}_{\phi,S_1}$. Intuitively, if Σ_{ϕ,S_1} is a "much larger" object than $\Sigma_{\phi,S}$ (although since they lie in different topoi, no direct comparison is available), the translation can only be "faithful" on some part of Σ_{ϕ,S_1} that can be identified with (the "image" of) $\Sigma_{\phi,S}$. A concrete example of this will show up in the treatment of composite quantum systems, see Sect. 13.13.3. As one might expect, a form of entanglement plays a role here.

13.11.4 Classical Physics in This Form

13.11.4.1 The Rules so Far

Constructing maps $\phi(\mathcal{L}(j)) : \mathrm{Hom}_{\tau_\phi(S)}\big(\Sigma_{\phi,S},\mathcal{R}_{\phi,S}\big) \to \mathrm{Hom}_{\tau_\phi(S_1)}\big(\Sigma_{\phi,S_1},\mathcal{R}_{\phi,S_1}\big)$ is likely to be complicated when $\tau_\phi(S)$ and $\tau_\phi(S_1)$ are different topoi, and so we begin with the example of classical physics, where the topos is always **Sets**.

In general, we are interested in the relation(s) between the representations $\rho_{\phi,S_1} : \mathcal{L}(S_1) \rightsquigarrow \tau_\phi(S_1)$ and $\rho_{\phi,S} : \mathcal{L}(S) \rightsquigarrow \tau_\phi(S)$ that is associated with an arrow $j : S_1 \to S$ in **Sys**. In classical physics, we only have to study the relation between the representations $\rho_{\sigma,S_1} : \mathcal{L}(S_1) \rightsquigarrow$ **Sets** and $\rho_{\sigma,S} : \mathcal{L}(S) \rightsquigarrow$ **Sets**.

Let us summarise what we have said so far (with σ denoting the **Sets**-realisation of classical physics):

1. For any system S in **Sys**, a representation $\rho_{\sigma,S} : \mathcal{L}(S) \rightsquigarrow$ **Sets** consists of the following ingredients.

    (a) The ground symbol Σ is represented by a symplectic manifold, $\Sigma_{\sigma,S} := \rho_{\sigma,S}(\Sigma)$, that serves as the classical state space.
    (b) For all systems S, the ground symbol $\mathcal{R}$ is represented by the real numbers $\mathbb{R}$, i.e., $\mathcal{R}_{\sigma,S} = \mathbb{R}$, where $\mathcal{R}_{\sigma,S} := \rho_{\sigma,S}(\mathcal{R})$.

(c) Each function symbol $A : \Sigma \to \mathcal{R}$ in $F_{\mathcal{L}(S)}(\Sigma, \mathcal{R})$ is represented by a function $A_{\sigma,S} = \rho_{\sigma,S}(A) : \Sigma_{\sigma,S} \to \mathbb{R}$ in the set of functions[112] $C(\Sigma_{\sigma,S}, \mathbb{R})$.

2. The trivial system is mapped to a singleton set $\{*\}$ (viewed as a zero-dimensional symplectic manifold):

$$\Sigma_{\sigma,1} := \{*\}. \tag{13.383}$$

The empty system is represented by the empty set:

$$\Sigma_{\sigma,0} := \emptyset. \tag{13.384}$$

3. Propositions about the system S are represented by (Borel) subsets of the state space $\Sigma_{\sigma,S}$.
4. The composite system $S_1 \diamond S_2$ is represented by the Cartesian product $\Sigma_{\sigma,S_1} \times \Sigma_{\sigma,S_2}$; i.e.,

$$\Sigma_{\sigma,\, S_1 \diamond S_2} \simeq \Sigma_{\sigma,S_1} \times \Sigma_{\sigma,S_2}. \tag{13.385}$$

The disjoint sum $S_1 \sqcup S_2$ is represented by the disjoint union $\Sigma_{\sigma,S_1} \coprod \Sigma_{\sigma,S_2}$; i.e.,

$$\Sigma_{\sigma,S_1 \sqcup S_2} \simeq \Sigma_{\sigma,S_1} \coprod \Sigma_{\sigma,S_2}. \tag{13.386}$$

5. Let $j : S_1 \to S$ be an arrow in **Sys**. Then

(a) There is a translation map $\mathcal{L}(j) : F_{\mathcal{L}(S)}(\Sigma, \mathcal{R}) \to F_{\mathcal{L}(S_1)}(\Sigma, \mathcal{R})$.
(b) There is a symplectic function $\sigma(j) : \Sigma_{\sigma,S_1} \to \Sigma_{\sigma,S}$ from the symplectic manifold Σ_{σ,S_1} to the symplectic manifold $\Sigma_{\sigma,S}$.

The existence of this function $\sigma(j) : \Sigma_{\sigma,S_1} \to \Sigma_{\sigma,S}$ follows directly from the properties of sub-systems and composite systems in classical physics. It is discussed in detail below in Sect. (13.11.4.2). As we shall see, it underpins the classical realisation of our axioms.

These properties of the arrows stem from the fact that the linguistic function symbols in $F_{\mathcal{L}(S)}(\Sigma, \mathcal{R})$ are represented by real-valued functions in $C(\Sigma_{\sigma,S}, \mathbb{R})$. Thus we can write $\rho_{\sigma,S} : F_{\mathcal{L}(S)}(\Sigma, \mathcal{R}) \to C(\Sigma_{\sigma,S}, \mathbb{R})$, and similarly $\rho_{\sigma,S_1} : F_{\mathcal{L}(S_1)}(\Sigma, \mathcal{R}) \to C(\Sigma_{\sigma,S_1}, \mathbb{R})$. The diagram in (13.382) now becomes

$$\begin{array}{ccc} F_{\mathcal{L}(S)}(\Sigma, \mathcal{R}) & \xrightarrow{\rho_{\sigma,S}} & C(\Sigma_{\sigma,S}, \mathbb{R}) \\ \Big\downarrow{\scriptstyle \mathcal{L}(j)} & & \Big\downarrow{\scriptstyle \sigma(\mathcal{L}(j))} \\ F_{\mathcal{L}(S_1)}(\Sigma, \mathcal{R}) & \xrightarrow[\rho_{\sigma,S_1}]{} & C(\Sigma_{\sigma,S_1}, \mathbb{R}) \end{array} \tag{13.387}$$

[112] In practice, these functions are required to be measurable with respect to the Borel structures on the symplectic manifold Σ_σ and $\mathbb{R}$. Many of the functions will also be smooth, but we will not go into such details here.

and, therefore, the question of interest is if there is a "translation representation" function $\sigma(\mathcal{L}(j)) : C(\Sigma_{\sigma,S}, \mathbb{R}) \to C(\Sigma_{\sigma,S_1}, \mathbb{R})$ so that this diagram commutes.

Now, as stated above, a physical quantity, A, for the system S is represented in classical physics by a real-valued function $A_{\sigma,S} = \rho_{\sigma,S}(A) : \Sigma_{\sigma,S} \to \mathbb{R}$. Similarly, the representation of $\mathcal{L}(j)(A)$ for S_1 is given by a function $A_{\sigma,S_1} := \rho_{\sigma,S_1}(A) : \Sigma_{\sigma,S_1} \to \mathbb{R}$. However, in this classical case we also have the function $\sigma(j) : \Sigma_{\sigma,S_1} \to \Sigma_{\sigma,S}$, and it is clear that we can use it to define $[\rho_{\sigma,S_1}(\mathcal{L}(j)(A)](s) := \rho_{\sigma,S}(A)\big(\sigma(j)(s)\big)$ for all $s \in \Sigma_{\sigma,S_1}$. In other words

$$\rho_{\sigma,S_1}\big(\mathcal{L}(j)(A)\big) = \rho_{\sigma,S}(A) \circ \sigma(j) \tag{13.388}$$

or, in simpler notation

$$\big((\mathcal{L}(j)(A)\big)_{\sigma,S_1} = A_{\sigma,S} \circ \sigma(j). \tag{13.389}$$

But then it is clear that a translation-representation function $\sigma(\mathcal{L}(j)) : C(\Sigma_{\sigma,S}, \mathbb{R}) \to C(\Sigma_{\sigma,S_1}, \mathbb{R})$ with the desired property of making (13.387) commute can be defined by

$$\sigma(\mathcal{L}(j))(f) := f \circ \sigma(j) \tag{13.390}$$

for all $f \in C(\Sigma_{\sigma,S}, \mathbb{R})$; i.e., the function $\sigma(\mathcal{L}(j))(f) : \Sigma_{\sigma,S_1} \to \mathbb{R}$ is the usual pull-back of the function $f : \Sigma_{\sigma,S} \to \mathbb{R}$ by the function $\sigma(j) : \Sigma_{\sigma,S_1} \to \Sigma_{\sigma,S}$. Thus, in the case of classical physics, the commutative diagram in (13.379) can be completed to give

$$\begin{array}{ccc} S_1 & \xrightarrow{\sigma} & \langle \rho_{\sigma,S_1}, \mathcal{L}(S_1), \textbf{\textit{Sets}} \rangle \\ {\scriptstyle j}\downarrow & & \downarrow {\scriptstyle \sigma(\mathcal{L}(j)) \times \mathcal{L}(j) \times \mathrm{id}} \\ S & \xrightarrow[\sigma]{} & \langle \rho_{\sigma,S}, \mathcal{L}(S), \textbf{\textit{Sets}} \rangle \end{array} \tag{13.391}$$

13.11.4.2 Details of the Translation Representation

The Translation Representation for a Disjoint Sum of Classical Systems

We first consider arrows of the form

$$S_1 \overset{i_1}{\to} S_1 \sqcup S_2 \overset{i_2}{\leftarrow} S_2 \tag{13.392}$$

from the components S_1, S_2 to the disjoint sum $S_1 \sqcup S_2$. The systems S_1, S_2 and $S_1 \sqcup S_2$ have symplectic manifolds Σ_{σ,S_1}, Σ_{σ,S_2} and $\Sigma_{\sigma,S_1 \sqcup S_2} = \Sigma_{\sigma,S_1} \coprod \Sigma_{\sigma,S_2}$. We write $i := i_1$.

Let S be a classical system. We assume that the function symbols $A \in F_{\mathcal{L}(S)}(\Sigma, \mathcal{R})$ in the language $\mathcal{L}(S)$ are in bijective correspondence with an appropriate subset of the functions $A_{\sigma,S} \in C(\Sigma_{\sigma,S}, \mathbb{R})$.[113]

There is an obvious translation representation. For if $A \in F_{\mathcal{L}(S_1 \sqcup S_2)}(\Sigma, \mathcal{R})$, then since $\Sigma_{\sigma,S_1 \sqcup S_2} = \Sigma_{\sigma,S_1} \coprod \Sigma_{\sigma,S_1}$, the associated function $A_{\sigma,S_1 \sqcup S_2} : \Sigma_{\sigma,S_1 \sqcup S_2} \to \mathbb{R}$ is given by a pair of functions $A_1 \in C(\Sigma_{\sigma,S_1}, \mathbb{R})$ and $A_2 \in C(\Sigma_{\sigma,S_2}, \mathbb{R})$; we write $A_{\sigma,S_1 \sqcup S_2} = \langle A_1, A_2 \rangle$. It is natural to demand that the translation representation $\sigma(\mathcal{L}(i))(A_{\sigma,S_1 \sqcup S_2})$ is A_1. Note that what is essentially being discussed here is the classical-physics representation of the relation (13.358).

The canonical choice for $\sigma(i)$ is

$$\sigma(i) : \Sigma_{\sigma,S_1} \to \Sigma_{\sigma,S_1 \sqcup S_2} = \Sigma_{\sigma,S_1} \coprod \Sigma_{\sigma,S_2} \tag{13.393}$$

$$s_1 \mapsto s_1. \tag{13.394}$$

Then the pull-back along $\sigma(i)$,

$$\sigma(i)^* : C(\Sigma_{\sigma,S_1 \sqcup S_2}, \mathbb{R}) \to C(\Sigma_{\sigma,S_1}, \mathbb{R}) \tag{13.395}$$

$$A_{\sigma,S_1 \sqcup S_2} \mapsto A_{\sigma,S_1 \sqcup S_2} \circ \sigma(i), \tag{13.396}$$

maps (or "translates") the topos representative $A_{\sigma,S_1 \sqcup S_2} = \langle A_1, A_2 \rangle$ of the function symbol $A \in F_{\mathcal{L}(S_1 \sqcup S_2)}(\Sigma, \mathcal{R})$ to a real-valued function $A_{\sigma,S_1 \sqcup S_2} \circ \sigma(i)$ on Σ_{σ,S_1}. This function is clearly equal to A_1.

The Translation in the Case of a Composite Classical System

We now consider arrows in **Sys** of the form

$$S_1 \overset{p_1}{\leftarrow} S_1 \diamond S_2 \overset{p_2}{\rightarrow} S_2 \tag{13.397}$$

from the composite classical system $S_1 \diamond S_2$ to the constituent systems S_1 and S_2. Here, p_1 signals that S_1 is a constituent of the composite system $S_1 \diamond S_2$, likewise p_2. The systems S_1, S_2 and $S_1 \diamond S_2$ have symplectic manifolds Σ_{σ,S_1}, Σ_{σ,S_2} and $\Sigma_{\sigma,S_1 \diamond S_2} = \Sigma_{\sigma,S_1} \times \Sigma_{\sigma,S_2}$, respectively; i.e., the state space of the composite system $S_1 \diamond S_2$ is the cartesian product of the state spaces of the components. For typographical simplicity in what follows we denote $p := p_1$.

There is a canonical translation $\mathcal{L}(p)$ between the languages $\mathcal{L}(S_1)$ and $\mathcal{L}(S_1 \diamond S_2)$ whose representation is the following. Namely, if A is in $F_{\mathcal{L}(S_1)}(\Sigma, \mathcal{R})$, then the corresponding function $A_{\sigma,S_1} \in C(\Sigma_{\sigma,S_1}, \mathbb{R})$ is translated to a function $\sigma(\mathcal{L}(p))(A_{\sigma,S_1}) \in C(\Sigma_{\sigma,S_1 \diamond S_2}, \mathbb{R})$ such that

$$\sigma(\mathcal{L}(p))(A_{\sigma,S_1})(s_1, s_2) = A_{\sigma,S_1}(s_1) \tag{13.398}$$

for all $(s_1, s_2) \in \Sigma_{\sigma,S_1} \times \Sigma_{\sigma,S_2}$.

---

[113] Depending on the setting, one can assume that $F_{\mathcal{L}(S)}(\Sigma, \mathcal{R})$ contains function symbols corresponding bijectively to measurable, continuous or smooth functions.

This natural translation representation is based on the fact that, for the symplectic manifold $\Sigma_{\sigma,S_1\diamond S_2} = \Sigma_{\sigma,S_1} \times \Sigma_{\sigma,S_2}$, each point $s \in \Sigma_{\sigma,S_1\diamond S_2}$ can be identified with a pair, (s_1, s_2), of points $s_1 \in \Sigma_{\sigma,S_1}$ and $s_2 \in \Sigma_{\sigma,S_2}$. This is possible since the cartesian product $\Sigma_{\sigma,S_1} \times \Sigma_{\sigma,S_2}$ is a product in the categorial sense and hence has projections $\Sigma_{\sigma,S_1} \leftarrow \Sigma_{\sigma,S_1} \times \Sigma_{\sigma,S_2} \rightarrow \Sigma_{\sigma,S_2}$. Then the translation representation of functions is constructed in a straightforward manner. Thus, let

$$\begin{aligned} \sigma(p) : \Sigma_{\sigma,S_1} \times \Sigma_{\sigma,S_2} &\rightarrow \Sigma_{\sigma,S_1} \\ (s_1, s_2) &\mapsto s_1 \end{aligned} \tag{13.399}$$

be the canonical projection. Then, if $A_{\sigma,S_1} \in C(\Sigma_{\sigma,S_1}, \mathbb{R})$, the function

$$A_{\sigma,S_1} \circ \sigma(p) \in C(\Sigma_{\sigma,S_1} \times \Sigma_{\sigma,S_2}, \mathbb{R}) \tag{13.400}$$

is such that, for all $(s_1, s_2) \in \Sigma_{\sigma,S_1} \times \Sigma_{\sigma,S_2}$,

$$A_{\sigma,S_1} \circ \sigma(p)(s_1, s_2) = A_{\sigma,S_1}(s_1). \tag{13.401}$$

Thus we can define

$$\sigma(\mathcal{L}(p))(A_{\sigma,S_1}) := A_{\sigma,S_1} \circ \sigma(p). \tag{13.402}$$

Clearly, $\sigma(\mathcal{L}(p))(A_{\sigma,S_1})$ can be seen as the representation of the function symbol $A \diamond 1 \in F_{\mathcal{L}(S_1\diamond S_2)}(\Sigma, \mathcal{R})$.

13.12 Theories of Physics in a General Topos

13.12.1 The Pull-Back Operations

13.12.1.1 The Pull-Back of Physical Quantities

Motivated by the above, let us try now to see what can be said about the scheme in general. Basically, what is involved is the topos representation of translations of languages. To be more precise, let $j : S_1 \rightarrow S$ be an arrow in **Sys**, so that there is a translation $\mathcal{L}(j) : \mathcal{L}(S) \rightarrow \mathcal{L}(S_1)$ defined by the translation function $\mathcal{L}(j) : F_{\mathcal{L}(S)}(\Sigma, \mathcal{R}) \rightarrow F_{\mathcal{L}(S_1)}(\Sigma, \mathcal{R})$. Now suppose that the systems S and S_1 are represented in the topoi $\tau_\phi(S)$ and $\tau_\phi(S_1)$ respectively. Then, in these representations, the function symbols of signature $\Sigma \rightarrow \mathcal{R}$ in $\mathcal{L}(S)$ and $\mathcal{L}(S_1)$ are represented by elements of $\text{Hom}_{\tau_\phi(S)}(\Sigma_{\phi,S}, \mathcal{R}_{\phi,S})$ and $\text{Hom}_{\tau_\phi(S_1)}(\Sigma_{\phi,S_1}, \mathcal{R}_{\phi,S_1})$ respectively.

Our task is to find a function

$$\phi(\mathcal{L}(j)) : \text{Hom}_{\tau_\phi(S)}(\Sigma_{\phi,S}, \mathcal{R}_{\phi,S}) \rightarrow \text{Hom}_{\tau_\phi(S_1)}(\Sigma_{\phi,S_1}, \mathcal{R}_{\phi,S_1}) \tag{13.403}$$

that can be construed as the topos representation of the translation $\mathcal{L}(j) : \mathcal{L}(S) \to \mathcal{L}(S_1)$, and hence of the arrow $j : S_1 \to S$ in **Sys**. We are particularly interested in seeing if $\phi(\mathcal{L}(j))$ can be chosen so that the following diagram, (see (13.382)) commutes:

$$
\begin{array}{ccc}
F_{\mathcal{L}(S)}(\Sigma, \mathcal{R}) & \xrightarrow{\rho_{\phi,S}} & \mathrm{Hom}_{\tau_\phi(S)}\big(\Sigma_{\phi,S}, \mathcal{R}_{\phi,S}\big) \\
\downarrow{\scriptstyle \mathcal{L}(j)} & & \downarrow{\scriptstyle \phi(\mathcal{L}(j))} \\
F_{\mathcal{L}(S_1)}(\Sigma, \mathcal{R}) & \xrightarrow[\rho_{\phi,S_1}]{} & \mathrm{Hom}_{\tau_\phi(S_1)}\big(\Sigma_{\phi,S_1}, \mathcal{R}_{\phi,S_1}\big)
\end{array}
\tag{13.404}
$$

However, as has been emphasised already, it is not clear that one *should* expect to find a function $\phi(\mathcal{L}(j)) : \mathrm{Hom}_{\tau_\phi(S)}\big(\Sigma_{\phi,S}, \mathcal{R}_{\phi,S}\big) \to \mathrm{Hom}_{\tau_\phi(S_1)}\big(\Sigma_{\phi,S_1}, \mathcal{R}_{\phi,S_1}\big)$ with this property. The existence and/or properties of such a function will be dependent on the theory-type , and it seems unlikely that much can be said in general about the diagram (13.404). Nevertheless, let us see how far we *can* get in discussing the existence of such a function in general.

Thus, if $\mu \in \mathrm{Hom}_{\tau_\phi(S)}\big(\Sigma_{\phi,S}, \mathcal{R}_{\phi,S}\big)$, the critical question is if there is some 'natural' way whereby this arrow can be 'pulled-back' to give an element $\phi(\mathcal{L}(j))(\mu) \in \mathrm{Hom}_{\tau_\phi(S_1)}\big(\Sigma_{\phi,S_1}, \mathcal{R}_{\phi,S_1}\big)$.

The first pertinent remark is that μ is an arrow in the topos $\tau_\phi(S)$, whereas the sought-for pull-back will be an arrow in the topos $\tau_\phi(S_1)$, and so we need a mechanism for getting from one topos to the other (this problem, of course, does not arise in classical physics since the topos of every representation is always **Sets**).

The obvious way of implementing this change of topos is via some *functor*, $\tau_\phi(j)$ from $\tau_\phi(S)$ to $\tau_\phi(S_1)$. Indeed, given such a functor, an arrow $\mu : \Sigma_{\phi,S} \to \mathcal{R}_{\phi,S}$ in $\tau_\phi(S)$ is transformed to the arrow

$$
\tau_\phi(j)(\mu) : \tau_\phi(j)(\Sigma_{\phi,S}) \to \tau_\phi(j)(\mathcal{R}_{\phi,S}) \tag{13.405}
$$

in $\tau_\phi(S_1)$.

To convert this to an arrow from Σ_{ϕ,S_1} to $\mathcal{R}_{\phi,S_1}$, we need to supplement (13.405) with a pair of arrows $\phi(j), \beta_\phi(j)$ in $\tau_\phi(S_1)$ to get the diagram:

$$
\begin{array}{ccc}
\tau_\phi(j)(\Sigma_{\phi,S}) & \xrightarrow{\tau_\phi(j)(\mu)} & \tau_\phi(j)(\mathcal{R}_{\phi,S}) \\
\uparrow{\scriptstyle \phi(j)} & & \downarrow{\scriptstyle \beta_\phi(j)} \\
\Sigma_{\phi,S_1} & & \mathcal{R}_{\phi,S_1}
\end{array}
\tag{13.406}
$$

The pull-back, $\phi(\mathcal{L}(j))(\mu) \in \mathrm{Hom}_{\tau_\phi(S_1)}\big(\Sigma_{\phi,S_1}, \mathcal{R}_{\phi,S_1}\big)$, with respect to these choices can then be defined as

$$
\phi(\mathcal{L}(j))(\mu) := \beta_\phi(j) \circ \tau_\phi(j)(\mu) \circ \phi(j). \tag{13.407}
$$

It follows that a key part of the construction of a topos representation, ϕ, of **Sys** will be to specify the functor $\tau_\phi(j)$ from $\tau_\phi(S)$ to $\tau_\phi(S_1)$, and the arrows $\phi(j) : \Sigma_{\phi,S_1} \to \tau_\phi(j)(\Sigma_{\phi,S})$ and $\beta_\phi(j) : \tau_\phi(j)(\mathcal{R}_{\phi,S}) \to \mathcal{R}_{\phi,S_1}$ in the topos $\tau_\phi(S_1)$. These need to be defined in such a way as to be consistent with a chain of arrows $S_2 \to S_1 \to S$.

When applied to the representative $A_{\phi,S} : \Sigma_{\phi,S} \to \mathcal{R}_{\phi,S}$ of a physical quantity $A \in F_{\mathcal{L}(S)}(\Sigma, \mathcal{R})$, the diagram (13.406) becomes (augmented with the upper half)

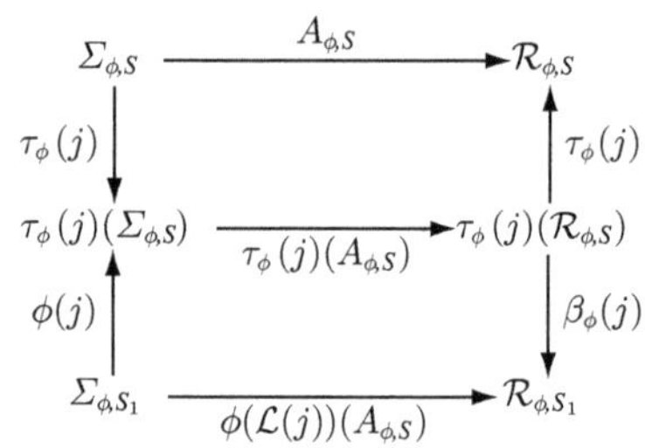

(13.408)

The commutativity of (13.404) would then require

$$\phi(\mathcal{L}(j))(A_{\phi,S}) = (\mathcal{L}(j)A)_{\phi,S_1} \tag{13.409}$$

or, in a more expanded notation,

$$\phi(\mathcal{L}(j)) \circ \rho_{\phi,S} = \rho_{\phi,S_1} \circ \mathcal{L}(j), \tag{13.410}$$

where both the left hand side and the right hand side of (13.410) are mappings from $F_{\mathcal{L}(S)}(\Sigma, \mathcal{R})$ to $\mathrm{Hom}_{\tau_\phi(S_1)}(\Sigma_{\phi,S_1}, \mathcal{R}_{\phi,S_1})$.

Note that the analogous diagram in classical physics is simply

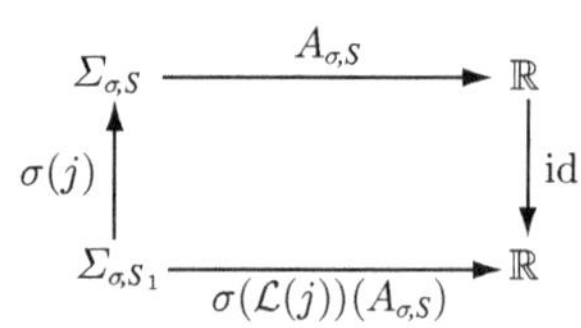

(13.411)

and the commutativity/pull-back condition (13.409) becomes

$$\sigma(\mathcal{L}(j))(A_{\sigma,S}) = (\mathcal{L}(j)A)_{\phi,S_1} \tag{13.412}$$

which is satisfied by virtue of (13.390).

It is clear from the above that the arrow $\phi(j) : \Sigma_{\phi,S_1} \to \tau_\phi(j)(\Sigma_{\phi,S})$ can be viewed as the topos analogue of the map $\sigma(j) : \Sigma_{\sigma,S_1} \to \Sigma_{\sigma,S}$ that arises in classical physics whenever there is an arrow $j : S_1 \to S$.

13.12.1.2 The Pull-Back of Propositions

More insight can be gained into the nature of the triple $\langle\tau_\phi(j), \phi(j), \beta_\phi(j)\rangle$ by considering the analogous operation for propositions. First, consider an arrow $j : S_1 \rightarrow S$ in **Sys** in classical physics. Associated with this there is (i) a translation $\mathcal{L}(j) : \mathcal{L}(S) \rightarrow \mathcal{L}(S_1)$; (ii) an associated translation mapping $\mathcal{L}(j) : F_{\mathcal{L}(S)}(\Sigma, \mathcal{R}) \rightarrow F_{\mathcal{L}(S_1)}(\Sigma, \mathcal{R})$; and (iii) a symplectic function $\sigma(j) : \Sigma_{\sigma,S_1} \rightarrow \Sigma_{\sigma,S}$.

Let K be a (Borel) subset of the state space, $\Sigma_{\sigma,S}$; hence K represents a proposition about the system S. Then $\sigma(j)^*(K) := \sigma(j)^{-1}(K)$ is a subset of Σ_{σ,S_1} and, as such, represents a proposition about the system S_1. We say that $\sigma(j)^*(K)$ is the *pull-back* to Σ_{σ,S_1} of the S-proposition represented by K. The existence of such pull-backs is part of the consistency of the representation of propositions in classical mechanics, and it is important to understand what the analogue of this is in our topos scheme.

Consider the general case with the two systems S_1, S as above. Then let K be a proposition, represented as a sub-object of $\Sigma_{\phi,S}$, with a monic arrow $i_K : K \hookrightarrow \Sigma_{\phi,S}$. The question now is if the triple $\langle\tau_\phi(j), \phi(j), \beta_\phi(j)\rangle$ can be used to pull K back to give a proposition in $\tau(S_1)$, i.e., a sub-object of Σ_{ϕ,S_1}?

The first requirement is that the functor $\tau_\phi(j) : \tau_\phi(S) \rightarrow \tau_\phi(S_1)$ should *preserve monics*. In this case, the monic arrow $i_K : K \hookrightarrow \Sigma_{\phi,S}$ in $\tau_\phi(S)$ is transformed to the monic arrow

$$\tau_\phi(j)(i_K) : \tau_\phi(j)(K) \hookrightarrow \tau_\phi(j)(\Sigma_{\phi,S}) \tag{13.413}$$

in $\tau_\phi(S_1)$; thus $\tau_\phi(j)(K)$ is a sub-object of $\tau_\phi(j)(\Sigma_{\phi,S})$ in $\tau_\phi(S_1)$. It is a property of a topos that the pull-back of a monic arrow is monic ; i.e., if $M \hookrightarrow Y$ is monic, and if $\psi : X \rightarrow Y$, then $\psi^{-1}(M)$ is a sub-object of X. Therefore, in the case of interest, the monic arrow $\tau_\phi(j)(i_K) : \tau_\phi(j)(K) \hookrightarrow \tau_\phi(j)(\Sigma_{\phi,S})$ can be pulled back along $\phi(j) : \Sigma_{\phi,S_1} \rightarrow \tau_\phi(j)(\Sigma_{\phi,S})$ (see diagram (13.408)) to give the monic $\phi(j)^{-1}(\tau_\phi(j)(K)) \subseteq \Sigma_{\phi,S_1}$. This is a candidate for the pull-back of the proposition represented by the sub-object $K \subseteq \Sigma_{\phi,S}$.

In conclusion, propositions can be pulled-back provided that the functor $\tau_\phi(j) : \tau_\phi(S) \rightarrow \tau_\phi(S_1)$ preserves monics. A sufficient way of satisfying this requirement is for $\tau_\phi(j)$ to be left-exact. However, this raises the question of "where do left-exact functors come from?".

13.12.1.3 The Idea of a Geometric Morphism

It transpires that there *is* a natural source of left-exact functors, via the idea of a *geometric morphism*. This fundamental concept in topos theory is defined as follows [66].

Definition 20 A *geometric morphism* $\phi : \mathcal{F} \rightarrow \mathcal{E}$ between topoi $\mathcal{F}$ and $\mathcal{E}$ is a pair of functors $\phi^* : \mathcal{E} \rightarrow \mathcal{F}$ and $\phi_* : \mathcal{F} \rightarrow \mathcal{E}$ such that

(i) $\phi^* \dashv \phi_*$, i.e., ϕ^* is left adjoint to ϕ_*;
(ii) ϕ^* is left exact, i.e., it preserves all finite limits.

The morphism $\phi^* : \mathcal{E} \to \mathcal{F}$ is called the inverse image part of the geometric morphism φ; $\phi_* : \mathcal{F} \to \mathcal{E}$ is called the direct image part.

Geometric morphisms are very important because they are the topos equivalent of continuous functions. More precisely, if X and Y are topological spaces, then any continuous function $f : X \to Y$ induces a geometric morphism between the topoi $\mathrm{Sh}(X)$ and $\mathrm{Sh}(Y)$ of sheaves on X and Y respectively. In practice, just as the arrows in the category of topological spaces are continuous functions, so in any category whose objects are topoi, the arrows are normally defined to be geometric morphisms. In our case, as we shall shortly see, all the examples of left-exact functors that arise in the quantum case do, in fact, come from geometric morphisms. For these reasons, from now on we will postulate that any arrows between our topoi arise from geometric morphisms.

One central property of a geometric morphism is that it preserves expressions written in terms of geometric logic. This greatly enhances the attractiveness of assuming from the outset that the internal logic of the system languages, $\mathcal{L}(S)$, is restricted to the sub-logic afforded by geometric logic.

En passant, another key result for us is the following theorem ([66] p359):

Theorem 14 *If $\varphi : \mathcal{C} \to \mathcal{D}$ is a functor between categories $\mathcal{C}$ and $\mathcal{D}$, then it induces a geometric morphism (also denoted φ)*

$$\varphi : \boldsymbol{Sets}^{\mathcal{C}^{op}} \to \boldsymbol{Sets}^{\mathcal{D}^{op}} \tag{13.414}$$

for which the functor $\varphi^* : \boldsymbol{Sets}^{\mathcal{D}^{op}} \to \boldsymbol{Sets}^{\mathcal{C}^{op}}$ takes a functor $\underline{F} : \mathcal{D} \to \boldsymbol{Sets}$ to the functor

$$\varphi^*(\underline{F}) := \underline{F} \circ \varphi^{op} \tag{13.415}$$

from $\mathcal{C}$ to ***Sets****.*

In addition, φ^* has a left adjoint $\varphi_!$; i.e., $\varphi_! \dashv \varphi^*$.

We will use this important theorem in several crucial places.

13.12.2 The Topos Rules for Theories of Physics

We will now present our general rules for using topos theory in the mathematical representation of physical systems and their theories.

Definition 21 The category $\mathcal{M}(\mathbf{Sys})$ is the following:

1. The objects of $\mathcal{M}(\mathbf{Sys})$ are the topoi that are to be used in representing the systems in **Sys**.

2. The arrows from τ_1 to τ_2 are defined to be the geometric morphisms from τ_2 to τ_1. Thus the inverse part, φ^*, of an arrow $\varphi^* : \tau_1 \to \tau_2$ is a left-exact functor from τ_1 to τ_2.

Definition 22 The rules for using topos theory are as follows:

1. A *topos realisation*, ϕ, of **Sys** in $\mathcal{M}(\mathbf{Sys})$ is an assignment, to each system S in **Sys**, of a triple $\phi(S) = \langle \rho_{\phi,S}, \mathcal{L}(S), \tau_\phi(S) \rangle$ where:
   (a) $\tau_\phi(S)$ is the topos in $\mathcal{M}(\mathbf{Sys})$ in which the physical theory of system S is to be realised.
   (b) $\mathcal{L}(S)$ is the local language that is associated with S. This is independent of the realisation, ϕ, of **Sys** in $\mathcal{M}(\mathbf{Sys})$.
   (c) $\rho_{\phi,S} : \mathcal{L}(S) \rightsquigarrow \tau_\phi(S)$ is a representation of the local language $\mathcal{L}(S)$ in the topos $\tau_\phi(S)$.
   (d) In addition, for each arrow $j : S_1 \to S$ in **Sys** there is a triple $\langle \tau_\phi(j), \phi(j), \beta_\phi(j) \rangle$ that interpolates between $\rho_{\phi,S} : \mathcal{L}(S) \rightsquigarrow \tau_\phi(S)$ and $\rho_{\phi,S_1} : \mathcal{L}(S_1) \rightsquigarrow \tau_\phi(S_1)$; for details see below.

2. (a) The representations, $\rho_{\phi,S}(\Sigma)$ and $\rho_{\phi,S}(\mathcal{R})$, of the ground symbols Σ and $\mathcal{R}$ in $\mathcal{L}(S)$ are denoted $\Sigma_{\phi,S}$ and $\mathcal{R}_{\phi,S}$, respectively. They are known as the "state object" and "quantity-value object" in $\tau_\phi(S)$.
   (b) The representation by $\rho_{\phi,S}$ of each function symbol $A : \Sigma \to \mathcal{R}$ of the system S is an arrow, $\rho_{\phi,S}(A) : \Sigma_{\phi,S} \to \mathcal{R}_{\phi,S}$ in $\tau_\phi(S)$; we will usually denote this arrow as $A_{\phi,S} : \Sigma_{\phi,S} \to \mathcal{R}_{\phi,S}$.
   (c) Propositions about the system S are represented by sub-objects of $\Sigma_{\phi,S}$. These will typically be of the form $A^{-1}_{\phi,S}(\Xi)$, where Ξ is a sub-object of $\mathcal{R}_{\phi,S}$.[114]

3. Generally, there are no "microstates" for the system S; i.e., no global elements (arrows $1 \to \Sigma_{\phi,S}$) of the state object $\Sigma_{\phi,S}$; or, if there are any, they may not be enough to determine $\Sigma_{\phi,S}$ as an object in $\tau_\phi(S)$.
   Instead, the role of a state is played by a "truth sub-object" $\mathbb{T}$ of $P\Sigma_{\phi,S}$.[115] If $J \in \mathrm{Sub}(\Sigma_{\phi,S}) \simeq \Gamma(P\Sigma_{\phi,S})$, the 'truth of the proposition represented by J' is defined to be

$$\nu\big(J \in \mathbb{T}\big) = [\![\tilde{J} \in \tilde{\mathbb{T}}]\!]_\phi \circ \langle \ulcorner J \urcorner, \ulcorner \mathbb{T} \urcorner \rangle \tag{13.416}$$

   See Sect. 13.6.2 for full information on the idea of a "truth object". Alternatively, one may use pseudo-states rather than truth objects, in which case the relevant truth values are of the form $\nu(\mathfrak{w} \subseteq J)$.

---

[114] Here, $A^{-1}_{\phi,S}(\Xi)$ denotes the sub-object of $\Sigma_{\phi,S}$ whose characteristic arrow is $\chi_\Xi \circ A_{\phi,S} : \Sigma_{\phi,S} \to \Omega_{\tau_\phi(S)}$, where $\chi_\Xi : \mathcal{R}_{\phi,S} \to \Omega_{\tau_\phi(S)}$ is the characteristic arrow of the sub-object Ξ.

[115] In classical physics, the truth object corresponding to a microstate s is the collection of all propositions that are true in the state s.

4. There is a "unit object" $1_{\mathcal{M}(\mathbf{Sys})}$ in $\mathcal{M}(\mathbf{Sys})$ such that if $1_{\mathbf{Sys}}$ denotes the trivial system in **Sys** then, for all topos realisations ϕ,

$$\tau_\phi(1_{\mathbf{Sys}}) = 1_{\mathcal{M}(\mathbf{Sys})}. \tag{13.417}$$

Motivated by the results for quantum theory (see Sect. 13.13.2), we postulate that the unit object $1_{\mathcal{M}(\mathbf{Sys})}$ in $\mathcal{M}(\mathbf{Sys})$ is the category of sets:

$$1_{\mathcal{M}(\mathbf{Sys})} = \mathbf{Sets}. \tag{13.418}$$

5. To each arrow $j : S_1 \to S$ in **Sys**, we have the following:

   (a) There is a translation $\mathcal{L}(j) : \mathcal{L}(S) \to \mathcal{L}(S_1)$. This is specified by a map between function symbols: $\mathcal{L}(j) : F_{\mathcal{L}(S)}(\Sigma, \mathcal{R}) \to F_{\mathcal{L}(S_1)}(\Sigma, \mathcal{R})$.

   (b) With the translation $\mathcal{L}(j) : F_{\mathcal{L}(S)}(\Sigma, \mathcal{R}) \to F_{\mathcal{L}(S_1)}(\Sigma, \mathcal{R})$ there is associated a corresponding function

$$\phi(\mathcal{L}(j)) : \mathrm{Hom}_{\tau_\phi(S)}(\Sigma_{\phi,S}, \mathcal{R}_{\phi,S}) \to \mathrm{Hom}_{\tau_\phi(S_1)}(\Sigma_{\phi,S_1}, \mathcal{R}_{\phi,S_1}). \tag{13.419}$$

   These may, or may not, fit together in the commutative diagram:

$$\begin{array}{ccc} F_{\mathcal{L}(S)}(\Sigma, \mathcal{R}) & \xrightarrow{\rho_{\phi,S}} & \mathrm{Hom}_{\tau_\phi(S)}(\Sigma_{\phi,\Sigma}, \mathcal{R}_{\phi,S}) \\ \Big\downarrow{\scriptstyle \mathcal{L}(j)} & & \Big\downarrow{\scriptstyle \phi(\mathcal{L}(j))} \\ F_{\mathcal{L}(S_1)}(\Sigma, \mathcal{R}) & \xrightarrow[\rho_{\phi,S_1}]{} & \mathrm{Hom}_{\tau_\phi(S_1)}(\Sigma_{\phi,S_1}, \mathcal{R}_{\phi,S_1}) \end{array} \tag{13.420}$$

   (c) The function $\phi(\mathcal{L}(j)) : \mathrm{Hom}_{\tau_\phi(S)}(\Sigma_{\phi,S}, \mathcal{R}_{\phi,S}) \to \mathrm{Hom}_{\tau_\phi(S_1)}(\Sigma_{\phi,S_1}, \mathcal{R}_{\phi,S_1})$ is built from the following ingredients. For each topos realisation ϕ, there is a triple $\langle \nu_\phi(j), \phi(j), \beta_\phi(j) \rangle$ where:

      (i) $\nu_\phi(j) : \tau_\phi(S_1) \to \tau_\phi(S)$ is a geometric morphism; i.e., an arrow in the category $\mathcal{M}(\mathbf{Sys})$ (thus $\nu_\phi(j)^* : \tau_\phi(S) \to \tau_\phi(S_1)$ is left exact).

      N.B. To simplify the notation a little we will denote $\nu_\phi(j)^*$ by $\tau_\phi(j)$. This is sensible in so far as, for the most part, only the inverse part of $\nu_\phi(j)$ will be used in our constructions.

      (ii) $\phi(j) : \Sigma_{\phi,S_1} \to \tau_\phi(j)(\Sigma_{\phi,S})$ is an arrow in the topos $\tau_\phi(S_1)$.

      (iii) $\beta_\phi(j) : \tau_\phi(j)(\mathcal{R}_{\phi,S}) \to \mathcal{R}_{\phi,S_1}$ is an arrow in the topos $\tau_\phi(S_1)$.

   These fit together in the diagram

$$\begin{array}{ccc}
\Sigma_{\phi,\Sigma} & \xrightarrow{\;A_{\phi,S}\;} & \mathcal{R}_{\phi,S} \\
\downarrow{\scriptstyle \tau_\phi(j)} & & \downarrow{\scriptstyle \tau_\phi(j)} \\
\tau_\phi(j)(\Sigma_{\phi,S}) & \xrightarrow[\tau_\phi(j)(A_{\phi,S})]{} & \tau_\phi(j)(\mathcal{R}_{\phi,S}) \\
\uparrow{\scriptstyle \phi(j)} & & \downarrow{\scriptstyle \beta_\phi(j)} \\
\Sigma_{\phi,S_1} & \xrightarrow[\phi(\mathcal{L}(j))(A_{\phi,S})]{} & \mathcal{R}_{\phi,S_1}
\end{array} \tag{13.421}$$

The arrows $\phi(j)$ and $\beta_\phi(j)$ should behave appropriately under composition of arrows in **Sys**.
The commutativity of the Diagram (13.420) is equivalent to the relation

$$\phi(\mathcal{L}(j))(A_{\phi,S}) = [\mathcal{L}(j)(A)]_{\phi,S_1} \tag{13.422}$$

for all $A \in F_{\mathcal{L}(\phi,S)}(\Sigma, \mathcal{R})$. As we keep emphasising, the satisfaction or otherwise of this relation will depend on the theory-type and, possibly, the representation ϕ.

(d) If a proposition in $\tau_\phi(S)$ is represented by the monic arrow, $K \hookrightarrow \Sigma_{\phi,S}$, the "pull-back" of this proposition to $\tau_\phi(S_1)$ is defined to be $\phi(j)^{-1}\big(\tau_\phi(j)(K)\big) \subseteq \Sigma_{\phi,S_1}$.

6.

(a) If S_1 is a sub-system of S, with an associated arrow $i : S_1 \to S$ in **Sys** then, in the diagram in (13.421), the arrow $\phi(j) : \Sigma_{\phi,S_1} \to \tau_\phi(j)(\Sigma_{\phi,S})$ is a monic arrow in $\tau_\phi(S_1)$.
In other words, Σ_{ϕ,S_1} is a sub-object of $\tau_\phi(j)(\Sigma_{\phi,S})$, which is denoted

$$\Sigma_{\phi,S_1} \subseteq \tau_\phi(j)(\Sigma_{\phi,S}). \tag{13.423}$$

We may also want to conjecture

$$\mathcal{R}_{\phi,S_1} \simeq \tau_\phi(j)\big(\mathcal{R}_{\phi,S}\big). \tag{13.424}$$

(b) Another possible conjecture is the following: if $j : S_1 \to S$ is an epic arrow in **Sys**, then, in the diagram in (13.421), the arrow $\phi(j) : \Sigma_{\phi,S_1} \to \tau_\phi(j)(\Sigma_{\phi,S})$ is an epic arrow in $\tau_\phi(S_1)$.
In particular, for the epic arrow $p_1 : S_1 \diamond S_2 \to S_1$, the arrow $\phi(p_1) : \Sigma_{\phi,S_1 \diamond S_2} \to \tau_\phi\big(\Sigma_{\phi,S_1}\big)$ is an epic arrow in the topos $\tau_\phi(S_1 \diamond S_2)$.

One should not read Rule 2. above as implying that the choice of the state object and quantity-value object are *unique* for any given system S. These objects would at best be selected only up to isomorphism in the topos $\tau(S)$. Such morphisms in the

$\tau(S)$[116] can be expected to play a key role in developing the topos analogue of the important idea of a *symmetry*, or *covariance* transformation of the theory.

In the example of classical physics, for all systems we have $\tau(S) = \mathbf{Sets}$ and $\Sigma_{\sigma,S}$ is a symplectic manifold, and the collection of all symplectic manifolds is a category. It would be elegant if we could assert that, in general, for a given theory-type the possible state objects in a given topos τ form the objects of an *internal* category in τ. However, to make such a statement would require a general theory of state objects and, at the moment, we do not have such a thing.

From a more conceptual viewpoint we note that the "similarity" of our axioms to those of standard classical physics is reflected in the fact that (i) physical quantities are represented by arrows $A_{\phi,S} : \Sigma_{\phi,S} \to \mathcal{R}_{\phi,S}$; (ii) propositions are represented by sub-objects of $\Sigma_{\phi,S}$; and (iii) propositions are assigned truth values. Thus any theory satisfying these axioms 'looks' like classical physics, and has an associated neo-realist interpretation.

13.13 The General Scheme Applied to Quantum Theory

13.13.1 Background Remarks

We now want to study the extent to which our "rules" apply to the topos representation of quantum theory.

For a quantum system with (separable) Hilbert space $\mathcal{H}$, the appropriate topos (what we earlier called $\tau_\phi(S)$) is $\mathbf{Sets}^{\mathcal{V}(\mathcal{H})^{\mathrm{op}}}$: the category of presheaves over the category (actually, partially-ordered set) $\mathcal{V}(\mathcal{H})$ of unital, abelian von Neumann subalgebras of the algebra, $B(\mathcal{H})$, of bounded operators on $\mathcal{H}$.

A particularly important object in $\mathbf{Sets}^{\mathcal{V}(\mathcal{H})^{\mathrm{op}}}$ is the *spectral presheaf* $\underline{\Sigma}$, where, for each V, $\underline{\Sigma}_V$ is defined to be the Gel'fand spectrum of the abelian algebra V. The sub-objects of $\underline{\Sigma}$ can be identified as the topos representations of propositions, just as the subsets of $\mathcal{S}$ represent propositions in classical physics.

In Sects. 13.8 and 13.9, several closely related choices for a quantity-value object $\mathcal{R}_\phi$ in $\mathbf{Sets}^{\mathcal{V}(\mathcal{H})^{\mathrm{op}}}$ were discussed. In order to keep the notation simpler, we concentrate here on the presheaf $\underline{\mathbb{R}^{\succeq}}$ of real-valued, order-reversing functions. All results hold analogously if the presheaf $\underline{\mathbb{R}^{\leftrightarrow}}$ (which we actually prefer for giving a better physical interpretation) is used.[117]

---

[116] Care is needed not to confuse morphisms in the topos $\tau(S)$ with morphisms in the category $\mathcal{M}(\mathbf{Sys})$ of topoi. An arrow from the object $\tau(S)$ to itself in the category $\mathcal{M}(\mathbf{Sys})$ is a geometric morphism in the topos $\tau(S)$. However, not every arrow in $\tau(S)$ need arise in this way, and an important role can be expected to be played by arrows of this second type. A good example is when $\tau(S)$ is the category of sets, $\mathbf{Sets}$. Typically, $\tau_\phi(j) : \mathbf{Sets} \to \mathbf{Sets}$ is the identity, but there are many morphisms from an object O in $\mathbf{Sets}$ to itself: they are just the functions from O to O.

[117] Since the construction of the arrows $\breve{\delta}(\hat{A}) : \underline{\Sigma} \to \underline{\mathbb{R}^{\leftrightarrow}}$ involves both inner and outer daseinisation, we would have double work with the notation, which we avoid here.

Hence, physical quantities $A : \Sigma \rightarrow \mathcal{R}$, which correspond to self-adjoint operators $\hat{A}$, are represented by natural transformations/arrows $\breve{\delta}^o(\hat{A}) : \underline{\Sigma} \rightarrow \underline{\mathbb{R}^{\succeq}}$. The mapping $\hat{A} \mapsto \breve{\delta}^o(\hat{A})$ is injective. For brevity, we write $\breve{\delta}(\hat{A}) := \breve{\delta}^o(\hat{A})$.[118]

13.13.2 The Translation Representation for a Disjoint Sum of Quantum Systems

Let **Sys** be a category whose objects are systems that can be treated using quantum theory. Let $\mathcal{L}(S)$ be the local language of a system S in **Sys** whose quantum Hilbert space is denoted $\mathcal{H}_S$. We assume that to each function symbol, $A : \Sigma \rightarrow \mathcal{R}$, in $\mathcal{L}(S)$ there is associated a self-adjoint operator $\hat{A} \in \mathcal{B}(\mathcal{H}_S)$,[119] and that the map

$$F_{\mathcal{L}(S)}(\Sigma, \mathcal{R}) \rightarrow B(\mathcal{H})_{\text{sa}} \tag{13.425}$$

$$A \mapsto \hat{A} \tag{13.426}$$

is injective (but not necessarily surjective, as we will see in the case of a disjoint sum of quantum systems).

We consider first arrows of the form

$$S_1 \stackrel{i_1}{\rightarrow} S_1 \sqcup S_2 \stackrel{i_2}{\leftarrow} S_2 \tag{13.427}$$

from the components S_1, S_2 to a disjoint sum $S_1 \sqcup S_2$; for convenience we write $i := i_1$. The systems S_1, S_2 and $S_1 \sqcup S_2$ have the Hilbert spaces $\mathcal{H}_1$, $\mathcal{H}_2$ and $\mathcal{H}_1 \oplus \mathcal{H}_2$, respectively.

As always, the translation $\mathcal{L}(i)$ goes in the opposite direction to the arrow i, so

$$\mathcal{L}(i) : F_{\mathcal{L}(S_1 \sqcup S_2)}(\Sigma, \mathcal{R}) \rightarrow F_{\mathcal{L}(S_1)}(\Sigma, \mathcal{R}). \tag{13.428}$$

Then our first step is find an "operator translation" from the relevant self-adjoint operators in $\mathcal{H}_1 \oplus \mathcal{H}_2$ to those in $\mathcal{H}_1$,

To do this, let A be a function symbol in $F_{\mathcal{L}(S_1 \sqcup S_2)}(\Sigma, \mathcal{R})$. In Sect. 13.11.2, we argued that $F_{\mathcal{L}(S_1 \sqcup S_2)}(\Sigma, \mathcal{R}) \simeq F_{\mathcal{L}(S_1)}(\Sigma, \mathcal{R}) \times F_{\mathcal{L}(S_2)}(\Sigma, \mathcal{R})$ (as in (13.358)), and hence we introduce the notation $A = \langle A_1, A_2 \rangle$, where $A_1 \in F_{\mathcal{L}(S_1)}(\Sigma, \mathcal{R})$ and $A_2 \in F_{\mathcal{L}(S_2)}(\Sigma, \mathcal{R})$. It is then natural to assume that the quantisation scheme is such that the operator, $\hat{A}$, on $\mathcal{H}_1 \oplus \mathcal{H}_2$ can be decomposed as $\hat{A} = \hat{A}_1 \oplus \hat{A}_2$, where the operators $\hat{A}_1$ and $\hat{A}_2$ are defined on $\mathcal{H}_1$ and $\mathcal{H}_2$ respectively, and correspond to the function symbols A_1 and A_2.[120] Then the obvious operator translation is $\hat{A} \mapsto \hat{A}_1 \in \mathcal{B}(\mathcal{H}_1)_{sa}$.

---

[118] Note that this is *not* the same as the convention used earlier, where $\breve{\delta}(\hat{A})$ denoted a different natural transformation!

[119] More specifically, one could postulate that the elements of $F_{\mathcal{L}(S)}(\Sigma, \mathcal{R})$ are associated with self-adjoint operators in some unital von Neumann sub-algebra of $\mathcal{B}(\mathcal{H}_S)$.

[120] It should be noted that our scheme does not use all the self-adjoint operators on the direct sum $\mathcal{H}_1 \oplus \mathcal{H}_2$: only the 'block diagonal' operators of the form $\hat{A} = \hat{A}_1 \oplus \hat{A}_2$ arise.

We now consider the general rules in the Definition 22 and see to what extent they apply in the example of quantum theory.

1. As we have stated several times, the topos $\tau_\phi(S)$ associated with a quantum system S is

$$\tau_\phi(S) = \mathbf{Sets}^{\mathcal{V}(\mathcal{H}_S)^{op}}. \tag{13.429}$$

Thus (i) the objects of the category $\mathcal{M}(\mathbf{Sys})$ are topoi of the form $\mathbf{Sets}^{\mathcal{V}(\mathcal{H}_S)^{op}}$, $S \in \mathrm{Ob}(\mathbf{Sys})$; and (ii) the arrows between two topoi are defined to be geometric morphisms. In particular, to each arrow $j : S_1 \to S$ in **Sys** there must correspond a geometric morphism $\nu_\phi(j) : \tau_\phi(S_1) \to \tau_\phi(S)$ with associated left-exact functor $\tau_\phi(j) := \nu_\phi(j)^* : \tau_\phi(S) \to \tau_\phi(S_1)$. Of course, the existence of these functors in the quantum case has yet to be shown.

2. The realisation $\rho_{\phi,S} : \mathcal{L}(S) \rightsquigarrow \tau_\phi(S)$ of the language $\mathcal{L}(S)$ in the topos $\tau_\phi(S)$ is given as follows. First, we define the state object $\Sigma_{\phi,S}$ to be the spectral presheaf, $\underline{\Sigma}^{\mathcal{V}(\mathcal{H}_S)}$, over $\mathcal{V}(\mathcal{H}_S)$, the context category of $\mathcal{B}(\mathcal{H}_S)$. To keep the notation brief, we will denote[121] $\underline{\Sigma}^{\mathcal{V}(\mathcal{H}_S)}$ as $\underline{\Sigma}^{\mathcal{H}_S}$.

Furthermore, we define the quantity-value object, $\mathcal{R}_{\phi,S}$, to be the presheaf $\underline{\mathbb{R}^{\succeq}}^{\mathcal{H}_S}$ that was defined in Sect. 13.8 . Finally, we define

$$A_{\phi,S} := \breve{\delta}(\hat{A}), \tag{13.430}$$

for all $A \in F_{\mathcal{L}(S)}(\Sigma, \mathcal{R})$. Here $\breve{\delta}(\hat{A}) : \underline{\Sigma}^{\mathcal{H}_S} \to \underline{\mathbb{R}^{\succeq}}^{\mathcal{H}_S}$ is constructed using the Gel'fand transforms of the (outer) daseinisation of $\hat{A}$, for details see later.

3. The truth object $\underline{\mathbb{T}}^{|\psi\rangle}$ corresponding to a pure state $|\psi\rangle$ was discussed in Sect. 13.6.3. Alternatively, we have the pseudo-state $\underline{\mathfrak{w}}^{|\psi\rangle}$.

4. Let $\mathcal{H} = \mathbb{C}$ be the one-dimensional Hilbert space, corresponding to the trivial quantum system 1. There is exactly one abelian sub-algebra of $\mathcal{B}(\mathbb{C}) \simeq \mathbb{C}$, namely $\mathbb{C}$ itself. This leads to

$$\tau_\phi(1_{\mathbf{Sys}}) = \mathbf{Sets}^{\{*\}} \simeq \mathbf{Sets} = 1_{\mathcal{M}(\mathbf{Sys})}. \tag{13.431}$$

5. Let $A \in F_{\mathcal{L}(S_1 \sqcup S_2)}(\Sigma, \mathcal{R})$ be a function symbol for the system $S_1 \sqcup S_2$. Then, as discussed above, A is of the form $A = \langle A_1, A_2\rangle$ (compare equation (13.358)), which corresponds to a self-adjoint operator $\hat{A}_1 \oplus \hat{A}_2 \in \mathcal{B}(\mathcal{H}_1 \oplus \mathcal{H}_2)_{\mathrm{sa}}$. The topos representation of A is the natural transformation $\breve{\delta}(\langle A_1, A_2\rangle) : \underline{\Sigma}^{\mathcal{H}_1 \oplus \mathcal{H}_2} \to \underline{\mathbb{R}^{\succeq}}^{\mathcal{H}_1 \oplus \mathcal{H}_2}$, which is defined at each stage $V \in \mathrm{Ob}(\mathcal{V}(\mathcal{H}_1 \oplus \mathcal{H}_2))$ as

$$\begin{aligned} \breve{\delta}(\langle A_1, A_2\rangle)_V : \underline{\Sigma}_V^{\mathcal{H}_1 \oplus \mathcal{H}_2} &\to \underline{\mathbb{R}^{\succeq}}_V^{\mathcal{H}_1 \oplus \mathcal{H}_2} \\ \lambda &\mapsto \{V' \mapsto \langle \lambda|_{V'}, \delta(\hat{A}_1 \oplus \hat{A}_2)_{V'}\rangle \mid V' \subseteq V\} \end{aligned} \tag{13.432}$$

where the right hand side (13.432) denotes an order-reversing function.

[121] Presheaves are always denoted by symbols that are underlined.

We will need the following:

Lemma 3 *Let* $\hat{A}_1 \oplus \hat{A}_2 \in \mathcal{B}(\mathcal{H}_1 \oplus \mathcal{H}_2)_{\text{sa}}$, *and let* $V = V_1 \oplus V_2 \in \text{Ob}(\mathcal{V}(\mathcal{H}_1 \oplus \mathcal{H}_2))$ *such that* $V_1 \in \text{Ob}(\mathcal{V}(\mathcal{H}_1))$ *and* $V_2 \in \text{Ob}(\mathcal{V}(\mathcal{H}_2))$. *Then*

$$\delta(\hat{A}_1 \oplus \hat{A}_2)_V = \delta(\hat{A}_1)_{V_1} \oplus \delta(\hat{A}_2)_{V_2}. \tag{13.433}$$

Proof Every projection $\hat{Q} \in V$ is of the form $\hat{Q} = \hat{Q}_1 \oplus \hat{Q}_2$ for unique projections $\hat{Q}_1 \in \mathcal{P}(\mathcal{H}_1)$ and $\hat{Q}_2 \in \mathcal{P}(\mathcal{H}_2)$. Let $\hat{P} \in \mathcal{P}(\mathcal{H})$ be of the form $\hat{P} = \hat{P}_1 \oplus \hat{P}_2$ such that $\hat{P}_1 \in \mathcal{P}(\mathcal{H}_1)$ and $\hat{P}_2 \in \mathcal{P}(\mathcal{H}_1)$. The largest projection in V smaller than or equal to $\hat{P}$, i.e., the inner daseinisation of $\hat{P}$ to V, is

$$\delta^i(\hat{P})_V = \hat{Q}_1 \oplus \hat{Q}_2, \tag{13.434}$$

where $\hat{Q}_1 \in \mathcal{P}(V_1)$ is the largest projection in V_1 smaller than or equal to $\hat{P}_1$, and $\hat{Q}_2 \in \mathcal{P}(V_2)$ is the largest projection in V_2 smaller than or equal to $\hat{P}_2$, so

$$\delta^i(\hat{P})_V = \delta(\hat{P}_1)_{V_1} \oplus \delta(\hat{P}_2)_{V_2}. \tag{13.435}$$

This implies $\delta(\hat{A} \oplus \hat{B})_V = \delta(\hat{A})_{V_1} \oplus \delta(\hat{B})_{V_2}$, since (outer) daseinisation of a self-adjoint operator just means inner daseinisation of the projections in its spectral family, and all the projections in the spectral family of $\hat{A} \oplus \hat{B}$ are of the form $\hat{P} = \hat{P}_1 \oplus \hat{P}_2$.

As discussed in Sect. 13.12, in order to mimic the construction that we have in the classical case, we need to pull back the arrow/natural transformation $\breve{\delta}(\langle A_1, A_2\rangle) : \underline{\Sigma}^{\mathcal{H}_1 \oplus \mathcal{H}_2} \to \underline{\mathbb{R}^{\succeq}}^{\mathcal{H}_1 \oplus \mathcal{H}_2}$ to obtain an arrow from $\underline{\Sigma}^{\mathcal{H}_1}$ to $\underline{\mathbb{R}^{\succeq}}^{\mathcal{H}_1}$. Since we decided that the translation on the level of operators sends $\hat{A}_1 \oplus \hat{A}_2$ to $\hat{A}_1$, we expect that this arrow from $\underline{\Sigma}^{\mathcal{H}_1}$ to $\underline{\mathbb{R}^{\succeq}}^{\mathcal{H}_1}$ is $\breve{\delta}(\hat{A}_1)$. We will now show how this works.

The presheaves $\underline{\Sigma}^{\mathcal{H}_1 \oplus \mathcal{H}_2}$ and $\underline{\Sigma}^{\mathcal{H}_1}$ lie in different topoi, and in order to "transform" between them we need we need a (left-exact) functor from the topos $\mathbf{Sets}^{\mathcal{V}(\mathcal{H}_1 \oplus \mathcal{H}_2)^{op}}$ to the topos $\mathbf{Sets}^{\mathcal{V}(\mathcal{H}_1)^{\text{op}}}$: this is the functor $\tau_\phi(j) : \tau_\phi(S) \to \tau_\phi(S_1)$ in (13.421). One natural place to look for such a functor is as the inverse-image part of a geometric morphism from $\mathbf{Sets}^{\mathcal{V}(\mathcal{H}_1)^{op}}$ to $\mathbf{Sets}^{\mathcal{V}(\mathcal{H}_1 \oplus \mathcal{H}_2)^{\text{op}}}$. According to Theorem 14, one source of such a geometric morphism, μ, is a functor

$$m : \mathcal{V}(\mathcal{H}_1) \to \mathcal{V}(\mathcal{H}_1 \oplus \mathcal{H}_2), \tag{13.436}$$

and the obvious choice for this is

$$m(V) := V \oplus \mathbb{C}\hat{1}_{\mathcal{H}_2} \tag{13.437}$$

for all $V \in \text{Ob}(\mathcal{V}(\mathcal{H}_1))$. This function from $\text{Ob}(\mathcal{V}(\mathcal{H}_1))$ to $\text{Ob}(\mathcal{V}(\mathcal{H}_1 \oplus \mathcal{H}_2))$ is clearly order preserving, and hence m is a genuine functor.

Let $\mu : \mathbf{Sets}^{\mathcal{V}(\mathcal{H}_1)^{\text{op}}} \to \mathbf{Sets}^{\mathcal{V}(\mathcal{H}_1 \oplus \mathcal{H}_2)^{\text{op}}}$ denote the geometric morphism induced by m. The inverse-image functor of μ is given by

$$\mu^* : \mathbf{Sets}^{\mathcal{V}(\mathcal{H}_1\oplus\mathcal{H}_2)^{\mathrm{op}}} \rightarrow \mathbf{Sets}^{\mathcal{V}(\mathcal{H}_1)^{\mathrm{op}}} \tag{13.438}$$

$$\underline{F} \mapsto \underline{F} \circ m^{\mathrm{op}}. \tag{13.439}$$

This means that, for all $V \in \mathrm{Ob}(\mathcal{V}(\mathcal{H}_1))$, we have

$$(\mu^* \underline{F}^{\mathcal{H}_1\oplus\mathcal{H}_2})_V = \underline{F}^{\mathcal{H}_1\oplus\mathcal{H}_2}_{m(V)} = \underline{F}^{\mathcal{H}_1\oplus\mathcal{H}_2}_{V\oplus\mathbb{C}\hat{1}_{\mathcal{H}_2}}. \tag{13.440}$$

For example, for the spectral presheaf we get

$$(\mu^* \underline{\Sigma}^{\mathcal{H}_1\oplus\mathcal{H}_2})_V = \underline{\Sigma}^{\mathcal{H}_1\oplus\mathcal{H}_2}_{m(V)} = \underline{\Sigma}^{\mathcal{H}_1\oplus\mathcal{H}_2}_{V\oplus\mathbb{C}\hat{1}_{\mathcal{H}_2}}. \tag{13.441}$$

This is the functor that is denoted $\tau_\phi(j) : \tau_\phi(S_1) \rightarrow \tau_\phi(S)$ in (13.421).

We next need to find an arrow $\phi(i) : \underline{\Sigma}^{\mathcal{H}_1} \rightarrow \mu^* \underline{\Sigma}^{\mathcal{H}_1\oplus\mathcal{H}_2}$ that is the analogue of the arrow $\phi(j) : \Sigma_{\phi,S_1} \rightarrow \tau_\phi(j)(3\Sigma_{\phi,S})$ in (13.421).

For each V, the set $(\mu^* \underline{\Sigma}^{\mathcal{H}_1\oplus\mathcal{H}_2})_V = \underline{\Sigma}^{\mathcal{H}_1\oplus\mathcal{H}_2}_{V\oplus\mathbb{C}\hat{1}_{\mathcal{H}_2}}$ contains two types of spectral elements λ: the first type are those λ such that $\langle\lambda, \hat{0}_{\mathcal{H}_1} \oplus \hat{1}_{\mathcal{H}_2}\rangle = 0$. Then, clearly, there is some $\tilde{\lambda} \in \underline{\Sigma}^{\mathcal{H}_1}_V$ such that $\langle\tilde{\lambda}, \hat{A}\rangle = \langle\lambda, \hat{A} \oplus \hat{0}_{\mathcal{H}_2}\rangle = \langle\lambda, \hat{A} \oplus \hat{1}_{\mathcal{H}_2}\rangle$ for all $\hat{A} \in V_{\mathrm{sa}}$. The second type of spectral elements $\lambda \in \underline{\Sigma}^{\mathcal{H}_1\oplus\mathcal{H}_2}_{V\oplus\mathbb{C}\hat{1}_{\mathcal{H}_2}}$ are such that $\langle\lambda, \hat{0}_{\mathcal{H}_1} \oplus \hat{1}_{\mathcal{H}_2}\rangle = 1$. In fact, there is exactly one such λ, and we denote it by λ_0. This shows that $\underline{\Sigma}^{\mathcal{H}_1\oplus\mathcal{H}_2}_{V\oplus\mathbb{C}\hat{1}_{\mathcal{H}_2}} \simeq \underline{\Sigma}^{\mathcal{H}_1}_V \cup \{\lambda_0\}$. Accordingly, at each stage V, the mapping $\phi(i)$ sends each $\tilde{\lambda} \in \underline{\Sigma}^{\mathcal{H}_1}_V$ to the corresponding $\lambda \in \underline{\Sigma}^{\mathcal{H}_1\oplus\mathcal{H}_2}_{V\oplus\mathbb{C}\hat{1}_{\mathcal{H}_2}}$.

The presheaf $\underline{\mathbb{R}^{\succeq}}^{\mathcal{H}_1\oplus\mathcal{H}_2}$ is given at each stage $W \in \mathrm{Ob}(\mathcal{V}(\mathcal{H}_1 \oplus \mathcal{H}_2))$ as the order-reversing functions $\nu : \downarrow\! W \rightarrow \mathbb{R}$, where $\downarrow\! W$ denotes the set of unital, abelian von Neumann sub-algebras of W. Let $W = V \oplus \mathbb{C}\hat{1}_{\mathcal{H}_2}$. Clearly, there is a bijection between the sets $\downarrow W \subset \mathrm{Ob}(\mathcal{V}(\mathcal{H}_1 \oplus \mathcal{H}_2))$ and $\downarrow V \subset \mathrm{Ob}(\mathcal{V}(\mathcal{H}))$. We can thus identify

$$(\mu^* \underline{\mathbb{R}^{\succeq}}^{\mathcal{H}_1\oplus\mathcal{H}_2})_V = \underline{\mathbb{R}^{\succeq}}^{\mathcal{H}_1\oplus\mathcal{H}_2}_{V\oplus\mathbb{C}\hat{1}_{\mathcal{H}_2}} \simeq \underline{\mathbb{R}^{\succeq}}^{\mathcal{H}_1}_V \tag{13.442}$$

for all $V \in \mathrm{Ob}(\mathcal{V}(\mathcal{H}))$. This gives an isomorphism $\beta_\phi(i) : \mu^* \underline{\mathbb{R}^{\succeq}}^{\mathcal{H}_1\oplus\mathcal{H}_2} \rightarrow \underline{\mathbb{R}^{\succeq}}^{\mathcal{H}_1}$, which corresponds to the arrow $\beta_\phi(j) : \tau_\phi(j)(\mathcal{R}_{\phi,S}) \rightarrow \mathcal{R}_{\phi,S_1}$ in (13.421).

Now consider the arrow $\breve{\delta}(\langle A_1, A_2\rangle) : \underline{\Sigma}^{\mathcal{H}_1\oplus\mathcal{H}_2} \rightarrow \underline{\mathbb{R}^{\succeq}}^{\mathcal{H}_1\oplus\mathcal{H}_2}$. This is the analogue of the arrow $A_{\phi,S} : \Sigma_{\phi,S} \rightarrow \mathcal{R}_{\phi,S}$ in (13.421). At each stage $W \in \mathrm{Ob}(\mathcal{V}(\mathcal{H}_1 \oplus \mathcal{H}_2))$, this arrow is given by the (outer) daseinisation $\delta(\hat{A}_1 \oplus \hat{A}_2)_{W'}$ for all $W' \in\downarrow\! W$. According to Lemma 3, we have

$$\delta(\hat{A}_1 \oplus \hat{A}_2)_{V\oplus\mathbb{C}\hat{1}_{\mathcal{H}_2}} = \delta(\hat{A}_1)_V \oplus \delta(A_2)_{\mathbb{C}\hat{1}_{\mathcal{H}_2}} = \delta(\hat{A}_1)_V \oplus \max(\mathrm{sp}(\hat{A}_2))\hat{1}_{\mathcal{H}_2} \tag{13.443}$$

for all $V \oplus \mathbb{C}\hat{1}_{\mathcal{H}_2} \in \mathrm{Ob}(\mathcal{V}(\mathcal{H}_1 \oplus \mathcal{H}_2))$. This makes clear how the arrow

$$\mu^*(\breve{\delta}(\langle A_1, A_2\rangle)) : \mu^*\underline{\Sigma}^{\mathcal{H}_1 \oplus \mathcal{H}_2} \to \mu^*\underline{\mathbb{R}^{\succeq}}^{\mathcal{H}_1 \oplus \mathcal{H}_2} \tag{13.444}$$

is defined. Our conjectured pull-back/translation representation is

$$\phi(\mathcal{L}(i))\big(\breve{\delta}(\langle A_1, A_2\rangle)\big) := \beta_\phi(i) \circ \mu^*(\breve{\delta}(\langle A_1, A_2\rangle)) \circ \phi(i) : \underline{\Sigma}^{\mathcal{H}_1} \to \underline{\mathbb{R}^{\succeq}}^{\mathcal{H}_1}. \tag{13.445}$$

Using the definitions of $\phi(i)$ and $\beta_\phi(i)$, it becomes clear that

$$\beta_\phi(i) \circ \mu^*(\breve{\delta}(\langle A_1, A_2\rangle)) \circ \phi(i) = \breve{\delta}(\hat{A}_1). \tag{13.446}$$

Hence, the commutativity condition in (13.422) is satisfied for arrows in **Sys** of the form $i_{1,2} : S_{1,2} \to S_1 \sqcup S_2$.

13.13.3 The Translation Representation for Composite Quantum Systems

We now consider arrows in **Sys** of the form

$$S_1 \overset{p_1}{\leftarrow} S_1 \diamond S_2 \overset{p_2}{\rightarrow} S_1, \tag{13.447}$$

where the quantum systems S_1, S_2 and $S_1 \diamond S_2$ have the Hilbert spaces $\mathcal{H}_1$, $\mathcal{H}_2$ and $\mathcal{H}_1 \otimes \mathcal{H}_2$, respectively.[122]

The canonical translation[123] $\mathcal{L}(p_1)$ between the languages $\mathcal{L}(S_1)$ and $\mathcal{L}(S_1 \diamond S_2)$ (see Sect. 13.11.2) is such that if A_1 is a function symbol in $F_{\mathcal{L}(S_1)}(\Sigma, \mathcal{R})$, then the corresponding operator $\hat{A}_1 \in \mathcal{B}(\mathcal{H}_1)_{\mathrm{sa}}$ will be 'translated' to the operator $\hat{A}_1 \otimes \hat{1}_{\mathcal{H}_2} \in \mathcal{B}(\mathcal{H}_1 \otimes \mathcal{H}_2)$. By assumption, this corresponds to the function symbol $A_1 \diamond 1$ in $F_{\mathcal{L}(S_1 \diamond S_2)}(\Sigma, \mathcal{R})$.

13.13.3.1 Operator Entanglement and Translations

We should be cautious about what to expect from this translation when we represent a physical quantity $A : \Sigma \to \mathcal{R}$ in $F_{\mathcal{L}(S_1)}(\Sigma, \mathcal{R})$ by an arrow between presheaves, since there are no canonical projections

$$\mathcal{H}_1 \leftarrow \mathcal{H}_1 \otimes \mathcal{H}_2 \to \mathcal{H}_2, \tag{13.448}$$

[122] As usual, the composite system $S_1 \diamond S_2$ has as its Hilbert space the tensor product of the Hilbert spaces of the components.

[123] As discussed in Sect. 13.11.2, this translation, $\mathcal{L}(p_1)$, transforms a physical quantity A_1 of system S_1 into a physical quantity $A_1 \diamond 1$, which is the "same" physical quantity but now seen as a part of the composite system $S_1 \diamond S_2$. The symbol 1 is the trivial physical quantity: it is represented by the operator $\hat{1}_{\mathcal{H}_2}$.

and hence no canonical projections

$$\underline{\Sigma}^{\mathcal{H}_1} \leftarrow \underline{\Sigma}^{\mathcal{H}_1 \otimes \mathcal{H}_2} \rightarrow \underline{\Sigma}^{\mathcal{H}_2} \tag{13.449}$$

from the spectral presheaf of the composite system to the spectral presheaves of the components.[124]

This is the point where a form of *entanglement* enters the picture. The spectral presheaf $\underline{\Sigma}^{\mathcal{H}_1 \otimes \mathcal{H}_2}$ is a presheaf over the context category $\mathcal{V}(\mathcal{H}_1 \otimes \mathcal{H}_2)$ of $\mathcal{H}_1 \otimes \mathcal{H}_2$. Clearly, the context category $\mathcal{V}(\mathcal{H}_1)$ can be embedded into $\mathcal{V}(\mathcal{H}_1 \otimes \mathcal{H}_2)$ by the mapping $V_1 \mapsto V_1 \otimes \mathbb{C}\hat{1}_{\mathcal{H}_2}$, and likewise $\mathcal{V}(\mathcal{H}_2)$ can be embedded into $\mathcal{V}(\mathcal{H}_1 \otimes \mathcal{H}_2)$. But not every $W \in \mathrm{Ob}(\mathcal{V}(\mathcal{H}_1 \otimes \mathcal{H}_2))$ is of the form $V_1 \otimes V_2$.

This comes from the fact that not all vectors in $\mathcal{H}_1 \otimes \mathcal{H}_2$ are of the form $\psi_1 \otimes \psi_2$, hence not all projections in $\mathcal{P}(\mathcal{H}_1 \otimes \mathcal{H}_2)$ are of the form $\hat{P}_{\psi_1} \otimes \hat{P}_{\psi_2}$, which in turn implies that not all $W \in \mathcal{V}(\mathcal{H}_1 \otimes \mathcal{H}_2)$ are of the form $V_1 \otimes V_2$. There are more contexts, or world-views, available in $\mathcal{V}(\mathcal{H}_1 \otimes \mathcal{H}_2)$ than those coming from $\mathcal{V}(\mathcal{H}_1)$ and $\mathcal{V}(\mathcal{H}_2)$. We call this "operator entanglement".

The topos representative of $\hat{A}_1$ is $\breve{\delta}(\hat{A}_1) : \underline{\Sigma}^{\mathcal{H}_1} \rightarrow \underline{\mathbb{R}^{\succeq}}^{\mathcal{H}_1}$, and the representative of $\hat{A}_1 \otimes \hat{1}_{\mathcal{H}_2}$ is $\breve{\delta}(A_1 \diamond 1) : \underline{\Sigma}^{\mathcal{H}_1 \otimes \mathcal{H}_2} \rightarrow \underline{\mathbb{R}^{\succeq}}^{\mathcal{H}_1 \otimes \mathcal{H}_2}$. At sub-algebras $W \in \mathrm{Ob}(\mathcal{V}(\mathcal{H}_1 \otimes \mathcal{H}_2))$ which are *not* of the form $W = V_1 \otimes V_2$ for any $V_1 \in \mathrm{Ob}(\mathcal{V}(\mathcal{H}_1))$ and $V_2 \in \mathrm{Ob}(\mathcal{V}(\mathcal{H}_2))$, the daseinised operator $\delta(\hat{A}_1 \otimes \hat{1}_{\mathcal{H}_2})_W \in W_{\mathrm{sa}}$ will not be of the form $\delta(\hat{A}_1)_V \otimes \hat{1}_{\mathcal{H}_2}$ for any V in $\mathrm{Ob}(\mathcal{V}(\mathcal{H}_1))$.[125] On the other hand, it is easy to see that $\delta(\hat{A}_1 \otimes \hat{1}_{\mathcal{H}_2})_W = \delta(\hat{A}_1)_{V_1} \otimes \hat{1}_{\mathcal{H}_2}$ if $W = V_1 \otimes \mathbb{C}\hat{1}_{\mathcal{H}_2}$.

Given a physical quantity A_1, represented by the arrow $\breve{\delta}(\hat{A}_1) : \underline{\Sigma}^{\mathcal{H}_1} \rightarrow \underline{\mathbb{R}^{\succeq}}^{\mathcal{H}_1}$, we can (at best) expect that the translation of this arrow into an arrow from $\underline{\Sigma}^{\mathcal{H}_1 \otimes \mathcal{H}_2}$ to $\underline{\mathbb{R}^{\succeq}}^{\mathcal{H}_1 \otimes \mathcal{H}_2}$ coincides with the arrow $\breve{\delta}(A_1 \hat{\diamond} 1)$ on the '*image*' of $\underline{\Sigma}^{\mathcal{H}_1}$ in $\underline{\Sigma}^{\mathcal{H}_1 \otimes \mathcal{H}_2}$. This image will be constructed below using a certain geometric morphism. As one might expect, the image of $\underline{\Sigma}^{\mathcal{H}_1}$ is a presheaf $\underline{P}$ on $\mathcal{V}(\mathcal{H}_1 \otimes \mathcal{H}_2)$ such that $\underline{P}_{V_1 \otimes \mathbb{C}\hat{1}_{\mathcal{H}_2}} \simeq \underline{\Sigma}^{\mathcal{H}_1}_{V_1}$ for all $V_1 \in \mathcal{V}(\mathcal{H}_1)$, i.e., the presheaf $\underline{P}$ can be identified with $\underline{\Sigma}^{\mathcal{H}_1}$ exactly on the image of $\mathcal{V}(\mathcal{H}_1)$ in $\mathcal{V}(\mathcal{H}_1 \otimes \mathcal{H}_2)$ under the embedding $V_1 \mapsto V_1 \otimes \mathbb{C}\hat{1}_{\mathcal{H}_2}$. At these stages, the translation of $\breve{\delta}(\hat{A}_1)$ will coincide with $\breve{\delta}(A_1 \hat{\diamond} 1)$. At other stages $W \in \mathcal{V}(\mathcal{H}_1 \otimes \mathcal{H}_2)$, the translation cannot be expected to be the same natural transformation as $\breve{\delta}(A \hat{\diamond} 1)$ in general.

13.13.3.2 A Geometrical Morphism and a Possible Translation

The most natural approach to a translation is the following. Let $W \in \mathrm{Ob}(\mathcal{V}(\mathcal{H}_1 \otimes \mathcal{H}_2))$, and define $V_W \in \mathrm{Ob}(\mathcal{V}(\mathcal{H}_1))$ to be the *largest* sub-algebra of $\mathcal{B}(\mathcal{H}_1)$ such that

---

[124] On the other hand, in the classical case, there *are* canonical projections

$$\Sigma_{\sigma,S_1} \leftarrow \Sigma_{\sigma,S_1 \diamond S_2} \rightarrow \Sigma_{\sigma,S_2} \tag{13.450}$$

because the symplectic manifold $\Sigma_{\sigma,S_1 \diamond S_2}$ that represents the composite system is the cartesian product $\Sigma_{\sigma,S_1 \diamond S_2} = \Sigma_{\sigma,S_1} \times \Sigma_{\sigma,S_2}$, which is a product in the categorial sense and hence comes with canonical projections.

[125] Currently, it is even an open question if $\delta(\hat{A}_1 \otimes \hat{1}_{\mathcal{H}_2})_W = \delta(\hat{A}_1)_{V_1} \otimes \hat{1}_{\mathcal{H}_2}$ if $W = V_1 \otimes V_2$ for a non-trivial algebra V_2.

$V_W \otimes \mathbb{C}\hat{1}_{\mathcal{H}_2}$ is a sub-algebra of W. Depending on W, V_W may, or may not, be the trivial sub-algebra $\mathbb{C}\hat{1}_{\mathcal{H}_1}$. We note that if $W' \subseteq W$, then

$$V_{W'} \subseteq V_W, \tag{13.451}$$

but $W' \subset W$ only implies $V_{W'} \subseteq V_W$.

The trivial algebra $\mathbb{C}\hat{1}_{\mathcal{H}_1}$ is not an object in the category $\mathcal{V}(\mathcal{H}_1)$. This is why we introduce the "augmented context category" $\mathcal{V}(\mathcal{H}_1)_*$, whose objects are those of $\mathcal{V}(\mathcal{H}_1)$ united with $\mathbb{C}\hat{1}_{\mathcal{H}_1}$, and with the obvious morphisms ($\mathbb{C}\hat{1}_{\mathcal{H}_1}$ is a sub-algebra of all $V \in \mathcal{V}(\mathcal{H}_1)$).

Then there is a functor $n : \mathcal{V}(\mathcal{H}_1 \otimes \mathcal{H}_2) \to \mathcal{V}(\mathcal{H}_1)_*$, defined as follows. On objects,

$$\begin{aligned} n : \mathrm{Ob}(\mathcal{V}(\mathcal{H}_1 \otimes \mathcal{H}_2)) &\to \mathrm{Ob}(\mathcal{V}(\mathcal{H}_1)_*) \\ W &\mapsto V_W, \end{aligned} \tag{13.452}$$

and if $i_{W'W} : W' \to W$ is an arrow in $\mathcal{V}(\mathcal{H}_1 \otimes \mathcal{H}_2)$, we define $n(i_{W'W}) := i_{V_{W'}V_W}$ (an arrow in $\mathcal{V}(\mathcal{H}_1)_*$); if $V_{W'} = V_W$, then $i_{V_{W'}V_W}$ is the identity arrow id_{V_W}.

Now let

$$\nu : \mathbf{Sets}^{\mathcal{V}(\mathcal{H}_1 \otimes \mathcal{H}_2)^{\mathrm{op}}} \to \mathbf{Sets}^{(\mathcal{V}(\mathcal{H}_1)_*)^{\mathrm{op}}} \tag{13.453}$$

denote the geometric morphism induced by π. Then the (left-exact) inverse-image functor

$$\nu^* : \mathbf{Sets}^{(\mathcal{V}(\mathcal{H}_1)_*)^{\mathrm{op}}} \to \mathbf{Sets}^{\mathcal{V}(\mathcal{H}_1 \otimes \mathcal{H}_2)^{\mathrm{op}}} \tag{13.454}$$

acts on a presheaf $\underline{F} \in \mathbf{Sets}^{(\mathcal{V}(\mathcal{H}_1)_*)^{\mathrm{op}}}$ in the following way. For all $W \in \mathrm{Ob}(\mathcal{V}(\mathcal{H}_1 \otimes \mathcal{H}_2))$, we have

$$(\nu^* \underline{F})_W = \underline{F}_{n^{\mathrm{op}}(W)} = \underline{F}_{V_W} \tag{13.455}$$

and

$$(\nu^* \underline{F})(i_{W'W}) = \underline{F}(i_{V_{W'}V_W}) \tag{13.456}$$

for all arrows $i_{W'W}$ in the category $\mathcal{V}(\mathcal{H}_1 \otimes \mathcal{H}_2)$.[126]

[126] We remark, although will not prove it here, that the inverse-image presheaf $\nu^* \underline{F}$ coincides with the direct image presheaf $\phi_* \underline{F}$ of $\underline{F}$ constructed from the geometric morphism ϕ induced by the functor

$$\begin{aligned} \kappa : \mathcal{V}(\mathcal{H}_1) &\to \mathcal{V}(\mathcal{H}_1 \otimes \mathcal{H}_2) \\ V &\mapsto V \otimes \mathbb{C}\hat{1}_{\mathcal{H}_2}. \end{aligned} \tag{13.457}$$

Of course, the inverse image presheaf $\beta^* \underline{F}$ is much easier to construct.

In particular, for all $W \in \mathcal{V}(\mathcal{H}_1 \otimes \mathcal{H}_2)$, we have

$$(\nu^* \underline{\Sigma}^{\mathcal{H}_1})_W = \underline{\Sigma}^{\mathcal{H}_1}_{V_W}, \tag{13.458}$$

$$(\nu^* \underline{\mathbb{R}^{\succeq}}^{\mathcal{H}_1})_W = \underline{\mathbb{R}^{\succeq}}^{\mathcal{H}_1}_{V_W}. \tag{13.459}$$

Since V_W can be $\mathbb{C}\hat{1}_{\mathcal{H}_1}$, we have to extend the definition of the spectral presheaf $\underline{\Sigma}^{\mathcal{H}_1}$ and the quantity-value presheaf $\underline{\mathbb{R}^{\succeq}}^{\mathcal{H}_1}$ such that they become presheaves over $\mathcal{V}(\mathcal{H}_1)_*$ (and not just $\mathcal{V}(\mathcal{H}_1)$). This can be done in a straightforward way: the Gel'fand spectrum $\underline{\Sigma}_{\mathbb{C}\hat{1}_{\mathcal{H}_1}}$ of $\mathbb{C}\hat{1}_{\mathcal{H}_1}$ consists of the single spectral element λ_1 such that $\langle \lambda_1, \hat{1}_{\mathcal{H}_1} \rangle = 1$. Moreover, $\mathbb{C}\hat{1}_{\mathcal{H}_1}$ has no sub-algebras, so the order-reversing functions on this algebra correspond bijectively to the real numbers $\mathbb{R}$.

Using these equations, we see that the arrow $\breve{\delta}(\hat{A}_1) : \underline{\Sigma}^{\mathcal{H}_1} \to \underline{\mathbb{R}^{\succeq}}^{\mathcal{H}_1}$ that corresponds to the self-adjoint operator $\hat{A}_1 \in \mathcal{B}(\mathcal{H}_1)_{\mathrm{sa}}$ gives rise to the arrow

$$\nu^*(\breve{\delta}(\hat{A}_1)) : \nu^* \underline{\Sigma}^{\mathcal{H}_1} \to \nu^* \underline{\mathbb{R}^{\succeq}}^{\mathcal{H}_1}. \tag{13.460}$$

In terms of our earlier notation, the functor $\tau_\phi(p_1) : \mathbf{Sets}^{\mathcal{V}(\mathcal{H}_1)^{\mathrm{op}}} \to \mathbf{Sets}^{\mathcal{V}(\mathcal{H}_1 \otimes \mathcal{H}_2)^{\mathrm{op}}}$ is ν^*, and the arrow in (13.460) is the arrow $\tau_\phi(j)(A_{\phi,S}) : \tau_\phi(j)(\Sigma_{\phi,S}) \to \tau_\phi(j)(\mathcal{R}_{\phi,S})$ in (13.421) with $j : S_1 \to S$ being replaced by $p : S_1 \diamond S_2 \to S_1$, which is the arrow in **Sys** whose translation representation we are trying to construct.

The next arrow we need is the one denoted $\beta_\phi(j) : \tau_\phi(j)(\mathcal{R}_{\phi,S}) \to \mathcal{R}_{\phi,S_1}$ in (13.421). In the present case, we define $\beta_\phi(p) : \nu^* \underline{\mathbb{R}^{\succeq}}^{\mathcal{H}_1} \to \underline{\mathbb{R}^{\succeq}}^{\mathcal{H}_1 \otimes \mathcal{H}_2}$ as follows. Let $\alpha \in (\nu^* \underline{\mathbb{R}^{\succeq}}^{\mathcal{H}_1})_W \simeq \underline{\mathbb{R}^{\succeq}}^{\mathcal{H}_1}_{V_W}$ be an order-reversing real-valued function on $\downarrow\! V_W$. Then we define an order-reversing function $\beta_\phi(p)(\alpha) \in \underline{\mathbb{R}^{\succeq}}^{\mathcal{H}_1 \otimes \mathcal{H}_2}_W$ as follows. For all $W' \subseteq W$, let

$$[\beta_\phi(p)(\alpha)](W') := \alpha(V_{W'}) \tag{13.461}$$

which, by virtue of (13.451), is an order-reversing function and hence a member of $\underline{\mathbb{R}^{\succeq}}^{\mathcal{H}_1 \otimes \mathcal{H}_2}_W$.

We also need an arrow in $\mathbf{Sets}^{\mathcal{V}(\mathcal{H}_1 \otimes \mathcal{H}_2)^{\mathrm{op}}}$ from $\underline{\Sigma}^{\mathcal{H}_1 \otimes \mathcal{H}_2}$ to $\nu^* \underline{\Sigma}^{\mathcal{H}_1}$, where $\nu^* \underline{\Sigma}^{\mathcal{H}_1}$ is defined in (13.458). This is the arrow denoted $\phi(j) : \Sigma_{\phi,S_1} \to \tau_\phi(j)(\Sigma_{\phi,S})$ in (13.421).

The obvious choice is to restrict $\lambda \in \underline{\Sigma}^{\mathcal{H}_1 \otimes \mathcal{H}_2}_W$ to the sub-algebra $V_W \otimes \mathbb{C}\hat{1}_{\mathcal{H}_2} \subseteq W$, and to identify $V_W \otimes \mathbb{C}\hat{1}_{\mathcal{H}_1} \simeq V_W \otimes \hat{1}_{\mathcal{H}_1} \simeq V_W$ as von Neumann algebras, which gives $\underline{\Sigma}^{\mathcal{H}_1 \otimes \mathcal{H}_2}_{V_W \otimes \mathbb{C}\hat{1}_{\mathcal{H}_2}} \simeq \underline{\Sigma}^{\mathcal{H}_1}_{V_W}$. Let

$$\begin{aligned} \phi(p)_W : \underline{\Sigma}^{\mathcal{H}_1 \otimes \mathcal{H}_2}_W &\to \underline{\Sigma}^{\mathcal{H}_1}_{V_W} \\ \lambda &\mapsto \lambda|_{V_W} \end{aligned} \tag{13.462}$$

denote this arrow at stage W. Then

$$\beta_\phi(p) \circ \nu^*(\breve{\delta}(\hat{A}_1)) \circ \phi(p) : \underline{\Sigma}^{\mathcal{H}_1 \otimes \mathcal{H}_2} \to \underline{\mathbb{R}^{\succeq}}^{\mathcal{H}_1 \otimes \mathcal{H}_2} \tag{13.463}$$

is a natural transformation which is defined for all $W \in \mathrm{Ob}(\mathcal{V}(\mathcal{H}_1 \otimes \mathcal{H}_2))$ and all $\lambda \in W$ by

$$\Big(\beta_\phi(p) \circ \nu^*(\breve{\delta}(\hat{A}_1)) \circ \phi(p)\Big)_W (\lambda) = \nu^*(\breve{\delta}(\hat{A}))(\lambda|_{V_W}) \tag{13.464}$$

$$= \{V' \mapsto \langle \lambda|_{V'}, \delta(\hat{A})_{V'}\rangle \mid V' \subseteq V_W\} \tag{13.465}$$

This is clearly an order-reversing real-valued function on the set $\downarrow W$ of sub-algebras of W, i.e., it is an element of $\underline{\mathbb{R}^{\succeq}}_W^{\mathcal{H}_1 \otimes \mathcal{H}_2}$. We define $\beta_\phi(p) \circ \nu^*(\breve{\delta}(\hat{A}_1)) \circ \phi(p)$ to be the translation representation, $\phi(\mathcal{L}(p))(\breve{\delta}(\hat{A}_1))$ of $\breve{\delta}(\hat{A}_1)$ for the composite system.

Note that, by construction, for each W, the arrow $(\beta_\phi(p) \circ \nu^*(\breve{\delta}(\hat{A}_1)) \circ \phi(p))_W$ corresponds to the self-adjoint operator $\delta(\hat{A}_1)_{V_W} \otimes \hat{1}_{\mathcal{H}_2} \in W_{\mathrm{sa}}$, since

$$\langle \lambda|_{V_W}, \delta(\hat{A}_1)_{V_W}\rangle = \langle \lambda, \delta(\hat{A}_1)_{V_W} \otimes \hat{1}_{\mathcal{H}_2}\rangle \tag{13.466}$$

for all $\lambda \in \underline{\Sigma}_W^{\mathcal{H}_1 \otimes \mathcal{H}_2}$.

Comments on These Results

This is about as far as we can get with the arrows associated with the composite of two quantum systems. The results above can be summarised in the equation

$$\phi(\mathcal{L}(p))(\breve{\delta}(\hat{A}_1))_W = \breve{\delta}(A_1)_{V_W} \otimes \hat{1}_{\mathcal{H}_2} \tag{13.467}$$

for all contexts $W \in \mathrm{Ob}(\mathcal{V}(\mathcal{H}_1 \otimes \mathcal{H}_2))$. If $W \in \mathrm{Ob}(\mathcal{V}(\mathcal{H}_1 \otimes \mathcal{H}_2))$ is of the form $W = V_1 \otimes \mathbb{C}\hat{1}_{\mathcal{H}_2}$, i.e., if W is in the image of the embedding of $\mathcal{V}(\mathcal{H}_1)$ into $\mathcal{V}(\mathcal{H}_1 \otimes \mathcal{H}_2)$, then $V_W = V_1$ and the translation formula gives just what one expects: the arrow $\breve{\delta}(\hat{A}_1)$ is translated into the arrow $\breve{\delta}(A_1 \hat{\diamond} 1)$ at these stages, since $\delta(\hat{A}_1 \otimes \hat{1}_{\mathcal{H}_2})_{V_1 \otimes \mathbb{C}\hat{1}_{\mathcal{H}_2}} = \delta(\hat{A}_1)_{V_1} \otimes \hat{1}_{\mathcal{H}_2}$.[127]

If $W \in \mathrm{Ob}(\mathcal{V}(\mathcal{H}_1 \otimes \mathcal{H}_2))$ is not of the form $W = V_1 \otimes \mathbb{C}\hat{1}_{\mathcal{H}_2}$, then it is relatively easy to show that

$$\delta(\hat{A}_1 \otimes \hat{1}_{\mathcal{H}_2})_W \neq \delta(\hat{A}_1)_{V_W} \otimes \hat{1}_{\mathcal{H}_2} \tag{13.468}$$

[127] To be precise, both the translation $\phi(\mathcal{L}(p))(\breve{\delta}(\hat{A}_1))_W$, given by (13.467), and $\breve{\delta}(A \hat{\diamond} 1)_W$ are mappings from $\underline{\Sigma}_W^{\mathcal{H}_1 \otimes \mathcal{H}_2}$ to $\underline{\mathbb{R}^{\succeq}}_W^{\mathcal{H}_1 \otimes \mathcal{H}_2}$. Each $\lambda \in \underline{\Sigma}_W^{\mathcal{H}_1 \otimes \mathcal{H}_2}$ is mapped to an order-reversing function on $\downarrow W$. The mappings $\phi(\mathcal{L}(p))(\breve{\delta}(\hat{A}_1))_W$ and $\breve{\delta}(A \hat{\diamond} 1)_W$ coincide at all $W' \in \downarrow W$ that are of the form $W' = V' \otimes \mathbb{C}\hat{1}_{\mathcal{H}_2}$.

in general. Hence

$$\phi(\mathcal{L}(p))(\breve{\delta}(\hat{A}_1)) \neq \breve{\delta}(A_1 \diamond 1), \tag{13.469}$$

whereas, intuitively, one might have expected equality. Thus the "commutativity" condition (13.409) is not satisfied.

In fact, there appears to be *no* operator $\hat{B} \in \mathcal{B}(\mathcal{H}_1 \otimes \mathcal{H}_2)$ such that $\phi(\mathcal{L}(p))(\breve{\delta}(\hat{A}_1)) = \breve{\delta}(\hat{B})$. Thus the quantity, $\beta_\phi(p) \circ \nu^*(\breve{\delta}(\hat{A}_1)) \circ \phi(p)$, that is our conjectured pull-back, is an arrow in $\mathrm{Hom}_{\mathbf{Sets}^{\mathcal{V}(\mathcal{H}_1 \otimes \mathcal{H}_2)^{\mathrm{op}}}}\big(\underline{\Sigma}^{\mathcal{H}_1 \otimes \mathcal{H}_2}, \underline{\mathbb{R}^{\succeq}}^{\mathcal{H}_1 \otimes \mathcal{H}_2}\big)$ that is not of the form $A_{\phi, S_1 \diamond S_2}$ for any physical quantity $A \in F_{\mathcal{L}(S_1 \diamond S_2)}\big(\Sigma, \mathcal{R}\big)$.

Our current understanding is that this translation is 'as good as possible': the arrow $\breve{\delta}(\hat{A}_1) : \underline{\Sigma}^{\mathcal{H}_1} \to \underline{\mathbb{R}^{\succeq}}^{\mathcal{H}_1}$ is translated into an arrow from $\underline{\Sigma}^{\mathcal{H}_1 \otimes \mathcal{H}_2}$ to $\underline{\mathbb{R}^{\succeq}}^{\mathcal{H}_1 \otimes \mathcal{H}_2}$ that coincides with $\breve{\delta}(\hat{A}_1)$ on those part of $\underline{\Sigma}^{\mathcal{H}_1 \otimes \mathcal{H}_2}$ that can be identified with $\underline{\Sigma}^{\mathcal{H}_1}$. But $\underline{\Sigma}^{\mathcal{H}_1 \otimes \mathcal{H}_2}$ is much larger, and it is not simply a product of $\underline{\Sigma}^{\mathcal{H}_1}$ and $\underline{\Sigma}^{\mathcal{H}_2}$. The context category $\mathcal{V}(\mathcal{H}_1 \otimes \mathcal{H}_2)$ underlying $\underline{\Sigma}^{\mathcal{H}_1 \otimes \mathcal{H}_2}$ is much richer than a simple product of $\mathcal{V}(\mathcal{H}_1)$ and $\mathcal{V}(\mathcal{H}_2)$. This is due to a kind of operator entanglement. A translation can at best give a faithful picture of an arrow, but it cannot possibly "know" about the more complicated contextual structure of the larger category.

Clearly, both technical and interpretational work remain to be done.

13.14 Characteristic Properties of Σ_ϕ, $\mathcal{R}_\phi$ and $\mathbb{T}$, $\mathfrak{w}$

13.14.1 The State Object Σ_ϕ

A major motivation for our work is the desire to find mathematical structures with whose aid genuinely new types of theory can be constructed. Consequently, however fascinating the "toposification" of quantum theory may be, this particular theory should not be allowed to divert us too much from the main goal. However, it is also important to see what general lessons can be learnt from what has been done so far. This is likely to be crucial in the construction of new theories.

In developing the topos version of quantum theory we have constructed concrete objects in the topos to function as the state object and quantity-value object. We have also seen how each quantum vector state gives a precise truth object, or "pseudo-state".

The challenging question now is what, if anything, can be said in general about these key ingredients in our scheme. Thus, ideally, we would be able to specify characteristic properties for Σ_ϕ, $\mathcal{R}_\phi$, and the truth objects/pseudo-states. A related problem is to understand if there is an object, $\mathbb{W}_\phi$, of all truth objects/pseudo-states, and, if so, what are its defining properties. Any such characteristic properties could be coded into the structure of the language, $\mathcal{L}(S)$, of the system, hence ensuing that they are present in all topos representations of S. In particular, should a symbol $\mathbb{W}$ be added to $\mathcal{L}(S)$ as the linguistic precursor of an object of pseudo-states?

So far, we know only two explicit examples of physically-relevant topos representations of a system language, $\mathcal{L}(S)$: (i) the representation of classical physics in **Sets**; and (ii) the representation of quantum physics in topoi of the form $\mathbf{Sets}^{\mathcal{V}(\mathcal{H})^{\mathrm{op}}}$. This does provide much guidance when it comes to speculating on characteristic properties of the key objects Σ_ϕ and $\mathcal{R}_\phi$. From this perspective, it would be helpful if there is an alternative way of finding the quantum objects $\underline{\Sigma}$ and $\underline{\mathbb{R}^{\leftrightarrow}}$ in addition to the one provided by the approach that we have adopted. Fortunately, this has been done recently by Heunen et al. [42]; as we shall see in Sect. 13.14.1, this does provide more insight into a possible generic structure for Σ_ϕ.

13.14.1.1 An Analogue of a Symplectic Structure or Cotangent Bundle?

Let us start with the state object Σ_ϕ. In classical physics, this is a symplectic manifold; in quantum theory it is the spectral presheaf $\underline{\Sigma}$ in the topos $\mathbf{Sets}^{\mathcal{V}(\mathcal{H})^{\mathrm{op}}}$. Does this suggest any properties for Σ_ϕ in general?

One possibility is that the state object, Σ_ϕ, has some sort of "symplectic structure". If taken literally, this phrase suggests synthetic differential geometry (SDG): a theory that is based on the existence in certain topoi (not **Sets**) of genuine "infinitesimals". However, this seems unlikely for the quantum topoi $\mathbf{Sets}^{\mathcal{V}(\mathcal{H})^{\mathrm{op}}}$ and we would probably need to extend these topoi considerably in order to incorporate SDG. Thus when we say "... some sort of symplectic structure", the phrase "some sort" has to be construed rather broadly.

We suspect that, with this caveat, the state object $\underline{\Sigma}$ may have such a structure, particularly for those quantum systems that come from quantising a given classical system. However, at the moment this is still a conjecture. We are currently studying systems whose classical state space is the cotangent bundle, T^*Q, of a configuration space Q. We think that the quantum analogue of this space is a certain presheaf, M_Q, that is associated with the *maximal* commutative sub-algebra, $M_Q \in \mathrm{Ob}(\mathcal{V}(\overline{\mathcal{H}}))$, generated by the smooth, real-valued functions on Q. This is currently work in progress.

But even if the quantum state object *does* have a remnant "symplectic structure", it is debatable if this should be axiomatised in general. Symplectic structures arise in classical physics because the underlying equations of motion are *second-order* in the "configuration" variables q, and hence first-order in the pair (q, p), where p are the "momentum variables".

However if, say, Newton's equations of gravity had been third-order in q, this would lead to triples (q, p, a) (a are 'acceleration' variables) and symplectic structure would not be appropriate.

13.14.1.2 Σ_ϕ as a Spectral Object: The Work of Heunen et al.

Another way of understanding the state object Σ_ϕ is suggested by the recent work of Heunen et al. [42]. They start with a non-commutative C^*-algebra, $\mathcal{A}$, of observables in some “ambient topos”, $\mathfrak{S}$—in our case, this is **Sets**—and then proceed with the following steps:

1. They construct the poset category[128] $\mathcal{V}(\mathcal{A})$ of commutative C^*-subalgebras of $\mathcal{A}$ that contain the identity operator $\hat{1}$.
2. One then constructs the topos, $\mathfrak{S}^{\mathcal{V}(\mathcal{A})}$ of covariant functors (i.e., *co*-presheaves) on the category/poset $\mathcal{V}(\mathcal{A})$.[129]
3. As a special object in the topos $\mathfrak{S}^{\mathcal{V}(\mathcal{A})}$, one constructs the “tautological” co-presheaf, i.e., covariant functor $\overline{\mathcal{A}}$ in which $\overline{\mathcal{A}}(V) := V$ for each commutative C^*-subalgebra, V, of $\mathcal{A}$. Then if $i_{V_1V_2} : V_1 \subseteq V_2$, the associated arrow $\overline{\mathcal{A}}(i_{V_1V_2}) : \overline{\mathcal{A}}(V_1) \to \overline{\mathcal{A}}(V_2)$ is just the inclusion map of $\overline{\mathcal{A}}(V_1)$ in $\overline{\mathcal{A}}(V_2)$.
4. Heunen et al. then show that $\overline{\mathcal{A}}$ is an internal *commutative* C^*-algebra in the topos $\mathfrak{S}^{\mathcal{V}(\mathcal{A})}$.
5. Using the recently published, very important, results of Banacheswski and Mulvey on constructive Gel'fand duality,[130] Heunen et al. show that the spectrum, $\overline{\Sigma}$, of the commutative algebra $\overline{\mathcal{A}}$ can be computed internally, and that it has the structure of an internal *locale* in $\mathfrak{S}^{\mathcal{V}(\mathcal{A})}$.

Thus Heunen et al. differ from us in that (i) they work in a general ambient topos $\mathfrak{S}$, whereas we use **Sets**; (ii) they use C^*-algebras rather than von Neumann algebras[131]; and (iii) they use covariant rather than contravariant functors.

It can be shown that their internal spectrum, the locale $\overline{\Sigma}$, is closely related to our spectral presheaf $\underline{\Sigma}$, which is a presheaf of topological spaces: namely, the Gel'fand spectra of the commutative subalgebras of the non-commutative algebra of physical quantities. We just want to mention briefly the natural way of relating the contravariant and the covariant situations.

The spectral presheaf can also be defined for C^*-algebras, and we assume for the moment that this has been done. For the spectral presheaf $\underline{\Sigma}$, we have the (continuous) restriction mappings $\underline{\Sigma}(i_{V'V}) : \underline{\Sigma}_V \to \underline{\Sigma}_{V'}$ (where $V' \subset V$) and these induce inverse-image mappings $\underline{\Sigma}(i_{V'V})^{-1} : \mathcal{O}\underline{\Sigma}_{V'} \to \mathcal{O}\underline{\Sigma}_V$ between the frames $\mathcal{O}\underline{\Sigma}_{V'}$, and $\mathcal{O}\underline{\Sigma}_V$ of open subsets (i.e.,the topologies) of $\underline{\Sigma}_{V'}$ and $\underline{\Sigma}_V$, respectively. The mapping $\underline{\Sigma}(i_{V'V})$ preserves finite meets and arbitrary joins and hence is a (local) frame morphism. Clearly, the frames $\mathcal{O}\underline{\Sigma}_V$, $V \in \mathcal{V}(\mathcal{H})$, together with the frame morphisms $\underline{\Sigma}(i_{V'V})^{-1}$, form a covariant functor $\overline{F}$. The sub-objects of this functor

[128] This notation has been chosen to suggest more clearly the analogues with our topos constructions that use the base category $\mathcal{V}(\mathcal{H})$. It is not that used by Heunen et al.

[129] They affirm that the operation $\mathcal{A} \mapsto \mathfrak{S}^{\mathcal{V}(\mathcal{A})}$ defines a functor from the category of C^*-algebras in $\mathfrak{S}$ to the category of elementary topoi and geometric morphisms.

[130] See [7–10]; the results partially go back to the early 1980s.

[131] One problem with C^*-algebras is that they very often do not contain enough projectors; and, of course, these are the entities that represent propositions. One might use AW^*-algebras, which are abstract von Neumann algebras, as mentioned in the first version of [42].

are the elements of an internal frame in $\mathfrak{S}^{\mathcal{V}(\mathcal{A})}$, and this internal frame is the internal Gel'fand spectrum.[132]

The concrete form of the opens in $\overline{\Sigma}$ as sub-objects of the functor $\overline{F}$ gives the external description of the Gel'fand spectrum, $\overline{\Sigma}$, of the internal commutative algebra as used by Heunen et al. This and further relations between the covariant and the contravariant approach to algebraic quantum theory in a topos are treated in detail in a forthcoming paper [26].

The fact that Heunen et al. arrive at essentially the same object as our spectral presheaf is striking. Amongst other things, it suggests a possible axiomatisation of the state object, Σ_ϕ. Namely, we could require that in *any* topos representation, ϕ, the state object is (i) the spectrum of some internal, commutative (pre-) C^*-algebra (or, perhaps, AW^*-algebra); and (ii) the spectrum has the structure of an internal locale in the topos τ_ϕ.

It is not currently clear whether or not it makes physical sense to always require Σ_ϕ to be the spectrum of an internal algebra. However, even in the contrary case it still meaningful to explore the possibility that Σ_ϕ has the 'topological' property of being an internal locale. This opens up many possibilities, including that of constructing the (internal) topos, $\mathrm{Sh}(\Sigma_\phi)$, of sheaves over Σ_ϕ.

13.14.1.3 Using Boolean Algebras as the Base Category

As remarked earlier, there are several possible choices for the base category over which the set-valued functors are defined. Most of our work has been based on the category, $\mathcal{V}(\mathcal{H})$, of commutative von Neumann sub-algebras of $B(\mathcal{H})$. As indicated above, the Heunen et al. constructions use the category of commutative C^*-algebras.

However, as discussed briefly in Sect. 13.5.5, another possible choice is the category, $\mathcal{B}l(\mathcal{H})$, of all Boolean sub-algebras of the lattice of projection operators on $\mathcal{H}$. The ensuing topos, $\mathbf{Sets}^{\mathcal{B}l(\mathcal{H})^{\mathrm{op}}}$, or $\mathbf{Sets}^{\mathcal{B}l(\mathcal{H})}$, is interesting in its own right, but particularly so when combined with the ideas of Heunen et al. As applied to the category $\mathcal{B}l(\mathcal{H})$, their work suggests that we first construct the tautological functor $\overline{\mathcal{B}l(\mathcal{H})}$ which associates to each $B \in \mathrm{Ob}(\mathcal{B}l(\mathcal{H}))$, the Boolean algebra B. Viewed internally in the topos $\mathbf{Sets}^{\mathcal{B}l(\mathcal{H})}$, this functor is a Boolean-algebra object. We conjecture that the spectrum of $\overline{\mathcal{B}l(\mathcal{H})}$ can be obtained in a constructive way using the internal logic of $\mathbf{Sets}^{\mathcal{B}l(\mathcal{H})}$. If so, it seems clear that, after using the locale trick of [42], this spectrum will essentially be the same as our dual presheaf $\underline{D}$.

Thus, in this approach, the state object is the spectrum of an internal *Boolean algebra*, and daseinisation maps the projection operators in $\mathcal{H}$ into elements of this algebra. This reinforces still further our claim that quantum theory looks like classical physics in an appropriate topos. This raises some fascinating possibilities. For example, we make the following:

---

[132] A frame is the same thing as a locale, and the elements of the internal frame are the opens in the locale $\overline{\Sigma}$.

Conjecture: The subject of quantum computation is equivalent to the study of 'classical' computation in the quantum topos $\mathbf{Sets}^{\mathcal{B}l(\mathcal{H})^{\mathrm{op}}}$.

13.14.1.4 Application to Other Branches of Algebra

It is clear that the scheme discussed above could fruitfully be extended to various branches of algebra. Thus, if $\mathfrak{A}$ is any algebraic structure,[133] we can consider the category $\mathcal{V}(\mathfrak{A})$ whose objects are the commutative sub-algebras of $\mathfrak{A}$, and whose arrows are algebra embeddings (or, slightly more generally, monomorphisms). One can then consider the topos, $\mathbf{Sets}^{\mathcal{V}(\mathfrak{A})^{\mathrm{op}}}$, of all set-valued, contravariant functors on $\mathcal{V}(\mathfrak{A})$; alternatively, one might look at the topos, $\mathbf{Sets}^{\mathcal{V}(\mathfrak{A})}$, of covariant functors.

For this structure to be mathematically interesting it is necessary that the abelian sub-objects of $\mathfrak{A}$ have a well-defined spectral structure. For example, let $\mathfrak{A}$ be any locally-compact topological group. Then the *spectrum* of any commutative (locally-compact) subgroup A is just the Pontryagin dual of A, which is itself a locally-compact, commutative group. The *spectral presheaf* of $\mathfrak{A}$ can then be defined as the object, $\Sigma_{\mathfrak{A}}$, in $\mathbf{Sets}^{\mathcal{V}(\mathfrak{A})^{\mathrm{op}}}$ that is constructed in the obvious way (i.e., analogous to the way in which $\underline{\Sigma}$ was constructed) from this collection of Pontryagin duals.

We conjecture that a careful analysis would show that, for at least some structures of this type:

1. There is a 'tautological' object, $\overline{\mathfrak{A}}$, in the topos $\mathbf{Sets}^{\mathcal{V}(\mathfrak{A})}$ that is associated with the category $\mathcal{V}(\mathfrak{A})$.
2. Viewed internally, this tautological object is a commutative algebra.
3. This object has a spectrum that can be constructed internally, and is essentially the spectral presheaf, $\Sigma_{\mathfrak{A}}$, of $\mathfrak{A}$.

It seems clear that, in general, the spectral presheaf, $\Sigma_{\mathfrak{A}}$, is a potential candidate for the basis of non-commutative spectral theory.

13.14.1.5 The Partial Existence of Points of Σ_ϕ

One of the many intriguing features of topos theory is that it makes sense to talk about entities that only "partially exist". One can only speculate on what would have been Heidegger's reaction had he been told that the answer to "What is a thing?" is "Something that partially exists". However, in the realm of topos theory the notion of "partial existence" lies easily with the concept of propositions that are only 'partly true'.

A particularly interesting example is the existence, or otherwise, of "points" (i.e., global elements) of the state object Σ_ϕ. If Σ_ϕ has no global elements (as is the case for the quantum spectral presheaf, $\underline{\Sigma}$) it may still have "partial elements". A partial element is defined to be an arrow $\xi : U \to \Sigma_\phi$ where the object U in the topos

[133] We are assuming that the ambient topos is **Sets**, but other choices could be considered.

τ_ϕ is a sub-object of the terminal object 1_{τ_ϕ}. Thus there is a monic $U \hookrightarrow 1_{\tau_\phi}$ with the property that the arrow $\xi : U \to \Sigma_\phi$ cannot be extended to an arrow $1_{\tau_\phi} \to \Sigma_\phi$. Studying the obstruction to such extensions could be another route to finding a cohomological expression of the Kochen-Specker theorem.

Pedagogically, it is attractive to say that the non-existence of a global element of $\underline{\Sigma}$ is analogous to the non-existence of a cross-section of the familiar "double-circle", helical covering of a single circle, S^1. This principal $\mathbb{Z}_2$-bundle over S^1 is non-trivial, and hence has no cross-sections.

However, *local* cross-sections do exist, these being defined as sections of the bundle restricted to any open subset of the base space S^1. In fact, this bundle is *locally trivial*; i.e., each point $s \in S^1$ has a neighbourhood U_s such that the restriction of the bundle to U_s is trivial, and hence sections of the bundle restricted to U_s exist.

There is an analogue of local triviality in the topos quantum theory where $\tau_\phi = \mathbf{Sets}^{\mathcal{V}(\mathcal{H})^{\mathrm{op}}}$. Thus, let V be any object in $\mathcal{V}(\mathcal{H})$ and define $\downarrow V := \{V_1 \in \mathrm{Ob}(\mathcal{V}(\mathcal{H})) \mid V_1 \subseteq V\}$. Then $\downarrow V$ is like a "neighbourhood" of V; indeed, that is precisely what it is if the poset $\mathrm{Ob}(\mathcal{V}(\mathcal{H}))$ is equipped with the topology generated by the lower sets. Furthermore, given any presheaf $\underline{F}$ in τ_ϕ, the restriction, $\underline{F}{\downarrow}V$, to V, can be defined as in Sect. 13.6.5. It is easy to see that, for all stages V, the presheaf $\underline{F}\downarrow V$ *does* have global elements. In this sense, every presheaf in $\mathbf{Sets}^{\mathcal{V}(\mathcal{H})^{\mathrm{op}}}$ is "locally trivial". Furthermore, to each V there is associated a sub-object $\underline{U}^V$ of $\underline{1}$ such that each global element of $\underline{F}\downarrow V$ corresponds to a partial element of $\underline{F}$.

Thus, for the topos $\mathbf{Sets}^{\mathcal{V}(\mathcal{H})^{\mathrm{op}}}$, there is a precise sense in which the spectral presheaf has "local elements", or "points that partially exist". However, it is not clear to what extent such an assertion can, or should, be made for a general topos τ_ϕ. Certainly, for any presheaf topos, $\mathbf{Sets}^{C^{\mathrm{op}}}$, one can talk about "localising" with respect to the objects in the base category C, but the situation for a more general topos is less clear.

13.14.2 The Quantity-Value Object $\mathcal{R}_\phi$

Let us turn now to the quantity-value object $\mathcal{R}_\phi$. This plays a key role in the representation of any physical quantity, A, by an arrow $A_\phi : \Sigma_\phi \to \mathcal{R}_\phi$. In so far as a "thing" is a bundle of properties, these properties refer to values of physical quantities, and so the nature of these "values" is of central importance.

We anticipate that $\mathcal{R}_\phi$ has many global elements $1_{\tau_\phi} \to \mathcal{R}_\phi$, and these can be interpreted as the possible "values" for physical quantities. If Σ_ϕ also has global elements/microstates $s : 1_{\tau_\phi} \to \Sigma_\phi$, then these combine with any arrow $A_\phi : \Sigma_\phi \to \mathcal{R}_\phi$ to give global elements of $\mathcal{R}_\phi$. It seems reasonable to refer to the element, $A_\phi \circ s : 1_{\tau_\phi} \to \mathcal{R}_\phi$ as the "value" of A when the microstate is s. However, our expectation is that, in general, Σ_ϕ may well have no global elements, in which case the interpretation of $A_\phi : \Sigma_\phi \to \mathcal{R}_\phi$ in terms of values is somewhat subtler. This has to be done internally using the language $\mathcal{L}(\tau_\phi)$ associated with the topos τ_ϕ: the overall logical structure is a nice example of a "coherence" theory of truth [35].

As far as axiomatic properties of $\mathcal{R}_\phi$ are concerned, the minimal requirement is presumably that it should have some ordering property that arises in all topos representations of the system S. This universal property could be coded into the internal language, $\mathcal{L}(S)$, of S. This implements our intuitive feeling that, in so far as the concept of "value" has any meaning, it must be possible to say that the value of one quantity is "larger" (or "smaller") than that of another. It seems reasonable to expect this relation to be transitive, but that is about all. In particular, we see no reason to suppose that this relation will always correspond to a *total* ordering: perhaps there are pairs of physical quantities whose "values" simply cannot be compared at all. Thus, tentatively, we can augment $\mathcal{L}(S)$ with the axioms for a poset structure on $\mathcal{R}$.

Beyond this simple ordering property, it becomes less clear what to assume about the quantity-value object. The example of quantum theory shows that it is wrong to automatically equate $\mathcal{R}_\phi$ with the real-number object $\mathbb{R}_\phi$ in the topos τ_ϕ. Indeed, we believe that this will typically be the case.

However, this makes it harder to know what to assume of $\mathcal{R}_\phi$. The quantum case shows that $\mathcal{R}_\phi$ may have considerably fewer algebraic properties than the real-number object $\mathbb{R}_\phi$. On a more "topological" front it is attractive to assume that $\mathcal{R}_\phi$ is an internal locale in the topos τ_ϕ. However, one should be cautious when conjecturing about $\mathcal{R}_\phi$ since our discussion of various possible quantity-value objects in quantum theory depended closely on the specific details of the spectral structure in this topos.

Heunen et al. brought up the idea of using the interval domain [42] (though, in their case, one obtains a *constant* functor, which is not capturing approximation processes). In general, the interval domain allows the systematic approximation of real numbers without assuming the continuum as given. Clearly, it is an interesting question if the quantity-value object $\mathcal{R}_\phi$ can be related to the interval domain. It remains to be seen if this has any generic use in practice.

In future applications, the quantity-value object $\mathcal{R}_\phi$ may well have a much weaker relation to the real numbers than in the topos form of quantum theory.

13.14.3 The Truth Objects $\mathbb{T}$, or Pseudo-State Object $\mathbb{W}_\phi$

The truth objects in a topos representation are certain sub-objects of $P\Sigma_\phi$. Their construction will be very theory-dependent, as are the pseudo-states, and the pseudo-state object, $\mathbb{W}_\phi$, if there is one. Each proposition about the physical system is represented by a sub-object $J \subseteq \Sigma_\phi$, and given a truth object $\mathbb{T} \subseteq P\Sigma_\phi$, the generalised truth value of the proposition is $[\![J \in \mathbb{T}]\!] \in \Gamma\Omega_\phi$; in terms of pseudo-states, $\mathfrak{w}$, the generalised truth values are of the form $[\![\mathfrak{w} \subseteq J]\!] \in \Gamma\Omega_\phi$.

The key properties of the quantum truth objects, $\underline{\mathbb{T}}^{|\psi\rangle}$, (or pseudo-states $\underline{\mathfrak{w}}^{|\psi\rangle}$) can easily be emulated if one is dealing with a more general base category, $\mathcal{C}$, so that the topos concerned is $\mathbf{Sets}^{\mathcal{C}^{\mathrm{op}}}$. However, it is not clear what, if anything, can be said about the structure of truth objects/pseudo-states in a more generic topos representation.

An attractive possibility is that there is a general analogue of (13.148) in the form

$$\begin{array}{ccc} \mathfrak{w} & \longrightarrow & \Sigma_\phi \\ & & \Big\downarrow \pi_\phi \\ 1_{\tau_\phi} & \xrightarrow[\ulcorner \mathfrak{w} \urcorner]{} & P\Sigma_\phi \end{array} \qquad (13.470)$$

and that obstructions to the existence of global elements of the state object Σ_ϕ can be studied with the aid of this diagram. If there is a pseudo-state object $\mathbb{W}_\phi$ then this could be a natural replacement for $P\Sigma_\phi$ in this diagram.

13.15 Conclusion

In this long article we have developed the idea that, for any given theory-type (classical physics, quantum physics, DI-physics,...) the theory of a particular physical system, S, is to be constructed in the framework of a certain, system-dependent, topos. The central idea is that a local language, $\mathcal{L}(S)$, is attached to each system S, and that the application of a given theory-type to S is equivalent to finding a representation, ϕ, of $\mathcal{L}(S)$ in a topos $\tau_\phi(S)$; this is equivalent to finding a translation of $\mathcal{L}(S)$ into the internal language associated with $\tau_\phi(S)$; or a functor to $\tau_\phi(S)$ from the topos associated with $\mathcal{L}(S)$.

Physical quantities are represented by arrows in the topos from the state object $\Sigma_{\phi,S}$ to the quantity-value object $\mathcal{R}_{\phi,S}$, and propositions are represented by sub-objects of the state object. The idea of a "truth sub-object" of $P\Sigma_{\phi,S}$ (or a "pseudo-state" sub-object of $\Sigma_{\phi,S}$) then leads to a neo-realist interpretation of propositions in which each proposition is assigned a truth value, given as a global element of the sub-object classifier $\Omega_{\tau_\phi(S)}$. In general, neo-realist statements about the world/system S are to be expressed in the internal language of the topos $\tau_\phi(S)$. Underlying this is the intuitionistic, deductive logic provided by the local language $\mathcal{L}(S)$.

These axioms are based on ideas from the topos representation of quantum theory, which we have discussed in depth. Here, the topos involved is $\mathbf{Sets}^{\mathcal{V}(\mathcal{H})^{\mathrm{op}}}$: the topos of presheaves over the base category $\mathcal{V}(\mathcal{H})$ of commutative, von Neumann sub-algebras of the algebra, $B(\mathcal{H})$, of all bounded operators on the quantum Hilbert space $\mathcal{H}$. Each such sub-algebra can be viewed as a context in which the theory can be viewed from a classical perspective. Thus a context can be described as a "classical snapshot", or "window on reality", or "world-view"/Weltanschauung. Mathematically, a context is a "stage of truth": a concept that goes back to Kripke's use of a presheaf topos to implement his intuitionistic view of time and process. However, for quantum theory, the contexts and their relations do not incorporate the idea of a process: there is no temporal aspect to them. Rather, what is encoded in the structure of the context category $\mathcal{V}(\mathcal{H})$ are the potential classical views on a quantum system and their relations (for a fixed time).

We have shown how the process of "daseinisation" maps projection operators (and hence equivalence classes of propositions) into sub-objects of the state object $\underline{\Sigma}$. We have also shown how this can be extended to an arbitrary, bounded self-adjoint operator, $\hat{A}$. This produces an arrow $\breve{A} : \underline{\Sigma} \to \underline{\mathcal{R}}$ where the minimal choice for the quantity-value object, $\underline{\mathcal{R}}$, is the object $\underline{\mathbb{R}^{\succeq}}$. We have also argued that, from a physical perspective, it is more attractive to choose $\underline{\mathbb{R}^{\leftrightarrow}}$ as the quantity-value presheaf. The significance of these results is enhanced considerably by the alternative, Heunen et al. derivation of $\underline{\Sigma}$ as the spectrum of an internal, commutative algebra.

These, and related, results all encourage the idea that quantum theory can be viewed as classical theory but in a topos other than the topos of sets, **Sets**.

Every classical system uses the same topos, **Sets**. However, in general, the topos will be system dependent as, for example, is the case with the quantum topoi of the form $\mathbf{Sets}^{\mathcal{V}(\mathcal{H})^{\mathrm{op}}}$, where $\mathcal{H}$ is the Hilbert space of the system. This leads to the problem of understanding how the topoi for a class of systems behave under the action of taking a sub-system, or combining a pair of systems to give a single composite system. We have presented a set of axioms that capture the general ideas we are trying to develop. Of course, these axioms are not cast in stone, and are still partly 'experimental' in nature. However, we have shown that classical physics exactly fits our suggested scheme, and that quantum physics "almost" does: "almost" because of the issues concerning the translation representation of the arrows associated with compositions of systems that were discussed in Sect. 13.13.3.

An important challenge for future work is to show that our general topos scheme can be used to develop genuinely new theories of physics, not just to rewrite old ones in a new way. Of particular interest is the problem with which we motivated the scheme in the first place: namely, to find tools for constructing theories that go beyond quantum theory and which do not use Hilbert spaces, path integrals, or any of the other familiar entities in which the continuum real and/or complex numbers play a fundamental role.

As we have discussed, the topoi for quantum systems are of the form $\mathbf{Sets}^{\mathcal{V}(\mathcal{H})^{\mathrm{op}}}$, and hence embody contextual logic in a fundamental way. One way of going "beyond" quantum theory, while escaping the a priori imposition of continuum concepts, is to use presheaves over a more general "category of contexts", $\mathcal{C}$, i.e., develop the theory in the topos $\mathbf{Sets}^{\mathcal{C}^{\mathrm{op}}}$. Such a structure embodies contextual, multi-valued logic in an intrinsic way, and in that sense might be said to encapsulate one of the fundamental insights of quantum theory. However, and unlike in quantum theory, there is no obligation to use the real or complex numbers in the construction of the category $\mathcal{C}$.

Indeed, early on in this work we noted that real numbers arise in theories of physics in three different (but related) ways: (i) as the values of physical quantities; (ii) as the values of probabilities; and (iii) as a fundamental ingredient in models of space and time. The first of these is now subsumed into the quantity-value object $\mathcal{R}_{\phi}$, and which now has no a priori relation to the real number object in τ_{ϕ}. The second source of real numbers has gone completely since we no longer have probabilities of propositions but rather generalised truth values whose values lie in $\Gamma\Omega_{\tau_{\phi}}$. The third source is also no longer binding since models of space and time in a topos could depend on many things: for example, infinitesimals.

Of course, although true, these remarks do not of themselves give a concrete example of a theory that is "beyond quantum theory". On the other hand, these ideas certainly point in a novel direction, and one at which, we almost certainly would not have arrived if the challenge to "go beyond quantum theory" had been construed only in terms of trying to generalise Hilbert spaces, path integrals, and the like.

From a more general perspective, other types of topoi are possible realms for the construction of physical theories. One simple, but mathematically rich example arises from the theory of M-sets. Here, M is a monoid and, like all monoids, can be viewed as a category with a single object, and whose arrows are the elements of M. Thought of as a category, a monoid is "complementary" to a partially-ordered set. In a monoid, there is only one object, but plenty of arrows from that object to itself; whereas in a partially-ordered set there are plenty of objects, but at most one arrow between any pair of objects. Thus a partially-ordered set is the most economical category with which to capture the concept of "contextual logic". On the other hand, the logic associated with a monoid is non-contextual as there is only one object in the category.

It is easy to see that a functor from M to **Sets** is just an "M-set": i.e., a set on which M acts as a monoid of transformations. An arrow between two such M-sets is an equivariant map between them. In physicists' language, one would say that the topos $\mathbf{Sets}^{M}$—usually denoted BM— is the category of the "non-linear realisations" of M.

The sub-object classifier, Ω_{BM}, in BM is the collection of left ideals in M; hence, many of the important constructions in the topos can be handled using the language of algebra. The topos BM is one of the simplest to define and work with and, for that reason, it is a popular source of examples in texts on topos theory. It would be intriguing to experiment with constructing model theories of physics using one of these simple topoi. One possible use of M-sets is discussed in [47] in the context of reduction of the state vector, but there will surely be others.

Is there "un gros topos"?

It is clear that there are many other topics for future research. A question of particular interest is if there is a *single* topos within which all systems of a given theory-type can be discussed. For example, in the case of quantum theory the relevant topoi are of the form $\mathbf{Sets}^{\mathcal{V}(\mathcal{H})^{\mathrm{op}}}$, where $\mathcal{H}$ is a Hilbert space, and the question is whether all such topoi can be gathered together to form a single topos (what Grothendieck termed "un gros topos") within which all quantum systems can be discussed.

There are well-known examples of such constructions in the mathematical literature. For example, the category, $\text{Sh}(X)$, of sheaves on a topological space X is a topos, and there are collections $\mathbf{T}$ of topological spaces which form a Grothendieck site, so that the topos $\text{Sh}(\mathbf{T})$ can be constructed. A particular object in $\text{Sh}(\mathbf{T})$ will then be a sheaf over $\mathbf{T}$ whose stalk over any object X in $\mathbf{T}$ will be the topos $\text{Sh}(X)$.

For our purposes, the ideal situation would be if the various categories of systems, **Sys**, can be chosen in such a way that $\mathcal{M}(\mathbf{Sys})$ is a site. Then the topos of sheaves, $\text{Sh}(\mathcal{M}(\mathbf{Sys}))$, over this site would provide a common topos in which all systems of this theory type—i.e., the objects of **Sys**—can be discussed. We do not know if this is possible, and it is a natural subject for future study.

Some more speculative lines of future research

At a conceptual level, one motivating desire for the entire research programme is to find a formalism that will always give some sort of "realist" interpretation, even in the case of quantum theory which is normally presented in an instrumentalist way. But this particular example raises an interesting point because the neo-realist interpretation takes place in the topos $\mathbf{Sets}^{\mathcal{V}(\mathcal{H})^{\text{op}}}$, whereas the instrumentalist interpretation works in the familiar topos, **Sets**, of sets, and one might wonder if the use of a pair of topoi in this way is a general feature of "topos-physics".

A related issue concerns the representation of the $\mathcal{PL}(S)$-propositions of the form "$A \varepsilon \Delta$" discussed in Sect. 13.5. This serves as a bridge between the "external" world of a background spatial structure, and the internal world of the topos. This link is not present with the $\mathcal{L}(S)$ language whose propositions are purely internal terms of type Ω of the form "$A(\tilde{s}) \in \tilde{\Delta}$". In a topos representation, ϕ, of $\mathcal{L}(S)$, these become propositions of the form "$A \in \Xi$", where Ξ is a sub-object of $\mathcal{R}_\phi$.

In general, if we have an example of our axioms working neo-realistically in a topos τ, one might wonder if there is an "instrumentalist" interpretation of the same theory in a different topos, τ_i, say? Of course, the word "instrumentalism" is used metaphorically here, and any serious consideration of such a pair (τ, τ_i) would require a lot of careful thought.

However, if a pair (τ, τ_i) does exist, the question then arises of whether there is a *categorial* way of linking the neo-realist and instrumentalist interpretations: for example, via a functor $I : \tau \to \tau_i$. If so, is this related to some analogue of the daseinisation operation that produced the representation of the $\mathcal{PL}(S)$-propositions, "$A \varepsilon \Delta$" in quantum theory? Care is needed in discussing such issues since informal set theory is used as a meta-language in constructing a topos, and one has to be careful not to confuse this with the existence, or otherwise, of an "instrumentalist" interpretation of any given representation.

If such a functor, $I : \tau \to \tau_i$, did exist then one could speculate on the possibility of finding an "interpolating chain" of functors

$$\tau \to \tau^1 \to \tau^2 \to \cdots \to \tau^n \to \tau_i \tag{13.471}$$

which could be interpreted conceptually as corresponding to an interpolation between the philosophical views of realism and instrumentalism!

Even more speculatively one might wonder if "one person's realism is another person's instrumentalism". More precisely, given a pair (τ, τ_i) in the sense above, could there be cases in which the topos τ carries a neo-realist interpretation of a theory with respect to an instrumentalist interpretation in τ_i, whilst being the carrier of an instrumentalist interpretation with respect to the neo-realism of a "higher" topos; and so on? For example, is there some theory whose 'instrumentalist manifestation' takes place in the topos $\mathbf{Sets}^{\mathcal{V}(\mathcal{H})^{\mathrm{op}}}$?

On the other hand, one might want to say that "instrumentalist" interpretations always take place in the world of classical set theory, so that τ_i should always be chosen to be **Sets**. In any event, it would be interesting to study the quantum case more closely to see if there are any categorial relations between the formulation in $\mathbf{Sets}^{\mathcal{V}(\mathcal{H})^{\mathrm{op}}}$ and the instrumentalism interpretation in **Sets**. It can be anticipated that the action of daseinisation will play an important role here.

All this is, perhaps,[134] rather speculative, but there is a more obvious situation in which a double-topos structure will be necessary, irrespective of philosophical musings on instrumentalism. This is if one wants to discuss the "classical limit" of some topos theory. In this case this limit will exist in the topos, **Sets**, and this must be used in addition to the topos of the basic theory. A good example of this, of course, is the topos of quantum theory discussed in this article.

Implications for quantum gravity

A serious claim stemming from our work is that a successful theory of quantum gravity should be constructed in some topos $\mathcal{U}$—the "topos of the universe"—that is *not* the topos of sets. All entities of physical interest will be represented in this topos, including models for space-time (if there are any at a fundamental level in quantum gravity) and, if relevant, loops, membranes etc. as well as incorporating the anticipated generalisation of quantum theory.

Such a theory of quantum gravity will have a neo-realist interpretation in the topos $\mathcal{U}$, and hence would be particularly useful in the context of quantum cosmology. However, in practice, physicists divide the world up into smaller, more easily handled, chunks, and each of them would correspond to what earlier we have called a "system" and, correspondingly, would have its own topos. Thus $\mathcal{U}$ is something like the "gros topos" of the theory, and would glue together the individual "sub-systems" in a categorial way. Of course, it is most unlikely that there is any preferred way of dividing the universe up into bite-sized chunks, but this is not problematic as the ensuing relativism is naturally incorporated into the idea of a Grothendieck site.

[134] To be honest, the "perhaps" should really be replaced by "highly".

Appendix 1: Some Theorems and Constructions Used in the Main Text

Results on Clopen Sub-Objects of $\underline{\Sigma}$

Theorem 15 *The collection, $Sub_{cl}(\underline{\Sigma})$, of all clopen sub-objects of $\underline{\Sigma}$ is a Heyting algebra.*

Proof First recall how a Heyting algebra structure is placed on the set, $\text{Sub}(\underline{\Sigma})$, of all sub-objects of $\underline{\Sigma}$.

The "$\vee$"- and "$\wedge$"-operations.

Let $\underline{S}$, $\underline{T}$ be two sub-objects of $\underline{\Sigma}$. Then the "$\vee$" and "$\wedge$" operations are defined by

$$(\underline{S} \vee \underline{T})_V := \underline{S}_V \cup \underline{T}_V \tag{13.472}$$
$$(\underline{S} \wedge \underline{T})_V := \underline{S}_V \cap \underline{T}_V \tag{13.473}$$

for all contexts V. It is easy to see that if $\underline{S}$ and $\underline{T}$ are clopen sub-objects of $\underline{\Sigma}$, then so are $\underline{S} \vee \underline{T}$ and $\underline{S} \wedge \underline{T}$.

The Zero and Unit Elements

The zero element in the Heyting algebra $\text{Sub}(\underline{\Sigma})$ is the empty sub-object $\underline{0} := \{\varnothing_V \mid V \in \text{Ob}(\mathcal{V}(\mathcal{H}))\}$, where $\varnothing_V$ is the empty subset of $\underline{\Sigma}_V$. The unit element in $\text{Sub}(\underline{\Sigma})$ is $\underline{\Sigma}$. It is clear that both $\underline{0}$ and $\underline{\Sigma}$ are clopen sub-objects of $\underline{\Sigma}$.

The "$\Rightarrow$"-operation.

The most interesting part is the definition of the implication $\underline{S} \Rightarrow \underline{T}$. For all $V \in \text{Ob}(\mathcal{V}(\mathcal{H}))$, it is given by

$$\begin{aligned}(\underline{S} \Rightarrow \underline{T})_V := \{\lambda \in \underline{\Sigma}_V \mid \forall\, V' \subseteq V,\ &\text{if} \\ &\underline{\Sigma}(i_{V'V})(\lambda) \in \underline{S}_{V'} \text{ then } \underline{\Sigma}(i_{V'V})(\lambda) \in \underline{T}_{V'}\} \quad (13.474)\\ = \{\lambda \in \underline{\Sigma}_V \mid \forall V' \subseteq V,\ &\text{if} \\ &\lambda|_{V'} \in \underline{S}_{V'} \text{ then } \lambda|_{V'} \in \underline{T}_{V'}\}. \quad (13.475)\end{aligned}$$

Since $\neg\underline{S} := \underline{S} \Rightarrow \underline{0}$, the expression for negation follows from the above as

$$(\neg\underline{S})_V = \{\lambda \in \underline{\Sigma}_V \mid \forall\, V' \subseteq V,\ \underline{\Sigma}(i_{V'V})(\lambda) \notin \underline{S}_{V'}\} \tag{13.476}$$
$$= \{\lambda \in \underline{\Sigma}_V \mid \forall\, V' \subseteq V,\ \lambda|_{V'} \notin \underline{S}_{V'}\}. \tag{13.477}$$

We rewrite the formula for negation as

$$(\neg\underline{S})_V = \bigcap_{V' \subseteq V} \{\lambda \in \underline{\Sigma}_V \mid \lambda|_{V'} \in \underline{S}^c_{V'}\} \tag{13.478}$$

where $\underline{S}_{V'}{}^c$ denotes the complement of $\underline{S}_{V'}$ in $\underline{\Sigma}_{V'}$. Clearly, $\underline{S}_{V'}{}^c$ is clopen in $\underline{\Sigma}_{V'}$ since $\underline{S}_{V'}$ is clopen. Since the restriction $\underline{\Sigma}(i_{V'V}) : \underline{\Sigma}_V \to \underline{\Sigma}_{V'}$ is continuous and surjective,[135] it is easy to see that the inverse image $\underline{\Sigma}(i_{V'V})^{-1}(\underline{S}_{V'}{}^c)$ is clopen in $\underline{\Sigma}_V$. Clearly,

$$\underline{\Sigma}(i_{V'V})^{-1}(\underline{S}_{V'}{}^c) = \{\lambda \in \underline{\Sigma}_V \mid \lambda|_{V'} \in \underline{S}_{V'}{}^c\} \tag{13.479}$$

and so, from (13.478) we have

$$(\neg\underline{S})_V = \bigcap_{V' \subseteq V} \underline{\Sigma}(i_{V'V})^{-1}(\underline{S}_{V'}{}^c) \tag{13.480}$$

The problem is that we want $(\neg\underline{S})_V$ to be a *clopen* subset of $\underline{\Sigma}_V$. Now the right hand side of (13.480) is the intersection of a family, parameterised by $\{V' \mid V' \subseteq V\}$, of clopen sets. Such an intersection is always closed, but it is only guaranteed to be open if $\{V' \mid V' \subseteq V\}$ is a finite set, which of course may not be the case.

If $V'' \subseteq V'$ and $\lambda|_{V''} \in \underline{S}_{V''}{}^c$, then $\lambda|_{V'} \in \underline{S}_{V'}{}^c$. Indeed, if we had $\lambda|_{V'} \in \underline{S}_{V'}$, then $(\lambda|_{V'})|_{V''} = \lambda|_{V''} \in \underline{S}_{V''}$ by the definition of a sub-object, so we would have a contradiction. This implies $\underline{\Sigma}(i_{V''V})^{-1}(\underline{S}_{V''}{}^c) \subseteq \underline{\Sigma}(i_{V'V})^{-1}(S_{V'}{}^c)$, and hence the right hand side of (13.480) is a decreasing net of clopen subsets of $\underline{\Sigma}_V$ which converges to something, which we take as the subset of $\underline{\Sigma}_V$ that is to be $(\neg\underline{S})_V$.

Here we have used the fact that the set of clopen subsets of $\underline{\Sigma}_V$ is a complete lattice, where the minimum of a family $(U_i)_{i\in I}$ of clopen subsets is defined as the interior of $\bigcap_{i\in I} U_i$. This leads us to define

$$(\neg\underline{S})_V := \operatorname{int} \bigcap_{V' \subseteq V} \underline{\Sigma}(i_{V'V})^{-1}(\underline{S}_{V'}{}^c) \tag{13.481}$$

$$= \operatorname{int} \bigcap_{V' \subseteq V} \{\lambda \in \underline{\Sigma}_V \mid \lambda|_{V'} \in (S_{V'}{}^c)\} \tag{13.482}$$

as the negation in $\mathrm{Sub}_{\mathrm{cl}}(\underline{\Sigma})$. This modified definition guarantees that $\neg\underline{S}$ is a *clopen* sub-object. A straightforward extension of this method gives a consistent definition of $\underline{S} \Rightarrow \underline{T}$.

This concludes the proof of the theorem.

The following theorem shows the relation between the restriction mappings of the outer presheaf $\underline{O}$ and those of the spectral presheaf $\underline{\Sigma}$. We basically follow de Groote's proof of Proposition 3.22 in [38] and show that this result, which uses quite a different terminology, actually gives the desired relation.

[135] See proof of Theorem 16 below.

Theorem 16 *Let $V, V' \in \mathrm{Ob}(\mathcal{V}(\mathcal{H}))$ such that $V' \subset V$. Then*

$$S_{\underline{O}(i_{V'V})(\delta^o(\hat{P})_V)} = \underline{\Sigma}(i_{V'V})(S_{\delta^o(\hat{P})_V}). \tag{13.483}$$

Proof First of all, to simplify notation, we can replace $\delta^o(\hat{P})_V$ by $\hat{P}$ (which amounts to the assumption that $\hat{P} \in \mathcal{P}(V)$. This does not play a role for the current argument). By definition, $\underline{O}(i_{V'V})(\hat{P}) = \delta^o(\hat{P})_{V'}$, so we have to show that $S_{\delta^o(\hat{P})_{V'}} = \underline{\Sigma}(i_{V'V})(S_{\hat{P}})$ holds.

If $\lambda \in S_{\hat{P}}$, then $\lambda(\hat{P}) = 1$, which implies $\lambda(\hat{Q}) = 1$ for all $\hat{Q} \geq \hat{P}$. In particular, $\lambda(\delta^o(\hat{P})_{V'}) = 1$, so $\underline{\Sigma}(i_{V'V})(\lambda) = \lambda|_{V'} \in S_{\delta^o(\hat{P})_{V'}}$. This shows that $\underline{\Sigma}(i_{V'V})(S_{\hat{P}}) \subseteq S_{\delta^o(\hat{P})_{V'}}$.

To show the converse inclusion, let $\lambda' \in S_{\delta^o(\hat{P})_{V'}}$, which means that $\lambda'(\delta^o(\hat{P})_{V'}) = 1$. We have $\hat{P} \in \underline{O}(i_{V'V})^{-1}(\delta^o(\hat{P})_{V'})$. Let

$$F_{\lambda'} := \{\hat{Q} \in \mathcal{P}(V') \mid \lambda'(\hat{Q}) = 1\} = \lambda'^{-1}(1) \cap \mathcal{P}(V'). \tag{13.484}$$

As shown in Sect. 13.8.3, $F_{\lambda'}$ is an ultrafilter in the projection lattice $\mathcal{P}(V')$.[136] The idea is to show that $F_{\lambda'} \cup \hat{P}$ is a filter base in $\mathcal{P}(V)$ that can be extended to an ultrafilter, which corresponds to an element of the Gel'fand spectrum of V.

Let us assume that $F_{\lambda'} \cup \hat{P}$ is not a filter base in $\mathcal{P}(V)$. Then there exists some $\hat{Q} \in F_{\lambda'}$ such that

$$\hat{Q} \wedge \hat{P} = \hat{Q}\hat{P} = \hat{0}, \tag{13.485}$$

which implies $\hat{P} \leq \hat{1} - \hat{Q}$, so

$$\underline{O}(i_{V'V})(\hat{P}) = \delta^o(\hat{P})_{V'} \leq \underline{O}(i_{V'V})(\hat{1} - \hat{Q}) = \hat{1} - \hat{Q} \tag{13.486}$$

and hence we get the contradiction

$$1 = \lambda'(\delta^o(\hat{P})_{V'}) \leq \lambda'(\hat{1} - \hat{Q}) = 0. \tag{13.487}$$

By Zorn's lemma, the filter base $F_{\lambda'} \cup \hat{P}$ is contained in some (not necessarily unique) maximal filter base in $\mathcal{P}(V)$. Such a maximal filter base is an ultrafilter and thus corresponds to an element λ of the Gel'fand spectrum $\underline{\Sigma}_V$ of V. Since $\hat{P}$ is contained in the ultrafilter, we have $\lambda(\hat{P}) = 1$, so $\lambda \in S_{\hat{P}}$. By construction, $\underline{\Sigma}(i_{V'V})(\lambda) = \lambda|_{V'} = \lambda' \in S_{\delta^o(\hat{P})_{V'}}$, the element of $\underline{\Sigma}_{V'}$ we started from. This shows that $S_{\delta^o(\hat{P})_{V'}} \subseteq \underline{\Sigma}(i_{V'V})(S_{\hat{P}})$, and we obtain

---

[136] In general, each ultrafilter F in the projection lattice of an abelian von Neumann algebra V corresponds to a unique element λ_F of the Gel'fand spectrum of V. The ultrafilter is the collection of all those projections that are mapped to 1 by λ, i.e., $F = \lambda_F^{-1}(1) \cap \mathcal{P}(V)$.

$$S_{\delta^o(\hat{P})_{V'}} = \underline{\Sigma}(i_{V'V})(S_{\hat{P}}). \tag{13.488}$$

It is well-known that every state $\lambda' \in \underline{\Sigma}_{V'}$ is of the form $\lambda' = \underline{\Sigma}(i_{V'V})(\lambda) = \lambda|_{V'}$ for some $\lambda \in \underline{\Sigma}_V$. This implies

$$\underline{\Sigma}(i_{V'V})^{-1}(S_{\delta^o(\hat{P})_{V'}}) = S_{\delta^o(\hat{P})_{V'}} \subseteq \underline{\Sigma}_V. \tag{13.489}$$

Note that on the right hand side, $S_{\delta^o(\hat{P})_{V'}}$ (and not $S_{\hat{P}}$, which is a smaller set in general) shows up.

De Groote has shown in [38] that for any unital abelian von Neumann algebra V, the clopen sets $S_{\hat{Q}}$, $\hat{Q} \in \mathcal{P}(V)$, form a base of the Gel'fand topology on $\underline{\Sigma}_V$. Formulas (13.488) and (13.489) hence show that the restriction mappings

$$\begin{aligned} \underline{\Sigma}(i_{V'V}) : \underline{\Sigma}_V &\to \underline{\Sigma}_{V'} \\ \lambda &\mapsto \lambda|_{V'} \end{aligned}$$

of the spectral presheaf are open and continuous. Using continuity, it is easy to see that $\underline{\Sigma}(i_{V'V})$ is also closed: let $C \subseteq \underline{\Sigma}_V$ be a closed subset. Since $\underline{\Sigma}_V$ is compact, C is compact, and since $\underline{\Sigma}(i_{V'V})$ is continuous, $\underline{\Sigma}(i_{V'V})(C) \subseteq \underline{\Sigma}_{V'}$ is compact, too. However, $\underline{\Sigma}_{V'}$ is Hausdorff, and so $\underline{\Sigma}(i_{V'V})(C)$ is closed in $\underline{\Sigma}_{V'}$.

The Grothendieck k-Construction for an Abelian Monoid

Let us briefly review the Grothendieck construction for an abelian monoid M.

Definition 23 A group completion of M is an abelian group $k(M)$ together with a monoid map $\theta : M \to k(M)$ that is universal. Namely, given any monoid morphism $\phi : M \to G$, where G is an abelian group, there exists a unique group morphism $\phi' : k(M) \to G$ such that ϕ factors through ϕ'; i.e., we have the commutative diagram

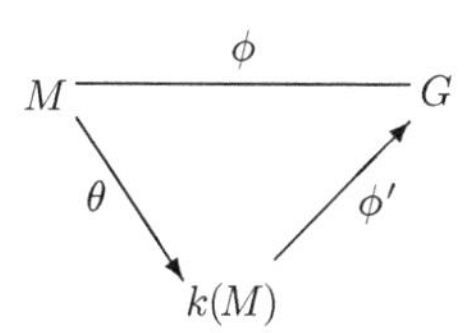

with $\phi = \phi' \circ \theta$.

It is easy to see that any such $k(M)$ is unique up to isomorphism.

To prove existence, first take the set of all pairs $(a, b) \in M \times M$, each of which is to be thought of heuristically as $a - b$. Then, note that *if* inverses existed in M, we would have $a - b = c - d$ if and only if $a + d = c + b$. This suggests defining an equivalence relation on $M \times M$ in the following way:

$$(a, b) \equiv (c, d) \text{ iff } \exists g \in M \text{ such that } a + d + g = b + c + g. \tag{13.490}$$

Definition 24 The Grothendieck completion of an abelian monoid M is the pair $(k(M), \theta)$ defined as follows:

(i) $k(M)$ is the set of equivalence classes $[a, b]$, where the equivalence relation is defined in (13.490). A group law on $k(M)$ is defined by

$$\text{(i) } [a,b] + [c,d] := [a+c, b+d], \tag{13.491}$$

$$\text{(ii) } 0_{k(M)} := [0_M, 0_M], \tag{13.492}$$

$$\text{(iii) } -[a,b] := [b,a], \tag{13.493}$$

where 0_M is the unit in the abelian monoid M.

(ii) The map $\theta : M \to k(M)$ is defined by

$$\theta(a) := [a, 0] \tag{13.494}$$

for all $a \in M$.

It is straightforward to show that (i) these definitions are independent of the representative elements in the equivalence classes; (ii) the axioms for a group are satisfied; and (iii) the map θ is universal in the sense mentioned above.

It is also clear that k is a *functor* from the category of abelian monoids to the category of abelian groups. For, if $f : M_1 \to M_2$ is a morphism between abelian monoids, define $k(f) : k(M_1) \to k(M_2)$ by $k(f)[a, b] := [f(a), f(b)]$ for all $a, b \in M_1$.

Functions of Bounded Variation and $\Gamma\underline{\mathbb{R}^{\succeq}}$

These techniques will now be applied to the set, $\Gamma\underline{\mathbb{R}^{\succeq}}$, of global elements of $\underline{\mathbb{R}^{\succeq}}$. We could equally well consider $\Gamma\underline{\mathbb{R}^{\leftrightarrow}}$ and its k-extension, but this would just make the notation more complex, so in this and the following Subsections, we will mainly concentrate on $\Gamma\underline{\mathbb{R}^{\succeq}}$ (resp. $\underline{\mathbb{R}^{\succeq}}$). The results can easily be extended to $\Gamma\underline{\mathbb{R}^{\leftrightarrow}}$ (resp. $\underline{\mathbb{R}^{\leftrightarrow}}$).

It was discussed in Sect. 13.8.2 how global elements of $\underline{\mathbb{R}^{\leftrightarrow}}$ are in one-to-one correspondence with pairs (μ, ν) consisting of an order-preserving and an order-reversing function on the category $\mathcal{V}(\mathcal{H})$; i.e., with functions $\mu : \text{Ob}(\mathcal{V}(\mathcal{H})) \to \mathbb{R}$ such that, for all $V_1, V_2 \in \text{Ob}(\mathcal{V}(\mathcal{H}))$, if $V_2 \subseteq V_1$ then $\mu(V_2) \leq \mu(V_1)$ and $\nu : \text{Ob}(\mathcal{V}(\mathcal{H})) \to \mathbb{R}$ such that, for all $V_1, V_2 \in \text{Ob}(\mathcal{V}(\mathcal{H}))$, if $V_2 \subseteq V_1$ then $\nu(V_2) \geq \nu(V_1)$; see (13.291). The monoid law on $\Gamma\underline{\mathbb{R}^{\leftrightarrow}}$ is given by (13.295).

Clearly, global elements of $\underline{\mathbb{R}^{\succeq}}$ are given by order-reversing functions $\nu : \mathcal{V}(\mathcal{H}) \to \mathbb{R}$, and $\Gamma\underline{\mathbb{R}^{\succeq}}$ is an abelian monoid in the obvious way. Hence the Grothendieck construction can be applied to give an abelian group $k(\Gamma\underline{\mathbb{R}^{\succeq}})$. This is defined to be the set of equivalence classes $[\nu, \kappa]$ where $\nu, \kappa \in \Gamma\underline{\mathbb{R}^{\succeq}}$, and where $(\nu_1, \kappa_1) \equiv (\nu_2, \kappa_2)$ if, and only if, there exists $\alpha \in \Gamma\underline{\mathbb{R}^{\succeq}}$, such that

$$\nu_1 + \kappa_2 + \alpha = \kappa_1 + \nu_2 + \alpha \tag{13.495}$$

Since $\Gamma\underline{\mathbb{R}^{\succeq}}$ has a cancellation law, we have $(\nu_1, \kappa_1) \equiv (\nu_2, \kappa_2)$ if, and only if,

$$\nu_1 + \kappa_2 = \kappa_1 + \nu_2. \tag{13.496}$$

Intuitively, we can think of $[\nu, \kappa]$ as being "$\nu - \kappa$", and embed $\Gamma\underline{\mathbb{R}^{\succeq}}$ in $k(\Gamma\underline{\mathbb{R}^{\succeq}})$ by $\nu \mapsto [\nu, 0]$. However, ν, κ are $\mathbb{R}$-valued functions on $\mathrm{Ob}(\mathcal{V}(\mathcal{H}))$ and hence, in this case, the expression "$\nu - \kappa$" also has a *literal* meaning: i.e., as the function $(\nu - \kappa)(V) := \nu(V) - \kappa(V)$ for all $V \in \mathrm{Ob}(\mathcal{V}(\mathcal{H}))$.

This is not just a coincidence of notation. Indeed, let $F\big(\mathrm{Ob}(\mathcal{V}(\mathcal{H})), \mathbb{R}\big)$ denote the set of all real-valued functions on $\mathrm{Ob}(\mathcal{V}(\mathcal{H}))$. Then we can construct the map,

$$\begin{aligned} j : k(\Gamma\underline{\mathbb{R}^{\succeq}}) &\to F\big(\mathrm{Ob}(\mathcal{V}(\mathcal{H})), \mathbb{R}\big) \\ [\nu, \kappa] &\mapsto \nu - \kappa \end{aligned} \tag{13.497}$$

which is well-defined on equivalence classes.

It is easy to see that the map in (13.497) is injective. This raises the question of the *image* in $F\big(\mathrm{Ob}(\mathcal{V}(\mathcal{H})), \mathbb{R}\big)$ of the map j: i.e., what types of real-valued function on $\mathrm{Ob}(\mathcal{V}(\mathcal{H}))$ can be written as the difference between two order-reversing functions?

For functions $f : \mathbb{R} \to \mathbb{R}$, it is a standard result that a function can be written as the difference between two monotonic functions if, and only if, it has bounded variation. The natural conjecture is that a similar result applies here. To show this, we proceed as follows.

Let $f : \mathrm{Ob}(\mathcal{V}(\mathcal{H})) \to \mathbb{R}$ be a real-valued function on the set of objects in the category $\mathcal{V}(\mathcal{H})$. At each $V \in \mathrm{Ob}(\mathcal{V}(\mathcal{H}))$, consider a finite chain

$$C := \{V_0, V_1, V_2, \ldots, V_{n-1}, V \mid V_0 \subset V_1 \subset V_2 \subset \cdots \subset V_{n-1} \subset V\} \tag{13.498}$$

of proper subsets, and define the *variation* of f on this chain to be

$$V_f(C) := \sum_{j=1}^{n} |f(V_j) - f(V_{j-1})| \tag{13.499}$$

where we set $V_n := V$. Now take the supremum of $V_f(C)$ for all such chains C. If this is finite, we say that f has a *bounded variation* and define

$$I_f(V) := \sup_C V_f(C) \tag{13.500}$$

Then it is clear that (i) $V \mapsto I_f(V)$ is an order-preserving function on $\mathrm{Ob}(\mathcal{V}(\mathcal{H}))$; (ii) $f - I_f$ is an order-reversing function on $\mathrm{Ob}(\mathcal{V}(\mathcal{H}))$; and (iii) $-I_f$

is an order-reversing function on $\mathrm{Ob}(\mathcal{V}(\mathcal{H}))$. Thus, any function, f, of bounded variation can be written as

$$f \equiv (f - I_f) - (-I_f) \tag{13.501}$$

which is the difference of two order-reversing functions; i.e., f can be expressed as the difference of two elements of $\Gamma\underline{\mathbb{R}^{\succeq}}$.

Conversely, it is a straightforward modification of the proof for functions $f : \mathbb{R} \to \mathbb{R}$, to show that if $f : \mathrm{Ob}(\mathcal{V}(\mathcal{H})) \to \mathbb{R}$ is the difference of two order-reversing functions, then f is of bounded variation. The conclusion is that $k(\Gamma\underline{\mathbb{R}^{\succeq}})$ is in bijective correspondence with the set, $\mathrm{BV}(\mathrm{Ob}(\mathcal{V}(\mathcal{H})), \mathbb{R})$, of functions $f : \mathrm{Ob}(\mathcal{V}(\mathcal{H})) \to \mathbb{R}$ of bounded variation.

Taking Squares in $k(\Gamma\underline{\mathbb{R}^{\succeq}})$

We can now think of $k(\Gamma\underline{\mathbb{R}^{\succeq}})$ in two ways: (i) as the set of equivalence classes $[\nu, \kappa]$, of elements $\nu, \kappa \in \Gamma\underline{\mathbb{R}^{\succeq}}$; and (ii) as the set, $\mathrm{BV}(\mathrm{Ob}(\mathcal{V}(\mathcal{H})), \mathbb{R})$, of differences $\nu - \kappa$ of such elements.

As expected, $\mathrm{BV}(\mathrm{Ob}(\mathcal{V}(\mathcal{H})), \mathbb{R})$ is an abelian group. Indeed: suppose $\alpha = \nu_1 - \kappa_1$ and $\beta = \nu_2 - \kappa_2$ with $\nu_1, \nu_2, \kappa_1, \kappa_2 \in \Gamma\underline{\mathbb{R}^{\succeq}}$, then

$$\alpha + \beta = (\nu_1 + \nu_2) - (\kappa_1 + \kappa_2) \tag{13.502}$$

Hence $\alpha + \beta$ belongs to $\mathrm{BV}(\mathrm{Ob}(\mathcal{V}(\mathcal{H})), \mathbb{R})$ since $\nu_1 + \nu_2$ and $\kappa_1 + \kappa_2$ belong to $\Gamma\underline{\mathbb{R}^{\succeq}}$.

The definition of $[\nu, 0]^2$.

We will now show how to take the square of elements of $k(\Gamma\underline{\mathbb{R}^{\succeq}})$ that are of the form $[\nu, 0]$. Clearly, ν^2 is well-defined as a function on $\mathrm{Ob}(\mathcal{V}(\mathcal{H}))$, but it may not belong to $\Gamma\underline{\mathbb{R}^{\succeq}}$. Indeed, if $\nu(V) < 0$ for any V, then the function $V \mapsto \nu^2(V)$ can get smaller as V gets smaller, so it is order-preserving instead of order-reversing.

This suggests the following strategy. First, define functions ν_+ and ν_- by

$$\nu_+(V) := \begin{cases} \nu(V) & \text{if } \nu(V) \geq 0 \\ 0 & \text{if } \nu(V) < 0 \end{cases} \tag{13.503}$$

and

$$\nu_-(V) := \begin{cases} 0 & \text{if } \nu(V) \geq 0 \\ \nu(V) & \text{if } \nu(V) < 0. \end{cases} \tag{13.504}$$

Clearly, $\nu(V) = \nu_+(V) + \nu_-(V)$ for all $V \in \mathrm{Ob}(\mathcal{V}(\mathcal{H}))$. Also, for all V, $\nu_+(V)\nu_-(V) = 0$, and hence

$$\nu(V)^2 = \nu_+(V)^2 + \nu_-(V)^2 \tag{13.505}$$

However, (i) the function $V \mapsto \nu_+(V)^2$ is order-reversing; and (ii) the function $V \mapsto \nu_-(V)^2$ is order-preserving. But then $V \mapsto -\nu_-(V)^2$ is order-reversing. Hence, by rewriting (13.505) as

$$\nu(V)^2 = \nu_+(V)^2 - (-\nu_-(V)^2) \tag{13.506}$$

we see that the function $V \mapsto \nu^2(V) := \nu(V)^2$ is an element of $\mathrm{BV}(\mathrm{Ob}(\mathcal{V}(\mathcal{H})), \mathbb{R})$.

In terms of $k(\Gamma\underline{\mathbb{R}^{\succeq}})$, we can define

$$[\nu, 0]^2 := [\nu_+^2, -\nu_-^2] \tag{13.507}$$

which belongs to $k(\Gamma\underline{\mathbb{R}^{\succeq}})$. Hence, although there exist $\nu \in \Gamma\underline{\mathbb{R}^{\succeq}}$ that have no square in $\Gamma\underline{\mathbb{R}^{\succeq}}$, such global elements of $\underline{\mathbb{R}^{\succeq}}$ *do* have squares in the k-completion, $k(\Gamma\underline{\mathbb{R}^{\succeq}})$. On the level of functions of bounded variation, we have shown that the square of a monotonic (order-reversing) function is a function of bounded variation.

On the other hand, we cannot take the square of an arbitrary element $[\nu, \kappa] \in \Gamma\underline{\mathbb{R}^{\succeq}}$, since the square of a function of bounded variation need not be a function of bounded variation.[137]

The Object $k(\underline{\mathbb{R}^{\succeq}})$ in the Topos $\mathbf{Sets}^{\mathcal{V}(\mathcal{H})^{op}}$

The Definition of $k(\underline{\mathbb{R}^{\succeq}})$

The next step is to translate these results about the set $k(\Gamma\underline{\mathbb{R}^{\succeq}})$ into the construction of an object $k(\underline{\mathbb{R}^{\succeq}})$ in the topos $\mathbf{Sets}^{\mathcal{V}(\mathcal{H})^{\mathrm{op}}}$. We anticipate that, if this can be done, then $k(\Gamma\underline{\mathbb{R}^{\succeq}}) \simeq \Gamma k(\underline{\mathbb{R}^{\succeq}})$.

As was discussed in Sect. (13.8.2), the presheaf $\underline{\mathbb{R}^{\succeq}}$ is defined at each stage V by

$$\underline{\mathbb{R}^{\succeq}}_V := \{\nu : \downarrow V \to \mathbb{R} \mid \nu \in \mathcal{OR}(\downarrow V, \mathbb{R})\}. \tag{13.508}$$

If $i_{V'V} : V' \subseteq V$, then the presheaf map from $\underline{\mathbb{R}^{\succeq}}_V$ to $\underline{\mathbb{R}^{\succeq}}_{V'}$ is just the restriction of the order-reversing functions from $\downarrow V$ to $\downarrow V'$.

The first step in constructing $k(\underline{\mathbb{R}^{\succeq}})$ is to define an equivalence relation on pairs of functions, $\nu, \kappa \in \underline{\mathbb{R}^{\succeq}}_V$, for each stage V, by saying that $(\nu_1, \kappa_1) \equiv (\nu_2, \kappa_2)$ if, and only, there exists $\alpha \in \underline{\mathbb{R}^{\succeq}}_V$ such that

$$\nu_1(V') + \kappa_2(V') + \alpha(V') = \kappa_1(V') + \nu_2(V') + \alpha(V') \tag{13.509}$$

for all $V' \subseteq V$.

---

[137] We have to consider functions like $(\nu_+ + \nu_- - (\kappa_+ + \kappa_-))^2$, which contains terms of the form $\nu_+\kappa_-$ and $\nu_-\kappa_+$: in general, these are neither order-preserving nor order-reversing.

Definition 25 The presheaf $k(\underline{\mathbb{R}^{\succeq}})$ is defined over the category $\mathcal{V}(\mathcal{H})$ in the following way.

(i) On objects $V \in \mathrm{Ob}(\mathcal{V}(\mathcal{H}))$:

$$k(\underline{\mathbb{R}^{\succeq}})_V := \{[\nu, \kappa] \mid \nu, \kappa \in \mathcal{OR}(\downarrow V, \mathbb{R})\}, \tag{13.510}$$

where $[\nu, \kappa]$ denotes the k-equivalence class of (ν, κ).

(ii) On morphisms $i_{V'V} : V' \subseteq V$: The arrow $k(\underline{\mathbb{R}^{\succeq}})(i_{V'V}) : k(\underline{\mathbb{R}^{\succeq}})_V \to k(\underline{\mathbb{R}^{\succeq}})_{V'}$ is given by $\big(k(\underline{\mathbb{R}^{\succeq}})(i_{V'V})\big)([\nu, \kappa]) := [\nu|_{V'}, \kappa|_{V'}]$ for all $[\nu, \kappa] \in k(\underline{\mathbb{R}^{\succeq}})_V$.

It is straightforward to show that $k(\underline{\mathbb{R}^{\succeq}})$ is an abelian group-object in the topos $\mathbf{Sets}^{\mathcal{V}(\mathcal{H})^{\mathrm{op}}}$. In particular, an arrow $+ : k(\underline{\mathbb{R}^{\succeq}}) \times k(\underline{\mathbb{R}^{\succeq}}) \to k(\underline{\mathbb{R}^{\succeq}})$ is defined at each stage V by

$$+_V \big([\nu_1, \kappa_1], [\nu_2, \kappa_2]\big) := [\nu_1 + \nu_2, \kappa_1 + \kappa_2] \tag{13.511}$$

for all $\big([\nu_1, \kappa_1], [\nu_2, \kappa_2]\big) \in k(\underline{\mathbb{R}^{\succeq}})_V \times k(\underline{\mathbb{R}^{\succeq}})_V$. It is easy to see that (i) $\Gamma k(\underline{\mathbb{R}^{\succeq}}) \simeq k(\Gamma \underline{\mathbb{R}^{\succeq}})$; and (ii) $\underline{\mathbb{R}^{\succeq}}$ is a sub-object of $k(\underline{\mathbb{R}^{\succeq}})$ in the topos $\mathbf{Sets}^{\mathcal{V}(\mathcal{H})^{\mathrm{op}}}$.

The Presheaf $k(\underline{\mathbb{R}^{\succeq}})$ as the Quantity-Value Object

We can now identify $k(\underline{\mathbb{R}^{\succeq}})$ as a possible quantity-value object in $\mathbf{Sets}^{\mathcal{V}(\mathcal{H})^{\mathrm{op}}}$. To each bounded, self-adjoint operator $\hat{A}$, there is an arrow $[\breve{\delta}^o(\hat{A})] : \underline{\Sigma} \to k(\underline{\mathbb{R}^{\succeq}})$, given by first sending $\hat{A} \in B(\mathcal{H})_{\mathrm{sa}}$ to $\breve{\delta}^o(\hat{A})$ and then taking k-equivalence classes. More precisely, one takes the monic $\iota : \underline{\mathbb{R}^{\succeq}} \hookrightarrow k(\underline{\mathbb{R}^{\succeq}})$ and then constructs $\iota \circ \breve{\delta}^o(\hat{A}) : \underline{\Sigma} \to k(\underline{\mathbb{R}^{\succeq}})$.

Since, for each stage V, the elements in the image of $[\breve{\delta}^o(\hat{A})]_V = (\iota \circ \breve{\delta}^o(\hat{A}))_V$ are of the form $[\nu, 0]$, $\nu \in \underline{\mathbb{R}^{\succeq}}_V$, their square is well-defined. From a physical perspective, the use of $k(\underline{\mathbb{R}^{\succeq}})$ rather than $\underline{\mathbb{R}^{\succeq}}$ renders possible the definition of things like the 'intrinsic dispersion', $\nabla(\hat{A}) := \breve{\delta}^o(\widehat{A^2}) - \breve{\delta}^o(\hat{A})^2$; see (13.306).

The Square of an Arrow $[\breve{\delta}^o(\hat{A})]$

An arrow $[\breve{\delta}^o(\hat{A})] : \underline{\Sigma} \to k(\underline{\mathbb{R}^{\succeq}})$ is constructed by first forming the outer daseinisation $\breve{\delta}^o(\hat{A})$ of $\hat{A}$, which is an arrow from $\underline{\Sigma}$ to $\underline{\mathbb{R}^{\succeq}}$, and then composing with the monic arrow from $\underline{\mathbb{R}^{\succeq}}$ to $k(\underline{\mathbb{R}^{\succeq}})$. Since only outer daseinisation is used, for each $V \in \mathcal{V}(\mathcal{H})$ and each $\lambda \in \underline{\Sigma}_V$ one obtains an element of $k(\underline{\mathbb{R}^{\succeq}})_V$ of the form $[\delta^o(\hat{A})_V(\lambda), 0]$. We saw how to take the square of these functions, and applying this to all $\lambda \in \underline{\Sigma}_V$ and all $V \in \mathcal{V}(\mathcal{H})$, we get the square $[\breve{\delta}^o(\hat{A})]^2$ of the arrow $[\breve{\delta}^o(\hat{A})]$.

If we consider an arrow of the form $\breve{\delta}(\hat{A}) : \underline{\Sigma} \to \underline{\mathbb{R}^{\leftrightarrow}}$, then the construction involves both inner and outer daseinisation, see Theorem 7. For each V and each $\lambda \in \underline{\Sigma}_V$, we obtain a pair of functions $(\delta^i(\hat{A})_V(\lambda), \delta^o(\hat{A})_V(\lambda))$, which are both not constantly 0 in general. There is no canonical way to take the square of these in

$\underline{\mathbb{R}^{\leftrightarrow}}_V$. Going to the k-extension $k(\underline{\mathbb{R}^{\leftrightarrow}})$ of $\underline{\mathbb{R}^{\leftrightarrow}}$ does not improve the situation, so we cannot define the square of an arrow $\breve{\delta}(\hat{A})$ (or $[\breve{\delta}(\hat{A})]$ in general.

Appendix 2: A Short Introduction to the Relevant Parts of Topos Theory

What is a Topos?

It is impossible to give here more than the briefest of introductions to topos theory. At the danger of being highly imprecise, we restrict ourselves to mentioning some aspects of this well-developed mathematical theory and give a number of pointers to the literature. The aim merely is to give a very rough idea of the structure and internal logic of a topos.

There are a number of excellent textbooks on topos theory, and the reader should consult at least one of them. We found the following books useful: [64, 34, 66, 55, 11, 59].

Topos theory is a remarkably rich branch of mathematics which can be approached from a variety of different viewpoints. The basic area of mathematics is category theory; where, we recall, a category consists of a collection of *objects* and a collection of *morphisms* (or *arrows*).

In the special case of the category of sets, the objects are sets, and a morphism is a function between a pair of sets. In general, each morphism f in a category is associated with a pair of objects.[138] known as its "domain" and "codomain", and is written as $f : B \to A$ where B and A are the domain and codomain respectively. Note that this arrow notation is used even if f is not a function in the normal set-theoretic sense. A key ingredient in the definition of a category is that if $f : B \to A$ and $g : C \to B$ (i.e., the codomain of g is equal to the domain of f) then f and g can be 'composed' to give an arrow $f \circ g : C \to A$; in the case of the category of sets, this is just the usual composition of functions.

A simple example of a category is given by any partially-ordered set ("poset") $\mathcal{C}$: (i) the objects are defined to be the elements of $\mathcal{C}$; and (ii) if $p, q \in \mathcal{C}$, a morphism from p to q is defined to exist if, and only if, $p \preceq q$ in the poset structure. Thus, in a poset regarded as a category, there is at most one morphism between any pair of objects $p, q \in \mathcal{C}$; if it exists, we shall write this morphism as $i_{pq} : p \to q$. This example is important for us in form of the "category of contexts", $\mathcal{V}(\mathcal{H})$, in quantum theory. The objects in $\mathcal{V}(\mathcal{H})$ are the commutative, unital[139] von Neumann sub-algebras of the algebra, $B(\mathcal{H})$, of all bounded operators on the Hilbert space $\mathcal{H}$.

---

[138] The collection of all objects in category, $\mathcal{C}$, is denoted $\text{Ob}(\mathcal{C})$. The collection of arrows from B to A is denoted $\text{Hom}_{\mathcal{C}}(B, A)$. We will only be interested in 'small' categories in which both these collections are sets (rather than the, more general, classes.)

[139] "Unital" means that all these algebras contain the identity operator $\hat{1} \in B(\mathcal{H})$.

Topoi as Mathematical Universes

Every (elementary) topos τ can be seen as a *mathematical universe*. As a category, a topos τ possesses a number of structures that generalise constructions that are possible in the category, **Sets**, of sets and functions.[140] Namely, in **Sets**, we can construct new sets from given ones in several ways. Specifically, let S, T be two sets, then we can form the cartesian product $S \times T$, the disjoint union $S \amalg T$ and the exponential S^T—the set of all functions from T to S.

These constructions turn out to be fundamental, and they can all be phrased in an abstract, categorical manner, where they are called the "product", "co-product" and "exponential", respectively. By definition, in a topos τ, these operations always exist. The first and second of these properties are called "finite completeness" and "finite co-completeness", respectively.

One consequence of the existence of finite limits is that each topos, τ, has a *terminal object*, denoted by 1_τ. This is characterised by the property that for any object A in the topos τ, there exists exactly one arrow from A to 1_τ. In **Sets**, any one-element set $1 = \{*\}$ is terminal.[141]

Of course, **Sets** is a topos, too, and it is precisely the topos which usually plays the role of our mathematical universe, since we construct our mathematical objects starting from sets and functions between them. As a slogan, we have: a topos τ is a category with "certain crucial" properties that are similar to those in **Sets**. A very nice and gentle introduction to these aspects of topos theory is the book [64]. Other good sources are [34, 65].

In order to "do mathematics", one must also have a logic, including a deductive system. Each topos comes equipped with an *internal logic*, which is of *intuitionistic* type. We will now very briefly sketch the main characteristics of intuitionistic logic and the mathematical structures in a topos that realise this logic.

The Sub-object Classifier

Let X be a set, and let $P(X)$ be the power set of X; i.e., the set of subsets of X. Given a subset $K \in P(X)$, one can ask for each point $x \in X$ whether or not it lies in K. Thus there is the *characteristic function* $\chi_K : X \to \{0, 1\}$ of K, which is defined as

$$\chi_K(x) := \begin{cases} 1 \text{ if } x \in K \\ 0 \text{ if } x \notin K \end{cases} \tag{13.512}$$

for all $x \in X$; cf. (13.95). The two-element set $\{0, 1\}$ plays the role of a set of *truth values* for propositions (of the form "$x \in K$"). Clearly, 1 corresponds to "true",

---

[140] More precisely, *small* sets and functions between them. Small means that we do not have proper classes. One must take care in these foundational issues to avoid problems like Russell's paradox.

[141] Like many categorical constructions, the terminal object is fixed only up to isomorphism: all one-element sets are isomorphic to each other, and any of them can serve as a terminal object. Nonetheless, one speaks of *the* terminal object.

0 corresponds to "false", and there are no other possibilities. This is an argument about sets, so it takes place in, and uses the logic of, the topos **Sets** of sets and functions. **Sets** is a *Boolean topos*, in which the familiar two-valued logic and the axiom (∗) hold. (This does not contradict the fact that the internal logic of topoi is intuitionistic, since Boolean logic is a special case of intuitionistic logic.)

In an arbitrary topos, τ, there is a special object Ω_τ, called the *sub-object classifier*, that takes the role of the set $\{0, 1\} \simeq \{\text{false}, \text{true}\}$ of truth values. Let B be an object in the topos, and let A be a sub-object of B. This means that there is a monic $A \to B$,[142] (this is the categorical generalisation of the inclusion of a subset K into a larger set X). As in the case of **Sets**, we can also characterise A as a sub-object of B by an arrow from B to the sub-object classifier Ω_τ; in **Sets**, this arrow is the characteristic function $\chi_K : X \to \{0, 1\}$ of (13.512). Intuitively, this 'characteristic arrow' from B to Ω_τ describes how A 'lies in' B. The textbook definition is:

Definition 26 In a category τ with finite limits, a sub-object classifier is an object Ω_τ, together with a monic true : $1_\tau \to \Omega_\tau$, such that to every monic $m : A \to B$ in τ there is a unique arrow $\chi_A : B \to \Omega_\tau$ which, with the given monic, forms a pullback square

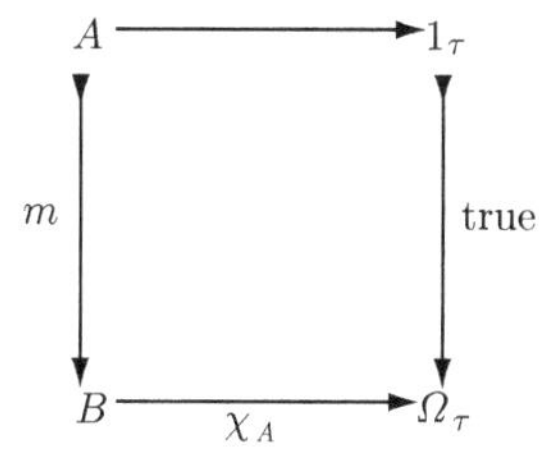

In **Sets**, the arrow true : $1 \to \{0, 1\}$ is given by true(∗) $= 1$. In general, the sub-object classifier, Ω_τ, need not be a set, since it is an object in the topos τ, and the objects of τ need not be sets. Nonetheless, there is an abstract notion of *elements* (or *points*) in category theory that we can use. Then the elements of Ω_τ are the truth values available in the internal logic of our topos τ, just like "false" and "true", the elements of {false, true}, are the truth values available in the topos **Sets**.

To understand the abstract notion of elements, let us consider sets for a moment. Let $1 = \{*\}$ be a one-element set, the terminal object in **Sets**. Let S be a set and consider an arrow e from 1 to S. Clearly, (i) $e(*) \in S$ is an element of S; and (ii) the set of all functions from 1 to S corresponds exactly to the set of all elements of S.

This idea can be generalised to any category that has a terminal object 1. More precisely, an *element* of an object A is defined to be an arrow from 1 to A in the category. For example, in the definition of the sub-object classifier the arrow "true : $1_\tau \to \Omega_\tau$" is an element of Ω_τ. It may happen that an object A has no elements, i.e., there are no arrows $1_\tau \to A$. It is common to consider arrows from sub-objects U of A to A as *generalised elements*.

---

[142] A *monic* is the categorical version of an injective function. In the topos **Sets**, monics exactly are injective functions.

As mentioned above, the elements of the sub-object classifier, understood as the arrows $1_\tau \rightarrow \Omega_\tau$, are the truth values. Moreover, the set of these arrows forms a Heyting algebra (see, for example, Sect. 8.3 in [34]). This is how (the algebraic representation of) intuitionistic logic manifests itself in a topos. Another, closely related fact is that the set, $\mathrm{Sub}(A)$, of sub-objects of any object A in a topos forms a Heyting algebra.

The Definition of a Topos

Let us pull together these various remarks and list the most important properties of a topos, τ, for our purposes:

1. There is a terminal object 1_τ in τ. Thus, given any object A in the topos, there is a unique arrow $A \rightarrow 1_\tau$.
   For any object A in the topos, an arrow $1_\tau \rightarrow A$ is called a *global element* of A. The set of all global elements of A is denoted ΓA.
   Given $A, B \in \mathrm{Ob}(\tau)$, there is a product $A \times B$ in τ. In fact, a topos always has *pull-backs*, and the product is just a special case of this.[143]
2. There is an initial object 0_τ in τ. This means that given any object A in the topos, there is a unique arrow $0_\tau \rightarrow A$.
   Given $A, B \in \mathrm{Ob}(\tau)$, there is a co-product $A \sqcup B$ in τ. In fact, a topos always has *push-outs*, and the co-product is just a special case of this.[144]
3. There is *exponentiation*: i.e., given objects A, B in τ we can form an object A^B, which is the topos analogue of the set of functions from B to A in set theory. The definitive property of exponentiation is that, given any object C, there is an isomorphism

$$\mathrm{Hom}_\tau\big(C, A^B\big) \simeq \mathrm{Hom}_\tau\big(C \times B, A\big) \tag{13.513}$$

   that is natural in A and C; i.e., it is 'well-behaved' under morphisms of the objects involved.
4. There is a sub-object classifier Ω_τ.

Presheaves on a Poset

To illustrate the main ideas, we will first give a few definitions from the theory of presheaves on a partially ordered set (or "poset"); in the case of quantum theory, this poset is the space of "contexts" in which propositions are asserted. We shall then use these ideas to motivate the definition of a presheaf on a general category. Only the briefest of treatments is given here, and the reader is referred to the standard literature for more information [34, 66].

---

143 The conditions in 1. above are equivalent to saying that τ is finitely complete.

144 The conditions in 2. above are equivalent to saying that τ is finitely co-complete.

A *presheaf* (also known as a *varying set*) $\underline{X}$ on a poset $\mathcal{C}$ is a function that assigns to each $p \in \mathcal{C}$, a set $\underline{X}_p$; and to each pair $p \preceq q$ (i.e., $i_{pq} : p \to q$), a map $\underline{X}_{qp} : \underline{X}_q \to \underline{X}_p$ such that (i) $\underline{X}_{pp} : \underline{X}_p \to \underline{X}_p$ is the identity map $\mathrm{id}_{\underline{X}_p}$ on $\underline{X}_p$, and (ii) whenever $p \preceq q \preceq r$, the composite map $\underline{X}_r \xrightarrow{\underline{X}_{rq}} \underline{X}_q \xrightarrow{\underline{X}_{qp}} \underline{X}_p$ is equal to $\underline{X}_r \xrightarrow{\underline{X}_{rp}} \underline{X}_p$, i.e.,

$$\underline{X}_{rp} = \underline{X}_{qp} \circ \underline{X}_{rq}. \tag{13.514}$$

The notation $\underline{X}_{qp}$ is shorthand for the more cumbersome $\underline{X}(i_{pq})$; see below in the definition of a functor.

An *arrow*, or *natural transformation* $\eta : \underline{X} \to \underline{Y}$ between two presheaves $\underline{X}, \underline{Y}$ on $\mathcal{C}$ is a family of maps $\eta_p : \underline{X}_p \to \underline{Y}_p$, $p \in \mathcal{C}$, that satisfy the intertwining conditions

$$\eta_p \circ \underline{X}_{qp} = \underline{Y}_{qp} \circ \eta_q \tag{13.515}$$

whenever $p \preceq q$. This is equivalent to the commutative diagram

$$\begin{array}{ccc} \underline{X}_q & \xrightarrow{\underline{X}_{qp}} & \underline{X}_p \\ {\scriptstyle \eta_q}\downarrow & & \downarrow{\scriptstyle \eta_p} \\ \underline{Y}_q & \xrightarrow[\underline{Y}_{qp}]{} & \underline{Y}_p \end{array} \tag{13.516}$$

It follows from these basic definitions, that a sub-object of a presheaf $\underline{X}$ is a presheaf $\underline{K}$, with an arrow $i : \underline{K} \to \underline{X}$ such that (i) $\underline{K}_p \subseteq \underline{X}_p$ for all $p \in \mathcal{C}$; and (ii) for all $p \preceq q$, the map $K_{qp} : \underline{K}_q \to \underline{K}_p$ is the restriction of $\underline{X}_{qp} : \underline{X}_q \to \underline{X}_p$ to the subset $\underline{K}_q \subseteq \underline{X}_q$. This is shown in the commutative diagram

$$\begin{array}{ccc} \underline{K}_q & \xrightarrow{\underline{K}_{qp}} & \underline{K}_p \\ \downarrow & & \downarrow \\ \underline{X}_q & \xrightarrow[\underline{X}_{qp}]{} & \underline{X}_p \end{array} \tag{13.517}$$

where the vertical arrows are subset inclusions.

The collection of all presheaves on a poset $\mathcal{C}$ forms a category, denoted $\mathbf{Sets}^{\mathcal{C}^{\mathrm{op}}}$. The arrows/morphisms between presheaves in this category the arrows (natural transformations) defined above.

Presheaves on a General Category

The ideas sketched above admit an immediate generalization to the theory of presheaves on an arbitrary "small" category $\mathcal{C}$ (the qualification "small" means that the collection of objects is a genuine set, as is the collection of all arrows/morphisms between any pair of objects). To make the necessary definition we first need the idea of a "functor":

The Idea of a Functor

A central concept is that of a "functor" between a pair of categories $\mathcal{C}$ and $\mathcal{D}$. Broadly speaking, this is an arrow-preserving function from one category to the other. The precise definition is as follows.

Definition 27 1. A covariant functor F from a category $\mathcal{C}$ to a category $\mathcal{D}$ is a function that assigns

(a)to each $\mathcal{C}$-object A, a $\mathcal{D}$-object F_A;
(b)to each $\mathcal{C}$-morphism $f : B \to A$, a $\mathcal{D}$-morphism $F(f) : F_B \to F_A$ such that $F(\mathrm{id}_A) = \mathrm{id}_{F_A}$; and, if $g : C \to B$, and $f : B \to A$ then

$$F(f \circ g) = F(f) \circ F(g). \tag{13.518}$$

2. A contravariant functor X from a category $\mathcal{C}$ to a category $\mathcal{D}$ is a function that assigns

(a)to each $\mathcal{C}$-object A, a $\mathcal{D}$-object X_A;
(b)to each $\mathcal{C}$-morphism $f : B \to A$, a $\mathcal{D}$-morphism $X(f) : X_A \to X_B$ such that $X(\mathrm{id}_A) = \mathrm{id}_{X_A}$; and, if $g : C \to B$, and $f : B \to A$ then

$$X(f \circ g) = X(g) \circ X(f). \tag{13.519}$$

The connection with the idea of a presheaf on a poset is straightforward. As mentioned above, a poset $\mathcal{C}$ can be regarded as a category in its own right, and it is clear that a presheaf on the poset $\mathcal{C}$ is the same thing as a contravariant functor $\underline{X}$ from the category $\mathcal{C}$ to the category **Sets** of normal sets. Equivalently, it is a covariant functor from the "opposite" category[145] $\mathcal{C}^{\mathrm{op}}$ to **Sets**. Clearly, (13.514) corresponds to the contravariant condition (13.519). Note that mathematicians usually call the objects in $\mathcal{C}$ "stages of truth", or just "stages". For us they are "contexts", "classical snap-shops", or "world views".

[145] The 'opposite' of a category $\mathcal{C}$ is a category, denoted $\mathcal{C}^{\mathrm{op}}$, whose objects are the same as those of $\mathcal{C}$, and whose morphisms are defined to be the opposite of those of $\mathcal{C}$; i.e., a morphism $f : A \to B$ in $\mathcal{C}^{\mathrm{op}}$ is said to exist if, and only if, there is a morphism $f : B \to A$ in $\mathcal{C}$.

Presheaves on an Arbitrary Category $\mathcal{C}$

These remarks motivate the definition of a presheaf on an arbitrary small category $\mathcal{C}$: namely, a *presheaf* on $\mathcal{C}$ is a covariant functor[146] $\underline{X} : \mathcal{C}^{\mathrm{op}} \to \mathbf{Sets}$ from $\mathcal{C}^{\mathrm{op}}$ to the category of sets. Equivalently, a presheaf is a contravariant functor from $\mathcal{C}$ to the category of sets.

We want to make the collection of presheaves on $\mathcal{C}$ into a category, and therefore we need to define what is meant by a "morphism" between two presheaves $\underline{X}$ and $\underline{Y}$. The intuitive idea is that such a morphism from $\underline{X}$ to $\underline{Y}$ must give a "picture" of $\underline{X}$ within $\underline{Y}$. Formally, such a morphism is defined to be a *natural transformation* $N : \underline{X} \to \underline{Y}$, by which is meant a family of maps (called the *components* of N) $N_A : \underline{X}_A \to \underline{Y}_A$, $A \in \mathrm{Ob}()\mathcal{C}$, such that if $f : B \to A$ is a morphism in $\mathcal{C}$, then the composite map $\underline{X}_A \xrightarrow{N_A} \underline{Y}_A \xrightarrow{\underline{Y}(f)} \underline{Y}_B$ is equal to $\underline{X}_A \xrightarrow{\underline{X}(f)} \underline{X}_B \xrightarrow{N_B} \underline{Y}_A$. In other words, we have the commutative diagram

$$\begin{array}{ccc} \underline{X}_A & \xrightarrow{\underline{X}(f)} & \underline{X}_B \\ {\scriptstyle N_A}\downarrow & & \downarrow{\scriptstyle N_B} \\ \underline{Y}_A & \xrightarrow[\underline{Y}(f)]{} & \underline{Y}_B \end{array} \tag{13.520}$$

of which (13.516) is clearly a special case. The category of presheaves on $\mathcal{C}$ equipped with these morphisms is denoted $\mathbf{Sets}^{\mathcal{C}^{\mathrm{op}}}$.

The idea of a sub-object generalizes in an obvious way. Thus we say that $\underline{K}$ is a *sub-object* of $\underline{X}$ if there is a morphism in the category of presheaves (i.e., a natural transformation) $\iota : \underline{K} \to \underline{X}$ with the property that, for each A, the component map $\iota_A : \underline{K}_A \to \underline{X}_A$ is a subset embedding, i.e., $\underline{K}_A \subseteq \underline{X}_A$. Thus, if $f : B \to A$ is any morphism in $\mathcal{C}$, we get the analogue of the commutative diagram (13.517):

$$\begin{array}{ccc} \underline{K}_A & \xrightarrow{\underline{K}(f)} & \underline{K}_B \\ \downarrow & & \downarrow \\ \underline{X}_A & \xrightarrow[X(f)]{} & \underline{X}_B \end{array} \tag{13.521}$$

where, once again, the vertical arrows are subset inclusions.

The category of presheaves on $\mathcal{C}$, $\mathbf{Sets}^{\mathcal{C}^{\mathrm{op}}}$, forms a topos. We do not need the full definition of a topos; but we do need the idea, mentioned in Sect. 13.17.2, that a topos has a sub-object classifier Ω, to which we now turn.

---

[146] Throughout this article, a presheaf is indicated by a letter that is underlined.

Sieves and The Sub-object Classifier $\underline{\Omega}$

Among the key concepts in presheaf theory is that of a "sieve", which plays a central role in the construction of the sub-object classifier in the topos of presheaves on a category $\mathcal{C}$.

A *sieve* on an object A in $\mathcal{C}$ is defined to be a collection S of morphisms $f : B \to A$ in $\mathcal{C}$ with the property that if $f : B \to A$ belongs to S, and if $g : C \to B$ is any morphism with co-domain B, then $f \circ g : C \to A$ also belongs to S. In the simple case where $\mathcal{C}$ is a poset, a sieve on $p \in \mathcal{C}$ is any subset S of $\mathcal{C}$ such that if $r \in S$ then (i) $r \preceq p$, and (ii) $r' \in S$ for all $r' \preceq r$; in other words, a sieve is nothing but a *lower* set in the poset.

The presheaf $\underline{\Omega} : \mathcal{C} \to$ **Sets** is now defined as follows. If A is an object in $\mathcal{C}$, then $\underline{\Omega}_A$ is defined to be the set of all sieves on A; and if $f : B \to A$, then $\underline{\Omega}(f) : \underline{\Omega}_A \to \underline{\Omega}_B$ is defined as

$$\underline{\Omega}(f)(S) := \{h : C \to B \mid f \circ h \in S\} \tag{13.522}$$

for all $S \in \underline{\Omega}_A$; the sieve $\underline{\Omega}(f)(S)$ is often written as $f^*(S)$, and is known as the *pull-back* to B of the sieve S on A by the morphism $f : B \to A$.

It should be noted that if S is a sieve on A, and if $f : B \to A$ belongs to S, then from the defining property of a sieve we have

$$f^*(S) := \{h : C \to B \mid f \circ h \in S\} = \{h : C \to B\} =: \downarrow\! B \tag{13.523}$$

where $\downarrow\! B$ denotes the *principal* sieve on B, defined to be the set of all morphisms in $\mathcal{C}$ whose codomain is B.

If $\mathcal{C}$ is a poset, the pull-back operation corresponds to a family of maps $\underline{\Omega}_{qp} : \underline{\Omega}_q \to \underline{\Omega}_p$ (where $\underline{\Omega}_p$ denotes the set of all sieves/lower sets on p in the poset) defined by $\underline{\Omega}_{qp} = \underline{\Omega}(i_{pq})$ if $i_{pq} : p \to q$ (i.e., $p \preceq q$). It is straightforward to check that if $S \in \underline{\Omega}_q$, then

$$\underline{\Omega}_{qp}(S) :=\downarrow\! p \cap S \tag{13.524}$$

where $\downarrow p := \{r \in \mathcal{C} \mid r \preceq p\}$.

A crucial property of sieves is that the set $\underline{\Omega}_A$ of sieves on A has the structure of a Heyting algebra. Specifically, the unit element $1_{\underline{\Omega}_A}$ in $\underline{\Omega}_A$ is the principal sieve $\downarrow\! A$, and the null element $0_{\underline{\Omega}_A}$ is the empty sieve $\emptyset$. The partial ordering in $\underline{\Omega}_A$ is defined by $S_1 \preceq S_2$ if, and only if, $S_1 \subseteq S_2$; and the logical connectives are defined as:

$$S_1 \wedge S_2 := S_1 \cap S_2 \tag{13.525}$$

$$S_1 \vee S_2 := S_1 \cup S_2 \tag{13.526}$$

$$S_1 \Rightarrow S_2 := \{f : B \to A \mid \forall g : C \to B \text{ if } f \circ g \in S_1 \text{ then } f \circ g \in S_2\} \tag{13.527}$$

As in any Heyting algebra, the negation of an element S (called the *pseudo-complement* of S) is defined as $\neg S := S \Rightarrow 0$; so that

$$\neg S := \{f : B \to A \mid \text{for all } g : C \to B,\ f \circ g \notin S\}. \tag{13.528}$$

It can be shown that the presheaf $\underline{\Omega}$ is a sub-object classifier for the topos $\mathbf{Sets}^{\mathcal{C}^{\text{op}}}$. That is to say, sub-objects of any object $\underline{X}$ in this topos (i.e., any presheaf on $\mathcal{C}$) are in one-to-one correspondence with morphisms $\chi : \underline{X} \to \underline{\Omega}$. This works as follows. First, let $\underline{K}$ be a sub-object of $\underline{X}$ with an associated characteristic arrow $\chi_{\underline{K}} : \underline{X} \to \underline{\Omega}$. Then, at any stage A in $\mathcal{C}$, the 'components' of this arrow, $\chi_{\underline{K}A} : \underline{X}_A \to \underline{\Omega}_A$, are defined as

$$\chi_{\underline{K}A}(x) := \{f : B \to A \mid \underline{X}(f)(x) \in \underline{K}_B\} \tag{13.529}$$

for all $x \in \underline{X}_A$. That the right hand side of (13.529) actually *is* a sieve on A follows from the defining properties of a sub-object.

Thus, in each "branch" of the category $\mathcal{C}$ going "down" from the stage A, $\chi_{\underline{K}A}(x)$ picks out the first member B in that branch for which $\underline{X}(f)(x)$ lies in the subset $\underline{K}_B$, and the commutative diagram (13.521) then guarantees that $\underline{X}(h \circ f)(x)$ will lie in $\underline{K}_C$ for all $h : C \to B$. Thus each stage A in $\mathcal{C}$ serves as a possible context for an assignment to each $x \in \underline{X}_A$ of a generalised truth value—a sieve belonging to the Heyting algebra $\underline{\Omega}_A$. This is the sense in which contextual, generalised truth values arise naturally in a topos of presheaves.

There is a converse to (13.529): namely, each morphism $\chi : \underline{X} \to \underline{\Omega}$ (i.e., a natural transformation between the presheaves $\underline{X}$ and $\underline{\Omega}$) defines a sub-object $\underline{K}^\chi$ of $\underline{X}$ via

$$\underline{K}^\chi_A := \chi_A^{-1}\{1_{\underline{\Omega}_A}\}. \tag{13.530}$$

at each stage A.

Global Elements of a Presheaf

For the category of presheaves on $\mathcal{C}$, a terminal object $\underline{1} : \mathcal{C} \to \mathbf{Sets}$ can be defined by $\underline{1}_A := \{*\}$ at all stages A in $\mathcal{C}$; if $f : B \to A$ is a morphism in $\mathcal{C}$ then $\underline{1}(f) : \{*\} \to \{*\}$ is defined to be the map $* \mapsto *$. This is indeed a terminal object since, for any presheaf $\underline{X}$, we can define a unique natural transformation $N : \underline{X} \to \underline{1}$ whose components $N_A : \underline{X}(A) \to \underline{1}_A = \{*\}$ are the constant maps $x \mapsto *$ for all $x \in \underline{X}_A$.

As a morphism $\gamma : \underline{1} \to \underline{X}$ in the topos $\mathbf{Sets}^{\mathcal{C}^{\text{op}}}$, a global element corresponds to a choice of an element $\gamma_A \in \underline{X}_A$ for each stage A in $\mathcal{C}$, such that, if $f : B \to A$, the 'matching condition'

$$\underline{X}(f)(\gamma_A) = \gamma_B \tag{13.531}$$

is satisfied.

Acknowledgments This research was supported by grant RFP1-06-04 from The Foundational Questions Institute (fqxi.org). AD gratefully acknowledges financial support from the DAAD.

This work is also supported in part by the EC Marie Curie Research and Training Network "ENRAGE" (European Network on Random GEometry) MRTN-CT-2004-005616.

We are both very grateful to the late Professor Hans de Groote for his detailed and insightful comments on our work. We also profited from discussions with Bas Spitter, Chris Heunen and Klaas Landsman, as well as with Samson Abramsky and Bob Coecke, and with Chris Mulvey and Steve Vickers. We thank Rick Kostecki and Bertfried Fauser for pointing out a number of typos in the first version of this paper.

CJI expresses his gratitude to Jeremy Butterfield for the lengthy, and most enjoyable, collaboration in which were formulated the early ideas about using topoi to study quantum theory.

References

1. Abramsky, S.: Domain Theory and the Logic of Observable Properties. PhD thesis, Queen Mary College, London University (1987)
2. Abramsky, S., Coecke, B.: A categorical semantics of quantum protocols. In: Proceedings of the 19th Annual IEEE Symposium on Logic in Computer Science, 415–425, IEEE Computer Science Press (2004). arXiv:quant-ph/0402130v5
3. Adelman, M., Corbett, J.V.: A sheaf model for intuitionistic quantum mechanics. Appl. Categ. Struct. **3**, 79–104 (1995)
4. Baez, J.C.: An Introduction to Spin Foam Models of Quantum Gravity and BF Theory. Lect. Notes Phys. **543**, 25–94 (2000)
5. Baez, J.: Quantum quandaries: A category-theoretic perspective. In: French, S., Rickles, D., Sahatsi, J. (eds.) The Structural Foundations of Quantum Gravity. Oxford University Press, Oxford (2006)
6. Baez, J., Schreiber, U.: Higher Gauge Theory. arXiv:math/0511.710v2 (2005)
7. Banaschewski, B., Mulvey, C.J.: A constructive proof of the Stone-Weierstrass theorem. J. Pure Appl. Algebra **116**, 25–40 (1997)
8. Banaschewski, B., Mulvey, C.J.: The spectral theory of commutative C_-algebras: the constructive spectrum. Quaest. Math. **23**, 425–464 (2000)
9. Banaschewski, B., Mulvey, C.J.: The spectral theory of commutative C_-algebras: the constructive Gelfand-Mazur theorem. Quaest. Math. **23**, 465–488 (2000)
10. Banaschewski, B., Mulvey, C.J.: A globalisation of the Gelfand duality theorem. Ann. Pure Appl. Logic **137**, 62–103 (2006)
11. Bell, J.L.: Toposes and Local Set Theories. Clarendon Press, Oxford (1988)
12. Bell, J.S.: On the problem of hidden variables in quantum mechanics. In: Speakable and Unspeakable in Quantum Mechanics, pp.1–13. Cambridge University Press, Cambridge (1987)
13. Beltrametti, E., Cassinelli, G.: The Logic of Quantum Mechanics. Addison-Wesley, Massachusetts (1981)
14. Birkhoff, G., von Neumann, J.: The logic of quantum mechanics. Ann. Math. **37**, 823–843 (1936)
15. Coecke, B.: Quantum Logic in Intuitionistic Perspective. Studia Logica **70**, 411–440 (2002)
16. Connes, A.: Noncommutative Geometry. Academic Press, San Diego (1994). Download from http://www.alainconnes.org/en/
17. Connes, A., Moscovici, H.: Background independent geometry and Hopf cyclic cohomology. arXiv:math.QA/0505475 (2005)
18. Corbett, J.V., Durt, T.: Collimation processes in quantum mechanics interpreted in quantum real numbers. Preprint (2007)
19. Crane, L.: What is the Mathematical Structure of Quantum Spacetime? arXiv:0706.4452 (2007)
20. Dalla Chiara, M.L., Giuntini, R.: Quantum logics. In: Gabbay, G., Guenthner, F. (eds.) Handbook of Philosophical Logic, vol. VI, pp. 129–228. Kluwer, Dordrecht (2002)

21. Döring, A.: Kochen-Specker theorem for von Neumann algebras. Int. J. Theor. Phys. **44**, 139–160 (2005)
22. Döring, A.: Observables as functions: Antonymous functions. arXiv:quant-ph/0510.102 (2005)
23. Döring, A.: Topos theory and 'neo-realist' quantum theory. In: Fauser, B., Tolksdorf, J., Zeidler, E. (eds.) Quantum Field Theory, Competitive Models. Birkhäuser, Basel, Boston, Berlin (2009). arXiv:0712.4003
24. A. Döring. Quantum States and Measures on the Spectral Presheaf. To appear in Special Issue of Adv. Sci. Lett. on Quantum Gravity, Cosmology and Black Holes, ed. Bojowald, M. (2009). arXiv:0809.4847
25. Döring, A.: The physical interpretation of daseinisation. In preparation (2009)
26. Döring, A.: Algebraic quantum theory in a topos: A comparison. In preparation (2009)
27. Döoring, A., Isham, C.J.: A topos foundation for theories of physics: I. Formal languages for physics. J. Math. Phys. **49**(5), 053515 (2008). arXiv:quant-ph/0703060
28. Döring, A., Isham, C.J.: A topos foundation for theories of physics: II. Daseinisation and the liberation of quantum theory. J. Math. Phys. **49**(5), 053516 (2008). arXiv:quantph/0703062
29. Döring, A., Isham, C.J.: A topos foundation for theories of physics: III. Quantum theory and the representation of physical quantities with arrows $\breve{\delta}(A) : \Sigma \rightarrow R^{\succeq}$. J. Math. Phys. **49**(5), 053517 (2008). arXiv:quant-ph/0703064
30. Döring, A., Isham, C.J.: A topos foundation for theories of physics: IV. Categories of systems. J. Math. Phys. **49**(5), 053518 (2008). arXiv:quant-ph/0703066
31. Döring, A., Lal, R.: Daseinisation and Galois connections. In preparation (2009)
32. Dowker, F.: Causal sets and the deep structure of space-time. In: Abhay A. (ed.)100 Years of Relativity—Space-time Structure: Einstein and Beyond. World Scientific, Singapore (2005)
33. Escard o, M.H.: PCF extended with real numbers: A domain-theoretic approach to higher-order exact real number computation. PhD thesis. Download from http://www.lfcs.inf.ed.ac.uk/reports/97/ECS-LFCS-97–374/index.html (2007)
34. Goldblatt, R.: Topoi: The Categorial Analysis of Logic. North-Holland, London (1984)
35. Grayling, A.C.: An Introduction to Philosophical Logic. Duckworth Press, London (1990)
36. de Groote, H.F.: On a canonical lattice structure on the effect algebra of a von Neumann algebra. arXiv:math-ph/0410.018v2 (2004)
37. de Groote, H.F.: Observables. arXiv:math-ph/0507.019 (2005)
38. de Groote, H.F.: Observables I: Stone spectra. arXiv:math-ph/0509.020 (2005)
39. de Groote, H.F.: Observables II: Quantum observables. arXiv:math-ph/0509.075 (2005)
40. Heidegger, M.: What is a Thing? Regenery/Gateway, Indiana (1967)
41. Heisenberg, W.: Philosophic Problems of Nuclear Science. Pantheon, New York (1952)
42. Heunen, C., Landsmann, K., Spitters, B.: A topos for algebraic quantum theory. arXiv:0709.4364v2 (second version 6. April 2008)
43. Isham, C.J.: Quantum logic and the histories approach to quantum theory. J. Math. Phys. **35**, 2157–2185 (1994)
44. Isham, C.J.: Topos theory and consistent histories: The internal logic of the set of all consistent sets. Int. J. Theor. Phys. **36**, 785–814 (1997)
45. Isham, C.J.: Some reflections on the status of conventional quantum theory when applied to quantum gravity. In: Gibbons, G. (ed.) Proceedings of the Conference in Honour of Stephen Hawking's 60'th Birthday. Cambridge University Press, Cambridge (2003)
46. Isham, C.J.: Is it true; or is it false; or somewhere in between? The logic of quantum theory. Contempory Phys. **46**, 207–219 (2005)
47. Isham, C.J.: A Topos Perspective on State-Vector Reduction. Int. J. Theor. Phys. **45**, 994–1020 (2006)
48. Isham, C.J., Butterfield, J.: A topos perspective on the Kochen-Specker theorem: I. Quantum states as generalised valuations. Int. J. Theor. Phys. **37**, 2669–2733 (1998)
49. Isham, C.J., Butterfield, J.: A topos perspective on the Kochen-Specker theorem: II. Conceptual aspects, and classical analogues. Int. J. Theor. Phys. **38**, 827–859 (1999)

50. Isham, C.J., Hamilton, J., Butterfield, J.: A topos perspective on the Kochen-Specker theorem: III. Von Neumann algebras as the base category. Int. J. Theor. Phys. **39**, 1413–1436 (2000)
51. Isham, C.J., Butterfield, J.: A topos perspective on the Kochen-Specker theorem: IV. Interval valuations. Int. J. Theor. Phys **41**, 613–639 (2002)
52. Isham, C.J., Butterfield, J.: Space-time and the philosophical challenge of quantum gravity. In: Callender, C., Huggett, N. (eds.) Physics Meets Philosophy at the Planck Scale, pp. 33–89. Cambridge University Press, Cambridge (2001)
53. Isham, C.J., Butterfield, J.: Some possible roles for topos theory in quantum theory and quantum gravity. Found. Phys. **30**, 1707–1735 (2000)
54. Jackson, M.: A sheaf theoretic approach to measure theory. PhD thesis, University of Pittsburgh (2006)
55. Johnstone, P.T.: Sketches of an Elephant: A Topos Theory Compendium I, II. Oxford Science Publications, Oxford (2002/03)
56. Kadison, R.V., Ringrose, J.R.: Fundamentals of the Theory of Operator Algebras, vol. 1: Elementary Theory. Academic Press, New York (1983)
57. Kochen, S., Specker, E.P.: The problem of hidden variables in quantum mechanics. J. Math. Mech. **17**, 59–87 (1967)
58. Kock, A.: Synthetic Differential Geometry (LMS lecture note series: 51). Cambridge University Press, Cambridge (1981)
59. Lambek, J., Scott, P.J.: Introduction to higher order categorical logic. Cambridge University Press, Cambridge (1986)
60. Landsman, K.: The principle of general tovariance. philsciarchive. pitt.edu/archive/00003931/ (2008)
61. Lawvere, F.W.: Functorial Semantics of Algebraic Theories. Ph.D. thesis, Columbia University, Columbia (1963). Available from www.tac.mta.ca/tac/reprints/articles/5/tr5abs.html
62. Lawvere, F.W.: Toward the Description in a Smooth Topos of the Dynamically Possible Motions and Deformations of a Continuous Body. Cahiers de Topologie et Geométrie Différentielle Catégorique **XXI**, 337–392 (1980)
63. Lawvere, F.W. (ed.): Introduction to Categories in Continuum Physics. Springer Lecture Notes in Mathematics No. 1174. Springer, New York (1986)
64. Lawvere, F.W., Rosebrugh, R.: Sets for Mathematics. Cambridge University Press, Cambridge (2003)
65. MacLane, S.: Categories for the Working Mathematician. Springer, New York, Berlin, Heidelberg (1971)
66. MacLane, S., Moerdijk, I.: Sheaves in Geometry and Logic: A First Introduction to Topos Theory. Springer, New York, Berlin, Heidelberg (1992)
67. Margenau, H.: Reality in quantum mechanics. Phil. Sci. **16**, 287–302 (1949)
68. Markopoulou, F.: The internal description of a causal set: What the universe looks like from the inside. Comm. Math. Phys. **211**, 559–583 (2002)
69. Mallios, A.: Modern Differential Geometry in Gauge Theories, vol 1: Maxwell Fields. Birkhäuser, Boston (2006)
70. Mallios, A.: Modern Differential Geometry in Gauge Theories, vol 2: Yang-Mills Fields. Birkhäuser, Boston (2006)
71. Mallios, A., Zafiris, E.: Topos-theoretic relativization of physical representability and quantum gravity. arXiv:gr-qc/0610.113 (2007)
72. Nakayama, K.: Topos-theoretic extension of a modal interpretation of quantum mechanics. arXiv:0711.2200 (2007)
73. Olson, M.P.: The Selfadjoint Operators of a von Neumann Algebra form a Conditionally Complete Lattice. Proc. AMS **28**, 537–544 (1971)
74. Ozawa, M.: Transfer principle in quantum set theory. J. Symbolic Logic **72**, 625–648 (2007). arXiv:math/0604349v2
75. Popper, K.: Quantum Theory and the Schism in Physics. Hutchinson, London (1982)
76. Raptis, I.: Categorical quantum gravity. Int. J. Theor. Phys. **45**, 1495–1523 (2006)

77. Smolin, L.: The case for background independence. In: French, S., Rickles, D., Sahatsi, J. (eds.) The Structural Foundations of Quantum Gravity. Oxford University Press, Oxford (2006)
78. Smolin, L.: Generic predictions of quantum theories of gravity. In: Oriti, D. (ed.) Approaches to Quantum Gravity—Toward a New Understanding of Space, Time, and Matter. Cambridge University Press, Cambridge (2006). arXiv:hep-th/0605.052
79. Smyth, M.: Powerdomains and predicate transformers: A topological view. In: Diaz, J. (ed.) Automata, Languages and Programming, Lecture Notes in Computer Science 154. Springer, Berlin (1983)
80. Sorkin, R.D.: Causal Sets: Discrete Gravity (Notes for the Valdivia Summer School). In: Gomberoff, A., Marolf, D. (eds.) Lectures on Quantum Gravity. Plenum, New York (2005)
81. Stone, M.H.: The Theory of Representations for Boolean algebras. Trans. Am. Math. Soc. **40**, 37–111 (1936)
82. Stubbe, I.: The Canonical Topology on a Meet-Semilattice. Int. J. Theor. Phys. **44**(12), 2283–2293 (2005)
83. Takeuti, G.: Quantum set theory. In: Beltrametti, E.G., van Fraasen, B.C. (eds.) Current Issues in Quantum Logic, pp. 303–322. Plenum, New York (1981)
84. Vickers, S.: Topology via Logic. Cambridge University Press, Cambridge (1989)
85. Vickers, S.: Issues of logic, algebra and topology in ontology. In: Theory and Applications of Ontology. In preparation (2008)
86. Vicary, J.: The quantum harmonic oscillator as an adjunction. Int. J. Theor. Phys. (2007). arXiv:0706.0711

Part VI
Geometry and Topology in Computation

Chapter 14
Can a Quantum Computer Run the von Neumann Architecture?

P. Hines

Abstract At the core of nearly every modern computer is a central processing unit running the von Neumann architecture. This computer architecture gives computationally universal machines, and non-trivial control structures arise naturally, leading to high-level programming constructs.

At the core of the von Neumann architecture is the notion that program code may be stored and manipulated in the same way as data. A datum describing an operation may be stored and processed in the same way as any other form of data, but may also be 'promoted' to an operation, and applied.

Classically, this is well-studied—particularly from a categorical point of view. We consider such operations in the quantum setting, including Nielsen and Chuang's orthonormal encoding, Abramsky and Coecke's categorical foundations, the BBC protocol, and the Choi-Jamiołkowsky correspondence.

Obstacles to a quantum analogue of the von Neumann architecture are also considered, including the no-cloning and no-deleting theorems, the "no-programming principle", and the Gottesman-Knill theorem.

14.1 Introduction

For impatient readers, the answer to the question posed in the title is, *No*. Readers familiar with quantum information and computation, may well think that the answer should be, *Of course not !*—although our intention is to prove that the question is more subtle than that. A more accurate, although less concise, title for this paper would therefore be, "Why is the von Neumann architecture so significant for classical computation, what are the differences between quantum and classical information that mean classical computers can implement the von Neumann architecture but quantum computers cannot, and what are the implications of this for models of quantum computation and information?"

P. Hines (✉)
Department of Computer Science, University of York, York, UK
e-mail: peter.hines@cs.york.ac.uk

Hines, P.: *Can a Quantum Computer Run the von Neumann Architecture?*. Lect. Notes Phys. **813**, 941–982 (2011)
DOI 10.1007/978-3-642-12821-9_14

In order to answer these questions, we analyse the von Neumann architecture from a classical point of view, in order to decide what features give it both practical utility and computational power, and then consider whether or not these essential features are shared by quantum systems.

In the classical world, the von Neumann architecture is ubiquitous. This is partly for practical reasons; binary values, logic gates, and a global clock are readily implementable via electronic circuits. However, it also bridges the gap between theoretical notions of computer science, and the underlying physical structures. Universality, programmability, compilation and higher-level control structures all arise in a natural way from the underlying architecture. All these are key concepts of computing—and are much less well-established in the quantum case.

Our claim is that the practical core of the von Neumann architecture is the interchangability of code and data. This is a fundamental concept of theoretical computer science, with close connections to formal logic and λ-calculus. From the categorical point of view, the code/data correspondence is exactly *Cartesian closure*—a special form of categorical closure. Quantum information also admits code / data correspondences—we consider similarities and differences, and their implications for machine architectures.

We emphasise that this paper is expository in the sense that results presented are generally well-known, or at least consequences of well-known theory. The aim is to view this categorical picture through the frame of the von Neumann architecture, and to consider implications from this very practical point of view.

14.2 The von Neumann Architecture

14.2.1 The Origins of the vN Architecture

In 1945, whilst on an extended stay at Los Alamos, J. von Neumann laid out in formal logical terms the basic operating principles of the EDVAC computer [77]. This was an (incomplete) draft report on the work of a team[1] on a project by the University of Philadelphia for the U.S. Army Ballistics Research Laboratory. Famously, this incomplete draft was widely distributed by H. Goldstine, an army mathematician who had originally introduced von Neumann to the project [19].

This distribution of an incomplete draft, listing von Neumann as sole author, later caused considerable bad feeling within the EDVAC team (see [10, 92, 71]), and it has been claimed [10] that von Neumann himself intended for a final completed version to be jointly co-authored by the entire team.

The EDVAC computer itself did not become fully operational until 1951, partly due to a dispute with the University of Philadelphia over intellectual property and patent rights resulting from this prior publication [70]. However, when finally

[1] The team was lead by J. Mauchley and J. P. Eckert, with J. von Neumann acting as a consultant. See [10] for details of the team members.

operational, the EDVAC computer was highly successful [10], and the the general principles outlined in [77] have become an almost universally accepted standard for processor design, known as the von Neuman, or vN, architecture.

14.2.2 The Fetch-Execute Cycle

Although technical details have changed beyond recognition (e.g. a significant advance of the EDVAC machine was the use of acoustic waves in mercury-filled tubes as a form of random-access memory [38]), the underlying principles of the von Neumann architecture were spectacularly successful, and remain in use today, in the form of the core operation of the Central Processing Unit of a computer. This *fetch-execute cycle* is a simple iterative step, performed on every clock cycle, that leads to the full range of behaviour of modern computers.

The fetch-execute step is as follows:

At the beginning of each cycle the *program counter* contains the value of a memory location.

1. The CPU *copies* the contents of the memory location referenced by program counter into the *instruction register*.
2. The data in the instruction register is *decoded* and the *control unit performs the action described*. This may be:

   a) *Copy a value* from memory into the *accumulator*.
   b) *Apply an instruction* (logic gate) to the accumulator.
   c) *Copy the contents* of the accumulator into a *memory location*.
   d) *Overwrite* the contents of the program counter with a *new value*.

3. The program counter is then *incremented* (in order to address the next instruction).

14.2.3 The Utility of the von Neumann Architecture

Practically, the vN architecture is significant for a number of reasons:

1. A *computationally universal* machine can be constructed. Up to memory constraints, any computation that may be performed by a Turing machine or the untyped lambda calculus may be performed on a von Neumann computer.
2. Computers may be *programmed*—the program executed is dependent of the contents of the computer memory, and no hardware reconfiguration is required in order to run a different computer program.
3. Manipulation of program code in a similar manner to data allows for branching, conditional execution, and subroutines. This opens the way to meaningful *control structures*.

4. Meaningful control structures allow high-level languages to be built on top of the basic machine code, and the equal treatment of code and data allows a von Neumann machine to run *compilers*, *interpreters* and *assemblers*.

Our interest in the vN architecture from a quantum-mechanical perspective is in order to seek analogues of 1.–4. above.

In terms of quantum computation, 1. has been intensively studied—we consider this further in Sect. 14.13.1. Also, despite the current paucity of quantum algorithms, 2. may become important at a later stage—we refer to [95] for some interesting ideas regarding stored-code quantum computers.

In terms of languages and control structures, several high-level languages for quantum computers have been proposed (see [34] for a survey). These are generally based on a "classical control, quantum data" paradigm, although purely quantum (i.e. superposition-preserving—see Sect. 14.3.3) conditionals have been proposed in [8, 40], and [52] considers control structures for conditional iteration based on purely quantum control. However, a comparison of [34] with any standard QM computation text (such as [41, 74, 79]) demonstrates that there are many more quantum programming languages than quantum algorithms.[2]

This brings us to 4. In the classical world, this feature has become so deeply ingrained into modern computing as to be almost invisible. The control structures and, to some extent, high-level languages, used in modern computation arise naturally from the underlying structure of the vN architecture. It is this feature that is of particular interest—that intuitive and useful structure arises from the underlying architecture of computer processors.

Finally, our interest in higher-level languages and control structures is in stark contrast to von Neumann's attitude [71]. With regard to his FORTRAN language, J. Backus recalls von Neumann as begin unimpressed, asking, "Why would you want more than machine code?". (See the title of [11] for a contrary view!) D. Gillies also recalls von Neumann's anger at his programming of the first assembler, on the grounds that this was, "using a sophisticated scientific tool to perform clerical tasks that could easily be carried out by graduate students" [71].

14.3 Relevant Quantum Information Theory

14.3.1 Basic Quantum Information

We briefly reprise some fundamentals of quantum information and computation. No attempt is made to give a full or consistent exposition—we concentrate on the

[2] This comment is deliberately unfair, in that languages presented in [34] have been designed for many purposes—including quantum communication protocols (of which there are many), proving correctness of both protocols and algorithms, formal proofs of security for quantum encryption and communication, &c.—and none of them have the creation of new algorithms as a stated objective. However, the point remains that going from quantum programming languages to quantum programs is a highly non-trivial exercise.

basic building blocks and properties relevant to this paper. For a full introduction, we refer to either [40, 64, 79], and [18] for comprehensive mathematical background on complex Hilbert spaces. We also use a pure state description, rather than considering mixed states, density matrices and completely positive maps—any of the previous references will also give a good exposition of this approach.

The atomic building-blocks of quantum information are *quantum bits*, or **qubits**, norm-1 vectors in a 2-dimensional complex Hilbert space Qu. Concatenation of qubits is given by the tensor product of Hilbert spaces, so n qubits are modelled by the 2^n-dimensional space $\otimes_{i=1}^{n} Qu$. Spaces of this form are called *quantum registers* of n qubits.

We will sometimes work in arbitrary finite-dimensional spaces, not just tensor product spaces of qubits. Norm-1 vectors in such spaces are sometimes known as *qudits*.

Operations on quantum registers are either *unitary maps*, or *measurements*. A unitary map, describing the evolution of an isolated quantum system, is simply an inner-product preserving linear isomorphism, and it is standard to talk about applying a unitary map to a quantum register. When given as matrices, unitaries are exactly those invertible matrices whose inverse is given by the complex conjugate.

A *measurement* is determined by a self-adjoint operator, or Hermitian matrix. By the spectral decomposition theorem, every finite Hermitian matrix has a unique decomposition as the sum of a complete set of projection operators, and the corresponding subspaces are taken to be the experimental outcomes of a measurement—we refer to [31] for details.

In quantum computation, as opposed to quantum mechanics generally, Hilbert spaces are equipped with a fixed orthnormal basis, known as the *computational basis*. Information-theoretically, this is a non-trivial step—the specification of a computational basis may be considered as *classical knowledge* about a quantum system.

A significant difference between quantum and classical information is the phenomenon of *entanglement*. Given Hilbert spaces H, K, a vector $\zeta \in H \otimes K$ is called *separable* when it may be written as $\phi \otimes \psi$, for some $\phi \in H$, $\psi \in K$, and *entangled* otherwise. Entanglement gives rise to many of the counter-intuitive and non-local effects of quantum mechanics, and is heavily studied. It is this phenomenon that is widely believed to provide a computational advantage in using quantum-mechanical rather than classical computing devices (We refer to [61] for analyses of the origins of speedup in quantum algorithms).

14.3.2 Dirac Notation, and Measurement Probabilities

An exceedingly useful formalism for manipulating quantum information is *Dirac notation*. This is based on the (very categorical—see [5]) notion we may work with linear maps only—instead of referring to a state vector $\psi \in \mathcal{H}$ we consider the linear map $|\psi\rangle : \mathbb{C} \rightarrow \mathcal{H}$, defined in the natural way as $|\psi\rangle(z) = z.\psi$. These linear maps are known as *Ket vectors*, and have duals, the *Bra vectors*, which are linear maps (functionals) $\langle\phi| : \mathcal{H} \to \mathbb{C}$ defined by the condition that the composite $\langle\phi| \circ |\psi\rangle$, as a linear endomap of $\mathbb{C}$, is the inner product of ϕ and ψ.

The physical interpretation of this composite is one of the key points of the Hilbert space formalism for quantum mechanics. Consider a state vector ψ, and a measurement specified by the Hermitian operator ζ, and a vector ψ which is an eigenstate of ζ. The probability of observing the state ϕ by the measurement ζ is exactly the norm square of the above inner product, so

$$\text{prob. of observing } \phi \quad = \quad |\langle\phi|\psi\rangle|^2$$

Strictly, $\langle\phi|\psi\rangle$ is a linear map from $\mathbb{C}$ to itself, given by multiplication with the inner product of ϕ and ψ. However, it is standard to abuse notation and refer to the complex number $\langle\phi|\psi\rangle \in \mathbb{C}$. Similarly, we refer to the state vector $|\psi\rangle \in \mathcal{H}$, with the understanding that it is in fact a linear map.

14.3.3 Superpositions and Coherent Operations

State vectors are norm-1 vectors in some Hilbert space H. Hence, given state vectors $|\phi\rangle$ and $|\psi\rangle$, the norm-1 vector $|\zeta\rangle = \alpha|\phi\rangle + \beta|\psi\rangle$ is also a state vector, for all $\|\alpha\|^2 + \|\beta\|^2 = 1$—we say that $|\zeta\rangle$ is a *superposition* of $|\phi\rangle$ and $|\psi\rangle$. The phenomenon of superposition is not uniquely quantum-mechanical (e.g. it is a feature of classical wave-mechanics and linear optics, although interpretations differ—see [27] for an early, but very readable account of superposition). However, *superposition-preserving* processes form an important part of quantum computation. From [72],

> Any quantum algorithm relies on the fact that if an arbitrary input state $|\Phi_i\rangle$ evolves to the final state $|\Psi_i\rangle$ then the superposition $\sum_{i\in I} \alpha_i|\Phi i\rangle$ evolves as $\sum_{i\in I} \alpha_i|\Phi i\rangle \mapsto \sum_{i\in I} \alpha_i|\Psi_i\rangle$.

This is also taken as the definition of *fully quantum* in both the original specification of a quantum Turing machine [28], and subsequent criticisms of this definition [76]. We refer to superposition-preserving processes as *coherent*. Note that unitary processes are by definition coherent. However, these are not the only coherent quantum processes; see [15, 16] for coherent processes involving unitaries, measurements, and classically-conditioned operations.

14.3.4 No-cloning, No-deleting, and Fan-Out

Two important constraints on quantum information are the *no-cloning* and *no-deleting* theorems.

The no-cloning theorem is due to [97]:

Theorem 1 *Let $|\phi\rangle$ be an arbitrary state vector in some Hilbert space H, and let $|e\rangle$ be a fixed state in the same space. There does not exist a quantum process that acts as $|\phi\rangle|e\rangle \mapsto |\phi\rangle|\phi\rangle$.* □

The no-deleting theorem is not simply the dual of the no-cloning theorem—rather, it states that an unknown quantum state cannot be deleted, *even in the presence of a copy*.[3] The significantly simpler statement that an unknown state cannot be deleted is known as no-erasure, and is a simple consequence of linearity.

The no-deleting theorem is due to [80]:

Theorem 2 *Let $\psi\rangle$ be an arbitrary state vector in some Hilbert space H, let $|e\rangle$ be a fixed state vector, and let $|A\rangle$ be some ancilla. Then any quantum process that acts as $|\psi\rangle|\psi\rangle|A\rangle \mapsto |\psi\rangle|e\rangle|A_\psi\rangle$, is simply (up to local unitary operations) a swap map on the second and third subspaces.* □

The presence of either copying or deleting operations in quantum systems would allow for superluminal (i.e. faster-than light) signalling [81]. Thus, these fundamental theorems of quantum information play an important rôle in ensuring the consistency of quantum physics with classical relativity.

The-no-cloning and no-deleting theorems above state that *arbitrary* quantum states cannot be copied or deleted. However, *computational basis states* may be copied using the *fan-out* operation. Consider an n-dimensional space H with computational basis $\{|0\rangle, \ldots, |n-1\rangle\}$. The (general) fan-out operation $F : H \otimes H \to H \otimes H$ is defined by its action on the computational basis states as:

$$F(|i\rangle|j\rangle) = |i\rangle|i + j \ (Mod\ n)\rangle$$

In particular, $F(|k\rangle|0\rangle) = |k\rangle|k\rangle$. However, this is *not* a general copying operation; given the superposition of two basis states, $|\phi\rangle = \alpha|j\rangle + \beta|k\rangle$, then

$$F(|\phi\rangle|0\rangle) = \alpha|j\rangle|j\rangle + \beta|k\rangle|k\rangle \neq |\phi\rangle|\phi\rangle$$

14.3.5 Resource-Sensitivity and the von Neumann Architecture

Resource-sensitivity will prove a key point in understanding why a quantum computer cannot implement the von Neumann architecture. On a very simple level, there are several explicit copying or deleting steps listed in Sect. 14.2.2, and these violate the no-cloning and no-deleting principles. A simple fix would be to replace the irreversible step

- *Copy a value* from memory location x into the accumulator.

by the reversible step

- *Swap the value* in memory location x with the contents of the accumulator.

---

[3] The no-deleting property is what logicians would refer to as the failure of the *contraction rule* (see [83] for an in-depth discussion of this), whereas the no-cloning property is a (special case of) failure of the *weakening rule*. These may be treated separately (i.e. we may consider logics with weakening, but not contraction, or vice versa), as in the field of *substructural logics* [84].

and a similar change may be made to the

- *Overwrite* the contents of the program counter with a new value.

Replacing "copy" and "overwrite" by "swap" is not the focus of this paper; embeddings of irreversible computation into reversible computation have been well-studied [14], including from a physical point of view [65]. Explicit reversible architectures based on the von Neumann architecture have also been studied in [94].

The objection is more fundamental, and has to do with the way that program code may be stored and manipulated in the same way as data—and in particular, the *decode the data in the instruction register, and perform the action described* step.

14.4 Data/Code Interchangeability, and Evaluation

Our claim is that the computational core of the von Neumann architecture is the "interpret a byte as an operation" step—and this is exactly the step that is problematic for quantum computation.

In the vN architecture, the *datum* in the arithmetic logic unit is interpreted as an *instruction* which may be applied to the contents of the data register. This is a physical implementation of the notion that an object may be "promoted" to a function, and functions may be stored and manipulated in the same way as any other data object (In Sect. 14.7.1 we show how this arises from *Currying*, and, in general, the related property of *monoidal closure*).

In simple terms, the von Neumann architecture implements an *Evaluation* operation: Consider a datum P of type $\mathcal{B}yte$ that specifies some operation $\Theta : \mathcal{B}yte \to \mathcal{B}yte$. We call P the *name* of Θ, and write $P = \ulcorner \Theta \urcorner$. An evaluation operation then takes the byte P, and another byte Q, and returns $\Theta(Q)$. At the core of the von Neumann architecture is a *physical* process that implements such an evaluation operation. Our aim is to consider how, or whether, such an operation may be physically implemented in a quantum setting—with all the implications this has for primitive machine architectures.

We will need to consider two variants of an evaluation operation:

1. *Resource-sensitive evaluation* $(P, Q) \mapsto \Theta(Q)$.
2. *Non- resource-sensitive evaluation* $(P, Q) \mapsto (P, \Theta(Q))$.

The most familiar form of evaulation is 2. above; in 1., the program code is "consumed" as the program is executed, which is certainly not a feature of classical computation. From either a logical or category-theoretic point of view, 1. is considered fundamental, and 2. arises from 1. in the presence of a primitive copying operation—we say that a map $(P, Q) \mapsto (P, \Theta(Q))$ contains an implicit copying step.[4] We thus need to bear in mind the essential resource-sensitivity of quantum information (as in Sect. 14.3.4), and its implications for evaluation operations.

14.4.1 Indirect Addressing, and Evaluation Operations

We have claimed that the core of the von Neumann architecture is the "Evaluation" operation that promotes data to code. An alternative point of view is that the power of the vN architecture arises from *indirect addressing*—i.e. operations such as the, *copy the contents of the memory location referenced by the program counter into the instruction register* step in the fetch-execute cycle. We now show that the existence of a suitable evaluation operation naturally allows for indirect addressing.

Consider a quantum computer with a suitable evaluation operation (which may, or may not exist !) Relative addressing may easily be implemented. Let us assume the hypothetical quantum computer has

- n **memory registers**, $\mathcal{M}_1, \ldots, \mathcal{M}_n$,
- a **program counter register** $\mathcal{P}$,
- an **instruction register** $\mathcal{I}$.

The complete "configuration space" of this computer is thus the space

$$\mathcal{C} = \mathcal{P} \otimes \mathcal{I} \otimes \mathcal{M}_1 \otimes \ldots \otimes \mathcal{M}_n$$

The first question is, given the value $|j\rangle$ in the program counter register, $\mathcal{P}$, can we "copy the contents of the memory register $\mathcal{M}_j$ into the instruction register $\mathcal{I}$"? Of course, (as per Sect. 14.3.5), we cannot copy the contents of $\mathcal{M}_j$ – nor can we irreversibly erase the contents of $\mathcal{I}$. However, we can swap the contents of $\mathcal{M}_j$ and $\mathcal{I}$ using a unitary operation.[5] Let us call this unitary $Load_j : \mathcal{C} \to \mathcal{C}$.

Let us now assume the existence of a suitable evaluation operation, and assume without loss of generality, that the name of the operation $Load_j$ is exactly the state $|j\rangle = \ulcorner Load_j \urcorner$. Applying the evaluation operation to the contents of the program counter register then provides exactly the relative addressing required.

---

[4] This statement is, of course, more general than formal. However, it may be formalised in a wide range of settings. In a logical setting, the observation that in his semantic models implication is not primitive but contains an implicit copying operation, famously motivated J.-Y. Girard's Linear Logic [36, 37]. The connection between evaluation and logical rules is beyond the scope of this paper—we refer to [29, 6] for logical interpretations of the particular structures presented, in terms of linear logic operations.

[5] The existence this swap is exactly the *symmetry* of the tensor product. For Hilbert spaces H and K, there is a natural isomorphism $\sigma_{H,K} : H \otimes K \to K \otimes H$. We shall also see in Section 14.9.1 that the existence of such a symmetry is required for a categorical treatment of evaluation—at least, in the quantum-mechanical setting.

The above discussion makes no mention of what would, or should, happen when a superposition of values is held in the program counter register, or when two registers of the quantum computer are entangled. These are questions that need to be answered in a discussion of the general properties of a quantum-mechanical evaluation operation.

We now consider, as a "toy example", the simplest possible classical case, and use this to motivate discussion of a quantum-mechanical form of evaluation due to [78].

14.5 Evaluation in the One-Bit Computer

As a starting point we consider evaluation in the simplest possible case: the one-bit classical computer. The data types are single bits, {0, 1}, and there are exactly two program instructions:

$$\begin{aligned} \text{The identity map } &: Id(b) = b \\ \text{Negation } &: Not(b) = b + 1 \ (Mod\ 2) \end{aligned}$$

We let 0 be the name of the identity map, and 1 be the name of the negation map, so the (non- resource-sensitive) evaluation map is an isomorphism that takes pairs of bits to pairs of bits $(Program, Data) \rightarrow (Program, newData)$ as shown in Fig. 14.1.

Before		**After**	
Prog.	Data	Prog.	Data
0	0	0	0
0	1	0	1
1	0	1	1
1	1	1	0

Fig. 14.1 Evaluation in the one-bit computer

Thus the *Eval* operation for the one-bit computer is simply the (classical) controlled-not logic gate. The quantum version of this logic gate is one of the basic building blocks of the quantum circuit model. It is therefore natural to consider whether a general such evaluation operation may be implemented by unitary maps.

14.6 Implementing *Evaluation* by Unitary Operations?

We now consider whether an "evaluation" operation may be implemented using unitary maps. However, as we are working in the finite-dimensional case, we are forced to consider non- resource-sensitive evaluation: Consider quantum registers **C**, **D** (the *code* and *data* registers). A resource-sensitive evaluation operation would have type $Eval_{rs} : C \otimes D \rightarrow D$ – but as $C \otimes D$ and D have different dimensions, no such unitary map can exist.

At this point, we should be suspicious. From a logical or category-theoretic point of view, a non- resource-sensitive evaluation operation involves an implicit copying step, and arbitrary quantum states cannot be copied.

This intuition is confirmed by the technique of "encoding unitary maps on an orthonormal basis", and the "no-programming theorem", presented in [78]. We will see that unitary evaluation may exist, but the names of maps must be computational basis vectors—recall that we cannot copy *arbitrary* quantum states, but a form of copying (i.e. the fan-out operation of Sect. 14.3.4) exists for *computational basis* vectors.

Definition 1 Unitary evaluation
Consider a family of unitary maps, $U_1, \ldots, U_k : \mathbf{S} \rightarrow \mathbf{S}$ that we wish to encode as members of the (sufficiently large) quantum register **R**. We may encode these as orthogonal vectors $\{\psi_1, \ldots, \psi_k\}$, and assume (without loss of generality) that this set of vectors is a subset of the computational basis of **R**.

The corresponding *unitary evaluation operator* $Eval_U$ is given by

$$Eval_U = \sum_{i=0}^{k} |\psi_i\rangle\langle\psi_i| \otimes U_i$$

and this satisfies the condition

$$Eval_U(\psi_i \otimes s) = \psi_i \otimes U(s) \;\; \forall s \in \mathbf{S} \, , \;\; \psi_i \in \{\psi_1, \psi_2, \ldots, \psi_k\}$$

When we take the quantum analogue of the 1-bit computer described above, both the identity and qubit negation (defined on the computational basis as $Not(|0\rangle) = |1\rangle$, $Not(|1\rangle) = |0\rangle$), may be implemented as unitary maps, so the above prescription gives the quantum **CNOT** gate.

$$Eval_U = CNOT = \begin{pmatrix} 1\,0\,0\,0 \\ 0\,1\,0\,0 \\ 0\,0\,0\,1 \\ 0\,0\,1\,0 \end{pmatrix}$$

14.6.1 A Limit to Unitary Evaluation Operations

It may be shown that at most n unitary maps may be encoded (with respect to a unitary evaluation map), as orthonormal vectors, on an n-dimensional space. The following theorem is based on that of [78]:

Theorem 3 *The Nielsen-Chuang "no-programming" theorem*

Consider n distinct unitary maps $U_1, \dots U_n : D \to D$. Let C be an n-dimensional space, and let Eval : $C \otimes D \to C \otimes D$ be a unitary map that satisfies

$$Eval(|c_i\rangle \otimes |d\rangle) = |c_i\rangle \otimes U_i(|d\rangle) \quad , \quad \{|c_i\rangle\}_{i=1}^n \subseteq C$$

Then

1. *no non-trivial superposition $\alpha|c_i\rangle + \beta|c_j\rangle$ encodes a unitary map.*
2. *the vectors $\{c_i\}_{i=1}^n$ are all orthogonal.*

Taken together, these imply that at most n unitary maps may be encoded in this way.

Proof Assume that, for some complex α, β satisfying $|\alpha|^2 + |\beta|^2 = 1$, the sum $\alpha|c_i\rangle + \beta|c_j\rangle$ encodes a unitary map V, so for arbitrary $|s\rangle \in \mathbf{S}$,

$$Eval((\alpha|c_i\rangle + \beta|c_j\rangle) \otimes |s\rangle = (\alpha|c_i\rangle + \beta|c_j\rangle)V(|s\rangle)$$

However, by linearity,

$$\begin{aligned} Eval((\alpha|c_i\rangle + \beta|c_j\rangle) \otimes |s\rangle) &= \alpha Eval(|c_i\rangle \otimes |s\rangle) + \beta Eval(|c_j\rangle \otimes |s\rangle) \\ &= \alpha|c_i\rangle \otimes U_i(|s\rangle) + \beta|c_j\rangle \otimes U_j(|s\rangle) \end{aligned}$$

This may be factorised as $(\alpha|c_i\rangle + \beta|c_j\rangle) \otimes V(|s\rangle)$ under any of the following conditions:

- $\alpha = 0$, in which case $V = U_j$
- $\beta = 0$, in which case $V = U_i$
- $U_j(|s\rangle) = V(|s\rangle) = U_i(|s\rangle)$, for all $|s\rangle \in \mathbf{S}$.

Either the superposition is trivial, or U_i, U_j, and V are all the same unitary operation. Using similar techniques, it may be shown that c_i is orthogonal to c_j, for all $i \neq j$. □

The above negative result states that (in an entirely unitary, finite-dimensional setting), operations may only be encoded on a fixed orthonormal basis—taken, for convenience, to be the computational basis.

This is sometimes interpreted as stating that quantum computers cannot operate on a "stored-code" principle, since an n-dimensional Hilbert space H can encode at most n unitary operations, whereas there are an infinite number of distinct unitaries from H to itself. In reality, practical proposals for quantum computation work with a small number of gates, up to a well-defined notion of approximation [85].

A stronger critique is that there is nothing particularly "quantum" about the way operations are encoded—computational basis vectors may be freely copied and deleted (via analogues of fan-out, Sect. 14.3.4), and are undisturbed by measurements in the computational basis. This is used to make the much stronger case that there is no advantage to storing program code as *quantum* rather than *classical* information.

There are, of course, two loopholes in the above interpretation, if not in the theorem itself. The first is that it only applies to finite-dimensional spaces. This feature will be the joker in the pack throughout this paper: many results presented (for both quantum and classical information) do not hold in the infinite-dimensional case—this becomes particularly relevant in Sect. 14.13.2. The second loophole is more practical: Theorem 3 above only applies to *unitary* evaluation operations.

As well as the orthonormal basis encoding technique, [78] considers using teleportation-like protocols to implement evaluation *probabilistically*, with reference to the Choi-Jamiołkowski correspondence [20, 55]. In [16] it is shown how, in certain cases, the probability of success may be increased to 1 by using unitary operations conditioned on the result of the measurement.

A post-selected form of teleportation, related to the Choi-Jamiołkowski correspondence, is at the core of the "categorical foundations" of quantum mechanics of [4]. We now take a categorical approach, and consider how evaluation arises from general principles, in both the classical and quantum worlds.

To illustrate the general category theory, and to make a link with the von Neumann architecture, we first present the theory of sets and functions. We then use this to motivate the general theory of categorical closure, and consider the particular form of categorical closure exhibited in the quantum world.

14.7 Evaluation as Currying

From a theoretical computer science perspective, evaluation operations arise from an abstract notion of Currying called *categorical closure*, and the theory of *(monoidal) closed categories*. We first present the classical motivation based on the theory of sets and functions (including logic gates and binary words as a special case).

14.7.1 Evaluation with Sets and Functions

The informal setting for an *Eval* operation is where we can find a representation of a function between two sets as a single element of another set, and can "promote" this to an operation to be applied. From either a category-theoretic or theoretical computer science point of view, this notion is not primitive but arises naturally as a consequence of the structure of sets and functions—notably the existence of *Currying*.

Definition 2 Cartesian products, Currying
Given sets A and B, their *Cartesian product* is the set defined by

$$A \times B = \{(a, b) : A \in A \ , \ b \in B\}$$

Given a function $f : X \times Y \to Z$, Currying is simply the process of, for each element $x \in X$, defining a function $f_x : Y \to Z$ by $f_x(y) = f(x, y)$.

Let us use categorical notation (formal definitions follow in Sect. 14.8). The category of sets, **Set** has the (proper class of) all sets as its *objects*, denoted $Ob(\mathbf{Set})$. Between any two objects $A, B \in Ob(\mathbf{Set})$ is the collection of *arrows*, $\mathbf{Set}(A, B)$. These are simply the functions from A to B.

For any sets $A, B \in Ob(\mathbf{Set})$,

1. the collection of all functions from A to B is itself a set, that we denote $[A \rightarrow B] \in Ob(\mathbf{Set})$.
2. the Cartesian product of A and B is itself a set $A \times B \in Ob(\mathbf{Set})$.

The existence of Currying can then be expressed succinctly, as

$$\mathbf{Set}(A \times B, C) \cong \mathbf{Set}(A, [B \rightarrow C])$$

Now consider the one-object set $I = \{*\}$. It is immediate that for all sets X,

$$I \times X \cong X \cong X \times I$$

since there is an obvious bijection between $\{(x, *) : x \in X\}$ and X itself.
Hence,

$$\mathbf{Set}(A, B) \cong \mathbf{Set}(I \times A, B)$$

and by Currying,

$$\mathbf{Set}(A, B) \cong \mathbf{Set}(I, [A \rightarrow B])$$

Finally, for all sets X,

$$[I \rightarrow X] \cong X$$

since for all $x \in X$, we have the function $\iota_x : \{*\} \rightarrow X$ defined by $\iota(*) = x$.

Hence, Currying and the properties of the one-object set give the bijection $\mathbf{Set}(A, B) \cong \mathbf{Set}(I, [A \rightarrow B])$ — that is, the existence of representations of functions from A to B as elements of some set. This formalises the intuitive notion of the "name" of an operation (from Sect. 14.4), at least in the category of sets and functions.

Definition 3 Given a function $g : A \rightarrow B$ we refer to the arrow $\ulcorner g \urcorner : \{*\} \rightarrow [A \rightarrow B]$ (or equivalently, the corresponding element of $[A \rightarrow B]$, considered as a set) as the *name* of the function g.

14.7.2 Is Any of this Non-trivial?

It may be objected that the above manipulations are trivialities—this is to some extent correct.[6] Of more interest is the strong similarity, between $I \times X \cong X \cong X \times I$ and the Hilbert space identity $\mathbb{C} \otimes H \cong H \cong H \otimes \mathbb{C}$. In particular (as formalised in [5]) the identity $[I \rightarrow X] \cong X$ is strongly reminiscent of Dirac notation, so $\iota_x : \{*\} \rightarrow X$ is the direct analogue of a Ket $|\psi\rangle : \mathbb{C} \rightarrow H$.

In order to consider similarities and differences more closely, we now to take the category-theoretic approach seriously, rather than simply as a form of notation.

14.8 Basic Category Theory

As with the section on basic quantum information (Sect. 14.3.1), we make no attempt to given anything approaching a comprehensive account of category theory—rather we pick and choose (and to some extent, simplify[7]) topics relevant to our discussion. A comprehensive account may be found in [73], with a physics-oriented approach given by [35]. We also refer to [66] for connections between category theory, logic and lambda calculus, and [13] for a computer science perspective.

Definition 4 Categories
A category $\mathbf{C}$ has a class of *objects*, denoted $Ob(\mathbf{C})$, and between any two objects $X, Y \in Ob(\mathbf{C})$ is a set of *arrows*, denoted $\mathbf{C}(X, Y)$. We often write $f : X \rightarrow Y$ for $f \in \mathbf{C}(X, Y)$, when the category $\mathbf{C}$ is clear from the context.

Arrows $f \in \mathbf{C}(X, Y)$ and $g \in \mathbf{C}(Y, Z)$ may be composed, giving $gf \in \mathbf{C}(X, Z)$, and composition is associative, so $h(gf) = (hg)f$. For each object $Y \in Ob(\mathbf{C})$ there is also an identity arrow $1_Y \in \mathbf{C}(Y, Y)$ satisfying $1_Y f = f$ and $g 1_Y = g$.

As a category $\mathbf{C}$ may have a *proper class* of objects, we cannot use set-theoretic operations on its objects. However, we may define—for example—the category of Sets to have all sets as objects, where the arrows $\mathbf{Set}(X, Y)$ are exactly the set-theoretic functions $f : X \rightarrow Y$.

[6] We refer to [57] for P. Freyd's perhaps controversial suggestion that the real function of category theory is to demonstrate that the trivial parts of mathematics are trivial for trivial reasons. Another point of view is that it allows us to formalise similarities and differences between the behaviour of mathematical structures—and we have a special interest in comparing the behaviour of Sets and Hilbert spaces.

[7] In particular, we will refer to indexed families of arrows in a category as "natural", without giving a formal definition. We refer to [73] for natural families as components of natural transformations, and [21] for an exposition without explicit reference to natural transformations.

14.8.1 New Categories from Old

The following operations on categories will be useful:

Definition 5 Opposite categories, product categories
Given a category $\mathbf{C}$, its *opposite category* $\mathbf{C^{op}}$ has the same objects, and the set of arrows $\mathbf{C^{op}}(X, Y)$ is exactly the set of arrows $\mathbf{C}(Y, X)$. Given $f \in \mathbf{C^{op}}(Y, X)$ and $g \in \mathbf{C^{op}}(Z, Y)$, the arrow $fg \in \mathbf{C^{op}}(Z, X)$ is exactly the composite $gf \in \mathbf{C}(X, Z)$.
Given categories $\mathbf{C}, \mathbf{D}$, the *product category* is defined to have, as objects, all pairs (X, A), where $X \in Ob(\mathbf{C})$ and $A \in Ob(\mathbf{D})$. Similarly, an arrow $f : (X, A) \to (Y, B)$ is just a pair $f = (f_1, f_2)$, where $f_1 : X \to Y$ and $f_2 : A \to B$.

Intuitively, the opposite category $\mathbf{C^{op}}$ may be thought of as "taking the category $\mathbf{C}$, and reversing all the arrows"—although a simple step, this gives many interesting dualities of mathematics (such as Stone dualities between topological spaces and lattices [56], and Pontryagin duality [86]). Also, for every definition or theorem in category theory, we derive a dual definition or theorem by moving to the dual category.

14.8.2 Structure-Preserving Maps Between Categories

As well as categories themselves, it is natural to define structure-preserving maps between categories.

Definition 6 Functors, adjoint pairs
A **functor** between categories $\Gamma : \mathbf{C} \to \mathbf{D}$ is a map that assigns

- to each object $X \in Ob(\mathbf{C})$, an object $\Gamma(X) \in Ob(\mathbf{D})$,
- to each arrow $f \in \mathbf{C}(X, Y)$, an arrow $\Gamma(f) \in \mathbf{D}(\Gamma(X), \Gamma(Y))$.

Functors are required to "preserve the categorical structure" in that for all $f \in \mathbf{C}(X, Y)$ and $g \in \mathbf{C}(Y, Z)$,

$$\Gamma(g)\Gamma(f) = \Gamma(gf) \in \mathbf{D}(\Gamma(X), \Gamma(Z)) \quad \text{and} \quad \Gamma(1_Y) = 1_{\Gamma(Y)}$$

Much of category theory is built on the notion of *adjointness*. Two functors $\Gamma : \mathbf{C} \to \mathbf{D}$ and $\Delta : \mathbf{D} \to \mathbf{C}$ are said to form an *adjoint pair* when, for all $X \in Ob(\mathbf{C})$ and $Y \in Ob(\mathbf{D})$, there exists a natural bijection

$$\mathbf{C}(X, \Delta(Y)) \cong \mathbf{D}(\Gamma(X), Y)$$

We say that Γ is *left adjoint* to Δ, or equivalently, that Δ is *right adjoint* to Γ.

Adjointness is a generalisation of the order-theoretic notion of *Galois connections*—indeed, partially ordered sets are themselves categories, and Galois connections are a special case of adjoint functors. The terminology "adjoint" comes from the notion of the adjoint of a continuous linear map of Hilbert spaces, defined by $\langle L(\phi)|\psi\rangle = \langle\phi|L^*(\psi)\rangle$.

A very practical tool in this field is Freyd's *adjoint functor theorem* that characterises when a given functor has a left (or, by working in the opposite category, a right) adjoint, e.g. in [73] it is demonstrated how the tensor product of Abelian groups may be derived.

For our purposes, adjoint functors will play a key rôle in defining *categorical closure*, giving the general category-theoretic approach to evaluation.

14.8.3 Monoidal Categories

A monoidal tensor for a category is a general notion covering operations such as the Cartesian product of sets and functions, the tensor product or direct sum of Hilbert spaces, disjoint union of Relations, &c. We follow the treatment given in in [73].

Definition 7 Symmetric monoidal categories
A *monoidal category* is defined to be a category $\mathbf{C}$, together with a functor $\otimes : \mathbf{C} \times \mathbf{C} \rightarrow \mathbf{C}$ that satisfies, for all $A, B, C \in Ob(\mathbf{C})$:

- *Unit objects* There exists $I \in Ob(\mathbf{C})$ satisfying $I \otimes A \cong A \cong A \otimes I$.
- *Associativity* $A \otimes (B \otimes C) \cong (A \otimes B) \otimes C$.

If a monoidal category satisfies the additional condition

- *symmetry* $A \otimes B \cong B \otimes A$.

it is called a *symmetric monoidal category.*

The isomorphisms above are required to be *natural*, and to satisfy various *coherence conditions*. However, MacLane's coherence theorems [73] mean that these conditions can generally be ignored with no harmful side-effects. In particular, we treat the associativity isomorphism $A \otimes (B \otimes C) \cong (A \otimes B) \otimes C$ 2. as though it were a strict identity.

From a strongly category-theoretic point of view, monoidal tensors may often be characterised by universal properties that they satisfy, without explicit reference to their behaviour on (for example) elements of some set of objects. For example, The Cartesian product of sets and functions may be characterised as a *categorical product*, as follows:

Definition 8 Products
Let $(\mathbf{C}, \otimes)$ be a monoidal category. The monoidal tensor $\otimes$ is a *product* when, for all objects $X_1, X_2 \in Ob(\mathcal{C})$ there exist arrows

$$X_1 \xleftarrow{\pi_1} X_1 \otimes X_2 \xrightarrow{\pi_2} X_2$$

such that, for all arrows $f_1 \in \mathcal{C}(Y, X_1)$ and $f_2 \in \mathcal{C}(Y, X_2)$ there exists a unique arrow $\langle f_1, f_2\rangle \in \mathcal{C}(Y, X_1 \otimes X_2)$ making the following diagram commute:

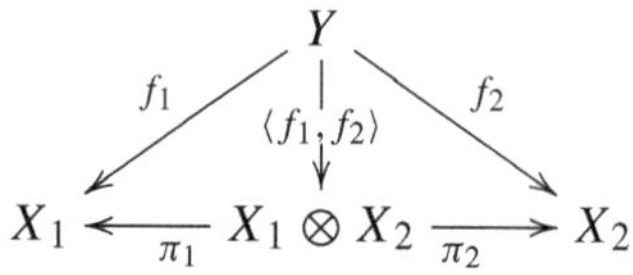

Recall that by considering the dual category, we may derive a dual definition. Reversing all the arrows in the diagram above gives the following:

Definition 9 Coproducts
Let **C**, ⊗ be a monoidal category. The monoidal tensor ⊗ is a *coproduct* when, for all objects $X_1, X_2 \in Ob(\mathcal{C})$ there exist arrows

$$X_1 \xrightarrow{\iota_1} X_1 \otimes X_2 \xleftarrow{\iota_2} X_2$$

such that, for all arrows $f_1 \in \mathcal{C}(X_1, Y)$ and $f_2 \in \mathcal{C}(X_2, Y)$ there exists a unique arrow $[f_1, f_2] \in \mathcal{C}(X_1 \otimes X_2, Y)$ that makes the following diagram commute:

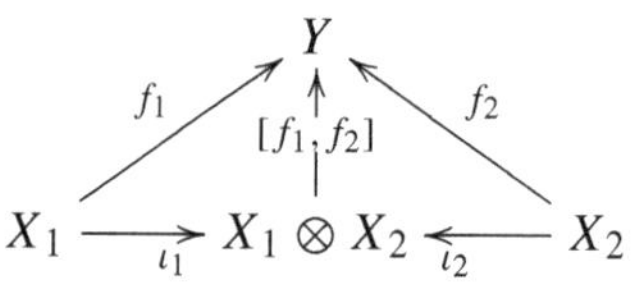

Examples

The Cartesian product of sets and functions is, as noted above, a categorical product. Dually, the disjoint union of sets and functions is a categorical coproduct. The direct sum of Hilbert spaces is both a product, and a coproduct (and hence a *biproduct*). The tensor product of Hilbert spaces is neither a product nor a coproduct—however, it may also be defined in terms of a universal property, as we now show.

14.8.4 (Monoidal) Closed Categories

The most general setting for a correspondence between objects and arrows in a category, and evaluation operations, is the field of *closed categories* [30, 67, 68]. We do not attempt an exposition of this, but work with the special case of *monoidal closed* categories—partly for simplicity, and partly because the specific examples we wish to consider are monoidal closed rather than simply closed.

Definition 10 Monoidal closed categories
Let (**C**, ⊗) be a monoidal category. We say that it is *monoidal closed* when there exists a functor

$$[\_,\_] : \mathbf{C}^{op} \times \mathbf{C} \rightarrow \mathbf{C}$$

called the *internal hom* functor, such that for fixed $B \in Ob(\mathbf{C})$, the functors given by

$$[B,\_] : \mathbf{C} \rightarrow \mathbf{C} \text{ and } \_ \otimes B : \mathbf{C} \rightarrow \mathbf{C}$$

form an adjoint pair.

This definition, although concise, is relatively abstract (and is only strictly accurate in the symmetric monoidal case). The following characterisations make a link to both an abstract form of Currying, and evaluation maps.

Theorem 4 *The following are equivalent to the definition of a monoidal closed category, in Definition 10 above:*

1. *There exists an internal hom. functor*

$$[\_,\_] : \mathbf{C}^{op} \times \mathbf{C} \rightarrow \mathbf{C}$$

satisfying

$$\mathbf{C}(A \otimes B, C) \cong \mathbf{C}(B, [A \rightarrow C])$$

2. *For every pair of objects $A, B \in Ob(\mathbf{C})$, there exists*
   - *an object $[A \rightarrow B]$,*
   - *an arrow $ev_{A,B} \in \mathbf{C}(A \otimes [A \rightarrow B], B)$*

   where, for all $f : A \otimes X \rightarrow B$, there exists unique $g \in \mathbf{C}(X, [A \rightarrow B])$ such that the following diagram commutes:

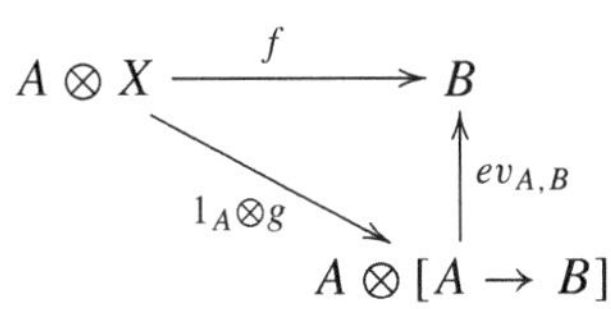

Proof Proofs may be found in any text on category theory or categorical logic (e.g. [73, 66]). □

Part 1. of Theorem 4 above provides the link with an abstract notion of Currying, and part 2. of the same theorem demonstrates that this is equivalent to the existence of an evaluation map satisfying the expected properties.

14.9 Categorical Closure and Hilbert Spaces

We now describe the particular form of compact closure exhibited by the category of (finite-dimensional) Hilbert spaces and linear maps, and its interpretation as quantum-mechancial protocols.

It has long been known that the collection of all linear maps between (finite-dimensional) Hilbert spaces H, K is itself a Hilbert space. This is a reflection of the categorical closure of the category of finite-dimensional Hilbert spaces—however, the categorical closure is of a particularly simple form. We first give an abstract exposition of this form of closure, including Hilbert spaces as a concrete example, and then give an overview of the 'categorical foundations' approach to quantum mechanics of [4].

14.9.1 Compact Closed Categories

Definition 11 Compact closure
A symmetric monoidal category $(\mathbf{C}, \otimes)$ is called *compact closed* when, for all $A \in Ob(\mathbf{C})$, there exists a *dual object* $A^* \in Ob(\mathbf{C})$ such that the functor $A \otimes \_ : \mathbf{C} \to \mathbf{C}$ is left adjoint to the functor $A^* \otimes \_$.

Although this definition is category-theoretically elegant—particularly when viewed as 2-category theory [62]—for our purposes it will be easier to work with the following characterisation, also given given in [62]:

Theorem 5 *A symmetric monoidal category $(\mathbf{C}, \otimes)$ is compact closed when, for every object $A \in Ob(\mathbf{C})$, there exists a dual object $A^* \in Ob(\mathbf{C})$ together with distinguished arrows*

- *The unit arrow $\epsilon_A : A \otimes A^* \to I$*
- *The counit arrow $\eta_A : I \to A^* \otimes A$*

that satisfy

$$\lambda(\epsilon_A \otimes 1_A)(1_A \otimes \eta_A)\rho^{-1} = 1_A \quad \textit{and dually,} \quad \rho_{A^*}(1_{A^*} \otimes \epsilon_A)(\eta_A \otimes 1_{A^*})\lambda_{A^*}^{-1} = 1_{A^*}$$

Using the diagrammatic notation of [58–60], this may be drawn as shown in Fig. 14.2.

The dual operation on objects $(\_)^*$, together with the unit and counit arrows may be used to define the *dual on arrows*. Given $f \in \mathcal{C}(A, B)$, then $F^* \in \mathcal{C}(B^*, A^*)$ is defined by

$$f^* = (1_{A^*} \otimes \epsilon_B)(1_{A^*} \otimes f \otimes 1_{B^*})(\eta_A \otimes 1_{B^*}) : B^* \to A^*$$

Diagrammatically, this is as shown in Fig. 14.3.

Note that in a compact closed category, the arrows η_A and μ_A are dual, so $\eta_A^* = \epsilon_A$.

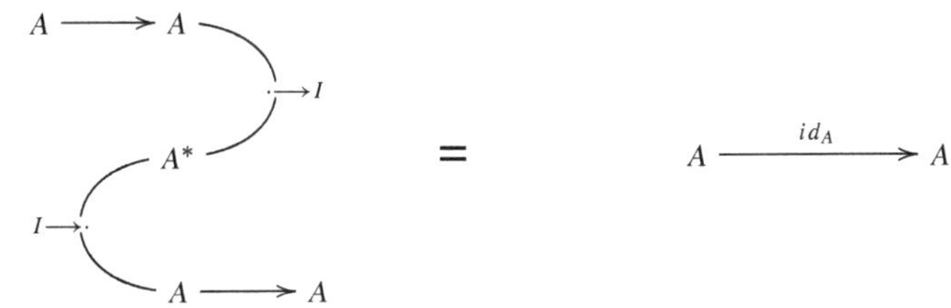

and dually,

Fig. 14.2 Axioms for compact closure
Note that (following the usual convention) the above diagrams omit the unit object isomorphisms $A \cong A \otimes I$, &c.

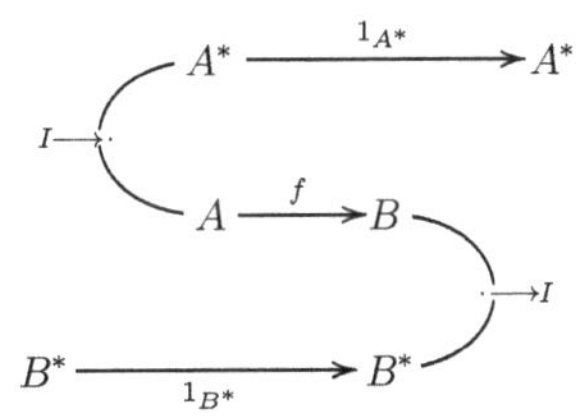

Fig. 14.3 The dual operation on arrows

14.9.2 The Internal Hom and Names in Compact Closed Categories

Compact closed categories have a particularly simple description of both the internal hom object, and names of arrows.

Theorem 6 *Let* $(\mathbf{C}, \otimes, \epsilon, \eta)$ *be a compact closed category, with* $A, B \in Ob(\mathbf{C})$. *Then*

- *The internal hom is given by* $[A \to B] = B \otimes A^*$.
- *Given an arrow* $f \in \mathbf{C}(A, B)$, *its name* $\ulcorner f \urcorner : I \to B \otimes A^*$ *is given, up to a symmetry map, by*

$$\ulcorner f \urcorner = (1_A^* \otimes f)\eta_A$$

Using the same diagrammatic notation as previously, the name of an arrow f is given by

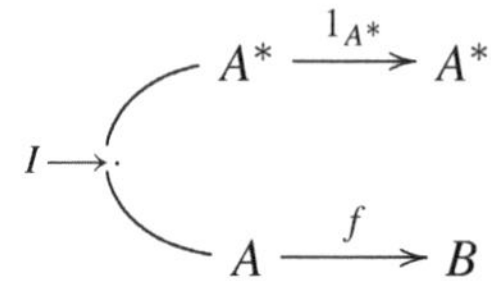

Note that both the ability to name arrows, and the adjunction giving monoidal closure, rely on the existence of symmetry isomorphisms $\sigma_{X,Y} : X \otimes Y \to Y \otimes X$. We refer to [90] for categories satisfying similar axioms, but with a weaker symmetry condition.

Definition 12 conames—the dual notion to names
In compact closed categories—by contrast to monoidal closed categories generally—there is a dual notion to *naming*; that of the *coname* of an arrow. Given an arrow $f \in \mathbf{C}(A, B)$, the *coname* is an arrow $\llcorner f \lrcorner \in \mathbf{C}([X, Y], I)$ given (dually to Theorem 6) by $\llcorner f \lrcorner = \epsilon_B(f \otimes 1_{B^*})$.

To see that conames do not exist in all monoidal closed categories, consider the category of sets and functions—here, for any set X, there exists exactly one function in $\mathbf{Set}(X, \{*\})$. This is a strong, and indeed significant, difference between the behaviour of evaluation for Sets and functions, and Hilbert spaces and linear maps. This difference accounts for the different behaviour of classical and quantum systems[8] given in Sect. 14.12.

14.9.3 Compact Closure and Hilbert Spaces

It has long been known that finite-dimensional Hilbert spaces are a canonical example of compact closed categories—and equally, Hilbert spaces in the general setting (i.e. allowing for infinite-dimensional spaces) are not compact closed [2].

Let us denote the category of finite-dimensional Hilbert spaces by $\mathbf{Hilb_{fd}}$. Arrows are linear maps (and hence, as we are working in the finite-dimensional case, both bounded and continuous), and we use the tensor product as the monoidal tensor.

Compact closure is easily exhibited. Consider a space H, with orthonormal basis $\{e_j\}_{j=1}^N$. Objects are self-dual, so $H^* = H$, and the unit and counit arrows $\epsilon : H \otimes H \to \mathbb{C}$ and $\eta_H : \mathbb{C} \to H \otimes H$ are given (using Dirac notation) by

---

[8] We emphasise that, although the category **Set** does not admit conames, they are by no means an exclusively quantum phenomenon—rather, they are simply a property associated with compact closure. For example, [46] uses compact closure in modelling classical Turing machines.

$$\epsilon_H = \sum_{j=1}^{N} \langle \mathbf{e}_j \mathbf{e}_j | : H \otimes H \to \mathbb{C} \text{ and } \eta_H = \sum_{j=1}^{N} | \mathbf{e}_j \mathbf{e}_j \rangle : \mathbb{C} \to H \otimes H$$

It is straightforward to check that the axioms of Definition 11 are satisfied.

The compact closure of Hilbert space, and its physical interpretation in terms of teleportation protocols, is at the core of Abramsky and Coecke's "categorical foundations" program for quantum mechanics [4]. We consider this in from Section 14.10 onwards, but first characterise states that are the names of unitary maps.

14.9.4 Naming Unitary Maps

Our ultimate aim is to describe what the state-map correspondence given by the categorical foundations program and the compact closure of finite-dimensional Hilbert space can tell us about the existence, or otherwise, of quantum-mechanical versions of evaluation. From the motivation given in Sects. 14.3.1 and 14.3.3, we are particularly interested in coherent quantum operations.

We now consider the state/map correspondence provided by the compact closure. We emphasise that this is *not* original to this paper. It is given explicitly in [4], and is implicit in the Choi-Jamiołkowsi correspondence between density matrices and completely positive maps [20, 55, 89].

Unwinding the definition of the name in a compact closed category gives a 1:1 correspondence between arrows $\mathbf{Hilb_{fd}}(\mathbb{C}, H \otimes H)$, and arrows $\mathbf{Hilb_{fd}}(H, H)$. Let us chose an orthonormal basis $\{\mathbf{e}_i\}_{i=1}^{N}$ for H, and consider a linear map described as a matrix $M = (m_{ij})_{i,j=1}^{N}$ on this basis. For any such matrix M we may define $\ulcorner M \urcorner \in \mathbf{Hilb_{fd}}(\mathbb{C}, H \otimes H)$ by

$$\ulcorner M \urcorner = \frac{1}{\sqrt{N}} \sum_{\alpha,\beta=1}^{N} m_{\alpha\beta}(|\mathbf{e}_\alpha\rangle \otimes |\mathbf{e}_\beta\rangle)$$

This naming operation is invertible; consider a vector $\psi \in H \otimes H$. Then $\psi = \ulcorner L \urcorner$, where

$$L = \begin{pmatrix} l_{11} & \ldots & l_{1N} \\ \ldots & & \ldots \\ l_{N1} & \ldots & l_{NN} \end{pmatrix}$$

satisfies

$$l_{ij} = \sqrt{N} \langle \mathbf{e}_i \otimes \mathbf{e}_j | \psi \rangle$$

Given this explicit description, it is clear that arbitrary linear maps on H may be named. We now characterise those states that name unitary maps:

Theorem 7 *The pure states that name unitary maps $U : H \to H$ are exactly the maximally entangled states of $H \otimes H$.*

Proof In order not to disrupt the expository flow of this paper, the proof of this result is given in Appendix.

14.9.5 The Choi-Jamiołkowski Correspondence

The correspondence between states and linear maps is a special case of the *Choi - Jamiołkowski correspondence* between completely positive maps and density matrices, discovered independently by M. Choi [20] and A. Jamiołkowsky [55]. As our exposition is in terms of *pure states* and *unitary maps*, rather than the more general *density matrices* and *completely positive maps*, so have given the correspondence in this restricted case, following [4].

It is demonstrated in [89] that the category of density matrices and completely positive maps is compact closed, and the state corresponding to a completely positive map under the Choi-Jamiołkowski correspondence is exactly the name of that map.

Finally, the connection between teleportation and the Choi-Jamiołkowski correspondence had been commented on (although not explored in detail) in [78]. The interpretation of teleportation as compact closure was developed in the *categorical foundations* program of [4].

14.10 Abramsky and Coecke's Categorical Foundations for Quantum Mechanics

As with the introductory sections on both quantum information and category theory, we make no attempt to give a coherent account or consistent history of the categorical foundations program—rather, we concentrate on those topics relevant to our interest in evaluation operations and the von Neumann architecture. We refer to [4, 5, 22, 23] for details, and other articles in this volumn for the current state of research.

14.10.1 Teleportation, Traditionally

The traditional description of teleportation is as follows: *Alice* has a quantum bit $|\psi\rangle = \alpha|0\rangle + \beta|1\rangle$ that she wishes to send to *Bob*. Alice is spatially separated from Bob, but they had previously shared a maximally entangled pair of particles $|\mathcal{B}ell\rangle = \frac{1}{\sqrt{2}}(|00\rangle + |11\rangle)$. This gives the overall state of the system as $|\psi\rangle \otimes |\mathcal{B}ell\rangle =$

$$\frac{1}{\sqrt{2}}(\alpha|0\rangle + \beta|1\rangle) \otimes (|00\rangle + |11\rangle) = \frac{\alpha}{\sqrt{2}}(|000\rangle + |011\rangle) + \frac{\beta}{\sqrt{2}}(|100\rangle + |111\rangle)$$

Alice then performs a measurement of her subsystem (i.e. the qubit $|\psi$, together with her half of the maximally entangled pair $|\mathcal{B}ell\rangle$) against a maximally entangled basis that contains the $\mathcal{B}ell$ state.

Let us assume that Alice observes the state $\mathcal{B}ell$. Bashing through the appropriate Hilbert space calculations will demonstrate that the system is now in the overall state $\frac{\alpha}{\sqrt{2}}(|000\rangle + |110\rangle) + \frac{\beta}{\sqrt{2}}(|001\rangle + |111\rangle$. Of course, this factorises as $\frac{1}{\sqrt{2}}(|00\rangle + |11\rangle)(\alpha|\rangle + \beta|1\rangle)$—that is, Bob now has the quantum bit $|\psi\rangle = \alpha|0\rangle + \beta|1\rangle$.

This is not the whole story—The probability of Alice observing the correct measurement result is $\frac{1}{4}$. If Alice observes a different state, she must inform Bob, via a classical communication channel, of the state she observed. Bob may then apply a 1-qubit unitary map to the state he holds, in order to recover $|\psi\rangle$.

We consider unitary corrections in Sect. 14.11.2. For the moment, we are satisfied with a *post-selected* version of the above protocol—that is, if Alice observe the "incorrect" outcome, the experiment is abandoned.[9]

14.10.2 Teleportation, Categorically

In the categorical foundations approach [4], the preparation of the Bell-state is modelled by the unit of a compact closed category, so the Bell state is simply the name of the identity map. Its dual, the counit, is a measurement where the Bell state is observed. The postselected teleportation protocol is then simply the defining axiom for a compact closed category:

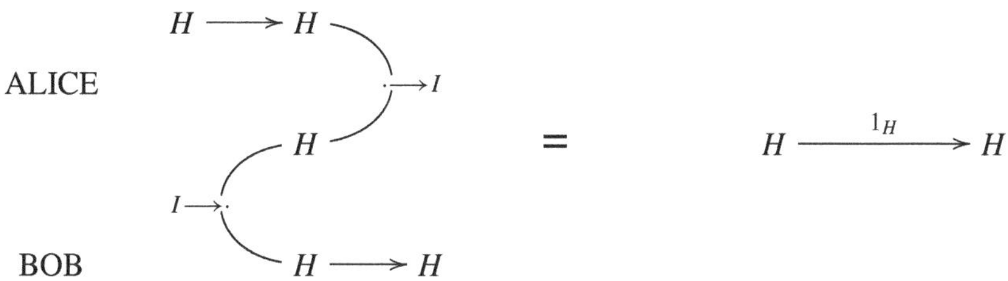

A natural question now is, "what happens when the state preparation, and measurement, are not based on names of the identity, but on names of other unitary maps?". It is demonstrated in [16] that when some other entangled resource is used in place of the Bell state, the teleportation protocol can, "apply a unitary to $|\psi\rangle$ on its way from Alice to Bob".

In the categorical foundations approach, this is the following: instead of using the unit $\epsilon_H = \ulcorner 1_H \urcorner$ and counit $\eta = \llcorner 1_H \lrcorner$, let us use the name and coname of some unitary maps $U, V : H \to H$, giving (via the definition of the name and coname) the result shown in Fig. 14.4.

---

[9] Even this post-selected version requires classical communication, in order to tell Bob when to give up on the experiment. The requirement for classical communication in teleportation protocols is important to prevent teleportation being used for superluminal signalling.

Fig. 14.4 Applying maps via teleporation (categorically)

Note (as emphasised in [22]) the apparently acausal order of application on the right hand side. Chronologically, the preparation of the state $\ulcorner U \urcorner$ happens first, then a measurement is made resulting in the state $\llcorner V \lrcorner$. However, when a state $|\psi\rangle$ is teleported using this arrangement, the result is $UV(|\psi\rangle)$—i.e. V is applied first, followed by U.

14.11 Evaluation by Teleportation, and the vN Architecture

14.11.1 The Story so Far

So far, we have seen that "evaluation" is the key part of the von Neumann architecture (Sect. 14.4). The only possible competition for this rôle is "relative addressing", and we have seen in Sect. 14.4.1 that this arises quite naturally from evaluation. We have also seen that evaluation is a categorical property that arises from *monoidal closure*—an abstract form of Currying, and the notion of *naming* an arrow.

The link with quantum mechanics follows from the categorical foundations program where compact closed categories are not only used, but are a key part of the program. Physically, compact closure is interpreted as the teleportation protocol of [15], and in general, the *implementing a logic gate by teleportation* of [16]. This strongly suggests that a (resource-sensitive) form of evaluation is available, and may be implemented in the quantum world. Modulo questions of reversibility and resource-sensitivity, can we therefore describe some form of von Neumann architecture for quantum computers?

The question that this section aims to answer is therefore:

Can we apply an unknown unitary map to a quantum state?

The "unknown unitary map" is given as a quantum resource—i.e. we are given its name[10]; a maximally entangled state vector $\ulcorner U \urcorner \in H \otimes H$. Our question now

[10] A natural question at this point is,'why not give the unknown map as a *coname*, rather than a name? From the physical interpretation of Sect. 14.10.2, a coname is interpreted as a (successful) measurement; that is, it is derived from a classically determined measurement apparatus. It is hard to see how we may go from an arbitrary quantum state to a measurement against some basis containing that state—thus the coname can only be given as classical information.

is, given $\ulcorner U \urcorner \in H \otimes H$, and a state vector $|\psi\rangle \in H$, can we reliably produce $U(|\psi\rangle) \in H$?

14.11.2 Postselection, and Unitary Corrections

Recall how, as shown in Fig. 14.4, unitary maps may be applied using a *teleportation protocol*. By letting the name in this diagram be our "unknown map", and taking the coname to be the coname of the identity, we are able to apply our unknown map to an arbitrary state $|\psi\rangle \in H$.

As it stands, this diagram describes a *post-selected* protocol; if the measurement at does not yield the required result (i.e the coname of the identity) we abandon the experiment and start again. Unfortunately, the experimenter has no control over the actual result of measurement—at best, he may specify a complete maximally entangled orthonormal basis set $\{\ulcorner V_j \urcorner\}_{j=1...n^2} \subseteq H \otimes H$, and measure against that. Thus, when working with a single qubit, we expect to observe the "correct" outcome with a probability of $\frac{1}{4}$.

This feature is why [78] refer to evaluation via teleportation protocols as "probabilistic evaluation". However, in the original teleportation protocol, [15], Bob applies a unitary operation to correct for this "incorrect experimental outcome". In [16], it is demonstrated how a similar technique can be used to implement certain quantum logic gates, with probability 1.

In the categorical foundations model presented in [4], the classical information flow and conditioning of a unitary correction on the result of a measurement are modelled using biproducts and canonical distributivity and associativity isomorphisms. We do not give an exposition of the categorical treatment of classical information here (see [23] for more details), but simply consider under what conditions a unitary correction may be applied, to give the desired result.

With a unitary correction, Fig. 14.4. becomes as shown in Fig. 14.5. Of course, what we wish to apply is the unitary $U : H \to H$, rather than the composite $CUV : H \to H$ — thus we need conditions for these two to be equal. The most general solution is $C = UV^{-1}U^{-1} : H \to H$ — however, C is a classically determined correction, and there is no classical information available about the operation U itself (from a von Neumann architecture point of view, it may have been loaded into

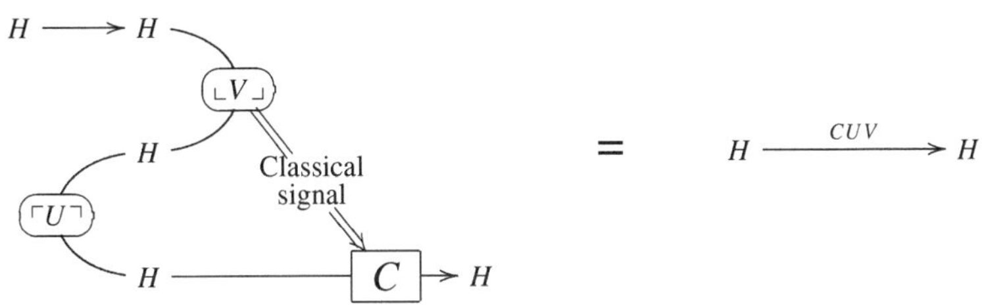

Fig. 14.5 Application by teleportation, with unitary correction

some "quantum instruction register" from memory, and be the outcome of some previous quantum computation).

Therefore, the unitary $C : H \to H$ may be conditioned on the measurement outcome $\llcorner V \lrcorner$, but cannot be dependent on U. This immediately imposes a restriction on the class of unitaries that may be implemented by teleportation.

It is immediate that operations that commute with all members of $\{V_j\}_{j=1}^{n^2}$ may be implemented deterministically, simply by taking $C = V^{-1}$. Slightly more generally, let us assume (without loss of generality — see Sect. 14.11.3 below) that $\{V_j\}_{j=1}^{n^2}$ form a group $\mathcal{G}$. In this case, we may implement the group $\mathcal{C}$ of unitary operations that satisfy $V_j^{-1} U V_j \in C$, for all $V_j \in \mathcal{G}$ and $U \in \mathcal{C}$.

Thus we cannot deterministically implement all unitary maps in using a teleportation protocol—the question now is: how severe is this restriction, and can we still do useful quantum computation?

14.11.3 Choosing a Measurement Basis

The only choice the experimenter has in the protocol shown in Fig. 14.5. is the choice of measurement basis—this must be a maximally entangled basis set for $H \otimes H$. As noted above, the basic assumption is that we have no information about the unitary map U, so in the general case there is no particular reason to favour one maximally entangled orthonormal basis over another. In any case, they are all equivalent up to some unitary isomorphism.

In a more explicitly computational setting, let us assume that H is the tensor product of a number of qubits. In this case, for both experimental and theoretical reasons, it is common to choose a basis based on the *Pauli group*.

Definition 13 Pauli groups, Bell basis
Consider the 2-dimensional Hilbert space of qubits Qu, with orthonormal basis $\{|0\rangle, |1\rangle\}$. The *(1-qubit) Pauli group* $\mathcal{G}_1$ that acts on Qu consists of the following unitary operations (given as matrices)

$$I = \begin{pmatrix} 1 & 0 \\ 0 & 1 \end{pmatrix}, \quad X = \begin{pmatrix} 0 & 1 \\ 1 & 0 \end{pmatrix}, \quad Y = \begin{pmatrix} 0 & -i \\ i & 0 \end{pmatrix}, \quad Z = \begin{pmatrix} 1 & 0 \\ 0 & -1 \end{pmatrix}$$

The names of these operations form a (maximally entangled) orthonormal basis for the space $Qu \otimes Qu$ called the *Bell basis*.

In general, the *Pauli group* G_n is the n-fold tensor product of G, so $\mathcal{G}_n = \{\otimes_{j=1}^n W_j : W_j \in \mathcal{G}_1\}$. Note that the names of the members of $\mathcal{G}_n$ also form a maximally entangled orthonormal basis for the space $\otimes j = 1^n Qu$

Physically, Pauli matrices correspond to the observables of spin $\frac{1}{2}$ particles—i.e. fermions such as protons, neutrons, &c. The names of Pauli group operations also form a very convenient maximally entangled basis for teleportation experiments, because of the following property:

Proposition 1 *Let U be a member of the Pauli group $\mathcal{G}_n$, and consider $|\psi\rangle \in \otimes_{j=1}^{n} Qu$. Then, experimentally, $U(|\psi\rangle)$ may be realised by 1-qubit operations.*

Proof This is immediate from the definition of $\mathcal{G}_n$ as the tensor product of a number of copies of $\mathcal{G}_1$. □

Corollary 1 *When using the names of the Pauli group $\mathcal{G}_n$ as the measurement basis for a teleportation protocol (as shown in Fig. 14.5), the required unitary corrections may be carried out as a series of 1-qubit operations.*

Corollary 2 *When using the names of the Pauli group $\mathcal{G}_n$ in a teleportation protocol, the operations that may be implemented deterministically are exactly the group $\mathcal{C}_n$ of operations that satisfy $\mathcal{C}_n = \mathcal{G}_n^{-1}\mathcal{C}_n\mathcal{G}_n$.*

14.11.4 The Clifford Group, and the Gottesman-Knill Theorem

The group specified in Corollary 2 above is well-known as the *Clifford group $\mathcal{C}_n$*—the stabiliser of the Pauli group $\mathcal{G}_n$. It is key to a number of different fields, including quantum error-correction, and measurement-based computation.

It is also the basis of one of the most significant recent results on quantum computation—the *Gottesman-Knill theorem*:

Theorem 8 *Any quantum circuit built up from:*

- *Computational basis preparations,*
- *Clifford group operations,*
- *Computational basis measurements*

may be efficiently simulated on a classical computer.

Proof This is proved in [39]—see also [79] for a good exposition.

Corollary 3 *If we wish to use teleportation to provide a deterministic evaluation operation, satisfying the conditions laid out in Sect. 14.11.1, we are restricted to quantum operations that can be efficiently simulated by a classical computer.*

A Comment: Measurement-Based Computing

For readers familiar with measurement-based computation, and the correspondence with implementing operations via teleportation given in [7], nothing in the above discussion should be interpreted as stating that measurement-based computation is restricted to the Clifford group. In particular, measurement-based computing is about applying *known* unitary operations, and the resulting feed-forward of classical information on experimental outcomes and corresponding unitary corrections is well-studied—see [25, 26] for a formal approach with a very theoretical computer science flavour.

A Question: Other Measurement Bases

A natural question at this point is whether choosing an alternative measurement basis (i.e. not based on the Pauli operations) will give us qualitatively different results. Recall that given two orthonormal basis sets B_1, B_2 for a space $H \otimes H$, we may give a unitary map $D : H \otimes H \to H \otimes H$ that maps one to the other—all orthonormal basis sets are equivalent up to isomorphism.

14.12 Naming an Unknown Arrow

After presenting such a negative result, it would be nice to discover something that quantum evaluation can do that classical evaluation cannot. So far, we have been considering how *data* may be interpreted as *code*, and applied to some other datum. Where the quantum setting wins out is in the dual process—given some implementation of a fragment of code, how may we *physically* produce the datum representing it (the *name* of the function)?

We treat a physical implementation of an instruction as a *black box* that we may give an input, and in return receive an output. In the classical case, we assume the black box is simply a function between the input and output sets, and in the quantum case, we assume a unitary map from the input to the output space.[11]

The question is then, given such a black box, how may we produce its name, as a physical resource?

14.12.1 The Classical Case

In the classical case, the situation is depressingly straightforward. We are given a black box that implements some function between finite sets, $f : X \to Y$. We have no additional information about the function f, but we wish to produce $\ulcorner f \urcorner \in Y \times X$, the "name" of $f : X \to Y$.

In the absence of other information about f, the only option is a brute-force investigation—we must feed into the black box every element of X, and record the output in each case.[12] Hence, the number of steps required to give the name of a function $f : X \to Y$ is exactly $|X|$. Note that the sucess of this procedure depends on

[11] In both cases, we ignore questions of timing and assume that the time spent processing is constant, regardless of the input. However, if the processing time is *not* constant, it maybe measured, and gives additional (classical) information about the operation of both quantum and classical black boxes. See [63] for applications of this in classical cryptography, and [17, 42, 72] for the key rôle that processing time plays in quantum computation.

[12] Even in the presence of a number of identical copies of the black box, and finite input/output sets, this procedure is at best tedious. In the infinite case, it is straightforwardly impossible—thus from a physical point of view, *arbitrary* functions cannot be named, even in the classical world. Under what conditions *computable* functions may be named is left as an interesting exercise.

- X is a finite set.
- the black box has no "internal states"—given an input $x \in X$ it returns the same output $f(x) \in Y$ regardless of the previous set of inputs to the black box.

14.12.2 The Quantum Case

The quantum case is also remarkably straightforward. Let us assume that the black box implements some unitary operation $U : H \to H$, and is subject to similar constraints to the classical case—i.e. a unitary map is implemented in constant time, and the input does not become entangled with some internal state of the black box.

Now recall the definition of the name of an arrow in a compact closed category, given diagramatically as

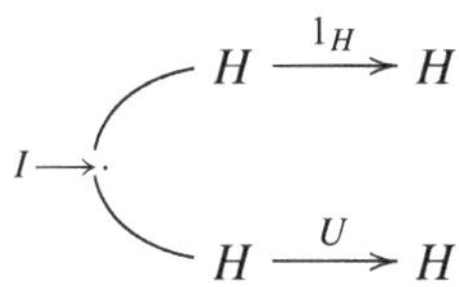

Interpreting the counit as the preparation of a maximally entangled pair (i.e. the name of the identity map), we simply create such an entangled pair, and pass one half of this pair through the black box, and do nothing at all to the other half. The resulting quantum state (taken as a whole) is then the name of the operation implemented by the black box.

Thus, creating the name of an unknown operation is a 1-step operation in the quantum case—compared to an arbitrarily long procedure in the classical case. Although this is arbitrarily more efficient for the quantum case, we may wonder what use it is . . .

14.12.3 A Fiction About Alice and Bob

Few papers on quantum computation or information are complete without a story about the QM information researchers, *Alice* and *Bob*. This paper is no exception, although the presented interpretation is more fanciful than most.

- Let us assume that Bob has developed a quantum computer $\mathcal{Q}$ that implements some interesting unitary map U on n-qubit registers. Alice, on the other hand, holds an n-qubit register R that she wishes to process using Bob's computer.
- Bob is happy for Alice to make this calculation, but does not wish to loan his computer to Alice—he may wish to prevent her reverse-engineering it by repeated applications to elements of the computational basis, or perhaps she still has not returned the textbooks she "borrowed" when they used to share an office.

To get around this impass, Bob works in the space of $2n$-qubit registers (and hence a space of size 2^{2n}, and produces the maximally entangled resource $\mathbf{B} = \ulcorner Id \urcorner$, where Id is the identity map. He then applies U to the final n registers of $\mathbf{B}$ to get some new resource $\mathbf{B}'$, where $\mathbf{B}'$ is the *name* of the unitary map U. Finally, he transmits the whole of $\mathbf{B}'$ to Alice.

Alice then treats the resource $\mathbf{B}'$ as the name of an unknown unitary map, and using the procedure described in Sect. 14.12.2 is able (albeit probabilistically) to produce $U(R)$, the result of applying Bob's quantum computer $\mathcal{Q}$ to her quantum register R.

In this manner, it seems that Bob may give out "samples" of his quantum computer $\mathcal{Q}$ that can be used exactly once, without Alice ever getting her hands on the precious machine. Perhaps the best conclusion to draw from this is that Digital Rights Management is much easier to enforce in the quantum world !

Unfortunately, if Alice is ever to implement Bob's unknown operation *reliably*, it must be a member of the Clifford group, and further classical communication with Bob will be required. So, it seems that the Quantum Rights Management™ protocol is in fact about distributing limited copies of classical programs.

However, if Bob can be persuaded to be a little less paranoid about Alice reverse-engineering is computer, he may find out from [85] that all he needs to perform universal quantum computation is *Clifford group operations*, and *single* $\frac{\pi}{8}$ *phase-shifts*. He may then split the above protocol into several parts, sending Alice either the name of some Clifford group operation, or a classical instruction to perform a $\frac{\pi}{8}$ phase-shift to a particular qubit, until the computation is complete.

Open question: If Bob performs the procedure described above, how much information can Alice deduce about the structure of his quantum computer?

14.13 Other Aspects

Our search for a quantum analogue of the von Neumann architecture has involved a lot of digressions into assorted, undeniably interesting, topics. Our perhaps unsurprising conclusion is that quantum computers cannot implement a suitable form of evaluation. However, our interest in quantum analogues of the vN architecture was not arbitrary—rather, we were interested in the computational features given in the list of Sect. 14.2.3. Although the behaviour of evaluation in the quantum setting prevents us from forming a direct analogue of the von Neumann architecture, it is worthwhile to consider topics related to this list directly.

14.13.1 Computational Universality and Quantum Computers

Much has been made of the fact that the von Neumann architecture allows for *computationally universal* machines. By *computationally universal* we mean, "capable of performing any computation that may be performed on a Turing machine". By the

Church-Turing thesis, this is believed to cover any computation that can be effiec-tively, or mechanically performed [43, 24].

The EDVAC machine, running the von Neumann architecture, was not the first computationally universal physical machine—this honour is believed to belong to the Z3 machine of Konrad Zuse machine [88] (also see [87] for an in-depth discussion, and the architecture of the Z3). Had the *analytic engine* of Charles Babbage [93] been completed, this would have claimed priority by at least 100 years.

Our interest in the von Neumann architecture has more been in the high-level control structures that arise naturally. However, it is worthwhile to consider whether a quantum computer can be computationally universal.[13] In one sense, this question is trivial; when restricted to a fixed computational basis, a quantum Turing machine [28] behaves exactly as a classical reversible Turing machine, and these are known to be computationally universal.

However, as discussed in Sect. 14.6.1, it is hard to see how a device restricted to a computational basis—for both "code" and "data"—may be considered in any way a *quantum* computer. A more interesting question is whether a computer may be both *coherent* and *computationally universal*—"fully quantum" in the terminology of [28, 76]. This question has been intensively studied, with no definite conclusion [76, 72, 42, 75]—and as always, the finite-dimensional case is much simpler [52].

Finally, the same features that make the von Neumann architecture universal contribute to its utility as a basis for high-level languages—although the connection between the ability to produce high-level languages, and computational universality, is far from clear. The papers [47, 49, 51] axiomatise the distinction between "high-level" and "low-level" languages in terms of domain theory, and make connections to computational universality. A quantum-mechanical version of this theory has also been presented in [50], but the situation there is even less clear.

14.13.2 Logical and Lambda-Calculii Interpretations of Monoidal Closure

Although we have emphasised the interchangeability of code and data as the key to the von Neumann architecture, it is more traditionally associated with other fields of theoretical computer science—Church's λ calculus, and categorical logic (and, of course, the connections between lambda calculii and logics given by the Curry-Howard isomorphism [91]). We refer to [69, 66, 13] for an introduction to categorical logic.

---

13 We strongly distinguish this from the idea of a *universal quantum gate* that can simulate (up to a reasonable approximation) any other quantum gate. In particular, universal computation requires the possibility of non-termination of an algorithm... and indeed the undecidability of termination is a fundamental theorem of theoretical computer science [54]. Both the usual quantum circuit model [17] and the (restricted forms of) quantum Turing machines contained in [17] require unconditional termination in exactly K steps, where K is some a priori fixed value.

The logical interpretation of categories used to model quantum protocols is given by Abramsky and Duncan, in [29, 6]. We do not give an exposition of this, but rather consider the conditions required for an *untyped* analogue, in the search for a computational system similar to the untyped lambda calculus.

Traditionally, the pure untyped lambda calculus is modelled by a *C-monoid*, or one-object Cartesian closed category[14] without a terminal object. (See also [3] for a computationally universal combinatory logic in terms of (untyped) *compact closure*, rather than *Cartesian closure*).

Given a Cartesian closed category $(\mathbf{C}, \times)$, a C-monoid is the endomorphism monoid of an object $D \in Ob(\mathbf{C})$ that satisfies

$$D \cong D \times D \cong [D \to D]$$

The identity $D \cong D \times D$ is relatively easy to satisfy: for sets and functions, and many other categories, this is satisfied by any countably infinite set (see [44, 45] for the general setting). However, $D \cong [D \to D]$ is less easy to satisfy—simple cardinality arguments demonstrate that no object in the category of sets and functions may satisfy this identity. The usual route to objects satisfying such identities is via *domain theory*. See [1] or [74] for a general perspective, and [66] for a categorical exposition related to C-monoids.

In compact closed categories, the situation is both simpler, and more subtle. The simple form of the internal hom object $[X \to Y] = Y^* \otimes X$ means that $N \cong [N \to N]$ is equivalent to self-duality $N = N^*$, together with the identity $N \cong N \otimes N$. Given any object $N \cong N \otimes N$, note also that $G = N \otimes N^*$ satisfies

$$G \cong G \otimes G \cong [G \to G]$$

We refer to [44, 48] for an explicit description of one-object analogues of compact closed categories.

In the particular case of Hilbert spaces, any separable infinite-dimensional Hilbert space H certainly satisfies $H \cong H \otimes H$. However, recall from [2] that only the category of finite-dimensional spaces is compact closed—thus, it is hard to see how evaluation in the quantum setting may be used to produce some form of (untyped) lambda calculus or combinatory logic. Infinitary versions of the Choi-Jamiołkowski correspondence have been explored in [82], in the context of order theory and C* algebras, but the categorical interpretation is not straightforward.

[14] The connection between C-monoids and Church's lambda calculus is not straightforward. As observed in [66], the product structure is equivalent to requiring *surjective pairing* in the lambda calculus. We also refer to [66] for a demonstration—via Occam's Razor—of how *combinatory logic* arises from C-monoids, without explicit reference to products, i.e. using the closed, but not monoidal closed, structure—giving what [66] refers to as "monstrous" coherence conditions. The author leaves it to those braver than himself to reason about quantum physics using the (admittedly elegant) language of such monstrosities!

14.13.3 Backus, Functional Languages, and Non- von Neumann Architectures

Throughout this paper, we have praised the von Neumann architecture as a significant advance in both theoretical and practical computer science. This is certainly the case; however, many programmers and theoreticians also see the near-universal reliance on it as an impediment, particularly with regard to either parallel or asynchronous computation. The *von Neumann bottleneck* is a common term first appearing in J. Backus' Turing award acceptance speech, "Can Programming be Liberated from the von Neumann Style?" [11]).

Thus, although we consider it unfortunate that quantum computers cannot run some analogue of the von Neumann architecture, it may instead be seen as an opportunity.

Backus' speech gives a strong defence of *functional programming* and functional programming languages. Functional programs are based on the notion of evaluating functions, rather than updating states. Unfortunately, they are often easier to characterise in terms of features they do *not* possess, such as

> Functional programming languages have no variables, no assignment statements, and no iterative constructs. This design is based on the concept of mathematical functions, which are often defined by separation into various cases, each of which is separately defined by appealing (possibly recursively) to function applications. — [32].

There is no space here for an exposition of the positive aspects of functional languages and programming—we refer to Backus' speech and subsequent works [11, 12, 53] for positive properties such as *referential transparency*, *lazy evaluation*, *freedom from side-effects* and *state-freeness*.[15]

As well as a plea for *programming languages* based on different principles, it is clear that Backus considered the von Neumann style of programming to be a direct consequence of the von Neumann architecture, and the persistence of the von Neumann architecture to be due to the universality of languages based on it. From [11],

> Our fixation on von Neumann languages has continued the primacy of the von Neumann computer . . . The absence of programming styles founded on non-von Neumann principles has deprived designers of an intellectual foundation for new computer architectures.

A stated aim of [11] is thus to provide programming concepts and languages that naturally lead to different underlying computer architectures—unfortunately, functional programs still tend to be executed on vN architecture machines!

Another key point of this program is that functional programming languages could, or should, come equipped with an "algebra of combining forms". This is an algebraic system that is intended to, *solve equations whose "unknowns" are*

[15] Side-effects and states are often considered essential for input/output, storage, exception-handling, &c. We refer to [96] for how such features are handled in functional programming using the very categorical idea of *monads.*

programs, in much the same way as one transforms programs in high school algebra—[11]. Whether such a system is possible for quantum algorithms remains open—although a step in this direction is [74], giving domain-theoretic analogues of differential equations with both classical and quantum search as their solutions.

Acknowledgments The author wishes to thank many individuals, and is very grateful for many discussions with: S. Abramsky and B. Coecke on the categorical foundations program and category theory generally, T. Altenkirch on quantum conditionals, iteration, and programming languages, V. Danos and R. Duncan for interest in an early draft of this paper, S. Braunstein on teleportation and the Choi-Jamiołkowski correspondence from a physicists perspective, and the software problem for quantum computers, K. Martin on the behaviour of quantum algorithms from a domain theory point of view, P. Selinger on functional programming and algebras of combining forms, and P. Scott on logical and categorical interpretations, and connections with lambda calculus.

Appendix

We consider the conditions required for a state vector to correspond to a unitary map, and show that this is intimately connected with questions of entanglement.

Definition 14 Let H denote a complex Hilbert space with orthonormal (computational) basis $\{\mathbf{e}_i\}_{i=1..n}$. Then for each basis vector $\mathbf{e}_i$ we define the *left-span* of $\mathbf{e}_i$ to be the subspace of $H \otimes H$ generated by the basis vectors $\{e_i \otimes e_j\}_{j=1}^n$. Dually, we define the *right-span* of $\mathbf{e}_j$ to be the subspace of $H \otimes H$ generated by the basis vectors $\{e_i \otimes e_j\}_{i=1}^n$. We denote these spaces by $lSpan(\mathbf{e_i})$ and $rSpan(\mathbf{e_j})$ respectively.

Our claim is that the vectors that are the names of unitary maps are exactly those that are equidistant to the left span and the right span of each basis vector $\mathbf{e}_i$, and thus maximally entangled.

Theorem 9 *Let H denote a complex Hilbert space with orthonormal basis $\{\mathbf{e}_i\}_{i=1..N}$, and let $M : H \to H$ denote a linear map. Then the following two conditions are equivalent:*

- **(i)** M is a unitary map.
- **(ii)** For each basis vector $\mathbf{e}_i$, the norm of the projection of $\ulcorner M \urcorner$ onto either $lSpan(\mathbf{e}_i)$ or $rSpan(\mathbf{e}_i)$ is $\frac{1}{n}$.

Proof (**(i)** ⇒ **(ii)**) By definition of a unitary map, M satisfies

$$MM^\dagger = I \text{ and } I = M^\dagger M$$

(It is easier to state that $M^\dagger = M^{-1}$. However, interesting and computationally important C^* algebras such as the Kuntz-Krieger algebras of [9] satisfy one-sided versions of these conditions, so we use them separately for future reference). Written in terms of matrix elements, we have that

$$(MM^{\dagger})_{ik} = \sum_{j=1}^{n} m_{ij}\overline{m_{kj}} = \begin{cases} 1 & i = k \\ 0 & i \neq k \end{cases}$$

$$(M^{\dagger}M)_{ik} = \sum_{j=1}^{n} m_{ji}\overline{m_{jk}} = \begin{cases} 1 & i = k \\ 0 & i \neq k \end{cases}$$

These conditions can also be characterised as "the sum of the norms of the entries in each row is 1, as is the sum of the norms of the entries in each column".

Moving to the name $\ulcorner M \urcorner \in H \otimes H$, we have

$$\langle \ulcorner M \urcorner | \ulcorner M \urcorner \rangle = \frac{1}{\sqrt{n}}\frac{1}{\sqrt{n}} = \left(\sum_{\alpha,\beta=1}^{n} \left(\sum_{i,j=1}^{n} \langle \overline{m_{\alpha\beta}}(\mathbf{e}_i \otimes \mathbf{e}_j) | m_{ij}(\mathbf{e}_i \otimes \mathbf{e}_j) \rangle \right) \right)$$

Using the Kroneker delta notation, $\langle \mathbf{e}_i \otimes \mathbf{e}_j | \mathbf{e}_\alpha \otimes \mathbf{e}_\beta \rangle = \delta_{\alpha i}\delta_{\beta j}$, so

$$\langle \ulcorner M \urcorner | \ulcorner M \urcorner \rangle = \frac{1}{\sqrt{n}}\frac{1}{\sqrt{n}} = \left(\sum_{\alpha,\beta,i,j=1}^{n} \overline{m_{\alpha\beta}} m_{ij} \delta_{\alpha i} \delta_{\beta j} \right)$$

Hence, by the condition imposed on the matrix elements by the unitarity requirement,

$$\langle \ulcorner M \urcorner | \ulcorner M \urcorner \rangle = \frac{1}{N} \sum_{\alpha,\beta=1}^{n} \overline{m_{\alpha\beta}} m_{\alpha\beta} = 1$$

So the unitarity of M implies that $\ulcorner M \urcorner$ has norm 1.

For the next step, observe that we may isolate the individual m_{ij} by

$$m_{ij} = \sqrt{n}.\langle \mathbf{e}_i \otimes \mathbf{e}_j | \ulcorner M \urcorner \rangle$$

and so the first unitarity condition gives that

$$1 = \sum_{j=1}^{n} \frac{1}{\sqrt{n}} \langle \ulcorner M \urcorner | \mathbf{e}_i \otimes \mathbf{e}_j \rangle \frac{1}{\sqrt{n}} \langle \mathbf{e}_i \otimes \mathbf{e}_j | \ulcorner M \urcorner \rangle$$

Hence

$$\frac{1}{n} = \sum_{j=1}^{n} \langle \ulcorner M \urcorner | \mathbf{e}_i \otimes \mathbf{e}_j \rangle \langle \mathbf{e}_i \otimes \mathbf{e}_j | \ulcorner M \urcorner \rangle$$

Using Dirac notation, for any orthnormal basis B the identity is given by $Id = \sum_{b \in B} |\mathbf{b}\rangle\langle\mathbf{b}|$, and so the inner product of vectors ϕ, ψ may be written as $\langle \phi | \psi \rangle =$

$\sum_{\mathbf{b}\in B}\langle\phi|\ \mathbf{b}\rangle\langle\mathbf{b}|\psi\rangle$. From the definition of the space $lSpan(\mathbf{e}_i)$ in terms of a basis set, we thus deduce that the projection of $\ulcorner M \urcorner$ onto the space $lSpan(\mathbf{e}_i)$ has norm $\frac{1}{N}$.

The dual condition about the right spans $\{rSpan(\mathbf{e}_i)\}_{i=1}^n$ follows from the second unitarity condition.

(ii) ⇒ (i) Let $\psi = \ulcorner M \urcorner$ satisfy the left and right span conditions. We may write these fully as

$$\frac{1}{n} = \sum_j \langle\psi|\mathbf{e}_i \otimes \mathbf{e}_j\rangle\langle\mathbf{e}_i \otimes \mathbf{e}_j|\psi\rangle$$

and

$$\frac{1}{n} = \sum_j \langle\psi|\mathbf{e}_i \otimes \mathbf{e}_j\rangle\langle\mathbf{e}_i \otimes \mathbf{e}_j|\psi\rangle$$

respectively. The definition of the naming operation $\ulcorner\ \urcorner$ gives that

$$[M]_{ij} = \sqrt{n}.\langle e_i \otimes e_j|\psi\rangle$$

and almost identical reasoning to above applied to the left span condition gives $\sum_{j=1}^n [M]_{ij}.\overline{[M]_{ij}} = 1$. Similarly, the right span condition gives that

$$\sum_{i=1}^n [M]_{ij}.\overline{[M]_{ij}} = 1.$$

From above, these are the two conditions required for unitarity, and hence our result follows. □

Interpretation

Although the above is presented abstractly, a quantum computational interpretation is immediate: given a quantum register $r \in qByte$, and an observation on a single arbitrary qubit with respect to *any* orthonormal basis $\{\mathbf{b}_1, \mathbf{b}_2\}$, then r is the name of a unitary map exactly when the observation of r gives either $\mathbf{b}_1$ or $\mathbf{b}_2$ with equal probability.

References

1. Abramsky, S.: Domain theory in logical form. Ann. Pure Appl. Logic **51**, 177 (1991)
2. Abramsky, S., Blute, R., Panangaden, P.: Nuclear and Trace Ideals in Tensored *-categories. J. Pure Appl. Algebra **143**(1) 3–47 (1999)
3. Abramsky, S., Haghverdi, E., Scott, P.: Geometry of interaction and linear combinatory algebras. Math. Struct. Comput. Sci. **12**(5) 625–665 (2002)

4. Abramsky, S., Coecke, B.: A categorical semantics of quantum protocols. In: Proceedings of the 19th Annual IEEE Symposium on Logic in Computer Science (LICS 2004), pp. 415–425. IEEE Computer Soc. Press (2004)
5. Abramsky, S.: Abstract scalars, loops, and free traced and strongly compact closed categories. In: Proceedings of CALCO 2005, Springer LNCS 3629, pp. 1–31 (2005)
6. Abramsky, S., Duncan, R.: A categorical quantum logic. Math. Struct. Comput. Sci. **16**(3) (2006)
7. Aliferis, P.: D Leung Computation by Measurements: A unifying picture. Phys. Rev. A **70**, 062314 (2004)
8. Altenkirch, T., Grattage, J., Vizzotto, J.: An Algebra of Pure Quantum Programming ENTCS, 3rd International Workshop on Quantum Programming Languages (2005) (to appear)
9. Anantharaman, C.: Delaroche C*– algèbres de Cuntz- Krieger et groupes Fuchsiens in Operator Theory, Operator Algebras and Related Topics (Timisoara 1996) The Theta Foundation, Bucharest, pp. 17–35 (1997)
10. Aspray, W.F.: Pioneer day '82: history of the stored program concept. Ann. Hist. Comp. **4**(4) 358–367 (1982)
11. Backus, J.: Can Programming Be Liberated from the von Neumann Style? A functional style and its algebra of programs. Commun. ACM **21**(8) 613–641 (1978)
12. Backus, J.: Function level programs as mathematical objects functional programming languages and computer architecture archive. In: Proceedings of 1981 Conference on Functional Programming Languages and Computer Architecture, pp. 1–10. Portsmouth, New Hampshire (1981)
13. Barr, M., Wells, C.: Category Theory for Computer Science, 3rd edn., Centre de Recherches Mathématique, Montreal (1999)
14. Bennett, C.: Logical reversibility of computation. IBM J. Res. Develop. **17** 525–532 (1973)
15. Bennett, H., Brassard, G., Crépeau, C., Jozsa, R., Peres, A., Wootters, W.K.: Teleporting an unknown quantum state via dual classical and einstein-podolsky-rosen channels. Phys. Rev. Lett. **70**, 1895–1899 (1993)
16. Brassard, G., Braunstein, S., Cleve, R.: Teleportation as a quantum computation. Physica D **120**, 43–47 (1998)
17. Bernstein, E., Vazirani, U.: Quantum complexity theory. SIAM J. Comput. **26**(5), 1411–1473 (1997)
18. Brown, A., Page, A.: Elements of Functional Analysis. Van Nostrand Reinhold Publishing, London (1970)
19. Burks, A.: Who invented the computer? The legal battle that changed computing history Promethius Books (2003)
20. Choi, M.: Completely Positive Linear Maps on Complex Matrices, Linear Algebra and Its Applications, pp. 285–290 (1975)
21. Coecke, B.: Introducing Categories to the practicisng physicist (Manuscript) Available as: http://web.comlab.ox.ac.uk/oucl/work/bob.coecke/Cats.pdf (2005)
22. Coecke, B.: Kindergarten quantum mechanics – Lecture notes. In: Quantum Theory: Reconsiderations of the Foundations III, p. 8198. AIP Press, New York (2006)
23. Coecke, B., Pavlovic, D.: Quantum measurements without sums. In: Chen, G., Kauffman, L., Lomonaco, S. (eds.) The Mathematics of Quantum Computation and Technology. Taylor and Francis (to appear)
24. Copeland, B.: The church-turing thesis. In: Zalta, E.N. (ed.) The Stanford Encyclopedia of Philosophy (Fall 2002 Edition), available as http://plato.stanford.edu/archives/fall2002/entries/church-turing/ The citation above refers to the version in the following archive edition (2002)
25. Danos, V., Kashefi, E., Panangaden, P.: The one way to quantum computation. In: Bugliesi, M. et al.: (eds.) ICALP 2006, Part II, Springer LNCS 4052 1321 (2006)
26. Danos, V., Kashefi, E., Panangaden, P.: The Measurement Calculus quant-ph: 0704.1263v1 (2007)

27. Darwin, C.G.: Free Motion in the wave mechanics. Proceedings of the Royal Society of London. Series A, Containing Papers of a Mathematical and Physical Character 117 (766) pp. 258–293 (1927)
28. Deutsch, D.: Quantum theory, the Church-Turing principle and the universal quantum computer Proceedings of the Royal Society of London, A 400, pp. 97–117 (1985)
29. Duncan, R.: Types for Quantum Computation, DPhil. Thesis, Oxford University (2007)
30. Eilenburg, S., Kelly, G.: Closed Categories Proceedings of the Conference on Categorical Algebra (La Jolla 1965), Springer 421562 (1966)
31. Feynman, R., Leighton, R., Sands, M.: The Feynman Lectures on Physics, vol. 3. Addison-Wesley, Reading, Mass (1965)
32. Finkel, R.: Advanced Programming Language Design. Addison-Wesley, Reading (1996)
33. Fouche, W., Heidema, J., Jones, G., Potgieter, P.: Halting in quantum Turing computation (2008)
34. Gay, S.: Quantum programming languages: survey and bibliography. Math. Struct. Comput. Sci. 16(4) (2006)
35. Geroch, R.: Mathematical Physics. University of Chicago Press, Chicago (1985)
36. Girard, J.-Y.: Linear logic. Theor. Comput. Sci. **50**(1), 1102 (1987)
37. Girard, J.-Y., Lafont, Y., Taylor, P.: Proofs and Types, Cambridge Tracts in Theoretical Computer Science, C.U.P (1989)
38. Godfrey, M., Hendry, D.: The computer as von Neumann planned it. IEEE Ann. History Comput. **15**(1), (1993)
39. Gottesman, D.: The Heisenberg representation of quantum computers. In: Corney, S.P., Delbourgo, R. Jarvis, P.D. (eds.), Group22: Proceedings of XXII International Colloquium on Group Theoretic Methods in Physics, pp. 32–43. Cambridge, MA. Extended version at arXiv:quant-ph/9807006 (1999)
40. Grattage, J.: QML – A Functional Quantum Programming Language, PhD Thesis, Univ. Nottingham (2006)
41. J. Gruska Quantum Computing. McGraw-Hill, New York (1999)
42. Hagar, A., Korolev, A.: What is quantum in quantum computing? A lesson from two halting problems,ESF Workshop on The Logic of Quantum Information 2004, available as http://www.webalice.it/shakush/ST1.pdf (2004)
43. Herken, R. (ed.): The universal Turing machine: A half-century survey, 2nd ed. Springer, New York (1995)
44. Hines, P.: The categorical theory of self-similarity. Theory Appl. Categories **6**, 33–46 (1999)
45. Hines, P.: A short note on coherence and self-similarity. J. Pure Appl. Algebra **175**(1), 135–139 (2002)
46. Hines, P.: A categorical framework for finite state machines. Math. Struct. Comput. Sci. **13**(3), 451–480 (2003)
47. Hines, P.: Physical systems as constructive logics. In: Calude, Dinneen, Paun, Rozenburg, Stepney (eds.) Springer LNCS, vol. 4135, pp. 101–112 (2006)
48. Hines, P.: Compact closed monoids – Definitions and Constructions GeoCal, CIRM Marseille (2006)
49. Hines, P.: Machine semantics. Theor. Comput. Sci. **409**, 1–23 (2008)
50. Hines, P.: Towards a quantum machine semantics. Mathematical Foundations of Programming Semantics (2007), invited talk
51. Hines, P.: From causality to computational models. Int. J. Unconventional Comput. **4**(3) 249–272 (2008)
52. Hines, P.: Quantum Circuit Oracles for Abstract Machine Computations. Available as: http://www.peterhines.net/downloads/papers/Circuits.pdf (submitted)
53. Hughes, J.: Why functional programming matters. In Turner, D. (ed.) Research Topics in Functional Programming. Addison Wesley, New York (1990)

54. Immerman, N.: Computability and complexity. In: Zalta, E.N. (ed.) The Stanford Encyclopedia of Philosophy (Fall 2006 Edition), available as: http://plato.stanford.edu/archives/fall2006/entries/computability/ (2002)
55. Jamiołkowski, A.: Linear transformations which preserve trace and positive semidefiniteness of operators. Rep. Math. Phys. **3**, 275–278 (1972)
56. Johnstone, P.: Stone Spaces, Cambridge Studies in Advanced Mathematics 3. Cambridge University Press, Cambridge (1986)
57. Johnstone, P.: Review of, 'Natural dualities for the working algebraist', by Clarke, D., Davey, B., American Mathematical Society (2000)
58. Joyal, A., Street, R.: The geometry of tensor calculus I. Adv. Math. **88**(1), 55–112 (1991)
59. Joyal, A., Street, R.: The geometry of tensor calculus II manuscript
60. Joyal, A., Street, R., Verity, D.: Traced Monoidal categories. Math. Proc. Camb. Phil. Soc. 425–446 (1996)
61. Jozsa, R., Linden, N.: On the rôle of entanglement in quantum speedup. Proceedings of the royal society. Math. Phys. Eng. Sci. **459**(2038), 2011–2032 (2003)
62. Kelly, M., Laplaza, M.: Coherence for compact closed categories. J. Pure Appl. Algebra **19**, 193–213 (1980)
63. Kocher, P.C.: Timing attacks on implementations of Diffie Hellman, RSA, DSS, and Other Systems. In: Koblitz, N. (ed.) Advances in Cryptology – CRYPTO 96, Springer-Verlag, LNCS, vol. 1109, pp.104–113 (1996)
64. Kitaev, A., Shen, A., Vyalyi, M.: Classical and Quantum Computation, Graduate Studies in Mathematics (47), American Mathematical Society, Providence (2002)
65. Landauer, R.: Irreversibility and heat generation in the computing process. IBM J. Res. Dev. **5**, 183–191 (1961)
66. Lambek, J., Scott, P.: An Introduction to Higher-Order Categorical Logic, Cambridge Studies in Advanced Mathematics 7. Cambridge University Press, Cambridge (1986)
67. Laplaza, M.: Coherence in non-monoidal closed categories. Trans. Am. Math. Soc. **230**, 293–311 (1977)
68. Laplaza, M.: Embedding of closed categories into monoidal closed categories. Trans. Am. Math. Soc. **233**, 85–91 (1977)
69. Lawvere, F.W.: Functorial semantics of algebraic theories. Proc. Nat. Acad. Sci. **50**(5) 869–872 (1963)
70. Lee, J.A.N.: Looking Back, IEEE Computer Magazine (Feb. 1996)
71. Lee, J.A.N.: John Louis von Neumann, The History of Computing Series (NSF project CDA-931261). available as http://ei.cs.vt.edu/_history/VonNeumann.html (2002)
72. Linden, N., Popsecue, L.: The halting problem for quantum computers, arXiv:quant-ph:/9806054 v2 1998 (1998)
73. MacLane, S.: Categories for the Working Mathematician, 2nd edn. Springer, New York (1998)
74. Martin, K.: Domain theory and Measurement This volume (2008)
75. Miyadera, T., Ohya, M.: Designing Quantum Turing Machine is uncomputable Proc. EQIS 02 (2002)
76. Myers, J.: Can a universal computer be fully quantum? Phys. Rev. Lett. **78**(9) 1823–1824 (1997)
77. von Neumann, J.: First Draft of a Report on the EDVAC U.S. Army report on Contract No. W-670-ORD-4926., University of Pennsylvania. Moore School of Electrical Engineering. See also [38] (1945)
78. Nielsen, M., Chuang, I.: Programmable quantum gate arrays. Phys. Rev. Lett. **79**, quant-ph/9703032 (1997)
79. Nielsen, M.: Chuang Quantum Computation and Quantum Information. Cambridge University Press, Cambridge (2000)
80. Pati, A., Braunstein, S.: Impossibility of deleting an unknown quantum state. Nature **404**, 164–165 (2000)
81. Pati, A., Braunstein, S.: Quantum deleting and signaling. Phys. Lett. A **315**, 208–212 (2003)

82. Raginsky, M.: Radon-Nikodym Derivatives of Quantum Operations arXiv:math-ph/0303056 (2003)
83. Renstall, G.: On logics without contraction, PhD Thesis, University of Queensland (1994)
84. Renstall, G.: An introduction to sub-structural logics. Routledge NY (1999)
85. Shi, Y.: Both Toffoli and Controlled-NOT need a little help to do universal quantum computation arXiv:quant-ph/0205115 v2 (2002)
86. Roeder, D.: Category theory applied to Pontryagin duality. Pacific J. Math. **52**(2), 519–527 (1974)
87. Rojas, R.: Konrad Zuse's legacy. IEEE Ann. History Comput. **19**(2), 5–16 (1997)
88. Rojas, R.: How to make Zuse's Z3 a universal computer. IEEE Ann. History Comput. **20**(3), 51–54 (1998)
89. Selinger, P.: Dagger compact closed categories and completely positive maps. In: Proceedings of the 3rd International Workshop on Quantum Programming Languages, ENTCS, vol. 170, pp. 139–163 (2007)
90. Shum, M.C.: Tortile tensor categories. J. Pure Appl. Algebra **93**, 57–110 (1994)
91. Sørensen, M., Urzyczyn, P.: Lectures on the Curry-Howard isomorphism, Studies In: Abramsky, Artemov, Gabbay, Kechris, Pillay, Shore (eds.) Logic and the Foundations of Mathematics, vol. 149. Elsevier (2006)
92. Stern, N.: From ENIAC to UNIVAC: An Appraisal of the Eckert-Mauchly Computers. Digital Press, Bedford, MA (1981)
93. Swade, D.: The difference engine: Charles Babbage and the quest to build the first computer, Penguin – Putnam (2001)
94. Thomsen, M., Gl¨uck, R., Axelsen, H.: Towards designing a reversible processor architecture. Proceedings of the Reversible Computation 2009, ENTCS (to appear)
95. Yu Vlasov, A.: Programmable quantum networks with pure states. In: Stones, J.E. (ed) Computer Science and Quantum Computing, pp. 33–61. Nova Science Publishers, Inc. (2005)
96. Wadler, P.: Monads for functional programming. In: Jeuring & Meijer (eds.) Advanced Functional Programming. Springer, LNCS 925 (1995)
97. Wootters, W., Zurek, W.: A single quantum cannot be cloned. Nature **299**, 802–803 (1982)

Chapter 15
A Categorical Presentation of Quantum Computation with Anyons

P. Panangaden and É.O. Paquette

Abstract In nature one observes that in *three space dimensions* particles are either symmetric under interchange (bosons) or antisymmetric (fermions). These phases give rise to the two possible "statistics" that one observes. In two dimensions, however, a whole continuum of phases is possible. "Anyon" is a term coined in by Frank Wilczek to describe particles in 2 dimensions that can acquire "any" phase when two or more of them are interchanged. The exchange of two such anyons can be expressed via representations of the braid group and hence, it permits one to encode information in *topological* features of a system composed of many anyons. Kitaev suggested the possibility that such topological excitations would be stable and could thus be used for robust quantum computation.

This paper aims to

1. give the categorical structure necessary to describe such a computing process;
2. illustrate this structure with a concrete example namely: Fibonacci anyons.

15.1 Introduction

The mathematics and physics of anyons probe the most fundamental principles of quantum mechanics. They involve a fascinating mix of experimental phenomena (the fractional quantum Hall effect), topology (braids), algebra (Temperley-Lieb algebra, braid group and category theory) and quantum field theory. Because of their topological nature, it is hoped that one can use them as *stable* realisations of qubits for quantum computation, as proposed originally by Kitaev [28]. In this article we

P. Panangaden (✉)
School of Computer Science, McGill University, Montreal, QC, Canada
e-mail: prakash@cs.mcgill.ca

É.O. Paquette (✉)
School of Computer Science, McGill University, Montreal, QC, Canada
e-mail: eopaquette@holon23.net

Panangaden, P., Paquette, É.O.: *A Categorical Presentation of Quantum Computation with Anyons*. Lect. Notes Phys. **813**, 983–1025 (2011)
DOI 10.1007/978-3-642-12821-9_15

review the mathematics of anyons and discuss the relations with braids, topology and modular tensor categories.

The spin-statistics theorem [34, 38, 44, 47] says, roughly speaking, that particles with $\frac{1}{2}$-integer spin satisfy Fermi-Dirac statistics (or are "fermions") while particles with integer spin satisfy Bose-Einstein statistics (or are "bosons"). This statement is one of the few actual *theorems* of relativistic quantum field theory. What this means is that it can be proved from very general assumptions of quantum mechanics and relativity and does not depend on particular models of particles.

The proofs traditionally given [47, 17, 18] involve quantum field theory and rest on assumptions about causality, invariance and positivity of energy. Nevertheless, the statement seems to have a compellingly topological flavor. Indeed such a topological reading has been given by Finkelstein and Rubinstein [20], but for extended objects rather than for elementary particles. The review article by Duck and Sudarshan [17] critiques this and other approaches to the spin-statistics theorem. Their essential point is that elementary particles are not extended objects and thus the topological arguments do not apply. Wightman points out that relativistic invariance is essential to the proof and argues that there is no spin-statistics connection when only euclidean invariance is assumed.

What is clear is that the proof does depend on the fact that relativistic quantum field theory is formulated on a $3+1$ dimensional spacetime. In a $2+1$ dimensional spacetime—which can be realised in the laboratory using surface phenomena—the usual argument for the existence of two kinds of statistics is not valid anymore. Nor is the argument for the existence of two kinds of spins. There is still a connection between spin and statistics but now a continuum of possibilities for each exists.

The experimental investigations of the relevant surface phenomena reveal many surprises. For example, the entities involved have flux tubes that *are* extended objects, so the topological intuitions underlying the spin-statistics theorem are no longer just heuristic. Furthermore there are collective excitations that behave like particles *but like particles with fractional electric charge*. Such fractional electric charges are never seen in nature and there are strong theoretical reasons (superselection rules) to think that they cannot occur free in nature.

In the mathematical physics literature, anyons seem to be intimately related to concepts such as *modular tensor categories* (MTC), a particular class of monoidal categories, *modular functors* (MF), *topological quantum field theory* (TQFT) and *conformal field theory* (CFT). There is, however, an order to this collection of ideas. Indeed, the preceding mathematical constructions may be related in the following way [5]:

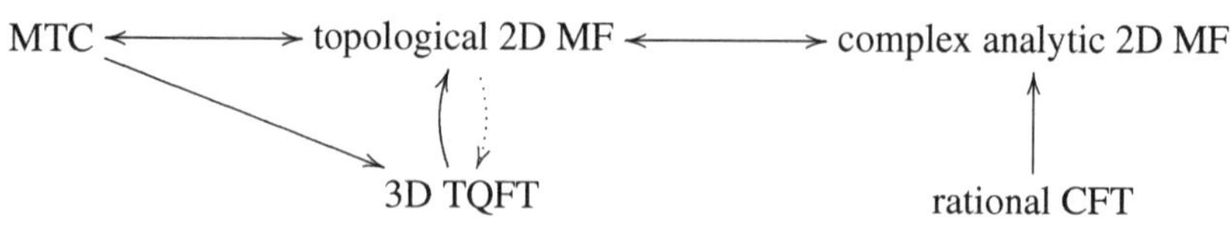

Definitions and an expository account of these relations—along with the precise assumptions needed to define them—are given by Bakalov and Kirilov in [5]. There, they essentially present a complete picture of the results found in [4, 13, 27, 41, 42] and [48]. The main point is that most of these structure are essentially equivalent. Using this and the fact that the theory of anyons is correctly described by semisimple MTCs, the categorical semantics for topological quantum computation with anyons that we present in Sect. 15.4 is based on these.

The prerequisites for this paper are relatively modest. We will assume that the reader is familiar with basic category theory and quantum computation. For an introduction to category theory, the reader is referred to [12] or, for a more technical introduction, to [31]. For quantum computation, we suggest [37] or, for a more categorical introduction to the subject [2].

15.1.1 Physical Background

The physical effect most associated with anyons is called the fractional quantum Hall effect (fqHe). In order to set the stage we first explain the classical Hall effect, then the (integer) quantum Hall effect and finally the remarkable features of the fqHe. The summary here is very brief and is no substitute for the many thorough papers that have been written in the physics literature. A particularly lucid presentation of the physical ideas appears in the 1998 Nobel lecture of Horst Störmer [46].

The Hall effect was discovered the same year that Einstein was born, 1879. The idea is very simple. Consider a fixed current flowing through a conductor, which we take to be a flat strip. The current flows along the long axis of the strip. If one measures the voltage at various points along the direction of current flow, one gets a drop in the voltage associated with the normal electrical resistance of the material. According to Ohm's law we have $R = V/I$, where V is the measured voltage drop between two points, I is the current and R is the resistance between the two points. One can also measure the voltage drop between two points lying *transverse* to the flow of the current. One does not expect to see a voltage drop, and indeed none is detected. However, if one applies a magnetic field in the direction perpendicular to the strip then there is a transverse force on the electrons flowing through the conductor: this is the well known Lorentz force law. This can be written as

$$\mathbf{F} = q\mathbf{v} \times \mathbf{B}$$

where $\mathbf{F}$ is the force, $\mathbf{v}$ is the velocity of the charged particle, q is its charge and $\mathbf{B}$ is the applied magnetic field. The $\times$ denotes the 3-vector cross product, so the force is perpendicular to both the direction of motion of the charge and the direction of the applied magnetic field. The upshot is that the electrons are pushed to one side and a voltage develops in the direction *across* the current flow. This is the Hall effect and is well understood in terms of classical electrodynamics.

The transverse voltage in the Hall effect V_H, yields an effective Hall "resistance" denoted by $R_H = V_H/I$. The resistance is now a *tensor*: the voltage and current are no longer in the same direction. Since the transverse force increases linearly with the applied magnetic field, one expects that R_H depends linearly on the applied magnetic field B; this is indeed what Hall found. Less obvious is the fact that the

Hall resistance decreases with increasing electron density. The reason for this is that, for a fixed current, the electrons have to travel faster to achieve the same current so according to the Lorentz force law the transverse force is greater.

A remarkable thing happens in two-dimensional systems at low temperatures. The simple linear behavior of the Hall resistance on the applied magnetic field is replaced by a complicated graph featuring plateaus followed by jumps. Furthermore, the value of the Hall resistance jumps according to a very simple law

$$R_H = \frac{h}{ie^2}$$

where, h is Planck's constant, e is the charge of an electron and i takes on positive integer values. The Hall resistance seems to be "quantised." In addition to this strange behavior of the Hall resistance, the ordinary resistance vanishes at the points corresponding to the plateaus of the Hall resistance. This behavior has been measured very accurately and seems to be universal, i.e. independent of the actual materials used. This phenomenon is called the *integer* quantum Hall effect. It was discovered in 1980 by Klaus von Klitzing [30].

It is worth emphasising that the two-dimensional nature of the system is crucial. In this case, by "two-dimensional" we mean that electrons are really confined to thin layers and can only move in two dimensions. The "thin" strips used by Hall are, of course, monstrously thick by these standards. Part of the brilliance of the experimentalists who made these discoveries is their skill in making ultra-thin and ultra-pure materials.

Roughly speaking, one can understand the iqHe as arising from the same mechanism that causes electron orbits in atoms to be quantised. From the wave mechanics point of view, electron orbits are quantised because the electron wave function for a confined electron has to obey periodic boundary conditions; this is the same reason that guitar strings have discrete spectra. In the case of the iqHe the Lorentz force does not cause the electrons to move in circular orbits but they tend that way. This is, of course, a very intuitive explanation. A rigorous understanding requires much more sophisticated arguments and detailed calculations. In 1981 Robert Laughlin [32] explained the iqHe as a manifestation of gauge invariance: a deep symmetry principle. However, despite the complexity of the phenomenon, the explanation of the iqHe can be given entirely in terms of the behavior of electrons at low temperature confined to two dimensions and interacting with impurities and with the applied magnetic field.

The fractional quantum Hall effect is like the iqHe except that instead of the integer values appearing in the formula

$$R_H = \frac{h}{ie^2}$$

we can have fractional values of i at which plateaus appear. Furthermore, these fractional values are simple fractions like $\frac{1}{3}$ or $\frac{2}{5}$. This discovery was made by

Horst Störmer and Daniel Tsui in 1981 and was a complete surprise [46]. The electrons formed complicated composite entities that behaved as if they had fractional charges! Free particles with fractions of the basic electron charge are never seen in nature. It is natural to think that composite entities could seem to have some multiple of the electronic charge but these fractionally charged "particles" were astonishing.

The crucial point is that one cannot explain the fqHe in terms of the behavior of electrons qua electrons. One has to think of the collection of electrons moving in a two-dimensional landscape as collectively forming a fluid with the excitations having a strikingly different character than individual electrons. They are effectively extended objects with nontrivial topological possibilities. The interaction with the magnetic field creates flux tubes that intersect the plane in which the electrons are confined and thus yield ribbon-like objects that can be wound around each other. These are called chargeon-fluxon composites.

15.2 Spin and Statistics

Consider a system of n identical particles in quantum mechanics. A permutation of these particles leaves the system physically unchanged in 3 space dimensions. Thus, the only way that the state of the system can change is to be multiplied by a phase. The Hilbert space of the system must carry a unitary representation ρ of the permutation group S_n.

Now if σ is a permutation, $\rho(\sigma) = e^{i\theta} I$ since all that can happen is a change of phase. Let τ be a transposition: $\tau^2 = id$, where id is the identity permutation. Thus, $\rho(\tau^2) = [\rho(\tau)]^2 = I$ so $\rho(\tau) = \pm I$. This holds for any transposition. Suppose that τ_1 and τ_2 share an element, i.e. τ_1 interchanges a and b while τ_2 interchanges b and c. This means that $\tau_1\tau_2\tau_1 = \tau_2\tau_1\tau_2$. It follows that $\rho(\tau_1) = \rho(\tau_2)$, for, if one of them, say τ_1, mapped to $-I$ and the other to I under ρ we would have $\rho(\tau_1\tau_2\tau_1) = I$ while $\rho(\tau_2\tau_1\tau_2) = -I$. We infer that, for a given ρ, either all transpositions map to I or all transpositions map to $-I$. Every permutation is the product of transpositions. Thus, there are just two possibilities for ρ, either all permutations map to I or all permutations σ map to $(-1)^{sign(\sigma)} I$. Particles obeying the former type of statistics are called *bosons* and those obeying the latter are called *fermions*.

We now consider the effect of transporting a particle in a loop. If we bring the particle back to its original position the physics must be unchanged. The Hilbert space must carry a representation of the rotation group $SO(3)$ which describes how the particle transforms under rotation. The group $SO(3)$ can be visualised as a sphere with antipodal points identified. Its homotopy group is $\mathbb{Z}_2$. This means that there are two kinds of loops: a closed loop on the surface of the sphere and a curve from a point on the sphere to its antipode. The first kind of loop is deformable to the trivial identity loop: that is, the loop that stays at a point. If one performs a 4π rotation the curve is homotopic to the identity. Thus a 4π rotation must correspond to the identity transformation and a 2π rotation to multiplication by ± 1.

The group $SO(3)$ has a covering group, that is, a group with trivial homotopy with continuous surjective homomorphism onto $SO(3)$; this is the group $SU(2)$ of unitary 2-by-2 matrices with determinant $+1$. The representations of $SU(2)$ are well understood: they are classified by a number s—called the *spin*—which is either an integer or a half-integer in natural units. The dimension of the representation is $2s + 1$: a qubit is often thought of as a spin $\frac{1}{2}$ particle. Particles in nature come in these two species: integer spin or half-integer spin.

According to the spin-statistics theorem, particles that obey Bose-Einstein statistics have integer spin while particles that obey Fermi-Dirac statistics have half-integer spin. This is not an assumption about particular models of elementary particles; it is one of the fundamental *theorems* of relativistic quantum field theory.

The spin-statistics theorem was first proved by Fierz [19] and Pauli [38] in 1940 for non-interacting quantum fields. Further proofs were given by Pauli himself [39] and deWet [15]. Almost 20 years after the original proof, the spin-statistics theorem was extended to interacting quantum fields by Lüders and Zumino [34] and by Burgoyne [11]. Later several new proofs were given.

The basic mathematical structure of some of the proofs was placed in the context of rigorous quantum field theory by Streater and Wightman [47] using the idea of describing a field theory in terms of the complex analytic properties of vacuum expectation values [50].

The proof in Streater and Wightman [47] is based on the following assumptions:

1. Poincaré invariance,
2. the vacuum is the lowest energy state, thus, the energy spectrum is bounded below,
3. particle annihilation operators annihilate the vacuum,
4. locality: fields at spacelike separation commute or anti-commute and
5. the metric on Hilbert space is positive definite.[1]

The proof given by Streater and Wightman depends heavily on properties of vacuum expectation values of monomials of field operators and their properties as holomorphic functions.

The proof goes by assuming the wrong statistics—e.g. by assuming commutators for spin $\frac{1}{2}$ particles—and then taking vacuum expectation values and finally using analytic continuation to deduce an equation from which it follows that the field vanishes. The analytic continuation process uses a complexified version of the Lorentz group. The essential reason why the proof works is that with the wrong sign in the commutation relations the energy is not bounded below.

This proof makes no mention of topology. Is there a topological reason for the spin-statistics theorem? There have been several papers on this topic: an interesting one is by Finkelstein and Rubinstein [20]. Their argument is based on the idea that

---

[1] To a mathematician this is part of the definition of Hilbert space. However, there have been proposals in physics to consider analogues of Hilbert spaces with an indefinite metric. In the physics literature these are sometimes also called "Hilbert spaces."

the exchange of two particles can be deformed to a rotation by 2π of one of them. They give a very appealing heuristic argument for this based on a "rubber band lemma." They make all this precise in the context of solitons—not for the elementary particles for which the original spin-statistics theorem was proved. Rafael Sorkin has been a major presence in the topological approach to spin-statistics theorems. Topological arguments were given by Balachandran et al. [6] for various situations; for example, in [7] they proved a topological spin-statistics theorem for strings. Many subsequent papers were written about a spin-statistics theorem for various kinds of "kinks", "geons" and other entities constructed from topological non-trivialities, see, for example [16]. In an important paper Berry and Robbins [8] showed that one could associate a geometric phase shift associated with exchanging two spin $\frac{1}{2}$ particles and obtain a spin-statistics theorem this way. However, this has not been extended to many particle systems.

There have been several critiques of the non-traditional proofs of the spin-statistics theorem. In a survey article Duck and Sudarshan [17] discuss these and several other proofs. The main point that they make is that the topological proofs apply to "extended objects" and that elementary particles cannot be assumed to have ribbons or other topologically nontrivial structures associated with them. Thus, the topological proofs go not give a substitute for the classical spin-statistics theorem. Wightman [51] pointed out that in the absence of Lorentz invariance there is no spin-statistics connection. Thus attempts to derive it from elementary principles based on euclidean invariance rather than invariance under the Poincaré group are doomed.

15.3 Anyons and Braids

The fact that in two dimensions there are more possibilities for the spin and the statistics is originally due to Leinaas and Myrheim [33], who said " In one and two dimensions a continuum of possible intermediate cases connects the bosons and fermion cases." The possibility was independently rediscovered a few years later by Wilczek [52, 53] and Sorkin [45].

Consider what happens when there are many particles. If they are all labeled as being distinct then arbitrary trajectories can be deformed into the identity transformation. However, in quantum mechanics particles are *indistinguishable*. Now when there is a trajectory it could correspond to an arbitrary permutation of the particles. However, the strands corresponding to each trajectory can be disentangled so all one has is the permutation.

When a particle in $2 + 1$ dimensional spacetime is wound around another twice a nontrivial winding occurs and there is no reason why the phase change should be ± 1. For a system of N particles the transformation in the wave function is given by a representation of the braid group.

The state-space of an n-particle system has to carry a representation of the braid group rather than of the permutation group. The representations are a much richer collection and we have the possibility of many more kinds of statistics in two dimensions: these particles are the *anyons*.

There is still a spin-statistics connection, however, it is now more complicated. As we have seen there are more than two possibilities for the "statistics": interchanging particles can cause arbitrary phase shifts. The rotation group in two dimensions is $SO(2)$. This group has the same homotopy group as a circle so it has an infinite family of types of "spin."

There is another new feature to be considered. As we have mentioned before, the physical quasi-particles that arise in the fractional quantum Hall effect are extended objects with charge and tubes of magnetic flux. Not only is there braiding but also *twisting*. Later, when we formalise the theory categorically we will introduce additional algebraic structure: the aptly named *ribbon structure* to capture this. For the moment we confine our attention to braiding.

The braid group can be described by giving generators and relations. We think of there being a fixed set of n points along a line segment and we visualise an element of the braid group as a set of strands connecting two such collections of n points. Each strand must go from one of the lower points to one of the upper points. The generators are interchanges of two adjacent strands: this can happen in two ways, the strand of particle i crosses *over* the strand of particle $i+1$ – we call this b_i – or it can cross under, we call this b_i^{-1}. For n points the generators are b_1 to b_{n-1} and their inverses. The generators obey the following equations:

$$b_i b_j = b_j b_i \quad \text{for } |i-j| \geq 2 \tag{15.1}$$

$$b_i b_{i+1} b_i = b_{i+1} b_i b_{i+1} \quad \text{for } 1 \leq i \leq n-1. \tag{15.2}$$

which respectively depicts as:

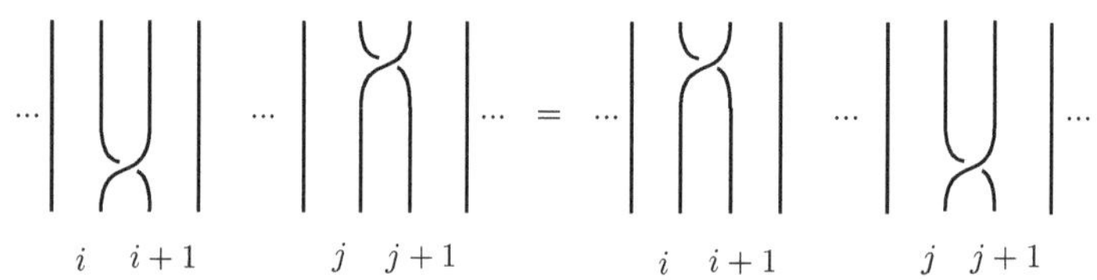

and

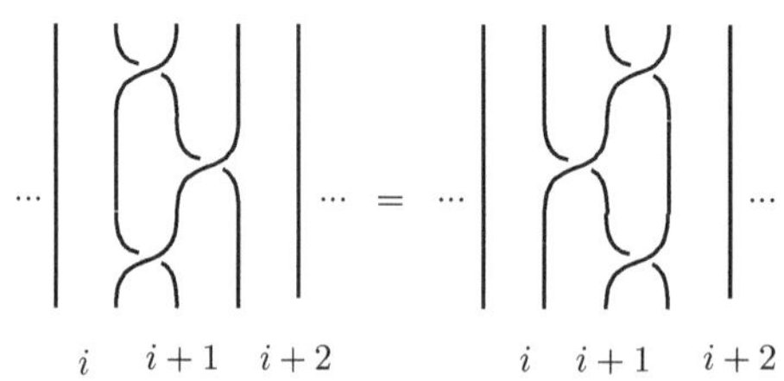

Now a collection of N anyons corresponds to a representation of the braid group on N particles. The simplest case is when they correspond to one-dimensional representations; higher dimensional representations are, in principle, also possible. In the case of one dimensional representations the wavefunction will transform in response to one particle being wound around another by acquiring a phase factor $\exp(i\theta)$; clearly one-dimensional representations of any group are always abelian. Thus, the generator b_j is represented by a $\exp(i\theta_j)$. Note that—unlike in the permutation group – $b_j^2 \neq Id$ so $\exp(2i\theta_j)$ need not be equal to 1. If one looks at the elements of the braid group in this representation they form an abelian group since, for each generator b_j occurring in a word, one gets a factor of $\exp(ik\theta_j)$ where k is the number of times that b_j appears in the word minus the number of times b_j^{-1} occurs. The order in which the generators appear is not important in this simple representation and for this reason anyons obeying these rules are called abelian anyons. In fact, the equations for the generators (the so-called Yang-Baxter equation) implies that the phase factors are the *same* for all the generators. To see this note that this equation implies that $\exp(i\theta_j)\exp(i\theta_{j+1})\exp(i\theta_j) = \exp(i\theta_{j+1})\exp(i\theta_j)\exp(i\theta_{j+1})$, whence the result follows immediately. Anyons transforming according to higher dimensional representations of the braid group will not have this simple abelian character: they are called nonabelian anyons.

Physically anyons are collective excitations rather than "elementary" particles. Thus when they are put together they form new excitations in complex ways. What is remarkable is that the complicated physics is summarised by simple algebraic rules called *fusion rules*. The best way to express the fusion rules is through the theory of semisimple monoidal categories which we will do in the next section.

In order to understand the meaning of fusion rules consider spin in quantum mechanics. Recall that elementary particles carry irreducible representations of $SU(2)$. When two particles are combined to form a composite entity one gets a system that transforms according to the tensor product of the representations. Such a tensor product need not be irreducible so one has to decompose it into irreducible representations: this is called "plethysm" in the mathematics literature. Thus, for example, when one combines a spin j particle with a spin k particle the one gets a system that can be in states of spin $|j-k|$ up to $j+k$ (see, for example, *Quantum Mechanics II* by A. Messiah [35]). Thus one can write heuristically

$$\mathbf{J} \otimes \mathbf{K} = \mathbf{J} - \mathbf{K} \oplus \ldots \oplus \mathbf{J} + \mathbf{K}.$$

The characterization of the irreducible representation of $SU(2)$ comes from the algebra of the infinitesimal generators (the Lie algebra $\mathfrak{su}(2)$). These are the familiar *angular momentum* operators obeying the following equations

$$[J_x, J_y] \stackrel{def}{=} J_x J_y - J_y J_x = i\hbar \varepsilon_{xyz} J_z.$$

In general, for a Lie algebra, one can have relations of the form

$$[K_\alpha, K_\beta] = C^\gamma_{\alpha\beta} K_\gamma$$

where the Ks are from the Lie algebra and the Cs are constants called structure constants. For more complicated groups like $SU(3)$ one can write similar relations: in the case of $SU(3)$ there are examples with more than one copy of a given representation.

In considering combinations of anyons of different types we associate with each type of anyon a *charge*, really this is just another name for "type." Then one will have rules for decomposing combinations of anyons of charges, say, A and B, which will take the form $A \otimes B = \ldots$. The theory of monoidal categories is just an abstract presentation of the notion of tensor products so it is the ideal setting to describe the combinatorics of fusion rules.

We can write the fusion rules in a form that looks like the defining equations for a Lie algebra. We write $\langle a, b\rangle$ for the fusion of anyons of type a and b. Then we can express fusion rules as

$$\langle a, b\rangle = N_{ab}^{c} c,$$

where a, b and c are anyon types and the Ns are just natural numbers. Thus a rule of the form

$$\langle a, b\rangle = 2a + b + 3c$$

would mean that fusing an a-type anyon with a b-type anyon yields either an a-type anyon, which can occur in two ways, or a b type anyon, or a c-type anyon, this last possibility can occur in three ways. For abelian anyons we have $N_{ij}^{k} = 0$ unless $k = i + j$, in which case it is 1. Formally, this looks like the rules for decomposing tensor products of irreducible representations of a Lie group into irreducible representations. However, this does not mean that the fusion rules correspond to the rules for combining irreducible representations of some Lie group. The resemblance is purely formal.

What is the connection between physical anyons and qubits? It is not an anyon by itself that forms a qubit, rather it is the set of fusion possibilities that forms a qubit. If we have a fusion rule with $N_{ab}^{c} = 2$ then when we fuse an "a" anyon with a "b" anyon to obtain a "c" anyon, we get a two dimensional space of fusion possibilities. This fusion space forms the qubit. In the case of Fibonacci anyons, to be discussed in detail in later sections, we have two types of anyons $\mathbf{1}$ and τ with fusion rule $\langle \tau, \tau\rangle = \mathbf{1} + \tau$. If we fuse τ and $\langle \tau, \tau\rangle$ we get $\mathbf{1} + 2 \cdot \tau$. Thus, if we look at the case when we have a τ as the result we get a 2-dimensional space of fusion possibilities and this simulates a qubit. An *operation* or a *one-qubit gate* on such a qubit consists of a braid connecting three τ anyons to three τ anyons where both triple have τ as global charge. Similarly, a two-qubit gate will be a braiding of two such triples.

15.4 The Algebra of Anyons: Modular Tensor Categories

From the discussion so far we can see that there are two aspects of anyonic behavior that need to be formalized. The first is the rather complicated kinematics involving braiding and the second is the dynamics of the anyons. Formalising the former entails having an algebraic structure rich enough to capture charges, braiding and fusion rules and the second requires a way of linking the kinematics with the dynamics according to the usual rules of quantum mechanics formulated in Hilbert spaces.

The kinematic side requires that we have a set of of algebraic rules that describes a system of anyons, their charge, their fusion rules and all this together with an action of the braid group, a formal system which embodies—at least partially—their trajectories in a 2+1 dimensional space.[2] Our aim is to give a categorical semantics that will take care of both the algebraic structures describing a system of anyons and that allows one to overlay the dynamics on top. Modular tensor categories provides such a language.

At this point it is worth reflecting on the use of categories. The first essential point is that there are different kinds of charges. Thus, an algebraic description must embody several types of objects. This is exactly what categories allow. Indeed, one can think of category theory as a kind of "higher-dimensional" algebra as advocated, for example, by John Baez. Second, the paths swept out by anyons are a crucial part of the physical description. One needs an algebraic formalism that allows these relations between anyons to be captured, the morphisms or arrows in a category give the right level of extra structure to express this. In particular we can have notions of objects being isomorphic without being identical.

Consider a category with tensor products, written $\otimes$. It could happen that $A \otimes (B \otimes C)$ is *identical* to $(A \otimes B) \otimes C$; if this is the case we say we have *strict associativity*. More commonly we have mere isomorphism between these two objects. In ordinary algebra equations like this are interpreted as identities and one would be forced to make everything strict. With categories one can have these equations holding non-strictly, or, as in the jargon of category theory, "holding up to isomorphism."

Let us first consider the basic properties of anyons and the algebra that is necessary to express these properties:

1. First, we have a system of labels, or types, that will represent the charges of our anyons.
2. We also need a way to express a compound system of anyons. This will be expressed by a monoidal (tensor) structure; this way, we will represent a compound system of anyons as the tensor product of their respective charges. The trivial charge is simply the tensor unit. Importantly, this category is not strict monoidal in general. This is physically important because, for instance, the bracketing of a compound system of charges will indicate in which order fusions occur.

[2] Indeed, the braid group is not sufficient as anyons are extended objects. We need to have ribbons (or framed) strands to adequately represent all movements such as, for instance, a rotation of an anyon by 2π.

3. As we saw in the preceding section, the worldlines of a system of anyons is described by representations of the braid group. We will require that our monoidal category has a braid structure as opposed to being symmetric.[3]
4. We need a way to express the notion of conjugate charge i.e. for a given charge A, its conjugate charge A^* is the unique charge that can fuse with A to yield the trivial charge. The structure that captures these notions is called a *rigid structure*.
5. The fact that the objects we are looking at are extended objects—flux tubes—means that, in general, representing their movements graphically with strands in $2+1$ dimensions is not enough; the correct graphical representation is realised by using ribbons, which can be twisted, instead of strands. The algebraic axiomatization of this has been given—long before mathematicians were aware of anyons—and is called a *ribbon structure* on our category. The axioms for a ribbon structure encapsulate correctly the algebraic rules imposed by the topological properties of ribbons including the possibility of twisting a ribbon.
6. We need a formal way to express the fusion rules and to map all the preceding algebraic formalism into the context of Hilbert spaces. This will be taken care of by an *semisimple* structure compatible with all the preceding structures.
7. Finally, we will consider a special class of semisimple ribbon categories called *modular tensor categories*. Such categories prohibit an infinite number of possible charges for an anyon of a given theory. Moreover, such a category contains within its defining data information about the fusion rules.

Note that most of the results that we present below are known and are discussed in from a physical standpoint in [9, 14, 22, 24] and [40]. Our presentation contrasts with these in the sense that we emphasise the link between the categorical structures and the physical phenomena.

We now give detailed definitions of the categorical structures we mentioned above. For a more detailed presentation of these notions, we refer the reader to [5] and [12].

15.4.1 Charges and Compound Systems: Monoidal Categories

Definition 1 A monoidal category is a category $\mathbf{C}$ equipped with a bifunctor

$$\otimes : \mathbf{C} \times \mathbf{C} \to \mathbf{C}$$

and three natural isomorphisms α, λ and ρ with components

$$\alpha_{A,B;C} : (A \otimes B) \otimes C \to A \otimes (B \otimes C) \ , \ \lambda_A : \mathbf{1} \otimes A \to A \text{ and } \rho_A : A \otimes \mathbf{1} \to A$$

[3] Being symmetric means that the braiding σ is such that $\sigma_{B,A}\sigma_{A,B} = 1_{A\otimes B}$ for all A and B.

such that for all A, B, C and $D \in |\mathbf{C}|$, both

(i) Pentagon axiom.

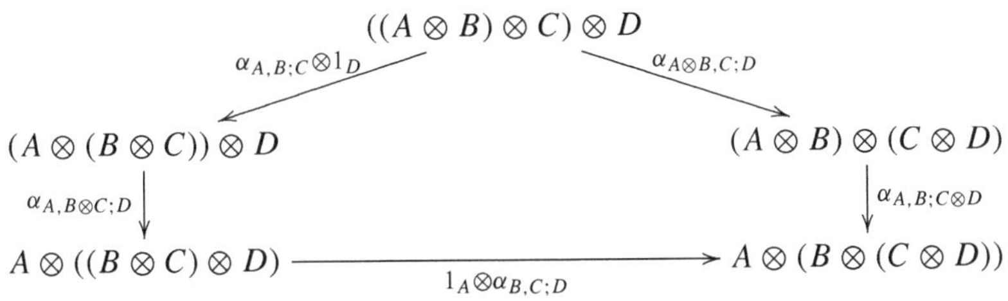

and

(ii) Triangle axiom.

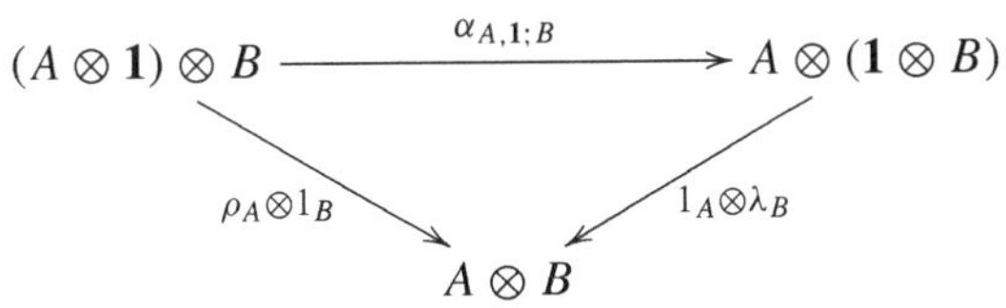

commute.

We interpret the components of this definition as follows:

- *Objects:* We will regard an $A \in |\mathbf{C}|$ as a label for a set of anyons. Note that this set might contain a single anyon. However, in that case, the object must satisfy some properties that we will consider below. Nonetheless, for simplicity purpose, in what follows, we will assume that an object A is the charge of a single anyon; this will make the explanations simpler.
- *Tensor:* Given a set of n charges $A_1, A_2, \ldots, A_n$, the compound charge of the system will be described by the n-fold tensor product $A_1 \otimes A_2 \otimes \ldots \otimes A_n$.
- *Unit:* The unit $\mathbf{1} \in |\mathbf{C}|$ is the label indicating the trivial charge.

The natural isomorphisms are interpreted as:

- α: The bracketing of an n-fold tensor product indicates the order of the fusions of the n components of the tensor product. The associativity isomorphism is used to change the pattern of fusion meaning that

$$\alpha_{A,B;C} : (A \otimes B) \otimes C \to A \otimes (B \otimes C)$$

changes the order of the fusions from $A \otimes B$ to $B \otimes C$ occurring first.

- λ *and* ρ*:* The natural isomorphisms $\lambda_A : \mathbf{1} \otimes A \to A$ and $\rho_A : A \otimes \mathbf{1} \to A$ simply tells us that combining an anyon with charge A with the trivial charge $\mathbf{1}$ changes nothing about the compound charge or even to the charge of an anyon obtained by fusing these two anyons together.

15.4.2 Worldlines: Braided Monoidal Categories

To correctly handle the movements of the anyons, our category needs at least a braid structure. As a compound system of charges is represented by their tensor product, the braiding will act on the components of such a tensor product of charges. Thus, the braid structure must behave coherently with the monoidal structure. What we need is called a *braided* monoidal category which is formally defined as follows:

Definition 2 A braided monoidal category $\mathbf{C}$ is a monoidal category equipped with a family of isomorphisms

$$\sigma_{A,B} : A \otimes B \xrightarrow{\sim} B \otimes A$$

natural in A and $B \in |\mathbf{C}|$, such that

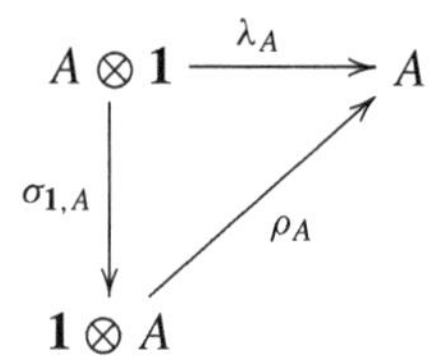

and both

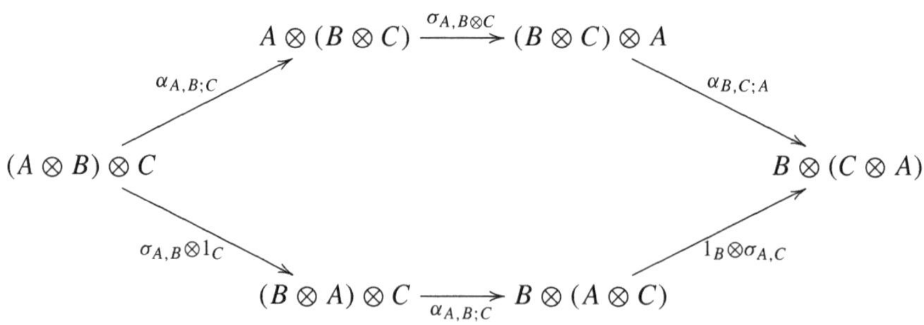

and the same diagram with σ^{-1} instead of σ, commute for all A, B and $C \in |\mathbf{C}|$.

Of course,

- σ: The natural isomorphism $\sigma_{A,B}$ is interpreted as the exchange of the charges A and B.

The reader might wonder if such a definition is enough for our purposes, in the sense that this is enough to adequately express the worldlines of a tuple of anyons. The remarkable answer is "yes!" There is a coherence theorem for braided monoidal categories, due to Joyal and Street, which says that given a natural isomorphism built from tensoring and composing identities and components of α, λ, ρ and σ from the braided monoidal structure, there is an associated braid. Moreover, two such isomorphisms are equal if their associated braids are equal. The proof of this theorem is quite technical and, for our purposes, the preceding comment is enough. We refer the reader to [25] for the details.

Remark 1 Despite the fact that the our category is not strict and since every monoidal category is equivalent to a strict monoidal category [31], we may omit bracketing and unit isomorphisms if these aren't necessary for the exposition.

15.4.3 Charge Conjugation: Rigidity

Now that we have some of the algebraic tools that describe the worldlines of anyons, we introduce another structure to express the conjugation of the charges which is similar to the notion of a compact structure on a monoidal category. This is given via the notion of duals within $\mathbf{C}$.

Definition 3 Let $\mathbf{C}$ be a braided monoidal category and $A \in |\mathbf{C}|$. A dual of A is an object $A^* \in |\mathbf{C}|$ together with two morphisms $i_A : \mathbf{1} \to A \otimes A^*$ and $e_A : A^* \otimes A \to \mathbf{1}$ that are such that

$$\begin{array}{ccc} A^* & \xrightarrow{1_{A^*} \otimes i} & A^* \otimes A \otimes A^* \\ & \searrow^{1_{A^*}} & \downarrow{\scriptstyle e \otimes 1_{A^*}} \\ & & A^* \end{array} \qquad \textit{and} \qquad \begin{array}{ccc} A & \xrightarrow{i \otimes 1_A} & A \otimes A^* \otimes A \\ & \searrow^{1_A} & \downarrow{\scriptstyle 1_A \otimes e} \\ & & A \end{array}$$

commute. A braided monoidal category $\mathbf{C}$ is rigid if each $A \in |\mathbf{C}|$ has a dual.

Physically speaking, we will interpret the structural morphisms of the previous definition in the following way:

- i: is interpreted as the creation of two quasi-particles with respective dual charges and
- e: as the annihilation of such a pair.

Now, given an $f : A \to B$ in a rigid braided monoidal category, we can define $f^* : B^* \to A^*$ as the composite

$$B^* \xrightarrow{1_{B^*} \otimes i_A} B^* \otimes A \otimes A^* \xrightarrow{1_{B^*} \otimes f \otimes 1_{A^*}} B^* \otimes B \otimes A^* \xrightarrow{e_B \otimes 1_{A^*}} A^*. \qquad (15.3)$$

It is easily verified this operation on morphisms together with $A \mapsto A^*$ on objects defines a functor.

Later, we will need the following

Proposition 1 *[5] Let* $\mathbf{C}$ *be a rigid braided monoidal category and* $B \in |\mathbf{C}|$ *together with its dual* $B^* \in |\mathbf{C}|$ *then, there are canonical isomorphisms*

$$\mathrm{Hom}(A \otimes B, C) \simeq \mathrm{Hom}(A, C \otimes B^*)$$
$$\mathrm{Hom}(A, B \otimes C) \simeq \mathrm{Hom}(B^* \otimes A, C).$$

15.4.4 Graphical Calculus for Rigid Braided Monoidal Categories

As we showed in Sect. 15.3, we can illustrate components of the braid group and their composition with pictures. In fact, we can do more: we will now give pictorial representation depicting completely the trajectories of anyons in 2+1 dimensions. Such a graphical calculus comes for free with rigid braided monoidal categories and adequately represents morphisms in such categories [5].

Let us now give the basic building blocs of such a graphical calculus:

- The *identity on* $\mathbf{1} \in |\mathbf{C}|$ is represented as the empty picture. This is not surprising: adding the trivial charge to the system is the same thing as adding nothing.
- The *identity on* a charge $A \in |\mathbf{C}|$ and its dual are respectively represented by

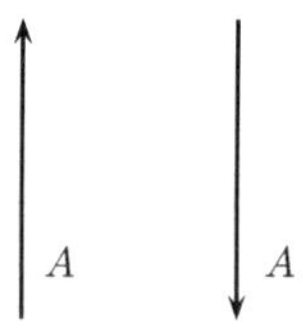

- A *morphism* $f : A \to B$ is depicted as

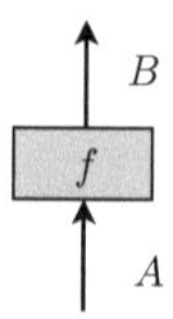

- The *composition* of morphisms $f : A \to B$ and $g : B \to C$ is given by stacking the graphical representations of and f and g and connecting the arrows labeled by B i.e.,

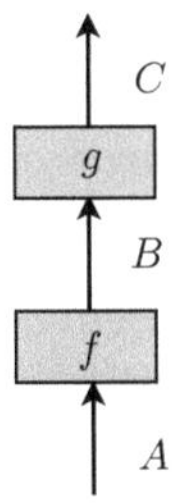

- The *tensor product* of morphisms $f : A \to B$ and $g : C \to D$ is given by aligning the graphical representations of and f and g side by side in the $f \otimes g$ order i.e.,

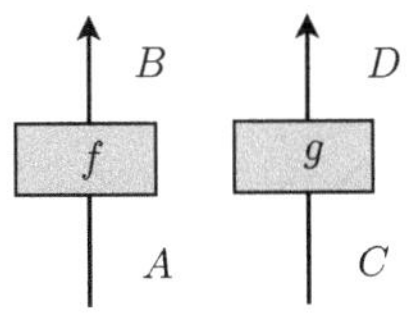

hence obtaining the representation for $f \otimes g : A \otimes C \to B \otimes D$.
- The morphisms $i_A : \mathbf{1} \to A \otimes A^*$ and $e_A : A^* \otimes A \to \mathbf{1}$ of the rigid structure on $A \in |\mathbf{C}|$ are represented as

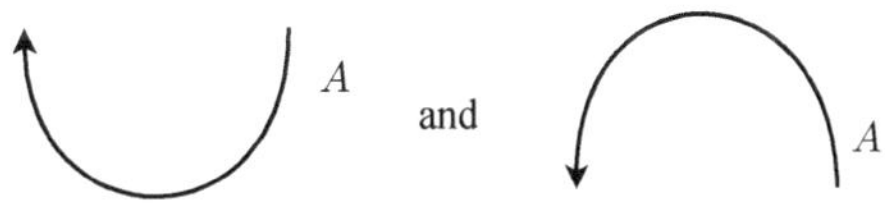

respectively.
- The *braiding* $\sigma_{AB} : A \otimes B \to B \otimes A$ and it's inverse are respectively depicted as

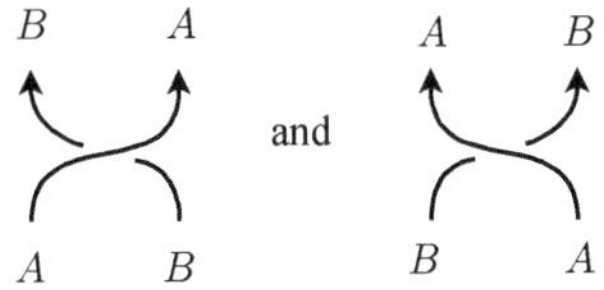

Remark 2 Note that the natural isomorphisms α, ρ and λ are not captured by this formalism. For the first, we will introduce a graphical notation for fusion later. For the latter two, it does not matter as already mentioned in our comment on the representation of $1_{\mathbf{1}}$.

15.4.5 A Twist in the Worldlines: Ribbon Categories

As noted in the introduction of this section, the braid group is not enough to capture completely the kinematics of the anyons. For instance, an anyon can revolve around some center by 2π and the change induced on the system is not an identity. Let us consider what this means.

Based on the language we already have from the theory of rigid braided monoidal categories, we can build following process [40]:

1. A pair of (quasi-)particles with respective charges A and A^* is created,
2. The two particles are swapped, the particle of charge A going behind the particle of charge A^* and, finally,
3. They annihilate.

Such a process is built from structural morphisms as

$$f = i_A \circ \sigma_{A,A^*}^{-1} \circ e_A.$$

From the graphical calculus on morphisms, $f : \mathbf{1} \to \mathbf{1}$ gets depicted as:

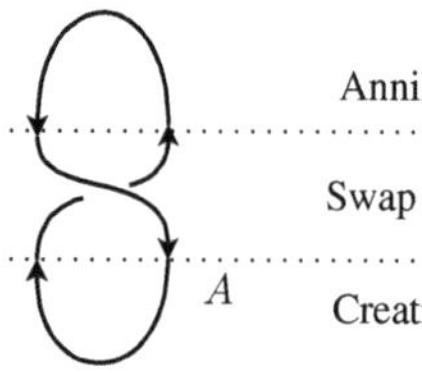

The key point here is that the amplitude of the process is non-trivial as there is an exchange occurring. Moreover, we have the following topological equivalences:

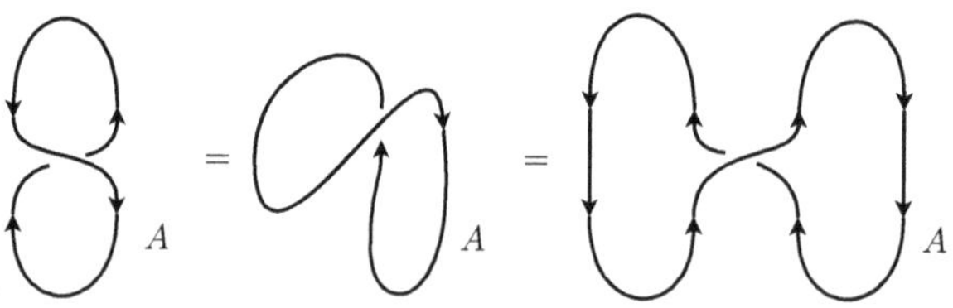

telling us that the amplitude for the first process is *equal* to the amplitude of the third which is, from bottom to top:

1. The creation of two pairs of particles with respective dual charges,
2. The exchange of the particle of charge A from the left pair with the one of charge A from the right pair and
3. The annihilation of the particle of charge A from the right pair with the particle of charge A^* from the left pair and the annihilation of the particle of charge A from the left pair with the particle of charge A^* the right pair.

Now, consider the following deformation:

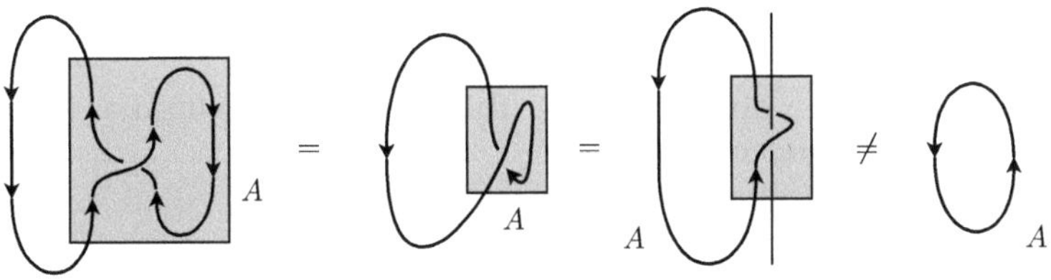

The third picture can be read as the creation of a pair of particles of charge A^* and A respectively and the particle of charge A gets rotated about 2π of some center which illustrated in the gray box. Now, as noted, this amplitude is different from the trivial amplitude depicted as the fourth picture. We conclude from this that illustrating the worldlines of our anyons with strands is not completely faithful to the process. Instead, we will use ribbons—or framed strands—so that:

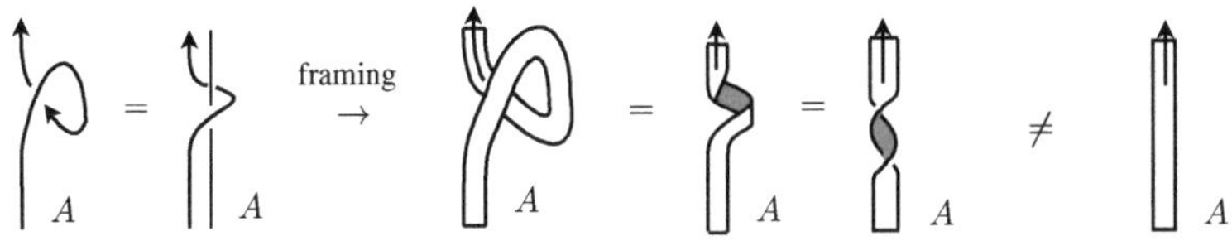

This "winding around some center" is almost given already by the rigid braided monoidal structure. Indeed, in *any* rigid braided monoidal category **C**, one can define

$$\gamma_A : A^{**} \to A \quad \text{as} \quad \gamma_A := (A \otimes e_{A^*}) \circ (A \otimes \sigma^{-1}_{A^* A^{**}}) \circ (i_A \otimes A^{**})$$

for any $A \in |\mathbf{C}|$ (the reader may check that γ_A is topologically equivalent to the framed ribbon of the preceding picture). However, note that we have a type mismatch if we compare with the twist depicted above as the later is of type $A \to A$. To complete the definition, we need a natural isomorphism of type $A \to A^{**}$ which will behave coherently with the rest of the structure. Formally:

Definition 4 A ribbon category[4] **C** is a rigid braided monoidal equipped with a natural isomorphism δ with components

$$\delta_A : A \to A^{**}$$

satisfying

(i) $\delta_{A\otimes B} = \delta_A \otimes \delta_B$;
(ii) $\delta_{A^*} = (\delta^*_A)^{-1}$ and,
(iii) $\delta_{\mathbf{1}} = 1$.

---

[4] Sometimes called a *tortile* category.

This is enough to formally define the "twist" that we discussed:

Definition 5 Let **C** be a ribbon category. The twist map is defined as the natural isomorphism θ with components

$$\theta_A := \gamma_A \circ \delta_A : A \to A.$$

Graphically, this is denoted as

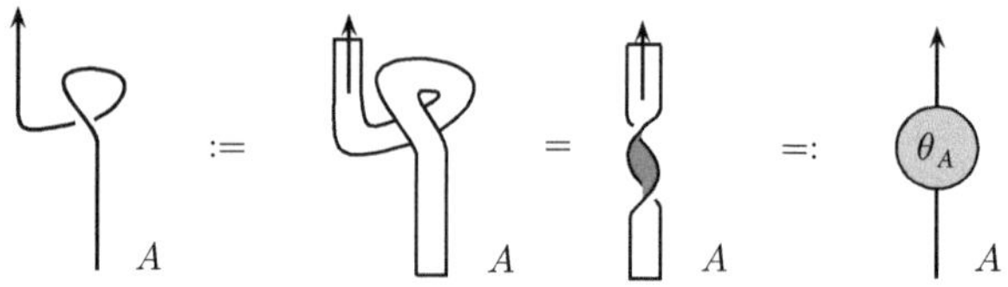

where we have deliberately omitted the isomorphism δ_A to avoid cluttering the picture. Note that we can (and did) get back to our strand notation and this is done without loss inasmuch as we use this notation for the twist, keeping in mind that this is a rotation of 2π of the strand around its center.

Interestingly, provided that we use these graphical conventions, *any* two processes that can obtained by obtained by continuously deforming one into the other will have the *same* amplitude! In fact, a theorem due to Reshetikhin and Tuarev tells us that for two such diagrams, the isomorphisms corresponding to each of these diagrams are equal. For an exact statement and a detailed proof, the reader is referred to [41].

This completes our discussion on the algebraic context describing the worldlines of anyons.

15.4.6 Towards Fusion: Semisimple Ribbon Categories

Now, we need fusion rules to build the fusion spaces where our quantum computational interpretation will take place but before, we need to introduce some new concepts:

Definition 6 A morphism $m : A \to B$ is a monomorphism (or is monic) when for any two $f, g : C \to A$, we have

$$m \circ f = m \circ g \quad \Rightarrow \quad f = g.$$

Conversely,

Definition 7 A morphism $h : A \to B$ is an epimorphism (or is an epi) when for any two $f, g : B \to C$,

$$f \circ h = g \circ h \quad \Rightarrow \quad f = g.$$

Because of its defining condition, a monomorphism (resp. epimorphism) is sometime called left-cancellable (resp. right-cancellable). The two previous definitions generalise the notions of injection and surjection in the usual sense. In fact, one may check that in **Set**, these concepts coincide i.e., monics are exactly injections and epis are exactly surjections.

Definition 8 A zero object in a category **C** is an object $0 \in |\mathbf{C}|$ which is both initial and terminal.

In particular, the presence of such an object enables us to define a *zero morphism*. Indeed, if **C** is a category with a zero object then, for any $A, B \in |\mathbf{C}|$, there exists a unique morphism $0 : A \to B$ defined as the composition $A{\to}0{\to}B$. Uniqueness of the zero morphism follows from the fact that 0 is simultaneously initial and terminal and hence, the set of arrows to and from it are singletons.

Such an object and its associated morphism enables in turn a generalisation of the notion of kernel as follows:

Definition 9 The kernel of a morphism $f : A \to B$ in **C**, a category with a zero object, is an arrow

$$Ker(f) := k : S \to A$$

such that if $f \circ k = 0$, the zero morphism, then for every $h : C \to A$ such that $h \circ f = 0$, h factors uniquely through k as $h = h' \circ k$. Diagrammatically,

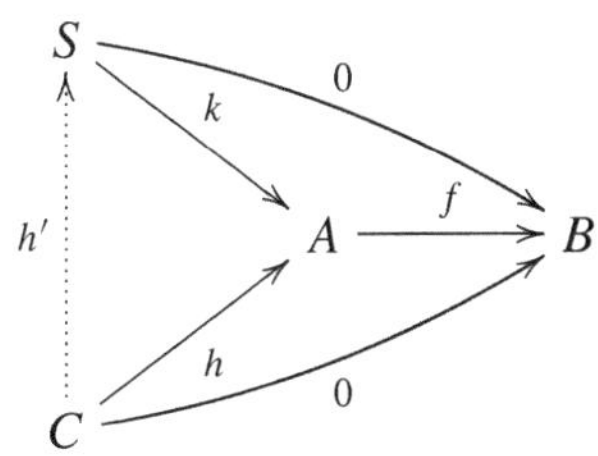

We can also define the dual notion:

Definition 10 The cokernel of a morphism $f : A \to B$ in **C**, a category with a zero object, is an arrow

$$CoKer(f) := u : B \to S$$

such that if $u \circ f = 0$ and if $h : B \to C$ is such that $h \circ f = 0$, then h factors uniquely through u as $h = u \circ h'$ i.e.:

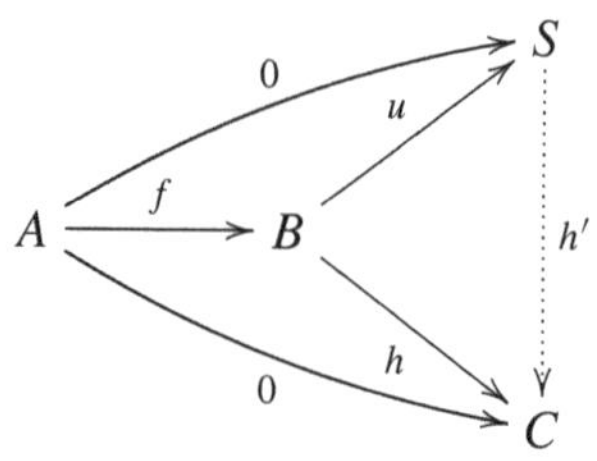

We now can state the central definition of this subsection[5]:

Definition 11 A category **C** is

a. Preadditive if its homsets are abelian groups (written additively) and the composition of morphism is bilinear over the integers;
b. It is additive if, in addition, every finite set of objects has a biproduct;
c. It is preabelian if it is additive and every morphism in **C** has a kernel and a cokernel and finally,
d. It is abelian if, in addition, every monomorphism is a kernel and epimorphism is a cokernel.

Let us emphasise the most important part of the previous definition from our point of view. The fact that a category **C** is abelian comes with a bonus: indeed, the previous definitions not only says that the kernels and the cokernels in **C** behave the same way as in vector spaces but also that its hom-sets are abelian groups and this can be seen as a first step toward an interpretation of our formalism within the context of complex vector spaces. To really get there, we need yet another notion inspired by the following fact: the charges of our basic anyons are irreducible in the sense that they cannot be decomposed into more elementary entities.[6] Such a property of the charges can be recast in categorical terms as follows:

Definition 12 Let **C** be an abelian category then, an $S \in |\mathbf{C}|$ such that $S \not\simeq 0$ is simple if for all $B \in |\mathbf{C}|$, $f : B \hookrightarrow S$ is either the zero morphism or an isomorphism.

This is the same as saying that A has no other subobject other than 0 and itself. From this, we have

Definition 13 An abelian category **C** is semisimple if any $A \in |\mathbf{C}|$ is such that

$$A \simeq \bigoplus_{j \in J} N_j S_j$$

where S_j is a simple object, J is the set of isomorphism classes of simple objects and $N_j \in \mathbb{N}$ are such that only a finite number of them are non-zero.

---

[5] This definition is equivalent [23] to the standard definition of an abelian category [31].

[6] Of course, this is an approximation in the effective field theory of these excitations.

This is enough to give our last definition in which there are now two distinct monoidal products; one from the abelian structure written as $\oplus$ and one from the ribbon structure denoted by $\otimes$.

Definition 14 A semisimple ribbon category $\mathbf{C}$ is a semisimple category endowed with a ribbon structure where the tensor unit $\mathbf{1} \in |\mathbf{C}|$ is simple, the tensor product is bilinear and where for each simple object $S \in |\mathbf{C}|$, $\mathrm{End}(S) \simeq \mathbb{K}$, a field of characteristic 0.

Remark 3 In what follows, we will assume that the field mentioned in the last definition is $\mathbb{C}$, the complex field.

Remark 4 To lighten the notation, we will often use the index of simple objects to identify morphisms involving these. For instance, 1_i is the identity on the simple object S_i and the natural isomorphism σ_{ij} is of type $S_i \otimes S_j \to S_j \otimes S_i$. Correspondingly, we will label the wires in the picture calculus only with the index i instead of the label S_i.

Now, having such a semisimple structure on $\mathbf{C}$ has many consequences. First, we can now handle fusions of anyons. Second, it is from this structure that complex vector spaces arise. Indeed, it is this fact that will enables us to define splitting spaces and unitary representations of the braid group therein. Such a structure also enables the following results:

Proposition 2 *In semisimple ribbon categories, there are natural isomorphisms*

$$A \otimes (B \oplus C) \simeq (A \otimes B) \oplus (A \otimes C) \quad \textit{and} \quad (A \oplus B) \otimes C \simeq (A \otimes C) \oplus (B \otimes C).$$

Moreover,

$$(A \oplus B)^* \simeq A^* \oplus B^* \quad \textit{and} \quad 0^* = 0.$$

15.4.7 Modular Tensor Categories

Semi-simple ribbon categories can already handle the kinematics of a system of anyons, our description of a particular model of anyon will take place within a particular class of such categories namely, modular tensor categories. First, such categories only allow a finite number of charges or rather, a finite set of isomorphism classes of simple objects. There is no reason to think that this is true in nature, however, all the applications that we have in mind for quantum computation can be done with a finite number of anyon types. A second point that we will inspect later is that the s-matrix defining the modularity condition has surprising connexions with the fusion rules.

More generally, these extra assumptions on semisimple ribbon categories lead to many interesting results; in particular, we can derive an equivalence between MTCs and modular functors. Since the goal of this paper is to give an introduction to computing with anyons, these results are not pursued here. The interested reader might want to look at [5] for a complete exposition of these subjects.

Definition 15 A modular tensor category is a semisimple ribbon category $\mathbf{C}$ such that

1. Finiteness condition: The index set J of isomorphism classes of simple objects is finite and $S_0 := \mathbf{1}$.
2. Modularity condition: For $i, j \in J$, the matrix s with entries

$$(s)_{ij} = \left[\lambda_{\mathbf{1}} \circ (e_i \otimes e_j) \circ (id_i \otimes \sigma_{ij} \otimes id_j) \circ (id_i \otimes \sigma_{ji} \otimes id_j) \circ (i_i \otimes i_j) \circ \lambda_{\mathbf{1}}^{-1}\right]_{ij}$$

which are depicted as

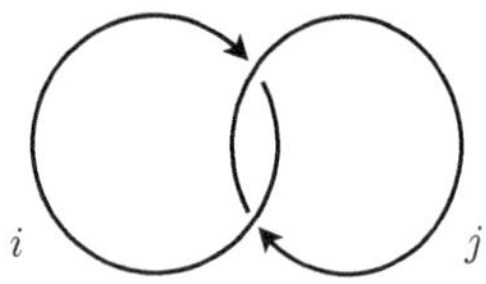

is invertible.

Remark 5 We indeed get a matrix for s since we have $\mathrm{End}(\mathbf{1}) \simeq \mathbb{C}$ – a field of characteristic 0—and the morphism depicted above is really of type $\mathbf{1} \to \mathbf{1}$ hence, a scalar.

The scalar components of the s-matrix form the so-called *Hopf link*. They can be thought of as the amplitude of the following process:

1. Two pairs of particles of respective charges A and A^* are created,
2. The particle of charge A^* from the left pair is wound around the particle of charge A from the other pair and
3. The two pairs annihilate.

We don't have yet the machinery to describe the surprising results that the modularity constraint entails, but we will do so in the subsection on the Verlinde Formula below.

15.4.8 Fusion Rules

For now on, we fix $\mathbf{C}$ to be a modular tensor category. As already mentioned, the charge of an anyon is represented by a simple object in $\mathbf{C}$. Suppose that two anyons with charges S_i and S_j fuse together into an anyon of charge S_k and that in N_{ij}^k ways, we will write this as

$$S_i \otimes S_j \simeq N_{ij}^k S_k.$$

There, the lower labels of N_{ij}^k then indicate which charges fuse together in order to yield the charge identified by the upper label.

Remark 6 Note that such an expression always makes sense, since by assumption the category is semisimple thus, each object is isomorphic to a direct sum of simple ones.

Now, the fusion process can produce different charges and this constitutes a generalisation of our description above. Taking in account this fact, we get the following

Definition 16 [Fusion rule] Let S_i and S_j be simple objects in **C** and J be the index set for the isomorphism classes of simple objects. The fusion rule of S_i and S_j is given by

$$S_i \otimes S_j \simeq \bigoplus_{k \in J} N_{ij}^k S_k.$$

There, the coefficients $N_{ij}^k = Dim(Hom((S_i \otimes S_j), S_k)$ are called the fusion coefficients of the fusion rule.

It is easy to verify that the fusion coefficients satisfy

$$N_{ij}^k = N_{ji}^k = N_{jk^*}^{i^*} = N_{i^*j^*}^{k^*} = N_{i^*j^*k}^0 = \dots \text{ and } N_{ij}^0 = \delta_{ij^*}.$$

and hence, in particular, we can lower and raise indices.

15.4.9 Translation to the Hilbert Space Context: Fusion and Splitting Spaces

We now come to the crux of the story. We now need to connect our rather abstract algebraic rules describing the kinematics of our anyons to dynamics which is expressed in the language of Hilbert spaces. Such a translation is already built-in a modular tensor category **C**. Indeed, we will use the facts that hom-sets of **C** are vector spaces over $\mathbb{C}$ and that for all simple objects S, $\mathrm{End}(S) \simeq \mathbb{C}$ to build the so-called splitting and fusion spaces in which we will simulate our qubits and coqubits.

Before carrying on to define such spaces for a family of anyons, we will need the following variant of Schur's lemma:

Lemma 1 *Let* $\boldsymbol{C}$ *be a semisimple abelian category and* $S_i, S_j \in |\boldsymbol{C}|$ *be simple objects such that* $i \neq j$*, then* $\mathrm{Hom}(S_i, S_j) = \{0\}$.

Proof Let $f : S_i \to S_j$ be arbitrary and consider $\mathrm{Ker}(f) : U \hookrightarrow S_i$ (the kernel is necessary monic) then, as S_i is simple by assumption, it follows that $\mathrm{Ker}(f)$ is either 0 or an isomorphism. If the kernel is 0, then f is an injection and hence, an isomorphism since S_j is simple, but this cannot be possible as $i \neq j$. Therefore, by simplicity of S_i we have $U \simeq S_i$ and $f = 0$ from which we conclude that $\mathrm{Hom}(S_i, S_j) = \{0\}$ as f is arbitrary. □

As a consequence of semisimplicity, the homset

$$\mathrm{Hom}(S_k, S_i \otimes S_j)$$

is really a complex vector space whose dimension is fixed by the fusion rule. Generally speaking, the passage from our modular tensor category to the category of finite dimensional complex vector spaces will be handled by this fact.

Definition 17 Let S_i, S_j and S_k be simple objects in **C**, a splitting space[7] is a complex vector space

$$V_k^{ij} \simeq Hom(S_k, S_i \otimes S_j)$$

of dimension $N_k^{ij} = dim(Hom(S_k, S_i \otimes S_j))$—the number of ways the charge S_k can split as the compound charge $S_i \otimes S_j$. Its states

$$Hom(S_k, S_k) \longrightarrow Hom(S_k, S_i \otimes S_j)$$

are called splitting states.

Note that the type of the splitting states makes sense. Indeed, if the splitting rule tells us that $\mathrm{Dim}(\mathrm{Hom}(S_k, S_i \otimes S_j)) = N_k^{ij}$, then through the fact homsets are complex vector spaces, a splitting state is a linear map of type

$$\mathbb{C} \longrightarrow \mathbb{C}^{N_k^{ij}}.$$

We now define the basis vectors of a splitting space. Let us introduce the following notation [40]: For S_i, S_j and S_k, some simple objects of **C**, if the following fusion rule $S_i \otimes S_j \simeq N_{ij}^k S_k$ holds, we will denote the basis state representing the corresponding to the ηth possible splitting of S_k into $S_i \otimes S_j$ as $|ij; k, \eta\rangle$ so that the set of basis vectors that spans the splitting space V_k^{ij} is

$$\{|ij; k, \eta\rangle \mid \eta \in \{1, 2, \ldots, N_k^{ij}\}\}.$$

In the category $\mathbf{FdVect}_{\mathbb{C}}$, the basis vector $|i-1\rangle : \mathbb{C} \to \mathbb{C}^n$ of $\mathbb{C}^n := \mathbb{C} \oplus \ldots \oplus \mathbb{C}$ n times, can be assimilated to the i-th canonical injection

[7] Such a space is also called a *fusion space* however, as the initialisation of a state takes place via a splitting, we prefer our proposed terminology.

$$\iota_i \,:\, \mathbb{C} \longrightarrow \mathbb{C}^n \;::\; 1 \;\mapsto\; \begin{pmatrix} 0 \\ \vdots \\ 1 \\ \vdots \\ 0 \end{pmatrix}$$

with the "1" on the i-th line and 0's elsewhere. Lifting this to our context, we can define the basis vector

$$|ij; k, \eta\rangle$$

for fixed k and η as

$$\begin{array}{ccc} \mathrm{Hom}(S_k, S_k) & \xrightarrow{\;|ij;k,\eta\rangle\;} & \mathrm{Hom}(S_k, S_i \otimes S_j) \\ & \searrow^{\mathrm{Hom}(1_{S_k}, \iota_\eta)} & \uparrow \simeq \\ & & \mathrm{Hom}(S_k, N_k^{ij} S_k) \end{array}$$

where ι_η is the ηth canonical injection into the N_k^{ij}-fold biproduct of S_k's.

According to the general form of the fusion rule, the compound charge $S_i \otimes S_j$ could be obtained via the splitting of different charges. In the light of lemma 1, each such splitting spaces are orthogonal one to the other. Thus, what we actually get is a tuple[8] of splitting spaces $\langle V_{k_1}^{ij}, \ldots, V_{k_n}^{ij}\rangle$ where $\{k_1, \ldots, k_n\} \subseteq J$, the index set of isomorphism classes of simple objects. Each component of this tuple carries a different possible branch of the computation labeled by the charge S_{k_l}. These spaces are mutually orthogonal in the *topological vector space* V^{ij} of dimension $\sum_{k_l \in J} N_{k_l}^{ij}$ spanned by the set of vectors

$$\{|ij; k_l, \eta\rangle \mid k_l \in \{k_1, \ldots, k_n\},\; \eta \in \{1, 2, \ldots, N_{k_l}^{ij}\}\}.$$

Remark 7 We can also define costates. Indeed, so far, we have spoken of a splitting state but the anyons can also—of course—fuse together thus yielding costates. As an example, for each basis vector $|ij; k, \eta\rangle$, we can define a bra $\langle ij; k, \eta|$ dualising the defining diagram for the basis vector and using canonical projections instead of the canonical injections.

Returning to our splitting spaces, we see that there is a morphism of type $S_k \to S_i \otimes S_j$ underlying each basis vector of a given splitting space. As it is determined by the ηth canonical injection and the splitting rule, we depict such a morphism as:

[8] Strictly speaking, it is a biproduct of fusion spaces but as the later is simultaneously a product and a coproduct, it also makes sense to speak of tuples.

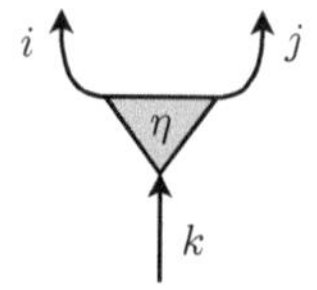

Of course, such a family of morphisms satisfies the following two relations:

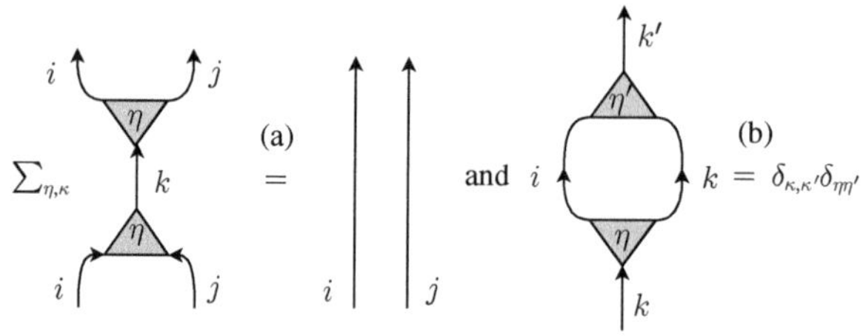

This is unsurprising as they define basis vectors. Indeed, the pictures above are simply the abstraction of the equations

$$\sum_{\eta,k} |ij;k,\eta\rangle\langle ij;k,\eta| = id \qquad \text{and} \qquad \langle ij;k,\eta|ij,k',\eta'\rangle = \delta_{k,k'}\delta_{\eta,\eta'}$$

satisfied by an orthonormal set of basis vectors.

More generally, a morphism

$$f : A \to B,$$

where A and B are understood as compound systems of anyons, is of the form

$$\bigoplus_{j\in J}\left(\text{Hom}(S_j, A) \xrightarrow{\text{Hom}(1_{S_j}, f)} \text{Hom}(S_j, B)\right).$$

Indeed, lemma 1 tells us that it is not possible to change the global charge from S_i to S_j for $i \neq j$ as the set of transformation between different simple objects is trivial. We will see some examples of such morphisms in the next section on Fibonacci anyons.

15.4.10 Quantum Dimension

In a modular tensor category, the *trace* of a map $f : A \to A$ is defined as

$$\text{Tr}(A) : \mathbf{1} \xrightarrow{i_A} A \otimes A^* \xrightarrow{f\otimes 1_A} A \otimes A^* \xrightarrow{\delta_A\otimes 1_{A^*}} A^{**} \otimes A^* \xrightarrow{e_{A^*}} \mathbf{1}. \qquad (15.4)$$

Now, each charge S_i has its own dimension. In fact, it is calculated much in the same way than in $\mathbf{FdVect}_{\mathbb{C}}$. Indeed, it is given by

$$d_i := \mathrm{Tr}(1_i).$$

Pictorially, this is

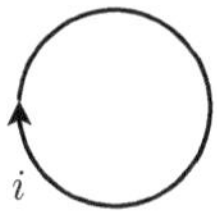

Such a number is called the *quantum dimension* of S_i. Before discussing the properties of this number consider the following

Lemma 2 *[40] Let S_i, S_j and $S_k \in$ **C** be simple objects then*

$$d_i d_j = N_k^{ij} d_k.$$

Proof We give a pictorial proof:

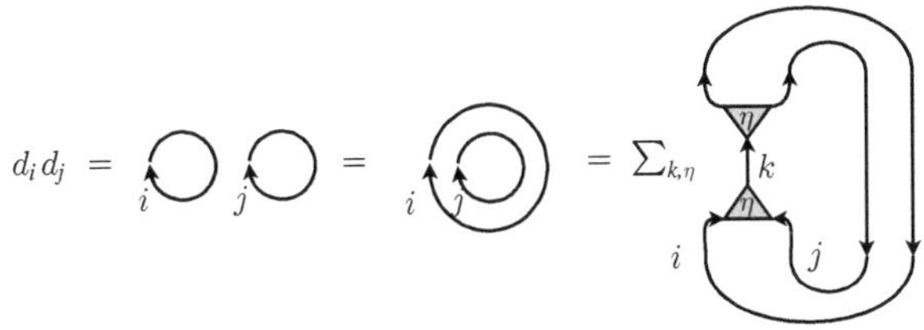

where for the last equality we used (a) from the relations for the morphism underlying the basis vectors. Now, using the topological invariance of the diagram, we get

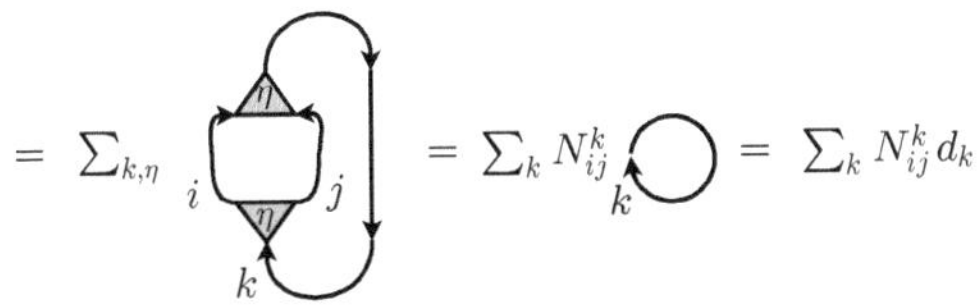

Where for the first equality we used (b) from the same pair of relations. □

From this, it is easy to see that the quantum dimension of abelian anyons is always 1 and that, independently of the charge. For a non-abelian anyon of charge S_i however, we have $d_i > 1$.

Example 1 We now give a specific example of quantum dimension via the Fibonacci anyons briefly introduced at the end of Sect. 15.3. Consider the splitting space

$$\mathrm{Hom}(I, \tau^{\otimes n})$$

there, $\tau^{\otimes n}$ is the n-fold tensor product of τ's all bracketed from the left. The dimension of this splitting space can be calculated using the fusion rule $\tau \otimes \tau \simeq \mathbf{1} \oplus \tau$. In doing so, one sees quite quickly that

$$\tau^{\otimes n} \simeq F_{n-2} \cdot I \oplus F_{n-1} \cdot \tau,$$

where F_m is the mth Fibonacci number. Now, using lemma 1, we have that

$$\mathrm{Hom}(I, \tau^{\otimes n}) \simeq \mathrm{Hom}(I, F_{n-2} \cdot I) \oplus \mathrm{Hom}(I, F_{n-1} \cdot \tau) \simeq \mathrm{Hom}(I, F_{n-2} \cdot I).$$

Thus, for the first few values of $n \geq 2$, we have

$$1, 1, 2, 3, 5, 8, 13, 21, \ldots$$

As this is the sequence of the Fibonacci numbers, the rate of growth must be given by the golden ratio ϕ.

On the other hand, using our calculations above, we find that $d_I = 1$ as two charges I fuse trivially. For d_τ, we have that

$$d_\tau^2 = 1 + d_\tau$$

from which one gets that, again, $d_\tau = \phi$, the golden ratio.

The fact that the quantum dimension is an irrational number illustrates that the splitting space obtained via such a set of τ anyons cannot be decomposed as a tensor product of smaller ones or, in other words, that the information is encoded into global degrees of freedom rather than local ones such as, for instance, the spin of an electron.

15.4.11 The Verlinde Formula

Let $\mathbf{C}$ be a semisimple modular tensor category, we can build the *fusion algebra* K of $\mathbf{C}$. Without getting into the technical details of such a construction[9] let us say that it has for basis the set of $x_j = \langle S_j \rangle$ for $j \in J$ and for unit $1 = x_0$. There, $\langle S_j \rangle$ denotes the isomorphism class of S_j. Of course, the multiplication in K is given by the fusion rules i.e.

$$x_i x_j = \sum_k N_{ij}^k x_k.$$

[9] This algebra is defined as $K(\mathbf{C}) \otimes_{\mathbb{Z}} \mathbb{K}$ where $K(\mathbf{C})$ is the Grothendieck ring of $\mathbf{C}$. See for instance [5] p. 32 and 53–54.

It turns out that this algebra can be diagonalised i.e. there exists a base in which the multiplication becomes diagonal:

$$x'_i x'_j = \delta_{ij} \alpha x'_j$$

where α is a scalar. The matrix that performs this diagonalisation is a renormalised s-matrix, the (modular) S-matrix which we now define.

Definition 18 Let d_j be the quantum dimension of $S_j \in |\mathbf{C}|$ and D be the scalar

$$\sqrt{\sum_{j \in J} d_j^2}.$$

The S-matrix is

$$S := \frac{1}{D} s.$$

We now have from [5] p. 52 the following

Proposition 3 *For a fixed $j \in J$, let N_j be the matrix of multiplication by x_j in the basis $\{x_j\}$ that is $(N_j)_{ab} = N^a_{jb}$ and also, let D_j be the diagonal matrix $(D_j)_{ab} = \delta_{ab} S_{ia}/S_{0a}$, then*

$$S N_a S^{-1} = D_a.$$

In fact, this proposition states that *the S-matrix diagonalises the fusion rules*. A more complete discussion along with proofs is given in the source of this proposition and [40].

Now, the previous proposition yields to the well-known result [49]:

Theorem 1 (Verlinde Formula)

$$N^k_{ij} = \sum_r \frac{S_{ij} S_{jr} S_{k^*r}}{S_{1r}}.$$

In turn, this theorem says that the S-matrix is not only related to the braids used to define it but also that given an S-matrix, one can calculate the fusion coefficients in $\mathbf{C}$.

15.4.12 Categorical Epilogue

This complete our categorical presentation of the algebra of a family of anyons. Note that even if, in what follows, we use only the semisimple ribbon structure in our description of topological quantum computation, specifying completely the modular tensor category structure was worth the work: indeed, specifying the simple objects, the fusion rules, the pentagon and hexagon axioms, the twist and the S-matrix completely determine the topological properties of a species of anyons!

15.5 An Example: Fibonacci Anyons

The strategy now will be to assume a set of fusion rules of a given species of anyons and solve the various algebraic constraints imposed by the semisimple modular structure. Our intended model to illustrate quantum computation with anyons is the formal semisimple modular tensor category **Fib** which captures the rules of Fibonacci anyons (see [9, 14, 22, 24, 40] from which the material of this section is derived).

- Fibonacci anyons have only two charges: $\mathbf{1}$ and τ, where $\mathbf{1}$ is the trivial charge,
- Both are their own anti-charge,
- They satisfies the following fusion rules:

$$\begin{aligned}
&\mathbf{1}\otimes\mathbf{1}\simeq\mathbf{1}\\
&\mathbf{1}\otimes\tau\simeq\tau\otimes\mathbf{1}\simeq\tau\\
&\tau\otimes\tau\simeq\mathbf{1}\oplus\tau
\end{aligned}$$

Categorically, this says that the semisimple modular tensor category **Fib** has

- Two simple objects $\mathbf{1}$ and τ where $\mathbf{1}$ is the tensor unit,
- That they are their own dual i.e.: $\mathbf{1}^*=\mathbf{1}$ and $\tau^*=\tau$ and,
- That $\mathbf{1}$ and τ satisfy the fusion rules given above.

Let us inspect the fusion rules. While the two first trivially hold, the third one says that the charge resulting from the fusion of two anyons of charge τ is either $\mathbf{1}$ or τ. It is precisely this third rule that tells us that our anyons are non-abelian as they can fuse in two distinct ways.

Now, back to our model, consider three anyons of charge τ all lined up $(\tau\otimes\tau)\otimes\tau$ and let them fuse in the order fixed by the bracketing. Such a process is algebraically described by:

$$\begin{aligned}
(\tau\otimes\tau)\otimes\tau &\simeq(\mathbf{1}\oplus\tau)\otimes\tau\\
&\simeq(\mathbf{1}\otimes\tau)\oplus(\tau\otimes\tau)\\
&\simeq\tau\oplus(\mathbf{1}\oplus\tau)\\
&\simeq\mathbf{1}\oplus 2\cdot\tau.
\end{aligned}$$

Hence, the fusion process for three τ anyons yields a final charge τ in 2 different ways or $\mathbf{1}$ in a single way. These three scenarios depict as

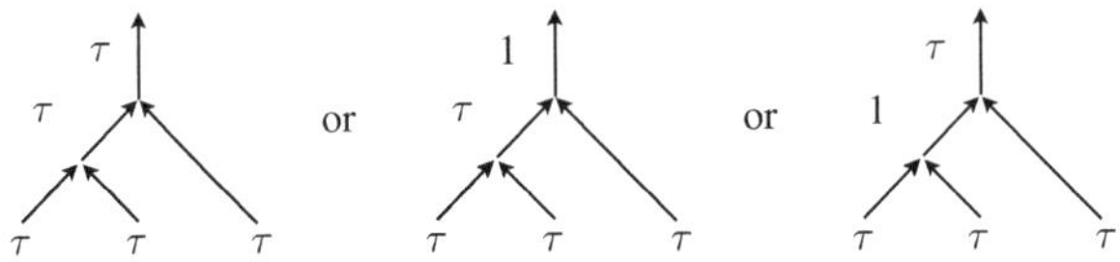

We now pass to the context of finite-dimensional complex vector spaces via the splitting spaces whose basis vectors are dual to the fusion states described above. Consider

$$\begin{aligned}\mathrm{Hom}(b, (\tau \otimes \tau) \otimes \tau) &\simeq \mathrm{Hom}(b, \mathbf{1} \oplus 2 \cdot \tau)\\ &\simeq \mathrm{Hom}(b, \mathbf{1}) \oplus \mathrm{Hom}(b, 2 \cdot \tau) \text{ and as } 2 \cdot \tau := \tau \oplus \tau \text{ this is}\\ &\simeq \mathrm{Hom}(b, \mathbf{1}) \oplus 2 \cdot \mathrm{Hom}(b, \tau). \end{aligned} \tag{15.5}$$

Now, using lemma 1 in conjunction with the property that for any $b \in \{\mathbf{1}, \tau\}$, $\mathrm{End}(b) \simeq \mathbb{C}$; if we set $b = \mathbf{1}$, then (15.5) is isomorphic to $\mathbb{C} \oplus 2 \cdot 0$. Conversely if $b = \tau$, then it is isomorphic to $0 \oplus 2 \cdot \mathbb{C}$.

From this, we conclude that considering the space of states with global charge $b \in \{\mathbf{1}, \tau\}$ is the same as considering

$$\mathrm{Hom}(b, (\tau \otimes \tau) \otimes \tau).$$

In its turn, such a consideration fixes either of the splitting spaces $\mathbb{C}$ or $2{\cdot}\mathbb{C} := \mathbb{C}^2$ as orthogonal subspaces of $\mathbb{C}^3$, the topological space representing our triple of anyons. It is within this two-dimensional complex vector space that we will simulate our qubit. Indeed, if $b = \tau$, we are left with two degrees of freedom which are the two possible outputs of the second splitting.

Remark 8 It is worth stressing that it takes *three* anyons of charge τ to simulate a *single* qubit. Moreover, we shall see later that braiding these anyons together simulates a unitary transformation on such a simulated qubit.

Remark 9 Since **Fib** is rigid, we can apply Proposition 1. We have

$$\begin{aligned}\mathrm{Hom}(\tau, (\tau \otimes \tau) \otimes \tau) &\simeq \mathrm{Hom}(\mathbf{1} \otimes \tau, (\tau \otimes \tau) \otimes \tau)\\ &\simeq \mathrm{Hom}(\mathbf{1}, ((\tau \otimes \tau) \otimes \tau) \otimes \tau).\end{aligned}$$

Comparing this fact with what we got in Example 1, we see that these two encodings are essentially the same. It is because of this that some authors, for instance J. Preskill in [40], prefer to encode their qubits within a quadruple of anyons of individual charge τ with global charge **1** instead. We choose the former to align with the work of Bonesteel et al. [9] that we will explain in Sect. 15.6.

Now that we have an expression for the topological spaces in **Fib**, it will be handy to fix a basis for them. Using the diagram given in the section on splitting spaces, we get

$$\begin{array}{ccc} \mathrm{Hom}(\tau,\tau) & \xrightarrow{|i-2\rangle} & \mathrm{Hom}(\tau,(\tau\otimes\tau)\otimes\tau) \\ & \searrow^{\mathrm{Hom}(1_\tau,\iota_i)} & \uparrow \simeq \\ & & \mathrm{Hom}(\tau,\mathbf{1}\oplus 2\cdot\tau) \end{array}$$

with $i \in \{2, 3\}$ and where the vertical isomorphism is built from the fusion rules.

Analogously, the basis vector $|NC\rangle$ spanning the one dimensional fusion space is defined as

$$\begin{array}{ccc} \mathrm{Hom}(\mathbf{1},\mathbf{1}) & \xrightarrow{|NC\rangle} & \mathrm{Hom}(\mathbf{1},(\tau\otimes\tau)\otimes\tau) \\ & \searrow^{\mathrm{Hom}(1_{\mathbf{1}},\iota_1)} & \uparrow \simeq \\ & & \mathrm{Hom}(\mathbf{1},\mathbf{1}\oplus 2\cdot\tau) \end{array}$$

It is labeled NC for *Non-Computational*. Indeed, the superposition of $|NC\rangle$ with $|0\rangle$ or $|1\rangle$ is prohibited.

15.5.1 The F-Matrix

In order to ensure consistency of the model **Fib**, splitting has to be associative as expressed categorically via the pentagon axiom from the monoidal structure [36].

There are two splitting spaces that can be obtained from a triple of anyons i.e.: $(\tau\otimes\tau)\otimes\tau$ and $\tau\otimes(\tau\otimes\tau)$. The basis vectors for these two splitting spaces are – of course – related by a unitary transformation called the F-*matrix* acting on the splitting spaces and defined via the natural transformation α in the following way:

$$F : \mathrm{Hom}(W,(S\otimes T)\otimes U) \xrightarrow{\mathrm{Hom}(1_W,\alpha_{S,T;U})} \mathrm{Hom}(W, S\otimes(T\otimes U)). \quad (15.6)$$

There, S, T, U and $W \in \{\mathbf{1}, \tau\}$.

Using splitting diagrams, we have:

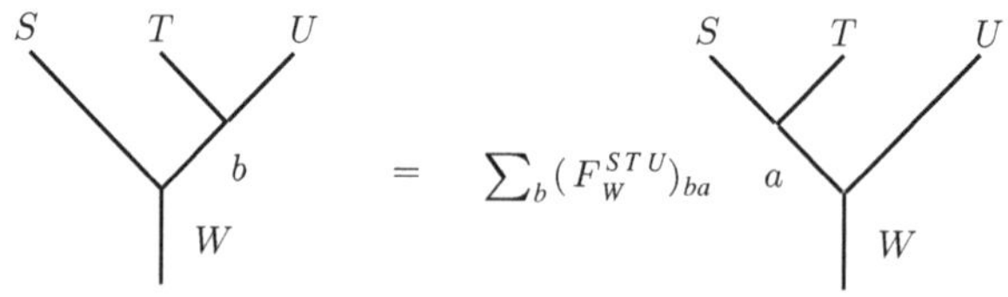

Considering the splitting diagram for fixed a and W as a basis vector, this is nothing but the matrix expression of F. In order to obtain a solution for the F-matrix, we need to recast the pentagon axiom from the monoidal structure in this context in such a way that we obtain a matrix equation. Consider

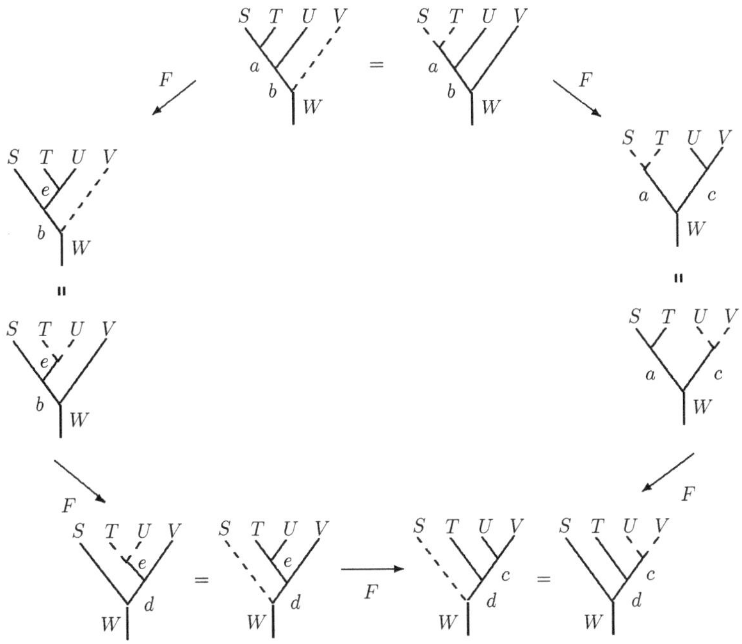

There, we explicitly expressed where F was acting on the splitting states via subdiagrams drawn in solid lines. Passing to the underlying pentagon axiom is made via the expression of F given in Eq. (15.6).

Now, equating both sides of the diagram yields

$$(F_W^{STc})_{da}(F_W^{aUV})_{cb} = \sum_e (F_d^{TUV})_{ce}(F_W^{SeV})_{db}(F_b^{STU})_{ea}. \tag{15.7}$$

Solving this in conjunction with a given set of fusion rules yields the F-matrix. To solve such an equation, one has to fix the labels for all the possible states in the splitting basis and solve the resulting system of equations.

In **Fib**, for a triple of anyons of charge τ, the trivial charge can split into such a triple in only one way. In this particular case, the F-matrix is

$$F_{\mathbf{1}}^{\tau\tau\tau} = [1]$$

as the first splitting must yields $\tau \otimes \tau$.

Conversely, if the initial charge is τ then, the splitting process can occur in two distinct manners. In order to get the F-matrix, we must use Eq. (15.7). For instance, a possible splitting scenario occurs when one fixes $a = \mathbf{1} = c$ and $d = \tau = b$. Using this with (15.7) gives:

$$(F_{\mathbf{1}}^{\tau\tau\mathbf{1}})_{\tau\mathbf{1}}(F_{\mathbf{1}}^{\mathbf{1}\tau\tau})_{\mathbf{1}\tau} = \sum_{e\in\{\mathbf{1},\tau\}} (F_{\tau}^{\tau\tau\tau})_{\mathbf{1}e}(F_{\mathbf{1}}^{\tau e\tau})_{\tau\tau}(F_{\tau}^{\tau\tau\tau})_{e\mathbf{1}}$$

$$1 = F_{\mathbf{11}}^2 + F_{\tau\mathbf{1}}F_{\mathbf{1}\tau}.$$

Using this, the other consistency relations and the fact that F is unitary, we find:

$$F_{\tau}^{\tau\tau\tau} = \begin{bmatrix} F_{\mathbf{11}} & F_{\mathbf{1}\tau} \\ F_{\tau\mathbf{1}} & F_{\tau\tau} \end{bmatrix} = \begin{bmatrix} \phi^{-1} & \sqrt{\phi^{-1}} \\ \sqrt{\phi^{-1}} & -\phi^{-1} \end{bmatrix}$$

where ϕ is the golden ratio.

Finally, combining the results for $F_{\tau}^{\tau\tau\tau}$ and $F_{\mathbf{1}}^{\tau\tau\tau}$ yields

$$F = \left[\begin{array}{c|cc} 1 & 0 & 0 \\ \hline 0 & \phi^{-1} & \sqrt{\phi^{-1}} \\ 0 & \sqrt{\phi^{-1}} & -\phi^{-1} \end{array}\right]$$

which is also unitary. The lower-right block induces a change of basis on the 2-dimensional splitting space while the upper-left block is the trivial transformation on the one-dimensional splitting space.

15.5.2 Braiding Anyons

We now express what will be the consequence of exchanging two anyons on the splitting space. As such an exchange is represented categorically by a braiding, this will yield a representation of the braid group in the splitting space.

15.5.2.1 The R-Matrix

The game here is very similar to the one for the F-matrix except that we use the hexagon axiom from the braided monoidal structure instead [36]. The R-matrix is a morphism

$$R : \mathrm{Hom}(W, (S \otimes T) \otimes U) \xrightarrow{\mathrm{Hom}(1_W, \sigma_{S,T} \otimes 1_U)} \mathrm{Hom}(W, (T \otimes S) \otimes U).$$

or, using splitting diagrams:

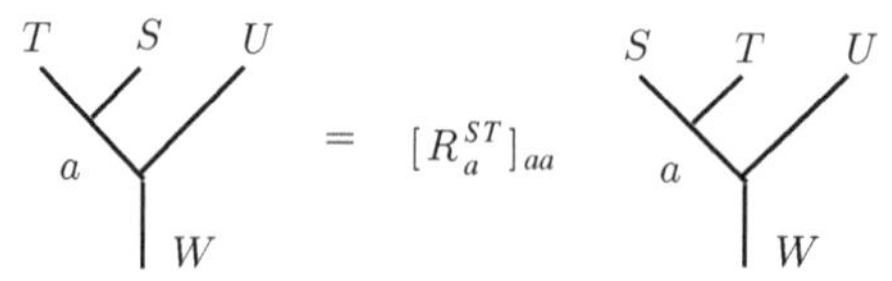

We already have the F-matrix thus, the hexagon needs to be solved only for the R-matrix. Recasted with splitting diagrams, the hexagon axiom becomes:

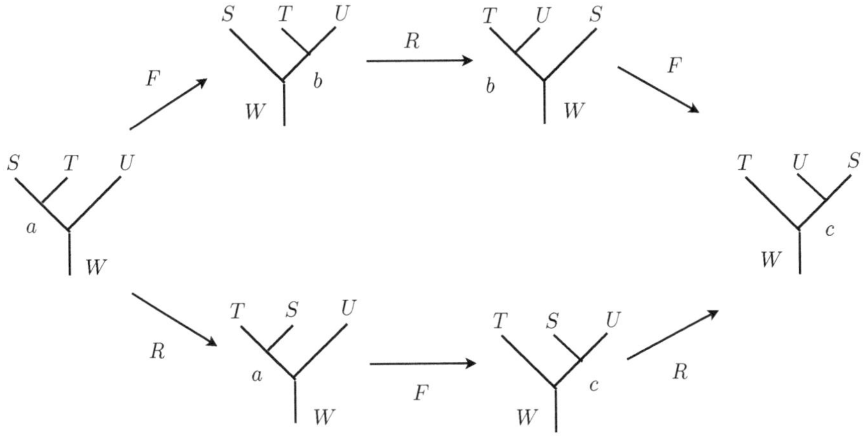

Writing it as a matrix equation yields

$$R_c^{SU}(F_W^{TSU})_{ca}R_a^{ST} = \sum_b (F_W^{TUS})_{bc}R_W^{Sb}(F_W^{STU})_{ba}.$$

For a triple of anyons with charge τ, explicit calculations of the R-matrix yields:

$$\left[\begin{array}{c|cc} -e^{-2i\pi/5} & 0 & 0 \\ \hline 0 & e^{-4i\pi/5} & 0 \\ 0 & 0 & -e^{-2i\pi/5} \end{array}\right]$$

Such a diagonal form is not surprising: whether the global charge of a couple is **1** (resp. τ), it must remain so even if we exchange the two components of the pair.

15.5.2.2 The B-Matrix

The R-matrix provided in the previous section give us a way to exchange the two leftmost anyons in a set of three. We now need a way to find the matrix that exchanges the two rightmost anyons, this will be the B-matrix and is defined as

$$\begin{array}{ccc} \mathrm{Hom}(W, (S \otimes T) \otimes U) & \xrightarrow{\quad B \quad} & \mathrm{Hom}(W, (S \otimes U) \otimes T) \\ \Big\downarrow{\scriptstyle \mathrm{Hom}(1_W, \alpha_{S,T;U})} & & \Big\uparrow{\scriptstyle \mathrm{Hom}(1_W, \alpha^{-1}_{S,U;T})} \\ \mathrm{Hom}(W, S \otimes (T \otimes U)) & \xrightarrow[\mathrm{Hom}(1_W, 1_S \otimes \sigma_{T,U})]{} & \mathrm{Hom}(W, S \otimes (U \otimes T)) \end{array}$$

As we found both the F and the R matrix in **Fib**, we can compute the B-matrix as

$$B := F^{-1}RF$$

15.6 Universal Quantum Computation with Fibonacci Anyons

The basic idea to simulate quantum computation with anyons is given by the following steps:

1. Consider a compound system of anyons. We initialise a state in the splitting space by fixing the charges of subsets of anyons according to the way they will fuse. This determines the basis state in which the computation starts.
2. We braid the anyons together, it will induce a unitary action on the chosen splitting space.
3. Finally, we let the anyons fuse together and the way they fuse determines which state is measured and this constitutes the output of our computation.

15.6.1 Simulating Qubits

First, the topological space for such a triple is a pair $\langle \mathbb{C}, \mathbb{C}^2 \rangle$ where the 2-dimensional space is spanned by the fusion states

$$|0\rangle := |(\tau \otimes \tau) \otimes \tau; \tau, 1\rangle \quad \text{and} \quad |1\rangle := |(\tau \otimes \tau) \otimes \tau; \tau, 2\rangle,$$

and the space of dimension one is spanned by:

$$|NC\rangle := |(\tau \otimes \tau) \otimes \tau; \mathbf{1}, 1\rangle.$$

Of course, the simulation of a qubit will occur on the 2-dimensional space spanned by $\{|0\rangle, |1\rangle\}$. Compound system of two or more qubits will be given by the compound system of such a triple of anyons. Note that even if we fix the global charge of the triple as τ, in the real world, it is possible that we may still measure $\mathbf{1}$. These errors are known as "leakage errors" as there is some unexpected "leaks" into another splitting space.

15.6.2 Quantum Computation

To perform actual quantum computation, it seems at first glance that we have two problems:

1. First, we would like to apply any gate on our simulated qubits but we have only the two braiding matrices and their inverses.

2. Second, even if we solve our first problem, it remains that this is not enough to quantum compute. Indeed, we also need a two-qubit gate.

We answer these. First, a composition of length l of R- and B-matrices and their inverses can get arbitrarily close to any element of $SU(2)$ and that, with l reasonably small. This is a consequence of the fact that our matrices together with their inverse satisfies the Solovay-Kitaev theorem, see [37] pp. 617–624 for a precise statement and a proof.

We now address the second problem. Following the works of Bonesteel et al. in [9], we explain how to build a **CNOT** gate for our anyons; for this we will need two triplets of τ anyons; one of them will act as our test qubit while the other will be the target qubit. The idea is relatively simple: we need to intertwine a pair of quasi-particles from the first triplet—the control pair—with the target triplet *without disturbing it*; as the braid operators are dense in $SU(2)$, we will arrange such an intertwining so that its representation in $SU(2)$ is close enough to to the identity. The next thing is to implement a **NOT**—actually a $i \cdot$ **NOT**—by braiding our two anyons of the control pair with those of the target triple. Finally, we extract the control pair from the second triplet—again—without disturbing it. Now, the key point is the following: a braiding involving the trivial charge **1** with an anyon of arbitrary charge does not change anything. Thus, when measuring the control pair, the $i \cdot$ **NOT** will occur if and only if the two anyons from the control pair fuse as an anyon of charge τ; otherwise the control pair only induces a trivial change on the system.

(a) Consider the following braiding:

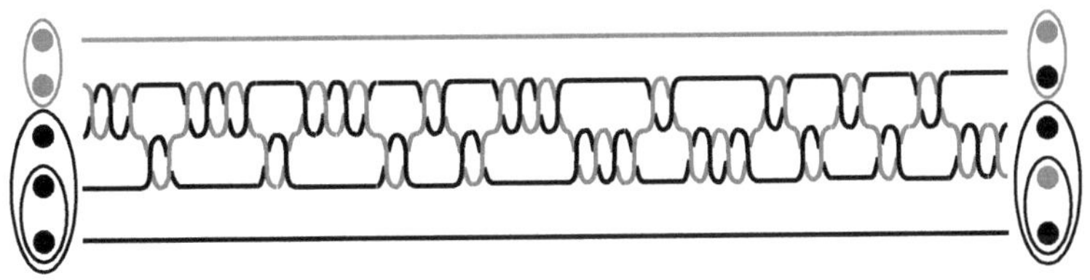

As an action on the splitting space of the three anyons involved, this is, in the same order as depicted in the picture:

$$B^3R^{-2}B^{-4}R^2B^4R^2B^{-2}R^{-2}B^{-4}R^{-4}B^{-2}R^4B^2R^{-2}B^2R^2B^{-2}R^3 \sim \left(\begin{array}{cc|c} 1 & 0 & 0 \\ 0 & 1 & 0 \\ \hline 0 & 0 & 1 \end{array}\right)$$

This tells us how the given braid insert an anyon within a triplet without disturbing it. In fact, this stresses the distinction between the dynamics of the anyons and the consequences on the splitting space. Indeed, even if we disturbed the initial configuration of anyons via multiple braidings, the effect on the splitting space is approximately the identity.

(b) Now, we implement an $i \cdot$ **NOT** as the following braid:

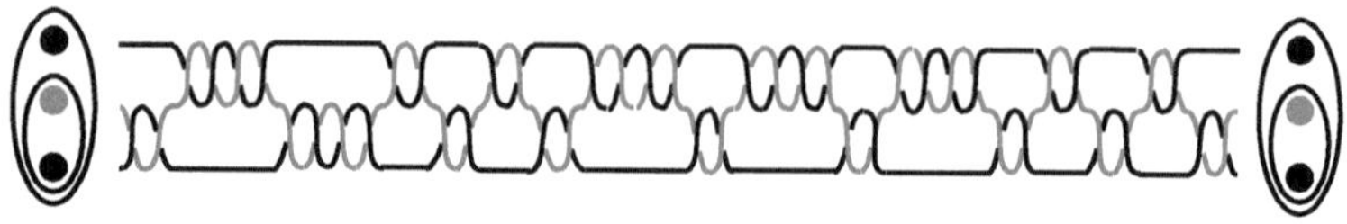

The unitary acting on the splitting space of the initial triple is given by:

$$R^{-2}B^{-4}R^{4}B^{-2}R^{2}B^{2}R^{-2}B^{4}R^{-2}B^{4}R^{2}B^{-4}R^{2}B^{-2}R^{2}B^{-2}R^{-2} \sim \left(\begin{array}{cc|c} 0 & i & 0 \\ i & 0 & 0 \\ \hline 0 & 0 & 1 \end{array}\right)$$

This combination of braids tells us how to implements a $i \cdot$ **NOT** gate on the two dimensional fusion space of our triple of anyons. Again, this gate is approximated.

(c) Finally, the $i \cdot$**CNOT** gate acting on two topological qubits is realised as follows:

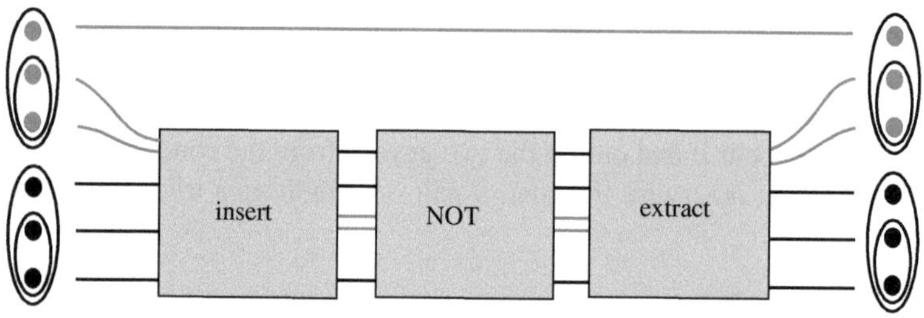

First, instead of inserting 1 anyon, we insert a couple that will be used as a test couple and that in the very same manner as described in (a) – as these two will fuse together yielding either **1** or τ, this is exactly what we want. Secondly, we apply the $i \cdot$ **NOT**-gate computed in (b). Finally, we extract the control pair returning it to its original position by applying the insertion procedure in reverse order. This is done, again, without disturbing the triple at stance here.

We claim that this implements a **CNOT**. Indeed, the test couple can fuse in two ways. If it fuse as **1**, then nothing happens as **1** is the trivial charge. If it fuse as τ, then we effectively apply the $i \cdot$ **NOT** gate computed in (b).

Interestingly, we may replace the $i \cdot$ **NOT** by any other braid thus obtaining a way to perform other controlled operations give such a pair of topological qubits. It turns out that this gate together with the R and B matrices form a universal set of quantum gates [10] as $i \cdot$ **CNOT** is entangling.

15.7 Conclusions

As we noted in the introduction of Sect. 15.4 the results that we presented here are not new. We gave an explicit description of the algebra of anyons in terms of modular categories, a description that is given in mathematical physics papers that may not be accessible to everyone.

The very rich nature of this subject makes the task of writing a complete introduction concisely almost impossible therefore, to complement our attempt, we now give some additional pointers to the litterature. On the physical side, Frank Wilczek edited *Fractional statistics and anyon superconductivity* [54], a book which comprises papers explaining central concepts related to anyons. This list wouldn't be complete without A. Kitaev's paper *Anyons in an exactly solved model and beyond* [29] where the reader can find another introduction on the algebra of anyons with a more physical flavour in appendix E.

On the quantum computing side, there is a growing interest in topological quantum computation due to the fact that such a model provides a more robust form of quantum computation. Some papers which emphasise topological quantum computation are [3, 43]. Perhaps closer to the essence of the present paper are the works of Freedman, Kitaev, Larsen and Wang presented in [21, 22].

Of course, apart from modular tensor categories, there are other links to draw between topology and quantum mechanics or quantum computing; for instance, the works of Louis H. Kaufman and Samuel J. Lomonaco Jr., in particular [26] in which they describe the role of braiding in quantum computing. There is also the work of S. Abramsky [1] that describes connections between knot theory, categorical quantum mechanics, logic and computation.

Finally, although the aim of this paper was to draw parallels between categorical structures and anyons, there is still work to do; for instance, our exposition was oriented towards **FdHilb**, the category of finite-dimensional Hilbert spaces. However, †-compact categories [2] constitute a correct framework to describe quantum mechanics and it would be interesting to see the benefits of describing topological quantum computation within this more abstract context.

Acknowledgments The authors would like to thank Samson Abramsky, Michel Boyer, Ross Duncan, Rafael Sorkin, Colin Stephen, Benoît Valiron and Jamie Vicary for useful discussions.

References

1. Abramsky, S.: Temperley-lieb agebra: From knot theory to logic and quantum computation via quantum mechanics. In: Chen, G., Kauffman, L., Lomonaco, S. (eds.) Mathematics of Quantum Computing and Technology, pp. 515–558. Taylor and Francis, New York (2007)
2. Abramsky, S., Coecke, B.: A categorical semantics of quantum protocols. In: Proceedings of the 19th Annual IEEE Symposium on Logic in Computer Science (LiCS'04), IEEE Computer Science Press (2004)
3. Aharonov, D., Jones, V., Landau, Z.: A polynomial quantum algorithm for approximating the jones polynomial. In: STOC '06: Proceedings of the 38th Annual ACM Symposium on Theory of Computing, pp. 427–436. ACM, New York (2006)
4. Bakalov, B., Kirilov, A. Jr.: On the Lego-teichmüller game. Transf. Groups **5**, 7–44 (2000)
5. Bakalov, B., Kirillov, A.: Lectures on tensor categories and modular functors. American Mathematical Society in University Lecture Series (2001)
6. Balachandran, A.P., Daughton, A., Gu, Z., Sorkin, R.D., Marmo, G., Srivastava, A.M.: Spin-statistics theorems without relativity or field-theory. Int. J. Mod. Phys. **A8**, 2993–3044 (1993)
7. Balachandran, A.P., McGlinn, W.D., O'Raifeartaigh, L., Sen, S., Sorkin, R.D., Srivastava, A.M.: Topological spin-statistics theorem for strings. Mod. Phys. Lett. **A7**, 1427–1442 (1992)

8. Berry, M.V., Robbins, J.M.: Indistinguishability for quantum particles: spin, statistics and the geometric phase. Proc. Roy. Soc. A **453**(1963), 1771–1790 (August 1997)
9. Bonesteel, N.E., Hormozi, L., Zikos, G.: Braid topologies for quantum computation. Phys. Rev. Lett. **95**, 140503 (2005)
10. Bremner, M.J., Dawson, C.M., Dodd, J.L., Gilchrist, A., Harrow, A.W., Mortimer, D., Nielsen, M.A., Osborne, T.J.: Practical scheme for quantum computation with any two-qubit entangling gate. Phys. Rev. Lett. **89**, 247902 (2002)
11. Burgoyne, N.: On the connection between spin and statistics. Nuovo Cimento **VIII**(4), 607–609 (1958)
12. Coecke, B., ´Paquette, E.O.: Categories for the practising physicist. In: Coecke, B. (ed.) New Structures in Physics. Springer Lecture Notes in Physics, New York (2008)
13. Crane, L.: 2-d physics and 3-d topology. Comm. Math. Phys. **135**, 615–640 (1991)
14. Das Sarma, S., Freedman, M., Nayak, C., Simon, S.H., Stern, A.: Non-abelian anyons and topological quantum computation. ArXiv e-prints, 707 (July 2007)
15. deWet, J.S.: On the connection between the spin and statistics of elementary particles. Phys. Rev. **57**, 646–652 (1940). Ph.D. thesis, Princeton University 1939
16. Dowker, H.F., Sorkin, R.D.: A spin-statistics theorem for certain topological geons. Class. Quantum Grav. **15**(5), 1153–1167 (1998)
17. Duck, I., Sudarshan, E.C.G.: Towards an understanding of the spin-statistics theorem. Am. J. Phys. **66**(4), 284–303 (1998)
18. Duck, I., Sudarshan, E.C.G., Arthur Wightman, S.: Pauli and the spin-statistics theorem. Am. J. Phys. **67**(8), 742–746 (1999). Review by Wightman of the book by Duck and Sudarshan
19. Fierz, M.: Uber die relativische Theorie kr¨aftfreir Teilchen mit beliebigem spin. Helv Phys. Acta **12**(3), 3–37 (1939)
20. Finkelstein, D., Rubinstein, J.: Connection between spin statistics and kinks. J. Math. Phys. 9(11), 1762–1779 (1968)
21. Freedman, M.H., Kitaev, A., Wang, Z.: Simulation of topological field theories by quantum computers. Comm. Math. Phys. **227**, 587–603 (2002)
22. Freedman, M.H., Larsen, M., Wang, Z.: A modular functor which is universal for quantum computation. Comm. Math. Phys. **227**, 605–622 (2002)
23. Freyd, P.: Abelian Categories. Harper and Row (1964)
24. Hormozi, L., Zikos, G., Bonesteel, N.E.: Topological quantum compiling. Phys. Rev. B **75**, 165310 (2007)
25. Joyal, A., Street, R.: Braided tensor categories. Adv. Math. **102**, 20–78 (1993)
26. Kauffman, L.H., Lomonaco, S.J. Jr.: Braiding operators are universal quantum gates. New J. Phys. **6**(134), (2004)
27. Khono, T.: Topological invariants for 3-manifolds using representations of the mapping class groups i. Topology **31**, 203–230 (1992)
28. Kitaev, A.: Fault-tolerant quantum computation by anyons. Ann. Phys. **303**(1), 3–20 (2003)
29. Kitaev, A.: Anyons in an exactly solved model and beyond (2006)
30. Klitzing, K.V., Dorda, G., Pepper, M.: New method for high-accuracy determination of the fine-structure constant based on quantized hall resistance. Phys. Rev. Lett. **45**(6), 494–497 (Aug 1980)
31. Mac Lane, S.: Categories for the working mathematician. Springer graduate text in mathemathics, vol. 5. Springer, New-York (1998)
32. Laughlin, R.B.: Quantized hall conductivity in two dimensions. Phys. Rev. B **23**(10), 5632–5633 (May 1981)
33. Leinaas, J.M., Myrheim, J.: On the theory of identical particles. Il Nuovo Cimento **37B**, 1–23 (1977)
34. Lüders, G., Zumino, B.: Connection between spin and statistics. Phys. Rev. **110**, 1450–1453 (1958)
35. Messiah, A.: Quantum Mechanics, vol. 1. Wiley, New York (1966)

36. Moore, G., Seiberg, N.: Polynomial equations for rational conformal field theories. Phys. Lett. B **212**(4), 451–460 (1988)
37. Nielsen, M.A., Chuang, I.L.: Quantum Computation and Quantum Information. Cambridge University Press, Cambridge (2000)
38. Pauli, W.: The connection between spin and statistics. Phys. Rev. **58**, 716–722 (1940)
39. Pauli, W.: On the connection between spin and statistics. Prog. Theor. Phys. **5**(4), 526–543 (1950)
40. Preskill, J.: Lectures notes physics 219/computer science 219 quantum computation (2004)
41. Reshetikhin, Y.N., Tuarev, V.G.: Ribbon graphs and their invariants derived from quantum groups. Comm. Math. Phys. **127**, 1–26 (1990)
42. Reshetikhin, Y.N., Tuarev, V.G.: Invariants of 3-manifolds via link polynomials and quantum groups. Invent. Math. **103**, 547–597 (1991)
43. Das Sarma, S., Freedman, M., Nayak, C.: Topologically-protected qubits from a possible non-abelian fractional quantum hall state. Phys. Rev. Lett. **94**, 166802 (2005)
44. Schwinger, J.: Spin, statistics and the TCP theorem and Addendum. Proc. Natl. Acad. Sci. USA **44**, 223–228, 617–619 (1958)
45. Sorkin, R.D.: Particle-statistics in three dimensions. Phys. Rev. D **27**, 1787–1797 (1983)
46. Störmer, H.: Nobel Lecture: Physics 1996–2000, Chapter The Fractional Quantum Hall Effect, pp. 295–326. World Scientific, Singapore (2002)
47. Streater, R.F., Wightman, A.S.: PCT, Spin, Statistics and All That. Benjamin, New York (1964)
48. Tuarev, V.G.: Quantum Invariant of Knots and 3-Manifolds. W. de Gruyter, Berlin (1994)
49. Verlinde, E.: Fusion rules and modular transformations in 2d conformal field theory. Nuc. Phys. B. **300**, 360–376 (1988)
50. Wightman, A.S.: Quantum field theory in terms of vacuum expectation values. Phys. Rev. **101**(2), 860–866 (January 1956)
51. Wightman, A.S.: The spin-statistics connection: Some pedagogical remarks in response to neuenschwander's question. Electro. J. Diff. Eq. 207–213 (2000). Conference on Mathematical Physics and Quantum Field Theory
52. Wilczek, F.: Magnetic flux, angular momentum, and statistics. Phys. Rev. Lett. **48**(17), 1144–1146 (April 1982)
53. Wilczek, F.: Quantum mechanics of fractional-spin particles. Phys. Rev. Lett. **49**(14), 957–959 (Oct 1982)
54. Wilczek, F. (ed.): Fractional Statistics and Anyon Superconductivity. World Scientific, New Jersey (1990)

Index

Note: 'f', 'n', 't' and 'b' followed by the locators refers to figures, note numbers, tables and boxes cited in the text

A

Abstract orthogonality, 398–403
Adjoint functor, 281–283, 362, 468–469, 956–957
Anyon, 167, 983–1023
Approximation, 38–39, 41, 439, 494–496, 498, 502–503, 506, 512, 514, 524, 526, 528, 541, 544–545, 547–548, 552, 575, 587, 589, 615, 618–622, 624, 650–652, 674–675, 687, 691, 696, 699, 720, 806, 828, 840, 842, 849–850, 871, 910, 952, 973n6
Autonomous categories, 307–322, 351, 363–364, 366, 368, 370–373, 412, 467–468

B

Bayesian order, 529–531, 587, 596, 630, 638–639, 643, 647–649, 655, 657, 662
Bicontinuous, 535–536, 688, 691, 693, 701
Bifunctor, 193, 196–198, 201, 209, 212, 214, 222, 236, 258, 271–272, 297–298, 482, 994
Bilinear logic, 475, 485–486
Biproduct, 250–253, 251n9, 256–257, 259–260, 262, 331–332, 335–340, 349–350, 353t, 383, 398, 400, 404, 406, 958, 967, 1004n8
Bohm momentum, 744
Boolean algebra, 50, 656–658, 758–759, 767–768, 773, 777, 790–791, 807–808, 855, 907–908
Braided monoidal category, 115–119, 122, 130, 138, 167, 303, 305–306, 315, 329–331, 342, 361, 996–998, 1001

C

λ-Calculus, 50, 53, 57–61, 64, 67, 71, 75, 77, 423, 437–438, 442, 445, 453, 456–457, 459–464, 713, 716, 722–723, 942
Cartesian category, 113–114, 118–119, 122, 140, 145, 154, 159, 233, 241, 333, 335, 361
Cartesian closed category (CCC), 50, 60–61, 80, 122, 128, 136, 154, 156–158, 974
2-Category, 36, 143, 154, 468–469, 471, 475–477, 479, 483–484, 960
Category of categories, 26–27
Causal set, 524, 694–695, 701, 755, 788
Causal structure, 187–188, 534–535, 688, 691–692, 702
Charge conjugation, 997–998
Choi-Jamiołkowski, 953, 963–964, 974
Classical state, 499–500, 504, 524, 528–531, 573, 578–580, 583, 593–596, 598, 600–609, 611–613, 615–616, 618–620, 623–624, 626, 628, 630, 632, 636–639, 643, 646–651, 653–655, 657, 665, 667–669, 673–674, 676, 753, 759, 762, 767, 769, 773, 802, 810, 814, 818, 855, 882, 905
Clifford algebra, 705–750
Clifford bundle, 725, 736–740, 749
Clifford group, 710–712, 720, 735, 744, 969–970, 972
Closed symmetric monoidal category, 138, 142–143, 154, 161–162, 164, 166
Cobordism, 96, 97t–98t, 99, 101–103, 166t, 213, 228–232, 240, 269, 279–281, 292
Coequaliser, 22t, 47
Colimit, 22, 37, 47–48, 309

Coecke, B. (Ed.): *Index*. Lect. Notes Phys. **813**, 1027–1031 (2011)
DOI 10.1007/978-3-642-12821-9

Commutative diagram, 22, 44, 124, 196, 201–202, 214, 216, 235, 244, 270, 291, 474, 823, 826, 870, 881, 884, 892, 919, 929, 931, 933
Comonad, 79–94, 283, 374
Compact (closed) category, 220–221, 223, 228, 262, 276
Continuous dcpo, 494, 496–497, 499–500, 506–508, 513, 519, 521, 523–524, 526, 539–540, 553, 560, 579, 586–587, 589, 615, 690, 696
Contraction, 68, 70, 86, 88–89, 135, 137, 140, 362, 368–369, 369t–370t, 373–376, 383, 390n3, 394, 401–402, 409, 443, 450, 455, 457–459, 462–463, 471–473, 477–478, 519–523, 527, 544–545, 586n3
Contravariant functor, 27, 34, 192–193, 202, 308, 340, 372, 404, 762, 770, 792–793, 807, 906, 908, 930–931
Coproduct, 14–20, 33, 42, 47, 72, 80–81, 91, 244–246, 248–252, 332, 335–340, 349, 353t, 361, 363, 366, 369f–370f, 371, 373–474, 380, 383, 836, 958n8
Curry-Howard correspondence, 51–68
Currying, 49, 59, 121–122, 127–129, 138–139, 141, 155, 948, 953–955, 959, 966
Cut-elimination, 70, 357–358, 366, 371, 373–375, 385–389, 395–397, 410–411, 443, 445, 461, 475

D

Dagger compact (closed) category, 223, 228, 231, 347–348
Dagger monoidal category, 202–203, 212–213, 231, 266, 341–343, 345, 347–348
Daseinisation, 753–754, 762, 770, 794n52, 800–808, 819, 828, 830–840, 837n80, 842–843, 845–846, 848–849, 853, 855–856, 859n96, 860, 863–871, 894, 896–898, 907, 912, 914–915, 924
Decoherence, 577, 674
Density operator, 531–532, 549–550, 575–577, 594, 632–634, 650, 652–654, 666, 671, 738
Dirac notation, 202–206, 211, 945–946, 955, 962, 977–978
Directed suprema, 501–502, 509–510, 556, 608, 612, 617, 621, 649, 689, 699
Direct sum, 16, 19, 110, 212–215, 249–253, 262, 361, 400, 404, 406, 818, 895, 957–958, 1007
Domain theory, 491–589, 594, 598–599, 621, 652–654, 674–682, 687–702, 850, 973–974, 976

E

Entanglement, 680–682, 882, 899–900, 904, 945, 976
Entropy, 493, 504, 524, 529–533, 549–550, 574, 579, 581–582, 584–587, 594–595, 598–600, 608–609, 615, 624, 626–628, 630, 652, 654–655, 672, 676–677, 679, 682
Epic (Epimorphism), 5, 10–13, 875, 893
Equaliser, 20–22, 22t, 47
Exponential, 48–50, 62, 66, 76–77, 86, 88–90, 137, 362, 366, 368–374, 406, 438, 447–459, 461–462, 464, 527, 589n25, 859n94, 926

F

Fan-out, 946–947, 951–952
Feynman diagram, 96–97, 99–100, 103, 108, 123, 437–486
Fibonacci anyons, 992, 1010–1011, 1014, 1020
Fixed point
  Banach theorem, 401
  least, 379, 491–493, 508, 524, 531, 579, 586
  of nonmonotonic mappings, 509–512
  unique, 401–403, 519–521, 528, 541, 556
Fixed point theorem, 401, 491–493, 511, 519, 579, 585–587
Formal distribution, 421–435, 445
Free compact 2-category, 475
Frobenius comonoid, 269, 280–281
Functional languages, 975–976
Functor category, 36, 47
Functor, *see specific functors*
Fusion rules, 991–994, 1002, 1005–1006, 1012–1014, 1016–1017

G

Galois connection, 835–836
General relativity, 98, 492–493, 524, 534, 687–702, 705, 708, 763, 871
Gentzen's proof theory, 366–371
Geometric logic, 774–775, 785, 890
Globally hyperbolic, 536–538, 688, 690, 692–695, 697–698, 700
GoI (geometry of interaction), 357–414
Gottesman-Knill theorem, 969–970
Grammar, 470–471, 672

Graph, 7, 36, 99, 223, 438, 529, 568, 572, 575, 599, 986
Groupoid, 10, 184, 193, 705–750

H
Hamiltonian, 442, 771, 774, 776n36, 783, 794, 864–865, 870
Heyting algebra, 134, 282, 759, 761, 766–769, 771, 773–774, 776, 797, 801, 803–805, 808–809, 813, 821, 823, 828, 916, 928, 932–933
Hilbert space, 95, 98t, 101–102, 105–106, 110, 122–123, 128, 131, 166t, 182, 184, 188, 192–194, 205, 215, 239, 246, 261, 263, 292, 297, 321, 340, 343, 348, 358, 361–362, 383, 386–387, 389, 398, 404–406, 421–425, 438–439, 531, 549, 587–588, 596, 630–631, 657–658, 661, 675, 681, 718n7, 721–722, 725, 743, 749, 754–755, 758–759, 777–778, 789–790, 793–794, 822, 848, 858, 865, 894–896, 899n122, 911–913, 925, 945–947, 949n5, 952, 955n6, 957–958, 960, 962–963, 965, 968, 974, 976, 987–988, 988n1, 993–994, 1007–1010, 1023

I
Incident algebra, 718–719
Inference rule, 43, 52t, 135–140, 142–146, 149
Informatic derivative, 493, 524, 539–559, 568, 572
Information order, 493–496, 500–501, 503–504, 513, 515, 560, 564, 566, 573, 671, 700
Initial object, 13–14, 20, 22, 33, 42, 185, 246, 251, 335, 877, 928
Interval domain, 491, 497, 504, 513, 522, 524, 537, 571, 688, 691, 695–698, 702, 850, 910
Intuitionistic logic, 51, 60, 133–134, 137, 145–146, 282, 753, 760, 768, 771–773, 778, 781, 926–928

K
Kauffman's calculus of distinctions, 716
Kleisli construction, 83–85
Kochen-Specker theorem, 757, 759, 762, 770, 788–790, 793b, 800, 807b, 814, 826–827, 839, 851, 909

L
Lambda-abstraction, 49, 53, 55, 150
Left dual, 125, 307, 309, 315
Limit, 47–48, 93, 134, 519, 524, 540–541, 552, 556, 559–560, 579, 584, 593–594, 607, 643–644, 675, 719–720, 747–749, 764–765, 915, 951–953
Linear logic, 70–76, 78, 85–86, 89, 97, 136–138, 140–141, 145, 187, 351–352, 357–360, 363–371, 373–375, 389, 402, 412, 437–438, 442, 444–445, 447, 450, 453, 456, 464, 475, 485n4
Linear realizability algebra, 442–445
Linguistics, 467–486
Liouville equation, 738
Lorentz distance, 538–539, 687, 689, 700–702

M
Modular tensor categories, 130, 166–167, 984, 992–1014, 1023
Monad, 79–94, 282–283, 374, 871n15
Monic (monomorphism), 5, 10–13, 29, 51, 708, 803, 817, 819–820, 823–826, 875, 889, 893, 909, 924, 927, 1002–1003, 1007–1008
Monoid, 8–9, 13–14, 24, 26, 29, 32, 42, 79, 183–184, 186–187, 214–215, 250, 253, 258, 261, 264–267, 281, 360, 362, 381, 388, 391, 430, 469, 500, 583, 754, 762, 854, 858–860, 863, 913, 919–920, 974
Monoidal category
  braided, 115, 119, 122, 130, 138, 167, 303, 305–306, 315, 329–331, 342, 361, 996–998, 1001
  closed, 121, 126, 129, 138–140, 154
  compact, 125–126, 128–129, 477, 486
  dagger, 212–213, 231, 266, 341–343, 345, 347–348
  strict, 186–187, 194–196, 211, 215, 264, 307, 359–360, 365, 430, 997
  strict dagger, 202–203
  symmetric, 306, 321, 331, 333, 335–336, 340, 342, 347, 349, 361–362, 365, 376–377, 385, 388, 390, 398–399, 402, 404, 957, 960
  traced, 290, 323, 377, 384–385, 402
Monoidal signature, 300–301, 322, 332, 339

N
Natural deduction system, 52t, 59b, 60t, 70
Natural transformation
  di, 147–149, 161
  monoidal, 87–89, 92, 275, 277, 280, 310–312, 316, 318, 345, 374, 385, 390
  unique, 933
No-cloning, 73, 115, 188, 232, 946–947, 947n3

No-deleting, 73, 188, 941, 946–947, 947n3
Nuclear ideal, 398, 423–425, 428, 431–433

P

Partiality, 400, 493–495, 524, 615, 619–620, 622, 624, 651–652
Partial order, 8, 13, 185, 187, 206, 281, 469–471, 476, 478–479, 494, 533, 539, 589, 593–682, 689, 788, 790, 798–799, 830, 836, 866, 932
Partial trace, 376–377, 398–403, 421
Pauli Clifford algebra, 714, 723–724, 732, 742
Pivotal categories, 312–315, 317–320, 322, 326–327, 330, 343, 346–348, 352
Postselection, 967–968
Pregroup, 467–471, 473, 475
Product
- bi, 250–253, 251n9, 256–257, 260, 262, 331–332, 335–340, 349–350, 353t, 383, 398, 400, 404, 406, 958, 967, 1004n8
- cartesian, 15–16, 20, 73, 92, 106, 110–113, 118, 121, 161, 208, 211, 213, 221, 224, 232, 242, 261, 293, 297, 306, 361, 371, 372f, 378–379, 404, 427, 875, 883, 885–886, 900, 926, 953–954, 957–958
- direct, 14, 485
- inner, 131, 192–193, 203–204, 214, 365, 383, 440, 531, 630, 648, 945–946, 977
- tensor, 73, 106–107, 110–111, 113–115, 118, 120, 122, 140–141, 155, 161, 163–164, 166t, 186, 188, 194, 198, 206, 212–213, 215, 261, 290, 296–297, 298t, 306, 328, 343, 351, 360–364, 367, 377, 380, 383, 387–388, 392, 404, 427, 476–477, 481–484, 878n122, 945n5, 957–958, 962, 968–969, 991–993, 995–996, 999, 1005, 1012

Proof net, 352, 388, 437–465
Pullback, 20–23, 47–48, 927

Q

QED (quantum electro-dynamics), 468, 476–481
Quantum computer, 167, 941–978
Quantum field theory, 96–97, 99, 132, 166, 232, 404, 423–424, 431, 437–442, 447, 464, 983–984, 988
Quantum information theory, 422, 493, 532–533, 550, 761, 944–948
Quantum logic, 249, 282, 524, 533–534, 755, 761n33, 803–808, 828, 967
Quantum measurement, 283, 596, 653
Quantum projection, 630, 657, 662
Quantum searching, 493, 572–575
Quantum state, 188, 239, 505, 524, 531–532, 587, 593–682, 817, 824, 849, 851, 905, 947, 951n10, 971
Quaternions, 486, 709–710

R

Relation
- approximation, 498, 526, 618, 620, 651, 675
- equivalence, 10, 37, 56, 103, 143–144, 157, 162–165, 321, 385, 792, 859, 919–920, 923
- symmetrical, 732
- transitive, 9, 58, 180–181, 537, 689, 694

Ribbon category, 320, 1001–1002, 1005–1006
Right dual, 125, 307, 310, 315

S

Schrödinger equation, 439, 739, 822
Scott continuous, 491–492, 501–503, 505–511, 520, 523–524, 526–528, 531, 533, 538–539, 545–546, 551, 576, 580, 582, 584, 586, 608–609, 613, 616, 622, 624–625, 628–629, 645, 649–651, 653, 687, 689, 697, 699, 701–702
Simply-typed λ-calculus, 57–61, 67, 71, 75
Spacetime geometry, 539, 701–702
Spacetime interval, 493, 505, 524, 534–539, 689, 702
Spacetime manifold, 537–539, 702, 705, 765
Spacial monoidal categories, 302, 353t
Spectral order, 532–534, 587, 630, 636, 638–639, 643, 647–649, 662, 665, 672, 828, 830, 832, 835, 837, 840, 842
Spectral presheaf, 753–754, 762, 783, 788, 792–794, 796, 800–803, 806–810, 812, 814, 818, 826–828, 838, 842, 850–851, 855, 860, 894, 896, 898, 900, 902, 905–909, 917, 919
Spinor, 439, 717–718, 720, 723–727, 730–733, 735
Spin statistics, 984, 988–990
Standard model, 98, 468, 479–481
Strict monoidal category, 186–187, 194–196, 202–203, 211, 215, 264, 307, 359–360, 365, 430, 997
Subcategory, 12, 37, 213, 278, 321, 337, 339–340, 361, 364, 383, 425, 432–433
Subobject classifier, 817, 821, 857, 911, 1004
Superposition, 463, 944, 946–947, 950, 952, 1016
Symmetric monoidal 2-category, 120t

Symmetric monoidal category, 74–75, 77, 92, 99, 119, 122, 130, 136, 138, 141–143, 154–155, 161–162, 164, 166, 196, 200, 203, 206, 213, 217, 223, 226, 256, 306, 321, 331, 333, 335–336, 340, 342, 347, 349, 361–362, 365, 376–377, 385, 388, 390, 398–399, 402, 404, 957, 960
Symplectic group, 744–749

T
Teleportation, 120, 128, 188, 221, 422, 953, 963–969
λ-Term, 54t, 56, 59, 65, 67, 78, 149
Terminal object, 13–14, 21, 22t, 32–33, 47–48, 50, 65, 85, 91, 113–114, 128, 133, 156, 185, 235, 246, 251–252, 265, 333, 335, 783, 786, 811, 825, 879, 909, 926–928, 933, 974
Topological quantum computing, 96, 99, 105, 130, 132, 166–167, 177, 188, 232, 271, 279, 984–985, 1013, 1023
Topology
  μ, 508, 539–543, 548, 551–552, 556, 560–561
  interval, 535–538, 687–688, 691, 693, 695, 697, 700–701
  manifold, 536–537, 688, 692–693, 695
  Scott, 501–503, 506–508, 512–513, 519, 521, 537, 541, 570, 589, 690–691, 694, 700, 702
Topos, 753–933
Tortile categories, 319–320, 322–323, 348
TQFT (topological quantum field theory), 96, 99, 132, 166, 188, 232, 271, 279–280, 984
Traced (monoidal) categories, 289, 302, 322–332, 348
Truth object, 754, 759, 762, 769, 776, 810, 812–822, 848–849, 891, 896, 904, 910–911
Twistor, 714, 731–734, 737
Types
  basic, 156, 162, 164, 445, 470, 873
  function, 59, 156, 162
  linear theory, 154–155, 159–165
  product, 59, 156, 162
  trivial, 162

U
Unitary operation, 947, 949–953, 967–969, 971
Universal arrow, 41–47, 49, 75

V
Vector space
  algebra, 3
  complex, 1004–1005, 1008, 1015
    finite dimensional, 29, 33, 131, 274, 290, 306, 361, 378, 399–400, 406,
    infinite dimensional, 428
  topological, 426, 1009
Vertex group, 424, 433–435
Von Neumann algebra, 358, 386, 404, 412, 793–794, 807–808, 818–819, 832, 835, 838n86, 844–847, 866, 902, 906, 918–919
Von Neumann architecture, 941–978

W
Weakening, 53, 58, 68, 70, 86, 88–89, 135, 137, 140, 282, 368–369, 370f, 373, 390, 395, 409, 443, 447, 452, 457–459, 463n3

MIX
Papier aus verantwortungsvollen Quellen
Paper from responsible sources
FSC® C105338

If you have any concerns about our products,
you can contact us on
ProductSafety@springernature.com

In case Publisher is established outside the EU,
the EU authorized representative is:
Springer Nature Customer Service Center GmbH
Europaplatz 3, 69115 Heidelberg, Germany

Printed by Libri Plureos GmbH
in Hamburg, Germany